Roland Traunmüller   Klaus Lenk (Eds.)

Electronic Government

First International Conference, EGOV 2002
Aix-en-Provence, France, September 2-6, 2002
Proceedings

 Springer

Series Editors

Gerhard Goos, Karlsruhe University, Germany
Juris Hartmanis, Cornell University, NY, USA
Jan van Leeuwen, Utrecht University, The Netherlands

Volume Editors

Roland Traunmüller
University of Linz, Institute of Applied Computer Science
Altenbergerstr. 69, 4040 Linz, Austria
E-mail: traunm@ifs.uni-linz.ac.at

Klaus Lenk
University of Oldenburg, Lehrstuhl für Verwaltungsinformatik
26111 Oldenburg, Germany
E-mail: lenk@uni-oldenburg.de

Cataloging-in-Publication Data applied for

Die Deutsche Bibliothek - CIP-Einheitsaufnahme

Electronic government : first international conference ; proceedings / EGOVS
2002, Aix-en-Provence, France, September 2 - 6, 2002. Roland Traunmüller ;
Klaus Lenk (ed.). - Berlin ; Heidelberg ; New York ; Barcelona ; Hong Kong ;
London ; Milan ; Paris ; Tokyo : Springer, 2002
(Lecture notes in computer science ; Vol. 2456)
ISBN 3-540-44121-2

CR Subject Classification (1998): K.4, K.6.5, K.5, K.3, C.2, H.5, H.4

ISSN 0302-9743
ISBN 3-540-44121-2 Springer-Verlag Berlin Heidelberg New York

Springer-Verlag Berlin Heidelberg New York
a member of BertelsmannSpringer Science+Business Media GmbH

http://www.springer.de

© Springer-Verlag Berlin Heidelberg 2002
Printed in Germany

Typesetting: Camera-ready by author, data conversion by Olgun Computergraphik
Printed on acid-free paper   SPIN: 10871160   06/3142   5 4 3 2 1 0

Preface

In defining the state of the art of E-Government, EGOV 2002 was aimed at breaking new ground in the development of innovative solutions in this important field of the emerging Information Society. To promote this aim, the EGOV conference brought together professionals from all over the globe. In order to obtain a rich picture of the state of the art, the subject matter was dealt with in various ways: drawing experiences from case studies, investigating the outcome from projects, and discussing frameworks and guidelines. The large number of contributions and their breadth testify to a particularly vivid discussion, in which many new and fascinating strands are only beginning to emerge. This begs the question where we are heading in the field of E-Government. It is the intention of the introduction provided by the editors to concentrate the wealth of expertise presented into some statements about the future development of E-Government.

The number of subject matters covered by the EGOV 2002 conference proceedings is best illustrated by listing some of them:

- Communication with citizens over the Net: One-Stop-Government, Single-Window-Access, and Seamless Government;
- Frameworks and guidelines for E-Government;
- International and regional projects, case studies, and international comparisons;
- Strategies, implementation policies, and best practice;
- Redesigning cooperation within and between agencies;
- Sustaining business processes, collaborative activities, legal interpretation, and administrative decision making;
- E-Democracy strategies, citizen participation in public affairs, and democratic deliberation;
- Technical implementation aspects (standards for information interchange and processes, digital signatures, platforms, security concepts, and provisions);
- Novel organizational answers and new forms of networks: adhoc cooperation and coalition between public agencies and public–private partnerships;
- Changing legal frameworks, and legal and social implications of new infrastructures and applications;
- Teaching of E-Government.

Many people cooperated over a long period of time to shape the conference and to prepare the program and the proceedings. Our thanks go to the members of the Program Committee (listed below) and to Gabriela Wagner, who heads the DEXA organization. The editors are particularly grateful to Ute Holler for her exceedingly engaged assistance in coordinating the preparation of the program and of the proceedings.

September 2002

Roland Traunmüller
Klaus Lenk

Program Committee

General Chair

Roland Traunmüller, University of Linz, Austria
Klaus Lenk, University of Oldenburg, Germany

Program Committee

Kim Viborg Andersen, Copenhagen Business School, Denmark
Chris Bellamy, Nottingham Trent University, UK
Trevor Bench-Capon, University of Liverpool, UK
Daniéle Bourcier, University of Paris 10, France
Jean-Loup Chappelet, IDHEAP Lausanne, Switzerland
Wichian Chutimaskul, King Mongkut's University of Technology Thonburi, Thailand
Nicolae Costake, Bucharest, Romania
Arthur Csetenyi, Budapest Univ. of Economic Sciences, Hungary
Christian S. Friis, Roskilde University, Denmark
Fernando Galindo, Universidad de Zaragoza, Spain
Michael Gisler, Bundesamt für Informatik und Telekommunikation (BIT), Switzerland
Dimitris Gouscos, eGovLab, University of Athens, Greece
Åke Grönlund, Umeå University, Sweden
Michel Klein, HEC Graduate School of Management, France
Sayeed Klewitz-Hommelsen, Fachhochschule Bonn-Rhein-Sieg, Germany
Friedrich Lachmayer, Republik Österreich BKA-Verfassungsdienst, Austria
Philippe Laluyaux, Clip Card, France
Alan Lovell, UK
Ann Macintosh, Napier University, Edinburgh, UK
Gregoris Mentzas, National Technical University of Athens, Greece
Thomas Menzel, University of Vienna, Austria
Jeremy Millard, Danish Technological Institute, Denmark
Javier Ossandon, Ancitel Spa, Italy
Reinhard Posch, Chief Information Officer, Graz University of Technology, Austria
Corien Prins, Tilburg University, The Netherlands
Gerald Quirchmayr, University of South Australia, Australia
Heinrich Reinermann, Deutsche Hochschule für Verwaltungswissenschaften Speyer, Germany
Reinhard Riedl, Fachbereich Informatik, Universität Rostock, Germany
Giovanni Sartor, CIRSFID Bologna, Italy
Erich Schweighofer, University of Vienna, Austria
Ignace Snellen, Erasmus University of Rotterdam, The Netherlands
Dieter Spahni, Institute for Business and Administration, Switzerland
Efthemis Tambouris, Archetypon S.A., Athens, Greece
Wim van de Donk, Tilburg University, The Netherlands
Mirko Vintar, University of Ljubljana, Slovenia
Francesco Virili, University of Cassino, Italy
Maria Wimmer, University of Linz, Austria

Table of Contents

Framework

Digital Olympics 2008: Creating the Digital Beijing

Knowledge Management

Requirements

Business Process Reengineering

Electronic Service Delivery

Designing Innovative Applications

Electronic Democracy

Information Society Technologies Programme (IST)

Implementing e-Government

Legal Issues

Technical Issues

Varied Contributions

Electronic Government: Where Are We Heading?

Klaus Lenk[1] and Roland Traunmüller[2]

[1] University of Oldenburg, Germany
lenk@uni-oldenburg.de
[2] University of Linz, Austria
traunm@ifs.uni-linz.ac.at

Abstract. In common understanding, Electronic Government focuses upon relatively simple transactions between identifiable customers (citizens, enterprises), on one side, and a multitude of government organisations in charge of particular activities, on the other. Attention is chiefly directed to Electronic Service Delivery. If the promise of e-Government as the principal key to modernising government is to be kept, this concept has to be broadened so as to include the full enabling potential of IT, as well as the complex reality of government and public governance. There is encouraging political support for e-Government, yet implementation problems could inhibit further success.

1   What Electronic Government Is About

Electronic Government (or e-Government) as an expression was coined after the example of Electronic Commerce. But it designates a field of activity which is with us for several decades yet. In many respects, e-Government is just a new name for the informatisation of the public sector, which has been going on for several decades now (Lenk 1994). The use of IT in public administration and in other branches of government (including parliaments and the judiciary) has attained a high level in many countries of the industrialised world. But there was hardly any political interest in this ongoing and almost invisible process of modernising government. Only academics and some far-sighted consultants insisted on the significance of informatisation for public governance and its modernisation (Snellen and van de Donk 1998).

For a long time, their message went unheard (Lenk 1998). Especially New Public Management as the most important explicit movement of government reform hardly recognised the enabling potential of IT for changing the work practices and the business processes in the public sector. Its image of IT was one of an auxiliary tool, to be used for supporting financial management and statistical information.

This situation changed fundamentally with the announcements of the then US Vice President Al Gore in 1993, heralding not only the potential for a renewal of society which an "Information Society" holds, but relating it directly to the need of improving the performance of the public sector. The lesson inherent in harnessing the propagation of an Information Society to public sector modernisation has been learned earlier in Asia than in the European Union, where the Action Plan, based on the Bangemann Report of 1994, concentrated one-sidedly on the private sector of the

R. Traunmüller and K. Lenk (Eds.): EGOV 2002, LNCS 2456, pp. 1–9, 2002.

economy. But the situation has now been redressed also in Europe. The American and Asian examples have greatly stimulated the interest in the potential of e-Government as one important facet of an Information Society.

Although the new impetus presents great opportunities for improving government processes, the underlying concepts of e-Government remain fairly vague. Moreover, they are still driven by analogies from E-Commerce. This may misdirect the attention of governments which are eager to innovate. Things are made to appear as if they were entirely new, and both the achievements and the lessons learned from several decades of using IT in the public sector are neglected by those who are discovering the enabling potential of IT for the first time.

Whilst in the past, IT-support was inward-looking and chiefly brought to bear on typical back office activities, the focus has now moved toward the external relationships of all branches of government. It is on electronic citizen service, on electronic procurement of goods, as well as on electronic democracy including democratic deliberations, citizen information, and electronic voting. Many early projects inspired by the Information Society rhetoric focussed on politically visible fields like online citizen services, without giving much consideration to whether the promised improvements were catering to the most pressing needs of citizens or enterprises. Also in the field of E-Commerce, early projects were launched without caring much about what the potential customers would actually need. Up to the "dot.com" crisis of the year 2000, market research and target group identification had been largely absent. But in the private sector, market forces quickly taught the right lessons.

The errors of e-Government are much harder to detect, and incentives to correct them do not always exist. A case in point is the assumption that online access "24 hours, 7 days a week" would meet the prime concern of most citizens when they have to approach a public organisation for services delivery or other reasons. The results of research on citizen-government relations, which were accumulated over decades, have been totally neglected in order to present things in such a way as if existing "online" solutions (which fall short even of involving interaction via videoconferencing) would hold the key to solving all problems. The political wish to announce serious actions and quick solutions has also led to focusing on transactions like registering a car or applying for an identity card, which citizens mostly do not consider as a service but rather as a nuisance. Many governments hoped to speed up the diffusion of Internet use within the population by offering relatively simple government "services" over this channel.

So it was less a desire to enhance back office productivity and service quality which prompted governments to embark on the type of Electronic Service Delivery, which is so prominent now. The overriding political concerns seemed to be of an economic nature, the state assuming a forerunner role in entering the Information Society. Administrative modernisation was piggybacked by economic policy. In federally funded projects in Germany it was at no time clear what the priorities actually were: propagating an electronic signature deemed necessary to make the country become a player in the world league of E-Commerce, or improving service quality and productivity in the public sector.

Among proponents of e-Government, this concentration on Electronic Service Delivery has contributed to a fairly distorted view of the whole machinery of government and of public governance. The system of public governance is now changing in many respects, and states as the former key players are re-positioning themselves with respect to global corporate players and a civil society which is discovering new ways of self-organisation. And again, quite like in the New Public Management wave of administrative modernisation, which is now ebbing off, the role which IT can play in this context is grossly underestimated.

Most of the present endeavours at promoting e-Government fall short of acknowledging two things, the complexity of government and of governance, and the potential of IT beyond what is cast on the market so far. But both are extremely important. Where these two circumstances will join, they will make e-Government a fascinating experience.

The convergence of new forms of governance and future ventures in IT will transform the ways in which we work, communicate, deliberate and negotiate. To set the stage, we have to broaden the concept of e-Government (Lenk and Traunmüller 2001). If we do not succeed in showing a way out of the narrow corridor of improving access to simple and highly automated business processes within a given institutional frame, e-Government might soon become another example of exaggerated "hype".

In order to broaden the concept of e-Government beyond Electronic Service Delivery, the German Society for Informatics (GI) published, together with the German Society of Electrical Engineering (VDE), in September 2000 a Memorandum on "Electronic Government as a key resource for modernising government. This memorandum dealt with the great prospects of a wider usage of information technology for a lasting modernisation of the state and public administration. With the same intention the OCG (Austrian Computer Society) established a "Forum e-Government" (www.gi-ev.de/informatik/presse/presse_memorandum.pdf and www.ocg.at).

If the promise of e-Government as the principal key to modernising government and governance in more than a superficial sense will ever materialise, a clearer view of the agenda of modernising public services should come to prevail. This view should not be tainted by considerations of applying readily available solutions to problems which are not sufficiently investigated. Knowledge of the administrative domain has to be harnessed to a good understanding of the opportunities opened up by technology. And the technological perspective should not be restricted to the present state of development of the technology and to what is on the market now.

In what follows we therefore deal, beyond the immediate perspectives of Electronic Service Delivery, with both the requirements for a pervasive e-Government flowing from the complexity of the public sector and the enabling potential of IT for e-Government. To conclude, we shortly address the significance of political support and successful change management, as critical factors for success. In so doing we draw on a theoretical foundation of e-Government which we published earlier (Lenk and Traunmüller 1999; see also Lenk and Traunmüller 2000 and 2001; Lenk, Traunmüller and Wimmer 2002).

2   Electronic Service Delivery: The Immediate Perspectives

The delivery of services over the Internet has attracted most of the attention devoted to e-Government. A "virtual administration" is more and more taking shape. Public administrations will eventually appear no longer as a set of independent agencies which have to be approached separately, but as a collective unit with which contact can be made via one and the same "portal", or "window". Such a common access structure will reduce neither their intrinsic complexity nor the required precision of their work. No institutional reform is required to make this happen. A One-Stop Government or "Single Window Service" will alleviate many burdens, for individuals and businesses alike. At the same time, it will make public administration more transparent.

One has to admit: this picture is still a more a vision than reality. Most administrative bodies are committed to first steps at improving their own services, without looking at what their neighbours do. Yet the integration of their business processes across organisational boundaries will become very attractive, at least from the addressees' standpoint. In order to make such a vision of "seamless government" come true, specifications of future service delivery arrangements have to be elaborated with great care. These have to take into account a multitude of dimensions:

- Addressing the needs of target groups (e.g. professionals, taxpayers, the elderly)
- Allowing for a multichannel access mix (one-stop shop, online, letter, fax etc.)
- Taking into account service complexity (which varies according to the categories of business processes supported)
- Establishing the required level of service integration (eventually single-window access to all services regardless of government level and organizational unit)
- Providing the required level of security (user identification; authentication, cancellation and non-repudiability of documents and communications)
- Implementing a data protection policy and transparency measures
- Making reliability and usability a prime concern (creating user interfaces which match existing skills, incentives, culture)

These requirements cannot be discussed in detail here (Lenk 2002). We will instead draw attention to the role of portals in an architecture for service delivery which is predicated on a clear separation between front offices and back offices. In such an architecture, front offices can be customer-centered, whilst back offices are task-driven. Some sort of middleware (or "mid office") is required to link front offices to back offices. Typically, front offices comprehend portals which may either directly be accessed by the citizens or used in a physical one-stop shop or a call center where citizens are served.

Portals cover a wider range of functions, and they can be designed to cater for specific demands of their users. Among their most prominent functions is the provision of information about services. This comprises basic information as to which services are available (or which duties citizens have to comply with), as well as information about how to get in contact, which evidence should be presented, etc. For a citizen, this may be of help in preparing a contact at a physical (front) office or in deciding about further steps on the Internet. Next, the download of forms is often of great help to citizens, even if such forms are filled in by hand and mailed back offline.

A great advantage of portals is the possibility of accessing information which is not confined to the range of services offered by a specific agency. At one single access point, citizens can obtain information about all levels of government, and it is of secondary importance which government level or which other (private or ssemi-public) actor runs the portal. Canada is perhaps the leading country with regard to single-window access to government. www.help.gv.at is the address of the Austrian national portal, where easy access to information concerning 55 life events (getting a passport, marriage, change of address etc) is provided. Similarly, the French Portal http://www.service-public.fr provides not only information but holds over 1,000 forms for download, covering the French national administration as well as local and regional government. Moreover, both portals have recently created specific entries for business situations, in addition to typical life events. The work on their extension into a gateway providing access to electronic transactions is in progress.

Since many contacts of ordinary citizens with public bodies are not so frequent, we predict that unassisted online transactions will fulfill only part of the needs of citizens. When it comes to more complex transactions, a citizen may need personalised help, or want personal contact and explanations. This should be made possible without the citizen having to go in person to the back offices where the service is produced or clerical work is done, and it can be achieved either indirectly via physical one-stop shops or call centres or directly through the portal.

Perhaps quite soon will we see the emergence of "telepresence" of a human agent based in a back office somewhere, via advanced forms of vidcoconferencing, in a front office service situation or directly on a screen at home. We also expect that more and more physical one-stop offices will be created, which rely on Internet-based (and Intranet-based) services. In an "administration à accès pluriel" (Rapport Lasserre 2000), multichannel access involving multifunctional front offices will bring the advantages of the Internet also to citizens who for one reason or another do not use the Internet personally. This option is not yet pursued in countries which still hope to encourage the diffusion of Internet access by making it attractive through online public service offerings. May it be recalled at this point that France, over 20 years ago, first embarked on administrative "Télé-services" in order to market the Télétel (Minitel) system. After some deceiving pilots, this strategy was quickly abandoned in favour of encouraging the (successful) diffusion of Minitel through zero-cost access to a nationwide phone directory. But in a world dominated by players from other world regions, European experiences obviously do not count, even not for the European Union.

To sum up, portals provide an ideal leverage for the modernisation of public services. Yet many problems still have to be solved. Opening a "window to the outside" will lead to considerable rearrangement inside government. This could be limited to re-arranging the interface of business processes so as to make them candidates for Web Services. But it can also amount to a momentous "E-Transformation" inside government, which would be attuned to the equally momentous changes in public governance which are already beginning to be felt worldwide. But, as we said before, narrow views about e-Government can prevent us from facing this challenge.

3   Broadening the View: Government and Public Administration from a Systemic Perspective

In order to apprehend the wide range of opportunities which e-Government now opens up for improving the public business, we have to recall that public governance is not just about delivering services. It includes democratic policy formulation, the execution of these policies, and the evaluation of their results so as to improve policy making in the future. The ways in which branches of government work: in policy-making and planning, in deciding cases and in settling conflicts, are manifold and often quite different from what can be found in the private sector.

The diversity of work practices, business processes, organisational structures and institutional settings in state, politics and administration cannot easily be described. It also depends on legal and political preconditions, which furthermore differ considerably from country to country. Especially at the level of the European Union, this diversity is seldom acknowledged. Moreover, vendors have an interest to downplay it in order to sell their products to more than just one country. But the governmental systems that have evolved on the European continent have a very complex structure. The French, the German and also the former Austro-Hungarian model of public administration are still very influential, and they are quite different from British or American governmental traditions. Many EU-financed European pilot projects have stumbled over these differences.

Most important is a look at what are the typical results, or products, of the executive branch of government, including local government and including also mixed forms of production like public-private partnerships. Only to a minor part are the products of administrative activity typical services, where individual "customers" can be identified. More often than not, they benefit a multitude of addressees, providing public goods e.g. in the form of common infrastructures. Examples include road construction and maintenance, or police patrols. Other kinds of government services consist in financial transfers, which aim at a redistribution and at social justice by giving to some and taking from others.

But the most important type of a administrative action is regulation through authoritative decision making, which takes place e.g. in granting licenses, in allocating rights and duties. Administrative decision-making has mostly to do with more or less complex situations where political and legal regulations are applied to distinct cases. To give an example, a building permit will not only benefit a houseowner, but compel the neighbours to tolerate building activities.

Finally, among the activities of the public sector we should not forget information management activities, especially in the large field of building up and maintaining fundamental data bases about inhabitants, land and economic activities. Such data bases include civic registers, geographic information systems, official statistics etc. They support administrative action and, at the same time, they provide services to the economy.

A closer look at the types of processes and products which are characteristic of the public sector is required for assessing which type of information system could support them. To complete the picture, we should also mention the policy making side, e.g. in the legislative branch of government. For many situations, there is no possibility of importing ready-made systems from the private sector. A case in point is "E-council": a system to support the deliberations and the work of local government council members (Schwabe 2000). Such systems are specific to the public sector, and there is not much willingness as yet to spend public money for their development.

4   Taking the Potential of Information Technology Seriously

Not only is government and public governance an uncharted field for many of those who presently pay lip service to e-Government. Also the vast enabling potential of IT, beyond what is to be found on the market so far, remains largely unacknowledged. It could be brought to improve many processes and structures in the public sector.

In our book, we conceive of the history of IT use in the public sector as a series of application generations, reflecting the respective state of advancement in hard- and software (Lenk and Traunmueller 1999, p.21ff.). e-Government is no exception to this. Here, the most relevant feature is communication and world wide information access over the Internet. Each generation of IT carried some general guiding ideas about what could be done with the technology. An example is provided by the idea of creating huge data banks (as well as that of regulating their use through data protection legislation), which took shape in the wake of disk storage devices. Another example is the "paperless office" as a guiding idea which was prompted by the advent of the PC. Each IT generation suggested new applications, and the practice of business was perceived principally in the light of what the latest generation of computers or information systems could do to support it. The general pattern is that problems always tended to be perceived in the light of available solutions.

The development of administrative informatics can thus be understood in a sort of dialectical movement. New applications suggested by new waves of technology seemed to arrive just in time so that problems besieging a field of practice could be tackled. The new generation of technology seemed to hold the ultimate solution for all problems. Yet when the new perspective was put into use it soon appeared that its promise was only partial. It became clear that under the spell of a central guiding idea its promise was overstated.

e-Government is firmly anchored within this trend. The current fashion of looking at administrative processes from a citizen-interaction perspective is just a continuation of the temptation to seek inspiration from technological progress and to derive from it guiding principles for good practice. To stress the positive side: now that, with the help of Internet technology, new forms of electronic service delivery appear possible, the problems of citizens in their dealings with administrative agencies seem to be taken seriously for the first time. But there is the other side of the coin as well: the interaction is interpreted in a way so as to make technology-driven solutions appear as valid solutions to them. It is seldom question of *social* innovations in administrative or political practice, which are IT-mediated or IT-enabled (Hoff et al. 2000). Not surprisingly, many truly important policy fields have not got yet advanced IT support. Providing services to handicapped persons, providing neighbourhood social services, or dealing with people with immigration status are hardly given a thought in e-Government strategies. A large part of the population seems to be simply absent in political statements about the E-Society.

5   Political Support – A Window of Opportunity

But nevertheless, the fast growing political interest in e-Government arouses great hopes. In order to prevent such hopes from dissipating, we now have to look for quick and tangible success for important groups of stakeholders. There are encouraging signs. Inter-organisational cooperation which is of vital importance to the innovation alliances has considerably increased. Even in a very complex polity like Germany where local governments compete with each other, and moreover are extremely jealous of anything the Land or the Federation does, cooperation is progressing. Still, the lacking willingness of many agencies to make investments in long-range projects, as well as the reluctance to spend money for qualifying staff, are points of distress.

Another point of concern is the lack of clear visions of what a modern public sector should look like. Among the central questions that have to be answered is the following one: Under which conditions do we want our public organisations to function in the future? Which products and services do we want them to provide? And should these be produced and/or delivered by public organisations themselves or from external sources or in partnership with others? The lack of well-founded visions of a modernised public sector becomes obvious when actors trying to promote e-Government find it difficult to figure out viable business models for new IT-based administrative services.

If such visions are not developed, the temptation will persist to look at daily practice only in the light of what the technology can do to improve it. In the end, therefore, strategic thinking will be required, Only if well-founded visions of the future work of state and administration will be developed, will e-Government become a lasting success.

6   Implementation – The Hidden Threat to e-Government

According to a recent management brief issued by OECD (OECD 2001), the inability of governments to manage large public IT projects threatens to undermine efforts to implement e-Government. In a climate of euphoria, it is easy to overlook hindrances on the way to a lasting improvement of governmental and administrative practice. Political discourse tends to lose contact with the reality of what can be achieved with given resources and in a reasonable period of time.

Action has to be taken to improve the conditions for successfully implementing e-Government projects. On one hand, thanks to the evaluation of past technical inventions we already have considerable knowledge about the success factors for projects and their diffusion. On the other hand, again and again we forget the lesson learned. One reason is that too many experiences made during implementation are generally scattered and not communicated. Also there is a widespread inclination to ascribe implementation difficulties to an immature state of technology.

Furthermore, there is a gap between those making concepts and those who have to implement them. The technical and logistic implementation of solutions is usually under the responsibility of field organisations and their management. In adapting software to the structures and working processes of the organisation they often miss adequate support for planning and implementing the required organisational changes. Software suppliers tend to provide technical solutions to complex socio-technical problems. Theirs is the role of an engineer, but there seems to be no architect in charge of the overall human-machine interaction system. Procedures of systems design will have to evolve toward holistic methodologies, balancing the technology package and the complex socio-technical work reality (Lenk and Traunmüller 1999, p.93ff.).

There is thus a real danger that e-Government will glide down the slope from a mountain of euphoria into a valley of deception. Only in broadening the concept and in recalling its basic tenets will we steer it toward lasting success.

References

1. Hoff, J., Horrocks, I, Tops, P. (eds.). Democratic Governance and New Technology. Technologically mediated innovations in political practice in Western Europe. London, New York: Routledge, 2000.
2. Lenk, K. Information systems in public administration: from research to design. In: Informatization in the Public Sector. 3, 1994, pp. 307-324.
3. Lenk, K. Reform Opportunities Missed: Will the innovative potential of information systems in public administration remain dormant forever? In: Information, Communication & Society. 1, 1998, pp. 163-181.
4. Lenk, K. Elektronische Bürgerdienste im Flächenland als staatlich-kommunale Gemeinschaftsaufgabe. In: Verwaltung & Management 8, 2002, pp.4-10.
5. Lenk, K., Traunmüller, R. Öffentliche Verwaltung und Informationstechnik (=Schriftenreihe Verwaltungsinformatik 20) Heidelberg: Decker, 1999.
6. Lenk, K., Traunmüller, R.. Perspectives on Electronic Government. In: Galindo and Quirchmayr (eds.). Advances in Electronic Government, Proceedings of the IFIP WG 8.5 Conference in Zaragoza, 2000, Zaragoza: University Press, 2000, pp. 11-27.
7. Lenk, K., Traunmüller, R. Broadening the Concept of Electronic Government. In: J.E.J. Prins (ed.), Designing E-Government, Amsterdam: Kluwer, 2001, pp.63-74.
8. Lenk, K., Traunmüller, R., Wimmer, M.A. The Significance of Law and Knowledge for Electronic Government. In A. Grönlund (ed.), "Electronic Government - Design, Applications and Management", Hershey (PA): Idea Group Publishing, 2002, pp. 61-77.
9. OECD PUMA Public Policy Brief No.8, March 2001. The Hidden threat to E-Government. Avoiding large government IT failures.
10. Rapport Lasserre. L'Etat et les technologies de l'information et de la communication: vers une administration à accès pluriel. Paris: La documentation francaise, 2000.
11. Schwabe, G. E-Councils – Systems, Experiences, Perspectives. In: A.M.Tjoa et al., Proceedings of the 11th International Workshop on Database and Expert Systems Applications, 4-8 September, 2000, Greenwich. Los Alamitos: IEEE Press, pp.384-388.
12. Snellen, I.Th.M.; van de Donk, W.B.H.J. (eds.). Public Administration in an Information Age. Amsterdam: IOS Press, 1998.

Centralization Revisited? Problems on Implementing Integrated Service Delivery in The Netherlands

Jeroen Kraaijenbrink

University of Twente, P.O. Box 217, 7500 AE Enschede, the Netherlands
j.kraaijenbrink@sms.utwente.nl

Abstract. In the Netherlands, the development of integrated public service delivery has been an important topic for over a decade. Despite the investments, the results are meager. In the literature, an overwhelming and contradictory amount of conceivable problems is mentioned that can explain these lagging results. Four case studies were carried out to find out which of these problems are most pressing in the particular context of integrated public service delivery. These are found to be: (1) indistinct and subdivided responsibilities, (2) focus on the autonomy of the own organization, and (3) insufficient scale. Given these problems, and given their different importance in the four cases, it is argued that the effective development of integrated public service delivery in the Netherlands requires more centralization.

1 Introduction

Enhancing the level of public service delivery has received much attention in the Netherlands but has not lead to substantial results yet [2], [4], [7], [9]. This applies in particular to the integration of service delivery (ISD). In short, ISD means that multiple public organizations cooperate to deliver their services in an integrated way, usually by means of an integrated counter. By doing so, these organizations try to offer a solution to the problem of fragmentation. Fragmentation of service delivery is seen as an important problem for at least thirty years in the Netherlands [8], [10], [11], [27]. It is considered problematic for both citizens and government. Whereas citizens cannot find their way in the bureaucracies, government does not reach its citizens, and public policy remains ineffective.

This paper argues that three characteristics of ISD explain why results are meager. First, to realize ISD, interorganizational *cooperation* is a necessity. Organizations must tune work processes, create new services together and mutually adapt their applications. This implies a major *change* for participating organizations, which can be problematic on its own. The third characteristic is the use of *information and communication technology* (ICT). Because public services are to a large extent information services [1], exchange of data and information is one of the crucial elements of ISD. It is unthinkable that this exchange of information can take place without the support of ICT.

R. Traunmüller and K. Lenk (Eds.): EGOV 2002, LNCS 2456, pp. 10–17, 2002.
© Springer-Verlag Berlin Heidelberg 2002

Given the literature, there are numerous theoretical reasons for the failure of interorganizational cooperation, organizational change and information system development, e.g. [5], [14], [15], [16], [19], [22], [23], [24], and [25]. Moreover, literature on success factors and solutions is innumerable as well, e.g. [3], [12], [13], [18], [19], [23], and [30]. Adding only the findings of this limited number of authors, at least 150 different reasons for success and failure appear [17]. However, it is not evident which of these are most important in the particular context of ISD in the Netherlands. Therefore this literature does not provide organizations with helpful insights in problems associated with ISD and situations in which these are more probable to occur. Because the literature is not well adapted to the particular situation of ISD the question remains:

Which are the problems hampering integrated service delivery in the Netherlands and why do these problems exist?

This paper reports about a research carried out to answer this question. It has the following structure. The next section discusses the research method. Section 3 presents the main problems and answers the first part of the central question. Section 4 answers the second part by discussing probable causes. Finally, Section 5 discusses the research and proposes topics for further research.

2   Research Method

To answer the research question, four case studies were conducted. Because the number of ISD projects that have passed the planning phase is limited in the Netherlands, the selection of cases was mainly based on convenience sampling. However, as Table 1 shows, a certain spread in domain, approach, results, and duration was achieved. The following cases were selected:

1. Counter for (starting) companies: cooperation of mainly chamber of commerce, tax office, and municipalities.
2. Counter for (starting) companies: combination of physical and virtual counter.
3. Health counter: cooperation of nursing and old people's homes and a municipality.
4. Counter for the unemployed: cooperation of municipal social service department, job centres, and social security.

Table 1. Short Description of the Cases

	Case 1	Case 2	Case 3	Case 4
Start	1999	1999	1996	1996
Approach	Mainly de-centralized	Mainly decentralized	Decentralized	Decentralized, then centralized
Integration	Little	Little	Virtually none	Some
Electronic communication	Little	Little	Virtually none	Much

The case studies consisted of an extensive documentation analysis and additional interviews with project leaders. A structured list of 150 theoretical reasons for success and failure was used to systematically check which of these were present and which were most pressing in each individual case. The period of data collection was September to December 2001.

3   Main Problems in the Cases

For answering the first part of the research question, this paper only presents cross-case results. The individual case studies are discussed in [17], but are not needed for the purpose of this paper. The main problems that were identified in the cases can be split up in three categories:

- Problems on indistinct and subdivided tasks and responsibilities
- Problems through a focus on the autonomy of the own organization
- Problems on scale

Problems on Indistinct and Subdivided Tasks and Responsibilities

These problems root in the legally defined organization structure of the public sector in the Netherlands. Within this structure, tasks and responsibilities are often not distinctly divided between organizations – both in a horizontal and a vertical way. As a result of this, the decision-making processes involve multiple organizations.

A. *Division of authority and responsibilities:* ISD asks for cooperation, tuning, and rethinking tasks and responsibilities. In the cases organizations emphasized their own – often legally based – responsibilities, and did not entrust them to other organizations. In Cases 1 and 3 parties tried to set up a foundation to overcome this problem. However they did not come to an agreement because none of the parties was willing or able to hand over responsibilities to this foundation.

B. *Confining role of national public organizations:* The cooperating local organizations are dependent on decisions of their national counterparts. Local organizations that try to cooperate perceive that these organizations often do not support them in their attempts. Therefore they are limited in their scope. In Case 2 the national public organizations were even called 'the common enemy' (translated from Dutch).

C. *Legal constraints and uncertainty:* A sizable part of public organizations' positions and tasks is legally defined. When parties cooperate, they must act within these constraints, which can lead to severe restrictions for ISD. In the Netherlands, tax offices are for example not allowed to make their systems accessible online. This makes electronic communication in Cases 1 and 2 difficult.

The frequent number of small policy changes was perceived as a problem as well because it causes uncertainty. In Case 4 the concerning project leader experienced the change that was promoted as major by government just as one out of many others.

Problems through a Focus on the Autonomy of the Own Organization

Each of the participating organizations has its own identity, culture, and way of working. When they cooperate, adaptations must be made on each of these. However, organizations experienced this as problematic and therefore kept their internal focus.

A. *Giving up identity:* Placing part of organization's services behind an integrated counter makes an organization less visible to its citizens. With losing this visibility, parties in the cases also feared they would loose part of their identity. This went hand in hand with a reserved attitude toward collective responsibilities and tasks. Case 1 illustrates this with its business cards: the logo of every participating organization is on it (see Fig. 1).

B. *Tuning of work processes:* Every participating organization has its own way of working. Tuning and adjusting work processes can streamline activities. However, the cases indicate that this asks for major efforts of all organizations, which lead to nearly no adaptations at all. The project leader of Case 2 illustrated this with stating that accumulating registration forms of the seven participating organizations would lead to a seventy centimeters' pile of paper. He remarked about this: 'when we interfere with that, we just have a very sour life and tiring discussions that lead to nothing' (translated from Dutch).

C. *Lack of standards:* Because of the large diversity of applications and formats in use in the participating organizations, electronic information exchange between them is far from easy. However, parties did not overcome this problem in other ways than by sending faxes, making phone calls, sending e-mails, and retyping data. There was also a lack of standards in definitions. In Cases 1 and 2 every participating organization had its own definition of entrepreneur on which they did not agree till date.

Fig. 1. Example of a Business Card

Problems on Scale
Because of the local level of the ISD, the projects have a relative small scale. Due to this small scale, organizations do not have sufficient resources and power to fulfill the needs for a sound integrated counter.

A. *Insufficient personnel:* During the ISD-projects organizations had to continue their regular activities. This lead to capacity problems for the projects. Moreover, because some important functions were only occupied by one or two persons, illness and dependence on single persons was reported as a problem. As a result of this, in Case 1 the communication about the project was stopped for half a year because of a long illness of only one of the project members.

B. *Insufficient financial capacity:* Parties indicated that the development of the necessary infrastructure and mid-office functionalities largely exceeded their bearing power. According to them these investments should be made at a higher aggregation level. The project leader of Case 2 indicates that the costs of building a suitable mid-office are estimated at about ten million Euro, with a budget of about one million Euro for this complete individual project.

C. *Ensuring privacy, authenticity, and security:* Electronic information exchange is an important enabler for ISD. Because of sensitivity of information however, exchange should be secure. This exchange is not limited to the local participating organizations because some databases are owned at a national level. Therefore, this can hardly be arranged at a local level. Case 4 illustrates that it is also difficult at a national level. Although electronic integrated information exchange was enabled, and security was claimed, the system was already hacked in its first few days of use.

Next to the nine problems in total, there were other problems mentioned as well. However, considering the four cases together, these nine appear to be the most pressing problems out of the 150 reasons for success and failure mentioned in theory. Yet, they do not appear in the same extent in each case. The next section discusses these differences.

4   Situations in Which They Are More Probable

Now that an answer is given to the first part of the research question, this section discusses the second part, which is about causes of the two types of problems. This section compares the approaches followed in each of the four cases.

Cases 1 and 2 seem to be most typical for the current way of developing public integrated counters in the Netherlands. As Table 1 shows, these cases have attained relative little integration and electronic communication. These two cases came also with most identified unsolved problems. The relative recent start of the projects could explain this. However, project leaders did not expect that the current approach would indeed lead to better integration and electronic exchange of information. According to them, solutions are needed on a higher organization level.

In the third case less problems were identified, but this project was also relatively unsuccessful in terms of level of integration and electronic communication. Because of the relative autonomous position of participating organizations, almost no problems were mentioned on tasks and responsibilities. Problems on scale were also scarcely mentioned, but this could be explained by the fact that ICT investments and attempts to really integrate were postponed.

In the fourth case all of the above-mentioned problems were identified during the project. This was particularly true in the stadium in which the project shifted from a decentralized to a centralized approach. However, after taking radical measures like changing laws and reorganization of the social security sector, this lead to relative good results on integration and electronic communication, as shown in Table 1.

Considering the different approaches of the four projects, it seems that the current most common approach in the Netherlands comes with most identified unsolved problems. This approach has centralized and decentralized elements in it, which means that none of the organizations on both national and local level has complete responsibility. It seems that as a result of this not much really happens. Although other causes exist, based on these limited number of cases I state that the current mixed approach is a major cause of the problems on integrating service delivery in the Netherlands.

Cases 3 and 4 support this statement. Both a centralized (Case 4) and a decentralized approach (Case 3) lead to fewer identified unsolved problems. In Case 3 few problems were reported, but as mentioned above, the results were relatively limited, partly because of limited resources. Considering the better results and solution of problems in Case 4, it seems that a centralized approach fits better to the needs of ISD. Conceivable reasons for this are that sufficient resources, capacity and authority are present to really change situations and reorganize towards a more integrated way of service delivery.

5   Towards More Centralization?

At first sight the suggestion that ISD requires a more top-down approach may seem outdated and in contrast with the current literature on organizational change and co-operation that favors participation and a bottom-up approach e.g. [5], [14], [23]. Considering the fourth case it appears indeed that forcing parties to cooperate just leads to frustration, stagnation, and a rigid attitude of parties and no results. However, after taking more radical measures like changing laws and redefining parties, it seems to lead to better results at the end. Therefore, it is not the obligation that leads to better results, but the reorganization that takes place. I state that problems exist because the current organization of the Dutch public sector does not fit the needs of ISD. ISD is more likely to succeed when this mismatch is eliminated. This is in accordance with interaction theory, which states that organizational problems should be fixed before introducing systems and that the organization should be in line with these systems [20], [21].

The preference for a top-down and discontinuous approach resembles much of what is said by Hammer and Davenport about Business Process Reengineering as well [6], [12]. Although BPR is about single private organizations and has not been very successful in the public sector [16], [28], it seems that at least some principles of this method are valuable in this particular context.

Although no cases were analyzed in which a centralized approach was taken from the beginning, some experiences in other public electronic information exchange projects indicate that it can be a successful approach. An example is the Kruispuntbank Belgium (enabling electronic information exchange in social security). Compared to the RINIS initiative in the Netherlands – which aims at similar results – it appears that in Belgium a centralized and drastic approach, involving changes in law, has lead to better results years ago than achieved with RINIS at this moment [30].

Of course these results are not a sufficient proof of the value of a centralized approach for ISD in the Netherlands. Therefore more and quantitative research is necessary to explore under which circumstances more centralization is advisable.

Although this paper suggests that a centralized approach has benefits for ISD, it is not a pleading for centralization in general. It is only due to the current relative decentralized approach of ISD in the Netherlands that more centralization is worth considering. This implies that the benefits that Simon relates to centralization – coordination, expertise, and responsibility – at this moment outweigh the costs, including higher workload of higher paid personnel, higher communication costs, and less available information [26].

Acknowledgements

The author would like to thank dr. Ronald Leenes, prof.dr. Robert Stegwee, dr. Jörgen Svensson, and dr. Fons Wijnhoven of the University of Twente for their comments, suggestions, and corrections, which greatly improved this paper.

References

1. Broek, M. van der: Informatievoorziening in de Excellente Gemeente. Elsevier Bedrijfsinformatie (2000)
2. Cap Gemini Ernst & Young: Web-based Survey on Electronic Public Services. Brussels: European Commission DG Information Society (2001)
3. Cavaye, A.L.M. & P.B. Cragg: Factors contributing to the success of customer oriented interorganizational systems. J. of Strategic Inf. Systems, vol. 4 (1995), no. 1, 13-30
4. CCTA (Central Computer and Telecommunications Agency): Benchmarking Electronic Service Delivery. A Report by the Central IT Unit (2000)
5. Daft, R.L.: Organization Theory & Design. West Publishing Company, Minneapolis/St. Paul, fifth edition (1995)
6. Davenport, Thomas, H.: Process Innovation: Reengineering Work through Information Technology. Boston, Massachusetts. Harvard Business School Press (1993)
7. Deloitte & Touche: Overheid oNLine. Trendonderzoek naar Gemeenten en Provincies op Internet. Onderzoeksrapport (2000)
8. Derksen, W. et al.: De Blik naar Buiten: Geïntegreerde Dienstverlening als Structuurprincipe. Wetenschappelijke Raad voor het Regeringsbeleid. Den Haag (1995)
9. Duivenboden, Hein van, & Mirjam Lips: Klantgericht Werken in de Publieke Sector. Inrichting van de Elektronische Overheid. Uitgeverij LEMMA BV, Utrecht (2001)
10. Eenmalige Adviescommissie ICT en Overheid: Burger en Overheid in de Informatiesamenleving: de Noodzaak van Institutionele Innovatie. Den Haag. September (2001)
11. Frissen, P.H.A.: De Virtuele Staat: Politiek, Bestuur, Technologie: een Postmodern Verhaal. Academic Service, Schoonhoven (1996)
12. Hammer, Michael: Reengineering Work: Don't Automate, Obliterate. Harvard Business Review (1990), July-August, 104-112
13. Hammer, Michael & James Champy: Reengineering the Corporation: a Manifesto for Business Revolution. New York. Harper Business (1993)

14. Jarrar, Y.F. & E.M. Aspinwall: Business Process Re-engineering: Learning from Organizational Experience. Total Quality Management, vol. 10 (1999), no. 2, 173-186
15. Kern, H. et al.: Building the New Enterprise: People, Processes, and Technology. Sun Microsystems Press. USA (1998)
16. Kock, N.F., Jr. & R.J. McQueen: Is Re-engineering Possible in the Public Sector? A Brazilian Case Study. Business Change and Re-engineering, vol. 3 (1996), no. 3, 3-12
17. Kraaijenbrink, Jeroen: De Lange Weg: een Onderzoek naar Knelpunten bij Interorganisationele Samenwerking rond Geïntegreerde Loketten. Graduation Essay. University of Twente, Faculty of Technology & Management. Enschede (2002)
18. Kurnia, S. & R.B. Johnston: The Need for a Processual View of Inter-Organizational Systems Adoption. Journal of Strategic Information Systems (2000) no. 9, 295-319
19. Lawrence, Peter: Workflow Handbook 1997. John Wiley & Sons. Chichester (1997)
20. Markus, M. Lynne: Power, Politics, and MIS Implementation. Communications of the ACM. Vol. 26 (1983), no. 6, 430-444
21. Markus, M.L.: Systems in Organizations: Bugs and Features. Pitman (1984)
22. Rose, G., H.Khoo & D.W. Straub: Current Technological Impediments to Business-to-Consumer Electronic Commerce, In: Communications of the Association for Information Systems (1999). http://members.aol.com/grose00000/cais/article.html. Visited. March 9, 2001
23. Sabherwal, R. & J. Elam: Overcoming the Problems in Information Systems Development by Building and Sustaining Commitment. Accounting Management & Information Technology, vol. 5 (1995), no. 3/4, 283-309
24. Senge, Peter et al.: De Dans der Verandering: Nieuwe Uitdagingen voor de Lerende Organisatie. Academic Service. Schoonhoven (1999)
25. Shaw, M. et al.: Handbook on Electronic Commerce. Springer-Verlag. Berlin (2000)
26. Simon, Herbert A.: Administrative Behavior. A Study of Decision-Making Processes in Administrative Organizations. Fourth edition, The Free Press, New York (1997)
27. SUWI: Structuur Uitvoering Werk en Inkomen: eerste Voortgangsrapportage. 13 Oktober (2000). http://www.suwi.nl, visited Oktober 18th 2001
28. Thaens, M., V.J.J.M. Bekkers & H.P.M. van Duivenboden: Business Process Redesign and Public Administration: a Perfect Match? Paper presented at the Conference of the European Group of Public Administration. Rotterdam. September 6-9 (1995)
29. Tidd, Joe, John Bessant & Keith Pavitt: Managing Innovation: Integrating Technological, Market and Organizational Change. John Wiley & Sons. Chichester (1997)
30. Zuurmond, A. et al.: Dienstverlening Centraal. De Uitdaging van ICT voor de Publieke Dienstverlening. Pre-advies Raad Openbaar Bestuur. (1998)

From Websites to e-Government in Germany

Dieter Klumpp

Alcatel SEL Stiftung, Lorenzstr. 10, 70435 Stuttgart, Germany
d.klumpp@alcatel.de

Abstract. European neighbors wondered at Germanys relatively slow start into Electronic Government. With a certain anxiety they had looked at this nation of eighty millions with its five million of public employees - the level of entire countries like Norway, Finland or Denmark. Consequently Europe was not surprised that Germany in 1997 took over a pole position presenting the first law on digital signature, an important element for Electronic Government purposes. But it turned out that this early law meant only a pre-dawn. It turned out that Germany - as it is undoubtedly big - needed much time and energy to get itself in motion. The presumed "giant" evolved to be a specter of mostly isolated actions and projects. Since two years - beginning with a memorandum on Electronic Government triggered by the two major IT associations - there are more and more actors striving for joint action. German Federal Government set up the program "BundOnline2005" putting comprehensive targets and respectable money into the necessary transformation of government processes, aiming at - as Chancellor Schroeder stated - "not citizens but data have to run". A public-private initiative "D21" is backing jointly strategies. Germany now appears to reach necessary pace to play the expected visionary role as well as to overcome the obstacles every innovative infrastructure at the beginning is confronted with.

1   Introduction

Germanys administration system is somewhat admired all over the world. But looking a bit closer there are rather the secondary virtues like "punctuality", "correctness", "quality", "thoughtfulness" and even "Prussian", which are connected with the image of Germany since times. Germans themselves of course hate their "bureaucracies" as all other nations do with theirs, but Germans are convinced that there exists some special kind of "reliability" within the German administration system. Without going deeper whether the German image is reflecting merely historical inheritance or today's reality, it appears to international observers that Germanys ranking in the Electronic Government sector has been on a quite surprisingly humble level for years. In early 2001 the French Prime Minister after presenting the first three years of "AdmiFrance" was dazzled and proud to learn from German Minister of the Interior that he felt Germany some years behind of France concerning the Electronic Government implementation.

Even after two years of a dedicated modernization program led by German Ministry of the Interior Germany has not found yet its specific role within the European movement towards Electronic Government. But this role is needed –

R. Traunmüller and K. Lenk (Eds.): EGOV 2002, LNCS 2456, pp. 18–25, 2002.
© Springer-Verlag Berlin Heidelberg 2002

Germanys contributions to structural modernization are estimated to be very valuable. The outstanding example was the development of the national rooted ("accredited") Digital Signature coming up with the first worldwide Federal law 1997 which became a pattern for implementation plans of many countries [1]. European countries in this early stage were somewhat reluctant with such German innovations because they feared that Germany was building up "too exhaustive" quality standards which could hamper the e-Business sector.

The "career" of the topic "Electronic Government" in Germany over the last five years shows some quite different roots, out of which some of the most important shall be outlined hereafter.

2   Cities "Internet Projects" As Website Collections

In mid Nineties the first worldwide parliamentarian commission on multimedia in the federal state of Baden-Wuerttemberg had the vision of a "tele-administration" being possible with the forthcoming broadband era. Visionary members of the group as Franz-Josef Radermacher predicted that Electronic Government would not arise out of technical possibilities but it would be compelled by the necessity "that we will have do 120% of public sector work with 80% of the clerks". Such forceful hints to Electronic Government procedures as a necessary means for rationalization vanished with the rising expectations of the public awaiting "millions of workplaces to be created by multimedia" (former EU commissioner Martin Bangemann).

German cities discovered Electronic Government predominantly participating in regional or EU programs [5]. As all over the world cities are confronted directly with the citizen some local administrations felt obliged to use the internet as an immediate opportunity to close an information gap between citizens and the governing authorities by offering electronic leaflets. The city projects tried to set up internet based "services for all" which in fact turned out to be not much more than the electronic access to the habitual paper-stuffed and always-occupied bureaucracy allowing to choose the license plate number for cars, asking for a paper form by mail up to clicking the Lord Mayor presented in cute colour photos. Cities, villages and their associations began to set up workshops on the new media "internet", mainly backed and funded by big players from the IT sector. The worldwide wave of public-private initiatives like the "Bay Area Initiative" slowly swapped over to German circles without raising the urge to modernize with all the power given [10].

With all respect to the early adopters one could witness the false step [7] towards Electronic Government: Simply "cloning" the procedures and patterns of e-Business for Electronic Government purposes leads to isolated public websites. This "experimental way" meant a tremendous loss of time and money for German cities. Each "successful" project overlapped with such of hundreds of 24.000 German cities leaving everywhere the problem of updating websites with too few people and resources. The most influential case for years was the city of Mannheim[1], where the two main modernizers, the Lord Mayor and his Communication Director succeeded in implementing a varied website with only a handful of contracted young IT specialists. There were very few cities following this "best principle" way - each one claimed to

---

[1] see www.mannheim.de and Blumenthal, J. in Alcatel SEL Stiftung [12]

have a "best principle" of its own. Similar leapfrog competitions[2] arose between the 16 German states [8].

3   Digital Signature and the Media@Komm Lead Projects

Another important root can be discovered in the broad German Science specter reaching from Public Administration Competence Centers like Speyer [9] to Knowledge Management and Organization Systems like the Fraunhofer Society Institute in Stuttgart. In 1996 one of Germanys most influential IT-specialized jurists, Alexander Rossnagel, hold the guest professorship funded by Alcatel Foundation at the Technical University of Darmstadt using this fruitful period to write down a first draft of the German Law on Digital Signature which was brought to final decision in 1997 [10]. It lined out an infrastructure based on public keys and long-term certified roots given by public or private trust centers. This innovation had to pass from the beginning a nightmare of misapprehension or better: a lack of understanding. The different actors in Germany and EU directories discussed the "security of the 128 kb key"[3], some businessmen confused signature with "a digital stamp imposed by authorities for every electron3ic move on the internet" and a few even suspected that the visible hand of the state was grasping for new tax revenues threatening to paralyze Internet economy".

German Ministry of Research meanwhile had built up a multimedia department promoting all kind of research for the "Information Society", which was after the 1999 election transferred to the Ministry of Economy and Technology. The "German" Digital signature was in worldwide discussions consequently milled to an "European" family of "electronic signatures", out of which the "self-determined user" should freely choose between worthless short-term software products and a long-term guaranteed infrastructure service[4]. One of the major tasks of the Media@Komm Lead Projects, funded with € 30bn should be the rollout of digital signature as infrastructure reaching the "critical mass". But from the beginning the "critical mass" could not be reached because the projects had to undergo a "full competition" way to find out ten winners out of 120 proposals. After many months of finding out of the ten the three winning proposals, even the winners (Bremen, Esslingen and Nuremberg-Erlangen) were not enabled to co-operate for another year. Consequently every project group had to begin from the bottom, EU project organization insisted even on the founding of private companies to deal with the private funding corresponding to the portion of public money.

The city of Bremen - which represents also the small state of Bremen - after two years climbed at the top of German communities with a broad interactive Electronic Government website and with a strong focus on standards. The impetus in Bremen emanates undoubtedly from scientific knowledge built up in years around the universitarian chair of Herbert Kubicek, supported by reform-oriented politicians.

---

[2] benchmarking is one of the favored marketing tools of the ever-so-present consulting firms

[3] a topic of kryptology

[4] countries like Japan and even European countries meanwhile discovered the charming of a nation-rooted Public Key Infrastructure and are settling a knowledge transfer with German Root Authority. Germany itself will need some time to join the way it opened years ago

4   The Memorandum "Electronic Government"

The lead projects on Electronic Government lacked cooperative structures. In early 2000 representatives of German Information Technology Society (ITG) and German Informatics Society (GI) met in Frankfurt to set up a Working Group on Electronic Government comprising more than 40 representatives of science, administrations, companies, unions and politics. After six months of bi-weekly meetings Klaus Lenk (Head of Administration Informatics of the Oldenburg University) presented the policy paper to State Secretary Brigitte Zypries and gathered experts in September 2000 stating: "With the pervasive success of the Internet, many opportunities are now emerging which together can be seen as leading to an "Electronic government". This expression is not meant to designate improved citizen services only. Instead, "government" is taken in a wide sense to denote governing and administrating in a democratic system including democratic processes of policy formulation. (...) The challenges addressed by the new motto are at least as great as those of electronic commerce. In view of the diversity of tasks involved in government and administration, they are even much broader and more varied" [2].

This vision came too early to prevent the specter of Electronic Government actors from falling into the e-Commerce depression a year later. The Memorandum quickly became the silent referential point in German discussion, it is still now present with hundreds of copies in all websites from associations, politics, science, cities foundations, trade unions and consulting companies. Being the presumably most important paper in German discussion it nevertheless has never been translated into another language. The latter is paradigmatic for the German way: the 42 who signed it up got a lot of applause, but none of those who applauded felt empowered to put it on the international scene simply translating it. All agreed to the recommendations inclusively the admonishing hint to begin immediately with reshaping of administration processes, "because there is open a time slot for Germany now".

The 1999 Memorandum established some important "Leitbilder" (models) to reshape the Electronic Government strategies: "From the point of view of the individual citizen, contacts in administrative offices can in the future be established via Internet portals and service shops, saving both time and effort. But the improvement of relations between citizens and the administration by means of new access structures constitutes just the tip of an iceberg. In order to get clearer view we have to add three further viewpoints which (...) are not looking at government and administration from the outside, but dealing directly with the machinery of public services and governance which constitute the larger part of the iceberg. These three viewpoints are:

- A reorganization viewpoint, consistently taking everyday (business) processes as the starting point.
- The perspective of telecooperation, which makes it clear that not only cooperation in routine affairs, but also complex and controversial negotiations may be carried out over distance, exhibiting ever less need for the persons involved to be present at one and the same place.
- A knowledge viewpoint, illustrating how much information technology can contribute to make processes more effective and transparent through support-ing knowledge as the most important asset of government and administration.

- In the triangle of relations between citizens (including companies), politics, and administration, the interplay between all of these perspectives is evident. But there is an urgent need to develop reference models and pilot projects in order to reveal the full extent of the possibilities at hand" [2].

Following Klaus Lenk the critical success factors include a strategic thinking instead of the "attitude of curious but indiscriminate trying out of different approaches". Without doubt a financing initiative is needed but very difficult to realize due to Germanys budget burdens during the still ongoing unification process. "People who drive the realization of electronic government are perhaps the most essential factor. In order to develop the requisite human resources, an unprecedented qualification effort has to be undertaken. A large number of the people working in the public sector need to become aware of the potential which information technology holds and also become able to estimate how they may better be able to structure their own working processes with this potential. This is closely related to a further requirement, namely a competent change management, which places people in the foreground"[2].

The Memorandum stated that a suitable IT infrastructure will be required providing the full range of availability and security. This clear request for "safer" infrastructures needed some time to trickle down from science and companies to the public sector including the politicians. One of the thresholds could be described as a traditional individualistic "non-cooperation attitude" and the incurable German passion to await visions from abroad, especially from overseas.

5   Thresholds Hampering Cooperation

To understand the slow start of Germany albeit being well equipped with all the necessary skills in "classic" administration, having leading scientists on Administration and well-trained professionals in all government sectors, one must examine the thresholds that hamper cooperation within Germany. As Federalism is an unchangeable constitutional basis of Germany, there are considerable difficulties to get along with common standards and procedures, which always mean a extensive degree of co-operation. After reunification German states still suffer several kinds of divides which could afford the necessary joint strategies towards Electronic Government. As even after 1989 there is a relatively sharp divide between the so-called "A states" (governed by Social Democrats) and "B states" (governed by Christian Democrats), there have grown more remarkable divides like between the "richer" (southern) and "poorer" (northern) states, between metropolitan states like Berlin versus the surrounding state of Brandenburg and last but not least the divide between the "old" (western) and the "new" (eastern) states after reunification. But all of them join immediately whenever Federal Government is suspected to interfere, because "all evil is central". Some German federal institutions and parties still try to maintain divides on such precedent topics like Electronic Government. Actors do have all types of explications why there is no need to extensively co-operate across the inner-German borders. As above mentioned, it is quite difficult to get a big mass to speed up.

Germany with its 80 Millions of people is not surprisingly the biggest public IT-market in Europe. KableNET [4] estimated Germany with (EUROS) 13,313 billions ahead of UK with 12,118 and France with 10,006 billions. With roughly 5 Millions of public employees, thereof 15% on federal level, 40% in the 16 states and 45% all over the communities, Germany counts on a big workforce in the public sector and can in future still count on it due to the mighty unionists of ver.di who clearly exclude major cutoffs, but are backing the modernization process with emphasis. Small and medium companies are discovering that Electronic Government business is much more than a niche and present products which are apt to transform paperwork to electronic processes. German SAP - spread over the globe - has reached to be No. 1 worldwide in offering comprehensive workflow systems which meet the demand of the public sector for reliability. German politicians - who are proud of their national car products - are still somewhat hesitant regarding and fostering export opportunities of German IT services, although the eastern countries are demanding for it in their phase of preparing EU membership. Russia dedicated its version of the D21 initiative, "Electronic Russia" exclusively to Electronic Government [12].

Electronic Government actors in Germany since 2001 begun looking forward to leaving individualistic approaches behind. Now it seems that after being despaired with the growing number of excellent projects that did not produce the necessary spark to the building up of common infrastructures, there are appearing new structures of cooperation and joint strategies.

6   Shared Strategies towards Implementation

In the 2002 Accenture research report, "e-Government Leadership - Realizing the Vision" [3] is perceived that "governments are, albeit slowly, realizing their visions. More importantly, there is a growing recognition that e-Government is not just about technology - but about harnessing technology as just one of the tools to transform the way governments operate". The report is finding an astonishing jump: "Germany significantly improved its overall ranking from 15th in the 2001 report to 9th this year, and was placed sixth among the Visionary Challengers. Its improved performance reflects a greater emphasis on bringing Government services online and further development of its strong base of mature services in the Revenue and Postal sectors. Following Chancellor Gerhard Schroeder's unveiling of Germany's e-Government vision, BundOnline 2005 (9/2000), the Government has implemented a range of measures to accelerate the implementation of its vision. The German Government 's focus is on the modernization of federal administrative structures that will deliver speedy, service-oriented, approachable and cheaper electronic administration by the year 2005. The Government identified 18 pilot projects over to lead its early efforts in delivering online services to citizens and businesses. They include the repayment of student loans, electronic tax declarations, and the processing of customs matters. The public procurement project, Öffentlicher Eink@uf Online is intended to combine the authorities' entire contract award process, from defining their requirements to delivering products".

The new strategic lines of Electronic Government in Germany are already visible. First of all, Electronic Government must be prevented from budget cuts that are threatening the smaller projects, most of them in communities. Second, the co-operations between cities and between German federal states must be fostered dramatically. The most likely move will be the creation of "Centers of Competence"[5]. Such centers are arising on the Federal Level, where the BSI (Federal Agency for IT Security) developed a "Handbook Electronic Government", which enables communities to develop secure administration processes, a valuable work which provides a manual for first steps and which will be equipped soon as a knowledge platform for best practices. The Mainz outlet of German Regulatory Authority presents its nation root signature organization every week to another international delegation weaving a worldwide network of partnership. In June 2002 Ministry of Interior launched SAGA[6] as important offer to the states and cities to shaping joint infrastructures [11].

Large states like North Rhine-Westphalia and Bavaria are constructing competence centers, the little (especially the eastern) ones intensify their talks about co-operation from technical infrastructure standards to "Shared Services Centers". Stuttgart Universities are creating modularized scientific entities to offer interdisciplinary approaches[7] to Electronic Government which will provide the necessary push for a Competence Center" funded by public and private partners offering Electronic Government training. Science already is acting along an European approach with partners.

"Shared Services Centers" (or as the author prefers: "Overlay Administration") could be the most important Shibboleth for Electronic Government in Germany. The discovery of co-operation is driven by the insight to concentrating the ever-too-small-budgets, to avoiding long trial-and-error, to creating new "highly interactive" and "automated" administration processes instead of passing on medieval structures to the 21st century Electronic Government. Like in other European countries the way to Electronic Government is not the easiest in Germany too. Modernizers in all sectors are confronted with their rivals who reject any change and are leaving the big tasks and burdens to the grandchilds. Local and regional politicians still prefer the "quick" internet project to sow up with "modernization", for their interests "rationalization" lies athwart popularity. Interaction between the European partners is the adequate remedy for the obstacles yet recognized, realizing a forceful push for innovative Electronic Government infrastructures.

References

1. Schwemmer, J., Elektronische Signaturen: praktische Erfahrungen, Beobachtungen, Schlussfolgerungen, in: Kubicek/ Klumpp/ Büllesbach et al. (Eds.), Innovation@ Infrastruktur. Zur Gestaltung der Informationsgesellschaft, Heidelberg 2002
2. Lenk, K./ Klumpp, D. (Eds.), Electronic Government als Schlüssel zur Modernisierung von Staat und Verwaltung, Memorandum des FA Verwaltungsinformatik der Gesellschaft für Informatik e.V. und des FB 1 der ITG im VDE, Bonn/Frankfurt September 2000
3. Accenture (Ed.), eGovernment Leadership - Realizing the Vision, April 2002
4. Annual Public Sector Information Technology Spending in Europe, 2000/2001, Kable Ltd., London 2001

---

[5] German term "Kompetenz-Zentrum" comprises "building up own skills" as well as "bundling administration tasks" and "knowledge transfer"

[6] The acronym for Standards and Architectures for eGovernment Applications reminds ironically to old Germanic very long stories, see: www.bund.de/saga

[7] see www.alcatel.de/stiftung (English version under construction)

5. Hagen, M./ Kubicek, H. et al.: One-Stop-Government in Europe - Results from 11 National Surveys, COST Action A 14 - Government and Democracy in the Information Age - Working Group "ICT in Public Administration", Bremen 2000
6. Klumpp, D., Von der Veränderung der Schnittstelle Mensch-Bürger, in: Kubicek et al. (Eds.), Multimedia@Verwaltung. Jahrbuch Telekommunikation und Gesellschaft, Heidelberg 1999,
7. Masser, K. / Gerhards, R., WEB-TEST II, Bundesländer im Vergleich, in: Innovative Verwaltung 5/97
8. Reinermann, H./ von Lucke, J. (Hrsg.): Portale in der öffentlichen Verwaltung, Forschungsbericht, Band 205, Forschungsinstitut für öffentliche Verwaltung, Speyer 2000, 1-6
9. Roßnagel, A.: Die digitale Signatur in der öffentlichen Verwaltung, in: Kubicek, H. et al.. Jahrbuch Telekommunikation und Gesellschaft 1999, Heidelberg 1999, 158
10. Klumpp, D./ Schwemmle, M., Wettlauf Informationsgesellschaft. Regierungsprogramme im internationalen Überblick, Berlin 2000
11. Bundesministerium des Innern (Ed.) SAGA - Standards und Architekturen für eGovernment Anwendungen im Rahmen der Initiative BundOnline 2005, Berlin, 04.06.2002 (draft v 0.9)
12. Alcatel SEL Stiftung (Ed.), Verwaltung und Region im Electronic Government, Tagungsdokumentation Brandenburg, Stuttgart 2002

BRAINCHILD, Building a Constituency
for Future Research in Knowledge Management
for Local Administrations*

Martin van Rossum[1], Daniele Chauvel[2], and Alasdair Mangham[3]

[1] The Hague University of Professional Education
vanrossum@thehague.nl
[2] European center for Knowledge Management
daniele.chauvel@free.fr
[3] London Borough of Camden
alasdair.mangham@camden.gov.uk

Abstract. The overall objective of this network of excellence is to train the BRAINCHILD change masters, champions and activists (called "Chief Knowledge Officers) in knowledge management for local administrations,. Applying the action learning method for a C.K.O. Graduate Course, a pilot group of C.K.O.'s will develop at the same time the strategic roadmaps for future applied research in Public Admin e-Work & Next Generation KM Systems. After validation the C.K.O. Graduate Course will be offered on a large European scale to the members of the founding BRAINCHILD networks (Telecities & Elanet) of advanced local authorities. The course will have official accreditation, from ecKM – Groupe ESC Marseille Provence and other participating academic institutes.

1 Relevance[1]

Today we are experiencing a transformation of the public administration into a demand driven public service organisation. Driving this is the citizen's need for information and services in their search for a better life as responsible and caring human beings. From this point of view, the way in which services are offered at the moment is often insufficient or difficult to access. In order to obtain the desired information /service, it is very common that several public departments and institutions must be contacted, that the information required is not always available and that consequently the process of collecting the information or requesting the service is very time-consuming.

---

* This article is a compilation of inputs from many contributors to the BRAINCHILD network of excellence, of which explicitly should be mentioned Daniele Chauvel - the Director of ecKM , the European center for Knowledge Management, at the Graduate School of Business Marseille Provence (France) and Alasdair Mangham - programme manager for the Camden Connect Team (London Borough of Camden, UK)
[1] Based on input from Alasdair Mangham

R. Traunmüller and K. Lenk (Eds.): EGOV 2002, LNCS 2456, pp. 26–32, 2002.
© Springer-Verlag Berlin Heidelberg 2002

At the forefront of applied research today is the development of demand driven information systems as a solution for the improvement of citizen services. This line of development includes a number of research challenges. Most important is the question of how to handle the dynamic aspects of demand driven information systems. The service offered as a response to specific questions is not static but operates within a dynamic growing service context. The citizens are looking for a service which will improve the quality of daily life. From the citizens perspective it is of no interest whether the information originates from the private or public sector in arrange of activities including: childcare, education, cultural activities, tourism, sports and health care. The citizen is interested in a service that will help in the specific situation, that is quick and responsive Current trends in Electronic Government are focused at delivering seamless services for citizens. This has involved Government agencies examining the service from a citizen centric point of view and then building electronic delivery platforms based around the service that the citizen requires. The most obvious manifestation of this approach is in the creation of portals where a number of services are joined together to give the citizen the impression of a holistic service.

What often lies behind the portals is a collection of Local and Central Government departments and/or private institutions / corporations who are feeding information out of their respective silos into the portal. The raised expectations of the citizen of the joined up-front end are very often not matched by the service that is delivered. Part of the solution is undoubtedly to re-engineer the back office to match the service paradigms being dictated by the new citizen focused front end. However, it will not be possible to re-engineer the back office to match every joined up service delivery scenario.

The creation of joined up service delivery models also raises expectations in the citizen of a joined up policy between the government actors. There is a common expectation that the joined up front end will be able to deal directly with up to 80% of routine transactions, with the 20% that require expert intervention being routed through to the back office.

The concentration of the tools tends to be on ways of managing the 80% of routine enquiries. Whilst these tools are important for the delivery of a seamless service to the citizens they do not enable either the experts or the policy makers to act in a joined up way. One important aspect of dealing with the more complex enquiries is the ability to join up data held in back office systems. However, data without context is meaningless. Data that can be viewed in context becomes information. Information that is analyzed and can be applied is knowledge. When this knowledge is distilled, organized, stored and redeployed according to specific user needs, then a corporation / organisation is employing Knowledge Management. For these reasons future applied research, for which BRAINCHILD network of excellence intends to design the strategic roadmaps, should target both the development of e-Work Systems as well as organisational knowledge management tools for accomplishing process improvements in public administrations. The alternative approach will be taken into account as well, that is to exploit artificial intelligence techniques to represent bureaucratic rules and procedures. Represented declaratively, bureaucratic rules and procedures, are not concealed in the black box of computer code, but made manageable by other (knowledge base) software tools.

The specific focus in that case will be to address the problem where a client must deal with the rules and procedures of several different agencies simultaneously to solve his/her problem. For political (etc) reasons, it is not always likely the diverse agencies will integrate their operations. Instead, we will alternatively also take into account a technical solution, in the form of *virtual integration* of procedures. This relies on the procedures being represented declaratively ,through utilising the new web services such as WSDL/UDDI and SOAP so other software can provide *on-the-fly re-engineering* to deliver a customised, integrated view of the procedure to the client.

To benefit from every customer or partner interaction, public administrations must give employees opportunities to record what was learned. Then, other employees must have access to the data and information and the means to understand it in context. Knowledge management solutions help an administration to gain insight and understanding from their own experiences. When employees use this KM system, best practices are perpetuated throughout the organization, and each employee has the tools to be as effective as the best employee. To the extent knowledge and information are the core assets of a public organisation, knowledge and information management become the means for ensuring success. From this perspective, the knowledge-related challenges facing the public and private sectors are similar. Public administration must develop a critical awareness of the value of its stock of knowledge, manage this resource optimally, and utilize it for more proficient service delivery (and policy making). However the role played by change masters, champions and activists is crucial for successful change in organisations. The overall objective of this network of excellence is to (develop training courses for) train the BRAINCHILD change masters, champions and activists (called "Chief Knowledge Officers) in knowledge management for local administrations,. Applying the action learning method for a C.K.O. Graduate Course, a pilot group of of to be trained C.K.O.'s will develop at the same time the strategic roadmaps for future applied research. After validation the C.K.O. Graduate Course will be offered on a large European scale to the members of the founding BRAINCHILD networks (Telecities & Elanet) of advanced local authorities. The course will have official accreditation, from e°KM – Groupe ESC Marseille Provence and other participating academic institutes.

The research should be based on results from earlier & current studies in European IST projects within the area of "smart government" In that sense BRAINCHILD will "cluster" ongoing IST project experiences. One of the requirements for the C.K.O. Graduate Course applicants will be that they play – on behalf of their local authorities – a key role in current "smart government" projects. It will be our intention to include from the outset local authorities from NAS-countries.

2   New Paradigm

During the past decade the source of economic competitiveness has shifted from the industrial paradigm to one that is knowledge-based. The value of know-how, knowledge and the intangible assets embraced by an organisation has become increasingly more important in comparison to tangible assets. The social reality within the industrial paradigm is conditioned by a high degree of formality, rule-based hierarchy, standardisation and an evasion of confrontation and criticism. The new paradigm, which is now emerging on the business landscape, calls instead for transversal and flat organisational structures, fast-innovation processes, managerial agility and interpersonal effectiveness.

3   Knowledge Management[2]

Knowledge Management (KM) is an emerging phenomenon variously at the centre of global economic transformation, organisational success, the eventual demise of private enterprise capitalism, new forms of work and the forthcoming paradigm shift from InfoWar to Knowledge Warfare (K Warfare). Competitive advantage is located in "learning organisations", "brain-based organisations", "intellectual capital" and the "economics of ideas." Knowledge has assumed this centrality in conjunction with sweeping changes in organisational forms and the dawning post-industrial and information revolutions. Definitions of and approaches to KM show that the meanings associated with "Knowledge Management" are multiple rather than singular, and the field has a set of intellectual roots which are neither incongruous, not consistent, but certainly different in their understanding of the matter. Nonetheless, actors and observers now agree that Knowledge Management is an important domain of study and application that has the ability to deliver significant business benefits.

The domain of Knowledge Management is becoming more sophisticated and extending its reach, reflecting the impact of emerging technologies as well as the influx of specialty domains (psychology, sociology, the organisational sciences) and new fields of endeavour (Public Administration, not-for-profit organisations). KM is no longer a systems issue, but rather a holistic initiative that understands organisations as socio-technical phenomena whatever their core activity.

As new-paradigm organisations become more knowledge-based, work becomes more knowledge-intensive. This truism applies to private and public sector organisations. It quickly becomes critical to understand the way people think, learn, use knowledge, and how these knowledge processes are embedded in the systems, structures and routines of an organisation. The knowledge assets in play include those that are explicit, easily accessed and codified, but especially those that remain implicit and locked in a psychosocial context. Knowledge Management is focused on the systematic development of all such assets.

Experience shows that successful organisational change requires (1) thorough diagnosis and scenario analysis, (2) clear change objectives, (3) organisational commitment that stems from the unfailing support of top management, (4) change masters or champions who provide leadership, and (5) careful, incremental reinforcement of any progress achieved. These principles apply to all types of organisations, whatever their sector of activity or change objective. A decade of experience with Knowledge Management change efforts similarly confirms these guidelines for action.

4   Public Administration

Public administration has pressures and constraints that are similar to those of any business enterprise. The private sector's primary objective is wealth creation, and the primary means for doing so increasingly centre on knowledge and knowledge processes within an organisation. The primary objective of public administration is social stability and effective public service, and historically the primary means for doing so has centred on legislation, procedural and information management.

---

[2] This and following paragraphs based on input from Daniele Chauvel

Public administration is now being called upon to act in an increasingly prosperous and knowledge-rich society while addressing the problems of the "digital divide." A radical transformation in the way public administration manages its knowledge is on the horizon, both within its organisations and in its external relationships with citizens. The knowledge-related objectives for today's public administrations include:

- Effective public services that address the public's issues and requirements relevantly, competently and timely, while consuming minimal resources.
- The preparation of citizens, organisations and public agencies to be effective policy partners: create sound public opinions; develop constructive public debates and policy formation; conceptualize, plan, decide and implement public actions
- Assure an acceptable quality of life through building, maintaining and leveraging commercial and public intellectual capital.
- Ensure a prosperous society by developing citizens to become competent knowledge workers, and institutions that are competitive.

5   C.K.O.s in Public Administration

To the extent knowledge and information are the core assets of a public organisation, knowledge and information management become the means for ensuring success. From this perspective, the knowledge-related challenges facing the public and private sectors are similar. Public administration must develop a critical awareness of the value of its (ever growing) stock of knowledge, manage this resource optimally, and utilize it for more proficient service delivery.

Change masters, champions and activists.

Peter Drucker has written that, "...unless an organisation sees that its task is to lead change, that organisation--whether a business, a university, or a hospital--will not survive.

Since the 1960's the discipline of Organisation Development has taught that organisational change requires steadfast leadership and management that is best materialised by dedicating a person or a team to the effort. In the late 1990's Nonaka, Krogh & Ichijo introduced the idea of a Knowledge Activist: "...someone, some group that takes on particular responsibility for energizing and coordinating knowledge creation efforts throughout the corporation."

The Knowledge Activist described by Nonaka et. al. fills the task description developed by Organisation Development some 50 years ago: a person who leads the knowledge-related change effort in an organisation. Knowledge Activists facilitate the creation and dissemination of knowledge, ensure interconnections between people and leverage innovation. This individual is a "merchant of foresight", developing a "bird's eye perspective" on the change required for knowledge development in a company.

6   The Need

The aim of BRAINCHILD Network of Excellence is to develop a program that addresses the theoretical foundations of knowledge management in public administrations, together with organisational change and development specific to public organisations, in parallel with the knowledge management system tools. This will lead to the development of best practices and technologies that allow public institutions to identify, share, disseminate and create or reuse knowledge & knowledge based systems.

This will form the pedagogical foundation on which a training program for public administration C.K.O.'s is developed. This educational program will benefit from the decade of Knowledge Management experience already acquired in the private sector, but will be specifically tailored to the particularities of the public sector.

BRAINCHILD will focus on the following priority thematic area:

Strategic Roadmaps for Applied Research Addressing Major Societal and Economic Challenges (1.1.2.i)

Objectives: To prepare the ground for RTD activities under FP6 by investigating future research challenges, roadmaps and associated implementation models in the domain of e-government/e-governance and e-work. A constituency of RTD stakeholders will be created by the joint networks of Telecities & Elanet.

Challenges will be identified in the following areas:

- Future e-work systems, "e-business" in public administrations, to facilitate seamless joint-up service delivery to the citizens;
- Organisational knowledge management including context- and location-sensitive solutions for acquisition, sharing, trading, and delivery of knowledge (including next generation knowledge management systems and artificial intelligence techniques [3]) to support public sector employees in their roles as seamless joint-up service delivery agents (**1.1.2.iv. Knowledge and interface technologies**);
- Models and scenarios to shape future policies for a knowledge-based economy in conjunction with the MUTEIS project (socio-economic research on the transition to a knowledge-based economy)

Focus: Activities of the C.K.O. Graduate Course participants will include studying the following aspects:

- Building and strengthening RTD communities that bring together research, business and user organisations with the aim of developing shared visions, scenarios and objectives and facilitating the integration of European research resources to address major future business and work challenges (clusters of IST research projects, national or industrial initiatives in which the C.K.O. Graduate Course participants play a key role themselves);
- Identifying research tasks for both objective-driven and exploratory research. Work should also help to identify and explore the set of complementary activities required to improve RTD impact. These include links to other research frameworks, innovation and take-up actions, training and mobility, standardisation, dissemination activities and the integration of international efforts.

---

[3] Actions to design & develop next generation knowledge management systems:
- Knowledge management of streaming and archival media resources for the open sharing of content"
- Digital media management, indexing / analysis and search / retrieval in the broadcast domain"
- km3 - Knowledge Monitoring, Maintaining, Mastering"
- Flexible Controls and Learning In Bureaucratic Systems"

- Identifying the key actors in the field, stimulating interest and achieving broad-based consensus on the way forward to meet the research challenges.

Addressing eEurope and eEurope+ Objectives

Objective: To support the broad adoption of IST solutions for "e-commerce" and e-work in public administrations, including the security of on-line transactions, greater flexibility in work organisation and better access to e-work facilities for local and virtual communities, and for SMEs, thus contributing to the realisation of eEurope and eEurope+ objectives.

Focus:

- Best practice actions for trust, security, e-work, organisational knowledge management and process improvement in public administrations (including next generation knowledge management systems);
- Encouraging the participation of Accession States (eEurope+), e-inclusion, implecations of European integration and enlargement for governance.
- Knowledge-based Society and social cohesion (**1.7.1.**).
- Citizenship, democracy and new forms of governance (**1.7.2**).

Clustering of Projects

Objectives: To facilitate synergy between existing projects that see an added value in working together on common objectives. The PACE, EDEN, PRELUDE, KEeLAN, EUSlanD and MUTEIS projects - in which Telecities & Elanet members play a key role - in particular will offer opportunities to share efforts with BRAINCHILD when designing strategic roadmaps for future applied research.

Organizing for Online Service Delivery: The Effects of Network Technology on the Organization of Transactional Service Delivery in Dutch Local Government

Marcel Hoogwout

Department of Public Administration and Public Policy (BSK) University Twente,
PO box 217, 7500 AE Enschede, The Netherlands
m.hoogwout@bsk.utwente.nl

1   Introduction

Dutch central government has, like many other governments, set high aims to offer government services on line. In 2002 about 25% of all services should be online. Because over 60% of all government services are provided by local governments, the challenge is to help these relatively less powerful local government organizations to realise this ambition. Local governments have to operate in an environment where investing in e-government is not evident. To overcome the problems they encounter in improving (online) government services new forms for organizing the service delivery emerge. This paper explores the problems local governments encounter in improving their transactional (online) service delivery and investigates the organizational solutions that arise to overcome these problems. The central question will be to what extent the new organizational forms contribute to the central government aims to realise the high e-government ambitions.

2   The Service Delivery Paradox

In spite of the relatively overwhelming political attention for client oriented government, Dutch local government is relatively slow in implementing electronic service delivery. There seems to be a structural contradiction, between the moral point of view in which it is generally accepted that citizens should be treated as clients of governments, and the practical point of view that makes local governments hesitant to invest in client convenience. Local politicians hesitate because there is little incentive for municipalities to improve their service delivery. Implementing improvements not only requires considerable financial resources, but also radical organizational changes and process redesign [1]. Legal problems have to be solved. Employee attitude has to change in order to promote citizen-oriented service delivery. Furthermore service integration requires that various organizations co-operate and share authority. This is difficult to achieve without a strong external incentive [2]. For most services, government not only has a monopoly, but also from an economic perspective, investing in service delivery improvement on a municipal level makes little sense. Costs will rise while turnover and income will not rise accordingly. The number of transactions for

R. Traunmüller and K. Lenk (Eds.): EGOV 2002, LNCS 2456, pp. 33–36, 2002.

most services is so low that most Dutch local governments simply are too small to warrant the investment. Pressing for improvements in service delivery is also not very rewarding for politicians. The relation between election behaviour and the experience of service delivery is, if it can be proven, very weak. On top of that Dutch citizens are relatively content with the quality of service delivery. Citizens award their local government a 7,2 on a 1-to-10-scale for service delivery. About 80% of all respondents affirm that they are more or less content about the transactional service delivery [3]. Besides these practical arguments also more fundamental dilemmas prevent local governments to embrace the citizens as client consumers. Transactional service delivery touches dilemmas like for example:

- Defending collective interest versus defending the interests of the individual: Sometimes government has to disappoint individuals on behalf of the society as a group.
- Maintaining rules and controlling citizens versus helping them.
- Privacy protection versus transparency through connecting governmental databases for optimal proactive service delivery: Big Brother versus Soft Sister [4];
- Neutral civil servants versus servants committed to the citizen's interest. The Weberian bureaucracy model which puts the chosen politicians forward as the sole clients of the civil servants collides with the concerned street level bureaucrat that cares for its clients and acts that way.
- Raising and educating citizens instead of spoiling them by treating them like a king client.
- 'Consumer democracy' versus 'representative democracy' [1]. Asking citizens for their needs through monitoring instruments and analysing consumer patterns compete with policy articulation based on chosen representatives.
- Hierarchy versus market allocation: by choosing for government as the sole entity to deliver a certain service and not leaving it to the market, one also chooses to accept the negative side effects of the hierarchy system of allocation.

In the light of the arguments and dilemmas shown above, it is understandable that most local politicians don't prioritise service delivery improvement. Given the efforts of central government of improving transactional service delivery this hesitation in local government is felt to be highly unsatisfying.

3   The Changing Organizational Landscape

Central and local governments are more and more aware of this service delivery paradox. To break through the hesitating attitude they are experimenting with new organization models for arranging transactional service delivery, models that avoid the paradox and benefit from the economies of scale and market incentives. An enabling force in this respect is the rapid development of network technology in recent years. Five dominant trends are visible in the changing landscape of the public sector in the Netherlands:

1. Back offices (production) and front offices (client contacts) are increasingly separated in time, space and organizational responsibility.
2. Back offices responsible for the production of the original legal task look for opportunities to co-operate and merge with other back offices to benefit from economies of scale.

3. Front offices are placed in competition. The incentives of the market economy are used to improve service delivery to clients, while avoiding the service delivery paradox.
4. Front office tasks are not only placed in a competitive environment with other governmental front offices but are also distributed over non-governmental organizations that are used to compete in the market.
5. Transactional service delivery tends to be modularised and standardised. The various transaction modules enable (physical and internet) front offices both in the public and private sector to combine a set of modules around the demand patterns (life events) of the client groups they want to serve. This also means that these front offices can shop among different governmental and non-governmental back offices to choose the optimal set of governmental and non-governmental transaction services for their own means. The modularisation eventually can lead to direct one-to-one transactions between the back office organizations and the citizens as government clients.

The five trends have in common that they all lead to an increasing variety of organization models and a fading of the organizational borders. Traditional government organizations exist next to more hybrid forms, while also powers and responsibilities are shared with competing private organizations. The Dutch central government recognizes the trends and the hazards that come with it. To a certain extent they embrace the change and are willing to experiment. The fading of the borders of governmental organizations creates however new challenges for the legislator to guarantee the legitimacy of governmental service transactions. The Dutch government has, for this reason, announced the constitution of a new law that will be designed to facilitate the great variety of new organizational arrangements (Wet Bestuurlijk Maatwerk).

4   Cases of New Organizational Forms in The Netherlands

Three examples of newly developed service delivery concepts in local governments illustrate the existence of the five trends mentioned above.

- **Rental Subsidy by the Ministry of Domestic Housing (VROM):** The department of collective housing is responsible for the allocation of rental subsidy to over one million households in The Netherlands. The development of a completely new modularised and automated transaction process has led to an experiment in which the intake is done not only by local governments alone, but also by real estate brokers and housing corporations as a natural complement to their services around rental housing. The administrative processing of the applications is done by the ministry of VROM. The local governments are put in a position that they have to compete with non-governmental organizations while the centralisation and modularisation of the transaction service enables to profit from economies of scale. (trends 1, 3, 4 and 5).
- **BV Woonnetwerk-Noord**: Woonnetwerk-Noord is an initiative of the City of Groningen with several housing corporations, real estate brokers, project developers, the utility company and the regional and central government. They intent to offer services both through the internet and through a walk-in singe window shop in the center of town (trends 1, 4 and 5).

- **WoningNet Utrecht**: This co-operation between the city of Utrecht and the 12 housing corporations in the Utrecht region has taken over the municipal tasks connected to the allocation and distribution of social housing in the City of Utrecht. Next to that it does the (on line) matching between supply and demand of social housing. The organization also provides the intake for financial rent support, all from the perspective of a single window for the citizen that is in need for social housing. The organization is also planning to facilitate both physical and digital front offices of local governments and housing corporations with their modularised intake and matching services (trends 1, 2, 3, 4 and 5).

5   Conclusion

The examples illustrate that the five organizational trends are increasingly redesigning the organizational landscape of service delivery in The Netherlands. They all rely heavily on the use of network technology to operate their service delivery concepts. For the citizens, as clients of governmental transaction services, dealing with government appears to improve. They enjoy the integration of associated services in a single window while the number of outlets and the individual choice increase. All examples show that mainly through the bundling of resources the necessary investments can be warranted. Especially the rental subsidy and the WoningNet example contribute in their new organizational form to the central government aims to realize the high e-government ambitions. Although the organizational models that were discussed above indeed seem to help to overcome the existing service delivery paradox, time will tell if these new models are viable enough to survive also in the long run.

References

1. Bellamy, C., and Taylor, J.A. (1998). *Governing in the information age*. Buckingham ; Bristol, PA: Open University Press
2. Kraayenbrink, J. Back to the Future: Centralization on its Revival? Problems in the Current Organization of Public Integrated Service Delivery. Paper presented at *DEXA 2002, submitted*.
3. Hoogwout, M. (2001). Leuker kunnen we het niet maken, maar willen we het wel makkelijker? Waarom overheden geen haast hebben met het verbeteren van de dienstverlening. In H.P.M. Van Duivenboden, and M. Lips (Eds.), *Klantgericht werken in de publieke sector: Inrichting van de elektronische overheid*. 149-166. Utrecht: Uitgeverij LEMMA BV.
4. Frissen, P., Koers, A.W., and Snellen, I.T.M. (1992). *Orwell of Athene? democratie en informatiesamenleving*. 's-Gravenhage: Sdu Juridische & Fiscale Uitgeverij.

Public Sector Process Rebuilding
Using Information Systems

Kim Viborg Andersen

Department of Informatics; Copenhagen Business School,
Howitzvej 60, DK – 2000 Frederiksberg
andersen@cbs.dk

Abstract. Ongoing modernization of the public sector, gray-zone/ semi-public organizations, and virtual/ teleworking/ Internet use are among the organizational features that need consideration for reorganizing the work processes using information systems. Although politics is not to be ignored, organizational and institutional changes alter the face of the public sector and pave the road for what we call Public Sector Process Rebuilding (PPR).

1   Introduction

With annual governmental expenditures on information systems (IS) sprouting, it is evident that the stakeholders seek a return from their efforts in implementing new IS in their organizations. For the government the motives can also include issues such as: did the R&D expenditures lead to higher economic growth, did the computerization of the public offices lead to savings in manpower, better service, and more services? Did interaction with citizens and companies improve? However, the impacts of using IS often do not live up to the expectations which is a continual source of vexation for the government.

Accordingly, one direction is to *include still more variables* to find valid explanations for *why IS does not always match the intentions, expectations, and needs*. The interest in factors leading to a successful implementation has stemmed from technical issues, to include concerns on environmental factors, organizational factors, and in turn political issues. Another route is to *refine the measurements of what constitute successful implementation*. Whereas, the managerial perspective and the workers' perspective have earlier dominated the literature on IS, recent studies have brought attention to the *internal and external work processes at both macro and micro-level* in the organizations. The process orientation of the studies of IS and organizational transformation is still a more popular route to follow.

There are indeed *various obstacles for reorganizing government*, such as the extreme *openness of the public sector organizations*. Also, public *organizations rarely change in any rapid nor top-down manner*. Rather they change in an incremental manner and only partly top-down. While this is true in a large part of the public sector, the increased use of teleworking, virtual organizations, and quangos/ semi-governmental organizations, only make us more optimistic on the possibility of reorganizing the "business processes" in the public sector.

R. Traunmüller and K. Lenk (Eds.): EGOV 2002, LNCS 2456, pp. 37–44, 2002.

2   Reorganizing the Processes

The very idea of reorganizing processes originates from the basic question: are we doing our business in the most optimal way? Are we doing our job well enough? Are we giving all we've got? (Osborne & Gaebler, 1993). This last question is just as crucial for the public sector as any other large organization, though it may be disputable whether this is contemplated enough by its members. For example, OECD (1995) emphasized the following significant areas in facilitating the exacting of maximum organizational benefits from IS:

- enhancing management, planning and control of the IS functions
- using technology to redesign and improve administrative processes
- providing better access to quality information
- harnessing the potential of new technologies
- developing and applying standards
- attracting and retaining high-caliber IS professionals
- increasing research into the economic, social, legal and political implications of new IS opportunities; and
- assessing experiences

Within the private sector, concepts of reorganization, redesigning and reengineering the processes have gained enormous popularity and – some argue –valuable impacts on the actual practices. Besides the initial articles and books (Davenport & Short, 1990; Hammer, 1990; Hammer & Champy, 1993), numerous books were published showing how IS was affecting dramatic and radical changes in an organization (e.g., Caudle, 1995; Champy, 1995; Davenport, 1993; Davenport & Stoddard, 1994). By contrast, the public administration community did not applause the reengineering concepts as breaking new ground. Rather, they jeered that, BPR was at best old wine in new bottles. At worst, BPR applied in the public sector would lead to misjudgment and actions inconsistent with the 'spirit' of the public sector.

The BPR approach emphasizes that changes in processes are to be drastic rather than incremental. Also, the approach points to broad, cross-functional processes and, if needed, a radical change towards such processes. This of course points to high risk for failure as well. Is risk part of the rationale of the public sector? Traditionally speaking, it would be a quick, resounding NO, but in 1997 we wouldn't be so fast to say taking risks is uncharacteristic of the public sector. First of all, one part of BPR tradition true enough emphasized short term efficiency and longer-term strategic advancement. However, as noted by Coombs and Hull (1996), during the 1990s a "Soft BPR" emerged emphasizing human costs and benefits rather than just organizational shape. Second, the face of the public sector is changing radically not only with rightsizing and downsizing, but also in organization and leadership. It is old news that public services do not inevitably have to be performed by the peers employed in the public sector. By all means, contract them out. Form quangos. Farm out, but keep a short leash on the administrative power. And don't forget to take steering and delegating seriously. These are just some of the central shifts in the public sector management during the 1980s and mid-1990s that makes it uncertain whether the public sector is overall risk wary.

Within the public sector, scholars affiliated with the *US National Academy of Public Administration* further define reengineering as: "...a radical improvement approach that critically examines, rethinks and redesigns mission product and service processes within a political environment. It achieves dramatic mission performance and gains from multiple customer and stakeholder perspectives. It is a key part of a process management approach for optimal performance that continually evaluates, adjusts, or removes processes."

The BPR concept argues likewise, that "researcher...and managers.. must begin to think of process change as a mediating factor between the IT initiative and economic return" (ibid.cit., p. 46). Thus, information technology is not seen as the sole factor that can lead to a miracle outcome. Instead, process reengineering is. However, IT is given the role as the almighty enabler connecting individuals, work groups and departments. Davenport formulated it as "..to suggest that process designs be developed independently of IS or other enablers is to ignore valuable tools for shaping processes" (p. 50). In Davenport's work, the impacts from IT on process innovation is grouped in nine categories: *automational* (eliminating human labor from a process), *informational* (capturing process information for purposes of understanding), *sequential* (changing process sequence), *tracking* (closely monitoring process status and objects), *analytical* (improving analysis of information and decision-making), *geographical* (coordinating processes across distances), *integrative* (coordination between tasks and processes), *intellectual* (capturing and distributing intellectual assets), and *disintermediating* (eliminating intermediaries from a process). Although these impacts differ slightly from the findings our research has identified for the public sector, these areas form a solid basis for addressing the use of IS in the service development, fulfillment and logistical functions of the public sector.

Using IS to increase the ongoing innovation of the work processes can for example include the use of computers to capture the work processes in case-handling within the social welfare administration and use these data to carefully examine the work processes. Within this area is also IS that enables one to track case status. It could mean that when Peggy Sue submits an application for housing and/ or subsidy to a public housing authority it will trigger actions by several public agencies. Also, to generate integrative organizations and have dialogue with agencies and various administrative levels of government spread over *different geographical locations*, IS is a powerful tool. In this area of IS applications, we have also placed forecasting and models. The use of the Internet or Intranet, can help co-working in designing new procedures, without long flights and wasteful commuting time. Finally, knowledge workers in the public sector often need to access same kind of data for designing new work procedures. IS can be used to rationalize the use of the *intellectual assets* in the organization by providing easy access to frequently used data. For example, easy access to the budget for the organization, the current account, documents describing the work processes, etc. can be granted.

IS can also be used to *reorganize the service fulfillment*. Meeting the citizens' needs is essential for the public sector, yet, quite often the employee has to choose between several solutions to match the needs specified by the citizen. E- procurement and the *one-stop services* for the citizens are examples of how IS reorganizes service fulfillment. Likewise, *voice communication* is essential to being responsive to the voice of partners. Computer-based voice response mail and answering machines are still rather uncommon in public offices. IS application here, that is virtual components and IS interorganizational communication tools, can eliminate the lack of transparency of each individual's work. Unfortunately, e-mail is still not an integrated part of the public sector, although studies show benefits in using e-mail in enhancing service quality and effectiveness.

The use of *logistical systems* is more than desirable in all parts of the public sector. For example, scheduling nurses' work week is a highly complicating matter, optimizing the financial issues, while considering labor market regulations etc. Logistical planning systems help keep a record of the laws, stock of workers etc. Also, IS allows remote monitoring of items/ processes. For instance, cars passing through a toll booth in Singapore's highways do not need to toss money in a receptor or in the hands of an attendant. Instead, smart cards with bar codes are read rapidly using telemetry. In addition, Singapore's central government is in charge of exchanging data on various import/ export issues, such as international trade bodies, traders, intermediaries, financial institutions, and port and airport authorities. Exchanging data involves for example collecting, manipulating and transmitting. By implementing new IS and reorganizing the work processes in the Trade Development Board, they were able to handle more cases, reduce the staff, and increase efficiency as well (Teo, Tan, & Wei, 1997).

3   Indicators of Change in Institutional Setting for IS Use

During the 1980s and 1990s governments have been committed to de-bureaucratization and have viewed privatization as a means to achieve this goal. The governments in most first world countries have behaved in line with this strategy. In fact, in recent years the concept of privatization has been used in different ways. Privatization can mean the sale of assets, contracting out, introduction of user charges, voucher schemes, deregulation and government withdrawal from public obligations.

The main aim of the modernization programs has been to improve public services for citizens and companies, to use resources more efficiently, and to halt the expansion of the public sector. Also, we have seen numerously budget reforms, the introduction of management techniques from the private sector, de-bureaucratization, decentralization, experiments with more independent local governments, and a more flexible job structure in the public sector.

For example, Andersen, Greve and Torfing (1996) found that *budget reforms* during the 1980s simplified many of the financial procedures within public institutions. Instead of practicing some highly complex procedures of budgeting and accounting, many public institutions are now free to act within a given financial limit.

Management techniques from the private sector are now more widely applied. Corporate management techniques are adopted, the skills of the chief executives upgraded, and focus on actual output receives more emphasis. *De-bureaucratization* has legitimized the task to minimize the number of public rules and regulations. This has been done by merging different public institutions, privatizing public tasks, and simplifying the remaining rules and regulations. However, the results have been viewed with some skepticism.

Decentralization has been a major component in the strategy to reorganize the public sector. Decision-making responsibility has been shifted downwards from the central government agencies to local government and public institutions. Within local governments the politicians have more power due to the extended use of framework laws, but this power is shared with street level bureaucrats and professionals, and they allow that users exerted more direct influence on the day-to-day operation of public institutions. Within public institutions overall, executive managers have been given greater freedom to formulate strategies for future development. A more *flexible job structure* in the public sector has been introduced. In fact, they allowed that salaries varied according to individual skills and performance.

These changes were all part of a responsive state strategy that aimed at increasing the efficiency of the public sector. A marketization strategy chaperoned this responsive state strategy which promised to intensify the use of market forces and competition within the public sector, hereby demonstrated by the current telecommunication industry, bus-services, airports, national train-companies. In most countries, we have noticed an increase in quarrels over who shall produce, organize, and finance the public services. The overall trend seems to be: democratic principles of the public sector will be left to the politicians. Outsourcing of producing and financing will stimulate the application of BPR in the public sector. The public sector's view on service information has accordingly been revised substantially. Thus, a recent OECD study on public administration arrived at the following objectives for information service made by the public sector (Arnberg, 1996):

- increase democratic legitimacy
- enable clients to claim entitlements
- improve clients' opportunities to influence service content and participate in the provision of services
- manage the expectations of clients about service levels and service quality according to the resources available
- facilitate and create conditions for client choice
- enforce performance on the providers
- restore the confidence of clients in the public sector and its agencies

Thus, IS is at the meso-political level seen as a tool that might help "to reinvent" democracy through various areas involving public contact and citizen access to public held data, one-stop shopping, interactive electronic service, point-of-contact data entry, expert systems, information storage, and revenue generation. Although we have seen this shift at the political agenda towards increased commitment in using IS in the public sector, we still have to see empirical studies that can demonstrate that such impacts are achieved.

3.1   Quangos and Outsourcing: Reorganizing the Incentives and Management

The use of new organizational forms for the public sector has taken its most radical form with the widespread use of quangos. In 1992, the quangos in the UK used about 30% of the total public budget, involving about 5,500 organizations. Similar importance is seen in the more socialist countries, such as the Netherlands and Denmark.

Also, outsourcing of computing activities has escalated during the past 10 years. In a study of computing in Japanese local government, Sekiguchi and Andersen (1999) found that the number of "outsourced" personnel has increased by 126% during the period 1985-1995. During the same period, the total number of computing personnel grew by about 88%. The expenses on computing for local government skyrocketed by about 175%. The high percentage of outsourced employees shows that the computer capability of the existing in-house staff lags far behind the increase in the need of computing skills.

3.2   Virtual Organizations, Teleworking, Internet: Reorganizing the Face of the Public Sector

Virtual organizations is a very popular term and has been connoted with some of the elements in rebuilding the organizations. Although there are varying definitions on what virtual organizations are, networking (e-mail, voice-mail, facsimile, instant messaging, M2M, etc.), restructuring of the organization (outsourcing jobs/ functions, downsizing, transformation), and team culture (geographically and functional), are three elements one finds in most examples (Dutton, 1997). Using such virtual construct is facing major challenges since:

- most users have so far rejected major innovations in telecommuting (or tele-access)
- managers are often unwilling to relinquish control and supervision
- trust and commitment can be undermined if IS is used as a substitute for person-to-person communication
- privacy of users and consumers are potentially threatened

Within the public sector, many teachers, researchers etc. work from home often in an arrangement that comes close to what can be termed virtual organizations. Although this group is important and numbers to a substantial group, few IS applications have aimed at stimulating this process and also to reorganize the work processes. Even less has been accomplished when we look at the much larger fraction of the public sector that is employed as case-workers (social security, housing, etc.).

Similarly, the use of the Internet to allow remote work and contact with the public sector's partners and citizens/ politicians, is an element that is rebuilding the public organization. So far, the knowledge of the impacts of this development is very limited. However, our studies on the organizational changes associated with the use of Internet is that they are driven not from top-down nor in a dramatically patterns. Our studies also show, however, that most Internet applications in government rarely allow two-way communication nor direct access to public employees.

4   Conclusions: Re-building the Work Processes

Above we have pointed to an overall modernization of the public sector, including quangos and a (semi-) virtual construct. We see this development as indicator of a public organization moving towards the technology based public organization. This does not mean that politics no longer is an issue. Nor does it imply that technocratic factors are no longer influencing the capability or opportunities more or less than before. Neither that information systems or new organizational forms is eroding power and politics as being an element of the public sector.

Thus, the overall message with our concept on PPR is to glean for some useful parts of BPR and process innovation ideas while giving ear to criticisms of the concept's application in the public sector. Though we in principle believe that governments should be as small as possible and contract out tasks as much as possible, the core of the public sector above all needs to be in optimal working order. The public organization of today and tomorrow is not easier to manage than older forms of organizations. On top of that, if all existing work procedures are merely transferred to remote workers without reorganizing the work processes, little would be accomplished and counter productive results may lurk.

Areas beyond the central government, but also local government, semi-governmental and other areas of governmental areas, can benefit from our concept. We must remember that political processes include a wide range of activities, such as limiting student enrollment at universities, hiring personnel, or setting the level of welfare service. These are essential parts of (implementing) general political decisions, and extremely important for the content of the public policy. The one million dollar question is: can IS be applied here along with reorganizing the processes? In other words, can we help graduates from vocational training schools get a job faster by using information technology, and yet not expand the number of state tasks? We believe the answer is yes.

In the public administration we face institutional powers with respect to checks and balances, power distribution and professional training. Whereas the low-risk automation has been ongoing in the public administration during the 1980s and 1990s, we believe that more high-risk re-engineering and paradigm-shifts will appear. Researchers at the US National Academy of Public Administration formulated six starting points for reengineering in the public administration. We have adopted their insights and adjusted their list of factors critical for a successful reengineering in the public sector.

Equally important to setting specific goals, is the rebuilding of the structures to support these goals along with implementing the new IS. This requires that we know the work processes. Although this is the case in a large part of the public sector, the flow of information, the share of information, and the manipulation of the information are just some of the items where our knowledge is in fact quite limited. However, without such knowledge prior to rebuilding the structures, the outcome will depend more on luck than professional responsibility, commitment, and involvement.

Also, the keywords "measurement" and "expectations" should be considered carefully. Within the public sector, it is difficult, but not impossible, to measure the processes (including the input and outcome of them). Likewise, the expectations from the "stakeholders" must be identified and tied to the performance management. This is naturally complicated by the change of a political cabinet after a national election and by the often rigid systems for the customers' to impose their influence on the content of the public service. Nevertheless, our message here is that rebuilding public organization is not successful if the only thing accomplished is increased satisfaction for the employees, or information systems that has a better user interface. The clue is that the expectations have to be known and that the important ones are not the expectations of the employees, whether they are shot-term or long-term.

References

1. Andersen, K. V., Greve, C., & Torfing, J. Reorganizing the Danish Welfare State 1982-93: A decade of Conservative rule. Scandinavian Studies, 68(2), 161-187. (1996)
2. Arnberg, M.: Informing clients: statements of service information and service standards. In OECD, Responsive government: service quality initiatives (pp. 245-64). Paris: OECD, PUMA. (1996)

3. Caudle, S. L.: Reengineering for results. Keys to success from government experience. Washington, DC: National Academy of Public Administration. (1995)
4. Champy, James: Reengineering management. The mandate for new leadership. New York: Harper Collins. (1995)
5. Coombs, Rod, & Hull, Richard: The politics of IT strategy and development in organizations. In William H. Dutton (Ed.), Information and Communication Technologies: Visions and Realities. Oxford: Oxford University Press. (1996)
6. Davenport, Thomas, & Short, J. E.: The new industrial engineering: Information technology and business process redesign. Sloan Management Review, 31, 11-27. (1990)
7. Davenport, Thomas: Process innovation. Reengineering work through information technology. Harvard Business School Press. (1993)
8. Donk, Snellen, & Tops: Orwell in Athens. A perspective on informatization and democracy. Amsterdam: IOS Press. (1995)
9. Dutton, W. H.: Virtual organizations. Unpublished manuscript. Seminar at CRITO. (1997)
10. Hammer, Michael: Reengineering work. Don't automate, obliterate. Harvard Business Review, 90, 104-12. (1990)
11. Hammer, Michael, & Champy, James: Re-engineering the corporation. A manifesto for business revolution. New York: Harper Business. (1993)
12. National Academy of Public Administration: http://www.clearlake.ibm. com/Alliance/regodata.html#concepts. (1996)
13. OECD: Governance in transition. Public management reforms in OECD countries. Paris: Author. (1995)
14. Osborne, David, & Gaebler, Ted: Reinventing government: How the entrepreneurial spirit is transforming the public sector. (1993)
15. Sekiguchi, Y., & Andersen, K.V. Information systems in Japanese government. Information Infrastructure and Policy, 6, 109-26. (1999)
16. Teo, Hock-Hai, Tan, Bernard C.Y., & Wei, Kwok-Kee: Organizational Transformation Using Electronic Data Interchange: The Case of TradeNet in Singapore. Journal of Management Information Systems, 13 (4), 139 - 166. (1997)
17. Thaens, Bekkers, & Van Duivenboden: BPR in the public administration. Conference for the European Group of Public Administration (EGPA). Permanent Study Group on ICT. Budapest, Hungary. (1995).

What Is Needed to Allow e-Citizenship?

Reinhard Posch

Secure Information Technology Center Austria
Inffeldgasse 16a, A-8010 Graz
Reinhard.Posch@a-sit.at

Abstract. e-citizenship is used as a term for participation of citizens in e-technologies. This includes all levels of e-government as well as e-commerce. However, e-government and e-commerce exhibit quite different characteristics. Whereas e-commerce has the clear target to intensify usage and turnover, e-government seeks its goals in availability and comfort not in augmented frequency. This paper discusses some aspects resulting from these facts using the Austrian approach as a model.

1   The Austrian View to Approach e-Citizenship

This report is presenting the Austrian view as to what is facilitating the use and the participation in e-government. E-government must be properly designed to allow for e-citizenship. Homogeneous approaches including all levels of administration are the key to acceptance. Such e-government consists of many elements that need to fit nicely together.

On one side such fitting together will be the basis for an efficient implementation, on the other hand efficiency is one of the effects. The targets resulting in this effect have to comprise increase of comfort and convenience of use as these two are the main factors yielding frequency. Therefore we have to agree on underlying principles that govern the implementation and use of information and communication technologies in the administration.

1. Citizens shall be encouraged to use e-technologies at their will.
2. Efficiency shall be looked at from a global rather than a local viewpoint.
3. We have to constitute the right for electronic proceedings for citizens.
4. Security and data protection have to be an overall principle.
5. Administration must be as transparent as possible.
6. We have to care for interoperability at all levels.
7. Open standards and freely available interface specifications have to be deployed wherever possible.
8. Appropriate change management has to enable further development at all times.
9. Competition, analogue to the private sector, shall enable long term efficiency.

E-government needs some basic elements to result in benefits adequate to the effort needed when implementing:

1. The structure of the e-government approach is a critical element. As the main effort is enabling change management, importance of such structure is often considered too late.

R. Traunmüller and K. Lenk (Eds.): EGOV 2002, LNCS 2456, pp. 45–51, 2002.
© Springer-Verlag Berlin Heidelberg 2002

2. The importance of open specifications of interfaces exceeds the importance of open source.
3. A further critical element is signature tokens. Speedy deployment of such tokens (the Bürgerkarte) is needed. As these tokens are placed in the hands of citizens, this is the element which should be introduced first. Thus, it can also act as a tool to raise awareness [1].
4. Applications have to form a structured back office that uses the specified interfaces.
5. Physical and virtual front offices that comply with the standard interfaces specified complement a modular e-government.
6. The whole system gains a European dimension both through electronic signatures [2] and through standards for forms that use XML/XSL [3] [4] and thus exhibit a potential of being language independent.

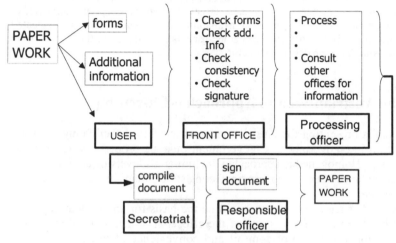

Fig. 1. The way paper works

Today many pieces of administration processes have electronic components. In general, these do not fit together automatically. A complicated chain of semi automated processing results from this situation. We have various roles and transitions that make the work difficult even if a set of tasks can be fully automated in many cases.

2   Structured e-Government

Many of the tasks shown in figure 1 are operations to transfer and transform information. In many cases this is the major part of the work to be done and exhibits a high potential of automation.

As a consequence e-government can be restructured (see figure 2). By introducing electronic signatures and thus avoiding all possible unwanted manipulation, there is a potential to let many administrative processes happen automatically. It is only necessary to interfere, when a decision is to be made which does not follow exact rules which are mapped into the workflow.

The process designed in figure 2 will be implemented in the financial administration (FinanzOnline [5]) by start of 2003.

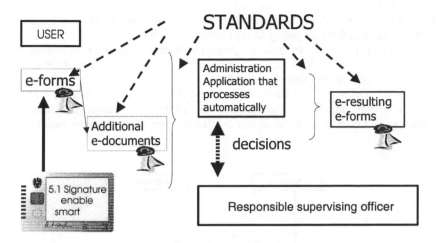

Fig. 2. The way e-government could work

Such modification of government processes is only acceptable for citizens, when it presents a uniform picture of the administration. To this end there is need for a well formed structure of the whole e-government process.

A well formed structure allows the administration to focus on the core business and to increasingly let private competition participate in the electronic and face to face services of the administration.

To achieve large acceptance help and information systems taking the special context into account for optimum comfort are crucial elements. These systems can integrate many elements of multimedia assistance. IP-telephony etc. for optimum assistance are just some of the examples.

For the resulting e-government strategy electronic signatures are the underlying principle. Today smart cards form the infrastructure so that citizens have a tool to "electronically present themselves" to the administration. Compatibility and inter-operability are to be assured on an as broad as possible level [6].

An appropriate structure allows for gradually introducing electronic procedures both, when accessing the e-government system and when accessing the back office. In some cases the system or back office can even be a simple workstation operated manually interfacing to the defined standards in the first approach.

3   Enabling Change Management

Efficiency is the effect that should be the result. Convenience and comfort as well as security are the tools to reach this goal. By implementing well defined structures and interfaces the tools available can be dynamically adopted.

For practical reasons there is a need for enabling a dynamical change of all parts within the system. This need is fulfilled by well defined interfaces. Security interfaces (signature [7], identification, etc.) and structural interfaces (person record [8], payment record, portal communication ...) have to form part of such systems.

Such a structure enables the implementation of semi automated or fully automated back office applications according to cost benefit conditions.

In all cases citizens can access government applications independent of the technology used for the back office application. This access can be face to face or fully electronically at the citizen's preference.

As a first step in this direction, Austria has decided to implement a web access through a generic applicable signed form for all federal ministries. Through this, proceedings can be managed that do not explicitly require special form handling.

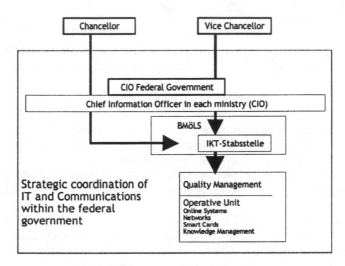

Fig. 3. Well structured e-government

4   Interoperable Documents

The use of appropriate formats is able to solve language and accessibility problems [9]. A form (document) can be prepared in Braille in English and it can be presented in German on the screen. The electronic signature is unchanged and valid in all cases by applying appropriate transformations.

New generations of browsers support such transformations directly. For older browsers automatic gateways can perform the transition in a web-based application for the administration. It is important that structure and formal description are interoperable among the various applications within the administration. This interoperability on a higher level can contribute to move from an e-member state to an e-Europe state in administration.

An "open interface initiative" is needed to allow interoperability on a transnational basis. Various levels of transformation can be used to cope both with standard formats and with individual approaches on a national level.

Not only is it important to provide standards for formats, electronic documents or at least their automated transformation must enable automated processing. There is a need for standard parsing mechanisms so that the source of a document need not be part of an application. Rather the document will itself tell what it can be used for, as it contains recognized tags that characterize the information.

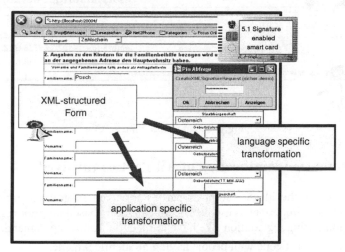

Fig. 4. Interoperability of formats and forms

Austria is currently building public accessible databases [10] that offer specifications, interoperability conditions as well as other information needed for the various institutions (public or private) involved in e-government.

Standard formats with formal description not only make it possible to process electronic forms automatically. By choosing XML documents it is also possible to include electronic signatures directly in the document [11] [12] – a benefit when implementing e-government applications. When designed properly, such forms can be identically rebuilt from the screen or from a printout. This makes it possible to include electronic signatures also on printed forms like tax bills etc. even when these are sent as printouts. A forgery is impossible.

Authentic, automatic reconstruction even from printout is crucial to cope with the various media and products that are and will be in place.

5   Integrated Signature

Figure 5 demonstrates such a form which is signed. In this case it constitutes a payment record issued by a bank for a specific administrative task. For tuition fees and many other applications such a technique makes the payment itself technologically independent from the administrative proceedings.

This same technique can be applied for issuing documents by the administration. More importantly on a European level this technique could enable independent interoperable administration still allowing fully electronic access.

With electronic signatures, with an appropriate structure of the e-government access, and with the necessary compatibility of forms citizen can govern their own data needed for administration and yield compatibility not only on a national but on a European level.

Fig. 5. Paying for administration

6   Automatic Identification of All Parties Involved

A crucial step is the identification of parties involved in government proceedings. This is especially important as e-government cannot be based on prior registration and identification during registration. We have to meet the following demands:

1. Identification must cope with people having an identical name.
2. Identification must be possible without prior contact with the administrative unit.
3. Identification must be based on adequate cryptography.
4. Identification must also comprise corporate bodies.
5. Identification must meet data protection requirements.
6. Identification must be capable of serving as a basis for confidentiality.

In the Austrian context these problems have been solved with a cryptographic binding of the ZMR (the number used in the registry of residents) to a certificate [13]. By application of a domain specific one way function this serves the above goals without infringing anonymity where appropriate.

7   Finally Structuring Organization

It is a key issue that the various appearances of e-government toward citizens present a coherent look and feel. This must be the case without respect to where the administrative application is located or whether it is federal or local. In a huge organization like governments this requires a clear organization and adherence to standards.

The information and communications technology board installed for the federal government in Austria is a means that serves as a tool to achieve such coordination which can also be communicated to and coordinated with the various levels of administration.

As a summary we can state that we need structures to enable e-citizenship. We need structures on the document level, structures on the security level comprising identification, signatures and confidentiality, and structures on the organizational level. Unless we reach such a structured approach it is unlikely that we meet the citizen's need in a mid and long term view.

References

1. Posch R. and Leitold H.: "White Book Bürgerkarte", first version by order of Ministry of Public Service and Sports, June 2001.
2. Directive 1999/93/EC of the European Parliament and of the Council on 13. December 1999 on a community framework for electronic signatures.
3. Bray T., Paoli J., Sperberg-McQueen C. M., Maler E.: "Extensible Markup Language (XML) 1.0 (Second Edition)", W3C Recommendation, October 2000.
4. Clark J.: "XSL Transformations (XSLT)", W3C Recommendation, 1999.
5. FinanzOnline, Ministry of Finance, http://www.bmf.gv.at/, preview demo available at http://fondemo.datakom.at/
6. eEurope Smart Card Initiative, Trailblazer 1, "Public Identity", http://eeurope-smartcards.org/
7. Hollosi A., Karlinger G., Posch R: "Security-Layer Specification", February 2002, http://www.buergerkarte.at/konzept/securitylayer/spezifikation/20020225/
8. Reichstädter P. and Hollosi A.: "PersonData – an XML Specification of Personal Data", April 2002, http://reference.e-government.gv.at/
9. Chisholm W., Vanderheiden G., Jacobs I.: "Web Content Accessibility Guidelines 1.0", W3C Recommendation, May 1999.
10. Austrian e-government reference portal of federal, provincial, and municipal administrations, http://reference.e-government.gv.at/
11. Eastlake D., Reagle J., and Solo D.: "XML-Signature Syntax and Processing", W3C Recommendation, 2002.
12. ETSI SEC: "XML Advanced Electronic Signatures (XAdES)", Technical Specification ETSI TS 101903, 2002.
13. Hollosi A.: "Identity Link – XML Specification", March 2002, http://www.buergerkarte.at/konzept/personenbindung/spezifikation/20020411/
14. Leitold H. and Hollosi A.: "An Open Interface Enabling Secure e-Government", forthcoming.
15. International Organization for Standardization: "Information technology - Security techniques - Evaluation criteria for IT security", ISO/IEC 15408-1 to 15408-3, 1999.
16. ETSI SEC: "Electronic Signature Formats, v.1.2.2", Technical Specification ETSI TS 103733, 2000.

Private Sanctity –
e-Practices Overriding Democratic Rigor in e-Voting

Åke Grönlund

Umeå University, Informatics
gron@informatik.umu.se

Abstract. The discussion on electronic voting has so far mostly focused on technical issues, mainly concerning security and privacy. This paper reports an empirical study on how the symbolic values of democracy, as manifested in the act of voting, are considered by e-voters. The study found that the voters in a student election in actions as well as in stated views gave priority to convenience over security and privacy. They voted electronically from home despite uncertainty about the security of the technical system. We argue that this is an indication that the view of the principles of democratic practices will change, and that what might be called an "e-practices mode of thinking" will to some extent prevail over a "rigid democracy mode".

1 Introduction

Electronic voting has been discussed for some time, most intensely so immediately after the US president election debacle in Fall 2000. The topic for the debate has mostly been technical issues relating to security, authentication, and privacy, but also turnout and, to some extent, economic considerations (CIVTF, 2000; Cranor, 2000; IPI, 2001; Rubin, 2001).

This paper reports a study that takes the issue a step further. Suppose we had perfectly secure systems for electronic voting – then what? What would they be used for? Voting is an act that is surrounded with a nimbus as it embodies the essence of democracy. The rules and procedures framing voting express the view of what a vote is, and hence what democracy is. Although different in details across countries and democratic variants (e.g. Barber, 1984; Åström, 2001), what is sometimes referred to as "the sanctity of the act of voting" (Statskontoret, 2001: 49) is usually described as coming from the following characteristics:

- The vote is an individual choice, and the result of individual deliberation. Although public debate is of course essential you should make up your mind yourself, at least without pressure from anyone else.
- The vote is at the same time an expression of a collective – you are part of a people who has decided to make decisions in public matters this way. Considering the alternatives, you should feel proud of the fact that you can indeed go to the polling booth. This means the way you view your voting is important – it is the expression of freedom of choice. The vote is a manifestation of the "sacred" mode of rule, democracy.

R. Traunmüller and K. Lenk (Eds.): EGOV 2002, LNCS 2456, pp. 52–60, 2002.
© Springer-Verlag Berlin Heidelberg 2002

- The content of the vote, or more precisely the link between the vote and you, is your private property. It is up to you whether or not you want to disclose your political preferences.
- The voting procedures, hence, must be considered as reliable in all aspects by everyone – even those who do not vote. This includes surrounding events such as the preceding and ensuing public debates. (CIVTF, 2000; Johansson, 2001; Statskontoret, 2001).

Examining current voting procedures we find that precincts are physically set up so that people are not too hard pressed to individually honor the above principles. For instance, when you actually cast your vote, usually by dropping a piece of paper in a ballot box, there is a curtain or some other arrangement to prevent others from seeing what you do.

The briefest thought about conditions in the home reveals that this physical guarantee for privacy cannot be arranged without Orwellian measures. So, if we had secure electronic voting from the home, privacy would have to be guaranteed by the individual rather than government. In fact, not only the voter herself but also, often, by her family who would have to agree not to shoudersurf. Would we do that? Or would we change or minds about the principles?

To gain at least some preliminary empirical findings about people's views of the act of voting in an e-voting context, we made an investigation of the first Swedish Internet election, arranged by a student union at Umeå University in May 2001.

2   The Election

Umeå Student Union, representing some 12 000 students in social sciences, humanities, and teacher education, is one of three unions at Umeå University, Sweden. Its Council is elected in annual public proportional elections. The procedures are quite similar to those in Swedish public elections. Voters vote for parties and are allowed to strike out candidates from the list provided by the party of their choice.

Turnout in student elections is generally very low across Sweden. In Umeå, the 2000 election attracted only 11,5 % of the electorate. The low turnout was the main reason for choosing to use Internet in 2001. Eventually sponsoring from the Department of Justice required broader aims, as the government was interested in a pilot test for voter behavior. This led to the goals eventually being the following:

- Increase turnout by about 50 % (from 11,5 % to ca 17 %).
- Investigate voters' attitudes to Internet elections (concerning privacy as well as symbolic values).
- Investigate strengths and weaknesses generally with Internet elections.
- Improve the election process.
- Get more people active in the democratic process.
- Make democratic debate more available to people.

It was hoped that improved marketing in combination with the "improved election process" would achieve this. The improvement of the election process included not only electronic voting, also an electronic discussion forum was set up. The voting technology used was developed by the US company Safevote (www.safevote.com).

The discussion forum was developed by Vivarto technologies (www.Vivarto.com), a Swedish partner to Safevote.

Voting could be done in three ways, by mail, at a precinct, or over the Internet. Voting by mail and by Internet was open from April 27 to May 10. Precinct voting was open May 10-11. It was possible to vote more than once over the Internet. It was also possible to override a mail vote by precinct voting. The voting periods and the option to re-vote were chosen so as to minimize the impact of technical failure, vote selling and changed conditions like some political scandal appearing during the early voting period (re-voting is allowed also in Swedish public elections –votes cast at post offices can be overridden by voting at a precinct).

These considerations led to that votes were given different priority: A precinct vote overrides an Internet vote, which overrides a mail vote.

3   The Investigation

On the initiative of the Department of Justice, an evaluation group was appointed. This group designed a set of measures for evaluations. The security of the technical system and of the logistics used was assessed by expert review. Voters' attitudes and behavior were investigated by a large telephone interview study designed by the evaluation group and executed by Sweden Statistics (http://www.scb.se/).

From the member file of Umeå Student Union (11 859 people), 2 500 names were randomly selected. Interviews were conducted from May 21 to June 6. 80 % responded, 2026 individuals.

3.1   The Findings

This investigation assessed voters' attitudes to critical issues of e-voting: how they view the "sacred act" of voting, and how they feel about privacy and integrity. As the possible effect of Internet voting on the rate of participation in elections has been on the agenda, we also tried to assess voters' reasons for participating or not, and their view of how practical it is to vote electronically from the home.

3.1.1   Participation

The marketing activities were considered successful. For instance, 87 % of those who did not vote knew that there was an Internet voting option. Despite that, the expected increase in turnout did not happen - on the contrary, it sank from an already low 10,4 % in the 2000 election to only 9,3 %.

3.1.2   Voting Method

Internet voting was by far the most popular method (Table 1). The low turnout may seem an indication that Internet voting does not help increase turnout, but the issue may be more complicated. There were some other factors that may have contributed to the decrease. One was that a rather big party (200 voters in the year 2000 election), also the only party directly targeting the students at the teacher education, did not run this time. Also, there were fewer precincts in the 2001 election. We therefore tried to find some further guidance by means of the interviews as to why people chose to vote or not.

Table 1. Voting by method

Voting method	Male %	Female %	Total %	Total number
Internet	66	60	63	678
Precinct	16	13	14	181
Mail	19	27	23	244

3.1.3   Reasons for Voting

Voters were asked about their reasons for choosing to vote or not. Table 3 below shows that the most common reason was that they saw it as natural to use their right to vote. Only 12 % were interested in the Union's activities. 24 % mentioned the Internet voting option as a reason. It should be noted that the voter population is very fluid. Only 28 % of those who voted in 2001 and were entitled to vote in the 2000 election also voted then. The Union's hypothesis was that Internet voting would make a positive difference - in a population where interest in the body to be elected is low and participation is not necessarily by routine, an easily accessible vote method may make a difference

There is indeed some support in the study that the Internet method appealed more to new voters. Dividing voters into two groups depending on whether or not they voted in the previous election shows that voting by the Internet was more common among the new voters, 67 % compared to 55 % (Table 2).

Table 2. Voting method 2001 compared to participation in the year 2000 election (%)

Voting method 2001	Voted 2000		Not entitled to vote 2000
	Yes	No	
Internet	**55**	**67**	63
Precinct	15	13	14
Mail	30	18	21

Further, 40 % of those who voted 2001 but not 2000 mentioned the Internet option as the reason for voting (Table 3). This should be compared with the corresponding figure for the total population, only 24 %.

Table 3. Main reason for voting among those who participated in the 2001 election (%)

	Interested in the Union	The Internet option	It is natural to vote	Other
All voters	12	**24**	62	4
Voted 2000	12	12	74	3
Did not vote 2000	10	**40**	43	7
Not entitled 2000	13	19	67	1

For one fourth of the voters, the main reason was that they felt attracted by the Internet voting alternative. For people who participated 2001 but not 2000, there was a greater share that felt this way. Viewed in this perspective, Internet voting could have increased the participation to a considerable extent, especially concerning attracting new voters to the booth.

3.1.4   Reasons for *Not* Voting

Some 90 % of the total population did not vote. Table 4 summarizes their stated reasons for not voting.

Table 4. Main reason for not voting (%)

	Not interested in the Union	Didn't get to doing it	Other
All non-voters (%)	48	24	27
Voted 2000	22	43	36
Did not vote 2000	54	21	25
Not entitled 2000	48	23	29

As Table 4 shows, close to half the population claims to have refrained from voting due to a lack of interest in the student union and its activities. About one fourth mentions that the main reason for not voting that "it just did not happen". Among those who participated in 2000 but not in 2001, there was a greater share who mentioned that they just did not make it and comparatively few, less than one fourth, that stated a lack of interest in the student union which caused them to not to vote.

The figures indicate a volatile electorate, with a high number "just not making it" and with a general democratic mindset ("it is natural to vote") by far overriding the interest for the object of the election, the student union. Even the attraction of the voting method was twice as high as the interest in the union. This seems to give further support for the hypothesis that the Internet option has indeed helped in preventing turnout to fall even more. Internet would then be an enabler of turnout, although not a determining factor.

3.1.5   The Sacred Act of Voting?

As discussed above, the "sanctity" of the act of voting can be expressed by several parameters: the *reasons* for voting, the *actual privacy* in the voting act and during the ensuing handling of the vote, the *integrity* of the vote itself after it has been cast, the *view* of just how this privacy should be implemented, and the general *credibility* of the election as a whole.

As for the *reasons for voting*, as we discussed above people seemed to be quite serious about it, as the most prominent reason for voting was that it is natural to use one's vote.

Privacy during the voting act when performed in the home obviously has to be arranged by each individual voter. 90 % said nobody else saw what they voted for. When someone did see it, it was a family member (8 %) or a friend (2 %). This indicates that the voters approached the privacy of voting much the same way as in a precinct voting. On the one hand, it might be claimed that 10 % seeing the voting is too much. On the other hand, probably more than 10 % of the population voluntarily reveals their political preferences to other people that are close to them.

It seems fair to say the voters in this election maintained what might be called a "personal sanctity" which is comparable to what the imposed privacy of the voting booth achieves.

It thus seems that Internet voting does not change voter's view of how the voting situation should be arranged.

3.2   Privacy and Integrity

The technology in the Umeå election was not completely safe. It used standard operating systems and standard web browsers. It did not use active components such as JavaScript, but it required this feature of the browser to be enabled[1]. There were also security problems in the handling of the codes used for identifying voters, as these were distributed by ordinary mail.

There was ambiguous information to voters, as the system was announced as simply "safe". As the setup clearly made use of standard technology, at least the somewhat knowledgeable user should raise some doubts about this statement. We were interested in how voters perceived the system's ability to guarantee privacy and integrity, and their view of how any (fears for) deficiencies in this respect would affect them.

During interviews, the respondents were asked to rate their agreement or disagreement with some statements relating to privacy and integrity matters by rating them on a 5-grade scale: strongly agree, partly agree, partly disagree, strongly disagree, and don't know. The results were as follows.

23 % agreed partly or strongly to the statement My vote may be disclosed to some unauthorized person:

	Strongly agree	Partly agree	Partly disagree	Strongly disagree	Don't know
All Internet voters (%)	7	16	18	48	10

Almost half of the Internet voters agreed partly or strongly to the statement *My opinion can be registered:*

	Strongly agree	Partly agree	Partly disagree	Strongly disagree	Don't know
All Internet voters (%)	24	22	15	31	8

44 % felt more or less strongly that My vote may disappear somehow:

	Strongly agree	Partly agree	Partly disagree	Strongly disagree	Don't know
All Internet voters (%)	15	29	16	32	7

Only 42 % agreed strongly to the statement *I feel confident that my vote will be counted.* This means a clear majority was not so sure about it:

	Strongly agree	Partly agree	Partly disagree	Strongly disagree	Don't know
All Internet voters (%)	42	33	12	3	8

Only 34 % agreed strongly to the statement Safety *is enough to protect voting secrecy:*

	Strongly agree	Partly agree	Partly disagree	Strongly disagree	Don't know
All Internet voters (%)	34	35	11	5	14

---

[1] More precisely, the actual voting system did not require this but the Web site from which the voting took place did. This meant in practice voters did in fact have to enable the JavaScript option

Altogether, it seems fair to say that the above answers reveal a great skepticism as to the privacy and integrity of Internet voting. Against this background it appears strange that they did in fact vote – note again that these are the views of those who actually cast their vote over the Internet.

A possible explanation could be that convenience is more important to them, so we asked a couple of questions about that.

3.3   Convenience

To assess the students' view of the convenience of Internet voting we asked them to assess fourstatements, again on a similar 5-grade scale as above. 90 % considered it very good or rather good not having to go to a precinct:

	Very good	Rather good	Rather bad	Very bad	Don't know
All Internet voters (%)	74	16	3	3	3

Although a solid majority of 81 % found Internet voting easy, it may come as something of a surprise that 6 % found it not so easy. After all, these were students with enough Internet experience to choose Internet as voting method. There were in fact a lot of technical problems - 33 % of the voters experienced such. This usually meant they had to reload the page as the connection timed out (instructions for this were on the screen as the possibility was foreseen). It is therefore likely that the 6 % who found it not so did so because of the actual system.

	Strongly agree	Partly agree	Partly disagree	Strongly disagree	Don't know
All Internet voters (%)	54	27	5	1	13

59 % felt that restriction of voting to precincts would indeed affect turnout by disagreeing partly or strongly to the statement *It does not affect turnout if voting is restricted to precincts with a higher degree of security:*

	Strongly agree	Partly agree	Partly disagree	Strongly disagree	Don't know
All Internet voters (%)	12	14	17	42	14

More speculatively 54 % thought that the Internet voting option had increased the turnout in this particular election: *What influence do you think the Internet voting option has on turnout in a student election?*

	Probably increased it	Probably not changed it	Probably decreased it	Don't know
All Internet voters (%)	54	33	5	7

Taken together it seems fair to say that these answers show that respondents found Internet voting practical. Indeed – because they actually voted – so practical that convenience overrides the perceived deficiencies in privacy and integrity.

4   Discussion

The figures presented above clearly tell that in this particular election, convenience was considered more important than security. Is this also valid more generally?

Obviously this investigation cannot give a clear answer to that. Students are a special group, a student election is not really considered as important as a national or municipal one, and e-voting is not yet an established practice so views may be tentative.

Still, when it comes to privacy issues, or the more particular right to keep your political opinion to yourself, if your political views are disclosed on the Internet it doesn't really matter if this happens in a small-scale and politically unimportant election or a national, important one. Therefore, the case should represent the views expressed on these issues.

A question that this investigation cannot answer is whether this is an expression of Swedish naiveté ("it doesn't happen to us in safe Sweden"), or if it is indeed a more profoundly considered active priority. To the support of the latter it may be argued that frequent Internet users do have to make some such prioritization if they are going to be able to use the Internet for shopping and such. Perhaps Internet voting is viewed as just a kind of Internet shopping? On the other hand it appears not unlikely that people in non-democratic countries, and probably young democracies, as well as Swedish immigrants from such countries, would think differently.

Another issue that is not clearly answered by this investigation is whether the answers reflect emerging opinions or residues of old ones. For instance, is the students' view of how the voting should be done – alone – a residue of the "manual" procedure, which is deep rooted? Are the views of privacy an example of technological naiveté leading to over-optimistic assumptions of the (low) probability that the kind of things that they state "might" happen actually will occur?

The students were experienced technology users - all used computers at work, and most probably also at home, most were their early 20s and thus from a generation which have used computers also in school. They were also experienced Internet users – all have student email addresses, and an estimated 80 % have broadband connection in their home. Altogether, it does not seem unfair to say that this indicates that the answers represent new views, those of "pragmatic Internet users" rather than anything else.

The implications of the findings, if they are valid also for a wider population, are that voting procedures will have to change:

1. Internet voting was the priority one voting method, yet turnout did not increase. Perhaps the optimistic view that e-voting will increase turnout should be reformulated: e-voting will become necessary so as to prevent turnout to fall even more.
2. If e-voting from the home is used on a large scale, the idea of the "voting day" will probably have to change, for convenience as well as because of technical issues. Attackers will have to go on for a longer period of time to make any difference, and this will increase the risk/chance of disclosure, and technical problems will not have such a disastrous effect.
3. If e-voting from the home is allowed, the task of guaranteeing privacy at the moment of voting will have to be delegated to the individual voter. The system will have to accept "private sanctity", that is ideologically and/or psychologically rather than physically enforced privacy. The Umeå experiences indicate that at least this population was up to the ethical standards necessary, but trusting this is probably the hardest challenge for the democratic system.

References

Barber, B (1984) *Strong Democracy*. Berkeley: University of California Press.
CIVTF (2000) *A Report on the Feasibility of Internet Voting*. January 2000. California Internet Voting Task Force (http://www.ss.ca.gov/executive/ivote).
Cranor, L.F. (2001) *Voting After Florida: No Easy Answers* (http://www.research.att.com/~lorrie/voting/essay.html).
IPI (2001) *Report from the National Workshop on Internet Voting*. Internet Policy Institute (http://www.netvoting.org/Resources/InternetVotingReport.pdf).
Johansson, S (2001) *Teknik och administration i valförfarandet*.(Technology and administraton in voting procedures)   Slutbetänkande från Valtekniska utredningen (SOU 2000:125). http://justitie.regeringen.se/propositionermm/sou/pdf/sou2000_125.pdf
Olsson, A.R. (2001) *E-röstning. En lägesbeskrivning*. (E-voting. State of the art) Stockholm: IT-kommissionen. Observatoriet för IT, Demokrati och medborgaskap, Observatorierapport 35/2001.
Rubin, A. (2001) *Security Considerations for Remote Electronic Voting over the Internet* AT&T Labs – Research, Florham Park, NJ (http://www.avirubin.com/e-voting.security.html).
Statskontoret (2001) *Utvärdering av kårvalet vid Umeå studentkår*. (Evaluation of Umeå Student Internet Election). Stockholm: Report 2001:26.
Åström, J. (2001) Should online Democracy be Strong, Thin or Quick? *Communications of the ACM*, Jan 2001.

Reconfiguring the Political Value Chain:
The Potential Role of Web Services

Francesco Virili[1] and Maddalena Sorrentino[2]

[1] Dipartimento Impresa e Lavoro, Università di Cassino
Via Mazzaroppi, 1 - 03043 Cassino (FR) - Italy
francesco.virili@eco.unicas.it
[2] Dipartimento Scienze dell'Economia e della Gestione Aziendale
Università Cattolica del Sacro Cuore di Milano
Via Necchi, 5 - 20123 Milano - Italy
mso@mi.unicatt.it

Abstract. A new technological standard, called 'Web services' has recently made its first appearance in the Web technologies arena. Our question here is: what is the role of Web services for eGovernment? In the present contribution, the concept of 'political value chain' is introduced and the process of value reconfiguration is illustrated, evidencing one of the potential roles of IT on administrative activities: the facilitation of 'citizen value' creation activities connection. A brief illustration of the Web services technology is then given, finally exploring its potential for e-Government activities and the related research issues.

1 Introduction

A new technological standard, called 'Web services' has recently made its first appearance in the Web technologies arena. Our question here is: what is the role of Web services for eGovernment?

In the following sections, the concept of 'political value chain' is introduced (Section 2), and the process of value reconfiguration is illustrated (Section 3), evidencing one of the potential roles of IT on administrative activities: the facilitation of 'citizen value' creation activities connection. In Section 4 the Web services technology is illustrated, finally exploring, in Section 5, its potential for e-Government activities and the related research issues.

2 The Political Value Chain

In a recent contribution about the use of IT for business process reengineering in the public administration, Anderson [1] maps on a value chain model, based on the classical Porterian concept [10], some examples of IT applications in the public sector. Anderson calls this scheme, inspired by [5] and depicted here in Figure 1, 'political

R. Traunmüller and K. Lenk (Eds.): EGOV 2002, LNCS 2456, pp. 61–68, 2002.

value chain'. Actually, the transposition of a typical concept of industrial business strategy (that of value creation) to the Public Administration, would require more caution: the author notices, a bit superficially, that 'in the public sector there is typically no financial margin of value to be added by innovation. Instead, the public sector can partly add value by shaping the business environment and helping companies be more efficient and effective. In part, too, the public sector is legitimised by its political actions in the democratic domain. So the margin of value in Figure 1 is cast as some combination of the economic, the democratic and the technical'. The idea of value production in the public administration would deserve a deeper analysis, and also the notion of purely 'financial margin' in the industrial business doesn't give justice to the more complex and comprehensive concept of 'customer value' discussed by Porter, that certainly includes many immaterial and qualitative aspects the customer is willing to pay for.

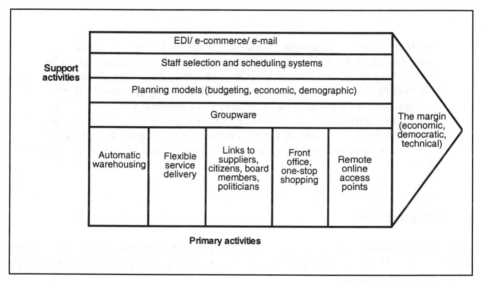

Fig. 1. IT opportunities within the political value chain. From [1], page 320

Moreover, the value chain is better used to represent the value creation process in the industrial production than in the service sector. From this point of view, the analysis of Stabell and Fjeldstad [14] certainly deserves attention: the two authors observe, not without reason, that there is a great difference between the value production processes of a typical industrial business, that inspired Porter's value chain proposition, and that of a service provider, like a bank or a hospital. For example, in Figure 1 the phase of 'inbound logistic', usually referred to the 'raw materials' of the production process, is somehow kept in the model to figure out an 'automatic warehousing' IT application. It is hard to say that inbound logistic is a value production phase in most of the typical public administration activities: immaterial services often constitute the main output, and there is no relevant inbound logistic process. In such cases, different value configurations may be considered as a starting point, like the 'value shops' and the 'value networks' discussed by Stabell and Fjeldstad.

Nevertheless, the political value chain is a simple and interesting starting point for our present work, and we would use it as it is, knowing that it may deserve some deeper analysis and (re)definition. For our purposes, we take for granted the existence, in the public sector, of a sequence of value creation activities, aimed to the production of what Anderson calls 'margin', that we would still call 'value'. We may better say 'citizen value', instead of 'customer value'. We won't investigate, in this contribution, on the specific characteristic of the 'citizen value' and on the details of the value creation activities description, categorization and sequence.

3   The Value Reconfiguration Process

Taken for granted the existence, in the public administration activities, of several value creation activities, we may point the attention to a process that is becoming increasingly common in several industrial and service sectors, generally called 'value reconfiguration', that is well described, for example, by Malone, Yates and Benjamin, in [8] using the framework of the transaction costs theory [17]. According to the authors, ICT may significantly reduce the overall transaction costs, inducing organisations to externalise some activities of the value production process without loosing control (value disaggregation). Moreover, an extensive use of Information and Communication Technologies would then allow the integration of third parties value production activities, creating new inter-organizational value configurations. By this value reconfiguration process, the organisation at the final end of the value creation system may develop and manage a wider and more articulated offer, integrating products and services from several other organisations. For example, Seifert and Wimmer [14], describe the value reconfiguration process focusing on the financial industry. They analyse the case study of a German mortgage bank, the Rheinische Hypothekenbank (Rheinhyp), that externalised the division of 'direct customers' (mortgages distributed via Internet) to a new joint venture company, 'Extrahyp'. Extrahyp was involved in the value production activities related to the new distribution channel; moreover, it was used to develop a richer product/service offering: in addition to the basic Rheinhyp mortgages other products and services issued by third parties were introduced. Finally, Extrahyp started issuing IT services to other banks.

Is this concept of value reconfiguration applicable to the public administration activities? The framework used by Malone, Yates and Benjamin [8] is based on Oliver Williamson's theories, that were later extended by the same author to the governance mechanisms [18], with some important modifications: a significant new concept is that of 'inefficiencies by design' (see also [16] for an application). Basically, we should now take into account, besides the classical transaction costs, also the cost of political consensus. In facts, some degree of governance inefficiency may be accepted (and even introduced on purpose) in order to 'buttress weak political property rights' ([18], page 199) extending consensus with compromising governance choices. The existence of this efficiency/consensus trade-off should not affect the potential role of ICT as transaction costs reducer and driver of value reconfiguration processes [2][8][10], though some research work should be devoted to deal with the enhanced complexity of the modified framework, with its peculiar aspects, and also to some recent criticisms like [3], based on the ambivalent effects of IT externalities on transaction costs.

In the next section we are pointing the attention to a new technology that may potentially play a central role in the value reconfiguration process.

4   Web Services: An Emerging Standard

In April 2001, some 52 IT companies and 'power users', participating in the W3C consortium (including Microsoft, IBM, HP, Sun, SAP, Boeing, …) conveyed to a workshop in San Jose (California) to advise the W3C on the further actions to be taken with regard to Web services. All of them published their 'position papers', (http://www.w3.org/2001/01/WSWS), discussing their peculiar view and means of implementation of the new technology.

Web services are self-contained, modular business process applications that Web users or Web connected programs can access over a network via a standardized XML-based interface, in a platform-independent and language-neutral way [4] [5]. This makes it possible to build bridges between systems that otherwise would require extensive development efforts. Web services are designed to be published, discovered, and invoked dynamically in a distributed computing environment. By facilitating real-time programmatic interaction between applications over the Internet, Web Services may allow companies to more easily exchange information, leverage information resources, and integrate business processes.

In practice, a Web service is a software reusable component (i.e. a small functionality, a little 'piece' of an application) that can be written by anybody (for example a software vendor), and published to be later retrieved and dynamically used within an existing application by anyone (for example an IS developer). Adopting this framework, companies in the future will be able to buy their information technologies as services provided over the Internet, rather than owning and maintaining all their hardware and software (Hagel, 2001). The functionalities that can be implemented by Web services have virtually no limits, ranging from major services as storage management and customer relationship management (CRM) down to much more limited services such as the furnishing of a stock quote and the checking of bids for an auction item.

Users can access some Web services through a peer-to-peer arrangement rather than by going to a central server. Some services can communicate with other services and this exchange of procedures and data is generally enabled by a class of software known as middleware. Services previously possible only with the older standardized service known as Electronic Data Interchange (EDI) are now likely to become Web services. Besides the standardization and wide availability to users and businesses of the Internet itself, Web services are also increasingly enabled by the use of the Extensible Markup Language (XML) as a means of standardizing data formats and exchanging data.

Through Web services systems can advertise the presence of business processes, information, or tasks to be consumed by other systems. Web services can be delivered to any customer device - e.g., cell phone, (PDA) and PC - and can be created or transformed from existing applications. More important, Web services use repositories of services that can be searched to locate the desired function to create a dynamic value chain. Web services go beyond software components, because they can describe their own functionality, look for, and dynamically interact with other Web services. They provide a means for different organizations to connect their applications with one another to conduct dynamic e-business across a network, no matter what their application, design or run-time environment.

By this new software layer, it's possible to build applications without having to know whom users are, where they are, or anything else about them. Users of these applications can source them as easily as they would be able to source static data on the Web, with complete freedom and no concern about the format, platform, or anything else.

So the revolutionary aspect of using Web services is that they are self-integrating with other similar applications. Until now, using traditional software tools to make two e-business technologies work together required lots of work and planning, to agree on the standards to pass data, the protocols, the platforms, etc. Thanks to Web services, applications will be able to automatically integrate with each other wherever they originate, with no additional work.

4.1   The Web Services Architecture

The Web Services architecture, depicted in Figure 2, is based upon the interactions between three roles: service provider, service registry and service requestor[4]. The interactions involve the 'publish', 'find' and 'bind' operations. Together, these roles and operations act upon the Web Service software module and its description. In a typical scenario, a service provider hosts a network-accessible software module (an implementation of a Web service).

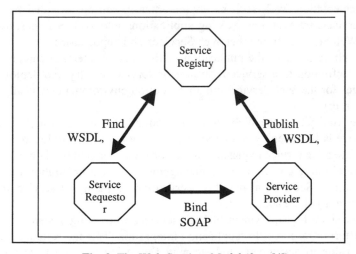

Fig. 2. The Web Services Model (from[4])

The service provider defines a service description for the Web service and 'publishes' it to a service requestor or service registry. The service requestor uses a 'find' operation to retrieve the service description locally or from the service registry; it uses the service description to 'bind' with the service provider and to invoke or interact with the Web service implementation. Service provider and service requestor roles are logical constructs and a service can exhibit characteristics of both.

The foundation of Web Services is represented by new standard technologies which meet the basic requirements for their implementation:

- SOAP (to communicate) a standard mechanism for sending requests to services and receiving responses;
- WSDL (to describe) a standard way to describe services, with a input/output interface specifications and some meta information (copyright, version, update URL, etc.);
- UDDI (to advertise and syndicate) a standard means of locating relevant services with the desired characteristics.

4.2   Building Web Services

Using Web services, as usual in componentized software applications, the system development process may be fractioned in two major parts: the standard components development and the integration into the target system.

Even if, given the novelty of the platform, a consolidated methodology does not yet exist, we may figure out, in the development process of the standard components, that approximately four major phases might be followed: building, deployment, running, management.

The 'build' phase includes the development and testing of the Web service implementation, and the definition of the descriptions for both the service interface and the service implementation. Web services implementation can be provided by creating new Web Services, transforming existing applications into Web Services, and composing new Web Services from other Web Services and applications.

The 'deploy' phase includes the publication of the service interface and service implementation definition to a service requestor or service registry and deployment of the executables for the Web service into an execution environment (typically, a Web application server).

During the 'run' phase, the Web service is available for invocation. At this point, the Web service is fully deployed, operational and network-accessible from the service provider. Then the service requestor can perform the find and bind operations.

The 'manage' phase covers ongoing management and administration of the Web service application. Security, availability, performance, quality of service and business processes must all be addressed.

On the user side, the deployment of Web services in existing systems should not require any effort or resources for application integration. This fact would surely have a significant value to developers, and we may figure out it could have a dramatic impact on the way of designing and implementing Information Systems that may be dynamically adapted to new business needs or organizational changes. The whole IS development process may be radically transformed, as foreseen by (Lyytinen et al, 1998): '... the distinctions between 'internal' and 'external' applications have greyed. The impact of this greying is both the altering and the broadening of design considerations such as availability, security, support and access for all applications. In response to these issues new mechanisms and methods of application assembly are emerging. [...] These are a far cry from the application-oriented, data flow diagramming, functional design and bespoke application days of yore. Against these changes, the role of the software developer necessarily changes. Some will manufacture components; the majority will facilitate their adaptation, choice, understanding and use'. (page 248).

5   Preliminary Conclusions: Exploring the Potential Role of Web Services

What is the potential role of Web services in the value reconfiguration process? John Hagel III, in [5], writes:

Two and a half years ago, Marc Singer and I wrote 'Unbundling the Corporation' [6]. In that article, we described [...] how the Internet would facilitate the unbundling [process], leading to much more tightly focused companies. The rise of the Web services architecture will not only speed this unbundling but will spur the growth of the new companies by letting them mobilize a greater range of resources to reach a broader set of customers (page 113).

Obviously, this statement is only a hypothesis that should be confirmed by evidence and better investigated. If we transpose this hypothesis to the Public Administration sector, the peculiar aspects of governance [10][11] and the higher complexity of the resulting framework would obviously require some additional efforts. The resulting research agenda would encompass theoretical aspects like transaction cost theory investigation and application, management aspects like the definition of the new organisational assets, and applicative aspects like the development of a security infrastructure to ensure the required level of trust.

References

1. Andersen, K.V. Reengineering public sector organisations using information technology, in 5, pp. 313-329.
2. Ciborra C., Teams, Markets and Systems, Cambridge University Press, 1993.
3. Cordella, A., Does Information Technology Always Lead to Lower Transaction Costs?, Proceedings of ECIS 2001, Bled (Slovenia).
4. Kreger, H., Web Services Conceptual Architecture, white paper, IBM Software Group, May 2001.
5. Hagel III, J. and J.S. Brown, Your Next IT Strategy, Harvard Business Review, October 2001, 106-113.

6. Hagel III, J. and M. Singer, Unbundling the Corporation, Harvard Business Review, March-April 1999, Vol.77(2) pp.133-141.
7. Heeks, R. (ed.), Reinventing Government in the Information Age: International Practice in IT-Enabled Public Sector Reform. Routledge, 2001.
8. Malone, T.W., Yates, J and R. Benjamin, Electronic Markets and Electronic Hierarchies. Communications of the ACM, 30(6), 1987, 484-497.
9. Moreton, R. and M. Chester, Transforming the Business: The IT Contribution, McGraw-Hill, 1996.
10. Lenk, K. and R. Traunmüller, Perspectives on Electronic Government, IFIP WG 8.5 IS in Public Administration Working Conference on Advances in Electronic Government, 10-11 February 2000, Zaragosa, Spain.
11. Lenk, K. and R. Traunmüller (eds.), Öffentliche Verwaltung und Informationstechnik - Perspectiven einer radikalen Neugestaltung mit Informationstechnik. Heidelberg Decker, 1999, German.
12. Picot A., Bortenlänger C. and H. Röhrl, Organization of electronic market: contributions from the new institutional economics, The Information Society, 13, (1997),107-123.
13. Porter, M.E. Competitive strategy: techniques for analyzing industries and competitors, NY: Free Press, 1998 first edition 1980.
14. Seifert, F. and A. Wimmer, Towards networked banking: the impact of IT on the financial industry value chain, Proceedings of the European Conference on Information Systems (ECIS), Bled, Slovenia, 2001, pp.474-84.
15. Stabell, C.B. and Ø. Fjeldstad, Configuring Value for Competitive Advantage: On Chains, Shops, and Networks, Strategic Management Journal 19 (1998), pp. 413-37.
16. Virili, F., The Italian e-Government Action Plan: from Gaining Efficiency to Rethinking Goverment. Proceedings of DEXA 2001 International Workshop on e-Government, Munich 2001.
17. Williamson, O.E., Markets and Hierarchies: Analysis and Antitrust Implications, Free Press, 1975.
18. Williamson, O.E., The Mechanisms of Governance, Oxford University Press, 1996.

The E-GOV Action Plan in Beijing

Xinxiang Chen

Capinfo Company Limited
Beijing Network & Multimedia Lab

1   IT Brings New Opportunities and Challenges

Information and Communication Technology (IT) is one of the most magic forces. It heavily affects people's daily life, learning and work, even the working style of the government in civil society. IT will make sure that each individual in the society can common share other's knowledge and ideas. By means of IT, people can collaborate with each other without limitation of time difference and geographic location. It will help people to exploit their potential and accelerate the development of the society.
Also IT can bring large challenges, such as innovation, productivity and efficiency, in front of enterprises, firms and government. They will face the competition not only in the special local area, but also all over the world.
IT is fast becoming a vital engine of growth for the world economy. So IT reforms the earth. The preparation of manuscripts, which are to be reproduced by photo-offset requires special care. Papers submitted in a technically unsuitable form will be returned for retyping, or canceled if the volume cannot otherwise be finished on time.

2   Internet in China

2.1   The Growth of Internet Users in China Shown in Figure 1

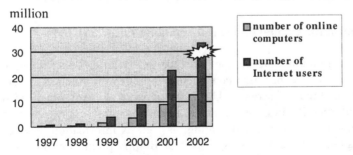

Fig. 1. The Growth of Internet user

2.2   Here Is the Statistics/Investigation from CNNIC in 01/2002

- Internet users in China: 33.7 million
- Online computers: 12.54 million

R. Traunmüller and K. Lenk (Eds.): EGOV 2002, LNCS 2456, pp. 69–74, 2002.

- Registered domain names in CNNIC: 127,319; Beijing accounting for 33.8%
- Registered Web Sites in CNNIC: 277,100; Beijing accounting for 20.6%
- Total capacity for international traffic: 7,597.5M

2.3   Penetration of IT in Beijing(01/2002)

- Internet users: 4.18 million, 12.39%
- PCs: 110.5 per thousand people, the ratio of the computers for family use has propagated to 28.4%
- Mobile phones: 5.95 million, 4.00%
- Telephones: 5.16 million, 2.84%
- TV sets: 148.9 per hundred houses
- 234 thousand employees in IT area

3   The E-GOV Action Plan in Beijing

- First Stage (1998 ~ 2000): Launching the supporting infrastructure, including information database and the portal website. Starting the E-GOV process in Beijing
- Second Stage (2001 ~ 2002): Developing, examining and managing the One-Stop, online interactive public services, available to the enterprises and citizens in Beijing
- Third Stage (2003 ~ 2005): to establish a systematic, well-structured, electronic service delivery system based on the high-speed, broadband municipal network, to build one of the best E-GOV in China

4   The E-GOV Action in Beijing: First Stage (1998-2000)

4.1   Establish the Capital Public Information Platform (CPIP)

Based on information transmission network and international standard protocol, CPIP, a public facility, has been constructed to realize information exchange and to provide various kinds of services. Under the support of many other existing public network platforms, it can create an environment for information gathering, processing, distribution, exchanging and sharing, as well as information service providing and information management. It is a pivot of Beijing public information gathering, and a gateway for inter-communication and information exchange of professional industry. Covering all the 10 districts in Beijing, the CPIP is interconnected with key information resources such as CHINANET, CSTNET, CERNET, and CHINAGBN. Today the CPIP, providing the fundamental services for cyber Beijing, is one of the largest IP-VPN based metropolitan area networks in Beijing and utilized by most of the government entities. It is also an important port for information exchange between Beijing and other cities both home and abroad. It also acts as a joint point, facilitating the inter-connection between cable TV network and multi-media broadband tele-communication network. The network structure of CPIP is shown in figure 2.

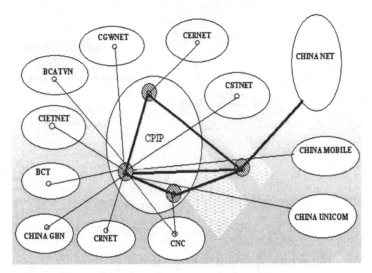

Fig. 2. Network structure of CPIP

4.2   Develop the Portal Website for the Municipal Government (http://www.beijing.gov.cn)

Beijing-China project - "the window of the capital" opened in 07/01/1998. As a key component of "Digitalization of Beijing" program, it is a uniform and unique portal for E-GOV, currently linking over 100 websites of different government departments in Beijing. The multi-language versions of the portal are shown as below.

Fig. 3. Chinese Simplified Version

4.3   Construct the Internal Network for the Government Departments and Computerize the Civil Services

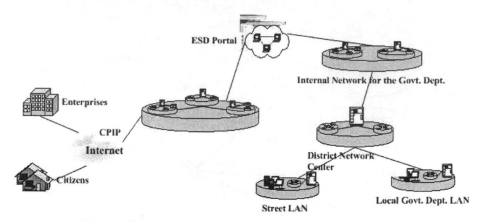

Fig. 4. Network architecture of E-GOV

4.4   Promote Important Application Projects

- Beijing Medical Insurance Information Subsystem & Beijing Citizen Card: the system will incorporate personal data of about six million insured people in Beijing by the end of 2002. Upon completion, it will be one of the most advanced social security information systems in China.
- Beijing Community Services Information System: the system is designed to cover all the 18 districts in Beijing and to set up a hotline call center system incorporating the 149 street-level service stations.
- Beijing Municipal Government e-Procurement System: it is an online bidding system enabling the Beijing Municipal Government, by inviting both domestic and foreign suppliers, to carry out procurement activities on a more open and transparent basis.

5   The E-GOV Action in Beijing: Second Stage (2001~2002)

- Upgrade the Organization, Reform the management modes and optimize the business process
- Integrate all information resources by building and updating a batch of administration database
- Launch the ESD (Electronic Services Delivery) schema to realize the electronic transaction between enterprises, citizens and the government. The ESD platform will provide a unified authentication and authorization service for government department, allowing the users to conduct their business with their counterparts in a secure manner, using any application, at anytime and from anywhere
- Continue to perfect the CPIP and speed up establishing the broadband network administration system

- Continue to improve the portal website – "the window of the capital", bringing innovative, interactive and efficient delivery to the Beijing government as well as high quality online services to the public

  1. Realize Single Sign-On for all departmental services, one identity across all departmental services.
  2. As part of EDS, an application integration platform was implemented, featuring the following aspects:
  3. Integration with the BackOffice Transaction System
  4. Reliable data transmission based on the XML integration
  5. Routing and dispatching of the workflow

- Carry out overall planning and security protection, formulate unique standard, relevant policies and regulations, and create a good soft-environment for implementation of E-GOV

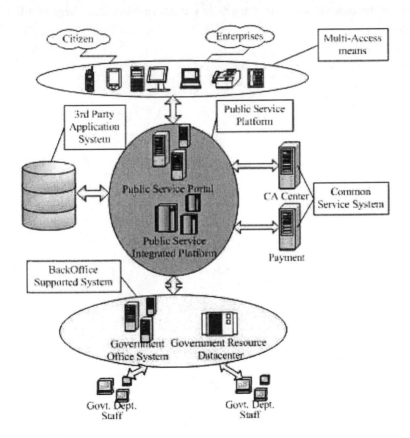

Fig. 5. Framework of ESD

6   The E-GOV Action in Beijing: Third Stage (2003-2005)

To establish a systematic, well-structured electronic services delivery system based on the high-speed, broadband municipal network, build one of the best E-GOV in China.

7   Conclusion

Over the past five years, the Beijing municipal government provided a lot of financial support for the programs related to the E-GOV Action plan, and will continue to offer help. But other than the money problem, the most important and difficult part involving the implementation of the action plan is the mindset change that is a requirement for all the citizens. Beijing will go ahead with its efforts to carry out E-GOV and hope to play a leading role in this particular field, providing more and better services for the industries and the citizens in the information era. There is still a long way to go.

The POWER-Light Version:
Improving Legal Quality under Time Pressure

Tom M. van Engers and Radboud A.W. Vanlerberghe

t.m.van.engers@acm.org, ravanl@oimp.nl

Abstract. The Dutch Tax and Customs Administration conducts a research program Program for an Ontology-based Working Environment for Rules and regulations (POWER). In this research program that was started in 1999 and is sponsored by the European Commission (E-POWER) since September 2001 an ICT-based methodology has been developed that enables the formalization of legal sources and finally the design of legal knowledge-based systems. The full-scale POWER-method however although much less time consuming than normal software design methodologies is still too elaborate especially if we want to apply this method in legal drafting or policy making processes. We therefore created the POWER-light version, a variant of the POWER-method that helps to improve legal quality and can be used with relatively little effort and in short time. Although the POWER-light version lacks many of the advantages of the regular POWER-method (e.g. its verification, simulation and knowledge-based component generation abilities) it offers a first step. The POWER-light approach offers the tools to get the best possible legal quality given the time restrictions.

1 Introduction

One of the goals of E-Government is providing citizens with means to access the governmental body of knowledge. This knowledge is based upon legislation, but also incorporates the business policy and interpretation that is added to the explicit knowledge corpus as it is reflected in the many legal documents like the different laws, regulations, case law etc.

In the POWER research program (Program for an Ontology-based Working Environment for Rules and regulations) a method and different supporting tools have been developed that support the chain of processes from drafting to implementation. Central to the POWER-method is a formalization process in which the legal knowledge sources are captured and translated into formal models, which we refer to as POWER-models (see Van Engers and Glassée 2001). These formal models are the basis for the systems development process(in which we create knowledge-based components) that in many cases follows the modeling process. The POWER-models are also used to detect (potential) defects in the knowledge sources e.g. inconsistencies and circularities (see Spreeuwenberg et al. 2001). The formal models can be used for simulation of the effects of (new) legislation as well.

It is obvious that the initial legal quality has great impact on the quality of the (e-) governmental services that are based upon it. Simulation of legal effects and verification of the (legal) knowledge sources helps to improve legal quality.

R. Traunmüller and K. Lenk (Eds.): EGOV 2002, LNCS 2456, pp. 75–83, 2002.
© Springer-Verlag Berlin Heidelberg 2002

In this paper we will give a brief description of the POWER-method.

While this method already reduces implementation time and improves legal quality, it is still too time consuming if the focus of the POWER-user is limited to the political decision making process. In this process legislation drafters, sometimes working closely together with knowledge groups in the public administrations, have to produce drafts under enormous time pressure. This leaves almost no time to integrally apply the POWER-method. Therefore we designed a POWER-light version. This method is specifically suited to conduct a quick scan of legal quality and can be applied with very little effort. That way even in the pressure cooker of the political decision-making process it is still possible to perform a quality check on draft legislation. In this paper we will explain the POWER-light method and show some experiences with this method in a recent legislation drafting process.

2   Managing Corporate Knowledge

If we want to model the knowledge of public administrations or other organizations that execute regulations, it seems best to focus on the existing documentary knowledge sources first before eliciting experts in order to model the knowledge of a certain domain. In most cases these documents contain the 'rules' in the form of an informal (or pseudo-formal) representation, e.g. the income tax law. Interpretation of these informal expressions is needed. This interpretation reflects the opinions of the public administration and consequently influences the business policy.

We consequently have to capture the expert knowledge to establish the correct interpretation of these documentary knowledge sources. The experts are also consulted to understand the processes in which the domain knowledge is used. This process knowledge is expressed in process models.

Usually experts from different disciplines and backgrounds are involved in the knowledge specification processes. Their knowledge is made explicit with help of the knowledge engineers, knowledge that would otherwise have remained implicit. The knowledge can furthermore be specified in a way that makes it easier to establish its validity. In addition to improving the efficiency of constituency treatment in its operational units, the knowledge-based systems serve primarily as a dissemination vehicle allowing the DTCA to make more effective use of the knowledge of its sparse experts, while improving the quality of law enforcement[1] in its operational units. The POWER program elaborates on that insight. The focus of the POWER program differs from 'traditional' knowledge engineering approaches (see for example Sudkamp 1988). In POWER we focus primarily on the knowledge specification. This specification can be used to create knowledge-based systems but this is just one application form. The POWER knowledge specifications are also used for enhancing the quality of legislation as well as for e.g. policy-impact analysis. The position of the POWER method is depicted in figure 1.

---

[1] Quality of law enforcement is defined as satisfaction of the constituency with the adoption of the principles of equality before law, predictability of law enforcement and proper use of authority by law enforcement agencies.

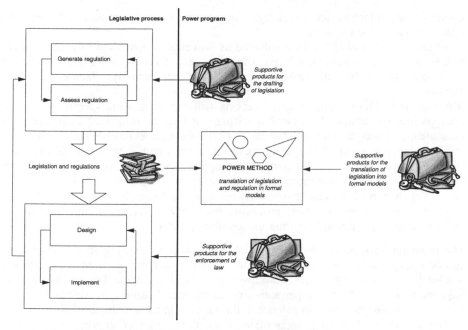

Fig. 1. The POWER method and tool in the legislative process

In common modeling approaches used in traditional information science process and domain knowledge are intermingled, if not in the different models or design documents that are used than in many cases in the software code. In the AI (expert systems or knowledge-based systems approach) the knowledge is separated from the inferences that are made upon it. But although a knowledge-based approach already has some advantages compared to traditional modeling methods we could even further improve the strength of knowledge modeling by besides distinguishing and separating process from domain knowledge also distinguishing the different knowledge sources and include explicit references to the original sources (and even specific distinct parts of these sources) in our models. By creating knowledge models this way, organizations get a means to manage their knowledge.

3   The POWER Modeling Process

The POWER modeling process first legal sources, such as law, executive measures both from the authorized ministry and the executive bodies, but also from expertise groups from such bodies that are active within the field of application, are translated into formal specifications. These formal specifications are termed conceptual models. These initial conceptual models closely reflect the structure and nature of the legal sources.

In most cases, however, this will result in weakly structured models due to the nature of the political processes that produced the legal source. Their emphasis is on reaching an agreement between the negotiating parties. Hence a second process that combines and refactors dependent legal sources into a consistent and well-structured conceptual model is required.

These processes need not be conducted as sequential steps. Typically, the modeler will select a legal source, most likely a few chapters of law, he/she knows is relevant to the domain, conduct the translation process and then partially refactor the specifications.

Verification of those resulting specifications will expose missing specifications, and perhaps inconsistencies. Additional specifications may be translated from executive measures, and inconsistencies may be resolved through executive decisions of the authorized expert group. Refactoring the specifications with these newly added specifications results in a more consistent and complete conceptual model. Eventually, the verification process demonstrates the consistency and completeness for the task at hand. The design of the enterprises software systems for enabling integration of the task into the executive business processes may be well underway by that time, in order to achieve the throughput time-to-execution required.

The modeling process is also likely to take advantage of existing specifications that can be reused for new executive tasks, and needs to take into account the process of reaching a political agreement, resulting in frequent changes. Tractability will be required both to verify the proper authority for the specifications, as well as for effectively maintaining this correspondence in the face of changes in the legal sources.

The legislation-drafting process often takes place under an enormous time pressure. This is due to the political consensus making process that leaves almost no time for the drafter to incorporate the outcomes of this process in a new draft. We therefore developed a POWER-light version that is based upon the original POWER-method, but consists of a more global approach. The consequence is that no formal models are produced and consequently the formal verification process, simulation of effects as well as the generation of knowledge-based components are not supported. The purpose of the POWER-light version is to support a global first scan of the draft's quality within a very short period of time. This can also be the first step to a full scale POWER-modeling and analysis process.

4   POWER-Light

The POWER-light method is comprised of two separate analysis steps; Structure analysis and Domain modeling.

The first step of the POWER-light method consists of a structural analysis of the legal knowledge sources (documents). Legal documents typically have an invariant structure that is used to construct meaning (e.g. lex specialis is constructed from the order of the members in the law). Another characteristic of legal texts is the use of references, both absolute references (e.g. 'article 3.11') or relative (e.g. 'the previous member'). Problems often occur during the drafting process due to the time pressure. Many drafters may be involved in a drafting process and even when special attention is paid to co-ordination legislative texts contain structural errors. Correcting these errors after a new law has become effective (i.e. after the political decision-making process has taken place) is costly (and in practice only happens when additional changes have to be made). Detecting these errors prior to subjecting them to a political decision may eliminate the need for these costly corrections.

The second part of the POWER-light method deals with possible defects regarding the system of the legislative text on a material level. Co-operation of different legislative drafters under time and pressure occasionally leads to inconsistencies or redundancies within the text. By trying to model (parts of) the legislative text some of these defects may be detected and consequently corrected before the legislative text is passed.

4.1   The Structural Analysis

Most legislative texts have a distinct (and invariant) structure (the invariance may be law-family dependent). Since legislation drafting is a dynamic process, often executed under time pressure with involvement of a group of people, it is prone to structural defects. These defects may cause serious problems in the operational units of the governments bodies. An example of such structural defect is a reference to a missing member (e.g. a reference to a concept that should be defined in a previous member).

We developed a software tool (Juridical Editor and Working Environment for Legislation drafters, in short JEWEL) that supports the structural analysis of legal documents.

JEWEL supports the detection of the most common structural defects:

Discontinuity in the numbering of structure blocks.

Legal texts are composed of structural elements (such as sections, article, members, sentences etc.). We call these structural elements structure blocks. To be able to refer to these structure blocks legislation drafters use a form of numbering. While creating and updating legislative texts structure-blocks are frequently edited, replaced and/or added. This may result in e.g. a numbering of the structure-blocks that is not continuous, structure-blocks that are placed out of sequence etc. Maintenance of the numbering is part of legislation drafting.

Incorrect use of references.

References in a text that have been heavily modified can be defective in two ways. The first possibility is that a reference refers to a structure-block that no longer exists. This type of reference defect is easy to detect and the automated tool will show that specific reference to be 'empty'. The other type of reference defect is harder to spot. This defect involves references to existing structure-blocks that do not have content relating to the issue addressed by the reference. Although it is possible for analysts to identify a few of these defects, only experts can truly expose these defects.

Inconsistent hierarchy.

Most legislative texts possess a structure that encompass a specific hierarchy. In the co-operation between different departments the use of two different hierarchy-types will diminish the readability of the legislative text and hampers future adjustments. Automated detection is possible, because as a result of the use of two different hierarchy-types, the structure-analysis will not function properly.

Defects found in this part of the analysis are, apart from the content-reference defect, self-apparent and have no need for validation by a legal expert.

4.2   Domain Analysis

During legislative drafting different people work on small parts of a legislative text. As a result it is difficult for the final editor to maintain an overall consistency throughout the entire text. In a first attempt to model a legislative text, many of these potential systematic defects come to light. By not only restricting oneself to the translation of legislative texts into conceptual models but also take into account integration and process-design issues that become apparent during translating, even more potential defects are identified in an early stage of the legislation design process.

From this analyses the following potential defects can be identified:

Definition of concepts and terms. Defects regarding the definition of concepts can take several forms.

First and foremost is ambiguity. Through attempting to model the definitions in a formal language, it often becomes apparent that more than one way of modeling the text is possible. By pointing out these options legislative drafters can be made aware of those passages in the draft that need further or different specification.

Conflicts in definitions may arise from having several people work on the same legislation. By trying to define the constraints regarding a single concept, these conflicts are brought to light.

Another important potential defect is incompleteness of the text. By defining the scope of each definition and combining these it is possible to check whether all cases that are to be addressed by the prospective legislation are accounted for.

Redundancies. By constructing a simple decision model it is possible to see whether all segments of the legislative text are used in enforcing the law.

Vagueness. Due to the time pressure legislative drafters often do not have the time to work out all legislative rules out in detail at first, but intend to do so in a later stage of the drafting process. Sometimes however certain of these "rough" rules are overlooked in these later stages. By using a simple decision model knowledge engineers check whether all conclusions enclosed in the legislative text can be reached through the use of common knowledge/sense.

5   Operating Procedure (Putting the Quick-Scan into Practice)

The first step in the POWER light procedure is to have the text parsed by an automated tool built for just that purpose called JEWEL. The parser brings to light structural defects in two different manners. It is able to recognize defects that pertain to incorrect numbering of structure-blocks. The other way that defects in the structure become apparent is when the parser is unable to recognize parts of the text and fails to mark these parts as structure-blocks. The defects in the latter category are usually the result of existing legislative texts being modified over the years.

Additional to the automated detection of structural defects the text is scanned manually to determine whether the structure-blocks that are referred to exist, and to check wether the concept that is linked to the reference can be found back in that structure-block. This ensures that whenever a reference refers to an existing structure-block, it refers to the correct structure-block.

The next step is to export the structured text to a case tool for further modeling. This is supported by another tool developed in POWER: OPAL (Ontology-based Parser for the Analysis of Legal texts). This tool automates the building of a prototypical conceptual model that is used in the regular POWER-method. OPAL analyses the text using a lexicon and grammar to identify concepts and their relation to one another. The concepts and relations are translated into types and associations. During this translation OPAL presents different modeling options to the knowledge analists. These options may indicate that in certain cases two or more different ways of reading the text are present.

As with the structure analysis the automated translation is followed by a manual one.

At this point the POWER light method starts to differ from the regular POWER method. Normally the normative expressions in the domain are translated into formal representations (in OCL-invariants). In the POWER-light version we stop when a basic conceptual model containing the concepts and relationships (associations) is produced. Consequently the entire text is scanned for issues that would pose difficulties in the translation of the normative expressions, integration of partial models and task mapping. This is done by attempting to construct both partial integrated models and constructing a simple decision model based upon the text and its intended meaning.

All potentials defects found in the entire procedure are then grouped by the structure blocks in which they occur (allowing us to refer to the original knowledge source). At the present time these potential defects are specified in two ways. Where the text poses problems for modeling the potential defect it is presented in the form of a question. (e.g. "Is it correct that with regard to this specific legislation investment costs may be deducted?). When potential defects do not qualify as ambiguity in the text these defects are presented using specific cases that address the defect. (e.g. In this legislation a boat would be considered to be a truck for the purpose of taxation. Is this the intended meaning?) The report is presented to the legislation drafters. They will than reflect on these issues and clarify possible misunderstandings. This results in a report with the actual defects instead of potential defects.

Depending on the time available several versions of draft legislation can be reviewed this way.

6   Benefits of POWER-Light

Knowledge engineers can add a fresh look at the same issues legislative drafters work on. By critical reading and asking for clarification and questioning decisions made by drafters, knowledge engineers force legislative drafters to reconsider and reweigh issues, possibly resulting in adjustments of that legislation. Not only will this clarify explicit choices that have been made, but more importantly it will also expose implicit assumptions used by the drafters, assumptions that might needlessly lead to restrictions on the set of possible implementation scenarios considered.

POWER-light offers a systematical approach that enables us to enhance the quality of the legislation. It provides drafters with an extra control that helps clearing defects.

Further more POWER-light helps to create legislation that will be more easily implemented by considering enforcement issues (in the task-mapping step) in an early stage of the legislative process.

7   Experiences with POWER-Light So Far

Although the POWER-light method has only recently been developed and is still in an early development stage, it has already been put into practice in two different legal domains. The first concerned the adaptation of an existing law (succession, in Dutch successie), the second concerned the development of new legislation (mileage taxation, in Dutch kilometerheffing).

In the first case we focused on three constitutes of the legal domain, 'partner begrip' (partnership), 'algemeen nut' (general profit) and 'bedrijfsopvolging' (business succesion). A new element was added; 'constructie van conserverende aanslagen' (construction of preliminary taxation). In this case the POWER-light method was applied when most of the legislation had got its shape and was already approved by the Ministry of Finance. However the application of the light version did yield several benefits. The analyses pointed out areas where further legislation was desirable. Furthermore it provided the enforcers of that legislation with an insight into the system of that legislation, thereby improving their ability to implement the legislation.

The second application of the POWER light method is still a continuing project (mileage taxation, in Dutch kilometerheffing). Even though the project has not been finished yet, benefits of POWER light can already be identified. In this case the POWER light method was involved in an early stage of new legislation. The major contribution of the analyses was the reconsidering of the basis assumptions of the legislation drafters. By basing themselves upon existing legislation the legislation drafters had overlooked the system of that legislation. They than proceeded to use parts of that legislation without including other parts that were crucial to the legislation as a whole.

Furthermore an implicit assumption was that by using parts of one legal text all relevant case law regarding that text would be included in the new legislation. By making that assumption explicit the accompanying text with the legislation was revised and as a result it was ensured that judges would support the assumption. As a final result several concepts within the text where elaborated upon in order to ensure unity of policy in the several units responsible for enforcing the legislation. At this point the legislation has not been passed yet and is expected to undergo several more changes. The legislation drafters involved have requested that the POWER-light method will be applied to the future versions of the legislation.

8   Conclusions

The POWER-light version of the POWER-method helps to improve legal quality and can be used with relatively little effort and in short time. Although the POWER-light version lacks many of the advantages of the regular POWER-method (e.g. its verification, simulation and knowledge-based component generation abilities) it offers a first step. Legislation drafters and people from the knowledge groups involved in drafting new legislation often have to work under enormous time pressure. It is evident that legal quality has great impact on the citizens' compliance. The POWER-light approach offers the tools to get the best possible legal quality given the time restrictions. By consequently applying the regular POWER-method when in the implementation stage the time pressure has been a little reduced, we can create knowledge-based applications that provide better E-services to the citizen.

References

1. Van Engers, T.M., Glassée, E., 2001, Facilitating the Legislation Process Using a Shared Conceptual Model, in IEEE Intelligent Systems January/February 2001 p 50-58.
2. Van Engers, T.M., Kordelaar, P.J.M., Ter Horst, E.A., POWER to the E-Government, in 2001 Knowledge Management in e-Government KMGov-2001, IFIP, ISBN 3 85487 246 1.
3. Spreeuwenberg, S., Van Engers, T.M., Gerrits, R.,2001, The Role of Verification in Improving the Quality of Legal Decision-Making, in Legal Knowledge and Information Systems, IOS press, ISSN 0922-6389.
4. Sudkamp,T. A., 1988, Languages and Machines; An Introduction to the Theory of Computer Science, Reading Mass, Addison-Wesley Publishing Company,

Intranet "Saarland*Plus*" – Enabling New Methods of Cooperation within the Ministerial Administration

Benedikt Gursch, Christian Seel, and Öner Güngöz

Institute for Information Systems (IWi)
at the German Research Center for Artificial Intelligence (DFKI),
Stuhlsatzenhausweg 3, Bd. 43.8, 66123 Saarbruecken, Germany
{Gursch,Seel,Guengoez}@iwi.uni-sb.de

Abstract. The potentials of the information and communication technology become more and more important for the optimization and support of the administrational work processes. The Saarland state government sets a special focus on the usage of the new technologies within the public administration and attaches value to it in the context of a global E-business strategy.
The presented document gives a short overview on the implementation of an intranet as a platform for ministerial comprehensive communication.

1   Introduction

The usage of the Information and communication technologies (ICT) creates beneficial potential of optimisation for the necessary restructuring of the public administration. In general, the digitalisation of information and communication as well as the rapid rise of the internet as a central resource for information exchange facilitates the integration of value-added chains independent from time and space. From the internal perspective, this enables new potentials of co-operation within and between administrative authorities, for example by creating virtual project rooms.[1]

According to this trend, an E-Business Strategy was developed by the state government of Saarland – one of the smaller federal states of Germany – as far back as the beginning of the year 2000. Included are different strategic projects, which give impulses for the efficient arrangement of administration procedures and which should set milestones for the modernisation of administration. The spreading of ICT, new forms of work and education, as well as the creation of jobs within the technical and service sector are in focus along with modernising administration [2].

The project "Intranet Saarland*Plus*", a subsidiary project of the strategic project "Networked State Government" aims at the optimization of the information and knowledge exchange within and between the different ministerial administrative authorities. This is achieved by the implementation of an intranet solution as a departmental comprehensive communication platform. Using this project as case-study, the presented paper provides in the following a short overview of the gained implementation polices and experiences.

R. Traunmüller and K. Lenk (Eds.): EGOV 2002, LNCS 2456, pp. 84–87, 2002.

2   Implementing the Intranet Saarland*Plus* – Experiences within the Ministerial Administration

2.1   Business Case and Challenges

The ministerial administration as the area of application for the intranet Saarland*Plus* is characterized by a dual role. On the one hand, duties of a supervising and intervening administration have to be fulfilled. On the other, the ministerial administration provides services for specific groups of citizen. As a result, the addressees can emerge at the same time as recipients of the administration's output and as subject to the supervision of the ministries. There is only a rare contact to the citizen, moreover the main target group of the services performed is built by commercial and social partners who play the role of a intermediary to the primary beneficiaries [3].

The state government of Saarland consists of eight departments and employs an overall amount of 25.000 people. One can see, just from the number of employees involved, that a radical reorganisation of work processes in the sense of a Business Process Reengineering (BPR) would be problematic, the more so as the existing procedures have to correspond to certain formal criteria according to the described duties. Furthermore, cultural barriers as well as resistance against organisational changes represent a vital obstacle. However, to improve work processes and the flow of information, a Continuous Process Improvement (CPI) approach was pursued to implement the system. Consequently, the Intranet "Saarland*Plus*" will enable that

- the knowledge is available at the location where it is needed, and
- the knowledge is stored in the place where it originated.

In addition, the Intranet "Saarland*Plus*" permits the adaptation of the internet technologies already used for the state government's web presence to the internal structures.

2.2   Conception and Implementation

The vision for Intranet "SaarlandPlus" was to provide an integrated communication platform that would incorporate the existing isolated intranet applications within the state government and that would support the co-operation within and between the various ministerial departments. To achieve this goal an integrated framework was developed that anticipates the various requirements within the state government. Considering the organizational structure described above it was necessary to realize the intranet in three expansion stages.

The first expansion stage is available for all state employees accessing the intranet. Information relevant to the whole workforce are stored there. Lists of telephone numbers and addresses, directories of legal regulations, work distribution plans, job advertisements, ICT related information as well as information concerning departmental vocational trainings fall under this category.

The second expansion stage contains the design of department specific sections. It is directed at the staff of each department and is only accessible for the particular employees. Announcements of the executive board, information of the staff council, important internal appointments or the internal departmental employee newspaper can be accessed here.

With *the third expansion stage*, cross-departmental project teams can use the intranet as a virtual project room in order to exchange relevant information for their team, e.g. appointments, protocols and other internal data.

Each of the described stages follows a procedural model containing four phases – requirement analysis, main and detailed concept, implementation and evaluation. During the *requirements analysis*, the requirements referring to the contents and functional range of the intranet are determined by the employees and the decision makers. This parallel top-down and bottom-up proceeding is necessary to detect any differences in perception at an early stage. The framework concept results from an iterative process, which has to take the results of the requirements analysis into consideration. It comprises a *main and a detailed concept* of hardware and software as well as the contents to be implemented. This is extended through a maintenance concept, a marketing strategy and a training concept. The *implementation* contains the technical and organisational implementation of the specifications which were compiled in the framework concept. The results of the implementation are then aligned with needs of the users during *evaluation*. As a consequence, this guarantees that the intranet is adapted to the user requirements and that possible barriers of use are anticipated.

The described phases were encapsulated by an *integrated framework* focusing on the change management process, the organizational structures, the technical infrastructure and the contents (information and knowledge). This ensures the active participation of the executives and employees during all phases.

The *change management process* was one of the most important parts of the project. The essential element of a successful change management is the commu-nication. Therefore several efforts were taken to enforce an open-minded information policy. The deployed communication tools ranged from kick-off-meetings in every department, e-mail-newsletters, circular letters and workshops to training, roadshows and marketing measures at the official starting date of the intranet. This integrated approach of treating the employees like customers and integrating them in every stage and phase of the development ensured a successful change management process.

Nevertheless, change management and communication are not sufficient to change given organizational structures. Therefore it was necessary to define *areas of responsibility* and assign them to specific employees. This areas of responsibility consisted in:

- *Organization:* Support for the intranet activities and promotion of the intranet in the departments and building an internal support-network for the intranet.
- *Content:* Support for the content provider and training of the employees. Designation of a contact person in every department for content-specific questions and a coordinator of the content providers.
- *Technique:* Support for technical issues and especially for the installation and support of the software needed to access the information via the intranet.

The process of assigning responsibilities was strongly enhanced by the fact that the project was residing in the staff office of innovation, research and technology that is assigned to the director of the state chancellery.

An important premise for the technical implementation of the intranet was represented by the existing content management system (CMS). RedDot was already in use for the state government's internet presence and had to be also utilised as the platform for the intranet. This led to several technical and organizational challenges:

- The chosen CMS is a very powerful tool, but it is not very useful for integrating content of different departments. As the integration was a main part of the conception, it was quite a challenge to solve this problem.
- The RedDot-technology which is used to maintain the content is quite simple, but requires a basic understanding of the difference of a web editor and a word processing software. It was another challenge to train the employees according to their knowledge in the use of the RedDot system.
- There was a bottle-neck in the RedDot implementation, caused by an internal service provider who was also responsible for the internet presence using the same platform.

To achieve a broad acceptance and usage of the intranet, the content has to be useful for the work process of the employees and, in addition to that, the installation of content has to be very simple. Using the existing CMS was one step in this process, another consisted in providing a general mandatory procedure for evaluating, converting and providing the contents for the intranet. Finally the accessibility was enhanced through usability-studies and corresponding measures.

3 Conclusion

The chosen procedural model has proven itself to be very successful. Due to the active integration of the employees, executives and stakeholders during the requirement analysis, it was possible to achieve consensus among all those involved. In addition, the knowledge resulting from the already existing isolated intranet applications was very useful. During the different phases, the integration of the employees and anticipation of cultural restraints were quite significant. The chosen expansion stage concept proved to be quite advantageous here. Through step-by-step expansion, employees had the opportunity to become familiar with the system and to actively participate in the formation of their own information and communication platform. The next steps of the projects will consist in the continuous expansion of the intranet's contents as well as the sustainable institutionalisation of intranet related activities within the ministerial administrations.

References

1. PwC: Die Zukunft heißt E-Government - Deutschlands Städte auf dem Weg zur virtuellen Verwaltung, Frankfurt (2000), p. 10.
2. Staatskanzlei des Saarlands: Szenario für eine eBusiness-Landschaft im Saarland, http://www.staatskanzlei.saarland.de/innovation_2121.htm
3. Breitling, M.; Heckmann, M.; Luzius, M.: Nüttgens, M.: Service Engineering in der Ministerialverwaltung. IM Information Management & Consulting, 13 (1998), pp. 91-98.

e-Learning for e-Government

Michel R. Klein and Jacques Dang

HEC School of management, 78350 Jouy-en-Josas, France
{Kleinm,Dang}@hec.fr

Abstract. The present paper presents some trends in educational activities for central and local government in France. It describes briefly the main characteristics of a brokerage platform used to import and export digitalized educational materials while protecting intellectual property rights.

1   Trends in Education for Central and Local Government

Central and local governments are facing the same kind of problems that other educational organisations, in particular private business organisations, are facing with respect to the continuous education of their personnel.

- the competence of the local government personnel is, as anywhere else, a prerequisite for improving the efficiency of administrative services.
- the persons in charge of providing continuous education are subject to budget constraint as anywhere else.
- there is a trend toward organising mix seminars where some teaching is done face to face but there are periods during which the participants interact with instructors from the distance.
- they have to help their personnel to receive training concerning new topics and software.

To these standard constraints are added others which are known to anybody who has been working for local government. Financial incentive used to reward improvement in competence is nearly non existent since promotion is essentially linked to age. Of course, there is always the reward of doing more interesting things, but in spite of the existence of this motivation for the personnel, it is unfortunately true, in our experience, that persons in authority , in our case elected members of the town council and the Mayor , are usually not very keen on professional education and put their energy on other matters.(like politics, for example!)

Another problem is that the public of potential learners in local government are spread all around the country and so it is natural that the educational centres are also spread over the country.

In France continuous education for local government is provided essentially by the Centre National de Formation des Personnels Territoriaux (CNFPT) which is offering seminars in about ten different educational centres and longer educational programs in four schools in Anger, Strasbourg, Montpellier and Nancy. Towns and other local governments can also use services of the local universities and private companies. The CNFPT is itself a public institution financed through a compulsory

R. Traunmüller and K. Lenk (Eds.): EGOV 2002, LNCS 2456, pp. 88–92, 2002.
© Springer-Verlag Berlin Heidelberg 2002

tax paid by towns its seminars are free and towns rely mostly on its service for their basic continuous educational needs.

As a consequence once content has been developed by an expert on a particular topic it is necessary to educate and support local instructors who are going to teach this content in decentralised centres.

This constraint is all the more important since the decentralised centres have a fairly high degree of autonomy. They select their own instructors who are usually local professionals. The standard procedure is to educate local instructors to the content and let them use it in courses or seminars run locally.

One example in France concerns the educational needs which were generated by the new accounting plan M14 for local government which became mandatory a few years ago in all French towns.

The content of the course to introduce accounting personnel to the new accounting plan was developed by a group of experts. Then it was neccssary to familiarise the local instructors with this content. These instructors were usually financial managers of medium size towns. For some specialised topics such as the impact of the change of the accounting plan on the financial management of towns it was a need to train instructors in the use of specialised software. In such seminars it is usually greatly appreciated to have an expert taking part in the seminar who can give the benefit of his experience.

In certain domains the trend is moving fairly fast. The development of training required with respect to office systems and Internet is a good example of such a domain. The type of competence to acquire is not different from that of any other organisation private or public. In such a situation it is often good strategy to re-use existing course material and learning resources which have been developed elsewhere. In such a situation it is useful to have outside support in order to:

-   import and export educational material in electronic form
-   support instructors in mastering the learning resources
-   make it possible for experts to participate from the distance under the form of a live lecture through video (Klein,1998,2001).

2   Learning Resources, Course Description and Learning Activities

A learning resource in our context is something which is used to transfer knowledge during a course. It can be a case study, a case studies guide, a set of slides to introduce basic concepts of a topic of teaching, a chapter of a book on a given topic, an article, a collection of exercises with solutions etc...

The basic descriptors of a learning resource (LR) are: discipline to which it belongs, its type (case studies, complete course, syllabus,...), the key words which helps to define its topic, the language used, the level of dealing with the subject (elementary, medium, advanced..), the author, the educational institution , etc...
Persons in charge of teaching a topic are usually looking for a complete set of LR to teach this topic or a specific LR.

A learning activity is something dynamic such a live lecture , the coaching of a group of students working on a project, a business game session , etc...

3   The Use of a Brokerage Platform for Learning Resources

We shall now describe briefly the characteristics of one particular brokerage platform, which can be use to help solve some of the problem we have mentioned above. This software was developed within the framework of the European Project UNIVERSAL For more detail the reader is referred to [Brantner et als,2001].

3.1   Main Function of the System

The main functions of the system can best be introduced by presenting the main window of the system.(see fig 1). These functions are:

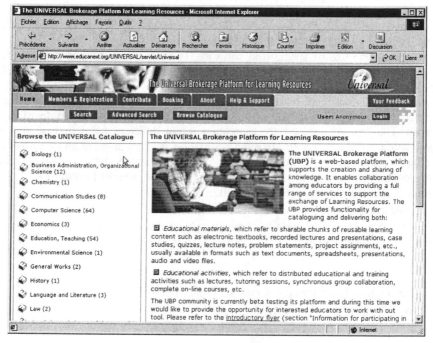

Fig. 1. Main Windows of the UNIVERSAL system

Contribute: this function allows a content provider (expert or a faculty member) who has been developing a course or just a given Learning Resource (LR) to add the **description** of the course or of the learning resource to the catalogue and to up-load in electronic form the learning resource itself: on the platform.

Booking: this function allows the instructor to download a LR once he has accepted the terms of use. These terms of use take the form of a contract. It is a this

Members and Registration. This function allow a potential user (content provider, instructor, student) to register to the system, to use its functions and to communicate with other authorised persons.

Search: allows a teacher or student to search for a given learning resource according to various criteria and to access the syllabus and document used in a seminars or course.

help & assistance: Provides access to a short introduction to the use of the system, in the future to its user manual.

3.2   Description of the Functions

3.2.1   Registering to the System

This function allows an institution to provide a list of persons belonging to the institution (faculties, students, ..) who are authorised to use the system. The registered user is provided with a user name and password. The system then knows to which institution the user belongs.

As a consequence a registered user of a given institution can be authorised to access the LR put together by a group of educational institutions. For example all Business Schools of a given alliance can decide to allow their faculties to access each others LR.

3.2.2   Searching for a Course Description or a Learning Resource

This function allows a registered user to search for a specific learning resource: for example an introduction to the town accounting plan or a complete course or seminar to a topic at a certain level such as an introductory seminar to the accounting plan of French town. In this latter case the user may wish to search for course descriptions on a given topic, ie their syllabus, describing the course with the topics treated at each class and the associated LR. Two types of search are allowed in UNIVERSAL simple and advanced.

3.2.3   Booking and Delivery of a Learning Resource

This function allows a registered user to book and import a learning resource. At the present stage two types of contracts are available to book a LR.. free access or restricted access to the members of a given organisation or association of organisations. The contract is there to provide protection of intellectual property rights and is a key element for decision by potential providers of LR.

3.2.4   Contribute

This function allows a content provider to:
- enter the description of the learning resources he wishes to make accessible.
- define the terms of the contract he wishes to have with potential users
- make the learning resource accessible from his own Universal interoperable server or upload it to the Universal Brokerage Platform (UBP) to make it accessible from the UBP directly.

4   Supporting Educational Activities

One important feature of UNIVERSAL is to provide a multi point IP- based video function called ISABEL. This function allows the teacher to teach to his students face to face and from the distance to other students through video. This service is presently running in the Linux environment. This function was, for example, used in a joint virtual course with students at HEC near Paris and Wirtschaft University in Vienna UNIVERSAL can automatically set up video conference sessions on demand and monitor the status of the live connection. It provides a convenient interface for managing on line class sessions.

5   Conclusions

The system we have presented is not specifically designed to support the exchange of learning resources for government personnel. This system was designed to support educational activities for higher and continuous education in general. It can be used for the education in the governmental sector. The fundamental goal of such a system is to make more readily available what is existing to interested persons and to protect intellectual property rights of authors.

Bibliography and Web Sites

1. For ISABEL see http://Isabel.dit.upm.es
2. For UNIVERSAL see http://www.ist.Universal.org
3. Brantner Stephan, Enzi Thomas, Guth Suzanne, Neumann Gustaf, Simon Bernd, UNIVERSAL-Design and Implementation of a highly Flexible E-Market Place for Learning Resources, Infonova and Vienna University of Economics and Business Administration, Information System Dept., 2001
4. Klein Michel, Borgman Hans, PC-based video as a tool for supporting collaborative work in teaching and research in Management, Proceedings 3 rd International Conference , Louvain- la –Vieille, May 7-9, 1998.
5. Klein Michel, Gauthier Valérie, Mayon-White William, Rajkovic Vladislav, Developing Synergies between Faculty and Students of European Business Schools through Telecommunication and Computer Supported Cooperative Tools , Proceedings Fourteenth Bled Electronic Commerce Conference , O'Keefe Bob, Gricar Jose et als (Eds), Moderna Organizacija, 2001.
6. Klein Michel, Keravel Alain, Impact des recherches sur les technologies de l'information et de la communication et leurs usages sur la formation à la gestion , Recherche et Enseignement de la Gestion, B.Moingeon (Editeur) , L'Harmattan, 2002.

Multi-level Information Modeling
and Preservation of eGOV Data*

Richard Marciano[1], Bertram Ludäscher[1], Ilya Zaslavsky[1],
Reagan Moore[1], and Keith Pezzoli[2]

University of California San Diego, 9500 Gilman Drive, San Diego, CA 92093 USA
[1] San Diego Supercomputer Center, MC 0505
{marciano,ludaesch,zaslavsk,moore}@sdsc.edu,
http://www.sdsc.edu
[2] Urban Studies and Planning Program, MC 0517
kpezzoli@sdsc.edu, http://regionalworkbench.org

Abstract. This paper addresses the issue of long-term preservation of and access to digital government information. We show how the preservation process is enhanced by storing an infrastructure-independent representation of the raw data, together with a model dependency graph (an executable graph of database view mappings). This allows for the design of decision-support tools and services for improving government transparency and promoting citizen access to eGOV data. A case-study, the *Florida Ballots Project,* is used to illustrate the approach.

1   Introduction and Approach

A common demand is that e-Government services promote citizen access to government information [1], such as official records kept at an archival institution [3]. Today, thanks to the ubiquitous Web, access to digital data is often less of a problem than actual information content. We argue that a multi-level information or "deep" modeling approach combined with an appropriate infrastructure independent representation mechanism can greatly enhance the value of eGOV data to the interested public, future researchers, and "digital archeologists/historians". We use the 2000 U.S. Presidential Election as an example of the deep modeling approach.

"On behalf of the State Elections Canvassing Commission and in accordance with the laws of the State of Florida, I hereby declare Governor George W. Bush the winner of Florida's 25 Electoral Votes," said Florida's Secretary of State, Katherine Harris, as she certified George W. Bush the winner over Al Gore, on November 26, 2000.   The National Archives and Records Administration (NARA), went on to record this 25-Vote result by collecting two documents for permanent retention:

- *Certificate of Ascertainment,* containing the proposed Electors:
  http://www.nara.gov/fedreg/elctcoll/2000/certafl.html

* Work partially supported by NSF/NPACI ACI-9619020 award (National Archives and Records Administration / NARA supplement) and National Historical Publications and Records Commission / NHPRC award ("Methodologies for Preservation and Access of Software-dependent Electronic Records").

R. Traunmüller and K. Lenk (Eds.): EGOV 2002, LNCS 2456, pp. 93–100, 2002.
© Springer-Verlag Berlin Heidelberg 2002

- *Certificate of Vote,* capturing the winning Electors:
  http://www.nara.gov/fedreg/elctcoll/2000/certvfl.html

More recently, two election media studies, started rethinking the entire process:

(1) *USA Today / the Miami Herald* on April 4, 2001,
   http://www.cnn.com/2001/ALLPOLITICS/04/04/florida.recount.01/
(2) The NORC *Florida Ballots Project*[1], on November 12, 2001,
   http://www.norc.org/fl, the results of which we use in our case study.

These studies present parameters under which either candidate could have won.[2] They suggest that with the 25 Votes, one should consider the retention of a parameter space that captures greater context. Fig. 1 depicts a model dependency graph we derived from examining NORC, and tries to formally define such a suitable parameter space as an example of our "deep modeling" approach.

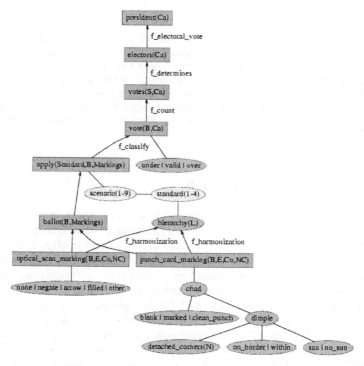

Fig. 1. 2000 Presidential Florida Election Model Dependency Graph, derived from NORC.

---

[1] At NORC, the National Opinion Research Center, at the University of Chicago, a consortium of news organizations including: the New York Times, the Wall Street Journal, the Washington Post Co. (The Washington Post, Newsweek), Tribune Publishing (LA Times, Chicago Tribune, etc.), CNN, Associated Press, and others. While the first study only looked at the *Undervote* (ballots rejected because no vote was recorded for the president), the second study looked at all 180,000 uncertified ballots in Florida's 67 counties, including the *Overvote* (ballots rejected because more than one presidential candidate was selected).

[2] Interactive analysis of the USA Today study at:
   http://usatoday.com/graphics/news/gra/gvote/frame.htm.

2   Multi-level Model Dependency Graphs

For several reasons, the depth of information modeling that corresponds to news reports and even official archival records is quite limited (e.g., the top three nodes in Fig.1), with NORC being a notable exception. One of the findings was that, depending on the specific scenario or applied standards, different outcomes can result.

The multiple possible outcomes can be made precise through "deep modeling" using a *graph of database mappings* as follows. The graph in Fig. 1 is an abstraction of such a graph (i.e., a network of "views" in the database sense). A view is a relational table that is defined by a query expression. Note that views can be layered and defined on top of other views, resulting in a graph of mappings. The overall graph itself defines a (complex) view, mapping the ("raw") input data to the final result (the President Elect in Fig.1). In general, a deep modeling approach using database views comprises:

1. *relational schemas* for all relevant entities and relationships (in the figure: parameterized entities and attributes)
2. *view definitions* (= database queries) precisely defining the mappings from one schema to another
3. *constraints* (logic formulas) over the relational schemas (to express, e.g., which standards can be applied to which ballot type)

Thus, in the actual graph, nodes correspond to relational views defined on top of other views or base tables. In our abstraction in Fig.1, nodes stand for *parameterized entities* (boxes) and *attributes* (ovals), while directed edges denote *database mappings*, i.e., functions between relational schemas.

Together with the raw data, the graph of database mappings can then be *executed* as a (complex) database query with a *verifiable* and non-controversial output. Of course this does not prevent a political controversy from happening, but it can be dealt with at a less superficial and more informed level: In Fig. 1, the *scenario* (see *Appendix E*) and/or the specific *standard* (see *Appendix D*) being applied to specific sets of ballots, uniquely determine the database tuples in the views above; in particular, the topmost tuple, i.e., which president should be named president elect. Thus, the only degree of freedom and non-determinism that such a graph of mappings allows is in the input data, in this case, the raw ballot data and the scenario/standard to apply.

NORC did everything to guarantee that the raw data was as objective as possible – in particular the coders did *not* compute the function *f_classify* themselves, i.e., they did not determine the votes. Instead every coder just described the markings seen and the *f_classify* determines the vote (under/valid/over) as a function of the standard and the markings on the ballot (see *Appendix B*). The crux is that those functions can be expressed and implemented as *database queries*. For example, the edge:

$$f\_electoral\_vote: electors(S,Ca) \rightarrow president(Ca)$$

means that whether candidate Ca is elected president is a function (called "electoral vote") of the electors (of all States S) of Ca. The latter is itself a function of *votes(S,Ca)*, i.e., the votes that candidate Ca received in state S. Clearly, given the corresponding relational tables, the result of *president(Ca)* or *electors(Ca)* can be represented as a database query on a table representing the votes per state and candidate (=*votes(S,Ca)*).

Some citizens may be interested in the top-most node only: who is the president elect? Others may choose to study the reports from the news agencies and study how many electors each candidate could win or how many certified votes per state each candidate had. The extremely close outcome of the presidential race (the differences in votes between candidates were below the statistical error margin in the state of Florida – the state which ultimately determined the election, 271 to 266 Electoral Votes for Gore) sparked an enormous controversy about the official outcome of the election and almost led to a constitutional crisis.

As a result NORC conducted a thorough study aimed at resolving the issues. Translated in our framework, this means that one can resolve the controversial issues in a precise and for the interested citizen, verifiable way (depending on the available raw data of course). The model dependency graph shows that at the lowest level, the raw data consists, e.g., of *optical_scan_markings* and *punch_card_markings*. For example, *B,E,Co,NC* means that on the ballot with identifier *B*, the coder *Co* has described the element *E* (e.g., a specific chad or specific box for a candidate) to have a marking NC (=NORC Code, e.g., *dimpled chad with two detached corners*). In the graph, the node *ballot(B,Markings)* then provides a convenient way to represent the information: the ballot *B* with all of its markings (including, for each element *E* and each coder *Co*, the observed markings encoded as *NC*). One point of the controversy was which *standard* should be applied to determine the intent of the voter.[3] Depending on the county or even precinct, and the type of ballot (see *Appendix A*) different standards could be applied. For convenience, NORC created *scenarios*, where each scenario explicitly states which standard is applied to which set of ballots. The markings coming from different ballot types (optical and punch card) were "harmonized" (see *Appendix C*) so that one could easily express standards even if applied *across* different ballot types. This "harmonization mapping" was itself documented but was added only as another convenience: one can still apply standards directly to the markings of a ballot (but without harmonization one needs to do this for each ballot type individually).

Thus, under the assumption that the raw data is uncontroversial, by using a graph of mappings, the dispute can be localized to the specific choice of scenarios/standards being applied. In this way, transparency and verifiability of the process can be guaranteed for every interested citizen.

3   Preservation Issues

The modeling of the study as a network of database transformations also has advantages for the preservation information. The NORC study provided all raw data online and precise descriptions of the mappings as part of the accompanying documentation. Moreover, a "Scenario Manager" (see Fig. 2) has been developed that allows the user to inspect the outcome of applying different scenarios. From an archival point of you, however, the specific choice of system (Microsoft Access)

---

[3] Of course if other more *reliable* technologies were available that would lead to unambiguous voting results and avoid the discussion about which standard to apply – however, this is not the point here: even if this specific controversy was caused by an anachronistic voting system, many other eGOV data issues (e.g., redistricting) will always present a "deep modeling challenge".

introduces an infrastructure dependency: *"Will a researcher be able to evaluate and experiment with the study 5, 10, or 50 years from now?"*[4]

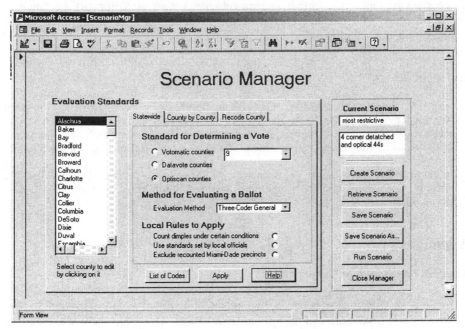

Fig. 2. Snapshot of a NORC tool created by Elliot Jaspin (Cox Newspapers)

A more robust and generic solution is to explicitly model the functionality of the Scenario Manager in an infrastructure independent way. First, mappings in a standard such as SQL. In this way, an "archival package" can contain the raw data together with all mappings and can be run on any future SQL engine, provided the standard is still supported.

One can go even one step further and create a *self-instantiating, self-validating archive* [2]. As in the case of SQL, we store an *executable* version of the constraints and mappings of a scenario. Moreover, we also package the *execution engine itself.* If the engine is SQL, this may not be a viable solution. Instead, for our running example, one could express all mappings as logic programs (Datalog/Prolog queries) and archive the complete execution engine (some complete Prolog system are smaller than the raw data of the NORC study) as part of the archive. Assuming that the logic engine is implemented in an infrastructure independent way (e.g., a Prolog engine in Java Byte Code), the complete analysis can be unrolled in the future by instantiating the graph of database mappings, and validating its integrity constraints. If the mappings satisfy certain properties, the analysis can in fact be *reversed,* i.e., one could try to solve the inverse problem and ask under which scenarios/standards a specific outcome is obtained.

---

[4] In fact, the problem occurs today: the Scenario Manager crashed several times during our experiments.

4   Conclusion

Multi-level (or "deep") information modeling provides a mechanism for capturing process information in a formal and unambiguous way as a network of database transformations. The characterization of the modeling process itself leads to the notion of self-instantiating, self-validating archives [2].

References

1. Cowell, E., Jacobs, J., Peterson, K.: Government Documents at the Crossroads. American Libraries. Infotrac. (Sep. 2001), 52–55
2. Ludäscher, B., Marciano, R., Moore, R.: Preservation of Digital Data with Self-Validating, Self-Instantiating Knowledge-based Archives, ACM SIGMOD Record, Vol. 30, No. 3 (2001) 54–63.
3. Moore, R., Baru, C., Rajasekar, A., Ludaescher, B., Marciano, R., Wan, M., Schroeder, W., Gupta, A.: Collection-based Persistent Digital Archives, D-Lib Magazine Vol. 6, No. 3 & 4, (Mar. 2000, Apr. 2000)

Appendices

Appendix A: Types of Ballots

Logically, we distinguish between two types of ballots: (1) **punch-card** ballots (*Votamatic* and *Datavote*), and (2) **optical-scan** ballots. However, there were really 5 types of voting systems in use in Florida. **Votomatic**, where a hand-held stylus was used to punch the pre-scored paper or *chad*, **Datavote**, where voters use a mechanical punching machine, **Optical Scan**, where ovals are filled in, or arrows connected, **Lever**, where the *Datavote* process was followed, and **Paper**, where the *Optical Scan* process was followed.

Voting system	Number of counties	Undervotes	Overvotes	Total number of uncertified ballots
Votomatic	15	53,215	84,822	138,037
Datavote	9	771	4,427	5,198
Lever	1			
Optical Scan	41	7,204	24,571	31,775
Paper	1			
Total	**67**	**61,190**	**113,820**	**175,010**

Appendix B: Visual Markings and NORC Codes Used

Most of the uncertified ballots were due to *overvotes*, and most of the problems came from the *punch-card* ballots. It is easy to see why, by looking at the following animation of a *Votomatic* voting machine (Doug Jones, University of Iowa): http://www.cs.uiowa.edu/~jones/voting/votomat/animate.html

Punch-card ballots

The punch-card ballots presented a number of possibilities for error, making the term "chad" a household word.

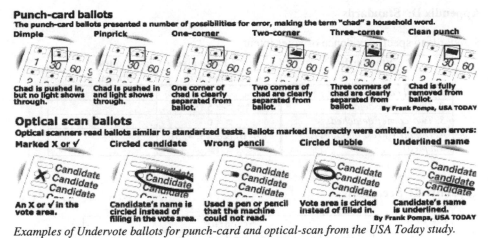

Dimple	Pinprick	One-corner	Two-corner	Three-corner	Clean punch
Chad is pushed in, but no light shows through.	Chad is pushed in and light shows through.	One corner of chad is clearly separated from ballot.	Two corners of chad are clearly separated from ballot.	Three corners of chad are clearly separated from ballot.	Chad is fully removed from ballot.

By Frank Pompa, USA TODAY

Optical scan ballots

Optical scanners read ballots similar to standarized tests. Ballots marked incorrectly were omitted. Common errors:

Marked X or √	Circled candidate	Wrong pencil	Circled bubble	Underlined name
An X or √ in the vote area.	Candidate's name is circled instead of filling in the vote area.	Used a pen or pencil that the machine could not read.	Vote area is circled instead of filled in.	Candidate's name is underlined.

By Frank Pompa, USA TODAY

Examples of Undervote ballots for punch-card and optical-scan from the USA Today study.

NORC Codes used to classify each mark on **punch-card** and **optical-scan** ballots.

Punch-card ballots		Optical-scan ballots	
Code	Meaning	Code	Meaning
0	blank, no mark seen	00	blank, no mark seen
1	1-corner of chad detached	11	circled party name
2	2-corners of chad detached	12	other mark on or near party name
3	3-corders of chad detached	21	circled candidate name
4	4-corners of chad detached, clean punch	22	other mark on or near candidate name
5	dimpled chad, no sunlight	31	arrow/oval mark other than fill: circle, x, /, check, scribble
6	dimpled chad, with sunlight	32	other mark near oval/arrow
7	dimple within chad area, off chad, with or without sunlight	44	arrow/oval filled
8	dimple on border of chad area, with or without sunlight	88	arrow/oval filled or marked other than fill, then erased or partially erased
9	chad marked with pencil or pen	99	negated mark: scribble-through, cross-out, "NO", and similar

Appendix C: Harmonized Codes

Equivalence Classes	NORC Codes	
	Punch-Card	Optical-Scan
0	0	00 / 99 / 88
1	8	
2	7	
3	5	11 / 12 / 21 / 22 / 31 / 32
4	6	
5	1	
6	2	
7	3	
8	4	44
9	9	

Appendix D: Standards

Standards to specify evidence of voter intent.

Standards	Equivalence Class Codes	
	Punch-Card	Optical-Scan
1. Dimple or better	>= 3	>= 3
2. One-corner detached	>= 5	>= 3
3. Two-corner detached	>= 6	>= 3
4. Dimple (if rest of ballot is dimpled)	(>= 6) && (3, 4, 5)	>= 3

Appendix E: Scenarios

Scenarios
1. Prevailing statewide standard
2. Supreme Court "simple"
3. Supreme Court "complex"
4. 67-county custom standards
5. Two-corners-detached statewide
6. "Most inclusive" statewide
7. "Most restrictive" statewide
8. The Gore 4-county recount strategy
9. "Dimples when other dimples present"

For example, *Scenario 5.*, *Two-corners-detached statewide*, is based on arguments made by George W. Bush's attorneys during the 36-day period following Election Day, where Standard 3. is applied statewide.

Also, *Scenario 8.*, *The Gore 4-county recount strategy*, is based on early post-election results, where the Gore camp requested hand counts in 4 heavily Democratic counties: Miami-Dade, Broward, Palm Beach and Volusia. Standard 2 is applied to some of Miami-Dade precincts.

e-Government and the Internet in the Caribbean: An Initial Assessment

Fay Durrant

Head Department of Library and Information Studies
University of the West Indies, Mona, Jamaica
claudette.durrant@uwimona.edu.jm

Abstract. Effective e-government requires cultural change, the incorporation of inter-organizational teams, identification and evaluation of knowledge management assets, and incorporation of facilitating information and communication technologies. Government services must harness this range of information resources. Several governments in the Caribbean have recognized the importance of consolidating and exploiting their dispersed knowledge resources. The objectives of e-government are being analysed with a view to determining the most appropriate means of delivering services via electronic means. The paper examines particularly communication with citizens over the Internet, the delivery of Internet based government information, and aids to the citizenry in using these new facilities. Telecentres located in libraries and community centres, in Jamaica and other parts of the Caribbean, demonstrate the early development of facilities for enhancing government communication with citizens over the Internet, and interaction between citizens and those providing services

In the context of policies for "access to information" and delivery of government services, Caribbean governments have initiated programmes for electronic communication with citizens and the delivery of some services. While the process is still at an early stage it is instructive to assess the advances which have been made and to identify and evaluate the opportunities for further exploitation of the existing human, material and technological resources.

Governments have adopted the electronic and telecommunications facilities which have become available in the region. The Internet, with special applications, is now one of the visible signs, which has allowed people to be connected, and to exchange information. It therefore provides one of the basic requirements - electronic networking - for delivery of information services.

The United Nations Public Administration Network (UNPAN) defines e-government as "a permanent commitment by government to improve the relationship between the private citizen and the public sector through enhanced, cost-effective and efficient delivery of services, information and knowledge."

Electronic government can begin with making specific pieces of information available, and can subsequently integrate information sources to make interactive services available electronically. Most governments in the Caribbean are at the first stage and have websites for ministries, departments, corporations, and agencies. These websites serve to communicate information to citizens, and in some cases to receive feedback. It is anticipated that these websites will provide the bases for government and organizational portals, and the full scale provision of electronic services.

R. Traunmüller and K. Lenk (Eds.): EGOV 2002, LNCS 2456, pp. 101–104, 2002.
© Springer-Verlag Berlin Heidelberg 2002

For e-government to be successfully implemented, an important step would be the identification of a single point of initial contact for the citizen to locate government services and or information on these services. Portals and gateways are being developed by governments to facilitate "one-stop shopping". and pathways and links to services and related websites. These are provided by the government information services, and in some cases the Office of the Prime Minister. These sites must of course be linked to the information held by other government organizations, so that a citizen's consultation is seamless and efficient.

As governments in the Caribbean continue to use the Internet to develop websites, to inform the public of policies, missions, and services, it seems opportune to undertake an assessment of these websites, and to identify their stages of development, and the incorporation of features considered essential for effective information services.

A pilot survey of government websites in the Caribbean shows that there are now sites of the central banks, the government information services, ministries of finance, ministries of agriculture, and the offices of disaster preparedness, statistical offices, and the Registrar General's Department. Of the twenty sites surveyed, all permitted or encouraged citizens and others to make contact with the organization, while fifty percent, provided mission statements of the organizations, and provided online publications.

Darrall West (2000) used twenty-seven features as the basis for undertaking a survey of government websites to determine their roles in providing information and services. These include:

Access features, design features such as tables of contents, site and subject indexes, frequently asked questions, interactive online services, multimedia, feedback, payment facilities, security and privacy. This researcher considers that additional features may include statements of government policies and statements of government services offered.

Of the twenty sites reviewed in this pilot survey, it was found that these sites can be classified as follows:

Statements of government policies and services	85%
Street Address, Phone, Fax, Email	75%
Comment and feedback	55%
Links to other sites	70%
Online publications	60%
Online databases – library or other	20%
Table of contents, site map	60%
Interactive online services	10%
Privacy and security policy statements	5%
Frequently asked questions	50%
Search engine, search engine	25%
Chat and instant messaging	0%
Date of last update	60%
Online payments	0%

While the sites do not necessarily state that they are part of an e-government strategy, the majority of the designers have clearly included statements of government policy and services. The majority are actively updated and permit comments and feedback.

There is however much further to go in meeting the demands of citizens and the challenges of knowledge management. Interactive access to information held by governments will require an integrated set of communications networks, which can result from interconnection of the communications channels and the content held on various databases and legacy systems. Offering online services such as the payment of taxes , passport services, or the issuing of certificates are important first steps. The Government of Jamaica, for example, has made some important advances in this area and the Registrar General's Department now enables citizens to download application forms or birth, death and marriage certificates.

From this preliminary survey there is also an indication that the features relating to presentation and delivery of services still need to be further developed and analyzed. The services identified so far include, for example, provision by the Central Banks of current information on the exchange rates. This information can be seen as very useful to the general population. Several of the central banks show the daily exchange rate with a range of currencies, running along the bottom of the screen. There can be further study as to how this type of information is accessed and used by a selection of business people and users from the general public.

Such features are the beginning of online government services, and to be effective, they must be based on the integration of the various sources of data held by different departments and agencies.

At this time the most important issue is assurance that integration of content and technology provide the basis of "one stop shopping " for government services. Our short or medium term vision shows citizens being able to access a portal to obtain information on government services, to transact business with the government, and to purchase licenses or permits, In the long term the objective will be to reduce the number of departments a citizen has to contact to obtain a required service. Interesting developments are taking place in the websites of the Gateway to the Dominican Republic, the Bahamas, St Kitts Nevis, Belize and the Jamaica Information Service.

The websites studied show that they are developed by individual ministries, and departments without always being linked to a "gateway" to government organizations and their services, which is generally recognizable by the general population. Some governments have used systems of naming URLs to provide some means of easy recognition by the public. The Country Gateway project being implemented by the World Bank globally has two instances in the Caribbean. The portal for the Dominican Republic has been established on the Internet, and in Jamaica a similar project is well advanced.

The development of e-government must involve each ministry department or corporation focusing on its own strategy within it priority areas. An important difference between business and government is that the number of citizens who are likely to use the services. Everyone citizen potentially has interest in accessing government information. The Registrar General's Department in Jamaica is an example of a site which all citizens are likely to access, as their need for certificates and other documents arises. The portal sites must therefore have the capacity for continuous use by multiple users.

In the content of e-government services in the Caribbean, there will be need for the citizenry to have public access to government portals. Telecentres such as those set up in the public libraries in Jamaica by the Jamaica Library Service and by the National Library of Jamaica provide examples of access points where information on government services can be accessed via the Internet. In addition to providing access to the

Internet these centres can provide training to the public and have begun providing access to local information resources.

The telecentres established by the Jamaica Sustainable Development Network in collaboration with the Government of Jamaica, the United Nations Development Programme and the University of the West Indies also provide opportunities for citizens to access government services via the Internet, and to take advantage of the services as they develop.

A survey of telecentres in Jamaica (Durrant 2001) showed that the network of public libraries and related information centers, are an example of facilities which citizens can use to access e-government websites. They indicate an area for further research to determine best practices in the development of the effective websites, as well as type of site and information service which satisfies the needs of citizens in certain contexts.

The data gathered so far shows that the Caribbean governments are making efforts to bridge the "digital divide". There is need, however, for more systematic study to inform and advise how further development would most effectively be done. Specific recommendations are for:

- the development by each government of a clearly identifiable website as an entry point for that government's presence on the Internet;
- use of URLs for the government websites based on a pattern which presents mnemonic features for easy recognition:
- incorporating information from the existing databases, to facilitate seamless access to a range of government services.
- development of access to online publications
- development of privacy and security policies and procedures
- implementation of online payment systems for government services.

References

United Nations Public Administration Network http://www.unpan.org

West, Darral M: Assessing E-Government: The Internet, Democracy, and Service Delivery, by State and Federal Governments. September 2000.

Durrant, Fay: Telecentres in Jamaica: report prepared for the Jamaica National Commission for Unesco. November 2001.

Durrant, Fay: Report on the establishment of two multipurpose telecentres: report prepared for the Jamaica National Commission for Unesco. February 2002.

Towards Interoperability
amongst European Public Administrations

Alejandro Fernández

Ibermatica, IT Services, Partenon 16, 28042 Madrid, Spain
a.fernandez.martinez@ibermatica.com

Abstract. In this paper we present the technical approach followed in the InfoCitizen project. Its novel approximation to Public Administration necessities lies in the combination of Web Services as providers and Intelligent Agents as consumers of public services. The distributed nature of a Web Service network requires a similarly distributed architecture, that in turn requires an assurance of availability and transparency achieved with the use of Intelligent Agents.

1   Interoperability in European Public Administrations

The goal of this paper is to present the technical solutions the Infocitizen [1] development team is implementing to cope with the goal of providing a software solution to the interoperability of European public services. InfoCitizen attempts to conduct electronic transactions in multi-agent settings – e.g. multi-country involvement – in a manner as transparent as possible for the citizen. In the following we will elaborate on the requirements posed by InfoCitizen regarding:

1. transparent public service provision for the citizen and
2. multi-agent setting of public service provision.

The conceptual model for the project has already been presented in [2]. This document is organized as follows: we will start by providing a brief overview of agent technology in the scope of the requirements of the project, followed by the generic architecture that will be implemented in order to achieve the goals of the project. Finally we will introduce the IberAgents architecture, which provides a generic multi-agent and distributed [3] framework, which does not only fulfill InfoCitizen requirements, but also provides a generic integration and distributed framework to be used in any kind of distributed scenario.

2   Agent Responsibilities in the Scope of InfoCitizen

As the delegate of a user or a Public Administration in the system, an agent is responsible for the negotiation on behalf of the entity it represents. Thus, it will identify the parties it communicates with, and will not disclose information marked as "private".

If some further information is required to perform a service, e.g. some user information, it will try to obtain it using the available resources. If this step fails, the

R. Traunmüller and K. Lenk (Eds.): EGOV 2002, LNCS 2456, pp. 105–110, 2002.
© Springer-Verlag Berlin Heidelberg 2002

Agent will then notify the requesting party, which must then decide which steps to take. Otherwise, the achieved result is communicated and the operation marked as successful.

In what follows, two kinds of agents will be discussed. The representative of the user in the system is the *Personal Agent*, and thus the only element in the system that communicates directly with the user, either on demand or on its own initiative. It can locate other components, but does not necessarily contain the know-how to fulfill a user necessity.

The second type is the *Interoperable Agent*. It is in these components that process information resides, and have the capacity to perform a task if the necessary input is provided.

Two main issues agents have to take into account are security [4] and anonymity of the citizens they will interact with. Regarding security, a Personal Agent must authenticate the citizen or PA employee requesting an operation. In the first case, where access to the system is public, a secure mechanism (such as Digital Signatures) must be used. In the special case where only PA employees can access Personal Agents, password authentication can be enough to enforce security.

An Interoperable Agent is often the unique entry point for administrative requests. It is thus its responsibility to keep access information private and anonymous, including access logs, identity of service consumers, and citizen profiles.

It is extremely important that all privacy concerns must be considered when implementing these components. One of the crucial goals of the project is to prevent third parties from keeping track of user activities in the system; in this regard, even the corresponding PA must see the system as an opaque layer.

At the same time, an audit trail must be stored, for the sake of data protection and integrity. Only if the user provides his or her consent, will the trail be revealed to an administrator, to undo an unwanted operation or correct any accidental errors.

3   InfoCitizen Architecture

In Infociticen [1], components are categorized in the following levels

Table 1. Top-down organization of component levels

Level	Component	Responsibilities
Personal	Personal Agent	presents a web interface to the user checks user authentication locates necessary services accesses interoperable agents contacts the user if necessary
Interoperability	Interoperable Agent	abstracts a process converts information between formats ensures anonymity of service consumer
Registry	Global Registry	registers components locates agents and services
Service	Service-supply Component	gives access to a legacy system
Legacy	Legacy System	performs the actual task

The topmost level is the only one visible to the user: Personal Agents provide a web interface, as the external access point to the system. Either accessed by an authenticated citizen or PA employee, it has the ability to perform any administrative task contained in the system.

If the Personal Agent cannot do the task on its own, it will delegate on an Interoperable Agent. On this level, the abstract information that describes a process is made specific (either for an administrative unit, such as a country, or for a special situation). Additionally, Interoperable Agents keep access data private, to ensure the anonymity of the users.

To locate available resources, Agents access the *Global Registry*. This special component contains a directory-like listing of Agents and Services, and can perform searches for specific components.

In the Service level, Agents access *Service-supply Components* in a uniform way. These are components that encapsulate access to legacy systems, and provide a SOAP interface within InfoCitizen. They should not access any other system components.

The lowest level is the Legacy level, where *Legacy Systems* reside. These include existing infrastructure in the PA, such as registration databases, and all other components external to InfoCitizen (e.g., an e-mail server). The PA can use the Legacy System in the traditional fashion, but the existence of a Service-supply Component ensures accessibility from within InfoCitizen.

Component Hierarchy

The top-down approach assures that components above can only access those below them; Service-supply Components, being on the lowest level, are just passive providers (they cannot access other system components, only external Legacy Systems).

Location of Components

In UML terminology, a *node* is the physical location of a component. It represents a computer or a cluster of computers.

The topology of the InfoCitizen system should be kept as flexible as possible, so that components can reside on different nodes.

In the Personal and Interoperability levels, there are no *a priori* requirements for node assignment. However, a good strategy is to locate all Interoperable Agents for the same PA in a common node. Personal Agents might thus reside in different nodes, depending on the number of users expected.

In the Service level, the usual approach is to place Service-supply Components in the same nodes as Legacy Systems, i.e. within the PA network. An external access point is provided for the rest of InfoCitizen nodes.

4   IberAgents

All the functionally and requirements highlighted in the previous paragraphs will be implemented by a multi-agent and distributed platform which is being developed by Ibermatica. It is called IberAgents.

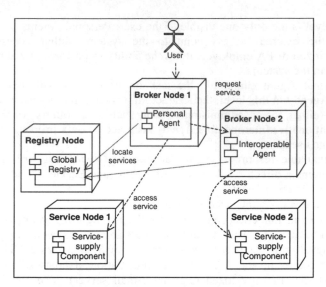

Fig. 1. Possible node organization of components. The Legacy Systems are not shown, since they usually reside in the same nodes as the Service-supply Components (Service Nodes)

The Agent Platform is organized in a distributed fashion, with several *components* (either *agents* or *services*) running on a number of *nodes*. At any given node there must be only one Java Virtual Machine (JVM) executing the agent platform.

Agents

In this architecture an *agent* has the following functionalities:

- *Interoperable*: agents can communicate with other agents.
- *Adaptive*: the path of execution of some tasks is not fixed, but depends on previous runs.
- *Delegated*: agents can make other processes run tasks assigned to them.
- *Proactive*: the agent is able to initiate operations on its own.

  The features below are desirable, but not essential for an agent.

- *Distributed*: the agent can execute code on several machines.
- *Customizable*: agents can represent users, and respond to their customization.
- *Interactive*: agents can communicate with users via a graphical interface.

Services

In InfoCitizen, a Service-supply Component wraps a Legacy System for SOAP [6] access.

Correspondingly, IberAgents uses the term *web service*. A web service is a component for creating open distributed systems. Their main advantage arises from their availability via lightweight protocols like HTTP, and XML standards like WSDL and SOAP.

Directory
Component discovery is performed using a directory of existing services and agents.

One of the design principles of IberAgents is that *developers do not have to know or care about what node a service is running in.* As exposed in [3], this poses a number of problems, which are being addressed with the use of Intelligent Agents.

Communication
Agents must be able to communicate with other agents, and with services located on any node. This communication among agents and services is performed via SOAP, being completely **asynchronous** whenever possible (i.e., when a service does not return anything).

A **broadcast** is also possible, in order to allow agents to communicate with multiple components at the same time; and to collect answers to any message from multiple sources.

A provision for communication with other agent platforms is also considered essential, to enable cooperation with other environments.

Supervised Running
In IberAgents, all the agents are proactive, which means that they can start their tasks without external calls. Therefore, they must keep running, even if an exception or error happens at runtime. The agent manager will restart the agent if necessary.

For system administration purposes, agents can be stopped and started from any node in the platform, via a web interface.

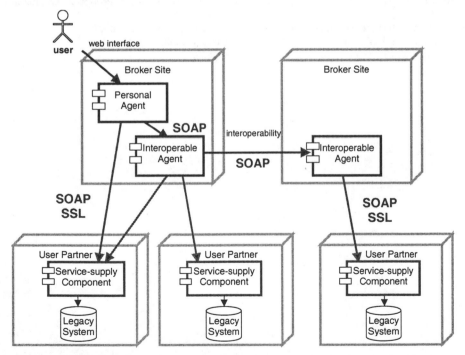

Fig. 2. InfoCitizen Integration Scheme. The *broker sites* can be developed using IberAgents, while the *user partner* sites might be implemented using other SOAP platforms

5   Integration of IberAgents with Other Modules

Just building a closed agent platform would hardly satisfy the needs of the InfoCitizen project. What is really needed is a way to interact with any external broker site, implemented using any underlying technologies.

In the spirit of web services, IberAgents publishes a SOAP interface for each of its accessible components. To the developer, there are no differences between an IberAgents site and any other InfoCitizen broker sites.

This openness leaves the door open to different implementations, and is essential to enable cooperative or competing solutions on other platforms.

As an example inside InfoCitizen, Service-supply Components will be developed using whatever technologies are suitable to the PA existing infrastructure. The standard SOAP interface will make Legacy Systems accessible from IberAgents.

References

1. InfoCitizen Consortium: IST-2000-28759 Annex 1
2. Tarabanis, K., Peristeras, V.: Towards Interoperability amongst European Public Administrations. Submitted to DEXA EGov (2002)
3. Waldo, J., Wyant, G., Wollrath, A., Kendall, S.: A Note on Distributed Computing. Sun Microsystems Laboratories, ref. SMLI TR-94-29 (1994)
4. IBM Corp., Microsoft Corp.: Security in a Web Services World: A Proposed Architecture and Roadmap (2002)
5. http://www-106.ibm.com/developerworks/webservices/library/ws-secmap/
6. W3C Note: SOAP: Simple Object Access Protocol 1.1 (2000).
   http://www.w3.org/TR/2000/NOTE-SOAP-20000508/

Assessing e-Government Implementation Processes: A Pan-European Survey of Administrations Officials

François Heinderyckx

Université Libre de Bruxelles, Dept of Information and Communication,
Av. F.D. Roosevelt 50/123, B-1050 Brussels, Belgium
Member of the Academic board of e-Forum
(Forum for European e-Public Services, www.eu-forum.org)
fheinder@ulb.ac.be

Abstract. A survey was conducted on behalf of e-Forum among 150 high ranking officials involved in e-government development in all 15 European Union countries. The results provide a unique pan-European examination of perceptions of officials driving the process of shifting towards what is generally referred to as 'e-government'. Issues covered in the survey include authentication techniques, financing e-government investments, benefits and fears among citizens, businesses, administrations and government, barriers and facilitators, priorities within the adminstrations.

1   Methodology

From December 2001 to May 2002, the Forum for European e-Public Services (e-Forum is a non-profit organization funded, for its start-up, by the European Commission) conducted a face to face survey among 150 key representatives of the public sector in all 15 EU countries. Interviewees were identified as key actors in e-government development and belonged mostly to general administration, be it at the national, regional or local level.

The questionnaire included closed and open questions covering various aspects of e-government with particular attention to perceived fears and expectations among citizens, businesses, administration and government as well barriers and potential catalysts in this area.

2   Results

The reader is reminded that respondents were all high ranking public servants, so that all indications collected are to be understood as these officials' perception of those indications.

One should also note that the sample size and the sampling method did not aim at providing a representative sample of the targeted population, but rather to collect information helping to assess broad trends in e-Government development in European administrations. As a result, breaking down the data even at the level of nations would

R. Traunmüller and K. Lenk (Eds.): EGOV 2002, LNCS 2456, pp. 111–115, 2002.

bear no relevance. Only two break-downs of the data will be considered, by groups of nations and by level of government (central versus local or regional).

2.1   Benchmarking by the European Commission

In 2001, EU member States agreed on a list of 20 basic public services (12 for the citizens, 8 for businesses) likely to offer e-government solutions. The Commission will monitor progress in implementation of those services on a half-yearly basis. The survey indicates that a significant number of people (about one third in our sample) in the administration are not fully aware of this new benchmarking initiative. Although about half of the respondents are uncertain whether the benchmarking will actually measure their progress towards e-Government, an overwhelming majority feel that it will increase their motivation to progress faster and will impact their plans or priorities.

2.2   Authentication

One of the central issues in e-Government projects remains authentication techniques. According to our survey, identification by user id and password is clearly the leading approach. Future developments include various means at medium and long term. PKI (Public Key Infrastructures, i.e. certificates sent by e-mail) are well represented, either currently or in future plans. Smart cards technology are significantly considered, but mostly within a few years time. Most administrations have no plans to resort to biometric recognition technologies. There is no spectacular difference in tackling authentication of citizens and businesses.

2.3   Financing e-Government Investments

Almost all respondents indicate that e-Government investments are included in their normal budget. A number of countries in Northern Europe, particularly in the British Isles and Ireland, do report special budgets on top of their usual departmental budgets. Very few countries report a possible co-financing by the private sector. Few respondents rely on benefit from cost reductions induced by e-Government implementation.

2.4   Priorities in Creating Benefits for the Citizens and the Businesses

The improvement of the quality of services was most often ranked as the top priority in developing e-government services for the citizens. Second in importance is to improve citizen's access to administrators and information (even more so in central administrations), followed by goals of improving efficiency, transparency and providing access 24 hours a day, 7 days a week (this is significantly more marked in Southern Europe). Improving cost-effectiveness is ranked higher in Northern Europe. Improving participation of citizens in democracy appears more crucial at the local and regional levels.

As for businesses, improving the quality of the services ranks at the top of priorities, particularly so in central administrations, and significantly more so in Northern Europe. Second to that priority come a group of 4 goals, namely continuous access to services (particularly in Southern Europe), enable services to be provided more cost-effectively (particularly in Northern Europe and in local administrations), improve the efficiency of administrative operations (particularly in Southern Europe and in local administrations) and improve business access to administrators and information (more so in central administrations and in Southern Europe).

2.5 Benefits for the Administration and the Government

As for the administration itself, the main goal is clearly to improve customer satisfaction, thus making the job easier (this is even more the case in Northern Europe). Also quite important is the expected increase in flexibility in working conditions, particularly so in local administrations, and even more in Northern Europe. Other high ranking priorities include personal development in new technologies (particularly in Southern Europe) and improved autonomy in the job (particularly in local administrations and in Southern Europe).

Respondents were also asked what they thought were their government's most important goals and objectives in developing e-government. It appears that administrations, quite homogeneously perceive their government's most important goals as seizing an opportunity to rationalize administrative procedures. Second in importance are improvement of citizens' well-being (particularly in local administrations and in Northern countries) and reduction of cost of administration (significantly more so in Northern Europe).

2.6 Fears Induced by the Development of e-Government
   (As Perceived by Administrations)

Administrations clearly (and with homogeneity) perceive 3 areas of concern among citizens: loss of information confidentiality, loss of human contact and digital divide (not all citizens will have access to the new services, and not all citizens will be able to use technologies properly). There is also a perceived concern following an increased control of citizens by the government (particularly in Southern Europe).

Regarding businesses' fears, as perceived by administrations, the major area of concern lies with the loss of information confidentiality and increased control by the government (particularly as perceived by local administrations for the latter). There is also concern about the fact that not all businesses will have access. But the loss of human contact and the issue of the ability to use the technologies is viewed as much less a concern for businesses than it is for citizens.

As regards the administration's perception of their own staff's fears, highest ranking concerns include inability to use new technologies properly, increased pressure from users/customers and inability to cope with increased speed (significantly more so in local administrations and in Northern Europe for the latter). Possible job cuts and increased control on individual performance are, to a lesser extent, other areas of concern. Loss of personal contact is perceived by administrations as much less a concern for their own staff than it is for citizens.

Finally, when asked about their perception of concerns induced within their government, administration officials clearly identify 4 areas of anxiety: failure of e-government projects (particularly in Northern Europe), digital divide, high cost of implementation (particularly among respondents working in central administrations) and risk of attacks and fraud by hackers (particularly in Southern Europe). The risk to end up with no real change ('window dressing') is also identified as a concern, particularly by respondents in Northern Europe and in local administrations.

2.7   Barriers and Facilitators

When asked to assess the importance of various barriers in the development of e-government, administration officials rank most often concerns about security and confidentiality as most prevalent. Second to that main concern, issues of lack of access among citizens, high set-up costs, lack of co-operation among administration departments and lack of political will and drive (particularly in Northern Europe for the latter) are most commonly identified as barriers in developing e-government.

When it comes to factors likely to facilitate implementation of e-government, one single factor stands out across Europe: strong leadership from the government. To a lesser extent, other factors also call upon political action: dedicated budgets, appropriate legal framework and availability of approved standards (particularly in Northern Europe). Also quoted are better internet penetration in households (particularly quoted within local administrations) and appropriate skills within the administration (significantly more quoted in Southern Europe).

Moreover, respondents were asked what their expectations were about an international association such as e-Forum. These questions provide valuable information as to what high ranking public servants involved in developing e-government are lacking in doing so. The strongest demand is, by far, on sharing experiences and best practices. This indicates that although each e-government project is clearly unique in its setting and constraints, administration officials in charge of their development are seeking experiences and practices elsewhere to feed into their set-up process. The prevalence of their demand in that respect indicates the absence of efficient structures in sharing such information at the European level.

Other salient expectations include: offer a repository for best e-government related documents in Europe (particularly among central administrations), provide an opportunity to develop informal network of colleagues (particularly in local administrations and in Northern Europe) and have a permanent up-to-date list of existing e-government services in European countries.

2.8   Priorities within Administrations

Face to face interviews allowed respondents for more spontaneous and open comments about the various issues related to the implementation of e-government. Some of these recurrent comments indicate patterns of opinions which appropriately supplement the main questionnaire.

A number of respondents believe that businesses as well as citizens expect e-government to provide a single point of access to administration and public services. Moreover, there is a recurrent view that e-government interfaces should be thought of as complements to existing, traditional systems rather than as substitutes, not only to accommodate those who can't access the new services, but also for those who do not want to. Regarding the issues of security and confidentiality, many consider that it is up to governments to build up people's trust and confidence. A number of respondents also stress the fact that efforts to develop e-government solutions should concentrate on back-office issues. It is also the case that many think that too much attention is focused on technical matters at the expense of considerations for a wide array of issues related to the more human aspects, i.e. the various problems to be solved regarding the people both as users and as administration employees. Many also expressed both their conviction that cross-departmental work was to be developed, and their skepticism that such change could really be achieved within the foreseeable future. Further along those lines, provided that e-government development is inseparable from administrative procedures' simplification and, broadly speaking, from a thorough business process reengineering, the transition can only be considered within a long term process which unfortunately exceeds usual political mandates and planning.

3   Perspectives

In spite of the necessary caution in using results from a survey conducted on a limited sample, converging views of these hand-picked key officials do provide some very relevant facts about the process at hand within administrations as perceived by those involved. Overall, it appears that officials managing the process of implementation have a well framed view of barriers, catalysts, fears and expectations associated with such process. The issues they raise clearly call for ample reforms which they appear dedicated to undertake. As for the way to achieve these e-government driven reforms, they seem to have developed a rather clear view of the priorities, although their agenda may appear to not necessarily match that of their political leaders, be it in nature or in timing. The shift towards e-government can be seen, to some extent, as the continuation of an on-going process which started with the implementation of computers and data-processing, so that   administrations' experience in that area should be seen as a real asset. However, given that e-government consists in developing automated tools to directly interface with citizens and businesses, there is little doubt that the reforms and transformations at hand significantly differ in nature and exceed in amplitude that which lead to computerization.

A One-Stop Government Prototype
Based on Use Cases and Scenarios

Olivier Glassey

IDHEAP and INFORGE, University of Lausanne,
1015 Lausanne, Switzerland
olivier.glassey@idheap.unil.ch

Abstract. In this paper we show the methodology we used to build a prototype
for One-stop Government. We started by defining ten simple use cases, and
then we developed scenarios, business rules and sequence diagrams for each of
them. This work was based on a conceptual model for One-stop Government
we developed in a previous research. We also explain why the use cases and the
scenarios proved very helpful for the conception and the development of the
prototype. Last we show the software architecture, based on distributed components, and the operation of the prototype with a few examples.

1 Introduction

In previous work we defined a conceptual model of a One-stop public administration,
putting the accent on both the structural and behavioral aspects of such a system. We
decided to work on such a model because we believe that there is a need for conceptual methodologies in that field, where various researches have already been conducted, such as the GAEL (Guichet Administratif En Ligne) project in Switzerland
[3] or the BTÖV (Bedarf für Telekooperation in der Öffentlichen Verwaltung)
method in Germany [6]. Indeed many researchers and public administrations need
stable models of processes that can help public administrations managing the complexity and the rapid evolution of new technologies. This is also what a Swiss working group on e-Government appointed by the federal government believes [5]. Furthermore this group identifies the definition of "patterns of applications" as one of
three essential action domains for e-Government.

During our modeling work, we took into account the internal processes of an administration, its relations with the customers and the expectations of the citizens: we
made a six month survey in the "Administration Cantonale Vaudoise" (ACV), a large
public administration at the cantonal level in Switzerland. There we met project managers, domain managers, departmental managers and users representatives for a total
of 28 formal interviews. We also had many casual talks with different civil servants.
Moreover we conducted two online surveys (1998 and 2000) in order to discover the
expectations of the citizens in the field of electronic administrative services and we
had close to 500 questionnaires to analyze. We also took into account the different
targets that a One-stop public administration can have (citizens, businesses, civil

R. Traunmüller and K. Lenk (Eds.): EGOV 2002, LNCS 2456, pp. 116–123, 2002.
© Springer-Verlag Berlin Heidelberg 2002

servants, other administrations, etc.) and different distribution canals (Internet, wireless, public kiosks, Intranet and so on). Using these field data we filled in "summary cards", inspired by CRC cards (Class-Responsibilities-Collaborators). For more information on this technique, we recommend [1]. First these cards helped us collecting the input from the various people we met and transforming it into a structural model through an iterative abstraction process. We defined this class model using the Unified Modeling Language (UML), developed by [2]. Then we used the cards to create use cases and scenarios following the methodology developed by [7]. We will not go into the details of this model because it is not the aim this paper where we want to explain how we built a prototype based on this model. In order to validate and to refine the conceptual model we decided to build a small system of a One-stop public administration and throughout this paper we will explain the different steps of this work. It has to be noted that it is only small prototype will limited functionalities because we lacked the resources to develop a full-scale system. However we think that the interest of this paper is focused on the methodology we used to build the prototype rather that on the final system. That is also why we will not show any implementation details and we will stay at the general software architecture level.

2   Use Cases and Scenarios

We wanted to illustrate rather typical Information-Communication-Transaction administrative services and we selected ten functionalities that we wanted to be part of the prototype. For each one of these we built use cases, scenarios and sequence diagrams. Here we shortly explained why we chose these particular functionalities even though they may seem quite simple:

– Content publication: this functionality demonstrates problems of content management, validation, update and access rights, as well as the ability of non-technical persons to publish content without specific knowledge.

– Directory: we found during our surveys of the citizens that it was one of the most expected services.

– Calendar: this tool illustrates typical problems of dynamic content management and of interfaces with databases.

– Appointments and personal agenda: the ability for a user to have his own agenda is a limited personalization capability in the prototype; a client can propose an appointment to a civil servant or receive reminders for various events.

– Validation and acceptation of an appointment: this shows limited workflow functionalities and interactions with a human actor; the system checks for conflicts in appointments and a person in charge validates it before the user receives a confirmation.

– Discussion forums: this service is an interesting feature because many forums are third party software and sometimes remotely hosted, so it allowed us to show the integration of commercial components or application in a One-stop public administration.

– **Chat:** basically we chose to add a chat to the prototype for the same reasons as above, with the idea of a future integration with the agenda, allowing the client to make an appointment for a official chat with a civil servant, which would provide advanced Communication services.

– **Search engine:** this is a very useful feature to find one's way in the mass of information provided by online government and it shows the ability to integrate distant functionalities within our prototype; a search engine is quite simple, but we can imagine to add a payment or a certification "engine" in a real system, which are necessary tools to support complex Transaction services.

– **Official forms:** official forms are the bases of a public administration's operations and they are at the convergence of Information, Communication and Transaction services; they are also the most requested feature that came out of our surveys.

– **Tax declaration:** this is the only really Transaction service of the prototype and we added it because it helped us illustrate the integration of legacy systems (we developed a prototype of tax declaration in 1997).

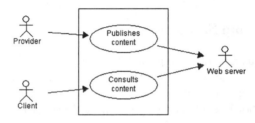

Fig. 1. Simple use case diagram

For each of these ten functionalities we created use case diagrams such as shown in Fig. 1 and we developed basic scenarios (Table 1), but here we will only show them for the first feature of the prototype. The use case diagrams and the scenarios are very useful to have a common language between technical persons, users, managers and leaders. They also provide a solid basis for a posteriori tests and validation of a system.

In order to complete the scenarios, which show in what order a procedure is accomplished, we defined business rules (an example is given in Table 2) to explain how a particular activity is conducted, using the methodology developed by [8].

When a scenario is a textual representation of a procedure or a service, UML also provides a graphical way of describing it. Indeed the sequence diagrams (Fig. 2) show the different steps of a procedure, but they also add information by introducing classes, actors and messages. They provide a solid basis for developers who can later on work on activity and state diagrams before they begin to actually work on the code. Furthermore there are many CASE tools that do code generation and reverse engineering, thus reducing greatly the workload of the software engineers.

Table 1. Basic scenario

Use case name	Content publication
Abstract	The public worker (or service provider in our terminology) publishes content on the Web: news, job offers, etc. The client can consult this content online after it as been validated by a person in charge.
Normal sequence of events	1. The provider publishes new content 2. The content is validated 3. The content is made available online 4. The client consults the content he needs
Alternate sequence	1. The client does not consult any content
Exception handling	At step 2, the content does not go online if it is not validated. A specific procedure begins.
Triggers	
Assumptions	
Preconditions	New content has to be published.
Postconditions	Content is accessible on line.
Authors and date	Olivier Glassey, February 12th, 2002

Table 2. Business rule

Business rule	Job offer publication
Short description	All job positions to be filled must be made publicly available, event if they are filled internally later on.
Source	Job regulation in public administrations
Type	Fact

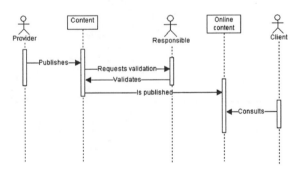

Fig. 2. Sequence diagram

As we mentioned above, we created such diagrams for each of the ten functionality of the prototype, then we used our structural model in conjunction with a CASE tool to create the class models of our prototype. These allowed us to generate the software "backbone", but we also needed to define logical software architecture before we started to really implement this system.

3   Architecture of the Prototype

We needed a software architecture that allowed us to take into account the following constraints, that we had found were very strong in public administrations:
- Ability to access heterogeneous applications and platforms
- Simple user interface and openness to the Internet
- Use of standard and robust technologies
- Modularity and scalability

To create the middleware layer we chose a distributed component architecture because it fitted our needs best and because there are really strong standards on the market. We will not present the characteristics of this type of architecture here, but we recommend [4] which is one of the foundation papers of this model. The three main components communication models are CORBA from the Object Management Group, COM+ (and now .NET) from Microsoft and Java/RMI for Sun Microsystems. Neither will we discuss these here, as there is an excellent comparison in [9]. Let us just say that for reasons of resources and simplicity we built our prototype using the COM+ model for the back-office applications. To be able to distribute these services to the clients we decided to build a dynamic Web interface based on the PHP/MySQL couple (respectively a server script language and a database, both of which are Open Source). The output displayed in the end-user interface is pure HTML, which means that it can be read by any browser such as Internet Explorer, Netscape or Opera.

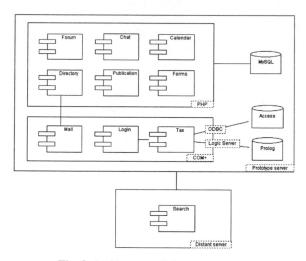

Fig. 3. Architecture of the prototype

Fig. 3 shows the general logical architecture of the prototype and of its functionalities. For most of the system, we adapted existing PHP applications (which can be seen also as components): forum, chat, calendar, directory, content publication tools and access to the official forms. On the other hand, the tax declaration application we built in 1997 was written in Delphi and used the ODBC gateway to connect to a Microsoft Access database and a Logic Server to retrieve calculation rules in Prolog. We simply wrote a COM component that encapsulates this application and is able to be distributed over the Internet using the ASP scripting language. We also added a commercial component called ASLogin that allowed us to secure the tax declaration pages. Finally we added the Google search engine in our prototype. We will not go into anymore details regarding the implementation of this prototype as this is rather a "toy" compared to real systems and because we think the interesting concept here is the general architecture: it allowed us to integrate various commercial and Open Source components and applications into what appears to be a unified and transparent system with a single interface.

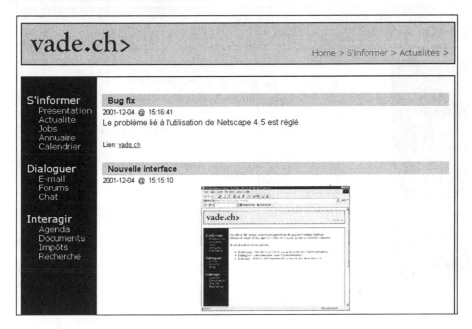

Fig. 4. Interface to access content

4   Operation of the Prototype

The prototype can be tested at http://uts.unil.ch/vade/, but it is only available in French. Here we will briefly describe its operation and show how users with different access rights can accomplish various tasks within a single interface. Let us mention that we made no particular work on the ergonomics of the prototype as it was not our intent. We used very classical navigation techniques with a left menu based on the Information-Communication-Transaction typology and an upper menu that shows the hierarchy of the pages. We realize that this is not entirely satisfying and we will work on a different interface in the future, most likely based on the "life events" metaphor.

The first screen (Fig. 4) shows the interface that a regular client (with no particular rights) sees when he chooses to see the latest news. The second one (Fig. 5) shows how a user with specific access right can publish new content within the same interface. Without any HTML knowledge he or she can publish text, links or images that will be validated before going online. Another example (not shown here) of clients with different access rights using a similar interface is the directory: a regular client can only browse or search the directory when selected civil servants have the access to add, edit or delete entries, in that case with additional links in the interface.

The prototype in itself does not show anything very impressive, but we think that its strong point is the concept of a single and universal interface that can be used via any distribution channel and by any type of client.

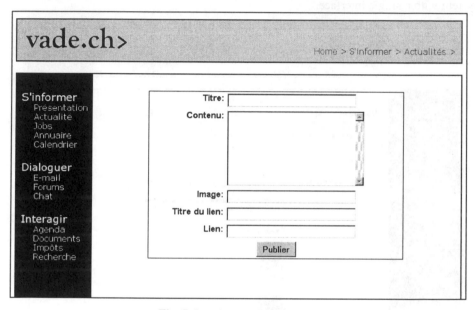

Fig. 5. Interface to publish content

5   Conclusions

What we have shown in this paper is the application of a conceptual model and methodology to a practical case of One-stop Government. The result of this work is a simple but functional prototype. We now want to emphasize what we believe are the strong points of this work. First we think that it is almost necessary to find a common language between technical persons, users and managers, and we found that the use of graphical models developed in UML provide a good way to achieve that. We also found that scenarios and sequence diagrams, although they are simple enough to be understood by anyone, are a great basis for software engineers to conceptualize and implement information systems that satisfy the needs of their users. We also realized that we needed a solid and standard logical architecture in order to integrate heterogeneous applications and software components into a seemingly unique One-stop Government system. Furthermore we think that, in order for this system to be accessible from anywhere and by anyone via a standard user interface, the best choice was to use a middleware layer interfaced with the Internet. In the future it is also very likely that this middleware will have to be interfaced with wireless technologies. Thus the architecture we chose allowed us to build a dynamic and universal interface that constitutes a single entry point to very heterogeneous electronic administrative services, providing true One-stop Government capabilities.

References

1. Bellin, D., Suchman Simone, S., Booch, G.: The CRC Card Book. Addison-Wesley, Massachusetts (1997)
2. Booch, G., Rumbaugh, J., Jacobson, I.: The Unified Modeling Language User Guide. Addison-Wesley, Massachusetts (1999)
3. Chappelet, J.-L., Le Grand, A.: Towards a Method to Design Web Sites for Public Administrations. In: Galindo, F., Quirchmayr, G. (eds.): Advances in Electronic Government, IFIP WG 8.5 in Zaragoza, Spain, 10-11 February (2000)
4. Fingar, P., Stikeleather, J.: Distributed objects for Business. Sunworld, USA (1996)
5. GCSI: 2e rapport du Groupe de Coordination Société de l'Information à l'intention du Conseil Fédéral. Suisse (2000)
6. Gräslund, K., Krcmar, H., Schwabe, G.: The BTÖEV Method for Needs-driven Design and Implementation of Telecooperation Systems in Public Administrations. Universität Hohenheim, Stuttgart, Germany (1996)
7. Kulak, D., Guiney, E.: Use Cases: Requirements in Contexts. Addison-Wesley, Reading, USA (2000)
8. Ross, R.: The Business Rule Book: Classifying, Defining and Modeling Rules, Version 4.0. Business Rules Solutions, Inc., USA (1997)
9. Suresh Raj, G.: A Detailed Comparison of CORBA, DCOM and Java/RMI. Wisconsin, USA (1998)

Reflections on the Requirements Gathering in an One-Stop Government Project

Johanna Krenner

University of Linz, Institute of Applied Computer Science,
Altenbergerstr. 69, 4040 Linz, Austria
krenner@ifs.uni-linz.ac.at

Abstract. This paper reports on the requirements analysis for one-stop government. It is focused on the work that was done in the analysis phase of the eGOV project - an EU project with the goal to develop an integrated platform for realizing online one-stop government - which is presented as a case study. Different types and sources of requirements are described. Furthermore the user-survey for the eGOV project is dealt. The paper concludes with some of the insights gained from the requirements analysis.

1   Introduction

At the moment e-Government is a highly popular topic. There are many projects running on communal, district, regional, national and European level. Thereby one-stop government is an important factor. The goal of one-stop government is to provide one point of access to the administration, which enables the user to take care of every public services s/he needs from there. Thereby it has to be taken into account, that most one-stop government solutions have three different target user groups: citizens, businesses and the public administrations themselves. As a consequence of this, there is a whole palette of different requirements.

To develop a successful one-stop government solution, careful requirements analysis has to be performed. In the EU-project "An Integrated Platform for Realizing Online One-Stop Government" – short: the eGOV project[1] (cf. [7], [9]) – a lot of effort was put into that phase.

The goal of the eGOV project is to develop a one-stop government platform. This platform includes a portal that can be used by citizens and businesses as the access point to all public administrations of a country. The user can do all of his/her public services there without having to know which one the responsible public authority is. Therefore the portal itself has to have some kind of "intelligence", so that it leads the user to the right public administration in the right region. Furthermore the eGOV platform includes elements through which public authorities can seamlessly connect to their back-office workflows.

The topic of requirements analysis for one-stop government is dealt with in this paper by first discussing different types of requirements. Then diverse sources of requirements will be presented. Based on this a closer look on user surveys will be taken exemplified by the eGOV project surveys. It will be concluded with insights and examples of requirements that were gained within the requirements analysis phase.

---

[1] eGOV Project homepage, http://www.egovproject.org/

R. Traunmüller and K. Lenk (Eds.): EGOV 2002, LNCS 2456, pp. 124–128, 2002.
© Springer-Verlag Berlin Heidelberg 2002

2   Basic Types of Requirements

When designing an integrated one-stop government system, one focus should be on public services. They are the central elements of the system from the point of view of the public administration as well as from the point of view of the citizens and businesses.

Wimmer [8] has developed the holistic reference framework, which covers three dimensions that have to be considered when dealing with public services: abstraction layers, progress of public services and different views. The dimension of special interest for the requirements analysis is the one of different views. Based on these views, categories of requirements - not only for the public services but for the whole one-stop government system - can be defined: process specific, technical, user, security related, law based, organizational, social and political as well as data and information specific requirements.

By taking a special point of view (the necessity of considering multiple viewpoints is also pointed out in [6]), the requirements of one category can be distinguished. Thereafter these requirements have to be analyzed, rated and reconciled within a category as well as category spanning.

3   Sources of Requirements

The following section presents different ways of gathering user requirements based on the experiences gained in the eGOV project. One advantage of that project is, that the project partners already represent different points of view, since they are representatives of different sectors: science, economy and public administration. Hence many requirements could be gained by exploiting the knowledge and know-how of the different project partners. The representatives of the businesses are from the IT sector and could therefore contribute much to the technical requirements, security related requirements and also data and information specific requirements. The scientific partners (partners from research institutions) have special knowledge concerning processes, organizational requirements and also data and information specific requirements. A decisive role is played by the public administration partners, also referred to as user partners, since they represent the field that the platform should finally be used in. Consequently they added not only to the user requirements, the organizational requirements and the social and political requirements but also made demands in the other categories.

Another source of requirements was the literature research. Besides the traditional way of gaining knowledge from books and papers on e-Government in general (e.g. [2], [4], [5]), as well as on project-task specific topics (e.g. process modeling, meta languages, etc.), the internet was used as a knowledge source too. Many public administrations already have their own web site. These sites differ a lot regarding the level of technical progress, implementation of progress of public services (cf. holistic reference framework [8]), usability, etc. Many requirements could be gained by evaluating the pros and cons of these features. To enable this, a state of the art report [1] was worked out. In this report, a closer look was taken on several existing e-Government web-sites / portals and their connection to the back office.

Special attention has to be paid to the acquisition of user requirements, which are especially crucial, since they also have an effect on requirements of other types. The needs and demands of the users can be gathered and evaluated best by asking them in an ordered and concrete manner. Therefore user surveys were performed in the eGOV project which included questionings of citizens, representatives of businesses and public administrations.

4   Surveys in the eGOV Project

The method utilized for gathering user requirement was based on questionnaires. Since the potential users of the eGOV system can be divided into three user groups - citizens, businesses, public administrations - the demands, expectations and fears of each group had to be figured out. Based on this the user requirements could be concluded.

In order to gather user requirements, three questionnaires have been developed, one for each target group (citizens, businesses, public administrations). The survey was mainly performed by handing/sending out questionnaires. In addition online-versions of the citizen questionnaire were created, which could be filled in via internet. In Austria and Switzerland the survey at the public administrations was done by performing interviews based on the public administration questionnaire. Switzerland also chose this way of gathering input for the businesses.

The persons questioned were divided into focus groups according to different attributes to ensure the coverage of the whole bandwidth of citizens, businesses respectively public administrations.

The focus groups for the citizens were categorized according to:

- Gender: male/female
- Internet usage: not used to use the Internet / used to use the Internet
- Age: <18, 18-34, 35-54, 55+

The businesses were divided into focus groups depending on their number of employees (<50 (Small/Medium Enterprise), >=50 (Large Business Unit)) as well as on the sector the business was acting in.

The public administrations were distinguished by their level (national, federal state, municipality, other).

5   Insights Gained from the Analysis

There is a wide range of detailed requirements that could be drawn from the analysis in the eGOV project. Only an overview on some of them can be presented as examples in this paper.

5.1   Citizens and Businesses

Even though citizens and business need a different content in the portal, the requirements they pose are similar. Most citizens and businesses considered a common one-stop portal for the whole public administration of a country as important and helpful and also explained their willingness to use such a portal for transactions. However certain conditions have to be met to reach a high acceptance.

Users, especially inexperienced ones, will need a good guidance through the portal. A well thought out structure, a comprehensible navigation as well as a consistent look of the web pages are a cornerstone for that. A good guideline to reach this can be found within [3].

It is also necessary to present the content in a way that is understandable for a normal user. This means that the public administrations must not describe things using their terminology, but have to make "translations" into everyday language. This also includes the representation of the structure of public services. Unlike public administrations, which often categorize public services by their affiliation to a public authority , citizens and businesses consider a structure according to life / business events understandable.

When navigating through the portal, it is of importance, that the user is aware at any time where s/he is and where s/he can go next. This makes it less likely that the user finds him/herself doing something s/he did not intend to and it prohibits that s/he gets lost.

Personalization is also very important to improve the acceptance of the portal. Since an authentification of a citizen is necessary for transactions anyway, this can also be used for personalization purpose. Then, when filling in an application form, the citizen would get a pre-filled form where the data known about him is already entered. It should also be possible to reuse former applications. This would be convenient as well as timesaving for the user and would reduce the probability of mistakes.

A business is likely to use the portal more often than a citizen, since they are obliged to do more public services. Therefore the demand of businesses for personalization is even higher than the one of the citizens. Especially the possibility to create shortcuts to often used services and to reuse former applications is of importance for businesses.

Users have concerns about data security especially in connection with personalization since they are aware that their data has to be stored therefore. Transparency can counter these concerns. The user needs to be informed what happens with his data, for whom it is accessible and how it is protected. This creates confidence in the portal.

It is also desired - especially by the citizens - that one can gather information without having to register. Any kind of registration should only be obligatory if the user wants to undertake transactions.

Another requirement for a one-stop government portal is that it needs to be intelligent in some way. If a user wants to use a specific public service, s/he should be automatically connected to the right public administration (e.g.: marriage - registry office) in the right town/municipality/region/... The location information should be drawn from the information known about the user if s/he is registered, or, if that is not the case, from the information given by the user.

Additional help should be available if it is needed. An element that might need further explanation can have a link added, which connects to the according help-file. This way, the clarity of a page is not impaired and experienced users do not have to deal with information they already know.

5.2 Public Administrations

Public administrations often have a strict hierarchical order and the way they work is fixed to a certain degree by laws. Hence, much attention has to be paid to legal aspects during the development of a one-stop government system, but since these aspects are very task specific, it cannot be gone into detail in this paper.

The way public administrations work can be reflected in models of their processes. Process models need to be part of the one-stop government system in order to enable a clear connection to the back office system. In contrast to e-Commerce processes, e-Government processes are not always well structured but often semi-structured and sometimes even unstructured. Therefore the process models representing the processes in the one-stop system must be flexible and allow nonlinear process flows. Also the processes themselves have to be thought over and changed with regard to the new medium to profit most of the new system.

At the moment many different back office systems are in use at the different public authorities. They are the backbone of the back office work and exchanging them would be extremely difficult and expensive if not even impossible. Therefore a new one-stop government system - as the one that is developed in the eGOV project - has to enable the integration of a whole range of different types of back-office systems to have a chance to be successful.

Furthermore, a one-stop government system needs to be easy to administrate and the communication between different public authorities should be facilitated by it. Therefore a standardized language for communication would be useful.

To profit most of a one-stop government system, the back office work at the public administrations must have a certain level of technical advancement. A workflow management system as well as an electronic record system are a good basis for the profitable usage of a one-stop government system.

References

1. eGOV (IST-2000-28471), Deliverable D121: Services and Process models functional specifications, January 2002
2. Lenk, Klaus, Traunmüller, Roland, Öffentliche Verwaltung und Informationstechnik, v. Decker, Heidelberg, 1999
3. Nielsen, Jakob, Usability Engineering, Morgan Kaufmann, Academic Press, San Diego, CA, 1993
4. Reinermann, Heinrich, von Lucke, Jörn (Hrsg.), Portale in der öffentlichen Verwaltung, Speyrer Forschungsberichte, Speyer 2000
5. Snellen, I. Th. M., van de Donk, W.B.H.J. (eds.), Public Administration in an Information Age, IOS Press Ohmsha, Amsterdam, 1998
6. Sommerville, Ian, Sawyer, Peter, Requirements Engineering, A good practice guide, John Wiley & Sons, Chichester, 1997
7. Tambouris, Efthimios, An Integrated Platform for Realising Online One-Stop Government: The eGov Project, in: Proceedings of the DEXA International Workshop "On the Way to Electronic Government", IEEE Computer Socity Press, Los Alamitos, CA, p. 359-363, 2001
8. Wimmer, Maria A., A European Perspective Towards Online One-stop Government: The eGOV Project. Electronic Commerce Research and Applications, Volume 1, 2002 (forthcoming)
9. Wimmer, Maria, Krenner, Johanna, Next Generation One-Stop Government Portale: das Projekt "eGOV", In Bauknecht, Brauer, Mück (eds.), Informatik 2001, Tagungsband der GI/OCG Jahrestagung, Band 1, OCG, Vienna, S. 277 - 284, 2001

Understanding and Modelling Flexibility
in Administrative Processes

Ralf Klischewski[1] and Klaus Lenk[2]

[1] Hamburg University, Department for Informatics, Software Engineering
Vogt-Koelln-Strasse 30, 22527 Hamburg, Germany
klischewski@informatik.uni-hamburg.de
http://swt-www.informatik.uni-hamburg.de
[2] Oldenburg University, Department of Economics and Law, Public Administration
26111 Oldenburg, Germany
lenk@uni-oldenburg.de
http://www.uni-oldenburg.de/verwaltungswissenschaft/

Abstract. Aiming to provide a platform for collaboration across agencies and to design appropriate IT support for the variety of administrative processes and decision making, concepts need to go beyond current approaches in business process modelling as well as workflow and record management. Drawing on these approaches, we suggest to focus on the unique tasks and activities of each actor involved and to present the relation of each individual contribution to the overall process as something tangible in order to support flexibility in the execution of administrative processes.

1 The Challenge to Support Administrative Processes

Electronic government is increasingly drawing attention to the need of reorganising many business processes within the public sector. In order to select appropriate forms of IT support for these processes, it is necessary to get a clear picture of their nature and of the purposes which they serve. Business processes in the public sector cover a wide range of tasks and of work arrangements. Whilst some of them can be fully automated, others rely on human agency and professional knowledge and require flexibility to a large extent. Unleashing the full enabling potential of IT for modernising the public sector requires a wider approach which presupposes a thorough familiarity with the "business" of the public sector and the characteristics of non-standardised work processes.

In this paper we re-examine approaches to understanding and modelling administrative processes and try to highlight the unique involvement of actors in administrative processes. The human element stands central in this approach, and we are looking for ways of modelling business processes which draw on the full range of the enabling potential of IT. We will propose a relational actor-oriented approach to modelling administrative processes and decision making across agencies, thus paving the way for a more appropriate IT support for public administration.

R. Traunmüller and K. Lenk (Eds.): EGOV 2002, LNCS 2456, pp. 129–136, 2002.

1.1   Characteristics of Administrative Processes

The characteristics of business processes in the public sector have much to do with the fact that most of the work there requires professional knowledge and experience. Mass-production of a type which can be fully automated does exist, but its scope is limited to simple processes of registering information, accounting and calculating. Of much greater importance are processes in which individual cases are dealt with, in more or less direct contact with the stakeholders. Legal rules and the explicit and implicit knowledge of administrators play an important role in such processes [4].

It is therefore adequate to say that the bulk of administrative processes in fields like assessing claims, granting licenses etc., is situated on a continuum which has on its one end fully standardised "production processes", and unstructured decision processes on the other. For processes of policy making, of legislating and of rendering justice, it is obvious that they depart to a large extent from the assumed model of production processes on which standard software in the private business sector is predicated. The same holds true for many processes occurring at the operative level of administrative agencies. Examples of such weakly structured decision processes include the granting of a license, assessing social benefit claims, issuing building permits, etc. When such processes start it is often not clear how long they will take, how much information is needed, and whether negotiations between the various agencies involved in the processes will take place.

Unfortunately, most computerised information systems in the public sector are still based on an understanding which takes well-structured and fully standardised processes as its starting point. These processes are recurrent in the private sector, e.g. in the field of accounting. Since many standardised processes can also be found in the public sector, e.g. in the fields of financial and personnel management, standard ERP (Enterprise Resource Planning) software such as SAP's R/3 is also making its way into institutions of the public sector. But besides such processes of an auxiliary nature, much of the work of public administration is characterised by primary processes at the operative level in which claims are processed, decisions made, and services rendered.

It is important to state that decision making in public administration occurs not only at the level of organisational management or policy, but is characteristic of its operative work. The officials in charge must remain flexible as to the workflow at stake. They must be able to ask for information, to ask a colleague for help, or to organise a meeting and insert the outcome of this meeting into the sequential work process.

1.2   The Quest for Flexibility

The options for standardising processes involving decision-making on individual cases or negotiations are very limited. Determination of some typical steps, which such processes should follow, may decrease service quality, effectiveness and efficiency. There are at least four reasons (which are not confined to the public sector) why more flexibility should be built into the execution of business processes where human agents collaborate using software of different types:

1. **Support of professional work:** from the perspective of an individual worker, an IT-supported workflow crosses a "workbench" which supports his or her work in all aspects, not confined to the work process to be acted upon. The official as a knowledge worker draws on many resources. He or she is used to invoke office tools, search for additional information, and uses platforms for collaboration with teams or groups, either on a steady basis or ad hoc. The interface between the flow of a process and the resources which this knowledge worker marshals can be construed as a situation where, according to the situated requirements of an ongoing process, he or she formulates demands toward the supportive environment which are met by "satisfiers" either available locally or brought in from elsewhere [5].
2. **Client's concern:** the occurrence of "service encounters" where an agent providing a service (or mediating it e.g. in a one-stop front office) is confronted with individual customer requirements flowing from a wide variety of life situations. These are difficult to specify in advance. Standardised service models should not preclude behaviour which caters to special wishes or needs of the customer. Rather, such models should serve as a resource, providing orientation in processes of service delivery which are adequate for a given situation [2].
3. **Unpredictable decision making processes:** decision-making is an important characteristics of its operative work in public administration. The officials in charge must be able to involve additional actors in decision making and to change the course of the process at stake.
4. **Limitations of cross-organisational feasibility:** actors co-operating across organisational borders have less or no possibility to discuss and commonly decide on details of their case-based collaboration during execution time. They frequently make assumptions or draw on commitments on what other agencies can contribute to the process execution. And it might turn out that actors in charge cannot act as planned and therefore must be able find other ways according to their available means and resources.

2   Modelling for Flexibility in Administrative Processes

Understanding the particular problems of different types of business processes in the public sector is a prerequisite for developing an adequate modelling approach. As pointed out above, many of the business processes in the public sector must not be predefined. We need to understand the details of how the actors involved bring in their expertise and how they collaborate and participate in decision making throughout the processes in order to choose or develop a modelling approach, which allows the design of appropriate IT support without losing sight of human agency and discretion in performing knowledge work, as well as of the collaborative aspect of such work.

Up to now, modelling of administrative processes is based on approaches known from business process reengineering, workflow management and/or record management. Those approaches and the current research in this areas do not focus on administrative processes. However, they do offer some support, but fall short of providing the flexibility required (see table 1).

Table 1. Major modelling approaches used to support processes in the public sector

approach	business process reengineering (BPR)	workflow management (WFM)	record management (RM)
original focus	business processes with the aim of reengineering, often based on event-process-chains (e.g. ARIS), usage of reference models for different domains	automatic management of data objects (e.g. documents) "flowing" through the work organisation while relating work items, work capacity and IT applications during run-time	processing administrative documents, i.e. creating, sharing (managing access authorisation), manipulating, registration/archive, retrieval etc.
current research	e-commerce	inter-organisational WFM	inter-organisational RM, semantic web
support for admin. processes	enables identification and overview of core processes of the organisations at stake in the process of modelling, actors are not taken as human agents working in a situated environment, but as attributes of process elements does not address workplace perspective, collaboration or flexibility (e.g. for officials the daily work is not triggered by "events")	adequate support of well-structured and standardised routine processes inclusion of independent subprocesses (e.g. across organisational boundaries) and other flexibility issues (e.g. exception handling) come increasingly into focus no support of officials or agencies in their way to organise or redirect processes according to situated needs	standardised IT support for record management throughout the organisation – but not beyond creates a more or less flexible collaboration environment, but no support for process management poor support for officials or agencies in their way to organise or redirect processes according to situated needs (e.g. to make adequate annotations)

2.1   Focussing on Actors and Relations within Processes

All of these approaches above can be applied successfully in public administration. But as they are inherently limited in supporting flexibility required for collaboration, we need to look for and/or develop modelling approaches which acknowledge the broad range of human work practices and take into account the understanding of human agency within administrative processes. Combining modelling practices from collaborative computing and process modelling (as in workflow and re-engineering methodologies), we try to identify the *unique involvement of actors* in administrative processes and present the *relation of each individual contribution to the overall process* as something tangible within the situated execution of processes and of related decision making. Focussing on actors and relations during modelling and IT implementation allows (even while a particular administrative process is ongoing) the official in charge to decide whom and what kind of contribution to enrol into the process.

From that point of view, the interconnection of process elements can be regarded as a form of individual contracting, framed by standards providing process patterns and rules for contracting. We radically depart from the assumption of a hierarchical world (in which process re-engineering is still caught). Instead of predefined processes being imposed and implemented from above, we assume contracting relationships – not only between external customers and an agency or an agent, but also within a process in which the results of each step performed should serve the next step of process execution. The actor in charge of this next step is thus considered as an internal customer. Such a contract model closely corresponds to the philosophy of New Public Management. One of its tenets is replacing hierarchical relationships with performance contracts, according to the principles of Management by Objectives. It is also related to a view which assigns tasks to units or agents not in a hierarchical way but by means of contracts, thus allowing for a wider range of institutional arrangements than classical administrative thinking.

2.2   Modelling Admin Points and Process Patterns

To support contracting and collaboration between officials / agencies as well as design of appropriate IT support, we suggest to model relational and actor oriented process patterns based on a repertoire of "admin points". The approach is a domain specific enlargement of serviceflow modelling ([1], [6]) which has been developed to model

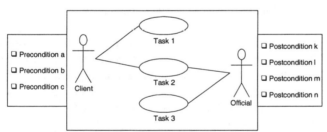

Fig. 1. Abstracted model of an admin point with three tasks (involving an official and a client, e.g. a citizen) and its pre- and postconditions

service processes in the field of tension between given standards and case-based reasoning. Within serviceflow modelling, the series of service points (denoted as a list) serves as the process plan or schedule (looking ahead) as well as the process history (looking back). Each of those points include a (UML-) specification of actors carrying out certain tasks/activities as well as the pre- and postconditions at each point (see abstracted model in figure 1) for "contracting" within process execution.

In principle, each of those points can be defined as needed. However, for communication and co-operation within domain specific processes and across organisational borders, it is most helpful to share a modelling "language", i.e. a common repertoire of premodelled admin points (see figure 2) in which each of these points are specified in terms of tasks/activities to be carried out, based on lists of pre- and postconditions.

Admin Points

Record point: official processes incoming record according to assigned tasks and given standards/rules, and documents results for further use within process

Record Point

Take-in Point: official or agent accepts application of citizen or other client of the administrative unit; if needed: official or agent evaluates situated concern and provides application assistance

Take-in Point

Approval/Decision Point: official approves or decides on a matter in relation to an incoming process/record and documents results for further use within process

Approval Point

Give-out Point: official or agent gives out result of application of citizen or other client ; if needed: evaluation of situated concern and assistance for next steps

Give-out Point

Negotiation Point (or start/end of negotiation process): several actors negotiate a matter in relation to an incoming process/record in order to decide on continuation of process (the responsible actor may also start a negotiation subprocess if necessary); results are documented for further use within main process

Negotiation Point

Start/End Negotiation Process

Legal Inspection Point: official inspects legal aspect of incoming process/record and documents inspection results for further use within process

Legal Inspection Point

Support Point (or start/end of support process): a responsible actor may call on a contribution not specified a priori to support processing (the actor may also start a support subprocess if necessary); results are documented for further use within main process

Support Point

Start/End Support Process

Research/Inquiry Point: official determines and/or inquires additional information in relation to an incoming process/record and documents results for further use within process

Research/Inquiry Point

Fig. 2. Example of an admin point repertoire

Based on these ideas, administrative processes may be predefined as process patterns, i.e. a series of admin points (e.g. figure 3 denotes an admin flow for the postal vote application through the web portal www.hamburg.de). In practice, actors involved in administrative processes may use these patterns as a general agreement for standardised co-operation and, in each particular case, as a template for the process and for the related individual documentation.

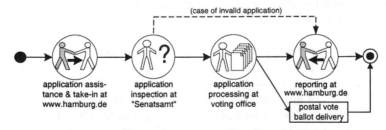

(case of invalid application)

application assistance & take-in at www.hamburg.de

application inspection at "Senatsamt"

application processing at voting office

reporting at www.hamburg.de

postal vote ballot delivery

Fig. 3. Process pattern for postal vote (through the web portal www.hamburg.de) based on the admin point repertoire

Here, we can only briefly indicate how modelling of admin points and process patterns improves flexibility. While processing an individual case it is possible that, e.g.,

- the admin point schedule is predefined, but may be changed if necessary
- the task list at each admin point is predefined, but may be changed if necessary
- skilled work is needed to compare preconditions at each admin point with the accumulated postconditions of process history, and to decide about action to take
- skilled work is needed to document results at each point, in particular to compare actual postconditions with those premodelled and/or expected by other points

In addition, there are various ways of integrating negotiations into process execution:

- the point schedule is suggested, but may be suspended for negotiation (point or subprocess) any time
- the task list at each admin point is suggested, but may be suspended for negotiation (point or subprocess) any time
- skilled work is needed to compare preconditions at each admin point with the accumulated postconditions of process history, and to decide about action to take
- skilled work is needed to document results at each point, in particular to compare actual postconditions with those premodelled and/or expected by other points
- skilled work and/or legitimated decision is needed to evaluate negotiation results and to decide about course of process continuation

In practice, processes may be of mixed character. E.g. the postal vote application is well structured except for handling individual cases with unexpected characteristics revealed at inspection or at application processing (in rare cases these might involve negotiations). For many kind of processes (e.g. when an application for a building permit is filed) it is not clear at the outset if they relatively straightforward or not, and how complex they will eventually become.

3   Discussion

Aiming to provide a platform for collaboration across agencies and to design appropriate IT support for the variety of administrative processes and decision making, we need new approaches to understand and model such processes. We suggest to focus on the unique tasks and activities of each actor involved and to present the relation of each individual contribution to the overall process as something tangible in order to support flexibility through case-based contracting of process elements. Prior to modelling it is essential to comprehensively understand the work situations which unfold every time when an individual process is started. In these processes, process patterns based on admin points should serve as a guide, allowing for departures from the patterns at the discretion of human actors.

Reviewing the different types of work encountered in public administration (mostly based on professional knowledge and processing information), we find that not all types can be adequately described as processes, especially those having to do with organisational learning [3]. There, approaches from collaborative computing might be a better choice for modelling. E.g. in the case of multilateral negotiations, such as in the area of house construction and issuance of building permits, the situation can best be rendered by assuming a platform for free collaboration where the architect, the owner and the issuing agency meet to discuss the relevant questions.

However, in most instances the process view is inherent in the nature of the work of public administration and also of the judiciary and of legislative bodies. It is always (except for individual actors trying to improve their knowledge without producing any tangible results) about delivering a product – mostly an informational product such as a legally binding administrative decision – to some actors in their environment, or to society at large. This implies that an input (demands, supports, legal constraints) is transformed through a conversion process into an output. It is therefore not advisable to model collaborative working situations without a process structure.

We do not expect that the modelling approach presented here will soon be adopted widely. It still needs further research and empirical evidence to (1) prove the feasibility of modelling admin points and admin flows in public administration, (2) provide guidelines as to the identification of process elements and the required granularity of modelling admin points and respective flow patterns, and (3) gain experience on the scope of the modelling approach and its (possible) impact.

Finally, we do not claim that the modelling approach exposed here is the only one possible permitting to escape from a view which implies the strict co-ordination of process steps and which treats human actors as simple executing agents. But we do argue that this approach offers more potentials for flexibility and for acknowledging the central role of human agency than any of the concepts applied up to now. A radical departure from standard workflow approaches is now required in order to achieve an integrative understanding of the work in the public sector and to avoid blocking opportunities for improving productivity, performance as well as working conditions.

References

1. Klischewski, R., Wetzel, I., Baharami, A.: Modeling Serviceflow. In: Godlevsky, M., Mayr, H. (ed.): Information Systems Technology and its Applications. Proceedings ISTA 2001. German Informatics Society, Bonn (2001) 261-272
2. Klischewski, R., Wetzel, I., Serviceflow Management: Caring for the Citizen's Concern in Designing E-Government Transaction Processes. In: Proceedings HICSS-35. IEEE (2002)
3. Lenk, K: Notwendige Revisionen des Geschäftsprozessdenkens. In: Maria A. Wimmer (ed.): Impulse für e-Government, Internationale Entwicklungen, Organisaton, Recht, Technik, Best Practices. Österr. Computer Gesellschaft, Wien (2002) 65-74
4. Lenk, K. Traunmüller, R., Wimmer, M.A.: The Significance of Law and Knowledge for Electronic Government. In: Grönlund, A. (ed.): Electronic Government – Design, Applications and Management. Idea Group Publishing, Hershey, London (2002) 61-77
5. Mowshowitz, A: Virtual Organization. Communications of the ACM, 40-9 (1997) 30-37
6. Wetzel, I., Klischewski, R.: Serviceflow beyond Workflow? Concepts and Architectures for Supporting Inter-Organizational Service Processes. In: Proceedings of 14th CAiSE (International Conference on Advanced Information Systems Engineering). Springer, Berlin (2002)

Business Process Management – As a Method of Governance

Margrit Falck

University of Applied Sciences for public administration and legal affairs Berlin,
Alt-Friedrichsfelde 60, D-10315 Berlin, Germany
margrit.falck@fhv.verwalt-berlin.de

Abstract. Practical examples of the public administration are discussed, in order to show the fact that business process management does not only serve the purposeful reorganization of administrative expirations but transports at the same time processes of organizational learning in the sense of Governance. Under the signs of e-Government there are above all the capability to cooperate and to act under on-line conditions, which must be learned by coworkers and administrative organizations. It is reported on a Virtual Community, which is used as instrument both for the business process management and for e-Governance.

1 It Does Not Go without Strategy and Examinable Goals

Business process management concerns the modelling of business processes with the goal of its gradual improvement (optimization). Which kind of improvement is aimed and in which measure improvement to be achieved, is to be defined before and specified on the basis examinable criteria.

With orientation on e-Government one goal is already set: the improvement of the process organization by means of the opportunities, which offer information and communication technologies on the state of the art. Decisive for kind and range of the use of technology however are such capability characteristics, those the organization of the process to be sufficient depending upon conditions are or the quality criteria, to which the result of the process correspond.

That has the consequence that the technological design concepts differ as a function of the goals, e.g. whether as improvement the balance of an administrative decision is the center of attention or the speed, with which it is present. In case of a development plan the employment of technology has to concentrate probably rather on transparency argumentation and decision making as well as on support forming of an opinion and bargaining processes, when in the case of a building permit, with which rather the automation of the processing and transportation logistics would have priority, after which the editors are supplied with information and documents for the quick task completion.

Administrative procedures are rarely in their capability characteristics homogeneous or in their quality criteria clearly. Rather many subprocesses are involved in a result, which concern one or more authorities and have in different segments different improvement goals.

R. Traunmüller and K. Lenk (Eds.): EGOV 2002, LNCS 2456, pp. 137–141, 2002.
© Springer-Verlag Berlin Heidelberg 2002

Example: The management of a building exists in a set of service processes, which include planning, decision, production and negotiation processes and in those at least two authorities are involved: the building-using authority ("tenant ") and the building-managing authority ("landlord"). The improvement goals extend of the quick accessibility of the responsible partners, over the rapid and economical completion from repairs to the flexibility of the demand services. Accordingly possible solutions for improvement will exist in a mosaic of single solutions, which must be developed by a systematic analysis and organization of the processes in the sense of administrative engineering. In the example of building management belonged to:

- the conclusion of a service agreement,
- the list with the telephone numbers of the responsible partners,
- making available forms and document patterns,
- the modernization of different workplaces in the equipment with means of communication as well as with hard- and software,
- authority-internal and authority-spreading solutions for information supply and collection,
- the access to the Intranet,
- the budgeting of financial ressources as well as
- new regulations in the contract design.

2   Organizational-Learning As an Important Goal

The selection of the realizable solutions depends primarily on it which possibilities the surrounding field of the business processes permits. In addition, it should be made dependent on it, in which measure with the modelling and optimization process possibilities of the organizational learning for the involved ones are created, which refer at the request of the future process organization in the sense of the e-Government desired.

What do I mean with it?

Under orientation e-Government the feasibility of an improvement solution essentially depends on the following factors:

- of available resources (finances, technology, personnel including their qualification),
- of the valid right and security requirements,
- of available standards, information and reliable knowledge,
- of the talents of the involved ones in handling new media as well as
- of preferential action standards and learned behavior.

Those are factors, which are to be searched in the surrounding field of the business process, i.e.. in the application environment of the planned process organization.

In addition, many of these factors belong to the success factors for the work on the project, i.e. for the joint work during the modelling and optimization procedure, in whose process the new process organization is agreed upon and established, which is to become after conclusion of the project the lived organization everyday.

What lies more near to already learn and learn as during the common work on the project exactly the qualifications and talents or train such behaviors and set standards, which are in the future necessary for the planned process organization and for e-Government?

In the example of the service processes between tenants and landlords the agreed upon process organization required of the involved ones:

- to deal with information openly and transparency; i.e.. to make and maintain information offers, to give the tenants insight into documents upon achievements agreed or planned measures of the landlord or inform about problems and complaints, in order to be able to react in time.
- to work and communicate service-oriented; i.e.. to consider desires appropriately or to make also demands appropriately, for confirming dates and inquiries, to inform about intermediate treatment states;
- to call up available information regularly or as required (pull-principle); i.e.. to use information offers and to inquire missing information.
- to act team-oriented and cooperatively; i.e.. to regulate deputy, to be able to give information, to use electronic communication efficiently, to share knowledge.

Many of these requirements refer to the fact that with the transition for e-Government competence in electronic communication and in the electronic task completion is needed, which could not be developed so far. Therefore we tried to set during the work on the project standards and promote behaviors, which contribute to the development of the demanded competences. So we have

- transparency and openness promoted, by informing in the Intranet about the course of the project continuously. In addition reports belonged to beginning and conclusion of the individual work packages, over the planned goals, proceedings and over the reached results.
- the communication ability promoted and time standards for reactions and feedbacks set, by answering electronic inquiries rapidly or Reactions to inquiries called consequently, to made calculable our accessibility by absence assistant, as well as, results of common meetings documented and distributed nearly in time, regular Reviews to the fulfilment of agreements and measures.
- authority-spreading co-operation develops, by restoring the common confidence basis with existing conflicts by appropriate presentation and were endeavored around mutual understanding.

Like that the business process management is not only a method administrative engineering but also an instrument to carry the culture change on the way to the e-Government.

3   Co-operation between Authorities As an Element of the Cultural Change

From our view the development of the ability for authority-spreading co-operation is a particularly important behavior for e-Government, because usually several authorities are involved at the production of a product or a decision for citizens, economics or the public. With the development of citizennear on-line services the contributions of the authorities involved must be integrated to a continuous process chain.

It becomes clear by the example of the registration of a trade (Fig. 1.), which will be possible for on-line over the Internet or off-line over an office for citizen or the district office for economy either. The expiration of request integrates office for citizen, call center, office for economy and senate administration to a process organization, whose optimal organization for citizens and administration is dependent on the close co-operation of all involved ones.

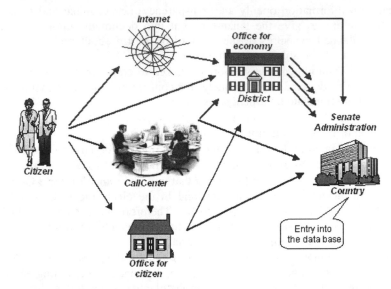

Fig. 1. Process variants „registration of a trade" Which process variant is going through, depends on different factors and decides in different places. For the citizen the choice of the entrance depends on its technical possibilities, on the use comfort, on the costs, on the quality of the consultation or other criteria. In the citizen office the run of the process depends on whether the case can be worked on finally or whether the case must be passed on to the economic office. The decision depends on which technical assistance and which know-how the coworkers have. In the Internet it decides whether the case can be worked on automatically or must be worked on manually. The decision depends on the fact whether the state of affairs is clear or must be interpreted. In the call center the further way decides likewise after the technical possibilities, after the existing qualifications and assistance as well as after the state of affairs of the case. The conditions at the different interfaces change with the time. Technical possibilities can be created, know-how grow, qualifications improve, assistance can be extended, etc.. Therefore the process optimization must be understood and realized as dynamic process, which presupposes the constant co-operation of the involved ones

An additional difficulty is that process organizations are subject to dynamics, with which the original optimization goals and criteria loses at validity. The dynamics can have their causes

- in the change of tasks,
- in the increase in experiences, competences and reliable knowledge,
- in the change of the division of responsibilities and the interfaces between the places involved and/or
- in the technology development.

The capabilities of cooperation of the authority involved, which was necessary for the development of the optimized process organization, is necessary thus also for the maintenance of the process organization.

In the case of the registration of a trade the representatives of the authorities were requested to the discussion in a virtual work space in the Intranet under the presentation of a neutral authority, in order to realize the difficult tuning processes between the numerous involved ones in a justifiable time. In this way the attempt is undertaken to establish the cooperative work forms necessary for the future process organization and to learn the appropriate abilities of the involved ones during the process design.

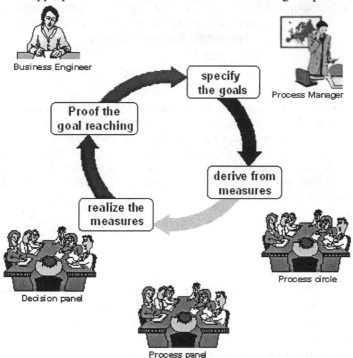

Fig. 2. Sustainability by process responsible person. The process manager observes the process organization, controls the process goals and suggests measures for further improvement. The process circle consists in theparticipants of the prozess, which collect information about the process, judges together, identifies weak points and looks for possibilities for the improvement. (similarly quality circles) The process panel consists of the process managers. They coordinates the suggested measures for process improvement. The Decision panel consists of the line managers of the organizational units taken part in the process. They decide on the realization of the suggested measures for process improvement. The business engineer supports the process responsible person in the process documentation. It knows GPM methodology and creates the connection to the information technology

4   Process Responsibility As Instrument of Sustainability

In order to reach the sustainability of an optimized process organization, it requires the continuous observation, analysis and improvement, in the sense of the process Controlling. As instruments of the process Controlling the function of the responsibility of process is currently tested. (Fig. 2.) About experiences can be reported at the time of the conference.

Proposal for a Dutch Legal XML Standard

Alexander Boer, Rinke Hoekstra, Radboud Winkels[1],
Tom M. van Engers[2], and Frederik Willaert[3]

[1] Dept. of Computer Science & Law, University of Amsterdam, Netherlands
[2] Dutch Tax and Customs Authority, Utrecht, Netherlands
[3] Application Engineers, Belgium

Abstract. This paper presents a proposal for an XML Standard for legal sources in the Netherlands. The standard intends to provide a generic and easily extensible framework for the XML encoding of the structure and contents of legal and paralegal documents. It differs from other existing metadata schemes for legal documents in two respects; It is language-independent and it aims to accommodate uses of XML beyond search and presentation services.

1 Introduction

This paper presents a proposal for an XML Standard for legal sources in the Netherlands. The research is carried out in the context of the E-POWER and e-COURT[1] projects and the results will be brought into the LeXML initiative[2]. The standard intends to provide a generic and easily extensible framework for the XML encoding of the structure and contents of legal and paralegal documents. This obviously includes legislation and case law, but also written public decisions, internal and external business regulations (for instance ship classification rules as in [7]), and contracts. XML elements and structure are defined in schemas that can be used to validate a document. Since there is a great variety of legal documents that cannot be covered by one normative standard, the standard consists of multiple schemas defining vocabularies that can be mixed in a document. While the standard aims to cover all possible legal sources, the focus of current work is on Dutch legislation: the 2001 Dutch law on income tax in the context of the E-POWER project, and the Dutch penal code of 1881 in the context of the e-COURT project. Later we will cover the structure of (Italian and Polish) court room transcripts (for e- COURT) and case law. The standard differs from other existing metadata schemes for legal documents in two respects; It is language-independent and it aims to accommodate uses of XML beyond search and presentation services.

---

[1] See Acknowledgements at the end of this paper.

[2] The LeXML initiative is the European equivalent of the Legal XML community in the Unites States.

R. Traunmüller and K. Lenk (Eds.): EGOV 2002, LNCS 2456, pp. 142–149, 2002.

1.1   Why a Legal XML Standard?

Clearly, legal documents serve a plethora of purposes, and for each purpose contain information that can be grouped in various general and specific categories. Many of the most common classification and reference systems even pre-date the storage of legal information on computers. XML promises dramatic improvements in the efficiency of managing and processing information in legal documents. XML and related technologies are essential parts of the more general idea of an integrated *semantic web*: A machine-readable and machine-understandable version of the current web coupled with new smart applications to exploit it. This machine-readable extension is called metadata: information about information.

Current XML schema and metadata definition efforts in the legal domain are concerned with government document locator services for general use like the British Legal and Advice sectors Metadata Scheme (LAMS) for 'Just Ask!'[3] and Australian Justice Sector Metadata Scheme (JSMS)[4]. In addition, there are XML schemes used by legal publishers for document locator services. These schemes focus mostly on classification of legal documents using traditional differentiae like:

- The author and legal status of the document
- Creation, modification, and promulgation dates
- The jurisdiction to which it applies, and the language(s) it is available in

These attributes are rather crude in meaning, and the resulting classification is superficial; It leaves out a lot of relevant detail and its quality is questionable for automated reasoning. Identification of documents by jurisdiction, for instance, assumes that the user of a search service knows what jurisdictions he is in – and that jurisdictions can be meaningfully delineated. Establishing jurisdiction is however a domain of legal research in itself. The meaning of the dates is clearer and can be directly used to establish validity of a document, but only relative to a known jurisdiction.

In addition to these traditional categories documents are usually classified in substantive terms with an attribute identifying a 'domain' of the document in a fixed classification. Viewing these attributes as metadata, or extra information about the document, creates a potential maintenance issue: The values of these attributes may change over the lifetime of a legal document, even if the document itself does not, as the concepts employed in the document change over time and become associated to (disassociated from) other concepts (see e.g. [5]). In reality, the agent of such a change in the web will usually be the creation or modification of a new legal document with certain attribute values. The information about the information is often not extra: It was in the document in the first place, or in another document referring to it. Classification at a 'heuristic level' always places the burden of maintenance on human domain experts.

---

[3] See e.g. http://www.lcd.gov.uk/consult/meta/metafr.htm
[4] See e.g. http://lawfoundation.net.au/olap/guidelines/metaintro.html

Furthermore, it is hard to conceive of any 'smart application' using the classifications as more than filters on search results. This is not because the meaning of 'jurisdiction' or 'appellate court filing' cannot be made understandable to the computer, it is because the level of the classification presupposes that the user of the classification system can read the document to find out why the classification was attached.

1.2   RDF and Representation of Meaning

The alternative approach to such a domain classification, is the direct identification of statements in the contents of documents. This includes statements about other documents and (fragments of) the document itself. For this purpose the Resource Description Framework[5] (RDF) was designed. In RDF, statements are encoded as as $(subject, predicate, object)$ triples. By describing legal concepts in different jurisdictions in an RDF 'dictionary'[6] as conceptual prototypes, the LexML initiative assumes that it is possible to identify and describe similarities and differences between legal concepts in different languages. Existing schemes (like JSMS and LAMS) rely on compatibility with HTML's META tag only allowing RDF-like statements about documents; In RDF terms that means that the subject of a statement is always a HTML document. The META tag can only be used to make statements about a document in the document itself.

The RDF data model is based on the concept of a statement and the concept of 'quotation' (or reification) of statements. Quotation is used to make statements about (reified) statements – the statement is treated as the subject or object of another statement. In addition to this data model additional RDF schema statements can be defined, whose meaningfulness depends on the extent to which they can be transformed to other schemas and whether there exist any applications that can validate statements against that schema with query or inference languages. Since RDF is serialized in XML documents, it is important to realize that the validity of an RDF statement is relative to an unspecified set of other statements in the memory of the validating application. There is no central authority guarding universal consistency or coherence of the semantic web and anyone can publish any truth, falsehood, opinion, or judgment.

The problem, obviously, is that of trust. Quotation allows one to create trust, because it allows one to express where something was stated, who stated it, who modeled a statement made by someone, and what level of guarantee they dare to associate with it. 'Syndication' of information is based on this notion of trust. You may be willing to pay for advice, for instance, if you trust the party that gives it even if advice on the same subject is also available for free from other parties. It is not merely about good intentions: Heuristic notions like 'domain' always remain open for disagreement. The quotation mechanism in RDF allows others to make qualifications about statements, solving another limitation of META tags.

---

[5] See http://www.w3.org/RDF/. RDF, like XML is an open standard from the World Wide Web Consortium (W3C) that is well-supported with free software.

[6] http://legalxml.org/Dictionary/

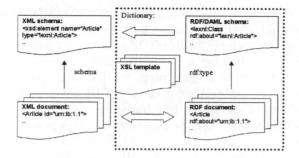

Fig. 1. RDF representation of legal documents.

Figure 1 shows the relationship we propose between the XML encoding of a document and the RDF encoding of the same document. If a document element is encoded as a set of RDF statements, any element of it can be subject, predicate, or object of statements *about* the document.

1.3   Aims of the Proposed Standard

This proposal is more complicated than comparable standards because it is motivated not just by concerns for 'smart' search, but also with document management and maintenance of existing and future decision support software used by public authorities like the fiscal applications developed by the POWER group in the Netherlands (cf. [1]). This proposal aims to standardize legal documents for the purposes of:

- Filtering
- Presentation
- Document Management
- Knowledge Representation
- Search
- Code Generation
- Rule generation
- Classification and Verification

Each meaningful element of the text up to a full sentence can be separately selected or changed by software through an XML API. Each element of the text can be marked with an ID attribute that is unique in the document. An element is easy to transform to (X)HTML (including hyperlinks) with familiar layout by application of XSL transformations. Because each element can have a unique identifier, it is possible to make external statements in RDF refer to the right part of the document. Only RDF, RDF Schema, and HTML support a widely accepted standard for reference anchors embedded in XML. The recent XLink specification released by the W3 Consortium creates a generic facility for referring unambiguously to parts of XML documents, but is not yet supported by most tools.

The standard consist of coupled schemas expressed in RDF Schema and XML Schema that are equivalent in meaning. A a two-way translator between the 'basic' XML standard and a more flexible RDF standard for software engineers is in development. RDF can be used to encode inverted file indices, association rules and, if time permits, self organizing maps (cf. [3]) to search for document elements using a concept index. A great variety of IR techniques requiring special indexing and active learning algorithms have been used for information retrieval on legal documents (see e.g. [6, 4] for an overview). RDF is syntactically also sufficient to represent description logics (in DAML+OIL) or UML. The standard thus allows use of existing UML and DAML+OIL editors for legal knowledge representation. DAML+OIL is an RDF Schema[7] that provides a number of standard 'logical' constructs for RDF descriptions (cf. [2]). The content of documents, e.g. norms and validity constraints, are expressed using the DAML+OIL vocabulary, because this vocabulary can be validated by free description logic theorem provers (FaCT and RACER) and translated to an expert system rule engine (JESS).

2   Description of Documents

We roughly distinguish three different viewpoints on how we look at legal documents:

Form In the Netherlands and in general a legal document can be 'recognized' and classifed by certain required phrases and formulas. Formal requirements on structure mostly reflect considerations of consistency of language and ease of access for the reader, but it also provides a context for the interpretation of the content of the document. This latter role is very specific for jurisdiction and timeframe and not part of the basic standard. Generic structural desiderata are defined in XML schemas.

Role Although we may look at the phrases and formulas in a written decision to classify a document as a law, we know that it is not the structure of the document that makes it a law, but the role the document plays in the activities of public bodies - most importantly the activities that produced the document. This is captured in RDF statements 'about' the document.

Content We also classify documents depending on what its content means: It represents a type of decision. If it is just a public decision its meaning is limited to a particular occurrence or case. If it is a norm or policy its meaning extends to general class of occurrences or cases and it postulates a value theory for making and judging decisions. This is captured in RDF statements 'about' this content: acts, norms , agents etc.

Figure 2 shows the relation between the various XML schemas that are part of the standard. An XML document on the bottom leftside that adheres to

---

[7] See http://www.w3.org/SW/ or http://www.daml.org for DAML+OIL or e.g. http://www.ontoknowledge.org/oil/ for its precursor OIL.

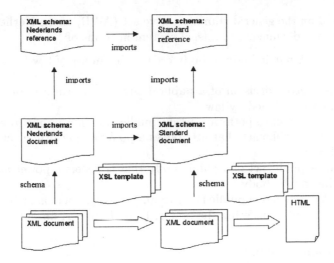

Fig. 2. Relations between components of the standard. The white arrows represent possible XSL transformations.

schema *Nederlands-document* uses the Dutch (Nederlands) XML element vocabulary for document structure and indirectly imports vocabulary *Nederlands-reference* and the structural requirements of *Standard-document* and *Standard-reference* – the 'language-independent' core vocabulary (in simple English). The language-specific schemas contain direct word-for-word translations (with the XML schema substitutionGroup element) to the standard vocabulary. An XML document is first transformed with an XSL transformation sheet to standard vocabulary, and the standard vocabulary can be transformed to HTML or RDF. RDF is used for further description of role and content.

Extensions built on top of the standard consist of relevant language-dependent vocabulary schemas, a standard schema that defines formal requirements, a simple XSL template that translates language-dependent schemas to standard vocabulary, an XSL template that translates the standard schema to standard-document, and optionally XSL templates for specialized presentations in HTML.

3   Terminology

Regardless of whether one sees the written law as an imperfect description of something else (*the* Law) or as a product of an social agreement that requires clear communication and predictable and foreseeable decisionmaking, it is clear that the written law refers to things 'in the world'. To regulate a world it must be described, after all. To be able to describe the meaning of the document the standard specifies a general vocabulary that can be used to describe it in RDF for specific purposes and domains. This work is in progress.

The conceptual core of these ontologies describes the relations between the various public documents published by public bodies in a fairly generic terminology, based on the general administrative act (AWB) of the Netherlands, that positions public documents in a legal system in terms of:

Public body A public body or body created by an act of law to serve a public interest.

Decision A written decision of a public body to perform a public act using a public power assigned by law.

Power A permission to perform certain public acts in a public interest.

Assignment A public act that attributes a power to perform a public act to a public body.

Delegation A public act of a public body that transfers a power assigned to it to another public body.

Mandate A decision of a public body to allow another person or body to use its power, without transfering the power, to perform certain public legal acts.

3.1   Time and Change

The importance of capturing the relations between public legal documents is made apparent by considering the requirements for correct updating of a collection of documents in time. Changes in laws are announced in separate documents and publishers must keep track of all documents from certain publication channels to be able to reconstruct what the form of an organic law is at some time point. Similarly, if you find a written decision on your doormat its validity status changes when a new written decision retracting it follows two days later.

To keep track of versions the standard provides a number of attributes for every structural XML element in the document that can be identified, selected, and thus changed; The *date-publication* of an element is the time the element is officially published or announced. The *date-enacted*, the time the content becomes applicable in decisionmaking, is always later than or the same as date-publication, but before *date-repealed*, the time the content becomes inapplicable in decisionmaking. Between date-enacted and date-repealed the element and its content is *active*, and outside this interval it is *inactive*. The *date-version* attribute represents the date the correctness of the content and other dates of the XML element was last verified. The XML document looses its value as a normative reference as time progresses and the time-interval between date-version and today increases.

If an XML element in a newly published document refers to another XML element the content of the element may repeal, enact, or change the other element. A XML element may also refer to another XML element to invoke its 'power'. This is the case if the official author of the document obtains the power to take the decision(s) communicated in the document from a specific assignment by law, delegation decision, or mandate decision. In such cases the element, or an ancestor element, becomes inactive if the element from which the power is obtained becomes inactive (at least in regulations from public bodies in the Netherlands). The standard aims to provide representational primitives that adequately capture legally relevant acts of this nature on legal documents without commitment to a specific 'model' for updating documents.

4   State of the Proposal

We currently have XML schema definitions of the structure (form) of Dutch legislation, Dutch documents in general and their standard counterparts, i.e. effectively all that is presented in Fig. 2. We have tried it out on the new Dutch income tax law and are in the process of applying it on other legislation. We have also created XSL translations from the XML format of a major dutch legal publisher and an XML format supported by the POWER tools (cf. [1]). We are building a large ontology for the legal domain, concentrating first on fiscal and penal law. By the end of June 2002 a substantial part will be finished.

Acknowledgements

E-POWER is partially funded by the EC as IST Project 2000-28125; partners are the Dutch Tax and Customs Administration, O&I Management Partners, LibRT, the University of Amsterdam (NL); Application Engineers, Fortis Bank Insurance (B); Mega International (F). e-COURT is partially funded by the EC as IST Project 2000-28199; partners are Project Automation, Ministry of Justice, CNR (I); Ministry of Justice (POL); Sema Group S.a.e. (SP); Intrasoft International (L); Universit Paul Sabatier (F); University of Amsterdam (NL).

References

1. Engers, T. van, Gerrits, R., Boekenoogen, M., Glass'ee, E., Kordelaar, P.: POWER: Using UML/OCL for Modeling Legislationn -an application report. In: *Proceedings of the 8th International Conference on Artificial Intelligence and Law (ICAIL 2001)*, pp. 157–167. ACM, New York, 2001.
2. Fensel, D., Horrocks, I., Harmelen, F. van, Decker, S., Erdmann, M. and Klein, M.: OIL in a nutshell. In: R. Dieng et al. (eds.) *Knowledge Acquisition, Modeling, and Management, Proceedings of the European Knowledge Acquisition Conference.* Lecture Notes in Artificial Intelligence, LNAI, Springer-Verlag, October 2000.
3. Kohonen, T.: Self-Organizing Maps. Springer Series in Information Sciences, Vol. 30, 1995. Springer, Berlin. Third edition 2001.
4. M-F. Moens, 'Innovative techniques for legal text retrieval', *Artificial Intelligence and Law*, **9**, 29–57, (2001).
5. E. Rissland and T. Friedman, 'Detecting change in legal concepts', in *Proceedings of the Fifth International Conference on Artificial Intelligence and Law (ICAIL-99)*, pp. 127–136, New York (NY), (1995). ACM.
6. H. Turtle, 'Text retrieval in the legal world', *Artificial Intelligence and Law*, **3**, 5–54, (1995).
7. R.G.F. Winkels, D. Bosscher, A. Boer, and J.A. Breuker, 'Generating Exception Structures for Legal Information Serving', in *Proceedings of the Seventh International Conference on Artificial Intelligence and Law (ICAIL-99)*, ed., Th.F. Gordon, pp. 182–195, New York (NY), (1999). ACM.

Size Matters – Electronic Service Delivery by Municipalities?

Ronald Leenes and Jörgen Svensson

University of Twente, P.O. Box 217, 7500 AE Enschede, Netherlands
{r.e.leenes,j.s.svensson}@bsk.utwente

Abstract. The development of e-government in the Netherlands shows two different worlds. The large national organisations implement Electronic Service Delivery (ESD) fairly successfully, while municipalities are slow to adopt ESD. This is a pity, since municipalities account for over 70% of the public services. They are expected to implement ESD on their own although they lack the necessary resources and distributed development is inefficient. In this paper we address the role of municipalities in the real and virtual world and argue that development of electronic (local) public services may be organized on a larger scale, depending on the type of service in question.

1   Introduction

In the international rat race for e-government, the Dutch government too has set ambitious targets. Twenty-five percent of the public services are to be delivered online by the end of this year. Although twenty-five percent does not seem much, and certainly is a long way from full-blown electronic government, even this goal is difficult to meet. In fact, in order to claim success, the Dutch government is already in the process of massaging the data. By redefining the phrase "twenty-five percent of services" into "twenty-five percent of service transactions", it is giving a disproportionate weight to the few large national service programmes, such as the Internal Revenue Service and the Dutch student bursary programme, which have succeeded in implementing electronic service delivery.

As a result of this redefinition, the twenty-five percent target may be reached. But, what about the next seventy-five percent, or even the next ten percent? With the large central programmes digitised, any future increase in electronic service delivery (ESD) will have to come from the Dutch municipalities. These municipalities deliver the vast majority of public services in the Netherlands, and they are currently considerably less successful in implementing ESD.

In this paper we discuss the reason why further ESD-development by the municipalities is problematic and we suggest some solutions to improve the chances of success.

The structure of the paper is as follows. In section 2 we compare the ESD track records of large national public service organisations on the one hand, and municipalities on the other. Then, in section 3 we provide the simple, but fundamental, problem of ESD-development by the municipalities: scale. Real ESD does not develop well in the context of small scale service delivery by municipalities. Then, building on this

R. Traunmüller and K. Lenk (Eds.): EGOV 2002, LNCS 2456, pp. 150–156, 2002.

conclusion, the central question in the remainder of the paper is whether the scaling up of ESD will be a viable option. To answer this question we first address the background of small scale service delivery in the Netherlands, the various arguments supporting the current arrangements for service delivery and the actual practices (section 4). Understanding this background provides insight in the conditions under which scaling-up will be desirable and possible, and in the methods that can be applied in specific circumstances (section 5). Section 6 provides some concluding remarks.

2   ESD, Fast and Slow

For some decades now, the Dutch government has had an eye on introducing ICT in public service delivery. Experiments with the use of legal expert systems in service delivery, for instance, date back to the late eighties [1]. The spectacular development of the Internet in recent years has boosted the expectations and prompted for even higher ambitions. Not only would ICT help to make service delivery more efficient, it would also improve service quality. Moreover, by giving the right examples, our government aims to propel Dutch society to the forefront of the information age [2–5].

The policy as outlined by the Dutch Government is aimed at both the quality and the quantity of electronic service delivery. All public service providers are to have a web site that not only provides basic information, but also allows for integrated service delivery (based on life events or demand patterns). Public service providers should also consider implementing pro-active service delivery. In 1998 the quantitative ambitions were expressed in measurable criteria. One of them was that the Dutch government aimed at bringing twenty-five percent of services on line by the year 2002 [3].

What progress have we made so far? Will the targets be reached this year? What can we say about quantity and quality of the services provided?

If we look at ESD in the Netherlands, we may distinguish two very different worlds. On the one hand, several organisations are indeed progressing on the road to electronic service delivery. Import duties in the port of Rotterdam are handled electronically. Most people's car license registrations are renewed by means of automatic bank transfers. The IRS offers its clients a computer programme, free of charge, for filing their tax returns. The data produced by the programme can either be sent to the IRS on-line or by means of a floppy disk. The student bursary system has been highly automated for over a decade and uses modern ICTs to communicate with its educated clientele. It not only offers general information to its clients, but also shows them the data that are stored about them and allows them to change certain data on-line.

Organisations such as these benefit immensely from electronic service delivery. The electronic communication with clients improves the speed and efficiency of communication and lowers the error rate in data entry. Also the clients benefit from ESD. They can interact with these organisations whenever and wherever they want.

Contrasted with the world of large organisations is the world of municipalities where progress in implementing electronic services is much slower.

The municipalities were supposed to be the driving forces in the development of electronic public services [3]. This idea was largely based on the fact that they account for some 70% of public services. Municipalities therefore have most citizen contacts, and they are also the biggest beneficiaries of a successful implementation of ESD. The quality of services may improve and also processes may become more efficient.

The strong focus on municipalities manifested itself in the Public Counter 2000 project (Overheidsloket 2000). Municipalities were encouraged to submit plans for funding local experiments. In 1996 the first phase of the Public Counter 2000 project started with 15 subsidized municipal pilots. Others were encouraged to follow these forerunners.

Some 5 years later we may conclude that municipalities have not come very far. Although many do their best, some still do not even have a website. In fact, in May 2001 only 282 of the 504 municipalities had one and the current aim is to have all 504 on-line in May 2002. If we look at the content of the websites, we may conclude that most of them only provide (sparse) basic information. Very few offer forms that can be downloaded. Only a few municipal websites offer the possibility of on-line transactions.

In sum, in 2002, we can conclude that some organisations indeed succeeded in realising advanced ESD, where at the same time others failed. The question is of course: why?

3   ESD, Large and Small

The fact that some organisations innovate and others do not can of course be related to many aspects [6]. Success in innovation depends, for instance, on the (correct) realisation of an actual necessity, on adequate management and on the organization's (work floor) culture. Indeed, with regard to ESD, all these arguments have been expressed, especially in explaining the lack of development in the municipalities. However, these explanations tend to overlook a simple factor. The organisations that have succeeded differ quite a lot from the organisations that did not.

The ESD champions in the Netherlands share a number of characteristics. First of all, each of these organisations is highly centralised, in the sense that the services are coordinated by a central administrative body. Second, each of them is only responsible for a limited number of related services: import duties, licences, taxes or bursaries. Third, the services are typically *high volume* services; they are offered to a larger audience (IRS, car-licences) and/or with a high frequency (import duties, bursaries). Finally, the organisations in question all have vast resources to develop electronic services.

The Dutch IRS (Belastingdienst) can serve as an example. It only administers the national tax programme. Of this programme, the income tax applies to six million citizens, who have to file their tax returns on a yearly basis. The tax office has a yearly ICT budget of roughly € 300 million and has an *IT* staff of over 2000 people, for a large part working in the special tax automation centre [7].

The municipalities contrast sharply with these ESD champions. Amsterdam, the largest city in the Netherlands, has some 750.000 inhabitants, followed in size by Rotterdam, The Hague and Utrecht. These large cities house 13% of the Dutch population. The rest of the population lives in the other 500 municipalities, resulting in an average number of inhabitants per city of about 30.000. Many towns of course, have even fewer inhabitants.

The size of the administration in municipalities depends on the size of the population. The same goes for the resources spent on ICT and ESD development. A town, such as Woudrichem (14.000 inhabitants) has an IT staff of 1.4 person and an annual IT budget of 160.000 euro.

In the meantime, every municipality, Woudrichem included, is expected to offer a very large number of services (300 to 400), ranging from garbage collection to education and from building permits to social assistance.

Can we expect these municipalities to develop the same advanced types of electronic service delivery as the Dutch IRS? Of course not!

Most municipalities, due to their size, lack the necessary resources to develop ESD for all of their services. But it is also questionable whether it makes sense to develop ESD locally for services that only target a limited population and generally have a low frequency [8].

The problem of developing ESD is clearly related to scale. For most service delivery the current development scale is that of the individual municipalities. This is not only inefficient, in many cases the scarcity of resources at the local level actually prohibits the development of ESD.

We may turn this argument around. If we really want to get ESD of the ground we have to develop electronic services on a larger scale (for groups of municipalities) thereby increasing the number of 'clients' and bundling the available resources.

Although this makes sense from an economic perspective, it also raises important questions from other perspectives. Economy of scale has always pleaded against local service delivery, so what is the rationale of providing services locally? Is the birth of an electronic government reason to change the existing arrangements?

4 The Rationale of Local Service Delivery and Local ESD Development

Public service delivery in the Netherlands is rooted in historical and legal grounds. Service delivery in most cases relates to a decision taken by some public body pertaining a right or an obligation of a citizen or an enterprise. Government is ultimately bound by law. The constitution and the laws based on the constitution determine the powers of the various government bodies. In the Netherlands most powers are distributed to the local level (art. 124 Dutch Constitution). The municipalities are therefore at the core of the public sector; the Netherlands are a decentralised unity state. In this the Netherlands differs from a country such as France, which is far more centralised.

There are various reasons for this distribution of powers. Among them are simpler democratic control, adaptability to local circumstances, better means for people to have their say and better integration of policy [9].

Besides the necessary powers to govern, municipalities also have the powers to provide public services. This of course makes sense, especially in the pre-internet era. Delivering services on a local level is practical (efficient) from the citizen's perspective. Having a service provider nearby saves time. Although municipalities formally are at the core of the public sector in the Netherlands, the system has become more complicated and obscured over the years. With the coming of the welfare state, central government became a more active player, intervening in the autonomous munici-palities. Policy no longer is developed and executed primarily on the local level. The higher levels of government (provincial and state level) nowadays have a much stronger role in policy formation and even in its execution.

In many cases policy is developed at the national level with municipalities only administering the national policy. This form of joint governance has serious conse-quences for the shaping of the public landscape and for public service delivery. In the current practice the powers and responsibilities are distributed over the various levels of government, sometimes making it unclear which level is responsible for a particu-lar task.

In this blurring of powers and responsibilities, municipalities are observed to pro-vide essentially three very different kinds of services:

- Truly local services: i.e. services which are provided based on local policy and lo-cal autonomy, concerning the management of the municipalities' own affairs free from interference by the State. Examples of such services are: street and commu-nity care and safety, local taxes, sports, recreation and culture.
- Joint governance services: i.e. services which are rooted in national legislation, but which are administered by the municipalities, with the municipalities having their own (additional) policy responsibilities and discretionary powers. An example in the Netherlands is the municipal social assistance, based on the General Assistance Act.
- Municipal delivery of national services: i.e. the administration of national policy by the municipalities, where the policy is completely defined at the national level and discretion is limited and the administration by the municipalities is simply a convenient means of bringing the service to the citizens. Clear examples of such services are the issuing of driver's licenses and passports.

In sum, municipalities deliver very different services, and there are clearly very dif-ferent reasons for service delivery on the local level. However, when we consider the strategy in developing ESD in the Netherlands, we see that these differences are not being considered. Initiatives like Public Service 2000 simply place municipalities at the centre of improving public service delivery, regardless of the type of services.
A rethinking of which services really have to be delivered by the individual munici-palities and which services might benefit form co-operation has not taken place.

The adage seems to be: services that are currently delivered at the municipal level should be informatized at the municipal level. But is this really the case? As we argue, the differentiation between truly local, joint and national services provides a basis to differentiate in the way ESD for the various services may be developed. While it un-derlines that some of the truly local services really require local ESD development, it also shows that some electronic services may be developed in co-operation and some may even be taken up by more central, national organizations.

5   Possibilities for Co-operation in, and Centralisation of ESD

Economies of scale provide the key to improve the speed at which ESD can be devel-oped, and as we argue, the type of service determines the possibilities for increasing this scale.

Truly local services, addressing local problems and based on local policies are indeed best dealt with at the level of the municipalities. The municipalities determine the content of these services and therefore should also determine and organize service delivery. When a service is really typical for the municipality in question, there seems to be little choice regarding ESD development: it will require the special development of this ESD for this municipality (either by the municipalities staff or by a commercial organization hired for this task). The possibilities of gaining economies of scale are very limited in this case, which implies that it may be wise not to develop ESD at all. Perhaps, however, solutions may be found in using general tools for developing simple service modules (JAVA applets, ASP code), or by co-operating with other municipalities with similar local services. However, this number of very specific local services is generally limited, and even differences in 'truly local' policies are not always as big as claimed. Many local bylaws are based on standard bylaws as produced by for instance the Association of Dutch Local Governments (logging permit, fire and safety measures), which makes co-operation between similar municipalities a viable option. In some circumstances it may be efficient to look for other, perhaps nongovernmental, organizations to actually deliver these truly municipal services, based on specifications from, and tailored to, the needs of the various municipalities [10].

For joint-governance services the core of the service delivery, the possibilities of co-operation and centralization of ESD development are far greater. Joint governance services, such as General Assistance, typically are based on a national core of regulation, which applies to all municipalities. This means that, in some cases, it may also be possible to partly centralize ESD development. Where the practice of municipal service delivery consists of combining pieces of national and local regulation, ESD development may be approached as a question of integrating local and national ESD-modules, thereby limiting the effort needed by each municipality. An important question concerning the feasibility of this approach is the amount of variation in local policy. For general assistance in the Netherlands, the viability of this approach has already been shown in the development of the MR-Expert systems, which contain the national legislation straight out of the box and to be supplemented with local rules [11].

For service based on national policy, the development of service modules on a national scale is an obvious choice. The responsible ministry could develop service modules, such as intelligent forms or expert system modules, and provide them to the municipalities to incorporate them in their websites. But, in this case also another obvious step can be taken: concentration (or centralization) of service delivery. An example where this already is possible, is the housing benefit. People fill in forms provided by the Ministry of Housing, Spatial Planning and the Environment (VROM), and send them to the Ministry, which takes care of the administrative process. The step to electronic service delivery by the ministry in this case is relatively small.

6   Conclusion

E-government dawns slowly in the Netherlands. One of the causes of the slow progress towards realizing true electronic service delivery is the choice for local ESD development. In the traditional service delivery model the central role for local government made sense. In the internet era the same choice hampers progress because it is inefficient to have all towns and cities develop ESD modules on their own with limited resources. An alternative is to move development to a different scale. This is possible for services that are the same for every town and city.

If we take e-government serious, we have to rethink the development process. We also have to rethink the way services are to be provided by the various levels of government. Which services should be provided by local councils, which may be provided by independent service providers (either public or private) and which services may be provided by the national government? ICT offers different means to establish better services: in the virtual world time and space lose importance. This opens the road to economies of scale by offering services on a supra municipal level. However, it is important that these advantages are weighed against other goals and functions of the public sector, such as universal service principles, and democratic control [12, 13]. The typology in types of services we have provided offers a starting point in this discussion.

References

1. Nieuwenhuis, M. A.: Tessec: Een expertsysteem voor de Algemene Bijstandswet. Kluwer, Deventer (1989)
2. Ministry of the Interior and Kingdom Relations (BZK): Terug naar de Toekomst: Over het gebruik van Informatie- en communicatietechnologie in de Openbare Sector, Ministerie van Binnenlandse Zaken en Koninkrijksrelaties. Den Haag (1995)
3. BZK: Actieprogramma Elektronische Overheid. Den Haag (1998)
4. BZK. Voorbij het Loket: Over de mogelijkheden en onmogelijkheden van pro-actieve dienstverlening voor de Nederlandse Overheidsorganisaties. Den Haag (1999)
5. BZK. Contract with the future: A vision on the electronic relationship between government and citizen. Den Haag (2000)
6. Levine, Arthur: Why Innovation Fails. State University of New York Press, Albany (1980)
7. Belastingdienst: Jaarverslag Belastingdienst 1999, (IRS, annual report) (1999)
8. Hoogwout, Marcel: Leuker kunnen we het niet maken, maar willen we het wel makkelijker? Waarom overheden geen haast hebben met het verbeteren van de dienstverlening. In H.P.M. van Duivenboden and M. Lips (eds.): Klantgericht werken in de publieke sector: inrichting van de elektronische overheid, Lemma, Utrecht (2001) 149-66
9. van Wijk, H.D, Konijnenbelt, W, van Male, R.M. et al.: Hoofdstukken van administratief recht. VUGA, 's-Gravenhage (1997)
10. Johnson, Peter: Knowledge management, knowledge based systems and the transformation of government. In: Leading People into 2000. Australian Human Resources Institute and the Public Service & Merit Protection Commission, Australia (1999)
11. Groothuis, M.M., Svensson, J.S.: Expert System Support and Juridical Quality. In: Breuker, Joost, Leenes, Ronald and Winkels, Radboud (eds.): Legal Knowledge and Information Systems (Jurix 2000) IOS-Press, Amsterdam (2000) 1-10
12. Bellamy, Christine, Taylor, John A.: Governing in the Information Age, Public Policy and Management. Open University Press, Buckingham; Bristol, PA (1998)
13. Stedman Jones, Daniel, Crowe, Ben: Transformation Not Automation the E-Government Challenge. Demos, London (2001)

Administration 2000 – Networking Municipal Front and Back Offices for One-Stop Government

Volker Jacumeit

Siemens Business Services GmbH & Co OHG,
Rohrdamm 85, 13629 Berlin, Germany
volker.jacumeit@siemens.com

Abstract. Administration 2000 is the first solution that combines legacy applications of different local administrations in order to offer a One-Stop Shop solution to citizens and private enterprises. The solution uses the Internet technology for a high secure and modern Intranet application and will be extend to a real Internet application later. The solution follows consequently the so called life event approach and also integrates services from private companies that belong to the specific life events. Besides the implementation of new technology the solution requires a paradigm within the public administrations and its business processes. In most cases the solution requires new contracts among the participating administrations and can lead to a change of the respective laws that define responsibility and process of the services.

1 One-Stop Government – New Services for Entrepreneurs, Public Administration and Citizens

Due to the growing competition and limited financial budgets Government, politics and public administration are forced to reform their Business. A new comprehension on public services hence changed and new tasks demand reengineering of business process that are consequently oriented by the needs of citizens, enterprises and other users of the public services.

Administration 2000 offers a complete access to services processed of public administration via the so called Citizen's Office (One-Stop-Government). Offerings for all situations of life in the field of e-Government fosters the connection with citizen, administration and private businesses.

A consortium of three counties and over 5 towns in German's state Schleswig-Holstein has run the project together with Siemens Business Services. To introduce a seamless One-Stop-Government for the citizens with the integration of local businesses and independent from the need of citizens to invest in equipment and technology, was the goal of the project. The services were on focus all the time and technology only used as a tool. Hence the reengineering of the business processes of the public authorities has consumed a big deal of time and money of the project. New ways of cooperation of the different departments as well as cooperation between different local administrations (municipalities and county council) have been identified and lead to organizational changes and proposed legal provisions.

Currently the new services are offered via Citizen's Offices or Walk-In Centers as these kind of public offices and administrations are called. In these offices civil ser-

R. Traunmüller and K. Lenk (Eds.): EGOV 2002, LNCS 2456, pp. 157–162, 2002.
© Springer-Verlag Berlin Heidelberg 2002

vants can offer the entire bunch of public services related to certain life events. The citizens do not have to walk from office to office in order to have their requests processed. "Instead of people we let move the data" is the concept of Germans Minister of the Interior and with the solution Administration 2000 we let this happen. Civil servants can now change data in databases of other municipalities and even of the county council, if this is necessary for the respective request. Also the very close cooperation with local businesses allow to extend the solution to real life event management.

1.1   Example of the Life Event Move House

People who move house today have to visit different offices of public administrations which are located often in different towns or districts of big cities. So they have to spent plenty of time standing in a queue in order to get to the right civil servant. They have to give redundant data several times and fill in many forms. In Germany the event Move House means that people have to deregister at municipality they live in before and then go to the municipality they have moved into for the entry in the new citizen register (update the ID-card and passport). A third administration must be visit for the change of the car-plate and the according car register. Additionally people have to inform service providers such as the gas- and power supplier, the waste collection company and more. In order to inform all those providers people fill in more forms (sometime via the internet) or send letters by traditional mail. But each time redundant date must be typed. Surely these boring tasks are more or less the same within all countries.

Now with the solution of Administration 2000 people can go to any municipality. The civil servant is the human life event manager for the citizen. The civil servant will first deregister the citizen at the former municipality, register the citizen at the municipality he has moved into, update ID-Card, passport and print necessary documents that certify the new address. Also the civil servant can process the data for the county in terms of changing the car-register and car plate. If the citizen wishes, the civil servant can even inform other service providers such as gas- and power supplier.

Technically spoken we have integrated all existing legacy applications of the public authorities and involved private companies in a new and unique user interface. This user interface is oriented to the respective services and works independent from the legacy applications with its proprietary interfaces.

With the growing number of households connected to the Internet, these kind of services will be offered also via the Web. But at this time we (Siemens Business Services as a solution provider and the public authorities and private businesses as the service providers) have gathered plenty of experiences and enhanced the solution to a mature version which can be used by everybody and that is comprehensive. An other advantage of this approach is, that due to the fact that the civil services are using the same interface as the people via the Internet, each civil servant can assist a citizen via telephone if any problem occurs with the application.

2   Demands of Administrations

Administrational decision maker intend to drive the modernization of the administration by using e-Government solutions.

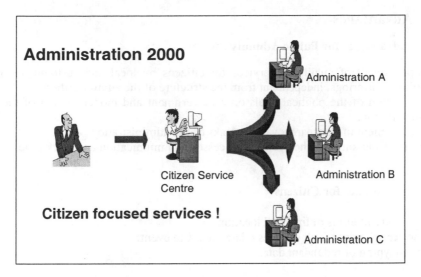

Fig. 1. Administration 2000

Table 1. Access Points for Processing Life Events

Portal	Citizen Office	Life Event Manager
• Information public and private sector • Search until you find the proper information and form if available • Fill in the form • Submit the form • No combination of data to other forms • No direct processing of data	• Any Citizen Office of the region / county • Civil servant collects data and processes these data online to all respective governmental processes and other entities • On-Stop Shop solution with human interface	• Integration of all governmental processes and of the private enterprises concerned by the life event • Intelligent and comprehensive dialog allows as much integration as the user wants • One-Stop Shop via Internet
All users via PC, Kiosk	Application for Citizen Offices and Call Center operated by civil servants	Citizens via PC, Kiosk Citizen Offices Call Center

Administrations like to offer integrated services to citizens, local enterprises and other administrations. These services must be secure, legally binding and optimized.

Using new Online-Services administrations like to improve the contact to its citizens. Services for local enterprises strengthen the competition of the region.

Administrations like to make use of experiences and the solution architecture of best practices.

Administrations like to built the new Online-Services based on standardized data and interfaces and keep already existing interfaces to legacy applications.

3   Advantages

3.1   Advantages for Public Administrations

- Improvement of quality of services for citizens by local and customer oriented solution offerings, independent from the structure of the administration.
- Realization of the political goals on e-Government and modernization of the administration.
- Improvement of transparency of the tasks of the administration
- Shortcut and simplify the business processes, communication and work time.

3.2   Advantages for Citizens

- Flexibility in terms of time and location.
- One access point for all services related to a life event.
- Avoid typing of redundant data.

3.3   Advantages for Enterprises

- Seamless interface to public administrations.
- Exchange of documents via e-mail.
- Online access to important data for planning.
- Online access to up-to-date laws and directions.
- Online Access to information on bid and tender process and monitoring data.
- Online procurement.

4   The Solution Components

Administration 2000 is a standardized and modular solution. Already existing best practice modules can be integrated on demand. Solutions for new business processes can easily be developed due to a tested an proven methodology approach and data descriptions of the new business processes.

4.1   Technical Overview

The technical concept is based on the usability of an internal networks of the administration and the legacy applications of the involved administrations. The local administrations and the county council are connected via a secure County-Backbone-Network. In a demilitarized zone an information server, application server and a communication server are operated. On the application server runs a web based application which is responsible for the processing of the integrated work flow of the administration. Using an unique user interface the different legacy applications of the participating administrations are operated. The communication server connects the user to the respective legacy applications because most of the legacy applications are operated at the local administrations. For this exchange of sensible and personalized data we are using Biz-Talk server from Microsoft. As the communication server. The communication server also takes care for the transfer of data, monitoring and reporting of the transactions via the network. The processing of data is done at the legacy applications on the local administrations. Security and integrity of data is guaranteed by the overall concept of the solution.

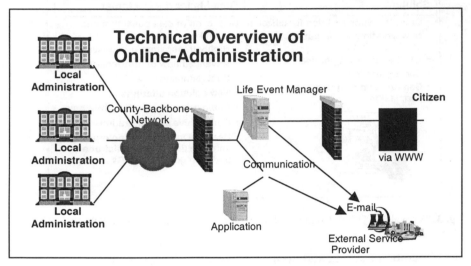

Fig. 2. Technical overview of Online Administration

Citizens will receive information about the services that the local administrations offer via the citizen offices. Hence the citizens can prepare themselves with all relevant documents and information necessary for the applications. They can arrange dates for visits of the citizen offices and thus avoid queuing.

4.2   Best Practice Modules

For the German market we have prepared business process descriptions and according XML-Schema. We have developed interfaces to main applications and a complete solution architecture available for the following processes:
- Solution package Citizen Register
- Car Register
- Construction Permission

Even if the solution of Germany can not be transferred without any changes to other countries, we can adopt these best practice modules.

4.3   The Methodology Approach

A tested methodology approach simplifies and accelerates the modeling of new business processes as well as the development of a solution architecture of new processes.

Existing definitions of interfaces, XML-Schema and a toolbox to customize the user interface are fostering the development and integration of new business processes.

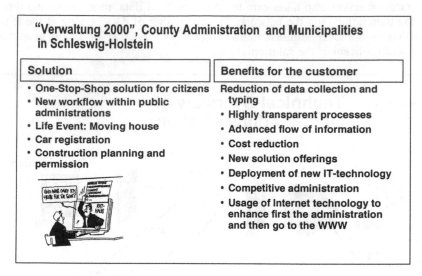

Fig. 3. "Verwaltung 2000", Country Administration and Municipalities in Schleswig-Holstein

4.4   Operation Concept and Security Concept

The operation concept defines the organizational tasks and the organizational environment as well as service and operation tasks for the operation of the solution Administration 2000. The security concept describes the threat to the system and possible defense activities. From this basis a step-by-step plan for the implementation of the new demands on security will be derived.

4.5   SBS Services

Siemens Business services fosters administration by the development of en e-Government strategy, Concept and Architecture of new Online-Services, Implementation and if demanded service providing, application support and operational services.

We develop for administration the appropriate e-Government solution and support administration in all phases of the development. Beginning with the vision and strategy up to the operation of the services.

4.6   Our Solution Partners

- The realization of the web application has been done by HEC GmbH, Bremen
- Standardized software products have been provided by Microsoft
- The County-Backbone-Network has been provided and is operate by Datenzenztale Schleswig-Holstein

The Experience of German Local Communities with e-Government – Results of the *MEDIA*@Komm Project

Tina Siegfried

MEDIA@Komm accompanying research , German Institute of Urban Affairs,
Strasse des 17. Juni 110, 10623 Berlin
siegfried@difu.de

Abstract. The largest multimedia project of the national government – *MEDIA*@Komm – has now run for almost two years. Over the course of time, the model municipalities and other German local communities have gained experience with the design of legally binding on-line transactions and the use of electronic signatures, and they have made progress with virtual town halls and the e-government project. The following article shows the results achieved in the *MEDIA*@Komm municipalities and considers the experience, obstacles and unsolved problems which have arisen in designing business processes without discontinuity of media in many German towns and cities. Also, as an important result of the accompanying research on *MEDIA*@Komm, it describes the central factors for success in the creation of virtual town halls.

1 The Multimedia Initiative of the National Government *MEDIA*@Komm

The *MEDIA*@Komm project is the largest multimedia-project of the national government to test legally secure electronic transactions between the administration, citizens and business companies. The aim is to design on-line transactions without discontinuity of media and to test the use of electronic signatures. A municipal competition was announced in 1998, and 3 municipalities were then selected to implement their concepts with the support of the national government. The winning projects were presented by the city of Bremen, the municipal association of Nürnberg, Erlangen, Fürth, Bayreuth and Schwabach and the town of Esslingen in partnership with Ostfildern. These projects have been implemented since 2000 with federal subsidies of 25 million euros. In addition to these funds the municipalities themselves, together with private partners, are raising about another 35 million euros for the implementation of the projects. This was originally scheduled to be completed by the end of 2002, but due to unforeseeable delays (cf. section 3) this is likely to be extended to about the middle of 2003.

The German Institute of Urban Affairs (Difu) was commissioned by the Federal Ministry of the Economy to act as the consortium leader and provide academic support and guidance to the federal government's largest multimedia initiative. Together wirth three other academic insitutions, each of which deals with specific aspects of the subject, Difu deals with economic issues and questions of adminstrative science. In addition, specialist events are held and a coopreation and communication network is being established which also includes the Internet presentation on *MEDIA*@Komm (www.mediakomm.net).

R. Traunmüller and K. Lenk (Eds.): EGOV 2002, LNCS 2456, pp. 163–168, 2002.
© Springer-Verlag Berlin Heidelberg 2002

2   Present State of Implementation in the Model Municipalities

2.1   Free Hanseatic City of Bremen

The Bremen *MEDIA*@Komm project "Legally binding multimedia services with digital signatures in the Free and Hanseatic City of Bremen" is being implemented by bremen online service GmbH & Co. KG and comprises three large core areas with further sub-projects: Access to secure and legally binding on-line services, platform and designing a standard for municipal online transactions and applications/life situations.

2.1.1 Successfully Implemented Local Community Applications

The first "visit" to a public authority carried out completely on-line, in which a marriage certificate was ordered and the fee was paid directly with a purse card via a secure Internet link, was presented in September 2000. Since then, the citizens of Bremen have been able to call on various on-line services of the administration and private service providers. In the life situation residence and change of address, for example, they include registration and cancellation of gas, electricity and water with the municipal utility companies, changing bank accounts at the Sparkasse and mail forwarding arrangements with the post office. Birth, marriage and death certificates can also be ordered on-line from the registry office in Bremen – and paid for electronically. Since June 2001 the residential registration office of the city Bremen has also offered on-line services. There, legally binding entries and deletions can be made in the residential register via the Internet by using the electronic signature. Since May 2001, applications for students have also been in place. A service for solicitors and business companies is currently in the test phase. They can call up information in electronic form free of charge from the register of companies database at the Local Court (Amtsgericht). A further process, which is also likely to interest small and medium-sized companies, is the on-line debt collection application, which has been developed for the Free Hanseatic City of Bremen. With this software, applicants can file debt collection applications electronically via the Internet, thus saving time, effort and costs. And finally, the prototype of a digital tender platform for the award of public orders in the construction industry is currently being developed in Bremen. The process is being developed in cooperation with Administration Intelligence (AI) and will make it possible in future to read announcements and download tender documents over the Internet. Since December 2001 citizens have been able to issue a direct debit authorisation with an electronic signature. This new possibility to pay any relevant charges can now be used to order a marriage certificate. Previously, marriage certificates ordered on-line were only sent out with an invoice by the registry office. Now the citizens no longer need to fill in the bank transfer form; instead, they simply issue the public authority with an electronically signed direct debit form for the respective transaction directly in the on-line application.

2.2   Nürnberg Municipal Association

The goal of the project "*MEDIA*@Komm in the Nürnberg region", implemented by the project sponsor, Curiavant Internet GmbH, is to offer legally binding multimedia services with digital signatures in the municipal association. A regional communication platform is therefore being created which will support secure communications and offer the citizens various communal and private services. The Nürnberg municipal association consists of five municipalities of different sizes in the region. The special challenge here is to develop on-line services and products which are equally "fitting" for all municipalities.

2.2.1 Successfully Implemented Local Community Applications

The citizens of Nürnberg have been able to use out municipal applications on-line since October 2000. With the multi-functional chip card from Curiavant Internet GmbH and a class 3 card scanner, they can apply and pay for their residents' parking permit from their PC. The provision of information from the register of residents is fully implemented, but it cannot yet be placed on-line due to legal regulations. Moreover, a municipal council information system has been developed which has been in use since April 2002. Another local community sub-project, the building site information system, has been in place since October 2001. The integration of geographical information systems and town maps in the Internet presence of Nürnberg enables citizens to call up information on traffic obstructions in the urban area via the Internet. Similarly, data and maps can be ordered on-line and paid for by invoice. Since the end of 2001, citizens in the Nürnberg urban region have also had access to a comprehensive information and booking system for regional cultural and educational facilities. Adult education centres, the municipal theatre and the municipal libraries already present their services on the Internet. The citizens can even enrol and receive a confirmation with their digital signature. The chip card also makes convenient payment possible – either on-line or by direct debit. On-line enrolment for an adult education course is possible; the payment (for now) is made by direct debit. In the next stage of development, this service will be extended to include access to data, e.g. lists of participants for the teachers. The municipal theatre not only presents its programme of performances, it also enables visitors to select, book and pay for their seat on-line. Tickets are then sent by post or deposited for collection at the evening ticket office. The libraries project, which offers citizens a number of library functions via the Internet, e.g. borrowing, reservations and research, has been in pilot operation in Erlangen since November 2001. Since December, the draft of the zoning plan for the city of Nürnberg has been available on-line in the framework of citizen participation.

2.3   Esslingen/Ostfildern

The central focus of the project in the towns of Esslingen and Ostfildern is on communication between citizens and the administration. With their community-based approach, the two towns aim to enable their citizens to participate actively in the life of the local community and to cooperate in municipal activities. The aim of the Esslingen approach is to create the necessary acceptance in the population for the use of signature cards to implement legally binding transactions and to make on-line "visits" to the municipal administration via the Internet. Six sub-projects make up Mediakomm Esslingen: Communal services, Education, E-commerce and e-business, Culture, Social affairs and Cross section.

The responsibility for each sub-project lies with a different institution or company; the sponsoring body Mediakomm e.V. in Esslingen is responsible for coordination.

2.3.1 Successfully Implemented Local Community Applications

In accordance with the principle of the local community of citizens, projects have now been implemented which provide information for the citizens and involve them in the process of discussion on developments in the town. Thus, citizen forums have been set up on the Internet pages which enable interested persons to discuss topics relevant to Esslingen. The zoning plan provides an even more practical example. The formal participation of the citizens in the development of a new construction area in Esslingen was also implemented via the Internet. To achieve goals such as greater transparency of the administration, customer orientation, better accessibility for the citizen and faster handling of administrative procedures, the Esslingen citizen information service ESSOS ("Esslingen On-line Service") was founded. It offers, or will soon offer, information and on-line services for various life situations. And to counteract the "digital divide", projects such as the supervised citizen PC (Bürger-PC™) have been implemented, which particularly aims to help population groups in Esslingen with little experience of new media to make use of the Internet and digital signatures. At central points in the town (e.g. in schools), PCs are available for the use of the public. Great attention was attracted by the world's first legally binding on-line election of a public body – the Municipal Youth Council in Esslingen – which was (partly) carried out via the Internet with the aid of digital signatures. The on-line election fulfills all legal requirements; the municipal regulations had to be adapted for this purpose. For construction projects, a prototype of the new service for on-line building permit applications is offered from the first quarter of 2002. In this service, all information (plans, drawings, correspondence) is available in digital form on an Internet platform after the user has identified himself with his digital signature. This can be used on-line by the administration, the citizens, contractors and architects. The aim is to make the construction process considerably more efficient because the approval of the digital original is also issued by an electronic signature.

3   Experience with the Use of Digital Signatures

The dependence of this development on the general framework is still underestimated. A number of factors can be identified which are major obstacles to the widespread dissemination and use of electronic signatures.

3.1   Development of the Legal Framework

The development and adjustment of legal requirements has taken a great deal of time, and for a while it caused great uncertainty among the users. The Digital Signature Act and the Digital Signature Ordinance were revised in 2001. And the adjustment of private law to enable the electronic signature to have equality of status with handwritten signatures also did not take place until 2001. The revision of administrative procedure law in the spring of 2002 is still to come, although the direction is now apparent and it is anticipated that the Administrative Procedure Act will pre-scribe the use of qualified signatures, with accredited signatures only required in exceptional cases.

3.2   Technology and Security

Users reported that the software which had to be installed for the use of signatures was often faulty or completely useless, and lay persons could only install it with great difficulty. Reports of complete and irrevocable system crashes during attempts to install it were no exception. These "teething troubles", which are so annoying for the user, are immense obstacles to widespread use. At least the problem of the lack of interoperability between digital signatures from different suppliers has been tackled since October 2001. TeleTrusT e.V. and the trust centre association T7 are working on behalf of the Federal Ministry of the Economy and Technology (BMWi) to develop a standard for electronic signatures which will draw together the present individual solutions.

The technical security requirements for the integration of signatures into the on-line processes between the administration and its clients and the requirements of the Data Protection Act were other stumbling blocks that needed to be overcome. For example, it is desirable that the client only enters his data into an on-line form once and can then assume that the administration knows his data, but the use of such existing data often collides with the provisions of the Data Protection Act.

3.3   Benefits and Acceptance

The high price of the necessary equipment is still an obstacle to the dissemination and use of digital signatures. The chip card and the associated software currently cost about 25 euros from the market suppliers, and there is also an annual fee of 25 euros for the administration of the certificates. The prices for the scanner, which is also needed, fluctuate depending on the level of security. And there is no specific application which leads citizens to feel that they absolutely need a signature card. The hopes of the *MEDIA*@Komm municipalities that the banks would include digital signatures as a standard feature on EC cards, and that a large proportion of the population would then automatically have a digital signature, have unfortunately not yet been fulfilled.

4   Practical Problems of Use
in the Local Community Administration

The design of on-line transactions should involve more than simply reflecting the existing processes and "converting" them into electronic form. Instead, consideration should be given to the question of whether processes can be redesigned in the interest of greater efficiency. To date, this point has been neglected because of the difficulties of integrating digital signatures into existing departmental procedures. The creation of public key infrastructures and concepts for key management within the local community are still central issues which remain to be solved. It must be considered whether every member of the administration should have a signature card, or whether there should be a central office to administer and process signatures. The present trend in the discussion among specialists seems to be towards a virtual postal centre which separates the incoming signatures and then passes the transaction itself on to the responsible person. But at the time of writing (spring of 2002) there is still no product which could take over these tasks. Another significant point in this connection is the training and integration of the staff; such projects cannot possibly succeed if the staff are not willing to cooperate.

5   Factors for the Success of e-Government

The *MEDIA*@Komm accompanying research has the task of advising and assisting the projects, but at the same time it must also evaluate them systematically. To ensure the comparability of the various approaches and enable the projects to be evaluated, a total of 9 dimensions were identified which play a central role in the implementation of e-government and virtual town halls. These dimensions in turn can be sub-divided into individual factors; the importance of each individual factor will be explained in greater detail. Criteria to assess the "maturity" of e-government and virtual town halls round off this measuring instrument for the evaluation of the *MEDIA*@Komm projects. In addition to evaluating the *MEDIA*@Komm municipalities, the factors for success can also be used as an instrument to measure the success of other local communities. The critical factors for success in the designing of virtual townhalls in Germany are: Vision and strategy, Organisation, Applications, Benefits, Use of Technology, Qualifications, Marketing, Cooperation and Resources. A more detailed description of the dimensions and sub-factors is expected to be published by the accompanying research in mid of 2002. For more information see www.mediakomm.net.

Electronic Public Service Delivery through Online Kiosks: The User's Perspective

Ruth Ashford[1], Jennifer Rowley[2], and Frances Slack[3]

[1] Department of Retail Marketing, Manchester Metropolitan University,
Manchester M15 6BH, UK,
R.Ashford@mmu.ac.uk

[2] School of Management and Social Sciences, Edge Hill University College,
Ormskirk, L39 4QP, UK,
rowleyj@edgehill.ac.uk

[3] School of Computing and Management Sciences, Sheffield Hallam University,
Sheffield S1 1WB, UK,
F.Slack@shu.ac.uk

Abstract. This paper reports a case study of Knowsley Metropolitan Borough's response to the UK Government's White Paper 'Modernising government' [1]. It provides unique data on user behaviour in relation to electronic public service delivery through public access kiosks and highlights some of the issues relating to the 'digital divide', the reduction of social exclusion. It offers a perspective on the uses for which customers perceive public access kiosks to be valuable and indicates barriers to kiosk use for other functions. Some of the messages reflect issues that have been debated in consumer responses to e-commerce and communication over the Internet. This is important because it suggests some consistency in the public reaction to IT-based service delivery, irrespective of the platform.

1 Introduction

In common with governments across the world, the British government has pledged to use information and communication technologies to increase participation in a greater part of local governance by 2005.

1.1 UK Government IT Initiatives

Both central and local government have been encouraged, through access to funding, to develop a co-ordinated approach to information technology procurement, working in partnership with the private sector. The main aim of this policy is to ensure that public services could be delivered 24 hours a day, seven days a week. This includes the use of interactive kiosks and call centres to enable the local community to access more convenient services. A recent survey of electronic service delivery by English District Councils (EDCs) found that 21% of councils surveyed currently use public access kiosks and 25% plan to use them in the future [2]. However, 54% of EDCs do not use them now and do not plan to use them in the future.

R. Traunmüller and K. Lenk (Eds.): EGOV 2002, LNCS 2456, pp. 169–172, 2002.

1.2   The IT Strategy at Regional Level

The roles of English local government are defined [3] as providing local democracy, local public policy making and local services to the community. The Government's White Paper [4] has had a dynamic effect on the way that UK local government now operates. At a regional level, the Mersey Wide Web (MWW) is a grouping of the public, private and voluntary sectors that aim to establish Merseyside, in the north-west of England, as a global player in the information society. MWW, established in 1997, aims to help the economic regeneration of Merseyside by ensuring that its workforce is geared to the future and raising the ICT presence in the region.

Knowsley Metropolitan Borough Council, an area of significant social deprivation in Merseyside, established the Knowsley Community Information Programme. The aims of the case study reported here were to investigate the behaviour and attitudes of users and non-users of the online kiosk located in the Knowsley Metropolitan Borough.

2   The Context of Online Kiosks

Online kiosks can be viewed as a medium through which it is possible to train, educate, inform, communicate, persuade, and relate. But, as with other public access systems, it is important that the kiosk is designed to support the task, the user profile and the environment in which the task is to be performed. A useful recent article in the context of public access kiosks is Maguire's [5] review of user interface design guidelines for public information kiosks. Rowley and Slack [6] argue that there are four components of public access systems that need to be considered: user characteristics, environment, task and technology, and they point to the limited attention that has been paid to environment or context.

Online kiosks may have both commercial and community functions. Early applications were designed to provide access to community or government information [7]. Integration of both community and commercial services and information is likely to characterise the kiosks of the future. Previous articles have tended to describe specific kiosks [8] or to develop taxonomies of the different types of kiosks [9]. This article offers unique data on user behaviour and attitudes.

3   Research Methodology

An I-plus information point created by Adshel and media technology company, City Space has been designed to equip the residents of Knowsley with access to information and learning systems and to promote personal and community development. This is a deprived area with an unemployment rate of 13.5% (the national average is 4%) and a high level of crime. The kiosk offered: free e-mail, business finder, child-care finder (in partnership with Mothercare), TV licence payment, access to local council services, transport information and entertainment information.

The multiple methods of data collection used a questionnaire survey with 1068 respondents, two focus groups and two in-depth interviews. The questionnaire comprised a mixture of behavioural, attitudinal and classification questions. In both the pilot study and the main study the questionnaire was administered by students who were trained in the process.

The sample for both the questionnaires and the focus groups were convenience samples; their composition was affected by the willingness of potential respondents to participate in the study. The individual interviews were used to supplement data collected in the focus groups and to develop some themes more fully. Usage statistics were also available to provide comparative statistics on usage levels, length of usage times and frequency of access to the local council web page.

Questionnaire data was entered into SPSS and analysed for descriptive statistics and relationships between variables. Content analysis was performed on the focus group and interview transcripts to identify recurrent themes that influence user behaviour and attitudes.

4   Results

Usage statistics revealed that the average number of users per day was 45. The average time per user was 3.05 minutes and one third of usage was to access the local council web page.

The majority (90%) of people questioned had not used the kiosk. This is a significant finding and suggests that for kiosks to fulfil their potential there is a need to understand their optimum role in communities. The main reason for non-use was a lack of awareness of the kiosk's existence. A significant group was aware that the kiosk existed but 'did not know what it did'.

Of kiosk users (slightly more than 100 respondents) half had only used it once, while 20% had used the kiosk five times or more, suggesting that they had found something of interest and value on the kiosk. It appeared that females were more likely to use the kiosk than males. Over 75% of users were under 35 years old and 45% of users were students. Users were generally quite positive about usefulness and ease of use. A correlation was evident between the perceived usefulness of the public access kiosk and the frequency of use.

Focus groups and interview respondents were asked to suggest solutions to the low level of awareness of the kiosk. An advert in a free local newspaper had generated some interest. It was felt that 'a lot of people would walk straight past it, because they don't want to talk to a robot'; and not be aware that there were 'things for them' on it. Suggestions ranged from creating a signpost, to point people in the right direction, to information leaflets, explaining the uses of the kiosk and a poster campaign in the surrounding shops.

Many kiosks are located under cover and where people are 'milling around' or waiting. This kiosk was located out of doors on a street and this provoked a number of negative comments, thus confirming the importance of location.

Internet access was available to 60% of respondents through work, home PC, or the local public library. The library is the only other free method of access available to the general public. On the other hand, 40% of respondents were disenfranchised. In addition, however, many of the kiosk services were available through other channels, such as the telephone or by making personal contact.

Focus group respondents expressed concern about service delivery through a kiosk. There was evident technophobia and a lack of trust in the technology (kiosks, Internet or telephone help lines) and the service exchange that it supported. Personal contact generates a sense of accountability and reassurance. Concern was also expressed about giving credit card details over the Internet and there was a preference for buying tickets in person.

5   Conclusion

Online kiosks have potential to offer Internet access and other information and services to people who might otherwise be excluded from participation in the 'information society'. This study has emphasised that local authorities and businesses need to treat a public access kiosk as a service to be promoted, particularly in contexts and communities where kiosks and other IT applications are unfamiliar. An enhanced understanding of the tasks that users are prepared to perform on a public access kiosk would also contribute to their success. Local government provision of electronic service delivery must be implemented with the needs and participation of the users in mind, not simply to comply with national Government initiatives. Without this level of engagement, online kiosks will remain a technology looking for a problem, rather than becoming a solution to people's problems.

References

1. 'Modernising government' (1999) HMSO,
   http://www.cabinet-office.gov.uk/moderngov/whtpaper/index.htm,
   [accessed 23 January 2002].
2. Phythian, M J and Taylor, W G K (2001) Progress in electronic service delivery by English District Councils. *International journal of public sector management*, 14 (7), 569-584.
3. Pratchett, L (1999) New technologies and the modernization of local government: an analysis of biases and constraints. *Public administration*, 77 (4), 731-750.
4. 'Modernising government' (1999) *loc cit.*
5. Maguire, M. C (1999) A review of user-interface design guidelines for public information kiosk systems. *International Journal of Human Computer Studies*, 50 (3), 253-286.
6. Rowley, J E and Slack, F (1998) *Designing public access systems.* Gower.
7. Ellis, B (1993) Kiosks handle employment queries. *Computerworld*, 27 (46), 88.
8. Miller, C (1996) New services for consumers without home page at home. *Marketing News*, 30 (9), 1-2.
9. Tung, L. L. and Tan, J. H. (1998) A model for the classification of information kiosks in Singapore. International Journal of Information Management, 18 (4), 255-265

FASME – From Smartcards to Holistic IT-Architectures for Interstate e-Government

Reinhard Riedl[1] and Nico Maibaum[2,*]

[1] University of Zurich
riedl@ifi.unizh.ch
[2] University of Rostock
maibaum@informatik.uni-rostock.de

Abstract. In this paper we shall present the results of the change management in the European research project FASME *(Facilitating Administrative Services for Mobile Europeans)*. First we shall compare the original objectives as they had been described in the Technical Annex with the achieved results. Second we shall highlight those issues of change, which were responsible for the success of the project. Only after the prototypical development of a product was abandoned and replaced by basic research on holistic solutions for interstate e-government, the already achieved progress became visible to the participants and a true interdisciplinary co-operation emerged. Concluding from this experience, we shall draw some conclusions for future e-government projects.

1   Introduction

A European citizen moving from city A in a European member state X to city B in a European member state Y has to deal with a lot of bureaucracy - such as the registration and deregistration of her living place, the change of her car license and of her insurance contracts, or the enrolment of children in school. This usually requires the following: First, the citizen has to investigate on administrative regulations and culture in Y and on the municipal services in B. Second, she has to collect personal documents from government agencies in X, or from the municipal administration in A. And third, she has to deliver these personal documents to agencies in Y, or to the municipal administration in B. In some cases this may fail because the authorities in her original place of living are unable to provide her with the documents needed; in other cases this may fail because the authorities in her new place of living are unable to interpret her foreign documents. As a consequence, even if the citizen manages to understand the procedures, it may become necessary that a tedious individual exception handling has to take place, which is expensive for both the citizen and the authorities, and whose outcome is unsure.

---

* Supported by the EU Fifth Framework Project FASME, http://www.fasme.org, and by a grant of the Heinz Nixdorf Foundation

R. Traunmüller and K. Lenk (Eds.): EGOV 2002, LNCS 2456, pp. 173–178, 2002.
© Springer-Verlag Berlin Heidelberg 2002

FASME *(Facilitating Administrative Services for Mobile Europeans)* was an inter-disciplinary R&D-project supported by the 5ᵗʰ Framework of the European Union[1], which intended to develop an intelligent Javacard for the authentication of citizens and the digital transport of personal data. That card was supposed to minimize the high, administrative migration efforts depicted above. During the project, we have found out that such an approach is indeed feasible, but it would reduce administrative migration efforts only marginally, and it would not alter the basic problems for the citizen significantly. In order to make life really easier for both citizens and civil servants, a much broader approach is required. In particular, a secure and trustworthy ad hoc procurement of so far not existing documents from remote authorities is needed, whereby it should be possible to tailor these documents to the needs of re-mote administrative processes, following different laws and guidelines formulated with the use of different administrative ontologies. This requires the following:

1. A **holistic solution for digital identity**, which flexibly realizes *the in effigy prin-ciple*, and which supports individual, group, and role identities, as well as anony-mous identities, and the delegation and the revocation of rights.
2. A **decoupling of inter-organizational processes** and **the provision of interoperability**, based on the concept of a loose C/S-coupling of local processes (i.e. no inter-organizational workflows, no global service point architectures), in-terface definitions for document services and hierarchical directory services.
3. The design of a **vertically and horizontally scaling, virtual information trans-fer space** for inter-organizational – and in particular international – exchange of personal information and other affiliated information, plus guidelines for the use and for the management of the space (e.g. the locality principle for relevance man-agement).
4. Joint provision of guaranteed and enforced **protection of privacy** and of **moni-toring facilities** for both civil servants and citizens – i.e. a fully-fledged transpar-ency management in accordance with European law and user requirements.

We have developed an architectural framework, which meets all these requirements, although with different degrees of detail, as we shall discuss below. Hereby the Java-card plays the role of a digital ID-card. After the card has verified the identity of the citizen with a biometric check, it speaks in effigy of her with local e-government services and remote document services. Thus, the Javacard is the enabling token for digital citizen-to-agency communication, while group and anonymous identities, as well as delegations of rights, are realized with soft identities [1]. The Javacard is used as a cache for digital documents, and as a key to a virtual memory for such docu-ments called Secure Card Extension (SCE) [2], but its main function goes beyond simple document transfers, it is the digital representative for the citizen interacting with e-government services. See also [3] for an illustration of the problem of hetero-geneity faced in FASME and [5] and [6] for a presentation of our solution.

We have implemented demonstration versions for the core IT-components in a simplified setting with the municipalities of Cologne (Germany), Grosseto (Italy),

---

[1] Approximate Figures: Total costs = close to 4 million , duration = 18 months plus extensions adding up to 21 months, head count = close to 60.

and Newcastle-upon-Tyne (England). Both the overall solution and the demos were successfully validated with user groups from six European member states.

In this paper, we shall depict how we have changed the main objectives during the project. Notably, this change was initiated by the work on the IT architecture. *Only after the prototypical development of a product was abandoned and replaced by basic research on solutions for interstate e-government, a true interdisciplinary co-operation emerged.* As long as the project was understood as a feasibility study based on a prototype close to a product, there was no common view of what the project was all about and the project partners did not realize its achievements. This changed only when the new focus on holistic solutions became accepted and thus leveraged inter-disciplinary work on the multiple dimensions of the solution. The observation that IT engineering activities contributed to the emerging interdisciplinary perspectives somewhat parallels the observation of the role of IT managers as brokers reported in [5].

Our bold statement above is confirmed by the results of the two project reviews carried out by the European Commission. The first took place, when the project was still run as an applied technology project. It severely criticized the progress of the project. Shortly afterwards we changed our goals and we directed activities towards the development of a holistic IT solution (for the price of providing demos which were less mature than the prototypes originally intended). The final review then was quite a success, as we presented the solution depicted above, which integrates the research results from the various disciplines with IT engineering R&D work to achieve a generic, customizable IT architecture for international e-government services. The reviewers' only point of criticism was the missing business plan for making money out of the project results.

2   Objectives versus Results

In the following, we shall compare the objectives (formulated according to the guidelines of the European Commission) with the achieved results.

Overall Plan: The main benefit of the FASME project will be a Javacard, which facilitates the mobility of citizens as well as the entailed administrative services. Resulting from the new possibilities offered by the Javacard, the subordinate benefits are reflected in the facilitation of procedures, concerning migrating citizens, municipalities and enterprises.	**Overall Achievement:** The main achieved benefits of the FASME project are a holistic understanding how to build scaling international e-government services, the identification of a next step of R&D topics in order to proceed systematically towards real world implementations, and a principal change of the engineering perspective for e-government project.
Overall Planned project goals: • The customizing of existing technology ... for novel administration procedures and the exploitation of potential benefits of interdisciplinary engineering processes ...	**Overall project goals achieved:** • Technological preferences shifted during the project; a new perspective on convergence management with so-called boundary objects for interdisciplinary R&D emerged.

• Simplification of the co-operation of municipalities, creation of public acceptance of innovative technologies (Javacards) ... • Facilitation of the European citizens' mobility ... • The early exploitation of Javacard technology ... to extend the European lead in Smartcard technology. • The reduction of financial and time efforts incurred by the migration of citizens.	• The way to more effective co-operation has been shown, public acceptance for Javacards is still an issue as its legal admissibility. • This goal was too far reaching for a small project like FASME. • The future success of Javacards is still unclear. • No empirical validation was provided (due to a lack of time) in the project, but the evaluation by civil servants has shown that it is indeed reasonable to believe it.
The **main planned project objective** is the development of a Javacard for the easy, safe, and secure transport of personal data needed for administrative procedures.	The **main result achieved** is the development of a holistic architecture for interstate e-government services.
The six **planned operational project objectives** are: • Development of business process models ... • Development of a social context model including social constraints for the application design process. • Development of a design framework (infrastructure, processes, and application software). • Implementation of a prototype. • Evaluation of the prototype by users (citizens, civil servants). • Development of plans for a European wide deployment.	The **achieved project results** are: • Development of citizen process models ... • Research has shown that problems with private service providers are hard to solve. They should be integrated in future R&D perspectives. Further, an analysis of the internal project development processes has suggested various guidelines for future R&D management (see below). • Done, but with a broader focus and a smaller degree of details than originally intended. This was vital for the project survival. • The same as above. • More standards for evaluation processes are needed. A lot of resources were spent on the discussion of the evaluation process. • The only point of criticism in the final review was that no plans for commercial use of project results were developed.

Measures of Success:	Reasons for Success:

There will be two main measures of success (cf. section on 'Innovations')
• The number of transport routes for documents …implemented… functionality, security, and ease of use.
• The quality of the interplay of all components:
  • … privacy and information self-determination granted.
  • … transparency of the usage processes.
	• Interdisciplinary results (as opposed to plain interdisciplinary work).
	• Feasibility, holistic nature, and scalability of the architecture.
	• Citizen control on what is done with her data.
	• The possibility to achieve proper transparency – although further work is required on this issue.

Milestones and expected results:	**Results achieved:**
Business process models of the administrative activities created by the migration of European citizens from one member state to another.	Change to citizen process models & transfer of service context for citizens from one municipality to another (if the citizen requests this).
Generic HW, SW and process design for Javacard enhanced administrative procedures.	Done on a conceptual level – too little time to compare the pros and cons of different technological solutions.
Prototypical implementations (infrastructure, applications, processes).	Demos instead of more complete prototypes as resources have been shifted from implementation work to basic research work.
Evaluations of the prototype under real world constraints.	Broadening the view and shortage of time resource did not allow enough real world testing (just the six cities).
Plans for a European wide deployment.	Business plans are still missing.

3   Divergence and Change

The positive result of the change management in FASME project may be traced back to the following key issues (compare [4]:

1. It changed from an applied research project to a basic research project. It delivered a validated concept for a solution rather than prototypes.
2. The planned floating multi-functionality was replaced by very few, easy to use, static functions, which can be used for different scenarios.
3. The intended international coverage could be resolved by local adaptation rather than by global standardization (except for the information transfer).
4. The intended main function of the Javacard as an information carrier was replaced by the implementation of digital identity as a main function.
5. The (traditional) transfer of documents with context specifications replaced the intended transfer of data and the idea of data consistency management was rejected.

6. The project team resisted the temptation to integrate payment functions, which are believed to be killer applications, but rather tend to kill projects.

The main difficulties of FASME were

1. It was too short and the efforts for interdisciplinary learning were underestimated.
2. We could not react to the finding that some major problems of migrating citizens arise from the need to consume private services.
3. The project plan relied on interfaces between different work-packages, none of which worked. Successful communication happened mostly during workshops.
4. Qualified human resources for research leadership had to be provided by the technical partners, who themselves faced serious staffing problems.

4   Conclusions

The diversity of public administration, reflecting the richness of socio-cultural differences, is one of the key challenges for a united Europe. This particularly refers to digital identity, data protection, and trust services. In the future, the main challenge for a European-wide implementation of interstate e-government services will be the heterogeneity of Europe with respect to administrative ontology and law, administrative processes and culture, the expectations of citizens, the existing legacy systems, and evolving new e-government legacies. Interdisciplinary R&D is required to further develop vertically and horizontally scaling, holistic solutions. We suggest focusing efforts on architecture in its holistic sense (rather than on prototypical products). Interdisciplinary projects should be provided with enough time, experienced convergence management, change management by scientists (or IT architects), and the possibility to integrate further partners at a later stage of the project.

References

1. C. H. Cap, N. Maibaum. *Digital Identity and it's Implications for Electronic Government.* In Towards the E-Society - E-Commerce, E-Business, and E-Government, Kluwer Academic Publishers, Boston, 2001.
2. C. Cap, N. Maibaum, and L. Heyden. *Extending the Data Storage Capabilities of a Java-based Smartcard.* In Proceedings ISCC 2001, IEEE Computer Society, 2001.
3. *Oostveen A.-M., van Besselaar P.*: Linking Databases and Linking Cultures, In Towards the E-Society - E-Commerce, E-Business, and E-Government, Kluwer Academic Publishers, Boston, 2001.
4. A.-M. Oostveen, P. v. Besselaar, and I. Hooijen. *Innovation as learning: three case studies of multi-functional chipcards.* In Proceedings EASST 2002, York 2002.
5. S.D. Pawlowski, D. Robey, and A. Raven. *Supporting Shared Information Systems: Boundary Objects, Communities, and Brokering.* In Proceedings ICIS 2001
6. R. Riedl R. *Interdisciplinary Engineering of Interstate E-Government Solutions,* Proceedings 4th International Conference on Cognition Technology: Instruments of Mind, Warwick 2001.
7. R. Riedl. Document-Based Inter-Organisational Information Exchange. In Proceeedings ACM SIGDOC 2001, Santa Fe, 2001.

The Local e-Government Best Practice in Italian Country: The Case of the Centralised Desk of "Area Berica"

Lara Gadda and Alberto Savoldelli

Politecnico di Milano, 20133 Milano, Piazza Leonardo da Vinci 32, Italy
Tel. +39 02 23992796; Fax. +39 02 23992720;
lara.gadda@polimi.it

Abstract. The reform impulse, which in the last years characterised the Public Administration of all the word, underlines the necessity to join legislative changes with process and change management. In more recent years, the attention to process oriented change management techniques has also emerged in the public sector, through attempts to draw from private sector, searching for new methodologies and managerial approaches that could satisfy the need of organisational innovation. The article aims to present a successful case related to the application of an e-government experience in Italy in a local context: the development of a centralised desk for issuing building permits, grouping twenty-two villages in north-eastern. This experience represents a best practice which could be transferred in many other context.

1 Introduction: The Context

In the last few years, industrialised Countries faced up the necessity of *reforming Public Administration*, a crucial problem since that context is quickly evolving. The change in progress is moving along two directions: on one hand, the users require a Public Sector's "product" risen in value and, on the other hand, there's the need to provide better services using the same resources (more trend towards efficiency) [1].

The progressive shift from a slow and inefficient bureaucracy toward a lean and dynamic organisation is one of the topic joining a number of interventions of the last decade, within the public sector administration all over the world. Though this deep change has been faced in each country in different ways [2], the fundamental characteristic is the presence, finally contemporary, of two separate trends: *the decentralisation and the modernisation*, intended as procedures and process simplification based on importing management models typical of the private sector and on the regulation of the relationship between politic and administrative responsibility in management of the public organisation.

After years of unsuccessful attempts also in the Italian context, the two fundamental reforming pressures are carried out and their most significant expression is the so called *Bassanini laws*[1], which on one hand determine the conferring of new functions and tasks to regions and local authorities (Bassanini 1) [3], on the other hand they

---

[1]The laws n°59/1997 and n°127/1997.

R. Traunmüller and K. Lenk (Eds.): EGOV 2002, LNCS 2456, pp. 179–186, 2002.
© Springer-Verlag Berlin Heidelberg 2002

enact the simplification of control and decision procedures (Bassanini BIS) [4]. In particular, the decentralisation laws transfer to regions, provinces and municipalities strategic tasks which have been performed up to now, by the central level, so including in the reforms the *sussidiarity principle* which states that "it's not due to the superior government all for which the inferior level is enough" [5].

The decentralisation represents a leverage for obtaining a more effective and efficient administrative management, nevertheless the functions delegation, up to now centralised, can be only a necessary condition, not certainly enough, for approaching of the public administration to the citizens. The enlargement of functions the regions should perform risks to undermine consolidated and, on the whole, efficient organisation, that could reveal inadequate to the "new decentralised system".

In order to shift the change from the legislative level to the organisational-functional one many models of change management have been experimented, as the *Total Quality Management* and the *Business Process Reengineering* [6] [7] [8]. Among the most important experiences, it's possible to remember:

- the necessity of PA to consider not only the services users, but also all the actors who, even if they aren't "clients", are interested or have influence on their allocation;
- the necessity to control equity, clearness and legitimacy principles of public processes;
- the presence of legislative constraints, the political commitment and many stakeholders

In this context the simple application of the business models from private to public sector creates several problems. So it's important the identification and evaluation of the actual processes and the definition of performance levels which should be reached by the change.

The article aims to present a successful case related to the application of an e-government experience in Italy in a local context: the development of a Centralised Desk for issuing building permits, grouping twenty-two villages in north-eastern ("Area Berica"). These administrations are bound by a Territorial Treaty which is finalised to reorganise the process of local development and to create new economic opportunities. Analysing the specific characteristics of the reality, this case is a best practice which could be transferred in many other contexts.

The next paragraph explains what Centralised Desk is, which are its aims and the involved actors. The third paragraph presents the definition of the case. At the beginning the "as is" analysis was realised on the base of an appropriate check list in order to identify the changes which allow the definition of the "to be" process. The fourth one is based on the chosen methodology, the change management approach and the introduction of IT solutions, and on the results of the implementation. Finally the article presents the conclusions: the relevant local changes and the possibility of extend the experience in other different local contexts.

2   The Centralised Desk

In the modernisation context of the local, regional and national Public Sector the Centralised Desk (CD) plays a very important role since it's a hinge between the local administration level and the economic-productive system.

The normative evolution of the CD is bounded to the DPR 447/98 and to its modification, the DPR 440/2000 [9]. Its subject is related to the localisation of productive system of goods and services, their realisation, restructuring, extension, closing-down, reaction and re-conversion of the productive activity, as well as the realisation of the internal works in building used as an enterprise.

The CD is the only one administrative enterprise interlocutor for all is related to the localisation, construction and restructuring of productive systems. It is the unique responsible of the whole administrative procedure and it's able to assure to everybody is interested the access to all information about the authorisation procedures and to the service and assistant activities to the enterprises. The creation of the CD transfers the collect of all needed authorisations from the enterprise to the Public Administration. The CD receives a schedule and its task consists in the production of all the documents useful to authorise the realisation of the required activity [10].

The Law establishes that the CD does either the administrative activities or the informative ones in order to help the contractors interested into the starting of an activity on a specific lot. So, on the administrative hand, the CD requires:

- the definition of the timing of the issue of an authorisation by PA;
- an internal organisation of the different offices;
- a change in the relationship among local PA which are involved in the process of the administrative authorisation related to different kind of productive activities;
- the necessity to develop forms of inter-institution co-operation which are based on the use of new technologies;
- a continuos process finalised to the administrative simplification

On the other hand, the informative activities could be related to:

- public information about who have already presented the requirement to the administration;
- the access to the knowledge of the state of art of each procedure by applicant;
- the facilities which the administration, manager of the CD, is able to assure to everything who wants to invest on its territory.

In conclusion it's possible to affirm that the CD is born in order to:

- simplify and speed up the administrative procedures;
- make administrative action clear and available to the citizens' participation;
- promote the local economic development also by the diffusion of information related to promotional and technical assistance activities.

The introduction of the CD finds its *problems in the change of the process.* it was difficult to convince the involved people to change their habits and to think that the new technologies could make the process better. This problem is due to the fact that in the little local administrations *the process is hindered by politic and parochial reasons.* Each administration is convinced that its working method is the best and they are worried to loose power and contact with the citizens/SMEs if the CD will take place[11].

3   The Definition of the Case

The realisation of the Centralised Desk started from a social/economic analysis. It was the result of several interviews made to the representatives of local administrations in order to identify the main characteristics of twenty-two villages which constitute the "Area Berica". The identified characteristics could be divided in two categories:

- specific elements of local administrations:

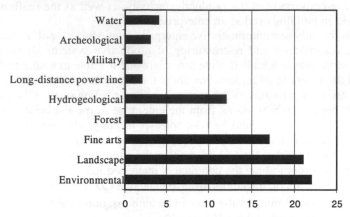

Fig. 1. The bonds of the protection of the territory

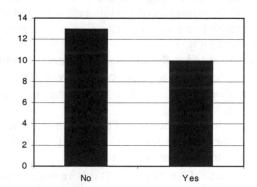

Fig. 2. IT in urban development plan

- specific elements of process of building permits:

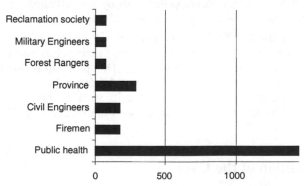

Fig. 3. The involved Third Administrations in procedures

4   The Identification of "As Is" Analysis to the Definition of "To Be" Process

The collected data through the social-economic analysis were the starting point to build the "to be process". First of all, in order to reach the best solution for the specific case, *weakness points & expected performances* were defined:

Table 1. Analysis "as is" – weakness points and expected performances

Weakness points	Expected performances
• Lack of operative and organisational structure; • Lack of accurate information for the citizens: all the procedures need to be integrated; • Lack of opportune technological tools; • Need to have a wide knowledge, but lack of available time; • Lack of standard procedures	• To standardise the procedures; • Major investments in technologic infrastructure; • Employment of high qualified people who carry out consulting activity; • Major information to users; • Simplification of procedures and reduction of timing of building permits; • Change of information between local and Third Administrations

This working approach let the drawing up of the *Centralised Desk Income Statement*: it was realised on the based of the collected data, the number of procedures and the involved people in "Area Berica".

Table 2. Analysis of the Income Statement

	"As is" analysis	"To be" process
Number of building permits	1440	1500
Number of involved people	25	30
Time dedicated to procedures by each person	94 days/year	75days/year
Costs (personnel, redemption and managerial expenditure) for each local administration	about 5.000 €	about 20.000 €
Covering of costs for each local administration	• a tax per head = 0,4 €	• a tax per head = 0,95 € or • a payment for the management of the relationship with the Third Administrations = 7.000 €

According with the introduction of Centralised Desk, the higher number of involved people is due to the presence of qualified personnel who requires an additional cost. On the other hand the Centralised Desk takes advantages to the final beneficiaries such as the possibility to have information in quick time, a major transparency in information exchange and the saving related with consulting costs.

5   Implementation Plan

Once the analysis and search phase was finished, the action plan was implemented:

- the *collected results were submitted to the twenty-two villages* in order to focus the attention on the common and the different characteristics of the "Area Berica";
- during the meetings with people involved in each local administration, some *proposals for the realisation of the Centralised Desk* were collected;
- in order to implement the CD it was important to open the dialogue towards the majors. It was possible to achieve the *signature of the treaty for the activation of the CD*;
- according with the characteristics of the villages, the project consortium decided *to divide the "Area Berica" in three parts called sub-areas.* In each sub-area, the villages referred to the biggest local administration which was in charge of co-ordinating activity;
- to realise the effective introduction of the CD *work teams* were organised. These teams convened directly people of different local administrations. The *training areas* dealt with:
    - *technological area.* Its programme was divided in three levels: presentation of CD web-site (www.sportellounico.net); guidelines to fill in the web-site the local administration rules and forms; simulation of procedure management by CD;
    - *administrative/managerial area.* Its programme dealt with the process analysis and the forms analysis. These training courses were firstly organised in three groups and then in only one group: each one expressed his/her opinion in order to define the standard form which satisfied all interests.

The main problems of the implementation plan dealt with:

- the *different opinions expressed by the involved representatives of local administrations.* Each of them was convinced his working methodology was better than anyone else and they were not in favour of the change;
- *the necessity to interact with the majors* who followed their political interests. In the twenty-two villages the majors belonged to different political currents and so it was very difficult to make them agreed;
- the importance to *find an agreement between the technicians and the majors* since their interests were different.

6   The Actualisation Status

According with the implementation plan and considering the emerged problems, the following decisions were taken:

- *An associated managerial model was created.* The three sub-areas were joined in a unique centralisation at the administration which heads the line of the Territorial Treaty. This model foresees the presence of two different administrations: ones which is responsible at local level and a co-ordinating ones. The activities of the local administration are:
    - to inform and help the citizens by information instruments;
    - to provide the standard and defined form;

- to receive the filled in form and the attached documentation;
- to send the form to the Centralised Desk and to sign building permits.

The activities of the co-ordinating administration are

- to record the form and to manage the unique procedure;
- to establish the relationships with the Third Administrations and to check it;
- to issue the authorisations for starting the productive installation;
- to transfer to the local administration the form which has completed the iter;
- to assure all the communications to the involved people.

- *The standard unique procedure was define. It refers to different documents.* Also these forms were standardised and they deal with building permits, starting activity declaration, habitability and feasibility certificate, and commerce;
  - *The standard form was filled in the web-desk (www.sportellounico.net).* The realisation of the on-line sharing of data simplifies the technicians' work, reduces the time of authorisation permits and makes the procedure monitoring and the data computerisation easier.

The project started on March 2001 and finished on October 2002. Now the administrations of the "Area Berica" carry out their building permits through the CD: all the local administrations take part in the project and they are enthusiastic for the achieved results.

7   Conclusions

The key elements of the CD project consist of the implementation method at local administrations. The outcomes are the result of acquiring relevant information from people involved in conducted analysis: that was allowed by the widely sharing of knowledge and strong co-operation of Majors. Specifically, public Authorities contributed to the work groups in a decisive way through their strong commitment to build widely accepted procedures and models, and on the basis of a sound experience in the field of knowledge and document management.

Moreover, the obtained model can easily be applied to complex information flows in local, but widely diffused, contexts. In fact, notwithstanding the regional characters of such procedures, the developed model provides a wide structure and every necessary feature for modelling CD. For that reason, the project constitutes a best practice to be applied in the European context.

The last element which should be considered is the possible future developments. The economic development area will become in the next future one of the most important and prestigious scope of local administration participation. Obviously, this is possible if the old competence "administrative" conception is overcome by politics characterised by local economic development. This passage requires an integrated vision of the different areas where it's possible to work.

References

1. AIPA, "La reingegnerizzazione dei processi nella Pubblica Amministrazione", pp. 1-24, 1999).
2. Klages, H., Loffler, E., "Administrative modernisation in Germany – a big qualitative jump in small steps", International Review of Administrative Sciences, pp. 373-384, (1995).

3. Legge 15 marzo 1997, n.59, "Delega al Governo per il conferimento di funzioni e compiti alle regioni ed enti locali, per la riforma della Pubblica Amministrazione e per la semplificazione amministrativa".
4. Legge 15 maggio 1997, n.127, "Misure urgenti per lo snellimento dell'attività amministrativa e dei procedimenti di decisione e di controllo".
5. Balboni, E., "I principi di innovazione del decreto Bassanini", Impresa&Stato, Aprile/ Giugno, pp.44-45, (1998).
6. Thompson, J.R., Jones V.D., "Reinventing the federal Government: the role of theory in reform implementation", American Review of Public Administration, June, (1995).
7. Kock, N.F., McQueen, R.J., Baker, M., "Re-engineering a public organisation: an analysis of a case of successful failure", The International Journal of Public Sector Management, Vol. 9, n. 4, (1996).
8. Pfiffner, J.P., "The National Performance Review in perspective", International Journal of Public Administration, 20(1), pp. 41-70, (1997).
9. D.P.R. 20 ottobre 1998, n.447 come modificato dal D.P.R. 7 dicembre 2000, n. 440,
10. Aa. Vv, "Realtà locali – periodico di informazione e cultura", n. 19, (2001).

The Immanent Fields of
Tension Associated with e-Government

Otto Petrovic

Karl Franzens University Graz and evolaris foundation,
Universitätsstrasse 15/E4, A-8010 Graz, Austria
otto.petrovic@uni-graz.at

Abstract. In the past few years the establishment of e-government has been given numerous new impulses. Totally new horizons have been opened both on the level of information and communication technologies, especially by the Internet, and on the level of administrative processes by new methods and tools applied in process design. But nevertheless especially e-government is faced with much more difficulties and opposition in its implementation than its counterpart in business, namely e-business, which in turn meets difficulties that are anything but small. E-government looks back upon more than 40 years of history [1,2,3,4], and the recipes for success that have been presented are often older than the people who today are in charge of implementing them. The present article tries to show that in approaches to e-government that are merely limited to the technological and the administrative process level many immanent fields of tensions remain unconsidered.

1 Many Unsolved Problems Despite Long Traditions

Many arguments such as the necessity to consider organization and technology in an integrated way (i.e. organizational reform taking place simultaneously to a technological reform) have similarly long traditions as e-government itself, date back even to the 1950ies and are thus anything but new [5]. This also applies in a very similar way to the demand to coordinate internal administrative changes and external services as well as to coordinate various decisions on technical investments within public administration. It seems rather unlikely that the two core innovation potentials of the Internet for e-government, namely the creation of uniform standards for data transmission and data representation and the penetration of the non-public sector with information and communication technologies, will be sufficient to make up for deficits in the implementation of e-government.

2 There Is Something That Gives a Meaning
to Administrative Processes

As soon as in the late 1980ies and early 1990ies both business and public administration realized that the organization of enterprises and public administration should not be oriented merely to individual functions but to a sequence of related activities – namely the business and administrative processes [6]. E-business – i.e. the counterpart

R. Traunmüller and K. Lenk (Eds.): EGOV 2002, LNCS 2456, pp. 187–190, 2002.

of e-government in business – however does not just concern the transformation of essential business processes but also a redesign of business models [7] with the aid of Internet technologies. Business models lend a more profound meaning to business processes. They determine which players are involved, which products and services these players offer and which revenues they are able to earn from which particular source of revenues. So business processes are always an implementation of a certain business model. Analogously there is a kind of 'administrative logic' that gives an intention-driven meaning to the underlying administrative processes. Merely considering the implementation of existing law as the only meaning in administrative processes would definitely be a drawback to a level that would roughly correspond to the sense of reality of the homo oeconomicus and would thus be of little pragmatic relevance for e-government. Activities in public administration and politics are based on limited rationality due to incomplete information, unsolved conflicts of interests and decisions that are characterized by emotions.

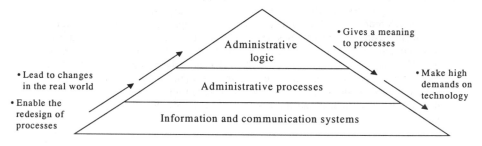

Fig. 1. The process levels of the e-government

As long as e-government primarily deals with technological matters on the level of the information and communication systems and with organizational matters on the level of administrative processes, citizens, public administration and political players will substantially object to changes and show their resistance to these changes either in an open or in a more subtle way. As a result the success of these changes will remain rather modest or at best just perceivable.

3   Understanding Public Administration As a Complex System

Public administration can be considered as an open system that internally is characterized by an extraordinarily high degree of functional differentiation, or to be more precise, of specialization. If such a complex system is faced with a highly dynamic change such as the development of the Internet, a field of tension is created because it is necessary to consider as many conflicting interests of the various parties involved in the process as possible (citizens, public administration, politics), as well as the costs incurred in terms of funds, time and contents.

While in the past ten years many highly efficient methods and tools for optimizing business processes were developed by means of information and communication technologies, the development of business models and administrative logics is still in its infancy. When developing business models, intuition, a 'funny feeling' or some knowledge gained through experience often play an important role. But intuition is very difficult to communicate and thus difficult to share with others so that decisions made by intuition very soon meet the limits of usefulness in a society that is increasingly based on division of work. An essential task of the e-government therefore should be to help develop methods and tools, show complex correlations within an administrative logic, make it communicable and analyze it from various points of view, in order to finally be able to make substantial use of the potential factors of the Internet when redesigning administrative processes.

4   True Administrative Logics Do Exist – Even Numerous Different Ones

Administrative processes could be managed rather easily in an efficient and effective way by means of the e-government if there was a consistent administrative logic underlying them. This administrative logic eventually determines which form of administrative process is considered as meaningful by the various players involved and is thus fostered by them either openly or in a more subtle way.

An example for such different administrative logics are e-government information systems for every day matters such as jobs, education, housing or health. The *citizen* expects that such services are provided by the public sector free of charge – and of course in a high quality and always up-to-date. The citizens' willingness to pay for such services is very low. From the point of view of *public administration* information systems for every day life represent a totally different logic. Due to the continuously growing cost-push, public administration is increasingly interested in e-government systems that will immediately contribute to cutting costs. Information systems for every day life, however, may well increase the quality of the service provided by public administration but also cause additional costs. *Political players* consider such information systems as highly valuable, additional tools of communication which can be used for self-expression, for communicating achieved goals, legitimizing public expenditure and above all for a very targeted formation of opinions.

5   But Who Is Entrepreneur, Bureaucrat and Politician at the Same Time?

The last field of tension associated with e-government is created by divergent requirements for planners, developers and operators of the e-government. Every-day applications such as tourist information systems, regional marketplaces for enterprises or information systems for businesses and culture require entrepreneurial thinking and actions. The ability to detect – or to generate – substantial gaps between demand and supply and bridge them with big economic success is in the center of interest. Thus searching for imbalances and balancing such divergences between demand and supply in a targeted way, as well as the variable 'generation of revenues' determine entrepreneurial thinking and actions.

Just the contrary applies to the demands made on e-government, especially when it comes to contacts with authorities, because in this field it is particularly important to fulfill specific requirements as efficiently as possible. Minimizing expenditure and optimally using existing funds as well as trying to establish a balance are in the center of interest. This means that the most important goal is to establish a balance between clearly defined demands made on public administration and the degree of fulfillment of these services.

Applications in the field of political participation in turn require an in-depth understanding of the impacts on the political structure of power, the functions of political processes and in many cases even political legitimization.

Acknowledgements

The author would like to acknowledge the funding of the evolaris foundation by contributions from the federal ministry of commerce and employment, the styrian business support association, the styrian provincial government's departments of commerce, finance and telecommunications and of innovation, infrastructure and energy, and the city of graz.

References

1. Kubicek, H., & Hagen, M., et al. (1999): Internet und Multimedia in der öffentlichen Verwaltung (Gutachten). Medien- und Technologiepolitik, Bonn: Friedrich-Eber-Stiftung, http://www.fes.de, retrieved on 6th August 2001.
2. Prins, J.E.J.(ed.) (2001): Designing E-Government: On the Crossroads of Technological Innovation and Institutional Change, Kluwer Law International.
3. Kraus, H. (1983): The Impact Of New Technologies On Information Systems In Public Administration In The 80, Verlag-North-Holland, Amsterdam - New York.
4. Traunmüller, R.; & Reichl, E.R. (1980): Die Entwicklung von Verwaltungsinformationssystemen, Proceedings of the 2nd Joint Technical Meeting of ÖGI/GI: Information Systems for the 1980ies.
5. Leavitt, H.; & Whisler T. (1958): Management in the 1980's, Harvard Business Review, Nov./Dec., S. 41-48..
6. Davenport, Th. H. (1993).: Process innovation: reengineering work through information technology.. Boston, Mass.
7. Petrovic, O.; Kittl, C.; & Teksten, R. D. (2001): Developing Business Models for eBusiness, Proceedings of the International Conference on Electronic Commerce, 31st October 2001 to 4th November 2001, Vienna, Austria.

VCRM – Vienna Citizen Request Management

Josef Wustinger[1], Gerhard Jakisch[2], Rolf Wohlmannstetter[3], and Rainer Riedel[2]

[1] Vienna City Administration, MKS, Rathausstraße 1, 1082 Vienna, Austria
wus@mks.magwien.gv.at
[2] Vienna City Administration, MD 14-EDP, Rathausstraße 1, 1082 Vienna, Austria
{jag,rie}@adv.magwien.gv.at; gerhard.jakisch@aon.at
[3] Vienna City Administration, MD 55, Niederhofstraße 23/304, 1120, Vienna, Austria
woh@bue.magwien.gv.at

Abstract. The vCRM is one of the eGovernment applications, which was awarded by the European Commission's eGovernance Competition 2001. [1 eGovernment] It is the only workflow application within the Vienna City Administration, which handles tasks administration wide and not like the standard procedures within a department. It deals with all comments, complaints, requests, which are not routine eGovernment processes and have a specific legally defined procedure. It deals not with requests that are designated to a specific department. One main advantage of this procedure is, that the officer in charge may easily adjust the workflow according to the specific need of the case. It is one of the finalists of the Global Awards for Excellence in Workflow [2 WARIA 2002]

1 Introduction

Vienna's Citizen Request Management (vCRM) is one of the most frequented applications of the implementation of the eEurope initiative [3 Liikanen] by the Vienna City Administration. The first pilots of the vCRM were the "street damage information system", which started 1989. It included the co-operation of 12 departments, the central citizens' matters system, (Zentrale Bürgeranliegen) which is in operation since 1985.

The vCRM formerly named Central Complaint management (CCM) was initiated 1998 by the CEO of Rapid Relieve to reduce the workload of those departments, citizen service centers and administrative units in the 23 districts dealing with Citizens' complaints. [4 Jakisch], to realize suggestions made during the organizational assessment and evaluation of the Vienna City Administration in 1998, to implement proposals made by administrative units with intensive client interaction and to meet citizen's expectations as described by the VCA's CEO in his inauguration speech 1997 [5 Theimer].

The vCRM may be described as one important action to build up CRM – Citizen Relationship Management – the adoption of Consumer Relationship Management concept in Vienna. It will play a very important role in the future of the interaction between Public Administrations and Citizens/Inhabitants to meet more successfully the citizens' expectations, requirements and needs.

R. Traunmüller and K. Lenk (Eds.): EGOV 2002, LNCS 2456, pp. 191–194, 2002.
© Springer-Verlag Berlin Heidelberg 2002

vCRM deals with those complaints, comments, requests, which are not well structured, need workflows that cannot be designed in advance, handle multiple complaints about one subject and which every inhabitant /citizen may ask.

It deals not with routine governmental activities as legally based requests, petitions, etc. that are based on a specific law and have a defined workflow and those have legally stated participants like building permissions etc.

2   Goal and Objectives

The goal of the vCRM is to improve substantially the administration's citizen orientation – the relationship between the citizens on the one hand and the administration on the other –. This will be achieved by implementing a new administration-wide tool – the vCRM – to record and process all citizens' requests, comments complaints and suggestions concerning all spheres of work of the Viennese administration.

The Vienna Citizens' Request Management has five objectives:

1. Coordinated workflow of comments, complaints and suggestions,
2. Recognition of parallel cases
3. Avoidance of different solutions for similar cases
4. 24/7 availability of information
5. Improvement of citizen's information

All incoming complaints, comments, informal request and suggestions are recorded administration-wide in one application the Vienna Central Request Management. It uses one specific basic workflow and predefined search tree of keywords including colloquial notions and synonyms.

The recognition of parallel and similar cases is performed via the keywords. New incoming cases are recorded by recognizing location (address and Geo-Reference System -code) and predefined keywords. Thus in future new cases concerning same subject and location will be added automatically to the file of already recorded and processed cases.

This will be achieved twofold by filing all cases concerning one matter in on file such already processed cases are the basis for a new one and free access to stored data for all administrative officers.

The information stored in vCRM is available for every user, the registered administrative officers. In future also the citizens will have a specific access to their cases. Only those cases dealing with citizen's privacy, other privacy or public security issues are secured via a special structured user ID and password.

The situation of the person sending a comment, informal request, complaint or suggestion will be improved by changing actual web form into a more interactive version, direct feed back of filing key and password and possibility for the citizen to access vCRM and read the actions performed and running.

3   The Procedures

The cases recorded in the vCRM [6 Resel] arrive via several modes of access: Email htttp://www.wien.gv.at/m55/b-bue.htm , fax or letter, Call Center or Face to face: in the Citizen Service Centers or all other participating administrative units.

At the moment all cases have to be recorded by a human intermediate – a civil servant –. The next version will offer an interactive web client too.

Every administrative unit participating accepts all interventions. This eases the procedure for the citizen because he needs no knowledge about the internal organization and responsibilities of the administration.

The procedures are divided into several steps, the recording, the take down, the different processes to finalize the complaint and the response to the requester.

4   vCRM Web Client

To overcome the obstacles caused by hard and software requirements of Fabasoft components running on each PC used to deal with vCRM on the one hand and to implement an interactive application offering advantages to administration by reducing the workload of the vCRM officers and the citizens by easier access and the overall goal to reduce costs led to the development of the Web Client solution.

It will be available by on all public terminals and may be accessed via Internet without having implemented Fabasoft Components. This application enables the VCA to spread the usage of vCRM faster to 120 Administrative Departments and 500 units because no additional hard and software is required and thus reduces their costs in future.

The main differences to the Version described in paragraph 3 are: the recording of the request is performed by the requester, the confirmation is accompanied by the creation of a specific User ID and PIN-Code presented to the requester, each requester may trace the current status of the processing of his request at any time by typing the ID and Pin in the Authentification form. This will reduce time of the vCRM officers spent on answering requesters' calls on status of current affairs.

5   Vision

Currently the Vienna City Administration is participating in a European Union Framework 5- IST project – EDEN [7 FP5-IST] – which deals with the citizen participation in administrative processes and eDemocracy. It develops several tools to improve interaction between citizens and administration.

These tools are based on NLP technologies (Natural Language Processing). One of them, the automatic routing system, is designed to analyze unstructured texts, find the relevant keywords and phrases, and pass the document automatically to relevant expert available. In case of vCRM such a tool may assist the vCRM officer recording requests by automatically filling in the record. It will be able to handle requests received by web client, email, fax and partially letters.

The necessary information about the experts may be taken from the advanced version of the VCA's electronic organization expert system "Überblick Wien" which may include not only the tasks and responsibilities of administrative units, job descriptions but also the expert knowledge of the administrative officers.

Thus matching the vCRM search results and expert information it could suggest next steps. Taking into account the developments of speech text conversion the authors will make no predictions how avatars may change administrative acting.

6   Conclusion

vCRM is the first city-wide document handling system introduced by the VCA. It is based on Faba Components workflow technologies allowing the Citizen Service department to document all incoming interventions.

It benefits as well Citizens as administration because it offers all involved parties actual information about the state of affairs. It reduces the processing time. The citizen may participate via email and public terminals. The incoming information is analyzed according the addressee, the location and content, thus allowing linking the information to the GIS maps.

The combination of this application with the EDEN Tools as described above will enable the administration to analyze the citizens complaints and comments multi modal. By giving all involved parties the possibility to interact, improvements will be achieved, which haven't been possible in the past.

Thus the (inter)-active Citizen will become an equal partner of the administration in the decision making process.

References

1. eGovernment, from policy to practice, 2001 Brussel
   http://europa.eu.int/information_society/eeurope/egovconf/projects_selected/austria/index_
   en.htm#CCM – Central Complaint Management
2. http://www.waria.com/awards/awards.htm
3. Liikanen, Making eEurope a reality, European IT Forum 2000 Monaco
4. G. Jakisch, eVienna living situation based eGovernment and eDemocracy, DEXA, Munich, 2001, 397 ff
5. E.Theimer "Electronic Services"
   http://wwws14.advge.magwien.gv.at/w4base/mdi/default.asp Vienna 1997
6. Resel , Eschner, ELAK ZBM, VCA MD 55 internal training document, Vienna 1999
7. Framework Programme 5 Contract IST-1999-20230, Brussels 2000

Public-Private Partnerships to Manage Local Taxes: Information Models and Software Tools

Mario A. Bochicchio and Antonella Longo

SET-Lab (Software Engineering & Telemedia Lab)
Department of Innovation Engineering, University of Lecce, Italy
{mario.bochicchio,antonella.longo }@unile.it

Abstract. The present work is about the first results of SOSECO, a public-private company operating in the southeast of Italy, created by Servizi Locali SpA and the Municipality of Castrignano, in cooperation with the University of Lecce (Engineering School and Law School). The goal of SOSECO is to improve the management of local taxes and to enhance the relationship between Citizens and Public Administration.

1 Introduction

From surveys [West] made on US federal and local agencies, e-government is still in its infancy: even if many governmental units are putting on-line a wide range of information and services for citizen and employees, a coherent and mature design of the underlying information infrastructure (software applications, standards, protocols, ...) is still at the down, far from the evolution already reached in the private world. Some argue that private-sector management techniques can be applied to government, to produce more efficient, effective and responsive public agencies. At the same time, new ideas about governance have also emerged, stressing collaborative relationships, network-like arrangements and hybrid public-private partnerships to enable a more effective problem solving and a greater citizen participation in public affairs [Mechling], [O'Toole], [Koppel].

In Italy, as in many other countries, we have interesting experiences about online services to citizens (e.g. in the "ICI On-Line Project" [Prato], or in "National Online Fiscal Service Project" [Finanze]), but the situation is not homogeneous. From an informal survey in the Southeast of Italy, for example, we found a scenario similar to the one depicted in [Vintar]. In this context we decided to support SOSECO in putting up a "Citizen Relationship Management" (CRM in the following) strategy, with the aim to create trust between Local Public Administrations (LPA in the following) and citizens, and to provide it with a modern, multichannel, personalized set of services.

The CRM strategy has 3 main parts:

- **Interaction CRM**, i.e. the ability to create new multichannel services (one-stop counters, call center, Web portal, ...) based on an integrated information resource.
- **Insight CRM**, i.e. the ability to continuously acquire updated info and to relate it to the existing ones. This is used to construct new, more effective user profiles.

R. Traunmüller and K. Lenk (Eds.): EGOV 2002, LNCS 2456, pp. 195–198, 2002.

- **Satisfaction CRM**: i.e. the ability to evaluate/correct the citizen satisfaction level, with respect to new services.

SOSECO information system has been designed and developed to fulfill all the main aspects of our CRM strategy.

2   The Proposed Model

Up to some years ago, the main model in management of local taxes has been its concession to private partners: so the concessionary (invested of public powers) replaced the Municipality in collecting local tributes and in checking for tax evasions. Moreover, municipalities were few interested in managing tributes efficiently, because the main source of incomes came from the Central Government.

After the last constitutional referendum about local autonomies (Oct. 2001), a new model of local government management came out, asking for management savings and for a more active role in the territorial development. So, many municipalities has been forced to find new founding ways and/or to better use all existing sources of incomes. Thus the management of tributes and entrances is becoming a core business for local administrations, and tax evasion is now a big issue at local government level.

As a consequence, a lot of municipalities start to manage entrances and local taxes and to double check evasion by themselves. The problem is that most LPAs, especially in the South of Italy, lack of the resources to design and to manage good fiscal services. To skip this problem they can adopt two possible solutions: acquiring resources (ICT infrastructure, technologies, skills, people, etc.) to directly manage fiscal entrances inside the municipality, or outsourcing the services necessary to support the core management of tributes.

A detailed comparison between the two solutions is out of the scope of this paper, but we can observe that, in general, the first approach is ineffective because of the lack of standards (in software, interoperability, data formats etc.) and because of the rapid evolution of technologies and skills. The second approach, instead, can be very effective if performed with the following constraints:

- it must be "transparent": data must be provided to the LPA with well documented formats, on open architectures (like standard DBMSs), ready to be reused for different purposes;
- the outsourcing must be on-line: each municipality must hold every time the actual ownership of its own data, on-line with all other municipal data bases;
- a sharp separation must be enforced, between the integrated data resource of the LPA and the services built over the data, so several service providers can concurrently offer services over the same data resources.

Thus public-private companies are the right component to create and to manage all the online LPA's data resources: the public part guarantees the respect of the previous constraints, while the private part provides the technological know-how. At the same time the role of the private partner changes: from substitute of the Administration (as in the case of the concessionary) to coach of innovation.

Basing on this model, in our project 3 partners cooperate to provide innovative fiscal services to citizens: the Municipality of Castrignano, the SOSECO (i.e. the data provider) and Servizi Locali (SL in the following), (i.e. the service provider); other private providers can cooperate with the Municipality and SOSECO to concurrently offer further services to citizens, business, and employees. At the same time with this approach the private partner can provide the same service to several Municipalities, without the need to manage the overall LPA's data resources.

3   The SO.SE.CO Project

SOSECO (Società Servizi Comunali – Society for Municipal Services) was born in Spring 2000 and is made up for 51% from SL as a private partner and for 49% from public partner (The Municipality of Castrignano del Capo, in the district of Lecce in the southeast part of Italy). The mission of SOSECO is:

- To adopt innovative tools and services to manage local tributes, like data quality assessment and citizen relationship management (CRM);
- To train Municipality's employees about the recent administrative reform;
- To technology support Municipality with data digitalization and integration;
- To support territory's research and development;

The relationship between the SOSECO and SL is based on a service agreement to supply innovative services and to manage fiscal entrances. SL provides innovative fiscal services but the whole management is still owned by the local Administration (through SOSECO).

The agreement between SL and SOSECO is about what they call "Introductory activities to tribute and entrances' collection", including:

- Census of real estates in the Municipal territory
- Estates data cleansing
- Cross checks among several data flows in order to check out evasions
- Services to citizens who can access their fiscal position and communicate with Administration and manage pre contentious, contentious and post contentious phase through different integrated channels

The Municipality has 15.000 inhabitants and last year, with this approach, it found 30% evasion level on a 2.5 M € approximate global entrance. The services provided by SL are paid by SOSECO with a percentage on plus evasion cashing assessed.

Major's Point of View is: "We are not outsourcing our fiscal services, but we are outsourcing innovation on fiscal services", with the following advantages:

- learning by doing for employees involved in the project, and knowledge transfer from the project team to the rest of employees;
- better aptitude to innovation;
- forcing the application provider to keep high the quality of service and innovate it not to be substituted by another provider.

Currently, SOSECO employs 3 people and 10 freelance consultants, but today we foresee the society growth like double over the next 12 months.

4   Implementation Issues

SETLab has supported both SL and SOSECO to define the model previously described, to build the innovative information infrastructures to the whole fiscal process, to design the multichannel applications for the CRM strategy and the communication with citizens.

To build, design and develop SOSECO's information system we used UWA
(Ubiquitous Web Application) [UWA] and W2000 methodology [Garzotto]. Moreover, we adopted a fast prototyping approach (based on MS Access and MS SQL
Server) to finely tune algorithms and filters for data cleansing and data management
for the SL side.

UWA helps us to reduce overtime costs in development, moving most of the issues
at the design phase, to improve effectiveness and ergonomics in functionality and
features, and to to dominate multichannel design and multichannel integration
through a proven methodology.

Four main software prototypes have been created in cooperation with SL to support its core processes (census taking, information cleansing and matching, tax as-
sessement and injunction of payment, citizen reception and advise).

The main research effort made by the SET-Lab has been to produce reusable tools
and standardized procedures for data cleansing, data integration and integrated multichannel services to citizens. In the first year of operation the information system
prototype we designed allowed SOSECO and SL to correctly manage the whole fiscal
process over 15,000 citizens with 3,5M€ of incomes, that is 30% more than the previous year. This result was achieved without contentious.

References

[Vintar] M. Vintar, "*Reengineering Administrative Districts in Slovenia*", Discussion Paper,
    No.11, Local Government and Public Service Reform Initiative, Open Society Institute,
    ISSN 1417 – 4855, 1999
[West] D.M.West, "*Assessing E-Government: The Internet, Democracy, and Service Delivery
    by State and Federal Governments, 2000*" Brown University, Sept., 2000
[Prato] http://www.comune.po.it/servcom/tributi/ici/htm/icionline.htm
[Finanze] http://www.finanze.it
[Mechling] J. Mechling, A customer service manifesto: using IT to improve government serv-
    ices. Government Technology, January, S27-S33, 1994
[Koppel] Koppel, J. G. S. (1999). The challenge of administration by regulation: preliminary
    findings regarding the U.S. government's venture capital funds. Journal of Public Admini-
    stration Resarch and Theory, 9, 641-666.
[O'Toole] O'Toole, L. J. (1997). Treating networks seriously: practical and research-based
    agendas in public administration. Public administration review, 57, 45-52.
[UWA] UWA (Ubiquitous Web Applications) Project: Deliverables D2 (General def. of the
    UWA framework), D7 (Hypermedia and Operation Design: Model, Notation, and Tools).
[Garzotto] F. Garzotto, "*Ubiquitous Web Applications*" - Invited Talk, In **Proceedings of
    ADBIS 2001**-5th East European Conference Advances in Databases and Information Sys-
    tems, A. Caplinskas, J. Eder (Eds.): Springer LNCS 5121, 2001

E-MuniS – Electronic Municipal Information Services – Best Practice Transfer and Improvement Project: Project Approach and Intermediary Results

Bojil Dobrev[1], Mechthild Stoewer[1], Lambros Makris[2], and Eleonora Getsova[3]

[1] Fraunhofer Institute for Secure Telecooperation, Schloss Birlinghoven
D-53754 Sankt Augustin, Germany
www.sit.fraunhofer.de
[2] Centre for Research and Technology – Hellas, 6th km Charilaou – Thermi Road
57001 Thermi – Thessaloniki
www.iti.gr
[3] Elisa Consult, 93, James Boucher Blvd., 1407 Sofia
Elisa@solo.bg

Abstract. The E-MuniS (Electronic Municipal Information Services – Best Practice Transfer and Improvement) Project aims to improve the best practices of the European Union municipalities regarding the use of information technology in municipal administration working processes and services to citizens and to transfer those results to South-Eastern European municipalities in particular from the Balkan region thus integrating it to the EU municipal network. The project consortium involves as participants couples of local municipality – IT-company partnerships from EU countries and from South East European countries. Within the project solutions for an e-municipality office (as back-office system) and prototypes of e-services to citizens and business (as front-office system) will be developed and implemented.

1 Project Main Goals and Expected Benefits

1.1 Project Background

Municipal services to citizens and municipal administration working processes have always posed challenges to both citizens and municipal employees. They were mostly associated with enormous paper work, procedural formalities, long queues in front of always busy administration offices, complete waste of time and a lot of nerves – again both for the citizens and administrative staff.

Although cities in the European Union have made rapid advancements in the field of implementation of innovative Information Technology for improving and facilitating the citizens' life, there are still deficits especially in providing e-services to citizens and integrating these services to internal applications. In the South-East European (SEE) region there are limited IT applications (mostly off-line) and a lack of use of internet technologies for municipal applications.

The E-MuniS (Electronic Municipal Information Services – Best Practice Transfer and Improvement) Project as a result of the successfully completed SEEmunIS pre-

R. Traunmüller and K. Lenk (Eds.): EGOV 2002, LNCS 2456, pp. 199–206, 2002.
© Springer-Verlag Berlin Heidelberg 2002

project (South East European Municipal Information Infrastructures, 1999/2000) provides an original contribution to the solution of the problem bringing together the efforts of technology providers and local authorities from EU and SEE based on public-private partnership relations.

Its results will be disseminated firstly to the project municipalities and following the dissemination strategy - to about 50 other municipalities from EU and the Balkan region, thus integrating the latter to the EU Community municipal network.

The project is realised with the financial support of the IST Programme of the European Union. It started in November 2001 and will end in October 2003.

1.2   Description of the Project Objectives

The E-MuniS ultimate goal is to provide opportunities for user-friendly implementation of the information technologies achievements in municipal administration working processes and services to citizens. The E-MuniS project main objective is to bridge the gap between EUMs and SEEMs regarding use of Information Technology in administration working procedures and services to citizens, thus facilitating the work of the municipal employees and making the life of he citizens easier.

E-MuniS project contributes to improving of the quality of life by:

- Better and more accessible services provided by local administrations, following the identification of the citizens' needs.
- Improving the working conditions of the municipal administration employees by reducing the time and efforts needed for information processing.
- Creation of new IT based jobs related to the latest achievements of the information technology usage in the municipal administration processes.
- Improving citizen's participation in all levels of local government life through a facilitated participation and interaction between citizens and local government.

2   Consortium

In compliance with the specifics of the priorities VIII.1.6-VIII-1.5 of the EU IST Programme, the project Consortium involves as participants couples of local municipality-IT company partnerships (municipalities and technology providers) from EU countries (Germany, Spain, Italy, Greece) and from South East European countries (Bulgaria, Former Yugoslav Republic of Macedonia, Croatia).

The E-MuniS tends to act as a driver for cooperation between public administrations of a number of cities in different countries, to contribute to public-private partnership development between the local administration and the IT industry, and to establish links and joint activities between IT companies from SEE and EU.

3   Key Issues

3.1   Project Aims

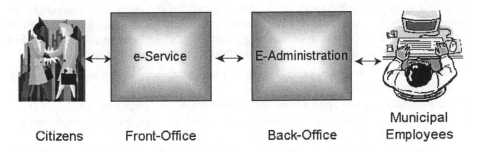

Fig. 1. Components of system development within the project E-MuniS

The project aims at development and implementation:

1. prototypes realizing an E-Municipality office (Back-office solution).
2. on-line services to citizens as front office solution with municipal and city websites and sets of interface tools allowing citizens access to municipal administration applications to ensure transparent information services to citizens. These solutions will be ported to an information kiosk system, too.

3.2   Project Approach

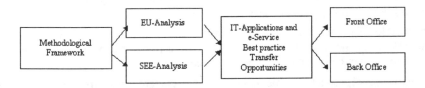

Fig. 2. Project approach

Basis of the project activities is the development of a methodological framework to analyze the situation in cities of the EU and South East European cities. The results will be used to identify best practice transfer opportunities for municipal IT-applications and e-services for front office and back office solutions.

   The overall project methodology will follow the top-down approach. All the tools for study, analysis, development and dissemination will be centrally elaborated. The responsibility for this will be delegated to the Work package leaders. The tools will be implemented in the project municipalities following a decentralized manner. All adapted or developed software tools will be completed to a prototype level with the relevant accompanying documentation. Pilot implementation will take place at two or three sites. All mid-term and final results will be immediately disseminated to all the project participants.

4   Results from the 1ˢᵗ Stage of the Project

4.1   Methodological Remarks

The first stage of the E-MuniS work plan focused on the study and analysis of EU municipalities' best practices and ICT infrastructure and the demand for improvement and ability for e-performance of working practices and services to citizens in 4 SEE municipalities. Therefore a set of questionnaires have been realized to:

- identify the current situation regarding implementation of Information Technology in EU municipalities
- identify best practice applications and elaborate on their improvement
- identify those applications that are better suited for transfer to the SEE municipalities based on certain criteria
- describe the most important services currently provided by SEE municipalities as well as information sources, IT applications and IT infrastructure.

For selecting best practice applications which are better suitable for transfer to SEE-municipalities a set of criteria has been used:

- is it relevant?
- is it feasible?
- what are the risks involved?

Is it relevant? Considering the scarcity of resources and the need to make an impact the E-MuniS project needs to carefully choose applications, which address a real need and whose results make a definite impact. The relevance of a given application depends on two considerations:

- is it relevant to the core activities and services of the municipality?
- would the solution make a measurable difference?

Is it feasible? The question about feasibility is more difficult to answer than the one about relevance, because there are a number of aspects that have to be considered. These aspects can be grouped into mainly three categories:

- organisational aspects - this concerns the current structures, legal restraints, communication mechanisms and procedures that are in place and which may be amenable to the introduction of IT systems or alternatively which require a complete reorganisation for their effective use.
- technological aspects - looking at the current state of technological development, especially concerning infrastructure availability, security technology, telecommunications and the Internet.
- human aspects - people are a significant factor in any organisation, and considering the need for change, there is a need for looking at the interest in new technology, trust in reliability, the need for training and potential cultural barriers.

What are the risks involved? In most cases the introduction or expansion of new technology like IT systems has a potentially dramatic effect on the organisation, especially when it is meant to lead to significant improvements in efficiency. There is a need for change in three different areas: in the organisational structure, in the organisational culture, in the information technology.

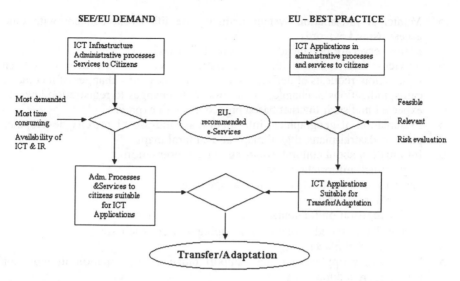

Fig. 3. Approach of selecting best practice applications which are suited for transfer

4.2   Intermediate Results

4.2.1   IT in EU Municipalities

The questionnaire, designed to gather information from EU municipalities, was disseminated to municipalities in Greece, Italy, Germany and Spain. In total, twenty-seven questionnaires from EU municipalities were selected and evaluated.
    Some results:

* All municipalities use MS Windows (98, 2000 or NT) as operating system while a large number (67%) have UNIX-based servers. Only 17% have Linux installed although a large number (83%) of municipalities consider the use of Open Software.
* All municipalities taking part in the survey have a LAN, 67% have PCs connected to a WAN. The majority of the municipalities have access to Internet (83%).
* Administrative municipality tasks (as citizens registration and financial administration) are in general supported by IT-systems. The use of various database systems is common.
* Geographical Information Systems are already introduced or will be implemented in near future.
* Beside this the prevailing trends within EU municipalities regarding administrative tasks are the use of modern technologies as electronic documents management systems, workflow components and the use of Internet technologies.
* Our survey shows that the municipalities already offer several e-services but there are still deficits concerning interactive systems using security technology and the integration of e-services to internal applications. The following applications seem to be proved:

o   Municipal information system realising the life event concept with quick access using keywords.
o   Forms server for downloading of forms (pdf files).
o   Services referring to citizens' registration such as extracts of citizens registration, requests of delivery of an income tax card - changes of income tax cards, indications of changes of address. With changes to registration laws it is expected that such Internet based procedures will expand.
o   Presentation of geographical information as land-use plans, city development plans, cadastral-plans, digital maps and cultural maps.
o   Information about cultural events and ticket reservation
o   City council information system
o   Services of the municipal waste management
o   Online library services
o   Online registration for courses at adult evening classes
- A few cities have started to realise challenging services such as:
o   Virtual market places
o   E-democracy applications that support citizens' participation in municipal planning procedures

4.2.2   IT in South East European Municipalities

Twenty-five questionnaires from SEE municipalities, districts and communities were answered. All of them were evaluated and summarized.

The situation and demand for Information Technology in SEE municipalities have been analyzed during the SEEmunis pre-project. According to the results of this project some problems are common to almost all municipalities:

- Lack of IT strategy as part of the strategy of the municipality
- Insufficient integration of IT applications (often using different tools and platforms)
- Insufficient computer equipment

This situation has not generally be changed.
Some results from the survey:

- The cities use personal computers which are connected to a LAN. All have Internet access.
- Microsoft technology is generally used (Windows as operating system, MS Office as personal productivity software). UNIX and Linux are not installed.
- The cities have websites which are used for provision of basic information.
- Administrative tasks as financial management, tax related administration, citizens registration, social services are supported by IT-applications, but the application related service to citizens and business is unsatisfying. The most important deficit is the lack of transparency of administrative procedures.
- The most demanded procedures are services to business and to issue permits and certificates for citizens.

4.2.3   Further Proceeding

Using the results of the survey and applying the selection criteria the following best practice applications will be realized within the E-MuniS project:

1. Information and E-Services Portal providing:
- Municipal information system – Information about all administrative issues of the municipality structured according to the life event concept
- Integration of forms (pdf-files) (one – way communication)
- Provision of services
  o For citizens: request of certificates (two-way interaction) (interactive forms are provided via portal).
  o For citizens: announcement of change of address (as all municipalities taking part in the project have already introduced a citizens' registration system, it will be possible to connect the e-service to this internal application).
  o For business: permits and certificates (using the example of starting or changing a trade). The e-service solution must be integrated in an internal application.

2. Electronic document management system with workflow components. This system will be connected to the administrative application and to the e-service application. This will ensure transparency and may be an example of multi channel approach, as there would be several ways of access to the municipal service: through internet, by post with the necessity to handle paper documents or via telephone.

5   Outlook

One of the most important issues of the project E-MuniS is to ensure on-going dissemination and exploitation of the project outcomes. The target is that about 50 municipalities of the Balkan Region should participate in the results.

At the Project level of dissemination workshops will be organized in each project municipality to demonstrate the applications, discuss the opportunities for their implementation in the particular municipality and conduct training for a designated group of staff to be working with them. The Project web site will contain a data base with information about all results achieved during the project stages and tools for analysis, modification and visualization. It is installed on a server and accessible via Internet.

The dissemination will be through professional societies and associations, NGOs, government bodies, supported by IT companies, based on public- private partnership.

At the International level of dissemination municipalities from other countries that have expressed interest in the E-MuniS outcomes will be potential end-users. The dissemination will take place through European associations and networks, such as Global Cities Dialogue, Tele-cities Network, Regional Innovation and Technology Strategies Networks, Association of the Balkan Municipalities, GISIG Network.

The project consortium will take great efforts to distribute the project outcomes aiming the integration of SEE-municipalities to the EU municipal network.

References

E-MuniS-Project Team:
1. Electronic Municipal Information Services – Best Practice Transfer and Improvement Project – a short project Presentation

2. Project Presentation and State-of-the-art of the ICT Infrastructure, IT Applications and Demand for new IT Projects in the SEE Municipalities
3. Methodological framework for study and analysis of the municipal administration working processes and services to citizens
4. IT applications in the municipal administration working processes and provision of e-services to citizens in the EUMs and SEEMs with best practice transfer opportunities

All documents are available on E-MuniS-Website: www.emunis-ist.org

Some Specific e-Government Management Problems in a Transforming Country

Nicolae Costake

Consultant Bucharest,
ncke@starnets.ro

Abstract. e-Government (e-G) raises specific technical, and also managerial problems. The managerial ones are particularly important for the transforming socio-economic systems. Recent experience of Romania in this field is shortly presented . Key managerial requirements are derived from the e-G business requirements. The proposal of orienting the back-office reengineering implied by e-G also on enforcing virtuous societal closed circuits and minimizing vicious ones is formulated.

1 Introduction. Statement of the Problem

E-Government (e-G) makes full use of ICT in the managerial subsystem of socio-economic systems (SES), implying the reengineering of its processes and links with its clients: citizens / organizations and internal users (managers, MPs, magistrates and civil servants). Last year, an EU Ministerial Declaration stated that e-G could improve the services, strengthen the European societies, raise the productivity and the well-being and also strengthen the democracy [1]. Technical aspects of e-G are studied in a vast literature.(e.g. [2]). Managerial aspects are, certainly, also important. The problem was studied by a number of authors. to name a few: Lenk and Traunmuller (e.g. split back-end from front-end activities, [tele-]cooperative work, organizational knowledge management.[3]).. Traunmuller and Wimmer (*i. a.* decisions and knowledge, including organizational memories [4], Lockenhoff (a general system theory for guiding the societal change.[5]). The managerial aspects of e-G are particularly important in the transforming countries, for assuring, and accelerating their trajectories towards the advanced Western SESs, as suggested e.g. in [2] and [6]. The present paper tries to answer the question what managerial e-G requirements (e-GMR) could result from its business requirements (e-GBR). Case study Romania is considered. The methodological approach is:e-G BR => e-GMR => comparison with the actual evolution =>consequences. The structure of the paper is:a) the challenge; b) e-G management requirements; c) recent developments for the information society in Romania: .d) conclusions.

2 The Challenge

The higher the decision level, the more adverse effects a mismanagement can induce. For the transforming countries, this conclusion is more evident. In the case of Romania, the difference in the interval 1990...2001 between the average evolution of the

R. Traunmüller and K. Lenk (Eds.): EGOV 2002, LNCS 2456, pp. 207–210, 2002.
© Springer-Verlag Berlin Heidelberg 2002

GDP of the former European communist countries and the actual evolution, represents a total value not far from that of the GDP in 1990. The 2001 Report of the European Commission on Romania's Accession [7] also identifies managerial problems, such as: the reform of the Public Administration, the reduced administrative capacity and the corruption.

3   Specific e-G Managerial Requirements. Case Study Romania

The experience in the field of e-G suggests a number of BR, which imply specific information system (IS) requirements [2]. The e-GBR can be structured in some categories: a) better services supplied to citizens, organizations and internal users: e.g. friendly interactive access to public information, minimization of the time and effort consumed in solving problems with the public institutions; b) better services for the users within the state institutions: e.g. adding flexibility and scalability; c) operational support: e.g. electronic documents and archives, centrally updated standard procedures, state and trend of resources, costs and revenues; d) better support for the executive management: feedback form the public, decision support based on data warehousing; e) general requirements: e.g. 24x7 availability, data security, and protection; protection against corruption and terrorism; possibility to build the coherent e-G information system in parallel by various state institutions. The general resulting key e-GMR is the need to assure the synergy within the SES. Following directions of managerial action can be enumerated: a) initial prerequisites support for the development of the communications infrastructure; political will to move towards the Information Society; readiness to fight bureaucratic resistance and corruption; .precise delegation of authority between central and local levels and. between / within state institutions; creation of the necessary legislative, regulatory and institutional infrastructure); b) support of the integrated informatisation of the public sector (including legislative and judiciary); c) creation of an empowered organizational entity for e-G; d) production of a national e-G strategy, support for : cooperative IS, e) identification of possible societal closed circuits, to be enforced if virtual, else minimized; f) training in basic ICT and e-G oriented applications, starting with high-level decision-makers. It is advisable to start with the creation of the informational coherence information system (as described in [2]), a shared informational resource for practically all information systems (IS) of the SESs.

Developments in Romania in 2001 and 2002 show the following: a) the mentioned managerial pre-requisites for e-G are now practically achieved.. In particular:, the telecommunications sector is growing,, the political will regarding the information society is shown by a specific chapter in the 2001..2004 government's action plan [8] and its update [9], the start for introducing networked PC's in all schools, growth of the national ICT industry; the adoption of a national anti- corruption strategy;. b) the essential specific legislation (copyright protection, electronic document and signature, data protection etc.) is in place. c) empowered authority for e-G exists. (A "Group for Promotion of the Information Technology", headed by the Prime Minister approves all major ICT projects. *I. a.,* it approved a Strategy for the Informatisation of the Public Administration [10]). The Ministry for Communications and Information Technology, quite dynamic, launched a bid for 19 IT projects, which became demonstrable prototypes produced by consortia of companies in less than one year.(end 2000). One of them, e- procurement, is now successfully experienced on a moderate scale and started to demonstrate the economic benefits for public acquisitions.. The scope and quality of the on-line access to public information was improved. The 2001-2004 action plan contains a number of provisions for e-G, many of them inspired by the eEurope+ plan of actions.. Its 2002-2003 update [9] foresees also the creation of unitary nomenclatures and informatised registers vital for achieving the coherence of the societal information. The national server for general interest nomenclatures (first defined in the strategic planning of 1992) started functioning.

However, from the point of view of the management science, some problems still exist:

a)   As a strategy based on the e-GBR (covering all the three Authorities of the state) is still under development, the action plans are not necessarily complete and do not necessarily represent the synergic trajectory towards the goal..

b)   The direct implementation of actions defined in the context of the advanced countries is not necessarily without difficulties. E.g., front-office oriented applications in general imply a back-office applications support. The version of the strategy [10] based on the Italian one [11] may encounter contextual problems. However, other experiences (e.g. [12]) could be directly implemented.

c)   Some of the action plans do not specify responsibilities and resources. and their financing source. Full use of advanced project planning and management methods and instruments can certainly be useful, also contributing to synergy.

d)   A general problem in societal management is the selection of the governance methods. Apart identifying invariant models to rely on, a rational choice is necessary between open- circuits commands (e.g. Ordinances)-or "hard governance (HG)" and regulations which create / enforce virtuous societal closed circuits, to automatically generate most of the actions provided otherwise in detailed action plans or "soft governance (SG)". This question is closer examined below.

A study [13] suggested the still great importance of the HG and also the need of closed informational circuits in the judicial field. A preliminary research suggested the existence of some tenths of societal closed circuits, some of them being switchable vicious / virtuous ones. [6]. One example, concerning the Legislative Authority follows:

(i) Examples of possible actions of the state are: a) precise delegation of the authority between the Parliament, the Government and the other state institutions able to generate regulations; b) minimization of modifications of the laws and other regulations, e.g. via the creation of instruments for simulating the likely consequences of their drafts and the standard structuring of their content; c) adoption of an efficient electoral law and friendly electoral interface; d) informatisation of the legislative processes, providing also transparency; e) public free access legislative database.

(ii) Example of a virtuous closed circuit: performant electoral law & implementation of the above exemplified actions of the state => elections => best choice of persons => improved legislation and better people in state managerial political positions =>good results => new better alternatives => elections.

(iii) Example of a vicious closed circuit: poor technological support => incomplete or incoherent legislation => possibilities to by-pass the law => growth of the underground economy => encouragement of incomplete or incoherent legislation.

4   Conclusions

Following conclusions are proposed: a) e-GMR derived from e-GBR suggest the importance and feasibility of creating synergy; b) for transitional SESs e-G presents a high economic potential; c) a rational governance mix must be found between HG and SG, with the accent on the latter.

References

1. EU Ministerial Conference on e-Government November 2001 http:// europa. eu. int / iin-foorma- tion_society
2. Costake, N., Jensen, F., H..:Towards an Architectural Framework of e-Government information systems   Paper prepared for KMGov 2002 Copenhagen, 2002
3. Lenk, K., Traunmuller , R.: Perspectives on Electronic Government IFIP WG8.5 Information Systems in Public Administration. Working Conference on Advances in Electronic Government Zaragoza, 2000
4. Traunmuller, R, Wimmer, M: Daten- Informtion – Wissen – Handeln: Management des Wissens.   *In: Reinermann, H (Hrsg): Regieren und Verwalten im Informationsalter v.Decker Heidelberg, 2000 482-498*
5. Lockenhoff, H.: Simulation for Decision Support in Societal Systems: Modelling for Guided Change to meet Complexity and the Future *In: Hofer, C., Chroust, G. (eds): Proceedings of the IDIMT- 2001 Interdisciplinary Information Management Talks   Trauner Linz, 2001*
6. Costake, N., Dragomirescu, H., Zahan, E: E- Governance: a Mandatory Reengineering for the Transforming Economies *In: Wimmer, M. (Ed): Proceedings KMGov Trauner Linz, 2001* 30-38
7. European Commission : 2001 Regular report on Romania's accession   SEC (2001) 1753 Brussels, 2001
8. Romanian Government's Plan of Actions 2002-2004 (in Romanian) http:// www. gov. ro
9. Plan of Action for the Social Democratic Party Governance in 2002 and 2003 (in Romanian) http:// www. gov. ro
10. National Strategy for the Informatization of the Public Administration (in Romanian) http:// www. gov. org
11. Presidenza del Consiglio dei Ministri, Dipartimento Funzione Pubblica: E- Government Action Plan 22 june 2000:// www. funzionepubblica. it/ download/ action plan. pdf (2000)
12. IDA Architectural Guidelines Technical Handbook Version 5.3 http:// www. ispo. cec. be (link to IDA)
13. Costake, N.:E- Governance and the Judicial System. A point of view. In:Tjoa, A., M., Wagner, R., R.: 12th International Workshop on database and Expert Systems Applications IEEE Computer Society Los Alamitos, 2001

Towards a Trustful and Flexible Environment for Secure Communications with Public Administrations

J. Lopez, A. Maña, J. Montenegro, J. Ortega, and J. Troya

Computer Science Department, E.T.S. Ingenieria Informatica
Universidad de Malaga, 29071 Malaga, Spain
{jlm,amg,monte,juanjose,troya}@lcc.uma.es

Abstract. Interaction of citizens and private organizations with Public Administrations can produce meaningful benefits in the accessibility, efficiency and availability of documents, regardless of time, location and quantity. Although there are some experiences in the field of e-government there are still some technological and legal difficulties that avoid a higher rate of communications with Public Administrations through Internet, not only from citizens, but also from private companies. We have studied two of the technological problems, the need to work in a trustful environment and the creation of tools to manage electronic versions of the paper-based forms.

Keywords: Public Administrations, Secure Communications, Electronic Forms, Certification Authorities, Public Key Infrastructure

1 Introduction

Approaches to electronic versions of many of the paper-based administrative procedures between Public Administrations and citizens can bring meaningful benefits. These benefits concern accessibility and availability of documents, regardless of time, location and quantity.

Although there are some experiences in the field of e-government there are still some technological and legal difficulties that avoid a higher rate of communications with Public Administrations through Internet, not only from citizens, but also from private companies. Any type of digital transaction is influenced by typical open networks risks. Agents involved (public organizations, private companies and citizens) need to work in a trustful environment. This environment must satisfy the required security levels in such a way that privacy and authentication of digital information is guaranteed to senders and receivers OGIT96 [4]. Also there is a lack of software tools that help to create, distribute and manage in an easy and flexible way the electronic versions of paper-based forms, which is the usual way of interaction with Public Administrations. Clearly, these tools must incorporate authenticity and integrity mechanisms that mimic those ones existing in the traditional paper-based documents [5].

R. Traunmüller and K. Lenk (Eds.): EGOV 2002, LNCS 2456, pp. 211–214, 2002.

In this paper we present the results of a research project that has focussed on the problems we have mentioned. We also show how the integration of the approaches produce a solution that enhance many of actual developments. Thus, the structure of the paper shows the two main works done in the project. Section 2 presents the design and development of a real hierarchical Public Key Infrastructure (PKI), which we consider the most convenient type of infrastructure for operation of any administrative procedure that involve a digital signature. Section 3 presents the design of a language for the description of electronic forms that allows the utilization of signed forms in all communications with Public Administrations. The paper finishes with conclusions in Section 4.

2   Development of a PKI Based on New Design Goals

Digital signatures schemes are based on the use of public-key cryptosystems [3]. The reasons are that these schemes offer the same functionality than handwritten signatures, and also a high protection against fraud. However, the global use of any of those cryptosystems needs a reliable and efficient mean to manage and distribute public keys, by using digital certificates. Such functionality is provided by a Public Key Infrastructure, which is formed by a diversity of Certification Authorities. A PKI becomes essential because without its use public key cryptography is marginally more useful than traditional symmetric one [2].

Although addition of certification capabilities in commercial electronic mail programs is a very helpful feature, a detailed analysis shows that these schemes result not satisfactory for e-government applications. Some design features that may compromise the security of the systems have been detected. We summarize some of the most important ones:

– It is common in most of Public Administrators that users share the same computer system. Therefore, private keys belonging to different users are not completely "isolated". This drawback does not allow the appropriate use of a very important security service for e-government applications, the non-repudiation service [7].
– Certificates needed for a verification of documents that have been digitally signed must be obtained from sources that are external to the electronic mail programs. Therefore, it is very possible that users do not verify them properly (as they are not forced to). Moreover, use of Certificate Revocation Lists (CRLs) is constrained.

These considerations has taken us to develop our own PKI [11], that has the following features:

– Adapted to the multi-hierarchical Internet structure because this is the operational environment.
– Provides secure means to identify users and distribute their public keys.
– Uses a CAs architecture that satisfies the needs of near-certification so the trust can be based on whatever criteria is used in real life.
– Eliminates problems of revocation procedures, particularly those associated with the use of Certificate Revocation Lists.

The main element in the hierarchy is the Keys Service Unit (KSU), which integrates certification and management functions. We use a scheme with various KSUs operating over disjoint groups of users, conforming a predefined hierarchy.

KSU hierarchy is parallel to the hierarchy of Internet domains. KSUs are associated to the corresponding e-mail offices. Every KSU is managed by a CA.

Additionally, it contains a portion of the certificates database to store the certified keys of its users. The third component is the key server, which receives requests and delivers the certificates. The key server manages a certificate proxy that keeps some of the recently received external certificates. The certified keys are managed solely by the corresponding CA; therefore, key updating and revocation are local operations that do not affect the rest of the system.

3 Description Language for Eforms: A New Design

Structured forms has been the traditional method of interaction with Public Administrations [6]. Moreover, the use of hand-written signatures in this type of documents has provided the necessary legal bindings for most of scenarios. Our previous study of common e-government applications has showed us that if paper-based forms have to be substituted by electronic forms, then these ones must have the following characteristics: integrity, or non-modification by external entities; non-repudiation, or non-deniability of agreements; and auditability.

Taking these features as a staring point, we have tried to design an appropriate language for the description of forms.

These reasons recommended us to try to design and develop our own Formal Description Language. Its name is FDL, and XML-like. To be more precise, it is based on XFDL [1].

The use of our own specific language, with its own tools, and completely adapted to XML [8,9], introduces many advantages in comparison with traditional use of HTML [10].

The most important advantages are briefly summarized next:

- *Regarding forms status*: It is easy to add useful components not included in standard HTML, and it provided automatic data validation without programming specific code. Also, the definition of the structure of the fields where signatures are contained simplify the (automatic) process of signature verification, and finally, the particular design of our language, together with the standard where it relies on, facilitate creation of parsers that automatically translate forms to any other language.
- *Regarding forms management:* The signer can store a copy of a partially filled document and one or several persons can sign these forms.
  FDL has two fundamentals concepts oppositely to HTML, the first notion is there are some extensions defined to distinguish different parts and formats in the same document, and the second one is the status of the form is preserved, thus our solution has been designed to organize any form in several pages while having data in memory continuously.

– *Regarding communication:* A proprietary format facilitates that the context of the signature is not lost.Moreover, the document is audited on its own. Oppositely to HTML, FDL provides a data structure and separates application, presentation and logic levels.

4   Conclusions

In this paper we have presented the results of a research project that studies the need of using security for communications over open networks, and the use of electronic versions of the paper-based forms to interact with Public Administrations.

We have shown the main features of the PKI specifically developed and the reasons for its design. Regarding the electronic forms we have designed a language for their description.

Modular design and development of those tools facilitates that the outcome of the work is integrated into e-government broader systems, and can be immediately applied to the social environment. These new solutions also help in establishing the basis for future design and development of schemes oriented to electronic forms signature in the communications with Public Administrations.

References

1. B. Blair, J. Boyer, "XFDL: Creating Electronic Commerce Transaction Records using XML", Computer Networks, n. 31, pp. 1611-1622, 1999.
2. W. Burr, "Public Key Infrastructure Technical Specifications (version 2.3). Part C: Concepts of Operations", Public Key Infrastructure Working Group, National Institute of Standards and Technology, November 1996.
3. W. Diffie, M. Hellman, "New Directions in Cryptography", IEEE Transactions on Information Theory. IT-22, n. 6. 1976, pp. 644-654.
4. European Commission, "Directive 1999/93 of the European Parliament and the Council on a Community Framework for Electronic Signatures", December 1999.
5. European Commission, "Directive on Certain Legal Aspects of Electronic Commerce in the Internal Market", February 2000.
6. "Improving Electronic Document Management", Guidelines For Australian Government Agencies, Australian Office of Government Information Technology, 1996.
7. "Non-Repudiation in Electronic Commerce", Artech House, 2001.
8. Extensible Markup Language (XML)" http://www.w3.org/XML
9. Canonical XML, Version 1.0, W3C Working Draft, September 2000. http://www.w3.org/TR/2000/WD-xml-c14n-20000907
10. HTML 4.01 Specification, W3C Recommendation, December 1999. http://www.w3.org/TR/html4/
11. J. Lopez, A. Mana, J. Ortega, J. Troya "Distributed Storage and Revocation in Digital Certificate Databases", Dexa 2000.

Supporting Efficient Multinational Disaster Response through a Web-Based System

Ignacio Aedo[1], Paloma Díaz[1], Camino Fernández[1], and Jorge de Castro[2]

[1] Laboratorio DEI. Departamento de Informática.Universidad Carlos III de Madrid
Avda de la Universidad 30, 28911 Leganés (Spain)
aedo@ia.uc3m.es, {pdp,camino}@inf.uc3m.es
[2] Dirección General de Protección Civil. Ministerio del Interior
Quintiliano 21, 28002 Madrid (Spain)
jcastro@procivil.mir.es

Abstract. The current process to deal with disaster mitigation has a number of drawbacks that can be solved using web technology. The basic problem is that there is a unidirectional and asynchronous flow of information among the different agents involved in a disaster mitigation procedure. This situation often results in a lack of coordination in the resources provision and in a useless assistance. In this paper we introduce ARCE, a web based system envisaged to cope with the lack of synchronism among assistance requests and responses in a multinational environment as the Latin-American Association of Governmental Organisms of Civil Defence and Protection is. ARCE makes uses of role-based access policies (RBAC) and information flow mechanisms to offer an efficient and reliable communication channel.

1 Introduction

In the last years, Internet is emerging as an essential platform to support cooperative tasks, as demonstrated in projects like the parametric earthquake catalogue [1], Arakne [2] or WebSplitter [3], particularly useful when the users of a system, who can be geographically scattered, do not access it in a synchronous way. This is the case of multinational disaster response, a process which involves different countries and organisms giving assistance to mitigate an emergency situation affecting to one or more countries. The use of a web system to coordinate and synchronize the efforts of each assistance supplier would make the emergency management quite more efficient avoiding duplicates and misunderstandings and providing reliable information on the actual needs. One of the main concerns of the Latin-American Association of Governmental Organisms of Civil Defence and Protection[1], involving 21 Latin-American countries, is to promote cooperation and mutual assistance in emergency situations. Nowadays, when one or more countries of this association are affected by a disaster they ask for cooperation to other countries or organisms in an unilateral way. In turn, each requested country or organism supplies assistance according to its possibilities but without taking into account how the others are contributing. As a consequence, an affected country can receive lots of perishable food, even more that what can be con-

---

[1] http://www.proteccioncivil.org/asociacion/aigo0.htm

R. Traunmüller and K. Lenk (Eds.): EGOV 2002, LNCS 2456, pp. 215–222, 2002.
© Springer-Verlag Berlin Heidelberg 2002

sumed by the victims, while the need for clothing is not been addressed by anyone. Moreover, most communication among countries is made through obsolete means like the phone, fax or telex that introduce delays and problems of flexibility, accessibility and reliability that can be easily overcome using the web. In this context, the Association in its IV Conference held in Azores in September 2000, committed to design a mechanism to make easier and more efficient the cooperation and mutual assistance tasks in emergency situations. As a consequence, the Spanish Civil Protection Department along with the DEI research group at Carlos III University of Madrid started to work on the development of ARCE, a web based system oriented towards enhancing the management of multinational disaster response within the scope of the Latin-American Association. ARCE is aimed at becoming a platform to share updated and reliable information among the associates in order to orchestrate an integrated and efficient response, respecting at the same time the peculiarities and autonomies of each member. In order to deal with different access rights (such as the ability to create an emergency, to ask for resources, to offer assistance and so on), the basic principles of role based access control (RBAC) [4] have been adopted and, therefore, a hierarchy of kinds of users makes up the basis to define how the system will be accessed. Moreover, not only the role assumed by a specific user will determine her capabilities to modify or browse the information but also the country she belongs to, since ARCE is implemented in a multinational environment. Finally, roles are also the basis to establish an information flow policy is used to push valuable information to each ARCE user.

2   Design Principles to Support Efficient Disaster Response

As mentioned before, the current process to ask for multinational assistance to mitigate the effects of a specific disaster has a number of drawbacks that can be solved thanks to the web technology. ARCE is a web application that provides mechanisms to notify an emergency, to ask for resources and to offer assistance. In a few words, when an emergency happens the country or countries affected, that we will call throughout this paper the emergency owner, can use the system to keep informed the other associates, called the assistance suppliers. If assistance is required, the emergency owner can ask for resources which are classified according to a multilingual glossary included in the system. Whenever an assistance request is received in ARCE the rest of the associates are notified by e-mail, so that they can access the application to see what the emergency owner is asking for and how they can help. For each emergency, the application will provide updated information about which resources are requested, which quantity was originally needed and how many items have been already supplied. Thus each associate can decide what to contribute taking into account what the others are providing. In order to develop a useful application a number of requirements were taken into account, including: accessibility; multi-user support; inter-state support; multi-purpose application; reliability; efficiency and maintainability.

Accessibility. The application has to be accessed anytime and anywhere to improve the communication among associates. With this purpose ARCE has been designed as a web system which can be used with any Internet browser (e.g. Internet Explorer, Netscape and Konqueror) and e-mail client.

Multi-user support. ARCE is accessed by different users with different responsibilities. This implies that each user has to be allowed to perform or not an operation (e.g. create an emergency, offer assistance, accept assistance) according to the role she plays in a specific organism. To deal with this issue ARCE assumes a role-based access policy (RBAC) described in [5]. Users are classified into roles identified in the Civil Defence and Protection departments as well as those of the invited organizations (e.g. EU, UN). Roles are structured in a hierarchy, so that general roles can be specialized into more concrete roles which inherit the access rules of their parents. Moreover roles can be aggregated into work teams. In this case, access rules are not propagated from the parent to the children. Figure 1 shows the roles hierarchy supported in the current version of ARCE. In the figure, aggregations denote the composition of teams, whereas generalizations represent the specializations of a role.

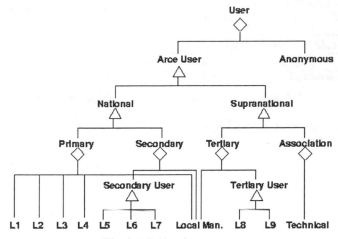

Fig. 1. ARCE roles structure

Each user of the application will be assigned one or more roles. This role along with the country or organism the user belongs to determines the information each user receives as well as the tasks she can perform on the system. For example, the Local Manager can add new users to her country structure while the Technical can perform maintenance operations such as create new Local Managers or new entities. In addition, different users have different views of the same information as shown in Figure 2 and Figure 3 where a request and assistance offer are shown respectively.

Inter-state support. ARCE has to provide each associate a certain degree of autonomy and control. For example, it is allowed to invite organisms or countries that do not belong to the Association to take part in a disaster mitigation. The access policy is then carefully designed in order to support enough roles as to provide control to the different associates while maintaining reliability and efficiency. This diversity of countries also affects to accessibility in terms of multilingual support. In the current version the two languages spoken in the Latin American community (Spanish and Portuguese) are supported both in the user interface and in the glossary of terms. The glossary was introduced in order to unify terminology concerning emergencies with a view to improving communication in a multicultural community. An on-line multilingual description of the technical terms is also provided.

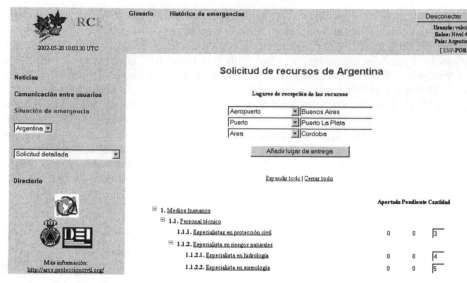

Fig. 2. Assistance from the point of view of an emergency owner

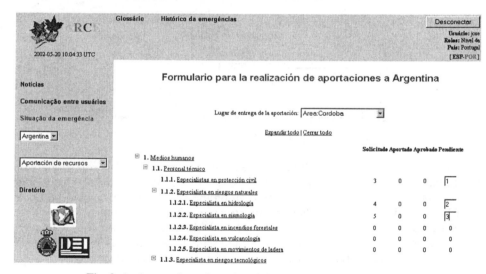

Fig. 3. Assistance from the point of view of an assistance supplier

Multi-purpose application. Even though the main goal of ARCE is to efficiently support emergency responses in a multinational context, it was considered that in order to improve its utility it was necessary to introduce the use of the application in other situations so that users can get used to it before an emergency occurs. Thus, ARCE has two operation modes: routine and emergency. In the routine mode, the application is used as a communication channel and to post news. There is a communication module that uses information flow policies to distribute messages among the different roles. These policies are aimed at ensuring that users only access the information for which they are authorized [6]. With this purpose, information is categorized into strategic, operational, technical, general and public. Information flows from one role to another according to the rules represented in Figure 4. Thus, when a user wants to send a message, first she has to decide which kind of information is going to send and then she will be able to select who is going to receive it from a list of potential targets. For instance, an L2 role will not be able to send strategic information although she could receive it from the upper level, and she can send Operational information to users holding an L2, L3 or L4 role.

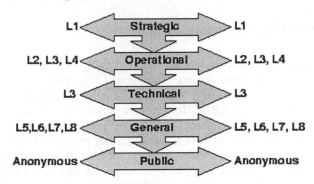

Fig. 4. Information flow policy in ARCE

In emergency mode, countries affected by a disaster can manage the emergency informing to the other associates, preparing a preliminary request for urgent resources, elaborating a more detailed request and coordinating the assistance offered by other countries. Since all the entities involved in an emergency, whether the owners or the assistance suppliers, have access to updated and reliable information about the real needs and how they are being solved, there are no problems of overlapping help. Indeed, before a supplier initiates the protocol to physically send any help, its assistance has to be approved by the emergency owner.

Reliability. This is a basic requirement in ARCE, since users have to trust the information and services offered by an application supporting a critical task. Indeed, users need to sure that information is updated, precise and accurate. All pages include the time of the last update in the system, so that users can have an idea of their "freshness". To avoid improper modification of data ARCE relies upon the use RBAC policies and authentication mechanisms, so that only authorized users can modify the information provided by the system. Moreover, the information flow policy ensures that messages received from the system are trustworthy, as far as only authorized users can send messages to the users who require or need that information.

Efficiency and maintainability. Since ARCE will be used in critical situations users have to receive enough information to plan their response and have mechanisms to provide an efficient response. All associates, through the appropriate role, will be able to create emergencies, ask for help and coordinate the multinational response to avoid redundancies. Moreover, they can invite external organisms or countries to take part in the mitigation of a specific disaster. They will also be able to access to the information concerning to a disaster, for which historical archives are maintained, and offer an assistance that is validated by the emergency owner. In order to increase the system usability, the user interface has been designed applying HCI principles paying special attention to usability and consistency issues [7]. Moreover, in order to improve this collaborative environment an iterative and user-centered development methodology has been adopted as suggested in [1]. Therefore, there is continuous evaluation process involving user representatives in order to empirically test the utility of the application.

3   ARCE Implementation and Evaluation

ARCE architecture relies upon an Apache server running over a number of modules developed on Zope, a platform to generate dynamic HTML pages. A number of modules have been built on Zope to deal with ARCE functions including: emergencies, messages, news, directory and RBAC. Information is held on a PostgreSQL relational database.

In order to assess the utility of ARCE, an empirical evaluation was carried out in the last meeting of the Latin-American Association held in Cartagena de Indias (Colombia) in February 2002. In this meeting representatives of 13 countries took part in a disaster simulation exercise and they used ARCE to coordinate a multinational response. In order to collect information on the system utility and usability, evaluators were asked to fill a questionnaire which was organized in three sections.

The first one ("Personal Information") was oriented towards gathering information on the evaluator command on technical aspects required to use ARCE, such us web browsers and e-mail clients.

The second one ("ARCE utility") gathered information concerning performance and usability, such as utility, speed, appropriateness of the roles structure implemented, reliability, legibility, quality of the user interface, and their degree of satisfaction after having used ARCE. They were asked to rate these features using a Likert-scale with five values, ranging from very good to really bad.

The last one ("ARCE operation") included questions about the different tasks they had performed during the exercise. In particular, evaluators expressed their opinion concerning five operation scenarios:

1. Creating a new emergency, that implies to create a preliminary report to inform the associates, to create a preliminary ask for assistance and to create a detailed ask for assistance. This task is performed by an emergency owner using ARCE in emergency mode.
2. Managing an emergency, that consists of modifying the ask for assistance; managing the proposals of assistance (accepting, denying or modifying them); accessing to the ask for assistance status and to the historical archive of an emergency. This task is performed by an emergency owner using ARCE in emergency mode.
3. Contributing to the disaster mitigation, that implies accessing to the ask for assistance (resources required and supplied), proposing a specific assistance and modifying a proposal of assistance. This task is performed by an assistance provider using ARCE in emergency mode.
4. Using the directory, that consists of accessing and modifying it. This task is performed by any user using ARCE in routine or emergency mode.

5. Using the communication through messages. This task is performed by any user using ARCE in routine or emergency mode. For each scenario, evaluators assessed the difficulty, utility and correctness of the different tasks they had to do to achieve their goals. They could also propose modifications to improve the system.

Evaluation results showed that users considered ARCE a quite useful tool. Thus, as it can be seen in Figure 5, most users were satisfied (80\%) or quite satisfied (20\%) with the application and all of them considered it as a useful tool to coordinate a multinational response. The most important conclusion of this experiment was that ARCE gave place to a positive attitude from the different organisms represented in the Association and reinforced their commitment to make use of web technology to orchestrate their efforts in an efficient way. Moreover, this simulation exercise was considered as a key activity in the iterative design of ARCE, since it has provided quite useful information to improve the application and, particularly, to take into account the specific needs of each entity of the Association.

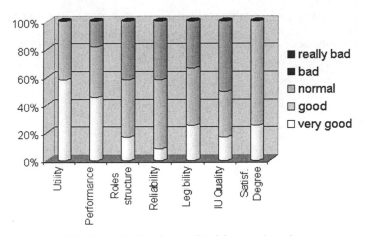

Fig. 5. Results for the part B of the questionnaire

4   Conclusions

In this paper, we have introduced ARCE a system using the web and technologies such as RBAC or information flow control policies in order to make more efficient the multinational response to a specific disaster. ARCE tears down distance and time barriers to make easier the management and coordination of mutual assistance in a multinational community as it is the Latin-American Association of Governmental Organisms of Civil Defence and Protection.

Acknowledgements. This work is supported by an agreement between Universidad Carlos III and "Dirección General de Protección Civil del Ministerio del Interior". RBAC module is based on the MARAH project funded by the "Dirección General de Investigación de la Comunidad Autónoma de Madrid y FSE" (07T/0012/2001). We'd like to thank Juan García and Miguel Àngel Hernández for their cooperation in the implementation of ARCE.

References

1. Padula, M, and Rinaldi, G.R. Mission-critical web applications. Interactions July 1999 Volume 6 Issue 4, 52 - 66
2. Bouvin , N.O., Designing user interfaces for collaborative web-based open hypermedia Proceedings of the eleventh ACM on Hypertext and hypermedia May 2000, 230 - 231
3. Han, R. Perret , V. and Naghshineh, M.. WebSplitter: a unified XML framework for multi-device collaborative Web browsing. Proceeding of the ACM 2000 Conference on Computer supported cooperative work. December 2000, 221 - 230
4. Ferraiolo, D.F., Barkley, J.F. and Kuhn, D.R.: A Role-Based Access Control Model and Reference Implementation within a Corporate Intranet. ACM Trans. on Information and Systems Security, 2(1), February (1999), 34-64.
5. Díaz, P., Aedo, I. and Panetsos, F.: Definition of integrity policies for web-based applications. Integrity and Internal Control in Information Systems: Strategic Views on the Need for Control. Ed.s. van Biene-Hershey, M.E. and Strous L. Kluwer Academic Publishers, USA 2000. 85-98.
6. Denning, D.E.. A lattice model of secure information flow. Comm. of the ACM, 19(5):236-243, 1976.
7. Rubin, J. Handbook of usability testing. New York: John Wiley & Sons. 1994.

KIWI: Building Innovative Knowledge Management Infrastructure within European Public Administration

Lara Gadda[1], Emilio Bugli Innocenti[2], and Alberto Savoldelli[1]

[1] Politecnico di Milano, 20133 Milano, Piazza Leonardo da Vinci 32, Italy
Tel. +39 02 23992796; Fax. +39 02 23992720;
lara.gadda@polimi.it
[2] Netxcalibur, 50123 Firenze, via Alamanni 25, Italy
Tel. +39 055 285859; Fax. +39 055 285760;
ebi@acm.org

Abstract. The paper is composed by two parts. The objective of the first part is to define a new approach to the innovation process in the Public Administration. In fact, in more recent years, the attention to process oriented change management techniques has also emerged in the public sector, through attempts to draw from private sector, searching for new methodologies and managerial approaches that can satisfy the need of organisational innovation. KIWI project analyses these techniques in one of the public management processes, the Knowledge Management. The second part of the paper is more related to the new IST tools that should be used to improve the efficiency and the effectiveness of non-profit organisations. In particular, the project aims at developing innovative, user-relevant, wireless technologies which make the relationship between PA and citizens easier.

1 Introduction

In the last few years, industrialised Countries faced up the necessity of reforming Public Administration, a crucial problem since that context is quickly evolving. The change in progress is moving along two directions: on one hand, the users require a Public Sector's "product" risen in value and, on the other hand, there's the need to provide better services using the same resources (more trend towards efficiency) [1].

Public Sector reform started with the adoption of a new set of rules. The progressive shift from a slow and inefficient bureaucracy toward a lean and dynamic organisation is one of the topic joining a number of interventions of the last decade, within the public sector administration all over the world. Though this deep change has been faced in each country in different ways [2], the fundamental characteristic is the presence, finally contemporary, of two separate trends: *the decentralisation and the modernisation*.

The mere political rules' transformation is not enough and it is necessary to develop a specific method in order to enhance organisations' performance, efficacy and efficiency [3]. The technological innovation and web oriented technology are the necessary starting point for improving PA performance. Therefore, there is the necessity to use a "change management" which should combine with information technology, change of organisation and human resources management [4].

R. Traunmüller and K. Lenk (Eds.): EGOV 2002, LNCS 2456, pp. 223–229, 2002.
© Springer-Verlag Berlin Heidelberg 2002

2   The Innovation Process in the Public Sector:
The e-Government

The change management is guided by four factors: the crisis of bureaucratic model, the monitoring of public expenditures, the pushes coming from the dissatisfaction of the citizen-user and the European integration. The advantages lead to a greater flexibility in activities management, to the possibility to eliminate the no added value activities and to an higher knowledge of the managed activity [5].

The main aim of PA is to be able to answer to the needs of citizens and enterprises. In the previous time the citizen was the receiver of services and public activities, now he is the client of a modern and efficient system which offers certain services [6].

In this context KIWI project is aimed at developing innovative knowledge management (KM) infrastructures able to transform *PA* at any level inside Europe into knowledge driven and dynamically *adaptive learning organisations* and empower *public employees* to be *fully knowledge workers*.

The expected results from KIWI project are:

- *Three re-designed Knowledge Management processes* related to large public administrations in Italy, Finland and France;
- the *KIWI Toolkit* designed to facilitate the implementation of Knowledge Management in large, multi-site, public administrations, based on the *Knowledge Warehousing tools* and the *Mobile Collaboration tools*.

Therefore, the project outcomes are to make a decision making faster and better informed, to improve the customer services, to increase the returns on investment as productivity, to save time and staff resources, to push to the innovation which is further stimulated by capitalising on knowledge and expertise, to reduce the costs.

3   The Key Process: The Knowledge Management

In order to be effective, the innovative process has to be funded on the existence, the development and the integration of the following four leavers related with the innovation key: the *policy*, the *culture*, the *organisation* and the *technology*. These four elements are so linked among them, that it's not possible to act on one without acting on all the others otherwise the effectiveness of the innovative actions is reduced. Along with the leavers, it's possible to identify four engines: *benchmarking*, *project management*, *total quality management* and *digital signature*. Combining the leavers with the engines, it's possible to identify the three main process which characterise the PA innovation: the *Change Management* (CM), the *Knowledge Management* (KM) and the *Citizen Relationship Management* (CRM) [8]. KIWI project is focused on the Knowledge Management (KM).

The knowledge is a combination of experience, values, information and specific competence which provides a framework for the assimilation of a new experience and new information. The knowledge comes from the information and, at the same time, the information comes from the data. So, the knowledge could be obtained by information through the instruments of the comparison, the consequences, the connections and the conversation [10]. But the knowledge is not an information enriched with content, it implies a judgement, involves the values, the emotions and the people perceptions, which generate a relevant impact on the available knowledge in the organisations [11]. The knowledge has to be correlated to the action because it is always a knowledge towards an "objective".

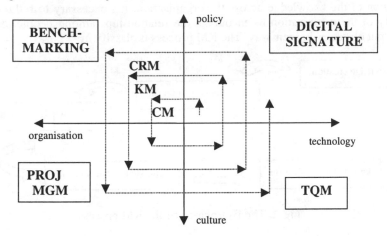

Fig. 1. The innovation sources for implementing e-government processes in public organisations

The knowledge could be classify in three main macro-typologies [12]:

- *declarative knowledge* (knowledge "about" something);
- *procedural knowledge* (knowledge of "how" something occurs);
- *causal knowledge* (knowledge of "why" something occurs).

There's a more general classification of the knowledge. The *explicit knowledge* is codified, expressed by formal and linguistic modalities, easily transmissible and conservable which can be expressed through words and algorithms. For Nonaka and Takeuchi (1995) this kind of knowledge is formalised, easily communicated and shared. The concept of *tacit knowledge* was introduced by Polanyi, who evidences the importance of a "personal" method to build the knowledge, influenced by the emotions and obtained at the end of a process of active creation and of organisation of the individual experiences [13]. It is difficult to define in a formalised way, it is linked to the reference context and it is personal. To be able to spread the tacit knowledge inside the organisation, it needs to convert it in words and numbers which could be understood by everybody. During this conversion, from tacit to explicit, the organisation knowledge is created.

Each organisation has a unique asset of knowledge and its internal problems. Each action of *knowledge management* has characteristics which are specific of the body for which it has been projected. The organisations could assure KM procedures oriented to the results and the strategic needs of the context where they operate, focusing the attention on the planning and the carrying out of the following areas:

- *process*: it assure the *KM* is aligned with the specific managerial processes.
- *organisational dynamics*: they over the barriers which obstacle the sharing of the knowledge and promote the innovation behaviour.
- *technology*: it allow people to share the activities using known instruments

The knowledge has also to be created. The mechanism of the knowledge creation is the condition and the "engine" of the innovation, in the two dimensions of the business innovation and the social innovation. The key point is the mobilisation and the migration of the knowledge across the organisation: it's necessary to rethink the capability of the organisation to encourage the relationships among people, or at least not to put obstacles in their way. The KM process is classify in:

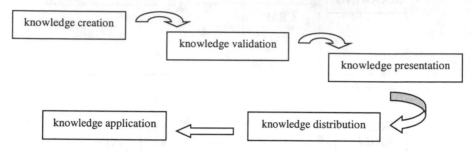

Fig. 2. The five phases of the KM process

The main aim of the organisation is to transfer the knowledge from tacit to explicit, from individual to collective, from collective to organisational:

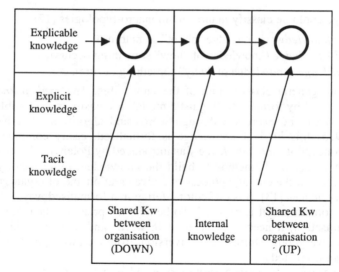

Fig. 3. The mapping of knowledge

4   The System Architecture

The KIWI platform will be structured around the following items:

- A web-based *Intranet Knowledge Warehousing* Toolset that will allow to build a wirelessly accessible knowledge warehouse. The knowledge warehousing will amass *internal* and *external knowledge*;

- *A Mobile Collaborative Environment* (built on Web-based Groupware tools), to support a realistic collaboration and knowledge sharing and transferring also among geographically distributed workforces, within and between public administrations. It will represent the convergence of technologies such as multimedia document/image management, videoconferencing, and mobile 3G technologies helping public administrations transcend all sorts of boundaries (geography, time and organisational structure) by *making available the right information to the right employee at the right place and at the right time.*

4.1   The Components

The KIWI platform is based on a *Intranet Knowledge Warehousing* Toolset. These tools will allow to build a wirelessly accessible knowledge warehouse (knowledge resources will includes manuals, letters, responses from citizens/companies, news, technological, organisational, legal and other relevant information from administrations, as well as knowledge derived from work processes) applications that support inter-organisational learning process.

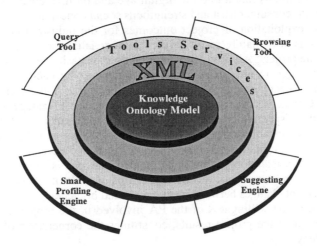

Fig. 4. The KIWI Knowledge Warehouse

The second Tool which composes the KIWI platform is the *KIWI Mobile Collaborative Environment*. The Groupware can help Public Administrations transcend all sorts of boundaries. Geography, time and organisational structure fade in importance in the groupware-enabled process. At the same time, mobile technologies are increasingly penetrating businesses, offering anytime/anyplace access to enterprise information. The novelty of the KIWI approach lies in the convergence of these two technologies.

The challenge of Mobile Groupware Tool development within KIWI is twofold:

- firstly, to allow for web-based, mobile groupware by supporting both *synchronous* (ie, real-time interaction such videoconferencing, chatting, electronic whiteboard, etc.) and *asynchronous* (ie, email, group discussions, etc.) communication, including live video/audio/text communications;

- secondly, to take full advantage of the Knowledge Warehousing tools, by integrating mobile groupware facilities within the KIWI Toolset. *This will lead to set up of a full Mobile Collaborative Environment.*

The project will look at the recent advancements in the mobile groupware standardisation such as the GroupDX (www.groupdx.org) initiative (linked to the WAP Forum), proposing an industrial grade XML Document Type Definitions (DTDs) and Object Schemata for Internet groupware applications. This new standard, called Groupware Mark-up Language (GML), is being established in order to facilitate data exchange among the various groupware applications, and facilitate data synchronisation among groupware applications and their individual counterparts.

5   Conclusions

The KM will result into a decisive improvement in inserting an information database. This allows the public employees to access easier to the needed data, independently from the place where they are. The aim is "to bring every citizens, home and school, every business and administration into the digital age and on line, creating a digitally literate Europe, build consumer trust and strengthens social cohesion".

Concerning the exploitation of project outcomes for the industrial component of the Consortium, this will mainly result in the commercialisation of the prototypes produced within the project. All prototypes will be used as basic elements to develop and produce marketable results: as in-house developments by each partner and in collaboration with project partners. A quick process of research transfer in production will assure to the Consortium partners an essential competitive advantage for a further consolidation of the respective positions on the market. The specific techniques implemented in the project will be used by most of the Partners to enhance the techniques already in use, contributing to consolidate a competitive advantage.

Once the innovative technologies are implemented, the idea is to realise a mobile groupware which let in-house functions be used at distance. It involves an organisation and a management changes between headquarter and branches.

Within this structure, the willingness of the PA involved in the project to provide a common exploitation of the project results, constitutes the cornerstone of the KIWI Exploitation Strategy.

References

1. AIPA, "La reingegnerizzazione dei processi nella pubblica amministrazione", (1999).
2. Klages, H., Loffler, E., "Administrative modernisation in Germany – a big qualitative jump in small steps", International Review of Administrative Sciences, pp. 373-384, (1995).
3. Hammer, M., Champy, J., "Reengineering The Corporation: A Manifesto for Business Revolution", HarperCollins, New York, (1993).
4. Osborne, D., Gaebler, T., "Reinventing Government. How the entrepreneurial spirit is transforming the public sector", New York, Plume, (1993).

5. D'Albergo, E., "Le sfide delle amministrazioni pubbliche al Change Management: una prospettiva socio-istituzionale", Rivista trimestrale di scienza dell'Amministrazione, n.1, (1998).
6. Halachmi, A., "Re-engineering and public management: some issues and conditions", International Review of Administrative Sciences, vol. 61, (1995), pp. 329-341.
7. Kettinger, W.J., Teng, J.T.C., Guha, S., "Business process change: A Study of Methodologies, Techniques, and Tools", MIS Quarterly, March, (1997).
8. Sbrana M., Torre T., "Conoscenza e gestione del capitale umano: la learning organization", FrancoAngeli, (1996).
9. Davenport T. H., Prusak L., "Working knowledge: how organizations manage what they know", Harvard Business School Press, (1998).
10. Takeuchi H., "Beyond Knowledge Management: lessons from Japan", Giugno, (1998), dal sito internet www.sveiby.com.au/LessonsJapan_it.html.
11. Zack M., "Managing Codified Knowledge", Sloan Management Review, Summer, (1999).
12. Polanyi M., "The tacit dimension", Routledge and Keagan, (1966).

Elektronische Steuer Erlass Dokumentation: A Documentation on Official Tax Guide Lines

Viktorija Kocman, Angela Stöger-Frank, and Simone Ulreich

Bundesrechenzentrum GmbH, Department A-VA-DX,
Hintere Zollamtsstr. 4, A-1030 Wien
{viktorija.kocman,angela.stoeger-frank,
simone.ulreich}@brz.gv.at

Abstract. E-Government can be defined as "carrying out government business transactions electronically". One aspect of e-government are online law documentation systems. In the Austrian Federal Dataprocessing Center (Bundesrechenzentrum), we have two online law documentation applications under development and support: Electronic custom-law documentation system (EZD) and electronic tax-law documentation system (ESED). EZD has been running since 1995, ESED is currently in a prototype state. The customer of both systems is the IT-section of the ministry of finance. These two applications have been developed especially for the use in the administration and are available for the staff in the Austrian government intranet. In the concrete paper, we will discuss technical and organisational aspects of the online tax-law documentation system ESED which in the first step will be used in the tax department of the ministry. The ministry of finance is responsible for ESED project management. Bundesrechenzentrum is responsible for the technical realisation of ESED.

1 History

The Federal Ministry of Finance, Austria's highest administrative office in tax affairs, is authorized in executing tax legislation. The applying tax rules are decided by the ministry. Further more it is necessary for all subordinated offices to know all those legal decisions. But various forms of information transfer are existing: paper, telefax and e-mail. So, no standardizised form exists to ensure that the information flow is done correctly and just in time. Another important fact of handling information is how to store them. Each office has to handle that by itself. Some of them collect the tax rules as paper sheets, others catalogue them or run an insufficiant electronic database in the office.

2 Declared Aims

Under these circumstances it was ordered to improve the information transfer to standardize the running out and, in addition, create a central archive for searching. Less paper, less time and certainly lower costs were demanded to be used as before. The usability of the new system should be extremly comfortable and trained easily. Additonally, knowledge management should be established within the complete finance administration throughout the whole country.

R. Traunmüller and K. Lenk (Eds.): EGOV 2002, LNCS 2456, pp. 230–233, 2002.

A working group was installed constisting of members of the ministry and tax officers of the subordinated offices. Just in case, BRZ supplied technical support. As a result a paper of demands was provided.

BRZ started the technical realization.

3   Technical and Organisational Aspects of ESED

Basically, there are 2 ESED-user groups: document writers and document readers. These two user groups can overlap. ESED writers work with a sophisticated Winword client (enriched with ESED Macros and ESED ActiveX components). ESED readers can query the documents with the ESED web client.

In general, ESED is a document database that is currently being updated with enactments. The amendments of enactments are also updated in the database, in order to acquire the relevance of the information.

ESED consists of a central database which stores the documents. We use a document database TRIP, which differs from relational databases. The emphasis in document databases lies on accurate (fragment) letter indexing to enable fast text search. The disadvantage of TRIP is that it cannot map relations that well as relational databases can.

ESED documents are organised in document collections which are referenced by a unique ID (Stammnummer). The documents are further divided into segments in order to give long text a structure. On each of these three levels (document collections, documents and segments) the ESED writer categorizes the document.

One of the used standards in the austrian administration is Winword. This was the guide value in designing the application. The input application had to be a Winword client. All of the ESED documents have to be written in the import ESED Winword client, and an easy import/export functionality has to be designed in order to raise the user acceptance and enable law experts to work on their private PCs outside of the ressort and the ministry of finance intranet.

The documents have a complex categorisation scheme. The categorisation scheme is specifical for the austrian tax law. The unique registration number (Geschäftszahl) is a unique ID for the file. Since ESED consists not only of tax enactments which have a registration number, but also of other document types like "announcements" which don't have a case number, we use another unique identifier for the database (Stammnummer). Another important attribute is the category "quota" (Norm), which is used to refer to citated statute. Each document can have more quotas. These quotes are later provided with hyperlinks which link to extern legal databases or to other ESED documents.

The TRIP database is located on UNIX, so there is a complex processing between the local PC and the database. After a document has been released for the ESED database, it is tranfered to a NT Server. NT Server has some conversion programmes running. The main assignement of the NT Server is to convert RTF documents to XML and to establish a communication to the unix database in order to fill the database with ESED text and categories.

This solution enables ESED documents to be processed immediately after the user has decided to release the document for the database. Seconds after the document release the text can be quieried with the web client.

4   Lessons Learned

E-government establishes special requirements on the information technology, especially when the application is designed to be used at the administration internally. There are more aspects to be considered as the mere technical concept.

Our experience has been that in order to gain acceptance by the administration staff, special emphasis has to be given to the *usability* of the application. It has to be considered that the administration staff is a special user group, less flexible in giving up their working habits than an "ordinary computer user". Much more, in the prototype stage it has to allow inaccuracy and individual user behaviour and find concept to cope with them. Once the user group got used to the process improvement enabled through the application and has lost their fears, it is also possible to require compromises from the users in order to win on processing efficiency.

The prototype stage of an e-government project designed for internal administration is crucial for the project success.

5   Prototype

Now a prototype already is running and it seems that the new system meets definetly the expectations: Each tax officer is registered by using ESED and gets therefore only the documents that are due for his field of work. In addition it is possible to search in the database in various ways. A catalogue of subject, law and catch phrases make it easy finding the right documents. Besides, a query mask with different searching fields offers a comfortable and efficient use. Novels are specially marked, long documents are splitted but could in case be joined together.

A lot of ESED-documents reference to other ESED-documents or to documents of another legal database (for instance RIS-Rechtsinformationssystem). That is why the documents are connected as hyperlinks to get the further information.

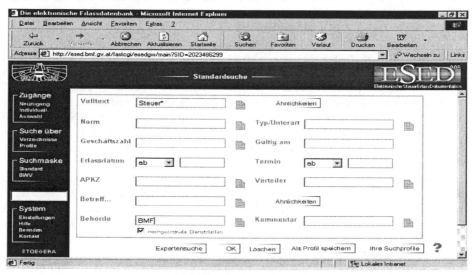

Fig. 1. Standard search mask

6   Prospects

Tax rules as well as important messages by the ministry will be recorded. To establish as a platform of communication the subordinated offices will offer their messages as well. Finally information-transfer will have no longer a one way direction.

Voting in the New Millennium: eVoting Holds the Promise to Expand Citizen Choice

Anthony Watson and Vincent Cordonnier

Professor, Edith Cowan University – Perth, Australia , Professor, Université des Sciences et Technologies de Lille – France and Adjunct Professor at Edith Cowan University.
a.watson@ecu.edu.au, vincent.cordonnier@univ-lille1.fr

Abstract. E-voting is not the same as e-democracy it is however a tool with the potential to help distribute rights in the voting process. The process may well prove to be a new 'killer application' for the Internet and a suitable authentication tool for voting security. A previous model outlined by the authors based on the use of voting smart cards and the Internet is expanded detailing concerns at the operational level and providing alternative solutions for security and the rigors required for voting scrutiny.

1   Introduction

The potential to invigorate the democratic process on a large scale through inclusivity and improved access based on adapted Internet technology may well change the way some social structures operate for the next generation and in particular the voting process. Accenture [1] suggests eDemocracy embodies these stages : 'Citizens access information ..., Decision making and influencing politicians, ...Voting electronically'. This paper focus on the latter aspect using a smart card based option.

Democracy exists when people who are concerned by a decision may participate in the making of that decision with many systems offering a mixture between direct and indirect voting alternatives (Watson and Cordonnier, [2]). In most cases the direct option would be preferred as the voter has their individual choice recorded as opposed to a delegated indirect vote. The use of information technology has the capacity to overcome some of the aspects of cost, time and distance in the voting process.

In Australia voting rates rose from 57.9% in the non compulsory era to 96.2% at the last election in the compulsory period [3]. In the US voting is by self direction and Kantor [4] suggests that the use of e-voting could have an impact and "help overcome the low voter turnout rate 44.9% in 1998". On the other hand it would appear that a postal voting option in Switzerland did little to improve the voting participation rate at a lowly 42.3% [3]. Voting from the comfort and security of ones home may have appeal in hot or cold climates or those areas with street violence or geographic issues.

Alexander [5] says "Casting a secret ballot in a fair and democratic election is, in fact, unlike any other kind of transaction" and it is true that the failure or abuse of the process can be more significant than a bank fraud. Recognising this it is suggested e-voting is feasible using a combination of smart card, cryptographic techniques and Internet technologies with a focus on authentication and security of data transport and storage.

R. Traunmüller and K. Lenk (Eds.): EGOV 2002, LNCS 2456, pp. 234–239, 2002.

The architecture of the e-voting system proposed is composed of five parts: i - Smart card for each citizen who may vote; ii - Terminal with card reading capacity and communications; iii - An acces to the Internet in some form; iv - A server to collect and manage the votes; v - A component which is the organisation process itself responsible for the voting delivery technology (pencil and paper or smart cards and the Internet).

2   Using a Smart Card As the Key to the Voting Process

Information technology designers recommend smart cards as a component of a system if the required service addresses a large and distributed population where each individual may require a personal treatment according to a specific profile and particularly so, if a high level of security is identified as a major issue of the application. Voting is an ideal domain of application for smarts cards in a technical sense but it is realised that any changes to an established voting practice is likely to encounter political and possibly social resistance.

Watson and Cordonnier [2] presented a scheme for what a digital voting process must offer and in [6] described security index options. The present goal it to address a limited number of issues including: authentication of the voter; verification of the right to vote; security of vote transmission; anonymity of the voter; guarantee of the vote counting process; possibility of reverse verification of some stages.

3   The Role of the Card

To operate effectively the smart card must authenticate itself and must verify in some way that it is used by the legitimate owner. The latter role is difficult .

Many applications already operational in the public area using a smart card use these important features of secure card technology: capacity to lead a mutual authentication with the server; identify the person who uses it (Authentication of the voter); sign an electronic document verifying the transaction; hide the message (encryption).

The way a card leads mutual authentication with the server is well known and not that different from what is used in many other applications.

It is more difficult to implement a satisfying solution to identify the voter. Many applications use a PIN code. This is possible if the card is used frequently however a voting card may be used only once or twice a year as an average. It is then likely that some voters will forget this code with, more abstentions as a consequence.

The most logical conclusion is to use biometrics where the user can be identified by a biological profile such as fingerprints, shape of the hand, characteristics of the eye. Most possibilities require an expensive special device to capture and analyse these biometric data. The argument is important if people are authorized to vote from their home. There is research suggesting Iris scanning technology, protected software and a Web camera could represent an effective and cheap solution.

A voice signature option is suggested because: voice identification techniques offer a reasonable level of security. Voice identification may be dynamic by asking the voter to repeat a word or a phrase he cannot prepare or copy on a recorder; most of the possible terminals as PC have a microphone and a loud speaker; it is even possible to implement these devices in the thickness of the card itself; voice is the simplest and most spontaneous means of expression of an individual; people who cannot easily read or write are not excluded by any other sophisticated identification process.

Another approach is to verify through the Internet that a voter is able to correctly answer a set of questions that relate to their life, family, job, holidays, etc. The method requires the data must be registered prior to the vote and stored in a secure environment, possibly the card itself. Some authors consider this solution as a good balance between security goals and access facilities, even for persons who do not have a high educational level. The system accepts a certain level of mistakes, can be personalised to any particular profile and is fast enough. As the proper responses are stored within the card, there is a limited risk for privacy, especially as any access performed is an internal comparison and then validated by a voting session.

4   Extended Roles of the Card

Microprocessor smart cards provide functionality beyond a static paper document, allowing for the overall architecture of a voting system which may offer new facilities or options. The card is responsible for driving the voting protocol. Voting at a distance implies the roles attributed to a voting officer must be delegated to another and only the card is secure enough to ensure this role. Furthermore as it holds most of the information to be protected, it is better to prevent this information leaving the card by giving the card itself a major role in the global management of the process.

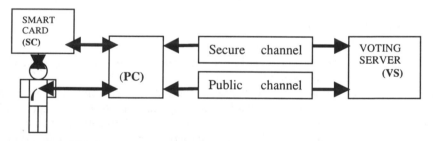

Fig. 1. representing both the voter and the voting bureau

Stage 1: The voter has to login to the server through the usual public channel importing a program that comprises interfaces with the voter, the card; and various electronic forms and documents to be used. The end of that stage is a request to the voter to present his card to the reader.

Stage 2: Is the identification process of the voter; it will use links from the PC that are not secure but the final decision to proceed will not allow any of the references (PIN code, prints of any profile) to be moved out of the card. The card responds through the PC in encrypted form allowing the voting process to continue. This is the point at which a particular token is identified.

Stage 3: May be considered as the informative time to propose to the voter the set of possible choices. It will use links from the voting server via the PC in a public environment. This stage may be personalised to the voter.

Stage 4: The vote is performed on the PC and data immediately moved to the card for encryption.

Stage 5: Is the transmission of the vote details to the voting server via a secure channel (meaning encrypted data).

There are two assumptions, which are crucial, to ensuring this model works:

1. each card uses a unique internal identification number determining that no 2 cards are the same;
2. the card itself is secure. This is an assumption we make with financial and health cards already and is usually a matter of degree.

5   The Voting Server (Verification)

The voting server may be local or global and it is responsible for managing, in real time, all the aspects of the voting session including server security, time stamping, dissemination of voting information and any voting variations.

The first role of the server is to replace the voting bureau with functionally to: verify the identity of the voter; provide voting material and information; receive the vote; date and time stamp votes and ensure the effective closure of the voting process for a given session; collect and aggregate the votes; and ensure annonymity.

The model proposed here should be seen as a framework and may be modified for national regulations There have been concerns expressed regarding the capacity to store individual voting transactions and 're-engineer' the vote to relate a particular vote to a particular individual. The model proposed here would prevent this. The Report from the National Workshop on Internet Voting (2001, [7]), suggests some specific design options for servers using protective firewalls and an archival system (P.7).

6   The User

Authentication of the user is a critical aspect and must be balance: prevention of another person using the card; simplicity of the associated device that allows multiple voting sessions; and capacity for any voter to easily provide the expected proof for identification.

Some countries authorise voters to use proxies. If a person A has to vote on behalf of a person B, the simplest solution is for B to temporarily give his card to A. Then the identification process would require B to give A the whole set of identification data. If A and B may attend together a trusted office with their cards, the transfer could be done prior to the vote. B will authenticate his right to vote by an appropriate identification and will transfer his voting capability to A. Then, at the voting time, A with his own identification process will be permitted to vote twice.

7   The Card and the Voting Session (Transaction)

It is possible to limit the role of the card to an authentication function but it could do more and it is these options which could distinguish the differences in e-voting systems. Firstly, the server will verify that the owner of the card is authorized to vote for that session; this is the result of a comparison of what is described in the card and the profile of the vote situation. For example, for a city council vote, the server will work on behalf of the city hall and verify that the owner is authorized to vote for that city for this particular voting instance.

Most of the voting models need to verify that a citizen has voted but nobody may know the content of that citizens vote. Manual systems are open to abuse with multiple votes recorded against the same name in different venues. This proposed model incorporates alternatives to ensure annonymity and a single vote check.

In addition to the authentication function the next requirement to be implememted in the card is a session token. Actually, as the card will be delivered for more than one session, it will be initially loaded with as many tokens as feasible and appropriate, thus ensuring a single card may last for several years. Each session will be given a number and the corresponding token will be used just once. This prevents an attempt to use the card more than one time for a session. The token will also provide a unique link between a card and a voting session.

The voting process will produce a voting array for that session and will be cyphered and transmitted to the voting server. It is not necessary to communicate the name or any id of the voter. However some countries, such as Australia, apply penalties on people who do not vote. It is preferable, then, to transmit some identifier to another server to ensure a token has been used and to make sure than there is no possible link between them. The single vote check is determined at the time of the sessiona and the token for that voting process erased preventing repettitive voting by a single card/voter.

8   Issues for Consideration

There are some technical matters for consideration in the actual moment of voting and although extreme need to be considered. The first must address the issue as to whether the person selecting a vote on a screen-based system is actually seeing the correct information. For example ensuring that the candidates are in the correct order on the screen. There are audit check mechanisms and the use of pre printed voting material, which should make it extremely difficult to fraudulently represent the information.

Another issue is to prevent 'dead people' voting and events of this nature have been recorded regularly in Italy, USA, France and Australia. The voting process is defrauded when a person pretends to be someone else and presents a vote in that name. It would be possible although difficult to obtain and use a deceased persons smartcard to register a vote. The smartcard option also allows for a defined life of a card. For example the card life could be set at 2 years which would limit the period of the fraud. There is also the argument of proxy vote by smartcard including voluntary and involuntary. The voluntary case could occur when an aged or infirmed person gave their voting card and authorization to a third party to vote on their behalf. The involuntary case occurs when a person is forced to hand over their voting smartcard and identification under duress to another person who uses that vote. This is a criminal activity and should be dealt with by law. In Australia there has been cases of electoral abuse through duress on individual voters and the law was changed to require a secret and scrutinized voting process in unions for example.

There must be a process to deal with lost or damaged voting cards. This is already the case with financial and health cards and is a systems management issue rather than a technical model matter.

9   Conclusion

One of the difficulties in accepting an e-voting approach may arise from social rather than technical issues including possible suspicion of the process and owners. There is still a test of time yet to take place which will look at the various brands of cards for security but the role of a smart card could lead a significant step forward in creating a remote eVoting secure system and the possible benefits that may bring.

References

1. Accenture, (2001), eDemocracy, http://www.accenture.com/xd/xd.asp.
2. Watson, A.C., Cordonnier, V., (2001), *Information Technology Improves Most of the Democratic Voting Processes*, Twelfth International Workshop on Database and Expert Systems Applications, IEEE Computer Society, Los Alamitos, California, p388-393.
3. Department of The Parliamentary Library, Research Note 29 1996-97, Parliament of Australia, http://www.aph.gov.au/library/pubs/rn/1996-97/97rn29.htm
4. Kantor, J., (1999), Obstacles to E-Voting, http://slate.msn.com//netelection/entries/99-11-02_44394.asp
5. Alexander, K. (2001), *Ten Things I Want People to Know About Voting Technology*, http://www.calvoter.org/publications/tenthings.html
6. Cordonnier, V., Watson, A.C.,(1997) *Access Control Determination of Multi-Application Smartcards Using a Security Index*, Third Australasian Security Research Symposium, July 1997, Brisbane.
7. Internet Policy Institute (2001), *Report of the National Workshop on Internet Voting : Issues and Research Agenda*, March 2001, National Science Foundation, University of Maryland.

e-Democracy Goes Ahead. The Internet As a Tool for Improving Deliberative Policies?

Hilmar Westholm

Center for Computing Technology, University of Bremen,
Postfach 330440, 28334 Bremen, Germany
westholm@tzi.de

Abstract. At first, important features of the current situation of citizens' participation in political procedures such as reasons for political apathy, political reactions with new ways of direct participation and the increase of communication-based means of participation are summarized. In the second part, "information", "consultation" and "active participation" are described as main components of non-organized citizen communication which must be embedded into users' and administrations' environment not only technically, but also economically, legally, organisationally, motivationally, and politically. Two case studies illustrate opportunities given by online-participation and underline the requirements to use online participation as a supplement, not as a replacement of traditional participation and to combine the advantages of various online and offline ways of citizens' involvement in a multi-channel approach.

1 Introduction

The situation of e-democracy usage has to be described against the background of formal as well as informal means of (offline-) participation. The current situation can be described with the following features:

Less involvement of citizens in traditional politics but increase in "instant participation". Factors like globalisation and individualisation cause more distance of citizens to the state and its institutions (e.g. administration, political parties). In many democracies, not just participation rates in elections are decreasing, but there is also a decline of membership in civil networks resulting in a precipitous drop in political engagement in general [1] with lower commitment to the political process and less trust in government. [2] Nowadays citizens prefer selective, focussed and limited involvement in political processes (mainly on the local level, often with NIMBY-(not in my backyard) character (i.e. individuals only participate politically if they feel a threat to their existing situation – e.g. if an industrial plant is to be built in their neighborhood) and transparency of political-administrative procedures. Traditional forms of political participation (e.g. elections, lobbying of interest-groups) mostly ignore these demands.

Thoughtful governments are now looking at the Internet not as a threat, but as a positive potential tool to re-engage the citizens in the business of governing. As the Internet has helped to empower a new generation of well informed and demanding consumers, some people argue that it will challenge the essentially passive relationship that the majority of people have with their government and politics.

R. Traunmüller and K. Lenk (Eds.): EGOV 2002, LNCS 2456, pp. 240–247, 2002.
© Springer-Verlag Berlin Heidelberg 2002

Increasing institutional possibilities for and use of direct participation. In some representative democracies in Europe direct-democratic tools are used more and more – sometimes combined with deliberative elements: Citizens' initiatives, referenda and ballots become part of the constitutional law and, mostly on the local level, normal democratic practice. "Legislatures are increasingly squeezed between the general public and the executive; the new technologies make plebiscite democracy more feasible and this possibility is putting pressure on representative democracy". [3] But these forms of votes are often criticised because they reduce complex political problems to simplifying yes/no-alternatives. [4] Besides, experts agree that ICT-supported voting by "pressing the button" has less importance than innovating the procedures of decisions. [5]

Part of such procedures (as well as indicators) are methods of gauging public opinion like polls, e-mails, or, more sophisticated, focus groups. The increasing use (and number) of these applied methods seem to be relieving the pressure on governments (and their representatives) to develop new and more effective means of public involvement. [6] But it is not finally decided if the main tool will be ICT; as representative experiences from the EU-supported DALI (Delivery and Access to Local Information and Services) project in Sweden showed: "Some politicians expressed doubts about participating in political debate via the Internet and preferred to use the phone when discussing issues with citizens. They felt that a written answer forced them to make a commitment on particular issues too hard, which thus made them more cautious. Also, some politicians felt that some questions from citizens could not be answered in a simple or public way. Such questions typically related to complex or politically sensitive issues, or involved a straightforward misunderstanding on the part of the citizen". [7]

Increase of deliberative involvement methods at various state levels. Most of the countries in Northern and Western Europe created a lot of elements of public involvement, especially in the area of public planning since the 1990s, like "public fora", "working groups", "citizen advisory groups", "caucuses", "visioning workshops" or "round tables", which go beyond legal minimum requirements. The Agenda 21, a document most countries of the world agreed on at the United Nations Conference on Environment and Development (UNCED) in 1992, gave a new input. A main emphasis of this document is to give citizens the opportunity to participate in issues dealing with transformation of societies (and local communities) to ensure sustainable development.

The challenge administrations are now facing is to establish appropriate approaches for engaging citizens as well as for (new) different interest groups in those new policy areas where IT permits the use of flexible and easily accessible means – not only for voting: For example, information can flow fast, it enables citizens to inform others about their opinion in an easy way or makes communication between representatives and voters easier. In regional planning processes, it becomes possible to involve interested citizens by providing shared workspaces with hyperlinks to relevant documents (e.g. maps) and to enable users to make their comments after using a structured filing via an Issue Based Information System (IBIS). Meanwhile interest groups are also better informed, better linked through networks, and, according to a survey in eight OECD-countries, better able to bring pressure to bear, especially on the middle level of bureaucracy. [8]

2   Steps of Participation in the Internet

There is a great hope that IT will improve political communication among citizens (or their interest groups), administrations, politicians and government. "E-democracy" is the keyword to bundle information and public participation and includes activities that follow these steps within the policy-making cycle from policy design through implementation to evaluation[1]:

2.1   Information

Most processes of political participation (should) start with intensive information. This "one-way relation" where government produces and delivers information for use by citizens covers "passive" access to information upon demand by citizens as well as "active" measures by government to disseminate information to citizens. On the state level it is now more or less self-evident that governments provide texts of laws and regulations and structures of their agencies via the Internet like political parties or NGO's do it with campaign-information. It is still a problem on the local level (often due to financial and human resource problems).

2.2   Consultation

This implies a "two-way relation in which citizens provide feedback to government - based on the prior definition by government of the issue on which citizens' views are being sought and requires the provision of information". [9] This happens in the form of opinion polling within city-marketing or as management of complaints. Managed electronically, these measures are cheaper than traditional means (surveys, household-questionnaires, see case study I below). Not representative, but more often used are online speaking hours of political party representatives or chats of politicians (who want to be re-elected). In Denmark, during a campaign in connection with the euro-referendum, cabinet ministers held online chat sessions with people. Resonance is normally rare – in rankings about usage of various online services, chats with politicians are in the last place beside the virtual visit of a museum.

Also for preparation of proposals of national acts or local planning procedures, governments are testing online applications. In Germany for instance, the federal government used the Web to publish the proposals of a Freedom of Information Act as well as another one of a Data Security Law and invited the citizens to comment on them in online forums. In these cases, government received high-quality contributions of the citizens who otherwise maybe would not have influenced the procedure.

A problem of Internet activities in traditional procedures is that they normally do not reach new citizen-groups. In an empirical research in the German town of Osnabrueck, researchers concluded that the function of organising or articulation via the Internet plays a secondary role and those parts of the population who normally do not participate in planning procedures will also not do so in Internet supported procedures. The Internet is a means for the communication-strong elite.

---

[1] Besides, the term e-democracy also includes e-voting, internal organisation among political parties and association, NGOs, etc. as well as new ways of participation like email-bombing.

Case Study I: Online Polling "Traffic Calming"
To gather the public opinion in a controversially discussed issue of traffic calming, both an online and offline opinion polling was launched with a questionnaire (mainly multiple choice-questions, one open question). Citizens were asked about quality of information by the district, if they support or reject the trial to reduce the traffic by installing barriers against non-resident traffic and what they think about the information and involvement policy of the district. Main results were:

- Politicians and administration had a quick feed-back of a large number of citizens' opinions to a controversial issue.
- Also those citizens participated who normally would not do that; the vocabulary was partially very rude.
- Users did not only make "clicks" to the multiple-choice questionnaire but every third user took the open question as an opportunity to express his own opinion.
- The combination of online and offline polling increased the representativity of participants compared to normal Internet-activities which are characterized by social exclusion (in this case, a large number of citizens elder than 60 years replied, only young citizens were represented worse than in reality; compared to other issues related to traffic, participation of women was high with 39%).
- Similar to offline ballot campaigns, interest groups mobilized their followers.
- The polling illustrated that an easy registration procedure to authenticate users is necessary because anonymous clicking enables participating several times.
- The polling process must be logically embedded in the political process (in this case, a citizens' interest group that fought for this issue for two years and had persuaded all the politicians, did not understand that political institutions launched an - even non-representative - polling when the trial started.
- Media were integrated into the polling process; the local newspapers reported about the (new) way of involvement many times thus increasing the number of participating citizens. [10]

2.3 "Active Participation"

A C2G-relation based on partnership with government, in which citizens actively engage in the policy-making process, is called "active participation" by the OECD. It acknowledges a role for citizens in proposing policy options and shaping the policy dialogue – although the responsibility for the final decision or policy formulation rests with the government. [11] Good experiences were made in the arctic Swedish town of Kalix where 7% of the population participated in an online discussion concerning the future development of the city centre. Other good experiences were gathered by the author with an online forum organized for five weeks during November and December which was especially related to the quality of citizens' contributions to the discussion (see case study II below). Also new forms of governance, e.g. the *delphi method*, a long-established approach for developing a consensual position statement through iterations of anonymous comments by a small invited panel, or alternative dispute resolution techniques like mediation can be very well supported by IT (esp. groupware applications) because the possible time-lag (between the contribution and the reading) enables the users to understand contributions and to react after collecting further information from other sources before contribute in a more qualified manner.

Case Study II: Online Forum Horn-Lehe

Within the EU-supported project E.D.E.N. (Electronic Democracy European Network) tools for better relationships between citizens and administrations are developed.[2] Two of them, a tool for online discussions and another one for visualizing maps, were applied first in a discussion about future development of the core district Horn-Lehe in Bremen. The discussion tool enabled anonymous and registered contributions with the right name as well as with a nickname to various issues from urban (district) planning to youth leisure activities. The forum combined information (masterplan and other map, minutes of district board meetings) with discussion, the problem of high expenses for moderation was met with the innovation of moderation by citizens (of the district); besides, it was a multi channel access while online discussion and visualizing maps was accompanied by physical committee meetings (where citizens could express their opinion), household-questionnaire, round-table. Some results were:

- About 6.100 visits to various discussions (not only hits to the homepage) were counted, 67 registered users and a unknown number of anonymous users (some 25-50) participated with 224 postings.
- A large amount of material was "produced", it was a big great effort for decision-makers to read them all (problem of information-handling capacity), an additional effect was that administrative experts became afraid to use the forum-tool for normal procedures.
- Surprisingly two third of the contributions were constructive (good and better) according to a five-point rating list, nevertheless most of the contributions were only expressive, some listening or responding, a small number emphatical and none of them persuasive (in both meanings: being persuaded as well as persuading). The quality of contributions made by registered authors was better.
- The quality of discussions profited by contributions of the head of the district administration and other experts.
- "Citizen-moderators" need to be trained intensively (technically but also "socio-technically" to summarize contributions, to find sensible titles for them (for the index-listing and better use) and to forward contributions to responsible experts for reply.
- A small number of users read the information provided additionally (such as the municipal masterplan, an explanation text to this plan, a zoning-code, minutes of the district committee, etc.).
- The project had a good press feed-back, importance of other PR-means (folder, poster) was small.

3   Adaptability of Applications

The few experiences existing demonstrate that the only technical ability to realize a means of citizen participation through the Internet has no effect on involvement from

---

[2] Within the E.D.E.N. project (duration until July 2003) tools are developed to improve communication between public administrations and citizens to increase knowledge and therefore support decision-making with urban planning as application domain through the use of natural language processing (NLP) and community tools (like polling and e-consultations) (www.edentool.org).

a content view. The great challenge is to embed these technical constructs into a specific technical, organisational, cultural, motivational, juridical and economical context. In other words: the electronic applications in the so-called virtual world need to become "adaptable" to processes and structures in the real world. [12] Adaptability must be related to the following factors:

From a *technical* point of view, online participation supplies must be integrated in the technical procedures in the back-offices. This is also true for planning procedures as well as for voter-registrations in online-elections. Besides, technical applications must not overstretch the users' equipment (e.g. in the Horn-Lehe forum end-users with modems needed too much time for downloading large files, e.g. for opening the maps).

Another issue are the *economic* framework conditions in an extended meaning of cost-benefit relations both of the public authorities and the users. How can governments both become "open" through practising more and better public involvement in the context of a financial crisis of public agencies and considering the necessity to decrease the role of governments trough outsourcing or privatisation of services? Public consultation is not for free but can be expensive when further requirements are considered, e.g. printing and distribution of popular brochures or the rent of venues.

Political participation (via the Internet or elsewhere) has only a chance of implementation if the target groups are motivated. *The readiness of citizens to participate* has often been over-estimated. According to the standard-model of political participation [13] it's no wonder – only citizens with a good financial as well as educational background and a good position in society believe that they can move something with their own political engagement. There is no reason to believe in a change due to the existence of new technologies. Young citizens will only be involved in planning issues that are related to their interests (such as planning of a youth centre) but not because there is a new attractive technology they use in everyday life. In other words: technically supported supplies must look for adaptability to existing discussions and motivations.

Motivation is also an issue from the point of view *of the internal users* in the administrations: If the personnel-situation is bad, employees will not be motivated to get more work by addressing planning issues via an Internet-forum.

Another precondition is the *legal admissibility* of existing or to be revised regulations. It is not only a question of implementation of the signature law but also one of procedure-related regulations in the circumstances described in this paper.

A further issue is the *political adaptability*: Participation supplies are less credible if a promise of implementation is missing. Online participation supplies must be integrated into the political process. Issues like legitimacy (elected committee, publicness of an online forum) must be weighed against the new input (knowledge of citizens) into the political process and the social demand for participation means other than elections.

4   Multi-channel Strategy

Under the precondition of a restricted budget, new public management models and budgeting within public administrations, cost-benefit relation becomes a decisive factor whether and how single experiments develop to widely used and sustainable applications. Positively influencing this relationship, resources have to be used not only for a single procedure but for many simultaneously and could be used from a larger number of operators. Therefore an economically driven online-participation platform should be operated that offers various functionalities (e.g. opinion polling, discussion fora, maybe also online-decision making – which can combine these ways and which is adaptable do the tasks of the clients) according to the example of commercially driven application service providers which could be rented by interested municipalities.

But nevertheless Internet-based participation will always be used as a supplement and support of offline tools but not alternatively. The social aspect of a meeting cannot be replaced, otherwise the possible anonymous use of Internet applications can be used by citizens who are not very trained in verbal communication. Similar to bank services which are provided physically in the branch as well as via telephone or Internet, we can speak of a multi channel strategy in which all channels relate to the same information sources: In the case of political participation, this includes the combination of public meetings and hearings, press releases, Internet-use and a citizen-hotline.

5   Conclusions

The Internet can be used to improve but not to replace traditional ways of citizens' involvement to political procedures. In a multi-channel application, the advantages of offline and online ways of involvement should be combined. eDemocracy does not implicate a jump ahead of democracy but can improve deliberation in many ways:

- It enables asynchronous communication: participants at virtual discussions have more time to think about the arguments of political opponents before they react.
- Discussions can be lead more rationally than emotionally.
- Internet-involvement-methods are more flexible related to time and location: users do not have to join evening meetings or visit the agencies at public opening hours.
- Planning can be visualized more easily.
- Citizens can prepare themselves before visiting an officer by looking at the planning via the Internet; communication can work on a qualitatively higher level.
- The anonymity of the Internet empowers people to participate who normally would not do that and enables discussion among citizens who usually don't have contacts (nevertheless, it seems to be a medium that favours literate and educated citizens).

If deliberative policies should be innovated by the Internet, it is necessary to take into account that the processes are embedded in an environment that is determined by the innovation not only technically but also politically, legally and economically.

References

1. Coleman, S, Goetze, J.: Bowling Together: Online Public Engagement in Policy Deliberation. Hansard Society, London (2001)
2. National Audit Office, 1999, Government on the web: a report by the comptroller and audit general, commissioned by: Commons, H. o., London (1999)
3. OECD, Impact of the Emerging Information Society on the Policy Development Process and Democratic Quality, PUMA Vol. 15. OECD, Paris (1998)

4. Bellamy, C., Taylor, J. A., 1998, Governing in the Information Age, Buckingham Philadelphia: Open Univ. Press (1998)
5. Lenk, K., Dienstleistungssysteme und elektronische Demokratie. Multimediale Anwendungen im Verhältnis von Bürger und Verwaltung. In: Schneidewind, U., Truscheit, A. Steingräber, G. (Eds): Nachhaltige Informationsgesellschaft. Analyse und Gestaltungsempfehlungen aus Management- und institutioneller Sicht, Metropolis, Münster, 135-153 (2000)
6. OECD 1998, ibid.
7. Ranerup, A., Internet-enabled applications from local government democratisation. Contradictions of the Swedish experience., in: Heeks, R. (Ed.): Reinventing Government in the Information Age. International practice in IT-enabled public sector reform. Routledge, London, New York (1999), 177-193
8. OECD 1998, ibid.
9. OECD, Engaging Citizens in Policy-making: Information, Consultation and Public Participation. f, PUMA Policy Brief Vol. 10, OECD, Paris (2001)
10. Westholm, H., (Mehr) Partizipation über Internet? Fallbeispiel einer Online-Meinungsumfrage. In: Stiftung Mitarbeit (eds.): Rundbrief Bürgerbeteiligung, Vol.. I (2002)
11. OECD 2001, ibid.
12. Kubicek, H., Westholm, H., E-democracy: Quo vadis? Stand und Perspektiven elektronischer Demokratie per Internet. In: Behörden Spiegel, Vol. 12 (2001) 44
13. Dalton, Russel J., Citizen Politics in Western Democracies. Public Opinion and Political Parties in the United States, Great Britain, West Germany, and France, Chatham House Publishers, Chatham N.J. (1988)

Discourse Support Systems for Deliberative Democracy

Thomas F. Gordon[1] and Gernot Richter[2]

[1] Fraunhofer FOKUS, Berlin, Germany
gordon@fokus.fhg.de
[2] Fraunhofer AIS, Sankt Augustin, Germany
gernot.richter@ais.fhg.de

Abstract. The idea of deliberative democracy is to facilitate broad and deep public participation in systematic, constructive discourses about legislation and policy issues, so as to enhance the legitimacy, efficiency, quality, acceptability and accountability of the political process. By **discourse support systems** we mean groupware designed to support structured, goal-directed discourses. The paper discusses the importance of discourse support systems for deliberative democracy, provides a brief overview of the Open Source Zeno system and mentions several e-democracy pilot applications of Zeno, including the DEMOS project of the European Union.

1 Introduction

E-government is about redesigning or **reengineering** the processes of government, taking into consideration the opportunities and risks of modern information and communications technology (ICT). **E-democracy** is a special case of e-government: using ICT to support the core political processes of government, sometimes called **governance**: policy debates, legislation, executive decisions, the resolution of legal and political conflicts, and the election of representatives.

There are various conceptions of how to best make use of ICT to "reinvent" democracy. (See [3] and [6] for an overview and case studies.) Some emphasize the potential of **e-voting** to facilitate processes of **direct democracy**, via referenda, where public interest groups can propose legislation which is put to a popular vote and decided by citizens directly, bypassing elected representatives. Direct democracy is controversial and we will not address its many issues here, except to point out that even proponents of direct democracy emphasize the importance of adequate information and deliberation, before putting issues up to a vote [4]. For us, the main potential of e-democracy is to enable greater citizen participation in political discourses, whether or not the citizens or elected representatives make the final decisions.

There is much talk about overcoming the problem of **digital divide**, to assure that all stakeholder groups have effective access to e-democracy processes. While this is important, we should not forget that only powerful special interest groups have access to the traditional print and broadcast media, creating an even greater **analog divide** which already has been severely detrimental to democratic ideals. What has greater influence on the outcome of an election: a substantial donation to a political party's "war chest", so as to be able to afford media events, or casting a vote at the polls? The

R. Traunmüller and K. Lenk (Eds.): EGOV 2002, LNCS 2456, pp. 248–255, 2002.
© Springer-Verlag Berlin Heidelberg 2002

new media of networked computers, especially the Internet, has given far more ordinary citizens an effective voice than any other technology in history. Consider, for example, the recent **weblog** phenomena, where thousands of ordinary citizens have begun publishing journals on the web [2,1].

In addition to weblogs, other more established Internet and web technologies can and have been used to facilitate political participation, including email, instant messaging, mailing lists, newsgroups, and web-based bulletin boards and discussion forums [3]. Each of these technologies has its advantages and appropriate uses, but due to a lack of space we cannot compare them here. Rather, the focus of the rest of this paper is on presenting a new kind of system, called **discourse support systems**, which unlike these other technologies are designed specially to support deliberation and other consensus building and conflict resolution processes, and discussing some experiences in applying such systems in e-democracy pilot projects.

2   Conceptual Model of Discourse Systems

Examples of discourse systems in politics and public administration are not difficult to find: 1) If a city plans to build a new airport, the applicable building codes may require the plan to be subjected to a public discussion with affected citizens and interest groups; 2) The cities of a region may work together to revise the zoning laws and plans to find a balance between growth and environmental protection; 3) A political party will need to discuss its political program and strategy for the next federal election; 4) Last but not least, parliaments, city councils and other law-making bodies deliberate about legislation in party factions, subcommittees and in plenary sessions, with input from various experts, lobbyists, professional staff and the public.

According to Walton in [13], a **dialog** is a goal-directed conventional framework in which two or more participants or parties "reason together in an orderly way, according to the rules of politeness or normal expectations of cooperative argumentation for the type of exchange they are engaged in." We define **discourse** as dialog, in Walton's sense, about some language artifact, such as draft legislation, project proposals, or city plans. We use the term **discourse system** to mean a "sociotechnical" system, consisting of human and technical "components", for performing particular discourse tasks within an organization, or between collaborating organizations. Finally, inspired among others by the work of Sumner and Shum [11], by **discourse support system** we mean the system of information and communications technology used as part of the infrastructure of a discourse system.

Conceptually, the main components of discourse systems are: the **actors** participating in the discourse, in their various roles; the **document** being discussed, including the history of changes made to the document; the **dialog** about the document, the subject matter of the document, or the dialog itself; and the **norms** which guide or regulate the dialog and modifications to the document.

Notice that the dialog can consist of many different kinds of speech acts: questions, motions, claims or assertions, arguments, offers, votes, and so on. In a more elaborate model, one might want to define separate components for different classes of speech acts. For example, there could be a component for managing claims and arguments and another for handling procedural issues.

Norms are of various kinds. They can provide mere guidance without imposing any obligations on the participants. Sources of norms are plentiful and varied, some general and some specific to the application. Example sources include social and linguistic conventions, rules of order, laws, regulations, administrative procedures, cases, principles, values, professional standards, and best practices. Norms may be conflicting and substantial reasoning may be necessary to decide which norms apply and how to resolve conflicts among them. Finally, norms are subject to change over time and may even change during and as a result of a particular discourse. For example, a participant might make an issue out of some rule of order.

3   Generic Use Cases

Having defined the relationship between discourse systems and discourse support systems, our next job is to consider what kinds of discourse tasks can be sensibly and usefully supported by modern information and communications technology. While a detailed requirements analysis would be possible only in the context of a specific application, we have been able to adduce some general requirements from our experience in several e-democracy projects. So-called "use cases" are a good starting point for identifying requirements. The use cases describe, at a very high level, the tasks and responsibilities of each role in the discourse.

We distinguish three main roles: **readers** browse the document and follow the dialog; **authors** write parts of the document or actively participate in the dialog; and **moderators** edit the document or moderate the discussion. Notice that we have used the same three roles for actors who interact with either the document or the dialog. This is because the protocol of a discussion can be conveniently viewed as a kind of complex, structured document.

Also, we have not distinguished between rights and obligations in these role definitions. For example, we leave it open whether authors have an obligation to make contributions to the document or only the right to do so if they want to. Here we are interested only in understanding the tasks which could be performed by each role, whether or not there is an obligation or even a right to perform such tasks in particular circumstances. This will ultimately depend on the norms appropriate for the particular application.

As usual, individuals may have several roles at once and several individuals may share the same role. For example, authors are typically also readers and several moderators may be responsible for some document or discussion. In a discourse, the moderators (i.e. editors) of the document being discussed need not be the same persons as the moderators of the discussion about the document.

Reader Tasks. Readers are interested in timely, relevant and accurate information about the participants and their roles, interests, background and activities; about the document and its subject matter; about the discussion; about the state of the proceedings in light of applicable procedures, about any other relevant norms; and about any background information helpful for understanding the issues. In particular, readers would like to be able to find information about similar past cases; to search for information in documents using metadata and the full text of the documents; to browse documents conveniently, using tables of contents, indexes and references (links); to find documents which are similar, in some sense, to a selected document; to cluster a set of documents and to categorize such clusters. Finally, readers would like to be able to keep informed about activity in the discourse, without having to regularly take the initiative (notification services).

Author Tasks. Authors are responsible for adding information to the system, to share them with readers. They need to be able add messages, articles and other kinds of documents to the system or to insert bibliographic information, abstracts and other data about these documents into catalogs and other databases. Authors need ways to refer to other documents or, ideally, parts of documents and make relationships between documents explicit. Finally, authors need ways to keep informed about tasks for which they are responsible and the status of these tasks, such as due dates, whether or not they have been completed, priorities and task dependencies.

Moderator Tasks. Moderators have final responsibility for the quality of the document or the discussion. They oversee and guide the entire process, helping other users to be aware of the applicable norms and thus their roles, tasks, rights and obligations. Their task is to assure that the discourse proceeds according to its purposes, so as to maximize the chance of achieving its goals. Moderators are responsible for applying appropriate moderation techniques to focus and guide the dialog. These include methods for broadening the dialog by gathering information about the problem and the interests of the parties, and brainstorming about possible solutions, and then narrowing the debate by clustering, categorizing and prioritizing options, and arguing about their relative merits. Relevant here is also the moderator's responsibility for opening and closing topics for discussion. Finally, moderators need tools for expressing, visualizing, presenting and analyzing relationships between parts of the dialog. Moderators need support in applying relevant norms to guide and regulate the process. Moderators require resources and methods for motivating other users to perform their tasks well in a timely manner. Moderators need to be able to conveniently monitor the activity in the discourse for which they are responsible. They need to be informed about new additions or changes to the document and new contributions to the discussion, without having to manually search for this information.

4   Overview of Zeno

The Zeno system, an Open Source groupware application for the Web written in Java, has been designed specifically for use as a discourse support system. This includes managing both the communities of users and groups who participate in the discourses and the content which is created and used in the discourses. A simple but powerful role-based access control scheme connects the two functional parts of Zeno: users and groups managed by the directory service are assigned access roles in journals where the content of discourses is stored. Discourse management functions for session management and event monitoring (logging, notification, discourse awareness, etc.) as well as communication services (messages to users and journals) provide the necessary support during a discourse.

Zeno's features are implemented in an extensible, object-oriented system architecture with easily customizable user interfaces, using the Velocity template engine and Cascaded Style Sheets.

4.1 Data Model

The design goal of Zeno led to a simple but general data model with a rich set of data structures and operations. The core of this data model is a persistent content store for **hyperthreads** of journals, articles, topics and links. **Journals** are container-like objects that can be used for many purposes, including shared workspaces, discussion forums, calendars, task management, and as a collaborative editing environment for complex, structured documents and content management. **Articles** are similar to email messages and support multiple MIME attachments. The contributions to a discourse are stored as articles. **Topics** are thematic collections of articles, that is, sets of articles which deal with the same subject. Topics and articles are contained in journals. When used as discussion forums, journals support both the threaded and the linear (topic-oriented) style of discussions. Journals, articles and topics, collectively known as Zeno **resources**, form a hierarchy or tree called the **compositional structure** of the content.

Typed links allow resources to be connected, which results in an arbitrary graph structure with labeled directed arcs called the **referential structure** of the content. A **link** connects a **source** resource with a **target** resource. A resource can be the source or target of any number of links. Thus, links can be used to create arbitrary directed graphs of resources. Links are typed with **labels** chosen from the set of **link labels** in the journal containing the source. The referential structure models non-compositional relationships between resources. Such graphs are much more general than the essentially hierarchical data structures typically used by file systems, shared workspaces, outliners or threaded discussion forums. We call the connected subgraphs of the referential structure **hyperthreads**, since they can be viewed as a generalization of the threads of discussion forums, replacing the reply relation by Zeno links.

Operations of the data model include full text search and powerful support tools for moderators: moving, copying, deleting, publishing and unpublishing articles, opening and closing topics, ranking or ordering articles and journals, and labeling articles and links to build conceptual graphs and visualize relationships. Automatic link management helps moderators to preserve the referential structure when they restructure the content of a discourse.

Journals are composed of a partially ordered sequence of any number of resources. Topics are composed of a partially ordered sequence of any number of articles. Articles are composed of any number of **attachments**. An attachment can be a file of any MIME type, such as word processing documents, spreadsheets, or image files.

Attributes describe the properties of resources, attachments and links which are relevant to the system or to the users. **System attributes**, such as the **creator, creation date** and **modification date** of a resource, are not modifiable by users, but rather set by the Zeno system. **Primary attributes** are standard attributes which may be modified by users, such as the **title, rank** and **note** of the resource. Finally, **secondary attributes** are ad hoc attributes defined by users for application-specific purposes.

All resources have the following system attributes: **id, creator, creation date, modifier** and **modification date**. All resources also have these primary attributes: **title, rank,** and **note**. The rank, an integer value, can be used for many purposes, such as prioritizing tasks or ordering the sections of a document. The note of a resource is a plain text string. Depending on the application, it can be used as the main part or body of a document, for example when journals are used as discussion forums, or as an abstract or description of files attached to the article, for example in journals used as content stores or shared workspaces.

Additional primary attributes of journals include, among others, **article labels** and **link labels**. They define the set of labels which can be used to tag articles and links. This feature enables journals to be used for **concept mapping, mind mapping, idea processing** and other approaches to modeling knowledge using labeled, directed graphs. For example, to model argumentation as in Issue-based Information Systems [7], one could define **issue, position** and **argument** labels for articles and **pro** and **con** labels for links.

Articles also have additional primary attributes, e.g., **label, keywords, begin date** and **end date**. The **label** is chosen from the set of the **article labels** of the journal containing the article. The **begin date** and **end date** attributes allow articles to be used to describe tasks, appointments or events, which can be used to generate reminders or displayed appropriately in calendar views.

4.2 User and Group Management

Zeno includes a directory service for managing users and groups of users. The directory maintains passwords, contact information, in particular email addresses, and user preferences. The directory can also be used for mailing lists.

For security and administration purposes, the directory is partitioned into a set of subdirectories, called **community directories**. A community directory can be configured so as to allow new users to register themselves in the community directory, without the assistance of an administrator. To allow self-registration, an administrator of the community directory gives **guest** users permission to register as new users if they meet the admission criteria stored in the community directory. The right to register new users is limited, and doesn't imply the right to view or modify existing records.

4.3 Role-Based Security Model

Access rights are controlled in Zeno by assigning the roles of **reader, author** or **moderator** to users and groups for each journal. That is, these roles are assigned for journals, but not directly for articles, topics or links. The access rights for an article or topic are those of the journal which immediately contains the article or topic. The access rights for a link depend on the access rights for the source of the link. Anyone with the right to view an article may also view the links from this article. Similarly, anyone with the right to modify an article may also modify the links from this article.

The rights of each role are fixed by the Zeno system. They cannot be redefined by users. Moderators have the most rights; with few exceptions they may do anything which can be done with a journal and its contents. Only moderators of a journal may create subjournals.

The readers of a journal may access and view every article in the journal which is published. Like moderators, readers may also access and view the identifiers and titles of subjournals. Further rights to a subjournal are controlled by the roles defined in the subjournal. The authors of a journal have the right to create new articles and topics in the journal. Participants in a discourse will often be both readers and authors.

4.4   Moderation and Editing Facilities

Based on feedback from users of prior versions of the system, Zeno provides a significantly richer set of features for moderating a discourse and editing its web of contributions. Only a few can be mentioned here.

Articles can be modified by editors at any time. The modification date and user id of the editor who made the changes are recorded in system attributes, to make it transparent to readers that the article has been modified, but Zeno does not currently keep a copy of the original version or provide any other version management services. Several articles, topics and journals can be selected and then, preserving their links, moved in a single transaction from one location to another in the compositional structure. Several articles and topics can be selected and then copied in a single transaction, in which case any links between the original articles are also copied. Resources can be deleted, recovered (undeleted) and permanently removed from the system.

Other features allow editors to close and re-open topics or journals, to publish and unpublish articles, to (partially) order direct components of journals (articles, topics, subjournals), and to define labels and qualifiers for articles.

5   e-Democracy Applications of Zeno

The first version of Zeno was developed as part of the European GeoMed project, which integrated Zeno with a Geographical Information Systems so as to enable citizens to discuss city plans on the Web [5,10]. This tradition has been continued; the current version of Zeno has been integrated with the CommonGIS system [12]. Zeno was recently used in an extensive e-democracy pilot application at the City of Esslingen, as part of the German Media@Komm project [9]. Finally, Zeno is being used as a part of the foundation of the DEMOS system [8]. DEMOS stands for Delphi Mediation Online System and is an e-democracy research and development project funded by the European Commission (IST-1999-20530). DEMOS offers innovative Internet services facilitating democratic discussions and participative public opinion formation. The goal is to reduce the distance between citizens and political institutions by providing a socio-technical system for moderated discourses involving thousands of participants about political issues at the local, national and European level. The vision and long-term goal of DEMOS is to motivate and enable all citizens, whatever their interests, technical skills or income, to participate effectively and actively in political processess which are both more democratic and more efficient than current practice. The DEMOS system is being validated in pilot applications in the cities of Bologna and Hamburg.

6   Conclusion

Many of the use cases we have identified for discourse support systems are not (yet) implemented by Zeno. There is a great deal of work remaining to be done. If there is one point we would like readers of this paper to remember, it would be that current tools only begin to scratch the surface of what is conceivable in the way of support for consensus building, conflict resolution and other core processes of democracy.

References

1. Rebecca Blood. The Weblog Handbook: Practical Advice on Creating and Maintaining your Blog. Perseus Pub., Cambridge, MA, 2002.
2. chromatic, Brian Aker, and Dave Krieger. Running Weblogs with Slash. O'Reilly, 2002.
3. Stephen Coleman and John Gøtze. Bowling together – online public engagement in policy deliberation, http://www.hansardsociety.org.uk, 2001.
4. UNI Unternehmerinstitute e.V. Für Effizienzstaat und Direktdemokratie. ASU Arbeitgemeinschaft Selbständiger Unternehmer e.V., Berlin, 2001.
5. Thomas F. Gordon. Zeno: A WWW system for geographical mediation. In P. J. Densham, Marc P. Armstrong, and Karen K. Kemp, editors, Collaborative Spatial Decision-Making, Scientific Report of the Intiative 17 Specialist Meeting, Technical Report, page 77–89. Santa Barbara, California, 1995.
6. Richard Heeks. Reinventing government in the information age: international practice in IT-enabled public sector reform. Routledge research in information technology and society. Routledge, London ; New York, 1999.
7. Werner Kunz and Horst W.J. Rittel. Issues as elements of information systems. Technical report, Institut für Grundlagen der Planung, Universität Stuttgart, 1970. also: Center for Planning and Development Research, Institute of Urban and Regional Development Research. Working Paper 131, University of California, Berkeley.
8. Rolf Lührs, Thomas Malsch, and Klaus Voss. Internet, discourses and democracy. In T. Terano et al., editors, New Frontiers in Artificial Intelligence. Joint JSAI 2001 Workshop Post-Proceedings. Springer, 2001.
9. Oliver Märker, Hans Hagedorn, Matthias Trénel, and Thomas F. Gordon. Internet-based citizen participation in the City of Esslingen. Relevance – Moderation – Software. In Manfred Schrenk, editor, CORP 2002 – "Who plans Europe's future?". Technical University of Vienna, 2002. 7th international symposion on information technology in urban and regional planning and impacts of ICT on physical space.
10. Barbara Schmidt-Belz, Claus Rinner, and Thomas F. Gordon. GeoMed for urban planning – first user experiences. In R. Laurini, K. Makki, and N. Pissinou, editors, Proceedings of 6th International Symposium on Advances in Geographic Information Systems, pages 82–87. 1998.
11. Tamara Sumner and Simon Buckingham Shum. From documents to discourse: Shifting conceptions of scholarly publishing. In Proceedings of CHI 98: Human Factors and Computing Systems, pages 95–102. ACM Press, Los Angeles, California, 1998.
12. Angi Voss, Stefanie Röder, Stefan Salz, and S. Hoppe. Group decision support for spatial planning and e-government. In Global Spatial Data Infrastructure (GSDI), Budapest, Hungary, 2002.
13. Douglas N. Walton. The New Dialectic: Conversational Contexts of Argument. University of Toronto Press, Toronto; Buffalo, 1998.

Citizen Participation in Public Affairs

Ann Macintosh and Ella Smith

International Teledemocracy Centre, University of Napier,
10 Colinton Rd, Edinburgh, EH10 5DT, UK
{A.Macintosh,E.Smith}@napier.ac.uk

Abstract. Reflecting on the European Commissions stated aim to broaden democracy this paper examines the nature of e-participation and considers concepts of democracy and issues surrounding citizen participation in pubic affairs. The paper describes how citizens are engaging with government and with each other about policy related issues that concern them, using technology specially designed for the purpose. The paper describes a case study of electronic participation developed for the Environment Group of the Scottish Executive in Summer 2001. Using the empirical data from this study the paper explores best practice guidelines for governments who wish to engage citizens in policy-making. The difficult task of addressing the requirements of all stakeholders, i.e. government, civil society organizations (CSOs) and citizens in designing the technology is discussed. The use and moderation of the electronic tools over the engagement period is assessed. Finally, the paper considers how the use of electronic tools can be monitored and their impact on citizen participation and the decision-making of government be assessed.

1 Introduction

Governments around the world are attempting to broaden democracy by providing an effective conduit between themselves and civil society organisations and between citizens themselves using innovative ICT to deliver more open and transparent democratic decision-making processes. It is argued that democratic political participation must involve both the means to be informed and the mechanisms to take part in the decision-making. ICTs have the potential to deliver e-democracy which addresses these joint perspectives of informing and participating. Over the last decade there has been a gradual awareness of the need to consider new tools for public engagement that enable a wider audience to contribute to the policy debate and where contributions themselves are both broader and deeper. A number of commentators have addressed this issue and at the same time highlighted the possible dangers of a technology-driven approach.

Barber [1] highlights the concept of strong democracy, creating active citizen participation where none had existed before. However he goes on to warn of the use of technology in that it could diminish the sense of face-to-face confrontation and increase the dangers of elite manipulation. Held [2] distinguishes nine different models of democracy. His participatory model reflects the need to engage both citizens and civil society organisations (CSOs) in the policy process. However, in order to engage citizens in policy-making, he and others recognise the need for informed and active

R. Traunmüller and K. Lenk (Eds.): EGOV 2002, LNCS 2456, pp. 256–263, 2002.

citizens. Fishkin [3] argues the need for 'mass' deliberation by citizens instead of 'elite' deliberation by elected representatives. Instant reactions to telephone surveys and television call-ins do not allow time to think through issues and hear the competing arguments. Van Dijk [4] addresses the role of information and communication technology with such participatory models of democracy in order to inform and activate the citizenry. However he also warns of the consequences of bad designs of technology.

The arrival of more sophisticated ICTs and the emergence of the Internet during the early 1990s coincided with widespread concern that politics and politicians had become increasingly irrelevant to ordinary people. It is clear from the increasingly low turnout at elections in the UK that traditional democratic processes do not effectively engage people. Other liberal democratic countries have also noted low voter turn-out. To start to address this problem, governments are beginning to consider the concept of e-voting and over the last year there has been a growing interest in internet voting. However, there remains a big question mark over whether this switch in methods of voting will actually change people's attitudes and address their growing disengagement from established political processes.

Several commentators discuss the use of technology to support the broader democratic process. Coleman and Gøtze [5] outline four possible scenarios for technology supporting democracy. The first e-democracy model is where the technology supports direct democracy. For example, Becker and Slaton [6] explore the current state and future of e-democracy initiatives that are designed specifically to move towards direct democracy. The second model is based on on-line communities, where technology is concerned with supporting civic communities. The work of Rheingold [7] on virtual communities assesses the potential impact of civic networks, questioning the relationship between virtual communities and the revitalization of democracy. Tsagarousianou et al., [8] give descriptions of a number of projects involved with e-democracy and civic networking. These authors suggest centrally designed government-led initiatives will clearly differ from grassroots civic developments, but argue also that "civic networking will not realize its objective unless it becomes more realistic in its goals and methods" (page 13). Coleman and Gøtze's third e-democracy model concerns the use of on-line techniques to gauge public opinion through surveys and opinion polls. However, Fishkin [3] questions whether opinion polls contribute to the complex issues of public policy. He argues that, as far as American citizens are concerned, they have the opportunity to be consulted on several occasions by opinion polls without prior warning or preparation, in order to find their views, even when the individual may have had no reason to develop any opinion on the subject being asked. He concludes that all that was gathered was an "attitude" that was created on the spot by the very process of participating in the survey (page 81). Finally, their fourth model focuses on the use of technology to engage citizens in policy deliberation, emphasizing the deliberative element within democracy.

The recent OECD report [9] emphasizes the need for: "a relation based on partnership with government, in which citizens actively engage in defining the process and content of policy-making." The OECD is addressing issues such as how to provide easier and wider access to government and parliamentary information, how to ensure that citizens have the ability to give their views on a range of policy related matters and also how to allow citizens to influence and participate in policy formulation. The OECD [9] report defines three types of interaction, namely one-way information provision, a two-way relationship where citizens are given the opportunity to give feedback on issues and, lastly, a relationship based on partnership where citizens are actively engaged in the policy-making process. Looking at these three types of inter-action, the OECD reports that the scope and quality of government provided informa-tion has increased greatly over the past decade. With regard to the two-way relation-ship, consultation is also on the increase, albeit with large differences between countries. For the final type of interaction, the OECD states that active participation and efforts to engage citizens in policy-making on a partnership basis are rare, under-taken on a pilot basis only and confined to a very few OECD countries. With regard to the application of ICT, the OECD reports that, while an increasing amount of gov-ernment information is obtainable on-line, the use of ICT for consultation is still in its infancy. The case study described in this paper addresses both the two-way relation-ship through consultation and the notion of active participation where citizens are helping to formulate policy.

2   Objectives of e-Consultation

One can consider three over-arching reasons for better engagement of citizens in the policy-making process:

- to produce better quality policy at the national level
- to build trust and gain acceptance of policy
- to share responsibility for policy-making.

All with the long term objective of strengthening representative democracy.

Given these overarching reasons, we argue that the objective of e-consultation is to improve the policy-making process through a range of devices to facilitate:

1. reaching a wider audience to enable broader consultation - in which case the par-ticipants themselves can be profiled to the extent that they have provided any iden-tifying details and they have given their informed consent to use these;
2. supporting participation through a range of technologies to cater for the diverse technical and communicative skills of citizens – in which case the 'ease of use' and 'appropriate design' of the e-consultation can be addressed;
3. providing relevant information in a format that is both more accessible and more understandable to the target audience to enable more informed consultation - in which case participants use of background information that is made available on-line can be analyzed, to give an indication of how relevant it has been;
4. engaging with a wider audience to enable deeper consultation and support delib-erative debate – in which content analysis and thread analysis of consultation dis-cussion forums can be considered;
5. analyzing contributions to support the policy-makers and to improve the policy – in which case analysis of what people have said in response to the consultation can be carried out more cost effectively since the responses are received in an electronic form (i.e. they do not need to be transcribed) and responses to closed questions can also be subjected to survey analysis techniques, again with cost-efficiency savings since they do not have to be transcribed from questionnaires;
6. providing relevant and appropriate feedback to citizens to ensure openness and transparency in the policy-making process.

Different forms of technology are appropriate for each of the above issues, but for all, the technology must be designed in ways that generate and support public trust in the process. However, Macintosh et al [10] describe how the very nature of governance means that the design of e-democracy systems becomes complex. For example, democratic needs for openness and transparency may conflict with needs for ease of use and simplicity of access. Whyte and Macintosh [11] discuss technology designed to enhance the transparency of public consultation practices, highlighting that many other problems also need to be addressed. For example, issues of unequal access to technology and the unequal technical and communicative capabilities of citizens demand systems that are simple to use and understand.

The web-based tool, e-consultant[1], which was used in the case study of electronic participation discussed in this paper begins to addresses each of the above issues. The tool is the Center's focus for e-participation research.

3 e-Consultation on Sustainable Development

The case study addresses the 'pre-policy document' stage of civic engagement i.e. where there are a number of pre-identified issues that government wishes to gather opinions on before drafting a policy document. The e-consultation allows government to invite discussion on issues - the aim is to get initial input from a wide audience so as to draft a more comprehensive policy document. The e-consultation was on behalf of the Environment Group of the Scottish Executive and was based around sustainable development issues facing Scotland. The aim was to equip Ministers with views to develop a policy document as input to the World Summit in South Africa in 2002. The study ran from 6th June to 8th October 2001. It aimed to inform people about the key issues facing a future Scotland and asked them to give their views on a range of issues from efficient use of resources to lifestyle and transport. The web site address for the study is: http://e-consultant.org.uk/sustainability/. It received a total of 392 contributions. These were made by 172 individuals and on behalf of 19 groups or organizations.

e-consultant is a dynamic website implemented in Active Server Pages. Scripts written in VBScript generate the HTML, and access and update the consultation data. Data is maintained in a SQL Server relational database, with the exception of the consultation's background information, which is held as static HTML. The e-consultant system resides on the ITC server, which in turn is on Napier University's network. In summary, the main sections of e-consultant for the sustainable development consultation were designed as follows:-

- *Overview:* A welcoming page outlining the purpose, target audience and timescales of the consultation, with links to the main websites of stakeholders.
- *Background Ideas:* Structured around the 7 key issues of sustainable development for Scotland; also included were links from the consultation issues to a comment page for each issue.
- *Have your say:* Here uses could enter their comment on an issue or respond to a previously made comment; also included were links from these comment pages back to the background ideas page for each issue.

---

[1] Developed with support from British Telecom, see internet site: www.e-consultant.org.uk

- **Tell a friend:** Here users could email people they felt would be interested in participating in the consultation.
- **Events and Links:** Details of the off-line seminars and links to other sites.
- **Feedback:** Space for a statement from the Scottish Executive as their response to the consultation once the consultation was complete and responses analyzed.
- **Review site:** An online questionnaire for users to complete to help the ITC evaluate the e-consultation.

As well as the above end-user sections, the management of the e-consultation process was facilitated by additional password-protected administrative services. These included functions to

- monitor comments on a 24hour post-moderation basis
- remove from view comments that breached the "conditions of use" statement
- view the entries to the on-line questionnaire
- view the most frequently read comments
- view the comments added in the last 24 hours
- view the comments received from postcode areas.

The ITC had access to all the above services to support managing the e-consultation. The Scottish Executive had access to all but the first 3 of the services. This helped them assess the consultation as it progressed.

In order to give some structure to the debate and encourage discussion topics to form, questions and ideas were divided into 7 issue based sections. The issues were also used to focus discussion on the less obvious aspects of the debate (economic and social) and help to explain how they fitted in. To encourage informed and deliberative debate the "Background Ideas" section was included in the site. This was divided into pages according to the various issues.

4   Analysis of Comments

All comments were read and categorised according to their content. The way this was done was inspired by Anthony Wilhelm's [12] analysis of 10 political newsgroups in October 1996 during the presidential election campaign in the United States. He used relatively simple content analysis categories to evaluate how far the participants provide and seek reasoned argument with evidence to support their contributions.

Our comments were analysed to judge the success of the consultation in terms of the way the e-consultation website was organised and the information given within it. This was done by first looking at the extent to which the comments answered the main questions set out in the e-consultation and second by an investigation of the relationship between the content initially provided on the website (the issues and background information) and the content of the comments.

The questions set by the e-consultation and the relevant categorises were:

- **Q1.** What sort of Scotland do we want to live in?
  - **Analysis categories:** Suggestion; no suggestion
- **Q2.** What could or should be done to achieve this by governing bodies, individuals and business?
  - **Analysis categories:** action-government, action-business, action-individuals, action-all

- **Q3.** What has been done well, so far, and what could be done better?
  - **Analysis categories:** Doing well; Improve

The results of the analysis showed that contributors engaged with the question "What sort of Scotland do we want to live in?". While most suggestions for improvement were directed towards government, there was an extensive realisation that everyone needed to be involved at some level. 97% of comments contained or agreed with suggestions about what could be done to improve our status in terms of sustainable development. Only 3% of comments made no suggestions and merely pointed out a current problem. Over half the comments (57%) included suggestions for government action. A reasonable proportion specified that businesses should take action (19%). A smaller number of comments pointed out where individuals should take action themselves (7%). A third of comments (33%) requested action from all parties (government, business and individuals).

The second part of the analysis was to discover whether the issues and background information provided in the e-consultation were clear and helpful. Here we assessed whether each comment appeared under the most appropriate issue and whether it reacted to or reflected the information provided as well as addressing the question. Almost all the contributors used the issues the way they were intended to be used (96%). Of those which could have perhaps more suitably appeared under a different issue (4%), most had some relevance to where they were placed. Comments that did not seem to fit any category tended to be about how policy was implemented. Some contributors did not divide their opinions into separate comments under specific issues, but grouped them all together under one issue. Most comments directly answered the questions that accompanied the issues or those posed in the "Background Information" section (91%), This is a good indication that the questions were successful in focussing the discussion. A further 7% of comments reflected themes from the "Background Ideas" section, without answering any of the questions posed within it.

5   Evaluation of the Process

The evaluation was based on past experience of conducting e-consultations, the meetings of a consultation Steering Group and the users' opinions of the site as given in response to the on-line "Review Site" questionnaire. Every page of the e-consultation had a link to the "Review Site" questionnaire. This aimed to gather information about the circumstances in which people used the site, how easy or difficult they found it to use and what they thought about using the Internet as part of a consultation process.

The Scottish Executive set up a Steering Committee drawn from key stakeholders across Scottish Society to guide the consultation. This Committee agreed the original design specification and the content of the consultation website through an iterative process. Best practice guidelines for "traditional" consultations were followed as closely as possible. As the target audience was anyone with an interest in sustainable development in Scotland, it was important that people using any Internet enabled computer at minimum connection rates and any web browsers could use the e-consultation. Both the structure and the content of the website were designed to cater for this diverse audience. It contained a clear statement on the conditions of use of the site and also a clear statement on privacy.

It was important to make it as easy as possible for any user to be able to add comments on the consultation, therefore the commenting process had to be as easy and attractive as possible. The ITC had found from previous e-consultations that users were put off by too intrusive a registration process. Therefore, it was agreed not to include a registration process. Instead, the users were asked to provide a minimum of personal information with each comment. Although this simplicity worked in the users' favour, it did lead to some difficulties during the evaluation process.

e-consultations provide a new mechanism to gather public opinion and as such they require new methods to promote them. Traditional promotional routes were augmented with more interactive "on-line" style promotion, "tell a friend" e-postcards and clickable logos advertising the consultation on related websites were used. The number of sites displaying the e-consultation link was disappointing, but they did have some success in involving people in the consultation. 40% of respondents to the on-line questionnaire stated they heard about the e-consultation through an electronic link. Almost half the respondents contributed to the e-consultation from their home, 30% from work, 11% from school or college, the remainder from a friend's house or community centre. Most felt that the site had been easy to use, although 7 people admitted to being confused in places and 4 found it quite difficult.

When one of the objectives of an e-consultation is to reach a wide target audience, the natural concern about the digital divide and hence the bias of an Internet based consultation has to be addressed. An e-consultation should be seen as just one route for participation and be supported by other opinion gathering events. All these participation routes should clearly link to one another. Importantly, an e-consultation provides the opportunity for deliberative participation. Well-structured background information on the website can serve to inform the debate. Allowing users to make their own comments and also to read and respond to comments from others provides a more transparent consultation process.

6   Conclusions

Although examples of e-democracy systems are often cited in the press and by politicians, there are few academic reports of these experiments. Few "real-world" examples of e-participation exist that provide sufficient empirical data on which to base sound research studies. It was because of this that we embarked on a programmed of active research.

We have attempted to show that while the evaluation issues are easy to state at this general level, their assessment has to deal with interdependencies between systems design, policy implementation, and the everyday politics and practice of communications between citizens, in all their variety, and public administrations in all their complexity. We suggest that evaluation of e-consultations has to take place within an analytical framework that takes into account the political, technical and social perspectives [13].

References

1. Barber, Benjamin. (1984). *Strong democracy: Participatory politics for a new age.* Berkeley: University of California Press.
2. Held, Anthony. (1996). *Models of democracy.* Cambridge: Blackwell Publishers.

3. Fishkin, James S. (1995) *The voice of the People. Public Opinion and Democracy.* Yale University Press
4. Van Dijk, Jan 'Models of Democracy and Concepts of Communication'. In Hacker, K.L and Jan van Dijk., (2000). (eds*) Digital Democracy issues of theory and practice.* Sage Publications
5. Coleman, Stephen. and Gøtze, John (2001) (2001). *Bowling Together: Online public engagement in policy deliberation.* Hansard Society and BT.
6. Becker, T. & Slaton, C. (2000). *The Future of Teledemocracy.* Westport, Conn. 2000. LC
7. Rheingold, H. (1993) *The Virtual Community.* Reading M.A. Addison and Wesley. Also see http://www.well.com/user/hlr/vcbook/vcbookintro.html (consulted February, 2002)
8. Tsagarousianou, Rosa. Tambini, Damian. & Bryan Cathy. (1998). (eds). *Cyberdemocracy: Technology, cities and civic networks.* London & New York: Routledge.
9. OECD (2001). *Citizens as Partners: Information, consultation and public participation in policy-making*: OECD.
10. Macintosh, A., Davenport, E., Malina, A.; and Whyte A.; Technology to Support Participatory Democracy*; in Electronic Government: Design, Applications, And Management*; edited by Åke Grönlund, Umeå University, Sweden; published by Idea Group Publishing January 2002; pp223-245.
11. Whyte A. and Macintosh A. Transparency and Teledemocracy: Issues from an 'E-Consultation.' In *Journal of Information Science.* July 2001.
12. Wilhelm, A.G. (1999) Virtual Sounding Boards: how deliberative is online political discussion? In Hague, B.N. and Loader, B. (Eds) (1999) *Digital Democracy: Discourse and Decision Making in the Information Age.* Routledge
13. Whyte, A. and Macintosh, A. 'Analysis and Evaluation of e-consultations; to appear in the e-Service Journal; Indiana University Press, 2002.

An Approach to Offering One-Stop
e-Government Services – Available Technologies
and Architectural Issues

Dimitris Gouscos[1], Giorgos Laskaridis[1], Dimitris Lioulias[2],
Gregoris Mentzas[2], and Panagiotis Georgiadis[1]

[1] eGovernment Laboratory, Dept. of Informatics and Telecommunications,
University of Athens
{d.gouscos,p.georgiadis}@e-gov.gr
glask@di.uoa.gr
[2] Dept. of Electrical and Computer Engineering, National Technical University of Athens
{dlioulias,gmentzas}@softlab.ntua.gr

Abstract. The right of citizens to high-quality e-Government services makes
one-stop service offerings an essential feature for e-Government. Offering one-
stop services presents many operational implications; an one-stop service provision (OSP) architecture is needed that, by means of a layered approach, provides facilities to refer to, invoke and combine e-Government services in a uniform way, in the context of cross-organisational workflows. Although enabling
technologies for all the layers of such an architecture are quickly evolving
(XML, WSDL, UDDI, WFMS et al) two major issues that need to be solved are
(a) abstracting the heterogeneity of the e-Government services that need to be
integrated and (b) identifying an appropriate style for cross-organisational
workflow control, somewhere in between the fully centralised and peer-to-peer
extremes. This paper presents an abstract layered OSP architecture, identifies
some major enabling technologies and briefly discusses those two issues.

1   Some Requirements for One-Stop e-Government

The delivery of e-Government services poses certain requirements inherent in the
mission of government itself, that differentiate e-Government from other e-service
application domains such as e-Business; customers have on *option* to choose the
services of a specific business, whereas citizens have a *right* to enjoy the services of
their own government. Therefore, e-Government service delivery must achieve (a)
maximal benefits (quality and performance of the service, added value of the content),
(b) maximal accessibility (simplicity of front-end logic, multi-linguality of user interfaces, anywhere/anytime availability, provision of alternative channels, no demands
for end-user IT skills) and (c) minimal costs (transportation, communication, document management); see, e.g. [1], [6], [10], [12], [15].

The idea of one-stop e-Government presents substantial promise for contributing to
all those optimality goals, whereas at the same time it is a natural fit to the concept of
web portals that by definition can bring together any number of arbitrarily heterogeneous web resources. It is well established, however, that the mere collection of links
to e-Government sites is a very superficial implementation of the one-stop concept;

R. Traunmüller and K. Lenk (Eds.): EGOV 2002, LNCS 2456, pp. 264–271, 2002.
© Springer-Verlag Berlin Heidelberg 2002

true one-stop e-Government calls for *compilation, presentation and delivery* of e-Government services in an one-stop fashion. This means that (a) on a conceptual level, services need to be compiled into some sort of *service bundles* around single real-world situations where they apply; these service bundles then must be (b) logically presented at the user interface level as responses to single real-world problems (the life-event approach) and (c) actually enacted and delivered at the operational level as if they were individual (in contrast to bundled), atomic services.

2   Operational Implications of One-Stop Service Offerings

Conceptual compilation of e-Government services in bundles corresponding to real-world situations must face problems like (a) how to categorize services by means of "ontological indexes" on their content (service ontologies), (b) how to refer to services (service naming schemes) and (c) how to store invocation pointers to service implementations (service repositories).

The issue of actually enacting and delivering these bundles as virtually atomic services sets outs some further requirements of its own. To achieve this sort of virtual atomicity, an one-stop service provider (OSP) must deliver a sense of seamlessness to service requestors. Consequently, an OSP needs to provide not only *bundling transparency*, i.e. hide that a "virtually atomic" service is actually enacted and delivered by invoking multiple lower-level services that may exchange data, synchronize their work/control flows and produce results finally composed into an atomic response, but also *bundling management*, i.e. the actual coordination mechanisms that take care of invocation, interaction and synchronization of the bundled services.

The issue of service bundling management, which is central to the provision of one-stop services, poses some hard problems on the operational level. Talking in an e-Government context, virtual one-stop e-Government services such as the issue of a business permit or the declaration of a change of address correspond, due to public sector functional disintegration, to bundles of multiple atomic services which, in the general case, are delivered by multiple providers. Therefore, it should be noted that (a) one-stop provision of bundled e-Government services entails multiple service providers and, what is more, (b) the technology that is used by individual providers for providing their share of bundled services cannot be assumed to be interoperable (in extreme cases, technology may not be used at all or only in an old-fashioned, merely esoteric, way). What is more, these providers being public agencies that are governed by complex or fuzzy regulatory frameworks in the general case, (c) their internal workflows may contain complex, multi-conditional portions, portions that cannot be modelled deterministically and many cases of exceptions.

One-stop e-Government services are required not only in an intra-border (i.e. local, regional or national), but also in a cross-border context; consider, for example, issuing work permits for immigrants or paying in country X a freelance worker based in country Y. Such as situation is additionally complicated since (d) public service providers from different countries will normally operate in different languages and be governed by non-harmonised regulatory frameworks and, apart from that, (e) it is quite risky to assume a given service level (e.g. response time, exactitude and completeness of response) from a public service provider in a country other than one's own; sometimes, a response cannot even be assumed at all within a reasonable amount of time.

3 Building Blocks
for an One-Stop Service Provision Architecture

All these operational level problems are posed as hard design constraints, when one tries to compile one-stop e-Government service bundles, and on the other hand as real-time quality and performance threats, during enactment of the service bundles and delivery of the bundled services. Therefore, work is needed in an architectural level, i.e. a level of systemic analysis and synthesis that considers *both* operational *and* technical issues, along the following directions:

- Service ontologies, service naming schemes and service repositories that offer capabilities to characterize, name, store and invoke a service by means of a well-defined set of "essential service characteristics" (e.g. application domain, content, inputs, outputs, behaviour, exceptions, invocation channel, delivery channel, information exchange format, quality levels, performance levels). Deployment of such ontologies, naming schemes and repositories would result in an *abstract service model*, on top of which it is possible to develop the service invocation and service coordination schemes mentioned below.

- Service invocation schemes that offer capabilities to uniformly invoke services of arbitrary information format and communication channel heterogeneity. Such schemes should be able to abstract from the caller any service-specific invocation and delivery details so that (a) information format heterogeneity could be abstracted by transparent transliteration from caller-native to service-native formats and vice versa (possibly to some standard format so that a star-like transliteration scheme is established), whereas (b) communication channel heterogeneity could be abstracted by having generic service invocation call that accept preferred channels as caller-defined parameters, consider channel capabilities retrieved from the service repository and then call channel-specific communication gateways.

- Service coordination schemes that offer capabilities to implement the atomic enactment of a service bundle and delivery of bundled results. Such a scheme should be able to handle increasing levels of co-ordination complexity, such as (a) invoking (serially or in parallel) *non-cooperating services* (i.e. services able to deliver results without any interim exchange of information or other synchronisation) and passing (possibly transliterated) information delivered from a service to the next invoked one, (b) invoking *co-operating services* with interim synchronization points (so that there are pending states where one service is waiting for another to reach some interim point of progress), (c) handling services (possibly co-operating) which are delivered through *fuzzy*, rather than deterministic, workflows and (d) handling *indeterminate situations* where a service actually does not (due to some service provider failure) or is not perceived to (due to some communication failure) reach expected progress within a reasonable amount of time.

- A service co-ordination scheme should also offer co-ordination management capabilities of increasing complexity, such as (a) enacting deterministic and *fuzzy* co-ordination scenaria, (b) enforcing *atomicity requirements* (i.e. try to complete the scenario or have it rollback as a whole, in a way reminiscent of database transactions), (c) reporting the current progress and status of an enacted co-ordination scenario and (d) mapping one-stop service requests to *pre-compiled* co-ordination scenaria, or *customising* pre-compiled scenaria according to request-specific characteristics, or *deducing optimal performance* co-ordination scenaria according to request-specific characteristics and current workload.

According to the above analysis, an one-stop e-Government service provision scheme should have a layered architecture like one depicted in Fig. 1.

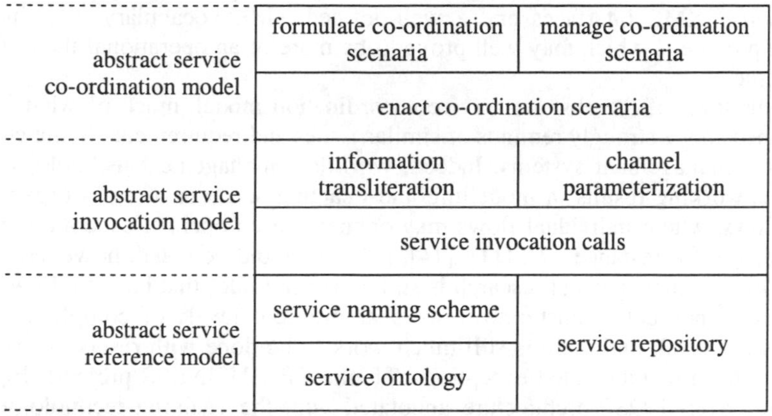

abstract service co-ordination model	formulate co-ordination scenaria	manage co-ordination scenaria
	enact co-ordination scenaria	
abstract service invocation model	information transliteration	channel parameterization
	service invocation calls	
abstract service reference model	service naming scheme	service repository
	service ontology	

Fig. 1. Layered architecture for an one-stop service provision scheme

4   Technologies That Can Serve One-Stop Services

As a complement to the layered OPS architecture a communication platform needs to be assumed that interconnects the OSP with individual service providers in a star-like topology. This platform incorporates all the service invocation and delivery channels that have been mentioned above; in a real-world setting, such a platform would include Internet (HTTP, email, WAP), telephony (voice, SMS, facsimile) or even paper-based connectivity between the OSP and individual service providers.

The current trends for e-Government certainly justify considering the Internet, and in particular the HTTP protocol, as the premium communication platform between an OSP and individual service providers; turning to HTTP, together with its accompanying technologies for deploying and accessing e-service content on the web, already offers a number of very promising, if not definitive, technical choices for implementing parts of this architecture.

Web service description languages, such as WSDL [5], as well as universal naming and access protocols, such as UDDI [13] and SOAP [2], are expected to solve many of the issues mentioned above for deploying service ontologies and naming schemes. The development of service repositories, on the other hand, is an active research issue (cf. the ideas of some IST e-Government cluster projects, such as EGOV and SMARTGOV), possibly looking at techniques from domains such as knowledge management. Not all technical problems are yet solved, but the fact that much research is active in many issues related to e-services, such as e-service composition and integration in heterogeneous environments (see, for example, [3], [4], [8], [9]) gives rise to a certain amount of optimism.

The need for abstraction of technical heterogeneity in information exchange between individual service providers can be effectively addressed by semantic and dynamic content exchange languages such as XML and DHTML; note that XML can also be used for dynamic control exchange in cross-organisational workflow architectures, as in [7], [11]. It should be noted, though, that adoption of a language such as XML just provides us with a syntax for formulating our messages, which is less than half of the solution. The remaining and most important part of the problem would be to establish XML interfaces and a well-defined XML vocabulary with individual service providers, which may well prove to be more of an operational than of a technical issue.

On the level of the abstract service co-ordination model, much of what has been mentioned above strongly reminds of similar issues and requirements in the context of workflow management systems. Indeed, workflow management technology already provides working results in modelling and enacting scenaria of cross-organisational work flows, where individual flows may or may not co-operate, exceptions may occur, etc; see, for instance, [7], [11], [14], [16]. It should be noted, however, that current workflow management research is still active in issues that have to do with management of non-deterministic workflows at different levels of complexity (fuzzy, weak workflows), and there is still much work to be done with respect to workflow recovery from unanticipated exceptions (cf. the CB-BUSINESS project). Fig. 2 depicts the layered OSP architecture annotated with the enabling technologies mentioned above.

As can be seen from this figure, many technological building blocks are there, but technologies still evolve and mature (WSDL, UDDI) and research is still active important open issues (WFMS). Apart from that, there is also a number of operational issues that have to be solved between OSPs and individual service providers in order for such architectures to operate. As a last, but not least, open-ended issue in the situation of Fig. 2, it is worth bringing into attention the underlying architectural paradigm on which this design is based.

abstract service co-ordination model	formulate co-ordination scenaria	manage co-ordination scenaria	WFMS technologies
	enact co-ordination scenaria		
Abstract service Invocation model	information transliteration	channel parameter-ization	XML, DHTML
	service invocation calls		
Abstract service Reference model	service naming scheme	service repository	WSDL, UDDI
	service ontology		
Communication platform	service invoca-tion messaging	result delivery mes-saging	Internet/HTTP

Fig. 2. Layered OSP architecture and some enabling technologies

5   A Discussion of Principles

In very general and abstract terms, this OSP architecture may be thought of as an architectural design that attempts to co-ordinate dynamic entities from multiple and independent information systems towards some common goal. Whether these entities are client-server *processes* in a distributed database, blue- or white-collar *workflows* in a manufacturing or office system or individual e-Government *services* to be bundled in one-stop offerings, from an architectural point of view their co-ordination towards a common goal entails some inevitable issues of integration. At some level these dynamic entities must communicate in order to exchange data and/or control. This need for communication and data or control exchange necessitates some degree of homogeneity, in the sense that a minimum amount of common conventions must be applied by all entities involved with respect to the communication channels that they use and the syntax and vocabulary of their messages.

As the technical conventions employed in an information system have mainly evolved out of business choices during the systems procurement or implementation, the technical homogeneity of different information systems is most often organisationally bounded. Systems within the same organisation follow (or are taken care of, in order to follow) compatible technologies, so that communication, information exchange and finally application integration is possible. In such intra-organisational contexts, therefore, one is usually (and justifiably) targeted at applying some standard conventions so that information systems are interconnected, at the networking and database levels, and interoperable, at the processing level.

The feasibility of this approach becomes quite questionable, though, in an inter-organisational setting like the one formed by multiple individual public service providers, whose e-Government services are to be bundled in true one-stop value propositions. In such a context, the diversity of existent IT platforms, as well as various business and political resistances to change leave very little room for assuming homogeneities at a technical or operational level (and this is even more true in the cross-border variant of this situation). Therefore, beyond the tight integration demanded by the interconnectivity or interoperability approaches, a new and more flexible architectural paradigm is sought to make dynamic co-ordination possible.

In the approach that we have presented in this paper, this is the paradigm of inter-mediation, namely the paradigm of establishing a third intermediator entity amidst all the dynamic entities co-ordinated, with a mission to manage the overall co-ordination scenario. This intermediator entity then (in our case the central OSP scheme), should first of all employ ontologies and naming schemes to acquaint with the dynamic entities that it co-ordinates, memorize their names and features in some sort of repository and isolate heterogeneities by means of specialised communication modules pluggable into an abstract co-ordination model. This, in more abstract terms, is the underlying approach that we have taken in designing the OSP scheme above.

The modelling part of this approach, namely the concept of applying some ontological modelling and some naming scheme in order to gain capabilities of acquainting and invoking multiple heterogeneous resources is already well-established in many computer science fields; this general principle can be found behind agent-based approaches in artificial intelligence, object-based approaches in software engineering and distributed architectures, let alone in the design of the Web itself. Unique Resource Identifiers (URIs) of web resources and Unique Resource Locators (URLs) of web pages are instances of this general concept, whereas description languages like WSDL and naming protocols like UDDI are explicitly based on similar principles to envision "web services networks", an idea which has some common grounds with the OSP architecture.

A substantial difference between these technologies and our own context, however, is that web technologies are completely open-ended as to the architectural location (probably absence) of control. In the true "uncontrolled" spirit of the Internet, these technologies are offered as building blocks that allow constructions of peer-to-peer structures, like the World-Wide Grid (WWG). Nevertheless, it is still architecturally possible to enforce, on top of these structures, higher-level layers that implement some sort of more or less strict control (federated, arbitrated, moderated, centralised styles) which is exactly the option that we exploit in our own OSP approach.

One of the major criticisms against this approach (and against architectural centralisation, in general) is that the center of control becomes at the same time a performance bottleneck and a single point of failure. This is indeed one of the open issues of our work and a major point for further investigation. The final choices on this are going to be made on architectural grounds, as well as on the operational requirements, constraints and capabilities which arise from particular application domain, that of offering one-stop e-Government services.

Acknowledgements

A major part of the work reported in this paper is performed in the context of the CB-BUSINESS (Cross-Border BUSiness INtermediation for Electronic Seamless Services, IST-2001-33147) project, which officially started in April 2002. The CB-BUSINESS consortium is led by Planet Ernst & Young S.A., and includes partners such as the eGovernment Laboratory of the University of Athens, SEMA Group sae, ComNet Media AG, public service providers and professional chambers.

References

1. Atkinson, R., and J. Ulevich (2000) Digital Government: The Next Step to Reengineering the Federal Government, Progressive Policy Institute Technology and New Economy Project, March 2000.
2. Box, D., D. Ehnebuske, G. Kakivaya, A. Layman, N. Mendelsohn, H.F. Nielsen, S. Thatte, and D. Winer (2000) Simple Object Access Protocol (SOAP) 1.1 Official Specification, W3C, May 8, 2000, http://www.w3.org/TR/SOAP/.
3. Boyer, K., R. Hallowell, and A. V. Roth (2002) E-services: operating strategy: a case study and a method for analyzing operational benefits, Journal of Operations Management, Vol. 20, No. 2, April 2002, pp. 175-188.
4. Casati, F. and M.-C. Shan (2001) Dynamic and adaptive composition of e-services, Information Systems, Vol. 26, No. 3, May 2001, pp. 143-163.
5. Christensen, E., F. Curbera, G. Meredith, and S. Weerawarana (2001) Web Services Description Language (WSDL) 1.1 Official Specification, W3C, March 15, 2001, http://www.w3.org/TR/wsdl.
6. Dawes, S., P. Bloniarz, K. Kelly, and P. Fletcher (1999) Some Assembly Required: Building a Digital Government for the 21st Century, Report of a Multidisciplinary Workshop, Center for Technology in Government, March 1999.

7. Lenz, K., and A. Oberweis (2001) Modeling Interorganizational Workflows with XML Nets, Proceedings Hawaian International Conference on System Sciences (HICSS), January 3-6, 2001, Maui, Hawaii.
8. Mecella, M. and B. (2001) Designing wrapper components for e-services in integrating heterogeneous systems, The VLDB Journal, Vol. 10, No. 1, 2001, pp. 2-15.
9. Sahai and V. Machiraju (2001) Enabling of the Ubiquitous e-Services Vision on the Internet, e-Service Journal, Vol. 1, No. 1, 2001, pp. 5-20.
10. Scheppach, R., and F. Shafroth (2000) Governance in the New Economy, Report of the National Governors' Association, Washington D.C., 2000.
11. Shegalov, G., M. Gillmann, and G. Weikum (2001) XML-enabled workflow management for e-services across heterogeneous platforms, The VLDB Journal, Vol. 10, No. 1, 2001, pp. 91-103.
12. Sprecher, M. (2000) Racing to e-government: Using the Internet for Citizen Service Delivery, Government Finance Review, October 2000, pp. 21-22.
13. uddi.org (2000) UDDI Technical White Paper, September 6, 2000, http://www.uddi.org/pubs/Iru_UDDI_Technical_White_Paper.pdf.
14. Vonk, J., W. Derks, P. Grefen and M. Koetsier (2000) Cross-organizational transaction support for virtual enterprises, O. Etzion, P. Scheuermann (Eds.), Proceedings of Cooperative Information Systems (COOPIS 2000), Springer, 2000.
15. West, D. (2000) Assessing E-government: The Internet, Democracy and Service Delivery by State and Federal Governments, Brown University, September 2000.
16. Workflow Management Coalition (1999) Interoperability Proving Framework 1.0, Technical Report WfMC-TC-1021, April 1999.

e-Governance for Local System:
A Plan and Implementation Experience

Cesare Maioli

Cirsfid, University of Bologna, Via Galliera 3, 40122 Bologna, Italy
maioli@cirfid.unibo.it

Abstract. In November 1999, the Regione Emilia-Romagna administration (RER) launched the Regional Telematic Plan (RTP), an initiative to increase the awareness of ICT at the different levels of the public sector, to foster their utilization and to allow better services for the small and medium sized enterprises and the citizens. The lead purpose of RTP is to increase the economic competitiveness of the regional enterprises having better served and ICT conscious citizens. We present the main findings after two years of experience.

1   Introduction

Regione Emilia-Romagna (RER) has 4 millions inhabitants, about 400.000 enterprises and, in year 2000, ranked 13th - on 190 regions in Europe - for gross per capita product.

At the end of 1999, RER launched the Regional Plan for Telematics [1], an initiative to foster the ICT insight and utilization and to allow better services for the enterprises and the citizens. The lead purpose of RTP was to increase the economic competitiveness of regional enterprise having better served and ICT conscious citizens, trying to balance and to integrate the efforts and investments in the ICT of more than 500 autonomous Local Administrations; nine are the Provinces, which actually are managing wide area networks and information services for other bodies.

RER has been a leader in several cases with regard to the ITC, at times even on a European level [2]. They were, however, almost always isolated cases: therefore we involved a "teamwork" of politicians, managers and technicians to support the identification of regional problems and interests in a regional process, integrating them in an inter-regional and European context so that the whole regional Public Administration, might improve itself and its ways of operating[3].

Cheaper and faster internet access, promotion of digital literacy, more advantage for the public sector from the digital technologies were the initiatives to bring on-line every citizen, every company, every administration [4].

RER planned to avoid the risks [5] that:

- its main cities or a few of its industrial districts would become increasingly estranged from their hinterlands leaving these areas excluded from global currents and unable, on their own, to chart a viable path towards participation in the information society
- cities, especially the university ones, would specialize in knowledge-based activities making wider the economic gulf between them and their hinterland.

R. Traunmüller and K. Lenk (Eds.): EGOV 2002, LNCS 2456, pp. 272–275, 2002.

The origin of the Plan dates back to 1997 when the regional Government identified computer networks and information services as the bases to leverage in order to increase the competitiveness of the whole regional system instead of considering them just as organizational services [6].

2   The Telematic Plan

The points of intervention proposed in the RTP to pursue the telematic development of the region in the perspective to reach an Agreement with the Local System, were the following:

1 - Innovation of services for citizens and enterprises
The objective is the development on the whole regional territory of high quality and efficient public services, integrated as much as possible, with easy and convenient access taking up the numerous opportunities offered by ICT.

2 – Strengthen and completion of the Unitary Network of Emilia-Romagna
To simplify and improve the public services for citizens and enterprises it is necessary to achieve a greater integration and a better internal communication in the whole regional public administration.

3 - Modernisation of the regional government
Innovation in the operational methods of the regional public administration fully concerns the Region administration itself.

4 - Diffusion of the "fourth knowledge" and public access to information society
Without a rapid and generalized development of competence on ICT the above-mentioned aims risk being unachievable..

5 - Promotion of electronic commerce and new media industry
A more intensive use of ICT by small and medium sized enterprises, in order to take advantage of the electronic commerce opportunities, and a reorganization of operative and managing processes represent a crucial challenge for the economic development of Emilia-Romagna

6 - Promotion of a competitive regional market of telecommunications and development of internet services
We considered points 2, 3, 4 as prerequisites for point 1 and points 1, 4, 6 as prerequisites for point 5, which constitutes the rationale of the entire RTP.

3   The Implementation of the Plan

During year 2000, RER made relevant investments to strengthen the regional telematic network: the co-operation with the Provinces in the design and financing made the connection between the offices of the different institutions more widespread.
Later the local telematic plans, compliant with RTP, were collected on the basis of a call issued by RER.

Since we wanted the political boards of the local authorities to consider the ICT not just a technical matter, we accepted the plans only in presence of a clear commitment of the political the boards. *We think that without a strong growth of interest and in-*

sight by the political decision makers in the ICT politics, the public sector risks a governance deficit about the processes reshaping everybody's life.

Also if we recognize the need for a constructive understanding between politics, industry and the public, we consider the ICT industry as a solution provider but we stress the need for politics to take the lead interpreting and anticipating the public requests and wishes.

In particular we asked the local authorities to contribute for about 50% of the cost in case of plan approval.

We financed the planning effort; often the plans were prepared by consortia of small bodies who never before had the chance to plan their information service development in a framework of political objectives.

We received 51 different plans, clustering about 300 different *projects* from more than 250 different authorities.

Finally, at the end of year 2000, RER refined its original PTR, and identified the fields of intervention to fund. They are of three different kinds:

A - Provision of a broad band telecommunication infrastructure for the regional territory

The solution will follow the final choice between the involvement of local utility companies, big ICT players, RER autonomous implementation through a project financing approach.

B - Design and implementation of ICT tools and services

They are developed together with Municipalities, utility companies and Provinces' administrations. The actual efforts are:

- integration of large archives dealing with health data, enterprise data, geographical information systems
- one-stop shop services in the fields of health services booking, managing of the tourist sector, monitoring of the natural environment data, information and procedures for the enterprises
- a unified regional Web portal on top of the numerous initiatives of Local Authorities' civic networks and internet information sites
- provision of tools for e-procurement and support to e-commerce initiatives via a regional electronic card.

C - Promotion and balance of the local administrations' information systems according to the RTP

The activities concern geographical information systems, statistical systems, electronic document management, interface usability and citizen-friendship, upgrading of technical skills of the personnel.

The financial investments are in the order of 200 million EUROs in five years for B and C, and of 70 million EUROs for item A.

4   Conclusions

We can summarize the main findings of our experience as follows.
From the organizational point of view, they are:

- the rate of involvement and insight of the political and regulatory bodies are of paramount importance; many times the decision making about apparently technical aspects given to information officers and technologists is a short minded and ineffective measure

- the appointment of the project ownership to local authorities make the inner energy grow.

From the technical assessment point of view, they are:

- the network and the internet paradigms are more and more accepted and assimilated and virtual communities of users of existing services spread the perception of the technologies as an enabler for sustainable communities
- in particular the one-stop approach and the availability of a bundle of transactional services behind a single query is felt as a target "at reach" by the wide public
- while great concern is detected about privacy and data transparency, some basic technical issues, such as data integrity, information redundancy and system consistence, are given for granted

From the management point of view, they are:

- the PTR experience provided a valid laboratory for experimenting new ideas, methods and new regulatory arrangements (e.g. area plans, consortia of bodies at various institutional levels)
- new and more effective partnership emerged among the local authorities
- the involvement of politicians has permitted a wider dissemination of best practice at local level, an increase of available resources and has encouraged the creation of the critical mass required for many projects.

References

1. Guidelines for the telematic development of Emilia-Romagna: a proposal to the local system, http://www.regione.emilia-romagna.it/pianotelematico/ Nov. 1999
2. E-Italia, Forum Società dell'informazione, Il Sole24Ore, June 2000
3. Sandri A., InRete: first action plan of the Regional Telematic Plan, RER, February 2000
4. Liikanen E.; Europe in the information age – accelerating the transition, Telecom Conference, Helsinki, October 1999
5. Gillespie A., Information technologies and integrated city-region development, TeleRegions Conference, Tanum, June 1999
6. La Forgia A., The Global Region, RER, July 1997

Transactional e-Government Services:
An Integrated Approach

C. Vassilakis, Giorgos Laskaridis, G. Lepouras, S. Rouvas, and Panagiotis Georgiadis

e-Gov Lab, Department of Informatics, University of Athens, 15784, Greece
{costas,glask,gl,rouvas,p.georgiadis}@e-gov.gr

Abstract. Although form-based services are fundamental to e-government activities, their widespread does neither meet the citizen's expectations, nor the offered technological potential. The main reason for this lag is that traditional software engineering approaches cannot satisfactorily handle all of electronic services lifecycle aspects. In this paper we present experiences from the Greek Ministry of Finance's e-services lifecycle, and propose a new approach for handling e-service projects. The proposed approach has been used successfully for extending existing services, as well as developing new ones.

1   Introduction

e-Government development may be quantified through a set of indicators to measure comparative progress. *eEurope* has published a list of 20 basic public services [1] which should be considered as the *first steps* towards "Electronic Government", along with a methodology for measuring government online services [2]. It is worth noting that among the basic public services listed in [1], 75% of them include e-forms filling and submission. Moreover, the citizen interface must be connected to the back-office in order to provide the transactional capability, which will allow it to offer the rich mix of services that customers want and governments have promised [3].

However, transactional services widespread currently lags behind the expected level, taking into account potential offered by technology. Besides structural reforms, and the adaptation of a customer-centric model [4], the development and maintenance processes for such services are quite complex: firstly, service requirements must be analysed; secondly, the service has to be designed, considering functional requirements, user interface aspects, and administrative issues. Implementation and deployment should then commence, and the e-service platform should be linked to some installed IT system, for exchanging data. Finally, when changes to the service are required, the whole process must be carried out, resulting in costs and delays.

2   Experiences from an Electronic Service's Life Cycle

The development of the electronic tax return form service for year 2001 followed a classic software engineering paradigm, starting off with the user requirements analysis. Four requirement dimensions were identified in this procedure (a) appropriate input forms (b) *input validation checks*, which verify that user input is conformant to tax legislation rules (c) forwarding of collected data to a back-end system for the tax computation process and (d) collection of the results of the tax computation

R. Traunmüller and K. Lenk (Eds.): EGOV 2002, LNCS 2456, pp. 276–279, 2002.
© Springer-Verlag Berlin Heidelberg 2002

process and user notification about the final result. These four requirement dimensions are in fact the "computerised" counterparts of the paper-based tax return form and tax computation process; some additional issues had also to be faced due to e-service operation (a) issuing of *identification credentials* service users (b) personal data management for authenticated users and (c) provision of a full service administration framework for the GmoF's administration team, including account management, database backup and recovery, statistics reports etc.

The design phase followed, producing detailed specifications of the various components and procedures, along with specification of the interfaces between the different stages of the information flow. Subsequently, each portion of the work was implemented "autonomously", and an integration step consolidated the different work dimensions into a single, operational platform. After the service became operational, maintenance tasks were performed, mainly for the purpose of correcting appearance problems and modifying or enhancing input validation checks.

During the life cycle of the project, a number of shortcomings of the followed approach were identified, which led to increased product delivery times and the need to involve more staff. These shortcomings are discussed in the following paragraphs.

1. Knowledge existed *implicitly* within the organisation, usually under the possession of experienced individuals, rather than stored in some publicly accessible repository in an explicit form. This affected all subtasks dealing with the organisation's *business logic*. During interviews, the analysis team often collected partial, or contradicting descriptions of the rules that applied to different cases.
2. User interface design did not include adequate domain expertise into forms. For instance, semantically related fields should be placed close together on the form for easier access. The user interface should also compensate the lack of expert assistance offered by tax officers in local tax offices.
3. Code reusability remained low, since code incorporated business rules, presentational elements, administrative issues and data repository accesses in an environment of increased cohesion. Thus changes to one of these issues triggered modifications to the other dimensions.
4. *Communication with back-end systems.* A full transaction processing cycle usually involves data exchange between the service platform and some organisational information system, either in an on-line or an off-line fashion. In both cases, building interfaces tailored for each case is not a good practice, since it requires substantial programming effort to address the peculiarities of the back-end system.

Regarding these issues, a number of technological solutions are available. Recording knowledge in an explicit and reusable format is addressed by knowledge management tools, while a number of research projects introduce novel methods to knowledge management ([5], [6]). For the service development and deploying phases both commercial products (e.g. XMLForms™ and Oracle E-Business Suite™) and Open Source solutions. In the standards area, the W3 consortium has published the *XForms* specification, an XML-based standard for specifying Web forms, while user-centric approaches for interfaces are discussed in the ISO 13407 standard.

3   Proposed Approach

Although the state-of-the-art provides sufficient tools for tackling the various phases of e-service lifecycle, these phases are still handled in isolation. To provide an alternative solution for the above-described problems one has to adopt a layered approach that introduces higher levels of abstraction (enhancing maintainability and re-usability), isolate knowledge from code, allow asynchronous development of modules and enable people with domain expertise to implement new services.

This approach shifts implementation focus from programmers to domain experts, who possess the knowledge needed to process the data gathered through the e-service. In this view, an e-service consists of the e-form and the processing of the submitted data. Processing has two phases: validation and information extraction. In the first phase submitted data is checked against a set of rules and, depending on the result, the user may be asked to change and re-submit the form. Although some simple error checks (e.g. date and number checking) are usually coded by programmers, for more complex dependency checks domain expertise is of the essence. For example, a check "if user selects field 170 and 180 and has less than two children, then field 190 has to be completed according to the law 1256/98" requires deep domain knowledge. So far, either the programmer worked closely with domain experts to program these checks, or the e-service gathered data that were processed manually, resulting in late error detections. The proposed approach enables experts to play a much more active role and carry out a much greater percentage of the implementation effort.

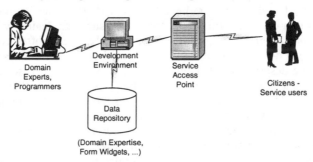

Domain
Experts,
Programmers

Development
Environment

Service
Access
Point

Citizens -
Service users

Data
Repository

(Domain Expertise,
Form Widgets, ...)

Fig. 1. System Architecture

As illustrated in Fig. 1., the development environment employs a data repository, where the ingredients of an e-service can be stored. These can be the widgets that compose the e-form along with their properties (type, multilingual labels, etc.) and the domain expertise needed to implement the service. Domain experts and/or programmers can use the development environment to implement and activate new services. The development environment has been designed to be friendly and intuitive, since a significant portion of its expected users (i.e. the domain experts) is not usually too familiar with the programming issues concerning the implementation of an e-service. To this end, an appropriate set of tools is provided.

Service development begins with the definition of the *forms* that comprise the service, and each form is then populated with the respective fields and labels. Fields and field groups may be directly drawn from the data repository, promoting thus uniformity between various services and enhancing reusability. For instance, the area displaying the user's personal data may be defined once, and then be used in all taxation e-forms. If no suitable component exists, the developer may create a new one, by entering information such as the description, its type (string, arithmetic, etc.), whether it is editable or not, its appearance on the screen etc. Additionally, a high-level specification of data interchange with the organisation's IT systems is provided, while *validation rules* may be attached to fields. A second level of validation rules may be attached at *form level* and *service level*, to cater for checks involving multiple components. Validation rules are defined using *semantically rich elements*, e.g. "the total husband's income from salaries", rather than implementation-oriented terminology, such as *table4_field22*. This enhances4 readability and facilitates maintenance, since the level of abstraction remains high and semantic information in retained in this representation. Finally, domain experts would also provide the test cases to verify the validation process.

The adoption of the proposed approach offers significant advantages: Firstly, development is faster since previously developed and tested e-service components can be directly used in new services; this allows for cost reduction as well. Secondly, maintenance is facilitated , since it is easy to locate the components and services affected by some change either in legislation or policy. Moreover, changes are now mainly conducted by domain experts. Thirdly, consistency is increased, through the introduction of a central repository for knowledge storage. Uniformity in the look and feel may be also achieved through usage of service templates. Finally, *multiple dissemination platforms* may be supported, through appropriate content generators that create content suitable for a variety of platforms (Web, WAP, etc).

4   Conclusions

In this paper we presented experiences from developing and maintaining a set of e-services for the Greek Ministry of Finance. The traditional software engineering approaches employed in the first development phases proved to be inadequate in handling all aspects related to the lifecycle for electronic services. In the second phase we used a new approach, together with appropriate software tools, which allowed for using higher levels of abstraction, enhancing thus the maintainability, portability and reusability of the project's results, and reducing overall development time.

References

1. eEurope, "Common list of basic public services"
2. eEurope, "eGovernment indicators for benchmarking eEurope".
3. Vivienne Jupp, "eGovernment – Lessons Learned, Challenges Ahead", eGovernment Conference: From Policy to Practice, 29-30 November 2001, Charlemagne, Brussels.
4. Frank Robben "(Re)-organising for better services", eGovernment Conference: From Policy to Practice, 29-30 November 2001, Charlemagne, Brussels.
5. Know-Net Consortium, "Manage Knowledge for Business Value"
6. DECOR Project, "Delivery of context-sensitive organisational knowledge"

Electronic Vote and Internet Campaigning: State of the Art in Europe and Remaining Questions

Laurence Monnoyer-Smith and Eric Maigret

University of Technology of Compiègne, CNRS, France
University of Paris 3- Sorbonne, CNRS, France
EVE partner

Abstract. Recent experiments shows that internet voting is not the political blessing that lots of politicians had hoped it would be to solve the non-ending legitimation crisis modern societies are going through. Our article focuses on four questions about internet voting to stress some experimental results and important remaining questions. We would like to insist on the lack of studies trying to apprehend how people trust these new electronic voting systems and how do they cope with the end of the voting rituals.

1   Introduction

The internet is quite commonly viewed as being able to serve as a new linkage institution enhancing voter information about candidates and elections in general, and as a consequence mobilizing an apathetic population and even, increasing voter turnout[1].

This is far from being a simple equation. Recent studies are showing interesting and mixed results about the ability of the internet to enhance political involvement: it is not quite obvious that internet is a political blessing. Rather, it seems to evolve toward a managerial model which concern "the 'efficient' delivery of government information to citizens and other groups of 'users' ; the use of ICT's to improve flows of information within and around the organs of government; a recognition of the importance of 'service delivery' to 'customers' ; the view that speeding up information provision is by itself 'opening up' government[2]..."

Our current EVE project, financed by the European Commission (DG IST), allows us to compare 11 different e-democracy experiments that permits us to stress a few points on which our knowledge is more or less salient and some remaining very important questions that have to be answered by those who are developping new processes aiming at voting, delibarating and participating in political life.

---

[1] See for exemple, the motivations developped by the UK government on their new on-line voting program: www.sheffieldvote.com,

[2] Chadwick A., *"Interaction between citizens in the age of internet: "e-goverment" in the United States, Britain and the European Union"*, Paper given to the APSA's Annual Meeting, San Francisco, August-September 2001, p. 2.

R. Traunmüller and K. Lenk (Eds.): EGOV 2002, LNCS 2456, pp. 280–283, 2002.

2   Do People Seek Political Information on the Internet during Election Campaign?

It is here very interesting to note how far from the expectations european and canadians users stand, especially compared to the USA. In the UK for exemple, James Crabtree[3] shows that during the june elections, although the UK is catching America's acess levels, british citizens lag far behind in understanding the real potential of internet.

Although we have to be cautious with extending the UK case to the other european countries, it seems that some important barriers prevent people for opting-in the internet contrary to the US citizens. The available information seems not to be interactive, attractive, and different enough from the one available off-line. During the last French presidential elections for exemple, none of the candidates has made specific efforts to present interactive informations.

The first contact with the information is crucial if the user is to come back for more: until more efforts are made, internet users won't pay enough attention to it.

3   Do the Use of Internet Enhance Voting Turnout?

Here again, we can't give straight answers considering the available datas. A few remarks can be made, concerning some specific populations studied during the past elections and the famous Arizona case.

There is no evident proof of an existing link between the act of voting and the use of internet whether it is to vote or to get political information. If there is a link, it might be very different from one election to another (from a presidential to a parlementary election): this might explain the differencies shown between the US and the UK and between the 2000 studies and the 1998 studies.

One has to consider as well that situations mights vary a lot from one country to another: in the UK last local elections for exemple (may 2002) electronic voting has attracted more people than usual: the publicity given to the Sheffield and Liverpoll experiments might be nevertheless a bias. Country in which elections are taking place on a week day might be more keen to vote via internet than those, like France, which organise their election on sundays.

Except from the Arizona study, which is very different from a national election, they all show that internet users are no more keen on voting than non users. It does not answer the question that if they could do so via the internet, people would vote more or less.

The Digital divide remains a fundamental issue for internet voting. The Arizona studies are congruent with each other underpinning that particular demographic groups already noted for being democratically disadvandaged are excluded even more than others with the introduction of internet voting.

---

[3] Crabtree J., « *Whatever happen to the e-lection* », a survey done by isociety, juin 2001. available on: www.ipf.co.uk/egovernment/posts/post99.htm

4   What Do We Know about the Young People and Politics on the Internet?

One of the most stable findings is that the young, are more likely to use the Internet than the elderly. The internet thus may represent an important venue for mobilizing young voters who have historically been underrepresented in the electorate, and traditionnaly less voting. Correlated with what have seen before, ie that internet users are not more likely to vote than non users, we can assume than this is partly caused by the importance of young people amoung users. Their political cynicism is spreading on the internet.

According to Henn and Weinstein (2001), the internet voting is not considered by young people to be a substancial reform within the political system and therefore will not more than marginnaly modify their behavior. –Which is better than nothing but will not in itself solve the legitimacy crisis that the political system suffers from-
Neverthless, some of the "substancial" problems could be partly solved by an interactive use of the internet (provide information, interact with voters...)[4].

This has been confirmed by the european project E-Poll which stressed that in Avellino where has been organised the first biometric recognition voting system for the Italian constitutional referendum on october 7th 2001, elderly people were more keen on trying this system that young people and they do trust the system more than under 30 years old people.

5   Trust, Rituals and Electronic Voting

No sociological research has been done on those specifics and yet fundamental aspects of the question of e-voting.

If one of the advantages of internet voting is certainly that it eliminates counting mistakes and saves a lot of time when giving the results, lots of security aspects remain uncertain for citizens. Security is not only a technical problem that might be solved by ingenieurs and improved as the technology is getting more complex and close to the 100% identification certitude (with fingerprint for ex), it is also a psychological issue. Then it is more appropriate to talk about "trust" than security from the citizen's point of view. If some research has been done on the e-business, none has tried to understand how to build a trust relationship via the internet with the organisation that will run the election on line. Time will probably be necessary to implement the system and to bring people to trust it as they do when they buy on line (Filser[5]). All e-business studies on that matter show as well the role played by intermediary actors in reinforcing the faith people have in a system (hot line, etc.): who could they be in the case of electronic voting and what kind of intervention could they have? These questions need to be studied or we might be disappointed by the low use of online voting systems.

---

[4]   Henn and Weinstein, « Youth and Voting behavior in Britain » ", Paper given to the APSA's Annual Meeting , San Francisco, August-September 2001.

[5]   Filser M, "Etat des recherches sur les canaux de distribution", in *Revue Française de Gestion*, Sept.-Oct 1998, pp. 66-76.

Another issue never analysed but stressed by some researchers[6] is whether the withdrawal of such an solemn event as voting, might loose his signification as a symbol which unite people in a common commitment toward democracy. In other words, electronic voting might remove from its pedestal such an important ritual from which generations have fought for. Does the move toward electronic voting correspond to this trend that tend to bring people closer to the elite or does it dilute democracy into a more vast consumer society were voting becomes a consumer act as buying a train ticket on line? One might underline that by introducing information technology at the heart of the political decision process, the society emphasises the passage from a culture of effort to a cultur of service in political matters. This has a lot of implications that should be studied and taken into account in future technological developments.

6   Conclusion

The potential for electronic governance is almost certain only in appropriate conditions. The lack of information that citizen often stress could be fullfiled by the new media. In concertation with political actors who have to be involved in the implementation of new technology, and in consideration of all the citizen's claim –not only access and practical questions[7]-, the development of e-governance processes could have then a positive influence on political participation. On its own, internet is not a blessing.

One recent hypothesis (Bucy et Gregson[8]) is that new media formats might satisfy this need for popular involvement by delivering a continuous stream of opportunities for civic engagement without overextending the government ability to respond. The multimedia thereby increases the likelyhood that citizen concerns will be heard. By making allowance for continuing mass involvement, new media formats serve the socially valuable purpose of bringing closer to reality to classical goal of full participation without over extending already burdened political institutions.

---

[6] Maigret E., Monnoyer-Smith L., « Le vote en ligne: Usages émergents et symboles républicains », Colloque ICUST, Juin 2001 ; Shnapper D., *Qu'est-ce que la citoyenneté*, Paris, Gallimard Folio, 2000.

[7] As C. Aterton already stresses it in 1987, the renewal of political life has very little to do with the technical aspects of voting. *Can technology protect democracy*, London, Sage.

[8] Bucy E.P., Gregson K.S., " Media participation: a legitimizing mechanism of mass democracy", *New Media and Society*, Vol.3(3)., pp. 357-380.

A Citizen Digital Assistant for e-Government

Nico Maibaum*, Igor Sedov**, and Clemens H. Cap

University of Rostock, Department of Computer Science,
Chair*** for Information and Communication Services
{maibaum,igor,cap}@informatik.uni-rostock.de

Abstract. In this short paper we describe the architectural concept of a Citizen Digital Assistant (CDA) and preliminary results of our implementation. A CDA is a mobile user device, similar to a Personal Digital Assistant (PDA). It supports the citizen when dealing with public authorities and proves his rights - if desired, even without revealing his identity. Requirements for secure and trusted interactions in e-Government solutions are presented and shortcomings of state of the art digital ID cards are considered. The Citizen Digital Assistant eliminates these shortcomings and enables a citizen-controlled communication providing the secure management of digital documents, identities, and credentials.

1 Introduction

'The quality of life is also depending on how good and how fast a government is doing its services to the public" said Germany's minister Otto Schily (Federal Ministry of the Interior) at the congress 'Efficient Government' in Berlin 2001 [1]. Today, every public authority is talking about a modern, efficient and citizen-friendly administration. The vision is to map the possibilities obtained through the new IT technologies, and demonstrated within the e-business sector, to the e-government sector. At the end, the citizen will communicate (via Internet) with a modern, efficient, cost-effective and transparent administration.

In order to achieve the above-mentioned goals research projects are dealing with different digital signature, national ID or citizen cards for usage within e-Government applications. Examples are the research project FASME[1], the FineID[2] and the DISTINCTID[3] project. The process of registering at a new place of living within Europe is facilitated by FASME. Administrative data, documents and profile infor-

---

*   Supported by the EU Fifth Framework Project FASME, http://www.fasme.org
** Supported by the National Research Foundation (DFG)
*** Supported by a grant of the Heinz Nixdorf Foundation
[1] Facilitating Administrative Services for Mobile Europeans (FASME), http://www.fasme.org
[2] Finnish Electronic Identification (FineID), http://www.sahkoinenhenkilokortti.fi
[3] Deployment and Integration of Smartcard Technology and Information Networks for Cross-Sector Telematics (DISTINCTID), http://distinct.org.uk

R. Traunmüller and K. Lenk (Eds.): EGOV 2002, LNCS 2456, pp. 284–287, 2002.
© Springer-Verlag Berlin Heidelberg 2002

mation of a citizen are stored on a JavaCard. One main problem within this project was the low memory of the JavaCard. The solution is a secure card extension using a kind of virtual memory for smart cards [2].

2   Requirements for Secure and Trusted Interactions

The following four requirements are necessary for a secure and trusted interaction between a citizen and public authorities. Presently, they are not satisfied by any consumer device (for example smart cards as digital ID cards).

The **Subscriber Identity Framework** is responsible for access control and controlled disclosure and release of identity (*real identity*, *pseudonymity* and *anonymity*). The on-card functionality of smart cards (e.g. digital signature, payment etc.) is normally restricted through a personal identification number (PIN) with a length of four up to eight digits, sometimes changeable through the cardholder. Today, the increasing number of different passwords and PIN codes is a serious problem for many people, because they are not able to remember all these codes. Furthermore, the verification of a PIN realises no real *proof of identity*, just *a proof of knowledge*.

The **Monitoring and Confidentiality Framework** is responsible for the management, supervision, and logging (security audit) of trustworthy transactions. Smart cards are the carrier devices for digital signatures. A resulting problem is obvious: *"How can it be guaranteed that the document to be signed by the smart card is indeed the document delivered to the smart card for signature?"* The smart card does not include a monitoring module. The cardholder must trust the card accepting device and the system behind it.

The **Secure Document Management Framework** is responsible for the secure storage, retrieval and administration of digital counterparts of paper-based documents (e.g. XML documents as e-docs), certificates, citizen profile information, and credentials [6]. At the moment everyone is talking about digitally signed documents but no one is talking about the secure storage of these e-docs. Smart cards are not utilisable because of their limited storage capacities.

The **Communication and Cryptographic Framework** is responsible for the communication between the citizen and public authorities. This includes unilateral, bilateral and multilateral security, secure key management, en- and decoding and the definition of different security levels. A smart card must rely on the card accepting device or the connected personal computer. The smart card is not able to connect to another device on its own.

3   Implementation Aspects and Application Scenarios

To fulfil the e-Government requirements described in Section 2, smart cards, PDAs, or mobile phones, are just partial solutions not fulfilling all these requirements. A Citizen Digital Assistant can be realised as a combination of a PDA and a mobile phone. We propose an architecture integrating a biometric component (*data storage*, *comparison process*, and *integrated biometric system*) for a real citizen authentication / identification. Biometric information never leaves the device (and therefore cannot be misused), comparable to private keys, and there is no need to rely on other external devices.

Every CDA is identical. Only after a personalization the CDA gets its own "citizen" identity. Therefore at least one card slot for a digital signature card or an electronic, national identity card is required in order to personalise the CDA. Problems, which still exist by using smart cards, can be solved with the CDA. The user does not need to insert her plastic card in any "unknown" device. She inserts the card in her trusted CDA and there is no danger that regular operations are in the sphere of influence of another instance or card-accepting device [4] with not checkable hard- and software.

The CDA is a device which *'talks"* and *"interacts"* vicariously for the citizen (see Figure 1). The user device is a combination of existing and widespread components. The costs continuously going down and the unbroken trend for miniaturisation of the components make the appropriate realisation not only thinkable but also possible.

The CDA can be used for e-voting and can be the carrier device for temporary credentials (see Fig. 1), any kind of digital documents or profile data.

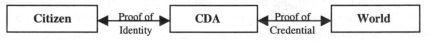

Fig. 1. Citizen remains anonymous via CDA

The citizen will no longer have to carry personal documents in paper form with them in order to obtain certain administrative services. Asynchronous data transfers (e.g. document verifications and document requests) with long latency can be supported, too. The citizen will not have to wait with her card inserted into a kiosk terminal until the complete administrative process is finished.

Such a user device can be used not only within the e-government sector but also within the e-business and internet sector, too. Everywhere the protection of privacy and identity is required the CDA can act.

4   Conclusion and Future Work

Our preliminary CDA implementation is based on the Compaq IPAQ 3660 running Linux 2.4.14[4]. To illustrate spontaneous networking in an e-government scenario the simple, low cost and low energy consuming Bluetooth technology (Ericson Bluetooth Application & Training Tool Kit) is utilised.

Enabling a personal identification without PINs, fingerprint technology based on the Siemens TopSec ID Module was selected. The advantage of this module is the

---

[4] Some parts are developed within the DFG research project "Security architecture and reference scenario for spontaneous networked mobile devices" at the University of Rostock [5].

secure enrolment, storage and verification inside the exchangeable module. The Service Discovery Protocol provides mechanisms for discovering and associating services within an ad-hoc community. To support multilateral security and the integration of security interest of communication partners an advanced security manager concept was developed, that is partly implemented [3] and is now evaluated.

Work is in progress to implement all the required interfaces. On top of the architecture will be a small Java-based e-government application, that was implemented within the European Research Project FASME including the usage of the Java-based FASME smart card.

The realization of such a CDA is the core module within an e-(government) system. Fixed IPs and unique processor serial numbers for the identification are avoided. Networks without user observability would facilitate a lot of problems [7].
In the future it will be possible that every citizen will have her own CDA including the desired management system in her pocket, possibly distributed by the authorities for a low fee like other official documents or different cards (e.g. passports and health insurance cards).

Smart cards are ideal for many application areas, but the missing monitoring functionality, the limited memory and the insufficient possibilities for biometric authentication requires new solutions.

References

1. O. Schily. Federal Ministry of the Interior, Germany. *Auf dem Weg zu einer modernen Verwaltung - BundOnline 2005*, 2001. Talk within the congress "Efficient Government".
2. C. Cap, N. Maibaum, and L. Heyden. *Extending the Data Storage Capabilities of a Java-based Smartcard*. In Proceedings of the Sixth IEEE Symposium on Computers and Communications ISCC 2001, pages 680 - 685, IEEE Computer Society, July 2001.
3. H. Buchholz. *Sicherheit in Ad-hoc Netzwerken*. Diploma Thesis, 2002.
4. H. Federrath and A. Pfitzmann. *Bausteine zur Realisierung mehrseitiger Sicherheit*, in G. Müller and A. Pfitzmann (Hrsg.): Mehrseitige Sicherheit in der Kommunikationstechnik, Addison-Wesley-Longmann, 1997.
5. Sedov, M. Haase, C. Cap, and D. Timmermann. *Hardware Security Concept for Spontaneous Network Integration of Mobile Devices*. In Proceedings of the Innovative Internet Computing Systems Conference, Ilmenau. Springer, June 2001.
6. Cap, Clemens H. and Maibaum, Nico. *Digital Identity and it's Implications for Electronic Government*. In Towards the E-Society - E-Commerce, E-Business, and E-Government, Kluwer Academic Publishers, Boston, 2001.
7. A.Pfitzmann and M.Waidner. *Network without User Observability*. Computers & Security, vol.2, no.6, pages 158-166, 1987.

A System to Support e-Democracy

Jan Paralic, Tomas Sabol, and Marian Mach

Technical University of Kosice, Letna 9, 041 20 Kosice, Slovakia
{Jan.Paralic,Tomas.Sabol,Marian.Mach}@tuke.sk
http://esprit.ekf.tuke.sk/webocracy/

Abstract. This paper briefly describes functionality and customisation support of the system called WEBOCRAT, which is being developed within the EU-funded project Webocracy (IST-1999-20364 "Web Technologies Supporting Direct Participation in democratic Processes"). The WEBOCRAT system represents a rich set of communication supporting tools that will bring public administration closer to citizens, making it more accessible and more accountable.

1   Introduction

The Webocracy project responds to an urgent need for establishment of efficient systems providing effective and secure user-friendly tools, working methods, and support mechanisms to ensure the efficient exchange of information between citizens and public administration (PA) [1]. This project addresses the problem of providing new types of communication flows and services from public PA institutions to citizens, and improves the access of citizens to PA services and information.

In [6] a three-phase strategy, for implementing e-democracy consisting of initiation, infusion and customisation phases, has been proposed. The first, *initiation phase*, starts with establishment of a portal that conveniently links citizens to PA. Next, *infusion phase*, means restructuring the organisation in order to accommodate innovation. And, finally, *customisation phase* of e-democracy system implements a one-to-one relationship between citizen and PA. The *WEBOCRAT* system focuses on support of infusion and mainly customisation phases by various specialised modules.

One of the main novelties of our approach is the knowledge-based support [4]. Information of all kinds produced by various modules or segments of these documents is linked to a shared ontology representing a domain of PA. The advantage of clear structuring and organising of information is more powerful search and retrieval engine, and more user-friendly content presentation [2].

2   WEBOCRAT System Functional Overview

From the point of view of functionality of the *WEBOCRAT* system it is possible to break down the system into several parts and/or modules [5]. They can be represented in a layered sandwich-like structure, which is depicted in Fig. 1.

R. Traunmüller and K. Lenk (Eds.): EGOV 2002, LNCS 2456, pp. 288–291, 2002.
© Springer-Verlag Berlin Heidelberg 2002

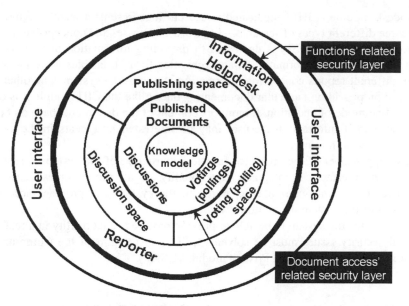

Fig. 1. *WEBOCRAT* system structure from the system's functionality point of view

A Knowledge Model module occupies the central part of this structure. This system component contains one or more ontological domain models providing a conceptual model of a domain. The purpose of this component is to index all information stored in the system in order to describe the context of this information (in terms of domain specific concepts).

Information stored within the system has the form of documents of different types. Since three main document types will be processed by the system, a document space can be divided into three subspaces – publishing space, discussion space, and opinion polling space. These areas contain published documents to be read by users, users' contributions to discussions on different topics of interest, and records of users' opinions about different issues, respectively.

Since each document subspace expects different way of manipulating with documents, three system's modules are dedicated to them. Web Content Management module (WCM) offers means to manage the publishing space. It enables to prepare documents in order to be published (e.g. to link them to elements of a domain model), to publish them, and to access them after they are published. Discussion space is managed by Discussion Forum module (DF). The module enables users to contribute to discussions they are interested in and/or to read contributions submitted by other users. Opinion Polling Room module (OPR) represents a tool for performing opinion polling on different topics. Users can express their opinions in the form of polling – selecting those alternatives they prefer.

In order to navigate among information stored in the system in an easy and effective way, one more layer has been added to the system. This layer is focused on retrieving relevant information from the system in various ways. Citizens' Information

Helpdesk module (CIH) is dedicated to search. It represents a search engine providing three different types of searches – keyword-, attribute- and concept-based.

Reporter module (REP) is dedicated to providing information of two types. The first type represents information in an aggregated form. It enables to define and generate different reports concerning information stored in the system. The other type is focused on providing particular documents – but unlike the CIH module it is oriented on off-line mode of operation. It monitors content of the document space on behalf of the user and if information the user may be interested in appears in the system, it sends an alert to him/her.

The upper layer of the presented functional structure of the system is represented by a user interface. It integrates functionality of all the modules accessible to a particular user into one coherent portal to the system and provides access to all functions of the system in a uniform way.

In order for the system to be able to provide required functionality in a real setting, several security issues must be solved [3]. This is the aim of the Communication, Security, Authentication and Privacy module (CSAP).

3   Customization Support in the WEBOCRAT System

Since the system can contain a lot of information in different formats (published information, discussion contributions, polling results), it may not be easy to find exactly the information user is looking for. Therefore he/she has the possibility to create his/her profile in which he/she can define his/her interests and/or preferred way of interacting with the system.

When defining an area of interest, user can select elements from a domain model (or subparts of this model). In this way user declares that he/she is interested in topics defined by the selected part of the domain model.

The definition of user's area of interest enables alerting – user can be alerted, e.g. on changes of the domain model, when a new discussion or opinion polling has been opened, new documents published etc. User has the possibility to set alerting policy in detail on which kind of information he/she wants to be alerted in what way (including extreme settings for no alerting or alerting on each event taking place in the system). The system compares each event (e.g. submission of a discussion contribution, publishing a document, etc.) to users' profiles. If result of this comparison is positive, i.e. the user may be interested in the event, then the user is alerted.

Alerting can have two basic forms. The first alternative is represented with notification using e-mail services. User can be notified on event-per-event basis, i.e. he/she receives an e-mail message for each event he/she is alerted on. Alternatively, it is possible to use an e-mail digest format – user receives e-mail message, which informs him/her about several events. The way of packaging several alerts into one e-mail message depends on user's setting. Basically, it can be based on time intervals and/or the size of e-mail messages.

The other alternative is a 'personal newsletter'. This does not disturb user at unpredictable time – user simply can access his/her newsletter when he/she desires to be informed what is on in the system. Moreover, he/she can access it from arbitrary gadget connected to the Internet. The personal newsletter has the form of a document published in the publishing space. This document is generated by the system and contains links to all those documents, which may be of interest for the user. Since the document is generated when user logs in, it can cover all information submitted and/or published since the last user's visit.

User registered in the system as an individual entity (i.e. not anonymous user) is provided with a personal access page ensuring him/her an individual access to the system. This page is built in an automatic way and can consist of several parts. Some of them can be general and the others are user-specific.

The personal access page hides division of the system into modules. Terms 'publishing space', 'discussion space', and opinion polling space' do not confuse users. The personal access page enables user to access all functionality of the system, which he/she is allowed to access in a uniform and coherent way.

Acknowledgements

This work is done within the Webocracy project, which is supported by European Commission DG INFSO under the IST program, contract No. IST-1999-20364 and within the VEGA project 1/8131/01 "Knowledge Technologies for Information Acquisition and Retrieval" of Scientific Grant Agency of Ministry of Education of the Slovak Republic. The content of this publication is the sole responsibility of the authors, and in no way represents the view of the European Commission or its services.

References

1. Becker, T. (2000). Rating the Impact of New Technologies on Democracy. *Communications of the ACM*, 44 (1), 39-43.
2. Dourish, P., et al (1999) Presto: an experimental architecture for fluid interactive document spaces, *ACM Transactions on Computer-Human Interaction*, 6 (2), 133–161.
3. Dridi, F., Pernul, G. and Unger, V: Security for the Electronic Government, *Proc. of the 14th Bled Electronic Commerce Conference*, "e-Everything: e-Commerce, e-Government, e-Household, e-Democracy", Bled, Slovenia, June 2001.
4. Dzbor, M., Paralic, J., and Paralic, M. (2000) Knowledge management in a distributed organization. In *Proceedings of the 4th IEEE/IFIP International Conference BASYS'2000*, Kluwer Academic Publishers, London, pp. 339-348.
5. Mach, M. and Sabol, T. (2001). Knowledge-based System for Support of e-Democracy. In *Proceedings of European Conference on e-Government ECEG'2001*, Trinity College, Dublin, pp. 269-278.
6. Watson, R.T., and Mundy, B. (2001). A Strategic Perspective of Electronic Democracy. *Communications of the ACM*, 44 (1), 27-30.

IST-Project:
AIDA – A Platform for Digital Administration

Anton Edl

anton.edl@infonova.com, http://aida.infonova.at/

1   Project Overview

1.1   The Consortium

The AIDA Consortium has a clear structure based on the objectives of the project. It includes:

- one leader in internet security: INFONOVA
- one big hardware equipment manufacturer: HP
- two universities: Politecnico di Torino working on the application side and Technical University of Graz, working in cryptography and software research
- An innovative telecommunication-applications company with great know how in market access: I&T
- Four organizations willing to support the design process and to implement and validate the services and technologies. Euro Info Correspondence Center Ljubliana, Ministry of economic affairs of Slovenia, Mestna Obcina Celje,

1.1.1   INFONOVA

INFONOVA is one of Austria's largest companies for the development of Internet-based telecommunication services and networks. Founded in 1989 by a small group of engineers, it has grown considerably and is now one of the biggest companies in telecommunication research and development in Austria and technology-leader for Public-Key-Infrastructure, E-Business and network integration.

Role in Project
Beside the project management, INFONOVA developed the WYSIWYS-software and deployed AIDA's core-platform.

1.2   General Objectives

AIDA's objective is to implement Advanced Interactive Digital Administrations by providing them with an infrastructure which supports administrations and other public bodies, to improve businesses' and citizens access to information and regulation and facilitate contacts, exchanges and feedback between administrations and between administrations and third parties, i.e. citizens, institutions and business.

The European Signature Guidelines as well as digital signature laws in different countries will lay the foundation for the issuing of electronic documents by public

R. Traunmüller and K. Lenk (Eds.): EGOV 2002, LNCS 2456, pp. 292–297, 2002.
© Springer-Verlag Berlin Heidelberg 2002

institutions such as administrative bodies, professional associations or universities, which – digitally signed – will be held equal to conventional paper-documents. This will enable these institutions e.g. to issue innovative forms of conventional documents, such as electronic certificates of birth, electronic trade licences, electronic diplomas and many more. Electronic documents like these can then be used instead of paper-documents.

1.2.1   Advantages of Electronic Documents

Using electronic documents has a lot of advantages - for the purpose of this proposal, documents providing certain qualities will be called e-documents :

- Electronic documents can be sent by means of the internet, thus enabling future applications for situations in which one would have to show up in person today – just for presenting paper-documents. Examples are: car-registration, registration of birth etc.
- Quite a lot of e-commerce activities require proof of identity, legitimization or authorization – which can easily be realized by presenting e-documents
- In combination with ITU-T X.509 digital certificates, the application and distribution of which are currently encouraged all over Europe, e-documents provide significant additional value to e-administration or, more general, to e-business environments.

1.2.2   Aims of the Project

Such electronic documents will only be valid if the digital signature they contain has been created by using a special environment. The European Signature Guidelines as well as national legislations recommend or even require the use of a trustworthy environment for the generation as well as for the verification of advanced digital signatures generally.

Now, this project has all these aims:

- to define and implement a trustworthy environment by means of a signature terminal, which does not only provide a secure solution but is also easy to handle by users,
- to define and develop machine-readable datastructures for electronic documents which can be used for national **and** international purposes,
- to define and develop electronic documents to replace conventional documents used at demonstrator sites and make proof of the usability of such documents in their electronic form for electronic administration environments, and, furthermore, to implement an environment and an infrastructure where such electronic documents can be used.

Summarising one can say, trustworthy digital signatures will be fundamental for various next generation services in the digital era. One of these services shall be an "e-Administration Service Provider" – and to achieve this, all   components, environments and techniques required shall be developed resp. integrated.

1.3   Technical Objectives

To enable Advanced Interactive Digital Administrations an infrastructure has to be developed, that consists of the following entities:

1.3.1   Signature Terminal

To create a trustworthy environment, all components have to be trustworthy and thus have to contribute to an overall secure solution.

- The signature creation device must make forging of signatures as difficult as possible by requiring an active user confirmation before signing anything. Entering a pin or fingerprint or any other authenticating information should ideally be done directly on the signature creation device.
- The signature creation device must be embedded into a secure environment, that also makes forging signatures difficult. This environment is responsible for preparing the data to be signed and to be sent to the signature creation device for hashing and signing. As the signature creation device might not be capable of displaying the data to be signed, the security of this environment and the binding to the signature creation device is crucial for the overall functionality of the system. Ideally, the signature creation device and the signature environment are integrated into one component.
- Displaying the data correctly is also very important. Correctly in this environment means that the user sees exactly the data he is going to sign and nothing else. A secure signature environment must therefore be able to parse and understand data and reject signing data that contain illegal parts which are unknown to the display unit. Examples for illegal contents are white text on white background or unknown HTML-tags etc. We call this module the WYSIWYS-module: What You See Is What You Sign.
- The software modules used must be authentic and therefore can only be run in an environment where such authenticated modules are supported and missing or wrong signatures would be detected reliably.

Theses project's objectives are to create and to incorporate all existing components needed to provide secure signature environment. It will provide a software-solution and an appropriate soft- and hardware solution, for example:

- Software-only solution for different platforms. To a large extent, Such a solution must rely on the environment, as it contributes largely to the security of the system. However, even if such a solution cannot provide perfect security, sometimes perfect security is not absolutely required. Features like the ability to display the data to be signed accurately, will still be very useful and appropriate for mass-market.
- Solution using small personal devices like subnotebooks or PDA's Integrating WYSiWYS-features into small portable computers like subnotebooks or PDAs equipped with a smart-card reader seem to be the ideal solution for secure signing equipment: the devices are small and can be carried around easily, while the display is usually still capable of displaying different forms of data to be signed.
- Solution using mobile phones. Mobile phones are also an ideal platform for a signing terminal: the devices are small and are carried around anyway and they already use smartcards. The main problem with traditional mobile phones is the limited display. For certain applications – like payment applications - this limitation causes no problem. Newer generation equipment integrates PDA-features with phones and should be the basis of an excellent future solution.

1.3.2   e-Documents

For e-documents, the first step will be to develop a general framework for such mutually acceptable documents for administrative purposes. This definition stage will concentrate on local administrations, e.g. cities and regions, but will have the European perspective in mind from the beginning. Documents like these need to be defined in a structured language, like XML, that easily allows translation into different languages and thus enables international deployment of this concept.

To provide e-documents for European citizens, the following techniques have to be integrated as well:

- Integration: signed XML. XML is of great importance for all data transactions in all domains of e-administration and/or e-business. In order to be able to warrant authenticity and integrity of transmitted data it is essential that digital signing is added to the features of XML.
- Specification: XML-data definitions. Appropriate XML-data definitions for of medical services have to be developed, especially taking into account the transnational network of correlations.

1.3.3   Management Platform for e-Documents

Around these document definitions, a basic universal infrastructure for e-administration services shall be implemented and/or integrated. An environment for generation and issuance of e-documents including a secure signing-infrastructure, a platform for citizens to enable them to display electronic documents and to verify their contents as well as the signatures of the authority, and a suitable directory infrastructure are required.

Special security devices in order to ensure authenticity and reliability as well as non-repudiation and retraceability have to be integrated into the fundamental platform for e-administration procedures.

Without any intention to anticipate the results of the corresponding evaluation task, well suitable security techniques applied in the field of public key infrastructures such as time stamping service, OCSP, chip cards (e.g. Java cards) etc. should be mentioned in this respect.

The management platform shall be a fundamental part of the "e-Administration Service Provider" strived for.

2   Case Studies

This is an overview of some Services, implemented during the AIDI Service Validation. More details and practical demonstrations will be shown at the conference live!

2.1   Exam Admission Service

The *exam admission service* is a widely and frequently used service inside any education institute and is a good candidate for e-administration due to the large spectrum of users and to the large number of paper documents usually involved.

Exam sessions at POLITO are programmed three times a year. Two exam sessions (in February and July) are ordinary sessions and one exam session (in September) is organized in order to help students pass the exams failed during the precedent sessions. Usually a student would have to sustain 6 exams/year. Starting with the academic year 2000/2001 the examinations for some courses will be split in 2 exams/course, so a student will have to sustain about 10-12 exams/year. Thus, he would usually need 12 paper EAC/year. Considering that a student could fail an exam, so he would need to obtain additionally another EAC. Thus, a student approximately needs $12 \times 1.5 = 18$ EACs per year.

POLITO has about 25.000 students, thus the total number of paper EACs released in a year would be approximately $25.000 \times 18 = \textbf{450.000}$.

2.2   Request for Extract from Birth-Book and Birth-Certificate

Extract from Birth-book (Birth-certificate) can be acquired by all physical persons for different purposes. It is a "personal birth-identification" of a person. There are different situations when an extract form Birth-book is needed:

- inscription of children to school,
- acquiring health card,
- marriage…

To get a Birth-certificate a request has to be put in to the competent public service. It can be written by hand or a pre-printed form can be used.

For passing the Birth-certificate the Administrative unit in the place of birth is in charge. Procedures for passing the Birth-certificate are regulated by different national laws, Paris and Vienna convention.

The Birth-certificate is handed on basis of correct fulfilled request. The request can be put in by:

- citizen for himself
- citizen for another person (in that case an authorisation is necessary)
- parents for children younger than 18 years.

The form of request is not prescribed. It can be also written by hand (e.g. MS Word). In that case the form has to contain all data needed for passing the certificate. Administrative unit Celje offers a standardised pre-printed form, which needs to be filled with required data. The pre-printed form is tax-free and available in the central office of Administrative unit, it can be also taken from Internet.

2.3   Permission for Lowering Accommodation Costs for Kindergarten Care

Duties for forfeiting the permission for the height of the payment for kindergarten care are according to several laws and directives valid in Slovenia in the working field of the local community. The process takes place in the municipality of Celje in the Department for social affairs, which is a part of the Ministry of work, family and social affairs of the Republic of Slovenia.

The right for lowering accommodation costs for kindergarten care is granted to any parent, on the basis of formulary, which is put in to the Department for social affairs.

In the next step the form is classified and a map with data is created. The application form is forwarded to the referent responsible for pre-school education in the Department for social affairs.

The referent is then in charge of the whole process and for informing parents about the result of the application. One copy is sent as information also to the kindergarten. Another copy is added to the application map and to the documentary files of the department.

City municipality of Celje has 49.000 citizens. The right for lowering accommodation costs for kindergarten care is taken in focus by 2000 parents. On a month basis we receive around 100 to 150 applications.

Main Achievements

The public demo platform at http://aida.infonova.at/demo/ proved to be a very stable one. All participants had numerous live-presentations via the public internet, which gave a lot of feedback for future enhancements of the system.

By the end of the project there are two **commercial products** - out of AIDA - on the market.

1. The Security Software Package – the so called "Crypto Library" is available on the market and a very stable and innovative product. See http://jcewww.iaik.tu-graz.ac.at/products/index.php for more information.
2. The WYSIWYS-Viewer was delivered to a customer in Austria by December 2001.The first certified Austrian Trust Center "A-Sign" http://www.a-sign.at/ recommends the SW-based version of the viewer for their strongest certificate (premium class certificate = equal to handwritten signature) to show signed XML-content.

So commercial success already started even before the project ended! This is actually one of **the** preferred outcomes of an EC-funded project.

For more information please contact: anton.edl@infonova.com

e-Government Strategies:
Best Practice Reports from the European Front Line

Jeremy Millard

Danish Technological Institute, Kongsvang Allé 29, 8000 Aarhus C, Denmark
Tel: +45 72 20 14 17; Fax: +45 72 20 14 14;
jeremy.millard@teknologisk.dk

Abstract. This paper reports on some of the recently completed work of the
EU-supported Prisma project examining the best of e-government experience
across Europe in relation to technology, organisational change and meeting the
needs of the user (citizens and business). Future work of Prisma involves developing scenarios of change over the next ten years, building future-oriented best
practice models and providing comprehensible and useful tools for practitioners
and researchers to guide their decision making and research priorities respectively. Apart from examining e-government and e-governance generally,
Prisma is also examining six service areas in detail: administrations, health,
persons with special needs (the disabled and elderly), environment, transport
and tourism.

1   Context and Drivers of Change
in Government and Governance

The importance of government is clear. Not only are we all dependent upon its services and the framework of law, peace and stability it provides, but in Europe it also
contributes 40% of GDP. Over the past few years, however, the concepts of government and governance have been dramatically transformed. Not only is this due to
increasing pressures and expectations that the way we are governed should reflect
modern methods of efficiency and effectiveness (like the best of business) but also
that governments should be more open to democratic accountability. This cauldron of
change is now finding itself once again brought to the boil by the impact of new
digital technologies on government. In many ways e-government enables the dual
goals of efficiency and democracy to be met more cheaply and easily than previously
envisaged, but the new technologies go much further than this. They are starting to
redefine the landscape of government by changing the relationships (power and responsibility) between players – between service providers and industry, between the
public and private sectors, and between government and citizen – by forging new
organisational and economic structures, by introducing new processes at work and in
the community, and above all by opening new opportunities as well as posing new
challenges, not least the threat of new digital divides.

Despite the power of the new technologies in providing global reach and interac-
tivity, models of e-government are cultural and political rather than technical. This

R. Traunmüller and K. Lenk (Eds.): EGOV 2002, LNCS 2456, pp. 298–306, 2002.

can be seen in Europe by comparing the very diverse levels of take up and of policy approaches, ranging from the Scandinavian and Anglo-Irish models of northern Europe, through the more statist, public sector driven responses of central Europe, to southern Europe's strong tradition for family, community and city-region driven approaches.

Across Europe, however, similar forces are acting out despite the variety of policy responses on the ground. The citizen is being treated more and more like a customer, the ICT literate are demanding the option of self-service through the increasing availability of on-line services, the roles of supplier and user of government services are blurring, as indeed are the distinctions between public and private sectors, for example as seen in the increasingly important role of the third sector and intermediaries and the establishment of many PPPs (public-private partnerships) to administer and deliver what have traditionally been seen as purely public services. Networks rather than hierarchies of interests, alliances and actors are increasingly determining how government, and especially e-government, is mapping out.

2   Lessons and Experiences from Europe

E-government is not just about a government portal with services offered electronically. It is also very much about the need to use ICT to support better quality "warm" human services, so that government 'on-line' complements rather than substitutes for government 'in-person'. These goals are simultaneously driven by and the result of:
1. both intra-and inter- governmental reengineering
2. the needs to strengthen efficiency, the public service ethic and democratic and open government.

E-government shares basic notions of business and work requirements with e-business, while at the same time it has its own distinctive features and problems manifest at European, national, regional, and local government levels. Thus, government, like business, requires greater efficiency, productivity, cost reductions, and treating citizens like customers. In this sense it shares the need for business process re-engineering. On the other hand, government, unlike business cannot choose its customers and, indeed, people are more than just customers, they relate to government as legal entities, e.g. as taxpayers, information users, hospital service customers and, generally, citizens who want to be aware, considered, recognised as participants in the democratic process, and free to express their opinions (e.g. through e-voting). Government also has stringent requirements such as:

- exemplary public service ethic with a focus on issues such as the welfare and health of the citizen and the equal treatment of all
- access for all
- caring for a sustainable environment, affordable public transport, etc.
- supply and demand for e-services, without a profit incentive, but subject to accountability and benchmarking
- provision of an institutional and a service framework for the wider economy.

In this sense, it is appropriate to talk of "government process re-engineering" (GPR), including not only services and relations with customers but also transparent, open and accountable government processes and relations with citizens in the digital democracy. This deep modernisation of government implies the development of a new culture for governance in public services, including new mindsets and sets of values, and new behaviours, as well as a re-engineering of structures.

2.1   Inter-governmental Process Re-engineering

A simple model of how the internal structures of government, typically when adopting e-government solutions, are being re-engineered in Europe, is provided in the accompanying four diagrams.

Diagram 1 shows the traditional structure of government before re-engineering. The first structural re-engineering tends to take place by the creation of a front-office (in diagram 2). This creates a one-stop shop enabling citizens and business to access government services through one point, regardless of their purpose, rather than through a multitude of points depending upon the organisational structure of government. Diagram 2 represents the state of art in many local and regional government agencies in Europe today.

In diagram 3, as the front-office logic (customer interface and service) takes hold and starts to determine the development of internal structures and processes, it is necessary for the back office to re-organise in order to reflect the needs of the front office and the customer. Traditional back-office departments, relationships and processes thus give way to ones determined by customer service through a front office.

Finally., in diagram 4, as demonstrated by a few European leaders, there is a decisive shift from "cold" administration (i.e. the back office) to "warm" ICT-supported human services (i.e. front office) in terms of personnel, and increasingly also in terms of resources. Small, ICT-automated back offices can serve and support very large front offices with more frontline ICT-supported human services based upon the improved cost-effectiveness and increased quality of administrative back office procedures.

In other words, the best of new e-governance in Europe is certainly not just about a government portal with services offered electronically. Rather, it is about this, plus internal reengineering of government, greater transparency, citizen access and involvement, and, above all, ICT support for better quality "warm" human services. For example, better direct health care of patients wherever they are, care of the elderly and disabled in their homes by government staff using interactive digital terminals to increase the timeliness and quality of information and support systems, improved responsiveness and efficiency of one-stop shop staff when dealing with direct enquiries, more efficient planning and control systems for delivering frontline services, etc. E-government could, and maybe should, lead to more, not less, human-orientated services.

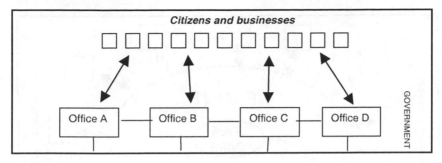

Fig. 1. Before re-engineering

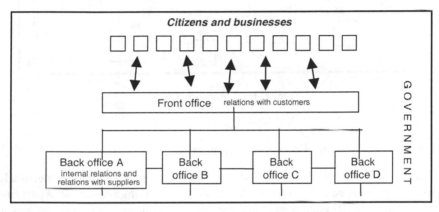

Fig. 2. Front office re-engineering

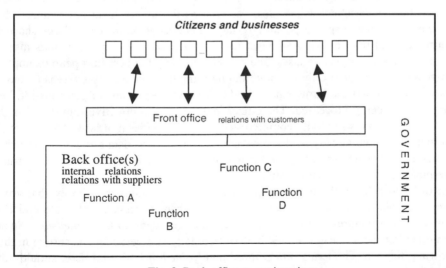

Fig. 3. Back office re-engineering

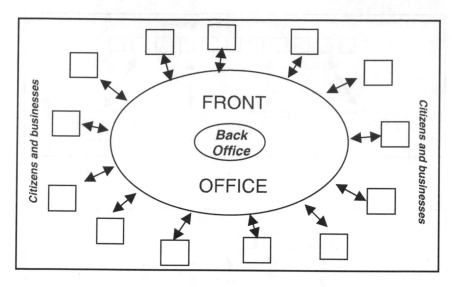

Fig. 4. Total re-engineering

2.2   Inter-governmental Process Re-engineering

Government in Europe is not only confronting massive internal changes but is also grappling with the need to re-examine the whole structure of government. Apart from the macro issues of EU enlargement and the need to specify more clearly the respective roles of European, national and regional/local institutions, is the need to look again at how government at the meso, as well as the micro, level is organised. E-government cannot reap the potentially large benefits it promises without sharing information and responsibility amongst its various parts – and this demands also a reconsideration of respective tasks. and powers. Already the best European examples of e-government require only one minimal data input or request by citizens or business who do not wish to know where, in the often vast apparatus of government, his or her case is being processed. The citizen only wants an effective, rapid and high quality service and is generally not interested in who provides it or how. The citizens' information is often already held somewhere, and as long as data protection, security and anonymity issues are safeguarded where warranted, Europeans generally trust government sufficiently on this point.

In order to achieve these objectives, electronic transactions thus imply that government agencies need to co-operate and even, where necessary, integrate and re-engineer their structures and processes. They certainly need to be *'joined-up'*. Such re-engineering should be considered both horizontally and vertically, as shown in the following diagram. The terms *'integration'* and *'joined-up'*, as used here, should not be confused with *'centralised'*. Indeed, government can be both highly integrated, joined-up and de-centralised. In fact, a third *'centralised-decentralised'* dimension could usefully be added to the diagram.

It is, of course, often the case that both vertical and horizontal integration, driven by ICT as well as other political and financial imperatives, are often part of the same policy, which is itself normally driven by a central government, top-down agenda (whether or not part of a centralising or de-centralising initiative).

2.3   Services Delivery

E-government enables many services to be delivered around the clock, throughout the year and ubiquitously across space. As shown above, it can put substantial power in the hands of the ICT-literate citizen so that the whole purpose, structure and mode of operation of government becomes re-engineered to the needs of those it serves. Many European government agencies and administrations are re-organising service access to take account of ICT, so that, in addition to face-to-face, new service channels are available which can either completely replace traditional channels or, as is more usual and desirable, supplement them, for example, phone and fax, Internet, kiosks, WAP, mobile, digital TV, etc. Not only are the number and types of channels proliferating, so too is the organisation of the service delivery itself. Especially in the case of Internet portals, the best of e-government in Europe now organises services:

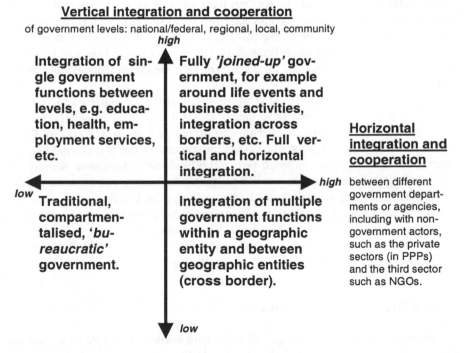

Fig. 5. Vertical and horizontal integration of e-government - inter-government process re-engineering

- for citizens, around life events, life episodes or the life cycle, such as birth, marriage, death, unemployment, education, living, home, working, sport and leisure, etc., etc.
- for business, around discrete business activities, such as VAT, tax, finance, employment, etc., typically supported by the business case approach where examples of typical situations are given and the service needs of businesses exemplified.

In this way, government re-organises to reflect citizen and business concerns rather than its own internal concerns and structure, and provides multiple interface channels.

2.4   Access for All

In many ways, the new technology, as mentioned above, does provide greater opportunities for access, for example by people in remote locations, those who work unsocial hours, or those who are immobile and thus cannot attend offices in person. However, there are important problems in determining precisely what citizens want and need, and how to provide e-government services in a user-friendly and effective way. User needs are always conditioned by what they already get, or imagine they can get. The best examples in Europe break down users into different categories, not just citizen and business, but also, for example, the elderly, the disabled, the visitor, the voter, professional users, etc., rather than treat users en bloc. The presentation of e-services through life events or business activities often supports such segmentation, although there can sometimes be problems of critical mass, i.e. a sufficient mass of a user segment to make the approach financially viable especially where special technologies like smart cards are needed for access and/or transactions like payment by or to the user.

The principle of access for all is also an important policy objective of the EU as well as of national and regional government authorities. This particularly means tackling the digital divide between citizens who do have access to ICT and are ICT-literate, on the one hand, and those who do not have access or the necessary skills, on the other. This is the biggest challenge facing the European Information Society in general, and e-government implementation in particular. Good examples to tackle digital exclusion include special training, the provision of PIAPs (public internet access points) and an acceptance that e-government does not only mean government-on-line but also a continuation of the public service ethic through government-in-person, though supported in terms of effectiveness and quality by ICT, both at the delivery point as used by government staff and in the back office.

2.5   Legal Issues, Security and Trust

Given both that governments are expected to be trustworthy and responsible guardians of the rights of citizens, and that they tend to hold vast amounts of data covering all aspects of the lives of citizens and businesses, systems and data security and trust in using them is vital. As a result, many European countries now have strong data protection laws, but these need constantly updating as new technologies come on stream and as new political and financial requirements manifest themselves. For example, much stress is now being placed upon trust, security and confidence, also in relation to fighting crime and terrorism against electronic, physical and personal property.

Security of transactions, for example using smart cards, is one of the biggest issues, although the barriers to widespread easy to use systems are often as much political and commercial than technical. Above all, there is a need for trust in interoperable systems, including data protection and privacy, as well as the need for anonymity where this is required, but balancing this with the legitimate needs of the authorities is not always easy. For the individual citizen or business, conducting electronic transactions requires cheap, easy to use and secure solutions in which information is easily authenticated and is reliable. The biggest challenge to interoperability across Europe is that legal systems between countries are highly incompatible. However, many trials and examples of roll-out have been successful, so there is confidence that solutions to these challenges are in sight.

2.6   Technology

Technical issues have been raised elsewhere in this paper, but all these need to be seen in the context of the speed and complexity of technological advance. This often makes it difficult for governments to keep pace and adapt their organisational structures, the skills of their staff and the actual e-services offered, in a timely, effective and financially responsible manner. European best practice in relation to technology shows that there is a need to adopt open standards, multiple access platforms, scalability, striking a balance between customised large scale turnkey solutions versus outsourcing and standardised solutions, and new technology models for eGovernment (e.g. mobile solutions, ambient intelligence, etc.)

2.7   Finance

Finance is, understandably, often the crunch factor. Like government the world over, European authorities tend to see their budgets squeezed in all directions, yet, at the same time, they are expected to deliver "more for less".

Experience shows that there is a trade-off between electronic and traditional services, in which e-services typically require heavy capital investments but can save substantial amounts in the longer term, both through improved efficiency and rationalisation, as well as because of greater service outreach and quality. Finance is needed not just for the technology but also for organisational change, as well as skills and competence development costs. Ultimately, government budgets are a political matter, especially as demand for government services tends to be infinite, particularly when offered free at the point of delivery. Good examples in Europe have shown how increased sources of investment and revenue can be generated, for example through greater efficiency of tax collection and the elimination of waste using ICT, well constructed agreements with private investors, the introduction of e-procurement particularly on a national or regional basis, etc. Getting the finance right, but not being driven only by a focus on costs whilst ignoring benefits, is a necessary condition for e-government.

References

1. OECD (2001) *Engaging Citizens in Policy-making: Information, Consultation and Public Participation*. PUMA Policy Brief No. 10
   http://www.oecd.org/pdf/M00007000/M00007815.pdf
2. PRISMA (2002) *Pan-European changes and trends in service delivery*, deliverable D2.2 of Prisma, a research action supported by the Information Society Technologies Programme of the European Union, 2000-2003, contact jeremy.millard@teknologisk.dk
3. PRISMA (2002) *Pan-European best practice in service delivery*, deliverable D3.2 of Prisma, a research action supported by the Information Society Technologies Programme of the European Union, 2000-2003, contact jeremy.millard@teknologisk.dk.

CITATION
Citizen Information Tool in Smart Administration

A. Anagnostakis, G.C. Sakellaris, M. Tzima, D.I. Fotiadis, and A. Likas

Unit of Medical Technology & Intelligent Information Systems, Dept of Computer Science,
University of Ioannina & Biomedical Research Institute - FORTH, PO BOX 1186, 451 10
Ioannina, Greece

Abstract. CITATION is an innovative software platform designed to facilitate
access to administrative information sources by providing effective information
structure, indexing and retrieval. CITATION improves electronic government
services, ensuring that citizens have easy and direct access to essential public
data and promoting online interaction between citizens and government. It
targets the improvement of administration services offered by governments and
therefore the creation of "smart" and flexible government structures that will be
able to provide their citizens with precise and personalised information.

1   Introduction

With the explosion of the Internet and communication technologies, the concept of
"information" transforms into a value per se. However, the overwhelming amount of
administrative information remains scattered and unstructured -due to an evident lack
of modeling and standardization-, stored in multi-site distributed repositories that
cannot be accessed, fused and delivered in a seamless manner [1-2].

CITATION is an innovative software brokering technological platform that helps
and supports information suppliers and consumers in their info-transactions. It
provides groundwork for the creation of "Smart" governmental structures that support
the provision of flexible and customised information services to the citizens.

CITATION focuses on the following:

- Effective access to a variety of information sources on the information global net.
- Transformation and representation of the retrieved information on a meta-level
- Matching operations between users' profiles and the information extracted from the
  various information sources
- Personalised information delivery.

1.1   Background

The first step in building a seamless information fusion environment such as
CITATION is the establishment of the "domain of discourse". Ontology is a formal
explicit description of concepts in a domain of discourse. Therefore, utilization of
domain-specific ontologies is the medium towards seamless information fusion [3-4].
Ontologies provide a compact, formal and conceptually adequate way of describing

R. Traunmüller and K. Lenk (Eds.): EGOV 2002, LNCS 2456, pp. 307–312, 2002.

the semantics of XML documents. By deriving DTDs from ontology the document structure is grounded on a true semantic basis and thus, XML documents become adequate input for semantics-based processing. The ontology provides a shared vocabulary that integrates the different XML sources, making the information uniformly acceptable and thus mediated between the conceptual terms and the actual mark-up used in XML documents [5].

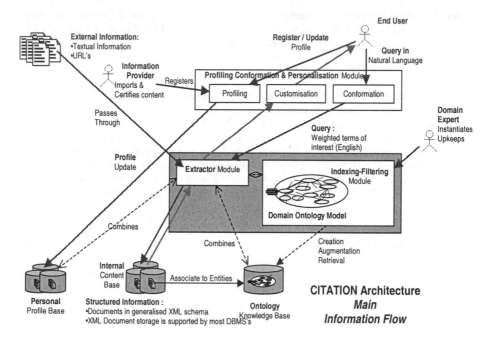

Fig. 1. CITATION architecture

2   Architecture

The CITATION architecture incorporates several technological advances, to deliver the desired innovative functional features.

CITATION internal Content Base holds assimilated external information, which is matched against the Internal Domain Ontology during the phase of import. The standardised DTD for administrative information allows for the uniform representation of the existing raw data, and provides a solid base for further processing and retrieval of information. The tokens from the user's query are extracted from the conformation model, augmented by external lexicons and dictionaries (UMLS, WordNet), filtered by the user's personal characteristics and matched against the semantic internal domain ontology. The domain ontology constitutes the conjunctive link among the actual data and the user's query. Fig. 1 captures the major modules of the system and depicts the data flow among them. The major modules identified are:

- The Profiling module: it is responsible for the generation and update of the user profiles.
- The Conformation module: it is responsible for preprocessing the user query (given in natural language) to facilitate the extraction of the tokens, and thus the internal mapping.
- The Customisation – Presentation module: it is responsible for the customised delivery of the final outcome of the user's query.
- The Extractor Module: it extracts the information from the external information sources and imports it into the CITATION Content Base.
- The Information Representation, Indexing and retrieval Module; it:
  - Does the initial matching of the imported data to the Ontology Model
  - Maintains and augments the Ontology model, delivering mechanisms for editing, allowing the Domain Expert to upkeep the model.
  - Facilitates the semi-automated creation of relationships among the newly inserted and pre-existing entities.
  - Checks and verifies the integrity of the internal domain model.
- The internal Domain Ontology base: it represents and stores the Domain Specific Knowledge in a Uniform manner, despite of the domain peculiarities.

2.1   Representative Objects

As mentioned above the main objective of the CITATION project is the design and implementation of a user-oriented knowledge discovery system, based on a semantic formalism. CITATION investigates the creation, collection and distribution of resource descriptions, to provide transparent means of searching for, and using resources. To improve searching, filtering and processing of information on the CITATION information repository, a common effort is made in the direction of "metadata". Metadata are defined literally as "data about data", but the term is normally understood to mean structured data about resources that can be used to help supporting a wide range of operations [11].

In CITATION we have defined a metadata structure based on implementations of the Extensible Markup Language (XML). In particular, XML aims at providing a common syntax to emerging metadata formats. Based on these lines we have proposed the CDRS (Citation Document Representation Schema) that is used to describe the indispensable information that CITATION handles and manipulates. Metadata in CITATION help administer and manage the resources, (e.g. keep information about their location and acquisition). It is also used to help managing user access and to help the versioning of the resources [6-8].

2.2   Extractor Module

The administrative information that the CITATION system provides either legacy textual information or links to Web-based information, has to be indexed through the internal Ontology Model. The action of linking the initial pieces of information to the Domain Ontologies and the Retrieval of the information in response to the User Query, is undertaken by the Extractor module.

Although the CITATION platform is highly open and capable of manipulating novel information formats, our preliminary search revealed that the administrative information is currently vastly textual, thus special care is being paid in the text manipulation methodologies. Research in NLP (Natural Language Processing) for concept extraction is an undergoing task for many years [9]. Although a reasonable level of performance and success has been achieved, the actual use and dissemination of data extracted from free text is still very limited. Standard NLP- techniques such as lexical scanning morphological analysis and parsing are important tasks in the process of automated analysis of natural language texts. However NLP techniques can only process a text syntactically. To capture, represent and understand the knowledge contained in the text it is necessary to have a semantic framework.

2.3   Ontology Model & Domain Ontology

Although CITATION selects "health" as its application domain, it manages to provide uniform ontology specification capable of representing various kinds of administrative terms and information. This facilitates the efficient communication and exchange of information among administrations in a variety of subjects of concern, since the introduced model is highly open and customisable.

Ontology provides conceptualisation (i.e. meta-information) that describes the semantics of the initial data [12]. CITATION comes to facilitate this exact common understanding, by introducing a "Universal" model that allow for adequate representation of the semantics of the existing informative administrative content [13-19]. For the core of the Domain Model the Protégé paradigm was adopted [17].

Ontology Model provides all the necessary functionality for the capture and storage of the domain knowledge items (entities) and the relationships among those in various fields of practice (i.e. the CITATION pilot cases). The most important step on the definition of the Ontology model is the determination of the elements constituting the model; during the study of the CITATION Ontology Model, we have identified three major characteristics to be modelled, namely:

- The abstract Entities (classes/concepts)
- The Supported Relationships
- The Metadata (class containing the meta-information for the instance-entities)

2.3.1   Entities & Relationships

Modelling entities is a straightforward task, including definition of a metadata distinct set plus additional information on its behavior in the ontology model.

However, modelling relationships have been proven to be rather twisted: On the one hand, simple and comprehensive mathematical rules had to be established in order to set an adequate set of integrity constraints, while on the other hand conceptual taxonomisation of the relationships is facilitated to allow for standardisation in the manipulation of the overall model.

2.3.2   Semantic Characterisation

Analysing the user requirements, revealed initially three large semantic categories, namely: Hierarchical, Generalisation & Sequential. Each of the imported relations is denoted accordingly, to allow for automated manipulation later in the information retrieval process. Handling of more categories is vastly supported, since the architecture of the knowledge base [14] is highly adaptive.

2.3.3   Meta-information

Based on legacy external lexicons to each of the entities a prime definition (in English) is assigned. A whole set of weighted meta-data terms (i.e. semantically equivalent definitions of the entity) are assigned in each term in a weighted manner. The terms of the Universal Translation Dictionary (EuroWordNet [17]) corresponding to the prime definition are assigned to the CITATION translation table allowing for multilingual semantic matching.

3   Results

Using XML based implementation for representing the various information objects we are able to control a well-formated object, based on the special requirements of the project. XML allows for the explicit declaration of element types and representation of document structure in the Document Type Definition (DTD) or XML Schema. The core elements of CDRS, *Resource, Unit, SubUnit,* are used to split the initial objects to sub-objects based on the concept of its content. The correlation between two objects is achieved by defining an identical number for these objects, but the correlation between CDRS object and the ontology model is accomplished by using special elements such as *Class, Concept* and *Metadata.* By using CDRS as a representative object any resource can be described and managed in a more efficient and flexible way. CITATION Ontology Model is highly adaptable, capable of representing domain ontologies over a wide spectrum of administrative information. The rules and constraints identified manage to capture and categorise a significant portion of the administrative knowledge, creating a solid ambient for semantic-based information retrieval. Enhanced capabilities in conceptual information retrieval are the outcome, and the indications so far are stimulating. The recall accuracy and the precision of the results rely heavily on the completeness of the domain ontology

4   Conclusions

CITATION meets the users' (citizens, business, public authorities) needs for flexible and ubiquitous access of administrative information. It provides a platform with multifunctional dialogue interfaces and multi-lingual features, being an intelligent information tool on administration issues.

CITATION offers effective and transparent access to governmental services through information indexing and retrieval using a semantic oriented, ontology based approach. It introduces models that allow for adequate representation of the semantics of the existing healthcare administrative information content.

Acknowledgments

The CITATION project is partially funded by the EU (IST-29379).

References

1. Green paper on public sector information: a key resource for Europe.
2. Eysenbach G:Consumer health informatics. BMJ 2000;320:1713-6.
3. Musen M. Dimension of knowledge sharing and reuse:Computers and Biomedical Research 25 (1992): 435-467
4. Gruninger M, Fox M.S:Methodology for the Design and Evaluation of Ontologies. Proceedings of the Workshop on Bsic Ontological issues in Knowledge Sharing, IJCAI-95, Montreal
5. Erdmann M, Studer R:How to structure and access XML documents with Ontologies. Data & Knowledge Engineering 36(2001) 317-335
6. ISO 639 - Codes for the representation of names of languages. http://www.oasis-open.org/cover/iso639a.html
7. ISO 3166 - Codes for the representation of names of countries. http://www.oasis-open.org/cover/country3166.html
8. Getty Thesaurus of Geographic Names. http://shiva.pub.getty.edu/tgn_browser
9. Bateman JA:Ontology Construction and Natural Language. Proc Formal Ontology &KR, 1993.
10. Eurodicautom http://europa.eu.int
11. Metadata Architecture http://www.w3.org/DesignIssues/Metadata.html
12. Bateman JA:Ontology Construction and Natural Language. Proc Formal Ontology KR, 1993.
13. Gruber T:A Translation Approach to Portable Ontology Specifications. Knowledge Acquisition 1993; 5(2):199-220.
14. Guarino N, Giaretta P:Ontologies and Knowledge Bases: Towards a Terminological Clarification, in Towards Very Large Knowledge Bases. N.J.I. Mars (ed); IOS Press, Amsterdam, 1995: 25-32.
15. Karp P, Chaudhri V:Thomere J. XOL: An XML-Based Ontology Exchange Language.1999.
16. Erdmann M, Studer R:How to Structure and Access XML Documents With Ontologies.Data and Knowledge Engineering, Special Issue on Intelligent Information Integration.
17. Cognitive science laboratory | Princeton University | 221 nassau st. |Princeton, nj08542, WordNet http://www.cogsci.princeton.edu/~wn/online/ 2002 Jan. 27
18. Vossen P, Department of Computational Linguistics University of Amsterdam, Euro WordNet http://www.hum.uva.nl/~ewn/ 1999.
19. Hahn U, Romacker M, Schultz S: How knowledge drives understanding- matching medical ontologies with the needs for medical language processing. Artificial Intelligence in Medicine 15 (1999) 25-51

Clip Card: Smart Card Based Traffic Tickets

Michel Frenkiel, Paul Grison, and Philippe Laluyaux

Clip Card, 1501-1503 Route des Dolines,
F-06560 Sophia-Antipolis

Abstract. The introduction of information society technologies (IST) in administrative processes poses a range of problems of technical, legal and human nature, that need to be properly addressed for the evolution to be successful. Some of these problems (and a few solutions) are reviewed in the scope of replacing paper-based traffic tickets by smart cards.

1   Project Objective

Traffic violations constitute the most common offence in modern societies. 150 million tickets are distributed every year in Europe. In the average, only 60% are eventually collected, because the administrations are unable to process unpaid tickets within the legal timeframe. This figure varies greatly between countries, even between regions, spreading the idea that citizen are not equal when facing the law.

The cost of collecting the money of a ticket also varies. The figures lay between 6€ and 12€ in the surveyed regions.

Clip Card results from in-depth discussions with public authorities: city officials, justice departments and law enforcement agencies, in France and Italy.

With Clip Card, the paper based traffic ticket is replaced by a smart card recorded on the spot of the offence by the traffic warden and clipped to the windshield wiper. **The innovation of Clip Card lies in the fact that the smart card is the traffic ticket itself, the payment means and the payment receipt.**

The offender may pay the fine at payment kiosks with his credit card or, as today by checks, in cash or by credit card, in tax offices, tobacco shops, post offices, etc. Back-office processing (matching paid tickets, issuing late penalties) is automated on a large scale, using up-to-date telecom technology (SMS, GPRS, secure communications) thus enabling considerable savings of public money.

Clip Card helps improve security and fight crime: dangerous vehicles are reported for being removed, stolen vehicles and vehicles with many unpaid tickets are detected and reported for further action.

To process traffic tickets, the traffic warden is equipped with pre-personalised smart cards and a mobile terminal (a hardened, communicating PDA).

All parties find advantages in Clip Card:

R. Traunmüller and K. Lenk (Eds.): EGOV 2002, LNCS 2456, pp. 313–318, 2002.
© Springer-Verlag Berlin Heidelberg 2002

- The traffic warden and the police officer who is released from administrative tasks and may concentrate on security missions.
- The offender who can pay his fine simply with his credit card at public kiosks. These kiosks could be combined with existing parking meters.
- The governments and the municipalities which save around 50% of the processing costs: in Europe, it means saving some 900 million € per year.

2   Project Status

- Clip Card has received the PROGSI label (French ministries of Finance, Justice, Industry) and go-ahead for trial. Clip Card is financially supported by the European Commission. It has also received the "Label Innovation" (Telecom Valley) and a support from ANVAR.
- System architecture is complete, back-office and terminals development are in progress with industrial partners
- Real life trials are scheduled to start during the Fall 2002 in Cannes, Ventimiglia and Torino.

3   Functional Overview

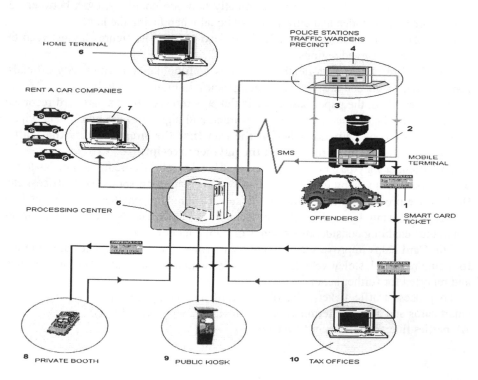

4   Interfaces with Administrations, etc.

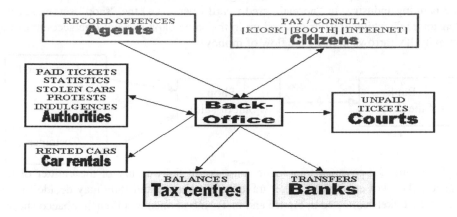

5   Key Questions Addressed

5.1   Technical

Availability, Reliability

- High availability (H24) required for the network
- Terminals on the network have availability and maintenance procedure comparable to payment terminals at points of sale
- Users and personnel are quite similar to those found at points of sale (customers-offenders are general public, sales clerks are administration personnel, tobacco shop personnel, etc.)
- Reliability must be very high for the mobile terminals and memory cards, as they are used in diverse, often rough conditions:

  - Indoor/Outdoor
  - In all weather (rain, sunshine, frost)
  - Day/night
  - Wardens are under stress, and constantly face a risk of aggression

Security Aspects

- payments: the same level of security as for self-service points of sales apply
- offence transmission requires a good level of security, but the network is entirely under control of Clip Card, and it presents a limited interest to hackers. As a full traceability of the transactions is required, mechanisms for detecting intrusion attempts are also present and freeze the most vital access to the system when an attempt is detected. There is a built-in high redundancy (as offences are both recorded in the mobile terminal and sent over the network), thus facilitating cross-checking and checkpoint/restart.

Cross-Border Payments
This may be the largest technical challenge, as it may require also an approval from
the banking industry. In "normal" credit card operations, the "shop" serves several
"customers", while all payments end up in the same bank account (the account of the
shop itself): (arrows indicate the flow of money)

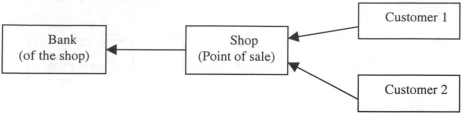

In Clip card, the payment transaction must credit the treasury of the emitter of the
traffic ticket. For example, a driver travelling in France and Italy may decide to pay
his traffic tickets collected in France and in Torino at once in a French tobacco shop:

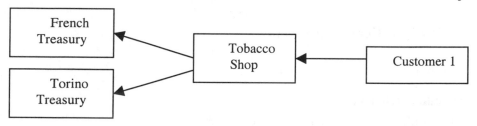

The point of sale terminals do not have the necessary capacity to hold all the programs needed to address the many treasuries (or banks). Regulations do not allow
Clip Card or a third party to cash in the money of transactions, and to pay back the
appropriate treasury. A solution has yet to be found.

5.2   Legal

Procedures of the Administration
The processing of traffic tickets may involve a number of different administrations.
For example, in France:

- law enforcement people (*Police Nationale, Police Municipale, Gendarmerie*), who
  may all issue tickets, report either to the Ministry of Interior, the Ministry of Defence, or the City Mayor.
- Payments are handled by the Treasury (*Trésor Public*) and the Ministry of Finance
- Late penalties are decided by the Ministry of Justice (and collected by the Treasury)
- Installation of public equipment is decided by the municipalities, and must be
  approved by the *Préfecture* (Min. of Interior)

Before considering the deployment of Clip Card, and the necessary changes in the
administrative procedures, the French administration requires a proof of the concept

via a significant trial, to be conducted in a large city. This means that a first version of Clip card must be developed, and used in a non-disruptive way regarding the Administrations: this requires no direct connection with any data base of the Administration.

Administration Data Bases

To store and process traffic tickets, a computer program was developed by the Administration, and instantiated by the Min. of Interior once for the *Police Nationale* (WinAF) and once for the *Police Municipale* (WinPM). These two instances are disjoint.

The data bases are used to match payments, identify late penalties, process indulgences. Interface with the Justice Department regarding late penalties is done via paper and diskettes.

Clip Card Trial

To trial the Clip Card concept without changing the existing procedures, we will produce diskettes that will be given to the *Préfecture* to feed in traffic violations into one of the systems. And we will also produce diskettes to feed in payments. Tickets and payments will be matched by the systems in place (WinAF, WinPM) under control of the administration. Late penalties and indulgences will not even be known by the Clip Card system.

However, as the most costly operation today consists of recording the handwritten traffic tickets onto one of the administration data bases, the Clip Card trial remains valid to demonstrate its advantages.

Similarly, our smart card "must contain the exact information found on the present forms, and the payment terminal must print the present forms exactly". This puts strong requirements on Clip Card!

In the future, rather than develop IST solutions that match existing administrative procedures, it would be more efficient for administrations to adapt their procedures to the new possibilities offered by IST. This is how private enterprises adopted IST.

The situation in Europe regarding legal constraints varies greatly between Member States. It took one year of discussions in Paris, at the various ministries, to obtain an authorisation for launching a Clip Card pilot operation. It took just a few days of discussions with the staff of the mayors of Torino and Ventimiglia, to obtain the same in those Italian cities, without the need to involve any central administration.

We are discovering now the requirements of the other Member States.

5.3   Human

Two aspects must be considered:
- Usability, by all people involved: offenders, wardens, administration personnel
- Quality of work, and future prospects, for the wardens.

Usability

- For the offenders, it is not more difficult to pay a traffic ticket than to use a pay telephone, a cash machine or a Proton terminal. These facilities are now well accepted.
- For the administration personnel, Clip Card is totally transparent.
- For the wardens, usability is a key aspect, which has not always received adequate answers in the past. For example, Malaysia adopted in 2000 the smart card to hold citizen's ID. Police forces received a hardened PDA to read those cards. The operation failed because the ergonomic characteristics were not adapted to the situation:

  - Too small, hard to read characters
  - Too small, too many buttons
  - Too complex, too many menus

Similarly, the police in Madrid was equipped with PDAs to record traffic violations, and print the tickets. But the printers are too often out of order for the system to be well accepted.

Learning from these experiences, Clip Card offers a very simple user interface, which can be activated either via voice or via buttons. Wardens will have plenty of time to practice, before they use it on the streets, so that their confidence is high.

Wardens Quality of Work and Future Prospects

Clip Card was presented first to the syndicates of police forces. It was well accepted, because it allows a better, more gratifying usage of the police forces. Instead of doing paperwork in their office, they will do their job of maintaining law and order with the people.

Clip Card was then presented to some of the wardens who have been selected to test it. Their reaction was also positive, because they feel that the quality of their equipment is a recognition of their value. Also, replacing the pen and paper by a PDA is a gratifying opportunity to enhance working conditions, and to enhance the image of their work. But they demand that this is accompanied by adequate education, and that the equipment is fully operational and tested at the start of the pilot.

For the future, mobile digital equipments are ubiquitous, with new applications popping up every day. Police wardens will not carry an array of devices. But one single, robust and flexible device should be available to accommodate smart ID cards, digital camera, fingerprints capture, mobile phone (for voice and data), GIS, and thus interfacing administration data bases in total security. Secure communications, but also security of the warden, with a Help button and geo-localisation will probably soon be offered. As Clip card is the first such application, there is a high stake (and considerable future opportunities) on its mobile terminal, as it may well become the de facto standard to process those future applications.

VISUAL ADMIN – Opening Administration Information Systems to Citizens

Benoît Drion[1] and Norbert Benamou[2]

[1] AIRIAL Conseil, 3 rue Bellini, F-92806 Puteaux Cedex, France
benoit.drion@airial.com
[2] BFC, 137 avenue du General Leclerc, F-92340 Bourg la Reine, France
norbert.benamou@business-flow.com

Abstract. The *VISUAL ADMIN* project aims at creating an eGovernment solution that makes easier for citizens and businesses to interact with administration. Considering administration activities from the perspectives of both "customers" (citizens and businesses) and administrations, *VISUAL ADMIN* intends to provide citizens and businesses with an global online view on information relevant to them. It also intends to organise the flow of information between administration services for handling customers' cases, and to act as a portal for getting access to relevant online information services. The current paper focuses on the requirements identified in the early months of the Project and presents the rationale of the platform under development.

1   Need for Simplifying Administration Processes

It is often difficult for administrations to avoid slowness and heaviness in dealing with citizens and businesses. These customers frequently encounter difficulties because a single need is often handled through a series of procedures sometimes distributed among several public bodies; for instance, in France, installing new offices in an industrial area might require interacting with up to five administrations.

Moreover the organisation of the public sector is also complex, due to the devolution of powers at several geographical levels (national, regional, intermediate and local levels). Therefore an administration service might also have to collaborate with other public bodies to process the customer's case, and specialised entities can have a role for some procedures (registration of patents, ...).

Such a complex organisation sometimes causes a huge delay for citizens and businesses, as well as for civil servants; and generally speaking implies many risks and discomforts for the customer:

- Infringing some regulations by forgetting asking authorisation to one administration service;
- Long delays for completing his case;
- Lack of transparency that makes difficult for him to understand what is happening and at what stage his case is.

On the other hand, inside and between administrations themselves, the processing of a customer's case is not always managed in an optimal way from start to finish, for front

R. Traunmüller and K. Lenk (Eds.): EGOV 2002, LNCS 2456, pp. 319–325, 2002.
© Springer-Verlag Berlin Heidelberg 2002

office services are sometimes structurally disconnected from back office information systems, databases or other services.

In this context, European governments have agreed on an initiative to simplify administrative procedures by going on line and by offering "eGovernment" services.

2   *VISUAL ADMIN* As a eGovernment Online Solution

The *VISUAL ADMIN* Project[1] has been launched in June 2001 by a European consortium gathering local governments and private companies with the objective of conceiving and experimenting an online eGovernment solution which could be set up by a local government as a Portal for efficiently supporting citizens and businesses in targeted policy areas.

Three local governments are involved in the project and will implement the *VISUAL ADMIN* solution in 2002:

*   In France, the *Communauté d'Agglomération d'Agen* (CAA) is a grouping of 6 municipalities in the Agen conurbation. CAA will use *VISUAL ADMIN* to better serve both citizens, with respect to its city policy; and businesses, in the frame of its economic development policy. Several types of actors will be implied in *VISUAL ADMIN* operations: territorial authorities, state services, non-profit organisations and consular bodies serving the targeted customers, particularly citizens in precariousness end enterprises.
*   In Italy, the *Comunità Montana Valle Maira* (CMVM) is a grouping of 14 municipalities in a rural area of the western Alps. CMVM will use Visual Admin to provide the "citizen" (in the broad sense of the term: resident and emigrated) and organisations representative of the civil society (associations, private organisations, trade unions) with information about natural research management and tourism in mountains areas, in the frame of the policies of water deployment and promotion of tourism.
*   In Poland, the *City of Lodz*, the second biggest city in the country, will use Visual Admin to manage the co-operation between public bodies of different geographical levels, social partners and non-governmental organisations, in settling and communicating around EU integration policy. Moreover it will support the partnership between state authorities, social organisations and private business units concerning EU structural funds (principles, objectives and applying procedures).

The common vision shared by the 3 pilots for the *VISUAL ADMIN* application is the one of a Portal set up by a territorial government to support one or several policies for which the territorial government has legal power. The Portal enables Citizens and Businesses concerned by that policy to:

*   Get access to relevant information and eProcedures either through the Web or Wireless information devices;
*   Be supported in having their case processed by the territorial government and through it by the whole set of relevant administrations.

---

[1] The *VISUAL ADMIN* Project is co-funded by the European Commission through the IST – "Information Society Technology" research programme

3   Identified Requirements

To guide the design and development of the *VISUAL ADMIN* solution, the Project relied on a study of the needs of three pilot territorial governments which was conducted during summer 2001 through interviews, surveys and focus groups. Actors that were contacted are:

- Territorial government staff: political managers and administrative-technical managers responsible for the policies addressed in the pilot, information system manager, ...;
- Managers of 3rd party public administrations with which the pilot should collaborate to implement *VISUAL ADMIN*;
- Citizens and enterprises being the administration's "customers".

Taking into account the series of needs expressed by the pilots and the regulatory environment for eProcedures in France, Italy and Poland, the current section presents a set of "abstracted" user requirements in relevant functional areas:

- Portal Organisation
- Case Processing
- Specialised eProcedures
- Case Monitoring and Supervision
- Security & Rights Management
- Accessibility
- Organisational Issues

3.1   Portal Organization

The Portal should be organized in areas that both contain ad hoc information managed by a large group of actors (the Territorial Government and 3rd parties such as other administrations); it could also embed information stored in 3rd party Web sites. The list of contact persons in all involved actors should be clearly identified in each Portal area and Citizens should be able reach them by *e-mail* through the Portal, with request for acknowledgement of receipt.

A first level of Portal areas concerns policies such as Economic Development, Social Welfare, etc..

Each Policy area should provide relevant *"Information"* such as a policy overview and the list of citizen needs that could be serviced within this policy. It should also support community development through an *"Open forum"* to discuss experiences and to comment on the existing initiatives, regulation changes, or other relevant contents.

A second level of Portal areas concerns specialized processes in the frame of a given policy, e.g. Enterprise Set Up Project, etc..

Each Process area should provide relevant *"Information"* such as a process overview (especially important for complex processes such as "Calls for Projects"), a Directory of actors, of their competencies and of contact persons, a Catalogue of all major document templates and forms to use (major formats such as PDF, Word, Excel, RTF or XML should be used), or a Guide about support measures and public aids that could concern citizens. It could also provide *"Self-Assessment"* facilities that could assist a citizen in positioning its need and identifying who to contact and which support measure to ask for. Finally, it should link enable Citizens to initiate a case through either eProcedures or traditional processes.

More generally *"Event-related advertisement"* such as announcement of a new Call for Projects, could be presented on the Portal Home page, in the related policy page and in the specific process sub-area.

3.2   Case Processing

The Portal should enable managing pilot specific workflows for processing Citizens cases which would gather Data forms, Files such as documents (Word, Excel, etc..), Maps (MapInfo format), Instantiation of document templates provided by the Portal, and Link with Web pages or components stored either on the Portal or on 3rd Party Web sites.

Depending on the specific workflow, the case initiator could be a Citizen or a civil servant. Actors involved in a workflow should be able to submit additional case information through either online data forms or formatted emails with attachments (filled in data forms or documents) and authenticated with digital signature. Workflows could also automate an eProcedure or the sending of e-mails with attachments such as data forms or documents.

Finally, a workflow could imply the creation of a private forum for supporting the communication between actors involved in the case processing.

3.3   "Official Tender" eProcedure

A generic type of eProcedures has been identified to support *"Official Tender"* as an alternative or complement to a paper-based request. The basic steps are:

- To package a digital proposal gathering an electronic request based on a "Request Template", the list of appendices (support information), and the set of appendices available in a digital form.
- To electronically send this proposal in a secure way (using digital signature)
- To receive an electronic acknowledgement of receipt with a registration number that must be used in any later communication, especially when sending "paper" appendices.
- To possibly receive an additional electronic acknowledgement of receipt when supplementary "paper" appendices are received by the administration
- To receive a "Eligibility Certificate" once the targeted administration has checked that all needed documents for a complete request have been delivered, mentioning also which organisations and which persons are in charge of processing the request.

Moreover, for official requests submitted in paper but prepared with electronic templates found on the Portal, the citizen may mention that (s)he authorizes the request processing to be monitored through the Portal.

3.4   Case Monitoring and Supervision

To monitor citizens' cases, a global "Case Status" indicator should be defined that reflects the workflow progress and is updated at each step of the workflow either automatically for steps performed through an information system or through follow-up forms or e-mails for steps performed externally.

A Citizen monitoring his/her own case should see both the "Case Status" and who is involved in its processing. Moreover, the citizen and all actors involved in the case could ask to be automatically notified via e-mail or fax of any change.

At a managerial level, Policy operators may use historical information about cases and workflows to assess needs for changes in work organization. Therefore, logs of cases handled through the Portal will be kept and will be used to create a data mart on which to base decision support facilities.

3.5   Security & Rights Management

Identification and authentication are critical to initiate or to get access to specific citizens' cases and associated forums. Adapted access rights should be granted by default to the case initiator and to the concerned citizen; they could also be granted to other users through a workflow template, when such a template exists, or by the initiator of the case.

Moreover official transactions with legal value must be secured with a digital signature carried out with a "secure signature creation device" and can be verified with "Qualified electronic certificates". For such purposes, the solution implemented must comply to domestic regulatory frameworks.

3.6   Accessibility

Accessibility first concerns the ease of adjusting the Portal to the Citizen. To achieve that, citizens using the Portal could *"Personalize"* their access by registering their interests and by selecting their preferred channel for later communication – email or fax, but perhaps also interactive chat or phone –. Voice recognition techniques could also support disabled people in getting access to the Portal.

Accessibility also concerns the ability to use the whole set of services from a single access point. Therefore the secure environment required for submitting an official request through an eProcedure should be easy to set up on a personal computer connected to the Internet, without the intervention of a specialist in information technologies. Moreover, to avoid the "digital divide" risk, this secure environment should also be made available to citizens in some public areas.

Finally accessibility concerns the possibility to involve users wherever they are. Therefore, to rapidly get in touch with a Citizen, civil servants should be able to capture on the Portal a text which would be automatically transformed into an SMS message. Furthermore, for civil servants intervening outside their office, mobile access will be supported through GPRS multimedia equipment.

3.7   Organizational Issues

An eGovernment project such as setting up a *VISUAL ADMIN* Portal goes beyond technology implementation and information management. It also implies setting up a set of support measures for both civil servants and citizens:

- Training and support about Internet, the Web and specific *VISUAL ADMIN* mechanisms should be offered to civil servants involved in the workflow processing, even outside the pilot itself.
- A Help Desk could be set up to help citizens in preparing requests. Minimal functionality should encompass Interaction management (email, interactive chat, etc...) and Case management (to follow up requests from candidates).

4   Rationale for the *VISUAL ADMIN* Platform

Taking into account the requirements presented above, the *VISUAL ADMIN* Portal is based on the combination of functionalities that significantly improve performance and quality of service:

- The Portal offers a single access point to the diversity of information systems available within the local government and in other related public bodies. It implies:
  - Information services for providing guidance to citizens and businesses on the targeted policy area;
  - Integration of existing databases and legacy applications of the administration and an online interaction between applications;
  - Interaction with existing eProcedures in public bodies.
- The portal includes a g-CRM (government's customer relationship management) module for handling the interaction of a customer with the administration workflow and for improving the quality and friendliness of the service. It implies:
  - Focusing on the customer's case rather than on constraints and needs internal to the administration.
  - Monitoring how each case is processed through the administration work flows in order to inform the customer on his case status.
- Mobile Access services are offered in order to provide citizens with a ubiquitous access to the administrative services through a multimedia mobile phone supporting GPRS or EDGE communication. The *VISUAL ADMIN* solution also offers an intuitive interface based on voice recognition for controlling access to administrative services and citizen's data.

5   Benefits Associated to *VISUAL ADMIN*

The *VISUAL ADMIN* Portal solution will thus enable a local government to:

- ***Better serve citizens:*** Relations between citizens and government services are often perceived as unbalanced with the citizen being more an anonymous item/case than a customer. By really involving citizens in their case processing in government services, the *VISUAL ADMIN* approach will improve people satisfaction with both the experience and the service quality.

Leveraging on local networks or communities of civil servants will also enable citizens to get a more comprehensive advice from the civil servants. Moreover, supporting civil servants that operate in nomadic work contexts will also bring closer together the citizens and the public service.

- *Settle better working conditions for civil servants:* By supporting the flow of information and work within and amongst government services, *VISUAL ADMIN* enables civil servants to decrease significantly time spent on low added value activities such as archiving or retrieving citizens' data, writing requests for additional information for other administration services, etc...
- *Reduce government expenditures:* By analysing bottlenecks in work organisation and supporting the redesign of processes thanks to workflow tools, and by reducing the time (and the cost) of data capture by civil servants.

e-Government Observatory

Freddie Dawkins

Senior Communications Manager, eGovernment Observatory,
GOPA Cartermill International, 6ᵗʰ Etage, 45, Rue de Trèves
B-1000 Brussels, Belgium

1 Description of Work in the e-Government Observatory

1.1 eEurope Context

The European Commission launched the eEurope initiative on 8th December 1999 with the adoption of the Communication 'eEurope – An Information Society for all'. The initiative aims at accelerating the uptake of digital technologies across Europe and ensuring that all Europeans have the necessary skills to use them.

The eEurope Action Plan sets ambitious goals concerning **"Government online: electronic access to public services"**.

The Lisbon European Council conclusions call for:

- efforts by public administrations at all levels to exploit new technologies to make information as accessible as possible.
- Member States to provide generalised electronic access to main basic public services by 2003.

The observatory takes as starting point one of the conclusions of the Sandhamn conference (Sweden on 13 and 14 June 2001):

"For eGovernment to be implemented successfully at the European level, policy should be developed that specifically addresses its European dimension. This is in addition to what is being achieved at the national level. Initially, such policy should be formulated to determine what European public administrations should do to make it easy for citizens and enterprises to transact business at the European level".

The Ministerial declaration produced as a conclusion of the eGovernment Conference in Brussels mandates IDA (Interchange of Data between Administrations) to set-up an *eGovernment Observatory.*

"Ministers agreed to encourage National Administrations and EU Institutions to establish a common view on which pan-European eServices are most essential on a European level, and to establishment of an eGovernment platform, building on the European Forum on eGovernment and the eGovernment Observatory"

The European Forum (www.eu-forum.org) on eGovernment and the eGovernment Observatory (www.europa.eu.int/ISPO/ida) work in a co-ordinated fashion and feedback to each other.

R. Traunmüller and K. Lenk (Eds.): EGOV 2002, LNCS 2456, pp. 326–329, 2002.
© Springer-Verlag Berlin Heidelberg 2002

1.2   IDA Context

The work described hereafter falls under the scope of Article 10 "Spread of Best Practice" as well as Article 9 "Interoperability with national and regional initiatives", of the IDA Interoperability decision [1]

IDA is well equipped to address this issue and has been consistently supporting work of relevance in the eGovernment domain.

1.3 Scope of Work

The subject matter of the Observatory is eGovernment domains, where IDA can add value focusing on the EU dimension. It helps to leverage national and regional initiatives of EU relevance, monitor other worldwide developments/trends and also promotes and learns from best practices.

Central to the Observatory is the gathering of information on state-of-the-art and trends concerning Information Society Technologies and their eGovernment applications being of relevance to IDA.

In this strategy, the IDA website plays a central role but the information is delivered using the most appropriate multimedia products and ensuring appropriate coordination. To date, the Observatory has used the website, electronic press releases, e-mail alerts, physical presentations in several EU Member States and wide-ranging electronic surveys, via Chgambers of Commerce, European Information Centres EICs) and others, to raise awareness of its work.

The information dissemination activities are coordinated with the promotion activities carried out in the context of the Information Society and of the eEurope/eGovernment initiatives.

An effective information network with the Member States Administrations is maintained, mainly through the National Government delegates to IDA, the Telematics for Administrations Committee.

The work addresses issues related to applications supporting the Internal Market, eDemocracy, public/private partnerships, and "pan-European eGovernment Services for European citizens and enterprises" at large.

The Observatory addresses both existing eGovernment initiatives and emerging trends in terms of applications, R&D technologies and commercial solutions which could have an impact to enable the effective deployment of the expectations raised by eGovernment.

2   Activities

The final objective of the Observatory is to proactively support IDA in the early identification and better understanding of emerging eGovernment areas where IDA shall play a role, complemented by Best Practice activities. Its outputs are so designed to contribute to the emergence of pan-European strategies.

---

[1] Decision no. 1720/1999/EC of the European Parliament and the Council of 12 July 1999 adopting a series of actions and measures in order to ensure interoperability of and access to trans-European networks for the electronic interchange of data between administrations (IDA) – OJ L203, 3-8-1999, p. 9.

The work involves the identification of relevant information, creating synergies with complementary initiatives in other DG's or Member States, packaging the information in the appropriate multimedia format(s), raising awareness and disseminating the information (web, information network with the Member States Administrations, meetings, documents, etc.). It also involves seeking feedback and spinning-off appropriate activities.

The information and its delivery has to combine clarity and wealth of content. It has to be appealing for people throughout Europe working on eGovernment or using its services. It supports the decision-making in Member States and European Institutions.

In order to achieve the objective above, the Observatory covers three major areas:

- Area 1: eGovernment Surveillance
- Area 2: In-depth information
- Area 3: Start-up

Area 1: eGovernment Surveillance

Central Repository and Continued Surveillance:
Being a virtual central repository where links with the most relevant eGovernment activities are kept. A "soft coordination" role is played. Surveillance of ongoing activities, trends, relevant technologies as well as developments in the private sector and information of relevance for the eGovernment domain. The information is disseminated through the IDA Web site[2] and a bi-monthly electronic newsletter is distributed along with regular meetings to exchange information and get feedback from Member States Administrations and involving as well key personalities in the domain.

An eGoverment "Tableau de bord" has been constructed on the website, and it monitors progress in national eGovernment plans as well as at the level of the Community.

National and Regional Initiatives and R&D Results:
Identify eGovernment national & regional initiatives in Member states as well as results from R&D projects of relevance for IDA. The strategy is to identify and give visibility to a few selected **"gemstones"** and then make proposals on appropriate actions in the different areas of activity of IDA (Best Practices, Generic Services, Common Tools, Security, content interoperability, etc.).

Area 2: In-depth Information

In-depth focused analysis is undertaken on a six monthly basis.
The first six monthly activities had as its focus:
On-line Services to Enterprises in Europe and the role of Public Private Partnerships.

---

[2] The Observatory provides the information but the actual delivery through the Website is subject of a separate project.

The objective is to identify the needs of enterprises concerning on-line services focusing on trans-border services. This involves analysing the electronic services already delivered as well as new services to be delivered. In order to understand those needs, both the offer side (administrations and others) and the demand side (enterprises) have to be analysed. Best Practices elsewhere world-wide will also be relevant.

The outcome will include possible activities to undertake at the Community level and proposed priorities.

The role Public Private Partnerships are already playing and the future role they can play in a Trans-European context need to be analysed. Other types of co-operation between the public and private sector are also relevant.

In the case of eProcurement, there is already some ongoing work on trans-border issues and Best Practices. The work will be a major input for the 2002 IDA Conference where all this work will be presented (September 19-20 in Brussels. It will also be an input for building IDA's policy for the future (a communication of the Commission is expected during 2002).

Each of the six monthly activities will ensure appropriate dissemination and exchange of information with national experts and will finish with a dissemination event.

The **Final Report** of the Observatory (after December 2002) will present a description of the work performed, the transborder catalogue and the recommendations issued by the project. The final report will be suitable for general dissemination, in particular via the IDA website.

References

For more information about eEurope see:
1. http://europa.eu.int/comm/information_society/eeurope/
2. http://europa.eu.int/comm/information_society/eeurope/documentation/index_en.htm

For more information about action line 3b of eEurope concerning "Government online: electronic access to public services" see:
3. http://europa.eu.int/comm/information_society/eeurope/actionplan/actline3b_en.htm)

Requirements for Transparent Public Services Provision amongst Public Administrations

Konstantinos Tarabanis[1] and Vassilios Peristeras[2]

[1] University of Macedonia, Thessaloniki, Greece
kat@uom.gr
[2] United Nations Thessaloniki Centre, Thessaloniki, Greece
per@untcentre.org

Abstract. In this paper we analyze the requirements posed by the Infocitizen project that attempts to make feasible the realization of a pan-European view for public service provision. The requirements are analyzed based on the project aims of conducting electronic transactions in multi-agent settings- e.g. multi-country involvement- in a transparent as possible manner for the citizen. Transparent public services provision for the citizen is posed as the requirement that both the inputs needed for the delivery of a service as well as the outputs produced by the service are respectively given and received in a transparent as possible manner for the citizen. That is, the citizen will only need to provide the input that cannot be automatically accessed from its relevant source and also the consequences of the delivered service will be automatically propagated to its relevant destinations. In order to achieve such intelligent provision of public services the forms of knowledge that need to be employed are also discussed.

1 The Problem Addressed by the InfoCitizen Project

As most other efforts in the area of electronic government, the InfoCitizen project addresses the "Service Provision" area [1][2] from those shown in Figure 1. The typical scheme for the provision of public services by a Public Administration (PA) involves a customer (citizen, business or other PA) who requests a service from the responsible PA. The PA then performs its internal tasks that may include interacting again with the customer during the various stages of process execution and at the end for service delivery.

The extent to which this simplified model of PA service provision is performed electronically has led to the conceptualization of the four levels of e-government reported in the literature [3]. The InfoCitizen project addresses the fourth level of e-government that of full service provision. Moreover, the project attempts to conduct electronic transactions in multi-agent settings- e.g. multi-country involvement- in a transparent as possible manner for the citizen. In the following we will elaborate on the requirements posed by InfoCitizen along its two main themes:

- transparent public service provision to the citizen and
- multi-agent setting of public service provision

R. Traunmüller and K. Lenk (Eds.): EGOV 2002, LNCS 2456, pp. 330–337, 2002.
© Springer-Verlag Berlin Heidelberg 2002

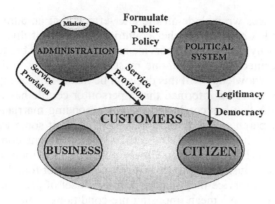

Fig. 1. InfoCitizen Scoping

1.1   Transparent Public Services Provision

Transparent public services provision for the citizen is addressed in InfoCitizen as the requirement that both the inputs needed for the delivery of the service as well as the outputs produced by the service are respectively given and received in a transparent as possible manner for the citizen.

Specifically, regarding the required input of a service, this should be automatically accessed when needed from its relevant source and the citizen will only need to provide anything that cannot be "pulled" in the above sense.

This "pull" strategy enables PA services with minimal input and precondition checks needed on behalf of the citizen-customer. That is, input that needs to be provided and preconditions that need to be checked in order to provide this service are, to the extent possible, automatically performed. However, we need to examine cases in which this "pull" strategy is difficult to apply and why.

In the other respect, the outputs of the service are defined in a broader sense than that of the delivered product or service. In this respect, as outputs of the service we also consider the consequences produced as a result of the service delivery itself. For example, when the PA service of a marriage is executed, the output of the service in a strict sense is the marriage certificate. However, outputs in a broader sense are all the by-products of the service. For example, the information that may need to be delivered to other agencies that document the change of marital status for the actors involved. The output may need to be delivered to the citizen, but its by-products (i.e. consequences) are required to be automatically "pushed" to the destinations (sinks) that are relevant.

The "push" strategy of the generalized output aside from alleviating the citizen from having to propagate the changes introduced by the public service that was just executed, also attempts to maintain consistency of the PA system as a whole across all countries considered. That is, any PA service that will be provided maintains consistency of all the individual PA systems, while any PA related question receives the same answer in all countries. For example, the marital status of a person receives the same answer anywhere posed.

This may mean that the consequences of the PA service provided are "propagated" to all relevant PAs in order to maintain a consistent state. As already described, under the term consequences we include all the after-effects that an administrative process could have on the PA systems of other countries. One part of this issue involves deciding when (and why) the local PA service has interest that is broader than the country where it is executed. In addition, one must determine how to decide the relevant PAs and with what knowledge will this decision be made. For example we could raise the question: who has to be informed that a person or couple has adopted a child in Greece? Is it the PA that has conducted the corresponding marriage? Is it the PA in which the child or parents were born in? Or both? Or is it some PAs in other countries? Also, perhaps the Adoption Act causes some different consequences in one country from another. For example, in Greece posting information regarding an adoption to the Ministry of Justice in Greece is a consequence not found elsewhere.

Furthermore, a PA service may require a different set of preconditions in order to be executed. This may also mean imposing pre-conditions to the PA service that are applicable in more than just the country of the PA that is responsible for its execution. For example: What if Greece does not have any age limit for adoption but Germany poses the age-limit of 21 for anyone interested in adopting a child? Or what will happen if a country gives the possibility of adopting a child to individuals, while another only to couples? In this case the overall PA system will be modeled in a "semi-consistent" state.

Taken together the InfoCitizen requirements of transparent PA service provision refer to transparent supply of input and delivery of output according to the knowledge management strategies of "pull" and "push" respectively.

As discussed an important issue in order to achieve transparent PA service provision involves the orchestration of the above "pull" and "push" strategies, that is, determining their applicability, the source of the "pull" and the destination of the "push" etc. These issues will be addressed by representing the aforementioned in a form of explicit knowledge. This knowledge has a control nature and must have a global scope. Such a role will be assumed by a "broker" or an "intermediate" that knows for example about the PA services in the countries considered. This "broker" will then route and facilitate all the information flows needed. It is important to emphasize at this point that this "broker" possesses mostly knowledge of a control nature and less that of data per se (e.g. personal data of the actors involved). It will contain control knowledge such as process descriptions and the relevant information that will enable the routing of the right piece of information to the right PA unit throughout the countries considered.

In order to build this "broker", schemes that model the control knowledge consisting of terms, pre-conditions, post-conditions, consequences, inputs, outputs etc. will be employed. Since such knowledge is expected to be related to a high-degree as PA systems have similarities amongst themselves, this knowledge will need to be modeled in a way that commonality in any of the above concepts is abstracted out. This "broker" is an important component of the InfoCitizen architecture.

This InfoCitizen requirement of transparent PA service provision may either define a new level of e-government above the four levels mentioned above or constitutes an advanced form of the fourth level. The fourth level of e-government, that is, conducting electronic transactions, can be achieved with a varying level of intelligence. For example, the simple exchange of the proper documents over a well-defined interface may lead to an electronic transaction. However, the InfoCitizen requirement of transparent PA service provision as analyzed above entails a high-degree of intelligence leading essentially to executable knowledge bases that support highly automated service provision.

1.2   Multi-agent Setting

The other InfoCitizen requirement, that of public service provision in a multi-agent setting also plays a central role for the project. This requirement has been motivated by the goal of the InfoCitizen project to facilitate a pan-European provision of PA services.

When looking at the internal tasks that are executed within PA in order to provide the service according to the previously discussed simple model for "Provide Service", it is often necessary for more PAs to participate other than just the PA responsible for the overall service. These associated PAs provide constituent services contributing to the main service. This gives rise to a process that has multi-PA agency participation and correspondingly multiple component processes. The form of participation of each PA agency and of its relevant component process is itself interesting to specify since it may vary to a large degree (e.g. simple provider of input, executer of part of the overall process, controller of part of the overall process etc.).

Multi-PA agency participation in the "Provide Service" process introduces more demanding coordination requirements for the responsible PA since the delivery depends in part on parties and processes external to the PA agency. Coordination requirements exist even in single PA agency case in order to orchestrate the steps of the process. However, in the multi-PA agency case the coordination requirements orchestrate the overall process both at the process step level as well as at the process-to-process level. The latter coordination requirements involve addressing issues such as "at what stage does another PA agency participate?", "which PA agency is responsible for this constituent service?", "what are its inputs and outputs?", "who controls this constituent process", etc. In order to address these coordination requirements knowledge again of a control nature is required.

In the previous discussion the implicit assumption is made that all such PAs (responsible and associated) are under the jurisdiction of the same country. InfoCitizen introduces the added complexity to a multi-agency "provide service" PA process in that the PA agencies involved are from the Public Administrations of more than one country. This added complexity affects the control knowledge discussed earlier in that:

- the control knowledge amongst Public Administration systems is different and may not even be compatible, and
- the terms used are different since on the one hand they are expressed in different languages while more importantly, differences amongst Public Administration systems lead to, among other things, slightly different definitions of linguistically identical terms.

These added requirements lead to the need of control knowledge at the meta-knowledge level in order to unify these distinct sets of control knowledge.

2   Knowledge-Based Intelligence for InfoCitizen e-Government

In order to achieve the intelligent provision of public services described previously in
the analysis of requirements of the InfoCitizen project, certain forms of control
knowledge and meta-knowledge need to be employed. Let us investigate the knowledge required as the level of intelligence of public service provision is gradually increased reaching the full level of InfoCitizen intelligence addressing transparent public service provision in a multi-agent setting.

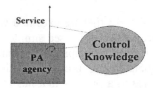

Fig. 2. Control Knowledge in a single PA agency setting

In the case of service provision by a single agency (Fig. 2), no other agencies are
required in order to deliver the service. That is, any "push" or "pull" that needs to be
executed lies within the domain of this agency. If the agency has achieved integration
in its information systems environment, the "push" and "pull" actions are already part
of its integrated operation. Otherwise, these "push" and "pull" actions need to be
implemented, for example, as middleware applications. Such integration is more
readily achievable since any organizational obstacles (e.g. who owns what information) are less pronounced. As a result, the knowledge that is required to implement the
"push" and "pull" actions in this setting lies implicitly in the application that achieves
integration and may not need to be explicitly represented.

In the case of service provision with the participation of multiple agencies all
within the PA system of one country (Fig. 3), achieving coordination of the "push"
and "pull" actions becomes a significant issue. Knowledge needs to be represented
regarding the following:

1. preconditions for provision of a service,
2. consequences of a service together with the affected party, both within the agency
   but also in other agencies,
3. controls that apply in any of the steps involved in the provision of the service together with the agency responsible for applying these controls,
4. inputs of a service together with their source,
5. outputs of a service together with their destination.

In the case of service provision with the participation of agencies from several
countries (Fig. 4), additional knowledge is needed in order to *unify* the distinct bodies
of knowledge that address the previous case of multiple agency participation (see
above list). Specifically, the additional knowledge that needs to be represented in this
case pertains to the following:

1. knowledge that can reason with term correspondence
   a) at a language level (i.e. corresponding terms in different languages) and
   b) at a conceptual level

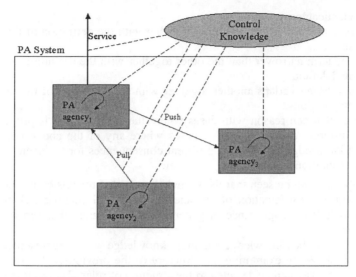

Fig. 3. Control Knowledge in a multiple PA agency setting

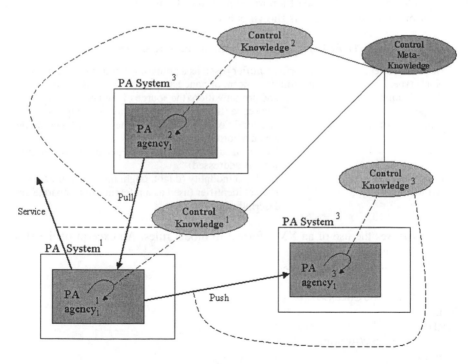

Fig. 4. Control Knowledge in a multiple PA agency setting corresponding to multiple PA systems

      i.   identical terms,

     ii. one term wider than the other together with the definition of the associated surplus,

   iii. one term narrower than the other together with the definition of the associated deficit,

   iv. one term overlaps another together with the definition of the non-overlapping regions,

2. knowledge that can reason with cases where the same service is provided differently in different countries, that is, cases where any of the above corresponding pieces of knowledge differ – e.g. different consequences for the same service, different preconditions for the same service etc.

From the above it can be seen that the knowledge needed for this case, that is service provision with the participation of agencies from several countries, corresponds in essence to meta-knowledge, since in general it is knowledge that can reason about other knowledge.

The above described knowledge and meta-knowledge will be represented in appropriate structures. When examining the structure of the previously described types of knowledge, similarity can be detected to the structure of rules. This is attributed to the fact that a great deal of the control knowledge employed in PA is based on laws enacted by the government, which by nature are structured as rules.

The representation of laws has been studied in the area of legal information systems. Kralingen [4] proposes the use of a norm frame in order to model laws. The structure of the norm frame of Kralingen is shown below:

Table 1. Structure of the norm frame of Kralingen

1 Norm identifier	The norm identifier (used as a point of reference for the norm).
2 Norm type	The norm type (norm of conduct or norm of competence).
3 Promulgation	The promulgation (the source of the norm).
4 Scope	The scope (the range of application of the norm).
5 Conditions of application	The conditions of application (the circumstances under which a norm is applicable).
6 Subject	The norm subject (the person or persons to whom the norm is addressed).
7 Legal modality	The legal modality (ought, ought not, may, or can).
8 Act identifier	The act identifier (used as a reference to a separate act description).

Leff [5] proposes the use of an XML format to model rules. An excerpt of the structure is shown below:

```
<Rule Name= "RuleOne"......>
<LHS>
     <Check></Check>
</LHS>
<RHS>
     <Action></Action>
</RHS>
</Rule>
```

For the meta-knowledge that needs to be represented appropriate structures will also be sought.

3   Conclusion

In this paper we analyze the requirements posed by the Infocitizen project that attempts to make feasible the realization of a pan-European view for public service provision. The requirements are analyzed based on the project aims of conducting electronic transactions in multi-agent settings- e.g. multi-country involvement- in a transparent as possible manner for the citizen. In order to achieve such intelligent provision of public services the forms of knowledge that need to be employed are also discussed. In future work we will investigate initial implementation directions, such as specializing on existing XML and derivative XML (e.g. ebXML) schemas [7] for the Public Administration domain and following for this the UN/CEFACT Modeling Methodology [6] in order to ensure compatibility.

References

1. Peristeras V., K., Tarabanis, "Towards an Enterprise Architecture for Public Administration: A Top Down Approach", European Journal of Information Systems, vol. 9, pp. 252-260. Dec. 2000. Also in proceedings of the 8th European Conference on Information Systems, Vienna, vol. 2, pp.1160-1167
2. Tarabanis K., Tsekos Th., and Peristeras V., "Analyzing e-Government as a Paradigm Shift", to be presented in the "Annual Conference of the International Association of Schools and Institutes of Administration (IASIA)" 17-20 June 2002, Istanbul, Turkey.
3. European Commission, "List of indicators for benchmarking eEurope as agreed by the Internal Market Council" at
http://europa.eu.int/information_society /eeurope/news_library/documents/text_en.htm.
4. van Kralingen, "A Conceptual Frame-based Ontology for the Law", Proceedings of the First International Workshop on Legal Ontologies, Melbourne, Australia, 1997
5. Lawrence L. Leff, "Automated Reasoning with Legal XML Documents", Proceedings of International Conference AI and Law – ICAIL 2001, St. Louis, Missouri, USA, 2001.
6. UN/CEFACT's Modeling Methodology, Draft, TMWG/N090R10, UN/CEFACT, November 2001
7. ebXML Business Process Specification Schema Version 1.01, Business Process Project Team, 11 May 2001 at www.ebxml.org/specs/ebBPSS.pdf

CB-BUSINESS: Cross-Border Business Intermediation through Electronic Seamless Services

Maria Legal[1], Gregoris Mentzas[2], Dimitris Gouscos[3], and Panagiotis Georgiadis[3]

[1] Planet Ernst & Young SA
m.legal@planetey.com
[2] Dept. of Electrical and Computer Engineering, National Technical University of Athens
gmentzas@softlab.ntua.gr
[3] eGovernment Laboratory, Dept. of Informatics and Telecommunications,
University of Athens
{d.gouscos,p.georgiadis}@e-gov.gr

Abstract. Business enterprises face significant obstacles in their quest to interact with public administrations and governments across Europe, such as bureaucracy, ambiguous procedures, functional disintegration, vague authority structures and information fragmentation. The recent trend towards the delivery of electronic services by governments (*"e-government"*) and the development of integrated and customer-oriented mechanisms (*"one-stop government"*) are efforts to overcome these problems. However, all related efforts focus on the national scene of each country and do not address the needs of businesses when they enter into cross-border processes. This paper presents the objectives, the overall approach and the architectural model of the CB-BUSINESS project, which aims to develop, test and validate an intermediation scheme that integrates the services offered by government, national and regional administration agencies as well as commerce and industry chambers of European Union and Enlargement countries in the context of cross-border processes.

1   Introduction

Business enterprises face significant obstacles in their quest to interact with public administrations and governments across Europe. The most common problems include bureaucracy, ambiguous procedures, functional disintegration, vague and/or overlapping authority structures and information fragmentation. The recent trend towards the delivery of electronic services by governments (*"e-government"*) and the development of integrated and customer-oriented mechanisms (*"one-stop government"*) are efforts to overcome these problems and radically revamp the services provided by European governments; see e.g. [1], [3], [5], [6] and [7].

At the European level, the *eEurope* initiative launched by the European Commission in December 1999 puts forward a concrete action plan for "Government On-Line (GOL)" with the aim to make public information more easily accessible and stimulate the development of new private sectors services based on the new data sources that become available; see [2].

However, related efforts both within and outside the European Union focus on the national scene of each country and do not address the needs of businesses when they

R. Traunmüller and K. Lenk (Eds.): EGOV 2002, LNCS 2456, pp. 338–343, 2002.
© Springer-Verlag Berlin Heidelberg 2002

enter into cross-border processes. Certain initiatives such as EUREGIO (see [4]) have been set up in this direction, but their emphasis is rather on facilitating transactions between neighboring border regions and, in this context, supporting cross-border processes at the local level, than on explicitly providing support for "anywhere-to-anywhere" cross-border transactions. The critical need is to support, for instance, a company (possibly without core business competencies in exports) that occasionally enters into export procedures or has to pay foreign subcontractors, requiring information and/or needing to make transactions with the public administration and government organisations of another (any other, possibly non-neighbouring) country. In this area the problems of bureaucracy, ambiguity, vagueness and disintegration that business enterprises have to face when interacting with foreign governments, get more sharp than in the national setting. Such problems are insurmountable for SMEs with limited resources.

The CB-BUSINESS project presented in this paper addresses directly this situation as its primary objective is to develop, test and validate an intermediation scheme that integrates the services offered by public administrations as well as chambers of commerce and industry in the context of cross-border issues. The CB-BUSINESS project is a 24-month project co-funded by the European Commission under the 'Information Society Technologies' program and involves as partners Planet Ernst & Young SA, SchlumbergerSema, University of Athens, ComNetMedia, Greek Ministry of Economy and Finance, Instituto Tecnologico de Canarias, Bulgarian Chamber of Commerce and Industry, Paris Chamber of Commerce and Industry, Athens Chamber of Commerce and Industry and Chamber of Commerce and Industry of Romania and Bucharest Municipality.

2   CB-BUSINESS Objectives

The CB-BUSINESS project has three specific objectives.

Objective 1. Design a unified true "one-stop shop" service model for "Business-to-Government" interactions. CB-BUSINESS aims to develop a service model based on an intermediation scheme that:

- extends the "first-shop" (i.e. information counter) and "convenience store" (i.e. one location for different transactions) service models to develop a true one-stop government model_(i.e. one location that integrates many services necessary to satisfy concerns of specific client groups in specific events) that is transparent to the end-user company;
- focuses on cross-border processes necessary for administrative support of cross-country business searches, contacts and transactions, falling under the general theme of "cross-border entrepreneurship" and mainly initiated by EU businesses targeted at Enlargement Country markets; and
- is structured around end-user needs (rather than provider services – i.e. governmental processes) by grouping services around "business life episodes".

Objective 2. Develop a WWW-based intermediation hub that implements this service model and will act as a pivotal point of contact for EU and enlargement country enterprises. CB-BUSINESS aims to develop a system that will:

- Have an "intermediation hub" technical infrastructure able to accept user requests, identify the cross-border processes that have to be enacted, trigger and dynamically coordinate process workflows of individual service providers (administrations and chambers) and integrate the final results for delivery to end-users.
- Be accessible over the Internet – but also open to future extensions, both for multi-channel service delivery (e.g. through mobile phones, digital TV, call centres, etc) and to complementary infrastructures (e.g. banking and postage services)
- Provide a "seamless service feeling" for end-users, who can have access to cross-border services through single-stop, single-session, single-sign-on procedures at a low cost and up to a standard quality, as well as to decrease operational costs and increase quality of service for both government/administration and chamber service providers.

Objective 3.Prove the validity of the service model and pilot-test the WWW-based intermediation hub in various specific cases that strengthen European integration and facilitate cross-border processes between EU-enterprises and businesses in enlargement countries.

3   CB-BUSINESS Approach

CB-BUSINESS establishes an intermediation scheme between service providers and end-users, with the aim of transforming the many-to-many service provider/end-user communication mesh (in which each end-user should communicate with as many service providers as are involved in serving his/her need) into a many-to-one-to-many star-like communication topology.

In this architectural scheme, an end-user communicates with the intermediation hub as a single point of reference, and the latter handles all complexity of triggering and co-ordinating service provider workflows. Therefore, end-users enjoy a "seamless service feeling" whereas service providers avoid the burden of inter-organisational communication, since the hub undertakes all co-ordination procedures. This results in more structured tasks for service providers, who are enabled to better define their operational interfaces towards the intermediation hub and concentrate on the establishment and improvement of quality and performance dimensions for their process workflows.

Upon submission of a user request (see Figure 1) the CB-BUSINESS intermediation hub shall identify involved services, competent service providers and user input requirements and ask for the latter as appropriate. Upon provision of required input data from end-users, the intermediation hub undertakes forwarding of user input, triggering and co-ordination of process workflows of individual service providers, rendering internal details regarding service execution procedures or workflows transparent for end-users; however, upon user demand, the processing status of end-users' requests may be monitored and presented. Therefore, the CB-BUSINESS intermediation hub employs overall workflows of cross-border processes to trigger and co-ordinate individual service provider workflows, thus being able to deduce the progress status of user requests and report it as appropriate.

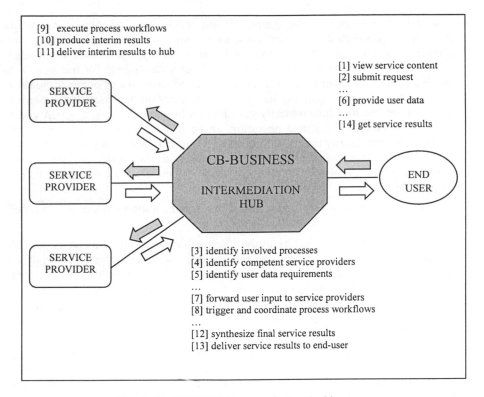

[9] execute process workflows
[10] produce interim results
[11] deliver interim results to hub

SERVICE
PROVIDER

[1] view service content
[2] submit request
...
[6] provide user data
...
[14] get service results

CB-BUSINESS

SERVICE
PROVIDER

INTERMEDIATION
HUB

END
USER

SERVICE
PROVIDER

[3] identify involved processes
[4] identify competent service providers
[5] identify user data requirements
...
[7] forward user input to service providers
[8] trigger and coordinate process workflows
...
[12] synthesize final service results
[13] deliver service results to end-user

Fig. 1. CB-BUSINESS Intermediation Architecture

From a structural point of view, the CB-BUSINESS intermediation hub is deployed as an added layer on top of existing government and business service schemes (which indeed corresponds to the idea that the CB-BUSINESS service-mediary constitutes an added-value layer on top of existing service provision schemes). The intermediation hub employs standardized modules for interfacing with individual service schemes, and such standardized modules are also deployed in the latter as interfaces to the CB-BUSINESS hub. These modules (referred to as CB-BUSINESS interfaces) encompass all the technical infrastructures and specifications, functional conventions and operational arrangements which are necessary in order to establish information and control flows between the intermediation hub and individual service schemes. Two points should be made here:

- This structural architecture is based on the concept of standardized interfaces between the intermediation hub and individual service schemes, and is not dependent upon internal details of service scheme implementation. In other words, although the CB-BUSINESS intermediation hub itself and the CB-BUSINESS interfaces are ICT-based, CB-BUSINESS service providers may internally implement their service provision procedures with IT- and/or paper-based work schemes. This degree of independence allows even non-IT-enabled service providers to integrate their services with CB-BUSINESS, which is considered an important strength of the architecture.

- The star-like topology of the intermediation architecture described here, where there is a single central point of co-ordination, may also be generalized into a hierarchical-star topology without affecting the basic principles. In such a topology, individual service providers are grouped in clusters (according, for instance, with geographical, sectoral or mission level criteria), and each service provider cluster is coordinated by a corresponding low-level intermediation hub. Low-level intermediation hubs are then hierarchically coordinated by the central hub, which undertakes responsibility for global operation. In such a multiple-hub topology, end-users do not necessarily refer to the central hub but may access any hub of the hierarchy (the one most suited to their needs according to the service providers that it clusters); all hubs, however, are able to provide the same service content so that no complexity is created on the end-user side.

As far as the implementation architecture of the intermediation scheme is concerned, the following basic principles apply:

- the overall technical architecture of the intermediation hub is as much as possible compliant to current standards (with respect to information formats, communication protocols, etc) and open to emerging ones, while at the same time getting the most benefit out of technologies for e-service provision and delivery platforms
- the overall technical and functional architecture of the intermediation hub is modular with the aim of facilitating incorporation of alternative and complementary e-service platforms.

With these two principles in mind, CB-BUSINESS architecture employs the notion of a "**communication gateway**" for referring to the implementation of communication and co-ordination links between the CB-BUSINESS intermediation hub, end-users and service providers. A CB-BUSINESS communication gateway represents a specific communication channel (such as the Internet or WWW, fixed or mobile telephony etc) together with all technical standards (communication protocols, information content or information medium formats etc) necessary for exploitation of the channel, together with all associated operational dimensions (e.g. quality-of-service, security-of-service etc). It is clearly desirable to support as many different communication ways as possible, so that prospective user communities are broadened and accessibility constraints are relaxed. With this objective in mind, the CB-BUSINESS implementation architecture has been conceived as comprising of certain indispensable core modules, implementing the basic intermediation hub functionality, complemented by a number of optional pluggable peripheral modules, each implementing one specific communication gateway employed by end-users and/or service providers In this way, the core architectural design can ignore communication gateway details, which are handled by dedicated pluggable modules, and assume an **abstract communication model**; even more importantly, additional communication gateways may be incorporated into the system as new e-service provision and delivery platforms emerge, which is a fundamental prerequisite for keeping up with technological state-of-the-art and exploiting the benefits of technological advancements. Last but not least, this architectural modularity also allows for provision of non-IT-based communication gateways (such as voice telephony, facsimile or paper mail) which can be valuable for printed document transfer and non-IT-enabled end-users.

4   Current Project Status

As the CB-BUSINESS project started just a month ago (April 2002), the present paper is based on material from the project proposal. Currently, the CB-BUSINESS consortium is working on the analysis of service provider processes and end-user requirements, as well as on the specifications of the functional and technical architecture of the CB-BUSINESS system. It is expected that by September 2002 the service scenarios of the CB-BUSINESS intermediation hub will have been defined and analyzed in detail, and all functional and technical aspects of the system demonstrator will have been specified, so as to proceed with the technical design and solution development.

It should be stated that the next steps of the project implementation involve, also, the formulation of a set of architectural provisions on CB-BUSINESS service delivery platforms, process workflow ontology and operational interfaces. This is an important process in order to ensure the wide market acceptance and acceptability of the CB-BUSINESS system. The architectural provision shall be publicized in the form of Request for Comments, so as to acquire external feedback and promote awareness and consensus.

References

1. Atkinson, R., and J. Ulevich, Digital Government: The Next Step to Reengineering the Federal Government, Progressive Policy Institute Technology and New Economy Project, March 2000.
2. Commission of the European Communities, eEurope: An Information Society For All, Communication on a Commission Initiative for the Special European Council of Lisbon, March 23-24, 2000.
3. Dawes, S., P. 3Bloniarz, K. Kelly, and P. Fletcher , Some Assembly Required: Building a Digital Government for the 21st Century, Report of a Multidisciplinary Workshop, Center for Technology in Government, March 1999.
4. EUREGIO, Das alltaegliche Europa, Gronau, 2000.
5. Scheppach, R., and F. Shafroth, Governance in the New Economy, Report of the National Governors' Association, Washington D.C., 2000.
6. M. Sprecher, "Racing to e-government: Using the Internet for Citizen Service Delivery", Government Finance Review, October 2000, pp. 21-22.
7. West, D. Assessing E-government: The Internet, Democracy and Service Delivery by State and Federal Governments, Brown University, September 2000.

Bridging the Digital Divide with AVANTI Technology

Antoinette Moussalli[1] and Christopher Stokes[2]

[1] London Borough of Lewisham, Lewisham Town Hall, Catford, London
antoinette.moussalli@lewisham.gov.uk
[2] Fujitsu Consulting, Lovelace Road, Bracknell, Berkshire RG12 8SN
chris.stokes@uk.consulting.fujitsu.com

Abstract. Whilst e-Government provides many opportunities for local authorities to serve citizens more effectively, it also runs the risk of widening the digital divide and making non-IT users second-class citizens. AVANTI aims to address this problem by focusing on people who cannot or think they do not want to be involved in the Information Society. To enable and encourage these citizens to interact electronically, we are developing a user friendly on-screen assistant to serve their needs and help turn the vision of universal Internet access for all into a reality.

1   Why Do We Need AVANTI?

Participants in the Information Society receive numerous benefits – better interest rates on banking facilities, special deals on utilities and insurance, cheaper goods and services and enhanced customer service options. People who cannot use technology or who are reluctant to do so miss out on these opportunities, and hence receive a lower standard of service from commercial organisations.

The introduction of e-Government, providing electronic access to services provided by local or national government, also presents many opportunities: it permits council departments to be available twenty-four hours per day, seven days per week to respond to citizen queries, it enables councillors and politicians to canvass the opinions of their constituents in their own homes, and it allows citizens to participate far more fully in the running of their communities or their country.

However, in the same way as the technology-poor are missing out on commercial offers and opportunities, the introduction of e-Government runs the risk of leaving these citizens behind in the information revolution. Some people are unable to use technology due to disabilities – visual impairment, manual dexterity problems or learning difficulties. Some do not possess the required technology and may not see the benefit of purchasing it. Others have difficulties with language, perhaps because they do not speak or understand the local language. Many people simply find technology too complex and confusing.

The eEurope Action Plan is designed to speed up and extend the use of the Internet to all sectors of European society, and special focus has been given to addressing the digital divide, the gap between people who currently use and are comfortable with technology and those who cannot or do not use it. For example, European Member States are encouraged to adopt and promote the Web Accessibility Initiative to make

R. Traunmüller and K. Lenk (Eds.): EGOV 2002, LNCS 2456, pp. 344–349, 2002.

public sector web pages more accessible, particularly to visually impaired users. However, in establishing the AVANTI programme we decided that a more radical approach to accessibility was required rather than simply enhancing standard web pages.

AVANTI seeks to bridge the digital divide by creating an on-screen intelligent assistant which takes over the interface, so that rather than dealing with conventional pages of text, the interaction is a more natural conversation between the citizen and the computer. In this way we are seeking to simplify the interface for people uncomfortable with technology, and by introducing sophisticated natural language processing technologies and speech generation we aim to address the needs of many of the citizens who are currently disenfranchised. AVANTI is a research project, part funded under the European Commission's fifth framework programme. As such, we recognise that some of the components we are working on may not be perfected during the life of the project; however, we acknowledge that AVANTI is not the whole answer to this difficult problem, but rather a first and key step along the road towards an inclusive information society for all.

2   How We Are Approaching the Problem

Before engaging with technologists, to identify the products and solutions which could assist in meeting the needs of these users, we identified the key stakeholders and established a programme to understand their needs.

The most important stakeholders are the target users of the applications, the citizens who currently cannot or think they do not want to use technology. Each of the participating cities in the AVANTI consortium has established a local user group comprised of representatives from the local community, including, as far as possible, people either within the target groupings for the project or those who can understand and represent their interests. Where applicable, we are also involving support agencies for the target users, including council staff who work with these groups and local and national organisations. Initially we sought the opinions of these people on the problems they have with current technologies, and how we could best address these. As the project has progressed, we have continued to involve the local users by demonstrating prototypes of our applications and gathering their feedback. The local users will have a key role throughout the life of the project, steering the development work and identifying or selecting enhancements to ensure the final product meets their needs as far as is possible.

The second key group of stakeholders is comprised of local service managers, the council employees who have responsibility for the services we are aiming to deliver. For the project to be successful, it has had to be integrated not just into the systems of the council but also into the thought processes of the council staff. We consider it vital that this project is not seen as another project led by technologists, but one which has been designed in partnership between the council departments, the users and the technologists. Whilst we are working primarily with service managers whose departments will be impacted directly by the project, we have also sought to involve the other service managers from the council as part of the preparation for a wider-scale deployment of this technology once it is proved.

Finally, we have been keen to ensure that the solutions we identify are as generic as possible. We have therefore established AVANTI as a trans-European consortium to research and demonstrate this technology, with representation from the United Kingdom (Lewisham and Edinburgh), Sweden (Stockholm) and Latvia (Ventspils). This is enabling us to research the technology with different cultures and languages, and develop a range of applications interfacing with local city systems.

3   Involving the Technology Partners

Whilst the project is led by the users and steered by the city service managers, it has also been necessary for us to research, develop, integrate and use state-of-the-art technology solutions, and two of the world's leading technology companies have been involved since the inception of the project.

The lead technology partner selected was Fujitsu, one of the world's largest IT solution providers, with many years of experience in designing, developing and deploying strategic solutions for local government. Fujitsu is one of the pioneers in online advice and conversation based applications, having been researching the area in the private sector for several years.

In addition, Microsoft has also been involved in the project as an associate partner. Microsoft has been actively researching and developing several of the components we are seeking to deploy, including Microsoft Agent technology, enterprise application integration products and speech technologies.

From the requirements identified by the users, we identified the core components of the solution, together with a range of additional facilities which we are currently researching. The consortium continues to work closely with the research functions of both Fujitsu and Microsoft to keep current with the state of the art. The city partners are also engaging with local technology suppliers to develop and enhance the product, for example integrating alternative languages.

4   Components of the Solution

The overall aim of the project is to make the interface with the computer as natural as possible, so as to make it usable by people who traditionally would be uncomfortable with technology. We therefore decided to replace a standard Internet interface, with long blocks of text, hyperlinks, frames, tables and buttons, with a far simpler one based around an on-screen character or avatar. This character is able to speak to the citizen in his or her own language, understands their responses in natural language and acts accordingly. When the project was conceived, and from the feedback received from user panels, it was apparent that an ideal solution would be for the citizen to be able to speak to the computer, and this is a major research area for the project; however, the limitations of current technologies have made this difficult to deploy in the short-term, and a keyboard entry is currently being utilised for the initial work around the development of the first applications. By the end of the project we hope to be able to implement a speech based solution.

The primary technology being implemented for the user interface is based around Microsoft Agent, which enables us to deploy a character on the screen, control its animations and gestures and send text to it which can be displayed in a speech bubble and converted into text using a text-to-speech engine. The toolset we have developed also allows for alternative delivery mechanisms such as a simpler text interface delivered to a personal computer, interactive television or mobile device, but the primary focus of the project is delivery via an animated on-screen character.

Once the citizen has given a response – by typing or ultimately by voice input – the system then has to understand what they have said in order to drive the conversation or interact with council systems as required. This 'understanding' is achieved by a natural language processing component. Various technologies for this have been researched by Fujitsu, ranging from simply looking for a specific phrase through to complex intent matching algorithms. The solution adopted is context sensitive, searching for key words or phrases within the response appropriate for the context of the question. In addition, the natural language processing component is able to identify and extract numbers, money, dates, times and ages as part of a response, whether they are in figures or in words. The aim is to enable the citizen to enter their request or response in as natural a way as possible, and for the system to understand and act on it.

The conversation is controlled by a conversation management component, which identifies the next action to take in response to the input from the citizen. This could be as simple as giving a reply back to them through the on-screen character. It could also involve accessing local databases to store or retrieve information, or interfacing with local city systems and databases.

The integration with city systems is managed by an enterprise integration component. This is highly configurable to enable it to access a selection of databases directly. To enhance the integration with city systems, this component can also interface with Microsoft's BizTalk Server, which then has the ability to integrate with an even wider range of databases, applications and messaging systems.

5   Applications for the Technology

The four cities in the AVANTI consortium are implementing a number of demonstrators, illustrating the range of electronic services which could be opened up to technologically challenged users.

The London Borough of Lewisham is part of a consortium of five UK local authorities developing a solution to enable electronic access to council processes, from social services and housing through to education and benefits. From these, an initial set of processes concerned with housing benefits has been selected to implement using the AVANTI technologies. Many of the citizens applying for such benefits are in the target groups for the project – elderly, disabled or socially disadvantaged – and without AVANTI many would not be able to use the electronic access mechanisms being developed. The AVANTI project is making access available to these services using the on-screen avatar, which can be deployed either into citizens' homes or via public access kiosks and cybercentres.

In addition, Lewisham is developing an electronic consultation application for their citizens. Electronic consultation – conducting surveys and opinion polls electronically for citizens – is a key way to make local government more accountable and democratic, as local residents can give their opinions on key issues as often as required rather than waiting for the next council elections. However, without technologies like AVANTI, citizens who cannot or do not wish to use computers will increasingly become disenfranchised in the democratic process; their views will simply not be recorded or represented. Using the friendly accessible approach of AVANTI enables these citizens to participate fully in the democratic process. This could also provide a model for electronic voting accessible to all citizens.

The City of Edinburgh is deploying AVANTI in its library service, providing an accessible interface to their existing information technology infrastructure in this area. Citizens will be able to interact with an avatar character to find details about local library facilities, ask questions about the service and carry out tasks like reserving books and extending book loans. The library service is already installing public access computers in the libraries, and in addition to enabling easier access to library applications, there is the potential to offer additional AVANTI services through the same infrastructure.

The borough of Kista in Stockholm, Sweden, is providing access to local information using the AVANTI toolset. This will enable citizens or visitors to the area to find answers to a range of frequently asked questions, such as the location of local amenities. Later in the project, they are seeking to interface the avatar to a range of municipal, regional and national databases.

In Ventspils, Latvia, the toolkit is being proved in a different culture, and with considerable language challenges. Within the Ventspils local project, citizens will initially be able to ask questions and be led through processes concerning local taxation.

Across the demonstration sites, the toolkit will be tested in six different domains of knowledge, using at least four languages and a variety of cultures. By drawing together the experiences of users in the four cities, we will aim to identify the features of the AVANTI toolkit which best address the needs of the technologically challenged citizens, any parts which do not work for these users, and areas for enhancement as we seek to develop and exploit the project.

6   Future Enhancements

The initial focus of the work has been to produce the toolkit to enable the cities to develop conversation-based applications for their local citizens. During this development work, and from the feedback we have received from our citizen users, we have identified a number of enhancements which are currently being researched by the cities and the two industrial partners.

A key technology for making electronic services accessible is voice recognition, enabling citizens to speak to the computer rather than having to type. This is an area where there have been major advances over the past five years, and two possible approaches have been identified. The first uses trained voice recognition, where the citizen has to spend approximately fifteen minutes reading text to enable the computer to understand their voice pattern; once this is completed, however, the citizen will be able to speak naturally to the system with the voice recognition module converting this speech into text. The second option is speaker independent voice recognition, which requires no training and is therefore more suitable for wide-scale deployment.

However, this can only be used to recognise a limited vocabulary of key words, making the conversation less natural. We will investigate both of these options towards the end of the project.

Another key area for e-Government is security, ensuring that where access is provided to confidential information in council systems, this is restricted to authorised citizens. Whereas regular computer users are familiar with passwords, these are less suited to citizens who are not comfortable with technology. We are investigating password-free options for AVANTI, including smart cards and biometric technologies, and will aim to integrate these later in the project.

We are also exploring new access devices for citizens, including interactive television and mobile devices. Almost every home has a television, and with the introduction of interactive digital capabilities, increasingly these sets are being used for more than just passive viewing. Likewise, many citizens now have mobile telephones, which are now being used for more than just voice calls. In both cases, we are keen to try to develop and deploy a more accessible interface to council services using these devices, whilst recognising the limitations of these channels.

7   An Integrated Service Offering

There are numerous initiatives being progressed in e-Government, and for a successful implementation it is vital that they present an integrated offering to the citizens, rather than a disjointed approach. We must also be careful not to restrict users to interacting with councils in the manner we decide is best for them, but rather provide options to enable them to decide how they want to access the services they require.

AVANTI is not designed to be the complete solution to e-Government. There will be many citizens for whom it is not suited, and in addition there will be applications which cannot be delivered using an on-screen character due to complexity or sensitivity. We would therefore expect to deploy it as part of a wider solution, which may include more conventional Internet technologies, assisted service capabilities either in council offices or remotely using call centres, and video conferencing links.

8   Summary

We recognise that there are many citizens who currently cannot participate in the Information Society, and as e-Government becomes more pervasive they will increasingly be left behind and become disenfranchised. Through the AVANTI project, working closely with this group of citizens, we are seeking to understand their needs and develop solutions for them.

Over the course of a two year research project, we do not expect to solve all of the problems of including these citizens in the Information Society. However, we believe that the research we are conducting and the solution we are developing will be a vital step in informing the debate, and will provide models for to delivering services to this key group of citizens. Only by understanding and addressing their needs will local and national government be able to realise the vision of universal electronic access for all citizens, and deliver true e-Government.

An Integrated Platform for Tele-voting and Tele-consulting within and across European Cities: The EURO-CITI Project

Efhimios Tambouris

Archetypon S.A., 236 Sygrou Ave., Kallithea, 176 72 Athens, Greece
tambouris@archetypon.gr
http://www.archetypon.gr

Abstract. Tele-democracy is becoming increasingly important for local authorities in Europe. The EURO-CITI project aims to specify, develop and evaluate an integrated platform for two tele-democracy services, namely tele-voting for opinion poll petitions and tele-consulting. The technical developments are divided into those for operators at local authorities and those for citizens. The platform empowers operators at local authorities to initiate a call-for-vote on a local problem, to dynamically set-up secure networks of cities and initiate a call-for-vote on common problems, to monitor voting results and extract statistical information, etc. Regarding security and privacy, authentication/authorization solutions are proposed and a Public Key Infrastructure is specified. The trial sites for the EURO-CITI platform are three European cities, namely Athens, Barcelona and London Borough of Brent.

1 Introduction

Electronic government and tele-democracy are high in the agenda of the European Commission [1][2]. The benefits of both e-government and tele-democracy are now well understood by local authorities worldwide that launch relevant initiatives [3][4].

In the case of tele-voting for realizing opinion polls petitions, the application of technology provides some straightforward advantages (such as increased convenience and accessibility and reduced costs) but also a historic opportunity to re-establish some form of direct democracy. The concept of direct democracy suggests that all citizens decide via voting on their problems. This concept was abandoned as local communities were growing in size.

The aim of this paper is to present an integrated tele-democracy platform for tele-voting and tele-consulting services within and across cities. The technical infrastructure is deployed in Athens, Barcelona and London and will enable the respective Local Authorities (LAs) to conduct "intra-city" or "local" as well as "inter-city" or "network" tele-voting and tele-consultations. This platform has been developed within EURO-CITI [5][6], a research project partially funded by the European Commission under the IST programme [7].

This paper is organized as follows. In section 2, a general overview of the EURO-CITI architecture and respective tele-democracy services is given. In section 3, the

R. Traunmüller and K. Lenk (Eds.): EGOV 2002, LNCS 2456, pp. 350–357, 2002.

characteristics of the services are outlined while in section 4 technical details are presented. In section 5 the approach to security/privacy issues is outlined. Finally, in section 6 the conclusions and future work are given.

2   The EURO-CITI Platform: Architecture and Services

The main objective of EURO-CITI is to exploit the potential of on-line democracy by developing and demonstrating new transaction services, namely tele-voting for realizing opinion poll petitions and tele-consulting. The development of these services calls for a common underlying architecture to facilitate their implementation and fully exploit their potential. In this section, the EURO-CITI technical architecture is presented and the EURO-CITI services are outlined.

2.1   EURO-CITI Architecture

The EURO-CITI architecture consists of a number of platforms (one per city) that communicate over the Internet. This architecture is depicted in figure 1 in the case of three cities, namely Athens, Barcelona and London Borough of Brent.

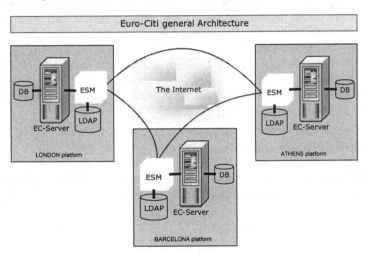

Fig. 1. EURO-CITI Architecture

The main components of the platform in each city are:
- The EURO-CITI (EC) Server where all applications for the operators and citizens reside.
- The EURO-CITI Security Manager (ESM) that is responsible for secure communications between platforms and authentication.
- The Lightweight Directory Access Protocol (LDAP) where all citizens and operators general information reside.
- The DataBase (DB) where all applications data reside.

Each local authority hosts one server. Those servers run the EURO-CITI services and are connected to sub-systems such as applications databases and LDAP repositories. EURO-CITI servers are able to communicate with each other thanks to the ESM component. Security requirements are fulfilled in order to provide citizens with trustworthy and secure services.

2.2   EURO-CITI Tele-voting

The EURO-CITI tele-voting application consists of two different tools:
- The *Tele-voting Administrative Tool* allows operators to initiate a call-for-vote, to invite other cities in a common call-for-vote, to determine the eligible voters, to initiate a call-for-vote in different languages etc.
- The *Tele-voting Service* allows citizens to vote, to request a call-for-vote, to extract statistics (if allowed by the operator), to switch between different languages etc.
- The EURO-CITI Tele-Voting service will be used for opinion poll petitions. In that context, three tele-voting scenarios have been identified by the participating local authorities as particularly important:
- "**Local** Voting". In this case, a voting issue is posted in one EURO-CITI server and eligible voters are citizens who are registered in that server.
- "Local Voting with European Scope". In this case, a voting issue is posted in one EURO-CITI server (termed initiator). Here, eligible voters consist of citizens who are registered in the initiator as well as citizens from other cities. These cities however must have been invited by the initiator and accepted that invitation.
- "Network Voting". In this case, a voting issue is proposed by one EURO-CITI server (termed initiator) and is posted in all servers (i.e. cities) that have accepted to participate in that voting. Here, eligible voters for each server are the citizens who are registered in that server.

2.3   EURO-CITI Tele-consulting

The Tele-Consulting module offers two types of services, Tele-Consultation and e-Forum. Each service is composed of two different tools:
- **Tele-consultation**
  - The *Tele-Consulting Administrative Tool* allows the operators to set up consultation campaigns.
  - The *Tele-Consultation Service* allows citizens to participate in consultation campaigns.
- **e-Forums**
  - The *e-Forums Administrative Tool* allows the operators to create new forums, to create new categories and to track the opinion given by the citizen in the different forums.
  - The *e-Forum Service* allows citizens to participate in the available forums by expressing their opinion or commenting on the opinion of other citizens.

In Tele-Consultation both "Local consultation" and "Local consultation with European Scope" scenarios are supported, where these scenarios have the same scope as in Tele-Voting. However, in e-Forums only "Local" scenarios are supported.

3   Characteristics

The main characteristics of the EURO-CITI integrated platform are:

- Intuitive, easy-to-use graphical interface for operators and citizens.
- Access from multiple devices for citizens.
- Authentication using multiple methods (login/password, smart cards, digital certificates).
- Security at the system level but also at the application level (in the case of tele-voting service).
- Ability to dynamically set up virtual private networks between cities in order to perform a common voting or consultation.
- Multilingual versions available for the operators to choose during installation.
- Multilingual content (e.g. postings) by operators are supported.
- Multilingual interface and content is available to citizens at any time.
- Archiving and auditing facilities are available to operators.
- Support of open standards e.g. Java, XML, WAP.

The specific characteristics of the tele-voting service are:

- Operators may create a new voting issue by inserting voting subject, options, duration, scope, category, keywords, URL for further information, multilingual information; by inviting other cities (in the case of network voting) and by determining eligible voters based on age, nationality and gender.
- The service supports multiple open voting issues at any time.
- Voting is secure and anonymous. No citizen is allowed to vote more than once for the same issue and no one is able to alter votes (democracy requirement). Also, citizens are able to verify their personal voting.
- Citizens are notified about forthcoming polls.
- Citizens are able to view the results of previous voting issues and the partial results of current voting issues (if the operator has enabled this option when creating the voting issue).
- Citizens are able to suggest a voting issue.

The specific characteristics of tele-consulting are similar with the relaxation of security constraints.

4   Development

The architecture used to develop the EURO-CITI platform is based on the J2EE standard. As an example, in figure 2 the software architecture for tele-voting is depicted. This architecture caters for a number of requirements (e.g. communication between services over ESM, communication of services with the Database and LDAP, access from multiple devices, support of multiple authentication methods etc.)

Each page of the resulting services is structured in three main parts (figure 3):

1. *Fixed part*: it includes the page head and the rest of components of the static design.
2. *The menu.*
3. *The page content.*

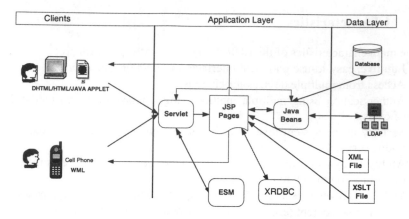

Fig. 2. Tele-voting Software Architecture

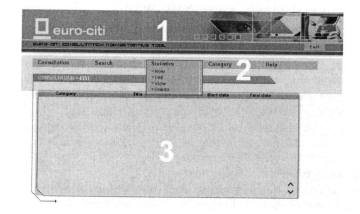

Fig. 3. Interface

5   EURO-CITI Security/Privacy Elements

Citizens access the EURO-CITI services through personal computers (Home PCs, kiosks) or WAP devices [8] using the Internet or wireless networks respectively. Links between the EURO-CITI nodes are protected by the following protocols (figure 4):

- From WAP devices to WAP gateway: **WTLS**
- From WAP gateway to EC servers: **SSL**
- From PC devices to EC servers: **SSL**
- From EC server to EC server: **IPSec**

Servers of different cities communicate with each other in the context of *network services*. A network service is launched by one city and is accessible from citizens of other cities. The participating cities can send their results to the city that has launched the service. In this context, citizens registered in a city and participating to a service of another city must be *remotely authenticated.*

For instance, if a citizen registered in London accesses a secure service proposed by Athens, the Athens server will have to ask to the London server if the citizen is authorized to access the service or not.

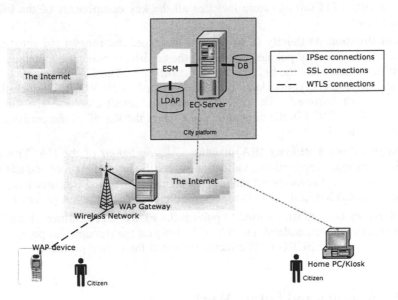

Fig. 4. EURO-CITI Security

With respect to authentication, two main methods are implemented: the login/password paradigm and the certificate-based authentication. These two solutions are combined with the use of Smart Card to realize four authentication solutions:

- **Simple login/password**: the citizen has to memorize his login/password pair. Citizens must use these credentials discretely in order to avoid their use by another person.
- **Login/password with smart card**: The smart card stores several login/password pairs. The citizen fills in the login/password window with a drag and drop application. The login/password pair can be provided to the citizen or stored in the smart card. Using the smart card is totally transparent for the EURO-CITI applications.
- **Certificates-based authentication**: certificates provide strong authentication with the use of complex cryptographic algorithms.
- **Certificate with smart card**: This is the strongest authentication method implemented. This authentication scheme is a two-levels authentication method. The use of the citizen's private key is protected by the card PIN code and it is never exported out from the smart card, thus enhancing a high security level.

In case a citizen owns a digital certificate, this certificate is stored either in a smart card or in the citizen's hard disk.

5.1   Public Key Infrastructure

The EURO-CITI architecture includes a Public Key Infrastructure (PKI) that manages digital certificates for citizens and web servers. A private PKI solution has been selected for managing citizen certificates while a public PKI solution handles EURO-CITI server certificates.

The EURO-CITI infrastructure includes all the key components of the following architectures:

The Certification Authority (CA) delivers, revokes, and renews the certificates. It implements the security policies that define the certificate content depending on both the certificate users and the future usages of certificates. The CA also archives certificates and private encryption keys (not implemented in EURO-CITI since data encryption is not required). The CA publishes the certificates and the *Certificate Revocation List* (CRL) in the directory. The CRL is the list of all the certificates that have been revoked.

The Registration Authority (RA) handles tasks *on behalf* of the CA. This mainly includes certificate applications, validation of certificate application, request of certificate suspension / revocation / renewal. In some cases, Local Registration Authorities can assist the RA in its task. These people handle locally the RA processes.

The Directory is a repository used to publish the EURO-CITI entities identities, like their name, first name, address, etc. The CA also uses the directory to publish certificates and CRLs. The EURO-CITI directory supports the LDAP protocol.

6   Conclusions and Future Work

The EURO-CITI platform equips local authorities with the necessary technical infrastructure in order to provide two important tele-democracy services: tele-voting for realizing opinion poll petitions and tele-consultations.

The trial sites for the evaluation of the EURO-CITI platform include three European cities, namely Athens, Barcelona and London Borough of Brent. For the evaluation, one hundred citizens of each city will be provided by smart cards while a significantly larger number will be provided by login/password credentials. The evaluation will include intra-city scenarios where, for example, citizens from one city will be able to vote on local issues. The evaluation will also include inter-city scenarios. In these scenarios, the operators at a city will propose a common call-for-vote and will invite other cities to join them. Upon acceptance, virtual private networks will be dynamically created and common votes will be possible for citizens across all participating cities.

Acknowledgments

The work presented in this paper was carried out as a part of the EURO-CITI project [6]. The EURO-CITI project (EURO-CITI IST-1999-21088) is partially funded by the European Commission under the IST programme [7]. The EURO-CITI consortium consists of the following partners: Archetypon S.A. (EL); University of Athens (EL); Schlumberger (F); T-Systems Nova (D); Indra Sistemas (E); Ajuntament de Barcelona (S); Municipality of Athens Development Agency (EL); London Borough of Brent (UK). The ideas expressed in this paper are those of the author and do not necessarily express the ideas of other partners.

References

1. European Commission: Public Sector Information: A Key Resource for Europe, Green paper on Public Sector Information in the Information Society, ftp.ccho.lu/pub/info2000/publicsector/gppublicen.doc (1999).
2. eEurope2002: An Information Society For All, Action Plan of the European Commission, available at http://europa.eu.int/information_society/international/candidate_countries/doc/eEurope_june2001.pdf, [Accessed 14 May 2002].
3. Caldow J.: Cinderella Cities, Institute for Electronic Government, IBM Report (2002).
4. Telecities home page, 2002, http://www.telecities.org, [Accessed 14 May 2002].
5. Tambouris E., Gorilas S., Spanos E., Ioanidis A. and Gomar G.I.L.: European Cities Platform for Online Transaction Services: The EURO-CITI project, Proceeding of the 14th Bled Electronic Commerce Conference, vol. 1 (2001) 198-214.
6. EURO-CITI project, http://www.euro-citi.org, [Accessed 14 May 2002].
7. IST Home page, http://www.cordis.lu/ist, [Accessed 14 May 2002].
8. Tambouris E. and Gorilas S.: Investigation of tele-voting over WAP, SoftCOM2000, vol. II (2000) 643-653.

EURO-CITI Security Manager: Supporting Transaction Services in the e-Government Domain

A. Ioannidis, M. Spanoudakis, G. Priggouris,
C. Eliopoulou, S. Hadjiefthymiades*, and L. Merakos

Communication Networks Laboratory
University of Athens, Dept. of Informatics and Telecommunications
Athens, Greece

Abstract. Transaction services that enable the on-line acquisition of information, the submission of forms and tele-voting, are currently perceived as the future of E-Government. The deployment of these services requires platform independent access and communications security as a basis. This paper presents the methodology, network infrastructure and software kernel, which are used to achieve these objectives in the context of the EURO-CITI project. Well-known and established technologies such as SSL/TLS and IPsec are used. The internal design of the EURO-CITI Security Manager (ESM) kernel is discussed. This kernel is an advanced software platform, residing within EURO-CITI hosts. ESM supports the transaction services discussed in this paper but also takes provision for future services in the E-Government domain

1 Introduction

The main objective of the EURO-CITI project (realised in the context of the EU IST Programme) is to assess the potential of on-line democracy by developing and demonstrating fully-fledged pilots on transaction services such as tele-voting, electronic submission of forms and tele-consulting. The unified EURO-CITI architecture has the following technical characteristics:

1. Access from different end-points (home or public PC, kiosks and GSM/WAP handsets).
2. Support of different user access levels using network security and authentication-authorisation mechanisms.
3. Dynamic configuration and management of secure trans-european networks (IP-based virtual private networks, VPN) of EURO-CITI nodes.
4. Facilitation of provision of added value network transaction services.

A basic element in this platform is the EURO-CITI server, fulfilling the requirement of providing transaction services not only on the local level (e.g., a voting issue with local scope) but also on the European level (i.e., allowing citizens from other cities to participate in a voting scenario).The EURO-CITI server allows Local

---

* Contact author. Tel: +301 7275362, Fax: +301 7275601, E-mail: shadj@di.uoa.gr, Mailing address: University of Athens, Dept. of Informatics and Telecommunications, TYPA Bldg, Panepistimioupolis, Ilisia, 15784, Athens, Greece

R. Traunmüller and K. Lenk (Eds.): EGOV 2002, LNCS 2456, pp. 358–361, 2002.
© Springer-Verlag Berlin Heidelberg 2002

Authorities to invite other Local Authorities and set-up a secure network, launch services on this network and, finally, drop the established associations.

EURO-CITI security services largely depend on network security mechanisms, authentication, authorisation and smart-card technology to meet the required level of privacy and security.

2   Basic Networking Requirements for the EURO-CITI Platform

Access to EURO-CITI services should be possible from home PCs or networked public PCs/kiosks via web browsers as well as through the industrially established Wireless Application Protocol (WAP). Additionally, the EURO-CITI platform should cater for network - level security and authentication using the login/password mechanism or smart cards.

Each EURO-CITI node is responsible for supporting and implementing services to local citizens but also to participate in a network[1] of EURO-CITI nodes which can be dynamically configured. Communication between servers should be performed through a well-defined interface providing a reliable and secure communications channel.

3   User Access Security Infrastructure

A basic requirement from the EURO-CITI platform is to provide support for citizens accessing the platform services using established and widely available tools. The selected medium for accessing the services was the web browser due to the high degree of penetration for almost all terminal platforms. Using a web browser to access services provides citizens with an interface they are most familiar and comfortable with. Other benefits include dynamic content and updates as well as platform independence.

Web browsers provide security by using the SSL and TLS protocols. SSL/TLS provides strong encryption and authentication at the application layer and is supported by most web browsers and also by some WAP-enabled mobile phones. SSL/TLS is considered a sufficiently strong security mechanism for the type of applications that are the targets of the EURO-CITI platform (sensitive but not critical).

Other options such as PPTP/L2TP and IPsec for citizen secure access have been considered. However, they are generally not available on as many platforms and using them requires special configuration.

Establishing a secure connection with a server requires trusting the server authenticity. This is accomplished through the use of server certificates, which are installed on the server (located within the EURO-CITI platform).

Citizen authentication is mainly supported through the use of username and password mechanisms. This is due to the lack of more secure yet abundant alternatives. However, for the few terminals equiped with smart card readers, the EURO-CITI server supports a special mode of the SSL/TLS protocol which provides simultaneous

---

[1] Here, the word network denotes a logical association between the involved parties

mutual authentication, instead of the more typical server-only authentication. This requires the presence of a user certificate to be installed on the smart card. The cards are issued by the Local Authorities to citizens wishing to use this authentication method and are prepared accordingly, using a Certification Authority (CA) present within the Local Authorities infrastructure. This CA is part of the EURO-CITI platform and is exclusively used for user authentication while accessing EURO-CITI services.

There is an additional issue concerning secure access through GSM terminals. WAP messages in GSM are secured using WTLS, which is translated to SSL/TLS at the WAP gateway before reaching the EURO-CITI server. During the translation, the data is momentarily decrypted and subsequently reencrypted. Fowever, it is considered that the security risks are acceptable for the applications supported by the platform.

4   Secure Communications between EURO-CITI Servers

For securing communications between EURO-CITI servers, the ESM (EURO-CITI Security Manager) component is used. ESM is responsible for creating and maintaining connections between Local Authorities wishing to cooperate. Connections are protected through the use of the IPsec protocol. IPsec provides a cross platform security mechanism, best suited for use by servers with sufficient processing resources. The security level offered by IPsec is generally agreed to be of the best available quality, with the potential of security every type of exchanged data whether it is application or connection control related.

ESM establishes IPsec associations between servers dynamically, for the duration of service availability (e.g. during periods where citizens are requested to vote for a particular issue). This is made possible through the use of an omnipresent channel (the P-Channel), responsible for controlling the establishment and termination of services as well as transferring additional control information. For the duration of a service between cooperating Local Authorities, a new separate channel is established, also protected with IPsec using a separate set of security parameters.

5   Services between Cooperating EURO-CITI Servers

The EURO-CITI platform supports two types of coperating services. The European Scope services allow users belonging to Local Authority A to use the service of Local Authority B. This requires that the Local Authorities have agreed to cooperate on a specific service and that Local Authority A wishes citizens registered to it, to participate to the service (e.g. a consultation issue) provided by Local Authority B. for this type of services, Local Authority A does not have any control over service execution and simply responds to requests from Local Authority B to authenticate its own citizens (Local Authorities are not capable of directly authenticating citizens registered elsewhere, such data remains private in each EURO-CITI server).

Network services are a second type of offered service. A networked service is essentially shared between multiple EURO-CITI servers and simultaneously provided to all citizens of participating Local Authorities. Information gathered during the service lifetime is also shared between all EURO-CITI servers. For example, several Local Authorities can start the same voting issue simultaneously. Votes which are collected are shared between the Local Authorities, so that at the end of the voting session everyone has the collected votes of all Authorities. This process is done ensuring that no citizen information is leaked and no citizen's vote can be revealed.

6   Conclusions

The use of widely available security technology such as SSL/TLS and IPsec allows for the creation of a flexible EURO-CITI base platform capable of reaching a wide audience and on top of which E-Government services can be deployed, improving upon the administrative procedures of Local Authorities and promoting direct democracy. Services can be offered supporting the cooperation between Local Authorities, while ensuring the security of transactions as well as citizen data.

The EURO-CITI platform has been developed in Java, ensuring platform and operating system independence. Where platform specific components must be accessed, as is the case with IPsec, abstraction layers and implementations have been created to facilitate integration with popular systems such as Windows.

References

1. T. Dierks, and C. Allen: "The TLS Protocol", IETF RFC 2246, January 1999.
2. S. Kent and R. Atkinson, "Security Architecture for the Internet Protocol", IETF RFC 2401, November 1998.
3. L. Barriga, R. Blom, C. Gehrmann, and M. Naslund, "Communications security in an all-IP world", Ericsson Review 2000, No 2, pp 96-107.
4. O. Kallstrým, "Business solutions for mobile e-commerce", Ericsson Review 2000, No 2, pp 80-92.
5. "WAP Architecture Specification, WAP Forum, April 30,1998.
6. Johan Hjelm, "Designing Wireless Information Services", Wiley, 2000.
7. IP Security for Microsoft Windows 2000 Server, http://www.microsoft.com/windows2000/docs/IPSecurity.doc, February 1999.

SmartGov*: A Knowledge-Based Platform
for Transactional Electronic Services

Panagiotis Georgiadis[1], G. Lepouras[1], C. Vassilakis[1], G. Boukis[2],
Efhimios Tambouris[2], S. Gorilas[2], E. Davenport[3], Ann Macintosh[3],
J. Fraser[3], and D. Lochhead[4]

[1] e-Gov Lab, Dept. of Informatics and Telecommunications, University of Athens
{p.georgiadis,gl,costas}@e-gov.gr
[2] Archetypon S.A.
{gboukis,tambouris,sgorilas}@archetypon.gr
[3] International Teledemocracy Centre, University of Napier
{A.Macintosh,E.Davenport,J.Fraser}@napier.ac.uk
[4] City of Edinburgh Council
dave.lochhead@edinburgh.gov.uk

Abstract. Public transaction services (such as e-forms) although perceived the future of e-government have not yet realised their full potential. E-forms have a significant role in e-government, as they are the basis for implementing most of the twenty public services that all member states have to provide to their citizens and businesses. The aim of the SmartGov project is to specify, develop, deploy and evaluate a knowledge-based platform to assist public sector employees to generate online transaction services by simplifying their development, maintenance and integration with already installed IT systems. This platform will be evaluated in two European countries (in one Ministry and one Local Authority). This paper outlines key issues in the development of the SmartGov system platform.

1   Introduction

According to the European Commission [1] *"transaction services, such as electronic forms, are perceived as the future of electronic government"*. Although a large number of initiatives have been undertaken at a local, regional or even national level, it is evident that these initiatives have not provided the expected results and in most cases public administration authorities have so far failed to exploit the benefits of using online transaction services, such as e-forms, in their processes. As stated in the eEurope initiative [2] *"eGovernment could transform old public sector organisation and provide faster, more responsive services. ... However this potential is not being realised."*

The SmartGov project suggests that an advanced knowledge-based platform for transaction services and particularly e-forms will allow realising the potential of these online services. The development of this platform however requires experience and

---

* Project partially funded by the European Community under the "Information Society Technologies" Programme (1998-2002) (Project Number IST-2001-35399).

R. Traunmüller and K. Lenk (Eds.): EGOV 2002, LNCS 2456, pp. 362–369, 2002.

expertise at different levels such as technical expertise in diverse areas (e.g. knowledge management, Internet, XML, networks, user-interfaces etc.), expertise in the operation of public authorities at all levels that aim to provide online services, expertise in process models and process improvement but also social aspects such as the fears of public sector employees when facing new technologies. As a result of the problem's complexity, the SmartGov project believes that a European synergy of public authorities, universities and industry is required in order to specify and develop a platform that will allow the potential of e-forms to be unleashed. By conducting that research at a European level not only the best players will be involved but also the results will be better evaluated and also disseminated and exploited.

The rest of the paper is structured as follows: The second section outlines the objectives of the SmartGov project with special emphasis on the issue of trust in electronic services, the third section depicts the technical issues concerning the development of the SmartGov platform and applications, the next section provides a summary of the two pilot applications of the project and the last section concludes with the future plans.

2   SmartGov Objectives

The aim of the SmartGov project is to specify, implement, deploy and evaluate a holistic approach for online transaction services specific to the public sector. It will achieve this by developing a **knowledge-based platform** to assist public sector employees to generate **online transaction services** by simplifying their development, maintenance and integration with installed IT systems. It will capitalise on emerging standards (such as XForms by W3C) to create an **open architecture** that ensures **interoperability** between installed IT systems and to **develop new applications** to exploit that architecture. It will derive a knowledge management framework to facilitate both the deployment and acceptance of the online transaction services. Applications will be user-friendly requiring only basic IT skills -besides the necessary domain knowledge- to deploy and manage electronic services and will be tested in selected public administration application areas.

Based upon a thorough investigation of the state-of-the-art in online transaction services technologies, a survey of the current situation at Public Administration Authorities and an analysis of the user requirements for each of the user groups involved, the project will generate detailed specifications for the knowledge-based core repository and the SmartGov services and applications. The initial analysis has determined a number of potential technologies to be used for the knowledge repository, such as Data Bases, Data Mining, Data Warehousing, XML, XSL and XForms. On the whole, the core repository will contain the basic Transaction Service Elements (TSE), used to build electronic services along with domain specific information and knowledge for each TSE.

Furthermore, based upon the user requirements the project will develop services and applications to support the involved user groups in carrying out their tasks. As depicted in the next figure, the SmartGov platform will include services (e.g. the SmartGov agent and the Information Interchange Gateway) to enable the communication with existing or new 3rd party Information Technology Systems.

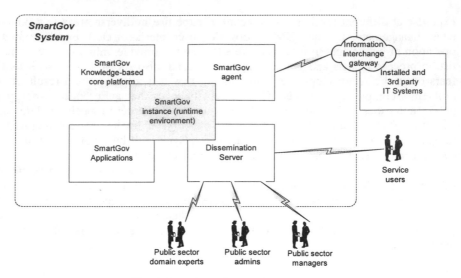

Fig. 1. Overview of *SmartGov* system

The platform will also include a dissemination service, to make available the Smart-Gov services to involved user groups, either internal to the public administration organisation (e.g. public sector domain experts, administrators and managers), or external user groups such as the end-users. Other envisaged SmartGov applications include administrative tools for capturing knowledge and for creating new transaction service elements.

User requirements analysis will also identify the end-user services that will be implemented and used to evaluate the SmartGov platform during the pilot application phase. In summary the SmartGov project will:

- Improve the working environment of public sector employees by equipping them with a knowledge-based platform, a set of relevant applications and a methodology allowing them to create and maintain e-services in an intuitive, user-friendly manner.
- Directly support staff involved in online transactions with citizens and businesses. The main features of the platform will include knowledge sharing and re-use, interoperability with installed IT systems, support of multiple access channels and full support of standards (such as XML and particularly XForms).
- Merge knowledge management principles with emerging standards on e-services (such as XForms) to enable administrators to capture and re-use their domain knowledge in the area of e-services.

A key issue in the development of electronic services is that of trust. Trust is an important resource in an e-services environment. While people dealing with commercial organisations are typically looking for financial integrity and confidentiality, when they deal with government agencies they expect not just integrity and confidentiality but also a level of transparency in the process that ensures trust in the service being provided. In SmartGov, representations of trust, trustingness and trustworthiness that take a more socially oriented approach required for public sector online transaction services, will be developed.

2.1   Trust in Electronic Services

To ensure that electronic services are designed, maintained, delivered and received effectively, it is important that people have trust in the components with which they interact.

Designers of services need to trust the procedures and tools that they use, particularly when redesigning existing services. They also need to trust the designers of other contributing or complementary services.

Deliverers of services have similar need of trust.

Clients (citizens and businesses) need to trust the behind-the-scenes people and procedures: trust that they are bona fide, trust that they will function as they are supposed to and trust that they will not misuse any information given by the client.

SmartGov, will endeavour to ensure that, in building models of electronic service delivery, models of trust will also be incorporated. Recent developments in social psychology suggest the value of studying situational trust, i.e. situations in which trust cues are provided by the situation or context as much as by the individual. This seems particularly relevant for services in which the various players may never come into direct contact with each other. This applies to many public authority services.

A comprehensive and informative analysis of *Trust formation in new organizational relationships* is offered by McKnight et al [3]. This report covers definitions of trust, the formation process, and the role of emotion in trust. Dibben [4] has decomposed business processes into a number of typical situations, and suggested what types of trust may apply in each of these.

There may also be particular relevance to SmartGov in studying "swift trust". The term "swift trust" was first used by Meyerson, Weick and Kramer [5] "to account for the emergence of trust relations in situations where the individuals have a limited history of working together". Relating this work to Dibben's examples of trust, learnt trust clearly does not exist in such scenarios and swift trust can arise as a result of situational trust.

3   Technical Aspects

Two main technology areas have been identified and will be addressed by the project: the knowledge-based core system and the applications and services.

3.1   The Knowledge-Based Core System

The project approach introduces and incorporates the key notion of the transaction service element (TSE), which is perceived as the main building block of transaction services. A TSE is the equivalent of a form field (such as the input space for a citizens id number or surname) but also contains metadata and domain knowledge that is attached by the form developer. Metadata may encompass the object's type, value range, multilingual labels, online help, while domain knowledge includes information about the relation of the object to other elements, legislation information, etc.

The knowledge-based platform provides a storage schema that is capable of storing and handling the services and the associated e-forms as well as the corresponding knowledge. The schema will be expandable and allow for the adoption of new services. This schema will be populated with Transaction Service Elements, forming thus the Transaction Service Elements Knowledge database (TSEKDB), which includes the essential elements for developing transaction forms along with all relevant information and knowledge. The domain knowledge embedded in installed systems will be used for the development of the TSEKDB.

Public sector employees interact with the TSEKDB through a user-friendly front-end (administrative) tool, which enables both the retrieval of already existing knowledge, as well as maintenance activities such as the addition of new knowledge, in an intuitive and user-friendly manner.

3.2   The Services and Applications

A Transaction Service (TS), within the SmartGov platform, is the equivalent of a form that contains a number of TSEs and some domain knowledge pertaining to the service as a whole. Under this scheme, development of a transaction service, consists of the following steps:

1. Selection of the appropriate TSEs to be included within the service
2. Decision of the layout that will be used to present the service to its users. This layout may be selected from within a standard template library (which may then be customised) or alternatively, any custom layout may be built from scratch.
3. Attachment of rules that govern the service, such as prerequisites for its usage, validation rules, triggering of other services etc.
4. Definition of MIS data and statistics to be captured for further processing.
5. When a transaction service has been developed, it may be deployed through service instantiation.

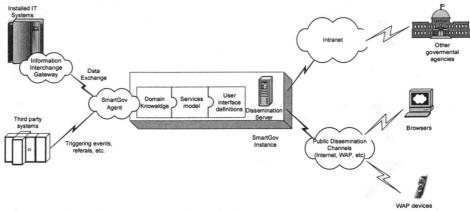

Fig. 2. An Operational SmartGov Instance

This procedure generates automatically a SmartGov instance, comprising of all web pages, forms, information repositories and programs needed to operate the service within the Web environment, wireless channels or any other supported service deployment infrastructure. The generated elements are installed on the Dissemination Server, which handles the presentation layer i.e. all interfaces with the applications users. The overall operation of an instantiated service is illustrated in Figure 2.

A service that has been deployed to the public may need to interact with an installed IT system in order to exchange data with it. All such communication is handled through the communication services, which include the SmartGov Agent and the Information Exchange Gateway. The Information Exchange Gateway is attached to the installed IT system and publishes an export schema, which contains all the data items that need to be accessed by services running within the SmartGov framework. The SmartGov Agent imports elements published within the Information Exchange Gateway's export schema within the SmartGov environment. Effectively, the Information Exchange Gateway encapsulates all peculiarities and idiosyncrasies of the installed IT systems, offering a uniform interface through which the SmartGov platform may communicate with virtually any IT system.

Besides providing the necessary link with the organisation's installed IT system, the SmartGov agent arranges for communication with third party systems the service should exchange data with, in order to access facilities that may complement or affect the running service. For instance, the SmartGov agent might provide linkage to document repositories where detailed instructions on form filling may be found, or support subscriptions to legislation databases, which emit alerts when legislation pertaining to the service operation is modified.

Service maintenance is also a major issue in operating transaction service environments that need to exchange data with installed IT systems. When a service undergoes modifications, for example due to legislation revisions, the electronic service published through the SMARTGOV instance must be 'in sync' with the organisation's private IT system, in order to carry out a full processing cycle for the service. In many cases, however, updating the private IT system may be quite cumbersome and time-consuming, while the 'front-end' part of the service, such as declaration submission, must resume operation rapidly. The SmartGov framework caters for these situations, by providing submission spooling mechanisms. These mechanisms allow for operating an electronic service and storing the submission data in a local information repository, until the organisation's back-end IT system is synchronised with the SmartGov instance. When the back-end IT system has been appropriately modified, the SmartGov instance may 'push' all collected submissions to the back-end, triggering thus the completion of the submission's processing cycle.

4   SmartGov Pilot Application

Pilot application will take place in two participating Public Administration Authorities: the General Secretariat for Information Systems, of the Ministry of Finance in Greece and the City of Edinburgh Council in Scotland.

4.1   General Secretariat of Information Systems

The General Secretariat of Information Systems (GSIS) is strategically oriented towards e-alignment of the services that it offers to citizens and businesses in taxation, customs and other application domains, exploiting the web as a major service delivery platform and interoperability technologies for integration with back-end IT infrastructures.

In this respect, exploitation of the SmartGov platform will present some substantial benefits for GSIS:

- GSIS employees at different levels will enjoy a user-friendly environment for implementing and maintaining e-services for the public (or even inter-organisational ones), as well as for transparently and seamlessly integrating these services with existing workflows and back-end IT infrastructures.
- Valuable domain knowledge associated with the e-services deployed through SmartGov will be preserved in a re-usable form that can be maintained as an organisational memory artefact.
- Various quality dimensions related to effectiveness, resource efficiency and reliability of the e-services deployed through SmartGov will be improved, whereas on the other hand the establishment and enactment of a performance management scheme is facilitated.

These benefits can substantially contribute to GSIS objectives for (a) achieving high quality of services towards citizens and businesses as a top-level strategic goal, as well as for (b) promoting Service Level Management (SLM) as a major operational policy.

4.2   City of Edinburgh Council

The City of Edinburgh Council (CEC) foresees a number of possible, different levels for the implementation of the SmartGov pilot application.

At the **national level**, with the establishment of the new Scottish Parliament and the Scottish Executive's commitment to modernising government through partnership working, there is an environment in Scotland that is conducive to testing new ways of sharing transaction knowledge and integrating service provision.

At the **city level**, the Smart City initiative in Edinburgh aims to provide a city portal that will be a single gateway to all relevant services and information.

At the **local community level**, there are many policy agendas and associated funding initiatives. The complexity of the mix is confusing and often leads to fragmented and disjointed efforts to take advantage of the opportunities presented.

The different conditions at city level and local level suggest that there is some value in running pilot applications at both levels, to compare the effectiveness of SmartGov in different situations. At the city level, policy is clear and SmartGov research can dovetail with the existing Smart City initiative. At the local level, there are particular challenges that SmartGov may find harder to address.

At the city level, the City of Edinburgh Council is in the process of establishing its Corporate Customer Service Model (CCSM), as a vital component of the Smart City. CCSM has many streams of activity, focused on people, process, technology and infrastructure. It presents an opportunity for SmartGov to make an impact on the Council's approach to designing and maintaining services efficiently and effectively. Staff at the International Teledemocracy Centre is working closely with Council staff to model existing processes, identify potential improvements and create a framework in which SmartGov principles can be applied. At the time of writing, several potential pilot applications have been suggested, such as citizens applying for housing benefits or businesses applying for licences to run bars.

At the local level, the West Edinburgh Community Planning Partnership area provides a challenging environment to test the SmartGov developments in conjunction with the local community. The area has:

- a strong infrastructure of community groups and local organisation with a history and experience of partnership working
- a local partnership organisation with multi-agency and cross sector representation
- an adopted Digital Inclusion Strategy
- a Community Learning Plan
- a number of access to employment initiatives

Key policy objectives of the Partnership are to increase access to learning, improve access to employment and enable social inclusion. These all require high levels of inter-agency trust and collaboration to be effectively delivered. At the time of writing, the Partnership is considering an appropriate application to do with access to learning and access to employment.

5   Current State and Future Work

SmartGov commenced on February 2002. So far, work has focussed on conducting a thorough investigation of the state of the art for e-services in the public sector, capturing user requirements and creating a high level set of system specifications. In the next phase work will continue to refine system specifications for each of the system components and to consequently implement them.

References

1. European Commission, 'Public Sector Information: A Key Resource for Europe', Green paper on Public Sector Information in the Information Society,
ftp.echo.lu/pub/info2000/publicsector/ gppublicen.doc
2. eEurope 2002 An information Society for All Action Plan, 19-20 June 2000,
http://europa.eu.int/information_society/eeurope/action_plan/actionplantext/ index_en.htm
3. McKnight, D.H., Cummings, L. and Chervany, N.L. (1995) Trust formation in new organizational relationships. Available at http://misrc.umn.edu/wpaper/WrkingPapers
4. Dibben, M.R. (2000). Exploring Interpersonal Trust in the Entrepreneurial Venture. London: Macmillan.
5. Meyerson, D., Weick, K. E., and Kramer, R. M. (1996). Swift trust and temporary groups. In: R. M. Kramer and T. R. Tyler, eds. Trust in organizations: Frontiers of theory and research. Thousand Oaks, CA: Sage Publications, 166-195.

Best Practice in e-Government

Josef Makolm

Austrian Ministry of Finance, Austrian Computer Society
josef.makolm@bmf.gv.at

Abstract. Evolution of e-Government is drifting. On the one hand, e-Government systems shift from information via communication and transaction systems to integrated systems. On the other hand, a partial trend from G2C and G2B systems towards G2G systems is observable. Different needs are to be met to build best practice solutions in the fields of G2C, G2B and G2G.

1 The Evolution of e-Government

1.1 From Information to Integration

As every system, e-Government is subject to the laws of evolution. Along the coordinate of time, a process of development becomes evident. Usually e-Government applications start as information applications, perhaps with components for communication. But this is just the starting point for further evolution because the availability of sufficient information creates the demand of interaction between citizens and authority. It is no longer sufficient to present application forms. Instead, it is necessary to enable the citizens to fill in these forms, to upload them to the authority and to receive an answer in electronic form, too. The e-Government system has to evolve to a transaction system. Starting from this point, the necessity to attach and upload documents comes up. These documents may be scanned documents coming from the paper world. Or they may be documents created by other electronic processes, e.g. a plan of a house or the actual balance-sheet of a company, which has to be transferred to the tax authority. This creates the demand for integration between the computer systems of applicants on the one hand, and of the authorities on the other hand, e.g. via XML structures. A "good practice" e-Government solution should meet these necessities.

A typical example for the evolution from an information system to a transaction system is the Austrian citizen information system "help.gv.at". It started as an information and communication system and brought life-situation related information to the citizens, supplemented by the ability to get answers to questions asked. As soon as the information in the system was more and more completed, the necessity to enable the citizens to submit their applications via the system came up. Now the system offers the ability to download several forms, to fill in these forms and to upload them to the authority. For example is it possible now to notify a dog or the migration of a dog to the authority, which is necessary for the matters of dog registration and dog taxation.

An example for the further evolution from transaction to integration is the possibility to upload a company balance-sheet to the Austrian company register. The

R. Traunmüller and K. Lenk (Eds.): EGOV 2002, LNCS 2456, pp. 370–374, 2002.

balance-sheet is created by the bookkeeping software and stored in an XML form. This XML data is checked by an applet, which can be downloaded from the authority's web site. If no error is detected by the applet, the XML form is uploaded to the authority.

1.2   From G2C and G2B to G2G

Citizens often have to accompany several documents when submitting their applications to the authorities. In many cases, these documents have been produced by other authorities and data relating to these documents is still available at the other authorities. Why should it be the duty of the citizens to run to several authorities for collecting their documents to complete their applications? Wouldn't it be a better way to let data run from one authority to the other? To make this offer to the citizens, it is the job of the authority to add necessary documents respectively data to the citizens' applications. And this collecting phase could be done by an electronic agent. This agent – located in a software driven workflow – could complete the citizens' applications before any official starts dealing with them. For example is it often necessary to add a birth or citizenship certificate to an application. These are typical cases to let data run and not citizens. The Austrian Citizen Card – available in 2003 – will help to realise this principle.

In any way, a broad social discussion – considering also the right of privacy – is a prerequisite for the described step of evolution. As a result of this discussion, political decisions are to be made.

2   Needs of G2C Systems

2.1   The Opinion Poll of the Austrian Computer Society

In the beginning of 2002, the conference "eGov Day" was organised in Vienna by the Austrian Computer Society[1]. As part of this conference, an opinion poll was done with the participants to find out the citizens' needs concerning G2C systems (a detailed analysis is planned but not yet published). To briefly sum up the main results of this survey, the following functions have been assessed and categorised as either "essential", "required", "useful" or "nice to have" (the order reflects the results of the opinion poll):

2.1.1   Essential Function
- Acknowledgement of successful receipt for the applicant (e.g. by e-mail).

2.1.2   Required Functions
- Indications pointing out whether a field has to be filled in mandatory or just optional;
- Choice between several paying modes for the user (e.g. credit card, internet banking, cash etc.);

---

[1] http://egov.ocg.at/egovday02.html

- Choice between several modes of delivery and the possibility to pick up the notice at the authority office;
- Pull-down menus, when certain content is necessary for form fields (e.g. a list of available documents in cases of document ordering);
- Possibility for the user to display the status of his application;
- Detailed information for the user concerning privacy, data protection and data security;
- Possibility of electronic submission of (scanned) plans or documents;
- After filling in the form and before submitting it, the possibility of a preview is necessary;
- All download forms have to be enabled as direct upload for electronic submission as well;
- After a form is accepted, it must be displayed with the possibility of printing;
- Possibility for the user to display his files kept by the authority;
- Online help for filling in the forms;
- The user can retrieve information concerning data transfer (encryption – technical security);
- At any time of an interaction (e.g. when filling in form), the user should be provided with adequate information on how many steps are finished and how many are still to be done (e g. page 1 of 5);
- Intelligent forms or help (form fields to be filled in change in context to typed input);
- Downloadable notice for further electronic processing by the user.

2.2.3   Useful Functions

- Personalization (e. g. after login the user's data and all his current proceedings are displayable);
- Possibility to log in as a test user to try out the system without any effect;
- Possibility of anonymous login (e.g. for submitting complaints and for displaying the status of ones anonymous complaints);
- Forms are designed in a way that these are not bigger than screen size.

2.2.4   "Nice to Have" Function

- Web information is displayable on the mobile phone.

3   Needs of G2B Systems

G2B systems are built for professional users. Satisfactory solutions have to meet the specific needs of this group of users:

- Transaction oriented system with login and logout function;
- Decentralised user and role administration;
- E-Application service;
- E-Delivery service.

Usually, professional users deal with more than one case. Therefore, they need transaction oriented systems with specific login and logout functions. Several roles have to be supported within one office. E.g. just one administrative key user is defined by the tax authority within the Austrian Tax Online System for a tax consultant's office. It is the duty of the tax consultant to assign the roles and authorisations within the tax online system to his employees: A secretary just may be authorised to submit e-applications to, or to receive e-notices from the tax authority while the tax consultant himself needs the authorisation to change his clients' booking within the accounts kept by the tax authority.

4   Needs of G2G Systems

G2G systems – like G2B systems – are built for professional users, too. For these inter-governmental systems, special needs are to be obeyed:

- Single login;
- Role based authorisation;
- Trusted inter-governmental authorisation;
- Workflow driven autonomous agents.

G2G respectively inter-governmental systems are used by government officials as part of their job. To avoid various logins and logouts when working with several systems, they need a single login function combined with trusted inter-governmental authorisation. Therefore, roles and authorisations for a person should be defined within the system the person logs in first. Other Systems just trust the authorisation handed over from this system. E.g. within the Austrian Ministry of Finance and its Tax Offices a Finance Portal offers the use of several systems: federal intranet, companies register database, real estates database, central registrations register, legal information system, intranet of the Ministry of Finance with the Tax Online System, the Online Tax Law Documentation System and several other internal systems.

At least, workflow driven autonomous software agents should be available for the routine checking and completing the incoming applications.

5   Technical Infrastructure – Outlook

Systems always evolve in layers, technical systems just as good as biological systems. After the process of evolution has formed a new layer, this layer works as starting base for further trials – and of course also errors – in a new evolutionary process. E-Government systems have their base in the existing internet infrastructure. Internet has exceeded the critical mass and is used as a matter of course. Secure data transfer and authentication techniques are available.

Partially governmental portals – connected by the Corporate Network Austria – are already active, Trusted Inter-Portal connectivity is starting. The Austrian Citizen Card with digital signature will be established in the beginning of 2003. A workflow system for the Austrian ministries is partially installed and will be completed in 2003. E-Payment methods, which meet the needs of the e-Government process, are discussed in a work group between banks and government. E-Application and e-Delivery service will be established in 2003.

As a conclusion can be said: There is a capable base and there are good promises for further evolution into effective e-Government services.

References

Maria A. Wimmer (ed.): Impulse für e-Government: Internationale Entwicklungen, Organisation, Recht, Technik, Best Practices; Tagungsband zum ersten eGov Day des Forums eGov.at, Band 158, Oesterreichische Computer Gesellschaft (books@ocg.at), Wien (2002)

e-Government Applied to Judicial Notices and Inter-registrar Communications in the European Union: The AEQUITAS Project

Carmen Diez[1] and Javier Prenafeta[2]

[1] Tools Banking Solutions, S.L. Paseo Independencia 32, 1. 50.004 Zaragoza. Spain.
diezc@tb-solutions.com
[2] Asociación para la Promoción de las Tecnologías de la Información
y el Comercio Electrónico (APTICE). María de Luna 11. 50.015 Zaragoza. Spain.
jprenafeta@aptice.org

Abstract. The new technological advances should be accessible to the citizen, achieving this purpose through the development of informatic tools that speed up services' rendering by the Administrations, taking into consideration the security aspects in these communications. The European AEQUITAS Project aims to develop an Informatic Tool that shall, via TCP/IP networks, allow secure communications and transmissions of electronic documents between juridical operators, using electronic signature and certification. The herein paper, describes in detail this Project, that can be encapsulated within the so-called *Networked Government*

1 Introduction

e-Government can be defined as a new technology information and communications system application used to enhance both relations between Public Entities and the citizen, as well as those internal procedures within these public entities.

The most significant advantages of this application are reduction in costs, better information accessibility, internal management improvement, data storage and processing, and generally a quicker service. These aforementioned advantages justify the fact that governments worldwide, even in those not so industrially developed countries, foster e-government applications development. Moreover, other benefits can be highlighted, such as security data processing tools that guarantee higher levels of confidentiality and integrity as compared to the traditional data management systems.

The development and implementation of these mechanisms, do not only require an upgrade in the existing computing equipment, infrastructure and communications facilities, but also provide the basis for new regulations that facilitate and promote the use of these new technologies in today's administrative procedures with full legal guarantee.

R. Traunmüller and K. Lenk (Eds.): EGOV 2002, LNCS 2456, pp. 375–382, 2002.
© Springer-Verlag Berlin Heidelberg 2002

Within the European Union, and mainly through the *e-Europe* initiative: *an Information Society for all*[1], the development of e-Government is fostered. The application of these new technologies in the public sector, as outlined during the European Council Summit held in Lisbon last 23rd and 24th March 2000, means that nowadays we are able not only to build the necessary mechanisms to render on-line services to the citizen, but also to realize those changes needed to use these tools in internal workflow processes.

Thus, within this frame, we can classify pursuant to the European Commission Information Society Project Office's point of view[2], a triple application of the information and communications technologies:

Open Government can be described as the development of government, ministry or other public organisms web sites whose aim is to make as much information as possible accessible to the general public.

Customer Orientated Government, aimed at offering, via the Internet, a series of services to the citizen through interactive platforms, adding authentication, integrity and confidentiality technical mechanisms needed in the different administrative procedures and processes. Generally speaking, we are referring to those mechanisms that can be applied to real estate registrars, tax payments, health and social security systems, ballot polls via Internet,... The objective is to offer 24X7 administrative services.

Networked Government. This application can be classified as the internal core part of the e-Government system. Being it's aim to integrate the new technologies in the Administrative management processes, to build intranets and to establish secure mechanisms with full guarantee when exchanging data and files amongst the different organizations and public entities at all levels, local, regional, central, or international.

The herein paper pretends to outline the TRUST FRAME FOR ELECTRONIC DOCUMENTS EXCHANGE BETWEEN EUROPEAN JUDICIAL OPERATORS (AEQUITAS) Project, set within the European Commission V Trust Frame Programme, located within the framework of the Key Action II, line II.4.2, called Large-scale trust and confidence and which envisages to extend, integrate, validate and prove technologies and architectures related to confidence in the context of large scale advanced settings for managerial or daily life. The validation should include, as a general rule, the evaluation of the legal consequences of the proposed solutions. More precisely, the objective of the AEQUITAS Project is to speed up procedures and increase the effectiveness of the quality service that the juridical operators, specifically Registrars and Attorneys, render to the society, through the use of secure and confidential Internet communications.

So as to reach this objective, the AEQUITAS Project pretends to develop a secure, confidential, scalable and interoperative system with all technical and juridical guarantees, as well as to transmit documents electronically between different European judicial operators. The following point gives more detail on the Project Participants, amongst which different Juridical Operators can be found.

---

[1] http://europa.eu.int/information_society/eeurope/index_en.htm

[2] *Public Strategies for the Information Society in the Member States of the European Union*, ESIS Report, September 2000. Available in http://www.irc-irene.org/documents/do-psismseu.html

2   Participants

2.1   Juridical Operators

- *Greffe du Tribunal de Commerce de Paris (France):*   The Greffiers fulfill two functions, on the one hand they are secretaries of Commerce Courts and, on the other, they act as Trade registrars (www.greffe-tc-paris.fr)
- *Consejo General de los Ilustres Colegios de Procuradores de los Tribunales de España (Spain).* The Procuradores are professionals whose principal function is to represent litigants before the Courts and, in particular, to transfer documents (notifications, writs, ...) with which parties and Courts communicate (www.cgpe.es)
- *Ilustre Colegio de Registradores de la Propiedad y Mercantiles de España (Spain).* The Registradores are responsible for registering properties (real estate transfer rights), mercantile operations and personal property (www.corpme.es)
- *Câmara dos Solicitadores (Portugal).* The Solicitadores are professionals who council individuals regarding juridical enquiries. Given that, in Portugal, Notaries are public officials whose functions are limited to legalizing Deeds, the Solicitadores have the mission to assist in the drafting of documents which need to attain the level of Public Deed (www.camara-solicitadores.pt)
- *Land Cadastre and Registry of Lithuania.* It is a governmental profit-seeking agency engaged in the following main activities: administration of Real Property Register and Cadastre and he Register of Legal Entities, the registration of real property objects and rights in them, the appraisal of real property, and cadastral surveying (http://www.kada.lt/imone.html).

The AEQUITAS Project, considering those aforementioned classifications, can be encapsulated under the *Networked Government* definition, to the extent that some of the participants comply with public functions (Greffiers and Registradores), whilst the activities of the others (Procuradores and Solicitadores) are specially linked to tribunals, or to Notaries, entities that undertake public functions.

These juridical agents that participate in the project are an example a many other juridical operators in the European Union. The electronic transmission to be found between these juridical agents are realized through the AEQUITAS Informatic Tool.

2.2   Other Participants

- *Tools Banking Solutions, S.L. (Spain).* A state-of-the-art Company that develops new information technologies. It is responsible for the coordination and technical management of the Project, as well as of the drafting and implementation of the software (AEQUITAS Informatic Tool) (www.tb-solutions.com)
- *Asociación para la Promoción de las Tecnologías de la Información y el Comercio Electrónico (APTICE, Spain).* This organism gives support in the coordination process and is responsible for the diffusion of the Project and its results (www.aptice.org)
- *Universidad de Zaragoza (Spain).* This institution is responsible for the validation of the results obtained in the project from a legal perspective (www.unizar.es)

- *Vilnius Law Faculty (Lithuania)*. This institution is in charge of investigating the current electronic signature and communications regulations in Lithuania. Moreover, it should give support during the trail phase of the AEQUITAS Informatic Tool, as well as in the diffusion of this tool (http://www.lta.lt/english.html).

3   Objective

The main aim of the AEQUITAS project is to achieve an improvement in the communications between European juridical operators, rendering more speed, security and confidentiality in these communications than those currently in use, and thus rendering a more effective and efficient service to the citizens.

So as to achieve this objective, a system has been designed to allow communications and telematic transmissions of documents between different participant juridical agents, in a secure, quick and confidentiality environment. This system is divided into the following elements:

- An **Association,** that groups, initially, the participant juridical operators in the Project. The members of this Association can be extended to the rest of participants in the Project, even though these are not juridical agents, as well as to third entities, not participants in the Project, namely, Judges, Secretaries and other juridical agents.   Statutes have been drafted, pending approval as of today, that regulate not only composition of the Association, but also all those aspects related to it: nature, function, governing bodies, object, etc...This Association acts as a third trust party in the telematic transmissions that arise between the members. These transmissions shall be performed via the AEQUITAS Informatic Tool. For this purpose, it is necessary that each Association member entity have a LDAP directory with the digital certificates of its members pursuant to the structure defined by the Association. The Association shall be based on these directories to accredit the identity and signatures of its members.
- A **Web** site, with the following URL http://www.euro-aequitas.net, has been developed to disseminate the Project, and thus make it known and accessible to other users.

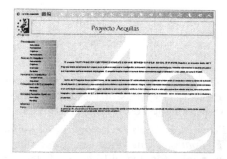

Fig. 1. Example of one of the AEQUITAS web pages

- A **Document Catalogue**, that shall allow the Informatic Tool to control the fulfilment of formal and signatory source requirements needed in order that these documents be accepted by the addressee. This document catalogue shall facilitate jurists from different countries to interpret these documents, described further on in more detail.
- An **Informatic Tool**, that allows for secure document transmissions, whether included or not in the aforementioned catalogue, and using for this purpose the so-called asymmetric key cryptography and electronic certification. A full description of this Tool is to be found further on.

4   Document Catalogue

The documents to be transmitted between the users can be either judicial or extrajudicial.

One of the tasks to realise within the AEQUITAS Project shall be the study of the existing communications flows between the different juridical operators, participants in the Project.

Nowadays, these information flows still prove to be scarce, but are bound to increase notably in the coming years as an effect of the more common enactment of substantive and procedural norms conceived to produce uniform effects throughout the European Union. Accordingly, those communications between the different European Registrars pursuant to the recently approved European Company Statute which is composed by a Regulation and a Directive that will come into force simultaneously on October 6th, 2004, have to be set; as well as those that take place pursuant to the European Council Regulation CE No 1348/2000, dated May 29th, regarding both civil and mercantile notices and transfers of judicial and extrajudicial documents in the Member States, enforced May 31st, 2001.

On the other hand, to be highlighted are the communications that exist between Proctors and Mercantile Registrars in the Spanish territory. These take place when the Proctors act as intermediaries and conduct Judicial Orders[3] issued by the Courts, handing these in at the Registrars, that once dealt with, return these to the Courts.

Once the communications flows have been studied, the documents sent in these communications shall be analysed. Work is performed using real documents, analysing its contents and so determine the data to be found in these, their signatories, as well as their nature and structure. In addition, formal and material requirements of these documents shall be determined, in compliance with the applicable legislation, and to be fulfilled so that these are accepted by the addressee and have juridical value.

Following the aforementioned document analysis, templates of these documents shall be developed in the XML language and integrated into the Informatic Tool. These templates shall show in the different fields that data that mandatory should be filled in the document. The users of the Tool should fill in and complete the templates that choose these data. In this way, there shall be a list of document templates from which to choose from and to be filled in by Tool users during the trial phase.

---

[3] Judicial Orders are used to order the issuing of certificates and testimonials and the practice of any action, whose execution corresponds to registrars of property, mercantile, naval, sales by instalments of moveable chattels, notaries, brokers or Court and Tribunal agents.

It is important to note that the objective is not to embrace all those documents currently transmitted between the Project's participant juridical operators, but to analyse a sufficient number of these so as to include them in the Tool and test them during the trial phase.

5   AEQUITAS Informatic Tool

5.1   Object

Tools Banking Solutions, S.L., within the AEQUITAS Project, shall develop a software application or informatic tool that shall allow, via TCP/IP networks, secure communications and transmission of electronic documents between the AEQUITAS Association members, using electronic signature and certification.

5.2   Users

The users of the Tool shall be the AEQUITAS Association members that shall use it to communicate with each other. Initially, therefore, the users shall be the juridical operators participants in the Project being that, for the moment, these are members of the Association.

Each of these juridical operators (Greffe du Commerce from Paris, Consejo General de los Ilustres Colegios de Procuradores de los Tribunales from España, Ilustre Colegio de Registradores de la Propiedad y Mercantiles from Spain, Câmara dos Solicitadores from Portugal and Land Cadastre and Registry from Lithuania) should be able to make available to their respective members (Greffiers, Procuradores, Registradores, Solicitadores, etc...) X-509 electronic certificates registered at their corresponding LDAP directories, which have to be adapted to the structure defined by the Association. These certificates shall be used in the Informatic Tool. There are two options: that the aforementioned users be constituted in a Public Key Infrastructure (hereinafter, PKI) and issue the electronic certificates to their members and publish them in the corresponding LDAP directories or else address an existing PKI so that it issues the electronic certificates and publish them in its LDAP directory. In both cases, these certificates should be accepted by the AEQUITAS Association.

5.3   Technical Structure

Architecture
The Tool has to main modules:

1. *Client Module*: that shall be installed in the Tool users' workstations (that, as we have already mentioned, shall be the juridical operators participating in the Project) to send documents between them. It a module that can be downloaded via Internet and compatible with Windows 98 or higher Operating Systems.

2. *Server Module*: there shall be a server at the Association and at each of the Association's member entities, in such a way that the internal communications in the same entity do not exit the corresponding entity's server. With regards to the communications between the members of the different participant entities in the Association, these shall use the Association server which acts as a third trust party, although it is possible to establish bilateral trust agreements between the organisms that are a part of the Association. This module is compatible with most frequently used Unix versions (Linux, Solaris, etc...) and with Windows NT or higher versions.

The servers used, that is, that of the Association and each of the others found in the different organizations, are STFIC servers, implemented by Tools Banking Solutions, S.L.

Certificates

The users shall be identified in the system via X-509 electronic certificates issued by each organization of which they are members of and accepted by the Association. Each organization shall administer the certificates issued by them, having to publish them in the LDAP directory that complies with the Association's defined structure, and revoke them when needed and publishing this act in their corresponding CRL.

The STFIC servers used by the system (that of the Association and those found at the Association member entities) shall have their corresponding server certificates.

5.4   General Workflow

1. The users shall certify via their electronic certificates so as to be able to access the Informatic Tool.
2. From the Informatic Tool, the document to be sent shall be selected. With this regards, there are two options:
   a)   There is a possibility that the user draft the document using that software application that he/she wants (for example, Microsoft Word, specific management programme, etc...). In this case, the document has to incorporate or import the Tool. The user himself/herself shall determine the security level that he/she wants, that is, the message shall be signed and ciphered or only signed.
   b)   It is also possible that from the Tool itself, one can choose, from a list of templates programmed in XML, a document drafted, complete and fill it in. In this case the security level is predetermined, setting as well as, the signatories needed. The Tool controls the compliance of all templates set requirements, preventing their transmission or warning the issuer and addressee of the irregularities detected.
3. Selecting the document to be sent, this document shall be signed electronically with the signatory's private key which shall be stored on a smartcard and duly protected with a password. This way, the identity and non-rejection of signatory, as well as document integrity, is guaranteed, being that once signed, any modification done to the document shall be detected. The documents can be signed by several signatories.
4. Furthermore, if required, the document shall be ciphered with a public key found in the addressee's certificate or in the organization's server of this addressee. This way, only the addressee shall be able to access the document contents guaranteeing thus confidentiality.

5. Once document is signed and, if required also, ciphered, it is sent to addressee and once delivered, a acknowledgement of receipt shall automatically be issued.
6. In any case, only those certificates accepted by the Association shall be accepted.
7. The servers validate the signatures of documents received by checking the certificate status through their corresponding CRL.
8. There is also the possibility that time stamping of signatures and communications acts be performed. Moreover, it is possible to make a copy or register of document contents sent be performed by a third trust party that could be the Association itself.

6   Conclusion

- The global accessibility of the new technologies to the citizens, can only be done through European projects, as the herein described, that in found within the so-called Networked Government.
- From a legislative point of view, it is necessary that there be a closer harmonization of the existing European State Member legislations.
- Moreover, from a technical perspective, it is totally necessary the setting of technical standards to achieve more compatibility between the different informatic systems used by the different European juridical operators.

The Concepts of an Active Life-Event Public Portal

Mirko Vintar and Anamarija Leben

University of Ljubljana, School of Public Administration, Gosarjeva 5, 1000 Ljubljana
{mirko.vintar,anamarija.leben}@vus.uni-lj.si

Abstract. Public Portals as common entry points to public services are becoming key elements of the future e-government infrastructure. In most countries, recent research in further development of public portals has been very intensive; however, approaches to the design of portal architecture and organization are still very diverse. In this paper, we will present current results of the research in progress aiming to develop prototype of an intelligent Life-Event Public Portal. We are focusing on the methodological aspects of the knowledge-based Life-Event Portals, which can provide much more efficient provision of e-services than conventional e-portals.

1   Introduction

In practically all European countries, numerous projects for development and implementation of e-government are in progress and Slovenia is no exception. The main objectives of these projects are to bring governments and citizens closer and to improve efficiency, effectiveness and transparency of its operation and increase the quality of its services. Through realization of these objectives, governments are starting to change their character from prevailingly power exercising institutions to pointedly more and more service providing and partnering institutions to the citizens and business community.

In Slovenia, a systematic approach to development of e-government started a few years ago. The basic information infrastructure for the implementation of electronic government is in place. The necessary legal framework was established by passing the Electronic Commerce and Electronic Signature Law in the year 2000. In the beginning of the year 2001, the Slovenian government accepted "The Strategy of e-Commerce in Public Administration of the Republic of Slovenia" for the development of e-government in Slovenia by 2004 [2] and the state's public web portal was introduced. This portal is designed to cover all three segments of e-government [8]:

- Government-to-Citizens (G2C): e-services for citizens provided by government,
- Government-to-Business (G2B): e-services for private sector provided by government,
- Government-to-Government (G2G): e-services for government provided by government.

Although a year has already passed since its introduction, at the time of writing this portal was still at the very initial state, meaning that it provides mainly information services and serves as a single entry point to the home pages of different administra-

R. Traunmüller and K. Lenk (Eds.): EGOV 2002, LNCS 2456, pp. 383–390, 2002.

tive bodies in Slovenia. It offers different information about the organization and functioning of public administration in Slovenia. In the first two segments (G2C and G2B), some e-services have already been introduced. The applications for the birth, marriage and death certificate are available for the public. For the businesses, some e-services concerning public procurement and public tenders are provided. For some other government services, downloadable application-forms are available, while some more sophisticated communication and transaction services still remain to be seen. Generally, communication with institutions and the relevant officials is available by e-mail.

2   Approaches to Provision of Services Using Web-Portals

A brief overview through the web-portals in different areas shows that we can define several levels of complexity of web-portals.

2.1   Simple 'Self-service' Portals

For these types of portals, it is characteristic that services are collected from different areas and administrative bodies and offered to the users via menus organised like shopping lists. Users are supposed to know exactly which services they need and which administrative bodies are responsible for their provision. Usually, most users don't have this information. In such situations, searching through endless lists of services and institutions may become a nightmare almost as unpleasant as the time-consuming chasing through offices in the traditional administration set up.

2.2   Life-Event Based Portals

Life-event based web portals are developed and organised according to the very realistic assumption that most users in a particular life situation do not know exactly which public services they need. For instance, the user only knows what he wants to achieve - to build a house, to start a business, to get married, etc. These situations are known as life-events. The system, i.e. the web portal, is supposed to have the necessary 'knowledge' to determine which services and administrative procedures are needed to be solved in order to assist the user in a particular life-event situation. To solve such a life-event, typically various administrative procedures at different administrative bodies usually have to be carried out.

Thereby, the system that guides the user trough the situation and helps him to identify the required services and their providers is needed. The web portal that includes such a system is called a *life-event portal* [5].

There are two types of life-event portals. The first is based on well-defined hierarchy of topics and life-events (*passive matrix of life-events*). The system allows user to select topics and subtopics and in this way guides him to particular life-event. When the life-event is selected, the information about required administrative procedures and the necessary assistance is offered. Examples of such portals are Austrian Internet

Service HELP [1] and Singapore e-Citizen [7]. These portals offer all information relevant to particular administrative procedure, e.g. competent administrative body, the documents to be presented, fees, and terms. In addition, adequate forms may be retrieved and filled. This information is manly offered in the form of a web page with links to other relevant web pages.

The second, user-friendlier type of life-event portal is based on so-called *active matrix of life-events*. The core system of such portals is a knowledge-based system. The knowledge-based system is a computer program based on inference mechanisms to solve a given problem employing the relevant knowledge [3,4]. The knowledge-based system in an active life-event portal (intelligent guide trough life-events) uses the pre-defined structure of particular life-event to form an active dialog with the user. In this way, the user is an active partner in the overall process of identifying and solving problems related to particular life situations.

3   The Architecture of an Active Life-Event Portal

Most of the 'supposed to be' active life-event portals available on the web today are still in the very initial state of development and are based on a very different architecture. In the next chapters, we will try to describe the basic building blocks of the prototype of the active life-event portal, which has been developed within our project [9]. An active life-event portal (portal based on active matrix of life-events) has the following main components [10] (Fig. 1):

- *Registry of procedures and forms* contains information required for implementation and execution of each type of administrative procedure (classification number, description, algorithm for execution, relevant administrative body, associated forms and documents – both input and output, the normative ground for its execution etc.)
- *Registry of normative regulations* contains all legal norms, which represent legal basis for the execution of administrative procedures.
- *Registry of life-events* contains data about topics and life-events requiring the governmental services. It also holds all data needed to define the decision aspect of life-events.
- *Electronic guide trough life-events* is a knowledge-based system that employs knowledge stored in above-mentioned registries. Together with the corresponding registries, it presents the communication interface of an active life-event portal.
- *Classification systems* are designed to classify administrative services procedures and forms for the entire public administration. The classification number of a procedure or form is its unique identifier.
- *Communication interface* is designed to enable an easy and user-friendly dialog between the user and the system through which all specific parameters of particular life-event of the user and his needs are defined.

The *communication interface* should meet three objectives of an active life-event portal. The first goal is to assist the user in selecting an adequate life-event. This can be achieved through the hierarchical structure of topics, which are supported by the portal. This structure helps user to identify the life-event that corresponds to his problem. The second goal is to identify the procedures needed to solve this life-event.

This could be achieved through the dialog with the user, based on the decision-making process, which is comprised in the structure of a life-event. This process results in the list of generic procedures. The third aspect is to identify an adequate variant of each generic procedure in this list. This is also a decision-making process, where the input parameters, needed to define the right version of the procedure, depend on the values obtained through the dialog with the user. For example, based on these parameters different supplements to the application form for the particular procedure are defined.

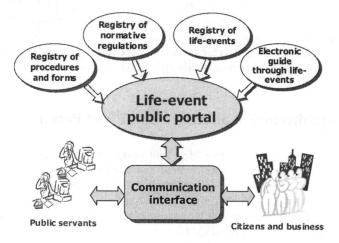

Fig. 1. The components of an active life-event portal

The decision tree through which communication interface of an active life-event portal should guide the users has, according to our architecture, three hierarchical levels (Fig. 2):

- Level of topics,
- Level of life-events,
- Level of administrative procedures.

Each hierarchical level of the decision tree is presented with specific model. For modelling of the decision making process related to the execution of a life-event, we decided to use selected concepts from eEPC (extended event-driven process chain) models [6].

3.1   Level of Topics

To establish hierarchical structure of topics, three types of topics are defined (Fig. 2):

- *Main topics* are topics at the highest level. When the communication interface is started, these topics are listed first. Main topics are always composed of at least two subtopics.
- *Subtopic* is composed of other topics (either subtopics or elementary topics). It can also include composed life-events.
- *Elementary topic* is composed only of composed life-events.

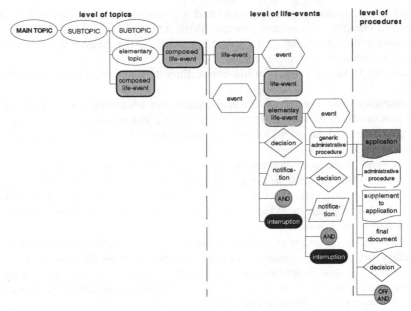

Fig. 2. Levels and main concepts representing decision tree of an active life-event portal

A model of a main topic is a tree. A root node of the tree is always a main topic, for which the model is designed. The leaves of the tree are always composed life-events, while other nodes are subtopics and elementary topics. The edges, connecting the nodes, are of the type 'is-composed-of'. They define the structure of the main topic.

3.2 Level of Life-Events

As this level includes a decision-making process, the models of life-events are more complex and more concepts are needed to define the structure of life-events (Fig. 2).

An *event* is a concept that helps to control the processing of life-event. Events occur in three different roles:

- As a start and end event: each life-event starts with start events (one or more) and ends with end events. The start event describes the impulse that initiates the life-event (e.g. 'I want to establish a business'); the end event describes the situation after the life-event is successfully finished (e.g. 'the business has been established').
- As a time event: time events identify the time at which a particular procedure or life-event is supposed to start. This time usually depends on some legal provision (e.g. 'in eight days after the applicant has been officially notified').
- As an event, describing in which state the life-event is at a particular point of time. With this type of events, the initiations of successive life-events or procedures are linked.

A *decision* is a basic concept in the life-event model as the decision-making process is defined by decisions and their *alternatives*. These two concepts define the active dialog with the user. A *notice* is another concept used in the dialog with user. It is usually connected with an *interruption*, which indicates, that processing of life-event is not successfully finished.

A *life-event*, an *elementary life-event* and a *generic procedure* define the actions, required to complete the overall process. With *AND-split* and *AND-join* points, the parallelism in the processing of life-event is modelled. The *control flow* indicates the direction in processing the life-event.

According to the structure of a life-event, three types of life-events are defined as follows:

- A *composed life-event* describes a sequence of life-events representing key steps in the processing of life-event. In the model of the composed life-event, only three concepts are used: an event, a life-event and control flow. It does not include a decision-making process.
- A *life-event* is the core element of the communication interface as it describes a decision-making process. In the model of the life-event, all above-described concepts may be used with the exception of composed life-event and generic procedure. Each life-event, included in the model of particular life-event, is further described with its own model.
- An *elementary life-event* is a special case of a life-event. In the model of an elementary life-event, all concepts may be used except other life-events. Instead of life-events, generic procedures are included. Consequently, the elementary life-events present the final step in the process of identifying the list of the generic procedures required to solve user's problem.

3.3   Level of Procedures

At the level of life-events, the generic procedures within life-events are identified. Variants of the same generic procedures are defined manly by the different documents that have to be presented with the application form to initiate a particular procedure. These documents (supplements to the application) can be understood as parameters of the generic procedure.

The concepts, used for modelling parameters of generic procedures, are shown in Fig. 2 and briefly described in the following.

An *administrative procedure* presents the generic procedure, for which the parameters are modelled. An *application* defines the document with which the procedure is initiated; the *final document* is the main output document (e.g. an official decision, a permit, etc.). An application and final document are always connected to an administrative procedure with the *data flow*.

Decisions and *alternatives* are used to define the structure of parameters: which *supplements to applications* are required depending on the chosen alternatives of the decisions. The structure of parameters is further defined with *AND/XOR-split* points. AND-split point is used to indicate that several supplements are needed simultaneously; XOR-split point indicates that one of stated supplements is required. The role of the *control flow* and *interruption* is the same as in the life-event model.

3.4   Data Model of Communication Interface

Key entities of the communication interface are shown in Fig. 3 and briefly described in the following.

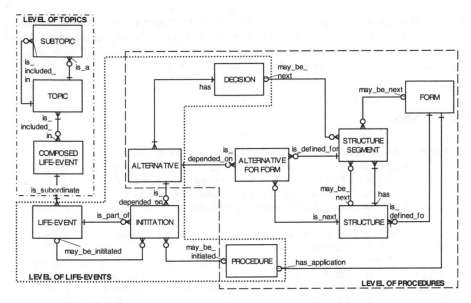

Fig. 3. Key entities of the communication interface

- Entities TOPIC and SUBTOPIC define the structure of topics, together with entity COMPOSED LIFE-EVENT they define the *level of topics* in the communication interface.
- Basic data of life-events are kept in entity LIFE-EVENT. Basic data about decisions and their alternatives and about events and notices are stored in entities DECISION and ALTERNATIVE. Data in entity INITIATION along with relationships define that part of the communication interface, which results in the list of generic procedures for particular life-event. In these relationships, the structure of life-event and basic life-event is described. The *level of life-events* in the communication interface is defined in this part of the data model.
- The basic data about administrative procedure and adequate forms are stored in entities PROCEDURE and FORM, which present the key entities in the registry of procedures and forms. Data in entities STRUCTURE, STRUCTURE SEGMENT and ALTERNATIVE FOR FORM along with the relationships define that part of the communication interface, which helps to identify the variant for generic procedure and to display relevant information for each procedure. The *level of procedures* in the communication interface is defined in this part of the data model.

4 Conclusions

On the basis of the presented concepts and mechanisms, we have developed a working prototype of an active life-event based web portal. The prototype was developed with the use of ASP (active server page) technology and ORACLE data-base management system. During further research and development, we will try to refine and generalize the modelling mechanisms and implement them for further development of the Slovenian public web portal. We believe that the concepts described in the paper can considerably contribute to the development of new solutions, and better ways of providing e-services to the citizens and companies. They are general enough to be easy implemented in developing other public web portals.

References

1. Austrian Internet Service HELP. http://www.help.gv.at. (February 2002)
2. Government Centre of the Republic of Slovenia for Informatics: The Strategy of e-Commerce in Public Administration of the Republic of Slovenia for the Period from 2001 until 2004. http://e-gov.gov.si/e-uprava/english/index.jsp. (February 2001)
3. Jackson P. Introduction to Expert Systems (third edition). Addison Wesley Longman Ltd., Harlow, (1999)
4. Klein M., Methlie L.B. Expert Systems: A Decision Support Approach. Addison-Wesley Publishers Ltd., Workingham, (1990)
5. von Lucke, J. Portale für die öffentliche Verwaltung: Governmental Portal, Departmental Portal in Life-Event Portal. In: Reinermann H., von Lucke J. (eds.): Portale in der öffentliche Verwaltung. Forschungsinstitute für öffentliche Verwaltung, Speyer (2000)
6. Scheer A.W.: Business Process Engineering, Springer-Verlag, Berlin-Heidelberg-New York-Tokyo, (1996)
7. Singapore e-Citizen. http://www.ecitizen.gov.sg. (February 2002)
8. Slovenian public web portal. http://e-gov.gov.si/e-uprava/english/index.jsp. February 2002.
9. Vintar M. et all. Report on the project: Development of public web portal (in Slovene). University of Ljubljana, School for Public Administration, Ljubljana, (December 2001)
10. Vintar M., Leben A. A Framework for Introducing e-Commerce Concepts in Public Administration (in Slovene). In: Proceedings of Days of Slovenian Administration, Portoro, (September 2000)

New Services through Integrated e-Government

Donovan Pfaff[1] and Bernd Simon[2]

[1] Goethe-University Frankfurt, Department of Electronic Commerce, Mertonstrasse 17,
60054 Frankfurt, Germany.
pfaff@wiwi.uni-frankfurt.de
http://www.ecommerce.wiwi.uni-frankfurt.de
[2] SAP Deutschland AG& Co. KG, Neurottstrasse 15a, 69190 Walldorf, Germany.
Bernd.Simon@sap.com
http://www.sap.com

1   Introduction

The New Public Management initiative in the 90's had a tremendous impact on the
principles of public administration. Cost transparency and customer orientation have
become strategic goals. The public sector is still in motion: eGovernment is a new
trend that also progresses the idea of customer orientation. International studies
document that eGovernment has become a well known phrase in many countries
worldwide. There are, however, significant differences in their respective development. Many administrations use the internet technology just to provide information[1].
The opportunity to generate additional revenues through integrative IT-solutions is
rarely used[2]. The challenge of performing additional tasks within a declining budget
forces governments to develop new ideas in order to increase revenues or to reduce
costs. This could be a future business for administrations. In the past, most administrations tried to realize e-government by establishing their own website. Current
initiatives concentrate on transactional aspects trying to connect specialized systems
with the web. Strategic concepts focusing on architecture and service portfolios are
becoming more and more important[3]. In addition to the political and administrative
part of eGovernment, there is also a commercial aspect of services. In particular those
commercial services associated with payment processes require and demand integrated transactions.

The first section of this paper documents the technical requirements for an integrated eGovernment framework. Legal aspects like data security or digital signature
will not be part of the discussion. Subsequent sections will describe the architecture of
an integrated solution. The last section presents proposals on new services of public
administrations.

---

[1] KPMG (2001), p. 14

[2] an Accenture Study on EGovernment in Germany in 2002 documents that only 31% of public
sector employees are convinced that EGovernment is a opportunity to receive additional revenues.

[3] e.g. BundOnline 2005 in Germany, uk.online or help.gv in Austria.

R. Traunmüller and K. Lenk (Eds.): EGOV 2002, LNCS 2456, pp. 391–394, 2002.
© Springer-Verlag Berlin Heidelberg 2002

2   Requirements for an Integrated Framework

A media break free communication between the individual IT-systems of one organization (like account system, reporting or document management) requires integration. In addition to this internal integration, special business scenarios having an effect on several institutions require an external or cross-authority integration as well. The best way to handle a business scenario "moving from city A to city B" is to realize a horizontal integration by connecting both systems directly. The web service of ordering a museum ticket for a special city in state government portal can serve as an example for a vertical integration.

Both kinds of process-oriented integration and the resulting collaboration scenarios are success factors for future business in the public sector. Business processes must be aligned to a common goal. It is not sufficient to provide an isolated web service. A common information base is needed in which web services are available when needed. All applications have to provide the functionality via interfaces to enable administrations to use the appropriate web services in any environment and to arrange innovative business processes using existing functions. In order to ensure interoperability between heterogeneous platforms, web services and interfaces must offer common, open technical internet standards enabling an cooperation between different technical platforms. These open technical standards are for example J2EE und Microsoft .NET. Usually these standards are HTTP (hyper text transfer Protocol), XML (Extensible Markup LANGUAGE), SOAP (Simple Object Access Protocol), WSDL (Web Services Description LANGUAGE) and UDDI (Universal Discovery, Description, and Integration).

A complete and perfect solution supports different standards to ensure a connection between applications and web services. An integrated eGovernment framework must be flexible in making changes and adding new process and components. A fast adjustment to rising volume of data requires a scalable architecture. The adding of new web services and processes, as well as compatible extensions of existing messages and patterns, must be possible without an interruption of business operations. This flexibility ensures the implementation of new services in a quick and efficient way.

Apart from the technical requirements, security plays an important role in an integrated e-government. Security means protecting and maintaining values like Integrity, Authenticity, Confidentiality and Availability. User and role administration, secure system management and digital signatures are keystones in maintaining these values.

3   Architecture

An ideal eGovernment landscape provides integrated business processes from the web portal to the backend systems. In order to implement such scenarios, it is necessary on the one hand to prevent media breaks and, on the other hand to provide basic functions such as security or workflow. It makes sense to provide these functions in a kind of middle office platform (e.g. mySAP CRM). The middle office should include functionality to connect and design specialized systems as well as tools to develop specific web services. Application Programming Interfaces (APIs) provide a data exchange between midoffice and specialized systems using XML. Irrespective of the relationship (Government-to-Consumer; Government-to-Business or Government-to-Government) it should also be possible to maintain central business partner master data for all kinds of customers as a prerequisite for a central customer account.

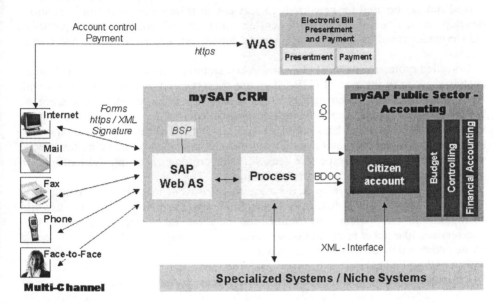

Fig. 1. Integrated eGovernment framework from SAP AG, Germany (www.sap.com)

When a customer uses the web service, the data coded by means of an electronic signature arrives via a secured Internet protocol (HTTPS) at the intermediate and goes from there over an interface into the correct specialized procedure.

When a fee occurs it must be passed on to the citizen's account. If the specialized procedure should not include invoice functionality, the invoice item in the customer account must be activated through the middle office by means of a billing engine. At that point an open item is created and presented to the customer through the web services (Electronic Bill Presentment and Payment) and made available for payment. The customer receives an overview of all open items and outstanding debits with the possibility of making the payment (for example invoice, debit or credit card). As current studies confirm the invoice is the most popular way of payment in the net. In order to also meet these demands within eGovernment, an integration of an Electronic Bill Presentment and Payment (EBPP) solution should be targeted. EBPP describes the electronic representation of the invoice, including a payment function.[4] Thus on the one hand a process without any media breaks is made possible for the administration and on the other hand the customers can pay using their preferred method. The open and balanced items are updated in the financial accounting of the administration. This scenario is supplemented by the documentation of the business transaction in an electronic records management system.

---

[4] Spann/Pfaff (2001), p. 509

4   New Services

The eGovernment framework presented above offers the advantage that central issues of e-government like security, electronic workflow, online payment are already resolved and can be used for new kinds of services. It is important that administrations develop creativity for new web services and extend their service portfolio. eGovernment could also have a commercial touch and help to refinance the framework.

So-called professionals (notaries, car dealers, architects etc.) represent an important customer group in the area of G2B. They are characterized by frequent contact with administration. A quite interesting business scenario for these customers could e.g. be the providing of geographical information for architect's offices. Instead of seeing the map material in the local land registries and making the necessary copies, the customer can download the required data from the administration portal using the special web service and pay the respective fees via their customer account. The official in charge no longer needs to deal with this business.

It is also possible to develop new business scenarios or to optimize existing services for customers (G2C), especially in the field of tourism. Big events e.g. a world championship offer various opportunities for local, federal or state government. Online services like ticket reservations, souvenirs, bar licenses or trading licenses could create additional value for customers and additional revenues for the administration.

The G2G relationship is a different one. The possibility of implementing collaborative scenarios between administrations (G2G) in the context of joint up government or one stop government is dependent on the level of networking and integration between administrations. For instance a central unit, such as the German "Bundeskasse" could offer their framework as a solution provider not only to the assigned state authorities but also to other public administration e-commerce-services. In order to exchange the data records in a secure way, a standard data exchange format (XML pattern) will have to be established.

Literature

1. **KPMG** (2001), "Verwaltung der Zukunft - Status quo und Perspektiven für eGovernment", Report.
2. **Spann, M. / Pfaff, D.** (2001), "Electronic Bill Presentment and Payment (EBPP)", Die Betriebswirtschaft (DBW), 61, 509-512

Risk Assessment & Success Factors
for e-Government in a UK Establishment

A. Evangelidis, J. Akomode, A. Taleb-Bendiab, and M. Taylor

School of Computing & Mathematical Sciences, Liverpool John Moores University,
Byrom Street, Liverpool, L3 3AF, England, UK
{cmsaevan,j.o.akomode,a.talebbendiab,m.j.taylor}@livjm.ac.uk
http://www.cms.livjm.ac.uk

Abstract. In a quest to modernise their activities and underpin their public-private partnerships, many governments around the globe have initiated their local eGovernment programmes. In this regard, best-practice, emerging Information Communications Technology (ICT) and e-business potential are leveraged to provide 24*7 access to online public services, ranging from online tax forms, to online voting. Whilst much may have been achieved towards developing and supporting one-stop shop to a range of online government services, more research is required, for instance, to provide a seamless integration and interoperation of these services, their integration with legacy systems, and risk management strategy. Based on an ongoing research focused on risk modelling and analysis of eGovernment web services, this paper introduces a categorisation of the main generic risk factors. The paper only elaborates on the first two categories of the risk factors and develops a set of potential success factors for eGovernment.

1 Motivation for the Research

Over the recent years, many governments around the globe have initiated their eGovernment strategies to exploit: ICT, e-business models and best-practice. The main interest is to improve operations and to support their citizens through the use of ubiquitous web access technology to public sector and eGovernment digital content. Currently, the way the public sector is implementing eGovernment may be viewed from two sides: (i) the development of eGovernment Web portals, providing various services to its customers, for example UK Online [1]; (ii) the development of Web sites 'dedicated' to a particular service by a single governmental department, for example TAXISnet ([2], pp.5-8).

At the moment, most governments around the world, including the UK government wish to take a step further and they have made (or are making) plans with tight deadlines to fully integrate all or at least most of the governmental services. An example could be made of the government of Canada ([3], p.5) and the UK government ([4], p.6). This 'next generation' of eGovernment is likely to be the result of a partnership between the public and private sector. The ongoing research seeks to identify risk factors in all aspects of eGovernment and develop a framework for risk assessment. The situation may serve as a guide to assist people working towards the next genera-

R. Traunmüller and K. Lenk (Eds.): EGOV 2002, LNCS 2456, pp. 395–402, 2002.
© Springer-Verlag Berlin Heidelberg 2002

tion of implementation and management of eGovernment projects to derive increased benefits.

Due to confidentiality the names and exact locations of the collaborating establishments are omitted but the quality of the material presented remains unaltered.

2   eGovernment Risk Factors

eGovernment projects are inherently complex, sharing similar risk factors with their e-business counterpart projects. Implementing eGovernment as a major development may not be easy since it may involve many factors of risk that could threaten the success of the project. Adequate risk assessment and management procedures may help in avoiding major pitfalls, though sometimes failures cannot always be predicted precisely.

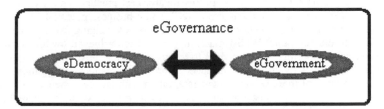

Fig. 1. eGovernance Perimeter

Before attempting to elicit the main generic eGovernment risk factors, it may be prudent to specify the environment in which eGovernment resides. Figure 1 shows that eGovernment finds itself within the perimeter (or boundary) of 'eGovernance' and interacts with the 'eDemocracy' concept. The latter may be defined broadly as the improvement of effectiveness and efficiency of democracy through the use of ICT [5].

Potential Areas of Risk in eGovernment				
Technological / Implementation	Social / Human	Security	Financial	Legal

Fig. 2. The Main Generic Risk Elements (or Factors) of eGovernment

The research indicates that the main generic eGovernment risk factors may be initially classified into the five categories [(i) technological/implementation; (ii) social/human; (iii) security; (iv) financial; (v) legal] shown in Figure 2.

A more detailed analysis of the first two categories of risk elements is presented below i.e. with regard to: (i) technological/implementation; (ii) social/human. Also, brief discussions are advanced on the latter three generic risk elements.

2.1 Technological and Implementation Risk Factors

Since eGovernment relies on the use of ICT, almost all eGovernment projects are ICT-centric. Thus, one of the main sources of risks in eGovernment-related projects lies on the technologies used and the way the technologies are implemented in order to serve the purpose a particular project is meant for. The most common technological and/or implementation risk factors identified are presented and discussed.

(i) *Risk of gaining quick political advantage:* eGovernment may be a rather new and trendy concept for most politicians, as such they often want to encourage the implementation of eGovernment projects quickly while they are still in office, in order to gain better political advantage because the alternative may be worse for them. The situation may lead to inadequate planning that may eventually result in the failure of the project. Experience from one UK establishment shows that such failure may be due to inadequate planning in linking technological capabilities with the requirements of the establishment, and it may be viewed as being propelled by an attempt to satisfy political request and instructions quickly.

(ii) *Risk of lack of adequate identification and classification of services:* The eGovernment concept is basically a way for the Governments (local or national) to offer their customers (citizens, businesses, other governments) better services. This research reveals that in this UK establishment there is a lack of: (a) adequate identification of service requirements; (b) adequate standard classification of the services the eGovernment system is meant to satisfy. This implies that the eGovernment (or IT) staff find it difficult to classify the services and therefore have problems in designing and developing the IT infrastructure. A clear example of this issue is the service of *Council tax* disregards, where – for instance – there is the case of students who are eligible for full exemption from paying the council tax and there is also the case of lone-parents who are eligible for a huge deduction from the council tax. In this case, it is obvious that without resolving the type of classification with city council officials it will be unclear to the IT staff on how to implement a workable eGovernment system to satisfy the need.

(iii) *Risk of lack of adequate methods for requirements identification*: Requirements identification, analysis and specification for a business or an organisation form an essential part that leads to the successful implementation of most ICT projects ([6]; [7]). There are various methods to achieve this goal of identification and analysis, which are based on the premise of Systems Analysis and Design (SAD) or Knowledge Elicitation (KE). In most ICT projects, this stage of requirements identification, analysis and specification are either ignored, shabbily carried out or inappropriate methods are employed. The situation often leads to project failures in ICT. In the case of the UK establishment investigated, it was found that no clear formal method of requirements identification and analysis or specification and classification was used before the next stage of technical implementation.

(iv) *Risk of lack of proper design and maintainability:* A design with risk elements employed in an ICT project can also lead to inappropriate implementation not minding the sophistication level of the hardware and software used. This situation often creates more risks, as the resulting system may not deliver the performance level expected from it. Eventually, maintainability of the system may become another added problem as the original 'product' fell short of its quality in performance. This represents one aspect of the main problem experienced by the public establishment(s) investigated in the UK with regard to eGovernment projects.

(v) *Risk of lack of resources or expertise by the public sector:* In most cases the public sector does not have all the resources or expertise to accomplish a full scheme of eGovernment and therefore collaboration with the private sector is of paramount significance. As such some parts of eGovernment projects may be outsourced to private establishments ([8], p.5). Experience from one public establishment (a city council) in the UK shows that such partnerships are not always healthy and can be susceptible to problems. In this case, there was a lack of mutual trust and understanding by both parties (public and private), as the private organisation did not honour all of its promises and the city council had to find a new business partner.

(vi) *Risk of integration with legacy systems:* A holistic approach towards eGovernment implies that everything relating to governmental services has to be integrated under one single eGovernment umbrella. The implication being that all governmental and non-governmental data, information, systems, services and other necessary items have to exist and interact in a common platform of communication. In the case of this UK establishment the IT staff found it difficult to achieve this goal when they attempt to integrate new IT systems with legacy systems. The main drawback was that in some cases, the whole system infrastructure (i.e. legacy system) in a locality (or a public establishment) may be so outdated that it may be quite difficult and frustrating or almost impossible to fit it into a modern technology platform of the desired eGovernment project.

(vii) *Lack of adequate business/project management skills:* The research divulges that lack of adequate leadership is a major eGovernment risk factor ([8], p.4; [9]), especially if such leadership is with regard to business or project management. Fundamentally, eGovernment is based on the use of information and communications technology as such technical experts of ICT are often viewed as the ones to lead and make the project a reality. On the other hand, eGovernment may be viewed as a grand business project that requires adequate business/project management skills. When an IT expert with little or no business management experience is placed in a leading managerial position of a huge eGovernment project, there may be an increased possibility of the risk of failure. Equally, someone with business and project management skills alone with no background in ICT may not be the appropriate person to head or lead an ICT (or an eGovernment) project. A knowledge of both sides (business/project management and ICT) by one person or group is likely to provide better leadership skills for successful eGovernment projects.

(viii) *Risk of large ICT projects:* Most eGovernment projects are huge information systems projects. The research reveals that the larger the project of IT, the more the likelihood of failure due to size and complexity. It may be advisable for eGovernment practitioners to opt for small projects ([8], p.2) or sub-divide a large project into manageable parts. This implies that in a large eGovernment project major activities of the project are to be identified clearly and appropriately grouped, in order to minimise the risk of failure. This situation enables a proper allocation of suitable staff to each of the major activities or group of activities as well as the use of inappropriate technology, which may lead to major implications or risks ([6]; [10]).

(ix) *Lack of formal risk management strategy:* Risk management involves a procedure for monitoring and controlling risks. It serves as an essential management tool (or strategy) for supporting project managers. This UK public establishment (city council) did not have a formal risk management strategy in place. The IT staff had to confront every newly discovered project threat on the basis of makeshift arrangements. The research discloses that the absence of adequate risk management strategies increases the chances of overall project failure.

(x) *Risk of lack of adequate security measures*: The issue of security in ICT projects is not only interesting, it is also vast and covers both: physical, technical and human aspects [11]. This section will only discuss briefly, the technical implications of security in ICT projects with regard to: confidentiality, integrity and availability ([12]; [13]). These three properties and their susceptibility form the core of technical security in ICT systems ([14]; [15]; [16]). Due to a combination of internal politics and expertise, security issues relating to the items discussed is an aspect of risk factors being experienced by the public establishment(s) investigated in the UK with regard to the issue of eGovernment project.

2.2 Social and Human Risk Factors

eGovernment projects are often huge and complex to implement. They usually affect the way people live and interact in a society. The belief is that there may be several social and human implications when implementing and using current available facilities of eGovernment or when developing the next generation of eGovernment. Major risk elements in this aspect are presented below.

(i) *Risk of lack of IT skills in the public sector:* The research findings indicate that social and human risk factors are capable of hindering progress in eGovernment projects, especially due to lacks of people with IT skills ([8], p.4; [9]). Also, experience from the UK establishment indicates that employees from other departments (not the IT department) do not understand the new systems. Therefore, they are either unwilling to cooperate with the IT staff or they simply cannot use the new systems. Similarly, people from the IT department of this city council complained of under-staffing (another major social risk factor), as they do not have enough human resource to carry out their tasks on time in the eGovernment project.

(ii) *Political risks*: Politicians are the people initially responsible for most projects of eGovernment implementation. The politicians decide whether or not a public establishment should proceed towards eGovernment, then they may provide the funding for the project. In most cases of eGovernment success or failure, the politicians may be in the position to receive the praise or blame ([8], p.2).

(iii) *Risk of lack of adequate collaboration arrangement*: Findings from the UK establishment have revealed a rather embarrassing issue. It may be reasonable to assume that within a national environment for accomplishing projects of eGovernment, the various local governments will and should collaborate and assist each other, in order to achieve a common goal of success. Unfortunately, this research reveals that this is not the case. Neighbouring local city councils do refuse to join forces due to political and cultural problems and beliefs. Certainly, this kind of behaviour from local city councils may damage any progress in eGovernment at a national level.

(iv) *Risk of digital divide and lack of adequate education*: Sadly for the officials or proponents of eGovernment, not all people have access to a computer or to the Internet [17]. Various reasons can be attributed to it, ranging from insufficient funds to lack of other resources but whatever the explanation may be, they form a modern social problem, for example, the so-called digital divide. As long as this digital divide exists and grows eGovernment projects may continue to be doomed to failure. Another factor of risk is that of the uneducated citizens. More specifically, many people may be able interested and willing to access eGovernment services, but they may not be sufficiently educated to be able to do so confidently unless they receive some level of training.

(v) *Risk of bureaucracy and fear*: Research findings from the UK establishment (city council) indicate that one major threat to eGovernment initiatives is the level of bureaucracy practised in the establishment. For instance, when staff from the IT department of the establishment approach employees from other departments for information to help them to develop the project further within the organisation, they often received unnecessary excuses as to why the required information cannot be made available. Consequently, it can be said that bureaucracy within an establishment - mostly due to fear of redundancy - can hinder eGovernment programmes.

(vi) *Risk of lack of citizens' understanding and privacy*: Another major social risk factor affecting eGovernment projects is the erroneous comprehension of the needs of customers (mainly the citizens) by the eGovernment developers ([8], p.5; [9]). There is no point in creating eGovernment programmes when the customers do not understand the delivered services or they are not delivered in the proper manner.

2.3   Financial Risk Factors

Financial risk factors in eGovernment are mainly related to lack of funding, especially lack of cross-agency project funding. Government funding models are not often set up to fund many of the eGovernment projects that are cross-agency or cross jurisdictional in nature [18]. The research in the UK establishments reveals further financial risks, but are out of the scope of this paper.

2.4   Legal Risk Factors

The legal aspects of eGovernment may be viewed as being still fluid. Such risks may include the issue that some transactions may not be processed electronically ([19], p.17), the enhancement of current laws ([20], p.23), the reluctance in reforming legislation ([21], p.213). Unfortunately, this discussion exceeds the scope of this article and may be presented in the future.

3   Potential Success Factors for eGovernment

The first two [(i) technological/implementation; (ii) social/human] generic risk factors may provide a clear explanation to the potential risks involved in the implementation of eGovernment projects and their use. Based on the discussion advanced, the more specific elements of risk can be turned around to create potential success factors, as shown in Table 1, for the improvement of eGovernment implementation and operation. The table can be used as a checklist (or guide) to support the successful implementation and operation of eGovernment projects.

Table 1. Checklist of Success Factors for eGovernment

Potential Success Factors for eGovernment	
Technological /Implementation	**Social / Human**
✓ Have clear targets for eGovernment projects	✓ Educate and train public sector employees
✓ Have standardised and classified services	✓ Invest in human resources for better ICT
✓ Improve relationship between public and private sector	✓ Assess any political cost/implication
✓ Address problem of integration	✓ Avoid any cultural and collaboration problems
✓ IT project management & Business management experts should lead an eGovernment project	✓ Address the issue of digital divide
✓ Sub-divide a large project	✓ Reduce bureaucracy and eliminate fear
✓ Employ standard method of requirements identification and systems analysis	✓ Understand the needs of the customer (citizen, private sector, other governments)
✓ Have a strategy for risk management	✓ Educate and train customers
✓ Consider all aspects of security for the system	✓ Employ standard methods of social intervention
✓ Educate and train staff	✓ Identify required services of the establishment
✓ Develop a plan for maintainability	

4   Conclusion

It should be recognised that eGovernment is not entirely a technology phenomenon. It is about re-inventing and re-organising the way service providers (public/private) and citizens (or customers) interact in the society. Consequently, an attempt has been made to give an insight into an ongoing risk assessment work carried out in some public establishments in the UK, regarding eGovernment implementation and management. From a holistic point of view a classification of the generic risk factors associated with eGovernment has been presented with a more elaborate discussion on the first two factors in Figure 2. The issue of security in eGovernment encompasses both physical, social and technical facets and can be enormous and quite exciting to discuss. A detailed presentation of the research findings on the components of security risks in eGovernment has been deliberately avoided in this paper, but an aspect of the associated technological security risk elements are discussed succinctly in section 2.1(x). Based on the results of the investigation carried out, Table 1 is presented as a checklist of potential success factors for eGovernment. It is hoped that the details in the table will be useful to practitioners, researchers and others interested in the domain of eGovernment implementation and management.

References

1. UK Online, http://www.ukonline.gov.uk. Crown copyright (2002)
2. Gouscos, D., Mentzas, G., Georgiadis, P., Planning and Implementing e-Government Service Delivery. Presented at the Workshop on e-Government in the context of the 8th Panhellenic Conference on Informatics, Nicosia, Cyprus, 8–10 November 2001
3. Government Of Canada, Government On-Line And Canadians. Canada (2002)
4. UK Prime Minister, Modernising Government. UK Cabinet Office (1999)
5. Watson, R., T., Mundy, B., A Strategic Perspective of Electronic Democracy. Communications of the ACM, vol. 44, no. 1, (2001) 27-30
6. Akomode, J., Moynihan, E., Employing Information Technology Systems to Minimise Risks in an Organisation. The 3rd International Conference of Business Information Systems BIS'99, 14-16, April 1999, University of Economics in Poznan, Department of Computer Science, Poland, (ed) Abramowicz, W., ISSN 1429-1851
7. Avison, D., Shah, H., The information Systems Development Life Cycle: a first course in information systems, McGraw-Hill (1997)
8. OECD, The Hidden Threat to E-Government. PUMA Policy Brief No.8 (2001) 5
9. West Sussex County Council, Implementing Electronic Government Statement – Risks. http://www.westsussex.gov.uk/e-government (2001)
10. Caldwell, F., Keller, B., Managing the Risk of US Voting Lawsuits. (ref. No. FT-12-8903), 11 January 2001
11. Akomode, J., Potential Risks in E-Business and Possible Measures for an Enterprise. The 10th Annual BIT Conference, 1st/2nd November 2000, (E-Futures), (ed) Hackney, R., (CD): ISBN 0 905304 33 0, No. 07
12. Mercuri, R., Voting Automation (Early and Often?). Communications of the ACM, vol. 43, no. 11 (2000) 176
13. Phillips, D., Von Spakovski, H., A., Gauging the Risks of Internet Elections. Communications of the ACM, vol.44, no.1 (2001) 73-85
14. Pfleeger, C. P., Security in Computing. Prentice-Hall (1997)
15. Briney, A, Got Security. http://www.infosecuritymag.com/articles/1999/julycover.shtml (1999)
16. Di Maio, P., Security Plans. http://www.mi2g.com/press/030400.htm (2000)
17. Stahl, B., C., Democracy, Responsibility, & Information Technology. Proceedings of the European Conference on e-Government, Trinity College Dublin (2001) 429-439
18. Keller, B., Baum, C., Identifying and Addressing Inhibitors to E-Government. (ref. No. COM-13-1123), Gartner Group, 8 March 2001
19. US Department of Labor, E-Government Strategic Plan. Office of the Chief Information Officer (2001)
20. Hagen, M., Kubicek, H., One-Stop-Government in Europe: Results from 11 National Surveys. Hagen, M., Kubicek, H., (eds.), Bremen, University of Bremen, ISBN: 3-88722-468-x (2000)
21. Klee-Kurse, G., One-Stop-Government in Finland. In One-Stop-Government in Europe: Results from 11 National Surveys, Hagen, M., Kubicek, H., (eds.), Bremen, University of Bremen, ISBN: 3-88722-468-x (2000)

Quo Vadis e-Government? – A Trap between Unsuitable Technologies and Deployment Strategies

Tamara Hoegler[1] and Thilo Schuster[2]

[1] Research Center for Information Technologies at the University of Karlsruhe (FZI),
Haid-und-Neu-Strasse 10-14, 76131 Karlsruhe, Germany
hoegler@fzi.de
[2] Cit GmbH, Kirchheimer Strasse 205, 73265 Dettingen/Teck, Germany
thilo.schuster@cit.de

Abstract. In Germany, eGovernment stagnates more than it progresses towards the electronic era of administrations. Lacking deployment strategies and acceptance problems concerning Smart Cards hinder its progress, resulting in a great gap between the targets and the actual state of eGovernment. This article describes a method for the stepwise introduction of electronic signatures placed on Smart Cards in public administrations and gives an approach for the deployment of eGovernment.

1 Introduction

According to the Resolution of the German Federal Cabinet[1], all transactions taking place between citizens and the Federal Administration should be available on-line up to the year 2005. Therefore, approximately 3.700 laws have to be adapted for the Internet[2], and public administrations have to hurry up with reducing the discrepancy between their targets and the actual state of eGovernment that seems to be more a standstill than a progress towards the electronic era of administrations.

Several problems accompany the deployment of eGovernment: On the one hand, all transactions between citizens and administrations need a legally valid signature that depends on rigid legal basic conditions. On the other hand, most of the administrative authorities do not have any strategy for the deployment of eGovernment solutions yet.

2 Digital Signature and Smart Cards – The Right Technological Pre-requisites for e-Government?

According to the German law, an electronic signature is valid legally if it meets the criteria described in the German signature decree[3]. Therefore, the electronic certificate is placed on a Smart Card and the user has to obtain the Smart Card from a trust cen-

---

[1] Date: 14th November 2001.
[2] Pricewaterhouse Coopers, PwC Deutsche Revision: Die Zukunft heißt E-Government, August 2000, p.18.
[3] The so called *Signaturverordnung*.

R. Traunmüller and K. Lenk (Eds.): EGOV 2002, LNCS 2456, pp. 403–406, 2002.
© Springer-Verlag Berlin Heidelberg 2002

ter in a highly secure, but lengthy ordering and identification process. Because of dis-advantages like the complicated and fussy application of Smart Cards, the need for special equipment[4] (mostly not included in today's hardware and software configura-tions) and an installation that can result in all kinds of technical compatibility issues, Smart Cards are bound up with acceptance problems within the population as well as administration[5].

A stepwise approach for the deployment of the digital signature is a solution for this dilemma, because experience in deploying such a system can be gained easily, employees of the administrative authorities can become accustomed to it slowly and the risks involved in using Smart Cards can be minimized.

Step 1: The Importance of Forms

The first step towards a fully integrated eGovernment system is to provide a down-loadable and printable version of forms, that are established in the administration and that guarantee a collection of data within the law. After the download, the user fills out a form (e.g. a PDF document), signs it and sends it to the administration by mail. After the arrival of the mail, an administration workflow begins.

A different representation of the data (as it could be the case when using a printed form, e.g. a PDF document) could result in the complete refusal of the request. There-fore, the employees have to be instructed to accept these self-produced or self-printed forms as well as the standard paper forms.

Step 2: Using an Intelligent Screen Dialog

The next step towards eGovernment is an electronic form that can be equipped with fields (that can be filled out on a client) and where the data can be validated by using an expression language. This process gives rise to the need of a well-known and sta-ble client environment that cannot always be guaranteed in the Internet application area. An interesting approach is an intelligent screen dialog, which can be flexible due to an advanced data validation mechanism. It is based on the user's input and offers online help facilities. In order to be independent from browser specific features, ser-ver based dialog technologies (like e.g. ASP or JSP) are suitable.

The user signs the (PDF) form electronically and sends it via email to the person in charge. The signature is verified and the files are stored safely for archive purposes. Afterwards, the (PDF) form can be transferred manually to the back-end processing system. Therefore, this process allows providing a large number of eGovernment ap-plications without waiting for a close integration with the back-end systems.

Step 3: The Electronic Inbox

The next step is to establish an electronic path from the public Internet dialog to the administrative employee. The user can sign the form electronically and send it to the person in charge. The data can be transferred either using e-mail or a browser-enabled

---

[4] Including a Smart Card reader, the corresponding device driver and certain application soft-ware that is necessary to be able to use an electronic signature.

[5] The focus of our deployment strategy lies on overcoming the acceptance problems of Smart Cards *in administrations*.

database application. This step allows providing a large number of eGovernment applications without waiting for a close integration with the back-end systems.

Therefore, the *electronic inbox* is a final step towards a fully integrated solution – an electronic workflow without a breach of the medias involved. Further steps are the creation of a signed response to the Internet user or extending the electronic workflow in order to include non-administrative organizations.

3 Strategies Concerning the Procedural Model for the Deployment of eGovernment

Aside from the digital signature, the deployment strategy for eGovernment solutions is a serious problem. A survey, carried out between May and November 2001 at the FZI[6], proves this fact. The following paragraphs describe the survey and the approach for an optimized eGovernment deployment strategy.

3.1 Basics and Methodology of the Survey

The survey's target was to get an overview of the current state of eGovernment in Germany. It was carried out by email in the form of a questionnaire and included 424 cities and municipalities[7]. A return ratio of 16.2% was achieved, cities with 10.000 to 50.000 inhabitants sent almost 80% of the returns.

3.2 Results

The survey points out, that apart from financial aspects a practice-oriented deployment strategy becomes a substantial factor of success for eGovernment solutions. A deployment strategy can reduce or even avoid financial risks and personnel bottlenecks, which were mentioned as main causes for the occurring problems.

Fact is, that most of the municipalities (79,7%) have neither a general eGovernment strategy nor concrete strategies concerning the deployment of online services (63,8%) or Municipal Information Systems (46,4%). Only 10% fall back on standardized methodologies of external providers like computer centers.

Containing the latest findings concerning critical and success factors for the deployment of eGovernment[8], this survey represents the basis of the procedural model, which is adapted to the needs of German administrations.

3.3 A Procedural Model for the Deployment of eGovernment

For a successful deployment of eGovernment a holistic strategy is needed, covering all aspects mentioned as essential during the survey. None of the analyzed models

---

[6] Reiter, Markus: Entwicklung eines Vorgehensmodells zur Einführung von E-Government in der Öffentlichen Verwaltung. Master Thesis at the Institute of Applied Informatics and Formal Description Methods, 2001.

[7] All located in the German states Baden-Wuerttemberg and Bavaria

[8] e.g. planning of applications, identified obstacles and priorities

(the Requirements Pyramid of Masser[9], the Classification Scheme of the BSI[10] and the Procedural Model of WIBERA[11]) covers all these aspects.

Covering all these aspects, our procedural model represents a holistic approach for the deployment of eGovernment by giving e.g. concrete contributions to the settling of targets during eGovernment projects as well as detailed guidelines for the definition of the strategy or the selection of suitable online services. Furthermore, the model includes support for the realization of workshops, the training of employees, the intern and extern marketing, for the security and for the financing of the project.

4   Summary and Conclusion

In this article, clear approaches regarding the deployment of eGovernment were given and a procedural model was presented, which is adapted to the needs of the German administrations. Practical requirements, established in 424 German cities and municipalities, form the basis of the procedural model.

The next few years and new technologies like the mobile signature will show if public administrations will overcome its lethargy or if the state of the art of eGovernment will stagnate at the status of its infancy.

Since it is independent of the Smart Cards in its first stages, our systematic approach to a fully integrated eGovernment system reduces project risks like mis-investments and acceptance problems because the employees are lead slowly to a new working style. The acceptance of such systems in public is reasonably higher since – in their initial stages – they can be used without having electronic certificates or a Smart Card equipment. This approach also reduces the time-to-market, which can be an important factor in satisfying the public expectations.

---

[9] Masser, K.: Kommunen im Internet. Neuwied: Luchterhand Verlag, 2000. Page 79.

[10] Bundesamt für Sicherheit und Informationstechnik, translated: Federal Agency for Security and Information Technology

[11] WIBERA is an agency for economic advice and auditing, situated in Germany.

A New Approach to the Phenomenon of e-Government: Analysis of the Public Discourse on e-Government in Switzerland*

Anne Yammine

fög – Forschungsbereich Öffentlichkeit und Gesellschaft, Universität Zürich,
Wiesenstrasse 9, 8008 Zürich, Switzerland
Anne.Yammine@access.unizh.ch

Abstract. EGovernment is commonly approached with a technical emphasis. In contrast to this perspective, we will take eGovernment into consideration as a phenomenon of communication. Our perspective is based on the assumption that the introduction of eGovernment has to be concomitant with a process of discussion in order to have lasting effects on a society. In contrast to the internet, eGovernment is approached more pragmatically through the public discourse. This line of argumentation will be highlighted by analysing the Swiss media discourse on eGovernment.

1 A New Perspective

EGovernment is commonly defined as "the use of information technology, in particular the internet, to deliver public services in a much more convenient, customer-oriented, cost-effective, and altogether different and better way[1]". In opposition to numerous technical and applied studies which are based on this definition, we would like to adopt a new perspective through which eGovernment is considered as a phenomenon of communication.

2 Assumptions and Basic Hypothesis

We assume that the introduction of a new technology is always concomitant with a process of reflection and discussion involving society as a whole and therefore postulate that eGovernment cannot be a factor of change until it causes a communicational process generating an overall positive attitude towards its implementation.

---

* This research is part of a doctoral thesis and of a research project conducted for the Swiss Federal Government, in which we will analyse the public discourses on eGovernment in Switzerland and on an international level.
[1] Holmes, D.: e.gov. e-business. Strategies for Government. Nicholas Brealley Publishing, London (2001) 2

R. Traunmüller and K. Lenk (Eds.): EGOV 2002, LNCS 2456, pp. 407–410, 2002.
© Springer-Verlag Berlin Heidelberg 2002

The introduction of the internet generated great expectations which were especially linked to the so-called dot-com economy flourishing within the internet hype. However, this hype collapsed in the fall of 2000, when the NASDAQ crashed and the dot-com sector slithered from a state of euphoria to one of despair[2]. Characterised by the context of the dot-com crash, the style of the eGovernment-discourse is more technique-orientated and overall precautious in its judgments and expectations.

Through combining these two assumptions, we can formulate a basic hypothesis which puts forward that, in comparison to the internet, eGovernment does not go through an initial phase of hype and euphoria and has not managed, at this point of the process, to create an overall positive reflective process within the society it affects.

3   Methodology and Research Question

Our analysis of the media discourse on eGovernment is based on the so-called methodology of issue-monitoring which allows us to capture a discussion systematically within a defined arena, to generate respective issues and to analyse these in regard to their relevance and their dynamics of communication[3]. The discussion on eGovernment represents a specific discourse, concentrated within one big issue which is part of the internet-discourse. In regard to the structures of attention of the media system, we are interested to know, under which circumstances, in which forms and with which contents the issue of eGovernment is generating attention in the Swiss media arena[4].

In order to grasp the resonance of the eGovernment issue, we have chosen a sample ranging from the most representative Swiss to the foreign newspapers[5].

---

[2] From that point onwards, the nature of the internet discourse changed into a more problem-focused and technically-orientated discussion, see: Imhof, K., Kamber, E.: Das Internet als Phänomen der massenmedialen Kommunikation. Vortrag anlässlich der Jahrestagung der Deutschen Gesellschaft für Publizistik- und Kommunikationswissenschaft (DGPuK) und der Österreichischen Gesellschaft für Publizistikwissenschaft (ÖGK). Wien (1st June 2000).

[3] see: Imhof, K., Eisenegger, M.: Issue Monitoring: Die Basis des Issues Managements. Zur Methode der Früherkennung organisationsrelevanter Umweltentwicklungen. In: Röttger, U. (ed.): Issues Management. Theoretische Konzepte und praktische Umsetzung. Eine Bestandesaufnahme. Westdeutscher Verlag, Opladen Wiesbaden (2001) 257-278.

[4] Our plan is to apply our analysis as well to the international media arena. For this publication however, we have chosen to focus on the Swiss media arena for reasons of publication terms.

[5] *Switzerland*: Neue Zürcher Zeitung, Tages-Anzeiger, Der Bund, Basler Zeitung, Neue Luzerner Zeitung, Berner Zeitung, Le Temps, Blick, Sonntagsblick, Sonntagszeitung, Facts, Cash, HandelsZeitung, Weltwoche, Mitteland-Zeitung, l'Hebdo, dimanche.ch; *France*: Le Monde, Le Figaro, Les Echos, L'illustré, La Tribune; *Austria*: Das Wirtschaftsblatt; *Germany*: Die Tageszeitung, Frankfurter Allgemeine Zeitung, Der Spiegel; *United Kingdom*: Financial Times, The Daily/ Sunday Telegraph, The Guardian, The Economist, The Times and Sunday Times; *USA*: The New York Times, The Los Angeles Times, The Wallstreet Journal, The Washington Post, Newsweek; *Canada*: The Standard, National Post, The Toronto Star, The Vancouver Sun, Calgary Herald, The Gazette.

4   Results of Analysis[6]

In 1999, the discourse on eGovernment in Switzerland generated significantly less resonance than it did in other national media arenas[7], such as in the USA, Canada, the United-Kingdom or Austria where the discussion on the internet spread more rapidly to the administrational and political system. Since the summer of 2000, Switzerland caught up with the international discourse dynamic however with less articles per newspaper. This tendency is overall maintained until the first quarter of the year 2002.

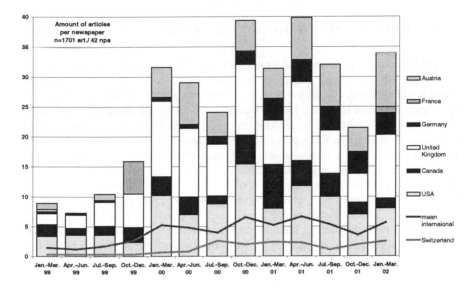

Fig. 1. Analysis of eGovernment-discourse dynamics 1999-2002. The figure shows the amount of articles per newspaper we found on eGovernment in Switzerland and within a few selected foreign media arenas. The mean value international represents all the foreign arenas summarised into one international arena which can be compared to the Swiss media arena.

For our categorical content analysis of the Swiss discourse, we have chosen to analyse all the articles focussing on eGovernment for the period of 2000 till the end of March 2002 which equates to a total of 276 relevant articles[8].

---

[6] In the presentation of our research results, we will focus on the Swiss discourse on eGovernment which can be situated into an international context through illustrating the relevance of articles on eGovernment issues in an international comparison.

[7] Previously to 1999, the internet discourse in Switzerland is related to the administrational system, however eGovernment is never mentioned as such.

[8] Our content analysis focuses on four categories which are applied to all the articles: per article we distinguish a maximum of six positions for each of which we determine the following categories: 1) *temporal perspective*: present-, past- or future-oriented; 2) *judgement/ evaluation*: positive, negative, neutral, ambivalent; 3) *focused problematic aspects* such as the social consequences of eGovernment and its technical and legal challenges; 4) *concrete applications* such as eVoting, eTaxes, eCensus and the "guichet virtuel".

From the point of view of the *temporal perspective*, the Swiss discourse on eGovernment focuses on the present process in 80% of all cases. In 50% of all cases, eGovernment is positively *judged*, but never euphorically. The overall discussion is lead in a controversial manner: most of the positive *judgements* are counterbalanced by negative or ambivalent *judgements* on the same aspect. Two third of the discourse is *focused* on the *problematic aspects* of the implementation of eGovernment, on its technical and legal challenges and on its effects on the administrational and political process. Social consequences are only evoked in 5% of all cases.

5   Conclusion and Prospect for the Continuation of the Research[9]

In regard to our first assumption concerning the implementation process of a new technology within a society, we can conclude provisionally that eGovernment has caused a controversial discussion which does not yet generate enough general positive expectations in order to have a lasting influence on the Swiss society as a whole.

Our second assumption about the nature of the eGovernment-discourse has been confirmed. Compared to the Swiss internet-discourse, the eGovernment-discussion is rooted much more in a technical and pragmatic context from its early beginnings. Though there are numerous positive judgements, they are never really euphoric and can be equalled in number by the sum of the negative and ambivalent positions which indicates the presence of a controversial and overall rather sceptical discourse.

In the continuation of our research process, we intend to broaden our categorical content analysis to the foreign media arenas mentioned in figure one and to integrate the political arenas into our sample as well as to incorporate expert discussions on eGovernment taking place within specific off- and online publications and circles.

References[10]

1. Donges, P., Jarren, O.: Politische Öffentlichkeit durch Netzkommunikation. In: Kamps, K. (ed.): Elektronische Demokratie? Perspektiven politischer Partizipation. Westdeutscher Verlag, Opladen Wiesbaden (1999) 85-108
2. Gisler, M., Spahni, D. (eds.): eGovernment. Eine Standortbestimmung (2nd edition). Verlag Paul Haupt, Bern Stuttgart Wien (2001)
3. Imhof, K.: Digitale Agora? Das Internet und die Demokratie. In: Standortbestimmung Internet. Bern (27. November 1997) 121-130
4. Jarren, O.: Internet − neue Chancen für die politische Kommunikation?. In: Politik und Zeitgeschichte, Vol. 40. Beilage zur Wochenzeitung "Das Parlament" (25. Sept. 1998) 13-21
5. Siedschlag, A., Bilgeri, A., Lamatsch, D. (eds.): Kursbuch Internet und Politik. Vol. 1: Elektronische Demokratie und virtuelles Regieren. Leske + Budrich, Opladen (2001)

---

[9] At this point of our research, we are not able to validate or reject our basic hypothesis definitively, because it would imply a comparison of the internet and the eGovernment discourse which we are unable to provide in this context of publication. However, we can give some indicators which support our two initial assumptions.

[10] This list of references only contains titles which are not already mentioned in the footnotes.

Self-regulation in e-Government: A Step More

Fernando Galindo

University of Zaragoza, 50009 Zaragoza
cfa@posta.unizar.es

Abstract. The paper presents an initial guide for the construction of codes of practice for e- government

1   Introduction

If it is necessary to create public-private Associations and frame regulations/codes of practice for electronic commerce[1], how much more so in the field of electronic government where services and functions are of vital importance and different to applications in electronic commerce[2]. Those functions may be briefly summarised in the following terms.

The field of electronic government[3] comprises a triple mission:

- To ensure open government and transparency in the activities of government agencies by designing web sites and portals to provide information and involve the general public in the doings of the administration.
- To provide on-line services enabling citizens to use the Internet to pay taxes, access registries, make applications or undertake procedures, elect their representatives, express their opinions, participate in administrative decision-making processes, and so on. All such online services and activities must of course guarantee the authenticity, integrity and confidentiality of the communications channels established by the State in a manner appropriate to official procedures in a democracy.
- To ensure the interconnection of government agencies. This means sharing workflows and infrastructure between agencies, regardless of the country where each is located, using modern information and communications technologies.

---

[1] See: GALINDO, F., "Public Key Certification Providers and E-government Assurance Agencies. An Appraisal of Trust on the Internet", in TJOA, A.M., WAGNER. R.R., (eds.), *12th International Workshop on Database and Expert Systems Applications*, Los Alamitos, IEEE, 2001, pp. 348-349

[2] On the particulatities of the applications in e government and public administrations see: LENK, K., TRAUNMÜLLER, R., WIMMER, M., "The Significance of Law and Knowledge for Electronic Government", in GRONLUND, A., *Electronic Government: Design, Applications & Management*, Hershey, Idea Group Publishing, 2002, pp. 61-77

[3] *Public Strategies for the Information Society in the Member States of the European Union*, ESIS Report, September 2000. Available at http://www.irc-irene.org/documents/do-psismseu.html. For a general discussion on electronic government, see: GRONLUND, A., *Electronic Government: Design, Applications & Management*, Hershey, Idea Group Publishing, 2002

R. Traunmüller and K. Lenk (Eds.): EGOV 2002, LNCS 2456, pp. 411–418, 2002.

On this view of the possible content of electronic government, there is little more that need be said in support of immediate regulation, even though actual implementation of e-government systems is still in its infancy. Having said this, however, we should note that the infrastructure required cannot be confined to the general proposal for e-commerce, but must be customised to fit the special conditions existing in the sphere of electronic government due to the services and functions involved.

In light of the above, in this paper we shall concentrate on the following issues:

- the presentation of a code of practice and an organization for electronic commerce as reference (2).
- an initiative launched to seek solutions for the practical problems arising from the spread of electronic government and establish appropriate regulatory frameworks (3).
- an initial proposal for the possible use of service charters to serve as a basis for the adaptation of electronic government applications. The contents and functions of these instruments are similar in the field of the public administration to the codes of practice used in electronic commerce (4).
- a short conclusion (5), finally.

2   The Code of Practice

2.1   Introduction

The internet's early history of develop in academic, military and industrial research carried out in the United States during the 1960s had a marked impact on the initial characteristics and common culture of the network. ICANN itself is a direct product of this tradition[4]. Now that internet use has spread to citizens all over the world, however, it has become necessary to progress further with the construction of a culture that takes the use of the network by a diverse community, rather than narrow academic, business or military interests into account. The whole of society now has a stake in the operation and governance of the internet. Naturally, domain names, one of the main fields of internet regulation, will remain an area of regulatory concern, but it has now become necessary to safeguard other rights such as data protection, ensure implementation of reasonable commercial practices, oversee internet content and establish dispute resolution mechanisms, to give but a few examples.

This is the context of the organisational and regulatory framework outlined in this section, which is representative of the will and opinion of citizens and social organisations. This organisation, which was established in April 2000 under the name APTICE (Asociación para la Promoción de las Tecnologías de la Información y el Comercio Electrónico – *Spanish Association for the Promotion of Information Technologies and E-Commerce*), has drawn up its own code of practice and implemented it through the creation of an independent institution, AGACE (Agencia para la Garantía del Comercio Electrónico – *Agency for the Guaranteeing of E-Commerce*).

---

[4] See on this history and regulation: GALINDO, F., "Autorregulación y Códigos de práctica en Internet", in CAYON GALIARDO, A., (ed.) *Internet y Derecho*, Zaragoza, Monografías de la Revista Aragonesa de Administración Pública, 2001, pp. 30-40

2.2   The APTICE Association: Membership

The Association for the Promotion of Information Technologies and E-Commerce, (APTICE: (www.aptice.org) was founded in Zaragoza, Spain, in April 2000.). The association currently has 83 members comprising private individuals, enterprise (telecommunications companies, banks, communications media, etc.) and public institutions, and is the fruit of a year-long period of debate and preparation by its founders (individuals, businesses and the Aragonese Development Institute, an independent government agency). APTICE reflects the conclusions reached from the joint R&D activities undertaken at the University of Zaragoza by companies and research teams, mainly associated with the Philosophy of Law Department.

2.3   The APTICE Code

APTICE has framed its code of practice/conduct for e-commerce in consultation with all of its members. The code was thus drafted in a spirit of consensus. The quality seal and guarantee infrastructure created in parallel with the code are charged with its implementation using the organisational machinery specifically designed for that purpose, which is embodied in the AGACE Agency.

The Code of Practice[5] is intended to provide a self-regulatory tool for the use of companies and public institutions in their relations with users, whether be citizens, other businesses or government agencies carrying out electronic transactions with subscribers. The code has been prepared on the basis of prevailing legislation in Spain and the European Union, taking into consideration the practices required by other similar codes worldwide, expert opinion on the issues involved and the experience of companies operating in the e-commerce industry.

To achieve its purpose, the APTICE Code of Practice contains seven general principles covering the key elements for building trust between the parties entering into on-line transactions over the internet and defining service quality and improvements needed in the activities and internal procedures of businesses and public institutions. These principles are as follows:

Principle #1: Identification of the Organisation.- In accordance with this principle, any organisation subscribing the APTICE code of practice must provide sufficient activities regarding its nature and activities in its web pages. The organisation must also comply with the domain name requirements established by the internet's central domain registries and with registration requirements established by legislation governing intellectual and industrial property. The future need for the use of advanced electronic signature systems and server authentication certificates is also provided for, as well as monitoring of legislation applicable to the establishment and its commercial activities.

Principle #2: Guarantees concerning claims and performance.- This principle requires that key commercial information (e.g. prices, delivery conditions, product descriptions, warranties, and many others) be displayed in the web site, together with instructions and procedures for carrying out on-line transactions, customer service details, and information concerning logistics, usability of web pages and contractual and extra-contractual liability.

---

[5] The full text is available in http://www.agace.org/en/index.html,

Principle #3: Security and technology infrastructure.- This establishes mandatory security policies for subscribing organisations.

Principle #4: Data protection.- This principle requires organisations subscribing the code of practice to comply fully with the Spanish Data Protection Act.

Principle #5: Content quality.- This centres on issues such as the organisation of illegal and offensive content, the protection of children and advertising practices.

Principle #6: Rules for out-of-court dispute resolution: The APTICE code of practice categorically requires subscribers to adhere to out-of-court dispute resolution systems. In principle, these would be bodies such as Consumer Arbitration Tribunals (business to citizen relations) and Chambers of Commerce (business to business relations). The intention is to ensure that any possible dispute that might arise between a company or public institution and customers (be they private individuals or other legal entities) is resolved as quickly and smoothly as possible, without the need for action in the courts, which are not sufficiently adapted to the exigencies of the new technology industries. APTICE will act as a mediator in disputes, which it will seek to resolve amicably or by referring cases to the most appropriate arbitration tribunal in the circumstances.

Principle #7: Requirements for the implementation of the APTICE Code of Practice.- This principle establishes the requirements that must be met by a company or institution intending to implement the code. These requirements refer, in particular, to the preparation of procedures manuals and records in accordance with the instructions provided by the institution or agency responsible for the performance of Code compliance audits. This principle will therefore allow auditors accredited by APTICE (to date only AGACE is an accredited auditor) to carry out appropriate procedures to examine the activities of an organisation and assess its compliance with the rules enshrined in the code of practice and, accordingly, its readiness to receive the award of the quality seal. Principle #7 also includes the sanctions mechanism established for cases of non-compliance by any subscribing organisation with its obligations upon accepting the code of practice as a guide for its own conduct.

The enumeration of these principles brings us to the need for a mechanism to implement the code of practice. This mechanism is now in place and comprises the quality seal and the organisational structure for its implementation. This infrastructure goes under the name of AGACE (*Agency for the Guaranteeing of E-Commerce*).

2.4   Guarantee Agencies

For some time now, it has been clear that the practicalities of the internet require the existence of specialised services, known as quality seals[6], to underpin the reliability of on-line transactions. Seals thus cover a different range of issues from guarantees referring to the identity of the parties entering into electronic transactions and the security of the messages exchanged between them. In view of the similarity of the goods

---

[6] It is common for standards agencies to award seals to companies manufacturing products or providing services subject to accepted quality standards. See: MOLES I PLAZA, R.J., *Derecho y calidad. El régimen jurídico de la normalización técnica*, Barcelona, Ariel, 2001, p. 29

affected, it seems appropriate in terms of the European legal tradition for such services to be able to guarantee both public and private operations.

Such services are already being implemented through initiatives such as TRUSTE[7], which concentrates on compliance with data protection regulations, BBB on line[8], which is concerned with on-line trading practices, and Web Trader[9], which designs codes of practice for on-line businesses.

The AGACE[10]e-commerce guarantee initiative bears a certain similarity to the above initiatives, but differs in that it is concerned with the various aspects of reliability and trust taken as a whole. Accordingly, the AGACE seal will be awarded only to those e-commerce activities that demonstrate compliance with the requirements of the APTICE code of practice, as described above.

AGACE has only recently commenced its activity, having carried out pilot consultancy work in the fields of e-commerce and e-government.

3   Organisational Initiatives

This point will present the main features of two different initiatives, all of which seek to safeguard the principles and values of public service in the design and implementation of electronic government applications. These are FESTE and AEQUITAS.

The first of these initiatives, FESTE (www.feste.org) was launched in November 1997 by the Spanish General Council of Notaries, the General Council of Legal Practitioners and the University of Zaragoza. The acronym FESTE is drawn from the name *Fundación para el Estudio de la Seguridad de las Telecomunicaciones* (Foundation for Secure Telecommunications Research). In accordance with its institutional object, FESTE provides security services for electronic communications under the provisions of prevailing legislation governing signatures and in light of the practices established in democratic legal systems for the activities of notaries public and commissioners of oaths[11].

The second initiative is AEQUITAS. This project is the result of a proposal made by the Spanish and French registrars, and the Spanish and Portuguese solicitors. The initiative has been awarded the status of a European Union project under the Information Society Technologies Programme (IST). The objective is to prepare the ground for the interconnection of the basic electronic government infrastructure (i.e. public key infrastructure and electronic communications certification services) that is already operational in participating organisations. The AEQUITAS initiative is also backed by APTICE.

Let us briefly describe the current status of AEQUITAS initiative.

The objective of the AEQUITAS project is to build the basic infrastructure for the mutual recognition of public key certificates in electronic government applications designed for the interconnection of government agencies.

---

[7] http://www.truste.com/
[8] http://www.bbbonline.com
[9] http://whichwebtrader.which.net/webtrader/
[10] http://www.agace.org/en/index.html
[11] See http://www.feste.org/.

The project's full name is *Trust Frame for Electronic Documents Exchange Between European Judicial Operators,* AEQUITAS project[12]. It is partially funded by a European Commission grant awarded within the V Framework Programme in the IST area, which is concerned with the information society[13]. The project is included in line II.4.2 (key action II), headed *Large-scale trust and confidence,* which is intended to extend, integrate, validate and pilot trust technologies and architectures in the context of advanced large-scale scenarios in business and everyday life.

From a legal point of view, the objective of the project is to enhance the quality of the services provided to society by judicial operators, and in particular by registrars, solicitors and their equivalents, by incorporating secure, confidential on-line communications into their activities.

AEQUITAS' specific objective is to ensure that e-mail messages exchanged between Spanish registrars and French *greffiers,* between Spanish and Portuguese solicitors, and between all such judicial operators are secure. The same infrastructure will subsequently be available for electronic communications with the citizens of the three countries involved. This security is sought not only in terms of the encryption programs and techniques employed in the exchange of messages, but also in terms of public trust and confidence in the institutions designated by the State to underpin the operation of the technology. This means laying the foundations for trust in law rather than relying on the *ad hoc* systems generated by the Internet itself following the ICANN philosophy practised by firms such as VeriSign, which we have already discussed in section 3.

In this context, one of AEQUITAS' key objectives is to create an Association formed by relevant stakeholders (e.g. professional organisations, companies, other associations, private individuals and legal entities). The future members of this Association hold that its mission should be to establish rules, certification practices and codes of practice in order to guarantee that the use of the Internet for electronic government purposes is in line with the practices and customs proper to a democratic society, in which the involvement of the citizen in public affairs and consensus-building in relation to issues with a profound impact on society are the ultimate gauge against which to measure the activity of government.

The Association will be created in the certainty that organisations of this kind are particularly well-suited for this and other tasks necessary for the proper functioning of the on-line world, such as involvement in the administration and running of high-level and territorial domains, the creation of voluntary accreditation systems for certification service providers, official recognition for firms specialising in the development and implementation of electronic government applications, clear technical standards and so on. In short, the Association's mission is to undertake tasks that are currently the province of multinational businesses.

---

[12] The use of the name AEQUITAS in connection with secure electronic communications in judicial matters goes back to 1997, when the first AEQUITAS project was promoted by the European Union as part of the INFOSEC programme. The contents of the 1997 project were, however, unrelated to the present undertaking. In fact the subject of the first AEQUITAS project was *The Admission as Evidence in Trials of Penal Character of Electronic Products Signed Digitally.* The content of the project is available in http://www.cordis.lu/infosec/src/study11.htm,.

[13] See http://www.tb-solutions.com/es/radff348.shtml. The project leader is the Spanish firm Tools Banking Solutions (www.tb-solutions.com).

4   Codes

The progressive regulation of electronic government by the institutions referred to and as a result of the initiatives described in the preceding section cannot, of course, take the place of general legislation in this area. As in the case of the APTICE Code, these are private rules or codes of conduct implemented by the same institutions as draw them up. They are intended to bind members and inform individuals or legal entities seeking the services of such institutions of the consequences of adhering.

The terminology used by these institutions to refer to the sets of rules they have created differs widely. FESTE, for example, uses the American-sounding expression "declaration of practices and certification policies" to refer to the basic regulations governing the certification service it has set up. This solution is similar to the terminology employed by other certification services. The AEQUITAS project, on the other hand, will resolve the issue of finding an appropriate title for its regulations in the future, in view of the action undertaken by the planned Association following its incorporation, which is currently in progress. The main concern at present is to draft the Association's statutes.

With a view to the direction that the initiatives described are likely to take in relation to the spread of electronic government applications, it may be of interest to explain here that an alternative to private regulations already exists to specify content in the field. This involves framing service charters, an instrument recognised in Spanish administrative law through Royal Decree 1259 of 16 July 1999 governing service charters and quality standards in government agencies.[14].

In accordance with article 3 of Royal Decree 1259/1999, "Service charters are defined as written documents comprising the instrument by which the agencies of the Spanish national government, Regional institutions and Social Security Management Entities and Services inform the public of the services entrusted to them and the quality standards established for the performance thereof, as well as the rights of citizens and users in relation to such services."

Pursuant to article 4 of Royal Decree 1259/1999, service charters must set out the following content:
1. General and legal content
   a) Identification and object of the agency or organisation providing the service
   b) Services provided
   c) Specific rights of citizens and users in relation to the services
   d) Collaboration or involvement of citizens and users in the improvement of services
   e) Up-to-date description of regulations governing each of the services and facilities provided.
   f) Availability and access to a complaints and suggestions book and, in particular, complaints procedures, response periods and effects of complaints
2. Quality standards
   a) Minimum quality standards, including at least the following information:
      1. Maximum periods for handling procedures and/or providing services
      2. Communication and reporting mechanisms for both general and personal information
      3. Public offices and opening hours

---

[14] For references to service charters see http://www.igsap.map.es/docs/cia/cartas/cartas.htm.

b) Information required to gain access to the service and under the best possible conditions.

c) Quality assurance, environmental protection and health and safety systems in place, where applicable

d) Quality assessment indicators

3. Other content

a) Postal addresses, telephone numbers and e-mail addresses of all offices where each service is provided, including a clear indication of routes and, where applicable, the nearest public transport

b) Postal address, telephone number and e-mail address of the unit responsible for framing the service charter

c) Other matters of interest concerning the services provided

Clearly, the general requirements for service charters contain basically the same items as those included in the APTICE Code, but with certain adaptations to fit them for their intended use as instruments by which government agencies undertake to perform public services in accordance with regulatory requirements. Specifically, charters must identify the government agency concerned, and refer to the quality of service content, the dispute resolution mechanism (complaints and claims procedures) and the procedure followed to frame and implement the charter in question (as provided in the remaining part of Royal Decree 1259/1999).

5   Conclusion

It would appear, then, that the contents and undertakings applied to the field of the public administration and electronic government are parallel and analogous to the APTICE Code of Practice described above.

The parallel solution to the APTICE e-commerce Code in the field of electronic government would therefore be to draw up a general or framework service charter setting out the basic requirements for any government agency or service entering into on-line transactions or relationships with citizens, other government agencies, companies and institutions.

This would not only be possible and a step forward in the self-regulation of electronic government, but would be in harmony with the democratic legal system as a whole.

UK Online: Forcing Citizen Involvement into a Technically-Oriented Framework?

Philip Leith and John Morison

School of Law, Queen's University
Belfast BT7 INN, N. Ireland, UK
{p.leith,j.morison}@qub.ac.uk

1 Introduction

UK Online is a centralised initiative which attempts to structure the nature of Government-citizen interaction, part of which is to expand notions of "citizen involvement" using technological approaches. The UK Online initiative lies within a general process of "modernisation" that is driven by the UK Government's White Paper *Modernising Government*[1]. We suggest that this project – along with other UK e-Government projects – which advertise a avowedly neutral strategy of developing ICT in government actually involves an attempt on the part of Government to structure and control a *new space* that is opened up.

The supposed "neutral" form of structured communication being planned by government is adversely affected by two forces: first, the desire of government to control the communication, but secondly, the technical challenges resulting from programming an interface to allow this communication - in particular the problems caused by "business model" communications when used by government.

Our critique of the e-Government project is not that it is not a useful – and perhaps efficient – way to move forward, but that it is inspired both by a mythology of the business model and the citizen as customer, rather than the citizen as citizen. We develop a conception of 'governmentality' to account for this re-ordering of relationships and urge a re-thinking of how UK on-line is developed.

2 Governmentality

The general movement from government to governance that is observable throughout the developed world stresses how today the state is involved more in "steering" than "rowing". The governmentality approach that is associated with the later work of Michel Foucault and other critics provides an important way of theorising about this general process.[2] This perspective helps to explain how government today is not so

---

[1] HMSO Cmd 4310 1999.

[2] See, for example, M. Foucault, *Power: The Essential Works 3* Harmondsworth: Penguin 2000) and "Governmentality" in *The Foucault Effect: Studies in Governmentality*, G. Burchell, C. Gordon and P. Miller (Hemel Hempstead, Harvester Wheatsheaf 1991) p.87-104 and *Technologies of the Self: A Seminar with Michel Foucault* edited by L. Martin, H. Gutman and P. Hutton (London, Tavistock, 1988). See also N. Rose *Powers of Freedom: Reframing Political Thought* (Cambridge, Cambridge University Press, 1999).

R. Traunmüller and K. Lenk (Eds.): EGOV 2002, LNCS 2456, pp. 419–423, 2002.

much as a set of institutions but as a domain of strategies and techniques through which different forces and groups attempt to render their particular programmes operable. The practice of government is seen as involving engagement with the many networks and alliances that make up a chain or network which translates power from one locale to another.

From within this theoretical framework, we study the UK government's "modernisation" programme and its emphasis on "information age government" where, in the words of the White Paper:

Government must modernise the business of government itself, achieving joined up working between different parts of government and providing new, efficient and convenient ways for citizens and businesses to communicate with government and receive services. [3]

The UK Online initiative seeks to put this into practice. We set this initiative into an continuum which represents the evolutionary process that changes the nature of communication between the state and the citizen and the very structure of the state bureaucracy itself. As Figure 1 suggests, the UK programme's ambitions are set at the highest level – stage 6 of the evolutionary move towards citizen/state interaction:

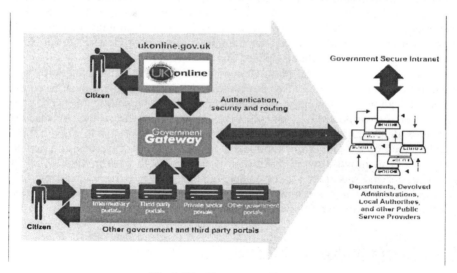

Fig. 1. The Government Gateway

We argue that the assumptions within this model must be challenged and its feasibility and desirability questioned. In particular, it is important to query the simple assumption that putting elements of government business on-line will necessarily enhance democracy by improving openness and participation. This, we assert, is subject to challenge.

Within UK Online the provision of services is actually quite limited: the promised revolution in government-citizen interaction is more equivalent to the transition from traditional shop to self-service supermarket rather than anything more fundamental. In relation to the promised transformation in consultation and democratic engagement

---

[3] Chapter 5, para. 5

which is intended to *"democratise the very institutions of democracy themselves"*, the reality can be seen as amounting to little more than providing an electronic suggestion box where government structures the interaction and dominates the process.   The governmentality framework allows us to see the UK Online revolution as an attempt to re-engineer Government-Citizen relations generally rather than as a merely neutral technical development as it is presented by its proponents in Government.

Indeed we take issue even with the idea of a neutral technology that is simply being harnessed by government in a general process of improvement. We argue instead that the whole idea of technicality (that is, simple use of business methodologies and techniques) is being used within a wider political process.

3   Technicality

The reader of any of the White Papers dealing with technology change within Government will find the same call to arms – that business methods demonstrate user-friendliness and efficiency and that the models of IT practice within business should be transferred over to government. The largest project to be so announced was that of the computerisation of the UK welfare department, the DSS, in the 1980s.   We have argued [4] that the project was problematic for a number of reasons:

- lack of technical understanding of the 'software crisis' (by which we mean problems found in large scale implementations)
- usage of a business model which was inappropriate for the kinds of processing being undertaken
- an attempt to force processing into a computer-suitable algorithm.

and later analysis demonstrated that our analysis was correct[5]. The DSS computerisation project was over-budget, under-performed and the technical framework imposed by programming and other considerations has led to the system being difficult to use and requiring various 'work arounds'. Also, as with current proposals, the claimed advantage that staff could be moved to more interesting work has not proven to be true; indeed, staff are usually deskilled by these implementations, serving the failures of the system more than the public.

The current e-Government project is also being pushed by IT models from business. Thus, the NAO report on e-Government to date[6] picks up on the optimism of business:

"The Oracle Corporation saving of £71 million through ... web-enabled ... functions ... and British Gas productivity improvements ..."[7]

This is despite the clear evidence that Government usage of technology is problematical. Reading the computing trade press demonstrates that projects are difficult

---

[4] "Law in a Changing Technological Society", in Law, Society and Change, Livingstone S. and Morison J. (eds), Gower, Aldershot, 1990. R.Geary & P.Leith.

[5] "From Operational Strategy to Serving the Customer: Technology and ethics in welfare law.", in International Review of Law Computers and Technology, 2001. R. Geary & P.Leith.

[6] "Better Public Services through e-government", Report by the Comptroller and Auditor General, HC 704-1 April 2002.

for government to implement. Rarely are IT projects successful – more usually they are late, over-budget and under perform. And they impose problems on future developments and transformations of government simply due to the expense of re-programming and re-engineering such expenses systems.

We are interested in why government is so poor at programming this face between the citizen and the state. Obviously, the software crisis is not unknown in the commercial world, but it seems to be particularly problematic in government usage of technology. We suggest that many of these problems arise from the difficult relationship of government to citizen and the difficulty of programming systems which properly effect this relationship: that is, systems which are flexible and not overtly formalist [7]. The glory of the business model is that complex transactions can be reduced to a relatively simple one of buy/sell. The IT systems developed have one usually one goal – to reduce costs in an economic transaction. When we move to government, though, the transaction is much more complicated – in ways which are not yet clear, but which do not support the view that the citizen is a customer. The rhetoric of e-government, though, remains that the citizen is a customer.

As one small example, we can look at the limitations on government use of information which are not present in the commercial world. In a commercial business relationship, that business would seek to reduce costs of storage and processing of information. A common database would be set up where information is verified, stored and processed – all with one aim in mind. In government, the amounts of information, the diversity of that information, the lack of funds to ensure it is up-to-date and verified[8], and arguments about 'big brother' and the collation of these diverse information sources, mean that the single pseudo-commercial relationship which the government want to implement via UK-online will always be technically problematical and difficult to overcome.

4   Connecting Governmentality and Technicality

If Government was simply a commercial relationship with the citizen, where taxes are collected, information provided, forms downloaded, then the strategy which is being outlined in e-government in the UK would be appropriate. However, a government has wider concerns. At a time when the health of orthodox, representative democracy is at best uncertain and voter apathy and general disenchantment with the answers offered by "big government" challenge us to widen and deepen democratic engagement it is important that we think more seriously about how to use this new space between citizen and government to enhance democratic engagement. While the approach of UK Online *may* bring some efficiency gains as it distributes services it misses an opportunity to use new technologies to facilitate more direct forms of democracy and reinvigorate traditional representative democracy. While technology cannot return us to the direct democracy of classical Athens, it may be that it can

---

[7] Formalism is the achilles heel of computing. See Formalism in AI and Computer Science, Ellis Horwood, 1990. Philip Leith.

[8] A complaint being made by the Information Commissioner with respect to many government data banks – including the police national computer system.

begin to supply an answer to Anthony Gidden's question of 'How can we democratize democracy[9]?'

5   Conclusion

We conclude by arguing the case for a much more limited but more realistic version of UK online which rather than seeking to structure and control the whole nature of government-citizen interaction, instead facilitates the more democratic possibilities that e-government can allow. We are endorsing a normative ideal of democracy as political communication and suggesting a model of democratic process where citizens must engage with one another and with government in a public space where they make proposals and criticise one another in an effort to persuade all parties of the best solutions to collective problems. This is active democracy where citizens are involved as much as is practical in making these fundamental decisions about how to live together as well as with simply transacting for services. Here it is important that the processes should be fair, open and, most of all, inclusive. This involves facilitating models of government and citizenship beyond those of service supplier and consumer to engineer both in technical terms and in a democratic sense a better approach to online government.

---

[9] *The Third Way: The Renewal of Social Democracy* London: Polity 1998 at p.72. This question keys into a variety of concerns about the nature of modern democracy and its ability to deliver the legitimacy that is required to underwrite government in a changing world. To take just one critic from a wide range proffering versions of this complaint, Barber argues that liberal democracy is a '"thin" theory of democracy, one whose democratic values are prudential and thus provisional, optional, and conditional——means to exclusively individualistic an private ends. From this precarious foundation, no firm theory of citizenship, participation, public goods, or civic virtue can be expected to arise'. *Strong Democracy: Participatory Politics for a New Age* (1984), p.4.

Data Security: A Fundamental Right in the e-Society?

Ahti Saarenpää

Professor of Private Law, Institute for Law and Informatics,
Faculty of Law University of Lapland, Box 122 , 96101 Rovaniemi Finland

Abstract. The birth of the modern network society and the strengthening of the idea of the constitutional state in Europe have occurred largely at the same time. This juxtaposition, although more accident than design, obligates us to examine from the legal perspective the tension between the array of opportunities (e.g., convergence) and new risks which the network society brings and the legal effectiveness of the constitutional state. The prevailing attitude towards data security offers an illuminating example of the new encounter between technology and law. A look at legislation and legal practice in this area reveals a variety of approaches. I present these and go on to argue for a position whereby data security can and should be assessed in terms of fundamental rights. We have a right to data security in the information infrastructure. At the same time, it must be pointed out that, if we are to avoid the potential risks involved, data security must quite literally be security, whereas legal regulation strives for certainty. When we attempt to forestall risks, the degrees of security and certainty needed at any given time should, in the final analysis, be assessed from the standpoint of fundamental rights.

1   A Changing Society and a Changing State

Society is changing profoundly. We have largely completed the transition to mass use of IT and data networks, and most of what at one time was the province of experts is now routine. The use of IT and networks is part of our daily lives. We can justifiably speak of "the network society" or, as it is already being dubbed, "the e-society." It is a society that has emerged rapidly, in the space of less than ten years. Few had the foresight to anticipate the change before it was upon us. (1)

A second significant change we are witnessing is *the rebirth of the constitutional* state and a concomitant decline of the administrative state. We now speak of the principle of *the constitutional state* as one of the guiding European legal principles. And, true to this commitment, we are now taking more seriously human rights, fundamental rights and the guarantees that these rights will be realized. At least this should be the case.

The changes in society and the state seem to have gone hand in hand for some time. After all, the approval of the *European Personal Data Directive* came at a time when both the development of IT and improved protection of fundamental human rights were prominent issues. While a superficial look at these two developments might suggest that the legislator had realized the significance of fundamental rights in the network society at the time, a broader historical perspective reveals that we are dealing with accident rather than design. The lengthy drafting process that preceded

R. Traunmüller and K. Lenk (Eds.): EGOV 2002, LNCS 2456, pp. 424–429, 2002.
© Springer-Verlag Berlin Heidelberg 2002

the enactment of the Directive was completed at a time when the use of open networks was already becoming extensive, so much so that even the experts were taken by surprise. We can hardly speak of these trends in IT and fundamental rights as an instance of joint or even coordinated development: chance has merely juxtaposed two different developments. Yet, these are changes that urge Law, in particular Legal Informatics, to assess afresh the relationship between law and technology.

Today we simply cannot any longer be content only to state that we live in an *e-society* and a *constitutional state*. We must probe deeper and identify the problems that impinge on - or might impinge on - the relationship between the network society and the constitutional state. We find ourselves faced with a complex cluster of legal issues: we must not only ponder the need for and potential of legislation that meets the requirements of the new network society but also analyze the extent to which the constitutional state needs to be renewed. For example, do we have legislative techniques that are sophisticated enough for the network society and does the network society require that we rethink what we mean by *the constitutional state*? (2)

In another context, I have gone so far as to assert that one of the distinctive legal features of our time is the scarcity of justice. For example, convergence, the commodification of knowledge, network communities, and globalization combine to make it harder than ever to achieve the effectiveness of rights through conventional legislative and procedural means. Promoting interests crucial to networks tends to restrict freedoms. Effectiveness is sought in the form of increased control and special procedural measures, which restrict even fundamental rights. The tension is plainly visible if only we want to see it. Moreover, many fundamental issues have been and continue to be overlooked, and we run an increased risk that the scarcity of justice will worsen. (3)

The assessment of *risks* has long traditions in different branches of Law, and we will do well to bear this in mind when considering the legal problems of the new information infrastructure. We also need legal risk analysis. We carried out such an analysis in the data security report written in 1997 at the *Institute for Law and Informatics* at the University of Lapland. The analysis identified 20 different risks, ranging from risks affecting currency and payment transactions to a *"drafting risk"*, which occurs when the legislator fails to realize the legal problems of the network society in time, or fails to notice changes taking place in the overall situation when implementing individual legislative measures. The report also dealt with the notion of an information war front and center. (4) Many found the issue bewildering at the time; fewer would find it as mystifying today.

2   Establishing a Position on Data Security

Of the many pivotal issues that emerge in this area, I have opted to focus in my presentation on *legal data security* in the network society characterized by a new *information infrastructure*. From the legal point of view, both data security and infrastructure are old as well as new issues. Indeed, the various provisions specifying the form and confidentiality of documents have promoted data security for centuries. Similarly, significant infrastructures have either been provided for in law or been under the direct control of public authority. The wealth of telecommunications directives in the European Union provides compelling evidence of this tradition.

It is appropriate at this point to provide a brief definition of *legal data security*. It can be described as the totality of laws, regulations, guidelines and practices whose purpose is to guide and assess the implementation of data security and to evaluate the various threats to it. Data security is very much - very much indeed - a legal issue.

This general description is of course only a basis for a closer examination of the issue. In looking at data security from the legal point of view, I see at least six perspectives that can be distinguished both in theory and in practice: these are data security as a tool, a technological development, a market, a system, an element of justice, and a fundamental right.

The tool perspective is the most traditional of the six positions: data security is seen as merely an aid in realizing more important objectives and, accordingly, is not given much weight when it comes to legislation. What we see here is essentially the distinction drawn by *Lord Snow* between humanistic and technical cultures: that which is less well known is less interesting. The consequence in the present case has been that the issue of data security has been addressed through unsystematic regulation of a general nature and, accordingly, has ended up being governed by sector-specific practices. The development curve of many a data security firm reflects this trend.

The technological perspective is a more modern one. After getting a start, the legislator trusts in favorable technological development. We even speak of the *technological imperative*. It is easy to find examples of this type of development in the network society, as the legislator hastens to come up with solutions dealing with network activities. This seems to be the case in many countries, as different states vie to be the frontrunners in the e-society. The Finnish electronic identity card, with no market for a lack of services and established standards, serves as an isolated but striking example.

Yes, the market. The market approach is to show confidence in the market and their ability to ensure adequate data security. Issues to be regulated are left to the market, which is expected to follow development so dependably that the legislator has no need for more detailed specifications. An illustrative example of this view in Finland is the Act on Privacy and Data Security in Telecommunications, which allows users to equip their connections with any of the security systems available. Although the provision also means that so-called strong encoding is permitted, it says even more about the view on the data security market I have mentioned here.

The system perspective represents another modest step forward legally. Adherents of the position see data security as an essential element of data systems. Data security must be taken into consideration in the planning of such systems. From the point of view of legislative technique, what we have here is a technical norm of sorts. A good example of this in Finland is Section 18 of the Act on Openness in Government Activities, which *deals with good information management practice*. The section obligates all units in the public sector to plan their information systems to take into account not only content but also data security. The complexly worded provision has as yet been applied very little in practice. On many occasions, I have referred to it as the most important provision in the Finnish public sector, where it would function as a central guiding provision in information-related work.

The legal perspective looks upon data security as a right of an individual or community - a right that must be rendered effective. While Section 18 of the Act on Openness in Government Activities clearly includes some of these features, the perspective is even more apparent in the data security provision of the Act on the Protection of Personal Data. Personal data must be duly protected in order for our *right to privacy to be safeguarded*. This principle, also seen in the Personal Data Directive, is significant in practice because it steers us towards applying the principle of proportionality in the name of the protection of privacy more so than the more conventional efficiency of economic production.

The fundamental rights perspective takes us a step further towards the new ways of thinking characteristic of the new constitutional state. It is not sufficient that we assess data security strictly as a legal issue. We must ask whether data security is one of the fundamental rights. To the best of my knowledge, the answer to this question – at least in Finland – would be affirmative. Three main arguments can be presented:

First, we must remain mindful of the link between the protection of personal data and data security. The two go hand in hand. The protection of personal data as a fundamental right requires sophisticated data security, thus making data security more than just a conventional right.

Second, it is also easy to see that the fundamental right to having one's affairs processed without undue delay by authorities and the courts as well as the fundamental right to *good government* require appropriate data security when working with information and making use of data systems in different ways. Here, too, data security is more than just a conventional right.

Third, we must take into account that we are changing over to electronic government and in fact have done so already to some extent. With this change and the development of electronic commerce, we increasingly exercise our fundamental rights - for example, and, in particular, privacy - in network environments. This observation, if no other, indicates that in the new information infrastructure data security has become both a substantial legal principle and a metaright, i.e., a fundamental right which is an ideological and actual precondition for the realization of statutory fundamental rights. Our right to self-determination can only be realized if we create the requisite data security information infrastructure.

What I have said here views the issue from the perspective of the individual. This of course is the primary approach we should take to data protection. But we cannot forget communities, private and public communities. There, too, the hierarchy of rights and metarights is relevant. The functionality of the free economy in the network society is largely dependent on the level of data protection and data security (information and commodity security). There is simply no way around this fact.

3 Security and Certainty

At this point, it will be appropriate to pause and consider a special issue - the relationship between security and certainty. On the one hand, we speak of data or information security. The key concept here is security, which is fairly transparent. On the other hand, we talk about legal certainty, which is a somewhat less accessible concept.

Yet, legal certainty is one of the pivotal elements of the constitutional state. It is often associated with the concept *rule of law*. The very expression *legal certainty* embodies a strong purpose - being able to anticipate the outcome of activities and legal decisions in legal life. Legal certainty refers to this predictability, which we pursue through an advanced legal culture.

Here I will bring in the concept of risk to illustrate the relationship between data security and legal certainty. Data security and legal certainty are the means we use to forestall the principal risks to information and information processing in the network society. The relationship among the three can be depicted in the form of a triangle:

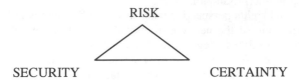

Fig. 1. The relationship among security, certainty and risk

The number of risks we face has not, to my knowledge, decreased. Quite the contrary. However, this is not the place to go into an extensive and detailed assessment of the different risks. Instead, I will focus on two risks that we covered in our report but which were not given the same scope. These are the risks affecting fundamental rights and the risk of excessive control.

The potential risk to *fundamental rights*, that is, the jeopardizing of our fundamental rights in making use of the information infrastructure, is a factor whose significance is growing as we exercise these rights to an increasing extent on networks, and open networks in particular. In *e-government* in the e-society, we are guided us towards exercising our fundamental rights on networks. We are increasingly network dependent. This is a situation which makes it essential that we assess the significance of the risk to our fundamental rights; this is not something we have spent much time doing. (5)

The risk of *excessive control* is crucially connected with the risk to fundamental rights. For example, the protection of personal data in Finland is nothing less than a fundamental right. The issue has to do not only with unforeseen risks that come to light when data security breaks down but also, and above all, with the increase in conscious control. Advances in IT and the resultant convergence enable control both on and of the network. Observations to this effect have been made in government in Finland. Efforts are made to increase control of the individual, accompanied by appeals to the need for heightened efficiency in bureaucracy and protection of the individual. The legal problems such developments entail are not always noticed or at least are not acknowledged. On balance, we have every reason to speak of a risk of excessive control.

Both the risk to our fundamental rights and the risk of excessive control illustrate that a tension - a growing tension - obtains between the developing network society and its *e-government,* on the one hand, and the constitutional state, on the other. If we are to rise to properly defend the rights of the individual, we must be prepared to reassess the constitutional state and its legal culture. (6)

4   Conclusion

For decades now, personal data protection has had the thankless task of being a buffer against many failings of knowledge, understanding and education that have accompanied the juridification of our increasingly technological society. We can still encounter those for whom data protection as a means for regulating the protection of personal data is either not a particularly important matter or is a downright curse word used to denounce what is seen as a hindrance to the smooth functioning of bureaucracy and the satisfaction of curiosity.

Data security became a pivotal tool in implementing the protection of personal data quite early on. Data protection and data security began to go hand in hand. For example, in Finland the 1987 Personal Data File Act had a provision on individual data security, albeit only one of the fifty or so provisions in the Act. The idea was to protect people through procedures that maintain and promote data security in the processing of data which someone has the right to process. The development of data protection legislation has in fact expressly linked data protection and data security. Appropriate data protection requires appropriate data security. Yet, this fact was not readily noticed as long as the new data security remained focused on technology.

Today, data protection and data security continue to go hand in hand, but their relationship has changed or at least is changing. Data protection still requires data security. But as a form of security data security has progressed to being a meta-level fundamental right. Realizing the significance of this metaright is becoming one of the cornerstones of the development and existence of the constitutional state in Europe. We have the right to expect that our fundamental rights will be realized in a network environment with the requisite data security. Similarly, in assessing the development of the constitutional state, we have every right to make the quality of the digital infrastructure the standard for the new legal culture. It is one of the basic pillars of certainty. And *certainty* is what we pursue in the constitutional state when undertaking to protect the individual.

References

1. Duff, Information Society Studies, 2000
2. Aarnio - Uusitupa (ed): Oikeusvaltio, 2002
3. Saarenpää: Law, Technology and Data Technology p 41 pp. Judicial Academy of Northern Finland Publications 4/1999
4. Saarenpää - Pöysti : (ed) Tietoturvallisuus ja laki, 1997
5. Saarenpää: Personal Data Protection and the Constitutional State p. 37 pp. In Judicial Academy of Northern Finland, Publications 3/2001.
6. Ferrajoli: Fundamental Rights p 1 pp in International Journal for the Semiotics of Law vol 14 (2001)

Legal Design and e-Government: Visualisations of Cost & Efficiency Accounting in the *wif!* e-Learning Environment of the Canton of Zurich (Switzerland)

Colette Brunschwig

University of Zurich, Department of Law, Freiestrasse 36,
8032 Zürich, Switzerland
colette.brunschwig@rwi.unizh.ch

Abstract. This paper applies Legal Design, a new field of inquiry, to discuss the form and contents of an E-Learning environment recently implemented by the Canton of Zurich (Switzerland) to enhance the training and development of public administration staff. It is argued that there is a need to visualise this environment more effectively. Working from basic notions of Legal Design and E-Government, the paper uses a set of clearly defined text visualisation rules and a multi-stage procedure adopted from visual communication to visualise one key module of the learning environment with a view to achieving a greater degree of iconicity and thus to meet established didactic and mnemotechnic criteria more successfully.

1 Legal Design and e-Government

Legal Design is still very much a new field. It deals with conceiving, creating and assessing visualisations of contents and materials that are either purely legal, or financial and economic ones, for example, that have a legal basis. Legal Design is applied in various contexts (research, teaching and practice) and shares a frame of reference with other disciplines, such as jurisprudence, economics, visual design and history of art.[1] In this paper, e-Government refers to those measures and actions public administration takes to employ modern information and communication technology, respectively the e-learning environments based on such new technologies, for the purposes of staff training and development.[2]

---

[1] On Legal Design, see Colette BRUNSCHWIG: Visualisierung von Rechtsnormen. Legal Design, Diss. Zürich 2001, Zürcher Studien zur Rechtsgeschichte, Vol. 45, ed. Marie Theres Fögen [et al.], Zürich 2001, p. 1ff.

[2] On the notion of e-learning, see, for example, Andrea BACK/Oliver BENDEL/Daniel STOLLER-SCHAI: E-Learning im Unternehmen. Grundlagen, Strategien, Methoden, Technologien, Zürich 2001, p. 28ff., and Michael KERRES: Multimediale und telemediale Lernumgebungen. Konzeption und Entwicklung, 2., completely rev. ed., München [et al.] 2001, p. 14.

R. Traunmüller and K. Lenk (Eds.): EGOV 2002, LNCS 2456, pp. 430–437, 2002.
© Springer-Verlag Berlin Heidelberg 2002

2   Starting Point

Since March 2002, the administration of the Canton of Zurich has been providing its staff with an online course on its reform of public administration (the reform bears the name *wif!*, short for „wirkungsorientierte Verwaltungsführung", i.e. ‚effect-oriented public administration'; cf. http://e-learning.wif.zh.ch; visited on 21 March 2002).[3] The learning environment consists of two modules: the first explains the principles of effect-oriented administration, while the second considers the Canton's steering instruments in action. This paper focuses on one of the submodules of the second module: Submodule 2e „Cost & Efficiency Accounting" (Kosten- und Leistungsrechnung or KLR in German; referred to hereafter as CEA). As regards the steering instruments, the principal learning objectives are that learners know how the various instruments are deployed and that they understand the instruments' mode of operation in the various phases of the controlling process. In terms of teaching methodology, the steering instruments are explained by way of reference to the Yearbook of Statistics of the Canton of Zurich, which serves as a model.

How has the *wif!* e-learning interface been designed? First, the *wif!*-logo appears at the top of the screen on the left. Beneath it, there is a vertical image bar, featuring a separate image for each submodule. The title bar „Wirkungsorientierte Verwaltungsführung" (i.e. ‚effect-oriented public administration') and the crest of the Canton of Zurich have been positioned to the right of the vertical image bar. Beneath the title, there is a navigation bar that learners can use to help them decide whether they wish to access the learning modules, any background information or the *wif!*-glossary, or whether they intend to order printed materials, mail an online competition entry, request help or call up links to online services of the Canton of Zurich. The central space on the site's screen features the syllabus and learning materials in multimedia format. Finally, there are additional navigation functions at the foot of the screen, enabling learners to move around in the learning environment.

The *wif!* e-learning environment contains quite a large number of different visualisations, such as charts, tables, diagrammes, images and animations. By contrast, Submodule 2e-1, which deals with what CEA is and for what purpose it is required by a local government office, and Submodule 2e-3, which considers the differences between CEA and general (financial) accounting, only feature a single image each in the vertical image bar to the left of the learning environment at the center of the screen: the fingers of a human hand holding the stem of a ring with a soap bubble sitting on it. Even at a second glance, this image has nothing to do with CEA. Therefore, in terms of didactics, the image is unnecessary: „Employing illustrations, animations and video sequences must not be an end in itself and should not distract [the

---

[3] In order to save space, I shall not present any screenshots of the *wif!* e-learning environment. I would like to refer the reader to the above mentioned Internet address. I would like to thank lic. rer. publ. HSG Sandra Vetsch, co-project leader of the *wif!* staff, for providing me with the information I needed to write this paper.

user] from the essential – the contents – (which is however often the case). Employing visuals must always help achieve pedagogic goals."[4]

3   Objective

The purpose of this paper is to visualise CEA.

4   Learner Features and Need Analysis

According to the introductory page of *wif!* e-learning, the learning environment and its materials are addressed to the staff of the public administration of the Canton of Zurich. Given the disparate nature of the members of this target group, an enormous effort would be required to identify what kind of learners these are and what their actual learning needs are.[5] It is beyond the scope of this paper to collect empirical data on these needs, particularly in view of visualising CEA. This means that the following visualisations have been conceived without taking the needs of all the various learners into either specific or detailed account.

5   Visualising Cost & Efficiency Accounting

5.1   Visualising Text As a Multi-stage Procedure

Visual communication has developed a multi-stage procedure to visualise text. The first stage involves identifying the topic of the text that is going to be visualised (receptive understanding). Then, the topic of the source text has to be processed (productive-creative procedure) through a heuristic phase, a conceptual phase and a production phase. I have discussed this procedure elsewhere in the context of visualising legal norms[6] and shall apply it here to conceive visualisations of CEA, respectively of what CEA is, not least since this has a legal basis in the public law of the Canton of Zurich.

5.2   Conceiving the Visualisation

Identifying the Topic: What Cost & Efficiency Accounting Is

In this phase, I shall set out to understand the text I intend to visualise by referring to the relevant literature. Thus, this phase involves consulting specialist texts on the topic treated in learning module 2e-1 (what is Cost & Efficiency Accounting).

---

[4] Egon DICK: Multimediale Lernprogramme und telematische Lernarrangements. Einführung in die didaktische Gestaltung, Nürnberg 2000, p. 91.

[5] On determining the addressees of an e-learning environment, see KERRES, see footnote 2, p. 52 and 135ff., and Ludwig J. ISSING: Instruktions-Design für Multimedia, in: Information und Lernen mit Multimedia und Internet. Lehrbuch für Studium und Praxis, ed. Ludwig J. Issing and Paul Klimsa, 3., completely rev. ed., Weinheim 2002, p. 159f.

[6] See BRUNSCHWIG, see footnote 1, p. 80ff. and 217ff.

Processing the Topic: What Cost & Efficiency Accounting Is

Heuristic Phase. The purpose of the heuristic phase is to collect materials well-suited to working out what CEA is. Due to the at times fairly abstract *wif!*-learning text, this involves spelling out its contents and meaning. On the one hand, this involves finding text material specifically on CEA; on the other hand, it means looking out for images that are somehow connected with CEA and that could be used as models for visualising what CEA is.

Textual concretisations of CEA. Although various cantonal decrees mention CEA (cf. Budget Act § 18 I, Fiscal Administration Ordinance § 16 and Global Budget Ordinance § 19), no further details are given. The legal regulations merely specify the purpose and preconditions of CEA, who is responsible for it and how CEA should be conducted.

„Cost and Efficiency Accounting consists of three main elements or principal areas: cost type accounting, cost centre accounting and cost unit accounting."[7] Cost type accounting concerns the costs incurred by particular services.[8] Cost centre accounting explains where costs are incurred.[9] In cost unit accounting, costs are attributed to the various services, respectively cost bearers, such as services, products etc.[10] In connection with CEA, NADIG uses visual metaphors, such as „**cash-river**",[11] „**value-drain**" and „**value-inflow**".[12]

Visual interfaces of CEA. Basically, the materials in learning module 2 are illustrated with activities of the Office of Statistics of the Canton of Zurich, in particular of the Yearbook of Statistics. This is also the case where CEA is explained. The Cost Accounting Sheet (hereafter CAS) in learning module 2e-2 visualises what CEA is in the form of a table by drawing on the Yearbook of Statistics.[13] The CAS breaks down the cost types and their amounts (cost type accounting) that arise in the context of the Yearbook of Statistics (book and CD-ROM format). Further, the CAS attributes these costs to the cost bearer, i.e. the Yearbook of Statistics of the Canton of Zurich (cost unit accounting). It also gives details of which costs have been incurred by which cost centre within the Office of Statistics of the Canton of Zurich. Finally, it details revenues from Yearbook sales, which in turn indicates the extent of cost recovery.

NADIG and SCHELLENBERG present charts or graphic overviews to visualise CEA.[14] To the best of my knowledge, there are no illustrations in the relevant literature that visualise Cost & Efficiency Accounting.

---

[7] Aldo C. SCHELLENBERG: Rechnungswesen. Grundlagen, Zusammenhänge, Interpretationen, 3., rev. and exp. ed., Zürich 2000, p. 267.

[8] Cf. Linard NADIG: Kostenrechnung als Führungsinstrument. Grundlagen, Zürich 2000, p. 26, and SCHELLENBERG, see footnote 7, p. 267.

[9] Cf. NADIG, see footnote 8, p. 26f., and SCHELLENBERG, see footnote 7, p. 267f.

[10] Cf. NADIG, see footnote 8, p. 27, and SCHELLENBERG, see footnote 7, p. 267f.

[11] NADIG, see footnote 8, p. 13.

[12] NADIG, see footnote 8, p. 18.

[13] On the features of tables, see Steffen-Peter BALLSTAEDT: Wissensvermittlung. Die Gestaltung von Lernmaterial, Weinheim 1997, p. 137ff.

[14] Cf. NADIG, see footnote 8, p. 28, Ill. 1/12, and SCHELLENBERG, see footnote 7, p. 268, Ill. 65; on charts, see BALLSTAEDT, see footnote 13, p. 107ff.

Conceptual phase. This phase involves selecting representative concrete illustrations of the text material that is going to be visualised from the materials (i.e. texts and visuals) collected in the heuristic phase. The materials that have been found are then ordered and attention is paid to identifying those texts and images that are typical of CEA.

From the textual material collected, I have chosen the visual metaphors „cash-river", „value-drain" and „value-inflow". As regards the visual sources of what CEA is, it is worth noting that the tabular CAS, contained in Submodule 2e-2, presents cost types and allocates them to the Yearbook of Statistics in book and CD-ROM format respectively. Besides, it is important that the CAS specifies which costs encumber which cost centre within the Office of Statistics. Moreover, it is quite decisive that the table and charts keep cost type accounting, cost centre accounting and cost unit accounting separate. Given the modest amount of text and visual material chosen, it is unneccessary to order it any further.

Production phase. During this phase, details have to be worked out and rules for visualising text have to be observed. As regards working out details, the concrete text material that has been found and selected is visualised, i.e. taken into account in designing the visualisation. Besides, the visual material that has been found and selected is used inasfar as it fits the subject matter of what CEA is. Finally, I shall also lay open the rules for visualising text that I have observed in designing the visualisation of CEA.

Before carrying out these steps, I should like to describe the visualisation of what CEA is. Essentially, these are sequences of moving images (animation). Within the scope of this paper, I shall limit myself to one animation.

Animation of the external costs of the Yearbook of Statistics in book form.

- Scene 1: Shows a printing press running at the printer's office of the Cantonal Stationery and Resources Office (KDMZ). The press is printing the Year*book* of Statistics. The letters „KDMZ" can be seen, indicating that the printing press is located at the KDMZ.
- Scene 2: Picture of the building where the Office of Statistics is located. The house is marked with the sign „Statistisches Amt des Kantons Zürich" (Office of Statistics of the Canton of Zurich). A river filled with banknotes and coins is flowing out of the building entrance to the right towards the open entrance of a building where the KDMZ (Stationery and Resources Office) is located. This building is clearly marked with the sign „KDMZ" and the sum of „CHF 80'880" (Swiss Francs) can be read in the „cash river".
- Scene 3: Shows a member of staff of Mendelin AG (private limited company) dispatching a sales promotion letter to Yearbook customers at a post office. Apart from a fictitious address, the text on the front of the letter reads „Yearbook Customer".
- Scene 4: Shows another cash river flowing out of the entrance of the Office of Statistics to the right straight towards the open entrance of Mendelin AG. The building is marked with a sign that reads „Mendelin AG" and the sum of „CHF 800" (Swiss Francs) is written in the cash river.

- Scene 5: Shows a post office worker of „Swiss Post" handing a list of potential customers to a member of staff of the Office of Statistics. The exchange takes place in a room at the Office of Statistics, marked as such.
- Scene 6: Shows a cash river flowing out of the entrance of the Office of Statistics to the right towards the entrance of a „Swiss Post" building. The sum of „CHF 1565" (Swiss Francs) is written in the cash river.

These details show that the metaphor of a river of cash – or „cash flow" as it is often called – has been integrated into designing the animations. Further, the animation will also consider concrete contents of the tabular CAS. More specifically, this means that some of these contents will be transferred directly into the animation, i.e. without designing any further images, whereas others require visual representation. Thus, the animation refers only indirectly to Submodule 2e-1. What remains to be done is to establish which text visualisation rules apply to the visualisation conceived so far. Given the limited scope of this paper, I shall only consider Scenes 1 - 4 of the animation by way of example.

Text Visualisation Rule with reference to Scenes 1 and 2: The rule of visual semantic repetition says that the designer repeats the contents and meaning of the text that is going to be visualised (source text) in the visualisation (target image).[15] In the CAS, it says, „KDMZ, Druckkosten Jahrbuch [...] CHF 80'880", i.e. ‚Stationery and Resources Office, Yearbook Printing Costs [...] 80'880 CHF (Swiss Francs)'. What this implies is that the printing office at the Stationery and Resources Office of the Canton of Zurich (KDMZ) prints the Yearbook and that printing costs amount to CHF 80'880. Scenes 1 and 2 provide a visual repetition of this fact. It follows that the text visualisation rule of visual semantic repetition has been observed. This rule also applies to the cash river flowing from the entrance of the Office of Statistics to the open entrance of the Stationery and Resources Office, since the visual metaphor of the cash river, which is significant for CEA, is repeated visually.

Text visualisation rules with reference to Scenes 3 and 4: If the rule of visual association is applied, this means that the source text is visualised by related image contents in that this association is accounted for either by experience, knowledge or meaning.[16] Even if the target image has different contents than the semantics of the source text, it is nonetheless related in terms of experience, knowledge or meaning to the semantics of the source text. According to the rule of experience-based visual association, visual representation occurs so that the components of the target image are related in time and/or space with those resulting either directly or indirectly from the textual source.[17] On the CAS, we read „Mendelin AG, Werbemailing [...] CHF 1'600", i.e. ‚Mendelin AG, Dispatch of Promotion Materials [...] CHF 1'600'. What this means is that Mendelin AG takes charge of dispatching promotion materials to Yearbook customers and that mailing costs for the book and CD-ROM version amount to CHF 1'600. Scene 3 (showing a member of staff of Mendelin AG dis-

---

[15] Cf. Werner GAEDE: Vom Wort zum Bild. Kreativ-Methoden der Visualisierung, 2., improved ed., München 1992, p. 92 and 101.

[16] Cf. GAEDE, see footnote 15, p. 34, 91 and 103.

[17] Cf. GAEDE, see footnote 15, p. 94f., 100 and 103.

patching sales promotion letters to Yearbook customers at a post office) is related to the contents of CAS in spatial terms in that CAS implies that the dispatch of promotion materials takes place at a post office and nowhere else. Experience shows that a member of the mailing company is responsible for doing this. It follows that the text visualisation rule of experience-based visual association has been observed. Further, the rule of visual repetition applies again with regard to the cash river flowing from the entrance of the Office of Statistics to the entrance of the building where Mendelin AG has ist premises (see explanation above). As the visualisation refers to the book version of the Yearbook of Statistics, I have reduced the costs to CHF 800, whereas I have considered the other half of the costs in the corresponding visualisation of the CD-ROM version (not covered in this paper); thus the two visualisations represent a sum total CHF 1'600.

6   Findings

The animation discussed above and that has been conceived with a view to establishing what CEA is, is an illustration. Its degree of iconicity is substantially greater than that of the tabular used in the *wif!* e-learning environment.[18] As such, the animation is much more concrete and graphic than the other, tabular mode of representation, and thus meets one of the criteria, among others, for assessing the didactic[19] and mnemotechnic[20] value of presentations and learning materials far more persuasively. It needs to be added that the weakness of the visualisation presented here is that it has been conceived separately from real learners' needs and that producing it requires a great deal of expense and labour.[21] In producing the animation, importance has to be attached that is does not run too fast and that it can be stopped, winded forward, rewinded and repeated. Like this, there is no danger that learners could not grasp the sequence of moving images.[22] Should producing the animation lead to the expenditure of too much effort, „the presentation of a sequence of single images, respectively illustrations"[23] would be possible, too.

---

[18] On the abstractness of diagrams compared to realistic images, cf. Wolfgang SCHNOTZ: Wissenserwerb mit Texten, Bildern und Diagrammen, in: Information und Lernen mit Multimedia und Internet. Lehrbuch für Studium und Praxis, ed. Ludwig J. Issing and Paul Klimsa, 3., completely rev. ed. Weinheim 2002, p. 66.

[19] On the didactic principle of vividness, cf. Beate BRUNS/Petra GAJEWSKI: Multimediales Lernen im Netz. Leitfaden für Entscheider und Planer, 3., completely rev. ed. Berlin [et al.] 2002, p. 22.

[20] Cf. Bernd WEIDENMANN: Multicodierung und Multimodalität im Lernprozess, in: Information und Lernen mit Multimedia und Internet. Lehrbuch für Studium und Praxis, ed. Ludwig J. Issing und Paul Klimsa, 3., completely rev. ed. Weinheim 2002, p. 52 and 61.

[21] Cf. ISSING, see footnote 5, p. 164.

[22] Cf. KERRES, see footnote 2, p. 229, and Bernd WEIDENMANN: Abbilder in Multimediaanwendungen, in: Information und Lernen mit Multimedia und Internet. Lehrbuch für Studium und Praxis, ed. Ludwig J. Issing und Paul Klimsa, 3., completely rev. ed. Weinheim 2002, p. 95.

[23] KERRES, see footnote 2, p. 181.

References

1. Andrea BACK/Oliver BENDEL/Daniel STOLLER-SCHAI: E-Learning im Unternehmen. Grundlagen, Strategien, Methoden, Technologien, Zürich 2001
2. Steffen-Peter BALLSTAEDT: Wissensvermittlung. Die Gestaltung von Lernmaterial, Weinheim 1997
3. Colette BRUNSCHWIG: Visualisierung von Rechtsnormen. Legal Design, Diss. Zürich 2001, Zürcher Studien zur Rechtsgeschichte, Vol. 45, ed. Marie Theres Fögen [et al.], Zürich 2001
4. Beate BRUNS/Petra GAJEWSKI: Multimediales Lernen im Netz. Leitfaden für Entscheider und Planer, 3., completely rev. ed. Berlin [et al.] 2002
5. Egon DICK: Multimediale Lernprogramme und telematische Lernarrangements. Einführung in die didaktische Gestaltung, Nürnberg 2000
6. Werner GAEDE: Vom Wort zum Bild. Kreativ-Methoden der Visualisierung, 2., improved ed., München 1992
7. Ludwig J. ISSING: Instruktions-Design für Multimedia, in: Information und Lernen mit Multimedia und Internet. Lehrbuch für Studium und Praxis, ed. Ludwig J. Issing und Paul Klimsa, 3., completely rev. ed. Weinheim 2002
8. Michael KERRES: Multimediale und telemediale Lernumgebungen. Konzeption und Entwicklung, 2., completely rev. ed., München [u.a.] 2001
9. Linard NADIG: Kostenrechnung als Führungsinstrument. Grundlagen, Zürich 2000
10. Aldo C. SCHELLENBERG: Rechnungswesen. Grundlagen, Zusammenhänge, Interpretationen, 3., rev. and exp. ed., Zürich 2000
11. Wolfgang SCHNOTZ: Wissenserwerb mit Texten, Bildern und Diagrammen, in: Information und Lernen mit Multimedia und Internet. Lehrbuch für Studium und Praxis, ed. Ludwig J. Issing und Paul Klimsa, 3., completely rev. ed. Weinheim 2002
12. Bernd WEIDENMANN: Multicodierung und Multimodalität im Lernprozess, in: Information und Lernen mit Multimedia und Internet. Lehrbuch für Studium und Praxis, ed. Ludwig J. Issing und Paul Klimsa, 3., completely rev. ed. Weinheim 2002
13. ID.: Abbilder in Multimediaanwendungen, in: Information und Lernen mit Multimedia und Internet. Lehrbuch für Studium und Praxis, ed. Ludwig J. Issing und Paul Klimsa, 3., completely rev. ed. Weinheim 2002

The First Steps of e-Governance in Lithuania: From Theory to Practice

Arūnas Augustinaitis and Rimantas Petrauskas

Law university of Lithuania
Ateities 20, LT-2057 Vilnius, Lithuania
{Araugust,rpetraus}@ltu.lt

Abstract. The article provides theoretical analysis of e-governance steps in Lithuania, based on comparative analysis of conceptual documents, strategies and plans, also draws conclusions on shortcommings thereof, and practical analysis of the issue. Existing two theoretical concepts of e-government in Lithuania are examined against selected theoretical and methodological criteria, while practical evaluation of e-governance steps in Lithuania is measured by experimental research of internet communication quality between citizens and government in Lithuania. In particular the research is targeted at websites of different public institutions and their feedback to citizens.

1   Introduction

"The new situation requires active work to meet new challenges in the field of high technologies, the internet and new communication services and to prepare a successful merge between the old and new economies" (Erki Liikanen). Seeking to improve the quality of public administration services provided to the citizens and increase their communication via the internet, a comprehensive research on different public institutions websites and their feedback to citizens in different countries has been carried out in the EU since 1999 [1]. Unfortunately, scientific research on similar topic in Lithuania has just started and has not preceded the drafting and enactment of the e-government foundation documents in Lithuania, the Lithuanian concept of e-government and the strategy of developing an information society in Lithuania. This article attempts a theoretical comparison of these two conceptual documents, explains different paradigms of e-government, which have resulted in the dualism of concepts of e-government, examines them against selected theoretical and methodological criteria. Practical measurement of e-government concepts is carried through the research of the quality of the Internet connection between the public institutions and the community.

2   Theoretical Perception of e-Government in Lithuania

The topic of e-government is one of the most discussed questions both at national and at European Union (EU) levels recently [2]. It is connected both with the paradigm-

R. Traunmüller and K. Lenk (Eds.): EGOV 2002, LNCS 2456, pp. 438–445, 2002.
© Springer-Verlag Berlin Heidelberg 2002

atic change of the governance institutions in the global context and with the impact of networked environment on the shift of living. The conception of e-government is not a simple or just a technological problem as it may seem from the first sight. This conception embodies several key structural complexes of analysis: *methodological, political, sociological and strategic*.

From the methodological point of view the most important thing is to gauge the general perception of e-government and the role it gets in the sphere of national governance and public administration. The role of e-government may be defined as:

- *Complementary factor*, which in one way or another organically complements the existing formal bureaucratic structure of governance.
- *Subsidiary factor*, which plays only subsidiary attendant role.
- *Synoptical vision* or a "big brother" - type political abstraction, the extreme case of which is understood as an absolute control of governance by technical means.
- *Substitute orientation*, when e-government increasingly replaces conventional forms of public administration and gives them new functional potential. It leads to absolute conceptions related to the technological rule of on-line democracy.
- *Management factor* as development of technologies and methods of public administration.
- *Social factor* as public service, which may be defined as a formula "on-line welfare for everyone".

From the political point of view e-government is some kind of democracy mechanism, because it is related with public information processes, dissemination of civic information and development of the public sphere. In a democratic society the shapes of e-government acquire value content through the mechanisms of communication. *In fact democracy coincides with the principles of organization and management of communication processes*. From the social point of view the development of e-government is related to the tendencies of knowledge society. This relation reflects the impact of e-government on the so-called *access to lifestyle*. It is the entrenchment and further development of networked life changes, where the key factors are the impact of Internet, problems of digital divide, governance attitudes towards global influences, political will and state information programmers.

Some kind of generalization may be made following the research in this field in Lithuania, but there are only the initial results, which require further systematic and deeper research projects [3].

It is not sufficient to make researches of direct social aspects of e-government. Civic information model is not conceptualized yet, nor is the model of the development of public sphere and there is a clear *deficit of social ideas, self-government and community conceptions* penetrating into the complicated process of entrenchment of legal environment and democratic values. Researches show a vast need for social ideas and social changes, but conservative stereotypes of public administration do not allow apply elements of e-government in the daily practice effectively and fast enough [4].

Another question is the practical realization of presumptions of e-government research. Is the idea of e-government acceptable to governance and citizens? In the first case one should bear in mind that in public administration there *prevail conservative governance methods*, based on systematic conception and the dominance of formal positions. At present, there are *two alternative e-government concep-*

tions prepared, which notwithstanding some from the first sight common formulation have different conceptual bases (Table 1).

Table 1. Comparison between e-government conceptions of Lithuania

Feature	Centralized model	E-government for citizens
Global context	-	+
Environment of changes	-	+
Integration into knowledge society	-	+
Education priorities	-	+
Accessability to information	+	+
Quality and development of citizens' interactions	-	+
Cohesion of citizens and governance institutions	+	+
Level of democratization	-	+
Quality of decisions	+	+
Optimization of public administration	+	+
Transparency of public administration	+	+
Individualization of public services	-	+
Governance depersonalization	-	+
Unification of public services	+	-
Departmentalism	+	-
Public spirit	+	+
Eurointegration	-	+
Integrated systems of state registers	+	+
Integration of state communication and data transmission networks	+	+
Common system for person identification	+	+
Social data system	+	+
Changes in legal environment	+	+
Data protection	+	+
Free general access to public information	+	+
Management of information processes	+	-
Development of e-business	+	-
Computer literacy	+	+
Compatibility of e-government projects	+	-
Standardization of e-government	+	-
Open formats	+	+

The first is traditional systematic approach which emphasizes high technological centralization and standardization. It continues the soviet regime principles which where implemented in such programmes as State automated control system or State automated scientific technical information system. E-government is understood as specialized activity or even a business sphere, which is governed by departmental principle. Centralized orientation is reflected in such projects as Single Governance Internet Portal, State Administrative Information System, infrastructure for electronic signature, integration of state registers and other similar projects, which present e-government as a cohesive normative system, hierarchically and functionally covering all the state ruling and public administration. The systematic approach is more or less tried out and better understood, and it leaves less possibilities for the unexpected.

The second approach towards e-government is based on tendencies of knowledge society and knowledge economy. It emphasizes diversity, decentralization, pluralism, personalization and oneness. Cultural and social priorities are more important than bureaucratic and normative centralization. This model is oriented towards future and is not void of uncertainty and certain elements of risk. Both approaches underline the need for society networking, but each treats this phenomenon in a totally different way.

There is the third approach towards e-government – a *pragmatic one*. The main feature of it is drift and some kind of deconceptualization. It deals only with partial tasks, which are offered by life. While deep in argument over the strategic dominance at the conceptual level, natural processes of formation of e-government in different Lithuanian governance and public administration spheres are going on, such as specialized public systems "Customs Rates in the Republic of Lithuania", "Vehicle Queues at Lithuanian Boarders", statistic data about the social and economic state of country, tourist information etc.

3 Quality of Communication by Internet between Citizens and Government

3.1 Experimental Methodology

The skills of the most government institutions are not sufficient for the applications of the new principles of communication and this to a certain extent hinders the development of the knowledge society [5]. The lack of application skills of separate institutions and individuals as well as the shortage of technical possibilities in the sphere of application of modern information technologies becomes obvious when assessing their application of the Internet to receive and disseminate the information. In order to see the extent of the Internet application in the offices of central and local public authority of Lithuania, starting January 2001 the Law University of Lithuania carried out an experimental research of the internet intelligence of some of the Lithuanian government institutions. The results of the research demonstrate the quality of communication by internet between citizens and governance. This research was carried out in two stages:

Research stage 1 (Jan of 2001). The research was carried out in two directions: by analyzing the websites of the highest rank central government institutions and examining how effectively the ministries' officials use the e-mail for communication with citizens.

Research stage 2 (Feb – March of 2002). The research was expanded to all ministries and municipalities. Citizens' e-mails with the some questions were sent to e-mail gates and personally to administration officers of these institution.

3.1.1 Research Stage One
The first part of the research was conducted according to the research methodology developed by Phil Noble (Sweden) and Bernhard Lehmann (Germany) to examine the Internet intelligence of the EU states' politicians and officials [1]. This methodology

is based on the specific analysis of the politicians' websites and the feedback from the community. The subject of their research was the four categories of the government institutions: offices of the presidents or prime ministers, offices of the parliament or the government, ministries of economy, social security or education. Depending on the category of the institution a questionnaire of 9-13 questions was developed for each of the four categories of institutions and according to the answers given to the submitted questions the institution of every category was assessed in the scale of 25 points. The research carried out by the *Law University of Lithuania* analysed the internet intelligence of the Seimas (The Parliament) of the Republic of Lithuania, the Office of the President and the Ministry of Economics, as well as the Ministry of Social Security and Labour.

The generalizing assessment of the Seimas was 19 points in the 25-point scale. If compared with the EU states [1], the internet intelligence of the Seimas is quite high. The Internet intelligence of the Office of the President assessment was good enough and equaled 17 points for the scale of 25 points. The results of the Ministry of Economy and the Ministry of Social Security and Labour was comparatively low and equaled 11 points.

The subject of the second part of the research of Stage 1 was the officers of different ministries of the Republic of Lithuania. In the process of this research, e-mails were sent to ten random officials of every ministry with a simple inquiry and then the results were analyzed. In order not to arouse the officials' suspicion about the research the question was simple. Moreover, the answer to this kind of message gives information about the way the ministry officials communicate with the ordinary people, the feedback style. Taking into consideration the fact that the ministers' and vice-ministers' e-correspondence is administrated by appointed people, they were not included into the list of addresses. Computer science specialists were also excluded.

Ten of thirteen of the Lithuanian ministries were examined. They were the ministries that indicated their officials' e-mail addresses. The answers were of different type: comprehensive (indicating the exact time and place of the reception, contact telephone numbers, fax numbers etc.), short, laconic or even of question type.

Information of officials' replies from the different ministries is illustrated in percent in Fig. 1. The findings show that only 11 percent of the respondents use e-communication. 82 percent of the ministries' officials did not respond to the inquiry.

3.1.2   Research Stage Two

The research was developed in the second phase (February-March, 2002). The websites of all ministries and local authorities were analyzed and the research was carried out in order to see how the ministry and local authority officials use the e-mail for the communication with citizens. It was decided to develop the research including all of the 13 Lithuanian ministries and 60 local authorities. The quality of the communication by the Internet between citizens and governance was examined by sending e-mails via e-mail gate or to administration officers of the institutions.

3.2   Analysis of the Websites

After reviewing the information on the Swedish, German, Finnish and Spanish local authorities' websites, 24 criteria, divided into 4 groups, were chosen. These 4 groups criteria help to assess the quality of the information displayed on the websites for the society:

I. Basic information about the local authority (7 criteria).
II. Important information for the citizens (10 criteria).
III. Entertainment information (4 criteria).
IV. Globalisation of the website and technical advantages (3 criteria).

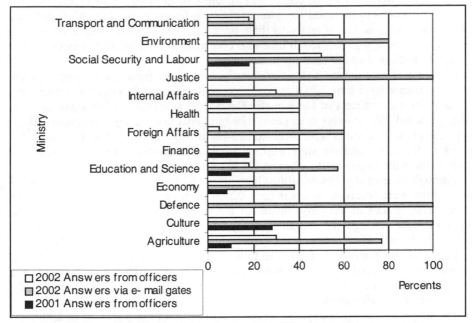

Fig. 1. Information of officials' replies from the different ministries in percentage

The ministries' websites of Belgium, Spain, Austria and Sweden were also reviewed. It should be mentioned that the ministries of the indicated countries do not have a uniform structure displaying the information on the Internet, due to the fact that the ministries are in charge of different spheres of activities. Therefore the following 14 basic criteria for the assessment of the information on ministries' websites may be distinguished: the structure of the Ministry; E-mail gate of the Ministry; personal e-mail address of the Minister; E-mail addresses of the Ministry staff; regulations of the Ministry; documents and drafts of legal acts; international relations; press releases; news; reception schedule; comments; FAQ, version in a foreign language; information on the website revision.

After reviewing the websites of the local authorities it was noted that 25 out of 60 Lithuanian local authorities still do not have their own websites designed. About 39% of the local authorities websites meet less than half of the indicated website information assessment criteria, 30% of the websites meet half of the criteria. Approximately 23% of the local authorities websites meet more than half of the criteria. Only 9 % of the analysed websites almost fully meet the criteria.

All the 13 ministries have their own websites that meet half or more of the basic criteria.

3.3   Research on the Internet Communication

In carrying out this research, typical citizens' questions were sent, thus imitating the inquiries of the citizens. 5 questions were designed for each ministry and local authority and were sent to the e-mail gate of these institutions. Also 1 universal and simple question was designed for administration officers of the ministries and local authorities and was sent directly to their personal e-mail boxes. While composing these questions, it was decided to take into consideration the list of frequently asked questions in the website of each institution.

613 e-mails with questions for the investigation of the feedback between citizens and governance were sent. 370 questions to e-mail gates and 243 questions were sent to administration officers of the ministries and local authorities. 47% answers via e-mail gates and 28% answers from personal e-mail boxes of the administration officers were received. 68 answers were received to 243 questions addressed to the personal e-mails of the administrative staff working in the Government, ministries and local authorities of the Republic of Lithuania (28 % of answers).

Through the e-mail gateway of the Government of the Republic of Lithuania 40 % of answers were received, whereas only 22 % of answers were received from the administrative staff of the Government of the Republic of Lithuania. 60 % of the answers came through the e-mail gateway of the ministries of the Republic of Lithuania (figure 1). Answers to all questions were received through the e-mail gateways of the ministries of Justice, Culture and National Defence. 33 % of the answers were received from the administrative staff of the Lithuanian ministries. The greatest number of answers (60 %) was received from the administrative staff of the Ministry of the Environment.

Through the e-mail gateways of the local authorities of the Republic of Lithuania 45 % of answers were received. Through the e-mail gateways, answers to all questions came from 6 authorities. 26 % of answers were received from the administrative staff of the local authorities.

The quality of the answers from e-mail gates was rather high. Only 10 % of the answers were not answered in essence, i.e. they referred to another person or institution. The rest of the answers were rather exhaustive and informative. The worse situation was with the answers from administration officers. About 69 % of the questions were not answered in essence i.e. they asked to indicate the position of the inquirer, to specify the details of the question or other. The rest of the answers were not very elaborate or informative.

4   Conclusions

1.   Lithuanian government makes a significant attempt to concentrate on the problems of e.government. Processes of looking for paradigmatic alternatives for two different e.government conceptions, which can be characterized on the basis of different theoretical criterions, is in progress. The first Lithuanian e-government conception is based on a centralized model of state administration. The second one is oriented towards citizens and social multiplicity (diversity).

2. The findings of the preliminary research show, that the highest Lithuanian government institutions, those of the Seimas of the Republic of Lithuania and the Office of the President, website Internet intelligence is close to the average level of the EU states.

3. The research carried out in 2002 demonstrated that the quality of the Lithuanian ministries' websites as well as the community feedback has greatly improved. Only 35 Lithuanian local authorities out of the total number of 60 have their own websites, but only one third of those satisfies the website information assessment criteria.

4. Many more replies, and of higher quality, were received by communicating via the e-mail gate of the ministries and local authorities than by sending e-mail directly to the institution officials.

References

1. Noble P., Lehmann, B.: Interactive Internet Study of EU Governments. Amsterdam - Maastricht Summer University http://www.amsu.edu/jac/ default3.htm. (2000).
2. Holmes Douglas. E-Gov. E-Business strategies for government. London: Nickolas Brealey Publishing, 330 p. (2001).
3. OSF - Lithuania. Public Policy Projects. http://www.politika.osf.lt/index.en.htm (2002).
4. Augustinaitis A. Informacijos visuomenės savivaldos tendencijos. Informacijos mokslai, 2000, (14), p.18-45.
5. Petrauskas R. Informacinių technologijų taikymas viešajame administravime. Vilnius: LTU, 65 p. (2001).

The Role of Citizen Cards in e-Government

Thomas Menzel and Peter Reichstädter

Chief Information Office Austria, Ministry of Public Services
{thomas.menzel,peter.reichstaedter}@cio.gv.at

Abstract. Citizen Cards serve as a central item in e-Government for the identification of the acting persons. Establishing the model of a Public-Private-Partnership, the Citizen Card will not be uniform, produced and issued by a public authority in Austria, but consist of diverse chip cards, issued by public or private organisations. These cards can be used for e-Commerce as well as in legally binding electronic communication with the administration. In this paper we discuss the basic requirements all these cards must meet, and present a typical e-Government session (a citizen applies to any authority using a signed XML form and web transport, and in return receives the official and legally binding decision of the public authority.)

1 Purpose of the Card

In the course of extension of functionality in e-Government-applications, direct communication between citizens and administration via Internet will be established. Because of the openness of this medium, special elements for secure identification and transaction, which are mandatory requirements for legally binding communication with public authorities, must be used in official filings with public authorities. Existing security breaches can be avoided by using electronic signatures for identification and authentication.

As also in conventional administrative proceedings, applicants must prove their identity once by showing, for instance, photo identification when filing their applications. Electronic filing requires proof of identity in an equally functional way. According to the European Signature Directive [SigDir01] and the Austrian Act on Electronic Signatures [SigAct99], electronic signatures, and in some sensitive cases secure electronic signatures[1], are required for official communication. Chip cards represent the state of the art technology for secure storage of signature creation data. If an electronic signature is used in e-Government, then the card to store the signature should be one based on "Concept Citizen Card". This concept is part of the Austrian e-Government-Strategy and covers items beyond the card itself, including specifications about issuance, applications and connected tools, like secure viewer, hash algorithm, Security Layer as interface, etc. The purpose of this concept is to develop an electronic identification card for the citizen's use on the data-highway by means of electronic signatures in a secure public-key-infrastructure.

---

[1] The Austrian Signature Act uses the term „secure electronic signature", which demands the same quality of an electronic signature as the term of the signature directive "advanced electronic signature based on a qualified certificate".

R. Traunmüller and K. Lenk (Eds.): EGOV 2002, LNCS 2456, pp. 446–455, 2002.
© Springer-Verlag Berlin Heidelberg 2002

2   Requirements for All Citizen Cards

The concept for a Citizen Card, prepared by A-SIT (in charge of examining technical standards for electronic signature products) and CIO Unit (responsible for e-Government strategy), does not expressly designate a certain card or type of card. Only minimum requirements are defined. All kind of cards which fulfil such minimum requirements can be used as a type of Citizen Card in accordance with the concept. The minimum requirements are based on legal guidelines provided in the Signature Act (as last amended) with respect to data security and the integration of the ZMR-number for accurate identification. Whereby Signature Act and Order determine here on one hand procedures and components, which are regarded as suitably for secure electronic signatures, and on the other hand confirmation bodies (A-SIT) evaluate the commercial products in this sector taking care of the fulfilment of all legal safety requirements. Citizen Cards must meet exactly the same criteria as all other cards, which are used in the framework of secure electronic signatures.

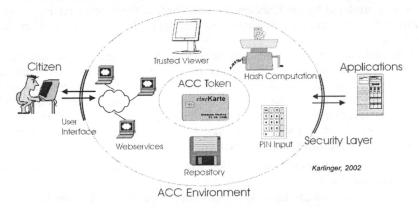

Fig. 1. Components for a Secure Signature Environment [KarG02]

Further the concept considers items that, while not obligatory are recommended since they substantially facilitate the use of the card. For example, info-boxes are recommended on the card to store user data, which is very often needed in connection with an e-Government session. The specification of a concept and the lack of a specific defined card should lead to a rapid and far spreading use of Citizen Cards, since a significant increase of chip cards with signature function, e.g. bank cards, is to be expected in the next years. All these cards should also be usable in e-Government and therefore be consistent with the Concept Citizen Card.

2.1   Secure Electronic Signature

The employment of secure electronic signatures in communication with the administration is the key element for the guarantee of security in e-Government. The Austrian Signature Act defines a secure electronic signature as an electronic signature that

a) is allocated solely to the signatory;
b) allows the signatory to be identified;
c) is created using devices under the signatory's sole control;
d) is linked with the data to which it refers to in a way which allows any subsequent change to the data to be identified; and
e) is based on a qualified certificate and is created using technical components and procedures which comply with the security requirements of the present federal law and the orders issued on the basis thereof.

A substantial security characteristic, especially important in meeting the requirement in (c), above, entails the suitable selection of the storage space for the signature creation data. The exclusive access of only the authorized signatory is best guaranteed by current technology by storing the signature creation data on a chip card with a crypto processor. In this manner, the encoding of the hash values for the signature is executed on the card, so the signature creation data never leaves the card; this is guaranteed by the architecture and operating system of the card processor. The development of other storage media, as for instance USB token or SIM cards for mobile phones, is likewise considered by the Concept Citizen Card. If these other media components feature sufficient security, they should also be applicable vis-à-vis the Concept Citizen Card.

Each card that achieves sufficiently secure electronic signatures is suitable for the Concept Citizen Card. The Austrian signature law equally applies to e-Commerce and e-Government, so that cards that are used for e-Commerce solutions and are consistent with the principles of the Concept Citizen Card in this context can likewise be used in e-Government as Citizen Cards. Although the legal basis is technology-neutrally formulated, at present only digital signatures fulfil all requirements. They are based on the following technical elements:

• Public key cryptography, also called asymmetric cryptography, represents the mathematical basis.
• Hash procedures ensure integrity of the data and permit an efficient signature creation process.
• Certificates bind the technical items like cryptographic codes to the identity of the signatory.
• Signature algorithms ensure the overall technical security of the electronic signature.

It is important to note that the legal framework permits both the application of RSA and Elliptic Curve Cryptography (ECC) as mathematical algorithm for signature creation. Therefore the exclusive use of ECC is not mandatory according to the Concept Citizen Card and RSA may also come into operation. But the longer period during which elliptic curves will be secure against brute force attacks, the relief of the chip hardware, and the shorter length of the value of a signature at the printout favor the use of ECC over RSA, in order to rollout future-oriented cards. [Certi98], [ZhIm98]

The private key, used to create a secure electronic signature, must be used exclusively for the signing process. This is necessary to ensure that only conscious declarations of will are signed. A clear separation between the process of signing and other key-based applications is essential. These card-based applications for authentication and content decryption also need access to a private key. Therefore, at least one further private key is stored on the card, which is used for these other applications in the Concept Citizen Card. Thus a separation of the signature process from all other processes, which are likewise based on asymmetrical cryptography, is ensured.

2.2 Personal Identity Link (PID)

Another feature is necessary to support a unique relationship between card and card holder. Each person living in Austria is listed in a Central Population Register and a unique, high-quality and life-constant identification number called a ZMR-number is assigned to every individual. This number is also used in the Concept Citizen Card for exact identification. During the initial registration process both public keys of the cardholder are sent to the Central Population Register. There these keys are bound to the card holder's ZMR-Number and the complete dataset is signed by the authority – this is the PID – and sent back to the registration authority, who stores it on the card for the cardholder's disposal. Therewith an exact identification of all persons living in Austria is guaranteed. The following diagram illustrates the structure of the PID:

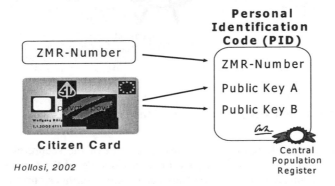

Fig. 2. Pattern of the Personal Identity Link

Due to data-protection demands, the PID in this form is not stored in the individual administrative files of a citizen. It is only used for identification purposes at the portal, but never stored in the back office databases, because storing this identification-number in databases would allow easy matching of unrelated databases (e.g. health care with income tax). Starting with the PID, a different context-dependent identification number (cd-id) is generated for each administrative procedure and only the hash-value of the cd-id is stored in the database. This procedure prevents the transmutation of cd-ids back to ZMR-numbers and PIDs in general and the conversion of different cd-ids of the same citizen. [HoLe02] The legal basis for the storage of the PID within the sphere of the card owner is outlined by the 2002 amendment to the administrative reform law 2001. The process is outlined in Figure 3.

2.3 Optional Info-Boxes

Elements needed during an e-Government session, apart from the keys and PID of the cardholder, consist mainly of rights and pointers to information. Supplements, which are needed in many different administrative procedures, as for instance for legal authorizations, birth certificates or similar documents, are at the citizen's disposal in electronic form. All these documents are also secured by an electronic signature of the issuing authority. They can be stored in info-boxes on the same card, on another card, on the PC of the card holder, or on other personal document safes anywhere on the Internet. For practical reasons the Citizen Card should supply the most essential documents directly, so that they can be used by the card holder anywhere without the need to connect to an external resource. Which information a card holder stores on his card is his decision.

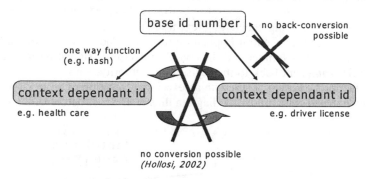

Fig. 3. System of context dependant id-numbers (cd-id)

3   Provision of Certification Services

The qualified certificate represents a substantial item of the secure electronic signature and so also of the Citizen Card, as it links the identity of the signatory to his key pairs. No public authority in Austria offers certification services currently or plans to do so in the near future. These services are provided entirely by private companies. In the context of a Public-Private Partnership, basic certification services (the generation and publication of a certificate in the directories and the execution of a revocation) are to be provided by the private certification service providers active in the market [RTR02], who offer qualified certificates. The registration process can be done by the public authority based on the principle of One-Stop-Shopping in e-Government. Figure 4 describes this process.

4   Possible Uses for Citizen Cards

The Concept Citizen Card, which must meet the minimum criteria mentioned above, should produce a scenario in which, within a short time, many different card types can also be used as a Citizen Card for communication with the administration. The trend in the context of the chip cards points towards the increasing integration of the electronic signatures on chip cards. So it is to be expected that most cards will also be signaturable, beginning in the year 2005, and therefore usable as Citizen Cards. Further the use of the secure electronic signatures is not limited on the Citizen Card to

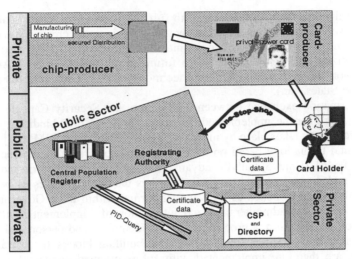

Fig. 4. Process of Citizen Card enrollment

communication with the administration. The signatures created with Citizen Cards can also be used for private legal transactions of the card holder, as a side effect increasing security in E-Commerce, E-Banking and other private sector electronic communication. An open list of possible Citizen Cards shows the versatility of the Concept Citizen Card:

- Electronic Identity Card
- Student-Service-Cards
- Social Insurance Cards (eCard)
- Cards for Chamber Members
- Service (ID) Cards for Officials in Public Administration
- Bank Cards
- Citizen Cards of other States

Fig. 5. Samples of Citizen Cards

5   Technology Independence

The versatility of the Concept Citizen Card also determines the definition of interfaces, which yield varying card types and specifications. The role of the Security Layers is seen as one of a standardized interface between various Citizen Cards and applications, which guarantees that applications can be developed without knowing or using the current technology. Therefore it is the responsibility of the Security Providers (e.g. certification service provider...) to fulfil the needs and requests of the applications, as well as to integrate a card or another safe signature creation device into the e-Government process. The interface definition of the Security Layers also allows the Concept Citizen Card to be adaptable to future developments, e.g. PDAs, mobiles, and to support the integration of foreign technologies (e.g. Citizen Cards from other EU Member States, etc.).

The information exchange between application and Security Capsule containing the Citizen Card can be implemented, for example by TCP/IP-bindings; coding the items between them takes place by means of XML [BPSM00], [TBMM01], [BiMa01] a fundamental technology based on an international specification and recommendation [ISO-10646], [Unic96] with the advantage of system- and platform independence as well as multilingual capability. XML also has the characteristics of being signaturable [EaRS02], established as a document format within the e-Government process, and used for the administration and integration of supplements, authorities or generally for the definition of procedure/process identifiers and personal identity links [Holl02], [Karl02]. Base services forming the building blocks for portals, market places, etc, are therefore implemented with future-oriented concepts and products such as Java, XML, TCP/IP or HTTP as well as open systems.

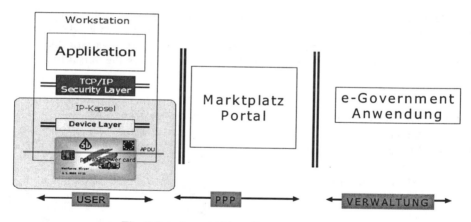

Fig. 6. Interfaces within e-Government process

6   Scenario of a Typical e-Government Session

There is no model of a 'typical' e-Government session; the session always depends on the e-Government process itself, whereby parts and phases have to be considered within a typical e-Government session. However, there are at least a few essential steps within the model that will always be found in e-Government.

The steps in the figure below are present in most e-Government procedures:

Authentification
OnLine dialog
HTML to XML conversion
Inclusion of enclosures
Creation of signed Data
Sending of signed data block

Gateway/Portal and
Back-Office process

Informing the applicant with
SMS, Mail, of delivery-process
Delivery of notifications/ records

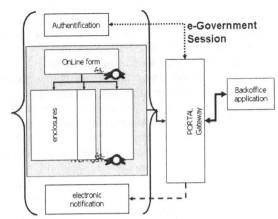

Fig. 7. Essential structure of typical e-Government session

After a (positive) authentification while using the citizen card and the functionality of the Security Capsule for determination of the authorizations, a TLS-secured session (including connection encoding) is established between the citizen and the Gateway/Portal. During the OnLine dialog, while completing HTML-forms with the WebServer application, there can be different types of help-systems activated, depending on problems and failures recognized when completing the e-forms. Some of the fields can either be completed with the information stored on the citizen card (to retrieve the stored data, a PIN may be required) or, when using personalized portals, with information from the personal data sheet. The e-form data is transmitted to the Web server, creating a specific XML-file, which file is then returned to the Web-Browser. This step is necessary due to the fact that the Security Capsule accepts only Input-data in a specific predefined XML format, before data can be signed through the capsule.

In some cases it may be necessary to electronically incorporate payment-files, mandates and other notifications/records (tax notification, criminal record) into an e-Government session. These enclosures can either be included as XML-structures or, until these enclosures are standardized and defined as XML structures, as .tif files. Finally, the XML e-form data and all necessary enclosures are put into an XML-Container, signed by using the electronic signature of the citizen stored on the citizen card together with the Security Capsule. Then they are sent (using the previously established TLS-session) to a virtual inbox of the portal, and forwarded to the relevant administrative office (i.e. interface to Backoffice applications) for further processing. The citizen receives a status message showing inbox-delivery.

Possibilities must exist for the citizen to place status queries regarding the executed e-Government procedures in order to further transparency in administrative proceedings. Further, there must be a possibility of communication with public authorities at key points with the goal of continuing incomplete administrative proceedings.

Depending on the preferences set during delivery service registration, the applicant will receive via some delivery service - after being informed through SMS, E-Mail, VoiceMail, fax – a signed notification in the form of an XML-Data Container (XML Data itself and XSL Style sheet). The XML structures underlying the notification from the public authorities must be standardized in respect of the XML-Tags that are employed (individual: family name, :given name, :occupation,...) [ReHo02],[HR-XML] in order to enable further automatic processing.

If the delivery of the notification does not occur within the TLS-session in which the e-form data is deposited, it will be necessary to consider a Delivery Service between administration units and citizens responsible for notification, logging and the delivery process itself. Such service will have to include the authentication of the administration signature within the notification XML record as well as the handling of receipts for delivery and decoding of the notification data. If the delivery process cannot take place electronically due to a technical or organizational reason, then it must take place by conventional means.

The individual steps depend on the e-Government procedure itself. In the sense of comprehensive administrative efforts however some steps toward standardized procedure definitions still need to be taken, including supplying documentation for the structures and groups of items as well as defining e-forms and electronic 'enclosures'. The free availability of these defined structures and documentation on (public) information servers (e.g. http://reference.e-government.gv.at) is absolutely necessary.

Together with the working group of the Austrian provinces in the context of the e-Austria initiative of the Austrian federal government, some progress has already been made. However, it will be important to look beyond national boundaries, and to define and standardize structures in the international context to prepare for and implement comprehensive administrative e-Government procedures throughout e-Europe.

7   Conclusion

Currently diverse new technical components, protocols and models are developed together with proven methods of administrative computer science to an efficient and constantly standardized IT system for modern administration. In addition to matching these new technologies with organizational concepts and legal bases (the signature law is a solid base for e-Government processes; other laws must be adapted for the sake of efficiency), a uniform concept must be considered for the systems used in the entire public administration in order to avoid interruption in communication between citizens and agencies or between the agencies and administrative units themselves.

The Citizen Card and the technologies of e-Government mentioned in this paper should be used to establish an efficient application of information technologies between the two ends of the interfaces - the citizen and the administration. By providing new methods of communication and information processing to support the work of the administration, computer science is able to do its part to contribute to the overall plan of administrative reform.

References

[BiMa01]: P. Biron, A. Malhotra: XML Schema Part 2: Datatypes, W3C Recommendation, 2001.

[BPSM00]: T. Bray, J. Paoli, C. M. Sperberg-McQueen, E. Maler: Extensible Markup Language (XML) 1.0 (Second Edition), W3C Recommendation, 2000.

[Certi98]: Certicom, The Elliptic Curve Cryptosystem for Smart Cards, http://www.certicom.com/resources/download/ECC_SC.pdf, 1998.

[EaRS02]]: D. Eastlake, J. Reagle und D. Solo: XML-Signature Syntax and Processing. W3C Recommendation, 2002. - http://www.w3.org/TR/2002/REC-xmldsig-core-20020212 .

[HoLe02]: Hollosi, A, Leitold, H, An Open Interface Enabling Secure e-Government - The Approach Followed with the Austrian Citizen Card, In: Proceedings of IFIP TC6 and TC11 Working Conference on Communications and Multimedia Security, CMS'2002, Portoroz, Slovenia, 26.-27. September 2002, in print.

[Holl02]: A. Hollosi, XML-Spezifikation der Personenbindung, März 2002.

[HR-XML]: HR-XML Consortium - http://www.hr-xml.org/.

[ISO-10646]: ISO/IEC 10646-1:2000. International Standard -- Information technology -- Universal Multiple-Octet Coded Character Set (UCS) -- Part 1: Architecture and Basic Multilingual Plane. UTF-8 is described in Annex R.

[Karl02]: G. Karlinger, Protokoll zur Abfrage der Personenbindung beim ZMR durch einen ZDA, März 2002.

[KarG02]: G. Karlinger, Anforderungen Bürgerkarten-Umgebung, http://www.buergerkarte.at/konzept/spezifikation/aktuell/Anforderungen%20Bürgerkarten-Umgebung.20020412.pdf, 2002.

[ReHo02]: P. Reichstädter, A. Hollosi, XML-Spezifikation der PersonData Struktur, Mai 2002.

[RTR02]: An official list of all Austrian Certification Service Providers is available at: http://www.signatur.rtr.at/en/directory/index.html.

[SigAct99]: (Österreichisches) Bundesgesetz über elektronische Signaturen (Signaturgesetz - SigG), öBGBl. I Nr. 190/1999, idF: BGBl. I Nr. 152/2001, download at: http://mailbox.univie.ac.at/thomas.menzel (follow the link "docs" in the left frame)

[SigDir01]: Directive 1999/93/EC of the European Parliament and of the Council of 13 December 1999 on a Community framework for electronic signatures, Official Journal L 013, 19/01/2000, p. 12 – 20.

[TBMM01]: H. Thompson, D. Beech, M. Maloney, N. Mendelsohn: XML Schema Part 1: Structures, W3C Recommendation, 2001.

[Unic96]: The Unicode Consortium. The Unicode Standard, Version 2.0. Reading, Mass.: Addison-Wesley Developers Press, 1996.

[ZhIm98]: Zeng, Y, Imai, H, Efficient Signcryption Schemes On Elliptic Curves. In: Papp, G/Posch, R, Global IT Security – Proceedings of the XV. IFIP World Computer Congress, OCG-Schriftenreihe, 1998, p. 75-84.

Indicators for Privacy Violation of Internet Sites

Sayeed Klewitz-Hommelsen

Fachhochschule Bonn-Rhein-Sieg, Grantham-Allee 20, D-53757 Sankt Augustin, Germany
Sayeed.klewitz-hommelsen@fh-bonn-rhein-sieg.de

Abstract. The purpose of the SAD (system for automated privacy check) project was, to scan public websites to explore, if there are any procedures or information found, which may potentially violate the privacy of the users. The criteria were to define, which information may indicate privacy relevance. To avoid misuse of the system, their where actions to be implemented. The system should support responsible persons with hints to potentially critical information.

1   Introduction

When the European Union started to harmonize the European privacy law, all member countries had to harmonize their privacy laws[1]. Germany was rather late in reforming its privacy law in 2001[2]. The following, however, will focus not on the specific regulations of privacy law, but rather on some aspects of the principles of privacy law.

Public administrations are searching for new ways to interact with their clients, especially with their citizens. This implies that the internet and particularly the world wide web are being used as a platform for new communication structures. Also the information issues of governments and administrations have changed. The trend is towards more, direct and complete information to the public. New processes are being established, new ways of participation and new partnerships are being developed. The number of web pages produced by public administrations and governments has become nearly uncountable.

Nevertheless, presenting content in the web and communicating with external partners always has an aspect of privacy protection for the involved persons. Every piece of personal data presented and every form on a web page gives or gathers information and often personal data. The idea of the "sad"-project[3] was to identify potentially

---

[1] Directive 95/46/EC (Directive 95/46) of the European Parliament and of the Council of 24 October 1995 on the protection of individuals with regard to the processing of personal data and on the free movement of such data (http://europa.eu.int/eur-lex/en/lif/dat/1995/en_395L0046.html), and the regulation (EC) No 45/2001 (regulation 45/2001) of the European parliament and of the council of 18 December 2000 on the protection of individuals with regard to the processing of personal data by the Community institutions and bodies and on the free movement of such data http://europa.eu.int/comm/internal_market/en/dataprot/news/reg45-2001en.pdf

[2] Bundesdatenschutzgesetz von 2001 (http://www.rewi.hu-berlin.de/Datenschutz/DSB/SH/material/recht/bdsg2001/bdsg2001.htm)

[3] SaD = **S**ystem zur **a**utomatisierten **D**atenschutzprüfung (system for automated privacy check), http://sad.inf.fh-bonn-rhein-sieg.de

R. Traunmüller and K. Lenk (Eds.): EGOV 2002, LNCS 2456, pp. 456–459, 2002.
© Springer-Verlag Berlin Heidelberg 2002

critical data or personal data and to give the author or the responsible person guidance in specifically checking his website for such relevant content. The idea included two separated questions to answer:

1. Would it be possible to identify such potential critical pages?
2. Would it be possible to find these pages using an automated procedure?

The objective was to test a special site, generate a report for the responsible person and explain for each item the reason why this item was marked.

2 What Should Be Identified?

European directive 95/46 on the protection of individuals with regard to the processing of personal data and on the free movement of such data and European Regulation 45/2001 for the processing of personal data by Community institutions and bodies and on the free movement of such data expect the controller of the processing of personal data to take responsibility of the following:

– The fair and lawful processing of personal data. Where not specified, this implies that explicit and legitimate purposes exist, that the subject has to consent to the processing of his data. (Art. 4 Regulation 45/2001, Art. 7 paragraph a) Regulation 45/2001)
– Informing the data subject about the purposes of the data collection (with exception in Art. 12 Paragraph 2 Regulation 45/2001)
– Providing appropriate security for the processing of personal data (Art. 17 Directive 95/46)
– The controller has to provide the data subject with information about his identity (Art. 10 and 11 Directive 95/46).

All these aspects offer at least the partial possibility of implementing automated procedures.

3 What Can Be Identified – Potential Indicators?

First, one must remember that all content-related exploration is language dependent. That means that there will be parts that must be designed to deal with different languages. For example, one can check every page on which we find a form, whether fieldnames indicate their potential use. These potential fieldnames are looked up in a table containing "suspicious" keywords[4]. This is a typical case of language dependency. As you can easily see, it is not really difficult to handle the language specific parts. So the method is language independent and can be easily reimplemented for each language you wish.

The idea, then, was to look for pages that use forms, because in that case information from the user is typically requested. Furthermore, we expect specials hints "near" the form about the privacy strategy and the intended use of the information. Will the gathered information be forwarded to any third party? When will the collected data be

---

[4] i.e. name, address, street, ZIP-Code etc.

erased? The method of scanning for the presence of such information leads via links to the area of forms and to keyword scans.

Will the transferred information be secure? Often forms request passwords without any security measures directly over the internet. So we tried to check if any kind of encryption was used during the information transfer. Secure sockets layer and transport layer security, shttp (secure hypertext transfer protocol) protocols are checked.

If the web server tries to establish cookies on the client computer, we check the timestamp of the cookie to determine if the duration makes sense with respect of the service used. As one can see, we reach here a barrier of context understanding, which a simple scan-machine can't overcome. At least the person who requested the report will be able to answer the question and so with this feedback method, a first approach to solving the problem can be found.

In addition to several information obligations, one must check for a correct address field somewhere on the web page. So the site is tested by keyword scan for typical links to the provider information[5] and similar expressions.

The possibility for the user to come into contact with the website provider is an important feature. So one action to prove this possibility was to scan the whole site for applied email addresses, rank them and send a test email to the most-used address. The scanner waits up to one week, to get an answer to his test mail. If no answer is received, it is marked in the report as a severe problem.

All of these aspects provide indicators for at least a threat on privacy protection. It does not mean, however, that the concrete page bears any danger for personal data. But in any case the offer of personal data (i.e. addresses of employees) or the input of personal data always carries the opportunity for violation of privacy within itself.

4   A Word About the Technical Details

The scanner was built on a standard Linux server. First the customer has to access a web server[6], where he enters a root web address and his email address (where the report should be sent). Then the scanning process is started. In the demonstration model we used a mix of standard tools. The entered website is completely downloaded by wget[7], a utility which retrieves files from the web. The limitation of this tool is that it doesn't interpret java script. So all links or menus generated by a java script are skipped.

After this part the result is analyzed by some shell scripts and in some special cases additional perl scripts came into use. So, for example, the entire email analysis is done by a perl script. The server was of moderate power (PC architecture, PIII 500 MHz, 128 MB, 60 GB) and nevertheless performed acceptably.

To reduce the download traffic, the system tracks whether a specific site was scanned the last time and starts a new scan only after a pause of several weeks. Should the same site be requested again, the system uses the earlier download.

---

[5] Art. 5 of the directive 2000/31/EC of the European Parliament and of the Council of 8 June 2000 on certain legal aspects of information society services, in particular electronic commerce, in the Internal Market ('Directive on electronic commerce')
http://europa.eu.int/eur-lex/en/lif/dat/2000/en_300L0031.html

[6] http://sad.inf.fh-rhein-sieg.de

[7] wget: http://www.gnu.org/software/wget/wget.html

5   Actions against Potential Misuse

One of the misuse possibilities we saw was that someone could scan the website of someone else. So we restricted the start of a check to people whose return email address had the same domain as the domain to be checked. This way, first the owner of the most-used email address gets the information on who had started a scan on his website. Furthermore, we plan to implement a kind of stop list for special well-known domains. So the famous free mail providers have thousands of users, with the correct domains in their email-addresses. Nevertheless, we don't expect most people to be entitled to request a scan for this domain. In such cases the scan starts when the operator sets the job to run level.

6   Final Remarks – Perspectives

During the project many additional aspects appeared and couldn't be integrated into the project. The most important was the idea of a kind of privacy/suspicion index. Such an index, if reasonable to define, could make web sites comparable on an abstract layer and motivate web site providers to optimize their sites under privacy aspects.

The project also showed that the service my become expensive, if many users use it. When the amount of text data we download from a site accumulates to a respectable volume, possibilities of financing such a service have to be evaluated.

A lot of technical details still haven't been solved. Java script and Java for example, can't be ignored nowadays. As special problem appeared with the typical use of several domains in conjunction with one domain. At this point, a change from www.domain.com to special.domain.com would not be followed. The implementation of more intelligence than just using a simple lookup table with keywords still seems to be challenging.

Verifiable Democracy:
A Protocol to Secure an Electronic Legislature

Yvo Desmedt[1] and Brian King[2]

[1] Florida State University,
desmedt@cs.fsu.edu
[2] Purdue School of Engineering and Technology, IUPUI campus,
briking@iupui.edu

Abstract. The manner in which a legislature votes is similar to a threshold signature scheme, and the power to sign legislation is similar to possessing shares to sign. The threshold k denotes the quorum number, the minimum number of legislators required to be present in order for legislature to be passed. Here we discuss techniques to ensure a secure electronic legislature.

1   Introduction

In democratic organizations, at a given time the number of legislators varies, while maintaining the relationship that a majority of the legislators can pass legislation. The integral part of democracy is the mechanism, which allows transfer of power from a set of n legislators to a subset of these n legislators. In a physical legislature, this mechanism is trivial to achieve. However, in an "electronic legislature", problems will arise. This paper will discuss how to achieve a verifiable democratic government using secret sharing techniques. We first proposed a verifiable democracy protocol in [1]. Subsequently, in [3], an alternate proposal was offered. However their proposal requires an administrator, which is not realistic. We will detail various problems which must be overcome, introduce the requirements for a verifiable democracy, discuss attempts at solving the verifiable democracy problem, and outline a protocol. Due to space limitations we omit technical details, such details will be included in an extended version of this paper.

The interest in developing an electronic government varies. One is that remote voting is desirable. Another is current events, in particular, terrorism attacks. As is often speculated in the media, in the September 11th terrorist attack, potential targets had included White House and/or the Capitol Building. Immediately following this attack, a second terrorism attack occurred, the mailing of anthrax spores to U.S. legislators. This attack successfully stopped the U.S. House of Representatives from meeting, and restricted the contact of the U.S. Senate. Fortunately, the House was able to meet within a few days and the contamination was limited to the Hart building, an office complex for senators. If the contamination had actually reached the congressional building, then the stoppage caused by the attack would have been much greater.

R. Traunmüller and K. Lenk (Eds.): EGOV 2002, LNCS 2456, pp. 460–463, 2002.

Many democratic organizations from legislatures to board of directors are susceptible to terrorist attacks. A solution to this problem of terrorism is to develop a distributed electronic legislature.

In an electronic government, the legislature's ability to pass or to not pass legislation should be thought of as the legislature digitally signing (with some secret key) the legislation or not signing the legislation. The power held by each legislator to vote on legislation will need to be a digital key, or rather should be thought of as a *share* of the legislature key (the one that will generate this legislature signature).

When considering an electronic government, we ask "will such a government be as representative as the physical government in place?" The danger of using a distributed electronic government is that the mechanisms for reigning-in legislative abuse is not necessarily in-place due to lack of the physical proximity of participants. The concern for the possibility of cheating among participants in an electronic government is warranted.

2   Background: Tools and Terminology

Suppose Alice wishes to send to Bob a *signature* of message M Alice applies a hash function $h()$ to M, so that $m=h(M)$. Alice sends to Bob M and $S=Signature(m)$, whereupon Bob can verify the signature. If the signature is verified then Bob accepts the message. Some examples of signature schemes that can be used in this protocol include the *RSA signature scheme* [4] and the *El Gamal signature scheme* [2].

In a *k out of n threshold sharing scheme* [5] the secret key K is shared out to n participants, so that any subset B of k participants can combine their shares and construct K while any subset of cardinality $\leq k-1$ gain no information about the K. In a *k out of n threshold signature scheme*, the signing key K is shared out to n participants so that any k participants can sign a message M. We denote S_i as participant P_i's *partial signature*. *Verifiable signature sharing* is a cryptographic sharing technique that allows a holder of document to distribute shares of the signature of the document to proxies (participants), so that the proxies can later reconstruct and sign the document (if they wish).

In an electronic voting scheme, if a voter leaves data/information such that this data allows others to verify that the voter's vote has been counted, we say that the voter has left a *receipt*. A voting scheme is said to be *receipt-free* provided that no information is left by the voter, which allows others to verify the voter's vote.

We represent the legislature by $A = \{P_1, ...,P_n\}$. We use A_t to represent the legislators present at time t, thus $A_t \subseteq A$. Here n is the size of the original legislature and n_t is the number of legislators present at time t. A *session* is a continuous period of time for which the legislators present A_t can vote on legislation and that the set of participants present remain fixed. The threshold k_t represents the threshold required to pass legislation at time t, for example in a legislature for which majority rules $k_t = \lfloor |A_t|/2 \rfloor + 1$. Every time the legislature A_t changes, some type of redistribution of shares will need to take place.

The following describes requirements for a verifiable democracy. Due to space limitation we have omitted the arguments as to why each is a requirement.

- The transfer of signature power needs to be temporary.
- The transfer of signature power needs to be done blindly.
- The participants from At, when given an opportunity to act on legislation must know that the outcome ("sign" or "not sign") is a result of their decision and not a result of bad faith on the part of the participants who had transferred them the power to sign.
- No set of participants should gain any information about a motion made during an illegal session.
- In a representative government it is the right of a citizen to know how their representative voted. In such a case we are referring to a voting scheme that requires a receipt.

3   Verifiable Democracy Protocol –
A Democratic Threshold Scheme

We start with a k out of n threshold scheme. At time t, n_t will represent the number of participants present, m_t will represent the message, and k_t will represent the dynamic threshold. A quorum exists provided $n_t \geq k$. Whenever $n_t \geq k$, we will naturally assume that $k_t \leq k$ and $k_t \leq n_t$. During the set-up, the legislature is empowered with a secret key so that any k out of n can compute the secret signing key. If $n_t \geq k$ we proceed with the protocol, if $n_t < k$ then there are not enough legislators to pass the legislation. At any time t, a message/law m_t, may be proposed. A_t represents the set of participants present at time t, $n_t = |A_t|$.

Legislative key generation. A secret key K is distributed to the n participants so that a "blinded message/law" can be signed in a k out of n threshold manner. In addition to distributing shares of K this distributor generates ancillary information. This ancillary information will be broadcasted to all, i.e. public record. The nature of the ancillary information is dependent on the verifiable sharing scheme that is used.

Blinding message. The participant P^* who proposes message m_t, blinds m_t before they present it to the legislative body A_t.

Transfer of Power -- Partial Signature Generation TPSG. As long as n_t exceeds (or equals) k the message will be considered for signing. If so k participants in A_t are chosen and they generate partial signatures for the blinded m_t.

Transfer of Power -- Partial Signature Distribution TPSD. Each of the k participants share out their partial signatures in a k_t out of n_t manner to A_t (we will refer to these k participants as *partial signature distributors*). Each participant in A_t has received k shares, whereupon they compress the k shares to one share. In addition to distributing partial signatures, the partial signature distributors will also distribute ancillary information that allows the legislative body A_t to verify the correctness of the partial signatures of the blinded m_t.

Transfer of Power -- Partial Signature Verification TPSV. The ancillary information provided in TPSD is first verified by each legislator in A_t. Upon verification the ancillary information is used by each legislator to verify the correctness of their "share of the partial signature of the blinded m_t". The verification procedure is devised so that with overwhelming probability it can be determined that a recipient has received a valid share this is achieved via a "verification and complaint" protocol. If a verification fails then a complaint will be raised, at that time a cheater has been detected, what remains is a protocol to determine whether the cheater is the "partial share distributor" or the "complainer". The consequence is that the completion of this stage with no complaints implies that the signature power for the message has been transferred to A_t such that any k_t can sign the message.

Unblind the message. The message is revealed to the legislature. Who reveals the message? P^* could. Or if one utilizes a trusted chairperson as in [3], then the trusted chairperson could reveal m_t. In [1], the protocol utilized RSA signatures and so the legislators themselves could unblind the message without the legislators revealing their partial signature of m_t.

Decision -- vote on m_t. The legislators decide whether to vote for or against m_t.

Partial Signatures Sent PSS. If any legislator wishes to vote for, the now unblinded, m_t they send their share of the partial signature of the blinded m_t.

Verification of the signature -- determining the passage of m_t. If k_t or more participants have sent their partial signatures then the message may be passed. If so, the combiner selects any k_t of the sent partial signatures and verifies the correctness of these partial signatures using the ancillary information provided within this protocol. For each one of these invalid partial signatures the combiner selects one of the remaining partial signatures sent and verifies it. If the number of valid partial signatures is less than k_t then the message m_t is automatically not passed. We have adopted a receipt-required version of the verifiable democracy protocol. The partial signature sends (PSS) together with the partial signature verification (PSV) implies k_t "valid votes". Who can play the role of the combiner? Any person, collection of people, or even the legislators can verify.

Message passed. The message is passed if a signature of m_t can be computed and there were k_t "valid votes'" sent and verified. A vote for m_t is a valid partial signature.

References

1. Y. Desmedt and B. King. Verifiable democracy. *IFIP TC6/TC11 Joint Working Conference on Communications and Multimedia Security (CMS'99)*, Kluwer Academic Publishers, 1999, pages 53-70.
2. Taher El Gamal: A Public Key Cryptosystem and a Signature Scheme Based on Discrete Logarithms. *CRYPTO 1984*, pages 10-18.
3. H. Ghodosi and J.Pieprzyk. Democratic Systems. *ACISP 2001*, pages 392-402
4. R. Rivest, A. Shamir, and L. Adelman. A method for obtaining digital signatures and public key cryptosystems. *Commun. ACM*, 21, pages 120-126, 1978.
5. A. Shamir. How to share a secret. *Commun. ACM*, 22, pages 612-613, Nov., 1979.

Arguments for a Holistic and Open Approach to Secure e-Government

Sonja Hof

University of Linz, Institute of Applied Computer Science,
Division: Business, Administration and Society, University of Linz, Austria
Eracom Technologies Switzerland AG,
sonja.hof@eracom-tech.com

Abstract. Security is widely acknowledged as one of the most important aspect for a successful e-Government implementation ([1], [2], [3]). Every month, new vulnerabilities and attacks are discovered and published. New security standards, patches, "fashionable" abbreviations pop up almost daily. How should a user trust the security and integrity of an e-Government portal in such an environment? This contribution starts with an overview of the technical as well as social aspects of security for an e-Government portal. They are introduced one by one and it is shown how to tackle them by emphasizing their drawbacks as well as their differences to the better-known world of e-Business. The paper finishes with by pointing out the requirements for the next generation e-Government platforms by proposing a simple and uniform approach security in its whole picture.

1 Introduction

Starting the implementation of a "secure" e-Government portal is an ambitious project. There are huge sets of different security aspects that have to be considered, solved and implemented ([1], [2], [4]). Some of them are still a matter of research, for which no "off-the-shelf" solutions exist. With each aspect, the paper shows their inherent problems and points out existing solutions or the lack of them. It also shows the individual drawbacks of them, as well as the drawback of this approach as it appears, if one puts all the different aspects and solutions together. Using different solutions for different aspects, instead of a complete solution that covers the full picture, hinders citizens to "identify" themselves with the system.

This paper gives an overview of the different aspects that have to be inspected for a secure e-Government portal. This is achieved by looking at the problem from two different angles: technical and social. As a third separate entity, this paper looks at the anonymity required for some processes, e.g. e-Voting. Finally, the paper concludes with a proposal for a uniform approach for a holistic view on e-Gov security.

R. Traunmüller and K. Lenk (Eds.): EGOV 2002, LNCS 2456, pp. 464–467, 2002.
© Springer-Verlag Berlin Heidelberg 2002

2   Technical Security Aspects

2.1   Authentication

Secure electronic authentication is of utmost importance for every e-Government project. The sureness to communicate with the appropriate partner, and not with an impersonator is an inherent feature of e-Government. On the technical side there are three different dimensions for authentication: how, who and when.

There exist many different techniques and approaches to securely authenticate a user electronically. They range from purely software solutions, e.g. passwords, to mixed solutions, e.g. transaction numbers, to hardware solutions, e.g. smartcards.

The second dimension to authentication defines who has to authenticate himself to the system. E-Government allows for communication between the government (one part of it) and a second party, e.g. citizen or a juristic person.

2.2   Privacy

Similar to the different authentication requirements, different tasks require different levels of privacy. Some tasks require no privacy, for others it is sufficient to ensure data privacy, and again other tasks require that even the act of using the e-Government portal has to be considered to be private.

Additionally, privacy has to be divided into short and long-term privacy. Short-term privacy has to ensure the privacy of the data while the task is active, e.g. trans-mission privacy. On the other side, long-term privacy deals with the time after the task has been completed. It deals with the data after it has been stored on the portal. Technical solutions in this area include SSL, file or disk encryption solutions.

2.3   Different Security Levels of Data

As mentioned above, it is important that data is only accessible to authorized users. Additionally to this restriction, there is a second dimension that restricts the access rights valid users actually have. This dimension defines the tasks that may be done with the data. Existing systems define multiple layers.

2.4   Data Integrity and Safety

Another important aspect of an e-Government portal is data integrity. The portal has to prevent tampering the data during transmission to and from the user, as well as to deny tampering the data once it is stored on the portal itself. The second aspect intro-duces a long-term security problem. Technical solutions in this area include digital signatures, unforgeable audits…

A further technical aspect regarding data is its safekeeping. Once transmitted and acknowledged, data should be stored in such a way that it is reconstructible that a data item was submitted, even if it lost for some reason. Technical solutions in this area include databases, transactions and digital signatures.

2.5   Quality of Service

For many tasks on an e-Government portal the quality of service (QoS) is of minor importance. Of course, responsiveness and availability are not to be neglected. However, compared to other requirement, they are of minor importance.

In contrast to that, there exist tasks, for which QoS is of utmost importance, e.g. e-Voting. In such cases it has to be guaranteed that a valid user is able to finish his task, e.g. to cast his vote. Technical solutions include firewalls or load balancer.

3   Social Security and Anonymity Aspects

E-Government portals have to cope with two different dimensions while coping with security. On the one hand there are technical security aspects as described in the previous chapter. On the other hand there are social aspects that influence security on quite a different level. Solutions to these issues are to be found on a social level.

3.1   Trust

Users tend to be careful in trusting the government to keep their data safe. Especially getting their trust for anonymous e-Voting is difficult. We think that the only way to gain their trust is to present them an open system, i.e. to allow other parties to give their opinion about the system, e.g. McKinsey or the Chaos Computer Club. In the end, all effort spent into an increased trust pays back.

3.2   Privacy and Simplicity

Privacy has, besides technical dimensions, also social aspects. Different users consider different data as private or public. The direct approach to always use private transmission is not feasible, as the necessary overhead is not always acceptable.

Simplicity seems to be an obvious aspect. Of course, an average user will never "understand" the security measurements and policies of an e-Government portal. However, he may know a "might-be" expert or try to inform himself with some literature. While the overall problems of course remain, one can decrease the "subjective" difficulty by trying to decrease the number of involved systems, e.g. a single authentication mechanism for all services.

3.3   Anonymity

E-Government portals have a peculiarity that distinguishes them from e-Business portals: Anonymity. For a set of tasks, the portal has to obey two contradicting rules: authentication and anonymity. Ready technical solutions in this area do not exist.

4   Conclusion

As shown in the previous sections, securing e-Government cannot be seen as a single process, but as a collection of hugely different tasks, systems, processes and requirements. In contrast to e-Business portals, e-Government has to go one step further regarding security. Additionally to technical security, it has to make this step also in the area of social security.

Seen from a technical side, an e-Government portal has, besides additional requirements (e.g. anonymity), to offer a higher level of security. The risks involved with a security leak are just too big. Most e-Business portals life with the fact of lost data and illegal transactions. E-Government portals just cannot accept this.

On the social side, the difference between e-Government and e-Business sites is even larger. For an e-Gov portal "trust" is of utmost importance, as a "user" has no way of circumventing the portal by not using it or by using an alternative one. Additionally, the user has also to cope with the fact, that even though he may not use the portal, the decisions made with its help, e.g. a vote, influence him directly.

Recapitulated, a secure portal does not only have to be secure, but also has to be "trusted" to be secure. To reach this goal, we propose to focus on global and open solutions to secure e-Government portals. As pointed out in the previous sections, we think that this goal can best be reached using an approach that has two distinct properties. First, it has to be open in the sense that third parties can evaluate its security. There should be as few restrictions as possible on what may be evaluated. Second, the security solution should consist of as few as possible building blocks, as each building block has to be "trusted", in order to trust the whole system.

References

1. J. von Lucke, H. Reinermann. Speyerer Definition von Electronic Government; http://foev.dhv-speyer.de/ruvii
2. Maria Wimmer and Bianca von Bredow. 2001. E-Government: Aspects of Security on Different Layers. In Proc. of the International Workshop "On the Way to Electronic Government", IEEE Computer Society Press, Los Alamitos, CA, pp. 350-358
3. Maria Wimmer and Bianca von Bredow. 2002. A Holistic Approach for Providing Security Solutions in e-Government. In Proc. of the HICSS 35. (ISBN 0-7695-1435-9)
4. Gerhard Weck. 2001. Zertifikate, Protokolle und Normen, Stolpersteine auf dem Weg zu einer PKI. In Patrick Horster (eds.). Elektronische Geschäftsprozesse, pp. 132-154 (ISBN 3-936052-00-X)

Supporting Administrative Knowledge Processes*

Witold Staniszkis

Rodan Systems S.A., Puławska 465, 02-844 Warszawa
witold.staniszkis@rodan.pl

Abstract. We present the general knowledge management model typical for central as well as local government agencies pre-requisite for successful implementation of knowledge management initiatives. We show that knowledge management systems comprising intelligent workflow management features are necessary to provide sufficient level of support for administrative knowledge processes. We conclude with a brief presentation of the ICONS project aiming at providing a KMS platform for e-government.

1 Introduction

The common fallacy of the IT side of the KM scene is focusing on the purely technological view of the field with the tendency to highlight features that are already available in advanced contend management systems. Such systems are commonly referred to as corporate portal platforms or, more to the point, as the knowledge portal platforms. From the KM perspective, as discussed in [McElroy1999], such claims may be justified only with respect to a narrow view of the field **focusing on distribution of existing knowledge throughout the organization**. The above views, called by some authors the "First Generation Knowledge Management (FGKM)" or "Supply-side KM", provide a natural link into the realm of currently used content management techniques, such as groupware, information indexing and retrieval systems, knowledge repositories, data warehousing, document management, and imaging systems. We shall briefly refer to existing content management technologies in the ensuing sections of the report to show that, within the above narrow view, the existing commercial technologies meet most of the user requirements.

With the growing maturity of the KM field the emerging opinions are that **IT support for accelerating the production of new knowledge** is a much more attractive proposition from the point of view of gaining the competitive advantage. Such focus, exemplified in stated feature requirements for so called "Second Generation Knowledge Management (SGKM)", is on enhancing the conditions in which innovation and creativity naturally occur. This does not mean that such FGKM required features as systems support for knowledge preservation and sharing are to be ignored. A host of new KM concepts, such as knowledge life cycle, knowledge processes, organizational learning and complex adaptive systems (CAS), provide the underlying conceptual base for the SGKM, thus challenging the architects of the new generation Knowledge Management Systems (KMS).

---

* This work has been supported by the project ICONS IST-2001-0324226

R. Traunmüller and K. Lenk (Eds.): EGOV 2002, LNCS 2456, pp. 468–471, 2002.
© Springer-Verlag Berlin Heidelberg 2002

Government agencies constitute a rather specific environment from the point of view of the knowledge management requirements. We attempt to identify the knowledge creation and dissemination processes in e-government, calling them administrative knowledge processes, and then, we discuss the KMS features required to provide IT support for administrative knowledge management initiatives.

2   Knowledge Management in Public Administration

The schematic view of the knowledge management cycle typical for a public administration agency is shown in figure 1. The view closely follows the Popper's three world model [Popper1971], with the bottom level corresponding to the realm of physical and abstract objects existing in the environment, the middle layer corresponding to perceptions, skills and attitudes of employees (tacit knowledge), and the upper layer representing the knowledge resources (explicit knowledge) maintained and disseminated in an organization.

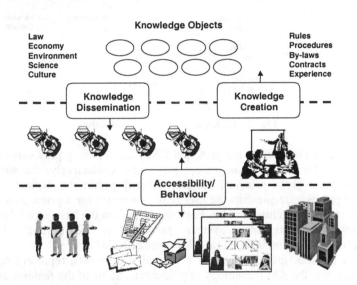

Fig. 1. The Knowledge management cycle in public administration

The accessibility and behavior are the principal characteristics of a public administration agency experienced by the environment (society, other organization), whereas the tacit knowledge [Nonaka1995] determines the actions of agency's employees. The knowledge management cycle is based on the one hand on externalization of tacit knowledge to create explicit knowledge artifacts to be accessible to others in the internalization process. Thus, although indirectly, the knowledge management cycle determines the quality of work in a public administration agency.

3   The Knowledge Management System Reference Architecture

A Knowledge Management System (KMS) is an IT platform supporting knowledge management processes taking place in an organization. A KMS reference architecture developed as the starting point of the IST ICONS project [ICONS2002] is presented in figure 2.

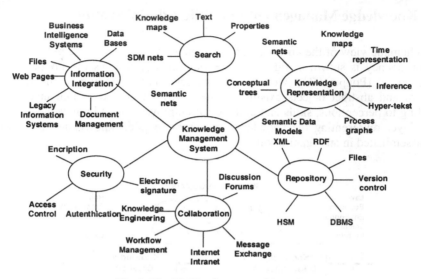

Fig. 2. The KMS reference architecture

The user functions clustered in the principal KMS features may play varying support roles within the knowledge management processes. Collectively, the sum of user requirements for a given principal feature, defined within the distinct knowledge management processes, represents the user requirement set for a given principal KMS feature. The reference architecture has been developed on the basis of a feature analysis of knowledge management systems presented in [KMForum2001, KMForum2001_D11, KMForum2001_D11a, KMForum2001_D12].

The KMS features, grouped into six principal feature sets, represent our current views pertaining to the KM technology requirements. Some of the features are already common in the advanced content management systems, referred to as the corporate portal platforms, some other are subject to the on-going KMS research efforts.

The Domain Ontology features pertain primarily to knowledge representation including the declarative knowledge representation features, such as taxonomies, conceptual trees, semantic nets, and semantic data models, as well as the procedural knowledge representation features exemplified by the process graphs. Time modeling and knowledge-based reasoning features pertain both to the declarative and the procedural knowledge representations. Hyper-text links are considered as a mechanism to create ad hoc relationships between content artifacts comprised in the repository.

Taxonomies provide means to categorize information objects stored in the content repository. Categorization classes may be arbitrary hierarchical structures grouping information objects selected by the class predicates. Class predicates are defined in the form of queries comprising information object property values or as full text queries comprising key word and/or phrases. Categorization classes are not necessarily disjoint.

Semantic networks provide means to represent binary 1:1 relationships, expressed usually as named arcs of a directed graph, where vertices are information objects belonging to any of the information object classes. Normally, the linked object classes are determined by the binary relationship semantics of the corresponding named arc. An example of a simple semantic net may be a binary relation Descendants defined as a subset of the Cartesian product of the set of Persons.

Business processes are usually represented by process graphs, typically by the Event-Condition Petri Nets or by directed graphs. Petri Net representation allows for expressing richer process semantics, in particular the pre-and post-conditions for process activities. The process specification must also be supplemented by the set of role definitions, one definition for each process activity, to enable the workflow management engine to properly assign tasks to KMS actors. The process graph representation should comprise a set of process metrics and, possibly, performance constraints and exception conditions.

References

ICONS2002, The ICONS Project Description, www.icons.rodan.pl, 2002.

KMForum2001, Weber, F., Kemp, J., Common Approaches and Standarisation in KM, EKMF Workshop on Standarisation, Brussels, June, 2001, www.knowledgeboard.com.

KMForum2001_D11, Kemp, J., Pudlatz, M., Perez, P., Ortega A.M., KM Technologies and Tools, European KM Forum, IST Project No 2000-26393, March, 2000, www.knowledgeboard.com.

KMForum2001_D11a, Kemp, J., Pudlatz, M., Perez, P., Ortega A.M., KM Terminology and Approaches, European KM Forum, IST Project No 2000-26393, March, 2000, ww.knowledgeboard.com.

KMForum2001_D12, Simpson, J., Aucland, M., Kemp, J., Pudlatz, M., Jenzowsky, S., Brederhorst, B., Toerek, E., Trends and visions in KM, European KM Forum, IST Project No 2000-26393, April, 2000, www.knowledgeboard.com.

KPMG1999, KPMG Consulting, Knowledge Management Research Report 2000, November, 1999, www.kpmg.co.uk.

McElroy1999, McElroy, M.W., Second-Generation KM, Knowledge Management, October 1999.

Nonaka1995, Nonaka, I., Takeuchi, H., The Knowledge Creating Company, Oxford University Press, 1995, New York, USA.

Popper1972, Popper, Karl R., Objective Knowledge, Oxford University Press, 1972, London, England.

IMPULSE: Interworkflow Model for e-Government

Aljosa Pasic, Sara Diez, and Jose Antonio Espinosa

SchlumbergerSema, Albarracin 25, 28037 Madrid
{aljosa.pasic,sara.diez,jose-antonio.espinosa}@sema.es

Abstract. New applications such as Virtual supply chains or E-government stress importance of application integration in heterogeneous environments. Workflow based approach is particularly interesting for public administrations, which traditionally tend to have hierarchic streamlined processes. We present a model for interworkflow complex service execution where control and co-ordination is performed by a supervisor workflow hub.

1 Workflow Tools in a Cross-Agency Process

Workflow Management Systems (WFMS) provide mechanisms to define businesses processes and to automate their execution in a way that isolates the business logic and the integration of different systems. In the past, WFMS have focused on homogeneous and centralised environment within the boundary of a single organisation. However, in newly created Internet paradigms such as B2B E-commerce or E-government, WFMS should support collaboration between different organisations. The challenge is therefore, to create new models for interaction of heterogeneous systems where internal workflow process details are abstract for other organisations and where enactment of different WF is performed in a co-ordinated way. Therefore, we will use interworkflow definition given in [1] to describe a tool that takes care of coordination of each workflow among different organisations.

On the other hand, a new concept closely related to enterprise application integration (EAI), has been built: BPI, Business Process Integration. BPI usually appears with another common acronym: BPM, Business Process Management, defined as: "The concept of shepherding work items through a multi-step process. The items are identified and tracked as they move through each step, with either specified people or applications processing the information. The process flow is determined by process logic and the applications (or processes) themselves play virtually no role in determining where the messages are sent." [2]. Our interworkflow model combines the best of a workflow tool with a powerful integration platform where CORBA and XML are the technological pillars under which the current model rests. The FORO Connector Architecture [3] defines a standard architecture for connecting the FORO Workflow Engine with external applications and legacy systems. These external applications will be invoked through common application drivers called connectors. The FORO Connector Architecture represents the implementation of the Interface 3 suggested by the Workflow Management Coalition [4] and [5].

The workflow interoperability standards define the mechanism that workflow product vendors are required to implement in order that a workflow engine makes

R. Traunmüller and K. Lenk (Eds.): EGOV 2002, LNCS 2456, pp. 472–479, 2002.

requests to another workflow engine to effect the selection, instantiation and enactment of known process definition by that other engine.

The requested workflow engine should also be able to receive back status information and the results of the enactment of the process definition. As far as possible this is to be done in a way that is "transparent to the user". Workflow enactment services provides the runtime environment in which process instantiation and activation occurs utilising one or more workflow management engines, responsible for interpreting and activating part, or all, of the process definition and interacting with the external resources necessary to process the various activities. Therefore, in our approach we distinguish the external resources such as humans or other software tools invoked to perform particular tasks from enactment services that are used to address other workflow engines.

Improving Public services (IMPULSE) is an IST programme project where two different workflow engines (Staffware and Foro-wf) are used in order to implement and validate this approach. The project will provide workflow enactment services and some of the external resources involved in sub-agencies will, in their turn, be implemented as other workflow enactment services (WES nesting).

2   e-Government: Transactions across Different Systems

The term E-Government, according to definition of World Bank [6] refers to the use by government agencies of information technologies (such as Wide Area Networks, the Internet, and mobile computing) that have the ability to transform relations with citizens, businesses, and other arms of government. Naturally, this interaction is implemented through web sites, but less obvious is the actual stage of e-government background implementation: the actual integration and transaction of procedures.

Therefore, some definitions of e-government are more restrictive, focusing on different stages, and making it the public sector equivalent of e-commerce.

In this paper we will limit our scope to what in [7] was described as Stage 2: Enabling inter-organizational and public access to information.

A number of governmental services is listed as promising candidates to be implemented as E-government services where workflow based integration forms the heart of back office system:

- claims processing and management;
- bid and proposal routing and tracking;
- handling of customer service and complaints;
- grant and scholarship award, approval, and processing; and
- human resource recruitment and hiring.

Any of above services contains processes that can be spread over various information systems, moreover they can be spread over various workflow systems. This type of services, which have been described in IMPULSE user scenarios [8] as "complex services", usually include several already available "simple" processes, which reside on the local (single agency managed) workflow (WF) or GroupWare (GW) systems. On the other hand, any service or sub-service of different governmental agencies that contains same variables (such as "change address" service), we defined as the "common services".

From the implementation point of view any service that requires the two or more WF and/or GW to be linked together, it is necessary to:

- Deal with the existing implementation problems where a physical boundary exists (Workflow and/or Groupware servers are not at the same site).
- Deal with the logical and organisational differences.

The end-to-end view describes the possible obstacles and differences that might exist between departments or organisations and offers an approach for dealing with them.

Furthermore, in our project each single agency WF/GW administration is responsible for publishing and maintaining its own service catalogue (with processes that have been solicited by interworkflow service model), depending on the policies and resources at its disposal. Global management and responsibilities of the IMPULSE interworkflow system, as well as responsibility for the design of the complex service, have also been left out of the scope of this project.

Finally, in figure 1, we present an excerpt from one of user cases in the project. The service "Apply for individual benefits" from Canary Island Government, spreads over three agencies and connects to Staffware and Foro-wf workflow management systems.

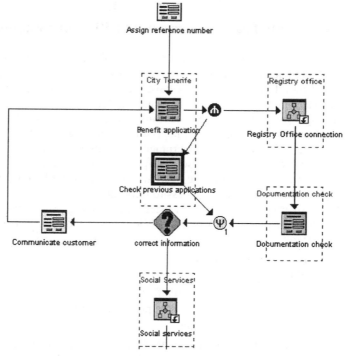

Fig. 1. A part of the complex service implemented with IMPULSE interworkflow tool

3   Interworkflow Modelling

Interworkflow process editor that we use is actually enhanced designer tool from one of the WFMS. The main difference is that sub-process, such as Registry office connection in the figure 1, can be inserted and defined as the external process that starts and ends with invocation of connector to an agency based workflow.

In [9] the modelling focuses on three main points: 1) start and end of work linkage; 2) structures for controlling the workflow and 3) exchanged data. Another model of coordination module between different WFMS is presented in [15] and has been named workflow mediation. However cooperative interworkflow models are more suitable for virtual enterprise and similar applications while public administrations are characterised by hierarchic structure of streamlined procedures. In the line with so called "one-stop-shop" idea, we propose supervising interworkflow hub, which will act as a control and co-ordination tier between single window for citizens or companies and separate agency managed workflow and groupware tools. Furthermore, control layer of this hub will also take care of connection, data translation and high level role mapping, while co-ordination layer is implemented by the existing workflow engine (with minimum modifications). This brings obvious investment savings.

IMPULSE model is presented in the figure 2.

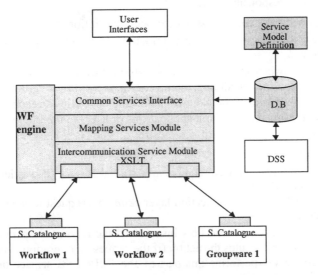

Fig. 2. IMPULSE model for interworkflow control and coordination in e-government applications

In our example from figure 1, we are making multiple data exchanges: data (citizen data request) is sent from the interworkflow to the WFMS1, from where it goes to the checking procedure that can take place in the same agency. In parallel, registry with previous applications is checked by IMPULSE interworkflow, which for this case acts as WFMS2. Except for execution of this task, IMPULSE is also responsible for running control agent that periodically checks control status and output data (if any) from the WFMS1. Finally, when both IMPULSE (WFMS1) and WFMS2 finished their sub-process, the first one will perform conditional step and will eventually pass the work to the WFMS3 (Social services).

Therefore, we are combining co-operative and hand-over types of linking.

Data exchange is based on XML-based interoperability specification [10] but it is able to interact with other XML markup vocabularies. The full overview of workflow interoperability and XML frameworks is given in [12]. In our interworkflow data model items may represent the properties of the process instance (workflow control or workflow relevant data), and/or any application related data associated with invoked applications during process enactment (application data) or they can represent process instance states. Also, various types of exceptions are defined, including temporary and fatal error types. However, the open problems remains, as also described in [13] where interworkflow approach was investigated in the healthcare domain, that each possible interleaving of steps that might occur at execution time, has to be predicted at process designing time, which leads to extreme complexity of design process.

The attractive characteristics of IMPULSE runtime server is that it will provide also runtime execution, based on existing WFMS engine. This reduces costs for companies that already have Staffware or Foro-Wf workflow engines. In IMPULSE architecture interworkflow environment is implemented as a layer around this existing engine that is responsible for:

- Interpretation of the complex service process definition
- Control of complex service process instances: creation, activation, suspension, termination...
- Navigation between process activities, which may involve sequential or parallel operations, deadline scheduling, interpretation of workflow relevant data
- Maintenance of workflow control data and workflow relevant data, passing workflow relevant data to /from applications or users to the interworkflow layer
- Supervision actions for control, administration and audit purposes

A typical sequence of events in cross-agency cases will be:

1. When the IMPULSE service requires an external service, it notifies the IMPULSE Intercommunication layer.
2. The IMPULSE Intercommunication layer *sends* the request to the Connector of the chosen WFMS or GW.
3. The Connector initiate a workflow instance (if the requested operation is a process in a workflow) or executes the action (if the requested operation is in GroupWare).
4. The domain server performs this part of the IMPULSE service on behalf of the IMPULSE server.
5. When the service part is completed, the Connector receives the results of the process. It forwards these to the Intercommunication layer.
6. The consumer Intercommunication layer hands the results to the corresponding IMPULSE workflow queue or it translate it (if necessary) and sends it to the following domain Connector.

4   Connector Agents with External WFMS/GW

IMPULSE tool accesses external applications for tasks and services execution, via standardised software applications named connector agents. These agents, positioned between the WFMS and the communication layer, can be suitable for a specific application or for a class of common applications, and they run on different platforms. Each agency WFMS must also include a connector agent (that can be easily installed) to support communication with IMPULSE server and other domains. This requires co-ordination logic that copes with invocation across different platforms and network environments, together with a means of transferring workflow relevant data in a common format or transferring it to the case instances in the individual application environments. In addition, the connector also takes care of commands for requesting the cancelling of a linkage between instances.

Connector agent also negotiates the information needed by elements, specifying name and type of the information element (variable or document). Once known the information required by IMPULSE engine, it sends the information elements and invokes external enactment services.

The basic protocol to follow is quite simple:

get_required_informationDef
set_input_information
need_moreInfo
get_returned_informationDef
get_output_information
release

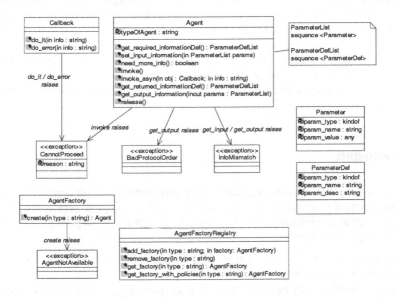

Fig. 3. IMPULSE connector agent architecture

IMPULSE connector agent architecture, presented in figure 3, is compliant to the WfMC specifications of workflow interfaces.

One of the common requirements for E-government applications is the citizens (in C2G services) or business (in B2G services) is the need for messaging of the case instance status. IMPULSE tool have several ways to interact with the case initiator: while the information is sent to public administration through web interface (Java server pages with special Tag library), the opposite way communication, that can be resolution data or interim status report, is either sent by an e-mail notification or it is placed in case initiators virtual tray. While resolution data sending is triggered by the case competition, interim status report is more dubious to design and implement because of the productivity trade-off. The proposed solution in connector agent is similar to a "courier" that periodically visits individual agencies involved in the case execution.

5   Further Work

In order to improve service design related communication between governmental agencies and their workflows which are handling data from multiple registries, a mechanism for representing semantic properties of governmental data is needed. Relationships between concepts and terms in the domain is crucial for designing complex services such those we mentioned in this paper. Some work on this subject, relying on a domain ontology that encapsulates the concepts in the small business initiation domain was done in [11].

As the public administration moves towards the competitive environment, adopting some of the business world paradigms, it is to expect a growing number of public-private partnership for performing a number of independent services that nowadays are completed by the public administration agencies.

This would introduce new needs into our model, such as the mechanism to compare different service offers and to derive and validate quality of service (QoS) parameters. The work done described in [15] could be exploited in order to meet this requirements.

6   Conclusion

From a functional point of view, the supervising interworkflow hub contain sufficient information necessary to establish, enact and control a complex service public administration exchange relationship. More specifically, the prototype we delivered in the IMPULSE project supports:

- connecting the relevant WFMS systems of different public administration agencies.
- designer interface for complex services
- Tag libraries for interface between "one-stop-shop" (single e-government window) and IMPULSE server
- the interactions needed to complete the complex services (mapping, translation...).
- the cross-agency status control.
- activities surrounding the service provision/consumption: e.g. remuneration, auditing.
- the termination mechanisms of the service enactment.

However, further research on ontology in public administrations and work on distributed knowledge representation is needed in order to design more effectively these complex cross-agency services.

References

1. K. Hiramatsu, Ken-ichi Okada, Yutaka Matsushita, H. Hayami: Interworkflow System: Coordination of Each Workflow among Multiple Organizations. CoopIS 1998: 354-363
2. ITQuadrant, Inc. ; "Glossary of terms"; November 2000.
3. Jose Antonio Espinosa, Antonio Sanz Pulido: IB-a workflow based integration approach, September 2001.
4. David Hollingsworth; "The Workflow Reference Model; Document Number TC00 – 1003"; Workflow Management Coalition; January 1995.
5. Workflow Management Coalition; "Workflow Management Application Programming Interface (Interface 2 & 3) Specification version 2.0"; Document Number WFMC-TC-1009; Workflow Management Coalition; July 1998.
6. http://www1.worldbank.org/publicsector/egov/definition.htm
7. Clay Wescot: E-government: enabling Asia-Pacific governments and citizens to do public business differently, july 2001
8. IMPULSE consortium: Improving public services requirements, April 2001
9. Interworkflow application model: WfMC-TC-2102 specification
10. Workflow standard interoperability Wf-XML binding: WfMC-TC-1023 specification
11. Nabil R. Adam, Francisco Artigas, Vijayalakshmi Atluri, Soon Ae Chun, Sue Colbert, Melania Degeratu, Adel Ebeid, Vasileios Hatzivassiloglou, Richard Holowczak, Odysseus Marcopolus, Pietro Mazzoleni, Wendy Rayner, Yelena Yesha: E-Government: Human-Centered Systems for Business Services
12. Michael zur Muehlen, Florian Klein: AFRICA: Workflow Interoperability based on XML-messages
13. Reichert M., Dadam P.: A framework for dynamic changes in WFMS,
14. Kamath Mohan and Krinith Ramamritharm: Failure handling and coordinated execution of concurrent workflows
15. Vassilis Cristophides, Richard Hull, Akhil Kumar and Jerome Simeon: Workflow mediation using VorteXML, March 2001
16. Justus Klingemann, Jurgen Wasch and Karl Aberer: Adaptive outsourcing in Cross-organisational workflows, ESPRIT project X-flow

Visualization of the Implications of a Component Based ICT Architecture for Service Provisioning

René Wagenaar and Marijn Janssen

School of Technology, Policy and Management, Delft University of Technology
Jaffalaan 5, NL-2600 GA, Delft, The Netherlands
Tel. +31 (15) 278 1140/8077, Fax. +31 15 278 3741
{Renew,MarijnJ}@tbm.tudelft.nl

Abstract. The planning and subsequent nationwide implementation of E-government service provisioning is faced with a number of challenges. Initiatives are confronted with a highly fragmented ICT-architecture that has been vertically organized around departments and with hardly any common horizontal functionality. It is anticipated that in the long run, an architecture based on generic, standardized components in the form of Web services will lead to a more flexible provision of government services over electronic channels. This paper reports on the use of a simulation environment for communicating the advantages of such a component based approach to ICT decision makers within local government.

1   Introduction

Public administrations should stay closer to citizens' every-day life, and act more proactively rather than reactively. Governmental organizations are challenged to provide more customer-oriented products and services [2,4]. Customers can be targeted through multiple channels, such as web-based, call centers and physical offices in the municipality hall. In order to exploit these channels in a coherent, efficient and effective way, the need to restructure administrative functions and processes is clearly felt to support coordination and cooperation between different departments. Legacy information systems within governmental organizations, however, often restrict the development towards new customer-oriented processes.

In general, the current situation is such that each governmental organization has developed its own information systems in rather isolation, and that for each product or service a separate information systems exists. No generic architecture is available that enables communication between front office and back-office applications, between back-office applications or with systems outside the own organization. Beneath being monolithic packages, enterprise information systems have been criticized that they often impose their own logic or business process view on an organization and lack flexibility and adaptability in today's dynamic environment [1]. Currently, pleas have been made for more open, flexible architectures constructed of relatively small

R. Traunmüller and K. Lenk (Eds.): EGOV 2002, LNCS 2456, pp. 480–483, 2002.

components, which can be configured to support a limited number of functions [3]. Governmental organizations are relatively slow in adapting such approaches as they lack sufficient insight into the pros and cons of such an approach. Gaining commitment is further complicated due to the large number of stakeholders that are involved such as politicians, process owners, information managers, ICT-departments, administrative departments etc. Some stakeholders might have a natural resistance against or not trust new initiatives; some might have too limited knowledge for decision-making or lack experience with ICT.

In this paper we will investigate the strength of simulation as a communication vehicle for the evaluation of a component based ICT architecture for E-Government service provisioning, as it allows to understand the essence of business systems, to identify opportunities for change, and to evaluate the effect of proposed changes on key performance indicators.

2   Project Description

Dutch municipalities are free to design their information architecture and to choose appropriate software vendors. Often there is no central management and departments can buy their own applications for each process. As a result, municipalities have a highly fragmented ICT-architecture, consisting of legacy systems for each product they offer. In short, there is an *interoperability* problem between applications within a municipality, but also between municipalities.

The VNG, the Dutch association of municipalities, has launched a number of initiatives to develop *communication standards*. The most important initiative in this respect is the creation of standards for the GBA, the Dutch authentic registration of data about all residents living in a particular city or village. Many applications, such as the passport or drivers license renewal request, and address changes need to use data of the GBA and consequently need to communicate with the GBA.

Apart from the interoperability problem, the Dutch software market for municipalities can be characterized by lock-in through a duopoly. Due to this dependency, municipalities have either to invest heavily for developing their own customized applications or pray that one of these companies will ship out software or customize an existing software package that will serve their needs. To counter this lock-in situation, the information managers of the Dutch cities with more than 100,000 inhabitants have joined forces in a cross-municipality Information Management Council (IMG) under supervision of the VNG. The goal of this council is to search for more open, flexible architectures that can support multiple municipalities. The IMG council initiated the *"AnalysePilot"* aimed at developing a reference architecture that should provide guidance for the development towards a component-based architecture. Such an architecture should not only bring online the 290 products currently provided in the municipalities' portfolio, but also support existing distribution channels. One of the goals of this project was to support management in their decision-making about the potential of a component-based ICT architecture.

Generally, a component based ICT-Architecture considers information systems as combinations and integrations of different software components. The manageability increases as large components can thus be constructed from smaller components. Each single component can be replaced by another component without affecting the others. Components can run from different computer platforms and interact with each other through standardized interfaces. Components can communicate with each other directly or through *middleware*. The advantages of using middleware over direct interaction are that fewer connections need to be established and maintained and that changes need only be made at one place in the overall ICT-architecture. Besides, middleware enables interoperability and portability between components; it prevents lock-in and allows users to select components based on criteria such as quality, costs, functionality etc. rather than to accept offerings from a few vendors.

Process analysis of the existing processes at a number of municipalities was used to determine which type of generic components could be suggested for the architecture. We focused on modeling the tasks for one particular process, the renewal of driver licenses.

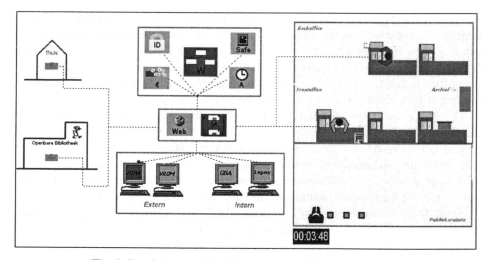

Fig. 1. Simulation model of Component based ICT architecture

3   Visualization Support

Simulation can be used as a communication instrument to stakeholders, since it allows to identify opportunities for change, and to evaluate the effect of proposed changes on key performance indicators without changing reality 5. The philosophy behind simulation is to develop a dynamic model of the problem situation, experiment with this model, and experiment with alternatives for the problem situation. In combination with visualization it can be used for creating understanding among non-experts for the impact of new process and ICT designs. As such, it was chosen by us in this project to show the Dutch municipalities which impact a generic component based ICT architecture could have on the execution of multi-channel service provisioning. Hitherto, two models were developed, an "as is" (current situation) and a "to be" situation (component-based and with multiple channels for service provisioning). Figure 1 shows a screenshot of the "to be" situation, with the new channels on the left side of the figure and the ICT-architecture based on components in the middle part of the figure. The functions of the components were simulated to ascertain that the model was independent of specific implementations. Municipalities should be able to replace any component with better ones when necessary, consequently only the functions of the components were simulated. Two new channels were added to the simulation, citizens using a computer at home and citizens using a terminal in the library, both connected to the Internet and requesting a driver's license. It shows that on a conceptual level a component-based approach is flexible and open enough to support a multi-channel approach. Future mobile channels can be supported using this approach as well.

4   Evaluation

Simulation models were presented to the information managers and management of the Dutch municipalities involved in this project. They agreed that a component-based approach created an open, flexible architecture because new components can be added to and generic components can be shared by multiple services and applications. They became also aware of the need for development of interface standards to counter potential drawbacks, such as time intensive wrapping of legacy systems, and issues concerning the maintenance of the ICT architecture.

Important *limitations* of the research are that we did not test the component based approach in practice, but limited the research to building simulation models to prove the concept. The purpose of this paper was to show that visualization based on simulation strongly supports in communicating new ICT concepts to the intended decision makers. It will be further used to assist the Dutch government in their efforts towards a more flexible, scalable and manageable ICT infrastructure.

References

1. F.J. Armour, S.H. Kaisler, and S.Y. Liu: A big-picture look at Enterprise Architectures. IEEE IT Professional, 1, 1, (1999) 35-42
2. R. van Boxtel: Actieprogamma Elektronische Overheid, The Hague, Ministerie van Binnenlandsezaken en Koninkrijksrelaties (1999) (in Dutch)
3. M. Fan, J. Stallaert, and A.B. Whinston: The adoption and design methodologies of component-based enterprise systems. European Journal of Information Systems, 9, 1, (2000) 25-35
4. K.J.L. Layne, Developing fully functional E-government: A four stage model. Government Information Quarterly, 18 (2001) 122-136
5. A.M. Law, and D.W. Kelton: Simulation Modeling and Analysis. New York: McGraw-Hill (1991)

Author Index

Lecture Notes in Computer Science

For information about Vols. 1–2371
please contact your bookseller or Springer-Verlag

Vol. 2407: A.C. Kakas, F. Sadri (Eds.), Computational Logic: Logic Programming and Beyond. Part I. XII, 678 pages. 2002. (Subseries LNAI).

Vol. 2408: A.C. Kakas, F. Sadri (Eds.), Computational Logic: Logic Programming and Beyond. Part II. XII, 628 pages. 2002. (Subseries LNAI).

Vol. 2409: D.M. Mount, C. Stein (Eds.), Algorithm Engineering and Experiments. Proceedings, 2002. VIII, 207 pages. 2002.

Vol. 2410: V.A. Carreño, C.A. Muñoz, S. Tahar (Eds.), Theorem Proving in Higher Order Logics. Proceedings, 2002. X, 349 pages. 2002.

Vol. 2412: H. Yin, N. Allinson, R. Freeman, J. Keane, S. Hubbard (Eds.), Intelligent Data Engineering and Automated Learning – IDEAL 2002. Proceedings, 2002. XV, 597 pages. 2002.

Vol. 2413: K. Kuwabara, J. Lee (Eds.), Intelligent Agents and Multi-Agent Systems. Proceedings, 2002. X, 221 pages. 2002. (Subseries LNAI).

Vol. 2414: F. Mattern, M. Naghshineh (Eds.), Pervasive Computing. Proceedings, 2002. XI, 298 pages. 2002.

Vol. 2415: J.R. Dorronsoro (Ed.), Artificial Neural Networks – ICANN 2002. Proceedings, 2002. XXVIII, 1382 pages. 2002.

Vol. 2416: S. Craw, A. Preece (Eds.), Advances in Case-Based Reasoning. Proceedings, 2002. XII, 656 pages. 2002. (Subseries LNAI).

Vol. 2417: M. Ishizuka, A. Sattar (Eds.), PRICAI 2002: Trends in Artificial Intelligence. Proceedings, 2002. XX, 623 pages. 2002. (Subseries LNAI).

Vol. 2418: D. Wells, L. Williams (Eds.), Extreme Programming and Agile Methods – XP/Agile Universe 2002. Proceedings, 2002. XII, 292 pages. 2002.

Vol. 2419: X. Meng, J. Su, Y. Wang (Eds.), Advances in Web-Age Information Management. Proceedings, 2002. XV, 446 pages. 2002.

Vol. 2420: K. Diks, W. Rytter (Eds.), Mathematical Foundations of Computer Science 2002. Proceedings, 2002. XII, 652 pages. 2002.

Vol. 2421: L. Brim, P. Jančar, M. Křetínský, A. Kučera (Eds.), CONCUR 2002 – Concurrency Theory. Proceedings, 2002. XII, 611 pages. 2002.

Vol. 2422: H. Kirchner, Ch. Ringeissen (Eds.), Algebraic Methodology and Software Technology. Proceedings, 2002. XI, 503 pages. 2002.

Vol. 2423: D. Lopresti, J. Hu, R. Kashi (Eds.), Document Analysis Systems V. Proceedings, 2002. XIII, 570 pages. 2002.

Vol. 2425: Z. Bellahsène, D. Patel, C. Rolland (Eds.), Object-Oriented Information Systems. Proceedings, 2002. XIII, 550 pages. 2002.

Vol. 2426: J.-M. Bruel, Z. Bellahsène (Eds.), Advances in Object-Oriented Information Systems.Proceedings, 2002. IX, 314 pages. 2002.

Vol. 2430: T. Elomaa, H. Mannila, H. Toivonen (Eds.), Machine Learning: ECML 2002. Proceedings, 2002. XIII, 532 pages. 2002. (Subseries LNAI).

Vol. 2431: T. Elomaa, H. Mannila, H. Toivonen (Eds.), Principles of Data Mining and Knowledge Discovery. Proceedings, 2002. XIV, 514 pages. 2002. (Subseries LNAI).

Vol. 2435: Y. Manolopoulos, P. Návrat (Eds.), Advances in Databases and Information Systems. Proceedings, 2002. XIII, 415 pages. 2002.

Vol. 2436: J. Fong, C.T. Cheung, H.V. Leong, Q. Li (Eds.), Advances in Web-Based Learning. Proceedings, 2002. XIII, 434 pages. 2002.

Vol. 2438: M. Glesner, P. Zipf, M. Renovell (Eds.), Field-Programmable Logic and Applications. Proceedings, 2002. XXII, 1187 pages. 2002.

Vol. 2439: J.J. Merelo Guervós, P. Adamidis, H.-G. Beyer, J.-L. Fernández-Villacañas, H.-P. Schwefel (Eds.), Parallel Problem Solving from Nature – PPSN VII. Proceedings, 2002. XXII, 947 pages. 2002.

Vol. 2440: J.M. Haake, J.A. Pino (Eds.), Groupware: Design, Implementation and Use. Proceedings, 2002. XII, 285 pages. 2002.

Vol. 2442: M. Yung (Ed.), Advances in Cryptology – CRYPTO 2002. Proceedings, 2002. XIV, 627 pages. 2002.

Vol. 2443: D. Scott (Ed.), Artificial Intelligence: Methodology, Systems, and Applications. Proceedings, 2002. X, 279 pages. 2002. (Subseries LNAI).

Vol. 2444: A. Buchmann, F. Casati, L. Fiege, M.-C. Hsu, M.-C. Shan (Eds.), Technologies for E-Services. Proceedings, 2002. X, 171 pages. 2002.

Vol. 2445: C. Anagnostopoulou, M. Ferrand, A. Smaill (Eds.), Music and Artificial Intelligence. Proceedings, 2002. VIII, 207 pages. 2002. (Subseries LNAI).

Vol. 2446: M. Klusch, S. Ossowski, O. Shehory (Eds.), Cooperative Information Agents VI. Proceedings, 2002. XI, 321 pages. 2002. (Subseries LNAI).

Vol. 2447: D.J. Hand, N.M. Adams, R.J. Bolton (Eds.), Pattern Detection and Discovery. Proceedings, 2002. XII, 227 pages. 2002. (Subseries LNAI).

Vol. 2448: P. Sojka, I. Kopeček, K. Pala (Eds.), Text, Speech and Dialogue. Proceedings, 2002. XII, 481 pages. 2002. (Subseries LNAI).

Vol. 2451: B. Hochet, A.J. Acosta, M.J. Bellido (Eds.), Integrated Circuit Design. Proceedings, 2002. XVI, 496 pages. 2002.

Vol. 2453: A. Hameurlain, R. Cicchetti, R. Traunmüller (Eds.), Database and Expert Systems Applications. Proceedings, 2002. XVIII, 951 pages. 2002.

Vol. 2454: Y. Kambayashi, W. Winiwarter, M. Arikawa (Eds.), Data Warehousing and Knowledge Discovery. Proceedings, 2002. XIII, 339 pages. 2002.

Vol. 2455: K. Bauknecht, A M. Tjoa, G. Quirchmayr (Eds.), E-Commerce and Web Technologies. Proceedings, 2002. XIV, 414 pages. 2002.

Vol. 2456: R. Traunmüller, K. Lenk (Eds.), Electronic Government. Proceedings, 2002. XIII, 486 pages. 2002.

Vol. 2469: W. Damm, E.-R. Olderog (Eds.), Formal Techniques in Real-Time and Fault-Tolerant Systems. Proceedings, 2002. X, 455 pages. 2002.

Vol. 2470: P. Van Hentenryck (Ed.), Principles and Practice of Constraint Programming – CP 2002. Proceedings, 2002. XVI, 794 pages. 2002.

Vol. 2483: J.D.P. Rolim, S. Vadhan (Eds.), Randomization and Approximation Techniques in Computer Science. Proceedings, 2002. VIII, 275 pages. 2002.

READINGS IN GLOBALIZATION
KEY CONCEPTS AND MAJOR DEBATES

Edited by
GEORGE RITZER
and
ZEYNEP ATALAY

WILEY-BLACKWELL

A John Wiley & Sons, Ltd., Publication

Readin

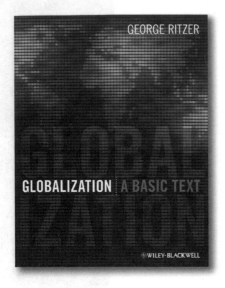

Globalization
A Basic Text

George Ritzer

This balanced introduction draws on academic and popular sources to examine the major issues and events in the history of globalization.

Globalization: A Basic Text is a substantial introductory textbook, designed to work either on its own or alongside *Readings in Globalization*. The books are cross-referenced and are both structured around the core concepts of globalization.

2009 • 608 pages • 978-1-4051-3271-8 • paperback

www.wiley.com/go/globalization

CONTENTS

INTRODUCTION TO THE BOOK

As the title makes clear, this anthology deals with globalization. We will operate with the following definition of globalization:

> Globalization is a transplanetary *process* or set of *processes* involving increasing *liquidity* and the growing multidirectional *flows* of people, objects, places, and information as well as the *structures* they encounter and create that are *barriers* to, or *expedite*, those flows.[1]

Globalization is, of course, a vast topic which cannot be covered completely in even as lengthy a volume as this one. The subtitle of this book makes clear what aspects of globalization will be dealt with here. Our first goal is to introduce the reader to at least some of the major *concepts* in the study of globalization. However, this introduction will be presented in the context of the *debates* that swirl in and around them. Indeed, the entire field of globalization studies is riddled with debates of all sorts and a secondary goal of this anthology is to introduce the reader to at least some of the major disputes in the field. These debates are important in themselves, but they also serve to clarify what we know about globalization. Furthermore, in many cases the debates also offer at least some examples of extensions to our knowledge of globalization that flow out of such debates. Such extensions are important because they illustrate that these debates are not merely exchanges of differing positions, but at times lead to advances in our understanding of globalization. All of the chapters illustrate this fact, but this is particularly the case where debates over a concept lead to new concepts. Examples of the latter include the debate over neoliberalism leading to such conceptual amplifications as neoliberalism as exception, exceptions to neoliberalism and graduated sovereignty (chapter 4); the debate about glocalization leading to the concept of grobalization (chapters 13, 14); and the debate over McDonaldization leading to the idea of glocommodification (chapter 15).

Chapter 1 of this volume stands alone and apart from the rest in the sense that it offers an overview of some of the major debates in the field. It constitutes an introduction to the volume, as well as to many of the debates to follow.

The remainder of the volume is divided into two broad parts. The first part deals with concepts, debates, and extensions in the *political economy* of globalization. As is explained further in the introduction to Part I, this heading was selected because in many cases it is difficult to clearly distinguish the political and economic aspects of globalization. The term "political economy" is rather old-fashioned; it was at one time coterminous with "economics" and related specifically to the economy of the state. However, it is now used more broadly to refer to the relationship between the state and the economy[2] and that is the way it will be deployed here. Some of the concepts (e.g. nation-state) covered in this part tend to be more political in nature, while others (e.g. neoliberalism) are more oriented to economics. Nonetheless, all deal in varying degrees with the relationship between politics and economics.

In Part II we turn to *culture* and its relationship to globalization. However, we hasten to point out that the issues covered in the two parts of the book overlap to some degree and the distinctions made are artificial, at least to a degree. All of the topics covered in Part I have cultural elements (and this is particularly true of such topics as civilizations, cosmopolitanism, McWorld, and Jihad). Further, there are certainly political and economic aspects to all the cultural concepts covered in Part II. For example, world culture encompasses the idea that the world's polities and economies have increasing cultural commonalities. However, all of the chapters in Part II deal with the issue of the degree to which it is possible to think in terms of a global culture or whether local culture inevitably retains its own distinctive character, even in the face of pressure from a globalized culture.

While the concepts covered in this volume are not exhaustive of the major ideas in the study of globalization, they are a good representation of those key ideas. The concepts covered are: *civilizations, Orientalism, colonialism, postcolonialism, neoliberalism, structural adjustment, nation-state, transnationalism, world systems, empire, network society and informationalism, world risk society, cosmopolitanism, McWorld and Jihad, creolization, hybridity, glocalization, McDonaldization,* and *world culture.* In addition, represented in the debates over these concepts are many of the major contributors to our understanding of globalization including Edward Said, Karl Polanyi, David Harvey, Manuel Castells, Samuel Huntington, Immanuel Wallerstein, Michael Hardt and Antonio Negri, Susan Strange, Linda Weiss, Leslie Sklair, William Robinson, Ulrich Beck, Benjamin Barber, Jan Nederveen Pieterse, Ulf Hannerz, Roland Robertson, George Ritzer, Malcolm Waters, and James L. Watson.

Overall, then, our belief is that the inclusion of so many major concepts and the work of so many leading contributors to the literature make this a worthy introduction to the field. However, it should be borne in mind that work on these concepts by these figures is presented in the highly dynamic context of the debates that rage about these ideas and thinkers. It is our opinion and hope that this serves to make this anthology not only highly informative, but also a dramatic and interesting introduction to the field of globalization studies.

NOTES

1 George Ritzer, *Globalization: A Basic Text.* Malden, MA: Wiley-Blackwell, 2010. For more on this definition see chapter 1 of that book.
2 See, for example, David Balaam and Michael Veseth, *Introduction to International Political Economy*, 4th edn. Englewood Cliffs, NJ: Prentice-Hall, 2007.

Introduction to Globalization Debates

The study of globalization is highly disputatious. Indeed, this entire volume is devoted to at least some of the major conceptual debates in the study of globalization. However, there are even more fundamental debates surrounding the whole issue of globalization. This first chapter of the book contains an essay by Mauro F. Guillén that examines five of the key debates in the field. While he does not include it as one of his debates, Guillén begins with the much discussed issue of just what is globalization. He reviews various definitions as well as proposing his own definition. He points out that globalization is not only a scientific concept but also an ideology with a multitude of meanings. In addition to disagreements over its definition, there is much dispute over just when globalization began.

Having in fact covered several debates in his introductory remarks, Guillén turns to what he considers the five key debates:

- Is globalization really happening?
- Does globalization produce convergence?
- Does globalization undermine the authority of the nation-state?
- Is globality different from modernity?
- Is a global culture in the making?

Guillén closes with some thoughts on what one of the fields covered in this book – sociology (others include political science, international relations, anthropology, economics, literary theory, geography) – has contributed to our understanding of globalization, as well as on the need for further research and more interdisciplinary work on the topic.

Several of the debates outlined by Guillén appear later in this book, but the highly disputatious nature of globalization is reflected in the fact that there are many other ongoing arguments in the field. Many of them appear in the following pages, but they represent only a small proportion of the large and growing number of debates in the study of globalization. While the fact of these exchanges does not promise any easy answers to the big issues in the field, it does reflect the field's enormous vibrancy.

 READING 1

Is Globalization Civilizing, Destructive or Feeble? A Critique of Five Key Debates in the Social Science Literature

Mauro F. Guillén

Introduction

Globalization is one of the most contested topics in the social sciences. Observers and theorists of globalization have variously argued that the rapid increase in cross-border economic, social, technological, and cultural exchange is civilizing, destructive, or feeble, to borrow Albert Hirschman's celebrated metaphors. Harold Levitt's "Globalization of Markets" or Kenichi Ohmae's *Borderless World* promise boundless prosperity and consumer joy as a result of globalization, i.e. the global as civilizing. In sharp contrast to this view, the historian Paul Kennedy warns in *Preparing for the Twenty-First Century* against our lack of structures to deal with a global world, while political economist Dani Rodrik rings a similar bell of alarm in *Has Globalization Gone Too Far?* concerning the increasingly free international economic and financial flows. As in the civilizing view, the destructive interpretation regards globalization as leading to convergence, albeit predicting harmful rather than beneficial consequences. Unlike the adherents to either the civilizing or the destructive views of globalization, other scholars, namely, Paul Hirst and Grahame Thompson in *Globalization in Question*, and Robert Wade in "Globalization and Its Limits", see it as a feeble process that has not yet challenged the nation-state and other fundamental features of the modern world.

In this chapter I first define globalization and its timing. Then, I review the main contributions of the various social sciences to research on globalization, with an emphasis on sociological perspectives. I organize the discussion and critique around five key debates or questions: Is globalization really happening? Does it produce convergence? Does it undermine the authority of nation-states? Is globality different from modernity? Is a global culture in the making?

What Is Globalization?

Intuitively, globalization is a process fueled by, and resulting in, increasing cross-border flows of goods, services, money, people, information, and culture. Sociologist Anthony Giddens proposes to regard globalization as a decoupling or "distanciation" between space and time, while geographer David Harvey and political scientist James Mittelman observe that globalization entails a "compression" of space and time, a shrinking of the world. Sociologist Manuel Castells emphasizes the informational aspects of the global economy when he defines it as "an economy with the capacity to work as a unit in real time on a planetary scale." In a similar vein, sociologist Gary Gereffi writes about global "commodity chains," whereby production is coordinated on a global scale. Management scholar Stephen Kobrin describes globalization as driven not by foreign trade and investment but by increasing technological scale and information flows. Political scientist Robert Gilpin defines globalization as the "increasing interdependence of national economies in trade, finance, and macroeconomic policy." Sociologist Roland Robertson argues that globalization "refers both to the compression of the world and the intensification of consciousness of the world as a whole." Also sociologist Martin Albrow defines globalization as the "diffusion of practices, values and technology that have an influence on people's lives worldwide." I propose to combine the perspectives of Robertson and Albrow, and so define globalization as a process leading to greater interdependence and mutual awareness (reflexivity) among economic, political, and social units in the world, and among actors in general.

Globalization, however, is also an ideology with multiple meanings and lineages. As Cox has observed, sometimes it appears loosely associated with neoliberalism and with technocratic solutions to economic development and reform. The term also appears linked to cross-border advocacy networks and organizations defending human rights, the environment, women's rights, or world peace. The environmental movement, in particular, has raised the banner of globalism in its struggle for a clean planet, as in its "Think Global, Act Local" slogan. Thus, globalization is often constructed as an impersonal and inevitable force in order to justify certain policies or behaviors, however praiseworthy some of them might be. In a broader historical sense, Mazlish and Robertson cogently argue that not only capitalism or advocacy movements but also Christianity, Islam, and Marxism have made global claims and harbored global pretensions. Hirsch and Fiss document that use of the term "globalization" in the press appears associated with multiple ideological frames of reference, including "financial market," "economic efficiency," "negative effect," and "culture."

The start of globalization is also a contested issue. One could argue that globalization begins with the dawn of history. The literature, however, has tended to date the start of globalization more recently in the experience of the West. At one end of the spectrum, historians have noted the importance of the first circumnavigation of the Earth in 1519–21. World-system theorists maintain that the expansion of European capitalism in the sixteenth century marks the start of globalization. Some economic historians point to the turn of the twentieth century as the heyday of international trade and investment before the convulsions of World War I and the Great Depression threw the world into spiraling protectionism. Robertson argues that globalization "took off" between 1875 and 1925 with the "time-zoning of the world and the establishment of the international dateline; the near-global adoption of the Gregorian calendar and the adjustable seven-day week; and the establishment of international telegraphic and signaling codes." Murphy recounts the history of international organizations to foster transportation and communication since 1850. Students of social movements for the abolition of slavery, woman suffrage, or the prohibition of female circumcision argue that the emergence of contemporary transnational advocacy networks can be traced back to the second half of the nineteenth century.

A third group of scholars starts the analysis of globalization at the end of World War II, with the coming of the nuclear age, the emancipation of colonies, the renewed expansion of trade and investment, and the economic rise of Northeast Asia. There is also justification for telling the story of globalization beginning with the unraveling of *pax americana* in the early 1970s or with the rise of neoliberal ideology in the late 1970s and early 1980s. In a more conceptually informed way, Kobrin distinguishes between the trade and investment linkages of nineteenth-century internationalization and the network and information ties of late twentieth-century globalization. Thus, there is no agreement as to whether it was with Magellan and Mercator, James Watt and Captain Cook, Nixon and Kissinger, or Thatcher and Reagan that globalization started or, to be more precise, that the narrative of globalization ought to begin. Lastly, it should be noted that the English term "globalization" was first used around 1960 in its world-wide sense as opposed to its much older meanings of the global as something spherical, total, or universal.

Definitions and timing aside, one of the persistent problems afflicting the study of globalization is that it is far from a uniform, irreversible, and inexorable trend. Rather, globalization is a fragmented, incomplete, discontinuous, contingent, and in many ways contradictory and puzzling process. Table 1 presents economic, financial, social, political, and bibliographical indicators of globalization. The measures are presented for the 1980–98 period not because globalization started in 1980 but rather because of data limitations. Foreign direct (excluding portfolio) investment as a percentage of GDP is 2.5 times greater today than twenty years ago – and nearly four times greater in the developing world. Trade has also grown, although not as fast as foreign investment. Financial globalization has grown fastest: foreign exchange turnover increased tenfold between 1979 and 1997 relative to world GDP, and both cross-border bank credit and assets have increased more than twofold as a percentage of world GDP.

Some key indicators of social exchange across borders are also increasing rapidly, including tourism and international telephone calls (see Table 1). International migration, though on the rise, has not reached important levels relative to world population. Also bucking the globalization trend is the growing number of nation-states – from 157 United Nations members in 1980 to 184 by 1998. And more ethnic groups than ever seem to be reasserting their identities and yearning to create their own state – Palestinians and Kurds, Basques and Catalans, Scots and Welsh, Tibetans and Kashmiris, Corsicans and Quebecois. Meanwhile, the number of international organizations has more than trebled. Among international advocacy groups, those concerned with human rights, the environment, Esperanto, women's rights, and world peace have grown fastest. And the internet has accelerated cross-border exchange during the 1990s, although less than

Table 1 Indicators of globalization, 1980–98

Indicators	1980	1985	1990	1995	1998
A. Economic					
Inward foreign direct investment stock, % world GDP	4.6	6.5	8.0	10.1	11.7[h]
Developed countries, % GDP	3.8	4.9	6.6	9.1	10.5[h]
Developing countries, % GDP	4.3	8.2	8.5	15.4	16.6[h]
Gross value added of foreign affiliates, % world GDP	—	5.2	6.4	6.3	7.8[h]
Exports of foreign affiliates, % total world exports	—	31.9	27.5	32.3	35.6
Exports + imports of goods, % world non-service GDP	72.7	68.1	76.0	87.5	92.1[h]
Developed countries, % non-service GDP	76.6	72.1	81.8	90.1	95.1[h]
Developing countries, % non-service GDP	60.9	54.6	55.0	77.3	83.2[h]
Exports + imports of goods and services, % world GDP	40.0	38.8	38.9	42.9	45.2[h]
Developed countries, % GDP	40.2	39.4	38.3	41.2	43.8[h]
Developing countries, % GDP	39.1	36.6	41.0	49.5	50.6[h]
B. Financial					
Daily currency exchange turnover, % world GDP[a]	0.7	1.3	3.8	5.6	6.8
Cross-border bank credit stock, % world GDP[b]	13.9	19.9	34.3	33.1	—
Cross-border banking assets, % world GDP[b]	13.7	19.9	28.1	28.5	—
C. Social and Political					
International tourist arrivals, % world population	3.5	6.7	8.6	9.9	—
Stock of international migrants, % world population[c]	1.5	1.8	2.0	2.2	—
International calls, minutes per million $ world GDP[d]	—	1,354	1,600	2,174	—
Internet hosts, number (thousands)[e]	—	5	617	12,881	19,459[h]
Nation-states with membership in the United Nations	157	157	159	184	184
International organizations, number	14,273[g]	24,180	26,656	41,722	48,350

Table 1 (*continued*)

Indicators	1980	1985	1990	1995	1998
D. Bibliographical					
Literature on globalization, annual entries:[f]					
Sociological Abstracts	89	142	301	1068	1009
Econlit	19	269	608	1044	924
PAIS (Politics and International Relations)	64	101	309	366	698
Historical Abstracts	69	81	103	166	157
Anthropological Literature	6	2	6	1	34
Books in Print	48	92	328	689	589

[a]Data are for 1979, 1984, 1989, 1995, and 1998.
[b]Data are for 1981, 1986, 1991, and 1995.
[c]Estimates.
[d]Excludes international calls using cellular phones or private networks.
[e]Data are for 1986, 1991, 1996, and 1997.
[f]Articles or books with the words "global" or "globalization" in the title, subject heading or abstract.
[g]1981.
[h]1997.
Sources: *World Investment Report; International Trade Statistics Yearbook; UN Statistical Yearbook*; Baldwin, R.E., Martin, P. (1999). Two waves of globalization: superficial similarities, fundamental differences. NBER Work. Pap. Ser. 6904. Cambridge, MA: Natl. Bur Econ. Res.; Tschoegl, A.E. (1998). Country and bank sources of international competitiveness: the case of the foreign exchange market. Work. Pap., Wharton School, Univ. Penn.; Vernon, R. (1998). *In the Hurricane's Eye: The Troubled Prospects of Multinational Enterprises*. Cambridge, MA: Harvard Univ. Press; Miguel Centeno, Dept. of Sociology, Princeton University; Yearbook of International Organizations; Penn Library Databases.

two or three percent of the population has access to it in most countries except the very rich ones.

It is perhaps ironic to observe that the fastest increase among the indicators included in Table 1 does not refer to globalization itself, but to the literature on globalization. As shown in Figure 1, there has been an explosion in the number of articles on globalization published in the economic, sociological, and political literatures. The number of books on globalization has also increased steeply. The historical and anthropological literatures, by contrast, have lagged behind. Among the social sciences, sociology was the first to pay attention to globalization. Sociology journals started to carry large numbers of articles on globalization during the early and mid 1970s, primarily induced by world-system theorizing. Some authors have attempted to summarize the literature, and several edited volumes have been compiled. Perhaps the most bewildering feature of the literature is not its sheer size but the remarkable diversity of authors that have contributed to it, ranging from postmodernist scholars or social theorists who rarely, if ever, engage in empirical research to number-crunching empiricists, politicians, and management consultants.

Five Key Debates

The five key debates that I identify in this chapter are not an exhaustive list of issues in the vast and rich literature on globalization. They capture, however, a broad spectrum of social, political, and cultural themes of interest to sociologists and other social scientists. [. . .] One should not assume those on the same side of the fence regarding a particular question actually agree with each other on other issues or that they approach the issue from exactly the same perspective.

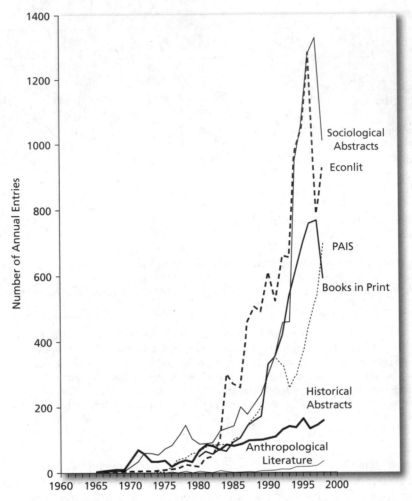

Figure 1 The literature of globalization

Is it really happening?

Most of the books and articles discussed in this chapter simply assume that the world is becoming more global, that is, more interrelated. Myriad policymakers, publicists, and academics take it as axiomatic that globalization is in fact happening without supporting their claim with data. Political economist and policymaker Robert Reich, for example, proclaims that "national economies" are disappearing and companies no longer have a nationality; only people do. There are, however, many skeptics.

Perhaps the best-documented case for the feeble argument against globalization has been made by Paul Hirst, an Oxford political scientist with ties to the Labour Party. In a recent book, Hirst and Thompson argue that the globalization trend of the last twenty years has been overstated as a process: it is not unprecedented in world history, they say, and foreign investment and trade are concentrated in the so-called triad – Western Europe, North America, and Japan. In sum, they argue that the economy is becoming more international but not more global. Political scientist Robert Wade echoes these criticisms: the volume of trade is small relative to the size of most economies; domestic investment is greater than foreign investment; multinationals locate most of their assets, owners, top managers, and R&D activities in their home countries;

and vast areas of the world have not been affected by globalization, namely, South and Central Asia, and the bulk of Africa.

The argument for the feebleness of globalization is useful in that it provides an important corrective to visions and myths of globalization assuming its inevitability and irreversibility. There are, however, two key counterarguments. Regarding the issue of the heterogeneous spread of globalization across the world, Castells correctly observes that the global economy is not meant to encompass the entire Earth. Rather, it comprises only certain segments of activity in both developed and developing countries. The second counterargument is that proponents of the feeble thesis focus almost exclusively on the economic and financial aspects of globalization to the detriment of political, social, and cultural ones. The literature offers and discusses evidence in support of political and cultural globalization that is, on the whole, quite persuasive. In addition, global warming, the AIDS pandemic, and the globalization of the media have heightened our awareness of living in an increasingly interconnected world. In sum, scholars arguing the feebleness of globalization have made a contribution in debunking certain myths and assumptions about a process that has all too often been uncritically reified. However, they are perhaps too wedded to a "monolithic" concept of globalization and oblivious to the notion that globality is a network of relationships that creates mutual awareness.

Does it produce convergence?

A second contested issue in the literature on globalization has to do with its consequences as to the convergence of societies toward a uniform pattern of economic, political, and even cultural organization. Most famously expressed in modernization theory, the spread of markets and technology is predicted to cause societies to converge from their preindustrial past, although total homogeneity is deemed unlikely. This line of thinking was advanced during the 1950s and 1960s by both economists and sociologists. Economic historians such as Jeffrey Williamson have documented convergence in income and labor markets during the nineteenth century and first decades of the twentieth. Sociologist Daniel Bell argued for a technologically driven convergence of postindustrial societies.

Further support for the convergence thesis comes from the world-society approach in sociology. In their summaries of an extensive empirical research program on the worldwide spread of educational systems and other forms of state activity, John Meyer and his associates and students argue that the expansion of rationalized state activities has acquired a momentum of its own, largely unaffected by cross-national differences in political structure or economic growth rates. Rather, the diffusion of rationalized systems follows the "exigencies of global social organization whose logic and purposes are built into almost all states." The result is that "the world as a whole shows increasing structural *similarities of form* among societies without, however, showing increasing *equalities of outcomes* among societies". Nation-states are seen as exhibiting convergent structural similarity, although there is a "decoupling between purposes and structure, intentions and results." World-society researchers argue that conformity comes both from the world-culture of rationalized modernity and from domestic groups that make claims on the state following the "consensus" over the formal acceptance of "matters such as citizen and human rights, the natural world and its scientific investigation, socioeconomic development, and education." They even present evidence to the effect that nationalism and religious fundamentalism "intensify isomorphism more than they resist it."

Social and political theorists as well as historians have elaborated a comprehensive critique of the presumed convergent consequences of globalization. Political historian Robert Cox writes that "the social and ethical content of the economy may be organized differently in various parts of the world." Historian Bruce Mazlish argues that "no *single* global history is anticipated." Sociologist Anthony Giddens adds an interesting twist when asserting that globalization "is a process of uneven development that fragments as it coordinates. [. . .] The outcome is not necessarily, or even usually, a generalized set of changes acting in a uniform direction, but consists in mutually opposed tendencies." In another book, Giddens elaborates: "Globalization has to be understood as a dialectical phenomenon, in which events at one pole of a distanciated relation often produce divergent or even contrary occurrences at another." In a similar vein, anthropologist Jonathan Friedman asserts that globalization is the product of cultural fragmentation as much as it is the result of modernist homogeneity, and that "what appears as disorganization and often real disorder is not any the less systemic and systematic."

These social and political theorists, however, have neither engaged in empirical testing of their propositions nor bothered to look for support in the existing literature. There is, though, a considerable body of empirical research backing the antithesis that globalization produces divergence and diversity or at least does not undermine national policies and institutions. Management scholar John Stopford and political economist Susan Strange document that the increasingly complex interaction between multinationals and states has produced a divergence in outcomes, while Doremus et al. show that differentiated national systems of innovation, trade, and investment remain firmly in place.

Political scientist Geoffrey Garrett has perhaps contributed the most extensive and solid body of empirical evidence, though it refers mostly to the experience of the advanced industrial democracies. He argues and demonstrates empirically that in the context of a global economy at least two paths are possible for national economic and social policymakers: adherence either to neoclassical economics or to social democratic corporatism. Garrett's analysis refutes simplistic views about convergence, proposing instead to view the balance of left–right political power and labor market institutions as the two key variables in a contingent analysis of economic performance. The best macroeconomic performance is obtained when the two variables are aligned with each other. For example, redistributive and interventionist policies combine with encompassing labor market institutions to produce macroeconomic performance in terms of growth and unemployment that matches or even surpasses the achievements of laissez-faire policies combined with weak labor market institutions. He concludes that there are "enduring cross-national differences" in economic policymaking and engagement of the global economy. In a broader study encompassing over one hundred countries during the 1985–95 period, Garrett finds no convergence in government expenditure patterns as a result of globalization. What has happened over the last decade is that many governments have pursued policies that buffer their citizens from the vagaries of global

markets and, in the presence of free capital mobility, willingly and knowingly accepted higher interest rates to keep capital at home.

Students of the varieties of capitalism, mostly political scientists, have long argued that firms and countries pursue different paths of incorporation into the global economy. Thus, German, French, Japanese, and American firms are competitive in the global economy, but rarely in the same industry and market segment. German firms excel at high-quality, engineering-intensive industries such as advanced machine tools, luxury automobiles, and specialty chemicals; French firms at large-scale technical undertakings such as high-speed trains, satellite-launching rockets, or nuclear power; Japanese firms at most categories of assembled goods, namely, household appliances, consumer electronics, and automobiles; and American firms at software, financial services, or biotechnology.

Comparative organizational sociologists have also presented qualitative and quantitative evidence to the effect that firms pursue different modes of economic action and adopt different organizational forms depending on the institutional and social structures of their home countries even as globalization increases. Moreover, they have collected data on newly industrialized countries in addition to the most advanced ones. Orrù et al. draw a number of systematic comparisons among East Asian and Western European countries, demonstrating that unique national patterns of organization not only persist over time but also contribute to the international competitiveness of firms. Guillén presents systematic case-study and quantitative evidence demonstrating that firms and labor unions in Argentina, South Korea, and Spain diverged in their patterns of behavior, organizational form, and growth even as their home countries became more integrated with the global economy during the post-World War II period.

Taken together, the empirical evidence provided by sociologists and political scientists supports well the case for diversity, or at least resilience, in cross-national patterns in the midst of globalization. It must be admitted, however, that world-society researchers also have a point, and one that is well supported by empirical evidence. The reason behind these seemingly irreconcilable empirical results might be that world-society research has made measurements at levels of analysis and abstraction higher than the finer-grained analysis of comparative sociologists and political scientists.

It should be noted that some sociologists reject the very terms of the convergence debate by arguing that globalization homogenizes without destroying the local and the particularistic. For example, Viviana Zelizer argues that "the economy [...] differentiates and proliferates culturally in much the same way as other spheres of social life do, without losing national and even international connectedness." Thus, globalization is not seen as precluding or contradicting diversity. Like Zelizer, Robertson sees the global as the "linking of localities."

Perhaps the most controversial aspect of the convergence debate has to do with the impact of globalization on inequality across and within countries. The evidence unambiguously indicates that there is today *more* inequality across countries than ten, twenty, fifty or even one hundred years ago. Stunningly, the gap in per capita income between rich and developing countries has grown five-fold between 1870 and 1990. There are, however, several noteworthy developing countries that have managed to close half or more of the gap since 1960, e.g. South Korea, Taiwan, and Ireland. Very few developing countries, though, have consistently grown faster than the most advanced ones since 1980. Thus, development levels across countries appear not to be converging as a result of globalization.

By contrast to cross-national inequality, it is not clear whether increased foreign trade and investment during the last twenty years have resulted in substantially higher wage inequality or unemployment *within* countries. Wage inequality has risen in most advanced countries during the last three decades. In a review essay, Kapstein presents several counterarguments to the claim that globalization has been the major cause of increased wage polarization, including that trade is too small a percentage of GDP to have a large impact, and that technological change is the ultimate cause of wage polarization. In agreement with Kapstein's reading of the evidence, Baldwin and Martin summarize the empirical literature as follows: "Virtually all studies find some impact of trade on the labor market in both the United States and Europe. The range of findings, however, is wide. Some find that trade accounted for virtually none of the wage gap, while others assigned 100 percent of the gap to trade. The consensus range is

perhaps 10–20 percent." As opposed to wage disparities, overall indicators of income inequality within countries have not increased during the last thirty years, and there is evidence indicating that when countries grow economically and become incorporated into the global economy poverty rates fall. Discussions and calculations of the impact of globalization on wage and income inequality within countries should take into account that while foreign trade and investment are powerful forces, domestic politics and processes still matter.

In sum, globalization does not seem to compel governments, firms, and individuals to converge in their patterns of behavior. While this may be regarded as a welcome aspect, it is important to bear in mind that increasing globalization has coincided in time with an exacerbation of income disparities across countries, and that at least part of the greater degree of income and wage inequality within countries is due to increased foreign trade and investment.

Does it undermine the authority of nation-states?

A third key issue surrounding the topic of globalization is whether this process has outgrown the governance structures of the international system of states and undermined the authority of the nation-state. For example, economist Raymond Vernon has long argued that the spread of multinational corporations creates "destructive political tensions," and that there is a "need to reestablish balance" between political and economic institutions. Historian Paul Kennedy asserts that governments are losing control, and that globalization erodes the position of labor and developing countries, and degrades the environment. "Today's global society," he writes, "confronts the task of reconciling technological change and economic integration with traditional political structures, national consciousness, social needs, institutional arrangements, and habitual ways of doing things." In a similar vein, Kobrin argues that globalization both challenges the autonomy or independent decision-making of the state and "raises questions about the meaning of sovereignty in its external sense of a system ordered in terms of mutually exclusive territoriality." And Mazlish argues that global history is an attempt to "transcend the nation-state as the focus of history."

International relations scholar Yoshikazu Sakamoto and political scientist Robert Cox concur in arguing that globalization generates problems of international governance and reduces the regulatory power of states. For Rodrik, globalization creates social and political tensions within and across nation-states. And political theorist Michael Mosher asks, "is there a successful way of reconciling the boundary transgressing character of markets with the boundary maintaining activities of nation-states?" He further notes that globalization has placed two liberal practices – the liberalism of the market and the liberalism of democratic citizenship – on a collision course, raising the dilemma of whether "moral concerns stop at the national border."

Sociologists have also joined the chorus of state doomsayers. For Waters, there is an "attenuation of the state," a rise of international organizations, and a trend toward more "fluid" international relations. McMichael also sees a decline of the state. For Albrow, "the nation-state has failed to confine sociality within its boundaries, both territorial and categorical. The sheer increase in cross-national ties, the diversification of modes of personal relationships and the multiplication of forms of social organization demonstrate the autogenic nature of the social and reveal the nation-state as just another timebound form." In a more empirically grounded way, Evans points out that globalization undermines the state because its associated neoliberal ideology is against the state and not because globalization is inextricably against the state. He further argues that the state may stage a comeback if there is a "return of the ideological pendulum," or a transformation of the state and a development of new elements of state-society synergy.

The analysis by British political economist Susan Strange is perhaps the most sophisticated articulation of the position that the international system of nation-states and the nation-state itself are coming under fire in a global world. She writes about the "declining authority of states" and preempts several possible criticisms. First, she notes that state interventionism is on the rise, although in relatively marginal matters. Second, she argues that there are more states in the world, especially after 1989, but that most of the new ones are weak and lack control. Third, she points out that the effectiveness of the East Asian state in orchestrating economic growth was only possible in a post-World War II order in which protectionism of the domestic market was acceptable and mature technologies were available. She further observes three power shifts in the global world, namely, from weak to strong states, from states to markets, and from labor markets to financial markets, with some power evaporating or dispersing.

Not surprisingly, those who argue that globalization is a feeble process also maintain that it can be easily handled by nation-states. For example, Hirst and Thompson and Wade assert that states can cope with globalization, although they have lost some freedom of action, especially concerning financial flows. Feeble proponents, however, are not alone challenging the notion that globalization undermines the nation-state.

Macrosociology has long maintained that the global arena is a "playground" for states, where they compete for economic, military, and political supremacy and survival. Thus, the world-system or the international arena, far from threatening states, actually fosters them. Neorealist international relations scholar Robert Gilpin points out that globalization reinforces the importance of domestic policies, as countries engage in regionalization, sectoral protectionism, and mercantilistic competition in response to changes in the international location of economic activities, resulting in a "mixed system," increasingly globalized and at the same time fragmented. A related, though distinct, argument against the presumed loss of state power in the wake of globalization comes from political scientist Leo Panitch. He rightly argues that "today's globalization is authored by states and is primarily about reorganizing rather than bypassing them." Moreover, as Cox observes, "power has shifted not away from the state but *within* the state, i.e. from industry or labor ministries towards economy ministries and central banks." And sociologist Seán O Riain sees states not as passive pawns but rather as "adapting, whether out of necessity or desire."

Another influential social scientist, Saskia Sassen, maintains that the state does not lose significance. Rather, there is a redefinition of the modern features of sovereignty and territoriality, a "denationalizing of national territory." Cox argues that globalization induces a transformation of the state, not its diminution. Stopford and Strange examine the new possibilities for state action in the global economy and conclude that its role has actually become magnified and more complex. According to most political scientists, therefore, the nation-state is alive and well, and the Westphalian order is unlikely to be replaced by a fragmented, medieval one. A key effect of globalization, however, has been the rise of global cities – New York, London, Miami, Singapore – whose role and stature transcend the nation-state in which they happen to be located.

Finally, the world-society view also rejects the claim that globalization undermines nation-states. Noting the expansion of state bureaucracies since World War II, Meyer et al. write that "globalization certainly poses new problems for states, but it also strengthens the world-cultural principle that nation-states are the primary actors charged with identifying and managing those problems on behalf of their societies." This argument is strikingly similar to the one offered by Panitch and Poulantzas. The modern nation-state, world-society scholars conclude, "may have less autonomy than earlier but it clearly has more to do."

The question of whether globalization undermines the authority of the nation-state comes best to life when examining the impact of globalization on the viability of the welfare state. Rodrik argues that globalization puts downward pressure on government spending for redistribution and welfare, and that the interaction of trade risk and openness calls for more welfare spending, but governments have trouble finding the money, an argument that Vernon finds persuasive. Stryker summarizes her assessment of the evidence in that globalization places limits on expansionary policies, represents a loss of power for the working class, and causes welfare state retrenchment. According to these social scientists, the challenge is "to engineer a new balance between market and society, one that will continue to unleash the creative energies of private entrepreneurship without eroding the social basis of cooperation". These arguments have become conventional wisdom among neoliberal policymakers and journalists. Gloomy, often unsubstantiated, forecasts about the inability of European welfare states to pay for generous social benefits have become commonplace since the early 1980s.

Other political scientists and sociologists, however, see things utterly differently. Political scientist Paul Pierson argues that the welfare state has declined not so much as a result of globalization but because of such indirect actions of conservative governments as reductions in the revenue base of the state and attacks on the strength of interest groups, especially labor. This is an argument that Fligstein and Gilpin endorse. Garrett empirically demonstrates the viability of social democratic corporatism even with increasing exposure to globalization in the forms of cross-border trade and capital mobility. He also proves that it is possible to win elections with redistributive and interventionist policies, and that better economic performance in terms of GDP growth and unemployment obtains, though with higher inflation than in the laissez-faire countries (United States, Britain). Garrett concludes that "big government is compatible with strong macroeconomic performance" and that markets do not dominate politics. In a direct rebuttal of Rodrik, Garrett analyzes data on more than 100 countries during the 1985–95 period to find that increasing exposure to globalization does not reduce government spending. Political scientist Evelyne Huber and sociologist John Stephens echo Garrett's conclusion that the welfare state is compatible with global capitalism, although they do admit that social democratic policies are today more constrained than in the so-called "golden age" of the 1950s and 1960s.

For Garrett, Huber, and Stephens and for Fligstein the welfare state is perfectly viable under conditions of globalization. Moreover, it may be able simultaneously to deliver social well-being and enhance national competitiveness. Thus, they reject the tradeoff that neoliberals see between welfare expenditures and economic competitiveness under conditions of globalization. In spite of the excellent, well-supported research by these authors, however, the debate in the media and among politicians throughout the world remains heavily tilted in favor of those blaming the welfare state for declining competitiveness and various social ills.

Is globality different from modernity?

Perhaps the most difficult debate surrounding globalization has to do with whether it is merely a continuation of the trend toward modernity or the beginning of a new era. On one side of the fence, Giddens argues that "modernity is inherently globalizing," and that "globalization [makes] the modes of connection between different social contexts or regions become networked across the earth's surface as a whole." This view follows directly from the concept of "disembedding" or "the lifting out" of social relations from local contexts of interaction and their restructuring across time and space," which Giddens considers a prerequisite for modernization. World-society scholarship takes sides with Giddens on this point: globalization results in a "sharing" of modernity across the world.

On the other side of the fence, British social theorist Martin Albrow argues that globalization is a "transformation, not a culmination," and the "transition to a new era rather than the apogee of the old." He proposes a stark distinction between modernity as the "imposition of practical rationality upon the rest of the world through the agency of the state and the mechanism of the market, the generation of universal ideas to encompass the diversity of the world," and globality as it restored "the boundlessness of culture and promotes the endless renewability and diversification of cultural expression rather than homogenization or hybridization." Other noted social theorists of globalization also support the same distinction, especially insofar as the modern nation-state is concerned: "The politics of identity substitutes for the politics of nation-building."

The debate over the relationship between modernity and globality is a central one for sociologists. If globality is merely the result of an intensification of modernizing trends, then the recent surge in the number of books and articles on this subject can hardly be justified. There is, however, a key theoretical argument to be made in favor of the view that globality is different from modernity. Modernity – like the distorting Mercator projection – is an outgrowth of the Western worldview. For reasons of theoretical consistency, one should reserve the terms "globalization," "global," and "globality" to denote, respectively, processes, qualities, and conditions that are not set into motion or dominated by any one model, paradigm, or worldview. In its broadest sense, globality is about a multiplicity of conceptions, not about cultural or paradigmatic hegemony; it is about the proliferation of cross-national network ties of an economic, political, social, and cultural nature. This criticism is especially germane in the case of authors who consider globalization to be an inevitable and sweeping process – neoliberals and Marxists in particular – as Fligstein has aptly pointed out.

Finally, Kobrin has proposed a distinction between globalization in the late twentieth century and the previous period of modern expansion of the world economy that is useful empirically. The international economy of the nineteenth century "*links* discrete, mutually exclusive, geographical national markets through cross-border flows of trade and investment." By contrast, the global economy of the late twentieth century is driven by the increasing scale of technology, the surge in cross-border collaboration of firms along the value-added chain, and the cross-border integration of information flows. Thus, globalization has "substantive meaning" because, this time around, "national markets are *fused* transnationally rather than linked across borders."

Is a global culture in the making?

Perhaps the most popular and controversial of the debates about globalization has to do with the rise of a global culture. Actually, there are only a few scholars who maintain that a global culture is in the making. The idea goes back to Marshall McLuhan's slippery concept of the "global village," later picked up by some influential marketing researchers who argued that the world was becoming increasingly populated by cosmopolitan consumers. Sociologist Leslie Sklair writes that a "culture-ideology of consumerism" – driven by symbols, images, and the aesthetic of the lifestyle and the self-image – has spread throughout the world and is having some momentous effects, including the standardization of tastes and desires, and even the fall of the Soviet order.

Other sociologists, however, argue against the homogenizing effects of mass consumerism. Zelizer writes that consumer differentiation should not be confused with segregation and posits that in the US economy differentiation is combined with connection: "the same consumer product can have at the same moment universal and local meaning." Zelizer urges sociologists to distinguish between the phenomenon of worldwide diffusion and the experience at the receiving end, which seems to be growing more diverse even as globalization intensifies. Similarly, anthropologist Arjun Appadurai argues that "individuals and groups seek to annex the global into their own practices of the modern," and that "consumption of the mass media worldwide provokes resistance, irony, selectivity, and, in general, *agency.*" Using cross-national attitudinal data over the 1981–98 period, Inglehart and Baker find that national cultures and values change over time, though in "path-dependent" rather than convergent ways. Even world-society arguments about the "world culture of educated individual choice and responsibility" stop short of announcing a global culture à la McLuhan. However, they do describe world-culture as binding society and individuals together "by rationalized systems of (imperfectly) egalitarian justice and participatory representation, in the economy, polity, culture, and social interaction." Other researchers have found that the spread of the mass media is not enough to account for the rise of cross-border advocacy groups, although "global governance" of major aspects of cross-border communication has been on the rise since 1850.

Political and social theorists and historians have noted the rise of what modernists would call "particularistic" identities as evidence against the rise of a global culture. Cox writes about globalization producing a "resurgent affirmation of identities," whereas Waters contrasts a cultural and "religious mosaic" with global cultural production and consumption of music, images, and information. Mazlish notes that "ethnic feeling is a powerful bond," and skeptically asks, "What counterpart can there be on the global level?" Political scientist Deborah Yashar rejects "global culture" and "global citizenship" concepts but also finds fault with the argument that globalization has induced the proliferation of ethnic movements. In her comparison of indigenous movements in Latin America, Yashar clearly demonstrates that no aspect of globalization – economic, political, social, or normative – can account for the rise of ethnic-based activism since the 1960s. Rather, globalization changes the characteristics of the state structures that activists face when making their claims.

Cross-border migration creates an unusually rich laboratory for assessing the rise of a global culture. Sociologist Alejandro Portes proposes the term "transnational communities" to refer to cross-border networks of immigrants that are "'neither here nor there' but in both places simultaneously." Different transnational communities, however, exhibit different origins, features, and problems, and they certainly do not form a monolithic global class of cosmopolitan citizens.

Similarly to Portes, Friedman accepts the basic notion of cultural fragmentation proposed by Appadurai, Smith, and Zelizer but argues that in today's world the existence of tribal societies cannot be correctly understood without explaining how they are embedded in global networks. In his view, cultural diversity must be seen in a global context.

Some of the most persuasive arguments against the idea of the emergence of a global culture come from anthropologist Clifford Geertz. He observes that the world is "growing both more global and more divided, more thoroughly interconnected and more intricately partitioned at the same time [. . .] Whatever it is that defines identity in borderless capitalism and the global village it is not deep going agreements on deep going matters, but something more like the recurrence of familiar divisions, persisting arguments, standing threats, the notion that whatever else may happen, the order of difference must be somehow maintained." Like Geertz, sociologist Anthony Smith is skeptical and notes an interesting "initial problem" with the concept of "global culture": "Can we speak of 'culture' in the singular? If by 'culture' is meant a collective mode of life, or a repertoire of beliefs, styles, values and symbols, then we can only speak of cultures, never just culture; for a collective mode of life [. . .] presupposes different modes and repertoires in a universe of modes and repertoires. Hence, the idea of a 'global culture' is a practical impossibility, except in interplanetary terms."

The ultimate question about the alleged rise of a global culture has to do with whether a global language is emerging. The diffusion of Esperanto has certainly not delivered on early expectations, and the "English-as-global-language" argument seems equally far-fetched and indefensible. As Mazlish observes, English "is becoming a sort of lingua franca [but] there are serious limitations to the use of English as the daily language of a global culture." Moreover, English is being challenged as the dominant language in parts of the United States and the United Kingdom. Even on the Internet, fewer than 50 percent of world users know English as a first language, and the proportion is dropping steadily as the new medium diffuses throughout the world. It is also instructive to recall that the most successful world language ever, Latin, evolved into a mosaic of Romance languages after spreading in its various vulgarized forms throughout the territory of the Roman Empire. Smith notes that, rather than the emergence of a "global" culture held together by the English language, what we are witnessing is the emergence of "culture areas" – not necessarily at odds or in conflict with each other, as Huntington would have it. Thus, Spanish, Russian, Arabic, French, Kiswahili, and Chinese have become the shared languages of certain groups, communities or population strata across countries located in specific regions of the world, namely, Latin America, the CIS, the Arab world, Subsaharan Africa, East Africa, and South East Asia, respectively.

Toward a Comparative Sociology of Globalization

The social science literature on globalization contains important theoretical and empirical disagreements. Scholars have provided very different answers to the five key debates discussed in this chapter. The balance of opinion appears to be tilted, however. Most research either assumes or documents that globalization is indeed happening, and most empirical studies – with the notable exception of the world-society approach – do not find convergence in political, social, or organizational patterns as a result of globalization. The most persuasive empirical work to date indicates that globalization per se neither undermines the nation-state nor erodes the viability of the welfare state. Some empirical evidence also documents that globality is different from modernity. Finally, it seems that no such thing as a global culture is emerging.

Relative to the other social sciences, sociology has contributed to the debate over globalization in three important ways. First, social theorists have developed an understanding of the nature and epochal implications of globalization. Although there is no agreement as to whether globalization is a continuation of modernity or not, there is an incipient body of work that outlines in detail what are the main theoretical perspectives and problems. Moreover, sociologists have called attention to the cultural, reflexive, and aesthetic aspects of globalization in addition to its economic and political dimensions. Second, world-society scholars have developed a macrophenomenological approach to globalization and the nation-state based on a sound

institutional theoretical foundation, and they have supported their view with systematic empirical evidence encompassing the entire world. Third, comparative sociologists have theorized about globalization's effects on cross-national difference and similarity. They have also offered empirical evidence in the forms of both rich case studies and quantitative analyses. Sociologists, however, need to continue reading the important contributions that economic historians, management scholars, political scientists, and anthropologists are making to the theoretical and empirical study of such a complex and multifaceted phenomenon as globalization.

The analysis and critique presented in this chapter indicate that globalization, far from being a feeble phenomenon, is changing the nature of the world. However, it is neither an invariably civilizing force nor a destructive one. Although further empirical investigation is warranted, there is already enough evidence available to reject either extreme. Globalization is neither a monolithic nor an inevitable phenomenon. Its impact varies across countries, societal sectors, and time. It is contradictory, discontinuous, even haphazard. Therefore, one needs to be open-minded about its unexpected and unintended consequences. One also needs to take into account the role that agency, interest, and resistance play in shaping it. As Pieterse has pointed out, globalization does not necessarily pose a choice between condemnation and celebration. Rather, it begs to be engaged, comprised, given form.

The complexity of globalization certainly invites additional research. We are in great need of further theoretical work to clarify the economic, political, cultural, and aesthetic dimensions of globalization and how they interact with each other. We also lack theoretical perspectives that bridge the micro–macro gap, i.e. that move across levels of analysis from the world-system to the nation-state, the industry, sector, community, organization, and group. Many of the empirical disagreements in the literature are primarily due to the various levels of analysis at which different researchers operate. Understanding globalization will require us to gather more and better data about its myriad manifestations, causes, and effects. We still know very little about what exactly causes it and what are its consequences on such key sociological variables as organizational patterns, authority structures, social inequality, and social movements, to name but a few. And sociologists need to work hard on government agencies and other data-gathering organizations so that they pay more attention in their surveys and censuses to relationships at various levels of aggregation.

Given the infancy of our efforts to understand globalization and the complexity of the phenomenon, it seems sensible to ask not only for an intensification of our interdisciplinary awareness but also for a comparative approach to the sociology of globalization. Comparing lies at the heart of the sociological enterprise.

We need to engage in comparative work in the dual sense of using multiple methods of data collection and analysis, and of applying our theoretical and empirical tools to a variety of research settings defined at various levels of analysis. The differences and similarities across such settings ought to give us a handle on the patterns according to which the causes and effects of globalization change from one setting to another. Without a comparative approach, the literature on globalization promises to remain as puzzling and contradictory as the phenomenon itself.

Political Economy

This first part of the book operates with the view that while the political and the economic aspects of globalization can be, and often are, separated for analytical purposes, it makes sense to combine them here under the heading of the political economy of globalization. As we will see, in global context, many seemingly political issues have economic implications, and the same is true in reverse.

We begin with *civilizations*; these are primarily cultural in nature, but all civilizations also have political and economic dimensions. Next, we deal with a set of interrelated ideas – *Orientalism*, *colonialism*, and *postcolonialism* – all of which have both political and economic dimensions and implications (as well as others, especially cultural). The highly interrelated ideas of *neoliberalism* and *structural adjustment* are generally thought of in economic terms, but all of them also have implications for the state and politics in general. The *nation-state* is obviously political, but from many points of view it is dominated by economic considerations, if not subordinated to economic interests (e.g. in Marxian theory, the state is part of the "superstructure" dominated by the economic "base"). *Transnationalism* encompasses a number of dimensions that bridge the political and economic including transnational corporations, the transnational capitalist class, the culture ideology of consumerism, and the transnational state. *World systems* involve the economic exploitation of the periphery by the core, but political entities are central to the world system. *Empire* is a new kind of postmodern global system that certainly involves economic exploitation of the multitude. It is not centered in the nation-state, but is controlled politically by a decentered constitutional system. The *network society* involves new global relationships based on *informationalism*, and this applies to both economic and political organizations and entities. The *world risk society* is one in which risks stem from both the economy and the polity and have an impact on both. *Cosmopolitanism* involves a broad outlook not limited to the nation-state and its particular political and economic interests. *McWorld* and the related idea of *Jihad* both pose a threat to democratic systems and implicitly, therefore, to successful economic systems, given the tendency to associate democracy and capitalism.

Civilizations

One of the most controversial of the theories developed in the post Cold War era is to be found in Samuel Huntington's (1993) *The Clash of Civilizations.* The central idea is that civilizations, the broadest cultural entities, are shaping patterns of cohesion, disintegration, and conflict in the post Cold War international system. Huntington identifies several major world civilizations: Western, Confucian, Japanese, Islamic, Hindu, Slavic-Orthodox, Latin American, and possibly African civilization. He states that "In this new world, local politics is the politics of ethnicity; global politics is the politics of civilizations."[1] The past intracivilizational clashes of political ideas, such as liberalism, socialism, anarchism, corporatism, communism, conservatism, Marxism, social democracy, and nationalism, are being replaced by intercivilizational clashes of *culture* and *religion.* In the new international order, culturally similar societies tend to cooperate, countries are prone to group themselves around the "core countries" of civilizations, and the relations between civilizations will not be close and will eventually lead to conflict, at least between some of them.

Huntington's thesis has been widely criticized for its conceptualization of "civilization"; for failing to differentiate between religion, culture, and civilization; for overlooking the integrative processes of capitalism, globalization, and modernization; for its lack of attention to the importance of nation-states and nationalism; and for its lack of scientific validity. Since it is impossible to present all of the critiques of Huntington's thesis in this part, we present three representative examples in this chapter.

Gray points out that Huntington's delineation of seven or eight civilizations is imprecise; this civilizational schema cannot accommodate certain cases. For instance, while Jewish culture is attached to Western civilization, Greek is not. Gray also identifies another major shortcoming of the civilizations thesis by showing that wars are not waged between civilizations. To the contrary, the twentieth century's history of conflicts demonstrates that there were several conflicts, clashes, and wars within the same civilization, as well as alliances between different civilizations. Gray argues that, contrary to Huntington's assumption that cultures create significant splits in international relations, culture by itself is not that powerful a factor. Differing cultural traditions rarely lead to major conflicts between states. It is their interactions with scarcities of resources, rival claims on territory, and conflicting agendas on trade that make cultural differences a source of war. Therefore, the whole idea of civilizational conflict is a "distorting lens" that prevents us from fully understanding "economic rivalries" and "military conflicts."[2]

In addition to the problems with the concept of civilization, Gray also criticizes the civilizations thesis for its neglect of globalization and modernization as integrating processes. Gray contends that there is a considerable connection between culture and political economy in that the global economic interdependence of world markets requires constant interaction among cultures.

The political climate in which the civilizations thesis was proposed (the end of the Cold War), as well as its political implications, are also underscored by the critics. It is argued that when the Cold War political taxonomy became obsolete, the civilizations thesis provided a convenient political ideology that, among other things, served to hold the Atlantic alliance together in spite of the demise of the threat posed by communism. This is related to the idea that Huntington identifies not only with the Atlantic alliance, but more specifically with the most important player in it, the United States. As a result, he is seen as offering a distinctly American perspective on the world's civilizations. According to Gray, Huntington's perspective "is an attempt to give a theoretical framework to American thinking about foreign policy in a context in which sustaining ideological enmities of the Cold War have vanished."[3]

Matlock agrees with Gray's criticisms of Huntington's thesis, arguing that the idea that civilizations are mutually exclusive is misleading. He states that it is difficult to accept the view that each civilization is somehow pure and harmonious when there are numerous examples of conflicts, clashes, and wars within the same civilization. Matlock also criticizes Huntington for endowing civilizations with a reality they do not have. He states that "civilization" is merely a convenient intellectual construct used to establish the boundaries of a field or topic of study.[4] In other words, "civilization" is an intellectual construct rather than an objective reality. Matlock specifically focuses on the difference between culture and civilizations. He argues that Huntington mistakes culture for civilization and lumps cultures into broader civilizations, and this serves to obscure the specifics of cultural differences and similarities.

Similarly, Brown questions Huntington's assumption that civilizations are self-contained and impermeable territories. Brown maintains that cultures are dynamic, living organisms that interpenetrate continually. Brown also argues that the physical "fault-lines" between civilizations are not preordained and eternal as Huntington assumes, but rather are man-made and of relatively recent origin.

No perspective on globalization has received more attention and more criticism than the clash of civilizations paradigm. Some consider it to be the fundamental view on the state of globalization in the late twentieth and early twenty-first centuries. Others see it as completely wrong-headed and even offensive. However, even its most ardent critics would acknowledge that it is an extremely useful perspective, if for no other reason than the fact that attacks on it serve to clarify much about contemporary globalization.

Much has been said about the criticisms of the clash of civilizations, but let us close with some thoughts by Huntington himself. Adopting a perspective based on Thomas Kuhn's philosophy of science, Huntington argues that what he has presented is a paradigm, or model, of global relations. As such, it is not enough to criticize his paradigm; it is incumbent on the critics to produce an alternative paradigm, one that better explains global realities today than does his model. When looked at in this way, it could be argued that while the critics may have wounded the clash of civilizations paradigm, they have not been able thus far to produce a better one. The challenge to Huntington's critics, indeed to all students of globalization, is to produce such a paradigm.

NOTES ••

1 Samuel Huntington, *The Clash of Civilizations.* New York: Simon and Schuster, 1993, 29.
2 John Gray, "Global Utopias and Clashing Civilizations: Misunderstanding the Present." *International Affairs* 74, 1, 1998: 159.
3 Ibid., 157.
4 Jack F. Matlock, "Can Civilizations Clash?" *Proceedings of the American Philosophical Society* 143, 3, 1999: 439.

 **READING 2**

The Clash of Civilizations?

Samuel P. Huntington

The Next Pattern of Conflict

[...]

It is my hypothesis that the fundamental source of conflict in this new world will not be primarily ideological or primarily economic. The great divisions among humankind and the dominating source of conflict will be cultural. Nation states will remain the most powerful actors in world affairs, but the principal conflicts of global politics will occur between nations and groups of different civilizations. The clash of civilizations will dominate global politics. The fault lines between civilizations will be the battle lines of the future.

[...]

The Nature of Civilizations

During the Cold War the world was divided into the First, Second and Third Worlds. Those divisions are no longer relevant. It is far more meaningful now to group countries not in terms of their political or economic systems or in terms of their level of economic development but rather in terms of their culture and civilization.

What do we mean when we talk of a civilization? A civilization is a cultural entity. Villages, regions, ethnic groups, nationalities, religious groups, all have distinct cultures at different levels of cultural heterogeneity. The culture of a village in southern Italy may be different from that of a village in northern Italy, but both will share in a common Italian culture that distinguishes them from German villages. European communities, in turn, will share cultural features that distinguish them from Arab or Chinese communities. Arabs, Chinese and Westerners, however, are not part of any broader cultural entity. They constitute civilizations.

A civilization is thus the highest cultural grouping of people and the broadest level of cultural identity people have short of that which distinguishes humans from other species. It is defined both by common objective elements, such as language, history, religion, customs, institutions, and by the subjective self-identification of people. People have levels of identity: a resident of Rome may define himself with varying degrees of intensity as a Roman, an Italian, a Catholic, a Christian, a European, a Westerner. The civilization to which he belongs is the broadest level of identification with which he intensely identifies. People can and do redefine their identities and, as a result, the composition and boundaries of civilizations change.

Civilizations may involve a large number of people, as with China ("a civilization pretending to be a state," as Lucian Pye put it), or a very small number of people, such as the Anglophone Caribbean. A civilization may include several nation states, as is the case with Western, Latin American and Arab civilizations, or only one, as is the case with Japanese civilization. Civilizations obviously blend and overlap, and may include subcivilizations. Western civilization has two major variants, European and North American, and Islam has its Arab, Turkic and Malay subdivisions. Civilizations are nonetheless meaningful entities, and while the lines between them are seldom sharp, they are real. Civilizations are dynamic; they rise and fall; they divide and merge. And, as any student of history knows, civilizations disappear and are buried in the sands of time.

Westerners tend to think of nation states as the principal actors in global affairs. They have been that, however, for only a few centuries. The broader reaches of human history have been the history of civilizations. In *A Study of History*, Arnold Toynbee identified 21 major civilizations; only six of them exist in the contemporary world.

Why Civilizations will Clash

Civilization identity will be increasingly important in the future, and the world will be shaped in large measure by the interactions among seven or eight major civilizations. These include Western, Confucian, Japanese, Islamic, Hindu, Slavic-Orthodox, Latin American and possibly African civilization. The most important conflicts of the future will occur along the cultural fault lines separating these civilizations from one another.

Why will this be the case?

First, differences among civilizations are not only real; they are basic. Civilizations are differentiated from each other by history, language, culture, tradition and, most important, religion. The people of different civilizations have different views on the relations between God and man, the individual and the group, the citizen and the state, parents and children, husband and wife, as well as differing views of the relative importance of rights and responsibilities, liberty and authority, equality and hierarchy. These differences are the product of centuries. They will not soon disappear. They are far more fundamental than differences among political ideologies and political regimes. Differences do not necessarily mean conflict, and conflict does not necessarily mean violence. Over the centuries, however, differences among civilizations have generated the most prolonged and the most violent conflicts.

Second, the world is becoming a smaller place. The interactions between peoples of different civilizations are increasing; these increasing interactions intensify civilization consciousness and awareness of differences between civilizations and commonalities within civilizations. North African immigration to France generates hostility among Frenchmen and at the same time increased receptivity to immigration by "good" European Catholic Poles. Americans react far more negatively to Japanese investment than to larger investments from Canada and European countries. Similarly, as Donald Horowitz has pointed out, "An Ibo may be [. . .] an Owerri Ibo or an Onitsha Ibo in what was the Eastern region of Nigeria. In Lagos, he is simply an Ibo. In London, he is a Nigerian. In New York, he is an African." The interactions among peoples of different civilizations enhance the civilization-consciousness of people that, in turn, invigorates

differences and animosities stretching or thought to stretch back deep into history.

Third, the processes of economic modernization and social change throughout the world are separating people from longstanding local identities. They also weaken the nation state as a source of identity. In much of the world religion has moved in to fill this gap, often in the form of movements that are labeled "fundamentalist." Such movements are found in Western Christianity, Judaism, Buddhism and Hinduism, as well as in Islam. In most countries and most religions the people active in fundamentalist movements are young, college-educated, middle-class technicians, professionals and business persons. The "unsecularization of the world," George Weigel has remarked, "is one of the dominant social facts of life in the late twentieth century." The revival of religion, "la revanche de Dieu," as Gilles Kepel labeled it, provides a basis for identity and commitment that transcends national boundaries and unites civilizations.

Fourth, the growth of civilization-consciousness is enhanced by the dual role of the West. On the one hand, the West is at a peak of power. At the same time, however, and perhaps as a result, a return to the roots phenomenon is occurring among non-Western civilizations. Increasingly one hears references to trends toward a turning inward and "Asianization" in Japan, the end of the Nehru legacy and the "Hinduization" of India, the failure of Western ideas of socialism and nationalism and hence "re-Islamization" of the Middle East, and now a debate over Westernization versus Russianization in Boris Yeltsin's country. A West at the peak of its power confronts non-Wests that increasingly have the desire, the will and the resources to shape the world in non-Western ways.

In the past, the elites of non-Western societies were usually the people who were most involved with the West, had been educated at Oxford, the Sorbonne or Sandhurst, and had absorbed Western attitudes and values. At the same time, the populace in non-Western countries often remained deeply imbued with the indigenous culture. Now, however, these relationships are being reversed. A de-Westernization and indigenization of elites is occurring in many non-Western countries at the same time that Western, usually American, cultures, styles and habits become more popular among the mass of the people.

Fifth, cultural characteristics and differences are less mutable and hence less easily compromised and resolved than political and economic ones. In the former Soviet Union, communists can become democrats, the rich can become poor and the poor rich, but Russians cannot become Estonians and Azeris cannot become Armenians. In class and ideological conflicts, the key question was "Which side are you on?" and people could and did choose sides and change sides. In conflicts between civilizations, the question is "What are you?" That is a given that cannot be changed. And as we know, from Bosnia to the Caucasus to the Sudan, the wrong answer to that question can mean a bullet in the head. Even more than ethnicity, religion discriminates sharply and exclusively among people. A person can be half-French and half-Arab and simultaneously even a citizen of two countries. It is more difficult to be half-Catholic and half-Muslim.

Finally, economic regionalism is increasing. The proportions of total trade that were intraregional rose between 1980 and 1989 from 51 percent to 59 percent in Europe, 33 percent to 37 percent in East Asia, and 32 percent to 36 percent in North America. The importance of regional economic blocs is likely to continue to increase in the future. On the one hand, successful economic regionalism will reinforce civilization-consciousness. On the other hand, economic regionalism may succeed only when it is rooted in a common civilization.

[. . .]

As people define their identity in ethnic and religious terms, they are likely to see an "us" versus "them" relation existing between themselves and people of different ethnicity or religion. The end of ideologically defined states in Eastern Europe and the former Soviet Union permits traditional ethnic identities and animosities to come to the fore. Differences in culture and religion create differences over policy issues, ranging from human rights to immigration to trade and commerce to the environment. Geographical propinquity gives rise to conflicting territorial claims from Bosnia to Mindanao. Most important, the efforts of the West to promote its values of democracy and liberalism as universal values, to maintain its military predominance and to advance its economic interests engender countering responses from other civilizations. Decreasingly able to mobilize support and form coalitions on the basis of ideology, governments and groups will increasingly attempt to mobilize support by appealing to common religion and civilization identity.

The clash of civilizations thus occurs at two levels. At the micro-level, adjacent groups along the fault lines between civilizations struggle, often violently, over the control of territory and each other. At the macro-level, states from different civilizations compete for relative military and economic power, struggle over the control of international institutions and third parties, and competitively promote their particular political and religious values.

The Fault Lines between Civilizations

The fault lines between civilizations are replacing the political and ideological boundaries of the Cold War as the flash points for crisis and bloodshed. The Cold War began when the Iron Curtain divided Europe politically and ideologically. The Cold War ended with the end of the Iron Curtain. As the ideological division of Europe has disappeared, the cultural division of Europe between Western Christianity, on the one hand, and Orthodox Christianity and Islam, on the other, has reemerged.

[. . .]

Conflict along the fault line between Western and Islamic civilizations has been going on for 1,300 years.

[. . .]

This centuries-old military interaction between the West and Islam is unlikely to decline. It could become more virulent. [. . .] Some openings in Arab political systems have already occurred. The principal beneficiaries of these openings have been Islamist movements. In the Arab world, in short, Western democracy strengthens anti-Western political forces. This may be a passing phenomenon, but it surely complicates relations between Islamic countries and the West.

Those relations are also complicated by demography. The spectacular population growth in Arab countries, particularly in North Africa, has led to increased migration to Western Europe. The movement within Western Europe toward minimizing internal boundaries has sharpened political sensitivities with respect to this development. In Italy, France and Germany, racism is increasingly open, and political reactions and violence against Arab and Turkish migrants have become more intense and more widespread since 1990.

On both sides the interaction between Islam and the West is seen as a clash of civilizations.

[. . .]

Historically, the other great antagonistic interaction of Arab Islamic civilization has been with the pagan, animist, and now increasingly Christian black peoples to the south. In the past, this antagonism was epitomized in the image of Arab slave dealers and black slaves. It has been reflected in the on-going civil war in the Sudan between Arabs and blacks, the fighting in Chad between Libyan-supported insurgents and the government, the tensions between Orthodox Christians and Muslims in the Horn of Africa, and the political conflicts, recurring riots and communal violence between Muslims and Christians in Nigeria. The modernization of Africa and the spread of Christianity are likely to enhance the probability of violence along this fault line. Symptomatic of the intensification of this conflict was the Pope John Paul II's speech in Khartoum in February 1993 attacking the actions of the Sudan's Islamist government against the Christian minority there.

On the northern border of Islam, conflict has increasingly erupted between Orthodox and Muslim peoples, including the carnage of Bosnia and Sarajevo, the simmering violence between Serb and Albanian, the tenuous relations between Bulgarians and their Turkish minority, the violence between Ossetians and Ingush, the unremitting slaughter of each other by Armenians and Azeris, the tense relations between Russians and Muslims in Central Asia, and the deployment of Russian troops to protect Russian interests in the Caucasus and Central Asia. Religion reinforces the revival of ethnic identities and restimulates Russian fears about the security of their southern borders. This concern is well captured by Archie Roosevelt:

Much of Russian history concerns the struggle between the Slavs and the Turkic peoples on their borders, which dates back to the foundation of the Russian state more than a thousand years ago. In the Slavs' millennium-long confrontation with their eastern neighbors lies the key to an understanding not only of Russian history, but Russian character. To understand Russian realities today one has to have a concept of the great Turkic ethnic group that has preoccupied Russians through the centuries.

The conflict of civilizations is deeply rooted elsewhere in Asia. The historic clash between Muslim and Hindu in the subcontinent manifests itself now not only in the rivalry between Pakistan and India but also in intensifying religious strife within India between increasingly militant Hindu groups and India's substantial Muslim minority. The destruction of the Ayodhya mosque in December 1992 brought to the fore the issue of whether India will remain a secular democratic state or become a Hindu one. In East Asia, China has outstanding territorial disputes with most of its neighbors. It has pursued a ruthless policy toward the Buddhist people of Tibet, and it is pursuing an increasingly ruthless policy toward its Turkic-Muslim minority. With the Cold War over, the underlying differences between China and the United States have reasserted themselves in areas such as human rights, trade and weapons proliferation. These differences are unlikely to moderate. A "new cold war," Deng Xaioping reportedly asserted in 1991, is under way between China and America.

The same phrase has been applied to the increasingly difficult relations between Japan and the United States. Here cultural difference exacerbates economic conflict. People on each side allege racism on the other, but at least on the American side the antipathies are not racial but cultural. The basic values, attitudes, behavioral patterns of the two societies could hardly be more different. The economic issues between the United States and Europe are no less serious than those between the United States and Japan, but they do not have the same political salience and emotional intensity because the differences between American culture and European culture are so much less than those between American civilization and Japanese civilization.

The interactions between civilizations vary greatly in the extent to which they are likely to be characterized by violence. Economic competition clearly predominates between the American and European subcivilizations of the West and between both of them and Japan. On the Eurasian continent, however, the proliferation of ethnic conflict, epitomized at the extreme in "ethnic cleansing," has not been totally random. It has been most frequent and most violent between groups belonging to different civilizations. In Eurasia the great historic fault lines between civilizations are once more aflame. This is particularly true along the boundaries of the crescent-shaped Islamic bloc of nations from the bulge of Africa to central Asia. Violence also occurs between Muslims, on the one hand, and Orthodox Serbs in the Balkans, Jews in Israel, Hindus in India, Buddhists in Burma and Catholics in the Philippines. Islam has bloody borders.

[...]

The West versus the Rest

The West is now at an extraordinary peak of power in relation to other civilizations. Its superpower opponent has disappeared from the map. Military conflict among Western states is unthinkable, and Western military power is unrivaled. Apart from Japan, the West faces no economic challenge. It dominates international political and security institutions and with Japan international economic institutions. Global political and security issues are effectively settled by a directorate of the United States, Britain and France, world economic issues by a directorate of the United States, Germany and Japan, all of which maintain extraordinarily close relations with each other to the exclusion of lesser and largely non-Western countries. Decisions made at the UN Security Council or in the International Monetary Fund that reflect the interests of the West are presented to the world as reflecting the desires of the world community. The very phrase "the world community" has become the euphemistic collective noun (replacing "the Free World") to give global legitimacy to actions reflecting the interests of the United States and other Western powers. Through the IMF and other international economic institutions, the West promotes its economic interests and imposes on other nations the economic policies it thinks appropriate.

[...]

The West in effect is using international institutions, military power and economic resources to run the world in ways that will maintain Western predominance, protect Western interests and promote Western political and economic values.

That at least is the way in which non-Westerners see the new world, and there is a significant element of truth in their view. Differences in power and struggles for military, economic and institutional power are thus one source of conflict between the West and other civilizations. Differences in culture, that is basic values and beliefs, are a second source of conflict. V. S. Naipaul has argued that Western civilization is the "universal civilization" that "fits all men." At a superficial level much of Western culture has indeed permeated the rest of the world. At a more basic level, however, Western concepts differ fundamentally from those prevalent in other civilizations. Western ideas of individualism, liberalism, constitutionalism, human rights, equality, liberty, the rule of law, democracy, free markets, the separation of church and state, often have little resonance in Islamic, Confucian, Japanese, Hindu, Buddhist or Orthodox cultures. Western efforts to propagate such ideas produce instead a reaction against "human rights imperialism" and a reaffirmation of indigenous values, as can be seen in the support for religious fundamentalism by the younger generation in non-Western cultures. The very notion that there could be a "universal civilization" is a Western idea, directly at odds with the particularism of most Asian societies and their emphasis on what distinguishes one people from another. Indeed, the author of a review of 100 comparative studies of values in different societies concluded that "the values that are most important in the West are least important worldwide." In the political realm, of course, these differences are most manifest in the efforts of the United States and other Western powers to induce other peoples to adopt Western ideas concerning democracy and human rights. Modern democratic government originated in the West. When it has developed in non-Western societies it has usually been the product of Western colonialism or imposition.

The central axis of world politics in the future is likely to be, in Kishore Mahbubani's phrase, the conflict between "the West and the Rest" and the responses of non-Western civilizations to Western power and values. Those responses generally take one or a combination of three forms. At one extreme, non-Western states can, like Burma and North Korea, attempt to pursue a course of isolation, to insulate their societies from penetration or "corruption" by the West, and, in effect, to opt out of participation in the Western-dominated global community. The costs of this course, however, are high, and few states have pursued it exclusively. A second alternative, the equivalent of "band-wagoning" in international relations theory, is to attempt to join the West and accept its values and institutions. The third alternative is to attempt to "balance" the West by developing economic and military power and cooperating with other non-Western societies against the West, while preserving indigenous values and institutions; in short, to modernize but not to Westernize.

[. . .]

Implications for the West

This article does not argue that civilization identities will replace all other identities, that nation states will disappear, that each civilization will become a single coherent political entity, that groups within a civilization will not conflict with and even fight each other. This paper does set forth the hypotheses that differences between civilizations are real and important; civilization-consciousness is increasing; conflict between civilizations will supplant ideological and other forms of conflict as the dominant global form of conflict; international relations, historically a game played out within Western civilization, will increasingly be de-Westernized and become a game in which non-Western civilizations are actors and not simply objects; successful political, security and economic international institutions are more likely to develop within civilizations than across civilizations; conflicts between groups in different civilizations will be more frequent, more sustained and more violent than conflicts between groups in the same civilization; violent conflicts between groups in different civilizations are the most likely and most dangerous source of escalation that could lead to global wars; the paramount axis of world politics will be the relations between "the West and the Rest"; the elites in some torn non-Western countries will try to make their countries part of the West, but in most cases face major obstacles to accomplishing this; a central focus of conflict for the immediate future will be between the West and several Islamic-Confucian states.

This is not to advocate the desirability of conflicts between civilizations. It is to set forth descriptive hypotheses as to what the future may be like. If these are plausible hypotheses, however, it is necessary to consider their implications for Western policy. These implications should be divided between short-term advantage and long-term accommodation. In the short term it is clearly in the interest of the West to promote greater cooperation and unity within its own civilization, particularly between its European and North American components; to incorporate into the West societies in Eastern Europe and Latin America whose cultures are close to those of the West; to promote and maintain cooperative relations with Russia and Japan; to prevent escalation of local inter-civilization conflicts into major inter-civilization wars; to limit the expansion of the military strength of Confucian and Islamic states; to moderate the reduction of Western military capabilities and maintain military superiority in East and Southwest Asia; to exploit differences and conflicts among Confucian and Islamic states; to support in other civilizations groups sympathetic to Western values and interests; to strengthen international institutions that reflect and legitimate Western interests and values and to promote the involvement of non-Western states in those institutions.

In the longer term other measures would be called for. Western civilization is both Western and modern. Non-Western civilizations have attempted to become modern without becoming Western. To date only Japan has fully succeeded in this quest. Non-Western civilizations will continue to attempt to acquire the wealth, technology, skills, machines and weapons that are part of being modern. They will also attempt to reconcile this modernity with their traditional culture and values. Their economic and military strength relative to the West will increase. Hence the West will increasingly have to accommodate these non-Western modern civilizations whose power approaches that of the West but whose values and interests differ significantly from those of the West. This will require the West to maintain the economic and military power necessary to protect its interests in relation to these civilizations. It will also, however, require the West to develop a more profound understanding of the basic religious and philosophical assumptions underlying other civilizations and the ways in which people in those civilizations see their interests. It will require an effort to identify elements of commonality between Western and other civilizations. For the relevant future, there will be no universal civilization, but instead a world of different civilizations, each of which will have to learn to coexist with the others.

 READING 3

Global Utopias and Clashing Civilizations: Misunderstanding the Present

John Gray

[. . .]

Why Wars Are Not Conflicts among Civilizations

Samuel Huntington's thesis of the clash of civilizations is a necessary corrective to a powerful recent trend in thinking about the international system. American foreign policy has long affirmed that the pursuit of peace is linked with the projection of human rights and support for democratic institutions. More recently, a similar view has been adopted by several other Western governments. Never more than one strand in the foreign policy of any country, it is often marginalized by other, more practically immediate considerations. But as an influence on thinking about international relations it is probably stronger today than at any other time.

Huntington makes some acute criticisms of this view. He is right to note that the individualist values embodied in Western understandings of liberal democracy do not command universal assent. They express the ethical life of a few Western societies. They are not authoritative for all cultures. Foreign policies which presuppose an eventual global consensus on liberal values will be ineffectual. This is an incisive criticism of Fukuyama's neo-Wilsonian certainty that Western values are universal; but in arguing that fault-lines between civilizations are the source of war Huntington misunderstands the present as grievously as Fukuyama does. As a result he gives a mistaken diagnosis of both the potential for tragedy and the opportunities for cooperation that our present circumstances contain.

Now, as in the past, wars are commonly waged between (and within) nationalities and ethnicities, not between different civilizations. Whether or not they are waged by the agents of sovereign states, the old,

familiar logic of territories and alliances often impels members of the same 'civilization' into enmity and members of different 'civilizations' into making common cause. In the Armenia–Azerbaijan conflict, Iran threw in its lot with Christian Armenia, not with Islamic Azerbaijan. The kaleidoscope of shifting alliances in the Balkans tells a similar story. Again, some of this century's decisive conflicts have been 'intra-civilizational'. The Iran–Iraq war and the genocide of Tutsis by Hutus occurred within what Huntington understands as single civilizations. The First World War is commonly, and not inaptly, described as a European civil war. The Korean war and the Vietnam war were conflicts among states all of which justified their claims by reference to 'Western' ideologies. Huntington's typology of civilizations does not map on to the history of twentieth-century conflict. Moreover, it is an imprecise, even arbitrary taxonomy. What is it that justifies the honorific appellation of 'civilization'? Huntington seems to believe that the world today contains somewhere between six and nine civilizations – Sinic (Chinese), Japanese, Hindu, Islamic, Latin American, Buddhist, Orthodox, African, and, of course, Western. Yet he is not altogether confident in this enumeration. He exhibits some doubt as to where Latin America should be placed; after some hesitation he includes the Jews in a sort of appendix to 'Western civilization', while concluding that Greece is no part of it. If one seeks for the criterion Huntington tacitly invokes for identifying a civilization, one soon discovers that it is an artefact of American multiculturalism: for Huntington, a community or a culture qualifies as a civilization if it has established itself as an American minority. Otherwise it does not.

The narrowly domestic perspective that informs much of Huntington's analysis gives a clue as to its historical provenance. It is an attempt to give a theoretical framework to American thinking about foreign policy in a context in which the sustaining ideological enmities of the Cold War have vanished. Unfortunately, Huntington's vision tells us more about contemporary American anxieties than it does about the late modern world. Huntington's watchword, 'Western civilization', is a familiar refrain in curricular debates in American universities. It has few points of contact with the world beyond American shores, in which 'Western' supremacy, and indeed the very idea of 'the West', are becoming anachronisms.

'The peoples of the West,' Huntington has warned, 'must hang together, or most assuredly they will hang separately.' This clarion call presupposes that Western civilization – 'the peoples of the West' – can be identified easily and unproblematically. Yet the old and familiar polarities of East and West never had a fixed or simple meaning. During the Cold War, 'the East' meant the Soviet bloc, which was animated by an unequivocally 'Western' ideology; in the Cold War's immediate aftermath, in former Yugoslavia and elsewhere, it came to refer to an older division between Eastern and Western Christianity; now it is being invoked, by Huntington and others, to capture America's relations with China and sections of the Arab world. When Huntington refers to 'Western civilization', he does not invoke an extended family of cultural traditions that has endured for centuries or millennia. He invokes a construction of the Cold War, with few points of leverage on the world that is taking shape around us.

Huntington is right to reject the view of the world, propagated by Fukuyama, in which modernization and westernization are one and the same. In many parts of the world, where countries are becoming modern by absorbing new technologies into their indigenous cultures, they are instead divergent developments. For some countries today, westernization of their economies and cultures would mean a step back from the late modern world: not modernization but a retreat from modernity.

The project of a global free market that is at present being advanced by many transnational organizations envisages reshaping economic life in every society so that it accords with the practices of a single type of capitalism – the Anglo-Saxon free market. But different kinds of capitalism reflect different cultures. There is no reason to think they will ever converge. Both the critics of capitalism and its supporters in Western countries have taken for granted that capitalist economies everywhere produce, or express, individualist values. This assumption was reasonable so long as developed market economies were confined to parts of western Europe, North America and the Antipodes. But the link it postulated was an historical accident, not a universal law. The capitalisms of East Asia are not the products of individualist cultures, and there is no reason to think that they will ever engender such cultures. Different patterns of family relations and different religious traditions are not facets of private life, like tastes in ethnic cuisines, without consequences for economic behaviour. They produce radically different market economies.

As global markets grow, the world is not being unified by a single economic civilization. It is becoming more plural. The increasing intensity of global competition is often noted; less often perceived is the fact that as competition between different cultures increases the comparative economic advantages of their family structures and religious traditions become more important. It is rather unlikely that the advantage in this competition lies always with highly individualist cultures. What are the economic costs of individualist patterns of family life, in which marriage is valued as a vehicle of self-realization? How does the cultural understanding of childhood as a phase of life exempt from obligations, which is strong in some Western countries, affect educational achievement? In the economic rivalries of the coming century such cultural differences will be central. Contrary to Huntington, however, this does not mean that the world can be divided up into well-defined, static civilizations. The emergence of genuine world markets in many areas of economic life makes continuing interaction among cultures an irreversible global condition.

What is new in our current circumstances is the worldwide spread of industrial production and its concomitant, the end of the global hegemony of any Western state. What is not new is conflict over territory, religion and commercial advantage between sovereign states. We must hope that wise policy can avert a rerun of the Great Game in which the world's powers struggled for geostrategic advantage in Central Asia and the Caucasus. But it is great power rivalries for control of oil, not cultural differences among the peoples that inhabit the eight nations of that region, that are likely to pose the most enduring risk to peace for its peoples.

Neither economic rivalries nor military conflicts can be understood when viewed through the distorting lens of civilizational conflict. Talk of clashing civilizations is supremely unsuited to a time when cultures – not least the extended family of peoples that Huntington loosely terms 'the West' – are in flux. In so far as such talk shapes the thinking of policy-makers it risks *making* cultural differences what they have been only rarely in the past – causes of war.

International Relations and Conflicts within Morality

Cultural differences can make international conflicts harder to resolve. They may make liberal democratic institutions of the kinds we are familiar with in Western countries unachievable, or even undesirable. That is one reason why I share Huntington's scepticism about foreign policies that aim to make liberal values universal. But the greatest obstacle to such foreign policies does not come from the evident fact of cultural variety. It comes from the awkward truth that even humanly universal values can be rivals in practice.

I put aside here the suggestion that all human values are entirely cultural constructions. This once fashionable doctrine of cultural relativism seems to me not worth extended consideration. It may well be true that some goods that are centrally important in Western societies are not universally valuable. That does not mean that all human goods and evils are culturally variable.

Personal autonomy, the authorship of one's life by one's own choices, is an urgent and pervasive demand in late modern Western cultures. At the same time, I am unpersuaded that it is a necessary feature of the good life for humans. Most human beings who have ever lived good lives did so without having much of it. Even where having a wide domain of personal options is one of the necessary ingredients of individual well-being, it is never the only ingredient. The worth of the options available matters as well. Nor am I convinced that as societies become more modern, personal autonomy is generally accorded a higher value. This seems to be true in the case of Britain, but it is a

mistake to take ourselves as a model for modernization everywhere. Perhaps, as economic and other risks multiply in late modern societies, people will be more willing to trade off portions of their autonomy if they can thereby achieve greater security.

To be sure, such trade-offs will sometimes enhance the 'on-balance' value that autonomous choice has for people. In other cases there will be a real conflict of values in which some autonomy is given up for the sake of another good. Compulsory saving for pensions may enhance the worth of personal autonomy on balance over a lifetime; but those who propose restricting freedom of divorce, say, because the stability of family life might thereby be promoted, must recognize that the personal autonomy of marriage partners is being curtailed for the sake of the well-being of children. Every human value has its price in other values with which it can conflict. Those who think, as I do, that the good for humans is not singular but plural, that human values are many not one, will find it hard to be convinced that this conflict should always be resolved in favour of autonomy. Liberal political philosophies that treat personal autonomy as a universal and overriding value are, or should be, controversial. The value of personal autonomy may well be a cultural construction, not something that is grounded in our common human nature. But, precisely because there is a common human nature, it cannot be true of all our values that they are cultural constructions.

Consider the chief evils to which human beings are vulnerable. Violent death is everywhere an evil. So is untimely death through malnutrition. Slavery, torture and genocide inflict injuries on their victims that block their chance of living any kind of worthwhile human life. The damage to human well-being wrought by these evils does not vary culturally to any significant extent. One of the central problems of ethical theory, in so far as it applies to international relations, is to determine which values are truly universal and which belong only to particular ways of life. Liberal values derive their hold on contemporary opinion partly from the fact that some of their injunctions – those forbidding torture, slavery and genocide, for example – are plausible components of a universal morality. Nonetheless, to identify the universal content of morality with the injunctions of recent Western liberal thought is a dangerous delusion. The difficult question is what is universal and what local in the morality of liberal regimes. This cannot be profitably discussed in the shop-soiled jargon of an incoherent debate about 'relativism'.

Cultural variations in political values do not generate the most serious of the ethical dilemmas that arise in international relations. The hardest question in the ethics of international relations is how to resolve conflicts among goods and bads that are indisputably universal. This is an issue that has been unduly neglected, partly owing to the revival of neo-Wilsonian ideas that attempt to deny its practical importance. Those who maintain that the foreign policies of liberal states should give a high priority to fostering democratic institutions throughout the world not only claim that liberal democracy has universal authority; they claim also that advancing democratic government promotes international stability. We are often reminded that liberal democracies rarely go to war with one another. As a natural, if tenuous inference from that fact, we are encouraged to believe that a world consisting only of liberal democratic regimes will be a world of perpetual peace. In this perspective promoting democracy can never conflict, save perhaps in the shortest term, with the pursuit of peace.

I do not think I have caricatured this conventional view. It marks a real correlation when it notes that wars sometimes arise from the domestic needs of tyrannies. Its cardinal defect is that the links that it affirms between peace and democracy are very far from being invariant. In the real world these two values are sometimes rivals. Nor are these conflicts so rare, or so trifling in their consequences, that they serve only to illustrate a limiting case. Consider a state in which populations of disparate nationalities and religious ancestries are held together in a dictatorial regime. Imagine that, for whatever combination of reasons, that regime begins to weaken, and demands for democratic institutions become politically irresistible. If the populations of such a dictatorial regime are territorially concentrated it is reasonable to expect the advance of democratic institutions to go in tandem with the fragmentation of the state.

We need not delve deeply into the literature of political science for an explanation. Functioning democracy requires high levels of trust. When populations are divided by memories of historical enmity trust is not easy to establish. When democratic deliberation concerns issues of life and death it is hard to begin. Where secession seems a real option it is likely to win support in the populations that most fear being overruled in such issues. If such fears predominate, the goal of secessionist movements will be to constitute a state sufficiently homogenous for trust – and thereby democracy – to be feasible.

I do not present this abstract scenario as a historical account of the break-up of any state that has ever actually existed. There is nothing inevitable in the process I have outlined, and in any actual historical context a multitude of accidents will play a large, often a decisive part. Yet without a reasonable level of trust democratic institutions cannot be sustained. Perhaps that is one of the reasons why tyrannies can endure: they are able to economize on trust in ways that democracies cannot. When tyrannous states that have in the past been able to economize on trust begin to move towards popular participation in government they tend – if they contain peoples that are geographically concentrated – to become fissiparous. In fortunate circumstances these tendencies may work themselves out peacefully. In many, perhaps most, contexts they incur a risk of war.

This is only one illustration of a truth of some practical importance. Even if liberal political morality is universal, applying its principles involves confronting fundamental conflicts of values. Some such conflicts are tragic in that wrong will be done however they are resolved. Advancing democracy does not always foster political stability. Preserving peace does not always coincide with the promotion of human rights. These are not transitory difficulties which we can expect someday to leave behind. They are permanent ethical dilemmas, deeply rooted in conflicts that states will always confront, which will never be fully resolved.

Liberal values cannot give definitive guidance in such cases. These are not conflicts between morality and expediency but *within morality itself*. It is a mistake to think that the most serious ethical conflicts in international relations are conflicts in which the demands of morality collide with considerations of expediency. Such conflicts are doubtless recurring and familiar. But the hardest dilemmas for sovereign states are not conflicts between observing moral principles to which they have committed themselves and promoting the economic interests of their citizens. They are conflicts among the moral principles to which they consider themselves committed. In confronting these inescapable ethical conflicts sovereign states are no different from any other moral agent.

Liberal political morality contains few solutions to the conflicts it generates. The goods that liberal principles protect are not always compatible. Promoting one often involves sacrificing others. We all know that the best foreign policies can have consequences that include significant collateral damage. I suggest that collateral damage is sometimes only another name for moral conflicts that are not wholly soluble. Consider the following examples. There is nothing in freedom of political association that is incompatible with strong government. Some states are fortunate enough to enjoy both. At the same time they are goods that do not always complement one another. Punctilious observation of the terms of its ultra-liberal constitution may have been one of the reasons why the Weimar Republic was short-lived. In that case, a weak democratic state was replaced by a genocidal totalitarian regime. Or consider a case from the world today. China has a long history of recurrent state disintegration. The evils flowing from anarchy are not hypothetical; they are a matter of common experience for hundreds of millions of Chinese now living. Memories of the interwar period and, even more, of the Cultural Revolution are widespread and vivid. Any regime which staves off the threat of anarchy in China has a potent source of political legitimacy in that achievement alone. Western opinion-formers who demand swift progress towards liberal democracy in China have not considered with sufficient seriousness the risks to freedom and security posed to ordinary Chinese by state disintegration. Yet preventing those evils of anarchy is a central feature of the liberal political morality that demands universal democracy. This is an ethical conflict that has no complete solution.

Conclusion

The Enlightenment thinkers who inspire contemporary liberal thought believed that the ethical conflicts that arise from the incompatibility of universal goods could be overcome: at some future point in human progress the species would be rid of the burden of such tragic dilemmas. That Enlightenment belief is an illusion with disabling effects on thought and policy today. Conflicts among the universal goods and evils recognized by liberal morality are not symptoms of backwardness we can hope someday to have transcended. They are perennial and universal.

Viewing the world today through the lens of apocalyptic beliefs about the end of history and 'the West versus the rest' conceals these universal and perennial conflicts. It encourages the hope that the difficult choices and unpleasant trade-offs that have always been necessary in the relations of states will someday be redundant. For that hope there is no rational warrant.

A more reasonable aspiration is that by understanding that some conflicts of values are intractable we will be better able to cope with them. There is much that is new in our present circumstances. What they do not contain is relief from the task of thinking our way through difficulties – conflicts of interests and ideals, incompatibilities among the values we hold most dear – that have always beset relations among states. For some, perhaps, this will seem a rather depressing result. Certainly there is nothing in it that is especially novel, or original; and it contains little that will gratify the commendable need for moral hope. But perhaps these are not quite the defects we commonly imagine them to be. The greatest liberal thinker of our time [Isaiah Berlin] was fond of quoting an observation by the American philosopher, C.I. Lewis: 'There is no *a priori* reason for supposing that the truth, when it is discovered, will necessarily prove interesting'. Nor, I would add, for thinking that it will be particularly comforting.

 **READING 4**

Can Civilizations Clash?

Jack F. Matlock, Jr

[. . .]

Questionable Points

If we examine Huntington's application of the concept of multiple civilizations (as distinct from his discussion of its definition), we find several features that, upon close examination, seem highly dubious.

First, his assumption that there is a high degree of coherence within the civilizations he postulates, which is pervasive in the book despite occasional caveats, is ill founded. The image of civilizations interacting to the point of conflict is that of entities sufficiently close-knit to be independent actors on the global stage. But civilizations, even as Huntington defines them, are not that at all. Pitirim Sorokin's criticism of Arnold Toynbee's concept is relevant.

> By "civilization" Toynbee means not a mere "field of historical study" but a united system, or the whole, whose parts are connected with one another by causal ties. Therefore, as in any causal system in his "civilization," parts must depend upon one another, upon the whole, and the whole upon its parts [. . .]

> Is Toynbee's assumption valid? I am afraid it is not: his *"civilizations" are not united systems but mere conglomerations of various civilizational objects and phenomena [. . .] united only by special adjacency but not by causal or meaningful bonds.*

In practice, Huntington makes the same error Toynbee did in assuming that the many disparate elements that make up his "civilizations" comprise a coherent, interdependent whole. They clearly do not, even if there are more causal relationships among the various elements than Sorokin was willing to admit.

Second, while he repeatedly refers to his civilizations as "the broadest level of cultural identity" or "the broadest cultural entities," he then assumes, without any real evidence, that breadth is correlated with intensity of loyalty. Why else would nations with similar cultures tend to cooperate, as he repeatedly asserts, while those with different cultures tend to fight? Why else should a state's "cultural identity" define its place in world politics?

Actually, there are at least as many conflicts within the civilizations Huntington postulates as there are between them, probably more, in fact. But even if this were not true, there is no reason to assume that a person's loyalty inevitably expands to encompass an area defined by some scholar as a civilization. Any attachment beyond the nation state is likely to be weak (if recognized at all) except in limited contexts, such as a feeling of religious solidarity.

Third, Huntington states repeatedly, without any convincing evidence, that cultural differentiation is increasing in today's world. This flies in the face of most observations of the impact of modernization, industrialization, and the communications revolution, all global phenomena. Huntington is surely correct when he argues that modernization should not be considered synonymous with "westernization," and also that its progress will not obliterate cultural differences. Let us hope and pray that this is the case, since cultural differences are not only sources of potential conflict; they are also the spice of life. Many differences are benign, even productive, and the variety they contribute to civilization in the singular enriches all mankind.

Nevertheless, while there is no reason to believe that we are rushing pell mell into some universal culture, it seems perverse to deny that present trends are creating cross-cultural ties and even uniformities that did not exist before. This is particularly true in those important areas of life such as the work people do, their access to information about the world beyond their locality, and the structure of institutions that shape their economic and civic life. Most human beings are in fact becoming more alike in some parts of their lives, even as they retain and sometimes accentuate their differences in others.

I was bemused by many statements in Huntington's book, but none puzzled me more than the following: "Politicians in non-Western societies do not win elections by demonstrating how Western they are. Electoral competition instead stimulates them to fashion what they believe will be the most popular appeals, and those are usually ethnic, nationalist, and religious in character."

I can only wonder how Huntington would characterize electoral competition in the West, and where he believes non-Western countries acquired the idea of electing political leaders. I can't find it in the Koran or Confucius.

Fourth, despite his extensive discussion of the difference between a culture and a civilization, in practice Huntington uses these words interchangeably in much of his discussion. This leads to repeated confusions, since a conflict sparked or exacerbated by cultural differences may or may not represent a "civilizational" divide. Many of the conflicts in which culture has played a role have been *within* the civilizations he postulates, and yet we often see a part cited as if it were the whole, an evident logical fault.

Furthermore, the concentration on "civilizational" conflict obscures and sometimes totally masks the elements of culture that contribute to conflict. Often, it is cultural similarity, not a difference, that nurtures conflict. Cultures that justify the use of force in disputes with people who are perceived as somehow different are obviously more likely to resort to violence than are those that value accommodation. If two of the first type live in close proximity, the likelihood of conflict would be higher whether or not they belong to different "civilizations." Attributing conflicts to a priori intellectual constructs such as "civilizations" can mislead the observer about the real causes.

[. . .]

Cultures, Not "Civilizations"

Huntington's thesis is not only deficient in predicting the most likely sources of conflict; by lumping cultures into broader civilizations, it obscures what we need to know if we are to understand the implications of cultural differences and similarities. Francis Fukuyama gives a striking example in his recent book, *Trust: The Social Virtues and the Creation of Prosperity*. Discussing a boom in small-scale industry in central Italy in the 1970s and 1980s, he points out some cultural similarities with Hong Kong and Taiwan:

> Though it may seem a stretch to compare Italy with the Confucian culture of Hong Kong and Taiwan, the nature of social capital is similar in certain respects. In parts of Italy and in the Chinese cases, family bonds tend to be stronger than other kinds of bonds not based on kinship, while the strength and number of intermediate associations between state and individual has been relatively low, reflecting a pervasive distrust of people outside the family. The consequences for industrial structure are similar: private sector firms tend to be relatively small and family controlled, while large-scale enterprises need the support of the state to be viable.

If we focus only on what Huntington calls "the broadest cultural entities," we lose the ability to detect and analyze specific cultural features that hold true across civilizations. And yet it is precisely such shared features that help us predict how rapidly specific institutions can spread from one culture to another, and what sort of modifications may result from their transplantation.

A Useful Concept Nevertheless

The faults I have described raise the question whether the analysis of "civilizations" has any utility at all. If one's goal is to understand the behavior of states and nations, it is clearly more important to understand the culture of these units than to presuppose behavior based on some broader cultural conglomerate. But if we define a "civilization" as simply the subject of an intellectual inquiry, it can be a useful term. As Fernand Braudel put it, "A civilization is first of all a space, a cultural area," and he goes on to say, "Whatever the label, there is a distinct French civilization, a German one, an Italian one, an English one, each with its own characteristics and internal contradictions. To study them all together under the heading of Western civilization seems to me to be too simple an approach."

Indeed, the broader the grouping, the more relevant detail is lost, and that which is lost may have a greater effect on behavior than traits held in common. Nevertheless, the extent of the cultural area to be studied is not the main point. There is nothing inherently wrong with looking at "Western civilization," however defined, for common cultural traits, studying how they developed, and examining how they are distributed within the area and how they interact with those of other societies. When used to define the scope of a study, the definition of a "civilization" can be based on any criteria the investigator chooses. Braudel, for example, wrote a magisterial work on the Mediterranean world at the time of Philip II. It does not matter that this work fuses parts of three civilizations as defined by Toynbee or Huntington, since the area had its own coherence, one based on geography rather than religion or politics. As Braudel put it in his preface to the English translation, "I retain the firm conviction that the Turkish Mediterranean lived and breathed with the same rhythms as the Christian, that the whole sea shared a common destiny, a heavy one indeed, with identical problems and general trends if not identical consequences."

It is a mistake, however, to treat a hypothetical "civilization" as anything other than a convenient intellectual construct used to establish the boundaries of a field or topic of study. Even Toynbee, who treated his "civilizations" virtually as organisms, noted in his volume of *Reconsiderations*, "[I]f the use of hypotheses is indispensable, it also has at least one besetting danger: 'the habit of treating a mental convenience as if it were an objective thing.'" Unfortunately, Huntington's application of his concept of civilizations is tainted by this habit.

A civilization by any definition is infinitely more complex than, say, a garden. Nevertheless, describing it is in principle no different. Each garden is unique, yet some will have common characteristics not shared by others. Some plants will grow well in some soils and poorly if at all in others. Some plants may take over if moved to a different environment. Some gardens are laid out in a strict geometry; others may be left, in places at least, to resemble wild growth. If the gardener is not careful, the colors of some flowers may clash. Observers can classify gardens, compare them, discuss whether elements harmonize or not.

Gardens, like civilizations, can be described, analyzed and interpreted. But one thing is certain. It would be absurd to speak of a "clash of gardens." It is equally absurd to speak of a "clash of civilizations." If the concept were valid, it would provide a useful shortcut to understanding the tensions and potential conflicts in the world. But it is not a shortcut to understanding. Rather, it is a diversion leading to confusion. If we are to understand where future conflict is most likely and how it can best be averted or contained, we must keep our attention on the actors on the international scene: the states, the organized movements, the international alliances and institutions. Their cultures are relevant, but so are other factors such as geographical position, economic and military strength, and membership in or exclusion from international institutions. We gain nothing by lumping cultures into broader conglomerates, and we can be seriously misled if we assume that difference inevitably means hostility. Life, and politics, are not so simple.

 READING 5

History Ends, Worlds Collide

Chris Brown

[. . .]

It is easy to pick holes in Huntington's work, especially the book-length version of his argument, which, precisely becomes it contains so much more detail is much more open to criticism – broad generalizations which pass muster in the enclosed context of a short

article are less tolerable when more space is available. Right from the outset his account of 'civilization' is *ad hoc* and muddled; civilizations are systems of ideas, and, as such, it is difficult to see how they could clash, although individuals and groups claiming to represent these ideas certainly can. Moreover, these systems of ideas are not now, nor have they ever been, self-contained or impermeable, a fact that Huntington acknowledges, but the significance of which he, perhaps, underplays. On the other hand, he deserves considerable credit for attempting to break up what was becoming in the early 1990s a rather sterile debate about the post-Cold War world. In his response to critics 'If Not Civilizations, What?', Huntington suggests that the only alternative models for what he is interested in are the old statist paradigm and a new 'un-real' vision of one world united by globalization; this is to put the matter rather starkly, but there is some justice to this claim. In effect, Huntington is providing a non-statist, but nonetheless realist, account of the world, which is an interesting addition to the conceptual toolkit of contemporary international relations theory. Part of the problem with Huntington's analysis, though, is that, although not statist, it remains spatial/territorial.

The prevailing metaphor in that book is that there are physical 'fault-lines' between civilizations. There are two problems with this notion; first, the analysis underplays the extent to which key dividing lines are man-made and recent – in former Yugoslavia, for example, the recurrent crises of the 1990s owe more to the success of Milosevic in mobilizing political support behind the nationalist cause of Greater Serbia than they do to largely spurious ethnic and religious differences, much less historical divides that go back to the Middle Ages or earlier. Such differences and divides certainly exist and have always existed, but their current political significance is the result of contingency rather than some inevitable process. Second, and rather more important, the 'tectonic' notion of civilizations does not recognise sufficiently the extent to which civilizations are already interpenetrated. The clash of civilizations, in so far as it exists at all, is more likely to take the form of the politics of multiculturalism and recognition in the major cities of the world than violent clashes on the so-called 'fault-lines'; policing problems in London are, thankfully, more characteristic of this politics than ethnic cleansing in Kosovo, horrifying though the latter may be.

[. . .]

This set of choices does indeed convey some sense of what is going on but on the whole it obscures more than it illuminates. What is particularly damaging about the way in which these oppositions are set up is that they tend to define the most important questions about the future in terms of a choice between universalism and particularism, with the underlying assumption that the former is the progressive option, while the latter, though possibly unavoidable, is regressive and not to be desired.

[. . .]

Equally, whether 'civilizations' clash along particular fault-lines is going to depend on how the inhabitants of those key areas, and their neighbours, near and far, choose to define themselves or allow political entrepreneurs to define them, and this is a political process, not one that follows a cultural recipe book. More generally, the future of globalization will be a product of political practice rather than cultural or economic theory. In short, one way or another, the major questions about the future of world order which this article has addressed will be answered in the years to come, but they will not necessarily be answered in their own terms; the contingencies of political power may have the last word, as so often in the past.

 READING 6

If Not Civilizations, What? Paradigms of the Post-Cold War World

Samuel P. Huntington

When people think seriously, they think abstractly; they conjure up simplified pictures of reality called concepts, theories, models, paradigms. Without such intellectual constructs, there is, William James said, only "a bloomin' buzzin' confusion." Intellectual and scientific advance, as Thomas Kuhn showed in his classic *The Structure of Scientific Revolutions*, consists of the displacement of one paradigm, which has become increasingly incapable of explaining new or newly discovered facts, by a new paradigm that accounts for those facts in a more satisfactory fashion. "To be accepted as a paradigm," Kuhn wrote, "a theory must seem better than its competitors, but it need not, and in fact never does, explain all the facts with which it can be confronted."

For 40 years students and practitioners of international relations thought and acted in terms of a highly simplified but very useful picture of world affairs, the Cold War paradigm. The world was divided between one group of relatively wealthy and mostly democratic societies, led by the United States, engaged in a pervasive ideological, political, economic, and, at times, military conflict with another group of somewhat poorer, communist societies led by the Soviet Union. Much of this conflict occurred in the Third World outside of these two camps, composed of countries which often were poor, lacked political stability, were recently independent and claimed to be nonaligned. The Cold War paradigm could not account for everything that went on in world politics. There were many anomalies, to use Kuhn's term, and at times the paradigm blinded scholars and statesmen to major developments, such as the Sino–Soviet split. Yet as a simple model of global politics, it accounted for more important phenomena than any of its rivals; it was an indispensable starting point for thinking about international affairs; it came to be almost universally accepted; and it shaped thinking about world politics for two generations.

The dramatic events of the past five years have made that paradigm intellectual history. There is clearly a need for a new model that will help us to order and to understand central developments in world politics. What is the best simple map of the post-Cold War world?

A Map of the New World

"The Clash of Civilizations?" is an effort to lay out elements of a post-Cold War paradigm. As with any paradigm, there is much the civilization paradigm does not account for, and critics will have no trouble citing events – even important events like Iraq's invasion of Kuwait – that it does not explain and would not have predicted (although it would have predicted the evaporation of the anti-Iraq coalition after March 1991). Yet, as Kuhn demonstrates, anomalous events do not falsify a paradigm. A paradigm is disproved only by the creation of an alternative paradigm that accounts for more crucial facts in equally simple or simpler terms (that is, at a comparable level of intellectual abstraction; a more complex theory can always account for more things than a more parsimonious theory). The debates the civilizational paradigm has generated around the world show that, in some measure, it strikes home; it either accords with reality as people see it or it comes close enough so that people who do not accept it have to attack it.

What groupings of countries will be most important in world affairs and most relevant to understanding and making sense of global politics? Countries no longer belong to the Free World, the communist bloc, or the Third World. Simple two-way divisions of countries into rich and poor or democratic and nondemocratic may help some but not all that much. Global politics are now too complex to be stuffed into two pigeonholes. For reasons outlined in the original article, civilizations are the natural successors to the three worlds of the Cold War. At the macro level world politics are likely to involve conflicts and shifting power balances of states from different civilizations, and at the micro level the most violent, prolonged and dangerous (because of the possibility of escalation) conflicts are likely to be between states and groups from different civilizations. As the article pointed out, this civilization paradigm accounts for many important developments in international affairs in recent years, including the breakup of the Soviet Union and Yugoslavia, the wars going on in their former territories, the rise of religious fundamentalism throughout the world, the struggles within Russia, Turkey and Mexico over their identity, the intensity of the trade conflicts between the United States and Japan, the resistance of Islamic states to Western pressure on Iraq and Libya, the efforts of Islamic and Confucian states to acquire nuclear weapons and the means to deliver them, China's continuing role as an "outsider" great power, the consolidation of new democratic regimes in some countries and not in others, and the escalating arms race in East Asia.

[. . .]

America Undone?

One function of a paradigm is to highlight what is important (e.g., the potential for escalation in clashes between groups from different civilizations); another is to place familiar phenomena in a new perspective. In this respect, the civilizational paradigm may have implications for the United States. Countries like the Soviet Union and Yugoslavia that bestride civilizational fault lines tend to come apart. The unity of the United States has historically rested on the twin bedrocks of European culture and political democracy. These have been essentials of America to which generations of immigrants have assimilated. The essence of the American creed has been equal rights for the individual, and historically immigrant and outcast groups have invoked and thereby reinvigorated the principles of the creed in their struggles for equal treatment in American society. The most notable and successful effort was the civil rights movement led by Martin Luther King, Jr, in the 1950s and 1960s. Subsequently, however, the demand shifted from equal rights for individuals to special rights (affirmative action and similar measures) for blacks and other groups. Such claims run directly counter to the underlying principles that have been the basis of American political unity; they reject the idea of a "color-blind" society of equal individuals and instead promote a "color-conscious" society with government-sanctioned privileges for some groups. In a parallel movement, intellectuals and politicians began to push the ideology of "multiculturalism," and to insist on the rewriting of American political, social, and literary history from the viewpoint of non-European groups. At the extreme, this movement tends to elevate obscure leaders of minority groups to a level of importance equal to that of the Founding Fathers. Both the demands for special group rights and for multiculturalism encourage a clash of civilizations within the United States and encourage what Arthur M. Schlesinger, Jr, terms "the disuniting of America."

The United States is becoming increasingly diverse ethnically and racially. The Census Bureau estimates that by 2050 the American population will be 23 percent Hispanic, 16 percent black and 10 percent Asian-American. In the past the United States has successfully absorbed millions of immigrants from scores of countries because they adapted to the prevailing European culture and enthusiastically embraced the American Creed of liberty, equality, individualism, democracy. Will this pattern continue to prevail as 50 percent of the population becomes Hispanic or nonwhite? Will the new immigrants be assimilated into the hitherto dominant European culture of the United States? If they are not, if the United States becomes truly multicultural and pervaded with an internal clash of civilizations, will it survive as a liberal democracy? The political identity of the United States is rooted in the principles articulated in its founding documents. Will the de-Westernization of the United States, if it occurs, also mean its de-Americanization? If it does and Americans cease to adhere to their liberal democratic and European-rooted political ideology, the United States as we have known it will cease to exist and will follow the other ideologically defined superpower onto the ash heap of history.

Got a Better Idea?

A civilizational approach explains much and orders much of the "bloomin' buzzin' confusion" of the post-Cold War world, which is why it has attracted so much attention and generated so much debate around the world. Can any other paradigm do better? If not civilizations, what? The responses in *Foreign Affairs* to my article did not provide any compelling alternative picture of the world. At best they suggested one pseudo-alternative and one unreal alternative.

The pseudo-alternative is a statist paradigm that constructs a totally irrelevant and artificial opposition between states and civilizations: "Civilizations do not control states," says Fouad Ajami, "states control civilizations." But it is meaningless to talk about states and civilizations in terms of "control." States, of course, try to balance power, but if that is all they did, West European countries would have coalesced with the Soviet Union against the United States in the late 1940s. States respond primarily to perceived threats, and the West European states then saw a political and ideological threat from the East. As my original article argued, civilizations are composed of one or more states, and "Nation states will remain the most powerful actors in world affairs." Just as nation states generally belonged to one of three worlds in the Cold War, they also belong to civilizations. With the demise of the three worlds, nation states increasingly define their identity and their interests in civilizational terms, and West European peoples and states now see a cultural threat from the South replacing the ideological threat from the East.

We do not live in a world of countries characterized by the "solitude of states" (to use Ajami's phrase) with no connections between them. Our world is one of overlapping groupings of states brought together in varying degrees by history, culture, religion, language, location and institutions. At the broadest level these groupings are civilizations. To deny their existence is to deny the basic realities of human existence.

The unreal alternative is the one-world paradigm that a universal civilization now exists or is likely to exist in the coming years. Obviously people now have and for millennia have had common characteristics that distinguish humans from other species. These characteristics have always been compatible with the existence of very different cultures. The argument that a universal culture or civilization is now emerging takes various forms, none of which withstands even passing scrutiny.

First, there is the argument that the collapse of Soviet communism means the end of history and the universal victory of liberal democracy throughout the world. This argument suffers from the Single Alternative Fallacy. It is rooted in the Cold War assumption that the only alternative to communism is liberal democracy and that the demise of the first produces the universality of the second. Obviously, however, there are many forms of authoritarianism, nationalism, corporatism and market communism (as in China) that are alive and well in today's world. More significantly, there are all the religious alternatives that lie outside the world that is perceived in terms of secular ideologies. In the modern world, religion is a central, perhaps *the* central, force that motivates and mobilizes people. It is sheer hubris to think that because Soviet communism has collapsed the West has won the world for all time.

Second, there is the assumption that increased interaction – greater communication and transportation – produces a common culture. In some circumstances this may be the case. But wars occur most frequently between societies with high levels of interaction, and interaction frequently reinforces existing identities and produces resistance, reaction and confrontation.

Third, there is the assumption that modernization and economic development have a homogenizing effect and produce a common modern culture closely resembling that which has existed in the West in this century. Clearly, modern urban, literate, wealthy, industrialized societies do share cultural traits that distinguish them from backward, rural, poor, undeveloped societies. In the contemporary world most modern societies have been Western societies. But modernization does not equal Westernization. Japan, Singapore and Saudi Arabia are modern, prosperous societies but they clearly are non-Western. The presumption of Westerners that other peoples who modernize must become "like us" is a bit of Western arrogance that in itself illustrates the clash of civilizations. To argue that Slovenes and Serbs, Arabs and Jews, Hindus and Muslims, Russians and Tajiks, Tamils and Sinhalese, Tibetans and Chinese, Japanese and Americans all belong to a single Western-defined universal civilization is to fly in the face of reality.

A universal civilization can only be the product of universal power. Roman power created a near-universal civilization within the limited confines of the ancient world. Western power in the form of European colonialism in the nineteenth century and American hegemony in the twentieth century extended Western culture throughout much of the contemporary world. European colonialism is over; American hegemony is receding. The erosion of Western culture follows, as indigenous, historically rooted mores, languages, beliefs and institutions reassert themselves.

Amazingly, Ajami cites India as evidence of the sweeping power of Western modernity. "India," he says, "will not become a Hindu state. The inheritance of Indian secularism will hold." Maybe it will, but certainly the overwhelming trend is away from Nehru's vision of a secular, socialist, Western, parliamentary democracy to a society shaped by Hindu fundamentalism. In India, Ajami goes on to say, "The vast middle class will defend it [secularism], keep the order intact to maintain India's – and its own – place in the modern world of nations." Really? A long *New York Times* (September 23, 1993) story on this subject begins: "Slowly, gradually, but with the relentlessness of floodwaters, a growing Hindu rage toward India's Muslim minority has been spreading among India's solid middle class Hindus – its merchants and accountants, its lawyers and engineers – creating uncertainty about the future ability of adherents of the two religions to get along." An op-ed piece in the *Times* (August 3, 1993) by an Indian journalist also highlights the role of the middle class: "The most disturbing development is the increasing number of senior civil servants, intellectuals, and journalists who have begun to talk the language of Hindu fundamentalism, protesting that religious minorities, particularly the Muslims, have pushed them beyond the limits of patience." This author, Khushwant Singh, concludes sadly that while India may retain a secular facade, India "will no longer be the India we have known over the past 47 years" and "the spirit within will be that of militant Hinduism." In India, as in other societies, fundamentalism is on the rise and is largely a middle class phenomenon.

The decline of Western power will be followed, and is beginning to be followed, by the retreat of Western culture. The rapidly increasing economic power of East Asian states will, as Kishore Mahbubani asserted, lead to increasing military power, political influence and cultural assertiveness. A colleague of his has elaborated this warning with respect to human rights:

> [E]fforts to promote human rights in Asia must also reckon with the altered distribution of power in the post-Cold War world [. . .] Western leverage over East and Southeast Asia has been greatly reduced [. . .] There is far less scope for conditionality and sanctions to force compliance with human rights [. . .]
>
> For the first time since the Universal Declaration [on Human Rights] was adopted in 1948, countries not thoroughly steeped in the Judeo-Christian and natural law traditions are in the first rank: That unprecedented situation will define the new international politics of human rights. It will also multiply the occasions for conflict [. . .]

Economic success has engendered a greater cultural self-confidence. Whatever their differences, East and Southeast Asian countries are increasingly conscious of their own civilizations and tend to locate the sources of their economic success in their own distinctive traditions and institutions. The self-congratulatory, simplistic, and sanctimonious tone of much Western commentary at the end of the Cold War and the current triumphalism of Western values grate on East and Southeast Asians.

Language is, of course, central to culture, and Ajami and Robert Bartley both cite the widespread use of English as evidence for the universality of Western culture (although Ajami's fictional example dates from 1900). Is, however, use of English increasing or decreasing in relation to other languages? In India, Africa and elsewhere, indigenous languages have been replacing those of the colonial rulers. Even as Ajami and Bartley were penning their comments, *Newsweek* ran an article entitled "English Not Spoken Here Much Anymore" on Chinese replacing English as the lingua franca of Hong Kong. In a parallel development, Serbs now call their language Serbian, not Serbo-Croatian, and write it in the Cyrillic script of their Russian kinsmen, not in the Western script of their Catholic enemies. At the same time, Azerbaijan, Turkmenistan, and Uzbekistan have shifted from the Cyrillic script of their former Russian masters to the Western script of their Turkish kinsmen. On the language front, Babelization prevails over universalization and further evidences the rise of civilization identity.

Culture Is To Die For

Wherever one turns, the world is at odds with itself. If differences in civilization are not responsible for these conflicts, what is? The critics of the civilization paradigm have not produced a better explanation for what is going on in the world. The civilizational paradigm, in contrast, strikes a responsive chord throughout the world. In Asia, as one US ambassador reported, it is "spreading like wildfire." In Europe, European Community President Jacques Delors explicitly endorsed its argument that "future conflicts will be sparked by cultural factors rather than economics or ideology" and warned, "The West needs to develop a deeper understanding of the religious and philosophical assumptions underlying other civilizations, and the way other nations see their interests, to identify what we have in common." Muslims, in turn, have seen "the clash" as providing recognition and, in some degree, legitimation for the distinctiveness of their own civilization and its independence from the West. That civilizations are meaningful entities accords with the way in which people see and experience reality.

History has not ended. The world is not one. Civilizations unite and divide humankind. The forces making for clashes between civilizations can be contained only if they are recognized. In a "world of different civilizations," as my article concluded, each "will have to learn to coexist with the others." What ultimately counts for people is not political ideology or economic interest. Faith and family, blood and belief, are what people identify with and what they will fight and die for. And that is why the clash of civilizations is replacing the Cold War as the central phenomenon of global politics, and why a civilizational paradigm provides, better than any alternative, a useful starting point for understanding and coping with the changes going on in the world.

Orientalism, Colonialism, and Postcolonialism

This chapter is concerned primarily with a topic, Orientalism, with roots in literary theory, but it also permits us to deal, at least briefly, with several other ideas closely related to globalization including colonialism and postcolonialism.

Literary theory involves, as its name suggests, studying, thinking about, and theorizing some body of literature. In the case of globalization, the most relevant body of literary theory involves the study of literature that was produced in, or is about, the experience of people who once lived in areas that were colonized, usually by the major Western powers (especially Britain). This literature is usually categorized under the heading of *postcolonialism*, or "a systematic discourse dedicated to investigating, analyzing, and deconstructing structures of knowledge, ideologies, power relations, and social identities that have been authored by and authorized by the imperial West in ruling and representing the non-West over the past 500 years."[1]

Edward Said's[2] *Orientalism* is "the founding document of post-colonial thought."[3] While it was not written with the idea of globalization in mind, and was written before the current era of globalization, it has powerful implications for contemporary thinking on globalization.

Orientalism has several interrelated meanings for Said. First, it is an area of academic interest (a discipline) with schools of "Oriental Studies." Thus, "the Orient was a scholar's word."[4] Second, it is a "style of thought based upon an ontological and epistemological distinction made between 'the Orient' and (most of the time) 'the Occident.'"[5] Third, and perhaps most importantly, Orientalism is a Western discourse "for dominating, restructuring, and having authority over the Orient."[6] It was the basis for the ways in which European culture "was able to manage – and even produce – the Orient politically, sociologically, militarily, ideologically, scientifically, and imaginatively."[7]

Orientalism was (and still is) a diverse cultural enterprise that included, among other things:

> The imagination itself, the whole of India and the Levant, the Biblical texts and the Biblical lands, the spice trade, colonial armies and the long tradition of colonial administrators, a formidable scholarly corpus, innumerable Oriental "experts" and "hands," an Oriental professorate, a complex array of "Oriental" ideas (Oriental despotism, Oriental splendor, cruelty, sensuality), many Eastern sects, philosophies and wisdoms domesticated for local European use.[8]

In spite of this diversity, and although it is far more than just ideas/discourse, Orientalism is primarily a set of ideas expressed in a specific discourse. Following

Michel Foucault (and Friedrich Nietzsche), knowledge cannot be divorced from power; and it was to a large degree as a result of Orientalism that Europe and the West more generally were able to exercise power over the East. To get at Orientalism as ideas/discourse, Said examines a variety of "texts" including not only scholarly works on the topic "but also works of literature, political tracts, journalistic texts, travel books, religious and philological studies."[9] The Orient that emerges from these texts "is less a place than a *topos*, a set of references, a congeries of characteristics, that seems to have its origin in a quotation, or a fragment of a text, or a citation from someone's work on the Orient, or some bit of previous imagining, or an amalgam of all of these."[10] The ideas associated with Orientalism are largely repeatedly reproduced fictions (although they are not totally false) that are rarely, if ever, based on observation, let alone careful empirical study.

Said's basic problem with Orientalism, aside from its disastrous effects on those labeled Orientals, is that it is an idea characterized by biases, ignorance, lack of knowledge, stereotypes, standardized views, and fictions. Orientalism reflects the power of the West and has little to do with the realities of life in the Orient. Negative stereotypes of Orientals abounded and they were shaped by Westerners' stereotypes of themselves. Westerners produced biased and limited "texts" about the Orient and it was those texts, and not life as it really existed in the Orient, which came to be considered the basis of the "truth" about the Orient.

There are a variety of intellectual problems with Orientalism that result from it "disregarding, essentializing, denuding the humanity of another culture."[11] People in the Orient were not discussed in individual or humanistic terms, but rather in collective or abstract terms. Furthermore, the view of the Orient has remained more or less the same in terms of both time and place for those in the West who think about, analyze, manage, and seek to subdue it. It is as if nothing has changed, or will ever change, in the Orient. More generally, Said argues that: "The West is the spectator, the judge and jury, of every facet of Oriental behavior."[12] Knowledge of the Orient, often unchanged over great stretches of time, was accumulated in the West, and this was closely related to the accumulation of both the people and the territories of the Orient by the West.

Said reserves his most scathing indictment for Orientalism as it relates to Islam. It is characterized by its "retrogressive position when compared with the other human sciences (and even with the other branches of Orientalism), its general methodological and ideological backwardness, and its comparative insularity from developments both in the other humanities and in the real world of historical, economic, social and political circumstances."[13]

Orientalism was, and still is, a highly influential book, but it is also one that has been subjected to many criticisms. Sadik Jalal al-'Azm offers several of the most important of these criticisms. For one thing, Said is seen as not restricting his analysis to the modern world, but tracing Orientalism back to the ancient Greeks and then up to, and including, the work of Karl Marx. The problem with this is that instead of being a product of a particular history, Orientalism tends to become *essentialistic*. That is, Said's work "simply lends strength to the essentialistic categories of 'Orient' and 'Occident,' representing the ineradicable distinction between East and West, which Edward's [Said's] book is ostensibly set on demolishing."[14]

Perhaps a more important criticism is that Said gives literature, and culture more generally, too much power. He seems to suggest that they are the "real source of the West's political interest in the Orient."[15] Downplayed in all of this are the political and material interests in the West in conquering and controlling the Orient. Thus, for example, France and Britain were interested in controlling the Suez Canal not because of "Orientalism," but because of the political, military, and economic advantages such control gave them. As al-'Azm puts it: "If Academic Orientalism transmutes the reality of the Orient into the stuff of texts . . . then it would seem that Said sublimates the earthly realities of the Occident's interaction with the Orient into the ethereal stuff of the spirit."[16]

Rattansi puts *Orientalism* in the context of the postcolonial studies that it played a central role in creating. On the one hand, postcolonialism refers to a time period after the period of *colonialism*, that is after the colonies of the Western imperial powers gained their independence. (Colonialism is the creation by a colonial power of an administrative apparatus in the country or geographic area that has been colonized in order to run its internal affairs, including its settlements.)

On the other hand, postcolonialism is a "*distinctive form of theorization and analysis*" that is not restricted to that time period or to those particular places.[17] Thus, Rattansi seeks to distinguish between *postcolonialism* as a type of intellectual inquiry and *postcoloniality* as historical epochs. What is crucial about postcolonialism, i.e., postcolonial studies, is that they involve "*the investigation of the mutually constitutive role played by colonizer and colonized . . . in forming . . . the identities of both the dominant power and the subalterns involved in the imperial and colonial projects of the 'West.'*"[18] It is in this context that Rattansi argues that *Orientalism* can be seen as "the founding text of modern postcolonialist studies."[19]

Rattansi examines some key works in postcolonial studies. A first set deals with the mutual constitution of identities between colonizer and colonized. A second is concerned with the ambivalence surrounding the relationship between colonizer and colonized, as well as the resistance that arises, at least in part, out of the instabilities in that relationship.

Rattansi also examines the relationship between the colonial/postcolonial and a series of related ideas such as the imperial/postimperial, the neocolonial, and the anticolonial. Most importantly for our purposes, Rattansi looks at the relationship between the idea of *globalization* (referring, in this case, to the general process of time–space compression) and postcolonialism. He concludes that the concept of postcolonialism remains useful because it reminds us that "imperial expansion and colonialism were key constitutive features, and indeed set both globalization and Western capitalism in motion and acted as continual fuelling forces."[20]

Rattansi closes with a rejection of the idea that postcolonial studies are restricted to those done by scholars associated with the former colonizers; instead he argues that such studies have become a truly international enterprise. He rejects the idea that postcolonial studies have ignored material forces such as Western capitalism. However, Rattansi also expresses reservations about postcolonial studies, including the work of Said. For example, he worries about the fact that this critical work fails to put forward an alternative vision of the future to that of the Orientalists and the colonialists. In spite of the fact that postcolonial studies have their weaknesses, they represent an important new body of work.

We close this chapter with Peter Marcuse's effort to relate Orientalism to today's world, especially globalization. More specifically, Marcuse seeks to relate Orientalism to what he calls "globalism." While "'Orientalism' was used to describe and categorize a specific geographic region, its people and its culture," "globalism" is employed "to suggest the way in which specific real processes at the international level, often lumped together under the term globalization, are discussed and portrayed in academic and popular circles."[21] Globalism is a specific view of globalization held by governments, scholars, and intellectuals. In this view, globalization tends to be seen as something new, dominant, involving a process free of individual choice, inevitable, and largely beneficial. As Marcuse puts it: "Globalism is to really existing globalization as Orientalism is to colonialism. Globalism is the hegemonic metaphor through which the actual process of globalization is seen/presented. It views development in the 'developing world' as inevitably following the superior path of development pursued by the 'developed world,' just as Orientalism sees the 'Orient' following (if it can) the superior form of development of the 'Occident.'"[22] Marcuse proceeds to iterate a number of other similarities between Orientalism and globalism. For example, just as Orientalism was a distorted lens through which to view the world, globalism is a distorted lens through which to view globalization. That is, globalization is seen as inevitable and is accepted unquestioningly. Such a view serves to defuse opposition to globalization. Said's work is seen as helpful here because it has been "a potent weapon on the side of social justice and the struggle for a humane world."[23] Marcuse sees a similar role for those who are critical of globalism such as those associated with the World Social Forum.

NOTES

1 Shaobao Xie, "Postcolonialism." In Jan Aart Scholte and Roland Robertson, eds., *Encyclopedia of Globalization*. New York: MTM, 2007, 986–90.

2 Among other key figures are Homi Bhabha (*The Location of Culture*. London: Routledge, 1994) and Gayatri Chakravorty Spivak (*In Other Worlds: Essays in Cultural Politics*. New York: Routledge, 1987; *A Critique of Postcolonial Reason: Toward a History of the Vanishing Present*. Cambridge, MA: Harvard University Press, 1999).

3 Joan Acocella, "A Better Place." *New Yorker* February 4, 2008: 68–9.

4 Edward W. Said, *Orientalism*. New York: Vintage, 1979/1994, 92.

5 Ibid., 2.

6 Ibid., 3.

7 Ibid., 3.

8 Ibid., 4.

9 Ibid., 23.

10 Ibid., 177.

11 Ibid., 108.

12 Ibid., 109.

13 Ibid., 261.

14 Sadik Jalal al-'Azm, "Orientalism and Orientalism in Reverse." In A. L. Macfie, ed., *Orientalism: A Reader*. New York: New York University Press, 219.

15 Ibid., 220.

16 Ibid., 221.

17 Ali Rattansi, "Postcolonialism and Its Discontents." *Economy and Society* 26, 4, 1997: 481, italics in original.

18 Ibid., 481, italics in original.

19 Ibid., 483.

20 Ibid., 492.

21 Peter Marcuse, "Said's Orientalism: A Vital Contribution Today." *Antipode* 2004: 809.

22 Ibid., 810.

23 Ibid., 816.

READING 7

Orientalism: Introduction

Edward W. Said

I

On a visit to Beirut during the terrible civil war of 1975–6 a French journalist wrote regretfully of the gutted downtown area that "it had once seemed to belong to [. . .] the Orient of Chateaubriand and Nerval." He was right about the place, of course, especially so far as a European was concerned. The Orient was almost a European invention, and had been since antiquity a place of romance, exotic beings, haunting memories and landscapes, remarkable experiences. Now it was disappearing; in a sense it had happened, its time was over. Perhaps it seemed irrelevant that Orientals themselves had something at stake in the process, that even in the time of Chateaubriand and Nerval Orientals had lived there, and that now it was they who were suffering; the main thing for the European visitor was a European representation of the Orient and its contemporary fate, both of which had a privileged communal significance for the journalist and his French readers.

Americans will not feel quite the same about the Orient, which for them is much more likely to be associated very differently with the Far East (China and Japan, mainly). Unlike the Americans, the French and the British – less so the Germans, Russians, Spanish, Portuguese, Italians, and Swiss – have had a long tradition of what I shall be calling *Orientalism*, a way of coming to terms with the Orient that is based on the Orient's special place in European Western experience. The Orient is not only adjacent to Europe; it is also the place of Europe's greatest and richest and oldest colonies, the source of its civilizations and languages, its cultural contestant, and one of its deepest and most recurring images of the Other. In addition, the Orient has helped to define Europe (or the West) as its contrasting image, idea, personality, experience. Yet none of this Orient is merely imaginative. The Orient

is an integral part of European *material* civilization and culture. Orientalism expresses and represents that part culturally and even ideologically as a mode of discourse with supporting institutions, vocabulary, scholarship, imagery, doctrines, even colonial bureaucracies and colonial styles. In contrast, the American understanding of the Orient will seem considerably less dense, although our recent Japanese, Korean, and Indochinese adventures ought now to be creating a more sober, more realistic "Oriental" awareness. Moreover, the vastly expanded American political and economic role in the Near East (the Middle East) makes great claims on our understanding of that Orient.

It will be clear to the reader (and will become clearer still throughout the many pages that follow) that by Orientalism I mean several things, all of them, in my opinion, interdependent. The most readily accepted designation for Orientalism is an academic one, and indeed the label still serves in a number of academic institutions. Anyone who teaches, writes about, or researches the Orient – and this applies whether the person is an anthropologist, sociologist, historian, or philologist – either in its specific or its general aspects, is an Orientalist, and what he or she does is Orientalism. Compared with *Oriental studies* or *area studies*, it is true that the term *Orientalism* is less preferred by specialists today, both because it is too vague and general and because it connotes the high-handed executive attitude of nineteenth-century and early-twentieth-century European colonialism. Nevertheless books are written and congresses held with "the Orient" as their main focus, with the Orientalist in his new or old guise as their main authority. The point is that even if it does not survive as it once did, Orientalism lives on academically through its doctrines and theses about the Orient and the Oriental.

Related to this academic tradition, whose fortunes, transmigrations, specializations, and transmissions are in part the subject of this study, is a more general meaning for Orientalism. Orientalism is a style of thought based upon an ontological and epistemological distinction made between "the Orient" and (most of the time) "the Occident." Thus a very large mass of writers, among whom are poets, novelists, philosophers, political theorists, economists, and imperial administrators, have accepted the basic distinction between East and West as the starting point for elaborate theories, epics, novels, social descriptions, and political accounts concerning the Orient, its people, customs, "mind," destiny, and so on. *This* Orientalism can accommodate Aeschylus, say, and Victor Hugo, Dante and Karl Marx. A little later in this introduction I shall deal with the methodological problems one encounters in so broadly construed a "field" as this.

The interchange between the academic and the more or less imaginative meanings of Orientalism is a constant one, and since the late eighteenth century there has been a considerable, quite disciplined – perhaps even regulated – traffic between the two. Here I come to the third meaning of Orientalism, which is something more historically and materially defined than either of the other two. Taking the late eighteenth century as a very roughly defined starting point Orientalism can be discussed and analyzed as the corporate institution for dealing with the Orient – dealing with it by making statements about it, authorizing views of it, describing it, by teaching it, settling it, ruling over it: in short, Orientalism as a Western style for dominating, restructuring, and having authority over the Orient. I have found it useful here to employ Michel Foucault's notion of a discourse, as described by him in *The Archaeology of Knowledge* and in *Discipline and Punish*, to identify Orientalism. My contention is that without examining Orientalism as a discourse one cannot possibly understand the enormously systematic discipline by which European culture was able to manage – and even produce – the Orient politically, sociologically, militarily, ideologically, scientifically, and imaginatively during the post-Enlightenment period. Moreover, so authoritative a position did Orientalism have that I believe no one writing, thinking, or acting on the Orient could do so without taking account of the limitations on thought and action imposed by Orientalism. In brief, because of Orientalism the Orient was not (and is not) a free subject of thought or action.

This is not to say that Orientalism unilaterally determines what can be said about the Orient, but that it is the whole network of interests inevitably brought to bear on (and therefore always involved in) any occasion when that peculiar entity "the Orient" is in question. How this happens is what this book tries to demonstrate. It also tries to show that European culture gained in strength and identity by setting itself off against the Orient as a sort of surrogate and even underground self.

Historically and culturally there is a quantitative as well as a qualitative difference between the Franco-British involvement in the Orient and – until the period of American ascendancy after World War II – the involvement of every other European and Atlantic power. To speak of Orientalism therefore is to speak mainly, although not exclusively, of a British and French cultural enterprise, a project whose dimensions take in such disparate realms as the imagination itself, the whole of India and the Levant, the Biblical texts and the Biblical lands, the spice trade, colonial armies and a long tradition of colonial administrators, a formidable scholarly corpus, innumerable Oriental "experts" and "hands," an Oriental professorate, a complex array of "Oriental" ideas (Oriental despotism, Oriental splendor, cruelty, sensuality), many Eastern sects, philosophies, and wisdoms domesticated for local European use – the list can be extended more or less indefinitely. My point is that Orientalism derives from a particular closeness experienced between Britain and France and the Orient, which until the early nineteenth century had really meant only India and the Bible lands. From the beginning of the nineteenth century until the end of World War II France and Britain dominated the Orient and Orientalism; since World War II America has dominated the Orient, and approaches it as France and Britain once did. Out of that closeness, whose dynamic is enormously productive even if it always demonstrates the comparatively greater strength of the Occident (British, French, or American), comes the large body of texts I call Orientalist.

It should be said at once that even with the generous number of books and authors that I examine, there is a much larger number that I simply have had to leave out. My argument, however, depends neither upon an exhaustive catalogue of texts dealing with the Orient nor upon a clearly delimited set of texts, authors, and ideas that together make up the Orientalist canon. I have depended instead upon a different methodological alternative – whose backbone in a sense is the set of historical generalizations I have so far been making in this Introduction – and it is these I want now to discuss in more analytical detail.

II

I have begun with the assumption that the Orient is not an inert fact of nature. It is not merely *there*, just as the Occident itself is not just *there* either. We must take seriously Vico's great observation that men make their own history, that what they can know is what they have made, and extend it to geography: as both geographical and cultural entities – to say nothing of historical entities – such locales, regions, geographical sectors as "Orient" and "Occident" are man-made. Therefore as much as the West itself, the Orient is an idea that has a history and a tradition of thought, imagery, and vocabulary that have given it reality and presence in and for the West. The two geographical entities thus support and to an extent reflect each other.

Having said that, one must go on to state a number of reasonable qualifications. In the first place, it would be wrong to conclude that the Orient was *essentially* an idea, or a creation with no corresponding reality. When Disraeli said in his novel *Tancred* that the East was a career, he meant that to be interested in the East was something bright young Westerners would find to be an all-consuming passion; he should not be interpreted as saying that the East was *only* a career for Westerners. There were – and are – cultures and nations whose location is in the East, and their lives, histories, and customs have a brute reality obviously greater than anything that could be said about them in the West. About that fact this study of Orientalism has very little to contribute, except to acknowledge it tacitly. But the phenomenon of Orientalism as I study it here deals principally, not with a correspondence between Orientalism and Orient, but with the internal consistency of Orientalism and its ideas about the Orient (the East as career) despite or beyond any correspondence, or lack thereof, with a "real" Orient. My point is that Disraeli's statement about the East refers mainly to that created consistency, that regular constellation of ideas as the pre-eminent thing about the Orient, and not to its mere being, as Wallace Stevens's phrase has it.

A second qualification is that ideas, cultures, and histories cannot seriously be understood or studied without their force, or more precisely their configurations of power, also being studied. To believe that the Orient was created – or, as I call it, "Orientalized" – and to believe that such things happen simply as a necessity of the imagination, is to be disingenuous. The relationship between Occident and Orient is a relationship of power, of domination, of varying degrees of a complex hegemony, and is quite accurately indicated in the title of K. M. Panikkar's classic *Asia and Western Dominance*. The Orient was Orientalized not only because it was discovered to be "Oriental" in all those ways considered commonplace by an average nineteenth-century European, but also because it *could be* – that is, submitted to being – *made* Oriental. There is very little consent to be found, for example, in the fact that Flaubert's encounter with an Egyptian courtesan produced a widely influential model of the Oriental woman; she never spoke of herself, she never represented her emotions, presence, or history. *He* spoke for and represented her. He was foreign, comparatively wealthy, male, and these were historical facts of domination that allowed him not only to possess Kuchuk Hanem physically but to speak for her and tell his readers in what way she was "typically Oriental." My argument is that Flaubert's situation of strength in relation to Kuchuk Hanem was not an isolated instance. It fairly stands for the pattern of relative strength between East and West, and the discourse about the Orient that it enabled.

This brings us to a third qualification. One ought never to assume that the structure of Orientalism is nothing more than a structure of lies or of myths which, were the truth about them to be told, would simply blow away. I myself believe that Orientalism is more particularly valuable as a sign of European-Atlantic power over the Orient than it is as a veridic discourse about the Orient (which is what, in its academic or scholarly form, it claims to be). Nevertheless, what we must respect and try to grasp is the sheer knitted-together strength of Orientalist discourse, its very close ties to the enabling socio-economic and political institutions, and its redoubtable durability. After all, any system of ideas that can remain unchanged as teachable wisdom (in academies, books, congresses, universities, foreign-service institutes) from the period of Ernest Renan in the late 1840s until the present in the United States must be something more formidable than a mere collection of lies. Orientalism, therefore, is not an airy European fantasy about the Orient, but a created body of theory and practice in which, for many generations, there has been a considerable material investment. Continued investment made Orientalism, as a system of knowledge about the Orient, an accepted grid for filtering through the Orient into Western consciousness, just as that same investment multiplied – indeed, made truly productive – the statements proliferating out from Orientalism into the general culture.

Gramsci has made the useful analytic distinction between civil and political society in which the former is made up of voluntary (or at least rational and non-coercive) affiliations like schools, families, and unions, the latter of state institutions (the army, the police, the central bureaucracy) whose role in the polity is direct domination. Culture, of course, is to be found operating within civil society, where the influence of ideas, of institutions, and of other persons works not through domination but by what Gramsci calls consent. In any society not totalitarian, then, certain cultural forms predominate over others, just as certain ideas are more influential than others; the form of this cultural leadership is what Gramsci has identified as *hegemony*, an indispensable concept for any understanding of cultural life in the industrial West. It is hegemony, or rather the result of cultural hegemony at work, that gives Orientalism the durability and the strength I have been speaking about so far. Orientalism is never far from what Denys Hay has called the idea of Europe, a collective notion identifying "us" Europeans as against all "those" non-Europeans, and indeed it can be argued that the major component in European culture is precisely what made that culture hegemonic both in and outside Europe: the idea of European identity as a superior one in comparison with all the non-European peoples and cultures. There is in addition the hegemony of European ideas about the Orient, themselves reiterating European superiority over Oriental backwardness, usually overriding the possibility that a

more independent, or more skeptical, thinker might have had different views on the matter.

In a quite constant way, Orientalism depends for its strategy on this flexible *positional* superiority, which puts the Westerner in a whole series of possible relationships with the Orient without ever losing him the relative upper hand. And why should it have been otherwise, especially during the period of extraordinary European ascendancy from the late Renaissance to the present? The scientist, the scholar, the missionary, the trader, or the soldier was in, or thought about, the Orient because he *could be there*, or could think about it, with very little resistance on the Orient's part. Under the general heading of knowledge of the Orient, and within the umbrella of Western hegemony over the Orient during the period from the end of the eighteenth century, there emerged a complex Orient suitable for study in the academy, for display in the museum, for reconstruction in the colonial office, for theoretical illustration in anthropological, biological, linguistic, racial, and historical theses about mankind and the universe, for instances of economic and sociological theories of development, revolution, cultural personality, national or religious character. Additionally, the imaginative examination of things Oriental was based more or less exclusively upon a sovereign Western consciousness out of whose unchallenged centrality an Oriental world emerged, first according to general ideas about who or what was an Oriental, then according to a detailed logic governed not simply by empirical reality but by a battery of desires, repressions, investments, and projections. If we can point to great Orientalist works of genuine scholarship like Silvestre de Sacy's *Chrestomathie arabe* or Edward William Lane's *Account of the Manners and Customs of the Modern Egyptians*, we need also to note that Renan's and Gobineau's racial ideas came out of the same impulse, as did a great many Victorian pornographic novels (see the analysis by Steven Marcus of "The Lustful Turk").

And yet, one must repeatedly ask oneself whether what matters in Orientalism is the general group of ideas overriding the mass of material – about which who could deny that they were shot through with doctrines of European superiority, various kinds of racism, imperialism, and the like, dogmatic views of "the Oriental" as a kind of ideal and unchanging abstraction? – or the much more varied work produced by almost uncountable individual writers, whom one would take up as individual instances of authors dealing with the Orient. In a sense the two alternatives, general and particular, are really two perspectives on the same material: in both instances one would have to deal with pioneers in the field like William Jones, with great artists like Nerval or Flaubert. And why would it not be possible to employ both perspectives together, or one after the other? Isn't there an obvious danger of distortion (of precisely the kind that academic Orientalism has always been prone to) if either too general or too specific a level of description is maintained systematically?

My two fears are distortion and inaccuracy, or rather the kind of inaccuracy produced by too dogmatic a generality and too positivistic a localized focus. In trying to deal with these problems I have tried to deal with three main aspects of my own contemporary reality that seem to me to point the way out of the methodological or perspectival difficulties I have been discussing, difficulties that might force one, in the first instance, into writing a coarse polemic on so unacceptably general a level of description as not to be worth the effort, or in the second instance, into writing so detailed and atomistic a series of analyses as to lose all track of the general lines of force informing the field, giving it its special cogency. How then to recognize individuality and to reconcile it with its intelligent, and by no means passive or merely dictatorial, general and hegemonic context?

[. . .]

My idea is that European and then American interest in the Orient was political according to some of the obvious historical accounts of it that I have given here, but that it was the culture that created that interest, that acted dynamically along with brute political, economic, and military rationales to make the Orient the varied and complicated place that it obviously was in the field I call Orientalism.

Therefore, Orientalism is not a mere political subject matter or field that is reflected passively by culture, scholarship, or institutions; nor is it a large and diffuse collection of texts about the Orient; nor is it representative and expressive of some nefarious "Western" imperialist plot to hold down the "Oriental" world. It is rather a *distribution* of geopolitical awareness into aesthetic, scholarly, economic, sociological, historical, and philological texts; it is an *elaboration* not only of a basic geographical distinction (the world is made up of two unequal halves, Orient and Occident) but also of a whole series of "interests" which, by such means as scholarly discovery, philological reconstruction, psychological analysis, landscape and sociological description, it not only creates but also maintains; it *is*, rather than expresses, a certain *will* or *intention* to understand, in some cases to control, manipulate, even to incorporate, what is a manifestly different (or alternative and novel) world; it is, above all, a discourse that is by no means in direct, corresponding relationship with political power in the raw, but rather is produced and exists in an uneven exchange with various kinds of power, shaped to a degree by the exchange with power political (as with a colonial or imperial establishment), power intellectual (as with reigning sciences like comparative linguistics or anatomy, or any of the modern policy sciences), power cultural (as with orthodoxies and canons of taste, texts, values), power moral (as with ideas about what "we" do and what "they" cannot do or understand as "we" do). Indeed, my real argument is that Orientalism is – and does not simply represent – a considerable dimension of modern political-intellectual culture, and as such has less to do with the Orient than it does with "our" world.

Because Orientalism is a cultural and a political fact, then, it does not exist in some archival vacuum; quite the contrary, I think it can be shown that what is thought, said, or even done about the Orient follows (perhaps occurs within) certain distinct and intellectually knowable lines. Here too a considerable degree of nuance and elaboration can be seen working as between the broad superstructural pressures and the details of composition, the facts of textuality. Most humanistic scholars are, I think, perfectly happy with the notion that texts exist in contexts, that there is such a thing as intertextuality, that the pressures of conventions, predecessors, and rhetorical styles limit what Walter Benjamin once called the "overtaxing of the productive person in the name of [. . .] the principle of 'creativity,'" in which the poet is believed on his own, and out of his pure mind, to have brought forth his work. Yet there is a reluctance to allow that political, institutional, and ideological constraints act in the same manner on the individual author. A humanist will believe it to be an interesting fact to any interpreter of Balzac that he was influenced in the *Comédie humaine* by the conflict between Geoffroy Saint-Hilaire and Cuvier, but the same sort of pressure on Balzac of deeply reactionary monarchism is felt in some vague way to demean his literary "genius" and therefore to be less worth serious study. Similarly – as Harry Bracken has been tirelessly showing – philosophers will conduct their discussions of Locke, Hume, and empiricism without ever taking into account that there is an explicit connection in these classic writers between their "philosophic" doctrines and racial theory, justifications of slavery, or arguments for colonial exploitation. These are common enough ways by which contemporary scholarship keeps itself pure.

Perhaps it is true that most attempts to rub culture's nose in the mud of politics have been crudely iconoclastic; perhaps also the social interpretation of literature in my own field has simply not kept up with the enormous technical advances in detailed textual analysis. But there is no getting away from the fact that literary studies in general, and American Marxist theorists in particular, have avoided the effort of seriously bridging the gap between the superstructural and the base levels in textual, historical scholarship; on another occasion I have gone so far as to say that the literary-cultural establishment as a whole has declared the serious study of imperialism and culture off limits. For Orientalism brings one up directly against that question – that is, to realizing that political imperialism governs an entire field of study, imagination, and scholarly institutions – in such a way as to make its avoidance an intellectual and historical impossibility. Yet there will always remain the perennial escape mechanism of saying that a literary scholar and a philosopher, for example, are trained in literature and philosophy respectively, not in politics or ideological analysis. In other words, the specialist argument can work quite effectively to block the larger and, in my opinion, the more intellectually serious perspective.

Here it seems to me there is a simple two-part answer to be given, at least so far as the study of imperialism and culture (or Orientalism) is concerned. In the first place, nearly every nineteenth-century writer (and the same is true enough of writers in earlier periods) was extraordinarily well aware of the fact of empire: this is a subject not very well studied, but it will not take a modern Victorian specialist long to admit that liberal cultural heroes like John Stuart Mill, Arnold, Carlyle, Newman, Macaulay, Ruskin, George Eliot, and even Dickens had definite views on race and imperialism, which are quite easily to be found at work in their writing. So even a specialist must deal with the knowledge that Mill, for example, made it clear in *On Liberty* and *Representative Government* that his views there could not be applied to India (he was an India Office functionary for a good deal of his life, after all) because the Indians were civilizationally, if not racially, inferior. The same kind of paradox is to be found in Marx, as I try to show in this book. In the second place, to believe that politics in the form of imperialism bears upon the production of literature, scholarship, social theory, and history writing is by no means equivalent to saying that culture is therefore a demeaned or denigrated thing. Quite the contrary: my whole point is to say that we can better understand the persistence and the durability of saturating hegemonic systems like culture when we realize that their internal constraints upon writers and thinkers were *productive*, not unilaterally inhibiting. It is this idea that Gramsci, certainly, and Foucault and Raymond Williams in their very different ways have been trying to illustrate. Even one or two pages by Williams on "the uses of the Empire" in *The Long Revolution* tell us more about nineteenth-century cultural richness than many volumes of hermetic textual analyses.

Therefore I study Orientalism as a dynamic exchange between individual authors and the large political concerns shaped by the three great empires – British, French, American – in whose intellectual and imaginative territory the writing was produced. What interests me most as a scholar is not the gross political verity but the detail, as indeed what interests us in someone like Lane or Flaubert or Renan is not the (to him) indisputable truth that Occidentals are superior to Orientals, but the profoundly worked over and modulated evidence of his detailed work within the very wide space opened up by that truth. One need only remember that Lane's *Manners and Customs of the Modern Egyptians* is a classic of historical and anthropological observation because of its style, its enormously intelligent and brilliant details, not because of its simple reflection of racial superiority, to understand what I am saying here.

The kind of political questions raised by Orientalism, then, are as follows: what other sorts of intellectual, aesthetic, scholarly, and cultural energies went into the making of an imperialist tradition like the Orientalist one? How did philology, lexicography, history, biology, political and economic theory, novel-writing, and lyric poetry come to the service of Orientalism's broadly imperialist view of the world? What changes, modulations, refinements, even revolutions take place within Orientalism? What is the meaning of originality, of continuity, of individuality, in this context? How does Orientalism transmit or reproduce itself from one epoch to another? In fine, how can we treat the cultural, historical phenomenon of Orientalism as a kind of *willed human work* – not of mere unconditioned ratiocination – in all its historical complexity, detail, and worth without at the same time losing sight of the alliance between cultural work, political tendencies, the state, and the specific realities of domination? Governed by such concerns a humanistic study can responsibly address itself to politics *and* culture. But this is not to say that such a study establishes a hard-and-fast rule about the relationship between knowledge and politics. My argument is that each humanistic investigation must formulate the nature of that connection in the specific context of the study, the subject matter, and its historical circumstances.

[. . .]

Much of the personal investment in this study derives from my awareness of being an "Oriental" as a child growing up in two British colonies. All of my education, in those colonies (Palestine and Egypt) and in the United States, has been Western, and yet that deep early awareness has persisted. In many ways my study of Orientalism has been an attempt to inventory the traces upon me, the Oriental subject, of the culture whose domination has been so powerful a factor in the life of all Orientals. This is why for me the Islamic Orient has had to be the center of attention. Whether what I have achieved is the inventory prescribed by Gramsci is not for me to judge, although I have felt it important to be conscious of trying to produce one. Along the way, as severely and as rationally as I have been able, I have tried to maintain a critical consciousness, as well as employing those instruments of historical, humanistic, and cultural research of which my education has made me the fortunate beneficiary. In

none of that, however, have I ever lost hold of the cultural reality of, the personal involvement in having been constituted as, "an Oriental."

The historical circumstances making such a study possible are fairly complex, and I can only list them schematically here. Anyone resident in the West since the 1950s, particularly in the United States, will have lived through an era of extraordinary turbulence in the relations of East and West. No one will have failed to note how "East" has always signified danger and threat during this period, even as it has meant the traditional Orient as well as Russia. In the universities a growing establishment of area-studies programs and institutes has made the scholarly study of the Orient a branch of national policy. Public affairs in this country include a healthy interest in the Orient, as much for its strategic and economic importance as for its traditional exoticism. If the world has become immediately accessible to a Western citizen living in the electronic age, the Orient too has drawn nearer to him, and is now less a myth perhaps than a place crisscrossed by Western, especially American, interests.

One aspect of the electronic, postmodern world is that there has been a reinforcement of the stereotypes by which the Orient is viewed. Television, the films, and all the media's resources have forced information into more and more standardized molds. So far as the Orient is concerned, standardization and cultural stereotyping have intensified the hold of the nineteenth-century academic and imaginative demonology of "the mysterious Orient." This is nowhere more true than in the ways by which the Near East is grasped. Three things have contributed to making even the simplest perception of the Arabs and Islam into a highly politicized, almost raucous matter: one, the history of popular anti-Arab and anti-Islamic prejudice in the West, which is immediately reflected in the history of Orientalism; two, the struggle between the Arabs and Israeli Zionism, and its effects upon American Jews as well as upon both the liberal culture and the population at large; three, the almost total absence of any cultural position making it possible either to identify with or dispassionately to discuss the Arabs or Islam. Furthermore, it hardly needs saying that because the Middle East is now so identified with Great Power politics, oil economics, and the simple-minded dichotomy of freedom-loving, democratic Israel and evil, totalitarian, and terroristic Arabs, the chances of anything like a clear view of what one talks about in talking about the Near East are depressingly small.

My own experiences of these matters are in part what made me write this book. The life of an Arab Palestinian in the West, particularly in America, is disheartening. There exists here an almost unanimous consensus that politically he does not exist, and when it is allowed that he does, it is either as a nuisance or as an Oriental. The web of racism, cultural stereotypes, political imperialism, dehumanizing ideology holding in the Arab or the Muslim is very strong indeed, and it is this web which every Palestinian has come to feel as his uniquely punishing destiny. It has made matters worse for him to remark that no person academically involved with the Near East – no Orientalist, that is – has ever in the United States culturally and politically identified himself wholeheartedly with the Arabs; certainly there have been identifications on some level, but they have never taken an "acceptable" form as has liberal American identification with Zionism, and all too frequently they have been radically flawed by their association either with discredited political and economic interests (oil-company and State Department Arabists, for example) or with religion.

The nexus of knowledge and power creating "the Oriental" and in a sense obliterating him as a human being is therefore not for me an exclusively academic matter. Yet it is an *intellectual* matter of some very obvious importance. I have been able to put to use my humanistic and political concerns for the analysis and description of a very worldly matter, the rise, development, and consolidation of Orientalism. Too often literature and culture are presumed to be politically, even historically innocent; it has regularly seemed otherwise to me, and certainly my study of Orientalism has convinced me (and I hope will convince my literary colleagues) that society and literary culture can only be understood and studied together. In addition, and by an almost inescapable logic, I have found myself writing the history of a strange, secret sharer of Western anti-Semitism. That anti-Semitism and, as I have discussed it in its Islamic branch, Orientalism resemble each other very closely is a historical, cultural, and political truth that needs only to be mentioned to an Arab Palestinian for its irony to be perfectly understood. But what I should like also to have contributed here is a better understanding of the way cultural domination has operated. If this stimulates a new kind of dealing with the Orient, indeed if it eliminates the "Orient" and "Occident" altogether, then we shall have advanced a little in the process of what Raymond Williams has called the "unlearning" of "the inherent dominative mode."

 READING 8

Orientalism and Orientalism in Reverse

Sadik Jalal al-'Azm

I Orientalism

In his sharply debated book, Edward Said introduces us to the subject of 'Orientalism' through a broadly historical perspective which situates Europe's interest in the Orient within the context of the general historical expansion of modern bourgeois Europe outside its traditional confines and at the expense of the rest of the world in the form of its subjugation, pillage, and exploitation. In this sense Orientalism may be seen as a complex and growing phenomenon deriving from the overall historical trend of modern European expansion and involving: a whole set of progressively expanding institutions, a created and cumulative body of theory and practice, a suitable ideological superstructure with an apparatus of complicated assumptions, beliefs, images, literary productions, and rationalisations (not to mention the underlying foundation of commercial, economic and strategic vital interests). I shall call this phenomenon *Institutional Orientalism*.

Edward Said also deals with Orientalism in the more restricted sense of a developing tradition of disciplined learning whose main function is to 'scientifically research' the Orient. Naturally, this *Cultural-Academic Orientalism* makes all the usual pious claims about its 'disinterested pursuit of the truth' concerning the Orient, and its efforts to apply impartial scientific methods and value-free techniques in studying the peoples, cultures, religions, and languages of the Orient. The bulk of Edward's book is not unexpectedly devoted to Cultural-Academic Orientalism in an attempt to expose the ties which wed it to Institutional Orientalism.

In this way Said deflates the self-righteous claims of Cultural-Academic Orientalism to such traits as scholarly independence, scientific detachment, political objectivity etc. It should be made clear, however, that the author at no point seeks to belittle the genuine scholarly achievements, scientific discoveries, and creative contributions made by orientalists and orientalism over the years, particularly at the technical level of accomplishment. His main concern is to convey the message that the overall image of the Orient constructed by Cultural-Academic Orientalism, from the viewpoint of its own technical achievements and scientific contributions to the field, is shot through and through with racist assumptions, barely camouflaged mercenary interests, reductionistic explanations and anti-human prejudices. It can easily be shown that this image, when properly scrutinised, can hardly be the product of genuinely objective scientific investigation and detached scholarly discipline.

Critique of orientalism

One of the most vicious aspects of this image, as carefully pointed out by Said, is the deep rooted belief – shared by Cultural-Academic and Institutional Orientalism – that a fundamental ontological difference exists between the essential natures of the Orient and Occident, to the decisive advantage of the latter. Western societies, cultures, languages and mentalities are supposed to be essentially and inherently superior to the Eastern ones. In Edward Said's words, 'the essence of Orientalism is the ineradicable distinction between Western superiority and Oriental inferiority'. According to this reading of Said's initial thesis, Orientalism

(both in its institutional and cultural-academic forms) can hardly be said to have existed, as a structured phenomenon and organised movement, prior to the rise, consolidation and expansion of modern bourgeois Europe. Accordingly, the author at one point dates the rise of Academic Orientalism with the European Renaissance. But unfortunately the stylist and polemicist in Edward Said very often runs away with the systematic thinker. As a result he does not consistently adhere to the above approach either in dating the phenomenon of Orientalism or in interpreting its historical origins and ascent.

In an act of retrospective historical projection we find Said tracing the origins of Orientalism all the way back to Homer, Aeschylus, Euripides and Dante. In other words, Orientalism is not really a thoroughly modern phenomenon, as we thought earlier, but is the natural product of an ancient and almost irresistible European bent of mind to misrepresent the realities of other cultures, peoples, and their languages, in favour of Occidental self-affirmation, domination and ascendency. Here the author seems to be saying that the 'European mind', from Homer to Karl Marx and A. H. R. Gibb, is inherently bent on distorting all human realities other than its own and for the sake of its own aggrandisement.

It seems to me that this manner of construing the origins of Orientalism simply lends strength to the essentialistic categories of 'Orient' and 'Occident', representing the ineradicable distinction between East and West, which Edward's book is ostensibly set on demolishing. Similarly, it lends the ontological distinction of Europe versus Asia, so characteristic of Orientalism, the kind of credibility and respectability normally associated with continuity, persistence, pervasiveness and distant historical roots. This sort of credibility and respectability is, of course, misplaced and undeserved. For Orientalism, like so many other characteristically modern European phenomena and movements (notably nationalism), is a genuinely recent creation – the product of modern European history – seeking to acquire legitimacy, credibility and support by claiming ancient roots and classical origins for itself. Certainly Homer, Euripides, Dante, St. Thomas and all the other authorities that one may care to mention held the more or less standard distorted views prevalent in their milieu about other cultures and peoples. However, it is equally certain that the two forms of Orientalism built their relatively modern repertoires of systematic conventional wisdom by calling upon the views and biases of such prestigious figures as well as by drawing on ancient myth, legend, imagery, folklore and plain prejudice. Although much of this is well documented (directly and indirectly) in Said's book, still his work remains dominated by a unilinear conception of 'Orientalism' as somehow flowing straight through from Homer to Grunebaum. Furthermore, this unilinear, almost essentialistic, presentation of the origins and development of Orientalism renders a great disservice to the vital concerns of Edward's book, namely, preparing the ground for approaching the difficult question of 'how one can study other cultures and peoples from a libertarian, or nonrepressive and nonmanipulative, perspective,' and for eliminating, in the name of a common humanity, both 'Orient' and 'Occident' as ontological categories and classificatory concepts bearing the marks of racial superiority and inferiority. It seems to me that as a logical consequence of Said's tendency to view the origins and development of Orientalism in terms of such unilinear constancy, the task of combating and transcending its essentialistic categories, in the name of this common humanity, is made all the more difficult.

Another important result of this approach bears on Said's interpretation of the relationship supposedly holding between Cultural-Academic Orientalism as representation and disciplined learning on the one hand, and Institutional Orientalism as expansionary movement and socio-economic force on the other. In other words, when Said is leaning heavily on his unilinear conception of 'Orientalism' he produces a picture which says that this cultural apparatus known as 'Orientalism' is the real source of the West's political interest in the Orient, ie, that it is the real source of modern Institutional Orientalism. Thus, for him European and later on American political interest in the Orient was really created by the sort of Western cultural tradition known as Orientalism. Furthermore, according to one of his renderings, Orientalism is a distribution of the awareness that the world is made up of two unequal halves – Orient and Occident – into aesthetic, scholarly, economic, sociological, historical and philosophical texts. This awareness not only created a whole series of Occidental 'interests' (political,

economic, strategic etc) in the Orient, but also helped to maintain them. Hence for Said the relationship between Academic Orientalism as a cultural apparatus and Institutional Orientalism as economic interest and political force is seen in terms of a '*preposterous* transition' from 'a merely textual apprehension, formulation or definition of the Orient to the putting of all this into practice in the Orient'. According to this interpretation Said's phrase 'Orientalism overrode the Orient' could mean only that the Institutional Orientalism which invaded and subjugated the East was really the legitimate child and product of that other kind of Orientalism, so intrinsic, it seems, to the minds, texts, aesthetics, representations, lore and imagery of Westerners as far back as Homer, Aeschylus and Euripides! To understand properly the subjugation of the East in modern times, Said keeps referring us back to earlier times when the Orient was no more than an awareness, a word, a representation, a piece of learning to the Occident:

> What we must reckon with is a large and slow process of appropriation by which Europe, or the European awareness of the Orient, transformed itself from being textual and contemplative into being administrative, economic, and even military.

Therefore Edward Said sees the 'Suez Canal idea' much more as 'the logical conclusion of Orientalist thought and effort' than as the result of Franco-British imperial interests and rivalries (although he does not ignore the latter).

One cannot escape the impression that for Said somehow the emergence of such observers, administrators and invaders of the Orient as Napoleon, Cromer and Balfour was made inevitable by 'Orientalism' and that the political orientations, careers and ambitions of these figures are better understood by reference to d'Herbelot and Dante than to more immediately relevant and mundane interests. Accordingly, it is hardly surprising to see Said, when touching on the role of the European Powers in deciding the history of the Near Orient in the early twentieth century, select for prominent notice the 'peculiar, epistemological framework through which the Powers saw the Orient', which was built by the long tradition of Orientalism. He then affirms that the Powers acted on the Orient the way they did because of that peculiar epistemological framework. Presumably, had the long tradition of Cultural-Academic Orientalism fashioned a less peculiar, more sympathetic and truthful epistemological framework, then the Powers would have acted on the Orient more charitably and viewed it in a more favourable light!

[. . .]

 READING 9

Postcolonialism and Its Discontents
Ali Rattansi

This paper's structure reflects its overall purpose: to provide a critical commentary on a fast-mushrooming area of research which I shall characterize as 'postcolonialism' or 'postcolonialist studies'. The paper is thus in two interrelated parts. The first will provide a provisional definition of the idea of the 'postcolonial' and explore some of the achievements of the field of 'postcolonialist' research as it has developed in cultural studies. The second part will highlight a number of key problematic areas in the field which have been the subject of considerable international debate.

1 In Praise of Postcolonialist Studies

Defining and theorizing the 'postcolonial'

Like all the 'posts' that are fashionable in current discourse, the idea of 'postcolonialism' faces formidable problems in mapping a terrain, an object of study, which is both coherent and can command consent among those supposedly working within the field. Many of the relevant problems are explored in the second half of this essay. For the present, a provisional set of delimiting boundaries and contents need to be defined.

Provisionally, *postcolonialism* may be marked out as a period in global time–space in which most of the former colonies of Western imperial powers have gained formal independence. It must be emphasized that there is no sense in which the 'postcolonial' is a singular moment. The reference must be to a *series* of transitions situated between and within the moments of colonization/decolonization. This emphasis on multiplicity is crucial. While a certain British or Northern European ethnocentrism has been tempted to conflate the postcolonial with the post (second world) war era, one has only to think of the 'Latin' American and indeed the North American context to appreciate the significance of the internal heterogeneity of the postcolonial period, spanning a time–space from the late eighteenth century in the 'North' or 'West' to the globality of the twentieth century. As Said has reminded us in *Culture and Imperialism*, the 'West' held something like 85 per cent of the world in the form of various possessions on the eve of the First World War in 1914.

One of the peculiarities that postcolonialism shares with that other ubiquitous 'post', postmodernism, is that *it marks out a supposed historical period as well as a distinctive form of theorization and analysis*. The similarities do not end there, for as fields of investigation both eschew traditional disciplinary boundaries and conventional conceptions of time, narrative and spatiality. In the case of postcolonialist studies, a heady, eclectic mix of poststructuralism, psychoanalysis, feminism, Marxism and postmodernism itself populates the field in varying combinations. Fanon, Freud and Lacan, Foucault, Derrida, Kristeva, Jameson and Gramsci jostle for position in the works of the major postcolonialist writers – Said, Spivak and Bhabha for example. Elsewhere, in discussing postmodernism, Boyne and I have suggested that it is useful to distinguish between *postmodernism*, as a set of cultural and intellectual currents, and *postmodernity* as an epoch in historical time–space which would include postmodernism as one of its elements. A similar conceptual discrimination would help here too: thus I propose to use *postcolonialism* and *postcolonialist studies* to refer to a particular form of intellectual inquiry and *postcoloniality* to index a set of historical epochs (the significance of the plural here will be clarified below).

It is my argument that the central defining theme of postcolonialism or postcolonialist studies is the investigation of the mutually constitutive role played by colonizer and colonized, centre and periphery, the metropolitan and the 'native', in forming, in part, the identities of both the dominant power and the subalterns involved in the imperial and colonial projects of the 'West'. Thus postcolonialism views the '"West" and the "Rest"' as mutually imbricated, although with due attention to the fundamental axis of inequality which defined the imperial process. The formation of nations and 'national cultures' in the centre and in the peripheries is therefore analysed as a series of outcomes of the imperial project *and the resistances to it*, which led to formal independence for the colonies and the inauguration of the 'postcolonial' by way of a variety of time–spaces of 'postcoloniality'.

In positing a certain mutuality to the processes of subject and identity formation as between colonizer and colonized, in effect the project of postcolonialist studies deconstructs the Manichean view of a binary opposition between the imperial and the subaltern, for there is a dismantling of the often-held conception of colonialism and imperialism as processes which wounded and scarred the psyches, the cultures and the economies of the colonized while leaving the metropolitan centres economically enriched, and culturally as a dominant, stable and indeed stronger set of formations. To put it differently, postcolonialist studies take as a premise that the cultures and psyches of the colonizer were not already defined, and only waiting, as it were, to be imposed, fully formed, on the hapless victims of the colonial project. The idea of the 'West' as white, Christian, rational, civilized, modern, sexually disciplined and indeed *masculine* was put into place in a protracted process in which the colonized Others were defined in opposition to these virtues. It was in constructing the 'natives' as black, pagan, irrational, uncivilized, pre-modern, libidinous, licentious, effeminate and childlike that the self-conception of the European as superior, and as not only *fit* to govern but as having the positive *duty* to govern and 'civilize' came into being.

However, as we shall see, the idea of mutual imbrication of identities in fact goes further than this. For the postcolonialist contention is that what was involved was an even more complex intertwining of identities-information, in which the Others against whom European identities were played off were not only outside but also *inside* the nation-states of the centre. The processes which led to the formation of Western modernity also involved an inferiorization and government or regulation and disciplining of internal Others such as women, children and the rapidly growing urban working class. Thus, 'internal' questions of the forms of incorporation of these subalterns into the national culture and polity became conflated with and superimposed onto issues involving the forms in which the 'natives' of the colonies were to be discursively comprehended and ruled.

Now, it is quite clear that viewed in *this* light, the imperial and colonial projects cannot be reductively analysed simply by reference to a decisive economic logic which narrates the formation of colonial cultures and polities as just another version of the familiar transition from feudalism to capitalism, except this time imposed from above by the metropolitan powers, and in which class formation, class interests and class conflicts remain the main engines of transformation. A properly 'postcolonialist' analysis, on the contrary, requires the acknowledgement of a set of processes in which cultural formation is dispersed along a number of axes of potentially *commensurate* importance – class, certainly, but also sexuality and gender, racism, familial relations, religious discourses, conceptions of childhood and child-rearing practices, and requiring therefore also an understanding of underlying processes of psychic development and 'deformation'. The societies that came into being through colonial encounters can no longer be discursively appropriated through a grid which reads them as re-runs of an oft-told linear narrative of the transition from one mode of production to another, whether in Marxist or Weberian vocabulary, and certainly not as an equally straightforward story of 'modernization' as functionalist, mostly American sociology would have it.

Very importantly, what is true of *colonial* formations seems to be true of the *metropolitan* societies as well, and by the same token, so to speak, for how could the seminal role of sexuality, gender, race, nation, the familial and so on be ignored as axes of cultural and political formation in the *centre* given the ever-growing understanding of the imperial project as involving *mutual imbrication* and *intertwining*? And there is, too, the question of how to understand the profound significance of the vast growth of 'knowledges' fostered in the processes of colonization and which appeared to have insinuated themselves at the heart of the forms of government through which colonial rule operated – anthropology, the systematizations of Oriental languages and histories, racial studies and eugenics for example – and which also appear to require a rethinking of received ideas of (material or economic) 'base' and (cultural and ideological) 'superstucture' which, even in the most sophisticated versions of the metaphor, cannot help but see such forms of knowledge, in the last instance, as epiphenomenal and thus miss their significance as *shaping* rather than merely *reflecting* the forms of colonial rule.

It is hardly surprising that in *Orientalism*, which can claim to be the founding text of modern postcolonialist studies, Said turned to the poststructuralism of Foucault to provide an alternative 'take' on questions of the relation between power and knowledge, given Foucault's attention to the imbrication between the formation of knowledges and their role in government, and also for the insights Foucault's work contains in analysing how European identities were formed in a process of what Foucault called 'normalization' which categorized and separated off a variety of internal figures that in the development of Western modernity came to be marked out as 'Other' – criminals, the supposedly insane, sections of the urban poor and so on. And it should be equally intelligible why, via Fanon especially, Freudian and Lacanian emphases have been prominent in a field of studies that has attempted to understand the profound psychological impact of colonial inferiorization on both the colonized and the colonizer. A variety of deployments of feminist approaches and appropriations of Freud, Lacan and Foucault, again, have quite understandably provided critical intellectual resources in attempts to unravel the complex relations between sexuality, class, race and relations of imperial and domestic domination and subordination, a task that has also been nourished by Gramscian insights on processes of hegemony and, in a different register, by Derridean theorizations of identity, alterity and *différance*.

However, it would be disingenuous, not to say naive, and certainly very un-poststructuralist to fail to register that all these theoretical resources have not simply been 'neutral' frames for the apprehension of the 'truth' of the effects of colonial encounters on colonizer and colonized. For the concepts that have structured the archive of postcolonialist studies have, of course, decisively influenced the distinctive manner in which the field has construed the nature of this relation between colonizer and colonized and in the way it has analysed postcolonial cultures as forms of displacement and postcolonial identities as particularly fragmented. There has not been, and there never can be, a simple relation of mirroring in which the 'truth' of colonial encounters can now be said to be *properly* narrated with the resources that had earlier not been used or – in the case of Foucault and to some extent even Freud – had simply not been available.

To put it differently, it should come as no surprise that the specificity of a postcolonial take on these issues has been the subject of sometimes quite acrimonious debate. I comment on these controversies in the second part of this paper. For the time being I point the reader to a significant exchange between O'Hanlan and Washbrook and Prakash where many of the issues around the legitimacy of specifically postcolonialist and Marxist 'takes' on narratives of colonialism and its aftermath are rehearsed in an illuminating manner.

Authority and identity

For the present, it is worth exploring, albeit very briefly, some 'typical' (post-*Orientalism*) postcolonialist investigations, to substantiate my claim that there are indeed elements worthy of praise in postcolonialist studies. The number of such studies is now extraordinarily large, in part because of the North American graduate studies machine which, with the participation of many students from the former colonies of Africa and India, has embraced the field and has begun to plough it with a not uncommon energy, enthusiasm and excellence. Some indication of the extent of the cultivation can be obtained from consulting the extensive, indeed daunting bibliography in a study such as Stoler's which refigures Foucault's work on sexuality in the light of postcolonialist studies and in Said's own sequel to *Orientalism, Culture and Imperialism*.

Investigations which exemplify the most general and fundamental theme of the field, that is, the complex ways in which aspects of the national cultures and identities of both the 'West' and the 'Rest' were formed by fateful colonial encounters, are an obvious starting point for (ap)praising postcolonialist studies.

Gauri Viswanathan's analysis of the formation of English literary studies in India and their subsequent growth in the academies of the imperial heartland is a particularly fruitful application and extension of Said's seminal arguments in *Orientalism*.

Viswanathan's research exemplifies the motifs of postcolonialism, for it demonstrates, among many other things, the following:

(a) That the project of teaching English literature to a certain class of Indians in India in the mid nineteenth century was part of a project to *govern* India by giving some Indians access to and insight into the greatness and supposed infinite moral superiority of English culture while at the same time creating a much-needed cadre of English speaking 'native' administrators and civil servants.

(b) That this was always also a self-conscious strategy to underwrite and mask the other British project of economically exploiting the subcontinent by giving it the veneer of a 'civilizing' mission.

(c) That in devising an education in English, the British were well aware of the significance of education in the creation of hegemony, for this was a project under way in Britain where a whole variety of strategies were being put into motion to contain the potential threat of the growing urban working class, schooling being one of the key planks, although in this case the attempt was infused with a Christian ethos which the British were aware had to be treated with caution in the Indian context. Here one can see the point about the interrelationship between one of the bourgeois West's internal Others – the urban working class – and the attempt to govern and exploit a set of external Other threatening subalterns, with strategies of containment being learnt and mutually transferred between the two widely separated territories of governance.

(d) That, ironically enough, it was the project of establishing English literary studies in *India*

which had a strong formative influence on the development of literary studies as a university subject in *Britain* in the last part of the nineteenth century, when English began to displace Latin and Greek languages and texts as the key medium for the education and disciplining of the middle and upper class English*man*.

This last point is particularly crucial. Given the manner in which English literature has functioned, and continues to work, to define Englishness, and given, too, the huge success of English literature as a university subject in India, and the significance of English literature in the education and Anglicization of contemporary middle-class Indians, the postcolonial point about the mutual imbrication of identities via the colonial encounter – although within the context of a fundamental asymmetry of power – seems thoroughly vindicated. And, of course, so too are the emphases on knowledge, power and governance, and their subject and identity-forming effects.

At various stages of the discussion so far I have alluded to the significance of both class and gender in postcolonial studies. For example, it is clear that, when one refers to the urban working class as an internal Other, the relation of alterity implies that the 'Otherness' operates dyadically *vis-à-vis* the dominant classes of Victorian Britain, or more generally of Europe. And that, given the gendered nature of educational access, the role of English, and education more generally, was of course of particular importance in the formation of imperial *masculinities* in the academy, although the way in which the imperial project shaped a particular conception of the role of *women* as reproducers of an imperial 'race' is also well documented. Moreover, the 'feminization' of the colonized male also of course occurred in the context of the masculinism of imperialism and the dominance of the male in the metropolitan order of things.

It is therefore appropriate to turn to another recent contribution to postcolonial literature in which many of these issues are particularly well highlighted. I refer here to the research of another Indian woman, Mrinalini Sinha, whose *Colonial Masculinity* offers a brilliant account of the changing configurations of Indian – and, more specifically, Bengali – masculinities and British imperial masculinities, set in the context of complex economic, social class and governmental transformations in this part of colonial India. From what is a complex and dense narrative, it is only possible here to extract a number of relevant arguments:

(a) That the conception of the 'effeminate' Bengali male in British colonial discourse in India – effeminization being a common enough, general discursive strategy of inferiorization in the imperial project – underwent, however, significant changes with the changing *class* structure of colonial Bengal. Effeminacy, from initially being attributed to all Indian men, then concentrated on Bengali men, and subsequently focused particularly on the Western-educated Bengali middle-class men who were beginning to make inconvenient political demands upon the colonial authorities. Interestingly enough, the Bengali male was not only ridiculed for his supposed lack of 'manliness', but also for his allegedly poor treatment of 'his' women! The combination was enough, in the eyes of the colonial authorities, to disqualify the hapless Bengali from participation in government.

(b) That the Bengalis developed complex classifications around their own sense of masculinity and emasculation. This too was related to class, with the petty clerks and then the declining rentiers conceiving of themselves as effeminized by the subservient nature of their work and their impoverishment respectively, the latter in an indigenous cultural context where masculinity was powerfully tied to the ownership of property.

(c) That, simultaneously, there was a process under way in the metropolis where English masculinity was being constructed around the public schools, Oxbridge, and so on, in deliberate contrast to what was regarded as the effeminacy of the colonial male. This was, to a significant degree, a specifically *English* rather than a British project, for there was considerable prejudice against recruiting civil servants from Scottish and Irish universities.

The research of Viswanathan and Sinha is only the tip of a veritable iceberg. A vast amount of other scholarship could be drawn upon to illustrate the interplay of class, gender, ethnicity, conceptions of the family, and so on, in the dynamics of the process which established crucial elements of identity for both the colonizer and the colonized, in a wide variety of geographical and national-imperial contexts. While it is somewhat invidious to pick out particular pieces of research from such a rich field, it is perhaps worth citing some other work which bears out the general themes of postcolonialist studies: for example, Catherine Hall's work on the formation of British national culture and citizenship in relation to the construction of colonial 'experiences'; David Arnold's research on the construction of 'Indianized' Western medical knowledges and practices in the context of the implantation of Western medicine in India, and also the manner in which this was implicated in the formation of conceptions of Oriental and Occidental bodies; Niranjana's discussion of the way in which English identities as well as those of Indians were formed by particular translations of key Indian traditional texts, the English being able to construct Indianness and, in alterity, Englishness from a selective reading of these texts, with Indians being similarly fed a version of themselves which conformed to English conceptions of their venality; Mudimbe's explorations of Western conceptions of Africa and the problems of 'recovering' and constituting an authentic African knowledge; Martin Bernal's *Black Athena* which attempts to contest the crucial element of Greek as opposed to Egyptian origins in the formation of the West's identity; and the important essays in the collections edited by Breckenbridge and van der Veer, Prakash and Chambers and Curti. The continuing durability of colonial discourses in Western scientific, sociological, anthropological and administrative knowledges and practices are investigated in, for example, essays by Mohanty on Western writings on 'Third World Women', Watney on the Western narrativization of AIDS and Rattansi on the sexualized racism which governed the British state's response to immigration from the colonies in the immediate aftermath of the Second World War.

Ambivalence and resistance

If the idea of the mutual constitution of identities provides one major set of themes for the architecture of postcolonialist studies, notions of ambivalence and resistance furnish another. The discussions are wide ranging, encompassing Bhabha's explorations of mimicry, speculations on the workings of imperialist sexual desire for the Other and readings of the specificities of white women's perceptions of the colonized.

It may seem inappropriate to bring together such a disparate body of work under the sign of ambivalence and resistance. But in my view what may be said to unite them is a specific element of postcolonialism which needs to be highlighted in a form that has not always been made explicit. That is, there is an aspect of this research which points to a chronic cultural and psychic instability at the heart of the colonial project, a sort of intrinsic dynamic of destabilization, whose mechanisms are formed around a complex interweaving of Self/Other relations as operationalized through sexuality and sexual difference. In the process of explicating this set of ideas in this way I shall be reading, or re-reading, some postcolonialist works, in particular those of Bhabha, in a form that is different from the explicit letter of the text or, to put it more accurately, I shall be adapting this work in ways more in keeping with my own interpretation of the psychic and sexual dynamics of colonization.

Elsewhere I have discussed what I call the sexualization of colonial discourses in much greater detail than is possible in this paper. Here I will only draw out the main lines of how sexuality and gender functioned to destabilize the relations between colonized and colonizer in ways which posed a constant threat to the strict division between the two on which the imperial project was inevitably premised.

Take, first, the forms of representation of 'primitive' sexuality among 'natives' of the lands of North America and Africa. The free and apparently natural sexual expressiveness that was supposedly 'observed' was a source of fascination, attraction, as well as fear and repulsion, with both male and female Africans and North American 'Indians' functioning as sexual Others, onto whom were projected the anxieties and *desires* of the European male. In pictorial representations the native land was often an attractive female, barely clothed, inviting European imperial penetration, while the native male was often depicted as effeminized – lacking bodily hair, in the case of the North American 'Indian', for example – and prey to the excessive sexuality of his woman.

This type of exotic eroticization of the native was an important element in the formation and reconstitution of sexualities and gender relations at 'home'. The white woman was seen as closer to the native than to the white male in many ways. She supposedly shared the lower intelligence, over-emotionality and potential sexual excess of the native – especially if she happened to be working class – therefore needing the same subordination and control, but by the same token requiring 'protection' from men, and the native in the colony, and her own sexual desire for other and Other males, allowing a legitimation of patriarchal gender relations at home and abroad.

Arguably, what are evident here are projections of white male – especially upper-class male – desires and anxieties which constantly threatened to breach the all-important binary between the colonizer and the native and which in practice, of course, were breached by widespread sexual liaisons between the two which have increasingly become the object of investigation in recent years. Note, too, the significance of homo-eroticism, sometimes under the surface, sometimes explicit as in the case of so many homosexuals who fled restrictions at home to fulfil their desires and fantasies in the Orient.

There is a sense in which the worst fears of the white colonial male were realized in the person of that curious creature, the white woman traveller who, in defiance of nineteenth-century expectations, decided to roam the colonies on her own, as it were, to 'see' for herself and then to commit the even greater transgression of writing about her 'experiences'. Women travellers to the colonies, imperial outposts and the 'virgin' territories soon to be colonized tended to write in a register different from that produced by the imperial gaze of the male 'discoverer' and adventurer. As Mills has pointed out, women's travel writing had more in common with that other tradition of travel writing which Pratt has categorized as deploying a 'sentimental' rhetoric in which the narrator is foregrounded and relationships with 'natives' become a crucial feature of the narrative. Women's travel writing, produced within the cracks of two conflicting subject positions – that relating to the private sphere of caring and emotional work and another which demanded a certain imperial authorial and authoritarian distance – was often that much more involved with and sympathetic to the 'natives'. As such it often functioned as a counter-discourse, and, although hegemonized by imperial assumptions about the 'civilizing' mission and subject to considerable ambivalence, especially when the pull of the suffragette movement came into conflict with the demands of the campaign for the abolition of slavery, may be regarded as subverting the colonizer/colonized binary in a potentially destabilizing manner.

Quite what the response of the 'memsahibs' was to the Western-educated middle- and upper-class native in India is not entirely clear. But it is time to return to the significance of the effeminization of such native males, the theme which organized Sinha's work discussed earlier. The effeminization may be seen as an inferiorizing device to a very particular threat posed by such natives. Both the threat and the response to it need to be seen in the context of women and natives as Others who chronically functioned as potential nightmares for the upper-class colonial male's desire for control not only over Self/Other relations, but also over the potential fragmentation of the internally riven male self.

It is here that an adaptation of Bhabha's brilliant insights on the effects of 'mimicry' provide an understanding of other mechanisms which destabilized the colonial project from within. Macaulay's famous Minute on Indian Education, which was the immediate catalyst for the development of a form of education for the formation of 'a class of persons, Indian in blood and colour, but English in taste, opinions, in morals, and in intellect' who would act as 'interpreters between us and the millions whom we govern', as Macaulay himself put it, succeeded only too well, but not necessarily with all the consequences that he intended. For the Anglicized Indian, in acquiring the tastes and scaling some of the heights of English intellectual accomplishment, also implicitly brought into question the innateness of the native's inferiority. He may not have mastered the nuances, and especially not the pronunciation, but this only served to produce in the colonies what Bhabha calls the 'forked tongue of English colonialism', a source and form of chronic ambivalence, for the colonialist was now constantly confronted with a sort of grotesque shadow, who returned the gaze of the colonizer in a partially displacing mode. Mimicry, then, is both a successful outcome of a technology of power and discipline, an 'English education', but also a 'menace', a threat, which — and this is a point that

Bhabha leaves implicit — was no mean influence in the production of intellectuals who demanded the liberty that many who had opposed the teaching of English had feared would be the consequence of English education in the colonies. The history of nationalisms in the colonies could, with a little exaggeration, be narrated as the history of the production of subjects who, by way of 'English' education both in the colonies and in the centre, acquired some of the cultural resources to contest and finally overthrow the state of subjection. This is mimicry as agency and empowerment, initiated by a process that was almost inevitable, given the exigencies of imperial government. This having been said, one might quite legitimately entertain doubts about the political significance and effectiveness of ambivalence as a form of resistance — Bhabha remains symptomatically silent on this question.

And at this point we are immediately confronted by other difficulties, paradoxes and ironies: for the continuing influence of 'English' education in the ex-colonies, the differentiation of these societies as nation-states, and so on, also poses acutely the question of what meaning can really be given to the idea of 'post*coloniality' or '*post*colonialism' and forms a convenient bridge into the second part of my discussion, which poses a number of questions which threaten to undo the whole idea of 'post'colonialist studies.

2 The 'Post' and the 'Colonial' in Postcolonialist Studies: Some Awkward Questions

If examined more rigorously, the idea of the 'postcolonial' reveals a number of chronic difficulties, often shared with other 'posts' fashionable today, especially 'postmodernism' (of which more later).

Take, first, a certain apparent confusion between the 'imperial' and the 'colonial'. In the paper so far I have used the terms almost interchangeably. But, arguably, an important distinction is thus being elided. Indeed, for some analytical purposes it would seem important to differentiate between *colonialism* as a particular form of direct rule and, more often than not, involving settlement, by a foreign power, and *imperialism* which could be reserved to denote a more diffuse expansionism. Despite the obvious overlaps, the two could be argued to have different dynamics and different consequences for the 'periphery' and the 'centre'. The usage would have to depend on particular contexts, for in many general discussions of course 'imperial' can be allowed to subsume the specificity of the 'colonial'.

But this only begs another question. Given that the term 'postcolonial' appears to have established itself over the more general 'postimperial', when does the *postcolonial* moment supposedly begin? Some authors argue that it begins *at the same moment* as the *beginning* of the imperial, used synonymously with the 'colonial', for the 'post-' signifies, above all, resistance to and active differentiation from imperial imposition. For others the term is basically an alternative to the ubiquitous Western designation of 'postwar' (referring to the Second World War). It does not seem helpful to argue that at the moment of resistance to the imperial encounter, in other words, almost at the very inception of the imperialist thrust, we are already in some sort of 'postcolonial' time–space. This is to homogenize very complex historical structures and periods. While all conceptual distinctions can only be provisional, and are related to specific analytical projects, I would want to maintain that, unless there are strong arguments for doing otherwise, the concept of the 'postcolonial' should, in terms of historical periodization, be restricted to time–spaces inaugurated by the formal independence of former colonies of Western powers.

This implies that specifically 'postcolonialist' writing may properly be said to emerge after the end of formal colonialism. Before that formal severance, what we have are forms of *anti*-colonial writing, which obviously cannot reflect upon the structures and events unleashed in the aftermath of independence. Arguably, even *historical* writings on colonialism undertaken after the end of formal colonialism will bear traces of the postcolonial experience and therefore may be said to be part of a postcolonial *oeuvre* although they may not always qualify as postcolonialist in the specific senses of 'postcolonialist studies' as delineated in the first part of this essay.

But is '*neo-colonial*' not a preferable term to postcolonial since it points up more explicitly the many forms of continuity between the periods of colonialism and formal independence? Moreover, the concept may have the advantage, as Young points out, of directing attention to the present, away from an endless restaging of the colonial encounters of the eighteenth and nineteenth centuries. There is something in this argument. Nevertheless, I do not think that the reasons offered are compelling enough for 'neo-colonial' to supplant rather than supplement 'postcolonial'. For one thing, we are still learning much from the novel manner in which the colonial encounter is being re-staged in postcolonialist studies. For another, the problem with the term neo-colonial is its connotation of a relationship between the ex-colony and the former colonizer that appears to posit a more conspiratorial role for the imperial power in the new period, and one which implies far too passive a role for those who govern the now independent states of Africa, Asia and so on.

At this stage of the discussion it becomes pertinent to effect a reversal of the argument just considered, and ask whether the description 'postcolonial' for particular periods and nation states does not actually *over-value* the impact of colonialism on the societies of both the colonized and the colonizer. The simple answer to this, in my view, is that the term may indeed imply effectivities which are exaggerated. Post*colonial*, as a designation, may draw attention away from the myriad other influences on the formation of these societies. This is an issue to which I will return in the conclusion to the paper. Incidentally, this is also an appropriate point at which to argue that, given the thesis of mutual imbrication of cultures and identities as developed in postcolonial studies, the term postcolonial has to be regarded as pertinent for the societies of ex-*colonial* powers as well as for the ex-*colonies*.

Now is the time to deal with another question that assumes relevance in this context. Is it helpful to lump together African, various Asian and Latin American societies/nation-states, and Australia, New Zealand and sometimes Canada and the USA as well, as 'postcolonial' when they have been formed by such very diverse histories and occupy such disparate time–spaces in the present cultural, economic and geo-political order of the world? *Patently, it is not. The concept of postcolonialism can provide only the most general framework of analysis.* Quite clearly, what is also required is an historical imagination and contemporary analysis which is aware, to take but one example, that the sense of marginality felt by white Australian and Canadian writers in relation to the metropolitan centres is not of the same order as that experienced by indigenous African and Asian writers, although it is undoubedly true that in being grouped together as part of something called 'Commonwealth Literature' they have all suffered a certain stigma of 'worthy but not quite' (to adapt a phrase from Bhabha's suggestive analysis of the status of the native 'mimic'). The point needs to be extended to any analysis of the general issue of comparative marginalization and peripheralization of the former White Dominions and those of black Africa and India in relation to Britain and the USA.

But the legitimacy of the term 'postcolonial' can still be seriously doubted when viewed from the perspective of the aboriginal populations of North and South America and Australia and New Zealand which are still fighting what they might see as *anti-colonial* struggles. This is where, again, extreme sensitivity to historical disjunctures and the specificities of time–spaces is crucial if the idea of the 'postcolonial' is to retain some analytical value. The point is to recognize the productiveness of the chronic ambivalence and potential destabilizations *within* the discourse of 'postcolonialism', an argument effectively explored in Prakash.

Finally, to bring this sort of conceptual ground clearing to a provisional close, one might ask whether *globalization* is a better concept than postcolonialism, especially given the fact that it is more inclusive, drawing into its ambit nation-states that have no significant formal recent history of colonialism but which nevertheless are participants in the present world order. However, in my view it would be a mistake to create this type of binary apposition. We need both concepts, one to signal a very general process of time–space compression – to borrow Harvey's inelegant but concise expression – and the other precisely to act as a reminder to those who insist on writing the narrative of globalization as if the process sprang from the internal dynamic of the transition from feudalism to capitalism in the West that imperial expansion and colonialism were key constitutive features, and indeed set both globalization and Western capitalism in motion and acted as continual fuelling forces.

The credentials of the idea of 'postcolonialism' have been provisionally established by the discussion so far, or at least so I would argue, although not without stripping it of certain pretensions which might otherwise render the concept vulnerable to dismissal. Nevertheless, a number of difficult questions remain, and will continue to perplex those who intend to work in the terrain opened up by the insights of postcolonialist studies.

[. . .]

Something to end with

Not that this is the end of the story. Ahmad, for instance, has charged that much of what passes as postcolonial scholarship, so often carried out by academics whose origins lie in the former colonies and who are now comfortably established in some of the most élite metropolitan universities, has merely used poststructuralism as a ruse to ally with some of the most fashionable but ultimately non-threatening fashions. In the process this has enabled the academics to land lucrative posts and obtain prestige in the centre while getting further divorced from the realities and political involvements of their countries of origin. Others (for example Dirlik) go even further, for they argue that postcolonialism involves a serious neglect of the role of global capitalism in perpetuating global inequalities in the present and that postcolonial studies may merely serve the cultural requirements of global capitalism.

There is probably a grain of truth in all this carping. But no more than a grain. Postcolonial scholarship is an international enterprise, and one of its most impressive aspects is the manner in which it has galvanized younger scholars in and from the former colonies, and also younger metropolitan researchers, to undertake investigations into the colonial encounter which are strikingly novel and profound, and these are undertakings in which the holy trinity of Said, Bhabha and Spivak, to cite the most prominent target of censure, has served as an admirable source of inspiration. While there are always dangers of academic co-option, Ahmad's Marxism has hardly escaped this insidious institutionalization either. Moreover, Ahmad seriously undervalues the long-term significance of critical intellectual work. There is an important point about the disappearance of the general intellectual of an earlier kind, for example, Sartre or C.L.R. James, which Said grieves over in *Culture and Imperialism*, but this is an altogether more sophisticated argument than that advanced by Ahmad.

And, to turn to Dirlik's source of discontent, it is simply untrue to say that global capitalism has been ignored in postcolonial research, although obviously what postcolonial studies has been about is finding non-reductionist ways of relating global capitalism to the cultural politics of colonialism, and indeed finding frameworks which allow the imperial and colonial enterprise to be seen not as external appendages of global capitalism, but as major constitutive elements (see Hall for a detailed critical reading of Dirlik which shows up some of its internal contradictions). Said's *Culture and Imperialism*, Viswanathan's *Masks of Conquest*, Spivak's *In Other Worlds*, Niranjana's *Siting Translation*, Sinha's *Colonial Masculinity*, Chatterjee's *Nationalist Thought and the Colonial World* – to take only some of the works cited in my paper – display varying but definite degrees of emphasis on the constitutive relation between imperialism, colonialism, class relations and global capitalism.

However, there are some reservations I do wish to enter. Although postcolonialist researchers have been scathing about the consequences of nationalist politics in the former colonies, it is not clear what kind of alternative vision they wish to advance. In the last part of *Culture and Imperialism* Said makes a brave but in my view an ultimately weak attempt to promote a sort of politics of cultural hybridity, which attempts to muster some optimism and finds an important role for postcolonial intellectual work, but which remains vague and perhaps naive. On the other hand, it is also naive to expect a politics, emancipatory or conservative, to be read off from the framework of postcolonialism in the same way that there is no necessary political belonging to postmodernism (an argument Boyne and I have elaborated elsewhere). Nevertheless, relatedly, and this may be because so much of postcolonialist studies is undertaken by historians and literary critics,

little attempt is made to connect with contemporary problems of 'development'. However, this is where sociologists, anthropologists, political theorists and economists with a postcolonialist sensiblility can make important contributions to a field, development studies, which is itself in crisis. Finally, it is worth pointing out that one must deflate any imperialistic mission on the part of postcolonial studies.

Postcolonialism is only one optic on the formation and dynamics of the contemporary world. It cannot be allowed to function as a totalizing perspective, indeed it is incapable of doing so, for it cannot remotely furnish all the intellectual frameworks required for any kind of cultural or any other kind of analysis. Take just one instance: there are only limited insights to be gained by designating contemporary Indian or African cinema 'postcolonial'. There is so much more to be said . . .

So, the enterprise of postcolonial studies is hardly unproblematic. But it is nevertheless the site of new, quite fundamental insights. At the risk of paradox, it might be said that it provides a non-essentialist but essential, non-foundationalist foundation on which to map the past, the present and the future in an age of transitions. The proliferation of so many 'posts' in the social sciences and the humanities is symptomatic of a widespread acknowledgement that the old categories will simply not do any more, even if the refusal to name anything positively, the tendency merely to gesture to the passing of something familiar bespeaks a deep uncertainty about how to map the future. And, to reinforce an argument made earlier, postcolonialism signals a more general de-centring of the West, from both within and without – an internal unravelling, as the Enlightenment project is questioned on several fronts, and an external transformation as the West's hegemony in the world order comes under severe pressure.

READING 10

Said's Orientalism: A Vital Contribution Today

Peter Marcuse

Edward Said's analysis of Orientalism was a powerful critique that showed how a concept, elaborated in

academic writings and popular discourse, achieved virtually hegemonic status although it was both wrong and supportive of relations of domination and exploitation on an international scale. His conclusion hardly needs demonstration today, when near Eastern policy at high ranks of United States decision-makers is challenged as being undertaken "to gain empirical evidence to test an assumption" that "the Arab-Islamic world is inherently allergic to democracy". Said's *Orientalism*, perhaps his most important book, is a striking model of engaged intellectual work, in which the link between deep scholarly effort and immediate political reality is ever present. We can learn much by trying to apply the same critical approach to other hegemonic concepts of our time. What follows is an initial attempt to do this with the concept of Globalism.

I want to argue that the richness of Said's approach can be extended quite directly to an analysis of the concept of Globalism, which in this sense is the inheritor of Orientalism's mantle. Just as "Orientalism" was used to describe and categorize a specific geographic region, its people and its culture, I want to use the term "Globalism" to suggest the way in which specific real processes at the international level, often lumped together under the term globalization, are discussed and portrayed in academic and popular circles.

Edward Said defined Orientalism as the hegemonic view in the "West" of the inferiority of the "East", a view both anticipating and justifying a colonial relation between dominant and subordinate, manifest in culture, language, ideology, social science, media, and political discourse. In Said's very influential book with that title, he lays out, in vibrant and often polemical prose, the minute details of the way in which Orientalism pervaded the world view of the leaders of European and United States societies, not as an intentionally malicious racism but rather as an often unconscious and sometimes benevolently intended set of attitudes and preconceptions arising out of relations of power. While Orientalism preceded nineteenth and twentieth century colonialism by several millennia, its earlier expressions fueled its later direct use in support of imperial policies in England, France, and finally the United States. Said begins his analysis with a devastating look at a parliamentary speech of Arthur Balfour in 1910, in which the condescending treatment of "Orientals" and the unquestioned belief in "Western" superiority is explicit. He then goes on to trace the manifestations of those same views in an implicit and even concealed but nonetheless pervasive form in literature, movies, public speeches, and works of art. Said's work is an outstanding example of what Pierre Bourdieu would call human capital in the service of power.

"Globalism" is an apt term for the latest manifestation of the infiltration of relations of power into the political and cultural understandings of our age. I use the term in a very specific and limited sense.

Globalism is the lens (trope, metaphor, set of implicit assumptions, world view, discourse) that underlies almost all current policies of most governments in the international arena. It sees the process of globalization as new, as the dominant feature of our time, a structural process independent of specific acts of choice, inevitable in its really existing form, and ultimately beneficial to all, although certain distributional inequities may be seen as needing correction. It is the lens through which a substantial portion of the scholarly and intellectual discussion of globalization sees its subject matter.

Globalization, in its really existing form, is the further internationalization of capital accompanied by and using substantial advances in communications and transportation technology, with identifiable consequences in cultural, internal and international political relations, changes in the capital/labor balance of power, work processes, roles of national government, urban patterns, etc.

Globalism is to really existing globalization as Orientalism is to colonialism. Globalism is the hegemonic metaphor through which the actual process of globalization is seen/presented. It views development in the "developing world" as inevitably following the superior path of development pursued by the "developed world", just as Orientalism sees the "Orient" following (if it can) the superior form of development of the "Occident". If we substitute the G7 for the Occident, and the Third World for the Orient, we can apply Said's insight with profit, keeping in mind the different roles of racism, geographic coverage, and cultural distortions involved in the parallels.

Globalism accepts as obviously true and not requiring proof the inevitable domination of global interests – specifically, globally organized capital – over all spheres of life and all countries of the world. As Orientalism paralleled and legitimated colonialism and imperialism and the domination of Western over "Third World" countries, so Globalism parallels and legitimates the priority of global capitalism over all forms of social organization, and the domination of capital over labor. As Said, in a nuanced discussion, concedes the significant contribution Orientalist scholars have made to accumulating facts and advancing knowledge about other little known societies to an audience in the West, so the contribution of globalist scholars to increasing the knowledge and understanding of the range and modes of operation of global capital must be conceded. Nevertheless, the underlying assumptions in both cases parallel the needs of established power. Orientalism and Globalism in fact overlap in critical ways: implicit racism/chauvinism and unquestioning acceptance of the value systems of the industrial and financial powerful nations (implicit in the acceptance of what "development" means) fuel both, and serve to buttress domination both within nations and among them.

Globalism, like Orientalism, is effective precisely because it pretends not to be an ideology, but just scholarship or description of the world as it is. As Pierre Bourdieu put it, "it goes without saying because it comes without saying".

Just as Said argues that "the Orient" is an artificial concept, one created, largely, by scholars and writers to describe a subject that does not exist in reality – or rather, to shape something that does exist in reality into a form that makes it manageable and manipulable by dominant powers located largely in the Western industrialized countries – so is "Globalism" an artificial concept, wrapping a set of developments whose real etiology is concealed into a single something that must be accepted as a "force", an actor, to which a whole range of results can then be attributed for which no one or group is responsible, which simply becomes part of reality, a given object to be studied and understood, described and quantified. But globalization is not *an* object, any more than eastern-located countries are *an* object; they are both names, concepts, artificially created in a particular social and political and historical context, and serving a particular social and political and historical purpose. There is no more a "force" of globalization than there is a "place" called the Orient.

The role that Balfour plays in Said's account is comparable to that played by Margaret Thatcher and Ronald Reagan in Globalism's ascendancy, with policy advisers such as the early Jeffrey Sachs and institutions such as the World Bank and International Monetary Fund, and discussions such as those at Davos, playing a leading role. In the social sciences, the lineage that Said painstakingly traces could be followed, in Globalism's case, with W. W. Rostow as an early representative and Manuel Castells, in his current work, or Anthony Giddens, today, as one of its latest and most sophisticated; Francis Fukuyama exposes the world view in cruder fashion, as does Thomas Friedman. The policies that Said tracks to the masters of the British Empire in the 19th century find their direct analogy in the masters of the Washington consensus at the end of the 20th and the beginning of the 21st.

But the real contribution of Edward Said is not to document the explicit biases and stereotypes of the colonialists, but to trace the more subtle but pervasive and hegemonic parallels of colonialism in the language, the metaphors, the discourse, and the cultural production of their times. Indeed, language, metaphors, discourse, are points along an increasingly comprehensive spectrum of representation that is Said's underlying theme: a lens through which the world, or parts of it, are seen, is the simile he himself uses. Globalism deserves the same attention today, as the lens through which globalization is seen and represented. The problem lies not in the scholarship that examines the operations of global capital, as it was not in the scholarship that examined the history or culture of colonial societies. It lies rather in the unquestioning acceptance of the appropriateness of what is being examined, of the pervasiveness of its reality, in short of its inevitability. Granting the inevitability of the increasing domination of global capital over all other forms of economic and social organization contributes to that domination, just as granting the inevitability of imperial relations contributes to the continued domination of those relations.

The uses of Globalism are legion; they support and legitimate globalization, and defuse the opposition to it. Globalism is the answer emanating from the World Economic Forum at Davos to the challenge from the World Social Forum at Porto Alegre; where Porto Alegre's slogan is, "Another World is Possible", Davos answers, "TINA, There Is No Alternative: really existing globalization is inevitable". Globalism is the understanding that undergirds the World Trade

Organization's response to Seattle and its successors, that frames the defense of NAFTA and the FTAA in the United States, that empowers employers in their bargaining with labor unions everywhere, that justifies low wages in developing countries. Globalism can also be used locally, supporting an odd coalition in which purely locally based interests, such as property owners, local political leaders, or locally attached residents, support a place marketing strategy that emphasizes a locality's key position in global exchanges. That global and local pressures as often complement each other as diverge has been often enough pointed out; they both rely on Globalism when it is to their advantage, and neither is internally homogeneous.

Said's nuanced discussion of Orientalism suggests similar caution in describing the scholarship having to do with globalization. In both cases, there is an underlying and important reality calling out for examination, and in both cases key figures in their exploration have contributed much to knowledge of the subject. As Said praises Vico and the Napoleonic expedition writers, so scholars like Friedmann and Sassen contributed much to an understanding of new developments on the world stage. The issue is not so much the worth of that endeavor, but rather whether the undercurrent within it, here categorized as Globalism, has not undermined the very utility of the term. One thinks of a similar situation with the concept "underclass", which William Wilson used to describe real developments in the inner cities of the United States. After significant criticism, e.g. by Herbert Gans, and reflection, Wilson has dropped the term completely, substituting the less catchy but more delimited term "ghetto poor". In the same way, the term "globalization" might, in the absence of a hegemonic Globalism, slowly be abandoned in favor of the more accurate if also less elegant "internationalization of capitalism".

A problem, in this account, both of Orientalism by Said and of Globalism here, is that both the world view being criticized and the material for its criticism come from similar, sometimes even identical, sources. Much of the material Said cites comes from Westerners, from the Western side of the lens of Orientalism. In the same way, much of the material that provides the most damning criticism of Globalism comes from writers and researchers and activists who are on the side of the victims of globalization. Their sympathies lie on the other side of the lens of Globalism, even as their "real" position is on the viewer side. So it is with Said: among the most trenchant material he cites is that which comes from acute Western observers, whose perspicacity he generously acknowledges. It is to be expected that the real representatives of the Orient would provide material for Said's indictment: why is so much that supports his position found in the work of Western scholars and leaders, from Christian writers of the eleventh century through Napoleon to the present? Franz Fanon one would expect; but the holders of endowed chairs at elite United States universities?

The answer perhaps lies in Said's use of the term "Orientalism" in some grammatical disjuncture with the term "Orientalist". It results from a differentiation I would wish to make explicit here. Much of the argument against Orientalism in fact comes from Orientalists; that term is rather used to denote those who study the discourse of Orientalism and the realities that are artificially subsumed under the term, rather than the exponents of the viewpoint of Orientalism. In the same way, many, including some of the most prominent writers on globalization, attack the implications of Globalism. One may, in both the Orientalist and Globalist case, distinguish three types of authors: (1) those who adopt the viewpoint of Orientalism or Globalism, the Balfours and the Rostows; while Said uses the term Orientalist more broadly, the term "Globalist" might be specifically applied to this group in the case of Globalism – the legitimators of globalization, the Globalists pure and simple; (2) those who study, describe, document, parse the processes going on in "the Orient" or in "globalization", who implicitly accept the tenets of the subject but may be critical of its results and may provide accurate and useful information for its understanding; also Orientalists in Said's usage, perhaps (a bit more awkwardly) the "scholars of globalization" here; and (3) students, writers, and activists on issues raised by Orientalism and Globalism who devote themselves to its critique – the critics of Globalism who however often move in circles overlapping those of the scholars. Said would certainly consider himself also an Orientalist, but in the sense of a critic of Orientalism, an Orientalist in the sense of (3), not (1), but moving in many of the same circles as (2), the scholars of the Orient. And certainly many dealing with globalization consider themselves concerned with the same issues and moving in the same circles as the scholars of globalization.

The dividing lines here are not sharp. Globalists celebrate globalization, and have no doubts as to its existence, but their work may involve scholarly examination of aspects of the underlying reality. Scholars of globalization may expose one or another of its negative realities, but largely do not question its fundamental tenets in their work; and critics of Globalism often contribute to its scholarly analysis. But at the extremes, the roles are clear.

Said speaks of Orientalism as a view of the colonies from the outside, as a Western lens shaped to meet Western needs. If there is a reality to the difference between "the West" and "the Orient" – and there is – is there any parallel with viewpoints on Globalism? It is Westerners that look through the lens of Orientalism from one side, seeing a distorted reality on the other; they are not on both sides of the lens. None of us, in "developed" or "developing" countries, are outside the reality of globalization that lies on the other side of the lens of Globalism, the reality of the internationalization of capital that does in fact infect all economies, all politics, all cultures, all languages, all ways of life, if in quite different forms. But the lens of Globalism is not a generalized one, created without actors, serving no particular purpose. It is a view from above, from those in power, able to dominate and exploit. They are active in "developing" countries as in "developed", just as Orientalists are as often found in the countries of the East as of the West. The purpose it serves is to distort the reality of those who are dominated and exploited, the oppressed, those below. Theirs is a reality the proponents of Globalism do not share, do not know. As with Said's Orientalism, this lens is one shaped well before the lens in its present form and use are perfected, well before the talk of some who are globalizers and some who are globalized. It builds on a view of the poor by the rich or their apologists that has evolved over centuries: on the distinction between the worthy and unworthy poor, the pictures of slum life that Jacob Riis described as depraved, the lumpen proletariat characterizations of Karl Marx, the culture of poverty thesis of Oscar Lewis, the descriptions of the poor and of criminals that Frances Piven and Michel Foucault so accurately describe and that Bertold Brecht so tellingly

limns. Were one as erudite as Said, one might go even further back and look at the representations of the poor in Victor Hugo, or in Shakespeare's Coriolanus, or perhaps even in Cicero; I do not believe it can be found in the classical Greeks, for here the poor, as slaves, were simply excluded from consideration. In any event, today, through the lens of Globalism, the representation of the poor is transformed into a discourse about the included and the excluded, the developed and the under-developed, the industrialized and the not yet industrialized, the rich and the poor – and thus, the global and the not-global or the globalizing.

It would seem churlish to press the parallel further, and to say, of the students of poverty, that their aim is to facilitate the control of "the poor", as the aim of the Orientalists (in sense 1) was to facilitate the control of "the Orient". But there are parallels. In the Manhattan Institute's attack on homelessness, the approach is to categorize the poor in order to bring them under control by addressing the disturbing characteristics of each separately; not even a bow in the direction of housing market inequities or desperate poverty is visible. The same may be said of some early studies of poverty, and even of some projects, such as the settlement houses (certainly the almshouses) of the past. Loic Wacquant makes a slashing attack on some current studies of poverty along the same lines, although he fails to discriminate between intent or motive and objective effect. But then the motivations of many Orientalists were also benevolent. To the extent that the poor are portrayed as exotic, studied as strange objects in the early British studies and the Pittsburgh study, the parallel holds. But of course the critical view is also strong; thus Barbara Ehrenreich's recent book is directly aimed precisely at de-exoticizing the poor.

The projects of Orientalism seem quite clear, from the Napoleonic expedition to Egypt to the British actions in the near East at the beginning of the century. So do the projects of Globalism, from the Bretton Woods agreements to the World Trade Organization, the International Monetary Fund, and the World Bank. Oddly enough, the actions of the Bush administration in Afghanistan and Iraq today seem closer to Said's Orientalism than to Davos' Globalism; Palestine policy even more so. Is the drive to Empire the successor to Globalism? Indeed, it seems in many ways to run counter to the earlier Globalist policies; its unilateralism, reliance on crude force, protectionism at home, contradict what Globalists have long advocated. Is the drive to Empire merely a temporary aberration, or does it now represent a new constellation of forces, and if so, one within or supplanting the relations of Globalism?

Since the process is one connected with real historical movements, it is also one of counter-movements, exposure of distortion, and presentation of alternate representations. Said also played a vital role in this counter-movement.

One of Edward Said's signal contributions was to clarify the intellectual substructure on which the colonial relations between the "West" and the "East", the imperial and the colonial powers, have been (and are being) built. The Orientalist world view continues in the period of globalization; it is not replaced by Globalism, but rather supplemented by it. In the ongoing conflict between the forces of exploitation and domination, Edward Said's many-faceted contributions have been a potent weapon on the side of social justice and the struggle for a humane world. The struggle against Globalism, exemplified by movements such as those represented in the World Social Forum, are not a replacement but a continuation of the struggle in which Said played such a prominent role. We miss him already.

Neoliberalism

While it came under severe attack in the global economic crisis that raged beginning in late 2007, neoliberalism has arguably been the most influential theory in globalization studies (and underlies the next chapter on structural adjustment). It has both strong adherents and vociferous critics. However, the critics have now gained the upper hand, at least for the moment, with much of that economic crisis being linked to the neoliberal belief in, and policies of, the free market and deregulation. It was the deregulation of the banks, financial institutions, and various markets that led to the high-risk ventures (subprime mortgages, credit default swaps, derivatives, etc.) that collapsed and led to the crisis. Nevertheless, one cannot understand globalization without understanding neoliberalism. It was a key factor in the emergence of the global age and the problems it created certainly had global implications. As the crisis deepened in Europe, French President Nicolas Sarkozy said: "It [neoliberalism] is a worldwide problem, and it should get a worldwide response."[1]

Neoliberalism is a theory that has implications for globalization in general, as well as for many of its elements. It is particularly applicable to the economics (especially the market and trade) and the politics (the nation-state and the need to limit its involvement in, and control over, the market and trade) of globalization.

Not only is it important in itself, but it has also strongly influenced other thinking and theorizing about both of those domains, as well as globalization in general.

A number of well-known scholars, especially economists, are associated with neoliberalism. We begin this chapter with some of the ideas of one neoliberal economist – William Easterly – in order to give the reader a sense of this perspective.

Easterly is opposed to any form of collectivism and state planning as they were espoused and practiced in the Soviet Union or are today by the UN, other economists, and so on.[2] Collectivism failed in the Soviet Union and, in Easterly's view, it will fail today. It will fail because it inhibits, if not destroys, freedom; and freedom, especially economic freedom, is highly correlated with economic success. This is the case because economic freedom "permits the decentralized search for success that is the hallmark of free markets."[3] Economic freedom and the free market are great favorites of neoliberal economists.

Easterly offers several reasons why economic freedom is related to economic success. First, it is extremely difficult to know in advance what will succeed and what will fail. Economic freedom permits a multitude of attempts and the failures are weeded out. Over time, what remains, in the main, are the successes and they serve to facilitate a high standard of living. Central planners can never have nearly as much knowledge as myriad individuals seeking success and learning from their failures and those of others. Second, markets offer continuous feedback on what is succeeding and failing; central planners lack such feedback. Third, economic freedom leads to the ruthless reallocation of resources to that which is succeeding; central planners often have vested interests that prevent such reallocation. Fourth, economic freedom permits large and rapid increases in scale by financial markets and corporate organizations; central planners lack the flexibility to make large-scale changes rapidly. Finally, because of sophisticated contractual protections, individuals and corporations are willing to take great risks; central planners are risk-averse because of their personal vulnerability if things go wrong.

Much of the contemporary critique of neoliberalism, especially as it relates to economics, is traceable to the work of Karl Polanyi and his 1944 book *The Great Transformation: the Political and Economic Origins of Our Time*. He is the great critic of a limited focus on the economy, especially the focus of economic liberalism on the self-regulating, or unregulated, market, as well as on basing all on self-interest. In his view, these are not universal principles, but rather were unprecedented developments associated with the advent of capitalism. Polanyi shows that the *laissez-faire* system came into existence with the help of the state and it was able continue to function as a result of state actions. Furthermore, if the *laissez-faire* system was left to itself, it threatened to destroy society. Indeed, it was such threats, as well as real dangers, that led to counter-reactions by society and the state (e.g. socialism, communism, the New Deal) to protect themselves from the problems of a free market, especially protection of its products and of those who labored in it. The expansion of the *laissez-faire* market and the reaction against it is called the *double movement*. While economic liberalism saw such counter-reactions (including any form of protectionism) as "mistakes" that disrupted the operation of the economic markets, Polanyi saw them as necessary and desirable reactions to the evils of the free market. Polanyi pointed to "the inherent absurdity of the idea of a self-regulating market."[4] He also described as mythical the liberal idea that socialists, communists, New Dealers, and so on were involved in a conspiracy against liberalism and the free market.

Rather than being a conspiracy, what took place was a natural, a "spontaneous," collective reaction by society and its various elements that were threatened by the free market. In his time, Polanyi sees a reversal of the tendency for the economic system to dominate society: "Within the nations we are witnessing a development under which the economic system ceases to lay down the law to society and the primacy of society over that system is secured."[5] This promised to end the evils produced by the dominance of the free market system, and also to produce *more*, rather than less, freedom. That is, Polanyi believed that collective planning and control would produce more freedom, more freedom for all, than was then available in the liberal economic system.

David Harvey argues that among the problems with neoliberalism as a theory is the fact that it assumes that everyone in the world wants very narrow and specific types of economic wellbeing (to be well-off economically, if not rich) and political freedom (democracy). The fact is that there are great cultural differences in the ways in which wellbeing (e.g. not to have to work very hard) and freedom (e.g. to be unfettered by the state even if it is not democratically chosen) are defined. Neoliberalism very often comes down to the North, the US, and/or global organizations (e.g. World Bank, International Monetary Fund) seeking to impose *their* definitions of wellbeing and freedom on other parts of the world. Furthermore, there is great variation on this among individuals in each of these societies, with the result that these definitions are different from at least some of theirs, but are nonetheless imposed on them.

Another problem lies in the fact that the theory conceals or obscures the social and material interests of those who push such an economic system with its associated technological, legal, and institutional systems. These are *not* being pursued because everyone in the world wants them or will benefit from them, but because *some*, usually in the North, are greatly advantaged by them and therefore push them.

Harvey offers a number of other criticisms of neoliberalism including the fact that it has produced financial crises in various countries throughout the world (e.g. Mexico, Argentina, and now globally); its economic record has been dismal since it has redistributed wealth (from poor to rich) rather than generating new wealth; it has commodified *everything*; it has helped to degrade the environment; and so on. Furthermore, there are signs that it is failing such as deficit financing in the US and China, symptoms of more immediate crisis (e.g. burgeoning budget deficits, the bailout of financial institutions, the current recession), and evidence that US global hegemony is crumbling.

Aiwha Ong makes an important contribution to our thinking about neoliberalism by distinguishing between neoliberalism as exception and exceptions to neoliberalism. One example of *neoliberalism as exception* involves the creation in various parts of the world of special economic zones which are largely separated from the rest of society and free from government control, and within which the market is given more-or-less free reign. These are "exceptions" because the market is not nearly as free elsewhere in society. For example, early in its move way from a communist economic system, China set up "special economic zones" and "special administrative regions" (as well as "urban development zones") characterized by "special spaces of labor markets, investment opportunities, and relative administrative freedom."[6] While the state retained formal control over these zones, *de facto* power rested with multinational corporations (MNCs) that set up shop within them. It was those corporations that controlled

migration into the zones as well as the ways in which people in the zones lived and worked.

Ong calls the political result of constructing these zones *graduated sovereignty*. That is, instead of governing the entire geographic area of the nation-state, the national government retains full control in some areas, but surrenders various degrees of control in others to corporations and other entities. While the creation of these zones may bring a series of economic advantages, it also can create problems for the nation-state that is no longer in full control of its own borders. (This is yet another indication of the decline of the nation-state: see chapter 6.)

Ong is primarily concerned with neoliberalism as exception, but she also deals with *exceptions to neoliberalism*. These can be double-edged. On the one hand, such exceptions can be used by the state to protect its citizens from the ravages of neoliberalism. For example, subsidized housing can be maintained even if a city's budgetary practices come to be dominated by neoliberal entities and processes. On the other hand, they can be used to worsen the effects of neoliberalism. For example, corporations can exclude certain groups (e.g. migrant workers) from improvements in the standard of living associated with a market-driven economy.

NOTES ···

1 Edward Cody, "No Joint European Strategy on Banks." *Washington Post* October 5, 2008: A20.
2 www.freetheworld.com/release_html.
3 William Easterly, "Chapter 2: Freedom versus Collectivism in Foreign Aid." *Economic Freedom of the World: 2006 Annual Report* 35.
4 Karl Polanyi, *The Great Transformation: the Political and Economic Origins of Our Time.* Boston, MA: Beacon, 1944, 145. This much quoted observation has been reworded in the edition excerpted in the present book.
5 Ibid., 251.
6 Aiwha Ong, *Neoliberalism as Exception: Mutations in Citizenship and Sovereignty.* Durham, NC: Duke University Press, 2006, 19.

READING 11

Freedom versus Collectivism in Foreign Aid

William Easterly

1 The New Collectivism

Marx was right about at least one thing: "History repeats itself, first as tragedy, second as farce." The 21st century has seen a farcical version of the collectivist utopian fantasies that led to such disasters in the 20th century. Fortunately, the new collectivism is far more tepid – less extreme, less powerful, and less coercive – than the ideologies that caused so much tragedy in the Communist bloc in the 20th century. The collapse of communism in Europe with the fall of the Berlin Wall, and the great success of the movement away from central planning towards markets in other places like China and Vietnam that remain nominally Communist (along with the poverty of the unrepentant Communist states in Cuba and North Korea) discredited the Communist notion of comprehensive central planning once and for all. Yet, by an irony that is not so amusing for its intended beneficiaries, the new farcical collectivism is still alive for the places that can afford it the least – the poorest nations in the world that receive foreign aid. Instead of the Berlin Wall, we have an "Aid Wall," behind which poor nations are supposed to achieve their escape from poverty through a collective, top-down plan. Instead of the individual freedom to prosper in markets, the successful approach of the nations that are now rich, the poor must let the international experts devise the collective solution to their miseries.

Jeffrey Sachs and *The End of Poverty*

Lest you think I exaggerate, consider some of the statements of the most prominent and extreme spokesman of the new collectivism for poor nations, Jeffrey Sachs. In his 2005 book, *The End of Poverty*, he says in the opening pages:

I have [. . .] gradually come to understand through my scientific research and on the ground advisory work the awesome power in our generation's hands to end the massive suffering of the extreme poor [. . .] Although introductory economics textbooks preach individualism and decentralized markets, our safety and prosperity depend at least as much on collective decisions to fight disease, promote good science and widespread education, provide critical infrastructure, and act in unison to help the poorest of the poor [. . .] Collective action, through effective government provision of health, education, infrastructure, as well as foreign assistance when needed, underpins economic success.

Sachs says that each poor country should have five plans, such as an "Investment Plan, which shows the size, timing, and costs of the required investments" and a "Financial Plan to fund the Investment Plan, including the calculation of the Millennium Development Goals Financing Gap, the portion of financial needs the donors will have to fill." These plans will be helpfully supported by the "international community":

each low income country should have the benefit of a united and effective United Nations country team, which coordinates in one place the work of the UN specialized agencies, the IMF, and the World Bank. In each country, the UN country team should be led by a single United Nations resident coordinator, who reports to the United Nations Development Program, who in turn reports to the UN secretary-general.

Everything will fit together in one great global plan run by "the UN Secretary General, [who] should ensure that the global compact is put into operation".

Like his collectivist predecessors, Sachs sees the achievement of prosperity as mostly a technical problem: "I believe the single most important reason why prosperity spread, and why it continues to spread, is the transmission of technologies and the ideas underlying them [. . .] science-based ideas to organize production". "Africa's problems [. . .] are [. . .] solvable with practical and proven technologies".

He sees one kind of scientific expert – the medical doctor – as the model for how to solve the problems of poverty:

> Development economics today is not like modern medicine, but it should strive to be so. It can improve dramatically if development economists take on some of the key lessons of modern medicine, both in the development of the underlying science and in the systematization of clinical practice, the point where science is brought to bear on a particular patient.

Of course, there are such things as public goods, which require solving a collective action problem to supply them. There is a role for government to supply such goods. However, Sachs (and the other collective approaches described below) seem to make little distinction between a lack of public goods and a lack of private goods, which is called poverty.

The United Nations' Millennium Development Goals

The United Nations is the main official sponsor of today's collectivist fantasies. These are called the Millennium Development Goals (MDGs), described on the United Nation's web site as follows:

> The eight Millennium Development Goals (MDGs) – which range from halving extreme poverty to halting the spread of HIV/AIDS and providing universal primary education, all by the target date of 2015 – form a blueprint agreed to by all the world's countries and all the world's leading development institutions. They have galvanized unprecedented efforts to meet the needs of the world's poorest.

Secretary-General Kofi Annan uses the collectivist "we":

> We will have time to reach the Millennium Development Goals – worldwide and in most, or even all, individual countries – but only if we break with business as usual. We cannot win overnight. Success will require sustained action across the entire decade between now and the deadline. It takes time to train the teachers, nurses and engineers; to build the roads, schools and hospitals; to grow the small and large businesses able to create the jobs and income needed. So we must start now. And we must more than double global development assistance over the next few years. Nothing less will help to achieve the Goals.

The Secretary-General uses "grow" as an active verb applied to business, something that "we must start now." Somehow collective action will create jobs and income, as opposed to the decentralized efforts of individual entrepreneurs and firms operating in free markets.

Insofar as the MDG campaign mentions private entrepreneurs, they are "partners" subject to "our" resolve:

> We resolve further: [. . .] To develop and implement strategies that give young people everywhere a real chance to find decent and productive work [. . .] To develop strong partnerships with the private sector and with civil society organizations in pursuit of development and poverty eradication.

Part of the reason for this campaign is not just to help the world's poor, but to help the UN, as Kofi Annan made clear at the September 2005 World Summit on the MDGs: "it is also a chance to revitalize the United Nations itself." In this it has been successful, at least at the World Bank and the IMF. These two organizations have long preached the virtues of free markets and ignored UN bureaucrats preaching statist rhetoric. Inexplicably, the World Bank and IMF have since 2000 embraced the UN MDG exercise and a lot of its planning. An OECD-DAC document explains this palace coup in favor of collectivist planning as follows.

> In the 1990s, the field of international development entered an era of reform and reformulation as the disparities between rich and poor countries increased. World leaders, in collaboration with the UN and other multilateral institutions, recognized the need for drastic measures to ensure that developing countries benefited from globalization and that development assistance funds were used equitably and effectively to achieve the global development aims embodied in the

Millennium Development Goals (MDGs) and other national development goals.

In their *Global Monitoring Report 2006: Millennium Development Goals: Strengthening Mutual Accountability, Aid, Trade, and Governance*, the IMF and World Bank make clear their embrace of the whole MDG planning exercise: "Donors and the international financial institutions must increase aid flows, improve aid quality, and better align their support with country strategies and systems." How would this be done? The World Bank and IMF reaffirm a commitment to "accountability for achieving results," which they note was already reaffirmed four years earlier in the UN Monterrey Summit. On the same page, the report notes without irony that "international financial institutions [such as the World Bank and IMF] still emphasize loans and reports rather than development outcomes." They are still having some difficulty, as a few pages later they cannot keep themselves from emphasizing loans, apologizing that "in 2005 lending through the concessional and non-concessional windows of the MDBs declined."

They plan to change their ways by "Implementing the results agenda":

> The 2004 Marrakech Roundtable on Results called for a monitoring system to assess the results orientation of the multilateral development banks (MDBs); that system is COMPAS, the Common Performance Assessment System, which draws on MDB frameworks and action plans to implement managing for development results (MfDR).

Managing for Development Results (MfDR)

Exactly what is MfDR? It is summed up in *Managing for Development Results Principles in Action: Sourcebook on Emerging Good Practice* (MfDR Sourcebook), prepared by the OECD and the World Bank. To clear up any confusion, the MfDR Sourcebook notes that "Performance management is a holistic, cultural change." When it does get a tad more concrete, MfDR seems to involve a lot of central planning, such as the following:

> *At the national level,* MfDR is used in the planning and implementation of results-based national plans, budgets,

and antipoverty strategies. International agencies may support this process with technical assistance.

> *In sector programs and projects,* partner countries and development agencies use MfDR in planning assistance programs or individual projects that are based on country outcomes and priorities defined in national or sector development plans.

It doesn't get any better reading the rest of the MfDR Sourcebook. In [a] table [in] the MfDR Sourcebook is the sensible principle: "Keep results measurement and reporting as simple, cost-effective, and user-friendly as possible." [Shown here] is an excerpt from the table giving the recipe for simple, cost-effective, and user-friendly results measurement and reporting. The old collectivists were lethal; the new collectivists just bury life and death issues under six layers of bureaucracy.

Examples of tools being used to manage for results in development agencies

---

M&E systems, plans and guidelines (incorporating MIS)

Audit and risk management frameworks

Performance measurement frameworks

Program/project monitoring frameworks

Audit guidelines and tools

Evaluation guidelines and tools

Risk analysis guidelines and tools

Training and guidelines for indicator design, data collection, and analysis

---

All the MDG planners use the word "accountability" frequently, but without understanding what "accountability" is. Unlike the individual accountability that each producer faces in free markets (you satisfy the customers or you go out of business), the MDG exercise has something called "mutual accountability." This murky notion appears to involve accountability, not to the intended beneficiaries, but to the other bureaucracies involved in the MDG plan, all of whom have a stake in the current system continuing regardless of results. Instead of individual accountability, we have collective responsibility: "Development agencies are creating results-based country assistance strategies in close dialogue with national governments [. . .] During this process, multiple agencies negotiate a process for working together to support country outcomes." A system in which everyone (multiple agencies and governments) are collectively responsible is equivalent to one in which nobody is individually responsible. If there are disappointing results, you can always blame someone else. Collective responsibility is to accountability what collective farms are to individual property rights.

2 Freedom versus Collectivism in Economic Development: the Empirical Record

The empirical record on the difference between the economic performance of freedom and that of collectivism is fairly clear to anybody following events of the last half century. There was a period from the 1930s through the 1950s when the rapid growth of the Soviet Union (since found to have been greatly exaggerated) made observers unsure as to which system delivered superior economic results. Unfortunately, these were the formative years of development economics and foreign aid policies, which led many of the early development economists to recommend that poor countries imitate the collectivist model, stressing forced saving and investment to achieve growth, and to advocate national economic planning (somewhere in the netherworld in between central planning and free markets). Although the World Bank and the IMF had abandoned central planning as the recommended approach to poor countries by the 1980s, foreign aid has never been able to shake its collectivist origins. For one thing, the World Bank and the IMF continued to function as large planning organizations; it was just that now the top-down expert-driven plans included adoption of freemarket liberalization (known as "structural adjustment"). The top-down planning by foreign experts and bureaucrats of how you should implement free markets did not lead to good results in the areas where it was most intensively practiced – Africa, the Middle East, Latin America, and (ironically) the former Soviet Union. This led to the unfortunate backlash against free markets that we are seeing today in many parts of those regions. The aid organizations retreated for self-protection into the MDG planning exercise described in the first section.

This is ironic, because the fall of the Berlin Wall and more access to information about the Soviet Union and its satellites made clear just how badly the most extreme version of collectivism had failed. Even prior to this, it was rather obvious that free societies were dramatically out-performing collectivist ones, as the most casual acquaintance with comparisons between East and West Germany, North and South Korea, or between the Soviet Union and the United States made clear.

The correlation between economic success and economic freedom

Today, long after the collapse of communism, there is still a huge amount of variation from free to unfree societies. To formalize the obvious, economic success is strongly correlated with economic freedom. I use the 2002 measure published in *Economic Freedom of the World: 2004 Annual Report* to match the last year for which a large sample of data on income is available.

Of course, there is a large problem of potential reverse causality – richer people might demand more economic freedom. Critics of the measures published in *Economic Freedom of the World* also might allege that they are constructed by those with strong prior beliefs that economic freedom is associated with prosperity and, hence, the indices might be unconsciously skewed to give higher scores to countries known to be success stories. (I don't know of any reason to doubt the Index published in *Economic Freedom of the World*, which uses only third-party data and includes no subjective judgments, but I bend over backwards to anticipate possible critiques.) Any such skewing would introduce a second kind of reverse causality. To address these possible objections, I show an instrumental variables regression in Table 1. Since the institutions of economic freedom originated in Europe and then spread to other temperate regions where Europeans settled (with some exceptions), I use distance from the equator as one instrument for economic freedom. Since different legal traditions (especially the British) favored economic freedom while others did not (obviously the socialist legal tradition), I use legal origin as another set of instruments for freedom. The test statistics on the validity of the instruments are mostly satisfactory, and we still show a very strong association between economic freedom and per-capita income.

Table 1 IV regression of log per-capita income (lpcy) in 2002 on economic freedom ratings

	lpcy2002
Economic freedom in the world, 2002 (from *Economic Freedom of the World: 2004 Annual Report*)	1.343 (8.48)**
Constant	−0.495 (−0.47)
Observations	86
Sargan over-identification test: *p*-value	0.0654
First-stage *F*-statistic on excluded instruments	8.25

Instruments for economic freedom: distance from equator, British, French, Socialist, or German legal origin.
* Significant at 5%; ** significant at 1%.

The "poverty trap" and the "big push"

Although economic freedom seems well established as a path to prosperity, advocates of collectivist solutions to world poverty allege that poor countries are in a "poverty trap." The poverty trap would prevent poor nations from experiencing economic growth even if they do have economic freedom, requiring a collectivist rescue operation. It is, again, Sachs who is the leading exponent of the "poverty trap" hypothesis. In *The End of Poverty*, he suggests three principal mechanisms. The first is that poor people do not save enough.

> When people are [...] utterly destitute, they need their entire income, or more, just to survive. There is no margin of income above survival that can be invested for the future. This is the main reason why the poorest of the poor are most prone to becoming trapped with low or negative economic growth rates. They are too poor to save for the future and thereby accumulate the capital that could pull them out of their current misery.

Sachs' second reason for a poverty trap "is a demographic trap, when impoverished families choose to have lots of children." Population growth is so high that it outpaces saving (which was already too low, according to the first reason).

The third element is increasing returns to capital at low initial capital per person (and low income per person):

An economy with twice the capital stock per person means an economy with roads that work the year round, rather than roads that are washed out each rainy season; electrical power that is reliable twenty-four hours each day, rather than electric power that is sporadic and unpredictable; workers who are healthy and at their jobs, rather than workers who are chronically absent with disease. The likelihood is that doubling the human and physical capital stock will actually more than double the income level, at least at very low levels of capital per person.

Sachs gives the example of a road with half of the road paved and half impassable due to missing bridges or washed out sections. Repairing the impassable sections would double the length of road but would much more than double the output from the road. "This is an example of a threshold effect, in which the capital stock becomes useful only when it meets a minimum standard."

The role of foreign aid is to increase the capital stock enough to cross the threshold level, in what became known as "the Big Push": "if the foreign assistance is substantial enough, and lasts long enough, the capital stock rises sufficiently to lift households above subsistence [...] Growth becomes self-sustaining through household savings and public investments supported by taxation of households." Without foreign aid, according to Sachs, "many reasonably well governed countries are too poor to make the investments to climb the first steps of the ladder."

Even before testing this hypothesis, it is worth noting that these ideas are not new. In fact, they were part of the founding ideas of development economics in the 1940s and 1950s and development economists used them to insist foreign aid was necessary for economic growth then, just as Sachs does now half a century later. After $568 billion in aid to Africa combined with the continent's economic stagnation over the past four decades, combined with the success of poor countries getting much smaller amounts of aid as a percent of their income in East Asia, one might have thought a little skepticism was in order before repeating the ideas of the 1950s.

Given the publicity that these revived, old ideas about foreign aid are receiving, let us test the hypothesis of the poverty trap and the necessity of the "Big

Table 2 Test of poverty-trap and economic-freedom hypotheses for economic growth

Instrumental variables regression	Dependent variable: per-capita growth, 1960–2002
Economic freedom in the world, averaged over 1970–2002	0.022
	(2.63)*
Log of initial per-capita income	−0.014
	(2.21)*
Constant	0.001
	(−0.05)
Observations	85
Sargan over-identification test: p-value	0.0542
First-stage F-statistic on excluded instruments	9.63

Instruments for economic freedom: distance from equator, British, French, Socialist, or German legal origin.
* Significant at 5%; ** significant at 1%.

Push" against the explanation that countries prosper because of economic freedom. The poverty-trap hypothesis would say that poor countries have low growth and rich countries have high growth, so there would be a positive association between initial income and growth. This positive association should hold up when we control for whether the country is "reasonably well governed" (such as whether the government facilitates economic freedom). So I do a regression combining economic freedom with initial income; as before I need to instrument for economic freedom to address possible reverse causality. A high value of (the average level of) economic freedom relative to initial income indicates that income potential is high (if the economic-freedom hypothesis is correct) compared to actual income and so would predict faster growth.

The results are shown in Table 2. The poverty-trap hypothesis loses out decisively to the economic-freedom explanation as to who prospers. Actually, initially poor countries grow *faster* than rich ones, once you control for economic freedom.

What about the role of foreign aid in launching the growth out of poverty? Does a "Big Push" of foreign aid lead to growth? There is a huge empirical literature on foreign aid and growth, with the latest verdicts being that foreign aid does *not* have any measurable impact on growth. I go back to the well one more time to see how aid flows affect the simple hypothesis testing introduced in Table 2.

In Table 3, I add foreign aid received as a ratio to Gross National Income of the recipient as an explanatory variable. Once again, there is the problem of reverse causality. I use the log of population size as an instrument for aid, taking advantage of a quirk in the aid system such that small countries receive large shares of their income as aid, unrelated to their economic performance or needs. Instrumenting for two right-hand-side variables at once leads to more complicated problems of identification and weak instruments, so let us treat this exercise as illustrative rather than definitive.

Controlling only for initial income and not for economic freedom, aid has no significant effect on economic growth. Once you control for economic freedom, aid has a negative and significant effect on growth. I am hesitant to stress this result too strongly, as the previous literature has generally found a zero effect of aid on growth, not negative. Much greater robustness testing is needed before the negative result can be taken too seriously, and the problem of weak instruments also needs much more examination. At the very least, however, this illustrative exercise is consistent with the previous literature that aid does not have a *positive* effect on growth.

Table 3 Per-capita growth 1960–2002 as function of aid, initial income, and economic freedom: instrumental variables regressions

	Regression 1	Regression 2
Aid/GNI 1960–2001	−0.001	−0.003
	(−1.43)	(3.32)**
Log of initial income, 1960	−0.001	−0.024
	(−0.29)	(2.68)**
Economic freedom in the world, averaged 1970–2002		0.024
		(2.09)*
Constant	0.025	0.081
	(−0.95)	(−1.95)
Observations	94	65
Sargan over-identification test: p-value		0.5718

Instrument for aid: log of population in 1980.
Instruments for economic freedom: distance from equator, British, French, Socialist, or German legal origin.
* Significant at 5%; ** significant at 1%.

3 Hayek and the iPod: Why a World of Uneven and Unpredictable Economic Success Needs Economic Freedom

What the collectivist vision always misses is that success is rare, failure is common. Economic success is always very uneven and unpredictable, across almost any possible unit of analysis one might consider. Economic freedom permits the decentralized search for success that is the hallmark of free markets. It is seldom known in advance what will succeed. Many thousands of searchers mount myriads of different trials as to what will please consumers. A free-market system gives rapid feedback as to which products are succeeding and which are not, and searchers adjust accordingly. Those activities that succeed attract more financing and more factors of production so that they can be scaled up enormously; those activities that fail to please consumers are discontinued. Planners don't have a search-and-feedback mentality; rather, they implement a preconceived notion of what will work and keep implementing it whether it is working or not.

Economic success stories are often unexpected and unpredicted. MP3 players were invented several years ago and seemed to offer great promise as a great new way for music lovers to listen to large amounts of their favorite music. Despite this promise, none of the early MP3 players caught consumers' fancy. (I was an "early adopter," buying one of these at a high price so I could see it die quickly.) Apple Computer, Inc., was known mainly for its strange failures in the PC market. It was a surprise when Apple Computer suddenly found a huge hit in the iPod mobile digital device, which as of March 2006 had 78% of the market for MP3 players. So far, Apple has sold 50 million iPods. The matching iTunes application program for selling songs on-line via download to an iPod accounts for 87% of the legal music downloads in the United States.

Ray Kroc was a salesman in the 1950s peddling Multimixers, a machine that mixed six milk shakes at a time. His original idea was to sell as many Multimixers as possible. In 1954, he visited a restaurant called "McDonald's" in San Bernadino, California. He noticed that the McDonald brothers kept eight Multimixers operating at full capacity around the clock. At first, he wanted to recommend their methods to his other clients, increasing the demand for his Multimixers. But then he changed his mind: he saw that preparing hamburgers, fries, and milk shakes on an assembly line was a way to run a successful chain of fast-food restaurants. He forgot all about Multimixers and the rest is

Golden Arches stretching as far as the eye can see. How many Ray Krocs has foreign aid lost by its emphasis on Plans?

Many consumer markets in the United States are similarly dominated by a small number of successful brands. The Coca-Cola and Pepsi-Cola companies together have 75% of the American market for carbonated soft drinks. Dr Pepper/Seven Up is in third place with another 15%. The remaining 10% of the market is split up among a large number of much smaller firms. Casual observation suggests many examples of brand dominance: Microsoft®, Starbucks®, Amazon.com®, Borders®, Barnes and Noble®, and so on. While brand dominance may reflect many factors about industrial organization, it also shows the incredible unevenness of product success associated with particular firms (as we will see in a minute), perhaps reflecting the kind of serendipity illustrated by the iPod and McDonald's®.

The uneven success of products is closely related to the uneven success of firms. Just 0.3% of firms in the United States accounted for 65% of all firm sales in 2002. Firm size is well known to follow Zipf's law (also known as a power law), in which the log of the size is a negative linear function of the frequency of this size occurring (or equivalently the rank). Power laws have generated a lot of hype; for the purposes of this paper, it is enough to point out how large-scale success is rare, while failure is common. In other words, the frequency distribution of firms (or whatever unit we are interested in) has a fat and long right-hand side tail, of which there are many special cases such as a log-normal distribution and a power law (Pareto distribution). In other words, most of the distribution is concentrated at some mediocre level, then there are a small number of firms that are just totally off the charts – way above what something like a standard bell curve would predict.

Even though large firms dominate the marketplace, it is not so easy to be a large firm. Of the world's largest 100 companies in 1912, some like Procter & Gamble® and British Petroleum were many times larger in 1995. However, they were the exception, as 1912's big 100 firms also included such dinosaurs as Central Leather and Cudahy Packing in the United States. Only 19 of the top 100 in 1912 were still in the top 100 in 1995, and 48 of 1912's big 100 had disappeared altogether by 1995. Business books lay out the secrets for success of a few large companies celebrated by the author, only

to see the firms fall upon hard times after the book is published. Business writers celebrated Enron® for its innovative approach right up to the last minute. Even the most successful business gurus have their embarrassments: Tom Peters' 1982 mega-best-seller, *In Search of Excellence*, included among its celebrated companies some that would later go bankrupt such as Atari Corporation, Wang Laboratories, and Delta Air Lines.

The difficulty of achieving and maintaining success is not peculiar to large firms. Every year about 10% of existing firms of all sizes go out of business. Not that it is so easy to start a new firm to replace the ones that go out of business. More than half of new firms fail within four years of the founding in the United States.

The economic success and failure of individuals is also well known to follow the same skewed tendencies. The distribution of individual income within countries generally follows a log-normal distribution for most of the range of income (covering 97–99 percent of individuals), with a power law covering the upper 1–3 percent of income earners.

Moving to international data, economic development is of course spectacularly uneven across countries, as well as across time. Observations of high average income are confined to a few countries in recent periods, with large parts of the world and large parts of human history bereft of this kind of success. A small minority of episodes attain very high income but this falls off almost vertically as we move down the ranks.

Manufacturing exports per capita

An indicator of development that shows even greater variation across countries is manufacturing exports per capita. This reflects many different factors: the transition from agriculture to manufacturing as countries develop, the many factors that influence openness to international trade and competitiveness in international markets, the gravity model of trade flows, and so on. At some more basic level than as a trade indicator, however, manufacturing exports reflects something that all countries can potentially do, and they are all competing in the same global marketplace. As an indicator, it also has the advantage of being evaluated at world market prices, unlike national incomes with different domestic prices, which are notoriously difficult to compare. Moreover, manufacturing exports are overwhelmingly dominated by the private sector and face a market test, unlike some of the components of GDP, such as a large government sector that is measured at cost rather than according to the value that individuals place on it. Success at exporting manufactures ranges all the way from Singapore's over $25,000 per capita to Burundi's 2¢ per capita (Table 4).

Whatever the advantages and disadvantages of manufacturing exports as a measure of success, manufacturing exports per capita are themselves highly correlated with the log of per-capita GDP. Not only is manufacturing export success itself spectacularly uneven across countries, it is also very uneven within each country across product categories. Data is available on manufacturing exports at the 6-digit product-classification level. Countries export as many as 2,236 different manufacturing products, with the average in the sample being 1,177. The top three out of this array of products account, on average, for 35% of export value, while the top 1% of products account for over half of exports. The distribution of export value across products is log normal, with the value within the top 20% of products (accounting for 94% of export value) following a power law.

In other words, the big difference between Ireland and Burundi (both small populations, with Burundi larger) is not that Ireland performs better on everything, but that it found three manufacturing export products (parts and accessories of data processing equipment, monolithic integrated circuits except digital, and sound recordings other than photographic products) that earned it $15 billion, while Burundi's top three (automobile spark ignition engine of 1,500–3,000 cc, sheet/tile and asbestos/cellulose fibre cement, and corrugated sheets of asbestos/cellulose fibre cement) earned it $151,000.

How do you achieve large-scale success on a few products? Again it is economic freedom that fosters success, finding the particular niche in international markets where the country can achieve enormous scale in exports. Burundi has one of the world's worst scores on economic freedom, while Ireland has one of the best. Economic freedom is highly correlated with manufacturing exports per capita. When we address causality by using the same instruments as above for economic freedom, we still find that economic freedom predicts success at manufacturing exports.

Why is economic freedom so conducive to large-scale manufacturing exports and to development in general? Why do planners fail so badly? In a world of great uncertainty and unpredictability, economic freedom succeeds for the following reasons:

1   There is a tremendous difficulty in knowing what will succeed. Economic freedom fosters competition and multiple attempts to find things that work, and weeds out the many failures. After a while, the economy consists mostly of the big successes, which facilitates a high standard of living. Planners cannot have enough knowledge of the complexities of success; moreover, they suffer from the delusion that they already know the answers.

2   Economic freedom gives markets, which are great feedback mechanisms for learning what is succeeding and what is failing. Central planning lacks feedback.

3   Economic freedom ruthlessly reallocates resources away from what is failing towards what is succeeding. Planning bureaucracies have departments that each constitute a vested interest resisting reallocation.

4   Economic freedom makes it possible to increase the scale of a successful activity rapidly and by a huge magnitude. Financial markets allocate funds to finance an expansion in scale and the organizational form of the corporation permits replication of the same activity that worked on a small scale on a much larger scale. Financial markets and corporations require economic freedom to function well. Planning bureaucracies seldom show much flexibility in expanding successful activities on a large scale.

5   Economic freedom makes possible sophisticated contracts that allow individuals and firms to deal with uncertainty. Given the rarity of success and the likelihood of failure, individuals and firms will only be willing to bet on finding a big hit if they have the ability to diversify risk and are protected against catastrophic consequences from failure. Limited liability in corporations, bankruptcy law, and financial markets help achieve these tasks in the world shaped by economic freedom. Risk-averse planning bureaucracies opt for low-risk, low-return activities.

Table 4 Ranking of countries by manufacturing exports per capita (Manfexppc)

Exporter	Manfexppc	Rank	Exporter	Manfexppc	Rank	Exporter	Manfexppc	Rank
Singapore	$25,335.56	1	Cyprus	$419.24	50	Venezuela	$37.76	99
Hong Kong	$23,345.09	2	Kuwait	$408.61	51	Bolivia	$37.00	100
Ireland	$11,714.59	3	Philippines	$401.97	52	Peru	$32.62	101
Belgium	$9,230.09	4	Tunisia	$387.58	53	Bangladesh	$30.22	102
Luxembourg	$7,687.62	5	Swaziland	$382.05	54	Rep. of Moldova	$30.19	103
Switzerland	$7,667.51	6	Greece	$369.04	55	Panama	$24.48	104
Netherlands	$6,331.30	7	Barbados	$362.31	56	Kazakhstan	$20.45	105
Sweden	$5,650.80	8	Belarus	$351.22	57	Madagascar	$19.92	106
Malta	$5,229.30	9	Romania	$284.20	58	Ecuador	$19.46	107
Macao	$4,954.83	10	Macedonia	$265.09	59	Egypt	$17.71	108
Denmark	$4,901.73	11	Latvia	$263.87	60	Armenia	$16.97	109
Finland	$4,813.37	12	Bulgaria	$243.14	61	Côte d'Ivoire	$16.83	110
Germany	$4,639.47	13	Fiji	$228.84	62	Zimbabwe	$16.22	111
Austria	$4,540.26	14	Antigua	$225.11	63	Georgia	$15.80	112
Canada	$4,451.37	15	Turkey	$212.77	64	Zambia	$15.09	113
France	$3,216.17	16	Polynesia	$179.70	65	Turkmenistan	$14.93	114
Japan	$3,128.05	17	Lesotho	$176.12	66	Gabon	$14.61	115
United Kingdom	$3,033.86	18	Trinidad	$168.23	67	India	$14.57	116
Slovenia	$2,953.41	19	South Africa	$148.02	68	Kyrgyzstan	$14.50	117
Italy	$2,821.06	20	Argentina	$147.09	69	Honduras	$12.55	118
Malaysia	$2,810.36	21	Jordan	$142.34	70	Nepal	$11.66	119
Rep. of Korea	$2,569.26	22	China	$135.91	71	Azerbaijan	$11.50	120
Israel	$2,529.26	23	Uruguay	$135.34	72	Suriname	$10.54	121
Hungary	$2,134.28	24	Morocco	$128.73	73	Iran	$9.71	122
USA	$1,924.84	25	Brazil	$123.48	74	Paraguay	$9.69	123
Czech Rep.	$1,828.87	26	Maldives	$117.37	75	Papua New Guinea	$9.10	124
Norway	$1,760.31	27	Indonesia	$105.25	76	Senegal	$8.67	125
Spain	$1,698.14	28	Saudi Arabia	$100.43	77	Kenya	$5.25	126
Estonia	$1,607.54	29	Botswana	$93.58	78	Cuba	$5.21	127
Portugal	$1,546.50	30	Belize	$88.85	79	Niger	$5.16	128
Slovakia	$1,270.32	31	Russia	$87.83	80	Nicaragua	$4.58	129
Mexico	$1,221.76	32	Serbia	$82.59	81	Ghana	$3.19	130
Qatar	$1,092.43	33	St Vincent	$80.61	82	Togo	$2.91	131
Mauritius	$855.41	34	Jamaica	$75.55	83	Sudan	$2.72	132
Bahamas	$782.86	35	Cambodia	$74.78	84	Algeria	$2.60	133
Costa Rica	$778.15	36	Ukraine	$71.13	85	Gambia	$1.69	134
New Zealand	$687.16	37	Chile	$70.73	86	Mali	$1.39	135
Thailand	$676.73	38	New Caledonia	$67.94	87	Burkina Faso	$1.34	136
Bahrain	$626.52	39	El Salvador	$64.44	88	Mozambique	$1.12	137
Australia	$594.48	40	Saint Lucia	$63.98	89	Comoros	$0.91	138
Croatia	$563.68	41	Greenland	$62.61	90	Uganda	$0.70	139
Iceland	$554.48	42	Colombia	$61.06	91	Guinea	$0.66	140
Lithuania	$534.74	43	Albania	$59.54	92	Benin	$0.62	141
Saint Kitts	$492.98	44	Lebanon	$50.10	93	Central Afr. Rep.	$0.59	142
Poland	$492.14	45	Cape Verde	$46.99	94	Tanzania	$0.54	143
Andorra	$474.97	46	Mongolia	$44.36	95	São Tomé	$0.44	144
Oman	$452.07	47	Guatemala	$43.93	96	Nigeria	$0.25	145
Grenada	$447.46	48	Namibia	$40.19	97	Ethiopia	$0.07	146
Dominica	$426.75	49	Guyana	$38.58	98	Burundi	$0.02	147

Individual freedom and progress

The idea that individual freedom leads to more progress than state planning is not new. It is part of a long intellectual tradition opposing top-down collectivist engineering in favor of bottom-up searching for solutions that goes back to Adam Smith and Edmund Burke. F.A. Hayek presciently noted more than 60 years ago how the complexity of knowledge required economic freedom and made planning impossible. A representative quotation is:

> The interaction of individuals, possessing different knowledge and different views, is what constitutes the life of thought. The growth of reason is a social process based on the existence of such differences. It is of essence that its results cannot be predicted, that we cannot know which views will assist this growth and which will not – in short, that this growth cannot be governed by any views which we now possess without at the same time limiting it. To "plan" or "organize" the growth of mind, or for that matter, progress in general, is a contradiction in terms [. . .] The tragedy of collectivist thought is that, while it starts out to make reason supreme, it ends by destroying reason because it misconceives the process on which the growth of reason depends [. . .] Individualism is thus an attitude of humility before this social process and of tolerance to other opinions and is the exact opposite of that intellectual hubris which is at the root of the demand for comprehensive direction of the social process.

This is not to say that economic freedom is easy to achieve. Even when such principles as private property, freedom of choice of occupation, protection against state expropriation, freedom of entry and competition in markets, prices determined by markets and not by state fiat are understood, it is difficult to implement the principles in practice. These principles rest upon a complex assortment of social norms, informal networks, formal laws, and effective institutions. To the extent that planners understand some of these principles, their characteristic mistake is to try to introduce everything at once from the top down in the self-contradictory combination of a "market plan." (Sachs, in an earlier incarnation, was the father of "shock therapy" for the ex-Communist countries, which tried to do exactly this.) Economic freedom is something that can only grow gradually within societies, with a lot of bottom-up searching for effective piecemeal reforms by political and economic actors – which helps explain why success at economic development is also relatively uncommon.

4 Conclusions

Alas, foreign aid has never been able to escape its collectivist origins. Today's collectivist fantasies such as the Big Push to achieve the Millennium Development Goals will fail just as badly as past varieties of collectivism. Indeed, the UN itself reports that they are already failing (it creatively sees this as a reason to solicit yet more funding for the Big Push). A peek inside the patterns of economic success shows the complexity of knowledge required to succeed, which dooms planning efforts and makes clear why economic freedom is so reliably associated with economic success.

Foreign aid could create new opportunities for the world's poorest people by getting them some of such essentials as medicines, education, and infrastructure, but only if foreign aid itself imitates the successful approach of economic freedom, by adopting a search and feedback approach with individual accountability instead of the current collectivist planning model. Even with these changes, outside aid cannot achieve the grandiose goal of transforming other societies to escape poverty into prosperity. Only home-grown gradual movements towards more economic freedom can accomplish that for the world's poor. Fortunately, that is already happening.

READING 12

The Great Transformation: The Political and Economic Origins of Our Time

Karl Polanyi

[. . .]

Birth of the Liberal Creed

Economic liberalism was the organizing principle of society engaged in creating a market system. Born as a mere penchant for nonbureaucratic methods, it evolved into a veritable faith in man's secular salvation through a self-regulating market. Such fanaticism was the result of the sudden aggravation of the task it found itself committed to: the magnitude of the sufferings that had to be inflicted on innocent persons as well as the vast scope of the interlocking changes involved in the establishment of the new order. The liberal creed assumed its evangelical fervor only in response to the needs of a fully deployed market economy.

To antedate the policy of laissez-faire, as is often done, to the time when this catchword was first used in France in the middle of the eighteenth century would be entirely unhistorical; it can be safely said that not until two generations later was economic liberalism more than a spasmodic tendency. Only by the 1820s did it stand for the three classical tenets: that labor should find its price on the market; that the creation of money should be subject to an automatic mechanism; that goods should be free to flow from country to country without hindrance or preference; in short, for a labor market, the gold standard, and free trade.

To credit François Quesnay with having envisaged such a state of affairs would be little short of fantastic. All that the Physiocrats demanded in a mercantilistic world was the free export of grain in order to ensure a better income to farmers, tenants, and landlords. For the rest their *ordre naturel* was no more than a directive principle for the regulation of industry and agriculture by a supposedly all-powerful and omniscient government. Quesnay's *Maximes* were intended to provide such a government with the viewpoints needed to translate into practical policy the principles of the *Tableaux* on the basis of statistical data which he offered to have furnished periodically. The idea of a self-regulating system of markets had never as much as entered his mind.

In England, too, laissez-faire was interpreted narrowly; it meant freedom from regulation in production; trade was not comprised. Cotton manufactures, the marvel of the time, had grown from insignificance into the leading export industry of the country – yet the import of printed cottons remained forbidden by positive statute. Notwithstanding the traditional monopoly of the home market an export bounty for calico or muslin was granted. Protectionism was so ingrained that Manchester cotton manufacturers demanded, in 1800, the prohibition of the export of yarn, though they were conscious of the fact that this meant loss of business to them. An act passed in 1791 extended the penalties for the export of tools used in manufacturing cotton goods to the export of models or specifications. The free-trade origins of the cotton industry are a myth. Freedom from regulation in the sphere of production was all the industry wanted; freedom in the sphere of exchange was still deemed a danger.

One might suppose that freedom of production would naturally spread from the purely technological field to that of the employment of labor. However, only comparatively late did Manchester raise the demand for free labor. The cotton industry had never been subject to the Statute of Artificers and was consequently not hampered either by yearly wage assessments or by rules of apprenticeship. The Old Poor Law, on the other hand, to which latter-day liberals so fiercely objected, was a help to the manufacturers; it not only supplied them with parish apprentices, but also permitted them to divest themselves of responsibility towards their dismissed employees, thus throwing much of the burden of unemployment on public funds. Not even the Speenhamland system was at first unpopular with the cotton manufacturers; as long as the moral effect of allowances did not reduce the productive capacity of the laborer, the industry might have well regarded family endowment as a help in sustaining that reserve army of labour which was urgently required to meet the tremendous fluctuations of trade. At a time when employment in agriculture was still on a year's term, it was of great importance that such a fund of mobile labor should be available to industry in periods of expansion. Hence the attacks of the manufacturers on the Act of Settlement which hampered the physical mobility of labor. Yet not before 1795 was the reform of that act carried – only to be replaced by more, not less, paternalism in regard to the Poor Law. Pauperism still remained the concern of squire and countryside; and even harsh critics of Speenhamland like Burke, Bentham, and Malthus regarded themselves less as representatives of industrial progress than as propounders of sound principles of rural administration.

Not until the 1830s did economic liberalism burst forth as a crusading passion and laissez-faire become a militant creed. The manufacturing class was pressing for the amendment of the Poor Law, since it prevented the rise of an industrial working class which depended for its income on achievement. The magnitude of the venture implied in the creation of a free labor market now became apparent, as well as the extent of the misery to be inflicted on the victims of improvement. Accordingly, by the early 1830s a sharp change of mood was manifest. An 1817 reprint of Townsend's *Dissertation* contained a preface in praise of the foresight with which the author had borne down on the Poor Laws and demanded their complete abandonment; but the editors warned of his "rash and precipitate" suggestion that outdoor relief to the poor should be abolished within so short a term as *ten* years. Ricardo's *Principles*, which appeared in the same year, insisted on the necessity of abolishing the allowance system, but urged strongly that this should be done only very gradually. Pitt, a disciple of Adam Smith, had rejected such a course on account of the innocent suffering it would entail. And as late as 1829, Peel "doubted whether the allowance system could be safely removed otherwise than gradually." Yet after the political victory of the middle class, in 1832, the Poor Law Amendment Bill was carried in its most extreme form and rushed into effect without any period of grace. Laissez-faire had been catalyzed into a drive of uncompromising ferocity.

A similar keying up of economic liberalism from academic interest to boundless activism occurred in the two other fields of industrial organization: *currency* and *trade*. In respect to both, laissez-faire waxed into a fervently held creed when the uselessness of any other but extreme solutions became apparent.

The currency issue was first brought home to the English community in the form of a general rise in the cost of living. Between 1790 and 1815 prices doubled. Real wages fell and business was hit by a slump in foreign exchanges. Yet not until the 1825 panic did sound currency become a tenet of economic liberalism, i.e., only when Ricardian principles were already so deeply impressed on the minds of politicians and businessmen alike that the "standard" was maintained in spite of the enormous number of financial casualties. This was the beginning of that unshakable belief in the automatic steering mechanism of the gold standard without which the market system could never have got under way.

International free trade involved no less an act of faith. Its implications were entirely extravagant. It meant that England would depend for her food supply upon overseas sources; would sacrifice her agriculture, if necessary, and enter on a new form of life under which she would be part and parcel of some vaguely conceived world unity of the future: that this planetary community would have to be a peaceful one, or, if not, would have to be made safe for Great Britain by the power of the Navy; and that the English nation would face the prospects of continuous industrial dislocations in the firm belief in its superior inventive and productive ability. However, it was believed that if only the grain of all the world could flow freely to Britain, then her factories would be able to undersell all the world. Again, the measure of the determination needed was set by the magnitude of the proposition and the vastness of the risks involved in complete acceptance. Yet less than complete acceptance spelled certain ruin.

The utopian springs of the dogma of laissez-faire are but incompletely understood as long as they are viewed separately. The three tenets – competitive labor market, automatic gold standard, and international free trade – formed one whole. The sacrifices involved in achieving any one of them were useless, if not worse, unless the other two were equally secured. It was everything or nothing.

Anybody could see that the gold standard, for instance, meant danger of deadly deflation and, maybe, of fatal monetary stringency in a panic. The manufacturer could, therefore, hope to hold his own only if he was assured of an increasing scale of production at remunerative prices (in other words, only if wages fell at least in proportion to the general fall in prices, so as to allow the exploitation of an everexpanding world market). Thus the Anti-Corn Law Bill of 1846 was the corollary of Peel's Bank Act of 1844, and both assumed a laboring class which, since the Poor Law Amendment Act of 1834, was forced to give its best under the threat of hunger, so that wages were regulated by the price of grain. The three great measures formed a coherent whole.

The true implications of economic liberalism can now be taken in at a glance. Nothing less than a self-regulating market on a world scale could ensure the functioning of this stupendous mechanism. Unless the price of labor was dependent upon the cheapest grain available, there was no guarantee that the unprotected industries would not succumb in the grip of the voluntarily accepted taskmaster, gold. The expansion of the market system in the nineteenth century was synonymous with the simultaneous spreading of international free trade, competitive labor market, and gold standard; they belonged together. No wonder that economic liberalism turned almost into a religion once the great perils of this venture were evident.

There was nothing natural about laissez-faire; free markets could never have come into being merely by allowing things to take their course. Just as cotton manufactures – the leading free trade industry – were created by the help of protective tariffs, export bounties, and indirect wage subsidies, laissez-faire itself was enforced by the state. The thirties and forties saw not only an outburst of legislation repealing restrictive regulations, but also an enormous increase in the administrative functions of the state, which was now being endowed with a central bureaucracy able to fulfil the tasks set by the adherents of liberalism. To the typical utilitarian, economic liberalism was a social project which should be put into effect for the greatest happiness of the greatest number; laissez-faire was not a method to achieve a thing, it was the thing to be achieved. True, legislation could do nothing directly, except by repealing harmful restrictions. But that did not mean that *government* could do nothing, especially indirectly. On the contrary, the utilitarian liberal saw in government the great agency for achieving happiness. In respect to material welfare, Bentham believed, the influence of legislation "is as nothing" in comparison with the unconscious contribution of the "minister of the police." Of the three things needed for economic success – inclination, knowledge, and power – the private person possessed only inclination. Knowledge and power, Bentham taught, can be administered much cheaper by government than by private persons. It was the task of the executive to collect statistics and information, to foster science and experiment, as well as to supply the innumerable instruments of final realization in the field of government. Benthamite liberalism meant the replacing of parliamentary action by action through administrative organs.

For this there was ample scope. Reaction in England had not governed – as it did in France – through administrative methods but used exclusively Parliamentary legislation to put political repression into effect. "The revolutionary movements of 1785 and of 1815–1820 were combated, not by departmental action, but by Parliamentary legislation. The suspension of the Habeas Corpus Act, the passing of the Libel Act, and of the 'Six Acts' of 1819, were severely coercive measures; but they contain no evidence of any attempt to give a Continental character to administration. In so far as individual liberty was destroyed, it was destroyed by and in pursuance of Acts of Parliament." Economic liberals had hardly gained influence on government, in 1832, when the position changed completely in favor of administrative methods. "The net result of the legislative activity which has characterized, though with different degrees of intensity, the period since 1832, has been the building up piecemeal of an administrative machine of great complexity which stands in as constant need of repair, renewal, reconstruction, and adaptation to new requirements as the plant of a modern manufactory." This growth of administration reflected the spirit of utilitarianism. Bentham's fabulous Panopticon, his most personal utopia, was a star-shaped building from the center of which prison wardens could keep the greatest number of jailbirds under the most effective supervision at the smallest cost to the public. Similarly, in the utilitarian state his favorite principle of "inspectability" ensured that the minister at the top should keep effective control over all local administration.

The road to the free market was opened and kept open by an enormous increase in continuous, centrally organized and controlled interventionism. To make Adam Smith's "simple and natural liberty" compatible with the needs of a human society was a most complicated affair. Witness the complexity of the provisions in the innumerable enclosure laws; the amount of bureaucratic control involved in the administration of the New Poor Laws which for the first time since Queen Elizabeth's reign were effectively supervised by central authority; or the increase in governmental administration entailed in the meritorious task of municipal reform. And yet all these strongholds of governmental interference were erected with a view to the organizing of some simple freedom – such as that of land, labor, or municipal administration. Just as, contrary to expectation, the invention of labor-saving machinery had not diminished but actually increased the uses of human labor, the introduction of free markets, far from doing away with the need for control, regulation, and intervention, enormously increased their range. Administrators had to be constantly on the watch to ensure the free working of the system. Thus even those who wished most ardently to free the state from all unnecessary duties, and whose whole philosophy demanded the restriction of state activities, could not but entrust the self-same state with the new powers, organs, and instruments required for the establishment of laissez-faire.

This paradox was topped by another. While laissez-faire economy was the product of deliberate State action, subsequent restrictions on laissez-faire started in a spontaneous way. Laissez-faire was planned; planning was not. The first half of this assertion was shown above to be true, if ever there was conscious use of the executive in the service of a deliberate government-controlled policy, it was on the part of the Benthamites in the heroic period of laissez-faire. The other half was

first mooted by that eminent Liberal, Dicey, who made it his task to inquire into the origins of the "anti-laissez-faire" or, as he called it, the "collectivist" trend in English public opinion, the existence of which was manifest since the late 1860s. He was surprised to find that no evidence of the existence of such a trend could be traced *save the acts of legislation themselves*. More exactly, no evidence of a "collectivist trend" in public opinion *prior* to the laws which appeared to represent such a trend could be found. As to later "collectivist" opinion, Dicey inferred that the "collectivist" legislation itself might have been its prime source. The upshot of his penetrating inquiry was that there had been complete absence of any deliberate intention to extend the functions of the state, or to restrict the freedom of the individual, on the part of those who were directly responsible for the restrictive enactments of the 1870s and 1880s. The legislative spearhead of the countermovement against a self-regulating market as it developed in the half century following 1860 turned out to be spontaneous, undirected by opinion, and actuated by a purely pragmatic spirit.

Economic liberals must strongly take exception to such a view. Their whole social philosophy hinges on the idea that laissez-faire was a natural development, while subsequent anti-laissez-faire legislation was the result of purposeful action on the part of the opponents of liberal principles. In these two mutually exclusive interpretations of the double movement, it is not too much to say, the truth or untruth of the liberal creed is involved today.

Liberal writers like Spencer and Sumner, Mises and Lippmann offer an account of the double movement substantially similar to our own, but they put an entirely different interpretation on it. While in our view the concept of a self-regulating market was utopian, and its progress was stopped by the realistic self-protection of society, in their view all protectionism was a mistake due to impatience, greed, and shortsightedness, but for which the market would have resolved its difficulties. The question as to which of these two views is correct is perhaps the most important problem of recent social history, involving as it does no less than a decision on the claim of economic liberalism to be the basic organizing principle in society. Before we turn to the testimony of the facts, a more precise formulation of the issue is needed.

Undoubtedly, our age will be credited with having seen the end of the self-regulating market. The 1920s saw the prestige of economic liberalism at its height. Hundreds of millions of people had been afflicted by the scourge of inflation; whole social classes, whole nations had been expropriated. Stabilization of currencies became the focal point in the political thought of peoples and governments; the restoration of the gold standard became the supreme aim of all organized effort in the economic field. The repayment of foreign loans and the return to stable currencies were recognized as the touchstone of rationality in politics; and no private suffering, no restriction of sovereignty, was deemed too great a sacrifice for the recovery of monetary integrity. The privations of the unemployed made jobless by deflation; the destitution of public servants dismissed without a pittance; even the relinquishment of national rights and the loss of constitutional liberties were judged a fair price to pay for the fulfillment of the requirement of sound budgets and sound currencies, these *a priori* of economic liberalism.

The 1930s lived to see the absolutes of the 1920s called in question. After several years during which currencies were practically restored and budgets balanced, the two most powerful countries, Great Britain and the United States, found themselves in difficulties, dismissed the gold standard, and started out on the management of their currencies. International debts were repudiated wholesale and the tenets of economic liberalism were disregarded by the wealthiest and most respectable. By the middle of the 1930s France and some other states still adhering to gold were actually forced off the standard by the Treasuries of Great Britain and the United States, formerly jealous guardians of the liberal creed.

In the 1940s economic liberalism suffered an even worse defeat. Although Great Britain and the United States departed from monetary orthodoxy, they retained the principles and methods of liberalism in industry and commerce, the general organization of their economic life. This was to prove a factor in precipitating the war and a handicap in fighting it, since economic liberalism had created and fostered the illusion that dictatorships were bound for economic catastrophe. By virtue of this creed, democratic governments were the last to understand the implications of managed currencies and directed trade, even when they happened by force of circumstances to be practicing these methods themselves; also, the legacy of economic liberalism barred the way to timely rearmament in the name of balanced budgets and stable exchanges, which were supposed to provide the only secure foundations of economic strength in war. In Great Britain budgetary and monetary orthodoxy induced adherence to the traditional strategic principle of limited commitments upon a country actually faced with total war; in the United States vested interests – such as oil and aluminium – entrenched themselves behind the taboos of liberal business and successfully resisted preparations for an industrial emergency. But for the stubborn and impassioned insistence of economic liberals on their fallacies, the leaders of the race as well as the masses of free men would have been better equipped for the ordeal of the age and might perhaps even have been able to avoid it altogether.

But secular tenets of social organization embracing the whole civilized world are not dislodged by the events of a decade. Both in Great Britain and in the United States millions of independent business units derived their existence from the principle of laissez-faire. Its spectacular failure in one field did not destroy its authority in all. Indeed, its partial eclipse may have even strengthened its hold since it enabled its defenders to argue that the incomplete application of its principles was the reason for every and any difficulty laid to its charge.

This, indeed, is the last remaining argument of economic liberalism today. Its apologists are repeating in endless variations that but for the policies advocated by its critics, liberalism would have delivered the goods; that not the competitive system and the self-regulating market, but interference with that system and interventions with that market are responsible for our ills. And this argument does not find support in innumerable recent infringements of economic freedom only, but also in the indubitable fact that the movement to spread the system of self-regulating markets was met in the second half of the nineteenth century by a persistent countermove obstructing the free working of such an economy.

The economic liberal is thus enabled to formulate a case which links the present with the past in one coherent whole. For who could deny that government intervention in business may undermine confidence?

Who could deny that unemployment would sometimes be less if it were not for out-of-work benefit provided by law? That private business is injured by the competition of public works? That deficit finance may endanger private investments? That paternalism tends to damp business initiative? This being so in the present, surely it was no different in the past. When around the 1870s a general protectionist movement – social and national – started in Europe, who can doubt that it hampered and restricted trade? Who can doubt that factory laws, social insurance, municipal trading, health services, public utilities, tariffs, bounties and subsidies, cartels and trusts, embargoes on immigration, on capital movements, on imports – not to speak of less-open restrictions on the movements of men, goods, and payments – must have acted as so many hindrances to the functioning of the competitive system, protracting business depressions, aggravating unemployment, deepening financial slumps, diminishing trade, and damaging severely the self-regulating mechanism of the market? The root of all evil, the liberal insists, was precisely this interference with the freedom of employment, trade and currencies practiced by the various schools of social, national, and monopolistic protectionism since the third quarter of the nineteenth century; but for the unholy alliance of trade unions and labor parties with monopolistic manufacturers and agrarian interests, which in their shortsighted greed joined forces to frustrate economic liberty, the world would be enjoying today the fruits of an almost automatic system of creating material welfare. Liberal leaders never weary of repeating that the tragedy of the nineteenth century sprang from the incapacity of man to remain faithful to the inspiration of the early liberals; that the generous initiative of our ancestors was frustrated by the passions of nationalism and class war, vested interests, and monopolists, and above all, by the blindness of the working people to the ultimate beneficence of unrestricted economic freedom to all human interests, including their own. A great intellectual and moral advance was thus, it is claimed, frustrated by the intellectual and moral weaknesses of the mass of the people; what the spirit of Enlightenment had achieved was put to nought by the forces of selfishness. In a nutshell this is the economic liberal's defense. Unless it is refuted, he will continue to hold the floor in the contest of arguments.

Let us focus the issue. It is agreed that the liberal movement, intent on the spreading of the market system, was met by a protective countermovement tending toward its restriction; such an assumption, indeed, underlies our own thesis of the double movement. But while we assert that the application of the absurd notion of a self-regulating market system would have inevitably destroyed society, the liberal accuses the most various elements of having wrecked a great initiative. Unable to adduce evidence of any such concerted effort to thwart the liberal movement, he falls back on the practically irrefutable hypothesis of covert action. This is the myth of the anti-liberal conspiracy which in one form or another is common to all liberal interpretations of the events of the 1870s and 1880s. Commonly the rise of nationalism and of socialism is credited with having been the chief agent in that shifting of the scene; manufacturers' associations and monopolists, agrarian interests and trade unions are the villains of the piece. Thus in its most spiritualized form the liberal doctrine hypostasizes the working of some dialectical law in modern society stultifying the endeavors of enlightened reason, while in its crudest version it reduces itself to an attack on political democracy, as the alleged mainspring of interventionism.

The testimony of the facts contradicts the liberal thesis decisively. The anti-liberal conspiracy is a pure invention. The great variety of forms in which the "collectivist" countermovement appeared was not due to any preference for socialism or nationalism on the part of concerted interests, but exclusively to the broad range of the vital social interests affected by the expanding market mechanism. This accounts for the all but universal reaction of predominantly practical character called forth by the expansion of that mechanism. Intellectual fashion played no role whatever in this process; there was, accordingly, no room for the prejudice which the liberal regards as the ideological force behind the anti-liberal development. Although it is true that the 1870s and 1880s saw the end of orthodox liberalism, and that all crucial problems of the present can be traced back to that period, it is incorrect to say that the change to social and national protectionism was due to any other cause than the manifestation of the weaknesses and perils inherent in a self-regulating market system. This can be shown in more than one way.

Firstly, there is the amazing diversity of the matters on which action was taken. This alone would exclude the possibility of concerted action. Let us cite from a list of interventions which Herbert Spencer compiled in 1884, when charging liberals with having deserted their principles for the sake of "restrictive legislation." The variety of the subjects could hardly be greater. In 1860 authority was given to provide "analysts of food and drink to be paid out of local rates"; there followed an Act providing "the inspection of gas works"; an extension of the Mines Act "making it penal to employ boys under twelve not attending schools and unable to read or write." In 1861 power was given "to poor law guardians to enforce vaccination"; local boards were authorized "to fix rates of hire for means of conveyance"; and certain locally formed bodies "had given them powers of taxing the locality for rural drainage and irrigation works, and for supplying water to cattle." In 1862 an act was passed making illegal "a coal-mine with a single shaft"; an act giving the Council of Medical Education exclusive right "to furnish a Pharmacopoeia, the price of which is to be fixed by the Treasury." Spencer, horror struck, filled several pages with an enumeration of these and similar measures. In 1863 came the "extension of compulsory vaccination to Scotland and Ireland." There was also an act appointing inspectors for the "wholesomeness, or unwholesomeness of food"; a Chimney-Sweeper's Act, to prevent the torture and eventual death of children set to sweep too narrow slots; a Contagious Diseases Act; a Public Libraries Act, giving local powers "by which a majority can tax a minority for their books." Spencer adduced them as so much irrefutable evidence of an anti-liberal conspiracy. And yet each of these acts dealt with some problem arising out of modern industrial conditions and was aimed at the safeguarding of some public interest against dangers inherent either in such conditions or, at any rate, in the market method of dealing with them. To an unbiased mind they proved the purely practical and pragmatic nature of the "collectivist" countermove. Most of those who carried these measures were convinced supporters of laissez-faire, and certainly did not wish their consent to the establishment of a fire brigade in London to imply a protest against the principles of economic liberalism. On the contrary, the sponsors of these legislative acts were as a rule uncompromising opponents of socialism, or any other form of collectivism.

Secondly, the change from liberal to "collectivist" solutions happened sometimes over night and without any consciousness on the part of those engaged in the process of legislative rumination. Dicey adduced the classic instance of the Workmen's Compensation Act dealing with the employers' liability for damage done to his workmen in the course of their employment. The history of the various acts embodying this idea, since 1880, showed consistent adherence to the individualist principle that the responsibility of the employer to his employee must be regulated in a manner strictly identical with that governing his responsibility to others, e.g., strangers. With hardly any change in opinion, in 1897, the employer was suddenly made the insurer of his workmen against any damage incurred in the course of their employment, a "thoroughly collectivistic legislation," as Dicey justly remarked. No better proof could be adduced that no change either in the type of interests involved, or in the tendency of the opinions brought to bear on the matter, caused the supplanting of a liberal principle by an anti-liberal one, but exclusively the evolving conditions under which the problem arose and a solution was sought.

Thirdly, there is the indirect, but most striking proof provided by a comparison of the development in various countries of a widely dissimilar political and ideological configuration. Victorian England and the Prussia of Bismarck were poles apart, and both were very much unlike the France of the Third Republic or the Empire of the Hapsburgs. Yet each of them passed through a period of free trade and laissez-faire, followed by a period of anti-liberal legislation in regard to public health, factory conditions, municipal trading, social insurance, shipping subsidies, public utilities, trade associations, and so on. It would be easy to produce a regular calendar setting out the years in which analogous changes occurred in the various countries. Workmen's compensation was enacted in England in 1880 and 1897, in Germany in 1879, in Austria in 1887, in France in 1899; factory inspection was introduced in England in 1833, in Prussia in 1853, in Austria in 1883, in France in 1874 and 1883; municipal trading, including the running of public utilities, was introduced by Joseph Chamberlain, a Dissenter and a capitalist, in Birmingham in the 1870s; by the Catholic "Socialist" and Jew-baiter, Karl Lueger, in the Imperial Vienna of the 1890s; in German and French municipalities by a variety of local coalitions. The supporting forces were in some cases violently reactionary and antisocialist as in Vienna, at other times "radical imperialist" as in Birmingham, or of the purest liberal hue as with the Frenchman, Edouard Herriot, Mayor of Lyons. In Protestant England, Conservative and Liberal cabinets labored intermittently at the completion of factory legislation. In Germany, Roman Catholics and Social Democrats took part in its achievement; in Austria, the Church and its most militant supporters; in France, enemies of the Church and ardent anticlericals were responsible for the enactment of almost identical laws. Thus under the most varied slogans, with very different motivations a multitude of parties and social strata put into effect almost exactly the same measures in a series of countries in respect of a large number of complicated subjects. There is, on the face of it, nothing more absurd than to infer that they were secretly actuated by the same ideological preconceptions or narrow group interests as the legend of the antiliberal conspiracy would have it. On the contrary, everything tends to support the assumption that objective reasons of a stringent nature forced the hands of the legislators.

Fourthly, there is the significant fact that at various times economic liberals themselves advocated restrictions on the freedom of contract and on laissez-faire in a number of well-defined cases of great theoretical and practical importance. Antiliberal prejudice could, naturally, not have been their motive. We have in mind the principle of the association of labor on the one hand, the law of business corporations on the other. The first refers to the right of workers to combine for the purpose of raising their wages; the latter, to the right of trusts, cartels, or other forms of capitalistic combines, to raise prices. It was justly charged in both cases that freedom of contract or laissez-faire was being used in restraint of trade. Whether workers' associations to raise wages, or trade associations to raise prices were in question, the principle of laissez-faire could be obviously employed by interested parties to narrow the market for labor or other commodities. It is highly significant that in either case consistent liberals from Lloyd George and Theodore Roosevelt to Thurman Arnold and Walter Lippmann subordinated laissez-faire to the demand for a free competitive market; they pressed for regulations and restrictions, for penal laws and compulsion, arguing as any "collectivist" would that the freedom of contract was being "abused" by trade unions, or corporations, whichever it was. Theoretically, laissez-faire or freedom of contract implied the freedom of workers to withhold their labor either individually or jointly, if they so decided; it implied also the freedom of businessmen to concert on selling prices irrespective of the wishes of the consumers. But in practice such freedom conflicted with the institution of a self-regulating market, and *in such a conflict the self-regulating market was invariably accorded precedence*. In other words, if the needs of a self-regulating market proved incompatible with the demands of laissez-faire, the economic liberal turned against laissez-faire and preferred – as any antiliberal would have done – the so-called collectivist methods of regulation and restriction. Trade union law as well as antitrust legislation sprang from this attitude. No more conclusive proof could be offered of the inevitability of antiliberal or "collectivist" methods under the conditions of modern industrial society than the fact that even economic liberals themselves regularly used such methods in decisively important fields of industrial organization.

Incidentally, this helps to clarify the true meaning of the term "interventionism" by which economic liberals like to denote the opposite of their own policy, but merely betray confusion of thought. The opposite of interventionism is laissez-faire, and we have just seen that economic liberalism cannot be identified with laissez-faire (although in common parlance there is no harm in using them interchangeably). Strictly, economic liberalism is the organizing principle of a society in which industry is based on the institution of a self-regulating market. True, once such a system is approximately achieved, less intervention of one type is needed. However, this is far from saying that market system and intervention are mutually exclusive terms. For as long as that system is not established, economic liberals must and will unhesitatingly call for the intervention of the state in order to establish it, and once established, in order to maintain it. The economic liberal can, therefore, without any inconsistency call upon the state to use the force of law; he can even appeal to the violent forces of civil war to set up the preconditions of a self-regulating market. In America the South appealed to the arguments of laissez-faire to justify slavery; the North appealed to the intervention of arms to establish a free labor market. The accusation of interventionism on the part of liberal writers is thus an empty slogan, implying the denunciation of one and the same set of actions according to whether they happen to approve of them or not. The only principle economic liberals can maintain without inconsistency is that of the self-regulating market, whether it involves them in interventions or not.

To sum up. The countermove against economic liberalism and laissez-faire possessed all the unmistakable characteristics of a spontaneous reaction. At innumerable disconnected points it set in without any traceable links between the interests directly affected or any ideological conformity between them. Even in the settlement of one of the same problem as in the case of workmen's compensation, solutions switched over from individualistic to "collectivistic," from liberal to antiliberal, from "laissez-faire" to interventionist forms without any change in the economic interest, the ideological influences or political forces in play, merely as a result of the increasing realization of the nature of the problem in question. Also it could be shown that a closely similar change from laissez-faire to "collectivism" took place in various countries at a definite stage of their industrial development, pointing to the depth and independence of the underlying causes of the process so superficially credited by economic liberals to changing moods or sundry interests. Finally, analysis reveals that not even radical adherents of economic liberalism could escape the rule which makes laissez-faire inapplicable to advanced industrial conditions; for in the critical case of trade union law and antitrust regulations extreme liberals themselves had to call for manifold interventions of the state, in order to secure against monopolistic compacts the preconditions for the working of a self-regulating market. Even free trade and competition required intervention to be workable. The liberal myth of the "collectivist" conspiracy of the 1870s and 1880s is contrary to all the facts.

Our own interpretation of the double movement on the other hand is borne out by the evidence. For if market economy was a threat to the human and natural components of the social fabric, as we insisted, what else would one expect than an urge on the part of a great variety of people to press for some sort of protection? This was what we found. Also, one would expect this to happen without any theoretical or intellectual preconceptions on their part, and irrespective of their attitudes toward the principles underlying a market economy. Again, this was the case. Moreover, we suggested that comparative history of government might offer quasi-experimental support of our thesis if particular interests could be shown to be independent of the specific ideologies present in a number of different countries. For this also we could adduce striking evidence. Finally, the behavior of liberals themselves proved that the maintenance of freedom of trade – in our terms, of a self-regulating market – far from excluding intervention, in effect, demanded such action, and that liberals themselves regularly called for compulsory action on the part of the state as in the case of trade union law and anti-trust laws. Thus nothing could be more decisive than the evidence of history as to which of the two contending interpretations of the double movement was correct: that of the economic liberal who maintained that his policy never had a chance, but was strangled by shortsighted trade unionists, Marxist intellectuals, greedy manufacturers, and reactionary landlords; or that of his critics, who can point to the universal "collectivist" reaction against the expansion of market economy in the second half of the nineteenth century as conclusive proof of the peril to society inherent in the utopian principle of a self-regulating market.

[. . .]

Freedom in a Complex Society

Nineteenth-century civilization was not destroyed by the external or internal attack of barbarians; its vitality was not sapped by the devastations of World War I nor by the revolt of a socialist proletariat or a fascist lower middle class. Its failure was not the outcome of some alleged laws of economics such as that of the falling rate of profit or of underconsumption or overproduction. It disintegrated as the result of an entirely different set of causes: the measures which society adopted in order not to be, in its turn, annihilated by the action of the self-regulating market. Apart from exceptional circumstances such as existed in North America in the age of the open frontier, the conflict between the market and the elementary requirements of an organized social life provided the century with its dynamics and produced the typical strains and stresses which ultimately destroyed that society. External wars merely hastened its destruction.

After a century of blind "improvement" man is restoring his "habitation." If industrialism is not to extinguish the race, it must be subordinated to the requirements of man's nature. The true criticism of market society is not that it was based on economics – in a sense, every and any society must be based on it – but that its economy was based on self-interest. Such an organization of economic life is entirely unnatural, in the strictly empirical sense of *exceptional*. Nineteenth-century thinkers assumed that in his economic activity man strove for profit, that his materialistic propensities would induce him to choose the lesser instead of the greater effort and to expect payment for his labor; in short, that in his economic activity he would tend to abide by what they described as economic rationality, and that all contrary behavior was the result of outside interference. It followed that markets were natural institutions, that they would spontaneously arise if only men were let alone. Thus, nothing could be more normal than an economic system consisting of markets and under the sole control of market prices, and a human society based on such markets appeared, therefore, as the goal of all progress. Whatever the desirability or undesirability of such a society on moral grounds, its practicability – this was axiomatic – was grounded in the immutable characteristics of the race.

Actually, as we now know, the behavior of man both in his primitive state and right through the course of history has been almost the opposite from that implied in this view. Frank H. Knight's "no specifically human motive is economic" applies not only to social life in general, but even to economic life itself. The tendency to barter, on which Adam Smith so confidently relied for his picture of primitive man, is not a common tendency of the human being in his economic activities, but a most infrequent one. Not only does the evidence of modern anthropology give the lie to these rationalistic constructs, but the history of trade and markets also has been completely different from that assumed in the harmonistic teachings of nineteenth century sociologists. Economic history reveals that the emergence of national markets was in no way the result of the gradual and spontaneous emancipation of the economic sphere from governmental control. On the contrary, the market has been the outcome of a conscious and often violent intervention on the part of government which imposed the market organization on society for noneconomic ends. And the self-regulating market of the nineteenth century turns out on closer inspection to be radically different from even its immediate predecessor in that it relied for its regulation on economic self-interest. *The congenital weakness of nineteenth-century society was not that it was industrial but that it was a market society.* Industrial civilization will continue to exist when the utopian experiment of a self-regulating market will be no more than a memory.

Yet the shifting of industrial civilization onto a new nonmarketing basis seems to many a task too desperate to contemplate. They fear an institutional vacuum or, even worse, the loss of freedom. Need these perils prevail?

Much of the massive suffering inseparable from a period of transition is already behind us. In the social and economic dislocation of our age, in the tragic vicissitudes of the depression, fluctuations of currency, mass unemployment, shiftings of social status, spectacular destruction of historical states, we have experienced the worst. Unwittingly we have been paying the price of the change. Far as mankind still is from having adapted itself to the use of machines, and great as the pending changes are, the restoration of the past is as impossible as the transferring of our troubles to another planet. Instead of eliminating the demonic forces of aggression and conquest, such a futile attempt would actually ensure the survival of those forces, even after their utter military defeat. The cause of evil would become endowed with the advantage, decisive in politics, of representing the possible, in opposition to that which is impossible of achievement however good it may be of intention.

Nor does the collapse of the traditional system leave us in the void. Not for the first time in history may makeshifts contain the germs of great and permanent institutions.

Within the nations we are witnessing a development under which the economic system ceases to lay down the law to society and the primacy of society over that system is secured. This may happen in a great variety of ways, democratic and aristocratic, constitutionalist and authoritarian, perhaps even in a fashion yet utterly unforeseen. The future in some countries may be already the present in others, while some may still embody the past of the rest. But the outcome is common with them all: the market system will no longer be self-regulating, even in principle, since it will not comprise labor, land, and money.

To take labor out of the market means a transformation as radical as was the establishment of a competitive labor market. The wage contract ceases to be a private contract except on subordinate and accessory points. Not only conditions in the factory, hours of work, and modalities of contract, but the basic wage itself, are determined outside the market; what role accrues thereby to trade unions, state, and other public bodies depends not only on the character of these institutions but also on the actual organization of the management of production. Though in the nature of things wage differentials must (and should) continue to play an essential part in the economic system, other motives than those directly involved in money incomes may outweigh by far the financial aspect of labor.

To remove land from the market is synonymous with the incorporation of land with definite institutions such as the homestead, the cooperative, the factory, the township, the school, the church, parks, wild life preserves, and so on. However widespread individual ownership of farms will continue to be, contracts in respect to land tenure need deal with accessories only, since the essentials are removed from the jurisdiction of the market. The same applies to staple foods and organic raw materials, since the fixing of prices in respect to them is not left to the market. That for an infinite variety of products competitive markets continue to function need not interfere with the constitution of society any more than the fixing of prices outside the market for labor, land, and money interferes with the costing-function of prices in respect to the various products. The nature of property, of course, undergoes a deep change in consequence of such measures since there is no longer any need to allow incomes from the title of property to grow without bounds, merely in order to ensure employment, production, and the use of resources in society.

The removal of the control of money from the market is being accomplished in all countries in our day. Unconsciously, the creation of deposits effected this to a large extent, but the crisis of the gold standard in the 1920s proved that the link between commodity money and token money had by no means been severed. Since the introduction of "functional finance" in all-important states, the directing of investments and the regulation of the rate of saving have become government tasks.

To remove the elements of production – land, labor, and money – from the market is thus a uniform act only from the viewpoint of the market, which was dealing with them as if they were commodities. From the viewpoint of human reality that which is restored by the disestablishment of the commodity fiction lies in all directions of the social compass. In effect, the disintegration of a uniform market economy is already giving rise to a variety of new societies. Also, the end of market society means in no way the absence of markets. These continue, in various fashions, to ensure the freedom of the consumer, to indicate the shifting of demand, to influence producers' income, and to serve as an instrument of accountancy, while ceasing altogether to be an organ of economic self-regulation.

In its international methods, as in these internal methods, nineteenth-century society was constricted by economics. The realm of fixed foreign exchanges was coincident with civilization. As long as the gold standard and – what became almost its corollary – constitutional regimes were in operation, the balance of power was a vehicle of peace. The system worked through the instrumentality of those Great Powers, first and foremost Great Britain, who were the center of world finance, and pressed for the establishment of representative government in less-advanced countries. This was required as a check on the finances and currencies of debtor countries with the consequent need for controlled budgets, such as only responsible bodies can provide. Though, as a rule, such considerations were not consciously present in the minds of statesmen, this was the case only because the requirements of the gold standard ranked as axiomatic. The uniform world pattern of monetary and representative institutions was the result of the rigid economy of the period.

Two principles of nineteenth-century international life derived their relevance from this situation: anarchistic sovereignty and "justified" intervention in the affairs of other countries. Though apparently contradictory, the two were interrelated. Sovereignty, of course, was a purely political term, for under unregulated foreign trade and the gold standard governments possessed no powers in respect to international economics. They neither could nor would bind their countries in respect to monetary matters – this was the legal position. Actually, only countries which possessed a monetary system controlled by central banks were reckoned sovereign states. With the powerful Western countries this unlimited and unrestricted national monetary sovereignty was combined with its complete opposite, an unrelenting pressure to spread the fabric of market economy and market society elsewhere. Consequently, by the end of the nineteenth century the peoples of the world were institutionally standardized to a degree unknown before.

This system was hampering both on account of its elaborateness *and* its universality. Anarchistic sovereignty was a hindrance to all effective forms of international cooperation, as the history of the League of Nations strikingly proved; and enforced uniformity of domestic systems hovered as a permanent threat over the freedom of national development, especially in backward countries and sometimes even in advanced, but financially weak countries. Economic cooperation was limited to private institutions as rambling and ineffective as free trade, while actual collaboration between peoples, that is, between governments, could never even be envisaged.

The situation may well make two apparently incompatible demands on foreign policy: it will require closer cooperation between friendly countries than could even be contemplated under nineteenth-century sovereignty, while at the same time the existence of regulated markets will make national governments more jealous of outside interference than ever before. However, with the disappearance of the automatic mechanism of the gold standard, governments will find it possible to drop the most obstructive feature of absolute sovereignty, the refusal to collaborate in international economics. At the same time it will become possible to tolerate willingly that other nations shape their domestic institutions according to their inclinations, thus transcending the pernicious nineteenth-century dogma of the necessary uniformity of domestic regimes within the orbit of world economy. Out of the ruins of the Old World, cornerstones of the New can be seen to emerge: economic collaboration of governments *and* the liberty to organize national life at will. Under the constrictive system of free trade neither of these possibilities could have been conceived of, thus excluding a variety of methods of cooperation between nations. While under

market economy and the gold standard the idea of federation was justly deemed a nightmare of centralization and uniformity, the end of market economy may well mean effective cooperation with domestic freedom.

The problem of freedom arises on two different levels: the institutional and the moral or religious. On the institutional level it is a matter of balancing increased against diminished freedoms; no radically new questions are encountered. On the more fundamental level the very possibility of freedom is in doubt. It appears that the means of maintaining freedom are themselves adulterating and destroying it. The key to the problem of freedom in our age must be sought on this latter plane. Institutions are embodiments of human meaning and purpose. We cannot achieve the freedom we seek, unless we comprehend the true significance of freedom in a complex society.

On the institutional level, regulation both extends and restricts freedom; only the balance of the freedoms lost and won is significant. This is true of juridical and actual freedoms alike. The comfortable classes enjoy the freedom provided by leisure in security; they are naturally less anxious to extend freedom in society than those who for lack of income must rest content with a minimum of it. This becomes apparent as soon as compulsion is suggested in order to more justly spread out income, leisure and security. Though restriction applies to all, the privileged tend to resent it, as if it were directed solely against themselves. They talk of slavery, while in effect only an extension to the others of the vested freedom they themselves enjoy is intended. Initially, there may have to be reduction in their own leisure and security, and, consequently, their freedom so that the level of freedom throughout the land shall be raised. But such a shifting, reshaping and enlarging of freedoms should offer no ground whatsoever for the assertion that the new condition must necessarily be less free than was the old.

Yet there are freedoms the maintenance of which is of paramount importance. They were, like peace, a by-product of nineteenth-century economy, and we have come to cherish them for their own sake. The institutional separation of politics and economics, which proved a deadly danger to the substance of society, almost automatically produced freedom at the cost of justice and security. Civic liberties, private enterprise and wage-system fused into a pattern of life which favored moral freedom and independence of mind. Here again, juridical and actual freedoms merged into a common fund, the elements of which cannot be neatly separated. Some were the corollary of evils like unemployment and speculator's profits; some belonged to the most precious traditions of Renaissance and Reformation. We must try to maintain by all means in our power these high values inherited from the market-economy which collapsed. This, assuredly, is a great task. Neither freedom nor peace could be institutionalized under that economy, since its purpose was to create profits and welfare, not peace and freedom. We will have consciously to strive for them in the future if we are to possess them at all; they must become chosen aims of the societies toward which we are moving. This may well be the true purport of the present world effort to make peace and freedom secure. How far the will to peace can assert itself once the interest in peace which sprang from nineteenth-century economy has ceased to operate will depend upon our success in establishing an international order. As to personal liberty, it will exist to the degree in which we will deliberately create new safeguards for its maintenance and, indeed, extension. In an established society the right to nonconformity must be institutionally protected. The individual must be free to follow his conscience without fear of the powers that happen to be entrusted with administrative tasks in some of the fields of social life. Science and the arts should always be under the guardianship of the republic of letters. Compulsion should never be absolute; the "objector" should be offered a niche to which he can retire, the choice of a "second-best" that leaves him a life to live. Thus will be secured the right to nonconformity as the hallmark of a free society.

Every move toward integration in society should thus be accompanied by an increase of freedom; moves toward planning should comprise the strengthening of the rights of the individual in society. His indefeasible rights must be enforceable under the law even against the supreme powers, whether they be personal or anonymous. The true answer to the threat of bureaucracy as a source of abuse of power is to create spheres of arbitrary freedom protected by unbreakable rules. For however generously devolution of power is practiced, there will be strengthening of power at the center, and, therefore, danger to individual freedom. This is true even in respect to the organs of democratic communities themselves, as well as the professional and trade unions whose function it is to protect the rights of each individual member. Their very size might make him feel helpless, even though he had no reason to suspect ill-will on their part. The more so, if his views or actions were such as to offend the susceptibilities of those who wield power. No mere declaration of rights can suffice: institutions are required to make the rights effective. Habeas corpus need not be the last constitutional device by which personal freedom was anchored in law. Rights of the citizen hitherto unacknowledged must be added to the Bill of Rights. They must be made to prevail against all authorities, whether state, municipal, or professional. The list should be headed by the right of the individual to a job under approved conditions, irrespective of his or her political or religious views, or of color and race. This implies guarantees against victimization however subtle it be. Industrial tribunals have been known to protect the individual member of the public even from such agglomerations of arbitrary power as were represented by the early railway companies. Another instance of possible abuse of power squarely met by tribunals was the Essential Works Order in England, or the "freezing of labor" in the United States, during the emergency, with their almost unlimited opportunities for discrimination. Wherever public opinion was solid in upholding civic liberties, tribunals or courts have always been found capable of vindicating personal freedom. It should be upheld at all cost – even that of efficiency in production, economy in consumption or rationality in administration. An industrial society can afford to be free.

The passing of market-economy can become the beginning of an era of unprecedented freedom. Juridical and actual freedom can be made wider and more general than ever before; regulation and control can achieve freedom not only for the few, but for all. Freedom not as an appurtenance of privilege, tainted at the source, but as a prescriptive right extending far beyond the narrow confines of the political sphere into the intimate organization of society itself. Thus will old freedoms and civic rights be added to the fund of new freedom generated by the leisure and security that industrial society offers to all. Such a society can afford to be both just and free.

Yet we find the path blocked by a moral obstacle. Planning and control are being attacked as a denial of freedom. Free enterprise and private ownership are declared to be essentials of freedom. No society built on other foundations is said to deserve to be called free. The freedom that regulation creates is denounced as unfreedom; the justice, liberty and welfare it offers are decried as a camouflage of slavery. In vain did socialists promise a realm of freedom, for means determine ends: the USSR, which used planning, regulation and control as its instruments, has not yet put the liberties promised in her Constitution into practice, and, probably, the critics add, never will. But to turn against regulation means to turn against reform. With the liberal the idea of freedom thus degenerates into a mere advocacy of free enterprise – which is today reduced to a fiction by the hard reality of giant trusts and princely monopolies. This means the fullness of freedom for those whose income, leisure, and security need no enhancing, and a mere pittance of liberty for the people, who may in vain attempt to make use of their democratic rights to gain shelter from the power of the owners of property. Nor is that all. Nowhere did the liberals in fact succeed in reestablishing free enterprise, which was doomed to fail for intrinsic reasons. It was as a result of their efforts that big business was installed in several European countries and, incidentally, also various brands of fascism, as in Austria. Planning, regulation, and control, which they wanted to see banned as dangers to freedom, were then employed by the confessed enemies of freedom to abolish it altogether. Yet the victory of fascism was made practically unavoidable by the liberals' obstruction of any reform involving planning, regulation, or control.

Freedom's utter frustration in fascism is, indeed, the inevitable result of the liberal philosophy, which claims that power and compulsion are evil, that freedom demands their absence from a human community. No such thing is possible; in a complex society this becomes apparent. This leaves no alternative but either to remain faithful to an illusionary idea of freedom and deny the reality of society, or to accept that reality and reject the idea of freedom. The first is the liberal's conclusion; the latter the fascist's. No other seems possible.

Inescapably we reach the conclusion that the very possibility of freedom is in question. If regulation is the only means of spreading and strengthening freedom in a complex society, and yet to make use of this means is contrary to freedom per se, then such a society cannot be free.

Clearly, at the root of the dilemma there is the meaning of freedom itself. Liberal economy gave a false direction to our ideals. It seemed to approximate the fulfillment of intrinsically utopian expectations. No society is possible in which power and compulsion are absent, nor a world in which force has no function. It was an illusion to assume a society shaped by man's will and wish alone. Yet this was the result of a market view of society which equated economics with contractual relationships, and contractual relations with freedom. The radical illusion was fostered that there is nothing in human society that is not derived from the volition of individuals and that could not, therefore, be removed again by their volition. Vision was limited by the market which "fragmentated" life into the producers' sector that ended when his product reached the market, and the sector of the consumer for whom all goods sprang from the market. The one derived his income "freely" from the market, the other spent it "freely" there. Society as a whole remained invisible. The power of the state was of no account, since the less its power, the smoother the market mechanism would function. Neither voters, nor owners, neither producers, nor consumers could be held responsible for such brutal restrictions of freedom as were involved in the occurrence of unemployment and destitution. Any decent individual could imagine himself free from all responsibility for acts of compulsion on the part of a state which he, personally, rejected; or for economic suffering in society from which he, personally, had not benefited. He was "paying his way," was "in nobody's debt," and was unentangled in the evil of power and economic value. His lack of responsibility for them seemed so evident that he denied their reality in the name of his freedom.

But power and economic value are a paradigm of social reality. They do not spring from human volition; noncooperation is impossible in regard to them. The function of power is to ensure that measure of conformity which is needed for the survival of the group; its ultimate source is opinion – and who could help holding opinions of some sort or other? Economic value ensures the usefulness of the goods produced; it must exist prior to the decision to produce them; it is a seal set on the division of labor. Its source is human wants and scarcity – and how could we be expected not to desire one thing more than another? Any opinion or desire will make us participants in the creation of power and in the constituting of economic value. No freedom to do otherwise is conceivable.

We have reached the final stage of our argument.

The discarding of the market utopia brings us face to face with the reality of society. It is the dividing line between liberalism on the one hand, fascism and socialism on the other. The difference between these two is not primarily economic. It is moral and religious. Even where they profess identical economics, they are not only different but are, indeed, embodiments of opposite principles. And the ultimate on which they separate is again freedom. By fascists and socialists alike the reality of society is accepted with the finality with which the knowledge of death has molded human consciousness. Power and compulsion are a part of that reality; an ideal that would ban them from society must be invalid. The issue on which they divide is whether in the light of this knowledge the idea of freedom can be upheld or not; is freedom an empty word, a temptation, designed to ruin man and his works, or can man reassert his freedom in the face of that knowledge and strive for its fulfillment in society without lapsing into moral illusionism?

This anxious question sums up the condition of man. The spirit and content of this study should indicate an answer.

We invoked what we believed to be the three constitutive facts in the consciousness of Western man: knowledge of death, knowledge of freedom, knowledge of society. The first, according to Jewish legend, was revealed in the Old Testament story. The second was revealed through the discovery of the uniqueness of the person in the teachings of Jesus as recorded in the New Testament. The third revelation came to us through living in an industrial society. No one great name attaches to it; perhaps Robert Owen came nearest to becoming its vehicle. It is the constitutive element in modern man's consciousness.

The fascist answer to the recognition of the reality of society is the rejection of the postulate of freedom.

The Christian discovery of the uniqueness of the individual and of the oneness of mankind is negated by fascism. Here lies the root of its degenerative bent.

Robert Owen was the first to recognize that the Gospels ignored the reality of society. He called this the "individualization" of man on the part of Christianity and appeared to believe that only in a cooperative commonwealth could "all that is truly valuable in Christianity" cease to be separated from man. Owen recognized that the freedom we gained through the teachings of Jesus was inapplicable to a complex society. His socialism was the upholding of man's claim to freedom *in such a society*. The post-Christian era of Western civilization had begun, in which the Gospels did not any more suffice, and yet remained the basis of our civilization.

The discovery of society is thus either the end or the rebirth of freedom. While the fascist resigns himself to relinquishing freedom and glorifies power which is the reality of society, the socialist resigns himself to that reality and upholds the claim to freedom, in spite of it. Man becomes mature and able to exist as a human being in a complex society. To quote once more Robert Owen's inspired words: "Should any causes of evil be irremovable by the new powers which men are about to acquire, they will know that they are necessary and unavoidable evils; and childish, unavailing complaints will cease to be made."

Resignation was ever the fount of man's strength and new hope. Man accepted the reality of death and built the meaning of his bodily life upon it. He resigned himself to the truth that he had a soul to lose and that there was worse than death, and founded his freedom upon it. He resigns himself, in our time, to the reality of society which means the end of that freedom. But, again, life springs from ultimate resignation. Uncomplaining acceptance of the reality of society gives man indomitable courage and strength to remove all removable injustice and unfreedom. As long as he is true to his task of creating more abundant freedom for all, he need not fear that either power or planning will turn against him and destroy the freedom he is building by their instrumentality. This is the meaning of freedom in a complex society; it gives us all the certainty that we need.

Freedom's Just Another Word . . .

David Harvey

For any way of thought to become dominant, a conceptual apparatus has to be advanced that appeals to our intuitions and instincts, to our values and our desires, as well as to the possibilities inherent in the social world we inhabit. If successful, this conceptual apparatus becomes so embedded in common sense as to be taken for granted and not open to question. The founding figures of neoliberal thought took political ideals of human dignity and individual freedom as fundamental, as 'the central values of civilization'. In so doing they chose wisely, for these are indeed compelling and seductive ideals. These values, they held, were threatened not only by fascism, dictatorships, and communism, but by all forms of state intervention that substituted collective judgements for those of individuals free to choose.

Concepts of dignity and individual freedom are powerful and appealing in their own right. Such ideals empowered the dissident movements in eastern Europe and the Soviet Union before the end of the Cold War as well as the students in Tiananmen Square. The student movements that swept the world in 1968 – from Paris and Chicago to Bangkok and Mexico City – were in part animated by the quest for greater freedoms of speech and of personal choice. More generally, these ideals appeal to anyone who values the ability to make decisions for themselves.

The idea of freedom, long embedded in the US tradition, has played a conspicuous role in the US in recent years. '9/11' was immediately interpreted by many as an attack on it. 'A peaceful world of growing freedom', wrote President Bush on the first anniversary of that awful day, 'serves American long-term interests, reflects enduring American ideals and unites America's allies.' 'Humanity', he concluded, 'holds in its hands the opportunity to offer freedom's triumph over all its age-old foes', and 'the United States welcomes its responsibilities to lead in this great mission'. This language was incorporated into the US National Defense Strategy document issued shortly thereafter. 'Freedom is the Almighty's gift to every man and woman in this world', he later said, adding that 'as the greatest power on earth we have an obligation to help the spread of freedom'.

When all of the other reasons for engaging in a preemptive war against Iraq were proven wanting, the president appealed to the idea that the freedom conferred on Iraq was in and of itself an adequate justification for the war. The Iraqis were free, and that was all that really mattered. But what sort of 'freedom' is envisaged here, since, as the cultural critic Matthew Arnold long ago thoughtfully observed, 'freedom is a very good horse to ride, but to ride somewhere'. To what destination, then, are the Iraqi people expected to ride the horse of freedom donated to them by force of arms?

The Bush administration's answer to this question was spelled out on 19 September 2003, when Paul Bremer, head of the Coalition Provisional Authority, promulgated four orders that included 'the full privatization of public enterprises, full ownership rights by foreign firms of Iraqi businesses, full repatriation of foreign profits [. . .] the opening of Iraq's banks to foreign control, national treatment for foreign companies and [. . .] the elimination of nearly all trade barriers'. The orders were to apply to all areas of the economy, including public services, the media, manufacturing, services, transportation, finance, and construction. Only oil was exempt (presumably because of its special status as revenue producer to pay for the war and its geopolitical significance). The labour market, on the other hand, was to be strictly regulated. Strikes were effectively forbidden in key sectors and the right to unionize restricted. A highly regressive 'flat tax' (an ambitious tax-reform plan long advocated for implementation by conservatives in the US) was also imposed.

These orders were, some argued, in violation of the Geneva and Hague Conventions, since an occupying power is mandated to guard the assets of an occupied country and not sell them off. Some Iraqis resisted the imposition of what the London *Economist* called a 'capitalist dream' regime upon Iraq. A member of the US-appointed Coalition Provisional Authority forcefully criticized the imposition of 'free market fundamentalism', calling it 'a flawed logic that ignores history'. Though Bremer's rules may have been illegal when imposed by an occupying power, they would become legal if confirmed by a 'sovereign' government. The interim government, appointed by the US, that took over at the end of June 2004 was declared 'sovereign'. But it only had the power to confirm existing laws. Before the handover, Bremer multiplied the number of laws to specify free-market and free-trade rules in minute detail (on detailed matters such as copyright laws and intellectual property rights), expressing the hope that these institutional arrangements would 'take on a life and momentum of their own' such that they would prove very difficult to reverse.

According to neoliberal theory, the sorts of measures that Bremer outlined were both necessary and sufficient for the creation of wealth and therefore for the improved well-being of the population at large. The assumption that individual freedoms are guaranteed by freedom of the market and of trade is a cardinal feature of neoliberal thinking, and it has long dominated the US stance towards the rest of the world. What the US evidently sought to impose by main force on Iraq was a state apparatus whose fundamental mission was to facilitate conditions for profitable capital accumulation on the part of both domestic and foreign capital. I call this kind of state apparatus a *neoliberal state*. The freedoms it embodies reflect the interests of private property owners, businesses, multinational corporations, and financial capital. Bremer invited the Iraqis, in short, to ride their horse of freedom straight into the neoliberal corral.

The first experiment with neoliberal state formation, it is worth recalling, occurred in Chile after Pinochet's coup on the 'little September 11th' of 1973 (almost thirty years to the day before Bremer's announcement of the regime to be installed in Iraq). The coup, against the democratically elected government of Salvador Allende, was promoted by domestic business elites threatened by Allende's drive towards socialism. It was backed by US corporations, the CIA, and US Secretary of State Henry Kissinger. It violently repressed all the social movements and political organizations of the left and dismantled all forms of popular organization (such as the community health centres in poorer neighbourhoods). The labour market was 'freed' from regulatory or institutional restraints (trade union power, for example). But how was the stalled economy to be revived? The policies of import substitution (fostering national industries by subsidies or tariff protections) that had dominated Latin American attempts at economic development had fallen into disrepute, particularly in Chile, where they had never worked that well. With the whole world in economic recession, a new approach was called for.

A group of economists known as 'the Chicago boys' because of their attachment to the neoliberal theories of Milton Friedman, then teaching at the University of Chicago, was summoned to help reconstruct the Chilean economy. The story of how they were chosen is an interesting one. The US had funded training of Chilean economists at the University of Chicago since the 1950s as part of a Cold War programme to counteract left-wing tendencies in Latin America. Chicago-trained economists came to dominate at the private Catholic University in Santiago. During the early 1970s, business elites organized their opposition to Allende through a group called 'the Monday Club' and developed a working relationship with these economists, funding their work through research institutes. After General Gustavo Leigh, Pinochet's rival for power and a Keynesian, was sidelined in 1975, Pinochet brought these economists into the government, where their first job was to negotiate loans with the International Monetary Fund. Working alongside the IMF, they restructured the economy according to their theories. They reversed the nationalizations and privatized public assets, opened up natural resources (fisheries, timber, etc.) to private and unregulated exploitation (in many cases riding roughshod over the claims of indigenous inhabitants), privatized social security, and facilitated foreign direct investment and freer trade. The right of foreign companies to repatriate profits from their Chilean operations was guaranteed. Export-led growth was favoured over import substitution. The only sector reserved for the state was the key resource of copper (rather like oil in Iraq). This proved crucial to the budgetary viability of the state since copper revenues flowed exclusively into its coffers. The immediate revival of the Chilean economy in terms of growth rates, capital accumulation, and high rates of return on foreign investments was short-lived. It all went sour in the Latin American debt crisis of 1982. The result was a much more pragmatic and less ideologically driven application of neoliberal policies in the years that followed. All of this, including the pragmatism, provided helpful evidence to support the subsequent turn to neoliberalism in both Britain (under Thatcher) and the US (under Reagan) in the 1980s. Not for the first time, a brutal experiment carried out in the periphery became a model for the formulation of policies in the centre (much as experimentation with the flat tax in Iraq has been proposed under Bremer's decrees).

The fact that two such obviously similar restructurings of the state apparatus occurred at such different times in quite different parts of the world under the coercive influence of the United States suggests that the grim reach of US imperial power might lie behind the rapid proliferation of neoliberal state forms throughout the world from the mid-1970s onwards. While this has undoubtedly occurred over the last thirty years, it by no means constitutes the whole story, as the domestic component of the neoliberal turn in Chile shows. It was not the US, furthermore, that forced Margaret Thatcher to take the pioneering neoliberal path she took in 1979. Nor was it the US that forced China in 1978 to set out on a path of liberalization. The partial moves towards neoliberalization in India in the 1980s and Sweden in the early 1990s cannot easily be attributed to the imperial reach of US power. The uneven geographical development of neoliberalism on the world stage has evidently been a very complex process entailing multiple determinations and not a little chaos and confusion. Why, then, did the neoliberal turn occur, and what were the forces that made it so hegemonic within global capitalism?

Why the Neoliberal Turn?

The restructuring of state forms and of international relations after the Second World War was designed to prevent a return to the catastrophic conditions that had so threatened the capitalist order in the great slump of the 1930s. It was also supposed to prevent the re-emergence of inter-state geopolitical rivalries that had led to the war. To ensure domestic peace and tranquillity, some sort of class compromise between capital and labour had to be constructed. The thinking at the time is perhaps best represented by an influential text by two eminent social scientists, Robert Dahl and Charles Lindblom, published in 1953. Both capitalism and communism in their raw forms had failed, they argued. The only way ahead was to construct the right blend of state, market, and democratic institutions to guarantee peace, inclusion, well-being, and stability. Internationally, a new world order was constructed through the Bretton Woods agreements, and various institutions, such as the United Nations, the World Bank, the IMF, and the Bank of International Settlements in Basel, were set up to help stabilize international relations. Free trade in goods was encouraged under a system of fixed exchange rates anchored by the US dollar's convertibility into gold at a fixed price. Fixed exchange rates were incompatible with free flows of capital that had to be controlled, but the US had to allow the free flow of the dollar beyond its borders if the dollar was to function as the global reserve currency. This system existed under the umbrella protection of US military power. Only the Soviet Union and the Cold War placed limits on its global reach.

A variety of social democratic, Christian democratic and dirigiste states emerged in Europe after the Second World War. The US itself turned towards a liberal democratic state form, and Japan, under the close supervision of the US, built a nominally democratic but in practice highly bureaucratic state apparatus empowered to oversee the reconstruction of that country. What all of these various state forms had in common was an acceptance that the state should focus on full employment, economic growth, and the welfare of its citizens, and that state power should be freely deployed, alongside of or, if necessary, intervening in or even substituting for market processes to achieve these ends. Fiscal and monetary policies usually dubbed 'Keynesian' were widely deployed to dampen business cycles and to ensure reasonably full employment. A 'class compromise' between capital and labour was generally advocated as the key guarantor of domestic peace and tranquillity. States actively intervened in industrial policy and moved to set standards for the social wage by constructing a variety of welfare systems (health care, education, and the like).

This form of political-economic organization is now usually referred to as 'embedded liberalism' to signal how market processes and entrepreneurial and corporate activities were surrounded by a web of social and political constraints and a regulatory environment that sometimes restrained but in other instances led the way in economic and industrial strategy. State-led planning and in some instances state ownership of key sectors (coal, steel, automobiles) were not uncommon (for example in Britain, France, and Italy). The neoliberal project is to disembed capital from these constraints.

Embedded liberalism delivered high rates of economic growth in the advanced capitalist countries during the 1950s and 1960s. In part this depended on the largesse of the US in being prepared to run deficits with the rest of the world and to absorb any excess product within its borders. This system conferred benefits such as expanding export markets (most obviously for Japan but also unevenly across South America and to some other countries of South-East Asia), but attempts to export 'development' to much of the rest of the world largely stalled. For much of the Third World, particularly Africa, embedded liberalism remained a pipe dream. The subsequent drive towards neoliberalization after 1980 entailed little material change in their impoverished condition. In the advanced capitalist countries, redistributive politics (including some degree of political integration of working-class trade union power and support for collective bargaining), controls over the free mobility of capital (some degree of financial repression through capital controls in particular), expanded public expenditures and welfare state-building, active state interventions in the economy, and some degree of planning of development went hand in hand with relatively high rates of growth. The business cycle was successfully controlled through the application of Keynesian fiscal and monetary policies. A social and moral economy (sometimes supported by a strong sense of national identity) was fostered through the activities of an interventionist state. The state in effect became a force field that internalized class relations. Working-class institutions such

as labour unions and political parties of the left had a very real influence within the state apparatus.

By the end of the 1960s embedded liberalism began to break down, both internationally and within domestic economies. Signs of a serious crisis of capital accumulation were everywhere apparent. Unemployment and inflation were both surging everywhere, ushering in a global phase of 'stagflation' that lasted throughout much of the 1970s. Fiscal crises of various states (Britain, for example, had to be bailed out by the IMF in 1975–6) resulted as tax revenues plunged and social expenditures soared. Keynesian policies were no longer working. Even before the Arab-Israeli War and the OPEC oil embargo of 1973, the Bretton Woods system of fixed exchange rates backed by gold reserves had fallen into disarray. The porosity of state boundaries with respect to capital flows put stress on the system of fixed exchange rates. US dollars had flooded the world and escaped US controls by being deposited in European banks. Fixed exchange rates were therefore abandoned in 1971. Gold could no longer function as the metallic base of international money; exchange rates were allowed to float, and attempts to control the float were soon abandoned. The embedded liberalism that had delivered high rates of growth to at least the advanced capitalist countries after 1945 was clearly exhausted and was no longer working. Some alternative was called for if the crisis was to be overcome.

One answer was to deepen state control and regulation of the economy through corporatist strategies (including, if necessary, curbing the aspirations of labour and popular movements through austerity measures, incomes policies, and even wage and price controls). This answer was advanced by socialist and communist parties in Europe, with hopes pinned on innovative experiments in governance in places such as communist-controlled 'Red Bologna' in Italy, on the revolutionary transformation of Portugal in the wake of the collapse of fascism, on the turn towards a more open market socialism and ideas of 'Eurocommunism', particularly in Italy (under the leadership of Berlinguer) and in Spain (under the influence of Carrillo), or on the expansion of the strong social democratic welfare state tradition in Scandinavia. The left assembled considerable popular power behind such programmes, coming close to power in Italy and actually acquiring state power in Portugal, France,

Spain, and Britain, while retaining power in Scandinavia. Even in the United States, a Congress controlled by the Democratic Party legislated a huge wave of regulatory reform in the early 1970s (signed into law by Richard Nixon, a Republican president, who in the process even went so far as to remark that 'we are all Keynesians now'), governing everything from environmental protection to occupational safety and health, civil rights, and consumer protection. But the left failed to go much beyond traditional social democratic and corporatist solutions and these had by the mid 1970s proven inconsistent with the requirements of capital accumulation. The effect was to polarize debate between those ranged behind social democracy and central planning on the one hand (who, when in power, as in the case of the British Labour Party, often ended up trying to curb, usually for pragmatic reasons, the aspirations of their own constituencies), and the interests of all those concerned with liberating corporate and business power and re-establishing market freedoms on the other. By the mid 1970s, the interests of the latter group came to the fore. But how were the conditions for the resumption of active capital accumulation to be restored?

How and why neoliberalism emerged victorious as the single answer to this question is the crux of the problem we have to solve. In retrospect it may seem as if the answer was both inevitable and obvious, but at the time, I think it is fair to say, no one really knew or understood with any certainty what kind of answer would work and how. The capitalist world stumbled towards neoliberalization as the answer through a series of gyrations and chaotic experiments that really only converged as a new orthodoxy with the articulation of what became known as the 'Washington Consensus' in the 1990s. By then, both Clinton and Blair could easily have reversed Nixon's earlier statement and simply said 'We are all neoliberals now.' The uneven geographical development of neoliberalism, its frequently partial and lop-sided application from one state and social formation to another, testifies to the tentativeness of neoliberal solutions and the complex ways in which political forces, historical traditions, and existing institutional arrangements all shaped why and how the process of neoliberalization actually occurred.

There is, however, one element within this transition that deserves specific attention. The crisis of capital accumulation in the 1970s affected everyone through the combination of rising unemployment and accelerating inflation. Discontent was widespread and the conjoining of labour and urban social movements throughout much of the advanced capitalist world appeared to point towards the emergence of a socialist alternative to the social compromise between capital and labour that had grounded capital accumulation so successfully in the post-war period. Communist and socialist parties were gaining ground, if not taking power, across much of Europe and even in the United States popular forces were agitating for widespread reforms and state interventions. There was, in this, a clear *political* threat to economic elites and ruling classes everywhere, both in the advanced capitalist countries (such as Italy, France, Spain, and Portugal) and in many developing countries (such as Chile, Mexico, and Argentina). In Sweden, for example, what was known as the Rehn–Meidner plan literally offered to gradually buy out the owners' share in their own businesses and turn the country into a worker/share-owner democracy. But, beyond this, the *economic* threat to the position of ruling elites and classes was now becoming palpable. One condition of the post-war settlement in almost all countries was that the economic power of the upper classes be restrained and that labour be accorded a much larger share of the economic pie. In the US, for example, the share of the national income taken by the top 1 per cent of income earners fell from a pre-war high of 16 per cent to less than 8 per cent by the end of the Second World War, and stayed close to that level for nearly three decades. While growth was strong this restraint seemed not to matter. To have a stable share of an increasing pie is one thing. But when growth collapsed in the 1970s, when real interest rates went negative and paltry dividends and profits were the norm, then upper classes everywhere felt threatened. In the US the control of wealth (as opposed to income) by the top 1 per cent of the population had remained fairly stable throughout the twentieth century. But in the 1970s it plunged precipitously as asset values (stocks, property, savings) collapsed. The upper classes had to move decisively if they were to protect themselves from political and economic annihilation.

The coup in Chile and the military takeover in Argentina, promoted internally by the upper classes with US support, provided one kind of solution. The subsequent Chilean experiment with neoliberalism demonstrated that the benefits of revived capital accumulation were highly skewed under forced privatization. The country and its ruling elites, along with foreign investors, did extremely well in the early stages. Redistributive effects and increasing social inequality have in fact been such a persistent feature of neoliberalization as to be regarded as structural to the whole project. Gérard Duménil and Dominique Lévy, after careful reconstruction of the data, have concluded that neoliberalization was from the very beginning a project to achieve the restoration of class power. After the implementation of neoliberal policies in the late 1970s, the share of national income of the top 1 per cent of income earners in the US soared, to reach 15 per cent (very close to its pre-Second World War share) by the end of the century. The top 0.1 per cent of income earners in the US increased their share of the national income from 2 per cent in 1978 to over 6 per cent by 1999, while the ratio of the median compensation of workers to the salaries of CEOs increased from just over 30 to 1 in 1970 to nearly 500 to 1 by 2000. Almost certainly, with the Bush administration's tax reforms now taking effect, the concentration of income and wealth in the upper echelons of society is continuing apace because the estate tax (a tax on wealth) is being phased out and taxation on income from investments and capital gains is being diminished, while taxation on wages and salaries is maintained.

The US is not alone in this: the top 1 per cent of income earners in Britain have doubled their share of the national income from 6.5 per cent to 13 per cent since 1982. And when we look further afield we see extraordinary concentrations of wealth and power emerging all over the place. A small and powerful oligarchy arose in Russia after neoliberal 'shock therapy' had been administered there in the 1990s. Extraordinary surges in income inequalities and wealth have occurred in China as it has adopted free-market-oriented practices. The wave of privatization in Mexico after 1992 catapulted a few individuals (such as Carlos Slim) almost overnight into Fortune's list of the world's wealthiest people. Globally, 'the countries of Eastern Europe and the CIS have registered some of the largest increases ever [. . .] in social inequality. OECD countries also registered big increases in inequality after the 1980s', while 'the income gap

between the fifth of the world's people living in the richest countries and the fifth in the poorest was 74 to 1 in 1997, up from 60 to 1 in 1990 and 30 to 1 in 1960'. While there are exceptions to this trend (several East and South-East Asian countries have so far contained income inequalities within reasonable bounds, as has France), the evidence strongly suggests that the neoliberal turn is in some way and to some degree associated with the restoration or reconstruction of the power of economic elites.

We can, therefore, interpret neoliberalization either as a *utopian* project to realize a theoretical design for the reorganization of international capitalism or as a *political* project to re-establish the conditions for capital accumulation and to restore the power of economic elites. In what follows I shall argue that the second of these objectives has in practice dominated. Neoliberalization has not been very effective in revitalizing global capital accumulation, but it has succeeded remarkably well in restoring, or in some instances (as in Russia and China) creating, the power of an economic elite. The theoretical utopianism of neoliberal argument has, I conclude, primarily worked as a system of justification and legitimation for whatever needed to be done to achieve this goal. The evidence suggests, moreover, that when neoliberal principles clash with the need to restore or sustain elite power, then the principles are either abandoned or become so twisted as to be unrecognizable. This in no way denies the power of ideas to act as a force for historical-geographical change. But it does point to a creative tension between the power of neoliberal ideas and the actual practices of neoliberalization that have transformed how global capitalism has been working over the last three decades.

The Rise of Neoliberal Theory

Neoliberalism as a potential antidote to threats to the capitalist social order and as a solution to capitalism's ills had long been lurking in the wings of public policy. A small and exclusive group of passionate advocates – mainly academic economists, historians, and philosophers – had gathered together around the renowned Austrian political philosopher Friedrich von Hayek to create the Mont Pelerin Society (named after the Swiss spa where they first met) in 1947 (the notables included Ludvig von Mises, the economist Milton Friedman, and even, for a time, the noted philosopher Karl Popper). The founding statement of the society read as follows:

> The central values of civilization are in danger. Over large stretches of the earth's surface the essential conditions of human dignity and freedom have already disappeared. In others they are under constant menace from the development of current tendencies of policy. The position of the individual and the voluntary group are progressively undermined by extensions of arbitrary power. Even that most precious possession of Western Man, freedom of thought and expression, is threatened by the spread of creeds which, claiming the privilege of tolerance when in the position of a minority, seek only to establish a position of power in which they can suppress and obliterate all views but their own.
>
> The group holds that these developments have been fostered by the growth of a view of history which denies all absolute moral standards and by the growth of theories which question the desirability of the rule of law. It holds further that they have been fostered by a decline of belief in private property and the competitive marker; for without the diffused power and initiative associated with these institutions it is difficult to imagine a society in which freedom may be effectively preserved.

The group's members depicted themselves as 'liberals' (in the traditional European sense) because of their fundamental commitment to ideals of personal freedom. The neoliberal label signalled their adherence to those free market principles of neoclassical economics that had emerged in the second half of the nineteenth century (thanks to the work of Alfred Marshall, William Stanley Jevons, and Leon Walras) to displace the classical theories of Adam Smith, David Ricardo, and, of course, Karl Marx. Yet they also held to Adam Smith's view that the hidden hand of the market was the best device for mobilizing even the basest of human instincts such as gluttony, greed, and the desire for wealth and power for the benefit of all. Neoliberal doctrine was therefore deeply opposed to state interventionist theories, such as those of John Maynard Keynes, which rose to prominence in the 1930s in response to the Great Depression. Many

policy-makers after the Second World War looked to Keynesian theory to guide them as they sought to keep the business cycle and recessions under control. The neoliberals were even more fiercely opposed to theories of centralized state planning, such as those advanced by Oscar Lange working close to the Marxist tradition. State decisions, they argued, were bound to be politically biased depending upon the strength of the interest groups involved (such as unions, environmentalists, or trade lobbies). State decisions on matters of investment and capital accumulation were bound to be wrong because the information available to the state could not rival that contained in market signals.

This theoretical framework is not, as several commentators have pointed out, entirely coherent. The scientific rigour of its neoclassical economics does not sit easily with its political commitment to ideals of individual freedom, nor does its supposed distrust of all state power fit with the need for a strong and if necessary coercive state that will defend the rights of private property, individual liberties, and entrepreneurial freedoms. The juridical trick of defining corporations as individuals before the law introduces its own biases, rendering ironic John D. Rockefeller's personal credo etched in stone in the Rockefeller Center in New York City, where he places 'the supreme worth of the individual' above all else. And there are, as we shall see, enough contradictions in the neoliberal position to render evolving neoliberal practices (vis-à-vis issues such as monopoly power and market failures) unrecognizable in relation to the seeming purity of neoliberal doctrine. We have to pay careful attention, therefore, to the tension between the theory of neoliberalism and the actual pragmatics of neoliberalization.

Hayek, author of key texts such as *The Constitution of Liberty*, presciently argued that the battle for ideas was key, and that it would probably take at least a generation for that battle to be won, not only against Marxism but against socialism, state planning, and Keynesian interventionism. The Mont Pelerin group garnered financial and political support. In the US in particular, a powerful group of wealthy individuals and corporate leaders who were viscerally opposed to all forms of state intervention and regulation, and even to internationalism sought to organize opposition to what they saw as an emerging consensus for pursuing a mixed economy. Fearful of how the alliance with the Soviet Union and the command economy constructed within the US during the Second World War might play out politically in a post-war setting, they were ready to embrace anything from McCarthyism to neoliberal think-tanks to protect and enhance their power. Yet this movement remained on the margins of both policy and academic influence until the troubled years of the 1970s. At that point it began to move centre-stage, particularly in the US and Britain, nurtured in various well-financed think-tanks (off-shoots of the Mont Pelerin Society, such as the Institute of Economic Affairs in London and the Heritage Foundation in Washington), as well as through its growing influence within the academy, particularly at the University of Chicago, where Milton Friedman dominated. Neoliberal theory gained in academic respectability by the award of the Nobel Prize in economics to Hayek in 1974 and Friedman in 1976. This particular prize, though it assumed the aura of Nobel, had nothing to do with the other prizes and was under the tight control of Sweden's banking elite. Neoliberal theory, particularly in its monetarist guise, began to exert practical influence in a variety of policy fields. During the Carter presidency, for example, deregulation of the economy emerged as one of the answers to the chronic state of stagflation that had prevailed in the US throughout the 1970s. But the dramatic consolidation of neoliberalism as a new economic orthodoxy regulating public policy at the state level in the advanced capitalist world occurred in the United States and Britain in 1979.

In May of that year Margaret Thatcher was elected in Britain with a strong mandate to reform the economy. Under the influence of Keith Joseph, a very active and committed publicist and polemicist with strong connections to the neoliberal Institute of Economic Affairs, she accepted that Keynesianism had to be abandoned and that monetarist 'supply-side' solutions were essential to cure the stagflation that had characterized the British economy during the 1970s. She recognized that this meant nothing short of a revolution in fiscal and social policies, and immediately signalled a fierce determination to have done with the institutions and political ways of the social democratic state that had been consolidated in Britain after 1945. This entailed confronting trade union power, attacking all

forms of social solidarity that hindered competitive flexibility (such as those expressed through municipal governance, and including the power of many professionals and their associations), dismantling or rolling back the commitments of the welfare state, the privatization of public enterprises (including social housing), reducing taxes, encouraging entrepreneurial initiative, and creating a favourable business climate to induce a strong inflow of foreign investment (particularly from Japan). There was, she famously declared, 'no such thing as society, only individual men and women' – and, she subsequently added, their families. All forms of social solidarity were to be dissolved in favour of individualism, private property, personal responsibility, and family values. The ideological assault along these lines that flowed from Thatcher's rhetoric was relentless. 'Economics are the method', she said, 'but the object is to change the soul.' And change it she did, though in ways that were by no means comprehensive and complete, let alone free of political costs.

In October 1979 Paul Volcker, chairman of the US Federal Reserve Bank under President Carter, engineered a draconian shift in US monetary policy. The long-standing commitment in the US liberal democratic state to the principles of the New Deal, which meant broadly Keynesian fiscal and monetary policies with full employment as the key objective, was abandoned in favour of a policy designed to quell inflation no matter what the consequences might be for employment. The real rate of interest, which had often been negative during the double-digit inflationary surge of the 1970s, was rendered positive by fiat of the Federal Reserve. The nominal rate of interest was raised overnight and, after a few ups and downs, by July 1981 stood close to 20 per cent. Thus began 'a long deep recession that would empty factories and break unions in the US and drive debtor countries to the brink of insolvency, beginning the long era of structural adjustment'. This, Volcker argued, was the only way out of the grumbling crisis of stagflation that had characterized the US and much of the global economy throughout the 1970s.

The Volcker shock, as it has since come to be known, has to be interpreted as a necessary but not sufficient condition for neoliberalization. Some central banks had long emphasized anti-inflationary fiscal responsibility and adopted policies that were closer to monetarism than to Keynesian orthodoxy. In the West German case this derived from historical memories of the runaway inflation that had destroyed the Weimar Republic in the 1920s (setting the stage for the rise of fascism) and the equally dangerous inflation that occurred at the end of the Second World War. The IMF had long set itself against excessive debt creation and urged, if not mandated, fiscal restraints and budgetary austerity on client states. But in all these cases this monetarism was paralleled by acceptance of strong union power and a political commitment to build a strong welfare state. The turn to neoliberalism thus depended not only on adopting monetarism but on the unfolding of government policies in many other arenas.

Ronald Reagan's victory over Carter in 1980 proved crucial, even though Carter had shifted uneasily towards deregulation (of airlines and trucking) as a partial solution to the crisis of stagflation. Reagan's advisers were convinced that Volcker's monetarist 'medicine' for a sick and stagnant economy was right on target. Volcker was supported in and reappointed to his position as chair of the Federal Reserve. The Reagan administration then provided the requisite political backing through further deregulation, tax cuts, budget cuts, and attacks on trade union and professional power. Reagan faced down PATCO, the air traffic controllers' union, in a lengthy and bitter strike in 1981. This signalled an all-out assault on the powers of organized labour at the very moment when the Volcker-inspired recession was generating high levels of unemployment (10 per cent or more). But PATCO was more than an ordinary union: it was a white-collar union which had the character of a skilled professional association. It was, therefore, an icon of middle-class rather than working-class unionism. The effect on the condition of labour across the board was dramatic – perhaps best captured by the fact that the Federal minimum wage, which stood on a par with the poverty level in 1980, had fallen to 30 per cent below that level by 1990. The long decline in real wage levels then began in earnest.

Reagan's appointments to positions of power on issues such as environmental regulation, occupational safety, and health, took the campaign against big government to ever higher levels. The deregulation of everything from airlines and telecommunications to finance opened up new zones of untrammelled market freedoms for powerful corporate interests. Tax breaks on investment effectively subsidized the movement of capital away from the unionized north-east and midwest and into the non-union and weakly regulated south and west. Finance capital increasingly looked abroad for higher rates of return. Deindustrialization at home and moves to take production abroad became much more common. The market, depicted ideologically as the way to foster competition and innovation, became a vehicle for the consolidation of monopoly power. Corporate taxes were reduced dramatically, and the top personal tax rate was reduced from 70 to 28 per cent in what was billed as 'the largest tax cut in history'.

And so began the momentous shift towards greater social inequality and the restoration of economic power to the upper class.

There was, however, one other concomitant shift that also impelled the movement towards neoliberalization during the 1970s. The OPEC oil price hike that came with the oil embargo of 1973 placed vast amounts of financial power at the disposal of the oil-producing states such as Saudi Arabia, Kuwait, and Abu Dhabi. We now know from British intelligence reports that the US was actively preparing to invade these countries in 1973 in order to restore the flow of oil and bring down oil prices. We also know that the Saudis agreed at that time, presumably under military pressure if not open threat from the US, to recycle all of their petrodollars through the New York investment banks. The latter suddenly found themselves in command of massive funds for which they needed to find profitable outlets. The options within the US, given the depressed economic conditions and low rates of return in the mid 1970s, were not good. More profitable opportunities had to be sought out abroad. Governments seemed the safest bet because, as Walter Wriston, head of Citibank, famously put it, governments can't move or disappear. And many governments in the developing world, hitherto starved of funds, were anxious enough to borrow. For this to occur required, however, open entry and reasonably secure conditions for lending. The New York investment banks looked to the US imperial tradition both to prise open new investment opportunities and to protect their foreign operations.

The US imperial tradition had been long in the making, and to a great degree defined itself against the imperial traditions of Britain, France, Holland, and other European powers. While the US had toyed with colonial conquest at the end of the nineteenth century, it evolved a more open system of imperialism without colonies during the twentieth century. The paradigm case was worked out in Nicaragua in the 1920s and 1930s, when US marines were deployed to protect US interests but found themselves embroiled in a lengthy and difficult guerrilla insurgency led by Sandino. The answer was to find a local strongman – in this case Somoza – and to provide economic and military assistance to him and his family and immediate allies so that they could repress or buy off opposition and accumulate considerable wealth and power for themselves. In return they would always keep their country open to the operations of US capital and support, and if necessary promote US interests, both in the country and in the region (in the Nicaraguan case, Central America) as a whole. This was the model that was deployed after the Second World War during the phase of global decolonization imposed upon the European powers at US insistence. For example, the CIA engineered the coup that overthrew the democratically elected Mosaddeq government in Iran in 1953 and installed the Shah of Iran, who gave the oil contracts to US companies (and did not return the assets to the British companies that Mossadeq had nationalized). The shah also became one of the key guardians of US interests in the Middle Eastern oil region.

In the post-war period, much of the non-communist world was opened up to US domination by tactics of this sort. This became the method of choice to fight off the threat of communist insurgencies and revolution, entailing an anti-democratic (and even more emphatically anti-populist and anti-socialist/communist) strategy on the part of the US that put the US more and more in alliance with repressive military dictatorships and authoritarian regimes (most spectacularly, of course, throughout Latin America). The stories told in John Perkins's *Confessions of an Economic Hit Man* are full of the ugly and unsavoury details of how this was all too often done. US interests consequently became more rather than less vulnerable in the struggle against international communism. While the consent of local ruling elites could be purchased easily enough, the need to coerce oppositional or social democratic movements (such as Allende's in Chile) associated the US with a

long history of largely covert violence against popular movements throughout much of the developing world.

It was in this context that the surplus funds being recycled through the New York investment banks were dispersed throughout the world. Before 1973, most US foreign investment was of the direct sort, mainly concerned with the exploitation of raw material resources (oil, minerals, raw materials, agricultural products) or the cultivation of specific markets (telecommunications, automobiles, etc.) in Europe and Latin America. The New York investment banks had always been active internationally, but after 1973 they became even more so, though now far more focused on lending capital to foreign governments. This required the liberalization of international credit and financial markets, and the US government began actively to promote and support this strategy globally during the 1970s. Hungry for credit, developing countries were encouraged to borrow heavily, though at rates that were advantageous to the New York bankers. Since the loans were designated in US dollars, however, any modest, let alone precipitous, rise in US interest rates could easily push vulnerable countries into default. The New York investment banks would then be exposed to serious losses.

The first major test case of this came in the wake of the Volcker shock that drove Mexico into default in 1982–4. The Reagan administration, which had seriously thought of withdrawing support for the IMF in its first year in office, found a way to put together the powers of the US Treasury and the IMF to resolve the difficulty by rolling over the debt, but did so in return for neoliberal reforms. This treatment became standard after what Stiglitz refers to as a 'purge' of all Keynesian influences from the IMF in 1982. The IMF and the World Bank thereafter became centres for the propagation and enforcement of 'free market fundamentalism' and neoliberal orthodoxy. In return for debt rescheduling, indebted countries were required to implement institutional reforms, such as cuts in welfare expenditures, more flexible labour market laws, and privatization. Thus was 'structural adjustment' invented. Mexico was one of the first states drawn into what was going to become a growing column of neoliberal state apparatuses worldwide.

What the Mexico case demonstrated, however, was a key difference between liberal and neoliberal practice: under the former, lenders take the losses that arise from bad investment decisions, while under the latter the borrowers are forced by state and international powers to take on board the cost of debt repayment no matter what the consequences for the livelihood and well-being of the local population. If this required the surrender of assets to foreign companies at fire-sale prices, then so be it. This, it turns out, is not consistent with neoliberal theory. One effect, as Duménil and Lévy show, was to permit US owners of capital to extract high rates of return from the rest of the world during the 1980s and 1990s. The restoration of power to an economic elite or upper class in the US and elsewhere in the advanced capitalist countries drew heavily on surpluses extracted from the rest of the world through international flows and structural adjustment practices.

[. . .]

 READING 14

Neoliberalism as Exception, Exception to Neoliberalism

Aihwa Ong

Neoliberalism seems to mean many different things depending on one's vantage point. In much of the world, it has become a code word for America's overweening power. Asian politicians and pundits view "American neoliberalism" as a strategy of market domination that uses intermediaries such as the International Monetary Fund (IMF) to pry open small economies and expose them to trade policies that play havoc with these nations' present and future economic welfare. For example, in the decade of the emerging Asian economies (1980s–90s), Asian leaders proclaimed that "Asia can say no" to American neoliberalism. Such rhetoric became more vociferous after the "Asian financial crisis" of 1997–8. In popular discourses, *neoliberalism* also represents unregulated financial flows that menaced national currencies and living conditions. South Korean anti-neoliberal protestors who lost their jobs due to imposed economic restructuring sported T-shirts that proclaimed, "IMF means I'M Fired!" In Latin America, the US drive for open markets and privatization is called "savage neoliberalism." Since the invasion of Iraq, critiques of neoliberalism have included the perception that America would stoop to conquest in order to grab oil resources for major corporations. Thus, in the global popular imagination, American neoliberalism is viewed as a radicalized capitalist imperialism that is increasingly tied to lawlessness and military action. As we shall see below, despite such widespread criticism, Asian governments have selectively adopted neoliberal forms in creating economic zones and imposing market criteria on citizenship.

Neoliberalism at Large

In the United States, in contrast, *neoliberalism* is seldom part of popular discourse outside the academy. Rather, *market-based policies* and *neoconservatism* are the native categories that code the ensemble of thinking and strategies seeking to eliminate social programs and promote the interests of big capital. *Liberty* has become a word that designates "free economic action" rather than political liberalism, which has become a dirty word. In rather broad terms, one can say that the Democratic Party promotes itself as the defender of individual rights and civil liberties against the excesses of an unfettered, market-driven ethos, while the Republican Party relies on a neoliberal (read neoconservative) discourse of individual solutions to myriad social problems. Both kinds of liberalism focus on free subjects as a basic rationale and target of government, but while the Democrats stress individual and civil freedoms, the Republicans underline individual obligations of self-reliance and self-management. For instance, the conservative columnist William Safire writes that "a Republican brain" chooses values that "include self-reliance over community dependence, intervention over isolation, self-discipline over society's regulation, finding pleasure in work rather than working to find pleasure." In political life, both kinds of liberal rationalities frequently overlap and fuse, but Republicans have strengthened neoliberalism's hold on America by casting (political) "liberalism" as "un-American." Such partisan debates in fact highlight the chasm that is opening up between political liberal ideals of democracy and the neoliberal rationality of individual responsibility and fate.

Upon his reelection to a second term, President George W. Bush claimed a political "mandate" to transform life in the United States. In a raft of proposed new "market-based policies," he has proposed to dismantle fundamental aspects of American liberal democracy institutionalized since the New Deal, from the privatization of Social Security and health care to the abolition of the progressive tax code. Bush calls his new vision the "ownership society," an explicit claim that American citizenship under his watch will shift toward a primitive, narrow vision of citizenship that includes only property owners, privileging "an independent and egoistical individual" in isolated pursuit of economic self-interest. In his second inaugural address, President Bush was explicit about "preparing our people for the challenges of life in a free society [. . .] by making every citizen an agent of his or her own destiny." This neoliberal view of citizenship also has the moral support of evangelical Christian groups.

But presidential attempts to marketize politics and reengineer citizenship have not gone unchallenged. Close to half the citizenry has opposed such policies of privatization. For decades, a plethora of protest movements have defended the steady erosion of the civil rights of prisoners, workers, women, homosexuals, minorities, and aliens, to name only a few. They promise to continue the fight to protect individual liberty and the national patrimony. But the Bush administration continues to seek to reverse antipoverty programs, health coverage, environmental protection, and food safety, among other policies, in the spirit if not in the name of neoliberal reason. This cluster of neoliberal logic, religion, rights, and ethics has become the problem-space of American citizenship, with outcomes as yet unknown. Nevertheless, as I mentioned above, since the 1970s, "American neoliberalism" has become a global phenomenon that has been variously received and critiqued overseas.

Neoliberalism and Exceptions

This book argues that as a new mode of political optimization, neoliberalism – with a small *n* – is reconfiguring relationships between governing and the governed, power and knowledge, and sovereignty and territoriality. Neoliberalism is often discussed as an economic doctrine with a negative relation to state power, a market ideology that seeks to limit the scope and activity of governing. But neoliberalism can also be conceptualized as a new relationship between government and knowledge through which governing activities are recast as nonpolitical and nonideological problems that need technical solutions. Indeed, neoliberalism considered as a technology of government is a profoundly active way of rationalizing governing and self-governing in order to "optimize." The spread of neoliberal calculation as a governing technology is thus a historical process that unevenly articulates situated political constellations. An ethnographic perspective reveals specific alignments of market rationality, sovereignty, and citizenship that mutually constitute distinctive milieus of labor and life at the edge of emergence.

I focus on the active, interventionist aspect of neoliberalism in non-Western contexts, where *neoliberalism as exception* articulates sovereign rule and regimes of citizenship. Of course, the difference between *neoliberalism as exception* and *exceptions to neoliberalism* hinges on what the "normative order" is in a particular milieu of investigation. This book focuses on the interplay of exceptions in emerging countries where neoliberalism itself is not the general characteristic of technologies of governing. We find neoliberal interventions in liberal democracies as well as in postcolonial, authoritarian, and post-socialist situations in East and Southeast Asia. Thus neoliberalism as exception is introduced in sites of transformation where market-driven calculations are being introduced in the management of populations and the administration of special spaces. The articulation of neoliberal exceptions, citizenship, and sovereignty produces a range of possible anthropological problems and outcomes.

At the same time, *exceptions to neoliberalism* are also invoked, in political decisions, to exclude populations and places from neoliberal calculations and choices. Exceptions to neoliberalism can be modes for protecting social safety nets or for stripping away all forms of political protection. In Russia, for instance, subsidized housing and social rights are preserved even when neoliberal techniques are introduced in urban budgetary practices. At the same time, in Southeast Asia, exceptions to neoliberalism exclude migrant workers from the living standards created by market-driven policies. In other words, exceptions to neoliberalism can both preserve welfare benefits for citizens and exclude noncitizens from the benefits of capitalist development.

But there is an overlap in the workings of neoliberal exceptions and exceptions to market calculations. Populations governed by neoliberal technologies are dependent on others who are excluded from neoliberal considerations. The articulation of populations and spaces subjected to neoliberal norms and those outside the purview of these norms crystallizes ethical dilemmas, threatening to displace basic values of social equality and shared fate. The chapters that follow present diverse ethnographic milieus where the interplay of exceptions, politics, and ethics constitutes a field of vibrant relationships. New forms of governing and being governed and new notions of what it means to be human are at the edge of emergence.

In this approach, I bring together two concepts – neoliberalism and exception – that others have dealt with separately. Neoliberalism as a technology of governing relies on calculative choices and techniques in the domains of citizenship and of governing. Following Foucault, "governmentality" refers to the array of knowledges and techniques that are concerned with the systematic and pragmatic guidance and regulation of everyday conduct. As Foucault puts it, governmentality covers a range of practices that "constitute, define, organize and instrumentalize the strategies that individuals in their freedom can use in dealing with each other." Neoliberal governmentality results from the infiltration of market-driven truths and calculations into the domain of politics. In contemporary times, neoliberal rationality informs action by many regimes and furnishes the concepts that inform the government of free individuals who are then induced

to self-manage according to market principles of discipline, efficiency, and competitiveness.

The political exception, in Carl Schmitt's formulation, is a political decision that is made outside the juridical order and general rule. Schmitt has argued that "the sovereign produces and guarantees the situation in its totality. He has monopoly over this last decision. Therein lies the essence of the state's sovereignty, which must be juridically defined correctly, not as the monopoly to coerce or to rule, but as the monopoly to decide." The condition of exception is thus a political liminality, an extraordinary decision to depart from a generalized political normativity, to intervene in the logics of ruling and of being ruled. The Schmittian exception is invoked to delineate friends and foes in a context of war. Giorgio Agamben has used the exception as a fundamental principle of sovereign rule that is predicated on the division between citizens in a juridical order and outsiders stripped of juridical-political protections.

In contrast, I conceptualize the exception more broadly, as an extraordinary departure in policy that can be deployed to include as well as to exclude. As conventionally understood, the sovereign exception marks out excludable subjects who are denied protections. But the exception can also be a positive decision to include selected populations and spaces as targets of "calculative choices and value-orientation" associated with neoliberal reform. In my formulation, we need to explore the hinge between neoliberalism as exception and exception to neoliberalism, the interplay among technologies of governing and of disciplining, of inclusion and exclusion, of giving value or denying value to human conduct. The politics of exception in an era of globalization has disquieting ethicopolitical implications for those who are included as well as those who are excluded in shifting technologies of governing and of demarcation. This book will explore how the market-driven logic of exception is deployed in a variety of ethnographic contexts and the ethical risks and interrogations set in motion, unsettling established practices of citizenship and sovereignty.

Interrelationships among exceptions, politics, and citizenship crystallize problems of contemporary living, and they also frame ethical debates over what it means to be human today. For instance, neoliberal exceptions have been variously invoked in Asian settings to recalculate social criteria of citizenship, to remoralize economic action, and to redefine spaces in relation to market-driven choices. These articulations have engendered a range of contingent and ambiguous outcomes that cannot be predicted beforehand. Neoliberal decisions have created new forms of inclusion, setting apart some citizen-subjects, and creating new spaces that enjoy extraordinary political benefits and economic gain. There is the Schmittian exception that abandons certain populations and places them outside political normativity. But articulations between neoliberal exceptions and exceptions to neoliberalism have multiplied possibilities for moral claims and values assigned to various human categories, so that different degrees of protection can be negotiated for the politically excluded.

The yoking of neoliberalism and exception, I suggest, has the following implications for our understanding of how citizenship and sovereignty are mutating in articulation and disarticulation with neoliberal reason and mechanisms. First, a focus on neoliberalism recasts our thinking about the connection between government and citizenship as a strictly juridical-legal relationship. It is important to trace neoliberal technology to a biopolitical mode of governing that centers on the capacity and potential of individuals and the population as living resources that may be harnessed and managed by governing regimes. *Neoliberalism* as used here applies to two kinds of optimizing technologies. *Technologies of subjectivity* rely on an array of knowledge and expert systems to induce self-animation and self-government so that citizens can optimize choices, efficiency, and competitiveness in turbulent market conditions. Such techniques of optimization include the adherence to health regimes, acquisition of skills, development of entrepreneurial ventures, and other techniques of self-engineering and capital accumulation. *Technologies of subjection* inform political strategies that differently regulate populations for optimal productivity, increasingly through spatial practices that engage market forces. Such regulations include the fortressization of urban space, the control of travel, and the recruitment of certain kinds of actors to growth hubs.

As an intervention of optimization, neoliberalism interacts with regimes of ruling and regimes of citizenship to produce conditions that change administrative strategies and citizenship practices. It follows that the infiltration of market logic into politics conceptually unsettles the notion of citizenship as a legal status rooted in a nation-state, and in stark opposition to a condition of statelessness. Furthermore, the neoliberal exception articulates citizenship elements in political spaces that may be less than the national territory in some cases, or exceed national borders in others.

The elements that we think of as coming together to create citizenship – rights, entitlements, territoriality, a nation – are becoming disarticulated and rearticulated with forces set into motion by market forces. On the one hand, citizenship elements such as entitlements and benefits are increasingly associated with neoliberal criteria, so that mobile individuals who possess human capital or expertise are highly valued and can exercise citizenship-like claims in diverse locations. Meanwhile, citizens who are judged not to have such tradable competence or potential become devalued and thus vulnerable to exclusionary practices. On the other hand, the territoriality of citizenship, that is, the national space of the homeland, has become partially embedded in the territoriality of global capitalism, as well as in spaces mapped by the interventions of nongovernmental organizations (NGOs). Such overlapping spaces of exception create conditions for diverse claims of human value that do not fit neatly into a conventional notion of citizenship, or of a universal regime of human rights. In short, components of citizenship have developed separate links to new spaces, becoming rearticulated, redefined, and reimagined in relation to diverse locations and ethical situations. Such de- and re-linking of citizenship elements, actors, and spaces have been occasioned by the dispersion and realignment of market strategies, resources, and actors.

Second, neoliberalism as exception refines the study of state sovereignty, long conceptualized as a political singularity. One view is of the state as a machine that steamrolls across the terrain of the nation, or that will eventually impose a uniform state bureaucracy. In actual practice, sovereignty is manifested in multiple, often contradictory strategies that encounter diverse claims and contestations, and produce diverse and contingent outcomes. In the course of interactions with global markets and regulatory institutions, I maintain, sovereign rule invokes the exception to create new economic possibilities, spaces, and techniques for governing the population. The neoliberal exception allows for a measure of sovereign flexibility in ways that both fragment and extend the space of the nation-state. For instance, in Southeast and East Asia, zoning technologies have carved special spaces in order to achieve strategic goals of regulating groups in relation to market forces. The spatial concentration of strategic political, economic, and social conditions attracts foreign investment, technology transfer, and international expertise to particular zones of high growth. Market-driven strategies of spatial fragmentation respond to the demands of global capital for diverse categories of human capital, thus engendering a pattern of noncontiguous, differently administered spaces of "graduated" or "variegated sovereignty." Furthermore, as corporations and NGOs exert indirect power over various populations at different political scales, we have an emergent situation of overlapping sovereignties.

For instance, technologies of optimization are repositioning the metropolis as a hub for enrolling networks of resources and actors, making the metropolis the hub of a distinctive ecosystem. Saskia Sassen has proposed an influential model of a few "global cities" – New York, London, and Tokyo – that control key functions and services that sustain global circuits. This transnational urban system dominates "cities in the global south which are mostly in the mid-range of the global hierarchy." The explosive growth of Shanghai, Hong Kong, and Singapore suggests the rise of a different kind of space – time synergy prompted by neoliberal exceptions. Market-driven calculations create novel possibilities for combining and recombining external and internal elements to reposition these cities as the sites of emergence and new circulations.

Situated mobilizations of strategic knowledge, resources, and actors configure vibrating webs of interaction, that is, space–time "ecosystems" that extend the scope of hypergrowth zones. This governmentality-as-ecology strategy does not seek to fit emerging Asian centers into a preexisting transnational urban system. Rather, the logic is to reposition the hometown (*oikos*) in its self-spun web of symbiotic relationships among diverse elements (ecosystem) for the strategic production of specific material and social values. This Microsoft-like approach creates "platforms" – "services, tools, or technologies – that other members of the ecosystem can use to enhance their own performance." It is a hub

strategy that uses capital not to perform conventional city functions but to leverage their relationships for innovative collaborations with global companies and research institutions that become intertwined with the future of the site.

Third, the calculative mechanisms of open markets articulate new arrangements and territorializations of capital, knowledge, and labor across national borders. Michael Hardt and Antonio Negri's influential book, *Empire*, contends that economic globalization has produced a uniform global labor regime. But the complex interactions between diverse zones and particular networks challenge sweeping claims about a unified landscape of labor regulation. Rather, I argue, different vectors of capital construct spaces of exception – "latitudes" – that coordinate different axes of labor regulation and of labor disciplining. Lateral production systems permit the stretching of governmentality as well as coercive labor regimes across multiple sites. Latitudinal spaces are thus formed by a hybrid mix of regulatory and incarceral labor regimes that can operate with little regard for labor rights across far-flung zones. Nevertheless, the latitudinal controls are subject to unexpected and unbidden challenges that rise intermittently from mobilities of labor among various sites.

Fourth, neoliberalism, as an ethos of self-governing, encounters and articulates other ethical regimes in particular contexts. Market rationality that promotes individualism and entrepreneurialism engenders debates about the norms of citizenship and the value of human life. For instance, in Southeast Asia, the neoliberal exception in an Islamic public sphere catalyzes debates over female virtue. *Ulamas* resist the new autonomy of working women, while feminists claim a kind of gender equality within the limits of Islam. Contrary to the perception that transnational humanitarianism replaces situated ethics, questions of status and morality are problematized and resolved in particular milieus shaped by economic rationality, religious norms, and citizenship values.

Indeed, different degrees of political and moral claims by the politically marginalized can be negotiated in the shifting nexus of logics and power. There are conceptual limits to models that pose a simple opposition between normalized citizenship and bare life. Giorgio Agamben draws a stark contrast between citizens who enjoy juridicallegal rights and excluded groups who dwell in "a zone of indistinction." But ethnographic study of particular situations reveals that negotiations on behalf of the politically excluded can produce indeterminate or ambiguous outcomes. Indeed, this is the complex work of NGOs everywhere, to identify and articulate moral problems and claims in particular milieus. At times, even business rationality may be invoked in seeking sheer survival for those bereft of citizenship or citizenship-like protections. Humanitarian interventions do not operate in a one-size-fits-all manner but must negotiate the shifting field of criss-crossing relationships.

Neoliberalism as exception articulates a constellation of mutually constitutive relationships that are not reducible to one or the other. Rather, ethnographic exploration reveals novel interactions between market-driven mechanisms and situated practices in space–time interrelationships through which problems are resolved. Technologies of self-governing articulate elements of citizenship, self-enterprising values are translated into movable social entitlements, and mobile entrepreneurial subjects can claim citizenship-like benefits in multiple locations. Meanwhile, the neoliberal exception in governing constructs political spaces that are differently regulated and linked to global circuits. Such reflexive techniques of social engineering and the reengineering of the self interact with diverse ethical regimes, crystallizing contemporary problems of citizenship and ethical living.

[. . .]

Structural Adjustment

In order to receive aid from global economic organizations such as the IMF and the World Bank, receiving nations have had to agree to restructure their economies and societies in line with neoliberal theory. Loans were given, but receiving nations had to agree to various economic reforms (e.g. cutting the size of government and its welfare system, privatization) that facilitated foreign investment and that led to free markets. This came to be known as "structural adjustment," a term first coined by the then World Bank President Robert McNamara in the late 1970s. It, like the closely associated "neoliberalism" (chapter 4), came to be despised by various academic critics as well as by those who lived in less developed nations and who were forced to undergo various structural adjustments in order to receive economic assistance.

Glassman and Carmody look at the economic impact of structural adjustment programs in Latin America in the late twentieth century. They associate structural adjustment with a number of negative economic consequences such as deindustrialization caused, at least in part, by high interest rates. Another negative effect was an increase in economic inequality as the rich grew richer while poverty increased. Control over local economies was increasingly in the hands of large multinational corporations (MNCs) and this served to weaken indigenous enterprise. Turning to Asia in

the 1990s, they find similar negative economic effects such as increased unemployment, declining wages, a weakening of labor unions, and increases in poverty. On the other hand, well-to-do domestic and foreign investors tended to prosper as a result of structural adjustment programs. For these reasons and others, Glassman and Carmody pull no punches in concluding that structural adjustment programs "are clearly wrong for Asia."[1]

Sarah Babb finds mixed conclusions in the literature on the economic impacts of structural adjustment programs (SAPs), but she focuses her attention on the evidence on, and debates about, the social consequences of structural adjustment for developing countries, especially in Latin America and the Caribbean.

Politically she finds that, when engaged in structural adjustment, states do less of some things (e.g. they are less directly involved in production) but more of other things (e.g. they strengthen private property and they make tax systems more regressive). They also encourage increased foreign direct investment (FDI). While these things tend to move those states in the direction of the American model of regulatory capitalism, there are important differences because markets have been transplanted to alien worlds, other societies often overshoot and go beyond the American model, and there is an erosion of social citizenship in many of these societies with a decline in the power of citizens and states *vis-à-vis* private investors.

A second set of impacts relates to the class structure. While differences between nations are less clear, it is the case that there is an increase in within-nation inequality. Among the causes of this increase in inequality are de-agriculturalization leading to less work for peasants, downsizing and layoffs resulting from the privatization of state-owned firms, high interest rates used to fight inflation, and an overall strain on labor markets.

Third, there is a rise in transnational networks. Most notable here is the rise of powerful global production networks involving, among others, MNCs. However, transnationalism (see chapter 7) is not restricted to the corporations as migrant workers also develop such networks (and use them, among other ways, to send remittances back home when they find work in other countries). Structural adjustment programs can also play a role in spawning border-spanning resistance movements.

Abouharb and Cingranelli look at the effects of SAPs undertaken by the World Bank between 1981 and 2000. They find that the overall effect of structural adjustment agreements (SAAs) "is to worsen government respect for physical integrity rights. Torture, political imprisonment, extra-judicial killing, and disappearances were all more likely to occur when a structural adjustment loan had been received and implemented."[2] They contrast their more critical orientation to a positive, neoliberal model of the direct effects of SAAs. In that model, rapid economic liberalization is seen as having a positive effect on human rights. They also look at various indirect effects of SAAs including less respect for economic rights, more domestic conflict, less democracy, and ultimately less respect for physical integrity rights. The authors then review other work on this topic and find that it generally confirms their critical perspective on SAAs.

In their conclusion, Abouharb and Cingranelli make it clear that the World Bank "probably" does not intend the negative outcomes they describe.[3] Among other things, the World Bank is publicly committed to good governance and sound human rights practices as ways to promote economic development; it is more likely to give loans to countries with positive records on such matters; and human rights practices improved in various nations in the early years of a loan, probably to impress the Bank.

Lloyd and Weissman found that both IMF and World Bank policies tend to undermine both labor power and the rights of labor. Among other things, these policies lead to a shrinking government labor force, privatization, greater labor flexibility including greater freedom to fire workers, wage reductions, and changes in pension programs that result in the need for people to work longer, to pay more for their pensions, but to get lesser amounts.

While there is much criticism of structural adjustment from many directions, Scott argues that at least in the case of Africa, failures there cannot be blamed on the IMF. Rather, he blames Africa's economic problems on its own corrupt leadership. He reviews various IMF programs – devaluation, reductions in government deficits, market prices, and privatization – and finds that in the main they have the potential to be helpful in Africa. Perhaps the most important thing that the IMF could do is to reduce corruption, but the problem of corruption is inadequately treated in IMF programs.

NOTES

1   Jim Glassman and Pádraig Carmody, "Structural Adjustment in East and Southeast Asia: Lessons from Latin America." *Geoforum* 32, 2001: 87.

2   M. Rodwan Abouharb and David L. Cingranelli,

"The Human Rights Effects of World Bank Structural Adjustment, 1981–2000." *International Studies Quarterly* 50, 2006: 234.

3   Ibid., 256.

Structural Adjustment in East and Southeast Asia: Lessons from Latin America

Jim Glassman and Pádraig Carmody

1 Introduction

The Asian economic crisis, which began in 1997, is a historical watershed. Should the crisis serve to derail the Asian "miracle" economies, it may usher in a new period in the geography of the global economy in which few, if any, developing countries can be optimistic about the prospects for rapid industrial growth – the Asian newly industrializing countries (NICs) having been the primary industrialization success stories in recent decades.

The crisis has also had important impacts on development theory and practice, for example with divisions emerging between the International Monetary Fund (IMF) and its supporters and the World Bank and various others who have argued that conventional structural adjustment programs (SAPs) are "the wrong medicine for Asia". While not questioning the general thrust of economic liberalization, these critics argue that the "demand reducing" elements of SAPs are designed for countries with large public sectors and substantial public debts such as those in sub-Saharan Africa, Eastern Europe, and Latin America, but that they are inappropriate for the Asian NICs, which for the most part have had relatively small states and debts which are largely held by the private sector.

We concur with the mainstream critics that the IMF's approach is the wrong medicine for Asia. However, this approach has also been inappropriate for countries elsewhere. Rather than the Asian economic crisis being the result of "cronyism" or corruption, which was then punished by international capital markets, we argue that it was the outcome of contradictions inherent in a globalized capitalist economy, and liberalization which exposed Asian countries to these contradictions.

Consequently, further marketization is likely to have systematically negative consequences for the Asian NICs. Seeing what these consequences are likely to be requires an examination of countries which have already implemented SAPs.

Not long ago, it was common to see work on development studies which ruminated on what Latin America could learn from the Asian "tigers". It is now appropriate to shift our geographic perspective and examine what East and Southeast Asia can learn from the experiences of structural adjustment in Latin America during the 1980s and 1990s, if more equitable and sustainable development strategies are to be implemented.

[...] In Section 2, we briefly describe the context and nature of SAPs. [...] In Section 5, we revisit the process of structural adjustment in Latin America, highlighting some of its major outcomes and relating these to broader crisis tendencies inherent in capitalist economies. In particular, we suggest that SAPs have interlocking core–periphery and class dimensions, as well as potential political consequences, which have negative implications for popular classes. In Section 6, we [...] [show] how the features of SAPs which exacerbated inequality and undermined industrial growth in Latin America are already having similar effects in the Asian NICs, and how these may increase the risks of future crises. We conclude by discussing alternatives to neoliberalism.

2 Global Structural Adjustment

Since the early 1970s the global economy has been in crisis. In the industrial countries this has been manifest in deindustrialization and falling real wages for the majority of the workforce. The breakdown of the Bretton Woods system of fixed exchange rates in the 1970s unleashed intense competitive pressures worldwide. Subsequently, the introduction of monetarist economic policies in the core countries in the late 1970s and early 1980s drove global interest rates dramatically higher and triggered a debt crisis in the developing world. Since that time developing countries have been called on to restructure their economies to correct resulting "disequilibrium" under the auspices of the world's two most powerful international financial institutions (IFIs) the World Bank and the IMF.

Structural adjustment is a policy package of "free market" economic reforms sponsored by the IFIs. Initially structural adjustment programs (SAPs) were introduced to offset what were seen as temporary balance of payments problems in developing countries resulting from increased oil prices and interest rates in the late 1970s. However, with the debt crisis, which broke in 1982, structural adjustment programs became more widespread and long-lived than was initially anticipated.

Structural adjustment consists of two distinct elements: macro-economic "stabilization" the purview of the IMF, and "structural adjustment" which entails the restructuring of the economy towards export-orientation under the auspices of the World Bank. Together the combined package is commonly known as "structural adjustment".

> The stabilization phase of adjustment focuses on demand restraint policies, usually effected by large reductions in government expenditure via measures such as subsidy removals, public sector employment cuts, and the introduction of user fees (for social services) [...] Structural adjustment involve(s) a realignment of the real exchange rate (through devaluation), privatization, liberalization of interest rates, and tax reform, including reductions in import/export barriers (removal/reduction of tariffs, quotas, and taxes) in order to improve the economy's relative trading position).

In the last 20 years, the vast majority of countries in the developing world have undergone a structural adjustment program. With the onset of the Asian economic crisis a number of countries there have also adopted them.

[...]

4 Embedding Structural Adjustment Programs in Place and Class: Theoretical Issues

The IMF and its structural adjustment policies have been criticized from a variety of perspectives. In particular we want to focus on the implications of SAPs for core/periphery and inter-class relations, as well as suggesting the gendering of some of their outcomes. The net results of structural adjustment are to subordinate peripheral economies to transnational corporations (TNCs), international banks, and core area governments; to generate greater inequality in the distribution of wealth and income between classes; and frequently to place a disproportionate share of the burden of adjustment on women. We also argue that in order to gain implementation against popular disapproval, structural adjustment frequently takes on politically authoritarian characteristics.

The global economic crisis which began in the 1970s has been worked out by the burden of adjustment being passed down from economically stronger areas and social forces to weaker and less politically organized ones. In the first instance, the IMF's insistence on currency convertibility and liberalization fosters domination of the periphery from the core by allowing relatively stronger capitals to dominate weaker capitals on an international level. This is largely so because liberalization of capital flows increases the power of international over domestic investors within a national economy. Meanwhile open trade and capital regimes help capital dominate labor by providing tools to resist working class demands for improved wages and social services. Thus SAPs typically allow local elites to pass the costs of adjustment onto the popular classes, because the participation of these elites is necessary in order for the IMF's agenda to be implemented. Within the popular classes, insofar as gender relations are already inegalitarian, women frequently end up taking on a disproportionate share of the burdens of adjustment. These tendencies are illustrated in Latin America.

5 Economic and Social Restructuring in Latin American NICs under Liberalization: Deindustrialization, Poverty, and Income Inequality

Two of the most pressing needs in developing countries are to reduce the level of unemployment and to diversify economies so that they are better able to withstand external shocks. One of the best ways to meet these needs is through the development of a competitive manufacturing sector that is labor-absorptive. However, structural adjustment causes deindustrialization in a number of ways. High interest rate policies detract from productive investment and negatively affect the balance sheets of companies already in debt. Simultaneously other "demand reducing policies" result in contraction of the domestic market, and trade liberalization may expose domestic producers to competitive displacement from overseas.

The experience of Chile is often invoked to justify the policies of the World Bank and the IMF. However, General Pinochet's post-1973 "stabilization" of the economy under IMF guidance resulted in deindustrialization, an absolute reduction in the number of manufacturing jobs, low investment and the reduction of productive capacity. The situation came to a head in 1981 as it was no longer possible "for firms to continue paying annual average real interest rates of 25–30%, while during the previous six years (1975–81) output had grown at an annual rate of only 7%". In some cases financial repayments rose to 50% of the total sales for firms. Consequently in 1982 there were record numbers of plant closures, capital flight and a "desubstitution of imports". From 1967 to 1982, total manufacturing employment fell from 327,013 to 223,138, with some sub-sectors, such as textiles, particularly hard hit.

In Chile, as in East Asia, economic liberalization was associated with the development of a financial "bubble". According to Barros external debt increased significantly after 1974, but much of this was not being used to finance domestic capital formation, but rather increasing amounts of non-traditional imports. This led to an appreciation of the real exchange rate, and a massive increase in the current account deficit. In the Southern Cone of Latin America, the IMF and World Bank have "repeatedly supported combinations of exchange rate appreciation and capital market liberalization which were doomed to fail".

SAPs also tend to be highly regressive in terms of their impact on income distribution. Indeed the motivation of such programs is partly to increase the profit share to "revive" the private sector economy. In response to the crisis which was driven by liberalization, Chile implemented the SAPs during all but one year between 1983 and 1990. Whereas the Asian NICs were noted for their "growth with equity", with often rapidly rising real wages, in Chile from 1981 to 1990 real wages dropped at an annual average of 5%, ending up 10% lower than they had been in 1970. Unemployment averaged 20% during 1974–87, compared to 6% in the 1960s, and by 1990 the richest 10% of Chileans had increased their share of the national income to 47%. Meanwhile, whereas only 17% of Chilean households lived below the poverty line in 1970, this had increased to 38% in 1986, declining only slightly (to 35%) by 1990. When the newly elected democratic government took over in 1990, it adopted a significantly less liberal policy regime.

Mexico's experience with SAPs was similar. The previous development strategy was one of import-substitution. However with the advent of the debt crisis, Mexico implemented SAPs in six out of eight years between 1983 and 1990. In contrast to predictions, however, this resulted in a shift not to export-oriented, but to import-oriented industrialization.

From the late 1980s the share of foreign direct to portfolio investment in Mexico declined dramatically. Mexico was able to attract substantial portfolio investment because it had previously met all the IMF conditions, and, as in many of the Asian NICs, the Mexican government pegged the peso to the US dollar. This led to an overvaluation of the exchange rate which hurt Mexico's export competitiveness and encouraged imports. Consequently the trade deficit increased from 0.51% of GDP in 1988 to 6.98% in 1992.

Given the over-valued exchange rate, and the consequent cheapness of imports, there were disincentives to invest in productive economic activity. Within the manufacturing sector dualism increased, as those subsectors associated with transnational investment or domestic oligopolies experienced rapid growth, whereas many more traditional domestically-oriented industries, such as textiles experienced a process of deindustrialization. According to Dussel Peters the main features of structural change in manufacturing in Mexico "are its heterogeneity, concentration and exclusion as well as a significant tendency to lose backward and forward linkages within the domestic economy". This may forebode the future trajectory of much of the manufacturing sector in East and Southeast Asia, as foreign investors have rushed in to buy up highly indebted companies at bargain prices after the devaluation of the region's currencies.

In terms of income distribution in Mexico: after the financial crisis of 1982–91, the purchasing power of the minimum wages dropped by 66%, in part the effect of repeated currency devaluations. This reduced the purchasing power of the minimum wages to just half of what it was during the years 1936–8.

While structural adjustment has been catastrophic for Mexico's popular classes, it has opened up new opportunities in trade and finance for the elite and increased the scale of concentration in the industrial sector. From 1988 to 1994 the number of billionaires in US dollar terms rose from 2 to 24 and by 1994, assets of the richest individual in Mexico exceeded the combined assets of the poorest 17 million. Moreover, the renewed financial crisis of 1994 forced another round of devaluation and pushed workers' wages down further yet. Falling incomes for working class families have forced many young women to find work in the burgeoning *maquiladora* sector at very low wages and under highly exploitative and patriarchal conditions.

While new inflows of capital to Latin America had, during the early 1990s dulled memories of previous crises, these have once again been rekindled by the financial crises of Mexico and more recently of Brazil, which have illustrated how tenuous are the putative gains from openness to international capital flows. Moreover, on each occasion where crisis has emerged, the core–periphery effects noted earlier by Payer have been prominent. For example, Mexico's bail-out package was accompanied by measures that gave the US Treasury de facto control over the proceeds of the Mexican national oil company, Pemex.

6 The Short-Term Consequences of Structural Adjustment in Asia

In Latin America, SAPs have had the effects described here because they altered neither the structural condi-

tions of dependence nor the class relations which led to or exacerbated the economic crisis – a situation of weak domestic demand (relative to market values produced) and heavy reliance on volatile global finance and increasingly competitive export markets. In fact SAPs exacerbated economic inequality and deepened poverty, thereby further weakening domestic markets. SAPs also increased the susceptibility of local economic processes to control by the most powerful international economic forces, particularly multi-national corporations and global finance, thus undermining much productive indigenous enterprise. In doing so, SAPs simultaneously serve the interests of the global economic core and certain fractions of international and domestic capital within the periphery.

While we recognize the specific differences between various Asian NICs and those of other regions, we do not believe that their successes exempt them from the broader dynamics we have described at work in the rest of the global capitalist economy. Though it is still too early to discern the medium and long-term effects of SAPs in Asia, we can note their results to date.

Along with mandating exchange rate flexibility, Thailand's SAP originally emphasized cuts in central budget expenditures (even though debts were overwhelmingly held by the private sector), with a targeted budget surplus equal to 1% of GDP for 1997/98. Capital inflows were initially to be encouraged through high interest rates and eased restrictions on equity participation in troubled financial institutions. Restructuring of the financial sector included the closure of fifty-eight insolvent finance companies. Wage increases were to be pegged to inflation, whereas in actuality the purchasing power of the minimum wage fell. The state also announced its intention to encourage privatization of state enterprises in the energy, transportation, utility and communications sectors.

As the economic situation in Thailand worsened throughout 1998, with GDP declining by more than 8%, some changes were negotiated with the IMF. High interest rates, that encouraged a continuing sense of crisis amongst foreign investors and had crushed many local businesses, were slowly lowered, and the state was allowed to run a budget deficit equivalent to 5% of the GDP during 1998/99. These reflationary measures helped the economy with the GDP growth for 1999 estimated at 3–4%.

More direct measures ensured that certain "private" interests would be bailed out with public money. The IMF funds were used to pay off the central bank's obligations and to indemnify foreign investors, as well as to restore currency reserves which had been depleted, in part, by efforts to bail out insolvent local finance companies. Overall however, these measures were insufficient to save many domestic capitalists, to the benefit of foreign investors who have been able to buy Thai assets at fire sale prices. Nonetheless, certain well-positioned Thai elites have also been able to take advantage of the opportunities presented by the crisis through activities ranging from arbitrage to new joint ventures with foreign investors.

While some of the edges were taken off Thailand's SAP to facilitate the restructuring of capital, there have been fewer efforts to directly rescue others. As the SAP took hold, unemployment more than doubled – from 1.9% of the workforce in 1997 to 4.2% in 1999. Other estimates place the 1999 rate even higher at 5.1%, while estimating a loss of 1.4 million construction and 140,000 manufacturing jobs between 1997 and 1999. Real wages for manufacturing workers fell from US$ 188/month in 1996 to US$ 133/month in 1999. Consequently the poverty rate is estimated by some to have doubled from around 10–20% of the population. Unemployment insurance has not yet been developed, and a major program of poverty alleviation was not put in place until 1999. Thus the Thai state has primarily relied on rural society to act as a shock absorber by finding work and residence for those laid off from urban-industrial occupations.

South Korea's SAP had similar outlines to the one implemented in Thailand, in spite of important differences in the industrial and political structures of the two countries. Again, reductions in government spending, increased foreign equity participation in ailing financial institutions, trade liberalization, and privatization measures were emphasized. The first letter of intent to the IMF (3 December, 1997) made restructuring of the financial sector the "centerpiece" of the SAP. Legal changes spurred by the SAP approved of hostile takeovers which will allow foreign investors to purchase up to a third of the shares of Korean companies as well as the establishment of subsidiaries of foreign banks and securities firms. Other changes eliminated the requirement for government approval of foreign takeovers involving Korean firms with more than 2 million won in assets, except in key industries and defense. As a consequence, Korea's recovery, which has so far been more robust than that of the other two Asian NICs undergoing adjustment, may result in the displacement of a significant number of formerly protected domestic producers by foreign firms through direct investment and imports from overseas. Deindustrialization would appear to be underway in some branches of manufacturing as textiles, motor vehicles, machinery and equipment, and particularly clothing production have been especially hard hit. Production of "wearing apparel" in South Korea was only 54.8% of its 1995 level in 1998.

Particularly important to the SAP was the attack on Korea's powerful labor unions under the guise of improved "labor market flexibility". The strong resistance of Korean labor to such demands was met by bringing it to the table in tri-partite (government, business, labor) bargaining sessions. While this represented a political gain for labor, it was used to impose concessions on it which increased the burden of unemployment, which rose precipitously from 2.6% in 1997 to a high of 8.6% in the February of 1999. Even with significant economic recovery by the end of the year, the number of workers still unemployed was twice what it had been before the crisis. During the crisis women were laid off at a rate seven times that of men, illustrating one of the ways in which class processes connected to restructuring are gendered. Income inequality has also jumped dramatically, with the richest 10% of urban households having incomes 8.5 times higher than the poorest 10%, up from 6.9 times two years previously.

As in Thailand, the economic situation in Korea deteriorated more rapidly than expected in 1998, with a nearly 7% decline in GDP. This forced some changes in the state budget, with the small surplus of 1997 turning into deficits equivalent to 5% of GDP during 1998 and 1999. Much of this deficit was the result of increased spending in support of financial sector restructuring, along with support for small- and medium-sized enterprises and export promotion. But the strength of Korea's labor unions and the need to try to limit their opposition to the SAP also helped produce an increase in spending on unemployment and social safety net programs. At the same time, however, spending on education and civil service salaries declined in both the 1998 and 1999 budgets.

In Indonesia the structural adjustment process has been even more difficult, helping to precipitate a continuing political crisis. Structural adjustment in Indonesia also followed an agenda of exchange rate flexibility, state expenditure reductions, financial sector restructuring, wage discipline, and privatization/liberalization. However, the Suharto regime, in spite of a general commitment to the SAP, vacillated during the key moments where the interests of powerful cronies were at stake, thus earning the distrust of much of the domestic and international investor community, precipitating the regime's violent downfall.

In spite of this political turmoil and the economic free-fall which accompanied it, the Indonesian state did in fact implement a comprehensive package of structural adjustment policies, including eliminating the foreign shareholding limit of 49% for financial firms, approving full foreign ownership of non-banking financial firms, lifting restrictions on foreign ownership of companies listed in the Jakarta stock exchange, cutting public spending (particularly on large infrastructure projects), eliminating a number of import monopolies, and cutting tariffs. The severity of the economic crisis, however, has made new opportunities for foreign investors less attractive, with the economy contracting by an estimated 13.7% in 1998.

As broad as the effects of the crisis and the SAP have been, there can be little doubt that workers and the poor have borne the brunt of the difficulties. Estimates of unemployment vary widely, but some place unemployment for 1998 as high as 15–20% of the workforce in Indonesia. Total reductions in the size of the formal workforce in 1998 have been estimated at over 5 million people, with manufacturing and service sector employment estimated to have contracted by 20% in 1998. Nominal wages were held constant between 1997 and 1998 and the Suharto regime canceled a planned 15% increase in civil service salaries and with dramatic inflation real wages declined between 30 and 50%, reducing them to their late 1980s level.

Poverty estimates in Indonesia are highly problematic, but there is a consensus that the crisis has increased poverty dramatically throughout the country. The International Labor Organization estimates poverty at 48.3%, and as elsewhere in Asia, weak or non-existent social safety net programs have exacerbated this. In spite of the severity of the crisis, it was not until September of 1998, that the Indonesian government announced the *possibility* of developing a social safety net program, with expenditures on food security, public works, health and education, and promotion of small and medium enterprises – equivalent in total to 6.5% of the federal budget. The development of these programs, however, is difficult in the environment of budgetary frugality which has prevailed under the SAP: even in the context of economic free-fall, the Indonesian state has limited reflationary expenditures and has held the deficit to less than 1% of GDP in 1998 and 1999.

6.1 Lessons and future prospects for Asia

Evidence of enhanced opportunities for powerful domestic and foreign investors and worsening short-term economic conditions for much of the population undergoing structural adjustment in Asia is incontrovertible. While it is not possible to determine precisely how much of this is due to the general crisis and how much is due to the specific measures undertaken as part of structural adjustment, it is clear that SAPs have, by design, pushed down wages and opened new investment opportunities for foreign capital. However, this is not an attempt to derail the Asian NICs general export-led growth drive (contrary to assertions by Malaysia's Prime Minister Mahathir), but to restructure it by enhancing the participation of foreign capital, and to open up the region to Western, particularly US exports. One US trade negotiator noted that the US had achieved more by way of opening the South Korean market for car parts in six months of bail-out talks than during ten years of bilateral trade negotiations.

In this context, what is important to the analysis of SAPs and their possible longer-term consequences is not merely the empirical evidence, but the explanation of the power relations that push in the direction of worsening income distribution and increased dependence. Some of the same kinds of general forces and outcomes which played out in Latin America under structural adjustment in the 1980s and 1990s are beginning to show in Asian countries undergoing SAPs. These forces, while instantiated in specific ways in different contexts, are likely to weaken the position of the popular classes while making the economies as a whole more dependent on Western capital flows.

Problems of dependency are likely to be the greatest in Thailand and Indonesia, which have relatively rudimentary levels of technology development and will be increasingly dominated by the decisions of TNCs, but even in South Korea this is an important issue. While the *chaebol* are far more technologically advanced and sophisticated than their counterparts in Southeast Asia, they retain a strong dependence on technology imports from Japan.

While the necessity is for further economic diversification and up-grading in Asia, SAPs will also reinforce an emphasis on competition through low-labor costs. In the short-term this will exacerbate underconsumptionist tendencies in the global economy and over-reliance on volatile export markets. In a global market, sustained competitive advantage is dependent on the introduction of new skills and technologies to raise productivity. Even if SAPs in Asia restore growth, and growth succeeds once again in raising wages, in a liberal environment, capital may respond by moving off-shore.

All of this points more generally to the dangers of a development strategy based on foreign capital inflows. Apart from its greater spatial mobility and the dependence this creates on decisions taken outside the national economy, foreign capital has other disadvantages. Many commentators now emphasize the importance of foreign direct over portfolio investment. Due to greater sunk costs, FDI has a longer-term commitment to an economy. It may also bring new skills and technology, but FDI is highly import-intensive and consequently current account deficits often tend to rise more than FDI inflows.

FDI is also meant to be a cheap form of finance, however the rate of profit remittances from FDI can also easily exceed international interest rates, implying a net loss for the national economy when comparing foreign debt to FDI as a source of capital for industrialization, at least on this score. Whereas South Korea made substantial use of foreign debt to finance its industrialization, it was channeled through the state and tied to performance standards on the part of firms. Once there was substantial capital account liberalization domestic firms in Korea accrued heavy foreign debts, making them vulnerable to devaluation in the context of a floating exchange rate regime. Structural adjustment will further reinforce this risk.

Foreign portfolio investment is particularly dangerous, because it flows mostly into stock markets and results in their appreciation in value, increases domestic stockholders' wealth. In the context of an open trade regime, this contributes to increased demand for imports, thereby worsening any trade deficit. Furthermore, inflows of foreign capital may contribute to an appreciation of the real exchange rate, thereby undermining export growth and making imports cheaper. Yawning current account deficits served as triggers for both the financial crises in Mexico in 1994 and Thailand in 1997 as international investors feared currency devaluations which would reduce the hard currency value of their investments. If Korean and other local firms are displaced by imports, as a result of trade liberalization, this may make the region's financial markets more fragile, while simultaneously increasing dependence on speculative and volatile portfolio capital as a source of growth. Mexico's previous experience is particularly instructive in this regard.

Structural adjustment also has wider political implications. The conventional wisdom has it that there is a mutually reinforcing relationship between "free markets" and electoral democracy. Thus, it could be argued that moves towards both political and economic liberalization took place concurrently in East Asia and Latin America in the 1980s and early 1990s. However, the timing of the relationship is important. In Latin America it was disillusionment with the structural adjustment policies of authoritarian governments which was partly responsible for the shift towards electoral democracy. In South Korea it was a militant workers' movement which pressed for political democratization.

In East and Southeast Asia the strictures on structural adjustment may lead to democratic reversals, rather than democratization, if the state responds to struggles against SAPs with repressive force, something which has already occurred to some extent in South Korea and Indonesia. Beyond this, there is a clear move in Asia to "insulate" economic policy-making from "political interference", with authoritarian macroeconomic governance by internationalist, neoliberal elites under an umbrella of formal political democracy. The hegemony of the neoliberal policies deployed by these internationalist state managers, along with the crucial practical support they garner from an unaccountable international investment community and the IMF, contradict the notion that liberalization is necessarily a move in the direction of democratization more broadly conceived. Rather, the rise of neoliberal hegemony, while helping to disable some of the more egregious military dictatorships (as in Indonesia) is supplanting this form of domination with more deeply entrenched practices of non-military domination; forcing unpopular policies on populations in the name of economic necessity and "competitiveness". SAPs are an integral component of this anti-democratic, neoliberal moment and have justifiably been a target of popular discontent. It is thus amongst anti-SAP coalitions that the struggle for genuine economic accountability and democratization is to be found.

7 Conclusion

In our view, SAPs are clearly wrong for Asia, not only because of their demand restraint elements, but also because of their more general emphasis on unrestricted trade liberalization and openness to international capital which have resulted in the "globalization of poverty". SAPs have well-documented and quite consistent outcomes across different countries, and this consistency reflects the relatively stable core–periphery and class characteristics of the structural adjustment process.

Specific SAPs do turn out somewhat differently, depending on the context. For example, in the Asian NICs undergoing adjustment, there have been substantial differences in social safety nets and other co-optive measures implemented by the state, and with their more highly developed technological capabilities, Korean firms are better placed to compete in the global market than their Thai or Indonesian counterparts. However all SAPs issue forth from the same kind of transnational class coalitions and have the same general purposes. They largely originate within the core and reflect the relative power of core and peripheral capitals. Given this, the fact that SAPs seem to consistently worsen income distribution, pose new burdens for working class women, and strengthen the position of core area investors is not surprising. To neglect these continuities would be, in our view, to undermine the political project of opposition to SAPs and the promotion of alternatives.

These alternatives are many. They range from nationalist initiatives such as those undertaken by the Malaysian state, which reintroduced capital controls to enable reflation of the economy, to more popularly based initiatives seeking a transformation in the structures of power, such as the activities of Thailand's Assembly of the Poor. In Asia, some scholars have called for policies and practices which reorient the region's economies towards the satisfaction of basic needs, empowerment and environmental sustainability by reducing the scale at which economic activity takes place. We do not know which combination of alternatives is likely to take root, but social forces committed to a more egalitarian and self-directed future will certainly resist the major features of the SAPs.

To be sure, the global prospects for the implementation of alternative approaches appear to be bleak. In terms of its core–periphery dimensions, the success of global structural adjustment in reasserting US economic dominance is evidenced by the economy growing rapidly at over 4% for 1999, with real wages rising for the first time in decades in the late-1990s. This gives great weight to the US governments' bullish adherence to the "Washington consensus" favoring global neoliberalism.

In order for local struggle to be effective it seems likely that there would need to be new international institutions which reduce the global power of finance capital. However, the US in particular has blocked recent attempts to reform the international economic system. This may change in the future as the US economy currently suffers from some of the same risks that brought about the crisis in East Asia. As portfolio investment has flowed in from overseas to the "safe haven" of the US stock market its value has risen. While the "new economy" in the US was partly built on the basis of a cheap dollar to revive exports in the late 1980s and early 1990s, the US is now dependent on a strong dollar to keep the confidence of international investors. The former Chairman of the Federal Reserve in the US, Paul Volker, argued recently that "the world economy was currently dependent on the US consumer, who was dependent on the stock market, which was dependent on about fifty stocks, half of which had not shown a profit". Much of the US consumer demand is being met by imports, with the US trade deficit at record levels. If domestic US producers are displaced by imports, and the connection between productive and financial returns "grounds" in the minds of investors, the US stock market may fall drastically in value, creating a global depression. Should that happen the US government may be forced to reconsider global neoliberalism. If and when it does, we are sure there will be no shortage of ideas for alternatives coming from the people and countries which have had to endure SAPs.

 READING 16

The Social Consequences of Structural Adjustment: Recent Evidence and Current Debates

Sarah Babb

Introduction

Once upon a time, intellectual debates around the relationship between wealthy and poor nations could be summed up under the rubric of modernization versus dependency. For modernization theorists, all good things went together: capitalist development, democratization, industrialization, urbanization, rational-legal administration, and increased well-being were assumed to be part of a single process that occurred in roughly the same way in all national contexts. In contrast, dependency theorists argued that the domination of rich over poor countries meant that modernization looked quite different at the periphery. Because of such relations of domination, foreign investment and national industrialization did not propel developing countries along the same trajectory as the wealthy democracies, but rather was compatible with manifold economic, political, and social distortions.

Beginning in the 1980s, however, this debate was rendered obsolete by a very different hegemonic order. Whereas modernization and dependency theorists alike had advocated for strong government involvement in promoting economic development, the new conventional wisdom demanded a dramatic downsizing of many government interventions. Associated with the structural adjustment lending programs of the World Bank and International Monetary Fund (IMF), and

neoliberal ideology, the new policy discourse suggested that it was only through thus liberating market forces that poor countries could grow and catch up to the developed world. Whereas modernization and dependency theorists were drawn from a range of social science disciplines, both the new model and its most prominent critics tend to be economists. Much of the recent work on the consequences of structural adjustment, therefore, has focused on its economic consequences. This essay, in contrast, seeks to revisit some of the older themes of modernization and dependency through looking at recent literature addressing the social dimensions of recent trends.

Structural Adjustment in Historical Perspective

Structural adjustment is a relatively recent phenomenon. In the decades following World War II, economic policy in the industrialized core reflected Keynesian economic ideas that prescribed the taming of markets through macroeconomic interventions. In poorer countries, much more direct state interventions in the economy were tolerated or even encouraged by the core. Even in nominally capitalist developing countries, state-owned enterprises played a significant role in national output during this period; indeed, they were actually encouraged and financed by the World Bank. Some other key elements of the postwar regime were controls on capital movements (which were explicitly condoned by the charter of the IMF) and systems of protection of domestic industries from foreign competition.

By the end of the 1970s, however, the seeds of a new regime had been sown. First coined by World Bank President Robert McNamara at the end of the 1970s, structural adjustment referred to a set of lending practices whereby governments would receive loans if they agreed to implement specific economic reforms. Although it was not clear what this meant at the time, only a few years later, World Bank and IMF lending arrangements had begun to aim at an ambitious agenda in keeping with the ascendant Reagan revolution: to encourage free markets and foreign investment.

The moment was precipitated by the outbreak of the Third World debt crisis in 1982. The indebtedness of LDC (least-developed country) governments can be traced back to the 1970s, when low interest rates, high inflation, and a glut of "petrodollars" led international banks to invest in the developing world. When global interest rates rose dramatically at the end of the 1970s, these debts became unsustainable. The debt crisis made persuading governments to implement policy reforms easier because such reforms could be required as preconditions to bailout funds. Privatization was particularly attractive because it both satisfied multilateral lenders and provided much-needed revenues. But there were also more subtle pressures: trapped under unwieldy debts and stagnating economies, governments were increasingly courting foreign portfolio investors, who were more likely to be attracted to governments that provided strong guarantees to property rights and did not interfere excessively in markets. Governments also came to rely on the advice of US-trained economists in high government posts, whose presence helped foster investor confidence – and who tended to be fervent believers in the need for market reforms. All these factors combined to create the conditions for the policy reforms of the following decades.

As a precise technical term, "structural adjustment" leaves a great deal to be desired: the policies associated with this term have shifted over time, and it is no longer associated with any particular lending program. In this review, therefore, I do not use structural adjust-ment as a technical term, but rather refer to its more interpretive and historical meaning – a term associated with a cluster of overlapping historical and conceptual associations in the same way as are the terms "modernity" or "democracy."

This review examines literature that reflects on the social characteristics of the era of structural adjust-ment. It is not designed to reflect on recent literature on economic development as measured by growth in national income and productivity. The relationship between globalization and development is at the center of an enormous, thriving, and complicated debate that would merit its own literature review. In contrast, this essay focuses on literature reflecting on the organizational, institutional, and class structures of national societies.

Even leaving aside issues of economic development, the consequences of structural adjustment are enormous, complex, and globe-spanning. I have therefore made several strategic decisions to pare this topic down to a more manageable size. First, I have opted to focus on the experience of developing countries – even though structural adjustment has contributed to the transformation of developed countries as well. Second, I have deliberately excluded literature on formerly state socialist economies, which have been subjected to most of the same policies but under very different historical circumstances. Third, my review focuses disproportionately on the experience of Latin America and the Caribbean, which is the focus of a great deal of the existing literature. Fortunately, in many respects, Latin America represents a relatively good laboratory for gauging the effects of structural adjustment in that it contains a range of incomes per capita, from the poorest of the poor (Haiti) to relatively well-off (Argentina).

What consequences has this shift in economic policy regimes had for underdeveloped societies? To what extent have the trends of the past two decades sharpened the distinctions between core and periphery – and to what extent have they brought them closer together? The following sections seek to answer these questions through examining three different social transformations: changes in the governance of economies, transformations in class structures, and the rise of transformational networks.

The Governance of Economies

Modernization theorists saw developed and developing countries coming together in an inevitable process of institutional convergence. As Marion Levy contended in 1967, "As time goes on, they and we will increasingly resemble one another [...] because the patterns of modernization are such that the more highly modernized societies become, the more they resemble one another". Has the era of structural adjustment made the periphery more structurally similar to the core? This section evaluates the extent of convergence in the governance of national economies.

Today, states in developing countries are doing a lot less of certain things. They are less directly involved in production: between 1988 and 1994, LDC governments transferred more than 3,000 entities from public to private hands. States are also decreasing their protection of domestic industries from foreign competition (through tariffs, licenses, etc.). They are putting fewer constraints on financial markets, fewer barriers on free movement of capital across their borders, and fewer regulations on labor markets. They are also operating with much tighter fiscal policy: even during recessions, they are refraining from using their central banks to finance deficit spending. To demonstrate their commitment to noninflationary monetary policies, many have adopted legislation making their central banks independent.

However, although states in developing countries have withdrawn from certain activities, they have simultaneously increased their involvement in others. To offset the revenues lost through removing tariffs, they have reformed their taxation systems to more effectively extract resources, commonly replacing taxes on income and wealth with more easily administered (but more regressive) value-added taxes. They have strengthened private property rights and expanded these rights for foreign firms – for example, by removing restrictions on foreign ownership of land and productive assets. They have joined the World Trade Organization (WTO), which promotes safeguards for property, including controversial intellectual property safeguards, in conjunction with trade opening. In addition to protecting property rights, LDC governments have recently been encouraged to adopt so-called governance reforms –

to construct institutional frameworks to help safeguard against market imperfections, such as bankruptcy legislation and judiciary independence.

The defining feature of the new regime is an increased role for private investment – particularly foreign private investment, in the economy. This trend represents both a continuation of and a break from the postwar governance regime. Foreign direct investment (e.g., Ford setting up a factory in São Paolo) was a staple of the "associated-dependent development" or "dependent development" system so sharply criticized by dependency theorists in the 1970s. Recently, however, foreign direct investment has become much more important to the economies of developing countries than it ever was during the heyday of dependency theory. Furthermore, the rise of private foreign portfolio investment (e.g., American investors buying stock in a Mexican telephone company, or buying Mexican government bonds) marks a qualitative historical break from the past. In 1970, portfolio investment in developing countries was, for all practical purposes, nonexistent; in 2000, there was a net inflow of \$47.9 billion.

These trends raise interesting theoretical issues about how to define the new institutional framework or organizing logic governing developing countries. In at least some respects, the sort of institutions that are emerging resemble the American model of regulatory capitalism. Under this model, the tasks prescribed for states include the enforcement of contracts, the regulation of natural monopolies, the administration of taxes, and the provision of infrastructure. Perhaps what we are witnessing throughout the developing world is a process of "institutional isomorphism," converging on the organizational patterns of the industrialized North in general, and the United States in particular. This interpretation, if true, would support the homogenizing predictions of modernization theorists.

Indeed, proponents of the new model unintentionally echo modernization theorists in asserting that opening to free trade and foreign investment will ultimately promote greater institutional convergence. Although opening to foreign competition may put inefficient local firms out of business, foreign investment brings improved technologies and management techniques, from which recipient nations will benefit. Because better management and technology increases productivity, more jobs will be available; over time, wages will rise, workers and citizens will demand more of governments and firms, and industrial, social, and environmental regimes will converge with those of the North.

In the remainder of this section, however, I argue that any strong claim that developing economies have been "Americanized" would be inaccurate – or at the very least, premature. On the one hand, it is true that state interventions have been replaced with a more uniform model reminiscent of the institutions of core capitalist powers. On the other hand, structural adjustment also illustrates the limits of convergence and has brought about the construction of institutions that depart, sometimes sharply, from the American model. Although there is insufficient space here to treat this issue completely, I focus on three divergent institutional outcomes: institutional mismatch, institutional overshooting, and the erosion of social citizenship.

One reason for divergent outcomes is that markets have been transplanted to alien worlds, governed by different norms and rules, and lacking the supporting institutions that took decades or even centuries to develop organically in their original contexts. As a result, there may be a mismatch between new and old institutions. In Mexico, for example, privatization and financial liberalization were conducted without a corresponding revision of bankruptcy legislation, which created the conditions for a $55 billion bailout of the banking system. Privatizations in developing nations have often been tainted by long-standing collusions between big business and government, which led to the consolidation of monopolies rather than the establishment of competitive markets. Although the governance reforms being promoted by multilateral lenders today are designed to prevent such undesirable outcomes, they are far more difficult to define and implement than the liberalizing reforms initiated in the 1980s.

In addition to institutional mismatch, there is also evidence of institutional overshooting – going beyond the American model. Such overshooting can often be traced to the extreme dependence of these governments on the resources of foreign investors and international financial institutions. Portfolio investors are known to conduct speculative attacks against these governments. Because of the perceived uncertainties of investing in emerging markets, portfolio investors hold the governments of developing countries to much higher standards of behavior than those of their developed counterparts. Third World governments must behave as unusually upstanding global citizens, or face the consequences of capital flight, destabilizing currency depreciations, and macroeconomic mayhem. Partly as a result of such pressures, many Third World governments have maintained very high interest rates and fiscal surpluses (a policy that stands in stark contrast, for example, to the policies of Federal Reserve Chairman Alan Greenspan since the mid 1990s), with negative consequences for growth.

Governments may also overshoot because of more direct pressures exerted by multilateral organizations, which condition their loans on policy reforms. At least some of the reforms promoted by these multilateral organizations seem to have surpassed the American model considerably in their degree of market friendliness. To return to the previous example, the IMF generally conditions its bailout funds on fiscal and monetary targets that are, by US standards, extraordinarily strict. To take another example, the World Bank, IMF, and US Agency for International Development (USAID) have been promoting the replacement of publicly funded social security systems with private, individual accounts; social security systems have been privatized throughout Latin America and the formerly communist world. In the United States, however, the privatization of social security has (as this article goes to press) been too politically controversial to implement, despite the best efforts of the current administration to rally support. The World Bank and the IMF have also promoted the implementation of "user fees" on primary education, thus interfering with the ability of poor families to send their children to school. One explanation for such overshooting is that multilateral (and certainly bilateral) organizations do not simply function as neutral transmitters of organizational templates, but are also subject to influence by vested economic, political, and organizational interests that influence which kinds of policies get promoted.

In evaluating the institutions being constructed in the new era, it is useful to distinguish between defining institutions of regulatory capitalism, on the one hand, and the institutions for promoting social welfare, on the other (we can think of these as corresponding very roughly to T.H. Marshall's civil citizenship and social citizenship). The enforcement of property rights and contracts, the regulation of monopolies, support for a standing army, etc., are examples of institutions without which competitive markets cannot function. It is mostly the American variety of these institutions that are being transferred to developing countries (even if with unexpected consequences).

Even in the market-friendly United States, however, these are not the only recognized functions of government. For example, with the exception of radical libertarians of the sort that populate the Cato Institute, most Americans consider it legitimate for governments to tax citizens to finance social programs and public education, regulate firms to guarantee worker safety, and protect citizens from environmental degradation. Overall, the institutions of social citizenship have been less consistently supported by multilateral organizations than the institutions of civil citizenship. The WTO has been criticized for failing to develop sanctions for governments that allow child labor and other practices considered abusive by the International Labor Organization. Recently, the World Bank and the IMF have begun to require that their most impoverished borrowers set aside a fixed percentage of their expenditures for "pro-poor" spending. However, because these lenders simultaneously require reductions in government spending, deflationary monetary policy, and the repayment of external debts, the effects of these poverty-reduction strategies may be cancelled out.

Leaving aside the influence of multilateral organizations, the institutions of social citizenship may be eroded simply because developing countries – unlike the United States – are burdened with external debt denominated in foreign currencies. Heavily indebted states have arguably adopted a role that diverges considerably from that adopted by core states: namely, the extraction of domestic resources and their export abroad. In some respects, this role is reminiscent of the colonial dependence of the nineteenth century. Resources spent on servicing debt are obviously resources that are not being spent on such recognized, basic functions of liberal capitalist government as the provision of public health, education, and infrastructure. A recent IMF study finds that external debt has a statistically significant negative impact on governments' ability to fund social programs.

Finally, structural adjustment may erode social citizenship by decreasing the bargaining power of states and citizens vis-à-vis private investors. To the extent that they cut into profits, the construction of social-welfare-governing institutions tends to be resisted by firms. The history of the industrialized democracies suggests that they are constructed in spite of resistance from firms, by states responding to the demands of organized social groups. Because Third World citizens and governments are in a disadvantaged bargaining position with respect to foreign investors, and even more so with respect to multilateral organizations like the IMF, they may be hampered in their ability to construct the institutions of social citizenship that developed countries take for granted. This is the premise underlying the famous "race to the bottom" so often cited by global justice activists: in their view, today governments are competing among themselves to attract foreign investors by providing the lowest taxes and the least stringent labor and environmental regulations (see http://www.aboutglobalization.com). Even standard neoclassical economic models provide some support for this idea.

However, there are at least two versions of the "race to the bottom" hypothesis. One version supposes that globalization subjects the workers and states of all regions – developed and developing alike – to such competition; the result should be institutional convergence of core and periphery toward uniformly low wages, standards, and social protections. Empirical analyses suggest that the overall trend toward reduced taxes on capital and declining unionization in OECD economies can be traced at least in part to economic globalization. There is also evidence that the North American Free Trade Agreement (NAFTA) has contributed to job losses among low-skilled workers in the United States. But although the wages of low-skilled workers and taxes on corporations may be declining in the wealthy North, nobody is yet claiming that social protections and environmental regulations in Germany and the United States are being downsized to resemble their counterparts in Zimbabwe and Bolivia. Most global trade and foreign direct investment occur among wealthy countries, rather than between wealthy and poor nations. Wealth and power continue to have their privileges, although there is no doubt that some of these privileges have been eroded for non-elites in developed countries.

What I examine here is the second version of the "race to the bottom," which focuses on competition among developing countries. In this view, structural adjustment puts developing countries in a particularly poor bargaining position. Heavily indebted, capital-poor countries with high levels of unemployment are desperate for foreign investment. However, in courting investors, they are flocking to a crowded market niche of similarly desperate countries, all selling low-wage, low-skilled work on the global marketplace. To make themselves look more competitive to investors shopping around for the best deal, they may offer lower levels of taxation, regulation, etc. If true, this pattern should lead to a polarization between developed and developing countries, with the latter converging among themselves on uniformly low regulatory standards and levels of social protection.

Anyone who has any experience with the antisweatshop movement on college campuses has seen an array of shocking facts; incredibly low wages, long hours, child labor, employer abuses, and wanton toxic dumping (see http://www.sweatshopwatch.org). However, to prove that there is a race to the bottom, we need evidence that the inhabitants of developing countries are worse than they would have been otherwise. Unfortunately, this presents manifold problems of measurement and controlling for extraneous factors. The removal of trade barriers and the opening to foreign investment occurred as part of a complex amalgam of social changes – external debt, increased pressures from multilateral organizations, privatization, vulnerability to balance-of-payments crises, etc. – that cannot be operationalized into a single variable. There are critical measurement problems with some of the most important elements of structural adjustment; "economic openness" itself is remarkably difficult to measure. Even assuming good measures for the independent variable, it is important to disaggregate the data to distinguish the impact on different social groups; but the demographic and labor market data from many developing countries are incomplete at best. Because disentangling and analyzing these different factors is so difficult, empirical evaluations of the race to the bottom hypothesis tend to be both partial and hotly contested.

Has state capacity to provide social welfare benefits declined? One circumstance that appears to support this idea is the rise of export-processing zones (EPZs) – special manufacturing areas where Third World governments offer investors exemption from taxation and regulation. According to the World Bank, whereas only a few such zones existed in 1970, by 1996 there were over 500 zones in 73 countries. This suggests that Third World governments are competing for foreign direct investment by lessening potentially welfare-enhancing interventions, such as the extraction of fiscal resources. Nevertheless, more optimistic observers would argue that existing taxes and regulations were too onerous to begin with, and that setting up EPZs is a necessary step in fostering economic development, which will ultimately increase human welfare. For reasons discussed in the following section, however, critics could reply that there is little evidence that such economic development is actually occurring.

What impact does structural adjustment have on the environment? There is little controversy over global environmentalists' assertion that external debt contributes to environmental degradation – after all, for a heavily indebted nation, the price of a clean and sustainable environment may be unaffordable. However, other assertions have been hotly contested. For example, the WTO has been accused by activists of systematically undermining national environmental standards by imposing sanctions on governments that try to enforce environmental standards in trade; other observers say these claims are exaggerated. Supporters of current policies suggest that liberalizing reforms generate economic development and that, in turn, such development increases respect for the environment: there is a strong correlation between environmental standards and GDP per capita. Once again, however, this argument rests on the contestable premise that development is occurring in the first place. It also overlooks the fact that not all indices of pollution decline with economic growth. The recent Carnegie Endowment report on the impact of NAFTA finds that it has not been as damaging to the environment as was originally feared, although there have been negative impacts in certain sectors, particularly in rural areas.

Has structural adjustment weakened labor unions in developing countries? In contrast to the literature on union decline in OECD nations, there has been little cross-national comparative research on trends in unions in the developing world. The partial accounts that exist paint an ambiguous picture that neither clearly supports nor refutes a race to the bottom in labor organizing. One national case that supports a pessimistic interpretation is that of Mexico, which lifted trade barriers and invited in foreign investment under the auspices of NAFTA. Since the implementation of NAFTA in 1993, real wages in Mexico have declined significantly, the minimum wage has been held down to foster international competitiveness, and unions have been weakened; in line with the predictions of neoclassical theory, unskilled workers appear to have been hurt the most. But it is not clear that the Mexican case can be generalized to the rest of the developing world. Frundt finds increased rates of unionization in Central America during the period of structural adjustment, although he suggests that the strength of unions may have declined. In a cross-national study, Mosley and Uno find that neither foreign direct investment nor trade openness are significant correlates of labor rights violations, although they do correlate with region and level of development. Murillo and Schrank observe that 13 of the 18 collective labor reforms implemented in Latin America between 1985 and 1998 enhanced rather than limited collective bargaining rights, an outcome they attribute partly to the strategies of traditional labor-backed parties and partly to transnational activism (discussed below).

This section has focused exclusively on the governance of national economies, broadly defined to include social-welfare-enhancing institutions. However, it is worth mentioning briefly another set of institutions that have been transformed in the era of structural adjustment: namely the rules of national politics. Existing literature on the topic of democratic transitions focuses on Latin America – arguably the continent in which the transformation has been most dramatic. Weyland argues that although the rise of market-friendly institutions has made Latin American democracy more sustainable, it has simultaneously limited the quality of this democracy. The end of the cold war and the opening of national economies to international markets led to increased pressures for minimal procedural democracy, both from the US government and from foreign investors in search of stable investment climates. It also weakened leftist parties and other proponents of radical reforms, decreasing elite groups' perception that dictatorship was the only solution. The net result has been that social groups and political parties are more likely to agree on the means (democratic elections), even if they disagree with the ends. However, Weyland

also points out that the changes associated with structural adjustment have also put severe constraints on the quality of democracy. Economic constraints and the threat of capital flight limit the latitude of possible policies. Such restrictions on policies have led to weakened political parties and depressed participation – eerily echoing the apathy of the US electorate. The accountability of elected leaders to their constituents has also declined.

Ultimately, what can we conclude about structural adjustment and institutional convergence? At the risk of sounding excessively conciliatory, I suggest that the available evidence echoes aspects of both modernization and dependency theories. On the one hand, institutions still work quite differently in the global South. States continue to service large and unsustainable debts; their policies must respond to the leverage of multilateral institutions and the need to maintain investor confidence. Now more than ever, dependency matters: there are fundamental differences between the roles of states in developed and developing countries that can be traced to large differences in bargaining power. On the other hand, we must concede that developing countries have adopted a model of governance that resembles, in its most general outlines, the sort of capitalism that is practiced in the United States. Whether this appears to have contributed to the further modernization of national societies is explored in the following section.

The Transformation of Class Structures

The two most hotly debated issues in the literature on liberalizing reforms are (a) whether they have promoted economic development, and (b) whether they have promoted equality. This section attempts to sort through some of the literature on changing national and global class structures.

Although there is not enough space in this review to address debates about economic development in more than a superficial way, we should briefly review some evidence on this point: economic growth, after all, has consequences for global social structure. The ostensible reason for implementing free-market reforms was that they would generate growth, development, and a convergence of the incomes of developed and developing countries. Twenty years later, the evidence in favor of these initial claims has been disappointing. For example, from 1960 to 1980, output per person grew 75% in Latin America and 36% in sub-Saharan Africa; in contrast, between 1980 and 2000, it grew by only 6% in Latin America and actually fell by 15% in sub-Saharan Africa.

These data, however, do not necessarily lead to the conclusion that market liberalization is bad for economic development. One counterargument is that national incomes have been dragged down by large external debts, which are the fault of governments, not market opening. Another counterargument is that market openings have not been carried far enough – if governments could remove remaining barriers to the functioning of markets, then there would be a more impressive rise in national incomes. A related argument is that development takes time, and that developing countries need to wait for the new model to bear fruit. Finally, the model's supporters point out that some countries have been doing very well: India and China, in particular, have been growing very rapidly.

What do the macroeconomic and demographic data tell us about trends in overall global inequality? First, it is important to distinguish between inequality within countries and inequality between countries. The question of inequalities between countries – whether countries like Mexico and India are catching up to countries like the United States and Japan – is quite controversial. Although some observers argue that inequalities across nations have declined, others have come to the opposite conclusion. To make sense of this apparent contradiction, Wade shows that the answer depends on how researchers measure and compare national wealth. One method is to compare the raw figures on national GDP converted into US dollar amounts, and compare across nations. According to these numbers, there is a clear pattern of rising inequality: some countries have been getting a lot wealthier, and others have been left behind. However, those claiming a convergence in national incomes use numbers that differ in two respects. First, they use numbers that are weighted by population: thus, the two largest developing countries (India and China) have an enormous impact on the final figures. Second, they use numbers that have been adjusted for purchasing-power parity (PPP), to control for the fact that a dollar in India, for example, will buy a great deal more than a dollar in the United States. The PPP-adjusted national GDPs, weighted for population, show a pattern of rising equality – but this effect disappears when India and China are subtracted from the calculations.

Thus, claims to rising equality across nations are based on the indisputable fact that India and China have been growing at a tremendous pace over the past two decades. What is extremely disputable, however, is whether this economic growth – and hence income convergence – can be attributed to structural adjustment. Neither India nor China is a particularly good representative of free market orthodoxy. Although it has used trade and foreign investment to its advantage, China continues to have an enormous state-owned sector and an inconvertible national currency. India's growth spurt began a decade before it began to implement liberalizing reforms, and protectionist tariffs actually increased during this first phase of growth. Meanwhile, Latin American economies in which market reforms have been implemented in a more orthodox manner have mostly suffered from stagnant levels of economic growth.

The data on global poverty have also generated a lively controversy. Basing their claims on in-depth knowledge of national case studies, a number of critics of structural adjustment have asserted that it has been pushing citizens of developing countries beneath the poverty line. But in 2002, World Bank Managing Director James Wolfensohn famously declared that the number of people living on less than $1.00 a day had fallen by 200 million. Does this mean that the global war on poverty is being won? Wade shows that in addition to a number of more minor problems, there is a fundamental methodological error in this claim: it compares figures from 1980 and 1998 that are not comparable because of a significant change in the World Bank's methodology for calculating the poverty line. An alternative is to look at demographic numbers on poverty, such as life expectancy at birth. Life expectancies at birth have increased among poor countries since the 1980s. However, during the 1980–98 period, the progress of poor countries in catching up to the life expectancy of wealthy ones slowed considerably compared with the previous 20 years.

The question of inequality within countries is less controversial than the question of between-country inequality, or the question of poverty; even optimistic observers, such as Firebaugh, concede that within national boundaries, income inequality has been increasing. To illuminate how these trends have played out in developing countries, there is a large and growing body of national case studies focusing on various indicators of social well-being and inequality. Two particularly useful studies are Portes and Hoffman's study of changing Latin American class structures and the recent Carnegie Endowment report on the impact of NAFTA on Mexico a decade after its ratification. Whereas the Portes and Hoffman study has the virtue of considering an entire continent's experience through the lens of a range of indicators of inequality and social welfare, the Carnegie study provides a detailed, in-depth account of the complexities of a single nation's experience with opening its economy to its wealthier and more powerful northern neighbors.

Both studies paint sobering portraits of the impact of structural adjustment on national class structures. During the 1980s and 1990s there was an increase in income inequality in Latin America, with a consistent concentration of wealth in the top decile of the population. Such income polarization has been particularly notable in Mexico. Meanwhile, the percentage of Mexicans beneath the poverty line is still greater than it was in the late 1970s, and real wages have actually declined.

Of course, the causes of these phenomena are complex, and we should not be too quick to jump to conclusions: the debt crisis in the 1980s and the peso devaluation in the 1990s played important parts in these trends, and it is not easy to disentangle these factors from market liberalization. Such ambiguities notwithstanding, the Mexican experience under NAFTA helps highlight some important processes that are contributing to qualitative changes in national class structures across the developing world. One such process is the movement of rural populations away from their native towns to urban centers or to places where they take jobs as low-wage agricultural workers. The mass movement off the land is part of a longer-term trend that predates the structural adjustment era by a many decades. However, structural adjustment has accelerated this trend by making traditional and small-scale agriculture even less viable. Under the new regime, small-scale farmers in the developing world receive fewer subsidies, face higher interest rates, and face competition with heavily subsidized and well-capitalized foreign agribusiness. Mexican government authorities estimate

a loss of 1.3 million jobs in the agricultural sector between 1993 and 2002.

This process of de-agriculturalization is only one of many simultaneous pressures on labor markets that may arise in the era of structural adjustment. A second source of pressure is the privatization of state-owned firms, which often leads to downsizing worker layoffs. Over the past two decades, there has been a significant contraction in formal sector employment in developing countries and a corresponding move toward employment in the informal economy. In other words, the labor force has come to be characterized less by employees and more by independent agents – from small business owners to ambulant chewing-gum sellers to garment pieceworkers. Although the rise of the informal economy is lauded by some observers as a necessary escape valve from cumbersome taxation and government regulations, other observers point out that it involves replacing stable, state-regulated jobs with a form of employment that tends to be precarious, poorly paid, and less productive.

A third source of pressure on labor markets in LDCs is the restrictive monetary policy that has become the norm under the structural adjustment regime. To foster the confidence of foreign investors and continue to receive financing from multilateral organizations (particularly the IMF), governments have prioritized the fight against inflation, often changing central bank legislation to take monetary policy out of the hands of the executive. However, reducing inflation to the levels preferred by the international financial community requires high interest rates – and high interest rates decrease domestic investment and increase unemployment.

Finally, there is evidence that labor markets are being strained by the bankruptcy of domestic firms that cannot compete with the flood of cheap imports from more open trade. Just as this job loss contributes to the informalization of the labor force, so it may be contributing to a restructuring of local bourgeoisies. A study by Silva on the fate of business during Chile's early experiment with liberalizing reforms under the Chicago Boys suggests that large, export-oriented businesses with access to international capital markets may be the hardiest, and that market concentration may result. Although there is evidence from various countries that smaller and domestic-oriented entrepreneurs may "wither away" in the face of foreign competition, there have also been unexpected adaptations to new conditions. Schrank documents the rise of a new class of indigenous investors in the EPZs in the Dominican Republic who have been able to profit from their combination of local connections and access to foreign capital. However, such firms have also suffered from high rates of bankruptcy, suggesting that we should not be too optimistic in our conclusions.

Although few observers are likely to shed sympathetic tears for the declining fortunes of formerly privileged industrialists, the fate of masses of unemployed workers and displaced peasants is cause for concern. In theory, foreign investment is supposed to compensate for labor shedding in inefficient sectors by creating jobs in more efficient, productive firms. Throughout the developing world, there is strong evidence that foreign-owned firms are indeed more efficient and productive than the domestic firms that they are replacing. But more productive plants have often translated into fewer rather than more jobs. Meanwhile, jobs created in EPZs may be vulnerable to capital flight to other low-wage regions. From 1994 to 2001, foreign direct investment from the United States to Mexico increased from about $5 billion per year to $16 billion per year. But most of the jobs created under NAFTA in the 1990s were in maquiladoras (EPZs), and about 30% of these jobs subsequently disappeared – many relocated to countries such as China where wages are even lower. Because foreign investment has not effectively compensated for the jobs lost through structural adjustment, many developing countries continue to be plagued with unemployment and poverty-level wages.

Although a number of studies suggest that structural adjustment has increased class inequality in many countries, the emerging evidence on gender inequality is more complex and ambiguous. In many places, structural adjustment has undermined traditional gendered divisions of labor, both by providing new opportunities for women to work for wages outside the home (e.g., in EPZs), and by contributing to male unemployment. However, whether this has led to a general empowerment of women with respect to men is a much more complicated question. Answering this question requires taking a number of other factors into account, such as the position women in developing countries adopt in the labor market. For example, they may come to rely on precarious and poorly paid work in the informal economy, keeping them dependent on male incomes. Gender roles may be slow to adapt to changing conditions (as "second shifters" in the United States know all too well), and multinational firms may actually encourage the reproduction of traditional roles. Furthermore, other circumstances related to structural adjustment, such as external debt and reduced government budgets, may undermine the position of women by eliminating resources such as access to education and healthcare. Thus, the impact of structural adjustment on gender inequalities is an area ripe for further research.

Overall, the consequences of structural adjustment for national and global class structures seem more suggestive of dependency than modernization. Under other circumstances, growing income inequality might be seen as compatible with the "Kuznets curve," in which rapid economic growth benefits upper- more than lower-strata groups; a rising tide may lift all boats, but in the early stages of development it may lift some boats more than others. But for the majority of developing countries in the past two decades, the tide has not risen at all, or only barely. Third World societies have undergone major transformations that are supposed to be the hallmarks of modernizing societies – mass movement off the land, urbanization, and industrialization. And yet, these transformations have not been consistently associated with economic growth and declining inequality across nations. This is precisely the sort of contradiction that interested dependency theorists – the emergence of social structures reminiscent of the core in some respects, but with very different underpinnings and consequences.

The Rise of Transnational Networks

Their numerous disagreements notwithstanding, a feature shared by modernization and dependency theorists alike was an emphasis on the nation-state as the unit of analysis. Both types of theorists focused on issues of national development and nation-level social transformations. The era of structural adjustment, however, has cast fundamental doubts on the utility of these postwar conceptual categories by contributing to the rise of social networks that span national borders. This section examines literature documenting transnationalism in three areas: business, labor markets, and policy.

Where business is concerned, it is not immediately apparent why transnationalism represents anything new – after all, during the 1970s, large foreign multinationals set up local branches in developing countries. What is new about current trends, however, is the spread of an organizational form characterized by networks rather than hierarchies. Today, global production increasingly relies on subcontractors and sub-subcontractors outside the scope of any single firm or nation. The computer one purchases at Best Buy, for example, contains components made and assembled by the workers of different firms in various different nations.

It is common to attribute this new production system to advances in technology. Advances in communications (e.g., the Internet) and transportation make it far easier for firms to subcontract to suppliers in faraway countries and to continually shop around for the suppliers that offer the most attractive prices. But the role of structural adjustment in creating these conditions should not be underestimated. Liberalizing reforms, combined with the setting up of EPZs in which regulations are reduced even further, facilitate global production networks by eliminating the friction of tariffs, taxes, complicated labor laws, and red tape. The global production networks that result are with "just-in-time" production: retailers order items from their suppliers (and the suppliers from their suppliers, and so on) as they are needed, rather than keeping large inventories in stock.

Among the virtues of this new system is that it is leaner: it eliminates bureaucratic inefficiencies and puts a premium on getting products to consumers quickly and at the lowest possible price. However, critics of the system point out that it is also meaner. In the garment and other industries, this system has been associated with an increased reliance on offshore sweatshop production. Whereas bureaucratic firms can be publicly criticized and sanctioned for unethical practices, holding them accountable for the practices of their suppliers, sub-suppliers, and so on down the food chain is much more difficult. Defenders of economic globalization often point out that the affiliates of foreign firms pay on average one third more than the prevailing national wages. But the pants one buys at Target are not produced by the Docker corporation; the company that

puts its label on a particular pair of pants may not provide – or even possess – information concerning the conditions under which it was produced.

However, firms are not alone in using networks that span national borders. A number of scholars have identified a trend toward "globalization from below" through the establishment of transnational migrant networks. Structural adjustment fosters the development of these communities at multiple levels. Most obviously, for the reasons enumerated above, structural adjustment puts pressure on national labor markets, leading to economic incentives to out-migrate. Meanwhile, foreign direct investment incorporates traditional segments of the population into the paid labor force and contributes to the Westernization of local cultures, making populations ripe for migration. High levels of foreign debt contribute to high interest rates in developing countries, which in turn cause their residents to work abroad to save up capital to invest in homes or small businesses. Under NAFTA, the illegal immigration of Mexicans to the United States has increased significantly, despite increases in border control. A recent United Nations report finds a 14% increase in the total world stock of migrants between 1990 and 2000 alone.

Transnational migration theorists suggest that new patterns of immigration differ from older waves in that they are not necessarily characterized by assimilation and permanent settlement. Many immigrants maintain strong social ties back home and travel back and forth between countries on a regular basis; others leave spouses and children behind in the expectation that they will return when enough money has been saved. Most recently, a more privileged class of transmigrant has emerged: well-paid, high-tech workers, often from India, with ties to both their receiving country and their country of origin. The experiences of transnational elites, which are obviously very different from those of the typical illegal Mexican factory worker, represent an underexplored and fascinating area of investigation.

One striking new trend linked to the rise of transnational migrant networks is the growing importance of remittances – cash sent home to the country of origin – in the economies of developing countries. According to a recent World Bank report, in 2001 the official total of remittances to developing economies was more than $70 billion, and contributed more than 10% to the GDP of nations that included Jordan, Lesotho, Albania,

Nicaragua, El Salvador, Cape Verde, and Jamaica. Although the World Bank tends to emphasize the beneficial effects of remittances for economic development, qualitative studies of the transmigrant experience emphasize the high human cost incurred by the people who work far from family and community, for low wages, and often without legal rights or protections.

Finally, structural adjustment has been met with a new kind of resistance that also relies on border-spanning social ties. Peter Evans identifies three kinds of transnational ties contributing to what he terms "counter-hegemonic globalization." First, there are transnational advocacy networks: globalization has created political openings that allow cross-border activists to leverage changes in state policies. For example, the Jubilee movement has drawn world attention to the issue of Third World debt and was arguably an important factor in pushing forward the Heavily Indebted Poor Countries initiative endorsed by the World Bank and IMF. Second, workers have strengthened contacts with their allies across borders to help compensate for the lack of bargaining power of workers faced with highly mobile capital. Third, there has been a proliferation of consumer-labor networks designed to help compensate for Third World states' inability or unwillingness to enforce fair labor practices, of the sort exemplified by the campus antisweatshop movement.

Final Thoughts

The era of structural adjustment has been associated with a number of fundamental and seemingly irrevers-

ible social transformations. Some of these changes, such as the rise of global networks, seem to have made the old modernization–dependency debates irrelevant. Others, such as the adoption of US-style patterns of economic governance around the world and the heightened salience of core pressures for policies in the periphery, echo the debates of the 1970s in ways that are interesting and potentially illuminating.

Over the past half-dozen years or so, there have been some signs that the intellectual and political underpinnings of the current order are being eroded, including a resurgence of Third World nationalist rhetoric, international social forums, and the rise and persistence of protests against multilateral organizations. Perhaps most interestingly, a number of prominent economists have begun to critique some of the fundamental tenets of the reigning model. However, although these trends have created space for debate, they have thus far coalesced into neither a school of thought nor a coherent set of policy alternatives.

This seems like a propitious time for sociologists to situate themselves within debates about what has happened, what went wrong, and what is to be done. Sociology lost considerable ground during the era of structural adjustment, which gave economists greater disciplinary dominance over discussions of the problems of poor countries. Consequently, many of the broader sociological, historical, and philosophical questions about the nature of modernity were thrust to the margins, as debates came to revolve around rational actors rather than the forces of history. A return to the big questions might be precisely what is needed to build a paradigmatic challenge, and a new terrain for debate.

 READING 17

The Human Rights Effects of World Bank Structural Adjustment, 1981–2000

M. Rodwan Abouharb and David L. Cingranelli

World Bank and International Monetary Fund (IMF) structural adjustment conditions require loan recipient governments to rapidly liberalize their economies.

According to previous research, these economic changes often cause at least short-term hardships for the poorest people in less developed countries. The

Bank and IMF justify the loan conditions as necessary stimuli for economic development. However, research has shown that implementation of structural adjustment conditions actually has a negative effect on economic growth. While there has been less research on the human rights effects of structural adjustment conditions, most studies agree that the imposition of structural adjustment agreements (SAAs) on less developed countries worsens government human rights practices. This study focuses on the effects of structural adjustment conditions on the extent to which governments protect their citizens from extra-judicial killing, torture, disappearances, and political imprisonment.

The results of this study suggest that existing theories of repression should be revised to take greater account of transnational causal forces. Previous studies examining variations in the human rights practices of governments have concentrated almost exclusively on state-level characteristics such as wealth, constitutional provisions, or level of democracy. The dominant theoretical framework underlying this research argues that, other things being equal, "repression will increase as regimes are faced with a domestic threat in the form of civil war or when a country is involved in international war". Other international factors besides involvement in international war such as the degree of integration into the global economy, sensitivity to international norms, and involvement with international financial institutions have received much less attention.

Empirically, this study advances our understanding of the human rights consequences of structural adjustment by correcting for the effects of selection. It is possible that the worsened human rights practices observed and reported in previous studies might have resulted from the poor economic conditions that led to the imposition of the structural adjustment conditions rather than the implementation of the structural adjustment conditions themselves. In other words, the human rights practices of loan recipient governments might have gotten worse whether or not a structural adjustment agreement (SAA) had been received and implemented. In addition, as our results will show, some of the factors that increase the probability of entering into a SAA, such as having a large population and being relatively poor, are also associated with an increased probability of human rights violations. For these reasons one must disentangle the effects of selection before estimating the human rights impacts of structural adjustment loans. In order to control for the effects of selection, a two-stage analysis was undertaken. In the first stage of the analysis, the factors affecting World Bank decisions concerning which governments receive SAAs were identified. In the second stage the impacts of entering into and implementing SAAs on government respect for human rights were examined.

The first-stage results demonstrate that the Bank does give SAAs to governments that are poor and experiencing economic trouble, but the Bank also employs a wide variety of non-economic loan selection criteria. The non-economic selection criteria examined in the first stage of the analysis build upon and extend selection models developed in previous research on the economic effects of structural adjustment. This research project is the first to demonstrate that the Bank prefers to give loans to governments that provide greater protection for worker rights and physical integrity rights of their citizens. Earlier research had shown that democracies were at a disadvantage when negotiating a SAA from the IMF, a finding consistent with expectations generated by Putnam's theory of two-level games. Our findings provide evidence that democracies also are at a disadvantage when negotiating with the World Bank.

After controlling for selection effects and other explanations of respect for physical integrity rights, the findings of the second-stage analysis show that the net effect of World Bank SAAs is to worsen government respect for physical integrity rights. Torture, political imprisonment, extra-judicial killing, and disappearances were all more likely to occur when a structural adjustment loan had been received and implemented. Governments that entered into SAAs with the World Bank actually improved their protection of physical integrity rights in the year the loan was received. Governments then reduced the level of respect for the physical integrity rights of their citizens during the years when structural adjustment conditions were imposed. This combination of findings suggests that governments seeking loans from the World Bank initially improved their human rights practices, possibly to impress Bank officials. However, the austerity measures required by the implementation of structural adjustment conditions led to a subsequent worsening of human rights practices by governments in loan recipient countries.

The theoretical argument is that there are both direct and indirect negative effects of the implementation of structural adjustment conditions on government respect for physical integrity rights. Structural adjustment conditions almost always cause hardships for the poorest people in a society, because they necessitate some combination of reductions in public employment, elimination of price subsidies for essential commodities or services, and cuts in expenditures for health, education and welfare programs. These hardships often cause increased levels of domestic conflict that present substantial challenges to government leaders. Some governments respond to these challenges by becoming less democratic as in the case of Peru under President Fujimori in the 1980s. The results presented here, like those of numerous other studies, have shown that increased domestic conflict and decreased democracy are associated with higher levels of repression. The case of Venezuela provides an illustration of the role of structural adjustment in producing increased domestic conflict, a weakened democratic system and repression. As Di John writes:

> A few weeks after the announcement of [structural adjustment] reforms, Venezuela experienced the bloodiest urban riots since the urban guerrilla warfare of the 1960s. The riots, known as the "Caracazo," occurred in late February 1989. A doubling of gasoline prices, which were passed on by private bus companies, induced the outburst [...] The riots that ensued were contained by a relatively undisciplined military response that left more than 350 dead in two days.

Although Venezuela's democratic system has been maintained, over the period of this study, dissatisfaction with economic policies has played a part in three attempted coups, multiple general strikes, two presidential assassination attempts, and has led to several states of emergency being imposed. Even today, debate over structural adjustment policies in Venezuela remains heated. President Hugo Chavez sustains his popularity largely based on his opposition to the kind of unregulated economic liberalization advocated by the IMF and the Bank.

The findings presented here have important policy implications. There is mounting evidence that national economies grow fastest when basic human rights are respected. SAAs place too much emphasis on instituting a freer market and too little emphasis on allowing the other human freedoms necessary for rapid economic growth to take root and grow. By undermining the human rights conditions necessary for economic development, the Bank is damaging its own mission.

Background

While each structural adjustment program is negotiated by representatives of the Bank and representatives of the potential loan recipient country, common provisions include privatization of the economy, maintaining a low rate of inflation and price stability, shrinking the size of its state bureaucracy, maintaining as close to a balanced budget as possible, eliminating and lowering tariffs on imported goods, getting rid of quotas and domestic monopolies, increasing exports, privatizing state-owned industries and utilities, deregulating capital markets, making its currency convertible, and opening its industries and stock and bond markets to direct foreign ownership and investment. Good governance emphases of the Bank include eliminating government corruption, subsidies, and kickbacks as much as possible, and encouraging greater government protections of human rights including some worker rights.

Most of the previous research has examined the IMF and its impacts, neglecting the role of the World Bank in promoting structural adjustment. Both are important actors, over the period examined in this study, the World Bank entered into 442 SAAs, while the IMF made 414. The remainder of the article briefly reviews previous work on the economic effects of structural adjustment, elaborates on the theory briefly outlined above, discusses the earlier research estimating the impact of structural adjustment on human rights; elaborates upon the need for a selection model, presents some specific hypotheses, and provides evidence supporting those hypotheses. Finally, the theoretical, methodological, and policy implications of these results are discussed.

The Economic Effects of Structural Adjustment

The purpose of structural adjustment programs is to encourage economic growth. According to neoliberal economic theory, structural adjustment programs reduce the size and role of government in the economy. A minimalist state produces and encourages economic growth, which promotes economic and social development. Limited government empowers individuals by giving them more personal freedom, making it more likely that all individuals will realize their potential. The ability to realize one's potential, according to this line of reasoning, leads to individual responsibility and self-reliance. Limited government maximizes individual opportunities, limits the opportunity for corruption and releases talented people into the more efficient private sector.

Many scholars have examined the link between structural adjustment policies and economic growth and the weight of the evidence so far is that structural adjustment is not effective. According to critics, the Fund and Bank use a conception of development that is too focused on economic growth, have misdiagnosed the obstacles to development in less developed countries, have failed to appreciate the value of government interventions into the private economy, and have insisted that structural adjustment reforms be implemented too quickly. It is possible that developing countries like China have been more successful, both in terms of aggregate economic growth and poverty reduction, because they have avoided SAAs from the IMF and World Bank. Unlike Russia, which has received a number of SAAs, China has avoided a rapid increase in economic inequality.

Theory: the Human Rights Effects of Structural Adjustment

Direct effects

Figure 1 depicts the main causal arguments of the conventional neoliberal and more critical views of the direct and indirect effects of structural adjustment on the human rights practices of governments. The direct effects may be theorized as positive or negative. The "positive" argument (linkage "a") is that a relatively limited government as required by SAAs is fundamental to all human freedoms. Limited government reduces barriers to the functioning of the free market, allowing people to enhance their opportunities and better pursue their own interests that are likely to be lost if human freedom is restricted. Consistent with this line of thought, Cranston has argued that respect for most human rights, including physical integrity rights (such as the right *not to be tortured*) only requires forbearance on the part of the state.

However, as linkage "h" of Figure 2 indicates, structural adjustment programs also may have the direct effect of worsening government human rights practices, because a substantial involvement of government in the economy is essential for the protection of all human rights. The historical record demonstrates, for example, that a reduced role of the state in capitalist economies has led to less protection of some human rights such as worker rights. From a principal–agent theoretical perspective, reducing the size of government also reduces the ability of principals (government leaders) to constrain the discretion of agents (police and soldiers). More administrative discretion is likely to lead to greater abuse of physical integrity rights. Also, in practice, the acceptance of structural adjustment conditions by the governments of less developed countries causes the adoption of new policies and practices. These new policies are designed to produce substantial behavioral changes in the affected populations. Evidence from literature about human learning suggests that people have a natural tendency to resist making substantial changes in their previous behavior. One of the tools government may use to overcome such resistance is coercion.

The idea that liberalization and economic development may conflict with respect for some human rights is an enduring theme in the debate over development policy and an implicit element of structural adjustment packages. Loan recipient governments are expected to reduce their efforts to protect the social and economic rights of their citizens in a variety of areas such as housing, health care, education, and jobs at least in the short run, with the expectation that they will be able to make much larger efforts toward these ends later. Civil and political liberties may have to be curtailed in order to ease the implementation of loan conditions.

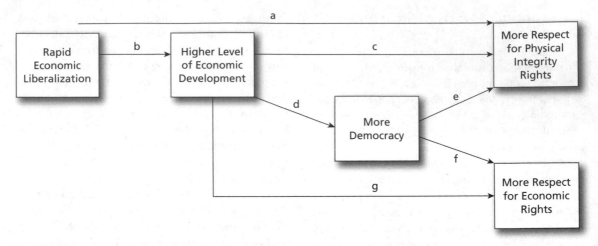

Figure 1 Structural adjustment and human rights: the neoliberal perspective

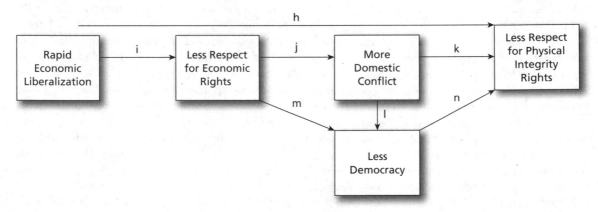

Figure 2 Structural adjustment and human rights: the critical perspective

People opposed to the policies of structural adjustment such as members of the press, trade unionists, leaders of opposition parties, clergy, social activists, and intellectuals may then be subjected to abuse of their physical integrity rights.

Indirect effects

Figure 1 also depicts the expected indirect effects of structural adjustment on the human rights practices of loan recipient governments. As noted, neoliberal economic theory suggests that structural adjustment will promote economic development (linkage "b" in Figure 1). Many previous studies have shown that

wealthier states have provided greater levels of respect for a wide variety of human rights including physical integrity rights (linkage "c"). Thus, if the imposition of a SAA increases the level of wealth in a less developed country, then the indirect effect of SAA implementation should be an improvement in the human rights practices of governments.

Despite findings showing that structural adjustment has not led to faster economic growth, the empirical debate over linkage "b" will continue. Thus, it is still important to understand the remainder of the neoliberal argument. As is indicated by linkages "d" and "e" in Figure 1, previous research has shown that wealthier states are more likely to be democratic, and relatively high levels of democracy are associated with a higher level of respect for most human rights including physical integrity rights. Therefore, if the imposition of a SAA promotes higher levels of democratic development through increased wealth, then an indirect consequence of SAA implementation should be an improvement in human rights practices.

Neoliberal defenders of the effects of SAAs on government respect for economic human rights have argued that higher levels of economic development caused by the implementation of a SAA will lead to improvements in government respect for economic rights (linkage "g") through what is now commonly referred to as the "trickle down" effect. That is, wealth will accumulate faster under a structural adjustment program, and, once accumulated, will trickle down to help the less fortunate in society. A number of studies have shown that the level of economic development has a strong, positive impact on basic human needs fulfillment. Moreover, as indicated by linkage "f," previous research has shown that democratic governments have been shown to make greater efforts to provide for the economic human rights of their citizens.

Unfortunately, all of indirect neoliberal arguments linking SAAs to better human rights practices depend upon supporting evidence for linkage "b" in Figure 1. Without linkage "b" all of the other indirect causal chains from rapid economic liberalization to better human rights practices by governments are broken. At an earlier point in time, one might have argued that it was too soon to conclude that there was no evidence that the implementation of SAAs led to the accumulation of more wealth by loan recipients, but SAAs were initiated by the World Bank in 1980 and the IMF has had conditionality associated with its loans as far back as 1952. If SAAs have had a stimulative effect on economic development, it should be observable by now.

The indirect effects posited by the critical perspective are summarized in Figure 2. There is a large body of research showing that implementation of a SAA has negative effects on government respect for economic human rights (linkage "i"). Rapid economic liberalization, according to many observers, forces loan recipient states to reduce or even stop making efforts to help their citizens enjoy internationally recognized rights to health care, education, food, decent work and shelter, because structural adjustment conditions almost always require reductions on government spending for social programs. Some studies have emphasized the disproportionate negative economic human rights consequences for women, for public sector employees and low-wage workers. The poor and those in the public sector have seen their wages fall in real terms, while at the same time they have faced increased living costs because of the removal of price controls and subsidies for essential commodities. The implementation of SAAs also has worsened the relative position of the worst off by increasing income inequality.

Less attention has been given to the relationships explicitly linking the implementation of SAAs to subsequent government respect for physical integrity rights. As shown in Figure 2, there are three indirect causal paths that should be considered (linkages "j–k," "j–l–n," and "m–n"). All lead to less respect for physical integrity rights, and all depend upon empirical support for linkage "i," which is plentiful. One line of thinking is that, by causing loan recipients to reduce their respect for the economic human rights of their most vulnerable citizens, externally "imposed" rapid economic liberalization of the type required by a SAA promotes domestic conflict (linkage "j"), which, in turn, leads loan recipient governments to become more repressive (linkage "k"). Acceptance of SAA conditions requires that decision makers in loan recipient countries enact unpopular policies. These policies cause hardships, especially among the poorest citizens, who are most dependent upon social programs. Citizens, often led by organized labor, protest against reductions in social welfare programs and public employment, commonly required in SAAs. Sometimes the protests become violent. The adjustment process also has intensified regional and ethnic conflicts as groups compete for a "dwindling share of the national cake". Increased repression (linkage "k") by the recipient government is one tool by which it can deal with violent protest. However, it is important to distinguish incremental economic liberalization that results from a societal choice without undue external interference and pressure from the kind of rapid economic liberalization required by SAA conditionality. Economic liberalization that is not required by the conditions found within a SAA may not affect or may actually reduce domestic conflict in societies. For example, Hegre, Gissinger, and Gleditsch examine the impact of economic liberalization and find no discernible impact on the probability of civil conflict.

Other critics of structural adjustment would like the Bank and Fund to give greater attention to the impacts of SAAs on issues such as democratic development. Increased domestic conflict caused by the implementation of SAAs presents serious challenges to democratic systems (linkage "l"). Also, as indicated by linkage "m," requiring democracies to enact unpopular policies, the Bank and Fund may be undermining democratic systems. The positive relationship between a state's level of democracy and its respect for all types of human rights (linkage "n"), as noted above, is well established in the literature. Thus, any policy that undermines democracy, undermines government respect for human rights.

Previous Research Linking Structural Adjustment to Human Rights Practices

The results of previous research explicitly focusing on the effects of SAAs on government respect for physical integrity rights are consistent with the expectations of the critical perspective. Camp Keith and Poe evaluated the human rights effects of getting a SAA from the IMF by comparing the human rights practices of governments with and without such loans while controlling for other factors reliably associated with good or bad human rights practices by governments. They focused on a global sample of countries between 1981 and 1987, and found some evidence indicating an increase in the level of repression of physical integrity rights during the implementation of a SAA. Using a cross-sectional analysis, Franklin also found some support for the argument that governments implementing IMF agreements were likely to become more repressive.

Furthermore, Camp Keith and Poe hypothesized that the very act of negotiating or entering into a loan with the IMF would have a temporary negative impact on the human rights practices of loan recipients. They were not clear about the rationale for this hypothesis, and their findings provided no statistically significant evidence for a "negotiations effect." Others have argued that the involvement of international actors has a moderating effect on domestic conflicts, which

should have the effect of improving government respect for physical integrity rights. There also is a specific reason to expect that negotiating a SAA from the World Bank would have at least a temporary positive impact on the human rights practices of loan recipient governments. The US International Financial Assistance Act in 1977 requires US government representatives on the decision making boards of the World Bank and IMF to use their voices and votes to advance the cause of human rights in loan recipient countries. The size of US contributions to the Bank gives it a strong voice in loan negotiations. Thus, one would expect the World Bank to make SAAs with countries that have good human rights practices.

Previous research has examined the effects of structural adjustment on the overall level of government respect for physical integrity rights but has not disaggregated the effects on torture, political imprisonment, extra-judicial killing, and disappearances. However, it is likely that the impacts of negotiating and implementing a structural adjustment program affect government respect for these kinds of physical integrity rights in different ways. In this early stage of the research program designed to develop theories explaining the human rights practices of governments, aggregate measures may mask theoretically important variations in how governments respect the human rights of their citizens. Disaggregating the measures of respect for physical integrity rights allows the investigation of whether governments improve or decrease their respect for different types of physical integrity rights to the same extent as a result of making and implementing a SAA from the World Bank.

Existing theories explaining why governments resort to violent forms of political repression conceive of repression as the result of conscious choices by rational, utility maximizing political leaders. Both the domestic and international costs and benefits of violating different types of physical integrity rights vary. Torture and political imprisonment are the most common forms of physical integrity rights abuse by governments. If government decision makers are rational, then policies allowing for the practice of torture and political imprisonment must offer higher net benefits than policies allowing the police or military to make citizens disappear or to kill them without a judicial process. If repression is a rational response to structural adjustment, then torture and political imprisonment should increase the most during the implementation of structural adjustment conditions. Since the end of the Cold War, however, there has been an increase in average worldwide government respect for the right against political imprisonment. This trend indicates that, over time, either the costs associated with this form of repression have increased, the benefits have declined or both.

[...]

Discussion

The most important substantive finding of this study is that receiving and implementing a SAA from the World Bank had the net effect of worsening government respect for all types of physical integrity rights. This finding is generally consistent with the findings of previous comparative and case study research on the human rights effects of IMF SAAs. It supports one of the main hypotheses in our research – that there would be a higher probability of physical integrity rights violations during the years a SAA was implemented. It is stronger, but generally supportive of the finding reported by Camp Keith and Poe regarding the effects of IMF structural adjustment conditions. The direction of our findings for political imprisonment were consistent with this hypothesis but were only statistically significant at the 0.11 level of confidence. It was hypothesized that the practices of torture and political imprisonment would be most affected by entering into and implementing SAAs. While the results did not provide strong support for this "differential effects hypothesis," the variation in the effects of SAAs across the four dependent variables examined did illustrate the usefulness of using disaggregated measures of physical integrity rights violations as advocated by McCormick and Mitchell. Consistent with Putnam, the findings also indicated that democratic governments had a disadvantage in negotiating SAAs with the Bank.

These findings concerning the effects of World Bank structural adjustment conditions on the human rights practices of loan recipients, with small differences, also pertain to the effects of negotiating and implementing a SAA with the IMF. In separate tests we have

examined the impact of IMF conditionality and the joint effects of structural adjustment loans by the IMF and/or the World Bank. No matter how the structural adjustment intervention is operationalized, the net effects on government human rights practices are found to be negative. We do not present all of those results in this paper mainly because of space limitations. However, there is also a void in the literature concerning the World Bank. While there have been numerous studies of the economic impacts of SAAs issued by the IMF, and Camp Keith and Poe and Franklin have conducted research on the human rights impacts of the IMF, there has been no previous global, comparative, cross-national research on the economic and human rights impacts of SAAs issued by the World Bank. As the number of SAAs issued by the World Bank and the IMF has been about the same over the period of this study, both international financial institutions have been about equally important in promulgating structural adjustment reforms. This paper, by focusing on the World Bank, begins to redress an unjustified imbalance in the literature.

Though it is clear that structural adjustment policies have negative human rights consequences for loan recipients, these bad outcomes probably have been unintended. First, the World Bank has been public in its commitment to good governance, including good human rights practices, as a way to promote economic development. Second, the selection stage findings indicated that the Bank has been more likely to give loans to governments with relatively good records of protection of physical integrity rights and worker rights. Third, the loan selection practices of the World Bank were not found to be strongly affected by the political interests of the major donors. Having an alliance with the United States or another major donor to the Bank had little effect on whether or not a country received a loan. Fourth, the findings showed that human rights practices improved during the years new SAAs were negotiated. One might infer that these improvements were designed to please Bank officials. Finally, there is no evidence that suggests that the Bank is aware of the negative human rights effects of structural adjustment.

In fact, in some very public ways, the World Bank has seemed concerned about advancing human rights, especially in recent years. James Wolfensohn, in speeches he gave as the former World Bank President, even came close to using a human rights framework in his discussion of the poverty reduction efforts of the Bank. This evidence of concern about human rights can be seen elsewhere in the Bank's activities. Since 1994, the World Bank's Governance Project has emphasized the role of good governance as a precondition for development. The Director of the Project has even argued that respect for human rights is a necessary condition for economic growth. However, despite this apparent concern about promoting good human rights practices, the World Bank continues to use the tool of structural adjustment as its principal way to promote economic development, and there is no evidence that the provisions of the SAAs negotiated by the World Bank have changed in recent years or are different from those negotiated by the IMF.

[. . .]

When coupled with the body of research showing that structural adjustment programs do not stimulate economic growth, the findings presented here cast serious doubt upon the wisdom of insisting upon rapid neoliberal structural adjustment as the main condition for providing loans. The Bank's structural adjustment policies were shown to lessen the four human freedoms examined in this study. Most likely, protecting these

and other human freedoms is critical to the promotion of economic growth. Thus, structural adjustment programs as presently conceived and implemented undermine the Bank's mission to alleviate poverty around the World, and instead generate conditions for its perpetuation. Besides expanding market freedom, the World Bank should insist upon improvements in respect for other human rights as a condition for receiving new structural adjustment loans.

Future research on the human rights effects of structural adjustment should examine the consequences for other types of human rights such as worker rights and women's rights. Future work also should focus on developing improved measures of structural adjustment loan implementation. New measures would allow for a closer examination of the direct and indirect effects of the speed and types of economic liberalization on democratization, domestic conflict and ultimately on government respect for human rights. Economic liberalization may not have inevitable negative consequences for the human rights practices of governments. However, the results of this research demonstrate that the rapid, externally imposed economic liberalization of the type insisted upon by the World Bank has led to increased government violations of physical integrity rights.

 READING 18

How International Monetary Fund and World Bank Policies Undermine Labor Power and Rights

Vincent Lloyd and Robert Weissman

After a decade of economic "reform" along lines advised by the International Monetary Fund (IMF) and World Bank, Argentina has plunged into a desperate economic crisis. The economy has been contracting for three years, unemployment is shooting up, and the country is on the brink of defaulting on its foreign debt payments. To avoid default, Argentina has negotiated for a new infusion of foreign funds to pay off the interest on old loans and obligations, and to forestall a pullout

by foreign investors. Traveling down that road took Argentina to the gatekeeper for such loans: the IMF. In August 2001, the IMF agreed to provide a new $8 billion loan for Argentina, intended to forestall default. That followed a nearly $40 billion January bailout package with a $14 billion IMF loan as its centerpiece.

But like the loans Argentina has negotiated with the IMF and World Bank over the last decade – and like all other such loans from the IMF and Bank – the new monies came with conditions. Among them are require-ments that Argentina: promote "labor flexibility" – removing legal protections that inhibit employers from firing workers; revamp its pension system to generate "new savings" by cutting back on benefits for retired workers; slash government worker salaries; and privatize financial and energy operations of the government. The requirements, and others, infuriated the Argentine labor movement, which responded in March 2001 with general strikes that stopped economic activity in the country. In August, with the latest loan package, tens of thousands of workers took to the streets in protest.

That the IMF would demand such terms is no surprise. A *Multinational Monitor* investigation shows that the IMF and World Bank have imposed nearly identical mandates on dozens of countries. Based on reviews of hundreds of loan and project documents from the IMF and World Bank, the *Multinational Monitor* investigation provides detailed evidentiary support for critics of the international financial institu-tions who have long claimed they require Third World countries to adopt cookie-cutter policies that harm the interests of working people.

Multinational Monitor reviewed loan documents between the IMF and World Bank and 26 countries. The review shows that the institutions' loan condition-alities include a variety of provisions that directly undermine labor rights, labor power, and tens of millions of workers' standard of living. These include:

- Civil service downsizing.
- Privatization of government-owned enterprises, with layoffs required in advance of privatization and frequently following privatization.
- Promotion of labor flexibility – regulatory changes to remove restrictions on the ability of government and private employers to fire or lay off workers.
- Mandated wage rate reductions, minimum-wage reductions or containment, and spreading the wage gap between government employees and managers.
- Pension reforms, including privatization, that cut social security benefits for workers.

The IMF and Bank say these policies may inflict some short-term pain but are necessary to create the condi-tions for long-term growth and job creation.

Critics respond that the measures inflict needless suffering, worsen poverty, and actually undermine prospects for economic growth. The policies reflect, they say, a bias against labor and in favor of corporate interests. They note as well that these labor-related policies take place in the context of the broader IMF and World Bank structural adjustment packages, which emphasize trade liberalization, orienting eco-nomies to exports and recessionary cuts in govern-ment spending – macroeconomic policies that further work to advance corporate interests at the expense of labor.

The Incredibly Shrinking Government Workforce

Perhaps the most consistent theme in the IMF/World Bank structural adjustment loans is that the size of government should be reduced. Typically, this means that the government should spin off certain functions to the private sector (by privatizing operations) and that it should cut back on spending and staffing in the areas of responsibility it does maintain.

The IMF/Bank support for government downsizing is premised, first, on the notion that the private sector generally performs more efficiently than government. In this view, government duties should be limited to a narrow band of activities that the private sector either cannot or does not perform better and to the few responsibilities that inherently belong to the public sector. In its June 2001 draft "Private Sector Develop-ment Strategy," the World Bank argues that the private sector does a better job even of delivering services to the very poor than does the public sector and that the poor prefer the private sector to government provision of services.

A second rationale for shrinking government is the IMF and Bank's priority concern with eliminating government deficits. The institutions seek to cut govern-ment spending as a way to close and eventually eliminate the shortfall between revenues and expenditures, even though basic Keynesian economics suggests that slow-growth developing nations should in fact run a deficit to spur economic expansion. In most countries, rich and poor, the government is the largest employer. In poor countries, with weakly developed private sectors, the government is frequently the dominant force in the nation's economy. Sudden and massive cuts in government spending can throw tens or hundreds of thousands out of work, and can contribute to a surge in unemployment and to a consequent reduction in the bargaining power of all workers.

[. . .]

Privatize, Privatize, Privatize

The civil service downsizing included in IMF and World Bank conditionalities is frequently bound up with privatization plans: under IMF and Bank instruction, governments agree to lay off thousands of workers to prepare enterprises for privatization. But privatization itself is frequently associated with new rounds of downsizing, as well as private employer assaults on unions and demands for wage reductions. Privatization is a core element of the structural adjustment policy package. Blanket support for privatization is an ideological article of faith at the IMF and Bank.

The range of IMF- and Bank-supported or -mandated privatizations is staggering. The institutions have overseen wholesale privatizations in economies that were previously state-sector dominated – including former Communist countries in Central and Eastern Europe, as well as many developing countries with heavy government involvement in the economy – and also privatization of services that are regularly maintained in the public sector in rich countries, such as water provision and sanitation, health care, roads, airports, and postal services.

[. . .]

Labor unions do not offer blanket opposition to all privatization. Particularly in the case of Central and Eastern Europe, but also in many developing countries, unions have agreed that privatization of some government operations may be appropriate. But they have insisted on safeguards to ensure that privatization enhances efficiency rather than the private plunder of public assets, and have insisted that basic worker rights and interests also be protected. But those safeguards by and large have not been put in place.

"Unfortunately, trade unions' proposals regarding the form of privatization, the regulatory framework and treatment of workers were usually not listened to during the massive privatization wave in Central and Eastern Europe," notes the International Confederation of Free Trade Unions (ICFTU) in a report published in advance of the fall 2001 IMF and World Bank meetings. The IMF and Bank acknowledge some of their mistakes in Central and Eastern Europe, the ICFTU notes, but "similar mistakes may well be repeated in Central and Eastern Europe and in other regions."

[. . .]

The Freedom to Fire

Another core tenet of IMF and Bank lending programs is the promotion of "labor flexibility" or "labor mobility," the notion that firms should be able to hire and fire workers, or change terms and conditions of work, with minimal regulatory restrictions. The theory behind labor flexibility is that, if labor is treated as a commodity like any other, with companies able to hire and fire workers just as they might a piece of machinery, then markets will function efficiently. Efficient-functioning markets will then facilitate economic growth.

Critics say the theory does not hold up. Former World Bank chief economist Joseph Stiglitz described the problem to *Multinational Monitor*: "As part of the doctrine of liberalization, the Washington Consensus said, 'make labor markets more flexible.' That greater flexibility was supposed to lead to lower unemployment. A side effect that people didn't want to talk about was that it would lead to lower wages. But the lower wages would generate more investment, more demand for labor. So there would be two beneficial effects: the unemployment rate would go down and job creation would go up because wages were lower."

"The evidence in Latin America is not supportive of those conclusions," Stiglitz told *Multinational Monitor*. "Wage flexibility has not been associated with lower unemployment. Nor has there been more job creation in general." Where "labor market flexibility was designed to move people from low productivity jobs to high productivity jobs," according to Stiglitz, "too often it moved people from low productivity jobs to unemployment, which is even lower productivity."

Indeed, some of the IMF and Bank documents treat labor flexibility almost as code for mass layoffs.

[. . .]

Spreading the Wage Gap

Few things more clearly run contrary to workers' interest than wage reductions. Wage freezes, wage cuts, and wage rollbacks are all commonplace in IMF and World Bank lending programs, as is "wage decompression" – increasing the ratio of highest to lowest paid worker. These initiatives usually occur in the public sector, where the government has authority to set wages and salaries, and where the rationale is to reduce government expenditures. (A different logic is applied to managers, however, where the assumption is that higher salaries are needed to attract quality personnel and to provide incentives for hard work.) Sometimes the IMF and World Bank-associated wage freezes or reductions do apply to the private sector, as in cases where the minimum wage is frozen or reduced. Sometimes the overarching policy is referred to as "wage flexibility" and is undertaken in connection with labor market reforms.

[...]

The institutions have elaborate justifications for opposing wage supports. An April 2001 World Bank policy working paper, for example, concludes that minimum wages have a larger effect in Latin America than in the United States – including by exerting more upward influence on wages above the minimum wage – and promote unemployment.

Pensions: Work Longer, Pay More, Get Less

Pension and social security reform has emerged as a high priority of the IMF and Bank in recent years, with the World Bank taking the lead. The thrust of the World Bank/IMF's proposals in this area has been for lower benefits provided at a later age, and for social security privatization.

[...]

The ICFTU reports that the World Bank has been involved in pension reform efforts, increasingly driving toward privatization, in over 60 countries during the past 15 years.

Dean Baker, co-director of the Washington, D.C.-based Center for Economic and Policy Research, says the Bank's support for social security privatization is not based on the evidence of what works efficiently for pension systems. "The single-mindedness of the World Bank in promoting privatized systems is peculiar," he says, "since the evidence – including data in World Bank publications – indicates that well-run public sector systems, like the Social Security system in the United States, are far more efficient than privatized systems. The administrative costs in privatized systems, such as the ones in England and Chile, are more than 1,500 percent higher than those of the US system."

Baker adds that "the extra administrative expenses of privatized systems comes directly out of the money that retirees would otherwise receive, lowering their retirement benefits by as much as one-third, compared with a well-run public social security system. The administrative expenses that are drained out of workers' savings in a privatized system are the fees and commissions of the financial industry, which explains its interest in promoting privatization in the United States and elsewhere."

Whither Labor Rights?

Few labor advocates argue that privatization should never occur or that no government layoff is ever necessary, though many would argue in almost all cases against certain IMF and Bank policies, such as reductions or mandated freezes on the minimum wage and privatization of Social Security. But among the most striking conclusions from the *Multinational Monitor* investigation of IMF and World Bank documents is the near-perfect consistency in the institutions' recommendations on matters of key concern to labor interests.

None of the documents reviewed by the *Monitor* show IMF or Bank support for government takeover of services or enterprises formerly in the private sector; they virtually never make the case for raising workers' wages (except for top management); they do not propose greater legal protections for workers. And on-the-ground experience in countries around the world shows little concern that implementation of policies sure to be harmful to at least some significant number of workers in the short term is done with an eye to ameliorating the pain. Worker safeguards under privatization, for example, repeatedly requested by labor unions around the world, are rarely put into force.

For former Bank chief economist Joseph Stiglitz, as well as unions and workers' advocates, the IMF/Bank record makes it imperative that basic workers' rights be protected. If there are to be diminished legal protections and guarantees for workers, and if IMF/Bank-pushed policies are going to run contrary to workers' interests, they say, then workers must at the very least be guaranteed the right to organize and defend their collective interests through unions, collective bargaining, and concerted activity.

But the Bank has stated that it cannot support workers' freedom of association and right to collective bargaining. Robert Holzmann, director of social programs at the World Bank, told a seminar in 1999 that the Bank could not support workers' right to freedom of association because of the "political dimension" and the Bank's policy of non-interference with national politics. Holzmann also raised a second "problem" with freedom of association. "While there are studies out – and we agree with them that trade union movements may have a strong and good role in economic development – there are studies out that also show that this depends. So the freedom by itself does not guarantee that the positive economic effects are achieved."

Shortly after the 1999 seminar, labor organizations met with the World Bank and IMF. According to a report from the ICFTU, World Bank President James Wolfensohn reiterated Holzmann's point, saying that while the Bank does respect three out of the five core labor rights (anti-slavery, anti-child labor, and anti-discrimination), it cannot respect the other two (freedom of association and collective bargaining) because it does "not get involved in national politics." The ICFTU reports that "this statement was greeted with stunned disbelief by many present."

 READING 19

Who Has Failed Africa?: IMF Measures or the African Leadership?

Gerald Scott

Introduction

Many writers have suggested that International Monetary Fund (IMF) Structural Adjustment Programs in Africa have not only damaged growth prospects for many countries, but have further worsened an already badly skewed income distribution. Some of these writers have claimed that IMF programs have ignored the domestic social and political objectives, economic priorities, and circumstances of members, in spite of commitments to do so. In a recent article, an African critic submitted that IMF measures have failed Africa. He claimed that "after adopting various structural adjustment programs, many [African] countries are actually worse off." Not unlike many, he seems to be suggesting that IMF programs have been somewhat responsible for the severe decline in economic conditions. Some critics of IMF programs have pointed out that the fact that economic conditions have deteriorated is not conclusive proof that conditions would be better without IMF programs. They do however stress that the developments associated with IMF programs have been extremely unsatisfactory. At the same time, this association does not necessarily imply that IMF programs cause economic decline in the region.

The main purpose of this paper is to argue that of all the feasible alternatives for solving Africa's current economic problems, IMF programs are the most promising. The paper will not contend that the panacea for the seemingly unsurmountable problems rest with the IMF. However, it will argue that IMF programs are better poised to help Africa reach its economic goals, or improve economic performance.

First, we will examine the main reasons for the region's dismal economic performance over several decades. Secondly, we will evaluate the evidence that has been used to reach the conclusion that IMF programs have had a deleterious effect on Africa's economic performance. Thirdly, we will present a case for the attractiveness of IMF programs, and a discussion of some specific prescriptions in IMF programs.

Why Has Sub-Saharan Africa (SSA) Performed So Poorly?

The problems of slow growth, high inflation, and chronic balance of payments problems continue to plague SSA well into the 1990s. In general these problems can be traced to international or domestic factors. During the last two decades a number of adverse events in the international economy have contributed to the economic decline in the region. These include oil crises, global recessions, deteriorating terms of trade, protectionism in the developed countries' markets, rising real interest rates, and the lack of symmetry in adjustment to payments problems. In addition a number of adverse developments in the domestic economy have inhibited productive capacity and thwarted the attempts to initiate and sustain economic growth.

No doubt, many countries lack appreciable amounts of essential resources and adequate infrastructure for sustained growth. It is also true that growth and development in many nations have been set back by droughts, civil wars, and political disturbances. It may even be true that colonial economic structures still account for many inflexibilities that inhibit economic growth. However, many countries could significantly improve economic performance and reduce poverty significantly if they managed their economies more efficiently, controlled population growth, and abandoned those policies that are so obviously anti-developmental.

The major setback has been gross mismanagement, which has largely resulted from corruption, rather than from incompetence and absence of skilled administrators. In many nations, national resources for investment, growth, and welfare have been consistently diverted into private hands and used largely for conspicuous consumption. Poor public sector management has resulted in large government budget deficits, which contribute to inflation, which in turn encourage undesirable import growth and serious balance of payments deficits. Quite simply African leaders, admin-

istrators, businesses, and political insiders have been engaged in corruption on a massive scale. The result has been almost complete destruction of the economic potential in many nations.

For the purpose of solving Africa's serious economic problems, there is need for the political will to attack the fundamental causes. If the present disquieting trends are not urgently tackled with the appropriate policies, then an even more somber future looms on the horizon for many Africans. Any package of measures should include policies designed to revitalize, expand, and transform the productive sectors into viable and self sustaining entities. In the absence of corruption public resources can be allocated efficiently to facilitate growth in the productive sectors. The microeconomic efficiency that results from efficient resource allocation, coupled with appropriate macroeconomic stabilization policies, would greatly enhance the prospects for economic growth and prosperity.

Assessing IMF Programs

Studies aimed at evaluating IMF programs in Africa conclude that the results are mixed, ranging from disappointing to marginally good. Inasmuch as it is difficult to assess the overall effect of IMF programs some studies have shown that they have been somewhat successful in terms of a number of key economic indicators. One main reason for the contention that IMF programs have been harmful is that many countries with programs have performed as badly as those without IMF programs. One must be cautious in examining the performance of key economic indicators following IMF programs because of the dynamics of the setting in which they are implemented. But let us suppose for the sake of argument that IMF programs actually result in deteriorating economic conditions immediately following the program. For example, suppose economic growth declines as a result of the program. Even though economic growth is perhaps the most important objective of national development policy, it is still reasonable to consider a program successful if it laid down the basis for future realization of economic growth, within some reasonable time period. In other words if it established the economic structure that promotes and facilitates long term growth, then it can still be regarded as successful. In addition it is possible that even though conditions did not improve, the program may have prevented economic conditions from deteriorating even more. It is not possible to subject IMF programs to controlled experiments. However, it seems more reasonable to argue that without IMF programs, in many countries, conditions would have been much worse, than one would argue that IMF programs cause conditions to worsen.

It has somewhat been fashionable, especially amongst those with very limited knowledge of the various economic rationales behind IMF recommendations, to reject those recommendations without presenting a feasible alternative. Many object to the IMF and some regard it not only as a representation of western economic interest, but as too uncompromising and arrogant in its relationship with nations in crisis. The indications are that IMF is usually anxious to intervene even before conditions deteriorate into a crisis. But like any prudent banker it has to be concerned about repayment prospect, which is essential for its very own survival and continuous provision of its service to other deserving members.

Why IMF Programs May Be the Answer for Africa

IMF programs in the 1990s should have a major attraction for Africans genuinely concerned with the welfare of the people for a number of reasons. First, the programs are no doubt based on sound theory, always a useful guideline for policy-formulation.

The peculiar social, political, and economic circumstances of African nations and the inability or refusal of the IMF to take them into account in the design and implementing of programs have been cited as reasons why IMF programs have "failed" in Africa, or are doomed to fail. On the contrary, these particular African circumstances are in fact another good reason why IMF programs may be the right answer to the problem. Because of the nature of African economic circumstances, particularly problems in economic administration, the conspicuous absence of commitment on the part of politicians and administrators to the development and welfare of the nations, the absence of institutional capacity, and the weak civic consciousness, the best policy

is to embrace IMF programs. IMF programs encourage the dismantling of controls and simplification of the bureaucratic process; emphasize the strengthening of institutional capacity; require public accountability and responsibility; emphasize efficiency and economic discipline; encourage private sector participation in the economy; foster coordination in economic decisions and promote macroeconomic stability; and emphasize measures designed to expand aggregate supply.

The optimal policy intervention for dealing with an inefficiency or distortion is to seek the source of the problem. IMF programs are attractive because they are designed to attack the problems at their source. In African nations there are many problems that are outside the control of the officials and administrators. However corruption is not one such problem and it need not be so pervasive and economically destructive. Although it is very important not to under-emphasize the importance of many other problems of development, corruption is an obstacle that can largely be controlled, if the top leadership is committed to that objective. It is not the same problem as say drought or poor resource endowment, or an absence of a skilled workforce, that is largely outside the control of officials.

One major attraction of IMF programs is that they tend to remove all opportunities for corruption, i.e., they seek the source of the problem. For example, the suggestion that controls should be dismantled is in recognition that the reliance on physical controls for resource allocation is inferior to the market mechanism, especially in the absence of an efficient administrative machinery for the effective administration of controls. But perhaps the more relevant point is that a proliferation of controls usually lays the foundation for corruption, which has continued to destroy economic life in the region.

In reality it is not the IMF who has failed Africa, but the African leadership. The politicians and public sector officials have conspired with private businessmen and firms to adopt and implement policies that benefit themselves at the expense of national development and welfare. The inability or unwillingness of Africans to demand more accountability and responsibility from both politicians and public servants ensures that violations of the public trust are not treated as illegal, immoral, unethical or non-nationalistic actions. If the IMF has failed Africa, it has done so by failing to vigorously condemn or expose corruption or even assign it the prominent place it deserves in the design of programs.

Many who oppose IMF programs have argued that they impose severe economic harm on the deprived peoples of Africa. Who are these deprived peoples and what is the evidence? The majority of them are rural inhabitants who have virtually been untouched by modernity. They are largely farmers, have limited participation in the modern economy, consume limited manufactured goods, and have very limited access to basic social services provided by governments. In the urban areas there are Africans of diverse economic circumstances ranging from those in abject poverty and squalor, to those of enormous wealth. The urban population is usually more politically powerful and its views have been the barometer used to measure or assess the political climate. On balance IMF programs will tend to harm the urban poor given the structure of their consumption basket and their production pattern. On the other hand the rural population could benefit immensely from IMF programs for similar reasons, and the efficiency gain to the nation would more than compensate for the loss experienced by the urban population. No convincing evidence has been advanced to support the claim of impoverishment of the majority of rural African peoples.

Even though the urban poor could face the most severe hardship as a result of IMF programs such adverse consequences could be mitigated even within the context of those same programs that supposedly impose such hardships. There is some empirical evidence that IMF reforms will improve the distribution of income and help the poor. There is also evidence that appropriate exchange rates and price incentives improve economic performance, and that private enterprises perform better than state enterprises.

What Africans need is a set of institutions that would enable them to effectively demand the very modest conditions the people deserve and subject all officials to full responsibility and accountability. Given the levels of ignorance, ethnic loyalties, poverty, disillusionment and despair, absence of strong nationalistic and patriotic attitudes, I shudder to imagine the difficulties associated with establishing such institutions. Notwithstanding, the task is possible if the leadership is committed to doing so. Based on the current structure

of African institutions, and the record of policy makers, IMF programs are more likely to be effective than other possible alternatives.

Let us examine some of the recommendations and issues in IMF programs and discuss their effects on national welfare.

Devaluation

A devaluation increases the prices of traded (relative to nontraded) goods and will induce changes in production and consumption. First, as imports become more expensive less will be demanded, thereby curbing excessive import demand which is a major source of balance of payments deficits. At the same time production of import substitutes will be encouraged. Secondly, exports will become expensive so that less will be consumed locally and more will be produced. Exports will also be cheaper in foreign countries, so that more will be demanded. Foreign firms that split production into several stages will find the country attractive for their investments, and tourism will also receive a boost. The devaluation will therefore stimulate the export and import substitution sectors. The political concern usually is that the urban consumers whose purchasing power has already been eroded by inflation partly from excessive government spending, will have to pay more for basic manufactured goods, the bulk of which are imported. Not surprisingly, there is usually an anti-devaluation sentiment in the main urban areas. It is very important to emphasize that the devaluation by itself will not correct the problem of macroeconomic instability. It must be accompanied by sound fiscal management that complements rather than counteracts the effects of the devaluation. For example, if the government continues to maintain significant fiscal deficits after devaluation, then the devaluation would soon be reversed as the exchange rate becomes overvalued again. An overvalued exchange rate is subversive to long-term growth and balance of payments adjustment.

The African rural population consumes imported manufactured goods only in limited amounts, but could potentially benefit from devaluation because it will increase the price of agricultural exports. A program that prescribes a devaluation so that exchange rates are competitive, should ensure that the producers of exports are not unreasonably exploited by middlemen (including government) to the extent that they have no incentive to expand production.

A legitimate concern is that devaluation will raise the price of essential inputs and stifle the supply response as the cost of production rises. In the first place, as long as cost of production lags behind prices, producers will find it profitable to expand production. In any case the appropriate supply response could be encouraged by an appropriate production subsidy. This of course involves an additional strain on the budget, and the IMF insists on fiscal restraint as we will see shortly. Fiscal reform involves maximizing tax revenue and ensuring that it is used to maximize macroeconomic performance. This means that those who have been avoiding their tax burden, especially the self employed, must be made to meet their tax obligations, and that frivolous and wasteful expenditures must be avoided.

Government budget deficit

When IMF programs recommend reductions in government expenditures, the concern is not only with the adverse effects of budget deficits on inflation and the balance of payments, but also with bogus budgetary appropriations that benefit private individuals and deprive the nation of developmental resources. As a result of the pervasiveness of corruption, many governments typically appropriate funds for the salaries of nonexistent civil servants or for goods and services that are not received. Similarly it is common for governments not only to pay highly inflated prices for goods and services, some of which are totally inessential, but also for governments to receive far less than market value for goods bought by some individuals or firms. IMF prescriptions on the budget can be viewed as perhaps a subtle way of telling African leaders that from their past record they cannot be trusted to appropriate the nation's resources in the national interest. This appears paternalistic, but should be acceptable to all concerned with the welfare of the mass of African peoples.

Government budget deficits as a percentage of GDP increased sharply after independence in many countries, as the states intervened ostensibly to correct the perceived flaws of a market economy. The evidence indicates that throughout the region the states have failed to perform the role of a prudent entrepreneur, and government investments have resulted largely in considerable inefficiency. Public enterprises have been inefficiently operated, as they have largely been used as a way of providing patronage to political insiders.

Government budget deficits financed largely through money creation, have contributed to serious inflation and balance of payments problem. These deficits have not been consistent with other macroeconomic objectives of the government. The control of the deficit usually requires reducing expenditure, including the elimination of subsidies to consumption, and increasing taxes. In many African nations it is common for the government to subsidize the consumption of essential food items, gasoline, electricity, public transportation etc. The major beneficiaries are the urban population and mostly political insiders who for example obtain goods at subsidized prices and resell at black market rates. The typical rural inhabitant, because of the structure of the consumption basket does not benefit much from government subsidies.

Market prices

IMF programs attempt to promote a strong link between work effort and reward. This involves appropriate prices of goods and services, and factors of production. Prices not only provide information to producers but serve as an incentive that facilitates efficient resource allocation. The major problem in African countries has been inadequate production. Production has been constrained by a large number of factors including inappropriate prices. In many African countries the tax system has turned the terms of trade against agriculture and has resulted in very slow or negative growth rates in this sector. Overvalued exchange rates are an implicit tax on exporters since exporters receive the official rate.

The imposition of market prices for agricultural commodities typically results in higher food prices. Rural farmers benefit as producers, but as consumers they lose. However as long as they can respond sufficiently as producers, their gains will be more than enough to compensate for their losses and the nation as a whole will benefit. The challenge of reforming prices is to ensure adequate production response, which may require other complementary policies.

The continuous proliferation of price controls will only continue to stifle production, worsen shortages, and reduce incentive for investment.

Privatization

African governments have argued that they have an obligation to provide goods and services usually provided by private enterprises in developed countries, because too often the market fails to do so. Thus they are compelled to invest in capital formation that will increase output, improve efficiency in resource allocation, and make the distribution of income more equitable. Those are desirable objectives and any government that achieves them deserves widespread commendation. Unfortunately the record of the public enterprises which are usually set up to pursue these objectives has been very disappointing. These public enterprises have been very inefficiently administered, and have been widely used by politicians as opportunities for patronage to their supporters.

In recommending privatization of certain public enterprises, IMF programs attempt to deal with two problems. The first is micro inefficiency in the productive sector, and the second is government budget deficits that result partly from the need to subsidize inefficiently run enterprises. Private enterprises that continuously make losses go out of business, but government enterprises with similar balance sheets receive political relief. By turning over certain enterprises to private institutions, the pressure on the budget eases, and there is a greater chance of increasing efficiency in production.

Corruption and rent-seeking

The issue of corruption is inadequately treated in IMF programs even though it is perhaps the most important cause of economic decline or stagnation. This issue has been left to the African peoples to deal with. Unfortunately they seem to lack the capacity to do so effectively. International institutions should adopt a more aggressive role in the process of eliminating corruption, rather than the somewhat lukewarm support for the establishment of democratic institutions.

In African nations in which corruption is acceptable and institutions are structured such that they can easily

be transformed into breeding grounds for corruption and rent seeking it is not surprising that corruption is so extensive.

The IMF way of dealing with corruption and rent-seeking is to destroy all opportunities for those activities. If all economic agents realize that prosperity can only be achieved through hard work, innovation or other legitimate means, then most people will become hard working, innovative or pursue other legitimate activities. But as long as public officials or businessmen can conspicuously display their enormous wealth that cannot be attributed to their innovation, business acumen, hard work, inheritance, winning a lottery, etc., without any fear of been asked by the appropriate authorities to account for their wealth, inefficiencies and corruption will continue to flourish. The average African must first realize that the luxury automobiles or the villas arrogantly displayed by a public servant, may be connected with his or her poverty and deplorable living conditions. Then the African must insist on full accountability of all public servants. Possessing a sense of nationalism that is much stronger than ethnic loyalties, in addition to strong leaders who are obsessed with the welfare of the people instead of an obsession with status, power, and wealth, will contribute immensely toward the elimination of corruption in the region.

There can be no doubt that if corruption is eliminated, or even controlled, a sizeable proportion of Africa's problems will disappear and the continent can then fully focus on utilizing its scarce resources for maximizing production and consumption.

Conclusion

African economic problems over the last two decades, can be traced mainly to a host of international and domestic factors. Many of the international factors and some of the domestic factors such as lack of suitable resource endowment, are outside the control of the governments and administrators. However, a significant part of the problems can be traced to corruption and other forms of inefficiencies. Instead of blaming the IMF for the dismal performance in Africa, we should focus on the African leaderships and their policies. The level of their commitment and the policies they have adopted and implemented increasingly seem to confirm only their deplorable lack of compassion for fellow Africans and a callous detachment from the people's welfare. The status quo must change to prevent further erosion of the economic base on the continent.

The best foreign assistance is one that has a lasting effect; it is one that would empower Africans to fully participate in the growth process, and provide them with the irrevocable ability to effectively demand the modest living conditions that they have been unjustly deprived of by their leaders for so long.

Nation-State

One of the most hotly debated issues in the study of globalization is the fate of the nation-state in the global age. There are so many statements on this issue, and so many different positions are staked out, that it is impossible to do them all justice in this chapter, even though the chapter includes a comparatively large number of essays. We devote much attention to this issue not only because it has been the subject of so much academic work, but also because it tells us a great deal about the process of globalization.

Donald Levine's work is useful in outlining some of the factors that have served to erode allegiance to the nation-state. Those factors are allegiances to the sub-national (the local or the primordial), the transnational (e.g. multinational organizations, MNCs), and the supranational (e.g. the EU and "Europeanness").

Susan Strange is concerned with the system which emerged from the Treaty of Westphalia in 1659 and involved the international focus on nation-states and their relationship with one another. In her view, however, that system is under great stress, leading her to the notion of the "Westfailure system." Thus her focus, at least here, is on the failure of the global system that accorded centrality to the nation-state and not on the failure of the nation-state *per se*. More generally, in other work, she feels that academics (especially those in the field of international relations, IR) accord too

much importance to the nation-state and in the process ignore other centers of political and economic power.

She focuses on problems created by the economic system that the nation-state has proven unable to handle. The first is global economic crises. Writing in the late 1990s, Strange deals with Asian financial crisis, but this failure has been demonstrated once again, and much more extremely, in the global recession beginning in late 2007. Here was a far greater economic crisis that was caused by the excesses in the economic system and the inability of the nation-states (especially the United States) to control and regulate economic agents within their borders. The plight of the nation-state is even clearer in the case of those agents that operated globally and were therefore beyond the control of any single nation-state. The second is the inability of the nation-state to deal with ecological problems – problems which are far worse today than they were a decade ago. Finally, there is the failure of the nation-state to do much of significance about the problem of global inequality which, like the other problems discussed by Strange, has grown far worse in the last decade.

Linda Weiss deals with the "myth of the power-less state." She argues that transnational movements and global flows are not new and they are not nearly as widespread as is believed by those who forecast the demise of the nation-state. Furthermore, the nation-state is more adaptable than its detractors believe. While nation-states have a differential ability to deal with transnational and global developments, there are at least some ("catalytic states") that are not only able to deal with them, but facilitate their development. Thus, she sees all of this as part of a general history of adaptation of the nation-state to external and internal changes. Over time, as globalization proceeds, nation-states will differentiate on the basis of their ability to adapt to globalization. The nation-states that adapt best will remain strong, while those that fail to adapt will risk being weakened and even overwhelmed by globalization.

Daniel Béland argues that "the role of the state is enduring – and even increasing – in advanced industrial societies."[1] He sees greater demands being placed on the state because of four major sources of collective insecurity: terrorism, economic globalization leading to problems such as outsourcing and pressures toward downsizing, threats to national identity due to immigration, and the spread of global diseases such as AIDs. Further, the state does not merely respond to these threats; it may actually find it in its interests to exaggerate or even create dangers and thereby make its citizens more insecure. A good example is the US and British governments' arguments prior to the 2003 war with Iraq that Saddam Hussein had weapons of mass destruction (WMDs) that posed a direct threat to them. The US even claimed that Iraq could kill millions by using offshore ships to lob canisters containing lethal chemical or biological material into American cities. The collective insecurity created by such outrageous claims helped foster public opinion in favor of the US invasion of Iraq and the overthrow of Saddam Hussein.

Rather than examining the nation-state *per se*, Robinson looks at the academic study of the nation-state within a global context. Because of the decline of the nation-state and the rise of the transnational, Robinson argues for the need for an epistemological shift to parallel the ontological change. That is, sociology and political science (especially IR), and other academic fields, should shift their focus from the national to the transnational (see chapter 7). In terms of sociology, Robinson argues that it should focus on the "study of *transnational social structure*."[2]

It is worth noting that Robinson, like many other analysts, reduces globalization to economic globalization. In Robinson's case, he goes further and reduces it to capitalism: "The essence of globalization is global capitalism."[3] This tendency to conflate globalization and economics, even when focusing on politics, greatly reduces the overall adequacy of analyses of the relationship between globalization and the nation-state.

NOTES ··

1   Daniel Béland, *States of Global Insecurity: Policy, Politics, and Society.* New York: Worth, 2008, 48.
2   William I. Robinson, "Beyond Nation-State Paradigms: Globalization, Sociology, and the Challenge of Transnational Studies." *Sociological Forum* 13, 4, 1998: 562.
3   Ibid., 563.

Sociology and the Nation-State in an Era of Shifting Boundaries

Donald N. Levine

[...]

Eroding Commitments to National Boundaries

Now. I find it fascinating to contemplate the fact that the new configurations that have come to compete with the authority of established disciplines resemble developments within the universe of nation-states. Just as allegiances to disciplinary fields have had to compete with intellectual alliances that I have glossed as *subdisciplinary*, *transdisciplinary*, and *supradisciplinary*, so over the past generation, commitments to national political entities have been weakened by the spread of allegiances that are *subnational*, *transnational*, and *supranational* in scope.

The locus classicus for formulating the dynamics of subnational loyalties is a paper written in 1962 by Geertz. "The Integrative Revolution." That paper identifies two powerful, interdependent, and often opposed motives: the desire to be recognized as a responsible person whose wishes, acts, hopes, and opinions "matter," and the desire to build an efficient, dynamic modern state. The one aim is to be noticed: it is a search for identity, and a demand that the identity be publicly acknowledged as having import, a social assertion of the self as "being somebody in the world." The other aim is a demand for progress – for a rising standard of living, more effective political order, greater social justice – and for playing a part in the larger arena of world polities.

As Geertz formulated the matter, tension between these two motives is a central driving force in the evolution of nations, yet one of the greatest obstacles to such evolution. The tension gets exacerbated in the new states, because of the accelerating importance of the sovereign state as a positive instrument for pursuing collective aims, when people's sense of self remains bound up with attachments based on blood, race, language, locality, religion, or tradition – attachments that Geertz designated generically (following Shils) as "primordial" ties.

A decade later Geertz was chagrined to note that the tensions among primordial groups that he had associated with the new states of Africa and Asia were by no means limited to those countries. When republishing his essay he confessed that in 1972

my passage about the declining role of primordial divisions in "modern" countries seems, to put it mildly, rather less convincing than [it] did in 1962, when this essay was originally written. But if events in Canada, Belgium, Ulster, and so on have made primordial definition seem less predominantly a "new state" phenomenon, they have made the general argument developed here seem even more germane.

Two decades after that, the assertion of subnational identities based on primordial ties had become one of the fastest-spreading social phenomena in the world, ranging from the benign move to legitimate a dozen provincial languages and dialects in France to conflagrations in Eastern Europe and Northeast Africa.

The remarkable thing is that precisely the opposite tendency has been increasing as well. Transnational organizations of many kinds have proliferated in recent decades. Corporations like General Motors, Royal Dutch Shell, and Goodyear Tire have emerged as multinational enterprises commanding vast resources outside the control of any national regulatory system. Some years ago, Elise Boulding reported a figure of about 10,000 transnational corporations with 90,000 affiliates, spread over all the continents. Comparable expansions took place with intergovernmental organizations (IGOs) and international nongovernmental organizations (INGOs). According to the 1995/1996 edition of the *Yearbook of International Organizations*, IGOs increased from a baseline of 37 in 1909 to nearly 4,000 in 1995. During the same period, INGOs exploded from a baseline of 176 to the current count of nearly 22,000.

Just as dramatic has been the expansion of supranational processes, processes that parallel what we have called the supradisciplinary spheres of discourse. In a colorful metaphor. Benjamin Barber designates this general phenomenon as "McWorld." Barber analyzes four imperatives that make up the dynamic of McWorld: a market imperative (all national markets now being vulnerable to the inroads of larger markets within which trade is free and currencies are convertible); a resource imperative; an information-technology imperative; and an ecological imperative. The trope – and reality – of cyberspace is a perfect manifestation of these new supranational realities.

Supranational forces manifest themselves even in the area of normative social controls. Yasemin Soysal has identified ways in which criteria of nationhood and citizenship have now accommodated to heavy streams of international migration. Her work, based on interviews in Britain, France, Germany, The Netherlands, Sweden, and Switzerland – and on data from an additional half dozen countries – documents the widely spread pattern of granting rights to guest workers and other non-nationals, rights that previously had been restricted to national citizens. These guests have been able to gain access to most kinds of employment, to enjoy social services such as health care, education, and social insurance schemes, and even to vote in local elections. Commenting on this new phenomenon, Soysal observes:

> The predominant conceptions of political sociology posit that populations are organized within nation-state boundaries by citizenship rules that acclaim national belonging as the legitimate basis of membership in modern states. This study, however, finds that [. . .] the state is no longer an autonomous and independent organization closed over a nationally defined population [. . .] My analysis of the incorporation of guest-workers in Europe reveals a shift in the main organizing principle of membership in contemporary polities: the logic of personhood supersedes the logic of national citizenship.

Soysal suggests that the principle of human rights has come to ascribe a universal status to individuals and their rights. She therefore theorizes these developments by construing them in terms of two contrasting principles of the global system: national sovereignty and universal human rights. Barber casts his analysis in terms of a different pair of principles. To the globalizing processes of McWorld he counterposes the principle of "jihad," the term he uses to designate a tendency to mobilize actors around more parochial kinds of allegiances. Thus, where Geertz found national loyalties in tension with subnational, primordial allegiances and Soysal finds national sovereignties in tension with global expectations about universal human rights, Barber removes nation-states from the equation altogether and conceives the great dualism of the contemporary world as a tension between tendencies toward global unification and subnational fragmentation.

In a searching critique of Barber's formulations, Roland Robertson rejects a conceptualization that counterposes local against global processes. While Robertson sees an inexorable trend toward globalization, he argues that it is incorrect to view this as taking place at the expense of or in opposition to local allegiances. Consequently, he urges us to adopt the term "glocalization" to symbolize a simultaneous expansion in both global and local directions, the universalization of particularisms and the particularization of universals.

However these processes get theorized, the growing consensus is that major developmental tendencies in the social world are weakening the claims of *national* boundaries. In Barber's words, the forces of jihad recreate ancient subnational borders from within, while the forces of McWorld make national borders porous from without. As Barber foresees it, the optimal development would be toward

> a confederal union of semi-autonomous communities smaller than nation-states, tied together into regional economic associations and markets larger than nation-states – participatory and self-determining in local matters at the bottom, representative and accountable at the top. The nation-state would play a diminished role, and sovereignty would lose some of its political potency. The Green movement adage "Think globally, act locally" would actually come to describe the conduct of politics.

[. . .]

The Westfailure System

Susan Strange

From a globalist, humanitarian and true political economy perspective, the system known as Westphalian has been an abject failure. Those of us engaged in international studies ought therefore to bend our future thinking and efforts to the consideration of ways in which it can be changed or superseded. That is the gist of my argument.

The system can be briefly defined as that in which prime political authority is conceded to those institutions, called states, claiming the monopoly of legitimate use of violence within their respective territorial borders. It is a system purporting to rest on mutual restraint (non-intervention); but it is also a system based on mutual recognition of each other's 'sovereignty' if that should be challenged from whatever quarter.

But while we constantly refer to the 'international political system' or to the 'security structure' this Westphalian system cannot realistically be isolated from – indeed is inseparable from – the market economy which the states of Europe, from the mid 17th century onwards, both nurtured and promoted. To the extent that the powers of these states over society and over economy grew through the 18th, 19th and 20th centuries, they did so both in response to the political system in which states competed with other states (for territory at first but later for industrial and financial power) and in response to the growing demands made on political authority as a result of the capitalist system of production and its social consequences. The label 'capitalist' applied to the market-driven economy is justified because the accumulation of capital, as the Marxists put it, or the creation and trading in credit as I would describe it, was the necessary condition for continued investment of resources in the new technologies of agriculture, manufacture and services. As I put it in *States and Markets*, the security structure and the production, financial and knowledge structures constantly interact with each other and cannot therefore be analysed in isolation. The point is 'kid's-stuff' to social and economic historians but is frequently overlooked by writers on international relations.

When I say that the system has failed, I do not mean to say that it is collapsing, only that it has failed to satisfy the long-term conditions of sustainability. Like the empires of old – Persian, Roman, Spanish, British or Tsarist Russian – the signs of decline and ultimate disintegration appear some while before the edifice itself collapses. These signs are to be seen already in the three areas in which the system's sustainability is in jeopardy. One area is ecological: the Westfailure system is unable by its nature to correct and reverse the processes of environmental damage that threaten the survival of not only our own but other species of animals and plants. Another is financial: the Westfailure system is unable – again, because of its very nature – to govern and control the institutions and markets that create and trade the credit instruments essential to the 'real economy'. The last area is social: the Westfailure system is unable to hold a sustainable balance between the constantly growing power of what the neo-Gramscians call the transnational capitalist class (TCC) and that of the 'have-nots', the social underclasses, the discontents that the French call *les exclus* – immigrants, unemployed, refugees, peasants, and all those who already feel that globalisation does nothing for them and are inclined to look to warlords, Mafias or extreme-right fascist politicians for protection. The point here is that until quite recently the state through its control over the national economy, and with the fiscal resources it derived from it, was able to act as an agent of economic and social redistribution, operating welfare systems that gave shelter to the old, the sick, the jobless and the disabled. This made up for the decline in its role – in Europe particularly – as defender of the realm against foreign invasion. Now, however, its ability to act as such a shield and protector of the underprivileged is being rapidly eroded – and for reasons to which I shall return in a while.

In short, the system is failing Nature – the planet Earth – which is being increasingly pillaged, perverted and polluted by economic enterprises which the state-system is unable to control or restrain. It is failing Capitalism in that the national and international institutions that are supposed to manage financial markets are progressively unable – as recent developments in east Asia demonstrate – to keep up with the accelerating pace of technological change in the private sectors, with potentially dire consequences for the whole world market economy. And it is failing world society by allowing a dangerously wide gap to develop between the rich and powerful and the weak and powerless.

The fact that the system survives despite its failures only shows the difficulty of finding and building an alternative. No one is keen to go back to the old colonialist empires. And though Islam and Christian fundamentalism make good sticks with which to beat the western capitalist model, the myriad divisions within both make any kind of theocratic-religious alternative highly improbable. So the old advice, 'Keep hold of nurse, for fear of worse' is still widely followed even while faith in her skill and competence is more than a little doubted.

[. . .]

The Three Failures

[The financial failure]

Let us start with the failure to manage this credit-creating system of finance. Up to summer 1997, the conventional wisdom was that states and their inter-governmental organisations between them were well able to supervise, regulate and control the banks and other institutions that created and traded in credit instruments – from government bonds to securitised corporate paper to derivatives. This was the message of a much-praised study by Ethan Kapstein. While national regulatory systems in each of the major developed economies functioned at the state level, the International Monetary Fund (IMF) and the Bank for International Settlements in Basle (BIS) functioned at the transnational level. This two-level, belt-and-braces system of governance could take care of any problems arising in the markets. But in the course of 1997, events in east Asia cast serious doubt on this comforting conclusion. The turmoil that hit the Malaysian, Indonesian and Thai currencies and stock exchange prices came out of a clear blue sky. Neither of those international regulatory institutions had foreseen or warned against such a contingency. As the turmoil spread and grew, the first rescue packages proved insufficient to restore even minimal confidence and had to be substantially increased. The common factor in all the stricken economies was an influx of mobile short-term capital, too much of which went in ill-considered speculative loans or in unproductive real-estate investments. Prime Minister Mahomed Mahathir of Malaysia blamed George Soros and other foreign speculators who had moved their funds out of the country as quickly as they had taken them in. But it was soon apparent that national regulations over the banks and over short-term capital movements in each of the east Asian countries (Taiwan excepted) had been totally inadequate. The admonitions to embrace financial liberalisation that came from Washington and the IMF had been taken altogether too literally.

But it is not just that the national systems and the international financial organisations were equally unprepared for the shocks of summer and autumn 1997. The case against Epstein's comfortable conclusions concern much more (a) the inadequacy of both the BIS and the IMF as global regulators; and (b) the inadequacy of *all* national systems of financial regulation. To be fair to Epstein, it only became apparent after he had done his study that the Basle system of capital-adequacy rules devised by the Cooke Committee in the 1980s and subsequently elaborated was not after all really effective. In its 1997 report the BIS more or less admitted as much and, making a virtue out of necessity, announced that in future the supervisory responsibility would rest with the banks themselves. Now, as the Barings story had shown, trusting the poachers to act as gamekeepers was an unconvincing strategy. The bosses at Barings neither knew nor wanted to know what Nick Leeson was up to. Barings' survival under acute international competition made them glad of the profits while discounting the risks he was taking. And even in the most prudent of banks these days, the complexities of derivative trading are often beyond the comprehension of elderly managers.

As for the IMF, its competence to coerce Asian governments into supervising and reforming their banking and financial systems is open to grave doubt. The IMF is used to negotiating with states (especially Latin American ones) over sovereign debts. Its officials – mostly economists – have no experience that helps them catch out wily and secretive bankers when they lie or cover up their business. Moreover, as the record in Kenya, for example, shows, IMF economists have no leverage when it comes to obdurate dictators protecting their corrupt and clientelist power structures. The problem with Suharto is above all political, not technical. The same is true of the African debt problem. Everyone, including the IMF, now agrees that rescheduling old debt in the Highly Indebted Poor Countries (HIPCs) is only making the problem worse, not better. But the IMF and World Bank are unable to force the creditor governments into the necessary agreement on whose debt should be wiped out and by how much.

As for the declining effectiveness of national systems of financial regulation and control, this may be less evident to Americans than it is to Europeans and Japanese. The German, French, British and Japanese systems function very differently. But all are currently being undermined by the technological innovations in financial dealing and the almost-instant mobility of capital across borders and currencies. A dangerous gap is therefore opening up between the international institutions that are unable and unwilling to discipline the banks, the hedge and pension fund managers and the markets, and the national systems of supervision and control whose reach is not long enough nor quick enough to prevent trouble. Eric Helleiner has argued that supervisors now have the technical know-how to trace funds as they move about the global financial system. True, but only far too slowly and with too much painstaking effort; not fast enough nor regularly enough to protect the system. So long as tax havens provide a refuge for wrongdoers, from drug dealers to corporate tax-evaders and heads of state who regard their country's aid funds as personal property, the national regulators' hands are tied.

The environmental failure

I have put the financial failures of the state-based system first because my recent research has convinced me that it is the most acute and urgent of the current threats-without-enemies. If we do not find ways to safeguard the world economy before a succession of stockmarket collapses and bank failures eventually lands us all in a 20-year economic recession – as the history of the 1930s suggests it might – then no one is going to be in a mood to worry overmuch about the long-term problems of the environment.

On the other hand the environmental danger is much the most serious. The planet – even the market economy – could survive 20 years of slow economic growth. But if nothing is done to stop the deterioration of the environment then the point might come with all these dangers when it was too late. The destructive trend might have become irreversible. Nothing anyone could do then would stop the vicious circle of environmental degradation, and it would be the Westfailure system that brought it about and prevented remedial and preventive action. Why? Because the territorial principle which lies at the heart of it proclaims that the territorial state is responsible for its own land – but not for anyone else's.

There are three distinct kinds of environmental danger. But for each, it is not the lack of technical knowledge, nor of appropriate policy measures that is lacking. It is the ability of the Westfailure system to generate the political will to use them. One is the destruction of the ozone layer. This is mainly attributed to the release of CFC gases from aerosols and other sources. As the 'hole' in the ozone layer grows larger, the protection from the sun given by the earth's atmosphere is weakened with serious atmospheric and climatic consequences. Another environmental problem is caused by carbon dioxide and sulphur pollution of the air. Some of this pollution comes from industry. But a lot comes from cars – cars that use petrol or diesel for fuel. Third, there is the depletion of the planet's resources – primarily of water, shrinking the acreage available for cultivation. Secondarily, there is the depletion of forests – not only rainforests – bringing unforeseeable climatic consequences, and also the depletion of species of plants, fish and animals, upsetting ecological balances that have existed for millennia.

With each of these environmental dangers, it is not hard to see that it is the state, with its authority reinforced by the mutual support provided by the Westfailure system, that is the roadblock, stopping remedial action. One consequence of the principle can be seen in the indifference of British governments to the acid rain carried by prevailing westerly winds to Scandinavian forests; or the indifference of US governments to the same kind of damage to Canadian forests. Another can be seen in the impasse reached at the Rio and Kyoto intergovernmental conferences on the environment. European and Japanese concerns left the United States substantially unmoved when it came to stricter controls over CFC gases. Nothing much has changed since. The agreements at the Kyoto conference in 1997 were more cosmetic than substantial. And when it comes to the pollution danger, the biggest *impasse* is between the developed countries and China. Pressure on Beijing from the United States and others to slow down the consumption of fossil fuels for the sake of the environment is met with the question, 'If we do, will you pay?' After all, they argue, the environmental dangers you perceive today were the result of your past industrialisation, not ours. Why should you expect us to be more environmentally aware today than you were yesterday? With our growing population, we cannot afford – unless, of course, you are prepared to pay – to slow down our growth to keep the air pure and the water unpolluted. Only rarely, as when Sweden offered to contribute funds to Poland to pay for tougher environmental rules on Polish coal and chemical plants, is the Westphalian territorial principle set aside. But Sweden is rich, was directly damaged by Polish pollution and could justify the transfer on grounds of self-interest. China and the rest of the developing countries are a far bigger nut to crack. So long as the Westfailure system persists, Nature will be its victim.

As Andrew Hurrell commented in a recent review, 'the pitfalls outweigh the promise by a very considerable margin' when it comes to transmuting short-term transfers into well-institutionalised long-term commitments on environmental matters. Hurrell also quotes one of the concluding chapters in the book, 'The studies of environmental aid in this volume paint a rather dark picture. Constraints on the effectiveness of environmental aid seem more pronounced than windows of opportunity'.

[The social failure]

The third Westphalian failure is social, or social and economic. The discrepant and divergent figures on infant mortality, on children without enough to eat, on the spread of AIDS in Africa and Asia, and on every other socio-economic indicator tell the story. The gap between rich countries and very poor ones is widening, and so is the gap between the rich and poor in the poor countries and the rich and poor in the rich countries. It is not that we do not know the answer to socio-economic inequalities; it is redistributive tax and welfare measures and what Galbraith called countervailing power to make good the tendency of capitalism to private affluence and public penury, and to booms followed by slumps. But applying that answer to world society is frustrated by the Westfailure system, so closely tied in as it is with the 'liberalised' market economy. If national Keynesian remedial policies are made difficult by the integrated financial system – as Mitterrand found out so painfully in 1983 – transnational Keynesian policies are practically inconceivable. We have had one demonstration of this in central Europe in the early 1990s. Here was a case, if ever there was one, for a second Marshall Plan to prime the pump for a rapid transition from state-planning to an open, competitive and therefore productive market economy. But the Reagan and Bush administrations were ideologically unsympathetic and the Germans too self-absorbed in their own unification to bother about the fate of their nearest neighbours. Indifference, whether to central Europe or to Africa, is not just a matter of the selfish, conservative mindsets that Gerald Helleiner recently parodied in verse:

> The poor complain. They always do. But that's just idle chatter. Our system brings rewards to all, at least to all that matter.

It is actually an inevitable result of the symbiosis between a world market economy and a state-based political system in which those with political authority are inherently unable to see that socio-economic polarisation is not in anyone's long-term interest. It is not just that the underprivileged may riot and loot as in Los Angeles in the 1980s or Jakarta today, or that they may pass their new epidemic diseases to the rich, or wage terrorist campaigns under the guise of religious *jehad*s. It is that socio-economic inequality becomes intolerable if people believe it will get worse, not better. They can bear deprivation and hardship if they believe that their children's lot will be better than theirs. Moreover, a flourishing market economy needs new customers, with money to spend, not homeless beggars and starving African farmers. America would not be what it is today without the millions of penniless immigrants that constantly expanded the mass market for its manufactures.

What Is To Be Done?

The two commonest reactions to the three failures of the system I have briefly described are either to deny the failures and to defend the dual capitalism–state system in panglossian fashion as the best of all possible post-Cold War worlds, or else fatalistically to conclude that, despite its shortcomings there is nothing that can be done to change things. Only quite recently has it been possible to detect the first tentative indications of a third response. It is to be heard more from sociologists than from international relations writers, perhaps because sociologists tend to think in terms of social classes and social movements rather than in terms of nation-states. As a recent collection of essays around the theme, 'The Direction of Contemporary Capitalism' shows, there is little consensus among them either about current trends or about possible outcomes. A good deal of this thinking has been inspired by the rediscovery of Antonio Gramsci and his concepts of hegemony, the historic bloc and social myths that permit effective political action. A common assumption is that the present system is sustained by the power of a transnational capitalist class (TCC).

I have no doubt that such a class exists and does exert its power over the market economy and the rules – such as they are – that govern it. Nearly a decade ago, I referred to it as the dominant 'business civilization'. I think Gill was mistaken in seeing evidence of its power in the Tripartite Commission, which was more a club of well-meaning has-beens than an effective political actor, a mirror rather than a driver. But he was right in spotlighting the emergence of a transnational interest group with powerful levers over national governments including that of the United States and members of the European Union. Recent research in telecommunications, trade negotiations concerning intellectual property rights and a number of other

spheres where international organisations have been penetrated and influenced by big-business lobbies all point to the existence of such a TCC. Yet to call it a class suggests far more solidarity and uniformity than in fact exists. The more I look into the politics of international business, the more I am struck by the growing divide between big business – the so-called multinationals – and the people running and employed by small and medium business enterprises. These enjoy few of the perks and privileges of the big corporations yet have to conform to the rules and agencies created by them. For them, globalization is something to be resisted, if only because it so blatantly tramples on the democratic principles of accountability and transparency.

The environmental issue area is a good example of the fissures in the TCC. On the one side are the big oil companies, the giant chemical combines, the vested interests of the car manufacturers and associated businesses. On the other are firms in the vanguard of waste disposal and clean-up technologies and interestingly – the transnational insurance business. Fear of the vast claims that might be made against their clients on environmental grounds is putting insurers increasingly in opposition to the polluters. Their opposition, of course, is predicated on legal systems that are sensitive to public opinion. The power of the latter meanwhile is also evident in the growing sensitivity of some elements in business to shareholders and consumers.

Thus, the notion tentatively posited by some of the neo-Gramscians that while there is some sort of TCC there is also an emerging global civil society is not lightly to be dismissed. To quote Leslie Sklair:

> No social movement appears even remotely likely to overthrow the three fundamental institutional supports of global capitalism [. . .] namely, the TNCs, the transnational capitalist class and the culture-ideology of consumerism. Nevertheless in each of these spheres there are resistances expressed by social movements.

Similarly, Rodolfo Stavenhagen, writing on 'People's movements, the antisystemic challenge' in the collection of essays edited by Bob Cox, finds the growth points of a nascent transnational opposition, or counterforce to Sklair's three institutional supports sustaining the Westfailure system. Not only, he says, are such social movements non-governmental, they are popular in the widest sense of that word; they are alternative to established political systems, and therefore often at odds with national governments and political parties and they seek 'to attain objectives that would entail alternative forms of economic development, political control and social organisation'.

In his introduction to this collection of essays, Cox does not predict the imminent demise of the 'fading Westphalian system'. The future world, he observes, 'will be determined by the relative strength of the bottom-up and top-down pressures'. The contest may be a long one and no one should underestimate the power of big business and big government interests behind these top-down pressures. Yet at the same time there is no denying that as Cox says, 'people have become alienated from existing regimes, states and political processes'. Witness the recent amazing, unforeseen turn-out – a quarter of a million in Paris and the same in London – in anti-government marches by country dwellers of every class and occupation. Everywhere, in fact, politicians are discredited and despised as never

before. The state is indeed in retreat from its core competences in security, finance and control over the economy; and this retreat is not inconsistent with its proliferating regulation of many trivial aspects of daily life. The new multilateralism Cox predicates 'will not be born from constitutional amendments to existing multilateral institutions but rather from a reconstitution of civil societies and political authorities on a global scale building a system of global governance from the bottom up'.

For international studies, and for those of us engaged in them, the implications are far-reaching. We have to escape and resist the state-centrism inherent in the analysis of conventional international relations. The study of globalisation has to embrace the study of the behaviour of firms no less than of other forms of political authority. International political economy has to be recombined with comparative political economy at the sub-state as well as the state level. It is not our job, in short, to defend or excuse the Westphalian system. We should be concerned as much with its significant failures as with its alleged successes.

 READING 22

Globalization and the Myth of the Powerless State

Linda Weiss

The new globalist orthodoxy posits the steady disintegration of national economies and the demise of the state's domestic power. This article, instead, seeks to show why the modern notion of the powerless state, with its accompanying reports about the demise of national diversity, is fundamentally misleading. It is undeniable that striking changes have taken place inside nation-states in recent times. On the social policy front, there has been a decisive move towards fiscal conservatism, whether from the Right or the Left, with reforms to taxation systems and the trimming of social programmes. In the economic sphere, governments have moved towards greater openness in matters of trade, investment and finance. These changes are often rep-

resented as prima facie evidence of the emergence of a new global 'logic of capitalism'. According to this logic, states are now virtually powerless to make real policy choices; transnational markets and footloose corporations have so narrowly constrained policy options that more and more states are being forced to adopt similar fiscal, economic and social policy regimes. Globalists therefore predict convergence on neoliberalism as an increasing number of states adopt the low-taxing, market-based ideals of the American model.

In contrast to the new orthodoxy, I argue that the novelty, magnitude and patterning of change in the world economy are insufficient to support the idea of a 'transnational' tendency: that is to say, the creation of genuinely global markets in which locational and institutional – and therefore national – constraints no longer matter. The changes are consistent, however, with a highly 'internationalized' economy in which economic integration is being advanced not only by corporations but also by national governments. Proponents of globalization overstate the extent and 'novelty' value of transnational movements; they also seriously underrate the variety and adaptability of state capacities, which build on historically framed national institutions. My argument therefore seeks not simply to highlight the empirical limits and counter-tendencies to global integration. More importantly, it seeks to elucidate theoretically what most of the literature has hitherto ignored: the adaptability of states, their differential capacity, and the enhanced importance of state power in the new international environment.

Given such variety, even where globalization has gone furthest, as in finance, we continue to find important differentials in national levels of savings and invest-ment, the price of capital, and even the type of capital inflows and outflows. This suggests that any significant 'weakening' in the capacity for macroeconomic management – to the extent that this has occurred – may owe at least as much to 'domestic' institutions as to global processes.

[. . .]

Different hypotheses of 'globalization'

While I have thus far alluded only to the 'strong globalization' hypothesis, there are in fact at least three hypotheses that can be identified in the literature:

(i)   strong globalization; state power erosion
(ii)   strong globalization; state power unchanged
(iii) weak globalization (strong internationalization); state power reduced in scope.

The findings of various studies summarized in the following section provide strong grounds for rejecting the first and second propositions in favour of the 'weak globalization' thesis. However, I find no compelling evidence for that part of the third proposition which claims that the state's role is now generally reduced to that of legitimating decisions initiated and implemented

elsewhere. Instead, I propose a fourth proposition that stresses the differential capacities of states and how the world economy, far from eliminating such differences, is more likely to sharpen and further emphasize their salience for national prosperity:

(iv)   weak globalization (strong internationalization); state power adaptability and differentiation emphasized.

The full development of this proposition rests on more extensive comparative material than can be mustered here. Nevertheless, I shall present a two-step argument, for the globalization thesis can be tackled in two different ways. The more common strategy to date has been to evaluate the extent of economic globalization: how far has it gone? What are its limits and counter-tendencies? Most of the literature has adopted this quantitative approach, often with considerable ability and finesse. Though such assessments are indispensable, they are also controversial. The controversy arises as much from the notorious inaccuracies of the available data as from the different uses to which the data are put. For these reasons, I shall confine this first part of my account to highlighting some of the main findings, and where relevant, the pitfalls. The second part of my argument is concerned with the impact of so-called globalization and its implications for the ability of states to pursue particular policy goals.

I Limits to Globalization

There is clearly some substance to the new globalist orthodoxy. The sheer volume of cross-border flows, of products, people, capital and, above all, of money is impossible to dispute. The important issue, however, turns on the meaning of these flows. Do they point to a clear globalization tendency? If such a tendency existed, one would expect to find evidence indicating that the changes in question conformed to at least three criteria: (i) novelty – is it unusual or without parallel, thus suggesting secular growth rather than oscillation? (ii) magnitude – how substantial is it in size? and (iii) distribution – to what extent is it world-wide in scope? I summarize the main counter-evidence under these three headings.

The novelty of global flows

Are contemporary international flows without historical precedent and therefore posing perhaps novel challenges? Are the post-war trends onward and upward? If the answer to both questions is in the affirmative, then we have clear evidence of a globalization tendency, of secular growth rather than oscillation. The answer, however, varies greatly according to when one starts to measure the changes, hence the often conflicting claims in the literature. At least two findings suggest room for caution.

First is the existence prior to 1913 of trade and capital flows not dissimilar in size to flows in the recent post-war period. [. . .]

The second finding is straightforward. The post-war trend towards greater trade integration, especially marked since the 1960s has been weakening. While world trade has grown much faster than output, this growth has actually been slowing over the 1980s and 1990s, the ratio declining from 1.65 in 1965–80 to 1.34 in 1980–90. Moreover, as Robert Wade has argued, there are not only cyclical but also structural reasons for expecting this slow-down to continue. Structurally, a gradual shift away from manufacturing within the OECD will mean less rather than more trade integration as the share of less trade-intensive services rises. Thus, from the perspective of our first criterion, evidence of an unprecedented tendency is not compelling.

The magnitude of global integration

How big are the changes? The answer depends not simply on when one starts measuring, but on what changes are measured. I will address this point with two examples commonly offered up by globalists as evidence of globalization: Foreign Direct Investment (FDI) and capital mobility.

FDI

Globalists identify the transnationalization of production as the driving mechanism of economic integration, drawing readily on aggregate FDI figures in support of that hypothesis. However, the use of aggregate FDI figures as proxies for the so-called 'globalization of production' seriously distorts reality.

[. . .]

Taking a more disaggregated approach to the investment figures, we can therefore see why FDI does not automatically extend economic linkages, especially in those areas of multinational economic activity that might have a direct bearing on state policies. If the level of FDI is indicative of a globalization tendency at work in the sphere of production, present trends do not point in that direction.

A more realistic indication of the extent to which the 'national' economy is being outflanked by transnational linkages can be gained by measuring inflows and outflows of FDI as a percentage of gross domestic investment. By this standard, the rates of FDI are actually quite modest. With certain notable exceptions – for instance, Britain and Sweden – gross domestic investment in Europe exceeds total FDI, both outbound and inbound, by at least 90 per cent.

Capital mobility

Globalists assume that the world economy is now so integrated that the constraints of location and of institutional frameworks are increasingly irrelevant; that corporations – whether satisfied or disgruntled with a particular national environment – can simply take a 'random walk' in the world market, escaping the confines of any one nation-state. It is this footloose quality of MNCs – above all the threat of exit – that is seen to pose the greatest threat to territorially constituted forms of governance. The reality, however, is at odds with this vision. For, as many studies report, the number of genuinely transnational companies is rather small.

On virtually all the important criteria – share of assets, ownership, management, employment, the location of R&D – 'the importance of a home base remains the rule, not the exception'.

Conventional wisdom nevertheless tells us that cost-reduction is the driving force compelling MNCs toward a footloose career, and that new transport and information technology liberates and encourages MNCs to exploit low-cost production sites, resulting in a globalization of production. Yet, if cost-reduction were the driving force behind the mobile MNC, we would expect to find most, or at least a very sizeable chunk of FDI going to the developing countries. However, the evidence firmly contradicts that expectation. As of 1991, a good 81 per cent of world stock of FDI was located in the high-wage – and relatively high-tax

– countries: principally the US, followed by the UK, Germany, and Canada. Moreover, this figure represents an increase of 12 points since 1967. Indeed the stock of FDI in the UK and the US exceeds the stock in Asia and the entire South.

Such figures underline the point that MNCs do not by and large invest where wages and taxes are lowest. Why not? Three considerations seem relevant. First, new technologies place a premium on fixed costs (equipment, machinery and so on), while reducing the importance of variable costs (such as wages and raw materials). While certain types of labour – especially knowledge-intensive labour – tend to be treated increasingly as a fixed cost, the general effect of this overall transformation is to reduce the cost savings to be gained by moving to low-income sites. Second, new production methods emphasize the growing importance of physical proximity between producers and suppliers – especially in non-assembly operations. These methods privilege local supplier networks, thus driving a trend towards the constitution of regional, not global, sourcing networks. A third factor, underscoring the critical importance of a home base, is the advantage firms derive from domestic linkages: national institutional frameworks which enmesh business in support relationships with trade associations, training and financial institutions, and national and local governments. In sustaining high-wage economies, one of the most important of these support systems is the relationship between government and business, which underpins the national innovation system. Being generally exclusive rather than open to all, support relationships of this kind constitute a competitive advantage.

These considerations suggest that the advantages of maintaining a firm 'home' or regional base may be stronger than ever, perhaps for most companies outweighing those to be gained from 'going global'. It would therefore appear that not only the incidence but also the advantages of mobility have been overstated. But the case against a strong globalization tendency does not rest here. We turn next to evidence concerning how the changes are distributed.

The distribution of trade and investment

Up to this point, my objective has been to show that the novelty and the magnitude of change has been overplayed. I have not sought to deny the existence of a more integrated world economy, a fact which I broadly acknowledge. My concern here is to draw attention to the way trade and investment are distributed. Three trends are inconsistent with a globalization tendency.

(i) *The national bases of production*   First, even if we accept that national economies are more integrated through trade and investment flows than in the recent past, it appears that in all but the smallest economies, trade constitutes quite a small share of GDP, with exports accounting for 12 per cent or less of GDP in Japan, the US and the EC. This means that in the main industrialized economies around 90 per cent of production is still undertaken for the domestic market. The national bases of production – and, as we saw, for investment – therefore seem as pronounced as ever.

(ii) *North–South divisions*   A second pattern runs counter to the idea of a globalizing tendency. Whereas globalization predicts more even diffusion between North and South, in fact world trade, production and investment remain highly concentrated in the OECD – that is, in the rich North. Over the 1970–89 period, the North's share of trade grew from 81 per cent to 84 per cent – though the decline of the South's share in world exports masks their changing composition, with largely negative growth of primary product exports, and a rising share of manufactured exports. Investment has followed a similar pattern, with around 90 per cent going to the North over the same period.

(iii) *Regionalization*   Finally, this predominantly Northern trade and investment is itself becoming more geographically concentrated in intra-regional patterns. For example, intra-European trade now accounts for some 62 per cent of its total export trade. Intra-regional trade within the American region – the US, Canada and Mexico – increased between 1980 and 1992 from 68 per cent to 79 per cent of total US–Japan and US–EU trade. Intra-regional trade has also become the dominant trend in Asia – China, ASEAN, Japan and the NICs – as the region has steadily enhanced its importance as export market and production site for Japan and the NICs. Intra-Asian trade in the period 1986–92 rose from 32.4 to 47.7 per cent of total exports, thus reversing the traditional dominance of trade with the US. In short, trade within Asia has been growing more rapidly than trade between Asia and the US.

[. . .]

Compelling evidence for a strong globalization tendency has thus far been wanting. In some respects, indeed, counter-tendencies seem more apparent. If we turn to the finance sector, however, the reality of a global market seems unassailable.

Since formal removal of the gold standard in 1971 and subsequent liberalization of exchange controls, international capital flows have reached truly spectacular levels. Whichever way we look, it is hard to escape the reality of global money markets where enormous sums are traded daily. This is the 'casino' face of capitalism, unleashed by national governments which now appear powerless to contain its destabilizing effects. It is this change which has given most life to the idea and reality of 'globalization'.

However, there is evidence of national diversity even in money markets. First, the price of capital has not converged. While studies disagree on whether real interest rates in different national markets continue to diverge, the price differential for both loan and equity capital remains considerable. Second, whereas globalization implies equalization, marked differences in savings and investment rates persist. For example, in 1992, the ratio of savings to GDP in eleven countries ranged from 0.5 to 25 per cent. In the lowest band (0.5–2 per cent) sat the US, the UK, Australia and Sweden; Germany and Austria occupied the middle band (10–15 per cent); and in the highest band (20–25 per cent) were Japan, Taiwan and Korea. The differentials in national investment rates tend to parallel those for savings. In 1992, investment as a percentage of GDP ranged from around 15 to 36 per cent, with the US, the UK, Australia and Sweden in the lowest band (15–19 per cent), Germany, Austria and Taiwan in the middle (22–25 per cent), and Japan and Korea in the highest (31–36 per cent).

This strong correlation between savings and investment rates has been interpreted to mean that countries do not draw freely on other countries' savings. Robert Wade, however, reports a fall in the OECD savings–investment correlation from 75 per cent in the mid 1970s to 60 per cent in the 1980s. Financial markets, he suggests, have therefore become more integrated, even if the mobility of capital is somewhat less than anticipated.

Finally, 'dualism' rather than 'transnationalism' seems to distinguish the operation of financial markets, most notably in the area of company shares. These tend to be fixed to specific national stock markets, thus contrasting dramatically with other parts of the financial market – for example, the bond, currency and futures markets – which are genuinely 'transnational'.

These qualifications to 'global' finance suggest that the relevance of national institutions is far from insignificant. Thus, the conclusion to the first part of my argument is that while national economies may in some ways be highly integrated with one another, the result – with the partial exception of money markets – is not so much a globalized world (where national differences virtually disappear), but rather a more internationalized world (where national and regional differences remain substantial and national institutions remain significant). What does this mean then for the power of governments to govern?

II The Extent of Government Powerlessness

For many commentators, the power of global finance – especially of the bond market – to undermine the monetary and fiscal policies of governments seems an incontrovertible truth. It is also viewed as the key constraining feature of a globalized economy: forcing all governments to adopt similar neoliberal – deflationary, fiscally conservative – policies. From this perspective, two conclusions follow. First, global money markets are all-powerful, forcing on governments fiscal conservatism – read 'powerlessness'. Second, it matters not whether a state is weak or strong; all national governments are impotent in the face of global finance. Here I will examine each of these claims in turn.

The problem with the 'powerlessness' argument is not that it is wrong about the new constraints on government capacity to make and implement policy. Rather, it is the assumption that such constraints are absolute rather than relative, and that they represent 'the end of state history' rather than an evolving history of state adaptation to both external and internal challenges. Three weaknesses in particular deserve highlighting.

Overstating earlier state powers

First, globalists tend to exaggerate state powers in the past in order to claim feebleness in the present. Whilst financial globalization is commonly identified as the factor undermining governments' ability to practise effective macroeconomic management – of the Keynesian reflationary variety – some commentators have recently questioned just how effective Keynesian demand management ever was. While in theory the fixed exchange rates guaranteed under the Bretton Woods system provided a more stable policy-making environment, in reality there is little compelling evidence that the state has ever had the sorts of powers that allegedly it has been forced to relinquish.

[. . .]

Overstating uniformity of state response

The fact that not all governments follow neoliberal dictums surely throws into question the central assumptions of the powerlessness argument.

[. . .]

Thus, if global finance has not exerted the uniformly debilitating effects so often claimed for it, why then, we may ask, has the idea of the powerless state seemed so persuasive to so many?

The political construction of helplessness

Perhaps more than anything, it has been the rise of monetarist policies in the 1980s, the emergence of fiscal retrenchment in bulwarks of social democracy like Sweden, and the various speculative attacks on national currencies that have led globalists to conclude that – while governments may reign – the global economy rules.

It must be said, however, that political leaders – especially in the English-speaking world dominated by neoliberal economic philosophy – have themselves played a large part in contributing to this view of government helplessness in the face of global trends. In canvassing support for policies lacking popular appeal, many OECD governments have sought to 'sell' their policies of retrenchment to the electorate as being somehow 'forced' on them by 'global economic trends' over which they have no control.

While it is true that governments are responding to similar pressures in the world economy – the long slump in world-wide demand, stagnant or falling living standards – it is quite misleading to conclude that these pressures derive solely or largely from 'globalization' tendencies, or that the latter produces a uniformity of response.

III Convergence versus Varieties of State Capacity

Globalists have not only overstated the degree of state powerlessness. They have also *over-generalized* it. It is to this final weakness in the globalist argument that we now turn.

The variety of 'national capitalisms' – continental European, East Asian, Anglo-American – finds a parallel in the variety of 'state capacities' for domestic adjustment strategies. In a different context, I will undertake to show how the two may be linked. At issue here, however, is the variety, as opposed to the convergence, of state capabilities. Contrary to globalist predictions, I propose that national differences are likely to become more rather than less pronounced in a highly internationalized environment, thus exacerbating rather than diminishing current differences between strong and weak states.

Yet even those who agree that 'globalization' has been highly exaggerated, nevertheless part company when considering the effects of economic internationalization on state capacity. While some conclude that the nation-state persists as an important locus of accumulation, and that national – and international – actors and institutions continue to structure economic space, others see state powers much more circumscribed through the shedding and shifting of traditional responsibilities.

In a comprehensive recent study, Hirst and Thompson propose that certain traditional powers are declining: 'The power of nation states as administrative and policy-making agencies has declined' while the state's role as an economic manager is 'lessening'. In this respect, they appear to overlap with the globalists. In a more nuanced approach, however, they insist on the enduring importance of the nation-state – not in traditional terms as sovereign power or as economic manager, but as the key source of legitimacy and the delegator of authority to powers above and below the national level. Its territorial centrality and constitutional legitimacy assure the nation-state a distinctive and continuing role in an internationalized world economy, even as conventional sovereignty and economic capacities lessen: 'Nation-states should be seen no longer as "governing" powers [. . .] Nation-states are now simply one class of powers and political agencies in a complex system of power from world to local levels.' According to this interpretation of current tendencies, state power is being reduced and redefined on a broad scale, stripped to the basics, becoming even a shell of its former self: still the supreme source of legitimacy and delegator of authority, but exercising no real capacity over its economic domain. The question is whether one can identify any clear cases which might fit this conception, and whether, having identified them, they represent not simply a group of traditionally 'weak' states, but a group where real power shifts are in train.

It is doubtful that the 'basic state' hypothesis fits even the EU experience, which appears to inform so much of this kind of reasoning. In the German case neither sub-national nor supranational agencies have supplanted the national state's coordinating capacities. Indeed, in a number of important respects – technological innovation and industrial investment – coordination has been growing, not declining, over the past two decades.

Although Hirst and Thompson do insist on the state's continuing importance as the source of legitimacy and the rule of law, and would therefore probably reject the 'weak state' characterization of their position, it is hard to see what kind of substantive powers the state would retain if it is no longer where the action is. If the state is increasingly becoming merely the place from which law is promulgated, authority delegated, powers devolved, then is that not simply a form of power shrinkage by stealth – somewhat akin to the centrifugal tendencies of feudalism? After all, their image of the evolving role of the state (as *Rechtsstaat*) has much in common with the role envisaged by eighteenth-century liberals: thus, not an eclipse of state power as some globalists are led to claim, but certainly a very narrowly defined power.

This seems to me mistaken. For it is blind to state variety and to adaptation. I, too, would emphasize change, but change is hardly novel to the state. Adaptation is the very essence of the modern state by virtue of the fact that it is embedded in a dynamic economic and inter-state system – even the evolving forms of warfare must be seen in that context. My argument is that nation-states will matter more rather than less – and, though not elaborated here, this will advance rather than retard development of the world economy. The argument is in three parts, emphasizing: (i) state adaptation rather than decline of functions; (ii) strong states as facilitators not victims of internationalization; and (iii) the emergence of 'catalytic' states consolidating national and regional networks of trade and investment.

Adaptiveness of the state

[. . .]

The major point to emphasize is that the capacity for domestic adjustment strategy does not stand or fall with macroeconomic capacity, whether of the reflationary or deflationary variety. It rests, perhaps more than ever, on industrial strategy, the ability of policy-making authorities to mobilize savings and investment and to promote their deployment for the generation of higher value-added activities.

This capacity for a coordinated and strategic response to economic change depends, in turn, not so much on specific policy 'instruments' or levels of 'integration into the world economy'. The contrasting cases of Singapore and Britain are testimony to this. Highly integrated Singapore – whose per capita GDP now exceeds that of Britain – maintains strong control over its savings and investment rates, thus engineering upward mobility in the international system. By contrast, highly integrated Britain, with little capacity for industrial adjustment, has failed to arrest its downward slide in the international order – Britain's traditional strength in promoting its financial sector being part of that drama. Thus, high integration does not necessarily mean the displacement of 'national' economies as the locus of accumulation, or the weakening of national economic management.

[. . .]

The state as victim or facilitator of 'globalization'?

In failing to differentiate state capacities, global enthusiasts have been blinded to an important possibility: that far from being victims, (strong) states may well be facilitators (at times perhaps perpetrators) of so-called 'globalization'. Although those researching in the field have yet to explore this possibility, there is sufficient evidence to suggest that this would be a promising line of enquiry. Such evidence as exists for Japan, Singapore, Korea, and Taiwan indicates that these states are acting increasingly as catalysts for the 'internationalization' strategies of corporate actors. As 'catalytic' states (see below), Japan and the NICs are taking the bull by the horns, providing a wide array of incentives to finance overseas investment, promote technology alliances between national and foreign firms, and encourage regional relocation of production networks.

[. . .]

The emergence of 'catalytic' states

The final strand in my argument is that we are witnessing changes in state power; but these changes have to do not with the diminution but with the reconstitution of power around the consolidation of domestic and international linkages.

As macroeconomic tools appear to lose their efficacy, as external pressures for homogenization of trade regimes increase, and as cross-border flows of people and finance threaten the domestic base, a growing number of states are seeking to increase their control over the external environment. State responses to these pressures have not been uniform. They have varied according to political and institutional differences. But, in general, one of two strategies has prevailed. Both involve building or strengthening power alliances: 'upwards', via inter-state coalitions at the regional and international level, and/or 'downwards', via state–business alliances in the domestic market.

To the extent that states are seeking to adapt and reconstitute themselves in these ways, they can perhaps best be seen as 'catalytic' states, to use Michael Lind's term. Catalytic states seek to achieve their goals less by relying on their own resources than by assuming a dominant role in coalitions of states, transnational institutions, and private-sector groups.

As a catalyst, this kind of state is one that seeks to be indispensable to the success or direction of particular strategic coalitions while remaining substantially independent from the other elements of the coalition, whether they are other governments, firms, or even foreign and domestic populations. Thus, far from relinquishing their distinctive goals and identity, states are increasingly using collaborative power arrangements to create more real control over their economies – and indeed over security. As such, these new coalitions should be seen as gambits for building rather than shedding state capacity.

There are many who would support the claim that we are witnessing the end of an era marked by the 'integral state', with assured territorial control over the means of legitimacy, security, and production. But at a time when serious analysis of 'state power' or the 'state's role' has become academically unfashionable, there will undoubtedly be less support for Lind's assertion that in place of the integral state we are now witnessing the rise of the catalytic state.

To what extent can the catalytic state be generalized? The first point to make is that 'catalytic' is being contrasted with 'integral'. It is a way of highlighting the tendency of states to seek adaptation to new challenges by forging or strengthening partnerships with other (state and non-state) power actors, rather than going it alone. Consolidation of such alliances is taking place primarily at regional and international level, between states, though also domestically, between states and corporate actors. The proliferation of regional agreements between nation-states – including the EU, APEC, and NAFTA – can be seen as one manifestation of this tendency. The evolving character of close domestic government–business cooperation, most notably in East Asia, is another.

The second point, however, is that even catalytic states have differential capabilities: some, like Japan and Germany, have both domestic and international clout, and hence are able to use their domestic leverage to position themselves advantageously, for example, in regional coalitions. Others, like the United States, exploit strong international leverage but at the expense of domestic adjustment capacity. Still others, like Russia, are so lacking in domestic capability that they are not even serious candidates for the kind of regional coalitions they otherwise might aspire to lead or join.

Recent examples of states using international agreements as a means of pursuing domestic economic goals include such initiatives as NAFTA and APEC. While both weak and strong states enter into such alliances, it is often the domestically weaker states which take the lead in seeking out this external path, aspiring to constrain others to adopt their own more 'hands off' approach to trade and industry. Australia's enthusiastic efforts in seeking to establish APEC, and the United States' leadership of NAFTA can be seen in this light. These states, with their traditional 'arms length' approach to the corporate sector, lack the more strategic capacities of their East Asian counterparts. In the absence of a normative and institutional base for strengthening developmental capabilities at home, both countries have sought instead to 'level the playing field' outside their domain. To this extent, one might agree with the conclusion that unlike the EU, such moves are driven not by a supranational vision but by 'insecure governments' seeking 'new tools to stimulate growth, employment, and a stable regional policy community'. To make the point in slightly different language, regionalism (inter-state coalitions) without domestic capacity (public–private coalitions) is only half the story, akin to conducting a war of movement without having established a war of position.

What this analysis suggests is that the most important power actors in these new inter-state coalitions will not be those initiating them – for instance, the US and Australia – but those who participate in them from a position of domestic strength. For the major solidity of Japan as a catalytic state in international coalitions is that it has developed robust capability at home via domestic (government–business) linkages. By contrast, the major weakness of the US is the underdevelopment of such linkages, reinforced by the overdevelopment of external strength.

If this reasoning is accepted, then we must enter a caveat to the notion of the rise of the catalytic state. Domestically strong states will more likely act in concert with others; while domestically weak states – especially large ones like the United States – will not completely lose their 'integral' character. In such cases, rather than a concentration on power-sharing we can expect to find an oscillation, as weak states shift between acting alone – through, for instance, defensive protectionism and bilateralism – and with others.

Thus, in this new era, the most successful states will be those which can augment their conventional power resources with collaborative power: engaging others – states, corporations and business associations – to form cooperative agreements and 'consortia' for action on this or that issue. But by far the most important of these coalitions will be partnerships of government and business, for this goes to the very heart of state capacity.

In contrast to Hirst and Thompson's conception discussed earlier, both domestic and regional coalitions imply that the state is not so much 'devolving' power – in a negative sum manner – to other power actors from whom it then maintains a passive distance. Rather, the state is constantly seeking power sharing arrangements which give it scope for remaining an active centre, hence being a 'catalytic' state.

Responses to globalization

Against the hypotheses of advancing globalization, diminishing state capabilities, and eroding institutional diversity, this paper has advanced three propositions. First, the world economy is an internationalized economy, increasingly a regionalized economy; but it is not genuinely a globalized economy in which territorial boundedness and geographic proximity have declining importance for economic accumulation. While money and finance have increasingly become 'global' in some – but not all – aspects of their operation, the same cannot be said of production, trade or corporate practice.

Second, convergence towards a neoliberal model of political economy is highly improbable. This is not simply because economic 'globalization' is rather more limited and subject to counter-tendencies than many accounts would suggest. It is also because nation-states themselves exhibit great adaptability and variety – both in their responses to change and in their capacity to mediate and manage international and domestic linkages, in particular the government–business relationship.

Finally, however, because domestic state capacities differ, so the ability to exploit the opportunities of international economic change – rather than simply succumb to its pressures – will be much more marked in some countries than in others. For while current tendencies in the world economy subject more and more national economies to similar challenges and opportunities, these are likely to solidify the institutional differences that separate the weaker from the stronger performers. Change is indeed occurring, but by the end of the millennium, one should be able to see more clearly that the changes in process in different national systems are those of adaptation rather than of convergence on a single neoliberal model.

The rise of East Asia, the national responses elicited by that challenge, together with the proliferation of regional agreements suggest that we can expect to see more and more of a different kind of state taking shape in the world arena, one that is reconstituting its power at the centre of alliances formed either within or outside the state. For these states, building state capacity, rather than discarding it, would seem to be the lesson of dynamic integration. As we move into the next century, the ability of nation-states to adapt to internationalization – so-called 'globalization' – will continue to heighten rather than diminish national differences in state capacity and the accompanying advantages of national economic coordination.

 READING 23

Globalization and the Resilience of State Power

Daniel Béland

Despite growing evidence that contradicts their claims, several prominent contemporary thinkers argue that globalization favors a decline of the national state. According to Michael Hardt and Antonio Negri, for example, the world is witnessing the emergence of a global capitalist "Empire" in which national states have a far less central position than before. Discourse on the decline of the national state, though shared across ideological lines, is especially popular on the far left, where Marxists and former Marxists have long promoted internationalism and the revolt of the "multitude" (i.e., ordinary people) against global capitalism.

For sociologist Manuel Castells, the planetary expansion of information networks like the Internet goes against national institutions and hierarchies:

> Networks dissolve centres, they disorganize hierarchy [. . .] Thus, contemporary information networks of capital, production, trade, science, communication, human rights, and crime, bypass the national state, which, by and large, has stopped being a sovereign entity.

From this perspective, the development of global capitalism and new communication technologies makes national states increasingly irrelevant: in a world of global communication, national boundaries lose their meaning.

However, these views oversimplify globalization's impact on the national state. Far from being passive in the process of globalization, policymakers in advanced industrial countries often promote free trade, economic integration, and foreign investment in order to gain electoral power and to push their own political agendas at home. These actors stress the domestic prosperity that can result from global exchange: economic openness may benefit countries and may even stimulate welfare state development and coordination. In Canada, for example, the Liberal Party in power between 1993 and 2006 promoted economic integration through the North American Free Trade Agreement (NAFTA) while stressing the need to preserve and even improve the country's welfare state. This welfare state was reframed as a competitive tool oriented toward the reproduction of a well-educated and competitive workforce.

The enactment of NAFTA and, more significant, the creation of the European Union (EU), are the most spectacular outcomes of the integration strategy many political leaders have initiated. Yet institutional and political integration remains limited even within the EU; for example, national states remain in charge of the large social insurance systems that protect workers and citizens against economic insecurity. As such, national states remain the primary source of economic, environmental, social, and military protection in advanced industrial societies.

Consequently, the role of the state is enduring – and even increasing – in advanced industrial societies, despite international variations in taxes and public spending levels. Since 2001, increased public awareness of terrorism in these societies has reinforced the state's legitimacy as the main source of security; when facing the threat of global terrorism, citizens and private businesses like the aviation industry turn to the state for protection. Considering this, as well as the long-term trends of public spending (see Table 1), economic globalization has not caused a massive decline of state power. These trends in public spending suggest that, in the advanced industrial world, national states remain massive actors involved in a number of complex and expensive tasks. Recent scholarship on the national state strengthens the claim that it can still implement policies that strongly affect the life of its citizens.

We should keep in mind three precautionary remarks about globalization and the resilience of state power. First, what is true of advanced industrial societies does not necessarily apply to other parts of the world. In many former socialist countries, for example, the departure from economic planning, widespread neo-liberal reforms, and the expansion of organized crime and the informal economic sector (i.e., activities that are neither taxed nor regulated by the legal system) have temporarily reduced the state's capacity to extract fiscal resources and protect citizens. Declining or insufficient state protection and growing collective insecurity stimulate the development of alternative providers of protection like militias and criminal organizations. This happened in Russia before and immediately after the collapse of the Soviet Union in 1991, suggesting that a strong decline in state power is possible, and that globalization is not necessarily the main factor behind it.

Second, state protection is not an institutional "status quo." Even in advanced industrial countries, state protection may not expand, or even maintain itself, indefinitely. Across countries and policy areas, neoliberal cutbacks and restructuring have reduced the concrete level of state protection offered without any influence from the global economy. The drastic 1996 American welfare reform [. . .] is an example of the decline of state protection in a specific policy area. Furthermore, in analyzing changes in the level of state protection, we must distinguish between political rhetoric about protection and the concrete reforms that have been enacted. For example, "social democratic" rhetoric may have hidden the true scope of cutbacks and restructuring that have significantly altered – and even reduced the level of – state protection in countries like Denmark and Sweden.

Third, domestic policy decisions may shrink a state's fiscal resources through the enactment of widespread income tax breaks, which in the long run may seriously reduce the state's capacity to protect citizens effectively. Because protection capacity is tied to fiscal revenues, and because income tax cuts are politically difficult to overturn, fiscal crises triggered by deep income tax breaks represent a potential menace to state protection. The deficits incurred by these tax breaks may legitimize budget cuts in social and environmental programs, and may lead to the multiplication of alternative,

Table 1 State real expenditure, 1937–95 (as a percentage of GDP)

	1937	1960	1980	1990	1995
Canada	10.1	13.4	19.2	19.8	19.6
France	15.0	14.2	18.1	18.0	19.3
Italy	. . .	12.0	14.7	17.4	16.3
Germany	21.0	13.4	20.2	18.4	19.5
Spain	10.7	8.3	12.5	15.5	16.6
Sweden	10.4	16.0	29.3	27.4	25.8
United Kingdom	11.7	16.4	21.6	20.6	21.4
United States	12.9	19.4	18.7	18.9	16.2

Source: Tanzi, V. and Schuknecht, L. (2000). *Public Spending in the 20th Century*. Cambridge: Cambridge University Press, p. 25.

market-based providers of protection. This situation may then lead to increased social inequality, as market-based protection tends to cover more affluent citizens.

The United States is probably the main advanced industrial country where a large-scale fiscal crisis is most likely to have a negative impact on state protection. The massive federal income tax cuts enacted in 2001 and 2003 have already led to the return of mammoth federal deficits. Though these tax cuts were ostensibly enacted as temporary measures, political pressure to make them permanent is strong, even in light of the new protection demands created by the 2005 Katrina catastrophe in New Orleans. In the future, fiscal crises related to these tax cuts could justify bolder budget cuts and reduce the federal state's capacity to effectively fight threats from economic insecurity and environmental hazards to international terrorism.

The United States faces a deepening contradiction between limited tax revenues and rising protection needs; elected officials who promote the economic interests of specific, frequently narrow, constituencies have significantly reduced the capacity of the state to raise revenues, while greatly increasing military spending and breeding fears concerning global terrorism. Fighting terrorism, environmental threats, and economic insecurity is increasingly expensive, and a growing number of citizens may soon discover that cutting income taxes – especially those of the wealthy – diminishes the state's capacity to protect society against the threats that concern them.

As we have discussed, state protection involves major tradeoffs and necessitates setting fiscal and policy priorities. Income tax cuts and budget deficits are thus a major aspect of the debate over the future of state protection, in the United States and abroad. The state must raise enough taxes to finance appropriate policy responses to growing protection demands. But who should pay for such expanding protection? This is a difficult *political* question, as setting fiscal priorities and tax levels is largely about power relations.

New Protection Needs

Globalization itself has not greatly reduced the capacity of the modern state to protect its citizens. However, globalization has definitely affected state protection and the politics of insecurity in two main ways.

Complicating the actions of the state

First, global trends may complicate or undermine the actions of the state regarding economic, social, or environmental insecurity. For example, pressures from global trade, capital markets, and production create fiscal constraints for policymakers who seek to attract foreign investment and to prevent companies from relocating to other countries. European monetary integration, an example of economic globalization, has forced EU member states to adopt strict budget policies that have reduced their capacity to enact new protection programs, or even to finance existing ones. Financial globalization and competition for foreign investment have also increased the political leverage of business interests, which tend to oppose constraining labor regulations and high corporate and payroll taxes. With the internationalization of protection, the increasingly common idea that firms could relocate to another country "gives more leverage to capital and thus puts downward pressures on employer contributions to welfare state programs and on corporate taxation." This increase in business power may also give more momentum to neoliberal campaigns aimed at privatizing – or downsizing – significant components of state protection. Where the welfare state is concerned, however, evidence shows that national actors and institutions still matter a great deal and that economic globalization has not favored strong institutional convergence.

Regarding the environment, the effects of air and water pollution are increasingly global in nature, and it is difficult for national states to act alone to fight these global environmental threats. Cooperation between states thus becomes necessary. [. . .] Environmental and economic globalization can reinforce each other, as in the field of food safety. In that policy area, global trade can facilitate the propagation of food hazards like BSE while creating new trade conflicts.

Growing protection demands

Second, globalization and the social, economic, and environmental fears it triggers can lead to new protection demands in society, as is clearly the case with global terrorism. In the past, local and regional terrorist networks challenged the authority of the national state in countries, as in Spain (Basque Country) and the United Kingdom (Northern Ireland). In recent decades, new communication technologies and increased transnational mobility have accelerated the development of global terrorist networks, making them difficult to detect and dismantle; to do so, national states rely on intelligence and international cooperation. The global nature of contemporary terrorist networks has complicated the role of the state, but has simultaneously reinforced its legitimacy as the main provider of protection. The strong reliance of the United States on the FBI, the CIA, and the recently created Department of Homeland Security to fight terrorist threats provides ground to this claim. Still, collaboration between these federal agencies and foreign state agencies like Britain's Secret Intelligence Service (SIS) is increasingly common.

Economic globalization can complicate or undermine the actions of the state, but it can also justify more comprehensive state protection. As workers from advanced industrial countries fear downsizing and international production relocation stemming from globalization, they depend more on the national state for economic security; paradoxically, global trade and finance aggravate economic insecurity, which in turn makes the national state the only stable source of protection against global insecurity.

In advanced industrial societies, political leaders can use the insecurity associated with economic globalization to appeal to voters and to justify the policy alternatives they champion. During the 2004 American presidential campaign, for example, Democratic candidate John Kerry referred often to economic globalization as a source of collective insecurity that the federal state should fight. For Kerry, this insecurity was a challenge that the state can confront, not an irremediable source of state decline. Despite the emergence of global social movements that challenge neoliberal globalization and promote alternative, transnational forms of governance and solidarity, many national political leaders still depict themselves as genuine defenders of ordinary citizens against the (perceived) negative effects of economic globalization.

Immigration, a symbol of globalization also remains an enduring source of concern and protection demands. In many advanced industrial countries, national identity is deeply rooted in common languages and culture. In these countries, immigrants may become scapegoats for the social and economic problems that citizens link to globalization. During the last two decades, far-right parties have exploited economic insecurity and urban delinquency to gain support from insecure voters who believe that immigrants are the source of these problems. In countries as varied as Austria, Belgium, Denmark, France, the Netherlands, and Switzerland, populist, far-right parties have depicted immigrants and their children as threatening national values and institutions; for these xenophobic parties, the state must protect national societies from "excessive levels" of immigration and globalization. When depicted as a threat to national identity, globalization can thus strengthen the protective mission of the state, but in this context protection applies only to native-born citizens, not to immigrants and their children. Although xenophobia is not a new phenomenon, politically manipulated fears associated with globalization and transnational migrations can legitimize a potentially repressive form of state protection.

The global spread of diseases such as BSE, SARS, and "bird flu" is another growing source of collective insecurity in advanced industrial countries. Because of their potential to sicken millions of citizens, these diseases receive much media attention, and as a result, global disease has become a key political issue.

Bird flu (avian influenza) presents a striking example of the relationship between global disease and collective insecurity. Especially since 2003, the public has been acutely aware of this potential pandemic threat. At the pinnacle of media buzz on this issue in late 2005, birds became the symbol of a health threat that propagates beyond national borders. As with BSE, bird flu has also presented political leaders with opportunities to depict themselves as competent risk fighters devoted to public safety and security. For example, in November 2005, President Bush used this issue to portray himself as a responsive politician able to cope with potential national and global emergencies. This helped the president to divert attention from the Hurricane Katrina catastrophe, during which many journalists and politicians accused his administration of responding slowly to emergencies.

Global issues like those discussed above create new anxieties and, for that reason, feed national debates about state protection and collective insecurity. Although globalization is clearly an important trend, variations between countries remain strong and the national states remain the enduring focal point of the politics of insecurity. Interestingly, however, global trends and the discourse about them are playing a growing political role within many of these national states.

[. . .]

Beyond Nation-State Paradigms: Globalization, Sociology, and the Challenge of Transnational Studies

William I. Robinson

Introduction

Sociology, and the social sciences in general, are attempting to come to terms with globalization as the world-historic context of events on the eve of the 21st century. Acknowledgment of the growing importance of studying the whole world "as a legitimate object of knowledge" has contributed to the emergence of multidisciplinary units dedicated to "global studies" or "transnational studies" in universities in the United States and elsewhere. Alongside this emergence is a proliferation of research institutes, nongovernmental and intergovernmental organizations dedicated to exploring the diverse dimensions of globalization, including its nature, consequences, and policy implications.

I do not propose in this essay a survey of the current state of transnational studies or a comprehensive review of recent literature, much less to elaborate a new transnational paradigm. Rather, my intent is twofold. First, I call for a break with the "nation-state framework of analysis" that continues to guide much macrosocial inquiry despite recognition among scholars that globalization involves fundamental change in our paradigmatic reference points. Even as the social sciences turn toward transnational studies, scholars often fail to recognize the truly *systemic* change represented by globalization, or what Ruggie terms an "epochal threshold". Consequently, research into transnationalism unfolds within the straightjacket of a nation-state framework. The nation-state is still taken as the basic unit of analysis, and transnationalism and globalization are seen as merely some new stage in *inter*national relations or in *cross*-national comparative studies. I suggest that much macrosocial inquiry has run up against certain cognitive and explanatory limitations in the face of globalization since nation-state conceptualizations are incapable of explaining phenomena that are transnational in character. The way out of this impasse is to shift our focus from the nation-state as the basic unit of analysis to the global system as the appropriate unit. Sociology's fundamental contribution to transnational studies should be the study of *transnational social structure* as the discipline's essential object of inquiry and as a key variable in the global system. I also will selectively examine some recent and promising lines of research into globalization, and suggest elements of an ongoing research agenda in transnational studies.

[. . .]

In sum, in its transnational stage, the national–international axis upon which the world capitalist system has been based has mutated into a qualitatively new global axis in which world zones (e.g., center, semiperiphery, periphery) and nation-states are no longer the central locus of social change. However, the supersession of the nation-state system will be drawn out over a lengthy period and checkered by all kinds of social conflicts played out along national lines and as clashes between nation-states. Social science should be less concerned with static snapshots of the momentary than with the dialect of historic *movement*, with capturing the central *dynamics* and *tendencies* in historic processes. The central dynamic of our epoch is globalization, and the central tendency is the *ascendance* of transnational capital, which brings with it the transnationalization of classes in general. In the long historic view, the nation-state system and all the frames of reference therein is in its *descendance*. However, capitalist globalization is a process, not so much consummated as in motion, and is unfolding in a multilayered world system. Determinacy on the structural side is shifting to new transnational space that is eroding, subsuming, and superseding national space as the locus of social life, even though this social life is still "filtered through" nation-state institutions. This situation underscores the *highly contradictory* nature of transnational relations as well as the *indeterminacy* of emergent transnational social structure.

One key disjuncture in the transnationalization process that has caused confusion in this regard is the internationalization of productive forces within an institutional system still centered around the nation-state. A full capitalist global society would mean the integration of all national markets into a single international market and division of labor and the disappearances of all national affiliations of capital. These economic tendencies are already well underway. What is lagging behind are the political and institutional concomitants – the globalization of the entire superstructure of legal, political, and other national institutions, and the transnationalization of social consciousness and cultural patterns.

[. . .]

While much neo-Gramscianism has emphasized the transformation of the nation-state system under globalizing dynamics, Sklair's "theory of the global system" proposes taking "the whole world" as the starting point [see chapter 7, reading 25] – that is, viewing the world not as an aggregate of nation-states but as a single unit and object of study, as "increasingly necessary for the analysis of a growing number of rapidly changing phenomena". Critiquing "state-centrism" in comparative and macrosociology, Sklair identifies transnational *practices* (TNPs) as operational categories for the analysis of transnational phenomena. The model involves TNPs at three levels: the economic, whose agent is transnational capital; the political, whose agent is a transnational capitalist class; and the

cultural, involving a "culture-ideology of consumerism": "The global system is made up of economic transnational practices and at the highest level of abstraction these are the building blocks of the system. The political practices are the principles of organization of the system. They have to work with the materials on hand, but by manipulating the design of the system they can build variations into it. The cultural-ideological practices are the nuts and bolts and the glue that hold the system together". Locating these practices in the field of a *trans*national global system, Sklair thus sets about to explain globalizing dynamics from outside of the logic of the nation-state system (indeed, he theorizes globalization at the *systemic* level). And Sklair, like the neo-Gramscians, is also concerned with the disjuncture between globalization and the continued institutional existence of the nation-state. "The nation-state [. . .] is the spatial reference point for most of the crucial transnational practices that go to make up the structures of the global system, in the sense that most transnational practices intersect in particular countries and come under the jurisdiction of particular nation-states". One result of this disjuncture is that "while capitalism is increasingly organized on a global basis, effective opposition to capitalist practices tends to be manifest locally".

Robinson attempts to synthesize neo-Gramscian insights with Sklair's theory of the global system in his analysis of an emergent global social structure of accumulation. A social structure of accumulation refers to a set of mutually reinforcing social, economic, and political institutions and cultural and ideological norms that fuse with and facilitate a successful pattern of capital accumulation over specific historic periods. A new global social structure of accumulation is becoming superimposed on, and transforming, existing national social structures of accumulation. Integration into the global system is the causal structural dynamic that underlies the events in nations and regions all around the world over the past few decades. The breakup of national economic, political, and social structures is reciprocal to the gradual breakup, starting some three decades ago, of a preglobalization nation-state based world order. New economic, political, and social structures emerge as each nation and region becomes integrated into emergent transnational structures and processes.

[. . .]

Concluding Remarks: a Research Agenda in Transnational Studies

If the picture I have painted here is incomplete (it is) by having not established any new transnational paradigm, then this essay will not have exceeded its intentions, which was to make a case for a break with nation-state analysis. To recapitulate by way of conclusion, a new multidisciplinary field of transnational studies should be predicated on a decisive break with the nation-state framework of analysis, and diverse *transnational phenomena and processes* should constitute its general subject matter. The "commanding heights" of transnational studies are economic globalization, the transnationalization of the state, classes, political processes, and culture, and the current integration processes taking place around the world (e.g., NAFTA, the European Union, etc.). In addition, transnational studies should interact with all area studies by helping to illuminate the changes globalization brings to each region as components of a global system. Perhaps the principal contribution of such a field, therefore, is less to open new avenues of research into the social universe than to *recast* numerous current social science research agendas in light of globalization, to expunge nation-state centrism in the process, and to explore the complex scenarios that emerge from the dialectic interaction of descendant nation-state and ascendant transnational spaces.

[. . .]

Transnationalism

Leslie Sklair distinguishes between two systems of globalization. The first – the neoliberal capitalist system of globalization – is the one that, as we have seen (chapter 4), is now predominant. The other is the socialist system that is not yet in existence, but is foreshadowed by current alter-globalization movements, especially those oriented toward greater human rights throughout the world. The alter-globalization movements, and the potentiality of socialism, are made possible by the problems in the current system of neoliberal globalization, especially class polarization and the increasingly ecologically unsustainable capitalist globalization.

While the nation-state remains important in his view, it is the case that Sklair focuses on transnational practices that are able to cut across boundaries – including those created by nation-states – with the implication that territorial boundaries are of declining importance in capitalist globalization. As a Marxist, Sklair accords priority to economic transnational practices, and it is in this context that one of the central aspects of his analysis – *transnational corporations* – predominates. Underlying this is the idea that capitalism has moved away from being an international system to a globalizing system that is decoupled from any specific geographic territory or nation-state.

The second transnational practice of great importance is political, and here the *transnational capitalist class* predominates. However, it is not made up of capitalists in the traditional Marxian sense of the term. That is, they do not necessarily own the means of production. Sklair differentiates among four "fractions" of the transnational capitalist class. The first is the *corporate fraction* made up of executives of transnational corporations and their local affiliates. Second, there is a *state fraction* composed of globalizing state and inter-state bureaucrats and politicians. The third, *technical fraction*, is made up of globalizing professionals. Finally, there is the *consumerist fraction* encompassing merchants and media executives. These four fractions are obviously very different from the capitalists conceptualized by Marx.

The transnational capitalist class may not be capitalist in a traditional sense, but it is transnational in various ways. First, its "members" tend to share global (as well as local) interests. Second, they seek to exert various types of control across nations. That is, they exert economic control in the workplace, political control in both domestic and international politics, and culture-ideological control in everyday life across international borders. Third, they tend to share a global rather than a local perspective on a wide range of issues. Fourth, they come from many different countries, but increasingly they see themselves as citizens of the world and not just of their place of birth. Finally, wherever they may be at any given time, they share similar lifestyles, especially in terms of the goods and services they consume.

The third transnational practice is culture-ideology, and here Sklair accords great importance to the *culture-ideology of consumerism* in capitalist globalization. While the focus is on culture and ideology, this ultimately involves the economy by adding an interest in consumption to the traditional concern with production (and the transnational corporations) in economic approaches in general, and Marxian theories in particular. It is in this realm that the ability to exert ideological control over people scattered widely throughout the globe has increased dramatically, primarily through the greater reach and sophistication of advertising, the media and the bewildering array of consumer goods that are marketed by and through them. Ultimately, they all serve to create a global desire to consume what benefits transnational corporations, as well as the advertising and media corporations that both provide examples of such corporations and profit from them.

Ultimately, Sklair examines the relationship among the transnational social practices and the institutions that dominate each by arguing that transnational corporations utilize the transnational capitalist class to develop and solidify the consumerist culture and ideology that is increasingly necessary to feed the demands of the capitalist system of production. Indeed,

it is this relationship that defines global capitalism today and it is the most important force in ongoing changes in the world.

William Robinson expands the idea of transnationalism by adding the concept of the *transnational state* (TNS). He accepts the notion that the nation-state, as well as the Westphalia system, have been superseded (see chapter 6), especially as they relate to capitalism. He looks at the TNS from the point of view of a neo-Marxian analysis of capitalism: "The TNS comprises those institutions and practices in global society that maintain, defend, and advance the emergent hegemony of a global bourgeoisie and its project of constructing a new global capitalist historical bloc. This TNS apparatus is an emerging network that comprises transformed and externally-integrated national states, *together with* the supranational economic [e.g. IMF] and political [e.g. UN] forums."[1]

Philip McMichael criticizes Robinson for developing his concept of the TNS abstractly and theoretically, rather than embedding it in "a conception of the contradictory historical relations within which it emerges."[2] Because he does not historicize his notion of TNS, Robinson is in danger of reifying the concept. In a way, Robinson is guilty of many of the same abuses as the "globalizers" who "impose a singular and abstracted logic on a culturally, ecologically, and politically diverse world."[3]

NOTES

1 William I. Robinson, "Social Theory and Globalization: The Rise of a Transnational State." *Theory and Society* 30, 2001: 165–6.
2 Philip McMichael, "Revisiting the Question of the Transnational State: A Comment on William Robinson's 'Social Theory and Globalization'." *Theory and Society* 30, 2001: 207.
3 Ibid., 208.

 READING 25

Transnational Practices
Leslie Sklair

The argument of this book is that we need to move to what we can term global systems theory if we are to understand the contemporary world and explain what is happening in it. We cannot ignore the nation-state, but this book attempts to offer in addition a conception of globalization based on transnational practices (TNPs). Globalization, therefore, is defined as a particular way of organizing social life across existing state borders. Research on small communities, global cities, border regions, groups of states, and virtual and mobile communities of various types provides strong evidence that existing territorial borders are becoming less important and that transnational practices are becoming more important. The balance of power between state and non-state actors and agencies is changing. This is what is meant by the transnational approach to globalization.

TNPs are analytically distinguished on three levels, economic, political, and culture-ideology, what I take to constitute the sociological totality. In the concrete conditions of the world as it is, a world largely structured by global capitalism, each of these TNPs is typically, but not exclusively, characterized by a major institutional form. The transnational corporation (TNC) is the major locus of transnational economic practices; the transnational capitalist class is the major locus of transnational political practices; and the major locus of transnational culture-ideology practices is to be found in the culture-ideology of consumerism. Not all culture is ideological, even in capitalist societies. The reason why I run culture and ideology together is that consumerism in the global system can only be fully understood as a culture-ideology practice. When we buy something that has been imported we are engaged in a typical economic transnational practice. When we are influenced to vote or support a cause by those whose interests are transnational we are engaged in a typical political transnational practice. When a global brand establishes a set of meanings for us and our friends and many others we do not know personally, we are engaged in a typical culture-ideology transnational practice.

The TNPs make sense only in the context of a global system. Global systems theory based on transnational practices is an attempt to escape from the limitations of state-centrism and to avoid the exaggerations of globalism. In order to do this, it is necessary to spell out exactly what these limitations and exaggerations are. The capitalist global system is marked by a very great asymmetry. The most important economic, political, and culture-ideology goods that circulate around the world tend to be owned and/or controlled by small groups in a relatively small number of places, mainly in and around global cities. Until recently it was both convenient and accurate to use the term Western to describe this asymmetry, and the idea of Western imperialism was widely acknowledged as a way of analysing the global system. Other terms, such as superpower, the triad of centre, semi-periphery, and periphery states, and hegemon state are also common. However, these terms appear to be losing their theoretical point as globalization threatens to displace state-centrism as the most fruitful approach for analysing the world today.

Nevertheless, the inter-state system has been the spatial reference point for most of the crucial transnational practices that go to make up the structures of the global system, in the sense that transnational practices intersect in particular places and these places usually come under the jurisdiction of particular nation-states. But it is not the only reference point, and some argue that it can distort the ways we try to understand the world today. The argument of this book is that the most important global force at the beginning of the twenty-first century is the capitalist global system. Transnational corporations provide the material base for a transnational capitalist class that unquestionably dictates economic transnational practices, and is the most important single force in the struggle to dominate political and culture-ideology transnational practices.

There are several other systems, regionally important, ethnically, culturally, and/or theologically based but none has, as yet, dominated the global system as capitalism did in the twentieth century. Resistances to capitalism, particularly in the form of radical social movements, have been and continue to be numerous and influential, though few offer genuine alternatives to capitalist society and none has had the pervasive success in state-building or the creation of institutions that capitalism enjoyed in the twentieth century. As I shall argue in the latter part of this book, this phase may be coming to an end.

The success of historical systems is often bound up with the success of the states that are their dominant powers. Britain in the nineteenth century was the leading power of the imperialist system, and the United States of America in the twentieth century was the leading power of the international capitalist system. Through their (respective) imperialist and neo-imperialist trajectories the ruling classes of these two countries etched the forms of home-grown capitalism onto what has become the capitalist global system. Mighty domestic economies, progressive ruling classes (in comparison with most others actually existing), and at least some desirable culture-ideology features particularly attractive to modernizing elites were combined with the willingness to use military force to open doors to them all over the world. This ensured the creation, persistence, and often aggrandisement of dominant social classes everywhere willing and eager to adopt their ways.

These dominant classes provided many members of what was to become the transnational capitalist class. The TCC consists of those people who see their own interests and/or the interests of their social and/or ethnic group, often transformed into an imagined national interest, as best served by an identification with the interests of the capitalist global system. In particular, the interests of those who own and control the major transnational corporations dictated the interests of the system as a whole. The fundamental in-built instability of the capitalist global system, and the most important contradiction with which any theory of the global system has to grapple, is that the dominant ideology of the system is under constant challenge. The substantive content of the theory, how those who own and control the transnational corporations harness the transnational capitalist class to solidify their hegemonic

control of consumerist culture and ideology, is the site of the many struggles for the global system. Who will win and who will lose these struggles is not a foregone conclusion.

The role of elites in Britain and the USA in the history of capitalism and the very existence of the TCC that the capitalist classes in Britain, the USA, and other places helped create, have historically built in the asymmetries and inequalities that now characterize capitalist globalization. Just as the leaders of dominant states (whether acting directly in the interests of the capitalist class or not) can call on superior economic, political, and culture-ideology resources in their dealings with those who challenge their interests, the transnational capitalist class enjoys similar dominance. The transnational approach to globalization that provides the framework for this book, therefore, is an attempt to replace the state-centrist paradigm of the social sciences with a paradigm of transnational practices, practices that cross state borders but do not originate with state actors, agencies, or institutions. It is not the state as such that drives globalization, but the transnational capitalist class (the institutional focus of political TNPs). The state, as we shall see, has a place in the transnational capitalist class via globalizing bureaucrats, politicians, and professionals. This class derives its material base from the transnational corporations (the institutional focus of economic TNPs) and the value-system of the culture-ideology of consumerism (culture-ideology TNPs).

[. . .]

Transnational Corporations and Capitalist Globalization

The impact of transnational corporations in the global system, especially *Fortune* Global 500 corporations, is plain for all to see. Tourists and business travellers will more often than not travel on a plane manufactured by one of the few corporations that dominate the aerospace industry, operated by one of the airlines that dominate the civil airline industry (nine airlines were big enough to make the FG500 in 2001). They will mostly occupy hotel rooms subcontracted to or owned or managed by the local affiliate of one of the few chains that dominate the global hotel industry. The cars they rent will be products of the few TNCs that dominate the global auto industry, and the agency may well be part of one of the small group of companies that dominate the car rental industry. And they will pay for some or all of this with one of the credit cards or travellers cheques issued by the few TNCs that control global personal finance.

The traveller will be able to watch television programmes and films produced and distributed by the major media conglomerates, will be able to buy globally branded products, at a price, and will usually be able to get around using English, the major global language. The traveller is also liable to be bombarded with advertisements for global consumer goods placed by the local affiliates of the transnational advertising agencies. While TNCs from the United States no longer dominate these sectors as they once did, they are still the leaders in a wide variety of fields and even when they are not the leaders it is often what are labelled American-style (a problematic idea) cultural products or local adaptations of them that are on offer.

This much is obvious at the level of perception. However, it would be simplistic to conclude that the two Mcs (McLuhan and McDonald's) have succeeded in shaping the global village in the form of a fast food outlet or that the real world is in the process of being reconstructed as a universal theme park along the lines of Disneyland. The reality is much more complex than this, though we would be well advised to remember the central insight of McLuhan, that the world is becoming a global village, and of McDonald's, that global packaging creates global desires.

In the previous chapters I dealt briefly with some of the major ways in which the global system has been categorized. Now is the time to act on the reasons why I find most of these approaches unsatisfactory. Lying behind my summary evaluations of these theories is the conviction that most of them are fixated around the unhelpful ideas that the state is the most appropriate unit of analysis and that First World states exploit Third World countries. The view that is propounded here is that it is more fruitful to conceptualize the global system in terms of transnational practices. Those who dominate in the realm of economic, political, and culture-ideology transnational practices in one community, one subnational region, country, supranational region or, indeed, globally, may exploit, ignore, or help those in other places. The state-centrist approach can lead to empirical

enlightenment in some cases [. . .], but at the expense of some theoretical confusion. The crux of the matter lies in moving beyond state-centrism to a theory of globalization based not in states and the inter-state system, but in transnational practices.

[. . .]

TNCs and foreign direct investment

The history of the TNC is, of course, bound up with the history of foreign direct investment (FDI). Although FDI had been substantial from the beginning of the twentieth century, it really took off in the 1950s, as a result of the flow of funds from the United States into Europe after the Second World War. US-based firms already had considerable sums invested in European subsidiaries since the second half of the nineteenth century, and post-1945 investments served both to rebuild what had been destroyed and to extend it. A political motive was clearly bound up with this economic activity. US foreign policy was based on the necessity of stopping the worldwide advance of communism in Europe and elsewhere through the economic development of areas under threat. US firms did not meekly follow the foreign policy line of their government against their own interests. There were large profits to be made from investing in a whole host of European industries and TNC executives and their local affiliates worked closely with globalizing politicians, bureaucrats, and professionals to make this happen.

In the 1950s and the 1960s many US firms grew so large so fast that Europeans began to speak of the American takeover of their economies. The widely read and influential book of the French politician and columnist, Servan-Schreiber, translated as *The American Challenge*, summed up these fears about the loss of economic independence. This and many other books and newspaper and magazine articles recommended that European industry and commerce should learn from the methods of the Americans and try to beat them at their own game. It is interesting to note that at the turn of the new millennium politicians, bureaucrats, and intellectuals in France were still displaying great suspicion about American influence in Europe. In the late 1990s France was the first government to break ranks on the OECD-sponsored Multilateral Agreement on Investment, and social movements against globalization, McDonald's, and *le fast food* were thinly veiled attacks on the Americanization of Europe. This populist rallying loses some credibility when it is discovered that the entrepreneur Ong Beng Seng from Singapore owned Planet Hollywood; Silas Chou of Hong Kong owned Tommy Hilfiger; Vincent Tan and Khoo Kay Peng, both from Malaysia, owned Kenny Rogers Roasters and Laura Ashley respectively. 'Western icons; Asian owners – such are the fruits of the global marketplace.'

American economic activity in the international arena (or American economic imperialism, as it was increasingly being labelled) began to be identified as a problem in urgent need of resolution. For many, the TNCs were the problem. Since the 1970s, almost all the major international agencies in the economic and trade fields have been producing recommendations on how to regulate the activities of the TNCs in recognition that both the rich countries in which the bulk of FDI was located and the poorer countries needed protection. TNC investments might appear minor relative to the total GNP of most large and rich countries, but they are extremely important in the context of specific economic sectors in poor countries as well as in struggling regions all over the world. The UN Department of Economic and Social Affairs took a special interest in these issues and a series of intensively researched reports in the 1970s led to the creation of a Commission on Transnational Corporations and a research centre. This eventually became institutionalized as the UN Centre on Transnational Corporations, with the difficult task of trying to reconcile the interests of TNCs, communities eager for their investments and those adversely affected. As part of a reorganization at the United Nations, the UNCTC was dissolved in the early 1990s and became the Transnational Corporations and Management Division of the UN Department of Economic and Social Development. Subsequently it was relocated from New York to Geneva and incorporated into the United Nations Conference on Trade and Development (UNCTAD) as the Division on Investment, Technology and Enterprise Development. The UNCTC influential quinquennial publication *Transnational Corporations in World Development* was replaced by an annual *World Investment Report*. While it still produces reviews of the place of TNCs in the global economy and a journal, *Transnational Corporations* (largely an outlet for conventional academic and policy-oriented articles), its role as an independent monitor of the practices of TNCs appears to be over.

Activity at the quasi-governmental level, like the UN and OECD, has been more than paralleled by a plethora of unofficial pressure groups that monitor the activities of the TNCs, wherever they may be. Church, consumer, and other campaigning groups frequently expose abuses of TNC power. The Amsterdam-based Transnational Information Exchange (TIE) was a pioneer of counter-strategies to combat the overwhelming resources that the TNCs can muster when they are attacked. These strategies are based on research to identify the interests behind the target companies 'to such an extent that their image, reputation and credibility are jeopardized by continued support of corporate denial of justice'. Some of the campaigns that TIE has been involved in with other networks have lasted many years, such as the Nestlé Infant Formula boycott, the campaign to force corporations to divest in South Africa, the struggles on behalf of Coca-Cola workers in Guatemala and Control Data workers in Korea, as well as several campaigns against TNC policies in the United States. The US-based International Labor Rights Fund is also very active in these areas. The Washington-based Public Citizen (part of the network founded by Ralph Nader), the New York based Inter-faith Center on Corporate Responsibility, the Boston-based INFACT, the Penang-based International Organization of Consumer Unions, and the Cambridge (UK)-based Baby Milk Action, have all also helped to organize successful campaigns, and there are thousands of similar small transnational networks now monitoring the TNCs in various parts of the world. Most of these organizations have regular newsletters, and many have influential magazines, for example Public Citizen's *Multinational Monitor*. The flood of environmentalist and consumer-advice literature that began in the 1980s often contains material critical of the TNCs.

The views of the TNCs can be found in a variety of sources, for example in their public interest advocacy advertising in the world's mass media, and in countless government sponsored settings. The contest between the TNCs and their critics is, however, very unequal. Mander noted, in all seriousness, that in the USA: 'During the early 1970s, all environmental groups together spent about $500,000 per year in advertising in order to offset an average of about $3 billion in corporate expenditures on the same subjects. This ratio was relatively small, only 6,000 to 1, which may help explain the early success of the environmental movement.' While some of the environmental and human rights organizations now have much greater budgets, the ratio is still weighted heavily in favour of the corporations and business in general.

These struggles pit the small people against the might of the transnational corporations, some of whom are richer than most countries. Nevertheless, even the poorest or smallest countries can, theoretically at least, frustrate the expansion plans of any one of these TNC giants by the simple, if often costly, expedients of refusing them permission to trade or manufacture within their territory or by nationalizing (expropriating) their property if they are already in business there. There is a large literature on this question, and this raises the thorny issue of the relations between TNCs and governments.

TNCs and governments

The theory of capitalist globalization presented here is a direct challenge to the conventional idea that there are different national styles of capitalism (Anglo-American, Japanese, German, French, Chinese, and so on) and that these are consequences of the relations between big business and governments, the historical trajectories of each country (path dependency) and styles of regulation and corporate governance. Obviously there is some truth in all this. There are some differences between big business and the organization of capitalism from place to place, between cities, regions, countries, areas settled by different ethnic groups, and so on, just as there are obviously differences between different industries, companies of different sizes, and companies operating under totally different systems of regulation, wherever they are located. The issue is not whether there are differences (of course there are) but what is the significance of these differences. Most theorists and researchers who accept the reality of globalization accept that there has been a fundamental change in the relations between transnational corporations and governments (or the state, not exactly the same thing). The globalizing challenge to the conventional view is that most governments and the states they

purport to govern have less power over domestic and foreign TNCs than they once had (this cannot be denied, in my view) and, more controversially, that most governments appear to be quite satisfied with this state of affairs and some even want to push it further. My explanation for this is bound up with the structure of the transnational capitalist class, and the role of the state fraction (globalizing politicians and bureaucrats) within it.

[. . .]

The Transnational Capitalist Class

The transnational capitalist class is not made up of capitalists in the traditional Marxist sense. Direct ownership or control of the means of production is no longer the exclusive criterion for serving the interests of capital, particularly not the global interests of capital.

The transnational capitalist class (TCC) is transnational in at least five senses. Its members tend to share global as well as local economic interests; they seek to exert economic control in the workplace, political control in domestic and international politics, and culture-ideology control in everyday life; they tend to have global rather than local perspectives on a variety of issues; they tend to be people from many countries, more and more of whom begin to consider themselves citizens of the world as well as of their places of birth; and they tend to share similar lifestyles, particularly patterns of luxury consumption of goods and services. In my formulation, the transnational capitalist class includes the following four fractions:

- TNC executives and their local affiliates (corporate fraction);
- globalizing state and inter-state bureaucrats and politicians (state fraction);
- globalizing professionals (technical fraction); and
- merchants and media (consumerist fraction).

This class sees its mission as organizing the conditions under which its interests and the interests of the global system (which usually but do not always coincide) can be furthered within the transnational, inter-state, national, and local contexts. The concept of the transnational capitalist class implies that there is one central transnational capitalist class that makes system-wide decisions, and that it connects with the TCC in each community, region, and country.

Political transnational practices are not primarily conducted within conventional political organizations. Neither the transnational capitalist class nor any other class operates primarily through transnational political parties. However, loose transnational political groupings do exist and they do have some effects on, and are affected by, the political practices of the TCC in most countries. There are no genuine transnational political parties, though there appears to be a growing interest in international associations of parties, which are sometimes mistaken for transnational parties. The post-Comintern Communist Movement, the Socialist International, international Fascist organizations, and various liberal and neo-liberal multi-state parties have never had much success.

There are, however, various transnational political organizations through which fractions of the TCC operate locally, for example, the Rotary Club and its offshoots and the network of American, European, and Japan-related Chambers of Commerce that straddles the globe. As Errington and Gewertz show in their study of a Rotary Club in Melanesia as well as my own research on AmCham in Mexico, these organizations work as crucial transmission belts and lines of communication between global capitalism and local business. For example, a visit to the website of BISNIS (Business Information Service for the Newly Independent States) of the USA Trade Center in the Russian Far East tells us that in addition to two International Business Associations there were eight Rotary Clubs operating in this remote region in 2001.

At a more elevated level are the Trilateral Commission of the great and good from the United States, Europe, and Japan whose business is 'Elite Planning for World Management'; the World Economic Forum which meets at Davos in Switzerland and the annual Global conferences organized by *Fortune* magazine that bring together the corporate and the state fractions of the TCC. Many other similar but less well-known networks for capitalist globalization exist, for example the Bilderberg Group and Caux Round Table of senior business leaders. There are few major cities in any First or Third World (and now New Second World) country that do not have members of or connections

with one or more of these organizations. They vary in strength from the major First World political and business capitals, through important Third World cities like Cairo, Singapore, and Mexico City, to nominal presences in some of the poorer countries in Africa, Asia, and Latin America. They are backed up by many powerful official bodies, such as foreign trade and economics departments of the major states. Specialized agencies of the World Bank and the IMF, WTO, US Agency for International Development (USAID), development banks, and the UN work with TNCs, local businesses, and NGOs (willing and not so willing) in projects that promote the agenda of capitalist globalization.

The political practices of the transnational capitalist class will be analysed in terms of two issues. First, how it operates to change the nature of the political struggle between capital and labour, and second, the downgrading of indigenous practices.

Labour and the transnational capitalist class

The relative strength of the transnational capitalist class can be understood in terms of the relative weakness of transnational labour. Labour is represented by some genuinely transnational trade unions. The World Federation of Trade Unions (WFTU) was founded in 1945, with 350 delegates representing 67 million workers in 56 countries. This immediately postwar show of labour unity included members from the CIO (Congress of Industrial Organizations), one of the two main union movements in the USA (but not the other, the AFL (American Federation of Labor)), Britain, the Soviet Union, China, and India. WFTU split under the pressure of the Cold War in 1949, when the British TUC and the CIO from the United States (followed by the AFL) set up in opposition the International Confederation of Free Trade Unions (ICFTU). ICFTU followed a strict international and national no-contact policy with the WFTU, which it saw as entirely Soviet-dominated. In the 1980s, the WFTU had over 200 million members in seventy countries (most of Eastern Europe and communist unions in Western Europe and Japan), though the Italian communist trade union had withdrawn and the French began to distance themselves in the mid 1970s, ostensibly to improve the climate for domestic solidarity. ICFTU had about 90 million members (in ninety-two countries, including Western Europe, the Americas, and most of the Third World). The World Council of Labour, a Christian-oriented movement, had about 15 million members. The collapse of the Soviet Union and communism in general in Eastern Europe in the 1990s led to the collapse of the WFTU, and splits that developed as a result of this in the ICFTU suggest that labour solidarity in opposition to capitalist globalization is an uncertain prospect.

In addition, there are some industrially based transnational union organizations, for example the International Metalworkers Federation, and the International Union of Food and Allied Workers' Associations. These have been involved in genuine transnational labour struggles, and have gained some short-term victories. However, they face substantial difficulties in their struggles against organized capital, locally and transnationally and they have little influence.

However, there is a good deal of research on how the labour movement reacts to globalization. The level of unionization to be found in TNC-owned industry in different countries varies widely as do the prospects for successful campaigns. Wills and Herod, in case studies from Europe and the USA respectively, both emphasize the need for strategic flexibility. In some circumstances organizing globally promises better prospects of success, in others organizing locally does. The question cannot be realistically discussed, however, unless there is some measure of the genuine independence of the union. We must distinguish at least three cases: first, where unions are prohibited or repressed; second, where unions are the creatures of governments or companies; third, where genuinely independent unions actually operate. While most TNCs in most countries will follow the local rules regarding the unions, host governments, particularly those promoting export-processing industries (not always under pressure from foreign investors), have often suspended national labour legislation in order to attract TNCs and/or to keep production going and foreign currency rolling in. Some cases will be discussed in the next chapter. With very few exceptions, most globalizing bureaucrats and politicians wanting to take advantage of the fruits of capitalist globalization will be unhelpful towards labour unions, if not downright hostile to them when they dare to challenge the transnational capitalist class.

Downgrading of indigenous practices

Even the most casual observer of transnational practices in the economic, political, and culture-ideology spheres cannot but be struck by the fact that indigenous practices are often unfavourably compared with foreign practices. Despite conceptual difficulties of the indigenous–foreign distinction (similar to traditional–modern), such comparisons are common between countries, cities, neighbourhoods, and regions. The downgrading of indigenous practices in many parts of the world is a subtle and circular process in which the newcomer has all the advantages and the incumbent all the handicaps. The necessity for and the presence of foreign companies, for example, are constant reminders of the deficiencies of the domestic economy. The new methods that TNCs bring are defined as more efficient (if not necessarily more desirable) than the traditional methods of production current in the host economy, and where entirely new products enter, this only underlines the inadequacies of the host. These can all have a depressing effect on local industry.

[. . .]

It is necessary to distinguish between economic, political, and culture-ideology practices here. In terms of economic logic, an indigenous enterprise may be fulfilling the needs of the local consumers through efficient use of domestic inputs, while in terms of political (transnational) logic it is perceived as quite inefficient because of its lack of international competitiveness. In more dramatic terms, the downgrading of local industries reflects the success of the transnational capitalist class in dragging them into the global economy and thereby transforming them, even in a rather minimal sense, into transnational industries.

The presence of expatriate managers and technicians in foreign firms in even the most industrially advanced economies serves to intensify the distinction between superior foreign and inferior indigenous industry. Recruitment of top management appears to be through two circuits, but with a predominantly one-way flow. Transnational companies, particularly those with global reputations, have less difficulty in recruiting the available staff, either from indigenous firms or from other foreign companies. Indeed, there is some evidence of a transnational staff circuit as random conversations in airports and more systematic interviews with TNC executives confirm. The larger transnationals commonly train key staff at headquarters (usually in the USA, Japan, and Europe) and for some a job with a major TNC is the first step in a global career. There is a good deal of evidence to suggest that managerial and technical talent flows from the indigenous sector to the transnational companies rather than vice versa, particularly but not exclusively in the Third World. Gershenberg argued this for Kenya as Okada did for Indonesia, though my own later research on Mexico suggests that this may be more of a two-way process in some industries. *Fortune* (18 August 1997) reported that Microsoft had subsidiaries in sixty countries employing 6,200 people of whom only five were expatriates! While this sounds exceptional, local economies may derive benefits from this type of brain drain, even sufficient to offset the costs, if there is seen to be fair competition between the TNCs and the indigenous firms for trained managerial, technical, and craft personnel. The optimum situation would be a policy that would encourage the TNCs to train young people rather than entice away those already trained and working in the indigenous sector. Some of these young people are, of course, tomorrow's transnational capitalist class.

The downgrading of indigenous industry may be compensated for by the more progressive business environment that foreign companies promote, and particularly the high-technology companies of US, European, or Japanese origin. Transnational corporations can give a competitive stimulation to existing local companies by demonstrating the business potential of new lines or products, and they can also directly influence the market for new indigenous firms, as Evans has shown for the computer industries in several countries. In general, higher expectations of transnational firms for business services and a better-educated workforce may provoke the state into public spending that might otherwise not have taken place. For example, some governments would probably not have spent as much on telecommunications and infrastructure as they have done without the stimulus of a foreign-dominated export sector that produces hard-currency earnings and the expectation that such facilities, however expensive, would attract even more companies. The managers and workers of those firms may well benefit from this in the long run, as well as the TNCs. It must be noted, however, that the managers and workers of those

indigenous firms that go under will not see this as an undiluted benefit and that state subsidies to attract FDI will not benefit the poor much.

There can also be a knock-on effect of the higher and more innovative technology that some foreign firms employ, all through society. This generates a climate for the technological upgrading of industry as a commercial proposition, and it also ensures that hardware and software are conveniently available, at a price, for those who wish to take advantage of them in any sphere. The presence of famous name globalizing firms undoubtedly encourages some enterprising local businesses to take opportunities that are offered for joint ventures and other forms of strategic alliances.

In these ways the transnational capitalist class downgrades certain indigenous practices by comparison with new and more glamorous transnational practices (some of which, paradoxically, might have originated locally as hybrid cultural practices). This creates what used to be termed a comprador mentality, the attitude that the best practices were invariably connected with foreigners who were the bearers of capitalist practices. Comprador mentality was either a cost or a benefit, depending on your position in the ideological struggle between those who believed that capitalism would inevitably damage Third World development prospects in the long run, and those who believed that there would be no development prospects without capitalism. This struggle revolved around the opposing material interests of competing classes and groups, and it still does.

Capitalist globalization has created new groups of what can be termed indigenous globalizers, aspiring members of the transnational capitalist class who have replaced the old compradors. They identify with global capitalism rather than any particular powerful country or corporation. Like all globalizers they are intellectually and geographically mobile. They make their connections with their countries of residence through the globalizing politicians and professionals who are officially responsible for regulating business, politics, and culture-ideology at the level of the national and local state.

The thesis that defines my approach to political transnational practices is that the state is a site of struggle between globalizers and localizers, principally between globalizing bureaucrats and politicians (indigenous globalizers) on the one hand and localizing bureaucrats and politicians on the other.

[. . .]

There are those who see the destiny of the world as bound up with the adoption of all that is modern, often embodied in the products and practices of the TNCs. On the other hand, there are those who are deeply suspicious of the modernization represented by the TNCs, particularly where this is perceived as Western or US dominance in culture, industry, warfare, science, and technology. A battery of concepts, some of which have migrated from social science jargon to the mass media, identify those on either side of the divide. The academically discredited distinction between traditional and modern is still common currency, while the notions of inward-oriented and outward-oriented describe those who look for guidance and sustenance to the resources of their own groups as opposed to those who look outside, usually to the West. Much the same idea is expressed by the distinction between local and cosmopolitan orientation. [. . .]

The price that the state will pay to sustain the costs of foreign investment will depend largely on the powers of indigenous globalizers, the local members of the transnational capitalist class. Whatever the price happens to be at a given time, and this can vary dramatically, it will be a price worth paying for some and not for others. What accounts for the complexity of the problem of evaluation is not only the economic and social costs involved themselves, but the interests, conflicting or in harmony, of those who pay the costs and those who reap the benefits. It may be an over-simplification to conceptualize all the different interests in terms of class struggle, particularly as some of the interest groups involved and some of the alliances of interests forged may defy analysis in conventional Marxist terms, particularly in the Third World. Nevertheless, there are class interests involved even though they may not always conveniently reduce to one labouring class versus one capitalist class.

The transnational capitalist class supported by the strata that the TNCs have created (globalizing bureaucrats, politicians, and professionals) and even in some circumstances privileged fractions of the labour force, will all increasingly identify their own interests with those of the capitalist global system. Those on the fringes of the TCC will often be forced to make a choice between acting on behalf of it against what many would define as the interests of their own communities, as the transnational practices of capitalist globalization penetrate ever deeper into the areas that most heavily impact on their daily lives. The specific function of those who are directly responsible for transnational political practices is to create and sustain the organizational forms within which this penetration takes place and to connect them organically with those indigenous practices that can be incorporated and mobilized in the interests of the capitalist global system. In order to do this the transnational capitalist class must promote, all over the world, a specific structure of culture-ideology transnational practices, namely the culture-ideology of consumerism. It is no accident that the age of capitalist globalization should have begun to flower in the second half of the twentieth century, just when the electronic revolution that heralded the age of the globalizing mass media took root.

[. . .]

The Culture-Ideology of Consumerism

The transformation of the culture-ideology of consumerism from a sectional preference of the rich to a globalizing phenomenon can be explained in terms of two central factors, factors that are historically unprecedented. First, capitalism entered a qualitatively new globalizing phase in the 1960s. As the electronic revolution got under way, the productivity of capitalist factories, systems of extraction and processing of raw materials, product design, marketing and distribution of goods and services began to be transformed in one sector after another. This golden age of capitalism began in the USA, but spread a little later to Japan and Western Europe and other parts of the First World, to the NICs, and to some cities and enclaves in the Third World. Second, the technical and social relations that structured the mass media all over the world made it very easy for new consumerist lifestyles to become the dominant motif for these media. Therefore, in the second half of the twentieth century, for the first time in human history, the dominant economic system, capitalism, was sufficiently productive to provide a basic package of material possessions and services to almost everyone in the First World and to privileged groups elsewhere. Capitalism, particularly in its neo-liberal phase from the 1980s, promised that eventually the rising tide would raise all boats, that is, everyone else in the world would get rich as long as they did what the transnational capitalist class told them to do. A rapidly globalizing system of mass media was also geared up to tell everyone what was available and, crucially, to persuade people that this culture-ideology of consumerism was what a happy and satisfying life was all about. In a powerful empirical study of the increasing hours and more intensive nature of work in the United States since the 1950s, Schor demonstrated how capitalist consumerism led North Americans (and, I would argue, other groups elsewhere) into a sort of Faustian bargain whereby those who can find work trade off their time for more and more consumer goods and services.

Mass media perform many functions for global capitalism. They speed up the circulation of material goods through advertising, which reduces the time between production and consumption. They begin to inculcate the dominant ideology into the minds of viewers, listeners, and readers from an early age, in the words of Esteinou Madrid, 'creating the political/ cultural demand for the survival of capitalism.' The systematic blurring of the lines between information, entertainment, and promotion of products lies at the heart of this practice. This has not in itself created consumerism, for consumer cultures have been in place for centuries. What it has created is a reformulation of consumerism that transforms all the mass media and their contents into opportunities to sell ideas, values, products, in short, a consumerist world-view. Elements of this are found in Boorstin's idea of the consumption community, integral to his thesis of American distinctiveness. Muniz and O'Guinn take this forward in the concept of brand community. Their ethnographic studies of owners of Macintosh computers and Saab and Ford Bronco cars illustrate the existence of three traditional markers of community, namely shared consciousness, shared rituals and traditions, and a sense of moral responsibility. They conclude, somewhat controversially: 'We believe brand communities to be real, significant, and generally a good thing, and evidence of the persistence of community in consumer culture.'

Contemporary consumer culture would not be possible without the shopping mall, both symbolically and substantively. As Crawford argued, the merging of the architecture of the mall with the culture of the theme park has become the key symbol and the key spatial reference point for consumer capitalism, not only in North America but increasingly all over the world. What Goss terms the magic of the mall has to be understood on several levels, how the consuming environment is carefully designed and controlled, the seductive nature of the consuming experience, the transformation of nominal public space into actual private terrain. Although there are certainly anomalies of decaying city districts interspersed with gleaming malls bursting with consumer goods in the First World, it is in the poorer parts of the Third World that these anomalies are at their most stark. Third World malls until quite recently catered mainly to the needs and wants of expatriate TNC executives and officials, and local members of the transnational capitalist class. The success of the culture-ideology of consumerism can be observed all over the world in these malls, where now large numbers of workers and their families flock to buy, usually with credit cards, thus locking themselves into the financial system of capitalist globalization. The integration of the medium of the mall and the message of the culture-ideology of consumerism had a formative influence on the trajectory of global capitalism. The medium looks like the message because the message, the culture-ideology of consumerism, has engulfed the medium. The problem, therefore, is not *Understanding Media* (the title of McLuhan's great if somewhat misconceived book) but understanding capitalist globalization, the system that produces and reproduces both the message and the media that incessantly transmit it.

A fundamental problem that has plagued media studies is the precise relationship between, on the one hand, the media and the messages they relay and, on the other, the audiences that receive these messages and the meanings they take from them and/or read into them. As we shall see below, it is naive to assume that most media messages actually do have the effects that their creators intend, even when the audiences are deemed to be unsophisticated and lacking in education. A growing body of theory and research has tested these ideas in a wide variety of social, cultural, and geographical settings. Ang, and Liebes and Katz, who carried out research projects on attitudes to the soap opera *Dallas*, discovered that different audiences read the same programmes very differently. While Ang's notion of a critical ethnography of reception and the social dynamics of meaning-making of Liebes and Katz problematize the message-reception issue very fruitfully, my contention here is that this research is mainly directed to a second order of meanings, no more and no less important than the first order of meaning of these media products. However, it is the first order of meanings, the culture-ideology of consumerism, with which I am concerned here. This provides the framework for the second order of meanings which raises different, more nuanced, and sometimes contradictory issues.

The connections between capitalist globalization and the culture-ideology of consumerism must be laid bare. In an attempt to do this, Featherstone develops a useful composite picture of contemporary consumer culture. He writes:

1 Goods are framed and displayed to entice the customer, and shopping becomes an overtly symbolic event.
2 Images play a central part, constantly created and circulated by the mass media.
3 Acquisition of goods leads to a 'greater aestheticisation of reality'.

The end result of these processes is a new concept of lifestyle, enhanced self-image. This 'glosses over the real distinctions in the capacity to consume and ignores the low paid, the unemployed, the old', though the ubiquity of the culture-ideology of consumerism actually does include everyone (or, at least, all those with the potential to buy) however poor, because no one can escape its images. And, it must be added, very few people would choose to escape its images and what they represent in terms of the good, or better, life. Monga insightfully analyses this issue through the stories of women from Africa who eventually found asylum in France and USA (and many more who did not). 'Though the perspectives of these women are in themselves of interest, what is of real import is their fundamental goal: survival in a rapidly changing world where the rhetoric of globalisation poorly conceals the reality of the increasing marginalisation of Africa and its inhabitants [including men].' This is concretely expressed in three strategies

for African women and their children mainly through migrating to the USA. The first is through the sale of beauty products for immediate income. The cosmetics industry in the USA, uniquely, has designed a range for black women, so women in Africa are keen to get hold of them, usually through high end informal sector locations. These locations are also socio-economic markers of a system based on credit in which authentic products straight from the USA are at a premium. The second strategy is education of children, a route to intermediate material well-being. The possibility of working through college in the USA makes this an attractive option. While France focuses on rhetoric for African women, the USA focuses on marketable skills, and the myth of America as the land of opportunity contrasts with the racism that black people often find in France. The third, long-term strategy, is the Americanization of children, through giving birth in the USA. This involves the rapid Americanization of names in Africa (usually taken from TV characters) and, Monga argues, illustrates a deeper desire to participate in the global village. She quotes Zhan to the effect that the 'success of American brand-name products abroad is due not to their "Americanism" per se but to their ability to match the demands of a diverse market throughout the world.' Monga is entirely on the mark when she argues: 'whereas women from Africa turn to American culture, some members of the African-American community look to African, or African-inspired culture as a means of expressing their need for self-affirmation and social recognition, often utilizing the same cultural markers as African women: first names, apparel, and art objects.'

The issue of Americanization is clearly a central dilemma of any critique of consumerism (and also of the politics of the consumer movement). Many scholars point up the distinctive role of the United States in the campaign to make consumer culture universal. Through Hollywood, and the globalization of the movies, via Madison Avenue, from where Ewen's captains of consciousness created the modern advertising industry, to the more geographically diffuse but ideologically monolithic television networking conceptualizers, the consumerist elites of the transnational capitalist class in the United States has assumed leadership of the culture-ideology of consumerism in the interests of global capitalism in the twentieth century.

A good illustration of this is in the origin of the soap opera, one of the most highly developed media forms through which mass consumerism is projected. It began in the 1920s when Glen Sample, an American advertising agent, had the idea of adapting a newspaper serial for the radio, a medium already dominated by commercial interests. The programme, *Betty and Bob*, was sponsored by a flour manufacturer, and Sample used the same idea to promote Oxydol washing powder for Procter & Gamble, under siege from Unilever's Rinso in the US market. Oxydol won out, and the so-called soap opera that was used to sell it gave its name to a genre, massively reinforced by its wholesale adoption by television all over the world since the 1950s.

The universal availability of the mass media has been rapidly achieved through relatively cheap transistor radios, cassette recorders, and televisions, which now totally penetrate the First World, almost totally penetrate the urban Second and Third Worlds, and are beginning to penetrate deeply into the countryside even in the poorest places. Thus, the potential of global exposure to global communication, the dream of every merchant in history, has arrived. The socialization process by which people learn what to want, which used to occur mainly in the home and the school, is increasingly taking place through what the theorists of the Frankfurt School had so acutely termed the culture industry.

[. . .]

 READING 26

Social Theory and Globalization: The Rise of a Transnational State

William I. Robinson

Globalization is a relatively new concept in the social sciences. What this concept exactly means, the nature, extent, and importance of the changes bound up with the process, is hotly debated. But few would doubt that it is acquiring a critical importance for the academic as well as the political agenda of the twenty-first century, or that it poses a distinctive challenge to theoretical work in the social sciences. The historic limitations of social theory, insofar as it has been informed by the study of "national" societies and the nation-state, are brought into focus by the universalizing tendencies and transnational structural transformations bound up with globalization. To what extent is the nation-state a historically specific form of world social organization now in the process of becoming transcended by capitalist globalization? This is the question that underlies the present essay, although the matter I intend to address is more circumscribed.

The debate on globalization has increasingly centered on the relation of the nation-state to economic globalization. But the issue of globalization and the state has been misframed. Either the nation-state (and the inter-state

system) is seen as retaining its primacy as the axis of international relations and world development – the "strong state" thesis – in a dualist construct that posits separate logics for a globalizing economic and a nation-state based political system, or the state is seen, as in the "weak state" or diverse "end of the nation-state" theses, as no longer important. Rejecting these frames, I intend here to clarify the relationship between globalization and the nation-state by critiquing and moving beyond this global-national dualism by developing the concept of a transnational state. I argue that the state and the nation-state are not coterminous. The conflation of the two in the globalization literature has impeded analysis of the increasing separation of state practices from those of the nation-state.

Specifically, I call for a return to a historical materialist conception of the state, and on this basis explore three interrelated propositions: (1) economic globalization has its counterpart in transnational class formation and in the emergence of a transnational state (henceforth, TNS) that has been brought into existence to function as the collective authority for a global ruling class; (2) the nation-state is neither retaining its primacy nor disappearing but becoming transformed and absorbed into this larger structure of a TNS; (3) this emergent TNS institutionalizes a new class relation between global capital and global labor.

[...]

A TNS apparatus is emerging under globalization *from within* the system of nation-states. The nation-state system, or inter-state system, is a historical outcome, the particular form in which capitalism came into being based on the complex relations among production, classes, political power, and territoriality. The material circumstances that gave rise to the nation-state are now being superseded by globalization. If capitalism's earlier development resulted in a geographic (spatial) location in the creation of the nation-state system, then its current globalizing thrust is resulting in a general geographic dislocation. What is required is a return to a historical-materialist theoretical conceptualization of the state, not as a "thing," or a fictional macro-agent, but as a specific social relation inserted into larger social structures that may take different, and historically determined, institutional forms, only one of which is the nation-state. Nothing in the current epoch suggests that the historic configuration of space and its institutionalization is immutable rather than itself subject to transformation.

This is to say that the political relations of capitalism are entirely historical, such that state forms can only be understood as historical forms of capitalism. Although the proposition cannot be explored here, I suggest that the explanation for the *particular* geographic expression in the nation-state system that world capitalism acquired is to be found in the historical uneven development of the system, including its gradual spread worldwide. Territorialized space came to house distinct market and capital accumulation conditions, often against one another, a process that tended to be self-reproducing as it deepened and became codified by the development of nation states, politics, and culture, and the agency of collective actors (e.g., Westphalia, nationalism, etc.). This particular spatial form of the uneven development of capitalism is being overcome by the globalization of capital and markets and the gradual equalization of accumulation conditions this involves.

To summarize and recapitulate: the state is the congealment of a particular and historically determined constellation of class forces and relations, and states are always embodied in sets of political institutions. Hence states are: (a) a moment of class power relations; (b) a set of political institutions (an "apparatus"). The state is not one or the other; it is both in their unity. The separation of these two dimensions is purely methodological (Weber's mistake is to reduce the state to "b"). National states arose as particular embodiments of the constellations of social groups and classes that developed within the system of nation-states in the earlier epochs of capitalism and became grounded in particular geographies. What then is a transnational state? Concretely, what is the "a" and the "b" of a TNS? It is a particular constellation of class forces and relations bound up with capitalist globalization and the rise of a transnational capitalist class, embodied in a diverse set of political institutions. These institutions are transformed national states and diverse supranational institutions that serve to institutionalize the domination of this class as the hegemonic fraction of capital worldwide.

Hence, I submit, the state as a class relation is becoming transnationalized. The class practices of a new global ruling class are becoming "condensed," to use Poulantzas's imagery, in an emergent TNS. In the process of the globalization of capital, class fractions from different countries are fusing together into new capitalist groups within transnational space. This new transnational bourgeosie or capitalist class is that segment of the world bourgeosie that represents transnational capital. It comprises the owners of the leading worldwide means of production as embodied principally in the transnational corporations and private financial institutions. What distinguishes the transnational capitalist class from national or local capitalist fractions is that it is involved in globalized production and manages global circuits of accumulation that give it an objective class existence and identity spatially and politically in the global system, above any local territories and polities.

The TNS comprises those institutions and practices in global society that maintain, defend, and advance the emergent hegemony of a global bourgeoisie and its project of constructing a new global capitalist historical bloc. This TNS apparatus is an emerging network that comprises transformed and externally-integrated national states, *together with* the supranational economic and political forums and that has not yet acquired any centralized institutional form. The rise of a TNS entails the reorganization of the state in each nation – I will henceforth refer to these states of each country as *national states* – and it involves simultaneously the rise of truly supranational economic and political institutions. These two processes – the transformation of nation-states and the rise of supranational institutions – are not separate or mutually exclusive. In fact, they are twin dimensions of the process of the transnationalization of the state. Central to my argument is that under globalization the national state does not "wither away" but becomes transformed with respect to its functions and becomes a functional component of a larger TNS.

The TNS apparatus is multilayered and multi-centered. It links together functionally institutions that exhibit distinct gradations of "stateness," which have different histories and trajectories, and which are linked backward and forward to distinct sets of institutions, structures, and regions. The supranational organizations are both economic and political, formal and informal. The economic forums include the International Monetary Fund (IMF), the World Bank (WB), the Bank for International Settlements (BIS), the World Trade Organization (WTO), the regional banks, and so on. Supranational political forums include the Group of 7 (G-7) and the recently formed Group of 22, among others, as well as more formal forums such as the United Nations (UN), the Organization of Economic Cooperation and Development (OECD), the European Union (EU), the Conference on Security and Cooperation in Europe (CSCE), and so on. They also include regional groupings such as the Association of South East Asian Nations (ASEAN), and the supranational juridical, administrative, and regulatory structures established through regional agreements such as the North American Free Trade Agreement (NAFTA) and the Asia-Pacific Economic Cooperation (APEC) forum. Here I wish to theorize this emerging configuration. These supranational planning institutes are gradually supplanting national institutions in policy development and global management and administration of the global economy. The function of the nation-state is shifting from the formulation of national policies to the administration of policies formulated through supranational institutions. However, it is essential to avoid the national–global duality: national states are not external to the TNS but are becoming incorporated into it as component

parts. The supranational organizations function in consonance with transformed national states. They are staffed by transnational functionaries that find their counterparts in transnational functionaries who staff transformed national states. These *transnational state cadres* act as midwives of capitalist globalization.

The TNS is attempting to fulfill the functions for world capitalism that in earlier periods were fulfilled by what world-system and international relations scholars refer to as a "hegemon," or a dominant capitalist power that has the resources and the structural position that allows it to organize world capitalism as a whole and impose the rules, regulatory environment, et cetera, that allows the system to function. We are witnessing the decline of US supremacy and the early stages of the creation of a transnational hegemony through supra-national structures that are not yet capable of providing the economic regulation and political conditions for the reproduction of global capitalism. Just as the national state played this role in the earlier period, I suggest, the TNS seeks to create and maintain the preconditions for the valorization and accumulation of capital in the global economy, which is not simply the sum of national economies and national class structures and requires a centralized authority to represent the whole of competing capitals, the major combinations of which are no longer "national" capitals. The nature of state practices in the emergent global system resides in the exercise of transnational economic and political authority through the TNS apparatus to reproduce the class relations embedded in the global valorization and accumulation of capital.

[. . .]

I have suggested here that the nation-state is a historically-specific form of world social organization in the process of becoming transcended by globalization. Historic structures may be transcended by their destruction and replacement. This is how, for instance, the historic structures of monarchy and feudalism in France were superseded. Such structures may also be superseded by transformation through incorporation into emergent new structures. This was the route through which monarchic and feudal structures were transcended in England. Hence there are monarchical and feudal residues in England that we do not find in France. I am suggesting here that a TNS is emerging through the latter route: the nation-state system is not being destroyed but transformed and incorporated through the process of globalization into the larger emergent structure of a TNS.

Let us recall that we study static structures for methodological purposes only, because there are abstractions from reality that can only be understood in relation to the dynamics of structural change. The static structure is of less concern than movement in structure. Social reality is best grasped in a synthesis of its synchronic and diachronic dimensions. Seen in this light, the nation-state and the inter-state system are not a constitutive component of world capitalism as an integral social system but a (the) historic form in which capitalism came into being. Temporally, the nation-state is penetrated from the past and the future as a disintegrating structure. The state, shed of its cotermination with the nation-state, may be seen as structure in motion whose form is changing under globalization. The emergent TNS as an unfinished and open-ended process is, as are all historic processes, subject to being pushed in new and unforseen directions and even to reversals. Beyond state theory, the globalization perspective presented in this essay may enhance our ability to comprehend the nature and direction of world social change in the new century and enrich the development of social theory more generally.

 READING 27

Revisiting the Question of the Transnational State: A Comment on William Robinson's "Social Theory and Globalization"

Philip McMichael

William Robinson's thoughtful and provocative essay calls for a recasting of the parameters of social theory in light of the structural shifts associated with globalization. In particular, he argues that the sociology of the state needs to acknowledge the growing "deterritorialization" of economic and political relationships at the turn of the twenty-first century. To accomplish this, he deploys the concept of the "transnational state" (TNS) as the embryonic political form of economic globalization. Robinson bases this conceptual intervention on a theoretical claim for "a 'deterritorialization' of the relationship of capital to the state," and "the 'pure' reproduction of social relations, that is, a process not mediated by fixed geo-political dynamics." This is a bold claim indeed. It suggests that we have arrived at a point where Marx's theory of capital now corresponds to reality. Alternatively, it reaffirms the claims made by the agents of capital that globalization is here to stay and there is no alternative. It is these implications that I wish to address in this comment.

While I support Robinson's position that social science is infected with an unhealthy dualism in state/market, and global/national terms, I question the way in which he advocates his position. Although he argues for a revival of a historical materialist conception of the state, the methodology employed tends toward an abstract formalism. The absence of a historical theory of capitalism is expressed in his unproblematized conception of globalization. Robinson views globalization as the "near culmination" of a process of capitalist expansion at the expense of "all pre-capitalist relations around the globe." The provocative telos here suspends the dialectic. This conception of globalization lacks contradiction and suppresses the fact that globalization is a relationship itself. Rather than viewing global/national tensions as immanent to globalization, Robinson presents, or dismisses, these as dualistic thinking. As a historical phenomenon, globalization contradicts national organization, local knowledge, self-sufficiency, and the like. Its proponents seek to deconstruct or appropriate these "obstacles," and in so doing they constitute the politics of globalization. My point is that a theorist may look for underlying tendencies that transcend such apparent "residuals," but "residuals" have a way of asserting themselves and conditioning the process under examination. The crisis of the Washington consensus, for example, expresses the global resistances and contradictions that constitute globalization.

Let me pursue the conception of "globalization" further. It seems to me that there are two ways to think about it. One way is to *theorize history* as a process of progressive commodification of social life, which allows one to state that "globalization is not a new process." This is Robinson's tack. The other way is to *historicize theory* and problematize globalization as a relation immanent in capitalism, but with quite distinct material (social, political, and environmental) relations across time and time–space. In this formulation, globalization assumes specific *historical* forms. These forms are not unrelated, in fact they can be theorized as either resolutions of prior, or preconditions of succeeding, forms of global arrangements. The current form of globalization, for example, can be viewed as a political counter-mobilization of capital to mid-twentieth-century state-protectionism, as a resolution of the crisis of the nineteenth-century self-regulating market institution, described in Karl Polanyi's *The Great Transformation*. In this view, the late-twentieth-century form of globalization is understood, via the method of incorporated comparison, as a repetition, but not a replication, of a prior globalization that conditioned its successor. Here, globalization is not simply the unfolding of capitalist tendencies, but a historically distinct project shaped, or complicated, by the contradictory relations of previous episodes of globalization.

The point is to develop a historical, rather than a theoretical, conception of capitalism, where the theory of capital is deployed methodologically and reflexively to interpret, rather than reveal, history. While Robinson notes that in "the historical materialist conception, the economic and the political are distinct moments of the same totality," his argument about globalization is that the political reorganization of world capitalism *lags* behind its economic reorganization. The implication is that globalization is essential to capitalist economic integration, which is currently outpacing its political form. That is, the political superstructure has yet to

complement its economic base. But "superstructures" are not distinct binary elements of capitalism with minds of their own. Certainly politics, law, and ideology, on the one hand, and economy, on the other, *appear* as independent binary elements, but these are fetishistic representations in thought. Political relations are economic relations, and vice versa. Even if there is an apparent mismatch between the scope of economic and political relations this is a theoretical, not a historical, observation. Contemporary globalization is a historical relation in which economic and political relations are necessarily in tension – in both historical and *ideological* terms. To suggest economic integration outpaces its political shell is to concede the definition of globalization to its ideologues. This mode of argument discounts the political moment, obscuring the political struggles that define the relations of globalization. It also encourages economic fetishism – attributing autonomy to the market, and eliminating a diverse array of social relations and lifestyles from consideration (especially among the roughly eighty percent of the world's population lacking consumer cash or credit to participate in the global market).

Robinson notes that historically "capitalism unfolded through a system of nation-states" whose boundaries are increasingly eroded by globalization, which supersedes the nation-state as "the organizing principle of capitalism, and with it, of the inter-state system as the institutional framework of capitalist development." The process of supersession involves the emergence of a "transnational state." I would not quarrel with this scenario other than with the image of supersession of the inter-state system. As David Myhre and I have argued, like Robinson, the concept of the "transnational state" speaks to the metamorphosis of the national state as much as it speaks to the elaboration of multilateral institutions to regulate global circuits of capital and commodities. But the multilaterals are extensions of their member states, some of which are more equal than others. Robinson's mode of argument is to map the trajectories of capitalism and the state together, shifting their scale in sequential moves from nation-state to transnational state. It is one thing to theorize a distinction between the national and the transnational state, but it is another to impose that theoretical distinction on the political history of capitalism, which has always been global.

Robinson claims that "there is nothing in the historical materialist conception of the state that necessarily *ties* it to territory or to nation-states," and that for Marx "the state gives a political form to economic institutions and production relations." Theoretically, there is no spatial specification in the movement of capital, beyond that social space governed by times of circulation and social reproduction, and composed of commodity circuits and class relations between the representatives of rent, profit, and wages. Geographic space only comes into play historically, but in the various theoretically posited social forms. For example, absolutism emerged out of medieval political domains, reformulating power as a politicized form of class rule by the European aristocracy. Capitalism emerged in the political alliance of "proto-capitalist" absolutist states and long-distance traders, where absolutist states "crystallized merchant wealth into capital by the political and legal regulation of commerce, thereby sponsoring the foundations of a world market which constituted the precondition of industrial capital." Absolutism integrated foreign and local commodity circuits, developing public authority and private property simultaneously through the recovery of Roman law and aristocratic power. That is, historically, capitalism emerged as a *political*-economic phenomenon.

The subsequent episodes of mercantilism, colonialism, and the movement toward "free-trade imperialism" of the nineteenth century all accompanied the maturation of the state as a national territorial, but world-historical, entity. Capitalist markets were never confined to the national territory – the colonies animated a global division of labor, incubated social labor in the form of slavery, and fueled state treasuries. The nation-state may have been the organizing principle of capitalist politics, but not of the composition and scope of markets. In fact, the nation-state was not only a world-historical product, but it was also the source of political expansion into the non-European world. It is not just that "territorialized space came to house distinct market and capital accumulation conditions, often against one another, a process that tended to be self-reproducing as it deepened and became codified by the development of national states, politics, and culture" as Robinson claims. Rather, territorial space was a vehicle of politics historically defined and redefined more by the claims made on states by merchants, industrialists, proletarians,

and eventually colonial subjects for certain (nationalist) political protections and entitlements within a global market, than by some underlying national economic logic. The nineteenth-century world market was organized by powerful (British) national capitalists, and, because of the dominant ideology of economic liberalism (backed by the force of the British state), the gold standard came to regulate national currencies. In this formulation, nineteenth-century globalization combined the international machinations of *haute finance* and the regime of gold, through which world market relations were embedded in states via the institution of central banking. Polanyi linked the rise of constitutionalism to the politics of currency adjustment under a gold regime, emphasizing that the nineteenth-century nation-state was an artifact of global monetary relations and their social consequences within states. Further, this was a political-economic arrangement orchestrated by British commercial hegemony vis-à-vis its rival states, and gunboat diplomacy vis-à-vis the non-European world.

In short, the nation-state is not simply an early spatial "protective cocoon" for capital, rather it is a historical product of specific global political-economic relations (an earlier "globalization"). Accordingly, it is questionable to argue, as Robinson does, that the "material circumstances that gave rise to the nation-state are now being superseded by globalization." Whether and to what extent this is so, the formulation loses sight of the significance of how global political-economic relations are embedded in the inter-state system. In particular, it obscures how they conditioned the "great transformation" toward the social-democratic, or developmentalist, state via early-twentieth-century class mobilizations that expressed the crisis of the international monetary regime based on gold. In my view, situating the "culmination" of the nation-state form in these conjunctural terms helps us to understand the complex interplay of global and national relations. And it suggests that the current "globalization project" is as much a counter-mobilization of capital against the constraints of social protectionism, as it is an expression of secular developments in the productive forces of capitalism. As such, globalization is not inevitable, rather it involves, again, a politically instituted world market privileging (rather than just expressing) "third wave" technologies.

Robinson's logic leads him to claim that the TNS "is attempting to fulfill the functions for world capitalism that in earlier periods were fulfilled by what world-system and international relations scholars refer to as a 'hegemon,' or a dominant capitalist power that has the resources and the structural position which allows it to organize world capitalism as a whole." This formulation reproduces the abstract formalism of state theory in positing a succession of state forms devoid of historical, geo-political content. Arguably, the mature nation-state form (as model) stemmed from the rise of the United States as a New World settler state challenging the nineteenth-century hegemonic model of the British state, which combined national with imperial relations of an international division of labor. The dynamic division of labor within the United States between agriculture and industry modelled coherence for the twentieth-century nation-state, and, following "the great transformation," came to embody the *ideal* of the developmentalist state universalized via US hegemony. In other words, the nation-state had definite geo-political lineages across time and space.

Just as the generalization of the nation-state form depended on specific geo-political relations, so the emergence of the TNS depends on specific geo-political relations, and is not simply a post-nation-state phenomenon. The transnationalization of state power embodies an attempt on the part of a declining hegemon (the United States) and (sometimes) its political allies in the G-7 to frame the institutions and political content of neo-liberalism in such a way as to preserve, or institutionalize, its power. Stephen Gill has captured the nuances of this process in his analysis of the deployment of neo-constitutionalism to lock neo-liberalism reforms into regional (NAFTA) and global (WTO) institutions, to prevent backsliding given neo-liberalism's relatively fragile status as a hegemonic ideology. Certainly Robinson rehearses the variety of impacts of neo-liberal restructuring of political and economic relations (states) in the process of constructing a TNS. And he includes an exemplary account of the multiple functions of the TNS as an expression of a global elite's attempt to reorganize regional, international, and multilateral institutions around the goal of sustaining the valorization and accumulation of capital on a global scale. However, his discussion of the construction of a global historic bloc by this elite lacks specific historical content,

discounting the contradictory initiatives taken by the United States via the G-7 and the World Economic Forum to build this power bloc.

It is no secret that the 1980s initiative for the GATT Uruguay Round, and for a free trade regime, came from the United States and its agribusiness lobby, which sought to institutionalize American "green power" to secure the United States as the "breadbasket of the world." It is also no secret that the United States and its corporate lobby subsequently initiated the establishment of the WTO in 1995. Washington favored the dispute-resolution and enforcement mechanism of the WTO as a decided improvement over the more diffuse rules and operation of the GATT. However, the enforcement of trade sanctions against member states violating "free trade" rules has had dramatically different impacts given the structural variation among states. The initial ambivalence of Japan regarding its industrial system and its rice culture, and of the EU regarding protection of agricultural policy, and the preference of southern states for plural institutions and negotiations rather than a single trade bureaucracy, still disrupts the attempt to establish a uniform set of rules. And, in Seattle, strong-arm measures orchestrated by the US Trade Representative, via the exclusive "green-room" negotiations, only confirmed Caribbean and African states' perceptions of the WTO as an instrument of the north. That is, the TNS is as much a tangle of geo-political relations as it is a political regulator of the global economy.

I have no argument with the concept of a TNS, but it is of questionable theoretical use if it is not derived from a conception of the contradictory historical relations within which it emerges. The historical conception tempers the tendency to transform a theoretical tendency into a trend, which in my view runs the risk of reification. Alliteration aside, we need to problematize globalization as a *historical project* rather than a *culminating process*. Certainly Robinson acknowledges that the TNS is a composite of "multiple centers and partial regulatory mechanisms," and the "diverse institutions that constitute a TNS have distinct histories and trajectories, are internally differentiated, and present numerous entry points as sites of contention." However, rather than using this insight to check an evolutionary conception of the TNS, Robinson goes on to detail the mechanisms of the TNS that replicate the functions (once) associated with the national state: compensation for market failure (bail-outs), money creation (the Euro), legal guarantees of property rights and market contracts, provision of public goods, transnational social policies, and global policing. Despite a disclaimer that these policies are not so much functions, as instruments, of a global elite, Robinson still disconnects the instrumentality from specific, historical geo-political relations.

My emphasis on geo-political relations is not a neo-realist reflex, instead it is a plea to historicize the social categories we deploy. It resolves, for example, Robinson's quandary concerning one function "that the TNS has not been able to assume, such as reining in speculation and excesses that so characterize the frenzied 'casino capitalism' of the global economy." This should not be a quandary, as it historicizes the politics of the global economy. In the first place, financial capital is the dominant fraction of the post-hegemonic era, empowered by the US-led monetarist counterrevolution of the 1980s and instrumentalized in the institutional politics of the globalization project. In the second place, in the wake of the US abandonment of the Bretton Woods regime and the deployment of the debt regime of the 1980s to impose financial liberalization, states have lost effective control of national currencies. Currency is an object of speculation and currencies (and hence countries) are brought directly into competition with one another. The effect is to force states to adopt competitive neo-liberal policies in order to defend their national currency. In moments of financial crisis, precipitated by currency speculation, the "currency hierarchy" among states asserts itself, allowing the dominant states to "export" the consequences to those states with weakened currencies, such as Mexico, South Korea, Indonesia, Malaysia, Thailand, Brazil, and Russia in the 1990s. Through the instrumentality of the IMF, such states cushion the global crisis with policies of enforced devaluation and financial liberalization. In short, the conditions and consequences of the speculative global economy express the power relations that stand behind the TNS.

In conclusion, Robinson's provocative and timely intervention problematizes many of the assumptions of *extant* social theory. My response stems from a long-held belief that all our social categories are historical and that we need to deploy them reflexively, that is, to capture the relations through which they are constituted. In this respect, Robinson's formulation is quite ambiguous: his positing of a "deterritorialization" of the relationship of capital to the state, where social relations are unmediated by fixed geo-political dynamics, is not consistent with his account of the transnational state emerging through the transformation and incorporation of the nation-state into a globalized economy. The theoretical image may be compelling, but the reality is even more compelling, namely that the project of globalization is riddled with contradictions. Not only is the world larger, more diverse, and more substantive than the horizons of globalization, but also it constitutes globalization in a profound sense because it expresses the material and discursive conditions that the corporate agents and bureaucratic functionaries of globalization seek to appropriate. The globalizers impose a singular and abstracted logic on a culturally, ecologically, and politically diverse world. As such, globalization must be conceived as a historical relationship that is continually undergoing reformulation – dramatized by the rising efficacy of a multitude of resistance movements. Because capital is a historically situated social relation, rather than a thing in circulation, it will always embody worldly relations in its innermost contradictions, and will not be reduced to the "pure" reproduction of social relations.

World Systems

The concept of the world system was created by a neo-Marxian thinker, Immanuel Wallerstein. He chose a unit of analysis very different from that of most Marxian thinkers. He did not look at workers, social classes, or even states because he found these too narrow for his purposes. Instead he looked at a broad economic entity with a division of labor not circumscribed by political or cultural boundaries. He found that unit in his concept of the world system, a largely self-contained social system with a set of boundaries and a definable lifespan (i.e. no world system lasts forever). It is composed internally of a variety of social structures and member groups. He viewed the system as held together by a variety of forces in inherent tension. These forces always have the possibility of tearing the system apart.

Wallerstein argues that thus far we have had only two types of world system. One was the world empire, of which ancient Rome was an example. The other is the modern capitalist world economy. A world empire was based on political (and military) domination, whereas a capitalist world economy relies on economic domination. A capitalist world economy is seen as more stable than a world empire for several reasons. It has a broader base because it encompasses many states, and it has a built-in process of economic stabilization. The separate political entities within the capitalist world economy absorb whatever losses occur, while economic gain is distributed to private hands. Wallerstein foresaw the possibility of a third world system, a socialist world government. Whereas the capitalist world economy separates the political from the economic sector, a socialist world economy reintegrates them.

Within the capitalist world economy, the *core* geographic area is dominant and exploits the rest of the system. The *periphery* consists of those areas that provide raw materials to the core and are heavily exploited by it. The *semiperiphery* is a residual category that encompasses a set of regions somewhere between the exploiting and the exploited. To Wallerstein, the international division of exploitation is defined not by state borders but by the economic division of labor in the world.

Leslie Sklair offers the oft-made critique of world systems theory that there is no "concept of the 'global' in most world-systems literature."[1] More specifically, any conception of the global in world system theory is "embedded in the world-economy based on the system of nation-states."[2] It is a much more an "inter-national" perspective than it is a global perspective. This is especially problematic for globalization theorists since most question the continuing importance of the nation-state. While he does not see it as a global perspective, Sklair is willing to acknowledge the fact that world systems theory helped to spread ideas about globalization in sociology.

NOTES ···

1 Leslie Sklair, "Competing Conceptions of Globalization." *Journal of World-Systems Research* 5, Summer 1999: 149.

2 Ibid., 151.

The Modern World-System: Theoretical Reprise

Immanuel Wallerstein

[. . .]

Theorizing is not an activity separate from the analysis of empirical data. Analyses can only be made in terms of theoretical schema and propositions. On the other hand, analyses of events or processes must include as a starting point a whole series of specific values of certain of the variables, on the basis of which one can explain how the final outcomes were arrived at. In order to convey the historical explanation with clarity, it is often the case that one has to assume or glide over the exposition of the formal interrelations between variables.

Consequently, it often makes sense to review the material a second time more briefly and abstractly at the conclusion. No doubt this should be useful to the reader. But it is even more important for the author, in forcing a degree of rigor in the analysis whose absence might readily pass unnoticed amidst the complexity of detail. The empirical material treated thus far has surely been complex – indeed, far more complex than it was possible to portray. Hence, I propose to review what I have been arguing in this book.

In order to describe the origins and initial workings of a world-system, I have had to argue a certain conception of a world-system. A world-system is a social system, one that has boundaries, structures, member groups, rules of legitimation, and coherence. Its life is made up of the conflicting forces which hold it together by tension, and tear it apart as each group seeks eternally to remold it to its advantage. It has the characteristics of an organism, in that it has a life-span over which its characteristics change in some respects and remain stable in others. One can define its structures as being at different times strong or weak in terms of the internal logic of its functioning.

What characterizes a social system in my view is the fact that life within it is largely self-contained, and that the dynamics of its development are largely internal. The reader may feel that the use of the term "largely" is a case of academic weaseling. I admit I cannot quantify it. Probably no one ever will be able to do so, as the definition is based on a counterfactual hypothesis: if the system, for any reason, were to be cut off from all external forces (which virtually never happens), the definition implies that the system would continue to function substantially in the same manner. Again, of course, substantially is difficult to convert into hard operational criteria. Nonetheless the point is an important one and key to many parts of the empirical analyses of this book. Perhaps we should think of self-containment as a theoretical absolute, a sort of social vacuum, rarely visible and even more implausible to create artificially, but still and all a socially-real asymptote, the distance from which is somehow measurable.

Using such a criterion, it is contended here that most entities usually described as social systems – "tribes," communities, nation-states – are not in fact total systems. Indeed, on the contrary, we are arguing that the only real social systems are, on the one hand, those relatively small, highly autonomous subsistence economies not part of some regular tribute-demanding system and, on the other hand, world-systems. These latter are to be sure distinguished from the former because they are relatively large; that is, they are in common parlance "worlds." More precisely, however, they are defined by the fact that their self-containment as an economic-material entity is based on extensive division of labor and that they contain within them a multiplicity of cultures.

It is further argued that thus far there have only existed two varieties of such world-systems: world-empires, in which there is a single political system over most of the area, however attenuated the degree of its effective control; and those systems in which such a single political system does not exist over all, or virtually all, of the space. For convenience and for want of a better term, we are using the term "world-economy" to describe the latter.

Finally, we have argued that prior to the modern era, world-economies were highly unstable structures which tended either to be converted into empires or to disintegrate. It is the peculiarity of the modern world-system that a world-economy has survived for 500 years and yet has not come to be transformed into a world-empire – a peculiarity that is the secret of its strength.

This peculiarity is the political side of the form of economic organization called capitalism. Capitalism has been able to flourish precisely because the world-economy has had within its bounds not one but a multiplicity of political systems.

I am not here arguing the classic case of capitalist ideology that capitalism is a system based on the non-interference of the state in economic affairs. Quite the contrary! Capitalism is based on the constant absorption of economic loss by political entities, while economic gain is distributed to "private" hands. What I am arguing rather is that capitalism as an economic mode is based on the fact that the economic factors operate within an arena larger than that which any political entity can totally control. This gives capitalists a freedom of maneuver that is structurally based. It has made possible the constant economic expansion of the world-system, albeit a very skewed distribution of its rewards. The only alternative world-system that could maintain a high level of productivity and change the system of distribution would involve the reintegration of the levels of political and economic decision-making. This would constitute a third possible form of world-system, a socialist world government. This is not a form that presently exists, and it was not even remotely conceivable in the sixteenth century.

The historical reasons why the European world-economy came into existence in the sixteenth century and resisted attempts to transform it into an empire have been expounded at length. We shall not review them here. It should however he noted that the size of a world-economy is a function of the state of technology, and in particular of the possibilities of transport and communication within its bounds. Since this is a constantly changing phenomenon, not always for the better, the boundaries of a world-economy are ever fluid.

We have defined a world-system as one in which there is extensive division of labor. This division is not merely functional – that is, occupational – but geographical. That is to say, the range of economic tasks is not evenly distributed throughout the world-system. In part this is the consequence of ecological considerations, to be sure. But for the most part, it is a function of the social organization of work, one which magnifies and legitimizes the ability of some groups within the system to exploit the labor of others, that is, to receive a larger share of the surplus.

While, in an empire, the political structure tends to link culture with occupation, in a world-economy the political structure tends to link culture with spatial location. The reason is that in a world-economy the first point of political pressure available to groups is the local (national) state structure. Cultural homogenization tends to serve the interests of key groups and the pressures build up to create cultural national identities.

This is particularly the case in the advantaged areas of the world-economy – what we have called the core-states. In such states, the creation of a strong state machinery coupled with a national culture, a phenomenon often referred to as integration, serves both as a mechanism to protect disparities that have arisen within the world-system, and as an ideological mask and justification for the maintenance of these disparities.

World-economies then are divided into core-states and peripheral areas. I do not say peripheral *states* because one characteristic of a peripheral area is that the indigenous state is weak, ranging from its non-existence (that is, a colonial situation) to one with a low degree of autonomy (that is, a neo-colonial situation).

There are also semiperipheral areas which are in between the core and the periphery on a series of dimensions, such as the complexity of economic activities, strength of the state machinery, cultural integrity, etc. Some of these areas had been core-areas of earlier versions of a given world-economy. Some had been peripheral areas that were later promoted, so to speak, as a result of the changing geopolitics of an expanding world-economy.

The semiperiphery, however, is not an artifice of statistical cutting points, nor is it a residual category. The semiperiphery is a necessary structural element in a world-economy. These areas play a role parallel to that played, *mutatis mutandis*, by middle trading groups in an empire. They are collection points of vital skills that are often politically unpopular. These middle areas (like middle groups in an empire) partially deflect the political pressures which groups primarily located in peripheral areas might otherwise direct against core-states and the groups which operate within and through their state machineries. On the other hand, the interests primarily located in the semiperiphery are located outside the political arena of the core-states, and find it difficult to pursue the ends in political coalitions that might be open to them were they in the same political arena.

The division of a world-economy involves a hierarchy of occupational tasks, in which tasks requiring higher levels of skill and greater capitalization are reserved for higher-ranking areas. Since a capitalist world-economy essentially rewards accumulated capital, including human capital, at a higher rate than "raw" labor power, the geographical maldistribution of these occupational skills involves a strong trend toward self-maintenance. The forces of the marketplace reinforce them rather than undermine them. And the absence of a central political mechanism for the world-economy makes it very difficult to intrude counteracting forces to the maldistribution of rewards.

Hence, the ongoing process of a world-economy tends to expand the economic and social gaps among its varying areas in the very process of its development. One factor that tends to mask this fact is that the process of development of a world-economy brings about technological advances which make it possible to expand the boundaries of a world-economy. In this case, particular regions of the world may change their structural role in the world-economy, to their advantage, even though the disparity of reward between different sectors of the world-economy as a whole may be simultaneously widening. It is in order to observe this crucial phenomenon clearly that we have insisted on the distinction between a peripheral area of a given world-economy and the external arena of the world-economy. The external arena of one century often becomes the periphery of the next – or its semiperiphery. But then too core-states can become semiperipheral and semiperipheral ones peripheral.

While the advantages of the core-states have not ceased to expand throughout the history of the modern world-system, the ability of a particular state to remain in the core sector is not beyond challenge. The hounds are ever to the hares for the position of top dog. Indeed, it may well be that in this kind of system it is not structurally possible to avoid, over a long period of historical time, a circulation of the elites in the sense that the particular country that is dominant at a given time tends to be replaced in this role sooner or later by another country.

We have insisted that the modern world-economy is, and only can be, a capitalist world-economy. It is for this reason that we have rejected the appellation of "feudalism" for the various forms of capitalist agriculture based on coerced labor which grow up in a world-economy. Furthermore, although this has not been discussed in this volume, it is for this same reason that we will, in future volumes, regard with great circumspection and prudence the claim that there exist in the twentieth century socialist national economies within the framework of the world-economy (as opposed to socialist movements controlling certain state-machineries within the world-economy).

If world-systems are the only real social systems (other than truly isolated subsistence economies), then it must follow that the emergence, consolidation, and political roles of classes and status groups must be appreciated as elements of this *world*-system. And in turn it follows that one of the key elements in analyzing a class or a status-group is not only the state of its self-consciousness but the geographical scope of its self-definition.

Classes always exist potentially (*an sich*). The issue is under what conditions they become class-conscious (*für sich*), that is, operate as a group in the politico-economic arenas and even to some extent as a cultural entity. Such self-consciousness is a function of conflict situations. But for upper strata open conflict, and hence overt consciousness, is always *faute de mieux*. To the extent that class boundaries are not made explicit, to that extent it is more likely that privileges be maintained.

Since in conflict situations, multiple factions tend to reduce to two by virtue of the forging of alliances, it is by definition not possible to have three or more (conscious) classes. There obviously can be a multitude of occupational interest groups which may organize themselves to operate within the social structure. But such groups are really one variety of status-groups, and indeed often overlap heavily with other kinds of status-groups such as those defined by ethnic, linguistic, or religious criteria.

To say that there cannot be three or more classes is not however to say that there are always two. There may be none, though this is rare and transitional. There may be one, and this is most common. There may be two, and this is most explosive.

We say there may be only one class, although we have also said that classes only actually exist in conflict situations, and conflicts presume two sides. There is no contradiction here. For a conflict may be defined as being between one class, which conceives of itself as the universal class, and all the other strata. This has in fact been the usual situation in the modern world-system. The capitalist class (the *bourgeoisie*) has claimed to be the universal class and sought to organize political life to pursue its objectives against two opponents. On the one hand, there were those who spoke for the maintenance of traditional rank distinctions despite the fact that these ranks might have lost their original correlation with economic function. Such elements preferred to define the social structure as a non-class structure. It was to counter this ideology that the bourgeoisie came to operate as a class conscious of itself.

[. . .]

The evolution of the state machineries reflected precisely this uncertainty. Strong states serve the interests of some groups and hurt those of others. From however the standpoint of the world-system as a whole, if there is to be a multitude of political entities (that is, if the system is not a world-empire), then it cannot be the case that all these entities be equally strong. For if they were, they would be in the position of blocking the effective operation of transnational economic entities whose locus were in another state. It would then follow that the world division of labor would be impeded, the world-economy decline, and eventually the world-system fall apart.

It also cannot be that *no* state machinery is strong. For in such a case, the capitalist strata would have no mechanisms to protect their interests, guaranteeing their property rights, assuring various monopolies, spreading losses among the larger population, etc.

It follows then that the world-economy develops a pattern where state structures are relatively strong in the core areas and relatively weak in the periphery.

Which areas play which roles is in many ways accidental. What is necessary is that in some areas the state machinery be far stronger than in others.

What do we mean by a strong state-machinery? We mean strength vis-à-vis other states within the world-economy including other core-states, and strong vis-à-vis local political units within the boundaries of the state. In effect, we mean a sovereignty that is *de facto* as well as *de jure*. We also mean a state that is strong vis-à-vis any particular social group within the state. Obviously, such groups vary in the amount of pressure they can bring to bear upon the state. And obviously certain combinations of these groups control the state. It is not that the state is a neutral arbiter. But the state is more than a simple vector of given forces, if only because many of these forces are situated in more than one state or are defined in terms that have little correlation with state boundaries.

A strong state then is a partially autonomous entity in the sense that it has a margin of action available to it wherein it reflects the compromises of multiple interests, even if the bounds of these margins are set by the existence of some groups of primordial strength. To be a partially autonomous entity, there must be a group of people whose direct interests are served by such an entity: state managers and a state bureaucracy.

Such groups emerge within the framework of a capitalist world-economy because a strong state is the best choice between difficult alternatives for the two groups that are strongest in political, economic, and military terms: the emergent capitalist strata, and the old aristocratic hierarchies.

For the former, the strong state in the form of the "absolute monarchies" was a prime customer, a guardian against local and international brigandage, a mode of social legitimation, a preemptive protection against the creation of strong state barriers elsewhere. For the latter, the strong state represented a brake on these same capitalist strata, an upholder of status conventions, a maintainer of order, a promoter of luxury.

No doubt both nobles and bourgeois found the state machineries to be a burdensome drain of funds, and a meddlesome unproductive bureaucracy. But what options did they have? Nonetheless they were always restive and the immediate politics of the world-system was made up of the pushes and pulls resulting from the efforts of both groups to insulate themselves from what seemed to them the negative effects of the state machinery.

A state machinery involves a tipping mechanism. There is a point where strength creates more strength. The tax revenue enables the state to have a larger and more efficient civil bureaucracy and army which in turn leads to greater tax revenue – a process that continues in spiral form. The tipping mechanism works in the other direction too – weakness leading to greater weakness. In between these two tipping points lies the politics of state-creation. It is in this arena that the skills of particular managerial groups make a difference. And it is because of the two tipping mechanisms that at certain points a small gap in the world-system can very rapidly become a large one.

In those states in which the state machinery is weak, the state managers do not play the role of coordinating a complex industrial–commercial–agricultural mechanism. Rather they simply become one set of landlords amidst others, with little claim to legitimate authority over the whole.

These tend to be called traditional rulers. The political struggle is often phrased in terms of tradition versus change. This is of course a grossly misleading and ideological terminology. It may in fact be taken as a general sociological principle that, at any given point of time, what is thought to be traditional is of more recent origin than people generally imagine it to be, and represents primarily the conservative instincts of some group threatened with declining social status. Indeed, there seems to be nothing which emerges and evolves as quickly as a "tradition" when the need presents itself.

In a one-class system, the "traditional" is that in the name of which the "others" fight the class-conscious group. If they can encrust their values by legitimating them widely, even better by enacting them into legislative barriers, they thereby change the system in a way favorable to them.

The traditionalists may win in some states, but if a world-economy is to survive, they must lose more or less in the others. Furthermore, the gain in one region is the counterpart of the loss in another.

This is not quite a zero-sum game, but it is also inconceivable that all elements in a capitalist world-economy shift their values in a given direction simultaneously. The social system is built on having a multiplicity of value systems within it, reflecting the specific functions groups and areas play in the world division of labor.

We have not exhausted here the theoretical problems relevant to the functioning of a world-economy. We have tried only to speak to those illustrated by the early period of the world-economy in creation, to wit, sixteenth-century Europe. Many other problems emerged at later stages and will be treated, both empirically and theoretically, in later volumes.

In the sixteenth century, Europe was like a bucking bronco. The attempt of some groups to establish a world-economy based on a particular division of labor, to create national states in the core areas as politico-economic guarantors of this system, and to get the workers to pay not only the profits but the costs of maintaining the system was not easy. It was to Europe's credit that it was done, since without the thrust of the sixteenth century the modern world would not have been born and, for all its cruelties, it is better that it was born than that it had not been.

It is also to Europe's credit that it was not easy, and particularly that it was not easy because the people who paid the short-run costs screamed lustily at the unfairness of it all. The peasants and workers in Poland and England and Brazil and Mexico were all rambunctious in their various ways. As R. H. Tawney says of the agrarian disturbances of sixteenth-century England: "Such movements are a proof of blood and sinew and of a high and gallant spirit [. . .] Happy the nation whose people has not forgotten how to rebel."

The mark of the modern world is the imagination of its profiteers and the counter-assertiveness of the oppressed. Exploitation and the refusal to accept exploitation as either inevitable or just constitute the continuing antinomy of the modern era, joined together in a dialectic which has far from reached its climax in the twentieth century.

 READING 29

Competing Conceptions of Globalization

Leslie Sklair

Introduction

Globalization is a relatively new idea in the social sciences, although people who work in and write about the mass media, transnational corporations and international business have been using it for some time. Jacques Maisonrouge, the French-born former President of IBM World Trade, was an early exponent of the view that the future lies with global corporations who operate as if the world had no real borders rather than organizations tied to a particular country. The influential US magazine *Business Week* (14 May 1990) summed this view up in the evocative phrase: 'The Stateless Corporation'. The purpose of this paper is to critically review the ways in which sociologists and other social scientists use ideas of globalization and to evaluate the fruitfulness of these competing conceptions.

The central feature of the idea of globalization is that many contemporary problems cannot be adequately studied at the level of nation-states, that is, in terms of each country and its inter-national relations, but instead need to be seen in terms of global processes. Some globalists have even gone so far as to predict that global forces, by which they usually mean transnational corporations and other global economic institutions, global culture or globalizing belief systems/ideologies of various types, or a combination of all of these, are becoming so powerful that the continuing existence of the nation-state is in serious doubt. This is not a necessary consequence of most theories of globalization, though many argue that the significance of the nation-state is declining (even if the ideology of nationalism is still strong in some places).

There is no single agreed definition of globalization, indeed, some argue that its significance has been much exaggerated, but as the ever-increasing numbers of books and articles discussing different aspects of it suggest, it appears to be an idea whose time has come

in sociology in particular and in the social sciences in general. The author of the first genuine textbook on globalization suggests that it may be '*the* concept of the 1990s'.

The argument of this paper is that the central problem in understanding much of the globalization literature is that not all those who use the term distinguish it clearly enough from internationalization, and some writers appear to use the two terms interchangeably. I argue that a clear distinction must be drawn between the inter-national and the global. The hyphen in inter-national is to signify confusing conceptions of globalization founded on the existing even if changing system of nation-states, while the global signifies the emergence of processes and a system of social relations not founded on the system of nation-states.

This difficulty is compounded by the fact that most theory and research in sociology is based on concepts of society that identify the unit of analysis with a particular country (for example, sociology of Britain, of Japan, of the USA, of Russia, of India, etc.), sub-systems within countries (British education, the Japanese economy, American culture, politics in Russia, religion in India) or comparisons between single countries and groups of them (modern Britain and traditional India, declining America and ascendent Japan, rich and poor countries, the West and the East). This general approach, usually called state-centrism, is still useful in many respects and there are clearly good reasons for it. Not the least of these is that most historical and contemporary sociological data sets have been collected on particular countries. However, most globalization theorists argue that the nation-state is no longer the only important unit of analysis. Some even argue that the nation-state is now less important in some fundamental respects than other global, forces; examples being the mass media and the corporations that own and control them, transnational corporations (some of which are richer than the majority of nation-states in the world today) and even social movements that spread ideas such as universal human rights, global environmental responsibility and the world-wide call for democracy and human dignity. Yearley identifies two main obstacles to making sociological sense of globalization, namely 'the tight connection between the discipline of sociology and the nation-state' and the fact that countries differ significantly in their geographies. Despite these difficulties (really elaborations of the local-global problem which will be discussed below) he makes the telling point that a focus on the environment encourages us to 'work down to the global' from the universal, a necessary corrective to state-centrist conceptions which work up to the global from the nation-state or even, as we shall see from individualistic notions of 'global consciousness'.

The study of globalization in sociology revolves primarily around two main classes of phenomena which have become increasingly significant in the last few decades. These are the emergence of a globalized economy based on new systems of production, finance and consumption; and the idea of 'global culture'. While not all globalization researchers entirely accept the existence of a global economy or a global culture, most accept that local, national and regional economies are undergoing important changes as a result of processes of globalization even where there are limits to globalization.

Researchers on globalization have focused on two phenomena, increasingly significant in the last few decades:

(i)   The ways in which transnational corporations (TNCs) have facilitated the globalization of capital and production.

(ii)   Transformations in the global scope of particular types of TNC, those who own and control the mass media, notably television channels and the transnational advertising agencies. This is often connected with the spread of particular patterns of consumption and a culture and ideology of consumerism at the global level.

The largest TNCs have assets and annual sales far in excess of the Gross National Products of most of the countries in the world. The World Bank annual publication *World Development Report* reports that in 1995 only about 70 countries out a total of around 200 for which there is data, had GNPs of more than ten billion US dollars. By contrast, the *Fortune* Global 500 list of the biggest TNCs by turnover in 1995 reports that over 440 TNCs had annual sales greater than $10 billion. Thus, in this important sense, such well-known names as General Motors, Shell, Toyota, Unilever, Volkswagen, Nestlé, Sony, Pepsico, Coca-Cola, Kodak, Xerox and the huge Japanese trading houses (and many other corporations most people have never heard of) have more economic power at their disposal than the majority of the countries in the world. These figures prove little in themselves, they simply indicate the *gigantism* of TNCs relative to most countries.

Not only have TNCs grown enormously in size in recent decades but their 'global reach' has expanded dramatically. Many companies, even from large rich countries, regularly earn a third or more of their revenues from 'foreign' sources. Not all *Fortune* Global 500 corporations are headquartered in the First World: some come from what was called the Third World or those parts of it known as the Newly Industrializing Countries (NICs). Examples of these are the 'national' oil companies of Brazil, India, Mexico, Taiwan and Venezuela (some owned by the state but most run like private corporations), banks in Brazil and China, an automobile company from Turkey, and the Korean manufacturing and trading conglomerates (*chaebol*), a few of which have attained global brand-name status (for example, Hyundai and Samsung).

Writers who are sceptical about economic globalization argue that the facts that most TNCs are legally domiciled in the USA, Japan and Europe and that they trade and invest mainly between themselves means that the world economy is still best analysed in terms of national corporations and that the global economy is a myth. But this deduction entirely ignores the well-established fact that an increasing number of corporations operating outside their 'home' countries see themselves as developing global strategies, as is obvious if we read their annual reports and other publications rather than focus exclusively on aggregate data on foreign investment. You cannot simply assume that all 'US', 'Japanese' and other 'national' TNCs somehow express a 'national interest'. They do not. They primarily express the interests of those who own and control them, even if historical patterns of TNC development have differed from place to place, country to country and region to region. Analysing globalization as a relatively recent phenomenon, originating from the 1960s, allows us to see more clearly the tensions between traditional 'national' patterns of TNC development and the new global corporate structures and dynamics. It is also important to realize that, even in state-centrist terms, a relatively small investment for a major TNC can result in a relatively large measure of economic presence in a small, poor country or a poor region or community in a larger and less poor country.

The second crucial phenomenon for globalization theorists is the global diffusion and increasingly concentrated ownership and control of the electronic mass media, particularly television. The number of TV sets per capita has grown so rapidly in Third World countries in recent years (from fewer than 10 per thousand population in 1970 to 60 per 1,000 in 1993, according to UNESCO) that many researchers argue that a 'globalizing effect' due to the mass media is taking place even in the Third World.

Ownership and control of television, including satellite and cable systems, and associated media like newspaper, magazine and book publishing, films, video, records, tapes, compact discs, and a wide variety of other marketing media, are concentrated in relatively few very large TNCs. The predominance of US-based corporations is being challenged by others based in Japan, Europe and Australia and even by 'Third World' corporations like the media empires of TV Globo, based in Brazil and Televisa, based in Mexico.

[. . .]

The World-Systems Approach

This approach is based on the distinction between core, semiperipheral and peripheral countries in terms of their changing roles in the international division of labour dominated by the capitalist world-system. World-systems as a model in social science research, inspired by the work of Immanuel Wallerstein, has been developed in a large and continually expanding body of literature since the 1970s.

The world-systems approach is, unlike the others to be discussed, not only a collection of academic writings but also a highly institutionalized academic enterprise. It is based at the Braudel Center at SUNY Binghamton, supports various international joint academic ventures, and publishes the journal, *Review*. Though the work of world-systems theorists cannot be said to be fully a part of the globalization literature as such, the institutionalization of the world-systems approach undoubtedly prepared the ground for globalization in the social sciences.

In some senses, Wallerstein and his school could rightly claim to have been 'global' all along – after all, what could be more global than the 'world-system'? However, there is no specific concept of the 'global' in most world-systems literature. Reference to the 'global' comes mainly from critics and, significantly, can be traced to the long-standing problems that the world-system model has had with 'cultural issues'. Wallerstein's essay on 'Culture as the Ideological Battleground of the Modern World-System', the critique by Boyne, and Wallerstein's attempt to rescue his position under the title of 'Culture is the World-System', illustrate the problem well.

Chase-Dunn, in his suggestively titled book *Global Formation*, does try to take the argument a stage further by arguing for a dual logic approach to economy and polity. At the economic level, he argues, a global logic of the world-economy prevails whereas at the level of politics a state-centred logic of the world-system prevails. However, as the world-economy is basically still explicable only in terms of national economies (countries of the core, semiperiphery and periphery), Chase-Dunn's formulation largely reproduces the problems of Wallerstein's state-centrist analysis.

There is, therefore, no distinctively 'global' dimension in the world-systems model apart from the inter-national focus that it has always emphasized. Wallerstein himself rarely uses the word 'globalization'. For him, the *economics* of the model rests on the inter-national division of labour that distinguishes core, semiperiphery and periphery countries. The *politics* are mostly bound up with antisystemic movements and 'superpower struggles'. And the *cultural*, insofar as it is dealt with at all, covers debates about the 'national' and the 'universal' and the concept of 'civilization(s)' in the social sciences. Many critics are not convinced that the world-systems model, usually considered to be 'economistic' (that is, too locked into economic factors) can deal with cultural issues adequately. Wolff tellingly comments on the way in which the concept of 'culture' has been inserted into Wallerstein's world-system model: 'An economism which gallantly switches its attentions to the operations of culture is still economism'. Wallerstein's attempts to theorize 'race', nationality and ethnicity in terms of what he refers to as different types of 'peoplehood' in the world-system might be seen as a move in the right direction, but few would argue that cultural factors are an important part of the analysis.

While it would be fair to say that there are various remarks and ideas that do try to take the world-systems model beyond state-centrism, any conceptions of the global that world-system theorists have tend to be embedded in the world-economy based on the system of nation-states. The 'global' and the 'inter-national' are generally used interchangeably by world-systems theorists. This is certainly one possible use of 'global' but it seems quite superfluous, given that the idea of the 'inter-national' is so common in the social science literature. Whatever the fate of the world-systems approach, it is unlikely that ideas of globalization would have spread so quickly and deeply in sociology without the impetus it gave to looking at the whole world.

[. . .]

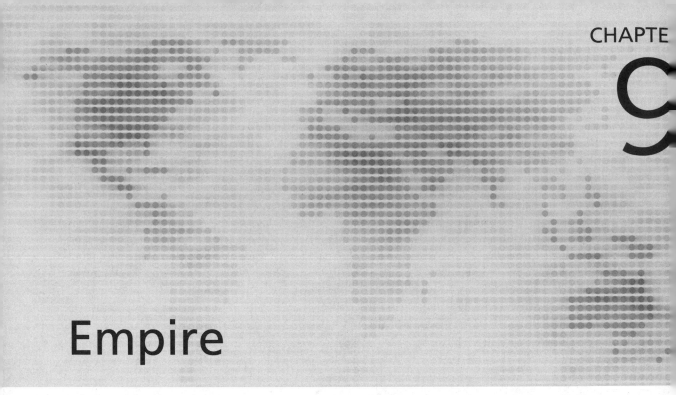

Empire

Michael Hardt and Antonio Negri's *Empire* presents a unique vision of globalization and contemporary, postmodern global realities. The authors associate modernity with a forerunner of globalization, *imperialism*, in which a given nation-state(s) stands at the center and controls and exploits, especially economically, a number of areas throughout the world. In contrast, *empire* is a decentered idea in which such dominance exists, but without any single nation-state (or any other entity) at its center. There is no center to empire: it is deterritorialized, it is virtual in the form of communication (especially through the media), and, as a result, empire is both everywhere and nowhere.

However, empire does not yet exist fully; it is in formation, but we can already get a sense of its nature and parameters. While there is no single power at its center, empire governs the world with a single logic of rule. Power is dispersed throughout society and the globe. Even the US, with its seeming global hegemony, is not an empire and does not even lie at its center. However, the sovereignty of the US does constitute an important precursor to empire and the US continues to occupy a privileged position in the world today. Nevertheless, it is in the process of being supplanted by empire.

Empire lacks not only territorial but also temporal boundaries, in the sense that it seeks (albeit unsuccess-

fully) to suspend history and to exist for all eternity. It also can be seen as lacking a lower boundary in that it seeks to expand its control down into the depths of the social world. That is, it seeks not only to control people's thought, action, and interaction, but also, via biopower, to control people's minds and bodies. All of this makes empire far more ambitious than imperialism.

The key to the global power of empire is that it is (or seeks to be) a new juridical power. That is, it is based on the constitution of order, norms, ethical truths, a common notion of what is right, and so on. This juridical formation is the source of empire's power. Thus, it can intervene in the name of what is "right" anywhere in the world to deal with what it considers humanitarian problems, to guarantee accords, and to impose peace. More specifically, it can engage in "just wars" in the name of the juridical formation; such wars are seen as legitimate. The enemy is anyone or anything the juridical formation sees as a threat to ethical order in the world. The right to engage in war against enemies is boundless in space and encompasses the entire globe. Empire is based on the ability to project force in the service of that which it regards as right.

Empire seeks to incorporate all that it can. It seeks to eliminate differences, resistance, and conflict. It also differentiates among people and uses that differentiation to hierarchize and to manage the hierarchy and the differences embedded in it. Hierarchization and its management are the real, day-to-day powers of empire.

Opposition to empire is found in the *multitude*, a collection of people throughout the world that sustains empire through labor, consumption, and so on. The multitude is the real productive force, but empire feeds off it like a parasite. It is the multitude that is the source of creativity in empire and it is a potentially revolutionary force. If it is successful, it will produce a similarly global counter-empire. Thus, while Hardt and Negri are critical of globalization, at least as it is practiced by empire, they also see utopian potential in globalization generated by the multitude. To Hardt and Negri, the sources of both our major problems and our liberation exist at the global level. Counter-empire must be global, it must be everywhere, and it must be opposed to empire. It is becoming more likely because empire is losing its ability to control the multitude. It is also more likely because while control is through communication and ideology, it is through that communication and ideology that the revolutionary potential of the multitude will be expressed, and it will be manifest globally. The key is that communication flows easily and effectively across the globe. This makes it easier for empire to exert control and to justify itself and its actions. Conversely, of course, it is also the mechanism by which the multitude can ultimately create counter-empire.

In excerpts from an interview, Hardt and Negri critique conventional thinking on globalization and offer their thoughts on it, including the importance of the use of biopower at the global level. They also point to the decline of the nation-state and the existence of new forms of sovereignty that require new forms of opposition and new alternatives to empire. In addition, they argue that the only effective way to oppose global imperial power is on an equally global scale. They admit that the idea of the multitude is vague, but they argue that it will be seen and emerge in its practices, especially those aimed globally at empire.

Barkawi and Laffey examine and critique *Empire* from the point of view of the field of international relations (IR), a field that focuses on the relationship among and between nation-states across the globe. They recognize that IR creates a kind of territorial trap – borders of sovereign states are relatively impermeable – from which it is difficult to extract oneself in order to get a fuller view of global relations. What is needed is a different conception of the global, such as the one offered by Hardt and Negri, "within which processes of mutual constitution are productive of the entities which populate the international system."[1] Another advantage of the Hardt and Negri approach is that it takes more seriously than does IR the position of the periphery or subalterns in global relationships.

While Barkawi and Laffey praise Hardt and Negri for offering a perspective that compensates for these weaknesses in IR, they also criticize Hardt and Negri on various grounds (as do many others). For one thing, they see a focus on empire as being too abstract and paying too little attention to "real relations of rule."[2] This is linked to Hardt and Negri's failure to see continuities between older imperial relations among nation-states and empire. Barkawi and Laffey are also critical of the idea of a break between US imperialism and empire; indeed, they argue that imperialism is still very much in evidence. Thus, they conclude that "globalisation and many of the phenomena Hardt and Negri describe are better understood by reference to an international state dominated by the US."[3] Nor do Barkawi and Laffey accept the idea that the era of interstate war is over, or that global relations are as "smooth" as Hardt and Negri suggest. The clearest evidence against this smoothness is the widespread global resistance to the US.

David Moore thinks about Africa from the point of view of the ideas developed in *Empire*. On this basis, he offers several critiques of that work. First, Hardt and Negri tend to focus on Europe and the US and tend to have less to say about, let alone to offer to, places like Africa. Hardt and Negri also seem to assume that everyone has passed through modernity en route to postmodernity, but this seems to exclude many in Africa who have yet to pass completely through modernity, let alone move on to postmodernity. Hardt and Negri's ideas do not reach far enough into, do not apply enough to, the nations of Africa (e.g. Zimbabwe, the Democratic Republic of Congo) that are struggling along the rocky roads to modernity. Hardt and Negri's postmodern orientation, while not irrelevant, does not offer nearly enough to the billions of people who continue to find themselves in this premodern reality. Finally, Moore finds Hardt and Negri ambivalent on humanitarian aid to those in Africa stuck in this reality. While he is critical of some forms of humanitarian aid based on biopolitical or military objectives, he is positive toward others coming from those concerned with civil rights and who fight for states that take citizenship seriously. Indeed, Moore argues that such humanitarians "have to be etched into the new wave of global solidarity Hardt and Negri assert as necessary."[4] Because they operate at such a general and abstract level, Hardt and Negri are unable to see clearly such a role for humanitarians and more generally to have much to say about the realities of Africa today. Moore concludes with the point that what African nations need today is democracy, but *Empire* has little directly to say about such a mundane matter and what it does say is not stated boldly enough.

Aronowitz also critiques Hardt and Negri for their abstractions. They fail to deal with such global organizations as the World Trade Organization, the International Monetary Fund, and the World Bank as concrete examples "of the repressive world government of Empire."[5] More importantly, they deal with resistance abstractly and theoretically rather than dealing with numerous real-world examples of resistance. Aronowitz argues that people continue to need to test the mettle of, and to resist, contemporary institutions (such as those mentioned above) and to force the still predominant nation-state into making reforms. They can do this while at the same time they can form the kinds of global alliances that Hardt and Negri associate with the multitude.

Finally, Hardt and Negri make it clear that they do not deny the reality of the nation-state, or argue for its end, but rather see its role as being transformed within empire. They defend their abstract sense of multitude, but recognize that they need to move toward a more concrete analysis of it as a revolutionary subject. To that end, they emphasize the "real transformative actions of the multitude" involving "resistance, insurrection, and constituent power."[6] Hardt and Negri recognize that the globe is not smooth and that there are differences among and between areas of the world. They also see their analysis as applying to areas usually considered outside it (such as the case of Africa mentioned above). Hardt and Negri close by acknowledging the fact that they have not provided all of the answers and they welcome the debate their work has stimulated. They see their work as contributing to a collective project and a collective (and emerging) body of knowledge.

NOTES

1 Tarak Barkawi and Mark Laffey, "Retrieving the Imperial: Empire and International Relations." *Millennium: Journal of International Studies* 31, 2002: 111.
2 Ibid., 122.
3 Ibid., 124.
4 David Moore, "Africa: the Black Hole at the Middle of *Empire*?" *Rethinking Marxism* 13, 3/4, 2001: 114.
5 Stanley Aronowitz, "The New World Order (They Mean It)." *The Nation* July 17, 2000: 27.
6 Michael Hardt and Antonio Negri, "Adventures of the Multitude: Response of the Authors." *Rethinking Marxism* 14, 3/4, 2001: 242.

READING 30

Empire

Michael Hardt and Antonio Negri

Preface

Empire is materializing before our very eyes. Over the past several decades, as colonial regimes were overthrown and then precipitously after the Soviet barriers to the capitalist world market finally collapsed, we have witnessed an irresistible and irreversible globalization of economic and cultural exchanges. Along with the global market and global circuits of production has emerged a global order, a new logic and structure of rule – in short, a new form of sovereignty. Empire is the political subject that effectively regulates these global exchanges, the sovereign power that governs the world.

Many argue that the globalization of capitalist production and exchange means that economic relations have become more autonomous from political controls, and consequently that political sovereignty has declined. Some celebrate this new era as the liberation of the capitalist economy from the restrictions and distortions that political forces have imposed on it; others lament it as the closing of the institutional channels through which workers and citizens can influence or contest the cold logic of capitalist profit. It is certainly true that, in step with the processes of globalization, the sovereignty of nation-states, while still effective, has progressively declined. The primary factors of production and exchange – money, technology, people, and goods – move with increasing ease across national boundaries; hence the nation-state has less and less power to regulate these flows and impose its authority over the economy. Even the most dominant nation-states should no longer be thought or as supreme and sovereign authorities, either outside or even within their own borders. *The decline in sovereignty of nation-states, however, does not mean that sovereignty as such has declined.* Throughout the contemporary transformations, political controls, state functions, and regulatory mechanisms have continued to rule the realm of economic and social production and exchange. Our basic hypothesis is that sovereignty has taken a new form, composed of a series of national and supranational organisms united under a single logic of rule. This new global form of sovereignty is what we call Empire.

The declining sovereignty of nation-states and their increasing inability to regulate economic and cultural exchanges is in fact one of the primary symptoms of the coming of Empire. The sovereignty of the nation-state was the cornerstone of the imperialisms that European powers constructed throughout the modern era. By "Empire," however, we understand something altogether different from "imperialism." The boundaries defined by the modern system of nation-states were fundamental to European colonialism and economic expansion: the territorial boundaries of the nation delimited the center of power from which rule was exerted over external foreign territories through a system of channels and barriers that alternately facilitated and obstructed the flows of production and circulation. Imperialism was really an extension of the sovereignty of the European nation-states beyond their own boundaries. Eventually nearly all the world's territories could be parceled out and the entire world map could be coded in European colors: red for British territory, blue for French, green for Portuguese, and so forth. Wherever modern sovereignty took root, it constructed a Leviathan that overarched its social domain and imposed hierarchical territorial boundaries, both to police the purity of its own identity and to exclude all that was other.

The passage to Empire emerges from the twilight of modern sovereignty. In contrast to imperialism, Empire

This reading comprises extracts taken from throughout the original book.

establishes no territorial center of power and does not rely on fixed boundaries or barriers. It is a *decentered* and *deterritorializing* apparatus of rule that progressively incorporates the entire global realm within its open, expanding frontiers. Empire manages hybrid identities, flexible hierarchies, and plural exchanges through modulating networks of command. The distinct national colors of the imperialist map of the world have merged and blended in the imperial global rainbow.

The transformation of the modern imperialist geography of the globe and the realization of the world market signal a passage within the capitalist mode of production. Most significant, the spatial divisions of the three Worlds (First, Second, and Third) have been scrambled so that we continually find the First World in the Third, the Third in the First, and the Second almost nowhere at all. Capital seems to be faced with a smooth world – or really, a world defined by new and complex regimes of differentiation and homogenization, deterritorialization and reterritorialization. The construction of the paths and limits of these new global flows has been accompanied by a transformation of the dominant productive processes themselves, with the result that the role of industrial factory labor has been reduced and priority given instead to communicative, cooperative, and affective labor. In the postmodernization of the global economy, the creation of wealth tends ever more toward what we will call biopolitical production, the production of social life itself, in which the economic, the political, and the cultural increasingly overlap and invest one another.

Many locate the ultimate authority that rules over the processes of globalization and the new world order in the United States. Proponents praise the United States as the world leader and sole superpower, and detractors denounce it as an imperialist oppressor. Both these views rest on the assumption that the United States has simply donned the mantle of global power that the European nations have now let fall. If the nineteenth century was a British century, then the twentieth century has been an American century; or really, if modernity was European, then postmodernity is American. The most damning charge critics can level, then, is that the United States is repeating the practices of old European imperialists, while proponents celebrate the United States as a more efficient and more benevolent world leader, getting right what the Europeans got wrong. Our basic hypothesis, however, that a new imperial form of sovereignty has emerged, contradicts both these views. *The United States does not, and indeed no nation-state can today, form the center of an imperialist project*. Imperialism is over. No nation will be world leader in the way modern European nations were.

The United States does indeed occupy a privileged position in Empire, but this privilege derives not from its similarities to the old European imperialist powers, but from its differences. These differences can be recognized most clearly by focusing on the properly imperial (not imperialist) foundations of the United States constitution, where by "constitution" we mean both the *formal constitution*, the written document along with its various amendments and legal apparatuses, and the *material constitution*, that is, the continuous formation and re-formation of the composition of social forces. Thomas Jefferson, the authors of the *Federalist*, and the other ideological founders of the United States were all inspired by the ancient imperial model; they believed they were creating on the other side of the Atlantic a new Empire with open, expanding frontiers, where power would be effectively distributed in networks. This imperial idea has survived and matured throughout the history of the United States constitution and has emerged now on a global scale in its fully realized form.

We should emphasize that we use "Empire" here not as a *metaphor*, which would require demonstration of the resemblances between today's world order and the Empires of Rome, China, the Americas, and so forth, but rather as a *concept*, which calls primarily for a theoretical approach. The concept of Empire is characterized fundamentally by a lack of boundaries: Empire's rule has no limits. First and foremost, then, the concept of Empire posits a regime that effectively encompasses the spatial totality, or really that rules over the entire "civilized" world. No territorial boundaries limit its reign. Second, the concept of Empire presents itself not as a historical regime originating in conquest, but rather as an order that effectively suspends history and thereby fixes the existing state of affairs for eternity. From the perspective of Empire, this is the way things will always be and the way they were always meant to be. In other words, Empire presents its rule not as a transitory moment in the movement of history, but as a regime with no temporal boundaries and in this sense outside of history or at the end of history. Third, the rule of Empire operates on all registers of the social order extending down to the depths of the social world. Empire not only manages a territory and a population but also creates the very world it inhabits. It not only regulates human interactions but also seeks directly to rule over human nature. The object of its rule is social life in its entirety, and thus Empire presents the paradigmatic form of biopower. Finally, although the practice of Empire is continually bathed in blood, the concept of Empire is always dedicated to peace – a perpetual and universal peace outside of history.

The Empire we are faced with wields enormous powers of oppression and destruction, but that fact should not make us nostalgic in any way for the old forms of domination. The passage to Empire and its processes of globalization offer new possibilities to the forces of liberation. Globalization, of course, is not one thing, and the multiple processes that we recognize as globalization are not unified or univocal. Our political task, we will argue, is not simply to resist these processes but to reorganize them and redirect them toward new ends. The creative forces of the multitude that sustain Empire are also capable of autonomously constructing a counter-Empire, an alternative political organization of global flows and exchanges. The struggles to contest and subvert Empire, as well as those to construct a real alternative, will thus take place on the imperial terrain itself – indeed, such new struggles have already begun to emerge. Through these struggles and many more like them, the multitude will have to invent new democratic forms and a new constituent power that will one day take us through and beyond Empire.

The genealogy we follow in our analysis of the passage from imperialism to Empire will be first European and then Euro-American, not because we believe that these regions are the exclusive or privileged source of new ideas and historical innovation, but simply because this was the dominant geographical path along which the concepts and practices that animate today's Empire developed – in step, as we will argue, with the development of the capitalist mode of production. Whereas the genealogy of Empire is in this sense Eurocentric, however, its present powers are not limited to any region. Logics of rule that in some sense originated in Europe and the United States now invest practices of domination throughout the globe. More important, the forces that contest Empire and effectively prefigure an alternative global society are themselves not limited to any geographical region. The geography of these alternative powers, the new cartography, is still waiting to be written – or really, it is being written today through the resistances, struggles, and desires of the multitude.

[. . .]

The Constitution of Empire

Many contemporary theorists are reluctant to recognize the globalization of capitalist production and its world market as a fundamentally new situation and a significant historical shift. The theorists associated with the world-systems perspective, for example, argue that from its inception, capitalism has always functioned as a world economy, and therefore those who clamor about the novelty of its globalization today have only misunderstood its history. Certainly, it is important to emphasize both capitalism's continuous foundational relationship to (or at least a tendency toward) the world market and capitalism's expanding cycles of development; but proper attention to the *ab origine* universal or universalizing dimensions of capitalist development should not blind us to the rupture or shift in contemporary capitalist production and global relations of power. We believe that this shift makes perfectly clear and possible today the capitalist project to bring together economic power and political power, to realize, in other words, a properly capitalist order. In constitutional terms, the processes of globalization are no longer merely a fact but also a source of juridical definitions that tends to project a single supranational figure of political power.

Other theorists are reluctant to recognize a major shift in global power relations because they see that the dominant capitalist nation-states have continued to exercise imperialist domination over the other nations and regions of the globe. From this perspective, the contemporary tendencies toward Empire would represent not a fundamentally new phenomenon but simply a perfecting of imperialism. Without underestimating these real and important lines of continuity, however, we think it is important to note that what used to be conflict or competition among several imperialist powers has in important respects been replaced by the idea of a single power that overdetermines them all, structures them in a unitary way, and treats them under one common notion of right that is decidedly postcolonial and postimperialist. This is really the point of departure for our study of Empire: a new notion of right, or rather, a new inscription of authority and a new design of the production of norms and legal instruments of coercion that guarantee contracts and resolve conflicts.

We should point out here that we accord special attention to the juridical figures of the constitution of Empire at the beginning of our study not out of any specialized disciplinary interest – as if right or law in itself, as an agent of regulation, were capable of representing the social world in its totality – but rather because they provide a good index of the processes of imperial constitution. New juridical figures reveal a first view of the tendency toward the centralized and unitary regulation of both the world market and global power relations, with all the difficulties presented by such a project. Juridical transformations effectively point toward changes in the material constitution of world power and order. The transition we are witnessing today from traditional international law, which was defined by contracts and treaties, to the definition and constitution of a new sovereign, supranational world power (and thus to an imperial notion of right), however incomplete, gives us a framework in which to read the totalizing social processes of Empire. In effect, the juridical transformation functions as a symptom of the modifications of the material biopolitical constitution of our societies. These changes regard not only international law and international relations but also the internal power relations of each country. While studying and critiquing the new forms of international and supranational law, then, we will at the same time be pushed to the heart of the political theory of Empire, where the problem of supranational sovereignty, its source of legitimacy, and its exercise bring into focus political, cultural, and finally ontological problems.

[. . .]

[We note] the renewed interest in and effectiveness of the concept of *bellum justum*, or "just war." This concept, which was organically linked to the ancient imperial orders and whose rich and complex genealogy goes back to the biblical tradition, has begun to reappear recently as a central narrative of political discussions, particularly in the wake of the Gulf War. Traditionally the concept rests primarily on the idea that when a state finds itself confronted with a threat of aggression that can endanger its territorial integrity or political independence, it has a *jus ad bellum* (right to make war). There is certainly something troubling in this renewed focus on the concept of *bellum justum*, which modernity, or rather modern secularism, had worked so hard to expunge from the medieval tradition. The traditional concept of just war involves the banalization of war and the celebration of it as an ethical instrument, both of which were ideas that modern political thought and the international community of nation-states had resolutely refused. These two traditional characteristics have reappeared in our postmodern world: on the one hand, war is reduced to the status of police action, and on the other, the new power that can legitimately exercise ethical functions through war is sacralized.

Far from merely repeating ancient or medieval notions, however, today's concept presents some truly fundamental innovations. Just war is no longer in any sense an activity of defense or resistance, as it was, for example, in the Christian tradition from Saint Augustine to the scholastics of the Counter-Reformation, as a necessity of the "worldly city" to guarantee its own survival. It has become rather an activity that is justified in itself. Two distinct elements are combined in this concept of just war: first, the legitimacy of the military apparatus insofar as it is ethically grounded, and second, the effectiveness of military action to achieve the desired order and peace. The synthesis of these two elements may indeed be a key factor determining the foundation and the new tradition of Empire. Today the enemy, just like the war itself, comes to be at once banalized (reduced to an object of routine police repression) and absolutized (as the Enemy, an absolute threat to the ethical order). The Gulf War gave us perhaps the first fully articulated example of this new epistemology of the concept. The resurrection of the concept of just war may be only a symptom of the emergence of Empire, but what a suggestive and powerful one!

[. . .]

There Is No More Outside

The domains conceived as inside and outside and the relationship between them are configured differently in a variety of modern discourses. The spatial configuration of inside and outside itself, however, seems to us a general and foundational characteristic of modern thought. In the passage from modern to postmodern and from imperialism to Empire there is progressively less distinction between inside and outside.

[. . .]

Finally, there is no longer an outside also in a military sense. When Francis Fukuyama claims that the contemporary historical passage is defined by the end of history, he means that the era of major conflicts has come to an end: sovereign power will no longer confront its Other and no longer face its outside, but rather will progressively expand its boundaries to envelop the entire globe as its proper domain. The history of imperialist, interimperialist, and anti-imperialist wars is over. The end of that history has ushered in the reign of peace. Or really, we have entered the era of minor and internal conflicts. Every imperial war is a civil war, a police action – from Los Angeles and Granada to Mogadishu and Sarajevo. In fact, the separation of tasks between the external and the internal arms of power (between the army and the police, the CIA and the FBI) is increasingly vague and indeterminate.

In our terms, the end of history that Fukuyama refers to is the end of the crisis at the center of modernity, the coherent and defining conflict that was the foundation and raison d'être for modern sovereignty. History has ended precisely and only to the extent that it is conceived in Hegelian terms – as the movement of a dialectic of contradictions, a play of absolute negations and subsumption. The binaries that defined modern conflict have become blurred. The Other that might delimit a modern sovereign Self has become fractured and indistinct, and there is no longer an outside that can bound the place of sovereignty. The outside is what gave the crisis its coherence. Today it is increasingly difficult for the ideologues of the United States to name a single, unified enemy; rather, there seem to be minor and elusive enemies everywhere. The end of the crisis of modernity has given rise to a proliferation of minor and indefinite crises, or, as we prefer, to an omni-crisis.

It is useful to remember here [. . .] that the capitalist market is one machine that has always run counter to any division between inside and outside. It is thwarted by barriers and exclusions; it thrives instead by including always more within its sphere. Profit can be generated only through contact, engagement, interchange, and commerce. The realization of the world market would constitute the point of arrival of this tendency. In its ideal form there is no outside to the world market: the entire globe is its domain. We might thus use the form of the world market as a model for understanding imperial sovereignty. Perhaps, just as Foucault recognized the panopticon as the diagram of modern power, the world market might serve adequately – even though it is not an architecture but really an anti-architecture – as the diagram of imperial power.

The striated space of modernity constructed *places* that were continually engaged in and founded on a dialectical play with their outsides. The space of imperial sovereignty, in contrast, is smooth. It might appear to be free of the binary divisions or striation of modern boundaries, but really it is crisscrossed by so many fault lines that it only appears as a continuous, uniform space. In this sense, the clearly defined crisis of modernity gives way to an omni-crisis in the imperial world. In this smooth space of Empire, there is no *place* of power – it is both everywhere and nowhere. Empire is an *ou-topia*, or really a *non-place*.

[. . .]

Counter-Empire

[. . .]

Being-against: nomadism, desertion, exodus

[. . .] One element we can put our finger on at the most basic and elemental level is *the will to be against*. In general, the will to be against does not seem to require much explanation. Disobedience to authority is one of the most natural and healthy acts. To us it seems completely obvious that those who are exploited will resist and – given the necessary conditions – rebel. Today, however, this may not be so obvious. [. . .] The identification of the enemy, however, is no small task given that exploitation tends no longer to have a specific place and that we are immersed in a system of power so deep and complex that we can no longer determine specific difference or measure. We suffer exploitation, alienation, and command as enemies, but we do not know where to locate the production of oppression. And yet we still resist and struggle.

[. . .] If there is no longer a place that can be recognized as outside, we must be against in every place. This being-against becomes the essential key to every active political position in the world, every desire that is effective – perhaps of democracy itself. The first anti-fascist partisans in Europe, armed deserters confronting their traitorous governments, were aptly called "against-men." Today the generalized being-against of the multitude must recognize imperial sovereignty as the enemy and discover the adequate means to subvert its power.

Here we see once again the republican principle in the very first instance: desertion, exodus, and nomadism. Whereas in the disciplinary era *sabotage* was the fundamental notion of resistance, in the era of imperial control it may be *desertion*. Whereas being-against in modernity often meant a direct and/or dialectical opposition of forces, in postmodernity being-against might well be most effective in an oblique or diagonal stance. Battles against the Empire might be won through subtraction and defection. This desertion does not have a place; it is the evacuation of the places of power.

Throughout the history of modernity, the mobility and migration of the labor force have disrupted the disciplinary conditions to which workers are constrained. And power has wielded the most extreme violence against this mobility. [. . .] Mobility and mass worker nomadism always express a refusal and a search for liberation: the resistance against the horrible conditions of exploitation and the search for freedom and new conditions of life. [. . .]

Today the mobility of labor power and migratory movements is extraordinarily diffuse and difficult to grasp. Even the most significant population movements of modernity (including the black and white Atlantic migrations) constitute lilliputian events with respect to the enormous population transfers of our times. A specter haunts the world and it is the specter of migration. All the powers of the old world are allied in a merciless operation against it, but the movement is irresistible. Along with the flight from the so-called Third World there are flows of political refugees and transfers of intellectual labor power, in addition to the massive movements of the agricultural, manufacturing, and service proletariat. The legal and documented movements are dwarfed by clandestine migrations: the borders of national sovereignty are sieves, and every attempt at complete regulation runs up against violent pressure. Economists attempt to explain this phenomenon by presenting their equations and models, which even if they were complete would not explain that irrepressible desire for free movement. In effect, what pushes from behind is, negatively, desertion from the miserable cultural and material conditions of imperial reproduction; but positively, what pulls forward is the wealth of desire and the accumulation of expressive and productive capacities that the processes of globalization have determined in the consciousness of every individual and social group – and thus a certain hope. Desertion and exodus are a powerful form of class struggle within and against imperial postmodernity. This mobility, however, still constitutes a spontaneous level of struggle, and, as we noted earlier, it most often leads today to a new rootless condition of poverty and misery.

A new nomad horde, a new race of barbarians, will arise to invade or evacuate Empire. Nietzsche was oddly prescient of their destiny in the nineteenth century. "Problem: where are the *barbarians* of the twentieth century? Obviously they will come into view and consolidate themselves only after tremendous socialist crises." We cannot say exactly what Nietzsche foresaw in his lucid delirium, but indeed what recent event could be a stronger example of the power of desertion and exodus, the power of the nomad horde, than the fall of the Berlin Wall and the collapse of the entire Soviet bloc? In the desertion from "socialist discipline," savage mobility and mass migration contributed substantially to the collapse of the system. In fact, the desertion of productive cadres disorganized and struck at the heart of the disciplinary system of the bureaucratic Soviet world. The mass exodus of highly trained workers from Eastern Europe played a central role in provoking the collapse of the Wall. Even though it refers to the particularities of the socialist state system, this example demonstrates that the mobility of the labor force can indeed express an open political conflict and contribute to the destruction of the regime. What we need, however, is more. We need a force capable of not only organizing the destructive capacities of the multitude, but also constituting through the desires of the multitude an alternative. The counter-Empire must also be a new global vision, a new way of living in the world.

[...]

New barbarians

Those who are against, while escaping from the local and particular constraints of their human condition, must also continually attempt to construct a new body and a new life. [...]

These barbaric deployments work on human relations in general, but we can recognize them today first and foremost in corporeal relations and configurations of gender and sexuality. Conventional norms of corporeal and sexual relations between and within genders are increasingly open to challenge and transformation. Bodies themselves transform and mutate to create new posthuman bodies. The first condition of this corporeal transformation is the recognition that human nature is in no way separate from nature as a whole, that there are no fixed and necessary boundaries between the human and the animal, the human and the machine, the male and the female, and so forth; it is the recognition that nature itself is an artificial terrain open to ever new mutations, mixtures, and hybridizations. Not only do we consciously subvert the traditional boundaries, dressing in drag, for example, but we also move in a creative, indeterminate zone *au milieu*, in between and without regard for those boundaries. Today's corporeal mutations constitute an *anthropological exodus* and represent an extraordinarily important, but still quite ambiguous, element of the configuration of republicanism "against" imperial civilization. The anthropological exodus is important primarily because here is where the positive, constructive face of the mutation begins to appear: an ontological mutation in action, the concrete invention of a first *new place in the non-place*. This creative evolution does not merely occupy any existing place, but rather invents a new place; it is a desire that creates a new body; a metamorphosis that breaks all the naturalistic homologies of modernity.

This notion of anthropological exodus is still very ambiguous, however, because its methods, hybridization and mutation, are themselves the very methods employed by imperial sovereignty. In the dark world of cyberpunk fiction, for example, the freedom of self-fashioning is often indistinguishable from the powers of an all-encompassing control. We certainly do need to change our bodies and ourselves, and in perhaps a much more radical way than the cyberpunk authors imagine. In our contemporary world, the now common aesthetic mutations of the body, such as piercings and tattoos, punk fashion and its various imitations, are all initial indications of this corporeal transformation, but in the end they do not hold a candle to the kind of radical mutation needed here. The will to be against really needs a body that is completely incapable of submitting to command. It needs a body that is incapable of adapting to family life, to factory discipline, to the regulations of a traditional sex life, and so forth. (If you find your body refusing these "normal" modes of life, don't despair – realize your gift!) In addition to being radically unprepared for normalization, however, the new body must also be able to create a new life. We must go much further to define that new place of the non-place, well beyond the simple experiences of mixture and hybridization, and the experiments that are conducted around them. We have to arrive at constituting a coherent political artifice, an *artificial becoming* in the sense that the humanists spoke of a *homohomo* produced by art and knowledge, and that Spinoza spoke of a powerful body produced by that highest consciousness that is infused with love. The infinite paths of the barbarians must form a new mode of life.

[...]

Now that we have dealt extensively with Empire, we should focus directly on the multitude and its potential political power.

The Two Cities

We need to investigate specifically how the multitude can become a *political subject* in the context of Empire.

[...]

How can the actions of the multitude become political? How can the multitude organize and concentrate its energies against the repression and incessant territorial segmentations of Empire? The only response that we can give to these questions is that the action of the multitude becomes political primarily when it begins to confront directly and with an adequate consciousness the central repressive operations of Empire. It is a matter of recognizing and engaging the imperial initiatives and not allowing them continually to reestablish order; it is a matter of crossing and breaking down the limits and segmentations that are imposed on the new collective labor power; it is a matter of gathering together these experiences of resistance and wielding them in concert against the nerve centers of imperial command.

This task for the multitude, however, although it is clear at a conceptual level, remains rather abstract. What specific and concrete practices will animate this political project? We cannot say at this point. What we can see nonetheless is a first element of a political program for the global multitude, a first political demand: *global citizenship*. During the 1996 demonstrations for the *sans papiers*, the undocumented aliens residing in France, the banners demanded "Papiers pour tous!" Residency papers for everyone means in the first place that all should have the full rights of citizenship in the country where they live and work. This is not a utopian or unrealistic political demand. The demand is simply that the juridical status of the population be reformed in step with the real economic transformations of recent years. Capital itself has demanded the increased mobility of labor power and continuous migrations across national boundaries. Capitalist production in the more dominant regions (in Europe, the United States, and Japan, but also in Singapore, Saudi Arabia, and elsewhere) is utterly dependent on the influx of workers from the subordinate regions of the world. Hence the political demand is that the existent fact of capitalist production be recognized juridically and that all workers be given the full rights of citizenship. In effect this political demand insists in postmodernity on the fundamental modern constitutional principle that links right and labor, and thus rewards with citizenship the worker who creates capital.

This demand can also be configured in a more general and more radical way with respect to the postmodern conditions of Empire. If in a first moment the multitude demands that each state recognize juridically the migrations that are necessary to capital, in a second moment it must demand control over the movements themselves. The multitude must be able to decide if, when, and where it moves. It must have the right also to stay still and enjoy one place rather than being forced constantly to be on the move. *The general right to control its own movement is the multitude's ultimate demand for global citizenship*. This demand is radical insofar as it challenges the fundamental apparatus of imperial control over the production and life of the multitude. Global citizenship is the multitude's power to reappropriate control over space and thus to design the new cartography.

Time and Body (the Right to a Social Wage)

[. . .]

This is a *new proletariat* and not a *new industrial working class*. The distinction is fundamental. As we explained earlier, "proletariat" is the general concept that defines all those whose labor is exploited by capital, the entire cooperating multitude. The industrial working class represented only a *partial* moment in the history of the proletariat and its revolutions, in the period when capital was able to reduce value to measure. In that period it seemed as if only the labor of waged workers was productive, and therefore all the other segments of labor appeared as merely reproductive or even unproductive. In the biopolitical context of Empire, however, the production of capital converges ever more with the production and reproduction of social life itself; it thus becomes ever more difficult to maintain distinctions among productive, reproductive, and unproductive labor. Labor – material or immaterial, intellectual or corporeal – produces and reproduces social life, and in the process is exploited by capital. This wide landscape of biopolitical production allows us finally to recognize the full generality of the concept of proletariat. The progressive indistinction between production and reproduction in the biopolitical context also highlights once again the immeasurability of time and value. As labor moves outside the factory walls, it is increasingly difficult to maintain the fiction of any measure of the working day and thus separate the time of production from the time of reproduction, or work time from leisure time. There are no time clocks to punch on the terrain of biopolitical production; the proletariat produces in all its generality everywhere all day long.

This generality of biopolitical production makes clear a second programmatic political demand of the multitude: *a social wage and a guaranteed income for all.* The social wage stands opposed first of all to the family wage, that fundamental weapon of the sexual division of labor by which the wage paid for the productive labor of the male worker is conceived also to pay for the unwaged reproductive labor of the worker's wife and dependents at home. The family wage keeps family control firmly in the hands of the male wage earner and perpetuates a false conception of what labor is productive and what is not. As the distinction between production and reproductive labor fades, so too fades the legitimation of the family wage. The social wage extends well beyond the family to the entire multitude, even those who are unemployed, because the entire multitude produces, and its production is necessary from the standpoint of total social capital. In the passage to postmodernity and biopolitical production, labor power has become increasingly collective and social. It is not even possible to support the old slogan "equal pay for equal work" when labor cannot be individualized and measured. The demand for a social wage extends to the entire population the demand that all activity necessary for the production of capital be recognized with an equal compensation such that a social wage is really a guaranteed income. Once citizenship is extended to all, we could call this guaranteed income a citizenship income, due each as a member of society.

Telos (the Right to Reappropriation)

[...]

Now we can formulate a third political demand of the multitude: *the right to reappropriation.* The right to reappropriation is first of all the right to the reappropriation of the means of production. Socialists and communists have long demanded that the proletariat have free access to and control over the machines and materials it uses to produce. In the context of immaterial and biopolitical production, however, this traditional demand takes on a new guise. The multitude not only uses machines to produce, but also becomes increasingly machinic itself, as the means of production are increasingly integrated into the minds and bodies of the multitude. In this context reappropriation means having free access to and control over knowledge, information, communication, and affects – because these are some of the primary means of biopolitical production. Just because these productive machines have been integrated into the multitude does not mean that the multitude has control over them. Rather, it makes more vicious and injurious their alienation. The right to reappropriation is really the multitude's right to self-control and autonomous self-production.

Posse

[...]

The name that we want to use to refer to the multitude in its political autonomy and its productive activity is the Latin term *posse* – power as a verb, as activity. [...] Posse refers to the power of the multitude and its telos, an embodied power of knowledge and being, always open to the possible.

[...]

As in all innovative processes, the mode of production that arises is posed against the conditions from which it has to be liberated. The mode of production of the multitude is posed against exploitation in the name of labor, against property in the name of cooperation, and against corruption in the name of freedom. It self-valorizes bodies in labor, reappropriates productive intelligence through cooperation, and transforms existence in freedom. The history of class composition and the history of labor militancy demonstrate the matrix of these ever new and yet determinate reconfigurations of self-valorization, cooperation, and political self-organization as an effective social project.

[...]

The posse produces the chromosomes of its future organization. Bodies are on the front lines in this battle, bodies that consolidate in an irreversible way the results of past struggles and incorporate a power that has been gained ontologically. Exploitation must be not only negated from the perspective of practice but also annulled in its premises, at its basis, stripped from the genesis of reality. Exploitation must be excluded from the bodies of immaterial labor-power just as it must be from the social knowledges and affects of reproduction (generation, love, the continuity of kinship and community relationships, and so forth) that bring value and affect together in the same power. The constitution of new bodies, outside of exploitation, is a fundamental basis of the new mode of production.

The mode of production of the multitude reappropriates wealth from capital and also constructs a new wealth, articulated with the powers of science and social knowledge through cooperation. Cooperation annuls the title of property. In modernity, private property was often legitimated by labor, but this equation, if it ever really made sense, today tends to be completely destroyed. Private property of the means of production today, in the era of the hegemony of cooperative and immaterial labor, is only a putrid and tyrannical obsolescence. The tools of production tend to be recomposed in collective subjectivity and in the collective intelligence and affect of the workers; entrepreneurship tends to be organized by the cooperation of subjects in general intellect. The organization of the multitude as political subject, as posse, thus begins to appear on the world scene. The multitude is biopolitical self-organization.

Certainly, there must be a moment when reappropriation and self-organization reach a threshold and configure a real event. This is when the political is really affirmed – when the genesis is complete and self-valorization, the cooperative convergence of subjects, and the proletarian management of production become a constituent power. This is the point when the modern republic ceases to exist and the postmodern posse arises. This is the founding moment of an earthly city that is strong and distinct from any divine city. The capacity to construct places, temporalities, migrations, and new bodies already affirms its hegemony through the actions of the multitude against Empire. Imperial corruption is already undermined by the productivity of bodies, by cooperation, and by the multitude's designs of productivity. The only event that we are still awaiting is the construction, or rather the insurgence, of a powerful organization. The genetic chain is formed and established in ontology, the scaffolding is continuously constructed and renewed by the new cooperative productivity, and thus we await only the maturation of the political development of the posse. We do not have any models to offer for this event. Only the multitude through its practical experimentation will offer the models and determine when and how the possible becomes real.

[. . .]

READING 31

The Global Coliseum: On *Empire*

Michael Hardt and Antonio Negri interviewed by Nicholas Brown and Imre Szeman

[. . .]

B&S: Your invention of the concept of 'Empire' itself would have to be the master example of this operation, and the older category it challenges is, of course, 'globalization'. The phenomenon that 'globalization' refers to has, for the most part, been treated as an empirico-historical event that requires intellectuals to consider how the speed of the present relates to the past, but which doesn't seem to require a wholesale invention of new concepts to make sense of it. Do you think you could encapsulate, briefly, what it is that 'Empire' allows us to think that 'globalization' is unable to encompass?

H&N: It may be right, as you imply, that globalization, especially in its economic guise, has often been conceived in quantitative terms – the increasing number, speed or distance of exchanges – rather than in qualitative terms and this has been an obstacle to understanding the real novelty of our contemporary situation. However, this may also be an indication of the limitation of the concept of globalization itself as the marker of our era. Many authors today, particularly on the Left, point out that globalization is nothing new or even that the quantity of global economic exchanges is lower than it was 50 or 100 years ago. This may be true from this limited perspective, but we think it is largely beside the point. We insist on the fact that what goes under the label 'globalization' is not merely an economic, financial or commercial phenomenon, but also and above all a *political* phenomenon. The political realm is where we most clearly recognize the qualitative shifts in contemporary history and where we are confronted by the need to invent new concepts. But, really, this distinction between the political and the economic (and the cultural) is no longer very satisfying either. We attempt to use the concept of biopower to name the zone characterized by the intersection of these old fields – an economy that is eminently cultural, a cultural field that is equally economic, a politics that comprehends the other two equally, and so forth. From this perspective, the concept 'globalization' is clearly too vague and imprecise. Empire seems to us a much more adequate concept for the new biopolitical order.

B&S: This vagueness or imprecision in the concept 'globalization' may explain why analyses based on it always seem to come down to the relatively banal question of periodization, that is, whether it indeed marks a genuine break with the past or whether it is merely the same old wolf in a new sheepskin. *Empire* insists on the need to abandon certain concepts and modes of critique in order to make sense of the present conjuncture. In particular, you point to the need to give up a form of critical thinking characteristic of Marxism and of postcolonial and postmodern critique – critique in general, for that matter – which was conceived as a challenge to a specific tradition of modern sovereignty that is tendentially extinct: the old wolf is in fact a dead horse. How easily can we give up our old habits of critical thought – not just concepts like 'globalization' but the very habits and structures of our current modes of thinking – and what are the consequences if we can't?

H&N: It does seem to us that posing the question in terms of sovereignty clarifies a variety of contemporary debates, such as those about the powers of nation-states in the age of globalization. There is no doubt that nation-states (at least the dominant nation-states) are still important political actors and exert significant powers. We argue, however, that the nation-state is

no longer the ultimate form of sovereignty as it was during the modern era and that nation-states now function within the imperial framework of sovereignty. The nature and locus of sovereignty have shifted and this, we believe, is the most significant fact that must be taken into account. This has a whole series of consequences that extend throughout the social field well beyond questions of the nation-state.

The consequences of recognizing this shift are indeed very high for both political thought and political action. Political arguments and strategies aimed against old forms of sovereignty may be ineffective against the new forms or they may even unwittingly contribute to its functioning. For example, propositions of hybrid identities or multiculturalism can seem like liberatory projects when one assumes that the power being confronted rests on pure notions of identity and stark oppositions of self and other. But when the sovereign power no longer resides on pure identities but rather works through hybridization and multicultural formations, as we claim it does in Empire, then those projects lose any necessary relation to liberation or even contestation. In fact, they could be complicit with imperial power itself. We do not mean to say because Empire works through multiculturalism and hybridity that we need to reject those strategies – rather we mean simply that they are not sufficient in themselves. In the face of the new forms of sovereignty, new strategies of contestation and new alternatives need to be invented.

[...]

B&S: Back to the notion of counter-Empire: you refuse categorically the now more or less accepted wisdom that globalization signals a crisis for agency and for politics. Instead, you suggest that Empire has produced the conditions of possibility for the production of new identities, collectivities and radically democratic polities – what you memorably describe as '*homohomo*, humanity squared, enriched by the collective intelligence and love of the community'. It is for this reason that you caution against a misplaced nostalgia for older forms, such as the nation-state, that might be imagined as protecting groups and individuals from the harsh winds of globalization. As we touched on earlier, this positive characterization of globalization might be resisted by many on the Left as a form of wishful thinking. Can you point us toward any situations or movements that exemplify the politics involved in the production of counter-Empire? It is tempting to see the protests against the WTO, the IMF and the World Bank as examples of such a politics. But even while these struggles are remarkable for the fact that they are directed precisely towards those institutions and organizations that help to 'structure global territories biopolitically', their politics still seem to be constructed around a modern idea of sovereignty insofar as it is built around the idea of an 'outside', a space or logic other than Empire.

H&N: Our primary point in the book is that a counter-Empire is *necessary*, even before considering how it is *possible*. In other words, our analysis leads us to the conclusion that the only effective contestation of global imperial power and the only real alternative to it must be posed on an equally global scale. Hence, the admittedly uncomfortable analogy, which runs throughout much of the book, with the rise of Christianity during the decline of the Roman Empire. Like then, a Catholic (that is, global) project is the only alternative.

Sometimes political theorizing runs up against obstacles that only practice can solve. Deleuze and Foucault, in their wonderful discussion on intellectuals and power, thought of this relation between theory and practice as a series of relays, passing back and forth the lead in the project. The example that strikes us as most significant in this regard is that way that Marx responded to the Paris Commune. Ever since his early writings he had been very sceptical of giving any positive content to the notion of communism, but suddenly the Parisian proletariat storms the heavens establishing its Commune and he learns from them more clearly in practical terms what communism can mean, how the state can be abolished, how democracy can be extended absolutely, and so forth. His thought could not move forward without the practical advances of the Parisian proletariat.

Well, we are not suggesting that we need today to wait for a new Paris Commune, but simply that practical experiences – like the protests against the global institutions of capital in Seattle, Washington, Prague, etc. – may suggest unexpected solutions. One of the great surprises in Seattle, for example, was that a variety of groups that we thought were irreconcilably antagonistic to one another suddenly appeared to have a common project: ecologists with labour unions, anarchists with human rights advocates, church groups with gays and lesbians. In our terms, we saw these developments as the construction of a *new place* within the dominant non-place of Empire, a new organization of the multitude. Or, at least, these events were allusions to that. It is very difficult to construct a new place of liberation within the non-place of Empire and nothing guarantees that it will not end up in a new kind of mystification. (Here is the negative side of our analogy to early Christianity.) Yet, the emergence of these struggles will undoubtedly contain the lessons for our moving forward both practically and theoretically.

One of our major criticisms of our book is that the concept of the multitude remained too indefinite, too poetic. In part, that is due to our primary focus on Empire and the length required to address its nature and structures. In any case, the multitude is the focus of our current work and we hope to be able to develop the concept more fully in the future.

[. . .]

 READING 32

Retrieving the Imperial: *Empire* and International Relations

Tarak Barkawi and Mark Laffey

For some, Michael Hardt and Antonio Negri's *Empire* is the 'most successful work of political theory to come from the Left for a generation'. It is certainly one of the most widely read analyses of international politics in recent years. Drawing on a combination of theoretic perspectives not found together in International Relations

(IR) – postmodernism, Marxism, and the communist and autonomist traditions of the Italian left – Hardt and Negri chart a new, unitary and global form of postmodern sovereignty which they term 'Empire', a 'logic of rule' worldwide in scope. Their project is twice removed from the discipline of IR, in its intellectual resources and in its object of analysis.

Born into a world of empires at war and amid contemporary processes of globalisation, IR remains centred on the logic of a modern system of sovereign states. Marxian analyses of the international, by contrast, concentrate on the interconnections between Europe, capitalism, and imperialism. Postmodern approaches, in a variety of disciplines, stress the encounter with the post-colonial and the inter-penetration of the European and non-European worlds. Hardt and Negri could only develop an approach to world politics that conceives the histories of the North and the South as common, shared and profoundly implicated in one another. This tension between a view of world politics based on the sovereign state and one that takes imperial relations seriously frames our engagement with Hardt and Negri. In common with *Empire*, we argue that understanding sovereignty requires locating it in histories of European expansion and engagement with the world outside the West.

[. . .]

Seeing through Sovereignty: *Empire* and the International

Although widely hailed as 'the Next Big Idea' in intellectual life, Hardt and Negri see *Empire* differently: 'Toni and I don't think of this as a very original book. We're putting together a variety of things that others have said. That's why it's been so well received. It's what people have been thinking but not really articulating'. Our interest in the book is less with questions of novelty than with the kind of analysis it represents. Engaging with and developing the long tradition of Marxian analyses of imperialism, *Empire* offers a 'total' analysis of world politics past and present. Core and periphery, North and South, East and West, inside and outside are treated as part of a single, increasingly global formation, structured and produced by imperial relations of diverse kinds. Following a brief exposi-

tion of their main argument, we focus on three themes central to the book: the role of the multitude in world politics, the transformation of sovereignty from a modern to a postmodern form, and the putative disappearance of imperialism.

Empire's thesis is a familiar one: sovereignty is not what it used to be. Under the pressure of capitalist globalisation, sovereignty's very nature is being transformed, from a modern to a postmodern form. In the process, a new global form of rule is emerging which Hardt and Negri term Empire. Imperialism is central to *Empire*'s account of world politics. Imperialism, they claim, operated through the modernist logic of inside/outside. Modern sovereignty and classical imperialism are thus inseparable: together they divided up the world and its population, in Europe and elsewhere. Imperialism was also 'a system designed to serve the needs and further the interests of capital in its phase of global conquest'. But from its inception, capital has tended toward world power in the form of the world market. The realisation of that power requires the remaking of modern sovereignty, which is a sovereignty of borders and limits. 'Imperialism is a machine of global striation, channelling, coding, and territorialising the flows of capital, blocking certain flows and facilitating others. The world market, in contrast, requires a smooth space of uncoded and deterritorialised flows'. It follows, on Hardt and Negri's account, that once the world market is achieved and there is no more outside, imperialism by definition is over. What remains is a new post-imperial and post-colonial world order.

Even though imperialism and modern sovereignty are in decline, capital still needs the state. From a Marxian perspective, the 'state-capital dialectic' is only conflictual from the point of view of the individual capitalist: '[w]ithout the state, social capital has no means to project and realise its collective interests'. The sovereign state and its powers may be undermined but state functions remain necessary and are 'effectively displaced to other levels and domains', local and transnational. The 'twilight of modern sovereignty' is also the dawn of Empire, a new 'decentered and deterritorializing apparatus of rule that progressively incorporates the entire global realm within its open, expanding frontiers'.

The model for understanding this new postmodern form of global capitalist sovereignty is the world market.

In contrast to imperialism, the new sovereignty is imperial but not imperialist, for the simple reason that 'its space is always open' rather than bounded: 'modern sovereignty resides precisely on the limit. In the imperial conception, by contrast, power finds the logics of its order always renewed and always re-created in expansion'. As modern sovereignty declines, the world is in fact becoming 'a smooth space' across which people, ideas and things move freely, albeit one cross-cut with new and old 'lines of segmentation', including class, that do not follow the boundaries of modern nation-states.

Most analyses of globalisation focus on the role of capital or the state in driving these changes. In contrast, Hardt and Negri stress the role of labour struggles, both in the emergence of globalisation as a capitalist strategy and in capitalist development more generally. Capital 'is not a thing but a social relationship, an antagonistic relationship, one side of which is animated by the productive life of the multitude', Hardt and Negri's term for what used to be called the proletariat. Successive stages in the evolution of capital and sovereignty are driven by this antagonism, with labour always the active subject. Significantly, the multitude is not located only in Europe but also outside. Hardt and Negri highlight the inter-related character of struggles across the globe and their role in driving capital forward, forcing it to respond to the multitude's essential creativity and plurality. Thus, the emergence of

> Empire and its global networks is a response to the various struggles against the modern machines of power, and specifically to class struggle driven by the multitude's desire for liberation. The multitude called Empire into being.

Indeed, *Empire*'s genealogy of the international functions as a grand narrative in which history is nothing but a series of struggles between the communism of the multitude and capitalist forces of reaction, the latter initially vested in modern sovereignty and the state and now located in Empire.

We stand here at some distance from a Westphalian view of the world and the disciplinary debates of IR. 'In the 1990s', observes Patomäki, 'after the short visit of Marxism in the mainstream of IR, there has been, perhaps more than ever, a tendency to reduce all problems of IR to an almost eternal dispute between political realism and liberalism'. In marked contrast to such disciplinary analyses, Hardt and Negri offer us a glimpse – albeit one that is sometimes partial, distorted or simply false – of what world politics looks like from a strikingly different angle of vision, one that takes both imperialism and Marxism after postmodernism seriously. As a result, they also help us see how attending to the imperial transforms our understanding of world politics. Nowhere is this more evident than *Empire*'s treatment of the multitude's struggles and their role in the historical development of sovereignty.

Putting the multitude at the centre of analysis is a major step forward in elaborating a 'thicker' conception of the international directly attentive to imperial relations. Focusing on labour grounds *Empire*'s analysis of the international in social forces and relations. A growing number of scholars have pointed to the everyday relations of power that underpin and enable the international system as conventionally understood, locating the international in the biopolitical. 'International politics', as E.H. Carr so famously observed, 'are always power politics'. But as Cynthia Enloe notes, 'it takes much more power to construct and perpetuate international [. . .] relations than we have been led to believe'. 'Ordinary people' have to be incorporated into the global social order so that their labour can sustain it.

Although seldom central to IR analyses, scholars in the interdisciplinary 'trading zone' of IPE regularly remind us of these relations. Aihwa Ong's analysis of the cultural politics of Chinese transnationalism shows how conceptions of national and ethnic identity are reworked and deployed, often in hybrid ways, in the service of capitalist entrepreneurialism and investment. Similarly, Jacqui True's discussion of post-socialist transformations in the Czech republic demonstrates the centrality of gender relations to capital's entry into new territories and construction of new markets. In these and other ways, the social relations of capital remake subjectivities. Beginning with the multitude, with people in the irreducible diversity of their daily lives, opens up space for a richer account of the international, one grounded in the everyday production of subjectivity and the intimate connections between and among the concrete struggles of peoples the world over.

An example helps draw out further some of the implications of a focus on the multitude for understanding world politics. We have already mentioned the role of Vietnamese peasants in producing the contemporary US. On 4 May 1970 Ohio National Guardsmen on the campus of Kent State University (KSU) opened fire on students protesting the US invasion of Cambodia. Thirteen students were shot, four of them fatally. That the students were white made the event all the more shocking to public opinion. 'Kent State' and 'May 4th' rapidly took on iconic status, as representative of an era wracked by imperial war in Southeast Asia and civil unrest in the US and elsewhere.

In the three decades since 1970, efforts to commemorate and memorialise the shootings at KSU have generated continuing controversy. As Scott Bills argues, 'the link between culture, narrative and empire is the key to examining post-1970 events at Kent State'. By their very nature, imperial adventures abroad and their consequences at home produce popular memories that contradict public or official histories. In representations of 'May 4th' dominant narratives and public myths of America confront both an event and memories of it that challenge and unsettle them. Similar struggles over memory and the nation are evident in the controversy surrounding the Smithsonian Institution's attempt to provide a historically accurate account of the US use of nuclear weapons at the end of the Second World War as well as in debates over the responsibility of past US policies for the strikes on the World Trade Center and the Pentagon on 11 September. In these and other ways, connections between widely dispersed populations are made manifest and translated into continuing struggles over history, memory and identity. The significance of such struggles for world politics is evident, for example, in the past and present impact of the American experience in Vietnam on US foreign policy. Seeing the multitude as central to what world politics is and how it changes over time directs our attention to a range of actors, locations and 'thick' relations all but invisible in contemporary IR.

A second theme from *Empire* that illuminates our larger argument about the significance of the imperial concerns the genealogy of sovereignty. Hardt and Negri offer a peripheral or subaltern re-reading of sovereignty. 'Modern sovereignty', they observe, may have 'emanated from Europe', but 'it was born and developed in large part through Europe's relationship with its outside, and particularly through its colonial project and the resistance of the colonized.' It follows that 'rule within Europe and European rule over the world' are 'two coextensive and complementary faces of one development'.

Critical scholarship in IR largely overlooks this integral relation. R.B.J. Walker's *Inside/Outside*, for example, has no index references to colony, empire or imperialism. Jens Bartelson's genealogy of sovereignty refers to empires and imperialism only in passing. David Held's writings on sovereignty also ignore or marginalise Europe's relations with its colonies. Even Hedley Bull and Adam Watson's *The Expansion of International Society* – explicitly addressed to the spread of sovereign recognition to formerly colonised territories – takes for granted that sovereignty emerges in Europe alone and then diffuses throughout the world.

In contrast to such views, Hardt and Negri force us to see that sovereignty, as a concept and an institution, developed in the encounter between Europe and the non-European world. The genealogy of sovereignty cannot be restricted to Europe itself but must include the imperial relations between Europe and its colonies: 'The colony stands in a dialectical opposition to European modernity, as its necessary double and irrepressible antagonist'. Inherent in sovereignty are racialised assumptions of European superiority and fitness for self-rule. Race, hitherto a marginal concern within the discipline, becomes central. As Gilroy argues in the case of modernity and slavery, Western political ideas and institutions cannot be separated out from their implication in the history of imperialism and its racialised terror and genocide. In these and other ways, understanding the West requires attention to its implication in world politics as a whole and to the 'thick' conception of the international outlined above.

While Hardt and Negri's re-reading of sovereignty is helpful in this regard, it must be supplemented with a more historically informed account of the relations between rule 'at home' and 'abroad'. Hardt and Negri take for granted that modern sovereignty in the form of imperialism functioned outside Europe in much the same way as it did inside, as a machinery of borders and limits. But sovereignty in the colonies was never what it was in the metropole. In purely juridical terms, at the height of the era of formal empire, one could speak of Belgian sovereignty over the Congo or British sovereignty over its Indian Empire. But often there was a considerable gap between the sharp lines and coloured spaces of imperial maps and the realities of colonial administration and rule. Large tracts were never adequately pacified, as on the Northwest Frontier of British India, while other areas were never brought under effective administration, as in much of Africa. Even at their height, European and other empires did not display the centralisation of authority taken for granted in discussions of the sovereign state. Relations between the formal apparatus of the 'home' state within an empire and the populations it ruled 'abroad' were multiple, diverse, and changing. Forms of rule were often overlapping and myriad arrangements were struck with local elites. Understanding world politics in terms of sovereignty – whether Westphalian or that of Hardt and Negri's *Empire* – too easily obscures real relations of rule.

Even after 1945, in the high noon of modern sovereignty, patterns of rule and power were often only contingently aligned with sovereign borders. In the wake of decolonisation, many new states were subject to high degrees of intervention by former imperial patrons and the superpowers, sometimes exceeding that experienced in formal empire when many areas were ruled more or less 'indirectly'. In the core too, the Cold War system led to high levels of superpower penetration of former great powers and other states as in Germany, Japan and Eastern Europe. Similar relations of international rule persist today in the policies and practices of the international financial institutions, the Western administered territories of Bosnia and Kòsovo, and the Anglo-American sanctions regime in Iraq. Modern sovereignty, even after decolonisation, was not a universal but at best only a regional practice of government and rule. This fact highlights the distorted and mystifying character of accounts of world politics that start with Westphalian sovereignty and its global diffusion. Attention to the everyday mechanics of rule also highlights difficulties with Hardt and Negri's account of Empire. Their claims for a sharp division between modern and postmodern forms of sovereignty founder in the face of the imperial continuities of international relations, past and present.

These reflections lead us to our third and final theme, the putative disappearance of imperialism. Hardt and Negri assert that imperialism is over for two reasons. First, the world market has been realised, at least tendentially. It is on this basis, as modern sovereignty collapses in the face of globalisation, that the world can now be characterised as a 'smooth space'. But Hardt and Negri's basic empirical claims about the decline of borders, as Petras and others have pointed out, are indefensible. Processes of liberalisation also have another side, namely, a massive effort to make it harder for undesirable flows – be they illegal economic migrants, asylum seekers, illegal drugs, crime, or contraband – to cross borders. As the European Union disassembles internal boundaries, for example, it simultaneously reinforces its external border.

The second reason imperialism is said no longer to exist stems from the unique character of the US. While many would agree with Edward Said's assertion that the US is replicating 'the tactics of the great empires', Hardt and Negri claim we are witnessing not a reinvigorated US imperialism but the birth of a post-imperial international system. They acknowledge US global hegemony over the use of force as well as its central role in controlling the international financial system. However, they argue that US policies are imperial not imperialist, in the sense that they are only ambiguously motivated by US national interests and do not seek to foster a world of closed spaces under US sovereignty. Indeed, US sovereignty was always postmodern according to Hardt and Negri and the US constitution provides the model for the network power that animates *Empire*. The validity of *Empire*'s argument for a sharp break between modern state sovereignty and postmodern global sovereignty rests in large measure on the plausibility of its analysis of the US in the world.

In this context, their claim that the Tet offensive of January 1968 marked the 'irreversible military defeat of the US imperialist adventures', takes on considerable importance. In fact, Tet resulted in a military stalemate. While it certainly was a political defeat for the Johnson Administration and its policy in Indochina, it was hardly irreversible in terms of the wider aims of US Cold War policy in the Third World. The US experience in Vietnam re-invigorated its efforts to find less costly and more effective ways to 'defeat communism', principally through the advising and supporting of Third World military and police forces, foreshadowed in the policy of 'Vietnamization' and codified in the Nixon Doctrine. Even in the depths of its Vietnam malaise, the US was able to sponsor covert operations in Chile, Angola, and elsewhere. Later, the so-called 'lessons of Vietnam' were crucial to the 'Second Cold War' launched in the latter half of the Carter Administration and pursued by President Reagan. The 1980s witnessed a renewal of US interventionism, including a war in Central America and US support for 'freedom fighters' in Afghanistan and elsewhere. The late 1980s saw the development of more effective forms of 'political' intervention, characterised by William Robinson as 'promoting polyarchy', which involved a careful combination of political, economic, military and covert intervention to produce 'stability' in Third World countries and open them up to US investment.

In all of this, it is hard to see how 1968 marks the 'irreversible' defeat of US imperialism. Not only is the inadequate nature of Hardt and Negri's historical analysis much in evidence here, it also becomes very difficult to locate the break at which US imperialism transforms into Empire. As we write, the US is establishing an arc of military bases across central Asia and developing patron–client relations with the authorities there. Such strategies of intervention and imperial control point to continuities not only with past US engagements in the Third World but also with older histories of imperialism. Now, as then, such engagements are also shaping the character of US democracy and society.

In our view, globalisation and many of the phenomena Hardt and Negri describe are better understood by reference to an international state dominated by the US. Immediately after the Second World War and in the decades since, state power was internationalised through a proliferating set of institutions and arrangements, with the US always at its core. In this respect, the categories and theories of classical imperialism, with the possible exception of Kautsky's ultra-imperialism, are a poor guide to the world in which we live. International state power is not reducible to the US alone. But in one domain after another, the concentration of US state power and its international reach is, if anything, greater now than in 1945. Hardt and Negri acknowledge that the main levers of world power

remain in the hands of US state agencies. Where then are we to locate the break between US imperialism and Empire?

These continuities and developments in US and international state power highlight additional difficulties with Hardt and Negri's account of political-military relations. In common with other analyses in the 1990s, they argue that the era of major inter-state war is over. This is due to the fact that nuclear weapons make

> war between state powers [. . .] increasingly unthinkable. The development of nuclear technologies and their imperial concentration have limited the sovereignty of most of the countries of the world insofar as it has taken away from them the power to make decisions over war and peace, which is a primary element of the traditional definition of sovereignty.

As a result, 'the imperial bomb has reduced every war to a limited conflict, a civil war, a dirty war, and so forth'. Military operations now take the form of police actions. These claims are fairly significant for Empire, as a world in which international war is alive and well is not one that is 'smooth' and subject to a single 'logic of rule'.

Unfortunately, Hardt and Negri's analysis of international security and the role of nuclear weapons overlooks significant political-military 'striations' in world politics. India and Pakistan directly contradict their assertions, as does the possibility of the use of weapons of mass destruction in the Arab–Israeli conflict. The end of the Cold War arguably made nuclear war more likely, especially given the fact that Soviet weapons, nuclear materials and technical personnel are far from being concentrated under imperial control and indeed may even be available for purchase on the open market. The buyers may well be non-state actors such as al Qaeda who, on the evidence of 11 September, would be far more willing to use weapons of mass destruction than the leadership of a state with a vulnerable homeland. If India and Pakistan, among other possibilities, indicate that inter-state and even nuclear war cannot so easily be assigned to the dustbin of history, al Qaeda and the 'War on Terror' are indicative of new forms of international and globalised war not reducible to the categories of police action. The possibility of US first use of nuclear weapons in such conflicts cannot be overlooked either, and may in fact be the most likely route to nuclear war other than accident. Hardt and Negri's claims regarding the 'smooth' and global nature of Empire's sovereignty are at best premature in the political-military domain.

Conclusion

A world composed of competing and potentially warring powers, whether states or other entities, is not the kind of world Hardt and Negri describe under the rubric of Empire. In direct contrast to the idea that the old imperialism is over, American policy analysts are resurrecting the language of empire and turning to Rome and Pax Britannica for inspiration. Charles Fairbanks of the Johns Hopkins University has announced that the US is 'an empire in formation' while Max Boot, editorial features editor of the *Wall Street Journal*, has called for the military occupation of Afghanistan and Iraq: 'Afghanistan and other troubled lands today cry out for the sort of enlightened foreign administration once provided by self-confident English-men in jodphurs and pith helmets'. As with Rome and Great Britain, American imperialism has and will continue to generate resistance. Within the conceptual categories of Hardt and Negri's *Empire*, these most recent developments in the history of imperial relations in world politics remain invisible.

'One of the central themes of American historiography', observes William Appleman Williams, 'is that there is no American Empire. Most historians will admit, if pressed, that the United States once had an empire. They then promptly insist that it was given away. But they also speak persistently of America as a World Power'. Perhaps the clearest evidence of the world's lack of 'smoothness' is the widespread resistance

to US power. In contrast, Hardt and Negri valorise the US. In a breath-taking lapse into American exceptionalism, they assert that US sovereignty is not like modern sovereignty; the US was postmodern from birth and US experience is 'truly new and original'. In times past, the US did sometimes act in imperialist ways but this was always an aberration, inconsistent with the defining essence of the US, the US constitution. In any case, with the realisation of the world market, US imperialism (indeed, all imperialism) is over. Marxism, postmodernism, and Italy notwithstanding, *Empire* is a deeply American book.

It has also been said that IR is a profoundly American social science. In important respects. *Empire* and IR represent world politics in distinctively American kinds of ways. From its inception, the US was figured as a 'city on a hill', one defined against European power politics and imperialism. This opposition between the new world and the old was reinforced after the Second World War as the US literally remade Europe. What kind of work does such an opposition do in these very different settings, in disciplinary IR and in a text hailed as 'a rewriting of *The Communist Manifesto* for our time'? In IR, the opposition between the US state and European empire is inscribed in post-war IR scholarship and reinforced by the development of area studies as a particular way of conceptualising the peripheral domains, a way tied more or less directly to US state interests and one which facilitated US imperial power. In *Empire*, the US is curiously abstracted from the blood-bespattered politics of the old world and returns only to remake the world as a whole in its own image, as Empire. In both cases, the trope of 'America' serves to obscure the imperial realities of world politics, past and present. We have sought to retrieve some of these realities for understanding world politics.

 READING 33

Africa: the Black Hole at the Middle of *Empire*?

David Moore

What does the magisterial sweep of *Empire* have to say about Africa? More important, what is there about

it that will, or should, affect the praxis of scholars and activists concerned with the struggles of Africa's "multitude?" How do the contradictions of postmodern, global informationalization, elucidated so eloquently in *Empire*, alter the continent ripped apart most severely by the two key crises on the way to modernism that its authors elaborate – primitive accumulation and nation-state formation (or sovereignty, in their language) – and the third one – democratization – that they don't?

Africa is caught on the cusps of what less severely avant-garde scholars than the Duke literature professor and the Italian political prisoner/writer-in-residence might call "civilizational" crises. The "dark continent" has not gone through all the blood and guts of the paths to *modernity* outlined in *Empire*, even if its (partial) incorporation in the global political economy has been accompanied by a catalogue of horrors of its own magnitude. Yet, there is clear indication of much more anguish to come as the information mode of production adds yet another level to the uneven articulations currently tearing the continent apart.

Africanists worry about what "paths" Africa is following – or wonder if there are any "paths" at all as opposed to meaningless meanderings of pain. How can we be sure that Africa's paths of "development" are *linear*, or that they are en route to something approximating "modernity" at all? The more pessimistic among them ask if "modernity" has not passed Africa by. Yet when Michael Hardt and Antonio Negri consider the continent at all, it is hoist by some petards of postmodernity and *its* hypercrises. *Empire* pours needed (although dialectical) scorn on those believing in the oxymoron of national liberation but then wishes its contemporary problems away. It is as if Africa and the rest of the "third world" had joined with the borderless multitude in advanced capitalist corners of the world.

One has only to come down to earth to remember that the millions of African refugees constitute a qualitatively different realm of existence than that lived by those rendered borderless by jets and cyberspace. The latter's subjectivities are formed in the merging of superstructure and structure occurring when communications become a means of production, and their differences are sublimated by Internet expertise. Yet in Africa, we see the deep, deep crises of modernity deferred – but now, perhaps, accelerated, thus more disruptive than ever in our post-cold war era. Borderlessness in Africa is due to poverty, war, and famine

and is subject to the mentalities of "tradition" (often invented, to be sure, but nevertheless counter to a strategy of Gramsci's "good sense") rather than a combination of supercool calculation and cyborgian connectivity. Does that mean that Africa (as always, we often end up thinking in spite of ourselves) is dependent on what the hyperadvanced multitude in the West decides for it? Is the discourse articulated in *Empire* yet another version, along with the various strands of development and underdevelopment theory over which we have pored in the past, yet another strand of academic "trickle-down"?

One wonders, then, if *Empire*, based on the European – let's face it, white – experience, can adequately recognize the African multitude? Can it outline the ways in which those at the peak of Empire (but whose radical nomadism contradicts it) can extend a difference-based solidarity with it? Or does the book, almost in spite of itself, place the continent on Fukuyama's wagon? "Sure," Hardt and Negri can almost be heard to say, "Africa's struggles bear the marks of nobility and tragedy, but their ends are almost predetermined, and maybe even farcical. Let's get back to Europe and America, where all the marches to Seattle or Prague *really* point to the heart of empire" (and remember, 747 flights and the Internet mobilization came first).

Are Africans' struggles – for affordable food in cities, for land in the country, to avoid war and famine all over – on the same plane as the multitude's "insurrectional event[s] that erupt within the order of the imperial system provok[ing] a shock to the system in its entirety"? How can they be, if Africans are not yet really "people" because the contradictions of nation-statehood have not yet been carried through on their soil? How can they be, if their struggles are not waged on the capitalist terrain furrowed by the ploughs of primitive accumulation, so that they can push capitalism forward – against its own will – into the heights of informatized productivity? Can the new mode of production propel Africa into these realms of efficiency and extraenlightened consciousness?

Do the intense manifestations of war-torn Africa's contradictions – the many wars about, battles within, and contests over primitive accumulation, sovereignty construction, and democratization so correctly identified in *Empire* as the building blocks of modernity – deserve the epitaph "been there, done that"? Or are they unique components of "the plural multitude of productive, creative subjectivities of globalisation that have learned to sail on this enormous sea"? Are there African "differences" with qualities that can add to Hardt and Negri's project of creating a world altered enough from the one we have now to warrant tearing one's self away from the pleasures of (digital) television and the World Wide Web? Hardt and Negri tend to assume that their subjectivities have all passed through modernity and are well ensconced in the age of informationalization: that they have learned the ropes on globalization's ships. They thus fall silent when confronted with the subjectivities still embedded in a contradictory mélange of productive and reproductive modes.

At one moment they condemn Marx for suggesting that the Indias of the world must follow the railroads to the antinomies of bourgeois freedom: otherwise, as the grandest old Eurocentric of them all said, the "passive [. . .] unresisting and unchanging societies [. . .] (with) no history at all" will remain with their superstitions and hierarchies. But Hardt and Negri can only hint at "the difference of Indian [and for the purposes of the point, African] society, the different potentials it contains." They do not refer to Marx and Engels's contemplation of Russian communes as a seedbed of alternatives: one wonders if that would be too populist, or a romanticization of backward-looking utopian socialism for them. Thus they soon move on to claim that the realm of autonomy to which they aspire can come only with the end of the world market. However, that market, with its "deterritorializing flows and the smooth space of capitalist development," can be realized only with the end of imperialism, which is a fetter on capital. But imperialism rests on competition between "nation-states," so they must go. This is an argument compatible, ironically, with the World Bank, the International Monetary Fund, and Hayek, and in making it, they, too, fall over as many contradictions as is humanly possible. More po-facedly, it is a thesis prescribed by the state-ensconced but neoliberal members of the South African Communist party who advise unbound globalization while the stalwarts of their party parry layoff after layoff and the rest of South Africa's poor negotiate the most unequal society in the world. But can this proposition be tested on societies not yet transcending the immanent planes of sovereignty, peoplehood, and disciplinary social

democratic compromises that have formed the centers of Empire's capitalism? Can it be tried on those social formations on the wrong side of the (post)colonial divide – a gap undergirding and enabling Empire while now inside it, the negative side of which has no contemporary equivalent other than the regressive ethno-national-tribal infinities which fool the warlords of nationhood into thinking they can replicate little empires of their own ad infinitum?

Or are Africa's laborers eligible for Hardt and Negri's global social wage, too? One would think that the only partially proletarianized poor of Africa are not quite the "new proletariat" of *Empire*'s teleology: most of Africa's exploited have not "moved outside the factory walls" because they have yet to be inside. Are the contract mineworkers, the sweatshop laborers, the sugar-plantation swathers, the cotton pickers, the teenage-girl silicon chip welders, the millions upon millions of "informal traders," the boychild soldiers, the part-market, part-subsistence agriculturists and, yes, the slaves all members of Hardt and Negri's "general intellect"? Do they, too, perform the "intellectual, immaterial, and communicative labor power" that has replaced factory-rooted labor-power as the source of surplus-value and that is at the "center of the mechanism of exploitation (and thus, perhaps, is at the center of potential revolt)"? Or are they dependent on the latter's largesse, on a new form of charity?

Even if the collective African laborer were fully proletarianized (if Hardt and Negri's partial appropriation of the dependency school's view of the already and always global factory had not been devastated by such scholars as Ernesto Laclau and Robert Brenner), would a global social wage stop the wars? Would a global social wage, in a package with a global parliament, global democracy, global currency (maybe even global unions that look like soviets were supposed to look), and all the rest in the panoply of the "cosmopolitical" thought mildly derided at the beginning of their book be the bedrock of a new identity superseding race, gender, and tribe? Would that new constitution of the (antihumanist) self allow all the other interpellations to sit under the mild and multicolored umbrella of mutually celebrated "difference"?

To cut the story short, if "*modernization has come to an end*" already, what about those whom it has not yet fully subsumed? What about those who are exploited through the formal subsumption of their noncapitalist labor, but who have yet to be "really" subsumed? Will they desire to be Taylorized, Fordized, and disciplined all the way to the full subsumption of their labor or will they resist? Will they choose primitive accumulation? Which process will be supported by the vanguard of the general intellect, or will it all be a matter of investment flows from proletarian core to proletarian periphery that somehow get conjured up by the affective biopolitical networks – perhaps run by those who manage their peers' pension funds?

Getting out of the Hole: Primitive Accumulation, Nation-State Formation, Democratization, and "Intervention"

The following words will demonstrate that some of the insights (and there is at least one on every page) in *Empire* are relevant in the "heart(s) of darkness" within and surrounding the bodies of over 700 million members of Hardt and Negri's (not quite?) multitude. *Empire* sheds much light on the contradictions of and struggles over primitive accumulation, sovereignty, and democratization in Africa. The rest of this essay will attempt to elucidate some of the ways in which Hardt and Negri assist the analysis of these processes. However, *Empire*'s impasse – a cul-de-sac born, ultimately, of too much damned philosophy and not enough empirical materialism – is shown at its blackest in Africa. Neither Empire nor the multitude's concomitant global solidarity (both productive of *and* resistant to Empire, and that combination may have unforeseen consequences for Africa beyond their sum!) may reach far enough into Africa to hasten the surmounting of the crises inherent on the road to the rocky reaches of modernity. Aside from the equivocal interventions of the humanitarian international, *Empire*'s map does not indicate any turns off the main road leading to alternative structures of political economy or identity. On both roads, Africa seems alone.

The next few pages will attempt to analyze some aspects of Africa's contemporary crisis and its potential resolution with the help of some of the concepts raised in *Empire*. The Democratic Republic of the Congo and Zimbabwe present two possible case studies of contradictory combinations of social relations sharing the label of "nation-state" on the continent which can "test" the heuristic power of Hardt and Negri's provocative text. The war in the Democratic Republic of the Congo brings to mind a turning point combining a crisis of nation-state formation and democratization. The crisis centering around Zimbabwe's "sovereign king" slipping from power and the intricately related land invasions heralds a crisis of democratization and primitive accumulation. Any study of Africa should also include South Africa, the continent's newest "democratic" social formation, which also happens to be Africa's most powerfully "advanced" capitalist outpost and is a prime candidate for Empire's loyal satellite or "regional hegemon." However, time and space preclude the "rainbow nation" on which much of Africa's whole fate may swing. Suffice it to note that its combination of pre-, present and post-phases of racially inscribed "modernity" may be Empire's future foretold. Crime, xenophobia, and the politics of privilege amidst poverty threaten to pull apart Mandela's magical "rainbow nation" in hypertime unless a classical social-democratic compromise (or something much more radical, of course) can ameliorate the crisis of basic needs faced by its multitude. Finally, the ambivalence of the humanitarian agenda as a means of the meeting of first and third world "multitudes" – through "intervention" – will be explored.

[...]

The Humanitarian Agenda: Vehicle for the Multitude or Empire's Masters?

In an interview with the Australian Broadcasting Corporation, broadcast 12 January 2000, the MDC-ML's Jacques Depelchin stated that Africans need solidarity from the Western multitudes on the scale of the antiapartheid movement. Perhaps that was the last moment of global solidarity for Africa's multitude. Perhaps, too, that solidarity was too easy: after all, its common denominator was a liberal belief in individual equality and the political means to negotiate that artefact of universal liberalism in an unequal arena constituted by economic liberalism. And who knows? It could pale in contrast with eighteenth-century antislavery campaigns and the nineteenth-century crusade against King Leopold. In any case, leftists in the west now fear encroaching on the barriers of sovereignty so rightly ridiculed in *Empire*, and are justifiably confused by the ever accumulating contradictions thrown up by the "dark" continent. Postcolonial relativism does not help, either. However, there may be positive indications in the "humanitarian international's" interventions in Africa. Is a "cosmopolitical democracy" augured in the humanitarian agenda the way ahead? Is it antithetical to *Empire*'s hopes for the multitude?

It is not surprising that when Hardt and Negri confront humanitarianism, they display their radical ambivalence at its height. Most devastating in *Empire* is the dualistic frame of mind it contains regarding what some people see as the nascent form of a humanitarian global order. When Hardt and Negri first approach the nongovernmental organization phenomenon, they impose closure upon it. They say that what could be seen as *solidarity* too easily becomes *intervention*. There, *Empire*'s deep negativity toward "intervention" seems *too* disapproving. They condemn "Amnesty International, Oxfam, and Médecins sans Frontières" as no more than the "most powerful pacific weapons of the new world order [. . .] the mendicant orders of Empire [. . .] blind[ing] th[eir] theorists to the brutal effects that moral intervention produces as a prefiguration of world order" and preparing the stage for military intervention – not to mention economic, social, and political control. With such a view, what global solidarity is available? Later, they lead us to wonder if the members of the humanitarian international are little more than hygienists, trying Sisyphus-like to keep AIDS-like scourges away from "us" in the "age of universal contagion." In this language, where does solidarity, and even the more paternalistic and passively revolutionary welfarist "assistance," end; and when does "intervention" begin? Are all the cosmopolitical democrats and humanitarians whom one might think are (and who think they are) on the positive side of globalization's double-edged sword actually the unwilling participants in Louis Althusser's "international of decent feelings"? As Althusser said about a different breed of internationalists on the morrow of World War II,

we are confronted with a phenomenon that is international in scope, and with a diffuse ideology which, though it has not been precisely defined, is capable of assuming a certain organisational form [. . .] one senses [. . .] a mentality in search of itself, an intention eager to embody itself in concrete form, an ideology seeking to define itself, entrench itself, and also furnish itself with a means of action. If this mentality is international, and in the process taking institutional form, then a new "International" is in the making.

Yet, caution should not transform into cynicism. In his acceptance speech for the 1999 Nobel Peace Prize, delivered 10 December in Oslo, Norway, James Obrinski, president of Médecins sans Frontières, said: "There is a confusion and inherent ambiguity in the development of so-called 'military humanitarian operations.' We must reaffirm with vigor and clarity the principle of an independent civilian humanitarianism. And we must criticize those interventions called 'military-humanitarian.' Humanitarian action exists only to preserve life, not to eliminate it."

Certainly, the convergence of moral and military intervention is a factor of empire, and that inspires the suspicion of those advancing the multitude's cause. One must be very careful to separate an intervention based on solidarity from bad faith and false pretenses. But in Africa, it seems as if even the interventionists of whom we are wary do not intervene! In mid-1999, the Congolese rebel groups fighting against Kabila asked the United Nations to enforce their Lusaka Peace Accord. Did they get any help? Not unless you call a very small, oddly Foucauldian "surveillance" force "enforcers"! Unless a quick fix can shove Empire's problems to the side, as in Kosovo and Kuwait, Empire's neoliberal organic intellectuals let laisser-faire do its tasks. That means war just as much as the "free" markets these global technocrats assume are always and already existent (even more so than Kelson and company presuppose universal juridical norms).

Yes, the vanguards of intervention who have transformed labor-education and agrarian-reform nongovernmental organizations into World Bank-funded human resources, industrial psychology, and genetically modified organism consultancies should be seen – and exposed – as the latest wave of biopolitical technicians, the disciplinarians of the global (but internalized) panopticon. So, too, the nascent military guards and, of course, their regional henchmen, for whom the state is not a poisoned bequest – as it is for most of the people under their rule. Rather, it is their tool of plunder so long as they attempt to follow Empire's dictates. But the human rights activists, especially those who blend the classical cornerstones of civil rights with their socio-economic sisters to fight for *states* taking citizenship seriously (the internationally coordinated Jubilee 2000, for example, which tirelessly campaigns to end the third world's colossal debt burden)? They have to be etched into the new wave of global solidarity Hardt and Negri assert as necessary. One asks – but could only ask at that point – how they fit into an analysis of this continent informed by *Empire*'s passions, principles, and perspectives. How do these itinerants of Empire fit in with *Empire*'s desire to free the (still only partially proletarianized, and *not* very powerful) poor?

Yet later in the book, the chroniclers of Empire forget their primary pessimism and change their position almost completely. Nearly three hundred pages on, the same nongovernmental organizations (although in the case of one, the first letters had been changed) and a plethora of others had become "the newest and perhaps most important forces in the global civil society." Hardt and Negri dismiss those who suggest that the nongovernmental organizations are naught but the nice face of neoliberalism (although they miss a few of the best critics of that phenomenon). They celebrate the view that a "subset" of the nongovernmental organizations "strive to represent the least among us, those who cannot represent themselves [. . .] [They] are in fact the ones that have come to be among the most powerful and prominent in the contemporary global order." The chroniclers of Empire's contradictions go even further: they assert that the humanitarian nongovernmental organizations do more than "represent the global People in its entirety" and "represent [. . .] the vital force that underlies the People and thus they transform politics into a question of generic life, life in all its generality." What one could call the "popular" as opposed to the technocratic nongovernmental organizations "extend far and wide in the humus of biopower; they are the capillary ends of the contemporary networks of power, or [. . .] they are the broad base of the triangle of global power."

How can the gap between these two perspectives be reconciled? Perhaps *Empire* contains a sleight of hand that allows one to go beyond what at first glance seems a big difference between the two statements on what might or might not be a new sphere of global solidarity. Indeed, the humanitarian international, in spite of its moralistic nature, may actually be closer to the globally nomadic labor force imagined by Hardt and Negri than the real thing. They come close to modern-day Francises of Assisi, too. They are the ascetics' international, rhizomed via the Internet to rouse the multitude.

Yet the problem might be that the nongovernmental organizations are still stuck in the mode of "representing" the "people." Thus they are trapped on the terrain of "democracy" which, for Hardt and Negri, is more about subjection to the nation-state's sovereignty than a medium of liberation, and as much about "discipline" as "redistribution." Nongovernmental organizations, then, do not go all the way. As long as they represent the "People" they actually "organize the multitude according to a representational schema so that the People can be brought under the rule of the regime and the regime can be constrained to satisfy the needs of the People." Thus the nongovernmental organizations become functional equivalents of the nation-state. Yet is that not better than being Trojan horses for Empire's military interventions, as Hardt and Negri presented originally? And anyway, Hardt and Negri's last confrontation with the nongovernmental organizations is not as harsh as its usual dismissal of democracy: it is positive. Perhaps their ambiguity about nongovernmental organizations is resolved in their perception of democracy but that, in turn, raises another problematic so long as it is confined within the boundaries of capital and nation-state. Whether a new form of global democracy and reciprocity can "develop" Africa without its going through the travails of primitive accumulation and nation-state construction is a very big question.

Africa and the Conclusion of Empire

Thus it can be concluded that some of Empire's merits trickle down to Africa, but one has the impression that if Hardt and Negri's multitude shake off the shackles of Empire, Africa still will have to pass through a lot of pain. Thus *Empire*'s messages for radical Africanists are mixed. There is no doubt it must be read. However, those who have not passed through the maelstroms of primitive accumulation (not fully subjected to the "ransacking (of) the whole world" of which Rosa Luxemburg spoke, the construction of state-nation sovereignty, and democratization, and then through the rhizomes of the Internet to the age of postmodernity probably cannot read it. And basic literacy training may be just too boring for Hardt and Negri's imagined multitude in the West. The problem is, it is too boring for most of the current rulers of the illiterate multitude, too, caught up as they are in the conspicuous and consequenceless consumption demanded by postmodern marketing and offered by the captains of Empire. These rulers are, of course, objectively illegitimate: that is, they are the ones who at best would turn the "multitude" into the "people," at worst exploit them to death – through war as well as the most primitive means of extracting their labor-power. Thus, the task for those who would help turn the "people" into the "multitude" is to make these rulers accountable – if not disappear. This would seem to be an issue of "democracy," but that concept is stated only between *Empire*'s lines, and not boldly enough.

 READING 34

The New World Order (They Mean It)

Stanley Aronowitz

[. . .]

Although *Empire* sometimes strays from its central theme, it is a bold move away from established doctrine. Hardt and Negri's insistence that there really is a new world is promulgated with energy and conviction. Especially striking is their renunciation of the tendency of many writers on globalization to focus exclusively on the top, leaving the impression that what happens down below, to ordinary people, follows automatically from what the great powers do. In the final chapters they try to craft a new theory of historical actors, but here they stumble, sometimes badly. The main problem is that they tend to overstate their case. From observations that the traditional forces of resistance have lost their punch, the authors conclude that there are no more institutional "mediations." Not so fast.

One of the serious omissions in *Empire*'s analysis is a discussion of the World Trade Organization, the International Monetary Fund and the World Bank, three of the concrete institutions of the repressive world government of Empire. Lacking an institutional perspective – except with respect to law – Hardt and Negri are unable to anticipate how the movement they would bring into being might actually mount effective resistance. Although not obliged to provide a program for a movement, the authors do offer indicators of which social forces may politically take on the colossus. Having argued that institutions such as trade unions and political parties are no longer reliable forces of combat, they are left with the postmodern equivalent of the nineteenth-century proletariat, the "insurgent multitude." In the final chapters of the book, incisive prose gives way to hyperbole, and the sharp delineation of historical actors melts into a vague politics of hope. Insisting that "resistance" precedes power, they advocate direct confrontation, "with an adequate consciousness of the central repressive operations of Empire" as it seeks to achieve "global citizenship." At the end, the authors celebrate the "nomadic revolutionary" as the most likely protagonist of the struggle.

The demonstrations against the WTO in Seattle last December and the subsequent anti-IMF and World Bank protests in Washington suggest a somewhat different story. The 40,000-plus demonstrators who disrupted the WTO meetings and virtually shut down the city consisted of definite social groups: a considerable fraction of the labor movement, including some of its top leaders, concerned that lower wages and human rights violations would both undermine their standards and intensify exploitation; students who have been protesting sweatshop labor for years and are forcing their universities to cease buying goods produced by it; and a still numerous, if battered, detachment of environmentalists – a burgeoning alliance that appears to have continued.

These developments shed light on the existence of resistance to Empire but also on the problem of theories that wax in high abstractions. Events argue that some of the traditional forces of opposition retain at least a measure of life. While direct confrontation is, in my view, one appropriate strategy of social struggle today, it does not relieve us of the obligation to continue to take the long march through institutions, to test their mettle. After all, "adequate consciousness" does not appear spontaneously; it emerges when people discover the limits of the old. And the only way they can understand the nature of the new Empire is to experience the frustrations associated with attempts to achieve reforms within the nation-state, even as the impulse to forge an international labor/environmentalist alliance proceeds.

 READING 35

Adventures of the Multitude: Response of the Authors

Michael Hardt and Antonio Negri

[. . .]

The Multitude inside Empire

[. . .] Our book does not provide a strong enough figure for the multitude, one that is able to support the legacy of the "revolutionary vocation of the proletariat." [. . .] We should point out, however, that our theorizing of the multitude up to this point has remained abstract but is nonetheless a necessary response that corresponds to a real condition. One can consider the multitude in the first instance as a logical hypothesis that follows from our analysis of the economic, political, and cultural structures of Empire. Along with our analysis of the contemporary forms of power, then, we have to develop the analyses of classes and their composition, contradictions and crises, the will to escape the yoke of capital and to express the power of liberation. This is a first step in an analysis of the multitude as a revolutionary subject.

The global condition of the multitude follows in part from our conception of Empire itself. Our contention,

This article was written by Hardt and Negri in response to essays in their book *Empire*.

expressed most generally, is that Empire is a global form of sovereignty that includes within its constitution supranational organizations, national structures (including nation-states), and local or regional organisms. In other words, our notion of Empire does not indicate an end of the nation-state. Nation-states remain extremely important but their functions have been transformed *within* the order of Empire. At the highest level, one could say that only Empire (and no longer any nation-state) is capable of sovereignty in a full sense.

The primary objection [by] some [. . .] with regard to this notion of imperial sovereignty has to do with the centrality or not of the United States as nation-state in the imperial global order. This can refer (negatively) to our claim that it is inaccurate to conceive contemporary global order in terms of US imperialism or (positively) to our notion of the genealogy of the imperial constitutional figure that has developed primarily through US constitutional history. The former, however, our argument against the notion of US imperialism, has inspired the most criticism [. . .]. This is clearly a delicate issue for the Left in many parts of the world. One way of understanding our hypothesis is to look at it from the perspective of capital and the critique of capital: capital has globalized the system of sovereignty without identifying itself with any single nation-state.

The imperial power of capital is exercised on the basis of a "non place." In other words, there is no center of imperial power and equally no outside to imperial power. It is interesting that this proposition is difficult to understand for political thinkers on the Left and the Right, whereas from the standpoint of any stock exchange or from the offices of any multinational corporation it is clear that capital has no country and in fact resists the control of nation-states.

To say that imperial sovereignty is global and that it has no outside does not mean in any way that conditions across the world have become the same or even tend toward homogeneity. The passage to Empire does indeed lessen some differences but it creates and magnifies others. Our world is just as uneven and hierarchical as the imperialist world was, but its lines of division cannot be adequately conceived along national boundaries. Perhaps we should say that our maps of global inequalities need to become much more complex. The concept of Empire does imply, however, that despite these differences we all share the common condition of being inside Empire. Even those regions that are sometimes thought to be excluded from the circuits of global capital (sub-Saharan Africa is often cited) are clearly inside when considered, for instance, from the perspective of debt. We are all within the domain of imperial control. Being inside is the common condition of Empire.

One consequence of this conception of global Empire is that it undermines the foundation of the concept of the people. In the modern tradition, the people (whether democratic or not) is founded on the nation and a real or imagined national sovereignty. As national sovereignty declines and the bounded national space dissolves, the people becomes unthinkable. What does it mean in our contemporary situation to pose the problem of a new subject that is not a people but is rather a multitude? Conceptually the difference should be clear: the notion of the people organizes the population into a bounded unity whereas the multitude conceives the population as an unbounded multiplicity. In *Empire* we allude to a variety of multitudes: the multitude in exodus, the multitude of barbarians, the multitude of the poor, and so forth. Some political consequences of these conceptions of the multitude are already clear. It is clear, for example, that we must move beyond the discourse of "the class that is made into a people" in which differences are made generic. This discourse has been hegemonic throughout the modern history of socialism and communism, but such a strategy will no longer work and is no longer desirable (if in fact it ever was). The multitude will not be made into a people. It is not a class properly speaking, despite the fact that it does contain, within its multiplicity, all the characteristics of the working class, the stigmata of exploitation, misery, and alienation. We must thus move beyond the discourse of the working class as people because it is no longer valid analytically or politically and we must abandon all nostalgia for that revolutionary figure. We must maintain, however, its amplitude, its powers, its will to resist exploitation, its spirit of revolt against the capitalist state, and its inventive force applied to the constitution of a future.

Where and who is this multitude, this new revolutionary subject? A host of authors [. . .] pose this question. This is indeed the right question to ask. What characterizes the real existence of the multitude today and what elements could help constitute it as a political subject?

[. . .]

The Power of Decision of the Multitude

The discourse of the multitude also must be developed with respect to its power for common political action. Earlier we emphasized the multiplicity of the concept of the multitude in contrast to the people, which tends to reduce multiplicity to unity. Now we must focus on the other aspect of the multitude – that is, how it is distinguished from the fundamentally passive conceptions of collective political subjectivity, such as the mob, the crowd, or the masses. How can the multitude make a "decision" and make itself a determinate force of transformation? [. . .] We should point out that this question itself goes against some of the fundamental assumptions of modern European political theory. According to that tradition only the one can decide and only a unity can act coherently; multiplicities are necessarily passive and incoherent. This is a basic axiom of the modern theory of sovereignty. We need to think, on the contrary, how the multitude, without denying its multiplicity, can take a decision and act effectively. We need to develop a political theory without sovereignty.

Our point of departure for beginning to address this question is the real transformative actions of the multitude. Three fundamental elements constitute the actions of the multitude: resistance, insurrection, and constituent power, or really, if one does not want to be so theoretical, micropolitical practices of insubordination and sabotage, collective instances of revolt, and finally utopian and alternative projects. These are the capacities of the multitude that are real and constantly present. Our hypothesis, then, is that in order for the multitude to act as a subject these three elements must coincide in a coherent project of counterpower. We need to discover a way that every micropolitical expression of resistance pushes on all the stages of the revolutionary process; we need to create a situation in which every act of insubordination is intimately linked to a project of collective revolt and the creation of a real political alternative. How can this be created, however, and who will organize it?

The obvious temptation here is to repeat, with regard to the multitude, the operation that (in his time) Rousseau operated on bourgeois society to make it into a political body. This is just the temptation, however, that we need to recognize and avoid, because for us the path leads in the opposite direction. It is not true that there can be no multiple agent without being unified. We have to overturn that line of reasoning: the multitude is not and will never be a single social body.

On the contrary, every body is a multitude of forces, subjects, and other multitudes. These multitudes assume power (and thus are capable of exercising counterpower) to the extent that they are enriched through this common productivity, that they are transformed through the force of invention they express, that they reveal and radically remake, through practices of commonality and mixture, their own multiple bodies. Self-valorization, revolution, and constitution: these become here the components of the capacity of decision of the multitude – a multitude of bodies that decides.

How can all this be organized? Or better, how can it adopt an organizational figure? How can we give to these movements of the multitude of bodies, which we recognize are real, a power of expression that can be shared? We still do not know how to respond to these questions. In the future, perhaps, we will have accumulated enough new experiences of struggle, movement, and reflection to allow us to address and surpass these difficulties – constituting not a new body but a multiplicity of bodies that come together, commonly, in action. We would like that the critiques of our book, *Empire*, be directed toward this incapacity of ours to give a complete response to these (and other) questions. We hoped that in writing *Empire* we would provide an argument that would stimulate debate. Risking being wrong is better than remaining silent. Ours is, after all, part of a collective project of all those who really think that the revolution of this world and the transformation of human nature are both necessary and possible.

Network Society and Informationalism

This chapter begins with an overview by Manuel Castells of his theory of the network society. Castells examines the emergence of a new global reality, society, culture, and economy in light of the revolution, begun in the United States in the 1970s, in informational technology (television, computers, PDAs, etc.). This revolution led, in turn, to a fundamental restructuring of the global capitalist system beginning in the 1980s and to the emergence of what Castells calls "informational capitalism" and "informational societies" (although with important differences among and between them). Both are based on "informationalism" which combines forces of production with knowledge and information. The information paradigm has five basic characteristics:

1   There exist technologies that act on information.
2   These technologies have a pervasive effect as information becomes a part of all human activity.
3   All systems using information technologies are defined by a "networking logic" that allows them to affect a wide variety of processes and organizations.
4   The new technologies are highly flexible, allowing them to adapt and change constantly.
5   The specific technologies associated with information are merging into a highly integrated system.

A new, increasingly profitable global informational economy has emerged. It is *informational* because the productivity of firms and nations depends on their ability to generate, process, and apply knowledge-based information efficiently. It is *global* because it has the "*capacity to work as a unit in real time on a planetary scale.*"[1] This globality is made possible for the first time by the new information and communication technologies. And it is "informational, not just information-based, because the cultural-institutional attributes of the whole social system must be included in the diffusion and implementation of the new technological paradigm."[2] While it is a global system, there are regional differences, even among those that are at the heart of the new global economy (North America, EU, Asian Pacific). Other regions (e.g. sub-Saharan Africa) are largely excluded, as are parts of the privileged regions (e.g. inner cities in the US).

Along with the rise of the new global informational economy is a new organizational form, the *network enterprise*, characterized by flexible (rather than mass) production, new management systems (often adopted from the Japanese), horizontal versus vertical models of organization, and the intertwining of large corporations in strategic alliances. Most important is the series of networks that make up the organization. The network organization is the materialization of the culture of the global informational economy; it makes possible the transformation of signals into commodities through the processing of knowledge. As a result, the nature of work is being transformed (e.g. flexitime), at least in developed nations.

Accompanying the development of multimedia out of the fusion of the mass media and computers has been the emergence of a culture of *real virtuality* "in which virtuality [e.g. the hypertext on the Internet] becomes a fundamental component of our symbolic environment, and thus of our experience as communicating beings."[3]

In contrast to the past dominated by "the spaces of places" (e.g. cities like New York or London), a new spatial logic, the "space of flows," has emerged. We have become a world dominated by processes rather than physical locations (which, of course, continue to exist). Similarly, we have entered an era of "timeless time" in which, for example, information is instantly available anywhere in the globe.

Beyond the network enterprise, the most important functions and processes in the information age are increasingly dominated by *networks* or "interconnected nodes" which are open, capable of unlimited expansion, dynamic, and able to innovate without disrupting the system. For the time being at least, capitalism has adopted such networks and created the "casino capitalism" (where money rather than production predominates) that allowed capitalism to become truly global and that dominated the 1990s and into the early twenty-first century. It was this, of course, that was responsible for the economic collapse beginning in 2007 which spread so rapidly around the globe because of networks. The state is rendered increasingly powerless in such a global system (the state becomes simply a node in a broader power network), and while counter-movements to the excesses of capitalism arise, they too are characterized by networks.

While a fan of Castells's earlier work (where Castells made important contributions to understanding the city), Peter Marcuse is a severe critic of his thoughts on the network society. He discusses a number of criticisms of Castells's more recent thinking under the following headings, among others:

- Human agency is eradicated in, for example, financial flows operated by electronic networks.
- Exclusion is a concern, but there is little discussion of those doing the excluding.
- The whole argument is presented in a passive voice (e.g. "relations of production have been transformed" rather than "capitalists have transformed those relations").
- Objects, things, structures appear to act.
- Even globalization seems to act; to be an actor; to be all-powerful.
- Conflict is bypassed or suppressed.
- Identities, and the social movements related to them, are reactive rather than active.
- Space is depoliticized.

Overall, in his later work, Castells comes off at the minimum as a disinterested observer and cataloguer of the contemporary world, and at the extreme as a supporter of the status quo.

NOTES

1   Manuel Castells, *The Rise of the Network Society.* Malden, MA: Blackwell Publishers, 1996, 92.
2   Ibid., 91.

3   Manuel Castells, "Toward a Sociology of the Network Society." *Contemporary Society* 29, 5, 2000: 694.

 READING 36

Toward a Sociology of the Network Society

Manuel Castells

The Call to Sociology

The twenty-first century of the Common Era did not necessarily have to usher in a new society. But it did. People around the world feel the winds of multi-dimensional social change without truly understanding it, let alone feeling a grasp upon the process of change. Thus the challenge to sociology, as the science of study of society. More than ever society needs sociology, but not just any kind of sociology. The sociology that people need is not a normative meta-discipline instructing them, from the authoritative towers of academia, about what is to be done. It is even less a pseudo-sociology made up of empty word games and intellectual narcissism, expressed in terms deliberately incomprehensible for anyone without access to a French–Greek dictionary.

Because we need to know, and because people need to know, more than ever we need a sociology rooted in its scientific endeavor. Of course, it must have the specificity of its object of study, and thus of its theories and methods, without mimicking the natural sciences in a futile search for respectability. And it must have a clear purpose of producing objective knowledge (yes! there is such a thing, always in relative terms), brought about by empirical observation, rigorous theorizing, and unequivocal communication. Then we can argue – and we will! – about the best way to proceed with observation, theory building, and formal expression of findings, depending on subject matter and methodological traditions. But without a consensus on sociology as science – indeed, as a specific social science – we sociologists will fail in our professional and intellectual duty at a time when we are needed most. We are needed because, individually and collectively, most people in the world are lost about the meaning of the whirlwind

Source: *Contemporary Sociology*, 29, 5, September 2000: 693–9.

we are going through. So they need to know which kind of society we are in, which kind of social processes are emerging, what is structural, and what can be changed through purposive social action. And we are needed because without understanding, people, rightly, will block change, and we may lose the extraordinary potential of creativity embedded into the values and technologies of the Information Age. We are needed because as would-be scientists of society we are positioned better than anyone else to produce knowledge about the new society, and to be credible – or at least more credible than the futurologists and ideologues that litter the interpretation of current historical changes, let alone politicians always jumping on the latest trendy word.

So, we are needed, but to do what? Well, to study the processes of constitution, organization, and change of a new society, probably starting with its social structure – what I provisionally call the network society.

A New Society

Except for a few stubborn academic economists, there is widespread consensus that we have entered a new economy. I contend we are also living in a new society, of which the new economy is only one component. Since this society will unfold, throughout the world, during the twenty-first century, the survival of sociology as a meaningful activity depends on its renewal, in accordance with the new phenomena to be studied and the new analytical issues to be tackled. But what is this new society? Since the focus of this article is on sociology, not society, I have no option but to be schematic and declarative, rather than analytical, taking the liberty to refer the reader to my trilogy on the matter (Castells [1996] 2000a). Here are, in my view, the main dimensions of social change that, together and in their interaction, constitute a new social structure, underlying the "new society."

First is a new technological paradigm, based on the deployment of new information technologies and including genetic engineering as the information technology of living matter. I understand technology, following Claude Fischer (1992), as material culture – that is, as a socially embedded process, not as an exogenous factor affecting society. Yet we must take seriously the material transformation of our social fabric, as new information technologies allow the formation of new forms of social organization and social interaction along electronically based information networks. In the same way that the industrial revolution, based upon generation and distribution of energy, could not be separated from the industrial society that characterized the last two centuries, the information technology revolution, still in its early stages, is a powerful component of multidimensional social change. While new information technologies are not causal factors of this social change, they are indispensable means for the actual manifestation of many current processes of social change, such as the emergence of new forms of production and management, of new communication media, or of the globalization of economy and culture.

The second dimension of social change is, precisely, globalization, understood as the technological, organizational, and institutional capacity of the core components of a given system (e.g., the economy) to work as a unit in real or chosen time on a planetary scale. This is historically new, in contrast with past forms of advanced internationalization, which could not benefit from information and communication technologies able to handle the current size, complexity, and speed, of the global system, as it has been documented by David Held et al. (1999).

The third dimension is the enclosing of dominant cultural manifestations in an interactive, electronic hypertext, which becomes the common frame of reference for symbolic processing from all sources and all messages. The Internet (248 million users currently, in 2000; 700 million projected by the end of 2001; 2 billion by 2007) will link individuals and groups among themselves and to the shared multimedia hypertext. This hypertext constitutes the backbone of a new culture, the culture of real virtuality, in which virtuality becomes a fundamental component of our symbolic environment, and thus of our experience as communicating beings.

The fourth axis of change, largely a consequence of the global networks of the economy, communication, and knowledge and information, is the demise of the sovereign nation-state. Not that current nation-states will disappear in their institutional existence, but their existence as power apparatuses is profoundly transformed, as they are either bypassed or rearranged in networks of shared sovereignty formed by national governments, supranational institutions, conational institutions (such as the European Union, NATO, or NAFTA), regional governments, local governments, and NGOs, all interacting in a negotiated process of decision making. As a result, the issue of political representation is redefined as well, since democracy was constituted in the national enclosure. The more key decisions have a global frame of reference, and the more people care about their local experience, the more political representation through the nation-state becomes devoid of meaning other than as a defensive device, a resource of last resort against would-be tyrants or blatantly corrupt politicians. In another axis of structural change, there is a fundamental crisis of patriarchy, brought about by women's insurgency and amplified by gay and lesbian social movements, challenging heterosexuality as a foundation of family. There will be other forms of family, as egalitarian values diffuse by the day, not without struggle and setbacks. But it is difficult to imagine, at least in industrialized societies, the persistence of patriarchal families as the norm. The real issue is how, at which speed, and with which human cost, the crisis of patriarchy will extend, with its own specific forms, into other areas around the world. The crisis of patriarchy, of course, redefines sexuality, socialization, and ultimately personality formation. Because the crisis of the state and of the family, in a world dominated by markets and networks, is creating an institutional void, there are (and increasingly will be) collective affirmations of primary identity around the key themes of religion, nation, ethnicity, locality, which will tend to break up societies based on negotiated institutions, in favor of value-founded communes.

Last, but not least, progress in scientific knowledge, and the use of science to correct its own one-sided development, are redefining the relationship between culture and nature that characterized the industrial era. A deep ecological consciousness is permeating the human mind and affecting the way we live, produce, consume, and perceive ourselves. We are just at the beginning of a most extraordinary cultural transformation that is reversing the course of thought that has prevailed among the world's dominant groups since the Enlightenment.

This new society was produced during the last quarter of the twentieth century, through the interaction among three independent processes that happened to coincide in time: the revolution in information technology; the socioeconomic restructuring of both capitalism and statism (with different fates for these antagonistic modes of production); and the cultural social movements that emerged in the 1960s in the United States and Western Europe. While this multi-dimensional social change induces a variety of social and cultural expressions in each specific institutional context, I propose the notion that there is some commonality in the outcome, if not in the process, at the level where new social forms are constituted – that is, in the social structure. At the roots of the new society, in all its diversity, is a new social structure, the network society.

The Network Society: the Social Structure of the Information Age

The new society is made up of networks. Global financial markets are built on electronic networks that process financial transactions in real time. The Internet is a network of computer networks. The electronic hypertext, linking different media in global/local connection, is made up of networks of communication – production studios, newsrooms, computerized information systems, mobile transmission units, and increasingly interactive senders and receivers. The global economy is a network of financial transactions, production sites, markets, and labor pools, powered by money, information, and business organization. The network enterprise, as a new form of business organization, is made of networks of firms or subunits of firms organized around the performance of a business project. Governance relies on the articulation among

different levels of institutional decision making linked by information networks. And the most dynamic social movements are connected via the Internet across the city, the country, and the world.

Networks are, however, a very old form of social organization. But throughout history, networks had major advantages and a major problem. Their advantages are flexibility and adaptability, characteristics essential for managing tasks in a world as volatile and mutable as ours. The problem was the embedded inability of networks to manage complexity beyond a critical size. Networks were historically useful for personal interaction, for solidarity, for reciprocal support. But they were bad performers in mobilizing resources and focusing these resources on the execution of a given task. Large, centralized apparatuses usually outperformed networks in the conduct of war, in the exercise of power, in symbolic domination, and in the organization of standardized, mass production. Yet this substantial limitation of networks' competitive capacity was overcome with the development of new information/communication technologies, epitomized by the Internet. Electronic communication systems give networks the capacity to decentralize and adapt the execution of tasks, while coordinating purpose and decision making. Therefore, flexibility can be achieved without sacrificing performance. Because of their superior performing capacity, networks, through competition, are gradually eliminating centered, hierarchical forms of organization in their specific realm of activity.

A network is a set of interconnected nodes. Networks are flexible, adaptive structures that, powered by information technology, can perform any task that has been programmed in the network. They can expand indefinitely, incorporating any new node by simply reconfiguring themselves, on the condition that these new nodes do not represent an obstacle to fulfilling key instructions in their program. For instance, all regions in the world may be linked into the global economy, but only to the point where they add value to the value-making function of this economy, by their contribution in human resources, markets, raw materials, or other components of production and distribution. If a region is not valuable to such a network, it will not be linked up; or if it ceases to be valuable, it will be switched off, without the network as a whole suffering major inconvenience. Naturally, networks based on alternative values also exist, and their social morphology is similar to that of dominant networks, so that social conflicts take the shape of network-based struggles to reprogram opposite networks from the outside. How? By scripting new codes (new values, for instance) in the goals organizing the performance of the network. This is why the main social struggles of the information age lie in the redefinition of cultural codes in the human mind.

The prevalence of networks in organizing social practice redefines social structure in our societies. By *social structure* I mean the organizational arrangements of humans in relationships of production/consumption, experience, and power, as expressed in meaningful interaction framed by culture. In the Information Age, these specific organizational arrangements are based on information networks powered by microelectronics-based information technologies (and in the near future by biologically based information technologies). Under the conditions of this new, emerging social structure, sociology must address several conceptual and methodological issues in order to be equipped to analyze core processes of social organization and social practice.

Theorizing Social Structure as Interactive Information Networks

The study of social networks is well established in sociological research, spearheaded in contemporary American sociology by Wellman (e.g., 1999), Fischer (e.g., 1992), and Granovetter (e.g., 1985). There is also an international association for the study of social networks, which constitutes a fruitful milieu of research. It can provide concepts and methods that will foster understanding of social networks as specific forms of organization and relationship, including electronic communication networks. Yet, while building on this tradition, I advance the notion that twenty-first-century sociology will have to expand the network-based perspective to the analysis of the entire social structure, in accordance with current trends of social evolution. This implies more than analyzing social networks. It will require reconceptualizing many social processes and institutions as expressions of networks, moving away from conceptual frameworks organized around the notion of centers and hierarchies.

For the sake of communication, I will use two illustrations to make my case, taking them from two different and very traditional sociological fields: industrial sociology and urban sociology. I will then draw some general theoretical implications from this change of perspective.

The prevailing form of business organization emerging in advanced societies and diffusing throughout the global economy is *the network enterprise*, which I define, in sociological terms, as the specific form of enterprise whose system of means is constituted by the intersection of segments of autonomous systems of goals. It follows a complete transformation of relationships of production and management, and thus of the occupational structure on which social structure is largely based. How can we conceptualize the role of producers of information in their differential position along an interactive network? How can we conceptualize the variable geometry of new industrial organizations, based on firms' permeable boundaries, bringing together workers, capital, and knowledge in specific projects that form, dissolve, and reform under a different configuration? Yes, work, workers, exploitation, cooperation, conflict, and negotiation do not disappear, but the ensuing individualization of the relationship between management and labor and the ephemeral character of project-based, industrial organizations require a new conceptual apparatus, focusing on networked relationships rather than on vertical hierarchies. In this perspective, I propose to conceptualize the new occupational structure around the interaction among three dimensions of production relationships: value making, relation making, and decision making.

For value making, in an information-based production process, we may differentiate various structural positions: the commanders (or strategists), the researchers, the designers, the integrators, the operators, and the human terminals. Relation making defines another set of positions: the networkers, the networked, and the switched-off. And the relative positioning in decision making differentiates among the deciders, the participants, and the executors. The three dimensions are analytically independent. Thus, the empirical observation of the various arrangements among different positions in the three dimensions built around the performance of a given project may yield some clues on the emergence of new social relationships of production, at the source of new social structure.

A second example: the transformation of spatial structure, a classic theme of urban sociology. With the diffusion of electronically based communication technologies, territorial contiguity ceases to be a precondition for the simultaneity of interactive social practices. But "the death of distance" is not the end of the spatial dimension of society. First, the "space of places," based in meaningful physical proximity, continues to be a major source of experience and function for many people and in many circumstances. And second, distant, interactive communication does not eliminate space; it transforms it. A new form of space emerges – "the space of flows." It is made of electronic circuits and information systems, but it is also made of territories, physical places, whose functional or symbolic meaning depends on their connection to a network, rather than on its specific characteristics as localities.

The space of flows is made of bits and pieces of places, connected by telecommunications, fast transportation, and information systems, and marked by symbols and spaces of intermediation (such as airports, international hotels, business centers, symbolized by de-localized architecture). For instance, in recent years there has been considerable debate about the emergence of "the global city." The global city is not just a major metropolitan center that ranks high in the worldwide geography of management of wealth and information. For such cities (New York, London, Tokyo, Paris, or São Paulo) we already had the descriptive notion of "world city," proposed 20 years ago. The global city, in the strict analytical sense, is not any particular city. And empirically it extends to spaces located in many cities around the world, some extralarge, others large, and still others not so large. The global city is made of territories that in different cities ensure the management of the global economy and of global information networks. Thus, a few blocks in Manhattan are part of the global city, but most of New York, in fact most of Manhattan, is very local, not global. These globalized segments of Manhattan are linked to other spaces around the world, which are connected in networks of global management, while being loosely connected to their territorial hinterlands.

So the global city is a network of noncontiguous territories, reunited around the task of managing globalism by networks that transcend locality (Graham and Simon 2000). From this theoretical perspective we can develop models to analyze the new spatial forms constituted around interterritorial networks, and then examine their differential relationship to their surrounding, local environments. Thus, it is the connection between local and global, rather than the "end of geography" in the age of globalization, that becomes the appropriate perspective for the new urban sociology (Borja and Castells 1997). Networks of discontiguous places in interaction with a diverse range of localities are the components of the new sociospatial structure. The central analytical question then becomes how shared social meaning is produced out of disjointed spatial units reunited in a purely instrumental, global logic (Castells 2000b). By redefining spatial structure on the basis of a networking logic, we open up a new frontier for one of the oldest sociological traditions, urban sociology.

The analysis of social structures as a multidimensional, evolving system of dynamic networks may help explain social evolution in the Information Age. Indeed, networks are dynamic, self-evolving structures, which, powered by information technology and communicating with the same digital language, can grow, and include all social expressions, compatible with each network's goals. Networks increase their value exponentially as they add nodes. In formal terms, as proposed years ago by computer scientist and Internet entrepreneur Bob Metcalfe, the value of a net increases as the square of the number of nodes on the net. (The precise formula is $V = n^{(n-1)}$, where V is the value of the network and n the number of nodes). Thus, a networked social structure is an open system that can expand indefinitely, as long as the networks included in the meta-network are compatible.

The issue arises, then, of the contradictions among networks, which lead to conflicts and social change. In fact, network theory could help solve one of the greatest difficulties in the explanation of social change. The history of sociology is dominated by the juxtaposition of and lack of integration between the analysis of social structure and the analysis of social change. Structuralism and subjectivism have rarely been integrated in the same theoretical framework. A perspective based on interactive networks as the common basis for social structure and social action may yield some theoretical results by ensuring the communication, within the same logic, between these two planes of human practice. A social structure made up of networks is an interactive system, constantly on the move. Social actors constituted as networks add and subtract components, which bring with them into the acting network new values and interests defined in terms of their matrix in the changing social structure. Structures make practices, and practices enact and change structures following the same networking logic and dealing in similar terms with the programming and reprogramming of networks' goals, by setting up these goals on the basis of cultural codes.

A theory based on the concept of a social structure built on dynamic networks breaks with the two reductionist metaphors on which sociology was based historically: the mechanical view of society as a machine made up of institutions and organizations; and the organicist view of society as a body, integrated with organs with specific bodily functions. Instead, if we need a new metaphor, the sociology of the network society would be built on the self-generating processes discovered by molecular biology, as cells evolve and develop through their interaction in a network of networks, within the body and with their environment. Interactive networks are the components of social structure, as well as the agencies of social change. The sociology of the network society may be able to bridge structure and practice in the same analytical grasp.

A New Methodology?

The renewal of the study of society cannot proceed just on theoretical grounds. Sociology is an empirical science, within all the limits inherent to the constraints of observation under non-experimental conditions. Thus, new issues, new concepts, new perspectives require new tools. The emergence of interactive information networks as the backbone of social structure makes even more acute the need to take up the greatest methodological challenge for empirical research in sociology. While most of our analytical tools are based on linear relationships, most social phenomena – even more so in the network society – are characterized by nonlinear dynamics. But in the last two decades, we have witnessed the development of numerous research tools able to deal with nonlinear relationships.

On one hand, we have an expanding field of the new mathematics of complexity based on notions such as fractals, emergent properties, autopoietic networks, and the like (Capra 1996). Most of these mathematical discoveries remain confined to formal exercises with slight relationship to empirical research. But they are tools ready to be used, transformed, and perfected by able researchers with both the knowledge of the tools and the substantive knowledge to make sense of this formal language.

On the other hand, enhanced power of computers, and new, flexible computer programming languages, enable us to handle the complexity of an interactive network structure in precise terms. Computer-based system analysis of dynamic networks may constitute a fruitful approach through which observation and theory can be reconciled without excessive social reductionism. Simulation models in the social sciences got off to a bad start in the 1960s because their underlying theories were utterly simplistic, and computer programs were technically constrained by their set of rigid assumptions. But new computing capacity, in dynamic interaction of alternative assumptions processed at high speed, may change everything – as is already happening in biological research. In this sense, computational literacy (that is, knowing how to interact with computers, rather than just run statistical programs) may be a fundamental learning requirement for the current generation of young sociologists – those who will analyze the network society.

In doing so, they will be fortunate enough to have access to a huge pool of information via the Internet. Given knowledge of languages (or automated translation programs), access to global sources may liberate sociology from the embedded ethnocentrism of its observation. Each study may be comparative or cross-cultural in its approach, by contrasting observation generated *ex novo* in a particular study to the accumulated knowledge on the matter from global sources. Naturally, critique of sources as well as problems of methodological integration of diverse data will be necessary requisites for use of this wealth of information. The practice of meta-analysis, in full development in other sciences, particularly economics, may become a standard tool of sociological research. This would also require proper training and methodological guidance for sociologists to benefit from expanded possibilities of information without being overwhelmed by it.

Overall, sociology should, and will, overcome the sterile, artificial opposition between quantitative and qualitative research, and between theory and empirical study. In the perspective of computational literacy, and with the formal integration of observations in a theory that conceives social structure as a network of interactive networks, it does not really matter what

comes from statistics or from ethnography. What matters is the accuracy of the observation, and its meaning. Thus, formal models scripted in the computer programs must be theoretically informed, yet able to be given information apt to answer the questions raised in the theory.

The sociology of the network society will develop through synergy among relevant theorizing, computational literacy, and sociological imagination.

REFERENCES

Borja, Jordi and Manuel Castells. 1997. *Local and Global: The Management of Cities in the Information Age.* London: Earthscan.

Capra, Fritjof. 1996. *The Web of Life: A New Scientific Understanding of Living Systems.* New York: Doubleday.

Carnoy, Martin. 2000. *Work, Family, and Community in the Information Age.* Cambridge, MA: Harvard University Press.

Castells, Manuel. [1996] 2000a. *The Information Age: Economy, Society, and Culture.* 3 vols. 2d Ed. Oxford and Malden, MA: Blackwell.

———. 2000b. "The Culture of Cities in the Information Age." Presented at conference on The Library of Congress, "Frontiers of the Mind in the 21st Century." Forthcoming (2001) in *The Castells Reader on Cities*

and Social Theory, edited by Ida Susser. Oxford and Malden, MA: Blackwell.

Fischer, Claude. 1992. *America Calling.* Berkeley: University of California Press.

Graham, Stephen and Marvin Simon. 2000. *Splintering Networks.* London: Routledge.

Granovetter, Mark. 1985. "Economic Action and Social Structure: The Problem of Embeddedness." *American Journal of Sociology* 19: 481–510.

Held, David, Anthony McGrew, David Goldblatt, and Jonathan Perraton. 1999. *Global Transformations.* Stanford, CA: Stanford University Press.

Wellman, Barry, ed. 1999. *Networks in the Global Village: Life in Contemporary Communities.* Boulder, CO: Westview Press.

 READING 37

Depoliticizing Globalization: From Neo-Marxism to the Network Society of Manuel Castells

Peter Marcuse

[...]

It is precisely the shift of focus away from the nature of, and the relationships among, social groups that marks Castells's trajectory. It is a move that suppresses the political, in the broad sense of the dynamic between the exercise of power and the resistance to it, and moves toward a determinism that undermines the relevance of political action. Power and conflicts over power disappear from view; classes, when they appear, have

a very subordinate role. Capitalism is conflated with globalization, but in an ambiguous and ahistorical fashion; technology, the media, demographic changes, the state appear as homogeneous, autonomous entities, actors themselves, behind whom actual actors are not to be seen. It is a classic case of reification, making the relations among human beings appear as a relationship among things, the relationships of social and economic position appear as relationships to or against technology, to or against the ascendance of "informa-tion." In place of the tensions, the contradictions, the conflicts among human actors and groups as the motor of change, there is a march of technology, of organizational forms, of their own accord, inexorably, globally. Human actors only react to these developments (some benefit from them, but not much attention is paid to them, and they are not seen as more than passive participants in the march). The critique of globalization implicit and often explicit in the books concludes with an appeal to "us" to understand, com-municate, become aware, together; any drawing of policy conclusions or indications for action is deliber-ately rejected. The discussion becomes depoliticized, both in its analysis and in its stance toward prescription: in Castells's words, "the power of flows takes precedence over the flows of power."

To be clear: by "depoliticized" I do not mean that Castells, or any other author, has an obligation to draw political conclusions and/or present political prescriptions as part of his or her work, although it may be desirable that more extend their work in these directions than now do. I mean rather that the political content present in the world Castells is analyzing is sup-pressed, played down, becomes incidental, in contrast to its role in reality. I take the political to be centered on relations of power among social actors; these play at best a secondary role in Castells's analysis, where they appear at all. The criticism is not that Castells fails to introduce a political analysis into the material he examines, but that he does not adequately deal with the content that is in fact in his material; not that he should politicize material that is nonpolitical, but that he has depoliticized material that is itself heavily political.

The problem is symbolized and encapsulated by the very title of Castells's *magnum opus*: *The Information Age*. What is central in the analysis is a technical devel-opment (and a somewhat mystified one at that [. . .]), not a social one. It makes the tools of production, rather than the relations of production, the characteristic of the age: thus the sequence might be: Stone Age, Bronze Age, Iron Age, Steam Age, Information Age, rather than Imperial Age, Feudal Age, Capitalist Age, Imperialist Age, Fordist Age, followed perhaps by various attempts at a further definition: Neo-Imperialist, Post-Fordist. The point is not the accuracy of any of these classifica-tion schemes, but what it is that is at the center of

them, what is taken as the indicative classificatory criterion. Even in traditional sociology and traditional economics, and certainly in Marx, it is the relations among and characteristics of groups within each society that are its defining characteristics. Not here.

The depoliticization of what would be, underneath it all, a sharp analysis of events can be traced in a number of areas. The language used systematically undermines the substance of the analysis and robs it of a political force it might otherwise have. A few examples highlight the issues here raised.

The Eradication of Human Agency

A key aspect of depoliticization is to make everything that happens anonymous, actor-less. It is not merely the old agency versus structure argument within Marxism, for in those discussions both sides always assume that structure refers to the pattern of relations among actors, among classes, and the issues invoke scale, proportion, relative weight, scope of human agency within struc-ture. With Castells, agency vanishes, actors disappear from sight. Both the language and the content of what he writes lead in this direction.

Castells does at times deal with the question of agency: "who are the capitalists?" he asks. He points out that there is no simple answer, that they are a "colorful array" of characters, and seems to open the door to a deeper discussion of class composition in advanced industrial societies and their global linkages. But then he proceeds: "above a diversity of human-flesh capital-ists and capitalist groups there is a faceless collective capitalist, *made up of financial flows operated by electronic networks.*" Important points do need to be made here as to the autonomy of individual capitalists, the differ-ence between a conspiracy and a class, how power is exercised, and so on. But the discussion does not go in this direction. Instead, the conclusion is the flat state-ment that "there is not [. . .] such a thing as a global capitalist class." Rather, "capitalist classes are [. . .] appendixes to a mighty whirlwind." "Who are the owners, who the producers, who the managers, and who the servants, becomes increasingly blurred." Maybe to Castells, but not to the majority of the world's peoples, I would guess. This is depoliticization with a vengeance: not power relations, but a "mighty whirlwind governs our actions [. . .] Power [. . .] is no longer concentrated in institutions (the state), organizations (capitalist firms), or symbolic controllers (corporate media, churches). It is diffused in global networks of wealth, power, information, and images [. . .] *The new power lies in the codes of information and in the images of representation around which societies organize their institutions* [. . .] *The sites of this power are people's minds.*" If power should be challenged, then, the entity responsible is the "society" which does the organizing; it does no good to criticize the state, or firms, or the media. The "realpolitik" of domination, to which Castells also refers elsewhere, is not the issue.

The Excluded Without the Excluders

In general there is much detail on those who are excluded, but not on those who exclude them. The process of exclusion is faceless, a world-historical process at the "end of millennium," not one for which any single group or class can be held accountable. In the substantial discussion of the exclusion of "the majority of the African population in the newest international division of labor," Castells concludes "that *structural irrelevance* (from the systems point of view) is a more threatening condition than dependency"; "a considerable number of humans [. . .] are irrelevant [. . .] from the perspective of the system's logic." Irrelevance is from "the system's" point of view, not from the point of view of those who can make no profit from the lives of the excluded. Some are excluded, but no one does the excluding. Actors disappear entirely in the blanket laid down by the language of sweeping phrases: "social forms and processes induced by the current process of historical change." (And one might raise the question of whether the excluded are really excluded from the system, or whether they are in fact quite useful for it but simply excluded from its benefits . . .)

In the conclusion to the third volume, Castells deals most explicitly with the question of who is responsible to the new informational/global economy. "The rule is still production for the sake of profit, and for the private appropriation of profit, on the basis of property rights – which is of the essence of capitalism. But [. . .] [w]ho are the capitalists?" The discussion then begins with a logical description: a "first level" which "concerns *the holders of property rights.*" The "second level [. . .] refers to *the managerial class.*" But here the reference to class ends; we get no closer than this to the flesh and blood of real actors. For "the third level [. . .] [has to do with] the nature of *global financial markets. Global financial markets, and their networks of management, are the actual collective capitalist* [. . .] *global financial networks are the nerve center of informational capitalism.*" So, in the end, the capitalists are not a "who" but a market; not those networking, but the network itself.

The Passive Voice

Castells uses the passive voice constantly, where an active grammar would raise the question of exactly who is responsible, or, if simple agency is not adequate to explain structural patterns, what forces, what relationships of power, what institutions or practices are involved and should be held accountable. The problem occurs from the opening to the closing of the three volumes. In the first chapter, "global networks of instrumental exchanges selectively switch on and off individuals, groups, regions, and even countries, according to their relevance in fulfilling the goals processed in the network, in a relentless flow of strategic decisions [. . .] **Our societies are increasingly structured around a bipolar opposition between the Net and the Self.**" "The" Net (capitalized?) and "the" Self (capitalized?). Just what does that mean? Networks among some groups are indeed in opposition to the self-development of other groups; there is "opposition" in the patterns Castells describes, but not conflict. In fact, it is not "global networks of instrumental exchanges" but networks of specific corporations, power blocs, states that "switch on and off" very specific individuals, groups, regions, and countries – and not any random individuals, countries, etc., all characterized by their concern with the "Self," but poor and working people, Third World countries, women.

In the last chapter, the passive voice continues to color the discussion of the transformations the three volumes describe. "*Relations of* production have been transformed." "[L]abor is redefined in its role as produced, and sharply differentiated according to workers' characteristics," and "generic labor is assigned a given task." "[C]apital is as transformed as labor is in this new economy," just as Castells elsewhere gives ample evidence of who benefits and who is hurt. But the presentation shifts the focus away from any person's or group's responsibility and on to the tools, the instruments, the "networks of instrumental exchanges" used by some to achieve their results at the cost of others.

The Imputation of Agency to Things

This is, in a sense, the mirror image of the disappearance of real actors from view: processes and relationships become reified, become actors themselves, autonomously, independently of human agency. Real actors disappear, and things become actors.

Technology becomes an independent actor, an autonomous force. We read sentences like: "technology has transformed the political role of the media." Not that political actors have taken advantage of technological developments to use media in a new role; the technology itself achieves the transformation. The new "techno-economic paradigm [. . .] is based primarily on [. . .] cheap inputs of information." The role of the media is indeed analyzed perceptively, even with an undertone of moral condemnation, but, since technology is to blame, there is no suggestion that calls for different ownership or control of the media would make a difference. The kind of media analysis undertaken by writers such as Herbert Gans, Noam Chomsky, or Douglas Kellner is not mentioned.

The opposite view is also to be found in Castells, with the contradictions unresolved. For instance, elsewhere Castells explicitly abjures technological determinism; he could hardly have said it more bluntly: "The Information Technology Revolution DID NOT create the network society." Yet, as is frequently the case, the language of the discussion constantly contradicts the broad theoretical statement. Technology is an independent process, independent both of economics and culture. At the same time, "Information technology bec[omes] the indispensable tool for the effective implementation of processes of socio-economic restructuring." The ambivalence as to the explanatory role of technology *vis-à-vis* socioeconomic restructuring runs throughout the discussion. For any analysis of the politics of the developments he describes, clarity

on that issue would seem vital, since if it is "technology that transforms," little can be done about it, absent Luddite initiatives, but if socioeconomic forces are involved, they can indeed be addressed, and with them the uses to which technology is put.

Globalization as "Actor," All-Powerful

It is treated as an entity, an active force; indeed, if the whirlwind has a name, it is globalization. Yet the precise meaning of globalization remains fuzzy. In volume 1 it appears primarily as a globalization of the economy, coupled necessarily with "informationalism," as a "historical discontinuity" from the past. In volume 2 its sweep is broader, and it assumes cultural and social forms as well. The issue of its newness "does not concern my inquiry." Yet we read that "globalization [. . .] dissolves the autonomy of institutions, organizations, and communication systems." If that is the case, just what globalization is, whether it is a new phenomenon or not, becomes critical, despite Castells's claim to the contrary. The picture suggests that not specific actors, not multinational corporations overriding national boundaries, not capital moving without effective restraint to and from wherever it wishes are at work, but the anonymous process of globalization. If globalization is not new, then we might well ask whether it is not capitalism as such, perhaps simply in a further advanced form, which is responsible for the developments Castells accurately describes. And if it is indeed capitalism, then capitalists might also bear some responsibility, and the political content of the conceptualization becomes clear. With the shift of focus to globalization, that political content disappears.

Nowhere is there an intimation that globalization is a process that can be altered or stopped, that really existing globalization is not the only form globalization might take. Globalization is presented as whirlwind, sweeping everything in its path.

Conflict Is Bypassed or Suppressed

The second volume, titled *The Power of Identity*, focuses on social movements, which are defined "as being:

purposive collective actions whose outcome, in victory as in defeat, transforms the values and institutions of society." The implication here is that conflict, victory or defeat, is the essence of what social movements are about, with those who support and represent the "values and institutions of society" as their clear antagonists. Conflict might thus be expected to be a critical element in the discussion of social movements, now discussed under the rubric of "identity." But in what follows "social actors [...] excluded from [...] the individualization of identity [...] in the global networks of power and wealth" are not engaged in conflict with those who have excluded them (nameless; see below), but rather these social actors are engaged in a search "for the construction of meaning." Their organizations, social movements, are not movements defined by conflict with those who have deprived them of meaning (and, presumably, of key material resources for living a decent life – the term "exploitation" does not feature in any of the three volumes). They are "cultural communes," "organized around a specific set of values [...] marked by specific codes of self-identification." As elsewhere, Castells has it both ways.

In the end, there need not be conflict; ultimately, the solution is for "all urban [sic] agents [to develop] a city project which impregnates civic culture and manages to achieve broad consensus." The earlier centrality of conflict has given way to the anticipation of consensus.

Identity (Social Movements) Becomes a Reactive Phenomenon

What identities react to, and indeed the definition of identity, is unclear. A formal definition is provided: "I mean by identity the process by which an actor [...] constructs meaning primarily on the basis of a given cultural attribute [...] to the exclusion of a broader reference to other social structures." Why an identity thus constructed cannot also have reference to other social structures is uncertain, and indeed in many examples in volume 2 they clearly do, for example, the feminist movement or the civil rights movement. And within a few pages fundamentalism, clearly taken as an identity movement, is put forward as a reaction to the exclusion of large segments of societies, presumably

a "reference to other social structures." And why is a working-class identity not an identity? And to what are "identities" reacting? In one place it is to "the logic of apparatuses and markets," in other words, to social circumstances; in other places it is to globalization; in others, to "excluders"; in another, to "the crisis of patriarchalism"; in still another, to "the unpredictability of the unknown." Granted that identities are indeed very diverse, in what sense can one then use the category as a meaningful single concept?

And yet, in the discussion, the functional differences among identities in the end disappear; all identities are treated as reactions, and reactions against generalized processes. Enemies do not appear; processes operate without operators or subjects. Although there is detailed and perceptive discussion of resistance movements in volume 2, the resistance is not against any one or any group in particular:

> Religious fundamentalism, cultural nationalism, territorial communes are [...] defensive reactions. Reactions against three fundamental threats [...] Reaction against globalization [...] Reaction against networking and flexibility [...] And reaction against the crisis of the patriarchal family [...] When the world becomes too large to be controlled [...] When networks dissolve time and space [...] when the patriarchal sustainment of personality breaks down [...] [people react.]

The reaction is not by people to other people doing things to them, but to faceless processes. True enough, people often do not see who is doing what to whom, and the descriptions Castells provides are often graphic and trenchant. But then is it not precisely the obligation of analysis to clarify who and what is involved, and are not formulations like those above in fact concealing what is happening, disarming more targeted resistance? In presenting identity movements as against faceless and actor-less processes, the movements themselves become similarly "soft"; they are not defined by their own interests, their own capacities, their own understandings, but only by that "process" which they are up against.

In fact, Castells also includes a much more analytic and political discussion of identities, differentiating between legitimizing identities, those which are introduced by dominant institutions and reinforce domination, resistance identities, those generated by the dominated to creates trenches of resistance, and project identities, those seeking to redefine positions in society and the transformation of the overall social structure. It is a useful categorization, harking back to the discussions of the 1960s as to the nature of social movements and their radical or system-maintaining roles. But it is a tool not then consistently carried forward in a discussion in which religious fundamentalism, the Zapatistas, the Patriot Movement in the United States, Japan's Aum Shinrikyo, the environmental movement, the women's movement, and the Lesbian and Gay Liberation movements are more or less given equal treatment under the uniform heading of "identity" movements.

Are there in fact any "project identities"? John Friedmann points out that the category of "project identity" into which Castells puts movements that "seek the transformation of overall social structure" is empty. Castells is a little ambiguous on the issue; at one point, he suggests that project identities may be involved in efforts at liberating women "through the realization of women's identity," or in movements, "under the guidance of God's law, be it Allah or Jesus." At another point he says that from "cultural communes" "new subjects [. . .] may emerge, thus constructing new meaning around *project identity*." And in the concluding chapter of the volume entitled *The Power of Identity*, he speaks merely of "project identities potentially emerging from these spaces [of resistance]." Identity, social movements built around identity, are not then today agents of political action; identity is not very powerful, according to Castells, despite the book's title.

The Independence of Key Phenomena

This is a part of the picture. At various times and places, Castells suggests the connections among the various phenomena he includes together under the various umbrella terms that frequently appear: the "information age," the "network society," the "global era." While these phenomena are discussed separately in the three volumes, Castells brings them together in a summary article: "The Information Technology Revolution [. . .]

The restructuring of capitalism [. . .] The cultural social movements." And he is explicit about the connection: "The network society [. . .] resulted from the historical convergence of [these] three **independent** processes, from whose interaction emerged the network society." The language is slippery: are they independent if they interact? To what extent does their interaction determine their nature and direction? Is the "historical convergence" just an accident? The detailed discussion of each suggests that they are indeed independent forces, each with an independent shape. Technological development, appearing independent, moves by its own laws, outside of political control, and social movements are not presented as efforts to control, redirect, or prevent the restructuring of capitalism. That a coherent set of actors is involved in each of the three phenomena drops out of sight. The evidence that "capitalist restructuring" molded the direction, extent, and nature of technological change, coming into conflict with, exacerbating, and highlighting cultural and social movements, is not taken up.

"The Depoliticization of Space"

This is a somewhat unexpected aspect of Castells's presentation. Castells has made a major contribution to the contemporary discussion of space in his evocation of the duality of the space of places and the space of flows; the terms have become an accepted part of the social science vocabulary. The space of places refers to that space to which some people are bound: perhaps unskilled workers, those without the means or the legal status for mobility, those to whom a particular location, city, territory, is a fundamental part of their identity, those who are tied to a particular space/place. The space of flows, by contrast, is used by those with unrestricted mobility and is the space in which capital moves, in which high-level financial transactions occur, in which decisions are made and control exercised, the space which the dominant networks of the advanced network society occupy. There is real meat here: the worlds of those who are location-bound and those with unrestricted mobility, both in their personal lives and in their transactions, are two different worlds; although, as Michael Peter Smith points out [. . .], to set the two up as a binary opposition hardly reflects their complex and overlapping nature: the users of the "space of flows" are also place-bound in many aspects of their activities, and many denizens of the "space of places" frequently move large distances and across borders, in increasingly frequent trans-national patterns.

Is it useful to convert the differences between these two worlds into a difference originating in/characterized by their use of space, rather than looking at the differences in the use of space as the outcome of differences in wealth, power, resources? Is the space of flows in any meaningful way really a space, or is it not rather a freedom from spatial constraints? Is the space of places really not also made up of flows as well as localities? What needs analysis, for political evaluation, is the extent to which those who use the "space of flows," the dominant groups in the global society, are or are not free of locational bounds. The difference between the occupants of the space of places and the users of the space of flows is a class difference, reflected in their relationship to space, reinforced but not created by it. Examining differences in the use of space without examining the differences in class, power, and wealth which produce those spatial differences is stripping social science analysis of its political relevance: depoliticization.

Worse, space itself becomes an actor, affirmatively displacing real persons and interests: "Function and power [. . .] are organized in the space of flows [. . .] the structural domination of its logic [. . .] alters the meaning and dynamic of places [. . .] a structural schizophrenia between two spatial logics [. . .] threatens to break down communication channels [. . .] a horizon of networked, ahistorical space of flows, aiming at imposing its logic over scattered, segmented places [. . .] Unless cultural *and physical bridges* are deliberately built between these two forms of space, we may be heading toward life in parallel universes whose times cannot meet." The logic of space becomes the cause, not the consequence, of social change. Just how do you build a "*physical*" bridge to a space of flows? An interesting conceptualization, with which Castells does not play; perhaps just an errant use of words. In any case, the insight has moved from a potentially striking and politically meaningful one into a play of metaphors, in which it is the "logic of space" that needs to be dealt with, not the relations among people using space. It

hardly helps to get a grip on industrial relations in a global age to be told that "the very notion of industrial location [has been transformed] from factory sites to manufacturing flows [. . .] [by] the logic of information technology manufacturing [and] the new spatial logic."

Playing with Time

As with the treatment of space, this is insightful and provocative in Castells's handling, but depoliticized; he fails to pursue his real insight to its logical conclusion. He points out, and illustrates, the differences in the "time-boundedness" of different actors and activities. To some extent the differentiation parallels longstanding Marxist and classical economists' distinction between those paid hourly wages and those on longer-term salary bases or making profit without regard to time spent, a distinction that then feeds into definitions of class and class relations. Castells deepens the differentiation: it is not just between those paid hourly and those paid in other ways, but between those for whom time itself is an important factor in determining the way their lives are lived and those independent of it, living in "timeless time." Time is thus a constraint on some much more than on others; it "means" different things to different people. Fine. But to different classes? No, the analysis does not go in that direction; it rather plays with the catchy phrase "timeless time" as characteristic of a type of person and activity, jetsetters, instant communicators and instant manipulators of capital, and instant and constant (time-independent) exercise of control. The truth is that some control the time of others but are free to determine their own time, while the time of others is controlled despite their will. Just as with "space of flows," the metaphor reflects a real truth, but the emphasis on the metaphor conceals the very real class differentiation it in fact only reflects. "Selected functions and individuals" do not "transcend time"; they simply have the power to control their own use of time, and that of others.

The Autonomy of the State

This is a complex subject. The intellectual and political tradition from which Castells comes had a central concern with the role of the state. Marx's classic formulation of the state as the "executive committee of the ruling class" was widely seen not as wrong but as incomplete. To explain contemporary developments, Castells's close friend Nicos Poulantzas produced a complex analysis of the subject that was at the heart of the intellectual ferment in which Castells first worked. But Castells opens his chapter on the state with a repudiation of Poulantzas's description as no longer applicable. Little of the earlier rich discussion survives, except as an echo. Instead, the state becomes an actor: "the state's effort to restore legitimacy," "the state's attempt to reassert its power." And there are sweeping statements such as "the nation-state [. . .] seems to be losing its power, although, and this is essential, *not its influence*." Or elsewhere: "the state does not disappear. It transforms itself. This transformation is induced not only by globalization, but by structural changes in the work process, and in the relationship between knowledge and power." There are outside pressures, but the state itself acts to transform itself.

What does that mean? Castells never returns to the formulation, but at the end of the chapter says that "in the 1990s, nation-states have been transformed from sovereign subjects into strategic actors." It is a muddled discussion. One possible interpretation might be that the nation-state remains important in the development of technology and in the support of "its" multinationals. Indeed, Castells emphasizes both points in various contexts in all three volumes. But why is that not a continuing source of power? The "nation-state" is used as a synonym for "state" in the global era, but the distinction between nation and state is never explored in the analysis; the capacity of the nation-state "is decisively undermined by globalization" but not by any specific actions of any specific actors, even though as a result multinationals can operate freely disregarding national borders. The nation-state has a "commitment to provide social benefits," although why that commitment should exist is not clear. There is a "destabilization of national states" through the globalization of crime and a "crisis of legitimacy" that is equally applicable to the Mexican and the United

States state, although both countries seem remarkably stable in almost every regard. Such an interpretation simply avoids the question of what the state is. Throwaway lines like "states are the expression of societies, not of economies" do nothing to help. Furthermore, Castells describes the state's activities as if it was or had been an independent, autonomous actor – precisely the conception that has been so systematically questioned in critical sociology over the last century and more. Yet there is also, in passing, the comment that "each nation-state continues to act on behalf of its own interests, or [*sic*] of the interests of constituencies it values most." That latter comment might be the beginning of a discussion of where power in and over states actually lies, a discussion opening up the political questions that are so little regarded in the books. But it is not a comment that is pursued. And its very formulation is already misleading: the question is posed as who "the state" autonomously values, the state as actor, the constituency as passive beneficiary, rather than as what active "constituencies" control or put pressure on the state. Remarkably, little of the current discussion about the state "losing control" ever specifies who is winning control.

And so we end with what appears a most ambiguous comment in the post-Seattle world: "the International Monetary Fund experts do not act under the guidance of governments [. . .] but as self-righteous surgeons skillfully removing the remnants of political controls over market forces." Of course, the International Monetary Fund and its related international bodies are deeply concerned with regulating, using the political power of governments and international transactions, and are critically dependent on governments for all of their activities – and particularly the one most powerful government in a one-superpower world. And in so doing they hardly act as independent experts or surgeons but are directly serving identifiable and very specific interests. Their actions are the subject of heated political discussion in countries around the world. Yet any discussion of those politics, however, is avoided.

[. . .]

World Risk Society and Cosmopolitanism

In this chapter we deal with two closely related ideas developed primarily by the contemporary German sociologist, Ulrich Beck. Beginning in the 1980s, Beck popularized the idea of "risk society" to describe the late modern era, in contrast to the "industrial society" that dominated the modern age. At the present moment, however, the industrial age lives on; both types of society coexist. The central issue in classical modernity was wealth and how it could be distributed more evenly. In advanced modernity the central issue is risk and how it can be prevented, minimized, or channeled. Safety has tended to replace equality as the central social issue. While people achieved solidarity in the past in the search for the positive goal of equality, in late modernity solidarity is achieved in the search for the largely negative and defensive goal of being spared from dangers.

Today's risks are largely traceable to industry and its side effects which are producing a wide range of hazardous, even deadly, consequences for society. Even in his early work Beck linked that activity to globalization, but this became more focal in his later work on "world risk society." Here Beck sees "all life on earth" endangered by such things as "nuclear energy . . . gene technology, human genetics, nanotechnology, etc. . . . unleashing unpredictable, uncontrollable and ultimately incommunicable consequences."[1] These are the result of "*un*natural, human-made, manufactured uncertainties and hazards."[2] It is not so much that risks have increased, but rather that they are less bounded by space (they are deterritorialized, that is nation-state borders do not restrict the flow of risks such as air pollution), time (e.g. nuclear waste will affect generations to come), and the social (e.g. who is affected by, and responsible for, a specific risk). In the essay presented here Beck focuses on the three examples of global risk: ecological, financial (made more relevant by the global recession beginning in late 2007), and terroristic. (In the essay with Sznaider he adds a fourth risk – moral.) There are, of course, differences among these risks: ecological risks are external, financial risks are internal, and terroristic risks are intentional. While these risks are global, they are not distributed equally throughout the world. Nevertheless, they require global solutions and global cooperation in order to achieve a solution. However, this leads to global conflicts, as well as to global solutions to these conflicts, including more global institutions and regulations.

Focusing on terrorism, specifically the events of September 11, 2001, Beck draws six lessons:

1 Humanity is able to form new bonds ("transnational cooperative networks") in reaction to terrorism and fear of further acts.

2   Internal security is no longer the exclusive province of the nation-state; indeed the borders that separated nation-states are overthrown. The nation-state itself is a "zombie concept"; it looks alive, but it's dead.

3   Neoliberalism's basic tenets (see chapter 4) – "that economics will supersede politics, that the role of the state will diminish"[3] – lost their force in the face of global risks like terrorism.

4   The only solution to global terror – as well as to other global problems – is transnational cooperation (see chapter 7). This is further evidence of the demise of the nation-state (see chapter 6).

5   We need to distinguish between global unilateralism as practiced by the US and multilateralism and the multilateral state, two examples of which are the "surveillance state" (where new powers of cooperation are used to build a new fortress state with a decline of freedom and democracy) and the "cosmopolitan state."

6   The cosmopolitan state, Beck's "new big idea," focuses on the "necessity of solidarity with foreigners both inside and outside the national borders" and the fact that global problems "cannot be solved by individual nations on their own."[4] The cosmopolitan state will form the "groundwork for international cooperation on the basis of human rights and global justice."[5]

Building on his work on the world risk society, Beck has made the cosmopolitan state, and more generally cosmopolitanism, the focus of much of his recent work.

Jarvis critiques the idea of a global risk society on several grounds. He devotes much attention to Beck's contention that global risks have increased because of the decline of the nation-state, especially in its ability to protect its citizens from these risks. Jarvis examines a variety of aspects of this contention and finds little support for the idea that the state and more generally the Westphalian system have declined. Among other things, he finds no evidence that globalization has caused declines in discretionary government spending on welfare, that states have been coerced into adopting neoliberal policies, or that there has been increased capital flight or capital scarcity which have made states more vulnerable to various risks. Overall, Jarvis concludes that rather than declining, the state is undergoing something of a renaissance. Furthermore, systems to control, distribute, and indemnify against risk continue to be in place and to function.

It seems clear that risk is not a new phenomenon. Further, rather than necessarily expanding in the current global age, there are many examples of declining risk (e.g. of nuclear war). In addition, even in the face of the kind of risks that concern Beck, it seems clear that social relationships will emerge in order to deal with them. There might be greater difficulties involved in dealing with these risks, but that does not negate the fact that there will be national and global efforts to manage them. It is already clear that such efforts have been undertaken and have succeeded, at least thus far, in dealing with many of the risks associated with SARS, AIDS, and global terrorism. Furthermore, we are witnessing such efforts to deal with the recession that began in late 2007. While it is not yet clear that these efforts will be successful, it is clear that national and global efforts have emerged to deal with the profound risks associated with the recession.

However, Jarvis points out that one of Beck's greatest weaknesses is his lack of attention to the risks associated with global finances and the global economy. That this has become a global crisis is a reflection of a serious weakness in Beck's theory. However, it is the case that in a broader sense this crisis tends to support Beck's thesis that we live in a new, different, and highly dangerous global risk society. Beck's thinking also serves to capture the increasing global awareness of, and concern about, global risks of all sorts.

Ulrich Beck and Natan Sznaider distinguish between cosmopolitanism as a moral position (it is to be preferred to nationalism) and *cosmopolitanization* involving "unintended . . . *side-effects* of actions which are not intended as 'cosmopolitan' in the normative sense."[6] As a result of the latter, the world is growing increasingly cosmopolitan whether we like it or not, whether or not we want it to be more cosmopolitan. Of course, cosmopolitanism also involves more conscious and normative undertakings such as "movements against global inequality or human rights violations."[7]

Beck and Sznaider distinguish between globalization (which takes place "out there") and cosmopolitanization (which happens "from within"). Cosmopolitanization involves "*really-existing relations of interdependence*."[8]

That is, it is people through their actions who create global relationships as well as the resulting risks. In focusing on actions and relations, a cosmopolitan sociology dissolves the distinctions – local, national, international, global – that lie at the core of most thinking about globalization. Above all, it means abandoning the traditional focus on the sovereign nation-state. Instead of viewing themselves as part of a nation-state, people are increasingly seeing themselves as part of the local *and* of the larger world. This is related to world risk in the sense that people come under pressure to cooperate globally because of these risks and threats.

Such a conclusion stems from "analytic-empirical cosmopolitanism" which should be distinguished from "normative-political cosmopolitanism"[9] (although the latter presupposes the former). Doing analytic and empirical work requires a shift from "methodological nationalism" (which is focused on the nation-state) to "methodological cosmopolitanism" (which abandons such a focus, as well as distinctions between local, national, international and global). In terms of the latter, one example would be focusing on "*transnational regimes of politics*" rather than the "state-centred distinction between national and international politics."[10] In this context, Beck and Sznaider critique other theories of globalization dealt with in this volume – world systems theory (chapter 8) and world polity (or culture) (chapter 16) – for maintaining and presupposing a national–international dualism. Cosmopolitanism encourages a multitude of perspectives: for example it is possible, maybe even necessary, to analyze a phenomenon like transnationality "locally and nationally and transnationally and trans-locally and globally".[11]

Craig Calhoun looks at Beck's thinking on cosmopolitanism within a broad view of that phenomenon. Most generally, cosmopolitanism involves a direct connection between the individual and the world. However, it can involve many things including a style of life, an ethos, a political project, any project beyond the local, a holistic view of the world, and so on. Calhoun critiques Beck's distinction between "cosmopolitanization" (the growing interdependence of the world) and "cosmopolitanism" (a moral responsibility on everyone) by questioning whether they are necessarily linked to one another. That is, does increasing interconnection necessarily mean that people will feel a greater moral responsibility for one another?

Beck sees cosmopolitanism as involving the freedom to choose whether or not to belong; that is, belonging becomes an option for the cosmopolitan. However, the freedom to make this choice is often restricted to societal elites; large numbers of people in the world, perhaps the great majority, are not able to make such a choice. Further, for elites, cosmopolitanism often serves as an escape from belonging to, for example, postcolonial societies. Created by this is the illusion that cosmopolitanism is able to transcend nation, culture, and place.

Beck is critiqued not only, at least implicitly, for being an elitist, but also for failing to see that it is impossible *not* to belong to something, if not many things. Thus, there is not as much freedom as Beck suggests. Furthermore, Beck's perspective tends to downplay the importance of national and local solidarity. Calhoun argues that in our rush to embrace cosmopolitanism we cannot and should not ignore, or wish away, such traditional forms of solidarity.

NOTES ··

1 Ulrich Beck, "The Terrorist Threat: World Risk Society Revisited." *Theory, Culture and Society* 19, 2002: 40.
2 Ibid., 41.
3 Ibid., 47.
4 Ibid., 50.
5 Ibid., 50.
6 Ulrich Beck and Natan Sznaider, "Unpacking Cosmopolitanism for the Social Sciences: a Research Agenda." *British Journal of Sociology* 57, 1, 2006: 7.
7 Ibid., 8.
8 Ibid., 9, italics in original.
9 Ibid., 13.
10 Ibid., 15, italics in original.
11 Ibid., 18.

The Terrorist Threat: World Risk Society Revisited

Ulrich Beck

Does 11th September stand for something new in history? There is one central aspect for which this is true: 11th September stands for the complete collapse of language. Ever since that moment, we've been living and thinking and acting using concepts that are incapable of grasping what happened then. The terrorist attack was not a war, not a crime, and not even terrorism in the familiar sense. It was not a little bit of each of them and it was not all of them at the same time. No one has yet offered a satisfying answer to the simple question of what really happened. The implosion of the Twin Towers has been followed by an explosion of silence. If we don't have the right concepts it might seem that silence is appropriate. But it isn't. Because silence won't stop the self-fulfilling prophecies of false ideas and concepts, for example, war. This is my thesis: the collapse of language that occurred on September 11th expresses our fundamental situation in the 21st century, of living in what I call 'world risk society'.

There are three questions I discuss in this article:

- First, what does 'world risk society' mean?
- Second, what about the *politics* of world risk society, especially linked to the terrorist threat?
- Third, what are the methodological consequences of world risk society for the social sciences?

What Does World Risk Society Mean?

What do events as different as Chernobyl, global warming, mad cow disease, the debate about the human genome, the Asian financial crisis and the September 11th terrorist attacks have in common? They signify different dimensions and dynamics of world risk society. Few things explain what I mean by global risk society more convincingly than something that took place in the USA just a few years ago. The US Congress appointed a commission with the assignment of developing a system of symbols that could properly express the dangers posed by American nuclear waste-disposal sites. The problem to be solved was: how can we communicate with the future about the dangers we have created? What concepts can we form, and what symbols can we invent to convey a message to people living 10,000 years from now?

The commission was composed of nuclear physicists, anthropologists, linguists, brain researchers, psychologists, molecular biologists, sociologists, artists and others. The immediate question, the unavoidable question was: will there still be a United States of America in 10,000 years time? As far as the government commission was concerned, the answer to that question was obvious: USA forever! But the key problem of how to conduct a conversation with the future turned out to be well nigh insoluble. The commission looked for precedents in the most ancient symbols of humankind. They studied Stonehenge and the pyramids; they studied the history of the diffusion of Homer's epics and the Bible. They had specialists explain to them the life-cycle of documents. But at most these only went back 2,000 or 3,000 years, never 10,000.

Anthropologists recommended using the symbol of the skull and cross-bones. But then a historian remembered that, for alchemists, the skull and bones stood for resurrection. So a psychologist conducted experiments with 3-year-olds to study their reactions. It turns out that if you stick a skull and crossbones on a bottle, children see it and immediately say 'Poison' in a fearful voice. But if you put it on a poster on a wall, they scream 'Pirates!' And they want to go exploring.

Other scientists suggested plastering the disposal sites with plaques made out of ceramic, metal and stone containing many different warnings in a great variety of languages. But the verdict of the linguists was uniformly the same: at best, the longest any of these languages would be understood was 2,000 years.

What is remarkable about this commission is not only its research question, that is, how to communicate across 10,000 years, but the scientific precision with which it answered it: it is not possible. This is exactly what world risk society is all about. The speeding up of modernization has produced a gulf between the world of quantifiable risk in which we think and act, and the world of non-quantifiable insecurities that we are creating. Past decisions about nuclear energy and present decisions about the use of gene technology, human genetics, nanotechnology, etc. are unleashing unpredictable, uncontrollable and ultimately incommunicable consequences that might ultimately endanger all life on earth.

'Risk' inherently contains the concept of control. Pre-modern dangers were attributed to nature, gods and demons. Risk is a modern concept. It presumes decision-making. As soon as we speak in terms of 'risk', we are talking about calculating the incalculable, colonizing the future.

In this sense, calculating risks is part of the master narrative of first modernity. In Europe, this victorious march culminates in the development and organization of the welfare state, which bases its legitimacy on its capacity to protect its citizens against dangers of all sorts. But what happens in world risk society is that we enter a world of *uncontrollable risk* and we don't even have a language to describe what we are facing. 'Uncontrollable risk' is a contradiction in terms. And yet it is the only apt description for the second-order, *un*natural, human-made, manufactured uncertainties and hazards beyond boundaries we are confronted with.

It is easy to misconstrue the theory of world risk society as Neo-Spenglerism, a new theory about the decline of the western world, or as an expression of typically German *Angst*. Instead I want to emphasize that world risk society does not arise from the fact that everyday life has generally become more dangerous. It is not a matter of the *increase*, but rather of the *de-bounding* of uncontrollable risks. This de-bounding is three-dimensional: spatial, temporal and social. In the spatial dimension we see ourselves confronted with risks that do not take nation-state boundaries, or any other boundaries for that matter, into account: climate change, air pollution and the ozone hole affect everyone (if not all in the same way). Similarly, in the temporal dimension, the long latency period of dangers, such as,

for example, in the elimination of nuclear waste or the consequences of genetically manipulated food, escapes the prevailing procedures used when dealing with industrial dangers. Finally, in the social dimension, the incorporation of both jeopardizing potentials and the related liability question lead to a problem, namely that it is difficult to determine, in a legally relevant manner, who 'causes' environmental pollution or a financial crisis and who is responsible, since these are mainly due to the combined effects of the actions of many individuals. 'Uncontrollable risks' must be understood as not being linked to place, that is they are difficult to impute to a particular agent and can hardly be controlled on the level of the nation state. This then also means that the boundaries of private insurability dissolve, since such insurance is based on the fundamental potential for compensation of damages and on the possibility of estimating their probability by means of quantitative risk calculation. So the hidden central issue in world risk society is *how to feign control over the uncontrollable* – in politics, law, science, technology, economy and everyday life.

We can differentiate between at least three different axes of conflict in world risk society. The first axis is that of *ecological* conflicts, which are by their very essence global. The second is *global financial* crises, which, in a first stage, can be individualized and nationalized. And the third, which suddenly broke upon us on September 11th, is the threat of global terror networks, which empower governments and states.

When we say these risks are global, this should not be equated with a homogenization of the world, that is, that all regions and cultures are now equally affected by a uniform set of non-quantifiable, uncontrollable risks in the areas of ecology, economy and power. On the contrary, global risks are per se unequally distributed. They unfold in different ways in every concrete formation, mediated by different historical backgrounds, cultural and political patterns. In the so-called periphery, world risk society appears *not* as an *endogenous* process, which can be fought by means of autonomous national decision-making, but rather as an *exogenous* process that is propelled by decisions made in other countries, especially in the so-called centre. People feel like the helpless hostages of this process insofar as corrections are virtually impossible at the national level. One area in which the difference is especially marked is in the experience of global financial crises, whereby entire regions on the periphery can be plunged into depressions that citizens of the centre do not even register as crises. Moreover, ecological and terrorist-network threats also flourish with particular virulence under the weak states that define the periphery.

There is a dialectical relation between the unequal experience of being victimized by global risks and the transborder nature of the problems. But it is the transnational aspect, which makes cooperation indispensable to their solution, that truly gives them their global nature. The collapse of global financial markets or climatic change affect regions quite differently. But that doesn't change the principle that everyone is affected, and everyone can potentially be affected in a much worse manner. Thus, in a way, these problems endow each country with a common global interest, which means that, to a certain extent, we can already talk about the basis of a global community of fate. Furthermore, it is also intellectually obvious that global problems only have global solutions, and demand global cooperation. So in that sense, we can say the principle of 'globality', which is a growing consciousness of global interconnections, is gaining ground. But between the potential of global cooperation and its realization lie a host of risk conflicts.

Some of these conflicts arise precisely because of the uneven way in which global risks are experienced. For example, global warming is certainly something that encourages a perception of the earth's inhabitants, both of this and future generations, as a community of fate. But the path to its solution also creates conflicts, as when industrial countries seek to protect the rainforest in developing countries, while at the same time appropriating the lion's share of the world's energy resources for themselves. And yet these conflicts still serve an *integrative* function, because they make it increasingly clear that global solutions must be found, and that these cannot be found through war, but only through negotiation and contract. In the 1970s the slogan was: 'Make love, not war.' What then is the slogan at the beginning of the new century? It certainly sounds more like 'Make *law*, not war.'

The quest for global solutions will in all probability lead to further global institutions and regulations. And it will no doubt achieve its aims through a host

of conflicts. The long-term anticipations of unknown, transnational risks call transnational risk communities into existence. But in the whirlpool of their formation, as in the whirlpool of modernity, they will also transform local cultures into new forms, destroying many central institutions that currently exist. But transformation and destruction are two inescapable sides of the necessary political process of experimentation with new solutions.

Ecological threats are only one axis of global risk conflict. Another lies in the risks of globalized financial markets. Crisis fluctuations in the securities and finance markets are as old as the markets themselves. And it was already clear during the world crisis of 1929 that financial upheavals can have catastrophic consequences – and that they can have huge political effects. The post-Second World War institutions of Bretton Woods were global political solutions to global economic problems, and their efficient functioning was an indispensable key to the rise of the western welfare state. But since the 1970s, those institutions have been largely dismantled and replaced by a series of ad hoc solutions. So we now have the paradoxical situation where global markets are more liberalized and globalized than ever, but the global institutions set up to control them have seen their power drastically reduced. In this context, the possibility of a 1929-size catastrophe certainly cannot be excluded.

Both ecological and financial risks incorporate several of the characteristics we have enumerated that make risks politically explosive. They go beyond rational calculation into the realm of unpredictable turbulence. Moreover, they embody the struggle over the distribution of 'goods' and 'bads', of positive and negative consequences of risky decisions. But above all, what they have in common is that their effects are deterritorialized. That is what makes them *global* risks. And that is what sets in motion the formation of global risk communities – and world risk society.

But while they show similarities, there are also important differences between the various kinds of global risk that significantly influence the resultant conflict. One is that environmental and technological risks come from the 'outside'. They have physical manifestations that then become socially relevant. Financial risks, on the other hand, originate in the heart of the social structure, in its central medium. This then leads to several other differences. Financial risks are more immediately apparent than ecological risks. A consciousness leap is not required to recognize them. By the same token, they are more individualized than ecological risks. A person and her/his next-door neighbour can be affected in very different ways. But, this aspect does not make financial threats potentially less risky. On the contrary, it increases their potential speed and reach. The economy is the central subsystem of modern society. And because all other subsystems depend on it, a failure of this type could be truly disastrous. So there are very compelling reasons to consider the world economy as another central axis of world risk society.

A further distinction can be made, however, between ecological and financial threats on the one hand, and the threat of global terrorist networks on the other. Ecological and financial conflicts fit the model of modernity's self-endangerment. They both clearly result from the accumulation and distribution of 'bads' that are tied up with the production of goods. They result from society's central decisions, but as unintentional side-effects of those decisions. Terrorist activity, on the other hand, is intentionally bad. It aims to produce the effects that the other crises produce unintentionally. Thus the principle of *intention* replaces the principle of *accident*, especially in the field of economics. Much of the literature on risk in economics treats risk as a positive element within investment decisions, and risk-taking as a dynamic aspect linked to the essence of markets. But investing in the face of risk presupposes trust. Trust, in turn, is about the binding of time and space, because trust implies committing to a person, group or institution over time.

This prerequisite of active trust, in the field of economics as well as in everyday life and democracy, is dissolving. The perception of terrorist threats replaces *active trust* with *active mistrust*. It therefore undermines the trust in fellow citizens, foreigners and governments all over the world. Since the dissolution of trust multiplies risks, the terrorist threat triggers a self-multiplication of risks by the de-bounding of risk perceptions and fantasies.

This, of course, has many implications. For example, it contradicts the images of the *homo economicus* as an autarkic human being and of the individual as a decider and risk taker. One of the consequences thereof

is that the principle of *private* insurance is partly being replaced by the principle of *state* insurance. In other words, in the terrorist risk society the world of *individual* risk is being challenged by a world of *systemic* risk, which contradicts the logic of economic risk calculation. Simultaneously, this opens up new questions and potential conflicts, namely how to negotiate and distribute the *costs* of terrorist threats and catastrophes between businesses, insurance companies and states.

Therefore, it becomes crucial to distinguish clearly between, on the one hand, the conventional enemy image between conflicting states and, on the other, the 'transnational terrorist enemy', which consists of individuals or groups but not states. It is the very transnational and hybrid character of the latter representation that ultimately reinforces the hegemony of already powerful states.

The main question is: who defines the identity of a 'transnational terrorist'? Neither judges, nor international courts, but powerful governments and states. They empower themselves by defining who is *their* terrorist enemy, *their* bin Laden. The fundamental distinctions between war and peace, attack and self-defence collapse. Terrorist enemy images are *deterritorialized, de-nationalized and flexible state constructions that legitimize the global intervention of military powers as 'self-defence'*. President George W. Bush painted a frightening picture of 'tens of thousands' of al-Qaida-trained terrorists 'in at least a dozen countries'. Bush uses the most expansive interpretation: 'They are to be destroyed.' Bush's alarmism has a paradoxical effect: it gives Islamic terrorists what they want most – a recognition of their power. Bush has encouraged the terrorists to believe that the United States really can be badly hurt by terrorist actions like these. So there is a hidden mutual enforcement between Bush's empowerment and the empowerment of the terrorists.

US intelligence agencies are increasingly concerned that future attempts by terrorists to attack the United States may involve Asian or African al-Qaida members, a tactic intended to elude the racial profiles developed by US security personnel. Thus the internal law enforcement and the external counter-threat of US intervention not only focus on Arab faces, but possibly on Indonesian, Filipino, Malaysian or African faces. In order to broaden terrorist enemy images, which, to a large extent, are a one-sided construction of the powerful US state, expanded parameters are being developed so as to include networks and individuals who may be connected to Asian and African terrorist organizations. This way, Washington constructs the threat as immense. Bush insists that permanent mobilization of the American nation is required, that the military budget be vastly increased, that civil liberties be restricted and that critics be chided as unpatriotic.

So there is another difference: the *pluralization* of experts and expert rationalities, which characterizes ecological and financial risks, is then replaced by the gross *simplification* of enemy images, constructed by governments and intelligence agencies without and beyond public discourse and democratic participation.

So there are huge differences between the external risks of ecological conflicts, the internal risks of financial conflicts and the intentional terrorist threat. Another big difference is the speed of acknowledgement. Global environmental and financial risks are still not truly recognized. But with the horrific images of New York and Washington, terrorist groups *instantly* established themselves as new global players competing with nations, the economy and civil society in the eyes of the world. The terrorist threat, of course, is reproduced by the global media.

To summarize the specific characteristics of terrorist threat: (bad) intention replaces accident, active trust becomes active mistrust, the context of individual risk is replaced by the context of systemic risks, private insurance is (partly) replaced by state insurance, the power of definition of experts has been replaced by that of states and intelligence agencies; and the pluralization of expert rationalities has turned into the simplification of enemy images.

Having outlined their differences, it should be no surprise that the three kinds of global risk, that is ecological, financial and terrorist threat, also interact. And terrorism again is the focal point. On the one hand, the dangers from terrorism increase exponentially with technical progress. Advances in financial and communication technology are what made global terrorism possible in the first place. And the same innovations that have individualized financial risks have also *individualized war*.

But the most horrifying connection is that all the risk conflicts that are stored away as potential could now be intentionally unleashed. Every advance from gene technology to nanotechnology opens a 'Pandora's box' that could be used as a terrorist's toolkit. Thus the terrorist threat has made everyone into a disaster movie scriptwriter, now condemned to imagine the effects of a home-made atomic bomb assembled with the help of gene or nanotechnology; or the collapse of global computer networks by the introduction of squads of viruses and so on.

Politics of World Risk Society

There is a sinister perspective for the world after September 11th. It is that uncontrollable risk is now irredeemable and deeply engineered into all the processes that sustain life in advanced societies. Pessimism then seems to be the only rational stance. But this is a one-sided and therefore truly misguided view. It ignores the new terrain. It is dwarfed by the sheer scale of the new opportunities opened up by today's threats, that is the axis of conflicts in world risk society.

People have often asked: 'What could unite the world?' And the answer sometimes given is: 'An attack from Mars.' In a sense, that was just what happened on September 11th: an attack from our 'inner Mars'. It worked as predicted. For some time, at least, the warring camps and nations of the world united against the common foe of global terrorism. I would like to suggest six lessons that can be drawn from this event.

The first lesson: in an age where trust and faith in God, class, nation and progress have largely disappeared, humanity's common fear has proved the last – ambivalent – resource for making new bonds. In his book *The Public and Its Problems*, John Dewey argues that it is not a decision, but its consequences and risk that create a public in the post-traditional world. So the theory of world risk society is *not* just another kind of 'end-of-history' idea; this time world history does not end with the resolution of political and social tensions, as Marx and Fukuyama believed, but with the end of the world itself. Nevertheless, what the global public discourse on global risks creates is a reason for hope, since the political explosiveness of world risk society displays a potential enlightenment function. The perceived risk of global terrorism has had exactly the opposite effect than that which was intended by the terrorists. It has pushed us into a new phase of globalization, the globalization of politics, the moulding of states into transnational cooperative networks. Once more, the rule has been confirmed that resistance to globalization only accelerates it. Anti-globalization activists operate on the basis of global rights, markets and networks. They both think and act in global terms, and use them to awaken global awareness and a global public. The term 'anti-globalization movement' is misleading. Many fight for an alternative globalization – global justice – rather than anti-globalization.

The second big lesson of the terrorist attack is: national security is no longer national security. Alliances are nothing new, but the decisive difference about this global alliance is that its purpose is to preserve *internal* and not external security. All the distinctions that make up our standard picture of the modern state – the borders that divide domestic from international, the police from the military, crime from war and war from peace – have been overthrown. It was precisely those distinctions that defined the nation state. Without them, it is a zombie idea. It still looks alive, but it is dead.

Foreign and domestic policy, national security and international cooperation are now all interlocked. The only way to deal with global terror is also the only way to deal with global warming, immigration, poison in the food chain, financial risks and organized crime. In all these cases, national security *is* transnational cooperation. Since September 11th, 'terrorist sleepers' have been identified in Hamburg, Germany, and many other places. Thus, German domestic policy is now an important part of US domestic and foreign policy. So are the domestic as well as foreign, security and defence policies of France, Pakistan, Great Britain, Russia and so on.

In the aftermath of the terrorist attack, the state is back, and for the old Hobbesian reason – the provision of security. Around the world we see governments becoming more powerful, and supranational institutions like NATO becoming less powerful. But at the same time, the two most dominant ideas about the state – the idea of the *national* state, and the idea of the *neoliberal* state – have both lost their reality and their necessity. When asked whether the $40 billion that the US government requested from Congress for the war against terrorism didn't contradict the neoliberal creed to which the Bush administration subscribes, its spokesman replied laconically: 'Security comes first.'

Here is the third lesson: September 11th exposed neoliberalism's shortcomings as a solution to the world's conflicts. The terrorist attacks on America were the Chernobyl of globalization. Just as the Russian disaster undermined our faith in nuclear energy, so September 11th exposed the false promise of neoliberalism.

The suicide bombers not only exposed the vulnerability of western civilization but also gave a foretaste of the conflicts that globalization can bring about. Suddenly, the seemingly irrefutable tenets of neoliberalism – that economics will supersede politics, that the role of the state will diminish – lose their force in a world of global risks.

The privatization of aviation security in the US provides just one example, albeit a highly symbolic one. America's vulnerability is indeed very much related to its political philosophy. It was long suspected that the US could be a possible target for terrorist attacks. But, unlike in Europe, aviation security was privatized and entrusted to highly flexible part-time workers who were paid even less than employees in fast-food restaurants.

It is America's political philosophy and self-image that creates its vulnerability. The horrible pictures of New York contain a message: a state can neoliberalize itself to death. Surprisingly, this has been recognized by the US itself: aviation has been transformed into a federal state service.

Neoliberalism has always been a fair-weather philosophy, one that works only when there are no serious conflicts and crises. It asserts that only globalized markets, freed from regulation and bureaucracy, can remedy the world's ills – unemployment, poverty, economic breakdown and the rest. Today, the capitalist fundamentalists' unswerving faith in the redeeming power of the market has proved to be a dangerous illusion.

This demonstrates that, in times of crises, neoliberalism has no solutions to offer. Fundamental truths that were pushed aside return to the fore. Without taxation, there can be no state. Without a public sphere, democracy and civil society, there can be no legitimacy. And without legitimacy, no security. From these premises, it follows that, without legitimate forums for settling national and global conflicts, there will be no world economy in any form whatsoever.

Neoliberalism insisted that economics should break free from national models and instead impose transnational rules of business conduct. But, at the same time, it assumed that governments would stick to national boundaries and the old way of doing things. Since September 11th, governments have rediscovered the possibilities and power of international cooperation – for example, in maintaining internal security. Suddenly, the necessity of statehood, the counter-principle of neoliberalism, is omnipresent. A European arrest warrant that supersedes national sovereignty in judicial and legal enforcement – unthinkable until recently – has suddenly become a possibility. We may soon see a similar convergence towards shared rules and frameworks in economics.

We need to combine economic integration with cosmopolitan politics. Human dignity, cultural identity and otherness must be taken more seriously in the future. Since September 11th, the gulf between the world of those who profit from globalization and the world of those who feel threatened by it has been closed. Helping those who have been excluded is no longer a humanitarian task. It is in the West's own interest: the key to its security. The West can no longer ignore the black holes of collapsed states and situations of despair.

To draw the fourth lesson I pick up my statement again that no nation, not even the most powerful, can ensure its national security by itself. World risk society is forcing the nation-state to admit that it cannot live up to its constitutional promise to protect its citizens' most precious asset, their security. The only solution to the problem of global terror – but also to the problems of financial risk, climate catastrophe and organized crime – is transnational cooperation. This leads to the paradoxical maxim that, in order to pursue their national interest, countries need to denationalize and transnationalize themselves. In other words, they need to surrender parts of their autonomy in order to cope with national problems in a globalized world. The zero-sum logic of mutual deterrence, which held true for both nation-states and empires, is losing its coherence.

In this context, then, a new central distinction emerges between sovereignty and autonomy. The nation-state is built on equating the two. So from the nation-state perspective, economic interdependence, cultural diversification and military, judicial and technological cooperation all lead to a loss of autonomy and thus sovereignty. But if sovereignty is measured in terms of political clout – that is, by the extent to which a country is capable of having an impact on the world stage, and of furthering the security and well-being of its people by bringing its judgements to bear – then it is possible to conceive the same situation very differently. In the latter framework, increasing interdependence and cooperation, that is, a *decrease* in autonomy, can lead to an increase in sovereignty. Thus, sharing sovereignty does not reduce it; on the contrary, sharing actually enhances it. This is what cosmopolitan sovereignty means in the era of world risk society.

Fifth lesson: I think it is necessary to distinguish clearly between on the one hand, not national, but *global unilateralism* – meaning the politics of the new American empire: *the Pax Americana* – and on the other hand, two concepts of multilateralism or the multilateral state: namely the *surveillance* state and *cosmopolitan* state. Before and after September 11th, US foreign policy changed rapidly from national unilateralism to the paradox of a 'global unilateralism'. In the aftermath of the Afghanistan war, the idea of a 'new world order' has taken shape in Washington's think-tanks and the US is supposed to both make and enforce its laws. The historian Paul Kennedy believes that the new American empire will be even more powerful than the classical imperial powers like Rome and Britain.

This is America's core problem today: a 'free society' is based on openness and on certain shared ethics and codes to maintain order, and Americans are now intimately connected to many societies that do not have governments that can maintain these ethics and order. Furthermore, America's internal security depends on peoples who are aggressively opposed to the American way of life. For America to stay America, a free and open society, intimately connected to the world, the world has to become – *Americanized*. And there are two ways to go about it: open societies either grow from the bottom up or freedom, democracy and capitalism are imposed from the outside by (the threat of) external intervention. Of course, there is the alternative: to affirm and value real international cooperation. Real cooperation will require the Bush administration to swallow a word that even September 11th didn't quite force down: 'multilateralism'. In effect, the message from Washington to Europe and the other allies is: 'We will do the cooking and prepare what people are going to eat, then you will wash the dirty dishes.'

On the other hand, we have to distinguish between two forms of multilateralism as well: surveillance states and cosmopolitan states. *Surveillance states* threaten to use the new power of cooperation to build themselves into fortress states, in which security and military concerns will loom large and freedom and democracy will shrink. Already we hear about how western societies have become so used to peace and well-being that they lack the necessary vigour to distinguish friends from enemies. And that priorities will have to change. And that some of our precious rights will have to be sacrificed for the sake of security. This attempt to construct a western citadel against the numinous Other has already sprung up in every country and will only increase in the years to come. It is the sort of phenomenon out of which a democratic authoritarianism might arise, a system in which maintaining flexibility towards the world market would be premised on increasing domestic rigidity. Globalization's winners would get neoliberalism, and globalization's losers would get the other side of the coin: a heightened fear of foreigners, born out of the apprehension of terrorism and bristling with the poison of racism.

This is my sixth and final lesson: if the world is to survive this century, it must find a way to civilize world risk society. A new big idea is wanted. I suggest the idea of the *cosmopolitan state*, founded upon the recognition of the otherness of the other.

National states present a threat to the inner complexity, the multiple loyalties, the social flows and fluids of risks and people that world risk society has caused to slosh across national borders. Conversely, nation states cannot but see such a fuzzing of borders as a threat to their existence. Cosmopolitan states, by contrast, emphasize the necessity of solidarity with foreigners both inside and outside the national borders. They do this by connecting self-determination with responsibility for (national and non-national) Others. It is not a matter of limiting or negating self-determination. On the contrary, it is a matter of freeing self-determination from its national cyclopean vision and connecting it to the world's concerns. Cosmopolitan states struggle not only against terror, but against the *causes* of terror. They seek to regain and renew the power of politics to shape and persuade, and they do this by seeking the solution of global problems that are even now burning humanity's fingertips but which cannot be solved by individual nations on their own. When we set out to revitalize and transform the state in a cosmopolitan state, we are laying the groundwork for international cooperation on the basis of human rights and global justice.

Cosmopolitan states can theoretically be founded on the principle of the national indifference of the state. This is a concept that is redolent of the way in which, during the 17th century, the Peace of Westphalia ended the religious civil war we call the '30 years war' through the separation of church and state. In a similar manner, the separation of state and nation could be the solution to some global problems and conflicts of the 21st century. For example: just as the a-religious state finally made possible the peaceful coexistence of multiple religions side by side, the cosmopolitan state could provide the conditions for multiple national and religious identities to coexist through the principle of constitutional tolerance.

We should seize this opportunity to reconceive the European political project as an experiment in the building of cosmopolitan states. And we could envision a cosmopolitan Europe, whose political force would emerge directly not only out of the worldwide struggle against terrorism, ecological and financial risks, but also out of both the affirmation and taming of European national complexity.

[. . .]

Risk, Globalisation and the State: A Critical Appraisal of Ulrich Beck and the World Risk Society Thesis

Darryl S. L. Jarvis

[. . .]

Global Risk Society

Global risk society is distinct from industrial modernity for Beck in one crucial respect: the "social compact" or risk contract is increasingly broken down. Risks are now incalculable and beyond the prospects for control, measurement, socialisation and compensation. "Nuclear power, many types of chemical and biotechnological production as well as continuing and threatening eco-logical destruction", argues Beck, are breaking down the "security pact" of industrial society, and thus the "foundations of the established risk logic are being subverted or suspended". This is the entry into global risk society and it occurs when

> the hazards which are now decided and consequently produced by society *undermine and/or cancel the established safety systems of the welfare state's existing risk calculations*. In contrast to early industrial risks, nuclear, chemical, ecological and genetic risk (a) can be limited in terms of neither time nor place, (b) are not accountable according to the established rules of causality, blame and liability, and (c) cannot be compensated for or insured against.

In the global risk society, no one any longer knows with certainty the extent of the risks we face through our collective technologies and innovations. Science now fails us, with conflicting reports, contradictory assessments and wide variance in risk calculations. Faith in the risk technocrats evaporates, the hegemony of experts dissolves and risk assessment becomes no more than a political game that advances sectional interests. The introduction of genetically modified food products in Western Europe, for example, has been mostly rejected by consumers not because of adverse findings by scientists in terms of prospective risks to human health, but because a wide spectrum of the population rejects the sanctity of the advice issued by risk experts who are seen as being influenced by big agrobusiness. Consumers now suspect the limited horizon of understanding that "experts" have about the unintended consequences of complex technologies and their risk externalities. The "social compact" of risk society thus breaks down under reflexive moder-nisation. Beck's portrayal of global risk society is a rather depressing one, increasingly dangerous and beyond meaningful control. Certainty and knowledge appear to break down, and the risk society seems more and more to engulf us all in a kind of cultural mindset of increasing fears, phobias, hyper-risks, and the pos-sibility of severe scientifically induced catastrophe. For Beck, the consequence of global risk society is the production of "organized irresponsibility" with expert division, contradiction and the limits of scientific knowledge paralysing political responses to emerging threats and risks.

Assessing the World Risk Society Thesis

The popularity of Beck's work is in part explained by its timing. Beck could not have foreseen that the publi-cation of his first work on world risk society in May 1986 would coincide with a catastrophe of monumental proportions, namely the explosion of the nuclear power plant at Chernobyl, Ukraine, on 25 April. Beck's concerns about reflexive modernity, his fears about the limits of science and technology and of the ability of human beings to control the consequences of the technologies they invented were all amply demonstrated when the number four reactor at Chernobyl suffered two fatal explosions allowing deadly radiation (30–40 times the radioactivity released by the atomic bombs over Hiroshima and Nagasaki) to escape into the atmosphere. In the days following the explosion the sight of men willingly sacrificing their lives as they were deployed by helicopter to crudely dump soil and concrete on the reactor in the hope of plugging any further radiation leakages only underscored the inability of science to respond meaningfully to the crisis it had unleashed. There was no crisis management, no response plan, no containment strategy other than to close down the facility, encase it in concrete, evacuate millions of people, seal off thousands of hectares of land and create a 30 km radius no-go zone around the reactor, later extended to a 4,300 km^2 exclusion zone. World risk society had, it seemed, arrived with a vengeance.

Yet, despite the timely publication of Beck's work and its resonance with the Chernobyl disaster, the broader contours of his thesis remain problematic and have attracted rigorous debate. Much of this debate has focused on the way Beck conceives of risk, but also the way he explains the process of individualisation and globalisation as antithetical to the logic of industrial modernity, the state and state-based mechanisms for risk control. Indeed, much of Beck's thesis rests on his observations about globalisation and what Beck sees as its negative effects upon state autonomy and institutional capacity. These, he believes, are challenged by complex interdependence, the globalisation of markets, heightened connectivity in media and opinion formation, capital mobility, as well as the advent of supranationalism. The leading patterns of political organisation that, since the Peace of Westphalia in 1648, have governed society in terms of its spatial-political and economic configuration are, for Beck, now eroded by activities (economic and political) that occur between states and by processes that are not state bound. The outcome is the transition from a Westphalian-based system of governance to a post-Westphalian system, where the bounds of the state and its capacity effectively to regulate and control all manner of processes, risks and externalities is

fatally compromised. States surrender parts of their sovereignty not willingly but surreptitiously, through cultural shifts, economic processes that bypass state regulatory regimes and political processes that ensnare states into complex regimes and transnational regulatory governance structures. The epicentre of society moves from a purely national setting to a worldwide community. Lorraine Eden and Stefanie Lenway capture the essence of this thesis:

> If we visualize the world of the 1970s and 1980s as a chessboard, then the immovable blocks were the national boundaries and trade walls behind which governments, firms and the citizens found shelter. Protected by politically made walls, countries could maintain their own cultures, traditions and ways of life, as well as their own choice of governance modes.

For Eden and Lenway, however, globalisation and the spate of neo-liberal policies that emerged during the 1980s have removed or "at least significantly reduced the impact of these immovable blocks between economies". In the process, the post-Westphalian system is born. Beck's reading of globalisation is a popular and widely held one; indeed, it has come to comprise the rationale for many of the anti-globalisation protest movements currently active all over the globe today. But what is the basis for the assumptions about the effects of globalisation on the state and the Westphalian system? If correct, we should be able to discern empirical variance and significant changes in, for example, the spread and distribution of wealth, foreign direct investment (FDI), the extent of multinational enterprise (MNE) relocation, perhaps increasing state failure as globalisation robs the state of its economic base and produces a fiscal crisis for the state. If, as Beck suggests, the state is now passing on to its citizens increasing burdens, offloading its welfare obligations as the tax base dwindles due to forced competition to reduce taxes and increase its attractiveness to highly mobile capital, then we should be able to track these changes and observe absolute reductions in government revenues and smaller government.

An examination of disparate empirical sources, however, reveals little to support Beck's thesis. First, there is little evidence of declining government tax receipts across a wide selection of OECD states. Nor is

Table 1 Government spending and tax revenue as a percentage of GDP: selected OECD states

	Government Spending			Tax Revenue		
	1960	1980	1998	1960[a]	1980	1997
Australia	21.2	31.4	32.9	22.4	28.4	30.3
Britain	32.2	43.0	40.2	28.5	35.1	35.3
Canada	28.6	38.8	42.1	23.8	32.0	36.8[c]
France	24.6	46.1	54.3	N/A	41.7	46.1
Germany	32.4[b]	47.9[b]	46.9	31.3[b]	38.2[b]	37.5
Italy	30.1	42.1	49.1	34.4	30.4	44.9
Japan	17.5	32.0	36.9	18.2	25.4	28.4[c]
Spain	N/A	32.2	41.8	14.0	23.9	35.3
Sweden	31.0	60.1	60.8	27.2	48.8	53.3
United States	26.8	31.4	32.8	26.5	26.9	28.5[c]
Averages[d]	28.3	40.5	43.8	25.1	33.1	37.6

[a]Estimated; [b]West Germany; [c]1996; [d]Unweighted.
Source: OECD figures as quoted in Raymond Vernon, "Big Business and National Governments: Reshaping the Compact in a Globalizing Economy", *Journal of International Business Studies*, Vol. 32, No. 3 (2001), p. 515.

there evidence of declining government spending. In fact, across the OECD government spending has increased in real terms as a percentage of GDP year on year (see Table 1) since 1960 – precisely when the effects of globalisation on Beck's account began to transform the international economy. As a percentage of gross domestic product (GDP), for example, government spending increased from 32.2% in Britain in 1960 to 40.2% in 1998, in Canada from 28.6% to 42.1%, in Italy from 30.1% to 49.1% and in the United States from 26.8% to 32.8%. Tax revenues have similarly shown significant growth trends, contrary to Beck's assertions. As a proportion of GDP, tax revenues increased in Britain from 28.5% of GDP in 1960 to 35.3% in 1998, in Canada from 23.8% to 36.8%, in Italy from 34.4% to 44.9% and even in the United States – an historically low-taxing state – increasing from 26.5% in 1960 to 28.5% in 1998. Rather than a fiscal crisis of the state or the retreat of the state in contemporary economic life, in OECD countries the state continues to be an integral part of the tapestry of modern economies.

The "hollowing out" of the welfare state thesis is also challenged by John Hobson, who notes that "reports of the death of taxation and the welfare state remain greatly exaggerated". Examining taxation policy in the OECD between 1965 and 1999, for example, Hobson

finds that rather than a downward trend of the tax burden there is, in fact, a clear upward trend – and not just for tax revenues but also for state expenditure (see Table 2). Indeed, as Hobson demonstrates, corporate tax rates in the OECD have actually increased at higher annual average rates than have government expenditure and aggregate tax burdens, with the average tax burdens applied specifically to capital increasing by more than 50% from 1960 to 1996–9 – the period typically identified with deepening and intensifying globalisation. As Hobson notes, "what *is* striking in an era of intensifying capital mobility, is the degree to which these broad fiscal indicators have *increased*, thereby suggesting a broadly positive rather than a negative relationship between globalisation and state fiscal capacity" – a finding diametrically opposite to the assertions of Beck and his characterisation of globalisation and its risk consequences for states and welfare societies.

The fiscal crisis of the state has thus not materialised, nor does it display any evidence of doing so in the near future. While, of course, the figures produced above are not indicative of discretionary government spending on welfare entitlements per se – of which there certainly might be evidence of reduced expenditure – they suggest that if this is the case it is *not* due to the forces of globalisation nor a compromised revenue base but ideational change among domestic constituencies and

Table 2 Tax and expenditure burdens, OECD, 1965–99

	1965–9	1970–4	1975–9	1980–4	1985–9	1990–4	1995–9
Aggregate tax burdens							
Average OECD	100	107	113	113	114	117	120
Average EU	100	106	114	118	119	122	125
Average expenditure burdens							
Average OECD	100	107	120	122	121	126	123
Average EU	100	106	121	125	126	129	128
Average tax burdens on capital							
Average OECD	100	117	143	141	148	148	152
Average corporate income tax burden							
Average OECD	100	105	109	116	126	117	131

Source: Adapted from John M. Hobson, "Disappearing Taxes or the 'Race to the Middle'? Fiscal Policy in the OECD", in Linda Weiss (ed.), *States in the Global Economy: Bringing Domestic Institutions back In* (Cambridge: Cambridge University Press, 2003), pp. 40, 44, 46.

the growth of new-right doctrines about the need for welfare reform. This is an entirely different set of issues, unrelated to induced fiscal austerity because of declining tax bases through capital mobility or globalisation.

As for the policy autonomy of states being "strait-jacketed" by globalising forces that *demand* conversion to neo-liberal policy agendas, fiscal conservativism and laissez-faire systems, there is little evidence of such homogenisation. Linda Weiss, for example, when examining policy autonomy and discretionary state manoeuvrability in emerging economies in Asia (Taiwan, South Korea) as well as developed states (Japan, Germany and Sweden), discovered greater latitude for state discretion than might be anticipated by mainstream globalisation theorists such as Beck. Rather than increasing institutional conformity between states or the loss of discretionary institutional capacity, divergence continues to be the order of the day. Indeed, Weiss's findings indicate that what she terms the "transformatory capacity" of the state remains robust, with states able to broker networks of domestic actors and innovate state policy to cultivate domestic industry transformation and engineer internationally competitive industry segments. Rather than globalisation being a "top-down" imposed process, as traditional globalisation theorists suggest, Weiss demonstrates the ways in which states and domestic policy innovation launch domestic actors into the international area

– effectively becoming catalysts of globalisation. By acting as "midwives", state institutions in Japan, Sweden, Germany, South Korea and Taiwan, Weiss demonstrates, but also in Australia, the United States, Britain and Singapore, have effectively launched overseas investment, regional relocation and global competitiveness. Globalisation, in others words, is a process utilised by states; it is an enabling strategy to mould policy goals and bring about nationally desirable developmental outcomes. Rather than "strait-jacketing" states, state–societal relations powerfully shape economic outcomes and harness globalisation. For Weiss, states remain powerful instrumentalities with strong institutional capacities which exhibit a high degree of institutional variation. Globalisation, in short, is what states make of it.

These findings contrast sharply with Beck's depiction of globalisation and its direct causal link with increased risk through the alleged reduction in the size of the welfare state. Beck tends to exaggerate the impact of globalisation, particularly in terms of capital mobility and his suggestion that capital mobility generates a systemic fiscal crisis for the state. If we look at FDI patterns in terms of its origins and destination, however, we observe little variance from historical patterns. In 1990 the triad regions of Western Europe, North America and Japan continued to account for the vast majority of FDI receipts – as they have done throughout the post-war period. In all, some 75% of the total accumulated stock of FDI and 60% of FDI flows in 1990 were concentrated in just three regions – North America, Western Europe and Japan. Globalisation has *not* changed this pattern other than to increase its volume. Capital might have become more mobile but it has not gone elsewhere and become more global or led to outright divestiture in the case of the triad economies.

Henry Wai-chung Yeung and Peter Dicken, among others, confirm the continuation of this trend for the 1990s. Rather than creating increased risk vulnerabilities because of capital mobility and its dispersal to cost-efficient havens in the far-flung corners of the earth, globalisation in fact displays a remarkable propensity to concentrate capital flows in developed economies, itself creating a problem for developing economies. Africa, for example, continues to attract less than 2% of global capital flows, while Latin America and the Caribbean are stalled at around 10–15% of global capital flows. Moreover, while about a third of FDI capital inflows find their way to developing countries as a whole, their dispersal tends to be predominantly to Asia (around 20%), while in Asia itself 90% of these flows concentrate in just 10 Asian countries, with the vast majority heading for China, Singapore and Hong Kong. Highly mobile capital, otherwise so often invoked as the nemesis of globalisation, in fact proves to be less mobile in terms of geographic spread than Beck suggests.

Similarly, if we look at the capitalisation of stock markets all over the world, which is indicative of the enormous growth in flows of portfolio foreign investment, we might expect to observe considerable leakage from triad stock exchanges and growth in the capitalisation of those in emerging economies consistent with mainstream globalisation theory. Yet little change is apparent, with the circulation of international portfolio investment seemingly content to stay in developed Western states and Japan (see Table 3). The United States, for example, still predominates with the vast bulk of the world's liquidity soaked up by the major US stock exchanges that in 2001 accounted for 48.5% of total global stock market capitalisation. By contrast, Latin America, all of Asia (excluding Japan), nondeveloped Europe, the Middle East and Africa, accounted for a mere 10.5% of global stock market capitalisation. Capital might now be mobile but it has certainly not gone global.

Table 3 Capitalisation of world stock markets, 2001

Country/region	Percentage of Global Total
United States	48.5
Developed Europe	31.3
Japan	9.7
Rest of Asia	5.3
Latin America	1.4
Rest of world	3.8

Source: Roger Lee, "The Marginalization of Everywhere: Emerging Geographies of Emerging Economies", in Jamie Peck and Henry Wai-chung Yeung (eds.), *Remaking the Global Economy: Economic Geographical Perspectives* (London: Sage, 2003), p. 73.

The point in highlighting these examples is to demonstrate that the beneficiaries of international capital flows continue overwhelmingly to be developed Western states and Japan. To be sure, this suggests the internationalisation of these economies and a structural transition in their economic composition, but does not suggest capital flight or capital scarcity and thus necessarily increased risk and vulnerability for industrial society as Beck insists. The process of financial liberalisation and capital mobility has thus been considerably more nuanced than Beck appreciates. Rather than an imposed condition foisted upon states, globalisation, at least in the sense of capital mobility, has been the result of deliberative state actions through capital account liberalisation, that is, state-sponsored initiatives generated by domestic actors. The causality of the globalisation process is thus opposite to that suggested by Beck and mainstream globalisation theorists. This might explain why many states have actually benefited from capital account liberalisation, developing highly successful financial service sectors and employment growth. At the very least, it demonstrates ample state capacity for adaptability, with most developed states structurally adjusting their economic composition to profit from the evolving forms of international capital circulation.

Finally, and perhaps most tellingly, Beck's much feared rise of reflexive modernity through the process of radicalised globalisation does not appear to have affected the durability of the state that, for all its weakness and supposed declining utility, appears to be enjoying something of a renaissance. At the very time when globalisation was accelerating, judged by increased flows of FDI and international trade, and at the same time as the state, according to Beck's account, was experiencing diminishing juridical authority through growing permeability and loss of political control, the number of states grew significantly – from 127 in 1970 to 191 in 2004. This, to say the least, is an oddity and suggests that rather than experiencing a transition to a post-Westphalian order, as postulated by Beck, we are in fact experiencing a deepening of the Westphalian system – evidence of the continuing utility of the state as a medium for economic and security protection – but it is losing in other domains. As Louise Pauly notes:

> If sovereignty is defined as policy autonomy, then increased international capital mobility seems necessarily to imply a loss of sovereignty. This old chestnut ignores, however, both an extensive literature on the evolution of the legal concept of sovereignty and a generation of research on the political trade-offs entailed by international economic interdependence. Furthermore, it downplays the stark historical lesson of 1914: Under conditions of crisis, the locus of ultimate political authority in the modern age – the state – is laid bare. Especially through its effects on domestic politics, capital mobility constrains states, but not in an absolute sense. If a crisis increases their willingness to bear the consequences, states can still defy markets. More broadly, the abrogation of the emergent regime of international capital mobility by the collectivity of states may be unlikely and undesirable, but it is certainly not inconceivable. As long as that remains the case, states retain their sovereignty. Nevertheless, in practical terms, it is undeniable that most states today do confront heightened pressures on their economic policies as a result of more freely flowing capital. The phenomenon itself, however, is not new. What is new is the widespread perception that all states and societies are now similarly affected.

Implications for Beck's Risk Society Thesis

Beck's use of globalisation as one of the principal determinants of risk under reflexive modernity makes his characterisation of globalisation central to validating the risk society thesis. As we have seen, however, it fails many empirical tests with relatively crude postulations. There is little empirical evidence to support Beck's suggestion that the state is in systematic retreat, that its fiscal base has been eroded, or its expenditure abilities reduced. If anything, among OECD countries, the institutional reach of the state, its fiscal base and expenditure commitments have all increased commensurate with deepening globalisation. Does this, then, invalidate Beck's world risk society thesis?

The answer to this question comes in many parts – much like Beck's thesis. As one of Beck's "five interrelated processes" that contribute to and generate increased risk, the extent of globalisation in terms of its dislocating impact upon the state, its political authority and ability to provide welfare has been overstated by Beck. While new historical precedents have been established through growing levels of interdependence, especially in terms of economic linkages (trade, finance, and investment), the suggestion that the state is withering away or that we are in a post-Westphalian system is premature, at least in these domains.

These observations, however, do not necessarily discount Beck's notion that individuals over the last few decades have been exposed to increasing personal vulnerabilities. Since the mid to late 1970s some OECD states (such as Britain, Australia and New Zealand) have witnessed a repudiation of social-democratic forms of governance such as a diminution of welfare entitlements combined with an increasing use of user-pays and fee-for-service systems in the provision of previously universally provided public goods (particularly in education, health, and transportation). Economic individualisation has thus undoubtedly exposed some groups to greater vulnerabilities and reduced the level of equitable access in relation to health and educational services. Indeed, the gulf between the rich and poor has been widening throughout the OECD. However, this widening gap is not a result of globalisation impoverishing disadvantaged strata of society but rather, as Timothy Smeeding notes, "by raising incomes at the top of the income distribution [spectrum.]" As he goes on to note:

> Notwithstanding [the influence of globalisation] domestic policies – labor market institutions, welfare policies, etc. – can act as a powerful countervailing

force to market driven inequality. Even in a globalized world, the overall distribution of income in a country remains very much a consequence of the domestic political, institutional and economic choices made by those individual countries – both rich and middle income ones.

Beck gives too little attention to the autonomous ideational changes that have championed the neo-liberal agenda – incorrectly ascribing these to structural forces endemic to radical modernisation. Of course, it is entirely conceivable that, depending on the prevailing political climate and the constellation of political forces, this agenda might be reversed, partially abandoned or modified. Thus the rise of the risk society, at least as it relates to the individualisation of risk through declining welfare provision or progressive taxation systems and globalisation, might not be as predetermined as Beck suggests.

Equally, some of Beck's other "interrelated processes" also appear problematic. For example, his assertion that rising and endemic underemployment will usurp the distributive function necessary to the reproduction of industrial modernity and transpose greater risks and vulnerabilities onto a growing segment of society does not appear empirically sustainable. To be sure, there has been a pronounced increase in the rate of casual and flexible employment practices, but the wholesale offshore movement of jobs has not taken place. Job redundancy and the replacement of "old economy" industries, for example, while a feature of the latter part of the 20th century and early part of the new millennium, have also been accompanied by job creation in the so-called "new economy" sectors (such as biotechnology, information technology, financial services, education, and the hospitality and tourism industries). Consequently, the fact that global unemployment stood at only 6.2% of the global workforce in 2003 (according to the International Labour Organisation – ILO) fails to indicate the emergence of a structural employment crisis. Indeed, this rate came off the back of a severe global economic slowdown (2000–3), the war on terror and disruptions to the global hospitality, tourism and aviation industries, and global panic associated with the outbreak of SARS in Asia. This rate, in other words, is cyclical not systemic and, according to the ILO, likely to trend downwards

as global economic activity picks up over the next couple of years.

What, then, might account for these premature assertions by Beck? The answer perhaps lies in appreciating the historical backdrop to his central thesis. Beck formulated many of his observations amid a period of tumultuous change in Germany. First, the rise of the Greens led to rapidly changing political affiliations in the 1980s, while the events surrounding the fall of the Berlin Wall and the problems of economic restructuring as a result of German reunification and post-reunification economic adaptation were tumultuous. The latter, in particular, have posed continuing challenges for Germany, especially in terms of labour market integration, economic equalisation and the modernisation of East German industry and infrastructure. Beck has undoubtedly been influenced by these events and the processes of accommodation and dislocation that naturally accompany them. At worst, Beck might thus be accused of a kind of "presentism" – a preoccupation with proximate current events and an assumption of both their ubiquity and universal validity as indices of a new risk civilisation. Robert Dingwall, for example, goes so far as to describe *Risk Society* as "a profoundly German book". As he notes, "most of the citations are to other German authors, the acknowledgements are to German colleagues and the book's drafting 'in the open hill above Starnberger See' is lovingly recorded". This is not, Dingwall insists, a xenophobic criticism but an observation of the milieu in which Beck's thoughts were influenced and the context in which his thesis has evolved – perhaps making Beck's concerns more local and parochial than he would care to admit. The point is a broader one, however. Anthony Elliott, for example, asks whether Beck's observations overstate the phenomena and relevance of risk. How, for example, should we compare risks in different historical periods? Are we really living in a unique historical epoch in which the calculus of risk is so extreme that it distinguishes itself from all previous epochs? As Brian Turner notes:

[A] serious criticism of Beck's arguments would be to suggest that risk has not changed so profoundly and significantly over the last three centuries. For example, were the epidemics of syphilis and bubonic plague in earlier periods any different from the modern environment illnesses to which Beck draws our attention? That is, do Beck's criteria of risk, such as their impersonal and unobservable nature, really stand up to historical scrutiny? The devastating plagues of earlier centuries were certainly global, democratic and general. Peasants and aristocrats died equally horrible deaths. In addition, with the spread of capitalist colonialism, it is clearly the case that in previous centuries many aboriginal peoples such as those of North America and Australia were engulfed by environmental, medical and political catastrophes which wiped out entire populations. If we take a broader view of the notion of risk as entailing at least a strong cultural element whereby risk is seen to be a necessary part of the human condition, then we could argue that the profound uncertainties about life, which occasionally overwhelmed earlier civilizations, were not unlike the anxieties of our own fin-de-siècle civilizations.

This goes to the core of Beck's thesis and questions its basic assumptions about the depth and extent of risk under reflexive modernity. Yet Turner fails to take his critique one step further and question whether, regardless of how extensive risk is, the regime of control and the social compact that distributes risk under industrial modernity is, in fact, breaking down as Beck asserts. Again, it seems highly problematic to suggest that the orderly distribution of risk or the ability to compensate or insure against risk are automatically mitigated on the basis of exceptionalism – the advent of nuclear weaponry, the prospects of nuclear mishap or the looming prospect of ecological disaster possibilities, and until they manifest themselves their possibility should not detract from the strength of existing regimes of control. Many states continue to display a high level of adeptness in indemnifying their constituents against natural disasters (floods, hurricanes, earthquakes, famine, humanitarian disaster). Indeed, the control regimes surrounding emergency management and response have probably never been so well formulated as they are today. The tsunami tragedy of 26 December 2004 in the Indian Ocean, for example, while representing one of the most devastating natural disasters of the last few centuries, inflicting cataclysmic destruction on multiple populations in several countries, was also one of the most well managed in terms of emergency response, humanitarian assistance and reconstructive aid efforts. Within hours of the disaster, emergency response teams were activated in Thailand, Sri Lanka and Indonesia, and within days international emergency and humanitarian assistance was deployed on a global scale, with these efforts redoubled as the calamity of the devastation became apparent. Perhaps only in terms of the immediate humanitarian emergency response in Western Europe at the end of the Second World War has the world witnessed such a massive mobilisation of resources, inter-agency effort and coordination, and global political coordination and response. Rather than a crisis of risk control and management, current crisis and emergency response systems represent an historical highpoint, having achieved greater levels of response effectiveness, early warning preparedness and crisis management than at any time before in history.

But for Beck, of course, this is not important, since all this would be swept away by the magnitude of looming, exceptional risks. But how accurate is this assumption? The Cold War has ended, the risk of nuclear confrontation has diminished (although proliferation may raise it), and so has the prospect of nuclear weapons accidents. Nuclear arsenals continue to be reduced and technical safety systems increased. Whilst there remains the prospect of weapons of mass destruction "falling into the wrong hands" and the development and deployment of so-called "dirty-bombs" based on the use of low-grade uranium, such a prospect scarcely matches the level of terror threatened during the Cold War. The consequences of risk exposure in these instances have traditionally been socialised, so why does Beck assume that such would not be the case again? The social compact would be stressed and challenged but not necessarily irreversibly broken. Likewise, even with recent events such as the BSE crisis in the United Kingdom, Europe and Canada, the outbreak of AIDS and SARS, the terrorist attacks in the United States, the ecological catastrophe of the cod crisis in Eastern Canada, the fish stock crisis in Europe, or any number of other events, the social compact has remained intact and subject to collective accommodation and response efforts. Imperfect though these may be, they have not yet led to systemic failure in the sense of realising the penultimate consequences of reflexive modernity. Nearly all have been addressed, most rectified or at the very least processes put in place to ameliorate their worst consequences and systemic causes.

Beck prefers to discount the success of these risk management efforts and tends to adopt, instead, a fatalistic view of the human condition, pointing to our inability to correct errors, an ineptitude when it comes to moderating risk-producing behaviour, and a collective inertia in the face of looming risk(s). Yet these assumptions seem to be less founded on empirical realities and more on a philosophy of fatalism, leading Beck to proffer a relatively simplistic prognosis that "institutions founder on their own success". But do they? Again, the empirical evidence for this is problematic. Beck, for example, invokes the case of the German crystal lead factory in Upper Palatinate in the Federal Republic of Germany:

> Flecks of lead and arsenic the size of a penny had fallen on the town, and fluoride vapours had turned leaves brown, etched windows and caused bricks to crumble away. Residents were suffering from skin rashes, nausea and headaches. There was no question where all of that originated. The white dust was pouring visibly from the smokestacks of the factory.

In terms of responsibility for the environmental risks produced by the factory, Beck is quite adamant that this was "a clear case". But, as he explains in disgust,

> on the tenth day of the trial the presiding judge offered to drop charges in return for a fine DM10,000, a result which is typical of environmental crimes in the Federal Republic (1985: 12,000 investigations, twenty-seven convictions with prison terms, twenty-four of those suspended, the rest dropped).

Science and the "organized irresponsibility" of the "security bureaucracies", Beck insists, increasingly dominate under reflexive modernity and, in the process, the apportionment of blame becomes obfuscated by an inept technocracy. In the case of the German crystal lead factory, Beck notes, "the commission of the crime could not and was not denied by anyone. A mitigating factor came into play for the culprits: there were three other glass factories in the vicinity which emitted the same pollutants". As a result, "the greater the number of smokestacks and discharge pipes through which pollutants and toxins are emitted, the lower the 'residual probability' that a culprit can be

made responsible". The limits of science and of the bureaucracy are revealed by their inability directly to connect the polluter with specific pollutants. The more pollution generated and the more polluters, for Beck, essentially dilutes the social compact and the ability to apportion blame, responsibility and thus secure compensation.

The example provided by Beck is meant to demonstrate the increasing failure of the social compact, of science and the technocracy to apportion blame and compensate for risk production. Eloquent though this example is, again its reification onto a universal plane seems premature. To what extent, for example, is the paucity of environmental law in the Federal Republic true, say, of the United States, Australia, Canada, or New Zealand? And in what sense should the example of the crystal lead factory be taken as a systemic condition of reflexive modernity? Surely it reflects little more than the paucity of outdated law in the German Federal Republic – a process that can be easily rectified by drafting better laws and by engaging political processes – much as Green movements throughout the world have done with increasing success. Beck, it seems, denies politics and the ability of political actors to change laws and respond to environmental damage. More generally, Beck fails to recognise that risk distribution and compensation have always been contentious affairs fraught with different legal opinions and with those responsible for the generation of risk keen to avoid the costs associated with it. Why, then, is this epoch distinctive from previous epochs where the same motifs have applied?

Unfortunately, for Beck, the point where his argument could be sustained empirically, and probably has greatest insight and utility, is precisely the point where he places too little investigative and analytical weight. The epochal distinctiveness of the current global economic order, for example, especially in terms of the risk posed by the constellation of opposing financial architectures, between semi-liberalised and non-liberalised state financial systems, the extraordinary growth in arbitrage instruments of various kinds, and the structural imbalances this creates in a global financial system now fiercely interdependent makes for an increasingly vexed global financial order posing greater risk to global wealth and the normal functioning of markets. While Beck refers to this phenomenon simply in terms of the structural changes foisted on FDI patterns by globalisation, he is left with little scope for exploring the fundamental changes in the global financial architecture and the increasingly precarious risk environment this generates and which, potentially, poses greater risk to global financial stability and the possibility of systemic global crisis. Beck, however, pays little heed to the basic difference between the movement of productive capital (FDI), short-term capital and the rise of the speculative or symbol economy. It is the latter, and the extraordinary growth in the volume of these transactions and the various arbitrage instruments engineered to secure them, where the emergence of the risk society thesis might be profitably applied but where Beck fails to do so.

Conclusion: Beck's Contribution to Risk Discourse

It is obvious that a purely empirical reading of Beck reveals serious shortcomings with the risk society thesis. To be fair to Beck, however, is this the correct way to read him? As Dirk Matten notes, "Beck's ideas are more of a provocative and conceptual nature rather than a minute empirical proof of certain social changes". They are perhaps better understood as a cultural and social commentary about the condition of late modernity and of its contradictions that both embody progress but also harm and risk. Like many of his contemporaries, Beck is alarmed by the fact of progress in almost every area of human endeavour amid a rampant disregard for ecological preservation, the use of technologies for nefarious purposes and the accelerated generation of unintended outcomes. Beck's

fixation with risk can thus perhaps be appreciated in an era in which all risk, no matter how finite, becomes ethically unacceptable and a bellwether of the social psyche. When Aaron Wildavsky asks "why are the healthiest, longest lived nations on earth so panicked about their health?" the answer must surely lie not in the empirical condition of longevity, the betterment of the human condition and the fact of medical advance. It is, perhaps, not so much a question about whether in fact there are more risks but how we perceive them and the adequacy of their management, compensation and mitigation. Read as a moment in the success of modernity, and at a time when risk tolerance has been reduced, risk aversion increased, and risk perception sensitised, Beck has undoubtedly captured the collective essence of a global society ill at ease. His greatest contribution perhaps lies in exposing these apparent paradoxes, capturing the essence of our collective angst about the limits of science, progress and rationality, about the sublimation of nature and the natural environment into ever more remote corners of our everyday experience, while at the same time we are still confronted by the limitation of knowledge, the fallibility of our existence, and the finitude of our mortality. Despite the success of science, technical knowledge, and the great leaps forward in our collective well-being, in the end each of us still faces the perils of everyday existence, the probabilities of meeting our fate through incurable illness, the uncertainty of our personal futures, or the possibility of accident and misfortune through exposure to the very products derived through scientific progress. Given the impossibility of transforming uncertainty, risk and harm into instruments amenable to total control and mitigation, Beck's work will surely resonate for generations to come.

 READING 40

Unpacking Cosmopolitanism for the Social Sciences: A Research Agenda

Ulrich Beck and Natan Sznaider

[. . .]

Cosmopolitanism is, of course, a contested term; there is no uniform interpretation of it in the growing

literature. The boundaries separating it from competitive terms like globalization, transnationalism, universalism, glocalization etc. are not distinct and internally it is traversed by all kind of fault lines. Yet we will argue that the neo-cosmopolitanism in the social sciences – 'realistic cosmopolitanism' or 'cosmopolitan realism' – is an identifiable intellectual movement united by at least three interconnected commitments:

First, the shared *critique of methodological nationalism* which blinds conventional sociology to the multi-dimensional process of change that has irreversibly transformed the very nature of the social world and the place of states within that world. Methodological nationalism does not mean (as the term 'methodological individualism' suggests) that one or many sociolo-gists have consciously created an *explicit* methodology (theory) based on an *explicit* nationalism. The argument rather goes that social scientists in doing research or theorizing take it for granted that society is equated with national society, as Durkheim does when he reflects on the integration of society. He, of course, has in mind the integration of the *national* society (France) without even mentioning, naming or thinking about it. In fact, not using the adjective 'national' as a universal language does not falsify but might sometimes even prove methodological nationalism. That is the case when the *practice* of the argument or the research pre-supposes that the unit of analysis is the national society or the national state or the combination of both. The concept of methodological nationalism is not a con-cept of methodology but of the sociology of sociology or the sociology of social theory.

Second, the shared diagnosis that the twenty-first century is becoming an age of cosmopolitanism. This could and should be compared with other historical moments of cosmopolitanism, such as those in ancient Greece, the Alexandrian empire and the Enlightenment. In the 1960s Hannah Arendt analysed the *Human Condition*, in the 1970s François Lyotard the *Post-modern Condition*. Now at the beginning of the twenty-first century we have to discover, map and understand the Cosmopolitan Condition.

Third, there is a shared assumption that for this purpose we need some kind of '*methodological cosmo-politanism*'. Of course, there is a lot of controversy about what this means. The main point for us lies in the fact that the dualities of the global and the local, the national and the international, us and them, have dis-solved and merged together in new forms that require conceptual and empirical analysis. The outcome of this is that the concept and phenomena of cosmopolitanism

are not spatially fixed; the term itself is not tied to the 'cosmos' or the 'globe', and it certainly does not encom-pass 'everything'. The principle of cosmopolitanism can be found in specific forms at every level and can be practiced in every field of social and political action: in international organizations, in bi-national families, in neighbourhoods, in global cities, in transnationalized military organizations, in the management of multi-national co-operations, in production networks, human rights organizations, among ecology activists and the paradoxical global opposition to globalization.

Critique of Methodological Nationalism

Methodological nationalism takes the following pre-mises for granted: it equates societies with nation-state societies and sees states and their governments as the primary focus of social-scientific analysis. It assumes that humanity is naturally divided into a limited number of nations, which organize themselves internally as nation-states and externally set boundaries to dis-tinguish themselves from other nation-states. And it goes further: this outer delimitation as well as the competition between nation-states, represent the most fundamental category of political organization.

The premises of the social sciences assume the collapse of social boundaries with state boundaries, believing that social action occurs primarily within and only secondarily across, these divisions:

> [Like] stamp collecting [. . .] social scientists collected distinctive national social forms. Japanese industrial relations, German national character, the American constitution, the British class system – not to mention the more exotic institutions of tribal societies – were the currency of social research. The core disciplines of the social sciences, whose intellectual traditions are reference points for each other and for other fields, were therefore *domesticated* – in the sense of being preoccupied not with Western and world civilization as wholes but with the 'domestic' forms of particular national societies.

The critique of methodological nationalism should not be confused with the thesis that the end of the nation-state has arrived. One does not criticize meth-odological individualism by proclaiming the end of the individual. Nation-states (as all the research shows) will continue to thrive or will be transformed into transnational states. What, then, is the main point of the critique of methodological nationalism? It adopts categories of practice as categories of analysis. The decisive point is that *national organization as a structuring principle of societal and political action can no longer serve as the orienting reference point for the social scientific observer*. One cannot even understand the re-nationalization or re-ethnification trend in Western or Eastern Europe without a cosmopolitan perspective. In this sense, the social sciences can only respond adequately to the challenge of globalization if they manage to overcome methodological nationalism and to raise empirically and theoretically fundamental questions within specialized fields of research, and thereby elaborate the foundations of a newly formulated *cosmopolitan* social science.

As many authors – including the ones in this volume – criticize, in the growing discourse on cosmopolitanism there is a danger of fusing the ideal with the real. What cosmopolitanism *is* cannot ultimately be separated from what cosmopolitanism *should be*. But the same is true of nationalism. The small, but important, difference is that in the case of nationalism the value judgment of the social scientists goes unnoticed because methodological nationalism includes a naturalized conception of nations as real communities. In the case of the cosmopolitan 'Wertbeziehung' (Max Weber, value relation), by contrast, this silent commitment to a nation-state centred outlook of sociology appears problematic.

In order to unpack the argument in the two cases it is necessary to distinguish between the *actor* perspective and the *observer* perspective. From this it follows that a sharp distinction should be made between *methodological* and *normative* nationalism. The former is linked to the social-scientific observer perspective, whereas the latter refers to the negotiation perspectives of political actors. In a normative sense, nationalism means that every nation has the right to self-determination within the context of its cultural, political and even geographical boundaries and distinctiveness. Methodological nationalism assumes this normative claim as a socio-ontological given and simultaneously links it to the most important conflict and organization orientations of society and politics.

These basic tenets have become the main perceptual grid of the social sciences. Indeed, this social-scientific stance is part of the nation-state's own self-understanding. A national view on society and politics, law, justice, memory and history governs the sociological imagination. To some extent, much of the social sciences has become a prisoner of the nation-state. That this was not always the case [has been shown by] Bryan Turner. [. . .] This does not mean, of course, that a cosmopolitan social science can and should ignore different national traditions of law, history, politics and memory. These traditions exist and become part of our cosmopolitan methodology. The comparative analyses of societies, international relations, political theory, and a significant part of history and law all essentially function on the basis of methodological nationalism. This is valid to the extent that the majority of positions in the contemporary debates in social and political science over globalization can be systematically interpreted as transdisciplinary reflexes linked to methodological nationalism.

These premises also structure empirical research, for example, in the choice of statistical indicators, which are almost always exclusively national. A refutation of methodological nationalism from a strictly empirical viewpoint is therefore difficult, indeed, almost impossible, because so many statistical categories and research procedures are based on it. It is therefore of historical importance for the future development of the social sciences that this methodological nationalism, as well as the related categories of perception and disciplinary organization, be theoretically, empirically, and organizationally re-assessed and reformed.

What is at stake here? Whereas in the case of the nation-state centred perspective there is an historical correspondence between normative and methodological nationalism (and for this reason this correspondence has mainly remained latent), this does not hold for the relationship between normative and methodological cosmopolitanism. In fact, the opposite is true: even the re-nationalization or re-ethnification of minds, cultures and institutions has to be analysed within a cosmopolitan frame of reference.

Cosmopolitan social science entails the systematic breaking up of the process through which the national perspective of politics and society, as well as the methodological nationalism of political science, sociology, history, and law, confirm and strengthen each other in their definitions of reality. Thus it also tackles (what had previously been *analytically* excluded as a sort of conspiracy of silence of conflicting basic convictions) the various developmental versions of de-bounded politics and society, corresponding research questions and programmes, the strategic expansions of the national and international political fields, as well as basic transformations in the domains of state, politics, and society.

This paradigmatic de-construction and re-construction of the social sciences from a national to a cosmopolitan outlook can be understood and methodologically justified as a 'positive problem shift', a broadening of horizons for social science research making visible new realities encouraging new research programmes. Against the background of cosmopolitan social science, it suddenly becomes obvious that it is neither possible to distinguish clearly between the national and the international, nor, correspondingly, to make a convincing contrast between homogeneous units. National spaces have become denationalized, so that the national is no longer national, just as the international is no longer international. New realities are arising: a new mapping of space and time, new co-ordinates for the social and the political are emerging which have to be theoretically and empirically researched and elaborated.

This entails a re-examination of the fundamental concepts of 'modern society'. *Household, family, class, social inequality, democracy, power, state, commerce, public, community, justice, law, history, memory and politics* must be released from the fetters of methodological nationalism, re-conceptualized, and empirically established within the framework of a new cosmopolitan social and political science. It would be hard to understate the scope of this task. But nevertheless it has to be taken up if the social sciences want to avoid becoming a museum of antiquated ideas.

Structure and Normativity: the Cosmopolitan Condition and the Cosmopolitan Moment

In order to unpack cosmopolitanism, we need to make another important distinction, namely that between normative-philosophical and empirical-analytical cosmopolitanism; or, to put it differently, between the cosmopolitan *condition* and the cosmopolitan *moment*. Up to now, much of the social scientific discourse has assumed the notion of cosmopolitanism as a moral and political standpoint, a shared normative-philosophical commitment to the primacy of world citizenship over all national, religious, cultural, ethnic and other parochial affiliations; added to this is the notion of cosmopolitanism as an attitude or biographical situation in which the cultural contradictions of the world are unequally distributed, not just out there but also at the centre of one's own life. A world of yesterday turned into an utopian future and reclaimed by social thinkers is elevating 'homelessness', 'fluidity', 'liquidity', 'mobility' to new heights. 'Cosmopolitanism' has a noble ring in a plebeian age, the nobility of a Kant in a postmodern age. This is the kind of cosmopolitanism familiar to philosophers since ancient times, but alien to social scientists. Here, cosmopolitanism is equated with *reflexive* cosmopolitanism. This idea of cosmopolitanism includes the idea that the self-reflexive global age offers space in which old cosmopolitan ideals could and should be translated and re-configured into concrete social realities and philosophy turned into sociology. Nevertheless, the question has to be asked and answered: *Why* is there a cosmopolitan moment now, at the beginning of the twenty-first century?

On the other hand the discourse on cosmopolitanism so far has not really paid attention to the fact that, besides the intended, there is an *unintended and lived* cosmopolitanism and this is of growing importance: the increase in interdependence among social actors across national borders (which can only be observed from the cosmopolitan outlook), whereby the peculiarity exists in the fact that this 'cosmopolitanization' occurs as unintended and unseen *side-effects* of actions which are not intended as 'cosmopolitan' in the normative sense. Only under certain circumstances does this latent cosmopolitanization lead to the emergence of global public spheres, global discussion forums, and global regimes concerned with transnational conflicts ('institutionalized cosmopolitanism'). Summarizing these aspects, we speak of the *Cosmopolitan Condition* as opposed to the *Postmodern Condition*.

The cosmopolitan condition

If we make a clear distinction between the actor perspective and the observer perspective, both in relation to the national outlook and the cosmopolitan outlook, we end up with four fields in a table representing the possible changes in perspectives and reality. It is at least conceivable (and this needs a lot of optimism!) that the shift in outlook from methodological nationalism to methodological cosmopolitanism will gain acceptance. But this need not have any implications for the prospect for realizing cosmopolitan ideals in society and politics. So, if one is an optimist regarding a cosmopolitan turn in the social sciences, one can certainly also be a pessimist regarding a cosmopolitan turn in the real world. It would be ridiculously naïve to think that a change in scientific paradigm might lead to a situation where people, organizations and governments are becoming more open to the ideals of cosmopolitanism. But again: if this is so why do we need a cosmopolitan outlook for the social sciences? Our answer is: in order to understand the really-existing process of cosmopolitanization of the world.

Like the distinction between 'modernity' and 'modernization', we have to distinguish between *cosmopolitanism* as a set of normative principles and (really existing) *cosmopolitanization*. This distinction turns on the rejection of the claim that cosmopolitanism is a conscious and voluntary choice, and all too often the choice of an elite. The notion 'cosmopolitanization' is designed to draw attention to the fact that the emerging cosmopolitan of reality is also, and even primarily, a function of coerced choices or a side-effect of unconscious decisions. The choice to become or remain an 'alien' or a 'non-national' is not as a general rule a voluntary one but a response to acute need, political repression or a threat of starvation. A 'banal' cosmopolitanism in this sense unfolds beneath the surface or behind the façades of persisting national spaces, jurisdiction and labelling, while national flags continue to be hoisted and national attitudes, identities and consciousness remain dominant. Judged by the lofty standards of ethical and academic morality, this latent character renders cosmopolitanism trivial, unworthy of comment, even suspect. An ideal that formerly strutted the stage of world history as an ornament of the elite cannot possibly slink into social

and political reality by the backdoor. Thus, we emphasize the centrality of emotional engagement and social integration and not only fragmentation as part of the cosmopolitan world. And this emphasizes that the process of cosmopolitanization is bound up with symbol and ritual, and not just with spoken ideas. And it is symbol and ritual that turns philosophy into personal and social identity and consequently relevant for social analysis. The more such rituals contribute to individuals' personal sense of conviction, the larger the critical mass available to be mobilized in cosmopolitan reform movements for instance, be they movements against global inequality or human rights violations. And the farther cosmopolitan rituals and symbols spread, the more chance there will be of someday achieving a cosmopolitan political order. This is where normative and empirical cosmopolitanism meet. At the same time, we must remember that a cosmopolitan morality is not the only historically important form of today's globalized world. Another one is nationalism. The nation-state was originally formed out of local units to which people were fiercely attached. They considered these local attachments 'natural' and the nation-state to be soulless and artificial – *Gesellschaft* compared to the local *Gemeinschaft*. But thanks to national rituals and symbols, that eventually changed completely. Now today many people consider national identity to be natural and cosmopolitan or world identity to be an artificial construct. They are right. It will be an artificial construct, if artificial means made by humans. But they are wrong if they think artificial origins prevent something from eventually being regarded as natural. It did not stop the nation-state. And there is no reason it has to stop cosmopolitan morality. However, the challenge will be to see these moral orders not as contradictory but as living side by side in the global world. Cosmopolitanism and nationalism are not mutually exclusive, neither methodologically nor normatively.

There can be no doubt that a cosmopolitanism that is passively and unwillingly suffered is a *deformed* cosmopolitanism. The fact that really-existing cosmopolitanization is not achieved through struggle, that it is not chosen, that it does not come into the world as progress with the reflected moral authority of the Enlightenment, but as something deformed and profane, cloaked in the anonymity of side-effects – this is an essential founding moment within cosmopolitan realism in the social sciences. Our main point is here to make a distinction between the moral ideal of cosmopolitanism (as expressed in Enlightenment philosophy) and the above mentioned cosmopolitan condition of real people. It's also the distinction between theory and praxis. This means, in our case, the distinction between a cosmopolitan philosophy and a cosmopolitan sociology.

Cosmopolitanism and globalization

But, one might object, isn't 'cosmopolitanization' simply a new word for what used to be called 'globalization'? The answer is 'no': globalization is something taking place 'out there', cosmopolitanization happens 'from within'. Whereas globalization presupposes, cosmopolitanization dissolves the 'onion model' of the world, where the local and the national form the core and inner layer and the international and the global form the outer layers. Cosmopolitanization thus points to the irreversible fact that people, from Moscow to Paris, from Rio to Tokyo, have long since been living in *really-existing relations of interdependence*; they are as much responsible for the intensification of these relations through their production and consumption as are the resulting global risks that impinge on their everyday lives. The question, then, is: how should we operationalize this conception of the world as a collection of different cultures and divergent modernities? Cosmopolitanization should be chiefly conceived of as globalization from *within*, as *internalized* cosmopolitanism. This is how we can suspend the assumption of the nation-state, and this is how we can make the empirical investigation of local–global phenomena possible. We can frame our questions so as to illuminate the transnationality that is arising inside nation-states. This is what a cosmopolitan sociology looks like.

[. . .]

 READING 41

Cosmopolitanism and Nationalism

Craig Calhoun

[. . .]

Cosmopolitanism has become an enormously popular rhetorical vehicle for claiming at once to be already global and to have the highest ethical aspirations for what globalisation can offer. It names a virtue of considerable importance. But, and these are my themes, it is not at all clear (a) that cosmopolitanism is quite so different from nationalism as sometimes supposed, (b) whether cosmopolitanism is really supplanting nationalism in global politics, and (c) whether cosmopolitanism is an ethical complement to politics, or in some usages a substitution of ethics for politics.

[. . .]

There are, however, three potential lines of confusion built into the idea of cosmopolitanism. We have noted two already. First, does it refer to what is common to the whole world and unites humanity? Or does it refer to appreciation of the differences among different groups and places? And second, does it refer to an individual attitude or ethical orientation, or does it refer to a condition of collective life? But confusion of the third sort is at least as common: cosmopolitanism is both description and normative program and the distinction is often unclear.

Indeed, part of the attraction of the idea of cosmopolitanism is that it seems to refer at once to a fact about the world – particularly in this era of globalisation – and to a desirable response to that fact. Ulrich Beck suggests that we should think of two linked processes. The growing interconnection of the world he calls 'cosmopolitanization'. He uses 'cosmopolitanism' for the attitude that treats these as a source of moral responsibility for everyone. But the very overlap in terminology suggests (despite occasional disclaimers) that one is automatically linked to the other. And this is not just an issue in Beck's writing but a wider feature of discourse about cosmopolitanism.

Clearly, neither the interconnectedness nor the diversity of the world brings pleasure to everyone. Growing global connections can become a source of fear and defensiveness rather than appreciation for diversity or sense of ethical responsibility for distant strangers. Globalisation can lead to renewed nationalism or strengthening of borders – as has often been the case since the 2001 terrorist attacks. But like many others Beck hopes that instead a cosmopolitan attitude will spread. He emphasises that risks such as environmental degradation turn the whole world into a 'community of fate'. Cosmopolitanism is, for him, the perspective on what humanity shares that will help us deal with this. Cosmopolitanism offers an ethics for globalisation.

[. . .]

It is impossible not to belong to social groups, relations, or culture. The idea of individuals abstract enough to be able to choose all their 'identifications' is deeply misleading. Versions of this idea are, however, widespread in liberal cosmopolitanism. They reflect the attractive illusion of escaping from social determinations into a realm of greater freedom, and from cultural particularity into greater universalism. But they are remarkably unrealistic, and so abstract as to provide little purchase on what the next steps of actual social action might be for real people who are necessarily situated in particular webs of belonging, with access to particular others but not to humanity in general. Treating ethnicity as *essentially* (rather than partially) a choice of identifications, they neglect the omnipresence of ascription (and discrimination) as determinations of social identities. They neglect the huge inequalities in the supports available to individuals to enter cosmopolitan intercourse as individuals (and also the ways in which certain socially distributed supports like wealth, education, and command of the English language are understood as personal achievements or attributes). And they neglect the extent to which people are implicated in social actions which they are not entirely free to choose (as, for example, I remain an American and share responsibility for the invasion of Iraq despite my opposition to it and distaste for the US administration that launched it). Whether blame or benefit follow from such implications, they are not altogether optional.

Cosmopolitanism seems to signal *both* the identity (and therefore unity) of all human beings despite their differences, *and* appreciation for and ability to feel at home among the actual differences among people and peoples. We focus sometimes on the essential similarity of people and sometimes on their diversity.

We should be careful not to imagine that either sort of cosmopolitanism is an immediately useful example for democracy. Modern democracy grew in close relationship to nationalism, as the ideal of self-determination demanded a strong notion of the collective self in question. Nationalism was also (at least often) an attempt to reconcile liberty and ethical universalism with felt community. This doesn't mean that we should not seek more cosmopolitan values, cultural knowledge, and styles of interpersonal relations in modern national democracy. It certainly doesn't mean that we should embrace reactionary versions of nationalism which have often been antidemocratic as well as anticosmopolitan. But it does mean that we need to ask some hard questions about how cosmopolitanism relates to the construction of political and social solidarities. Does cosmopolitanism actually underpin effective political solidarity, or only offer an attractive counterbalance to nationalism? How can we reconcile the important potential of multiple and hybrid cultural and social identities with political participation and rights? What is the relationship between valuing difference and having a strong enough commitment to specific others to sacrifice in collective struggle or accept democracy's difficult challenge of living in a minority and attempting only to persuade and not simply dominate others with whom one does not agree? It will not do simply to substitute ethics for politics, no matter how cosmopolitan and otherwise attractive the ethics. It will not do to imagine democratic politics without paying serious attention to the production of strong solidarity among the subjects of struggles for greater self-determination.

Many forms and visions of belonging are also responses to globalisation, not merely inheritances from time immemorial. Nations and national identities, for example, have been forged in international relations from wars to trade, in international migrations and among those who traveled as well as those who feared their arrival, and in pursuit of popular sovereignty against traditional rulers. Nationalism has often grown stronger when globalisation has intensified. Islam, Christianity, Buddhism and other religions arose in the contexts of empires and conflicts but also have been remade as frames of identity crossing nations and yet locating believers in a multireligious world. Religion has shaped globalisation not only as a source of conflict but of peacemaking. The significance of local community has repeatedly been changed by incorporation into broader structures of trade and association. And communal values have been articulated both to defend havens in a seemingly heartless world and to set examples for global imitation. While structures of belonging may be shaped by tradition, thus, we need to understand them not merely as traditional alternatives to modernity or cosmopolitanism but as important ways in which ordinary people have tried to take hold of modernity and to locate themselves in a globalising world.

In a broad, general sense cosmopolitanism is unexceptionable. Who – at least what sophisticated intellectual – could argue for parochialism over a broader perspective, for narrow sectarian loyalties over recognition of global responsibilities? Who could be against citizenship of the world? But the word 'citizenship' is a clue to the difficulty. Cosmopolitanism means something very different as a political project – or as the project of substituting universalistic ethics for politics – from what it means as a general orientation to difference in the world. And a central strand of political theory is now invested in hopes for cosmopolitan democracy, democracy not limited by nation-states. In the spirit of Kant as well as Diogenes, many say, people should see themselves as citizens of the world, not just of their countries. This requires escape from the dominance of a nationalist social imaginary (that is, a nationalist way of understanding what society is and constituting new political communities).

It is an escape that carries the risk of throwing the baby out with the bathwater. We should, I think, join in recognising the importance of transnational relations and therefore transnational politics, movements, and ethics. We should try to belong to the world as a whole and help it thrive, and be more just and better organised. But we should not imagine we can do so very well by ignoring or wishing away national and local solidarities. This is something I think the work of Ernest Gellner affirms. We need to be global in part through how we are national. And we need to recognise the ways national – and ethnic and religious – solidarities work for others. If we are among those privileged to transcend national identities and limits in our travel and academic conferences and reading and friendships we should nonetheless be attentive to the social conditions of our outlook and the situations of those who do not share our privileges.

McWorld and Jihad

In an argument first published as an article in *Atlantic Monthly* in 1992 and later in a book, *Jihad vs McWorld*, Benjamin Barber offered a hotly debated thesis on globalization. As the title suggests, Barber focuses on the antagonism between Jihad and McWorld. Whereas McWorld refers to the expansion of a unified political and cultural process that is progressively becoming omnipresent throughout the world, Jihad, as Barber defines it, is the localized, reactionary force that bolsters cultural parochialism, but which also leads to greater political heterogeneity throughout the world. However, he argues that Jihad must, in the end, yield to McWorld.

Jihad is usually associated with Islam, where it means the spiritual effort, struggle, and striving of the heart against vice, passion, and ignorance. It should be noted, however, that Barber does not restrict the use of the term to Islam; rather he uses it rhetorically to refer to groups that aim to redraw boundaries and reassert ethnic, racial, tribal, and/or religious identities. Thus, non-Islamic examples of Jihad include the Basque separatists, the Catholics of Northern Ireland, Québécois, and Puerto Ricans, among others.

By McWorld, on the other hand, Barber means the global businesses – McDonald's, Microsoft, Disney, Coca-Cola, Panasonic, Kentucky Fried Chicken, etc. – that are producing global homogeneity. He reduces all of them to "Mc" in order to highlight and to critique the culturally homogeneous nature of their products. McWorld suggests the threatening aspects of unregulated capitalism as well as the damage caused by the international corporations, which are committed to large and fast profits through their aggressive global expansion. Taken together, McWorld and Jihad symbolize a planet simultaneously being drawn together by communications and commerce and being split apart by the "Jihad" of fanatics and terrorists.

In Barber's view, McWorld is the true driving force in globalization and ultimately the more powerful of the two processes on which he focuses. A combination of economic expansionism and the spread of popular culture, McWorld produces a global marketplace in which the sovereignty of the nation-state is surrendered to transnational commerce. Barber argues that the requirements of the market lead McWorld to support international peace and stability, and to reduce the possibility of isolation and war, in order to make for greater efficiency, productivity, and profitability. Nonetheless, he believes that McWorld faces a series of challenges. Barber contends that market freedom does not necessarily mean democracy and that markets can only be a means, not an end. The economic efficiency of the market does not translate into democracy, full employment, dignity at work, environmental protection, and so on.

Barber contends that McWorld has not always delivered on its promises. Without regulation, markets can eliminate competition in many areas. This has various cultural and ideological implications. Of greatest importance is the possibility of the emergence of monopolies in the media which result in uniformity and censorship and which might help lead to totalitarianism. Further, it brings together diverse cultures and ethnicities under the heading of consumer culture, with the result that "consumer" and "person" become practically synonymous. For that reason, McWorld can be culturally manipulative and coercive; it replaces culturally unique societies with a global consumer society in which people consume the same goods and have the same symbols, lifestyles, and "so-called" necessities.

In the face of struggle between Jihad and McWorld, Barber does not foresee Jihad withstanding McWorld in the long run, especially Americanization, McDonaldization (see chapter 15), and Hollywoodization. The culture of McWorld is, to Barber, enmeshed in capitalism and this is more likely to mean the defeat of liberal democracy than its victory. Barber argues that although American consumerism is more democratic than most traditionalisms, a strong state has been the only sponsor of democratic freedoms and social equality.

Zakaria criticizes Barber's thesis on several grounds. He argues that Barber understands Islam as monolithic and inherently hostile to democracy. Zakaria also contends that Barber is ambivalent about nationalism. On the one hand, Barber argues that nationalism has an important role in history as it gives people a sense of belonging and self-determination, and he criticizes those who distort the concept. On the other hand, Zakaria states that Barber himself uses "a crude, xenophobic and inaccurate term to describe this complex phenomenon."[1] Zakaria maintains that the Jihad vs McWorld distinction is an oversimplification of a very complex world. According to Zakaria, Jihad is not discussed thoroughly, is used incorrectly, and is employed only as the antithesis of McWorld. Similarly, Zakaria argues that "McWorld" as a concept does not provide a substantial analysis of the global economy, international financial markets, or national governments.

Turner shares Zakaria's concerns about the use of Jihad and McWorld as all-encompassing concepts. Turner argues that Barber's presentation of Jihad and McWorld as uniform entities obscures the differences that exist within each. Furthermore, Turner criticizes the validity of "Jihad" as an explanatory category. He argues that by treating Islam as monolithic and inherently hostile to democracy, Barber fails to recognize the similarities between Islamic and Christian fundamentalism, equates fundamentalism with traditionalism in order to argue that fundamentalism is hostile to modernity, overlooks Islam's historical development with the West, and recreates a "friend or foe" perspective.[2]

Barber returns to his Jihad vs McWorld thesis in the post-9/11 world. He rejects the "war on terrorism" at least as it is being waged in light of the 9/11 attacks. Instead of a war against the jihadists, Barber urges a war, or at least a struggle, on behalf of democracy. The true enemy is what Barber calls "radical nihilists" and he believes they can be defeated by a variety of democratic movements already under way. In order to play a role in this, the US needs to change from a society dedicated to "wild capitalism and an aggressive secularism" and riddled with "social injustice";[3] it needs to tame its capitalism, ameliorate social injustice, and accept religion and civil society. He believes that 9/11 and its aftermath have given global democracy its moment.

Despite its shortcomings, Barber's Jihad vs McWorld thesis is, and continues to be, one of the noteworthy endeavors to explain the contemporary world in the light of globalization. By focusing on the role of culture, and discussing the links between consumption, markets, and democracy, Barber delivers one of the most hotly debated accounts of the globalization process.

NOTES

1 Fareed Zakaria, "Paris is Burning." *The New Republic*, January 22, 1996: 28.
2 Bryan S. Turner, "Sovereignty and Emergency: Political Theology, Islam and American Conservatism." *Theory, Culture and Society* 19, 4, 2002: 112.
3 Benjamin R. Barber, "On Terrorism and the New Democratic Realism." *The Nation*, January 21, 2002: 18.

Jihad vs McWorld

Benjamin R. Barber

Just beyond the horizon of current events lie two possible political futures – both bleak, neither democratic. The first is a retribalization of large swaths of humankind by war and bloodshed: a threatened Lebanonization of national states in which culture is pitted against culture, people against people, tribe against tribe – a Jihad in the name of a hundred narrowly conceived faiths against every kind of interdependence, every kind of artificial social cooperation and civic mutuality. The second is being borne in on us by the onrush of economic and ecological forces that demand integration and uniformity and that mesmerize the world with fast music, fast computers, and fast food – with MTV, Macintosh, and McDonald's, pressing nations into one commercially homogeneous global network: one McWorld tied together by technology, ecology, communications, and commerce. The planet is falling precipitantly apart *AND* coming reluctantly together at the very same moment.

These two tendencies are sometimes visible in the same countries at the same instant: thus Yugoslavia, clamoring just recently to join the New Europe, is exploding into fragments; India is trying to live up to its reputation as the world's largest integral democracy while powerful new fundamentalist parties like the Hindu nationalist Bharatiya Janata Party, along with nationalist assassins, are imperiling its hard-won unity. States are breaking up or joining up: the Soviet Union has disappeared almost overnight, its parts forming new unions with one another or with like-minded nationalities in neighboring states. The old interwar national state based on territory and political sovereignty looks to be a mere transitional development.

The tendencies of what I am here calling the forces of Jihad and the forces of McWorld operate with equal strength in opposite directions, the one driven by parochial hatreds, the other by universalizing markets, the one re-creating ancient subnational and ethnic borders from within, the other making national borders porous from without. They have one thing in common: neither offers much hope to citizens looking for practical ways to govern themselves democratically. If the global future is to pit Jihad's centrifugal whirlwind against McWorld's centripetal black hole, the outcome is unlikely to be democratic – or so I will argue.

McWorld, or the Globalization of Politics

Four imperatives make up the dynamic of McWorld: a market imperative, a resource imperative, an information-technology imperative, and an ecological imperative. By shrinking the world and diminishing the salience of national borders, these imperatives have in combination achieved a considerable victory over factiousness and particularism, and not least of all over their most virulent traditional form – nationalism. It is the realists who are now Europeans, the utopians who dream nostalgically of a resurgent England or Germany, perhaps even a resurgent Wales or Saxony. Yesterday's wishful cry for one world has yielded to the reality of McWorld.

The market imperative

Marxist and Leninist theories of imperialism assumed that the quest for ever-expanding markets would in time compel nation-based capitalist economies to push against national boundaries in search of an international economic imperium. Whatever else has happened to the scientist predictions of Marxism, in this domain they have proved farsighted. All national economies are now vulnerable to the inroads of larger, transnational markets within which trade is free, currencies are convertible, access to banking is open, and contracts are enforceable under law. In Europe,

Asia, Africa, the South Pacific, and the Americas such markets are eroding national sovereignty and giving rise to entities – international banks, trade associations, transnational lobbies like OPEC and Greenpeace, world news services like CNN and the BBC, and multinational corporations that increasingly lack a meaningful national identity – that neither reflect nor respect nationhood as an organizing or regulative principle.

The market imperative has also reinforced the quest for international peace and stability, requisites of an efficient international economy. Markets are enemies of parochialism, isolation, fractiousness, war. Market psychology attenuates the psychology of ideological and religious cleavages and assumes a concord among producers and consumers – categories that ill fit narrowly conceived national or religious cultures. Shopping has little tolerance for blue laws, whether dictated by pub-closing British paternalism, Sabbath-observing Jewish Orthodox fundamentalism, or no-Sunday-liquor-sales Massachusetts puritanism. In the context of common markets, international law ceases to be a vision of justice and becomes a workaday framework for getting things done – enforcing contracts, ensuring that governments abide by deals, regulating trade and currency relations, and so forth.

Common markets demand a common language, as well as a common currency, and they produce common behaviors of the kind bred by cosmopolitan city life everywhere. Commercial pilots, computer programmers, international bankers, media specialists, oil riggers, entertainment celebrities, ecology experts, demographers, accountants, professors, athletes – these compose a new breed of men and women for whom religion, culture, and nationality can seem only marginal elements in a working identity. Although sociologists of everyday life will no doubt continue to distinguish a Japanese from an American mode, shopping has a common signature throughout the world. Cynics might even say that some of the recent revolutions in Eastern Europe have had as their true goal not liberty and the right to vote but well-paying jobs and the right to shop (although the vote is proving easier to acquire than consumer goods). The market imperative is, then, plenty powerful; but, notwithstanding some of the claims made for "democratic capitalism," it is not identical with the democratic imperative.

The resource imperative

Democrats once dreamed of societies whose political autonomy rested firmly on economic independence. The Athenians idealized what they called autarky, and tried for a while to create a way of life simple and austere enough to make the polis genuinely self-sufficient. To be free meant to be independent of any other community or polis. Not even the Athenians were able to achieve autarky, however: human nature, it turns out, is dependency. By the time of Pericles, Athenian politics was inextricably bound up with a flowering empire held together by naval power and commerce – an empire that, even as it appeared to enhance Athenian might, ate away at Athenian independence and autarky. Master and slave, it turned out, were bound together by mutual insufficiency.

The dream of autarky briefly engrossed nineteenth-century America as well, for the underpopulated, endlessly bountiful land, the cornucopia of natural resources, and the natural barriers of a continent walled in by two great seas led many to believe that America could be a world unto itself. Given this past, it has been harder for Americans than for most to accept the inevitability of interdependence. But the rapid depletion of resources even in a country like ours, where they once seemed inexhaustible, and the maldistribution of arable soil and mineral resources on the planet, leave even the wealthiest societies ever more resource-dependent and many other nations in permanently desperate straits.

Every nation, it turns out, needs something another nation has; some nations have almost nothing they need.

The information-technology imperative

Enlightenment science and the technologies derived from it are inherently universalizing. They entail a quest for descriptive principles of general application, a search for universal solutions to particular problems, and an unswerving embrace of objectivity and impartiality.

Scientific progress embodies and depends on open communication, a common discourse rooted in rationality, collaboration, and an easy and regular flow and exchange of information. Such ideals can be hypocritical covers for power-mongering by elites, and they may be shown to be wanting in many other ways, but they are entailed by the very idea of science and they make science and globalization practical allies.

Business, banking, and commerce all depend on information flow and are facilitated by new communication technologies. The hardware of these technologies tends to be systemic and integrated – computer, television, cable, satellite, laser, fiber-optic, and microchip technologies combining to create a vast interactive communications and information network that can potentially give every person on earth access to every other person, and make every datum, every byte, available to every set of eyes. If the automobile was, as George Ball once said (when he gave his blessing to a Fiat factory in the Soviet Union during the Cold War), "an ideology on four wheels," then electronic telecommunication and information systems are an ideology at 186,000 miles per second – which makes for a very small planet in a very big hurry. Individual cultures speak particular languages; commerce and science increasingly speak English; the whole world speaks logarithms and binary mathematics.

Moreover, the pursuit of science and technology asks for, even compels, open societies. Satellite footprints do not respect national borders; telephone wires penetrate the most closed societies. With photocopying and then fax machines having infiltrated Soviet universities and *samizdat* literary circles in the eighties, and computer modems having multiplied like rabbits in communism's bureaucratic warrens thereafter, *glasnost* could not be far behind. In their social requisites, secrecy and science are enemies.

The new technology's software is perhaps even more globalizing than its hardware. The information arm of international commerce's sprawling body reaches out and touches distinct nations and parochial cultures, and gives them a common face chiseled in Hollywood, on Madison Avenue, and in Silicon Valley. Throughout the 1980s one of the most-watched television programs in South Africa was *The Cosby Show*. The demise of apartheid was already in production. Exhibitors at the 1991 Cannes film festival expressed growing anxiety over the "homogenization" and "Americanization" of the global film industry when, for the third year running, American films dominated the awards ceremonies. America has dominated the world's popular culture for much longer, and much more decisively.

In November of 1991 Switzerland's once insular culture boasted best-seller lists featuring *Terminator 2* as the No. 1 movie, *Scarlett* as the No. 1 book, and Prince's *Diamonds and Pearls* as the No. 1 record album. No wonder the Japanese are buying Hollywood film studios even faster than Americans are buying Japanese television sets. This kind of software supremacy may in the long term be far more important than hardware superiority, because culture has become more potent than armaments. What is the power of the Pentagon compared with Disneyland? Can the Sixth Fleet keep up with CNN? McDonald's in Moscow and Coke in China will do more to create a global culture than military colonization ever could. It is less the goods than the brand names that do the work, for they convey life-style images that alter perception and challenge behavior. They make up the seductive software of McWorld's common (at times much too common) soul.

Yet in all this high-tech commercial world there is nothing that looks particularly democratic. It lends itself to surveillance as well as liberty, to new forms of manipulation and covert control as well as new kinds of participation, to skewed, unjust market outcomes as well as greater productivity. The consumer society and the open society are not quite synonymous. Capitalism and democracy have a relationship, but it is something less than a marriage. An efficient free market after all requires that consumers be free to vote their dollars on competing goods, not that citizens be free to vote their values and beliefs on competing political candidates and programs. The free market flourished in junta-run Chile, in military-governed Taiwan and Korea, and, earlier, in a variety of autocratic European empires as well as their colonial possessions.

The ecological imperative

The impact of globalization on ecology is a cliche even to world leaders who ignore it. We know well enough that the German forests can be destroyed by Swiss and Italians driving gas-guzzlers fueled by leaded gas. We also know that the planet can be asphyxiated by greenhouse gases because Brazilian farmers want to be part of the twentieth century and are burning down tropical rain forests to clear a little land to plough, and because Indonesians make a living out of converting their lush jungle into toothpicks for fastidious Japanese diners, upsetting the delicate oxygen balance and in effect puncturing our global lungs. Yet this ecological consciousness has meant not only greater awareness but also greater inequality, as modernized nations try to slam the door behind them, saying to developing nations, "The world cannot afford your modernization; ours has wrung it dry!"

Each of the four imperatives just cited is transnational, transideological, and transcultural. Each applies impartially to Catholics, Jews, Muslims, Hindus, and Buddhists; to democrats and totalitarians; to capitalists and socialists. The Enlightenment dream of a universal rational society has to a remarkable degree been realized – but in a form that is commercialized, homogenized, depoliticized, bureaucratized, and, of course, radically incomplete, for the movement toward McWorld is in competition with forces of global breakdown, national dissolution, and centrifugal corruption. These forces, working in the opposite direction, are the essence of what I call Jihad.

Jihad, or the Lebanonization of the World

OPEC, the World Bank, the United Nations, the International Red Cross, the multinational corporation ... there are scores of institutions that reflect globalization. But they often appear as ineffective reactors to the world's real actors: national states and, to an ever greater degree, subnational factions in permanent rebellion against uniformity and integration – even the kind represented by universal law and justice. The headlines feature these players regularly: they are cultures, not countries; parts, not wholes; sects, not religions; rebellious factions and dissenting minorities at war not just with globalism but with the traditional nation-state. Kurds, Basques, Puerto Ricans, Ossetians, East Timoreans, Quebecois, the Catholics of Northern Ireland, Abkhasians, Kurile Islander Japanese, the Zulus of Inkatha, Catalonians, Tamils, and, of course, Palestinians – people without countries, inhabiting nations not their own, seeking smaller worlds within borders that will seal them off from modernity.

A powerful irony is at work here. Nationalism was once a force of integration and unification, a movement aimed at bringing together disparate clans, tribes, and cultural fragments under new, assimilationist flags. But as Ortega y Gasset noted more than sixty years ago, having won its victories, nationalism changed its strategy. In the 1920s, and again today, it is more often a reactionary and divisive force, pulverizing the very nations it once helped cement together. The force that creates nations is "inclusive," Ortega wrote in *The Revolt of the Masses*. "In periods of consolidation, nationalism has a positive value, and is a lofty standard. But in Europe everything is more than consolidated, and nationalism is nothing but a mania [...]"

This mania has left the post-Cold War world smoldering with hot wars; the international scene is little more unified than it was at the end of the Great War, in Ortega's own time. There were more than thirty wars in progress last year, most of them ethnic, racial, tribal, or religious in character, and the list of unsafe regions doesn't seem to be getting any shorter. Some new world order!

The aim of many of these small-scale wars is to redraw boundaries, to implode states and resecure parochial identities: to escape McWorld's dully insistent imperatives. The mood is that of Jihad: war not as an instrument of policy but as an emblem of identity, an expression of community, an end in itself. Even where there is no shooting war, there is fractiousness, secession, and the quest for ever smaller communities. Add to the list of dangerous countries those at risk: in Switzerland and Spain, Jurassian and Basque separatists still argue the virtues of ancient identities, sometimes in the language of bombs. Hyperdisintegration in the former Soviet Union may well continue unabated – not just a Ukraine independent from the Soviet Union but a Bessarabian Ukraine independent from the Ukrainian republic; not just Russia severed from the defunct union but Tatarstan severed from Russia. Yugoslavia makes even the disunited, ex-Soviet, nonsocialist republics that were once the Soviet Union look integrated, its sectarian fatherlands springing up within factional motherlands like weeds within weeds within weeds. Kurdish independence would threaten the territorial integrity of four Middle Eastern nations. Well before the current cataclysm Soviet Georgia made a claim for autonomy from the Soviet Union, only to be faced with its Ossetians (164,000 in a republic of 5.5 million) demanding their own self-determination within Georgia. The Abkhasian minority in Georgia has followed suit. Even the good will established by Canada's once promising Meech Lake protocols is in danger, with Francophone Quebec again threatening the dissolution of the federation. In South Africa the emergence from apartheid was hardly achieved when friction between Inkatha's Zulus and the African National Congress's tribally identified members threatened to replace Europeans' racism with an indigenous tribal war. After thirty years of attempted integration using the colonial language (English) as a unifier, Nigeria is now playing with the idea of linguistic multiculturalism – which could mean the cultural breakup of the nation into hundreds of tribal fragments. Even Saddam Hussein has benefited from the threat of internal Jihad, having used renewed tribal and religious warfare to turn last season's mortal enemies into reluctant allies of an Iraqi nationhood that he nearly destroyed.

The passing of communism has torn away the thin veneer of internationalism (workers of the world unite!) to reveal ethnic prejudices that are not only ugly and deep-seated but increasingly murderous. Europe's old scourge, anti-Semitism, is back with a vengeance, but it is only one of many antagonisms. It appears all too easy to throw the historical gears into reverse and pass from a Communist dictatorship back into a tribal state.

Among the tribes, religion is also a battlefield. ("Jihad" is a rich word whose generic meaning is "struggle" – usually the struggle of the soul to avert evil. Strictly applied to religious war, it is used only in reference to battles where the faith is under assault, or battles against a government that denies the practice of Islam. My use here is rhetorical, but does follow both journalistic practice and history.) Remember the Thirty Years War? Whatever forms of Enlightenment universalism might once have come to grace such historically related forms of monotheism as Judaism, Christianity, and Islam, in many of their modern incarnations they are parochial rather than cosmopolitan, angry rather than loving, proselytizing rather than ecumenical, zealous rather than rationalist, sectarian rather than deistic, ethnocentric rather than universalizing. As a result, like the new forms of hypernationalism, the new expressions of religious fundamentalism are fractious and pulverizing, never integrating. This is religion as the Crusaders knew it: a battle to the death for souls that if not saved will be forever lost.

The atmospherics of Jihad have resulted in a breakdown of civility in the name of identity, of comity in the name of community. International relations have sometimes taken on the aspect of gang war – cultural turf battles featuring tribal factions that were supposed to be sublimated as integral parts of large national, economic, postcolonial, and constitutional entities.

The Darkening Future of Democracy

These rather melodramatic tableaux vivants do not tell the whole story, however. For all their defects, Jihad and McWorld have their attractions. Yet, to repeat and insist, the attractions are unrelated to democracy. Neither McWorld nor Jihad is remotely democratic in impulse. Neither needs democracy; neither promotes democracy.

McWorld does manage to look pretty seductive in a world obsessed with Jihad. It delivers peace, prosperity, and relative unity – if at the cost of independence, community, and identity (which is generally based on difference). The primary political values required by the global market are order and tranquillity, and freedom – as in the phrases "free trade," "free press," and "free love." Human rights are needed to a degree, but not citizenship or participation – and no more social justice and equality than are necessary to promote efficient economic production and consumption. Multinational corporations sometimes seem to prefer doing business with local oligarchs, inasmuch as they can take confidence from dealing with the boss on all crucial matters. Despots who slaughter their own populations are no problem, so long as they leave markets in place and refrain from making war on their neighbors (Saddam Hussein's fatal mistake). In trading partners, predictability is of more value than justice.

The Eastern European revolutions that seemed to arise out of concern for global democratic values quickly deteriorated into a stampede in the general direction of free markets and their ubiquitous, television-promoted shopping malls. East Germany's Neues Forum, that courageous gathering of intellectuals, students, and workers which overturned the Stalinist regime in Berlin in 1989, lasted only six months in Germany's mini-version of McWorld. Then it gave way to money and markets and monopolies from the

West. By the time of the first all-German elections, it could scarcely manage to secure three percent of the vote. Elsewhere there is growing evidence that glasnost will go and perestroika – defined as privatization and an opening of markets to Western bidders – will stay. So understandably anxious are the new rulers of Eastern Europe and whatever entities are forged from the residues of the Soviet Union to gain access to credit and markets and technology – McWorld's flourishing new currencies – that they have shown themselves willing to trade away democratic prospects in pursuit of them: not just old totalitarian ideologies and command-economy production models but some possible indigenous experiments with a third way between capitalism and socialism, such as economic cooperatives and employee stock-ownership plans, both of which have their ardent supporters in the East.

Jihad delivers a different set of virtues: a vibrant local identity, a sense of community, solidarity among kinsmen, neighbors, and countrymen, narrowly conceived. But it also guarantees parochialism and is grounded in exclusion. Solidarity is secured through war against outsiders. And solidarity often means obedience to a hierarchy in governance, fanaticism in beliefs, and the obliteration of individual selves in the name of the group. Deference to leaders and intolerance toward outsiders (and toward "enemies within") are hallmarks of tribalism – hardly the attitudes required for the cultivation of new democratic women and men capable of governing themselves. Where new democratic experiments have been conducted in retribalizing societies, in both Europe and the Third World, the result has often been anarchy, repression, persecution, and the coming of new, noncommunist forms of very old kinds of despotism. During the past year, Havel's velvet revolution in Czechoslovakia was imperiled by partisans of "Czechland" and of Slovakia as independent entities. India seemed little less rent by Sikh, Hindu, Muslim, and Tamil infighting than it was immediately after the British pulled out, more than forty years ago.

To the extent that either McWorld or Jihad has a NATURAL politics, it has turned out to be more of an antipolitics. For McWorld, it is the antipolitics of globalism: bureaucratic, technocratic, and meritocratic, focused (as Marx predicted it would be) on the administration of things – with people, however, among the chief things to be administered. In its politico-economic imperatives McWorld has been guided by laissez-faire market principles that privilege efficiency, productivity, and beneficence at the expense of civic liberty and self-government.

For Jihad, the antipolitics of tribalization has been explicitly antidemocratic: one-party dictatorship, government by military junta, theocratic fundamentalism – often associated with a version of the *Führerprinzip* that empowers an individual to rule on behalf of a people. Even the government of India, struggling for decades to model democracy for a people who will soon number a billion, longs for great leaders; and for every Mahatma Gandhi, Indira Gandhi, or Rajiv Gandhi taken from them by zealous assassins, the Indians appear to seek a replacement who will deliver them from the lengthy travail of their freedom.

The Confederal Option

How can democracy be secured and spread in a world whose primary tendencies are at best indifferent to it (McWorld) and at worst deeply antithetical to it (Jihad)? My guess is that globalization will eventually vanquish retribalization. The ethos of material "civilization" has not yet encountered an obstacle it has been unable to thrust aside. Ortega may have grasped in the 1920s a clue to our own future in the coming millennium.

"Everyone sees the need of a new principle of life. But as always happens in similar crises – some people attempt to save the situation by an artificial intensification of the very principle which has led to decay. This is the meaning of the 'nationalist' outburst of recent years [...] things have always gone that way. The last flare, the longest; the last sigh, the deepest. On the very eve of their disappearance there is an intensification of frontiers – military and economic."

Jihad may be a last deep sigh before the eternal yawn of McWorld. On the other hand, Ortega was not exactly prescient; his prophecy of peace and internationalism came just before blitzkrieg, world war, and the Holocaust tore the old order to bits. Yet democracy is how we remonstrate with reality, the rebuke our aspirations offer to history. And if retribalization is inhospitable to democracy, there is nonetheless a form of democratic government that can accommodate parochialism and communitarianism, one that can even save them from their defects and make them more tolerant and participatory: decentralized participatory democracy. And if McWorld is indifferent to democracy, there is nonetheless a form of democratic government that suits global markets passably well – representative government in its federal or, better still, confederal variation.

With its concern for accountability, the protection of minorities, and the universal rule of law, a confederalized representative system would serve the political needs of McWorld as well as oligarchic bureaucratism or meritocratic elitism is currently doing. As we are already beginning to see, many nations may survive in the long term only as confederations that afford local regions smaller than "nations" extensive jurisdiction. Recommended reading for democrats of the twenty-first century is not the US Constitution or the French Declaration of Rights of Man and Citizen but the Articles of Confederation, that suddenly pertinent document that stitched together the thirteen American colonies into what then seemed a too loose confederation of independent states but now appears a new form of political realism, as veterans of Yeltsin's new Russia and the new Europe created at Maastricht will attest.

By the same token, the participatory and direct form of democracy that engages citizens in civic activity and civic judgment and goes well beyond just voting and accountability – the system I have called "strong democracy" – suits the political needs of decentralized communities as well as theocratic and nationalist party dictatorships have done. Local neighborhoods need not be democratic, but they can be. Real democracy has flourished in diminutive settings: the spirit of liberty, Tocqueville said, is local. Participatory democracy, if not naturally apposite to tribalism, has an undeniable attractiveness under conditions of parochialism.

Democracy in any of these variations will, however, continue to be obstructed by the undemocratic and antidemocratic trends toward uniformitarian globalism and intolerant retribalization which I have portrayed here. For democracy to persist in our brave new McWorld, we will have to commit acts of conscious political will – a possibility, but hardly a probability, under these conditions. Political will requires much more than the quick fix of the transfer of institutions. Like technology transfer, institution transfer rests on

foolish assumptions about a uniform world of the kind that once fired the imagination of colonial administrators. Spread English justice to the colonies by exporting wigs. Let an East Indian trading company act as the vanguard to Britain's free parliamentary institutions. Today's well-intentioned quick-fixers in the National Endowment for Democracy and the Kennedy School of Government, in the unions and foundations and universities zealously nurturing contacts in Eastern Europe and the Third World, are hoping to democratize by long distance. Post Bulgaria a parliament by first-class mail. Fed Ex the Bill of Rights to Sri Lanka. Cable Cambodia some common law.

Yet Eastern Europe has already demonstrated that importing free political parties, parliaments, and presses cannot establish a democratic civil society; imposing a free market may even have the opposite effect. Democracy grows from the bottom up and cannot be imposed from the top down. Civil society has to be built from the inside out. The institutional superstructure comes last. Poland may become democratic, but then again it may heed the Pope, and prefer to found its politics on its Catholicism, with uncertain consequences for democracy. Bulgaria may become democratic, but it may prefer tribal war. The former Soviet Union may become a democratic confederation, or it may just grow into an anarchic and weak conglomeration of markets for other nations' goods and services.

Democrats need to seek out indigenous democratic impulses. There is always a desire for self-government, always some expression of participation, accountability, consent, and representation, even in traditional hierarchical societies. These need to be identified, tapped, modified, and incorporated into new democratic practices with an indigenous flavor. The tortoises among the democratizers may ultimately outlive or outpace the hares, for they will have the time and patience to explore conditions along the way, and to adapt their gait to changing circumstances. Tragically, democracy in a hurry often looks something like France in 1794 or China in 1989.

It certainly seems possible that the most attractive democratic ideal in the face of the brutal realities of Jihad and the dull realities of McWorld will be a confederal union of semi-autonomous communities smaller than nation-states, tied together into regional economic associations and markets larger than nation-states – participatory and self-determining in local matters at the bottom, representative and accountable at the top. The nation-state would play a diminished role, and sovereignty would lose some of its political potency. The Green movement adage "Think globally, act locally" would actually come to describe the conduct of politics.

This vision reflects only an ideal, however – one that is not terribly likely to be realized. Freedom, Jean-Jacques Rousseau once wrote, is a food easy to eat but hard to digest. Still, democracy has always played itself out against the odds. And democracy remains both a form of coherence as binding as McWorld and a secular faith potentially as inspiriting as Jihad.

 READING 43

Paris Is Burning: *Jihad vs McWorld* by Benjamin R. Barber

Fareed Zakaria

[. . .]

Benjamin Barber is a professor of political philosophy at Rutgers University who has often written on big subjects. He is best known for his advocacy of "strong democracy." His book of that name, which appeared in 1984, was an argument for unmediated democratic politics. It advocated greater participation of all citizens in all aspects of social and political life; criticized communitarianism for its intolerance of individual choice and autonomy; and extolled civic education. But the book's animating purpose was an attack on America's distinctive political theory, liberal constitutionalism. As developed perhaps most importantly by James Madison, liberal constitutionalism seeks to tame the passions of direct democracy through various mediating mechanisms – delegated powers, deliberative representation, federal structures, and so on. For Barber, this was thinly veiled oligarchy. He rejected the very notion of mediation, dismissing – in the tradition of the American pragmatists John Dewey and Charles Pierce – all knowledge not grounded in experience. Again and again he quoted Rousseau's cry, "Once a people permits itself to be represented, it is no longer free."

Barber's new book could be read as a continuation of these themes. It, too, is deeply concerned about the fate of democracy. It, too, is littered with approving references to participation, civic education and (that most trendy Eden of all) civil society. It, too, quotes Rousseau often. On closer reading, however, *Jihad vs McWorld* is a wholesale refutation, unacknowledged or unwitting, of Barber's longstanding public philosophy. The most interesting and original parts of the new book comprise, at heart, a diatribe against the effects of unchecked participation by the masses.

More importantly, the book reflects a certain kind of unyielding leftism's final argument against the rise of liberal democratic capitalism. With political and economic critiques exhausted, what remains is an aesthetic case against capitalism, a strange exercise in the politics of taste.

The starting point of Barber's book is reasonable enough: the simultaneous rise of economic globalization and communal loyalties threaten the nation-state, from above and from below. Barber goes on to link the fate of the nation-state to the fate of democracy, which is his chief concern. "The modern nation-state has actually acted as a cultural integrator and has adapted well to pluralist ideas; civic ideologies and constitutional faiths around which their many clans and tribes can rally." If the state gets overwhelmed in its struggle with "Jihad" and "McWorld," Barber argues, our "post-industrial, post-national [. . .] [epoch] is likely also to be terminally postdemocratic."

Thus the book has two villains, who are the infelicitous entities of its title. "Jihad" is a metaphor, referring here not simply to the Islamic idea of a holy war, but to "dogmatic and violent particularism of a kind known to Christians no less than Muslims, to Germans and Hindis [sic] as well as to Arabs." Barber does discuss Islam, but he betrays more prejudice than knowledge. He seems to equate Islam with the Arabs (as the quotation above implies), imputing that region's political dysfunctions to that religion. The reader of Barber's book would not know, for example, that the four largest Muslim populations in the world are all outside the Middle East – in Indonesia, India, Pakistan and Bangladesh. Islam is a monolith, according to Barber, and one that is intrinsically inhospitable to democracy and "nurtures conditions favorable to parochialism, antimodernism, exclusiveness, and hostility to 'others.'" I guess Clinton didn't get to this part of the book.

Barber is quite ambivalent about nationalism. (And it is odd that he discusses nationalism in the context of "Jihad.") He recognizes nationalism's historical role in giving people a sense of belonging and self-determination. He scorns people who "use nationalism as a scathing pejorative," thus distorting "a far more dialectical concept." Yet he himself uses a crude, xenophobic and inaccurate term to describe this complex phenomenon. What on earth does "Jihad" have to do with the mood that he discerns in Occitan France, Spanish Catalonia, French Canada and German Switzerland?

Mercifully, the discussion of "Jihad" is short: a mere fifty of the book's 300 pages. This is partly because bashing "Jihad" isn't complicated. It does not take long to convince people in the West today that communal militancy is bad and a danger for democracy. But the real reason for the brevity of Barber's analysis of "Jihad" is that he seems to have decided that it is not really the problem after all. "Jihad" is simply a frightened reaction to the onslaught of "McWorld." It "tends the soul that McWorld abjures and strives for the moral well-being that McWorld [. . .] disdains." It becomes clear now that Barber's real enemy, his real obsession, is "McWorld."

Barber's discussion of "McWorld" is tough going. It is written in the breathless style of a futurologist, complete with invented words and obscure logic. Chapter four, for example, concludes: "This infotainment telesector is supported by hard goods, which in fact have soft entailments that help obliterate the hard/soft distinction itself." The book is studded with impressive-sounding, hollow lines such as this one:

"The dynamics of the Jihad–McWorld linkage are deeply dialectical."

Barber makes several arguments in the sections on "McWorld," not all of them consistent. "McWorld" itself is, variously, the global market, multinational business, rampant consumerism and global pop culture. Barber sometimes speaks in the gloomy tones of the declinists, suggesting that America barely survived the cold war, and then only by taking on a crippling national debt. Other times he speaks of an America poised to dominate the world economically and culturally. In some places he argues that the globalization of economics has created a world of multinational corporations that have no national character, but he also declares incessantly that "McWorld" is pervasively American – "'international' is just another way of saying global American." So what, exactly, is the problem: a weakening America in the midst of a nationless world, or American global hegemony?

The expansion of the global marketplace and its consequences is an important subject. It has spawned vigorous debates among political scientists and economists for decades. Which is to say, it is not as novel as Barber or some of the wide-eyed management consultants he cited think. The power of multinationals, for example, is not exactly a new phenomenon. India was colonized, in the seventeenth and eighteenth centuries, not by a country but by a multinational corporation, the British East India Company, which wielded financial, political and military powers that no modern-day corporation could ever have. Imagine Coca-Cola with its own army, its own courts, its own laws.

Scholarly studies with careful collections of data abound on topics such as foreign direct investment, home country controls and outsourcing – all of which complicate the simple picture of nation-states in decline in the face of global markets. International financial markets function smoothly, for example, owing to an elaborate regulatory structure created and sustained by national governments; but Barber pays little attention to such matters, filling his pages instead with a blizzard of anecdotes taken from the pages of newspapers and magazines. Many of his assertions about the frightening power of "McWorld" and its relentless thrust across the globe are supported by the evidence of [. . .] advertisements. Ralph Lauren's "Living Without Space" campaign to sell Safari perfume and Reebok's

"Planet Reebok" theme are illustrations of how "advertising colonizes space." By this method, I suppose, the gooey 1980s song "We Are the World" is proof of a new imperialism, with Quincy Jones its mastermind. In fact, pop music does worry Barber. In a chapter on MTV called "McWorld's Noisy Soul" there is an ominous two-page world map, reminiscent of the geopolitical primers of yore, that shows most of the globe shaded in grey. It turns out that all these countries receive music television. Where is Samuel Huntington when you need him?

Amid the din, however, one note can be heard throughout Barber's discussion: a distaste for "McWorld" in all its manifestations. Barber clearly abhors McDonald's, the evil empire itself, with its day-glo arches, plastic decor, factory food and tacky advertising. And McDonald's is merely the symbol for all large, consumer-based multinationals such as Coca-Cola, Pepsi, Nike, Reebok and Disney. Barber's vision is the sophisticated urbanite's suburban nightmare: "McWorld is an entertainment shopping experience that brings together malls, multiplex movie theaters, theme parks, spectator sports arenas, fast-food chains (with their endless movie tie-ins), and television (with its burgeoning shopping networks) into a single vast enterprise."

Barber hates the fact that the global consumer companies are destroying the delightful and quaint and individual cultures that one expects to see when one travels abroad. And the sinister new globalism has even hit France, the country he cherishes most. In the good old days, Barber lovingly recounts, "one ate nonpasteurised Brie and drank *vin de Provence* in cafés and brasseries that were archetypically French; one listened to Edith Piaf and Jacqueline Françoise on French national radio stations and drove 2CV Citroens and Renault sedans without ever leaving French roadways [. . .] An American in Paris crossed the waters to get away from TasteeFreez, White Castle and Chevrolet pickup trucks and once in France could be certain they would vanish." It is a novel objection to imperialism that it is ruining tourism.

I like many of the things that Barber likes – neighborhood stores, bistros, good food, good wine – but I try not to confuse my tastes with my politics. Barber misunderstands the phenomenon that he deplores. McDonald's and Coca-Cola and Nike and Disney have become so dominant because during the last hundred years, and especially during the last forty years, the industrialized world has seen a staggering rise in the standard of living of the average person. This means that vast numbers of people now have the time and the money to indulge in what used to be upper-class styles of life and leisure, most notably eating out and shopping. True, they eat and shop at places that Barber would not, but surely that is not the point.

The explosion of wealth and the rise of living standards, in what the Marxist historian E.J. Hobsbawm has called capitalism's "golden age," is among the most important social transformations in history. After thousands of years, more than a tiny percent of the population of these countries have some degree of material well-being. The recent debate over the very real problem of stagnating wages has made us forget how far we have come. A half century of peace and economic growth has created a new revolution of rising expectations. The average American family now consumes twice as many goods and services as in 1950. Then, less than 10 percent of Americans went to college; now, almost 60 percent do. The poorest fifth of the population of the United States consumes more today than the middle fifth did in 1955.

It is easy to demean the rise of mass consumption, as Barber does, mocking the individual "choice" that is reflected in the range of toppings on a baked potato or the variety of cereals in a supermarket. But this trivializes a remarkable phenomenon. Rising standards of living mean rising levels of hygiene, health and comfort. John Kenneth Galbraith, hardly a free-market ideologue, explained in 1967 that "no hungry man who is also sober can be persuaded to use his last dollar for anything but food. But a well-fed, well-clad, well-sheltered and otherwise well-tended person can be persuaded as between an electric razor and an electric toothbrush." When a middle-class person thinks of a house today, it has two bathrooms with heat and air conditioning in every room. This would have been considered prohibitively luxurious in 1950. Even measuring from 1973, when real wages began stagnating, standards of living have kept moving up. And the benefits are not mainly in the variety of cereal brands available. The number of cases of measles in America in 1974 was 22,094; it is now 312. A rising standard of living is not a form of corruption. It often represents an increase in the dignity of daily life.

McDonald's does look tawdry when compared to a Parisian bistro, but most of the customers at McDonald's, even at McDonald's in Paris, probably did not eat much in bistros before cheap fast-food restaurants appeared. Two generations ago, eating out was a luxury; today the average American eats out four times a week. (A weekly Big Mac is a fine expression of family values.) McDonald's and its look-alikes became successful because they offered ordinary people the convenience of eating out often and cheaply in sanitary (OK, antiseptic) conditions. And the rise of fast food has not exactly brought about the demise of fine dining. The world Barber likes is alive and well, but it is no longer central to society. Madonna looms larger in the general culture than Jessye Norman because more people listen to her sing: and in a democratic society it matters more how many listen than who listens. Indeed, *Jihad vs McWorld* can be read as a compendium of the social changes that a rising middle class has wrought on national cultures that were heretofore shaped by upper-class rituals and symbols.

A large part of Barber's discomfort with "McWorld" seems to stem from his discomfort with capitalism. This shows itself in two interesting ways. First, he litters his book with the usual paeans to civil society, by which he (like everyone else) means not all private groups, but the private groups he likes. Thus, conspicuously absent from his account of civil society are private firms. In fact, he sees corporations as actively hostile to civil society. "Who will get the private sector off the backs of civil society?"

As a professor of political theory, Barber must know that the concept of civil society emerged in Europe in the eighteenth century in part to describe private business activity. From Adam Ferguson and Adam Smith to Bernard Mandeville and David Hume, the philosophers who developed this idea spoke of the unintended good to society that results from selfish economic activity. In Mandeville's famous phrase, "private vice is public virtue." And leaving aside the matter of intellectual pedigree, how can one speak about organizations that provide individuals with personal autonomy and personal dignity, and shield them from the whims of the state, without mentioning private enterprise?

Second, individual choice that is exercised in a private economic sphere is, for Barber, somehow false. He celebrates the average person's hasty choice at the ballot box as genuine, but he scorns the careful decisions that the same person makes about where to work and live, what house or car to buy. These latter decisions, he implies, are forced on the unsuspecting consumer by omnipotent corporations. His book is sprinkled with calls for "real choice" and "genuine choice," but this is patronizing and unconvincing: he really means *choices like his*. The truth is that companies usually succeed when they cater to people's choices; and when they try to create people's choices, they often fail. Remember New Coke? Barber has his own nightmare backwards: it is the people of France, not the evil multinationals, who are abandoning French culture. If more Frenchmen ate in bistros and watched Louis Malle than eat at McDonald's and watch Arnold Schwarzenegger, there would be no slippage of French culture.

There is no denying that "McWorld" is not a pretty sight. The rise of a mass consumption society produces political, economic and cultural side effects that are troubling. But surely the criticism of this world, and of the liberal capitalism which created it, must first recognize its accomplishments. The political and economic changes that have created McWorld are, on the whole, admirable ones. Giving people the ability to live longer, to move where they want, own a house, to enjoy such pleasures as vacations and restaurants and shopping is good, even noble. And there is something distinctly unbecoming about an American intellectual disparaging the spread of American blandishments across the world. We like higher standards of living for ourselves, but we worry about their effects on others.

It is particularly strange to find that Barber, a man of the left, is so worried. After all, the left has been in favor of the goal of rising standards of living for the average person for centuries. In the pursuit of this goal it often made serious arguments against capitalism, questioning whether it was the right path. Some of these criticisms have proven wrong. (Communism, Lenin explained, would outproduce capitalism.) And some of these criticisms – relating to, say, income inequality – are important questions to this day. But Barber's book reflects a stubborn kind of leftism that has despaired of political and economic argument and, as a last resort, takes refuge in an aesthetic criticism of the market. This is the same leftism that produced Norman Mailer's one specific political position: tax plastic.

Barber's own work has been filled with paeans to ordinary people. He has championed measures that would give them greater autonomy and freedom of choice. But now that he is confronted by "the people's" actual – that is to say, tacky – choices, Barber wants them to choose differently. Indeed, he seems to desire what he has always denounced: elites, and the mediating institutions that try to cushion society from the direct effects of democracy while at the same time working to elevate people's judgments. Maybe Barber has discovered that there is something to be said, after all, for cultural leadership, for constraints on individual choice.

Sixty-three years ago, Ortega y Gasset, in *The Revolt of the Masses*, made a more intellectually honest argument against democratic capitalism, against the consequences of rampant and unfettered choice. He spoke directly to the Barbers of his time. "You want the ordinary man to be master," he wrote. "Well, do not be surprised if he acts for himself, if he demands all form of enjoyment, if he firmly asserts his will [. . .] if he considers his own person and his own leisure, if he is careful as to dress [. . .] Was it not this that it was hoped to do, namely, that the average man should feel himself master, lord and ruler of himself and of his life? Well, that is now accomplished. Why then these complaints of the liberals, the democrats, the progressives of thirty years ago? Or is it that like children they want something, but not the consequences of that something?"

 READING 44

Sovereignty and Emergency: Political Theology, Islam and American Conservatism

Bryan S. Turner

[. . .]

In 1995 Benjamin Barber presciently published *Jihad vs McWorld*, and republished it in 2001 [. . .] the book is essentially about the problem of democracy in the modern world, but the overt theme is the clash between the universal consumer world (McWorld) and the tribal world of identity politics and particularities (Jihad). The 'essential Jihad' is the fundamentalist movement of Islam, and the essence of McWorld is McDonald's. These two realities produce two radically different forms of politics. 'Jihad pursues a bloody politics of identity, McWorld a bloodless economics of politics'. Although in his 'Afterword' Barber retreats somewhat from an exclusive identification of Jihad with Islamic fundamentalism, the Islamic world does provide Barber with his most striking illustration of a committed or hot politics. By contrast, McWorld could be used to illustrate an argument from Oliver Wendell Holmes who, reflecting on the violence of the American civil war, came to the conclusion that conviction and certitude breed violence. McWorld is not a place for hot political emotions, but for steady pragmatic adjustments to contingencies.

The terrorist attack on New York can be taken as an illustration of Barber's dichotomy of politics in *Jihad vs McWorld*, because symbolically the World Trade Center towers perfectly embodied the cool systems of economic exchange that advanced capitalism had promoted against the hot politics of diasporic people and their cultures. Barber's typology is not merely a reproduction of Huntington's clash of civilizations in which once more Islam is chosen as a compelling illustration of an inevitable conflict between the West and the rest, or Fukuyama's reconstruction of the end-of-ideology thesis. Barber's analysis is in fact more complex and more interesting than Huntington's dichotomous model of endless conflict or Fukuyama's model of inevitable evolution towards liberal capitalism. For Barber, Jihad and McWorld stand in a dialectical relationship of mutual reinforcement. McWorld needs Jihad as its negative Other, while Jihad requires capitalism, or more specifically the United States, as its negative contrast. The cool universalism of McDonald's stands in a productive dialectic with the hot politics of Islamic Jihad, and yet at times they also interpenetrate each other. Jihad utilizes global technologies for its communication requirements and broadcasts its global message through modern media. Furthermore, Barber recognizes that American culture also produces jihadic politics in the form of radical Christian fundamentalism and violent militia men. McWorld and Jihad constantly intermingle and fuse with each other. He notes for example that Japan, in which national identity and national politics have been deeply preserved and fostered in the post-war period, has also embraced many components of western consumerism. In 1992, the number one restaurant in Japan as measured by the volume of consumers was McDonald's. Finally, his argument is constructed as a defence of democratic politics against both McWorld and Jihad. McDonald's undermines community and social capital, and thus erodes and corrodes the trust and communal membership that are essential foundations of secular democracy. The particularistic tribal mentality of Jihad is difficult to reconcile with democratic politics that requires compromise and cooperation between groups and communities that do not share the same ethnic identities. Democratic politics require a social space that has evolved beyond both McWorld and Jihad.

Although there are important political differences between Huntington, Fukuyama and Barber, their characterizations of Islam share a common set of assumptions. The result is the recreation of Orientalism. For example, Barber strongly identifies himself with democratic politics and his approach to the dialectic of McWorld and Jihad is summarized in the acknowledgements to his book when he says of Judith Shklar that she feared Jihad, distrusted McWorld and worked to make democracy possible. However, Barber's use of the term 'Jihad' is unfortunate because, unlike McWorld, it does refer to a specific institution within a given religious culture. Jihad in Islamic theology refers primarily to an internal spiritual struggle for self-mastery or self-overcoming, and its secondary meaning is a struggle against any threat to the integrity of Islam as a surrender to God. In the radical movement of the Muslim Brethren, Jihad was given a definite meaning of anti-colonial struggle by the Muslim teacher Sheikh Hassan al-Banna in the 1930s. While Barber struggles to separate 'Jihad' from 'Islam', the separation never quite works. Similarly, the attempts of western governments to separate 'Islam' from 'terrorism' are never

quite successful, partly because the media image of Muslims is now dominated by militarized images of armed Taliban.

Let us consider five criticisms of these American accounts of Islam. First, they fail typically to recognize the affinities between, for example, Protestant and Islamic fundamentalism. Second, they mistakenly identify fundamentalism with traditionalism in order to argue that fundamentalism is hostile to modernity. Third, western commentaries on fundamentalist Islam typically fail to consider the heterogeneity of contemporary Islamic belief. For example, the apparent triumph of fundamentalism has been challenged by many prominent liberal intellectuals in Islam and there is considerable opposition from radical Muslim women who are Islamist but reject the traditional seclusion of women, veiling and arranged marriages. Fourth, they treat Islam as an external and foreign religion without recognizing its historical development with the West. Finally, the creation of endless dichotomies between Islam and the West, or between Islam and modernity produces the division between foe and friend that follows directly from the political theology of Schmitt.

The first criticism is that in these American academic accounts 'fundamentalism' is undifferentiated and equated with 'militant Islam' of which the Taliban are the principal example. However, the arguments developed by Huntington and Fukuyama would apply equally to Jewish and Christian fundamentalism; they might also apply to Hindu nationalism, and to a range of socio-political movements where political and religious imagery are interwoven. Barber, by contrast, recognizes the global relationship between various types of fundamentalism and Jihad. There exists an 'American Jihad' of the Radical Right and American fundamentalist preachers like Jerry Falwell interpreted the attacks on New York and Washington as 'the wrath of God being vented on abortionists, homosexuals and the American Civil Liberties Union'. Barber's recognition that fundamentalism as a critique of modernity (McWorld) is also shared by Protestant fundamentalism in America represents a useful criticism of Huntington and Fukuyama. Huntington's account of the alienation of young men from modern society as a result of unemployment and under-achievement would be a powerful explanation of alienation and generational conflict in America and Europe.

Second, there is an assumption in these arguments that modernity is singular and uniform rather than plural and diverse. Contemporary ethnographic research, by contrast, has shown how modernity, postmodernity and tradition are completely interconnected in everyday life. For example, in contemporary ethnographic studies we find that Turkish women routinely integrate the Qur'an and *tefsir* collections from famous Sufi sheikhs with textbooks by Foucault, Habermas and Sontag. Because their daily activities combine intense prayer with political topics and pop music, they are not confronted by the inexorable dialectic of Jihad versus McWorld. In his brilliant study of social change in Morocco and Lebanon, Michael Gilsenan showed how the religious orders of Sufism, in adapting to urbanization and nationalist politics, have combined both traditional folk religiosity and modern cultural themes. These cultural hybrid systems lend support to S.N. Eisenstadt's argument that there are multiple modernities, and in particular that westernization is not identical with modernization.

Third, it is odd to regard Islam as a tribal or particularistic social movement, given the fact that Islam as a political movement challenges existing political structures precisely because it does not sit easily within nation-state boundaries. The case of Shi'ite Islam is in this respect important, since it does not recognize state authority but only the final authority of the hidden Imamate. We should not regard Islamic fundamentalism as anti-modern, because the implication of this opposition is to equate fundamentalism with traditionalism. Islamic fundamentalism has specifically criticized and rejected traditional Islam, which is seen as a principal source of weakness in the face of modernization. There are sociological arguments in favour of regarding puritanical forms of biblical fundamentalism – whether Christian or Islamic or Jewish – as sources of modernity in opposition to traditional patterns of spiritual mysticism. If we take [Weber's] *The Protestant Ethic and the Spirit of Capitalism* at all seriously, then fundamentalism stands at the roots of the ethos of McWorld in a much more generic and intimate fashion than Barber recognizes. Religious fundamentalism is the (often unintended) harbinger of austere modernity, and fundamentalism is the principal foundation of hostility to tradition. Islamic fundamentalism has been as much opposed to traditional religiosity, such as

Sufi mysticism, as it has been to the corruption of western consumerism. Fundamentalism as an ideal type involves a return to cultural roots in order to reform the present against the aberrations of the immediate past. Its cultural genre involves literalism towards (biblical) texts and typically an antagonism towards (baroque) decoration. Fundamentalism promotes personal asceticism against both mysticism and consumerist hedonism. The thrust of Islam has not been missionary in the same way that Christianity spread throughout the world, but any recognition of Islam as a 'world religion' pays tribute to its universalistic message. Indeed, many accounts of Islam, both internal and external, would recognize its primary commitments to equality and justice rather than to individual salvation. These theological distinctions should not be exaggerated, but one aspect of the dynamic nature of Islam in the modern world is this sense of universal justice.

A fourth criticism would be to argue that McWorld and Jihad should be treated technically as ideal types rather than descriptions of actual patterns of economic and religious organization. One problem with Barber's account is that it obscures the heterogeneity of Islam in actual societies. With globalization, Islam has itself become a diverse and complex cluster of cultures rather than a monolithic religious system. The differences between Shi'ism and Sunni Islam are well known, but the dispersion of Muslim cultures through migration has created a variety of diasporic forms of Islamic culture. The Muslim diaspora has resulted in a significant internal debate about, for example, the authority of traditional religious leadership and sacred texts, and similarly a variety of women's movements in Muslim societies have produced both internal debate and cultural change. The combined effect of these movements has been internal heterogeneity. Defining Islam as the foe has to deny or mask the wide range of distinctive cultural movements within Muslim societies. In military terms, there is a specific problem in Afghanistan where political alliances and networks between warlords change so rapidly and consistently that constituting a consistent foe is impossible.

Finally, the distinction between 'cool' McWorld and 'hot' Jihad is merely a re-description of Weber's ideal type distinction between open-universal (associative) and closed-particularistic (communal) social relationships.

The examples of associative and communal relationships in *Economy and Society* were indeed market and family. The sociological issue is that, whereas familial relations are affective, committed and particular, exchange relationships between strangers in the market place require neutrality, coolness and generality if trade is to prosper. The theoretical implication is that capitalism or McWorld, unlike other social systems, is not bellicose, but it is not the case that McWorld has no connection with state violence. Although under most circumstances McWorld does not fight wars directly, capitalist enterprises typically enter into military relations indirectly through the state or occasionally through mercenary forces. Capitalism is deeply involved in financing and profiting from wars, and indeed one obvious consequence of 9/11 was an improvement in share values of companies that are involved in military production. It is also the case that, at least in places such as Colombia, mercenary troops are financed directly by American business. The relationship between bloody Jihad and bloodless economics is more complex and more dialectical than Barber allows. This problem of what we might call the pragmatics of war and trade is equally at issue with respect to the Taliban, whose military machine has been largely financed, on the one hand, by western governments who wanted them to topple the Russian invasion, and by the global trade in heroin, on the other.

These comments on the distinction between markets and tribes raise a more important and serious problem in the debate about Islam and the West, namely the issue concerning the sovereignty of nation-states and the impact of globalization on state politics. Before 11 September, there was a consensus that globalization constrained the political autonomy of the state and that, in the long term, globalization might result in the decline of nation-states and the rise of global governance. The modern state might eventually give way to 'cosmopolitan governance'. The terrorist attack on New York and the offensive of the Northern Alliance in Afghanistan have raised an important question around the optimistic view of the development of global politics after the erosion of the nation-state. The current military conflict has clearly demonstrated that the United States is the only power that can wage a global war, and that American foreign policy needs will largely dictate the development of global governance. The consequence is to underscore the fact that international relations are primarily defined by conflicts between nation-states, and that America can shape global developments through its superior military and economic powers. The growth of the Russian oil economy, with the support of the US administration, indicates that America will not remain dependent on either the Saudi government or OPEC. Finally, the dominance of America over media systems (despite the importance of 'glocalism', alternative media systems and competition from al-Jazeera TV) has given the American government powerful control over the global presentation of the conflict.

The identification of militant Islam with the foe assumes, amongst other things, that Islam is the Other and that it is an external alien force. However, this view of Islam has to ignore the fact that Islam, partly through global migration, has become an important part of the cultural and economic life of the West. These negative images of Islam can be interpreted as aspects of a revival of Orientalism, but the paradox of these negative images is that, as a result of migration and globalization, Muslim communities have settled and evolved in most western industrial societies, where they constitute an important element of the labour force. It is estimated that there are 16 million Muslims in Europe, and Islam as a faith is also well established in the United States, where it is estimated that its adherents number between 1 and 6 million. The proximity and interpenetration of cultures are suppressed by the dependence on exclusionary dichotomies in the Orientalist vision of global divisions. In his *Representations of the Intellectual*, Edward Said complained that 'cultures are too intermingled, their contents and histories too interdependent and hybrid, for surgical separation into large and mostly ideological oppositions like Orient and Occident'. The foe/friend distinction has to remain largely oblivious or indifferent to Said's modest observations about cultural hybridity, and the futility of the Orient/Occident separation.

[. . .]

 READING 45

On Terrorism and the New Democratic Realism

Benjamin R. Barber

[. . .]

Can Asian tea, with its religious and family "tea culture," survive the onslaught of the global merchandising of cola beverages? Can the family sit-down meal survive fast food, with its focus on individualized consumers, fuel-pit stop eating habits and nourishment construed as snacking? Can national film cultures in Mexico, France or India survive Hollywood's juggernaut movies geared to universal teen tastes rooted in hard violence and easy sentiment? Where is the space for prayer, for common religious worship or for spiritual and cultural goods in a world in which the 24/7 merchandising of material commodities makes the global economy go round? Are the millions of American Christian families who home-school their children because they are so intimidated by the violent commercial culture awaiting the kids as soon as they leave home nothing but an American Taliban? Do even those secular cosmopolitans in America's coastal cities want nothing more than the screen diet fed them by the ubiquitous computers, TVs and multiplexes?

Terror obviously is not an answer, but the truly desperate may settle for terror as a response to our failure even to ask such questions. The issue for *jihad*'s warriors of annihilation is of course far beyond such anxieties: it entails absolute devotion to absolute values. Yet for many who are appalled by terrorism but unimpressed by America, there may seem to be an absolutist dimension to the materialist aspirations of our markets. Our global market culture appears to us as both voluntary and wholesome; but it can appear to others as both compelling (in the sense of compulsory) and corrupt – not exactly coercive, but capable of seducing children into a willed but corrosive secular materialism. What's wrong with Disneyland or Nikes or the Whopper? We just "give people what they want." But this merchandiser's dream is a form of romanticism, the idealism of neoliberal markets, the convenient idyll that material plenty can satisfy spiritual longing so that fishing for profits can be thought of as synonymous with trolling for liberty.

It is the new democratic realist who sees that if the only choice we have is between the mullahs and the mall, between the hegemony of religious absolutism and the hegemony of market determinism, neither liberty nor the human spirit is likely to flourish. As we face up to the costs both of fundamentalist terrorism and of fighting it, must we not ask ourselves how it is that when we see religion colonize every other realm of human life we call it theocracy and turn up our noses at the odor of tyranny; and when we see politics colonize every other realm of human life we call it absolutism and tremble at the prospect of totalitarianism; but when we see market relations and commercial consumerism try to colonize every other realm of human life we call it liberty and celebrate its triumph? There are too many John Walkers who begin by seeking a refuge from the aggressive secularist materialism of their suburban lives and end up slipping into someone else's dark conspiracy to rid the earth of materialism's infidels. If such men are impoverished and without hope as well, they become prime recruits for *jihad*.

The war on terrorism must be fought, but not as the war of McWorld against *jihad*. The only war worth winning is the struggle for democracy. What the new realism teaches is that only such a struggle is likely to defeat the radical nihilists. That is good news for progressives. For there are real options for democratic realists in search of civic strategies that address the ills of globalization and the insecurities of the millions of fundamentalist believers who are neither willing consumers of Western commercial culture nor willing advocates of jihadic terror. Well before the calamities of September 11, a significant movement in the direction of constructive and realistic interdependence was discernible, beginning with the Green and human rights movements of the 1960s and 1970s, and continuing into the NGO and "antiglobalization" movements of the past few years. Jubilee 2000 managed to reduce Third World debt-service payments for some nations by up to 30 percent, while the Community of Democracies initiated by the State Department under Madeleine Albright has been embraced by the Bush Administration and will continue to sponsor meetings of democratic governments and democratic NGOs. International economic reform lobbies like the Millennium Summit's development goals project, established by the UN to provide responses to global poverty, illiteracy and disease; Inter Action, devoted to increasing foreign aid; Global Leadership, a start-up alliance of corporations and grassroots organizations; and the Zedillo Commission, which calls on the rich countries to devote 0.7 percent of their GNP to development assistance (as compared to an average of 0.2 percent today and under 0.1 percent for the United States), are making serious economic reform an issue for governments. Moreover, and more important, they are insisting with Amartya Sen and his new disciple Jeffrey Sachs that development requires democratization first if it is to succeed.

George Soros's Open Society Institute and Civicus, the transnational umbrella organization for NGOs, continue to serve the global agenda of civil society.

Even corporations are taking an interest: hundreds are collaborating in a Global Compact, under the aegis of UN Secretary General Kofi Annan, to seek a response to issues of global governance, while the World Economic Forum plans to include fifty religious leaders in a summit at its winter meeting in New York in late January.

This is only a start, and without the explicit support of a more multilateralist and civic-minded American government, such institutions are unlikely to change the shape of global relations. Nonetheless, in closing the door on the era of sovereign independence and American security, anarchic terrorism has opened a window for those who believe that social injustice, unregulated wild capitalism and an aggressive secularism that leaves no space for religion and civil society not only create conditions on which terrorism feeds but invite violence in the name of rectification. As a consequence, we are at a seminal moment in our history – one in which trauma opens up the possibility of new forms of action. Yesterday's utopia is today's realism; yesterday's realism, a recipe for catastrophe tomorrow. If ever there was one, this is democracy's moment. Whether our government seizes it will depend not just on George Bush but on us.

Culture

Part II of this book opens with an excerpt from the work of Jan Nederveen Pieterse in which he differentiates among three theories of cultural globalization: *differentialism*, *hybridization*, and *convergence*. While these theories are treated here under the heading of culture, they have much broader applicability to many topics covered in this book. It could be argued, for example, that nation-states throughout the world remain stubbornly different ("differentialism"), are growing increasingly alike ("convergence"), or involve more and more combinations of various political forms drawn from many different parts of the world ("hybridization"). In fact, differentialism has already been covered in chapter 2 in a discussion of Huntington's work on civilizations. While this work is discussed earlier because of its political aspects and implications, it could also have been discussed here because civilization can be seen as "culture writ large." In spite of their broader applicability, the focus here will be on the other two types of cultural theory and the ways in which they relate to global culture.

What makes all of these theories particularly attractive is that they relate to the focal concern in the definition of globalization offered in the introduction to this book with flows and barriers. However, they take very different positions on them and their relationship to one another. In *differentialism*, the focus is much more on barriers that prevent flows that would serve to make cultures (and much else) more alike. In this view, cultures tend to remain stubbornly different from one another. In the *convergence* perspective, the barriers are much weaker and the global flows stronger, with the result that cultures are subject to many of the same flows and tend to grow more alike. In its extreme form, convergence suggests the possibility that local cultures can be overwhelmed by other, more powerful, cultures, or even a globally homogeneous culture. Finally, in the *hybridization* perspective, external flows interact with internal flows in order to produce a unique cultural hybrid that combines elements of the two. Barriers to external cultural flows exist in the hybridization perspective, and while they are strong enough to prevent those flows from overwhelming local culture, they are not strong enough to block all external cultural flows entirely. That which does succeed in gaining entry mixes with local culture to produce unique cultural hybrids.

While differentialism has already been covered, this part of the book deals with perspectives on hybridization and convergence, as well as the debates which surround them. We begin with hybridization which encompasses work on three different but closely related ideas: creolization, hybridity, and glocalization (chapter 13). This is followed by a selection of works (chapter 14) devoted to critiques of these ideas. This part of the book closes with two perspectives – McDonaldization (chapter 15) and world culture (chapter 16) – that represent the convergence perspective ("McWorld", covered in chapter 12, could also be included in this category).

Globalization and Culture: Three Paradigms

Jan Nederveen Pieterse

Globalization or the trend of growing worldwide interconnectedness has been accompanied by several clashing notions of cultural difference. The awareness of the world "becoming smaller" and cultural difference receding coincides with a growing sensitivity to cultural difference. The increasing salience of cultural difference forms part of a general cultural turn, which involves a wider self-reflexivity of modernity. Modernization has been advancing like a steamroller, erasing cultural and biological diversity in its way, and now not only the gains (rationalization, standardization, control) but also the losses (alienation, disenchantment, displacement) are becoming apparent. Stamping out cultural diversity has been a form of disenchantment of the world.

Yet it is interesting to note how the notion of cultural difference itself has changed form. It used to take the form of *national* differences, as in familiar discussions of national character or identity. Now different forms of difference have come to the foreground, such as gender and identity politics, ethnic and religious movements, minority rights, and indigenous peoples. Another argument is that we are experiencing a "clash of civilizations." In this view, cultural differences are regarded as immutable and generating rivalry and conflict. At the same time, there is a widespread understanding that growing global interconnectedness leads toward increasing cultural standardization and uniformization, as in the global sweep of consumerism. A shorthand version of this momentum is McDonaldization. A third position, altogether different from both these models of intercultural relations, is that what is taking place is a process of cultural mixing or hybridization across locations and identities.

This is a meta-theoretical reflection on cultural difference that argues that there are three, and only three, perspectives on cultural difference: cultural differentialism or lasting difference, cultural convergence or growing sameness, and cultural hybridization or ongoing mixing. Each of these positions involves particular theoretical precepts and as such they are paradigms. Each represents a particular *politics of difference* – as lasting and immutable, as erasable and being erased, and as mixing and in the process generating new translocal forms of difference. Each involves different subjectivities and larger perspectives. The first view, according to which cultural difference is immutable, may be the oldest perspective on cultural difference. The second, the thesis of cultural convergence, is as old as the earliest forms of universalism, as in the world religions. Both have been revived and renewed as varieties of modernism, respectively in its romantic and Enlightenment versions, while the third perspective, hybridization, refers to a postmodern sensibility of traveling culture. This chapter discusses the claims of these perspectives, their wider theoretical assumptions, and asks what kind of futures they evoke. Arguably there may be other takes on cultural difference, such as indifference, but none have the scope and depth of the three perspectives outlined here.

Clash of Civilizations

In 1993 Samuel Huntington, as president of the Institute for Strategic Studies at Harvard University, published a controversial paper in which he argued that "a crucial, indeed a central, aspect of what global politics is likely to be in the coming years [. . .] will be the clash of civilizations [. . .] With the end of the Cold War, international politics moves out of its Western phase, and its centerpiece becomes the interaction between the West and non-Western civilizations and among non-Western civilizations."

The imagery is that of civilizational spheres as tectonic plates at whose fault lines conflict, no longer subsumed under ideology, is increasingly likely. The argument centers on Islam: the "centuries-old military interaction between the West and Islam is unlikely to decline." "Islam has bloody borders." The fault lines include Islam's borders in Europe (as in former Yugoslavia), Africa (animist or Christian cultures to the south and west), and Asia (India, China). Huntington warns against a "Confucian-Islamic military connection" that has come into being in the form of arms flows between East Asia and the Middle East. Thus "the paramount axis of world politics will be the relations between 'the West and the Rest'" and "a central focus of conflict for the immediate future will be between the West and several Islamic-Confucian states." He therefore recommends greater cooperation and unity in the West, between Europe and North America; the inclusion of Eastern Europe and Latin America in the West; cooperative relations with Russia and Japan; exploiting differences and conflicts among Confucian and Islamic states; and for the West to maintain its economic and military power to protect its interests.

The idea of dividing the world into civilizations has a long lineage. In Europe, it goes back to the medieval understanding of a tripartite world of descendants of the three sons of Noah. Arnold Toynbee's world history divided the world into civilizational spheres. It informs the approach of the "Teen Murti" school of Contemporary Studies in Delhi. Kavolis divides the world into seven incommensurable civilizational systems based on religion: Christian, Chinese (Confucian-Taoist-Buddhist), Islamic, Hindu, Japanese (Shinto-Buddhist-Confucian), Latin American syncretism, and non-Islamic African. Galtung argues that each civilization has different ways of knowing the world. Dividing the world into civilizations is a cliché that echoes in every encyclopedia of world history; but it is also old fashioned and overtaken by new historiography and the emergence of "world history."

Huntington's position stands out for its blatant admixture of security interests and a crude rendition of civilizational difference. In view of its demagogic character it obviously belongs to the genre of "new enemy" discourse. In fact, it merges two existing enemy discourses, the "fundamentalist threat" of Islam and the "yellow peril," and its novelty lies in combining them.

Huntington recycles the Cold War: "The fault lines between civilizations are replacing the political and ideological boundaries of the Cold War as the flash points for crisis and bloodshed." "The Velvet Curtain of culture has replaced the Iron Curtain of ideology as the most significant dividing line in Europe." Hence there will be no "peace dividend." The Cold War is over but war is everlasting. This has been referred to as a new politics of containment and a new round of hegemonic rivalry, which is translated from an ideological into a civilizational idiom. Huntington's thesis has given rise to extensive debate and his argument has been widely rejected while acknowledging that its contribution has been to present culture as a significant variable in international relations. Huntington has developed his thesis in a book and followed up with a wider treatment of culture. I will not reiterate the debate here but bring up key points that show Huntington's view as one of three paradigms of cultural difference.

Huntington constructs the West as a "universal civilization," "directly at odds with the particularism of most Asian societies and their emphasis on what distinguishes one people from another." The charge against "the Rest" is that they attempt modernization without westernization. This may be the actual danger: the specter of *different modernities* and thus the breakdown of western civilizational hegemony. By now, multiple modernities are an accepted theme.

The geopolitics is odd. Significant arms flows between the Middle East and East Asia do not involve Islamic countries but Israel and its arms sales to China, which have been of particular concern to the US because they re-export high-tech equipment of US origin. Another instance, which Huntington does cite, exchanges of military technology between Pakistan and China, also involves an American angle. Major concerns from an American security point of view, such as military relations between China and Iran (and more recently, arms exports from North Korea), are not mentioned.

What is overlooked in this geopolitical construction are the dialectics of the Cold War and the role the United States has been playing. It's not so much a matter of civilizational conflict as the unraveling of geopolitical security games most of which have been initiated by the US in the first place, which the hegemon in its latter days can no longer control, so it calls on allied states to help channel them in a desirable direction. At the turn of the century, the British Empire in its latter days of waning economic and military power did the same, calling on the United States to "police" the Pacific, the Caribbean, and Latin America, on Japan to play a naval role in the China Sea, and to contain the Russian empire, and seeking allies in the European concert of powers. Then as now, the waning hegemon calls on "civilizational" affinities: the White Man's Burden and his civilizing mission, and now "democracy," freedom, and the virtues of the free market.

The sociologist Malcolm Waters formulates an interesting theorem according to which "material exchanges localize, political exchanges international-ize and symbolic exchanges globalize." This is difficult to maintain because it ignores how microeconomic dynamics at the level of firms propel the macroeco-nomic process of globalization; but interesting in this context is the view that the cultural, symbolic sphere is the first to globalize; a perspective diametrically opposed to Huntington's thesis. This shows the oddity of Huntington's view: it is a *political* perspective on culture coined in conventional national security lan-guage. Culture is politicized, wrapped in civilizational packages that just happen to coincide with geopolitical entities. Obviously, there is much slippage along the way and all along one wonders: what is national security doctrine doing in a world of globalization and in the sphere of cultural representations? While Huntington focuses on fault lines between civilizations, his pes-simism is matched by gloomy views on growing ethnic conflict.

Indeed the most remarkable element of the thesis is its surface claim of a clash of *civilizations*. Why is *cul-ture* being presented as the new fault line of conflict? Huntington's framework is a fine specimen of what he blames Asian societies for: "Their emphasis on what distinguishes one people from another." At a general level, this involves a very particular way of reading culture. Compare Immanuel Wallerstein on "Culture as the ideological battleground of the modern world-system": note that culture and ideology are being merged in a single frame, and that culture is defined as "the set of characteristics which distinguish one group from another." Anthony King uses a similar concept of culture as "collective articulations of human diversity."

If we would take this to its ultimate consequence then, for instance, bilingualism cannot be "cultural"

because "it does not distinguish one group from another." Indeed any bicultural, intercultural, multi-cultural, or transcultural practices could not according to this definition be "cultural." Whichever mode of communication or intercourse different groups would develop to interact with one another would not be cultural for culture refers only to intergroup diversity. We have thus defined any form of intergroup or transnational culture out of existence for such per definition cannot exist. Intercultural diffusion through trade and migration, a lingua franca between cultures, returnees from abroad with bicultural experience, children of mixed parentage, travelers with multi-cultural experience, professionals interacting cross-culturally, the fields of cyberspace – all of these fall outside "culture."

Obviously, this notion of culture is one-sided to the point of absurdity. Diversity is one side of the picture but only one, and interaction, commonality or the possibility of commonality is another. In anthro-pology this is cultural relativism and Ruth Benedict's view of cultures as single wholes – a Gestalt or configuration that can only be understood from within and in its own terms. It implies a kind of "billiard ball" model of cultures as separate, impene-trable units (similar to the way states have been repre-sented in the realist view of international relations). Over time, this generated ethnomethodology, eth-nosociology, and a trend toward the indigenization of knowledge. This is an anomalous definition of culture. More common a definition in anthropology is that culture refers to behavior and beliefs that are learned and shared: learned so it is not "instinctual" and shared so it is not individual. Sharing refers to social sharing but there is no limitation as to the boundaries of this sociality. No territorial or historical boundaries are implied as part of the definition. This understanding of culture is open-ended. Learning is always ongoing as a function of changing circum-stances and therefore culture is always *open*. To sharing there are no fixed boundaries other than those of common social experience, therefore there are no territorial limitations to culture. Accordingly culture refers as much to commonality as to diversity. [. . .] I refer to these fundamentally different notions of culture as territorial culture and translocal culture.

Cultural relativism represents an angle on culture that may be characterized as *culturalist differentialism*. Its lineages are ancient. They are as old as the Greeks who deemed non-Greek speakers "barbarians." Next, this took the form of immutable cultural difference based on religion, separating the faithful from heathens, unbelievers and heretics. The romantics such as Johann Gottfried Herder revived this view of strong cultural boundaries, now in the form of language as the key to nationhood. Both nationalism and race thinking bear the stamp of cultural differentialism, one emphasizing territory and language, and the other biology as destiny. Nation and race have long been twin and at times indistinguishable discourses. During the era of nationalism, all nations claimed cultural distinction for their own nation and inferiority for others, usually in racial terms. "Jewishness," "Germanness," "Japaneseness," "Englishness," "Turkishness," "Greekness," and so forth, all imply an inward-looking take on culture and identity. They are creation myths of modern times. They all share the problem of boundaries: who belongs, and since when?

Cultural differentialism can serve as a defense of cultural diversity. It may be evoked by local groups resisting the steamroller of assorted "developers," by ecological networks, anthropologists, and artists, as well as travel agencies and advertisers promoting local authenticity. Culture and development, a growing preoccupation in development thinking, may turn "culture" into an asset. It calls to mind the idea of the "human mosaic." An upside of this perspective may be local empowerment; the downside may be a politics of nostalgia, a conservationist posture that ultimately leads to the promotion of open-air museums. Either way the fallacy is the reification of the local, sidelining the interplay between the local and the global. The image of the mosaic is biased, as the anthropologist Ulf Hannerz points out, because a mosaic consists of fixed, discrete pieces whereas human experience, claims and postures notwithstanding, is fluid and open-ended. Accordingly critical anthropology opts for deterritorialized notions of culture such as flows and "traveling culture."

Huntington's thesis is at odds with the common self-understandings of East and Southeast Asian societies, which run along the lines of East–West fusion, as in "Western technology, Asian values." The Confucian ethic may carry overtones of East Asian chauvinism but also represents an East–West nexus of a kind because the neo-Confucianism it refers to owes its status to its reinterpretation as an "Asian Protestant ethic." While Confucianism used to be the reason why East Asian countries were stagnating, by the late twentieth century it has become the reason why the "Tigers" have been progressing. In the process, Confucianism has been recoded as a cross-cultural translation of the Weberian thesis of the Protestant ethic as the "spirit of modern capitalism." The Confucian ethic carries some weight in the "Sinic" circle of Singapore, Taiwan, China, and Korea; it carries less weight in Japan and no weight among the advocates of an "Asian way" such as Prime Minister Mahathir Mohamad of Malaysia and his "Look East" program. Given the tensions between the ethnic Chinese and the "bumiputra" Malays in Malaysia, just as in Indonesia, here an Islamic-Confucian alliance is the least likely option.

While Huntington reproduces standard enemy images of "the Rest," he also rehearses a standard self-image of the West. "The West" is a notion conditioned by and emerging from two historical polarities: the North–South polarity of the colonizing and colonized world, and the East–West polarity of capitalism–communism and the Cold War. These were such overriding fields of tension that differences *within* the West/North, *among* imperialist countries and *within* capitalism faded into the background, subsiding in relation to the bigger issue, the seeming unity of imperialist or neocolonial countries and of the "free world" led by the US. In view of this expansionist history, we might as well turn the tables and say: the West has bloody borders. Thus, Huntington practices both Orientalism and Occidentalism. In reinvoking "the West," the differences between North America and Europe are papered over. In fact, historical revision may well show that there are much greater historical affinities, in particular similar feudal histories with their attendant consequences for the character of capitalisms, between Europe and Asia than between Europe and North America.

In his usual capacity as a comparative political scientist, Huntington observes a worldwide "third wave" of democratization. Apparently, at this level of discourse civilizational differences *are* receding. In this domain, Huntington follows the familiar thesis of convergence, that is, the usual modernization paradigm of growing worldwide standardization around the model of the "most advanced country," and his position matches Fukuyama's argument of the universal triumph of the idea of liberal democracy.

McDonaldization

The McDonaldization thesis is a version of the recent idea of the worldwide homogenization of societies through the impact of multinational corporations. McDonaldization, according to the sociologist George Ritzer, is "the process whereby the principles of the fast-food restaurant are coming to dominate more and more sectors of American society as well as the rest of the world." The expression "the rest of the world" bears contemplating. The process through which this takes place is rationalization in Weber's sense, that is, through formal rationality laid down in rules and regulations. McDonald's formula is successful because it is efficient (rapid service), calculable (fast and inexpensive), predictable (no surprises), and controls labor and customers.

McDonaldization is a variation on a theme: on the classical theme of universalism and its modern forms of modernization and the global spread of capitalist relations. Diffusionism, if cultural diffusion is taken as emanating from a single center (e.g., Egypt), has been a general form of this line of thinking. From the 1950s, this has been held to take the form of Americanization. Since the 1960s, multinational corporations have been viewed as harbingers of American modernization. In Latin America in the 1970s, this effect was known as Coca-colonization. These are variations on the theme of cultural imperialism, in the form of consumerist universalism or global media influence. This line of thinking has been prominent in media studies according to which the influence of American media makes for global cultural synchronization.

Modernization and Americanization are the latest versions of westernization. If colonialism delivered Europeanization, neocolonialism under US hegemony delivers Americanization. Common to both is the modernization thesis, of which Marx and Weber have been the most influential proponents. Marx's thesis was the worldwide spread of capitalism. World-system theory is a current version of this perspective. With Weber, the emphasis is on rationalization, in the form of bureaucratization and other rational social technologies. Both perspectives fall within the general framework of evolutionism, a single-track universal process of evolution through which all societies, some faster than others, are progressing – a vision of universal progress such as befits an imperial world. A twentieth-century version of this line of thinking is Teilhard de Chardin's evolutionary convergence towards the noosphere.

Shannon Peters Talbott examines the McDonaldization thesis through an ethnography of McDonald's in Moscow and finds the argument inaccurate on every score. Instead of efficiency, queuing (up to several hours) and lingering are commonplace. Instead of being inexpensive, an average McDonald's meal costs more than a third of a Russian worker's average daily wage. Instead of predictability, difference and uniqueness attract Russian customers, while many standard menu items are not served in Moscow. Instead of uniform management control, McDonald's Moscow introduces variations in labor control ("extra fun motivations," fast service competitions, special hours for workers to bring their families to eat in the restaurant) and in customer control by allowing customers to linger, often for more than an hour on a cup of tea, to "soak up the atmosphere."

She concludes that McDonald's in Moscow does not represent cultural homogenization but should rather be understood along the lines of *global localization*. This matches the argument in business studies that corporations, also when they seek to represent "world products," only succeed if and to the extent that they adapt themselves to local cultures and markets. They should become insiders; this is the principle of "insiderization" for which the late Sony chairman Akio Morita coined the term "*glocalization*," or "looking in both directions." Firms may be multinational but "all business is local."

This can lead to counterintuitive consequences, as in the case of the international advertising firm McCann

Erickson, whose Trinidad branch to justify a local presence promotes Trinidadian cultural specificity. "The irony is, of course, that [. . .] it is advertising including transnational agencies which have become the major investors in preserving and promoting images of local specificity, retaining if not creating the idea that Trinidad is different, and inculcating this belief within the population at large." The profitability of the transnational firm hinges on the profitability of the branch office whose interest lies in persuading the firm that only local advertising sells.

So far, this only considers the angle of the corporation. The other side of global localization is the attitude of customers. The McDonald's Moscow experience compares with adaptations of American fast food principles elsewhere, for instance in East Asia. Here fast food restaurants though outwardly the same as the American models serve quite different tastes and needs. They are not down-market junk food but cater to middle class tastes. They are sought out for their "modern" aesthetics, are appreciated for food variation rather than uniformity, and generate "mixed" offspring, such as "Chinglish" or "Chamerican" restaurants in China. They offer a public space, a meeting place – in a sense culturally neutral because of its novelty – for new types of consumers, such as the consumer market of the young, of working women, and of middle class families. They function in similar ways in southern Europe and the Middle East. In wintry Tokyo, upstairs in Wendy's young students spend hours doing their homework, smoking and chatting with friends, because Japanese houses are small.

Thus, rather than cultural homogenization McDonald's and others in the family of western fast food restaurants (Burger King, KFC, Pizza Hut, Wendy's) usher in difference and variety, giving rise to and reflecting new, mixed social forms. Where they are imported, they serve different social, cultural, and economic functions than in their place of origin, and their formula is accordingly adapted to local conditions. In western metropoles, we now see oriental fast food restaurants and chains along with Latino, Middle Eastern, Turkish, and French eateries. Fast food may well have originated outside the West, in the street side food stalls of the Middle East, Asia, and Africa. American fast-food restaurants serve German food (hamburgers, frankfurters) with French (fries, dressing) and Italian elements

(pizza) in American management style. American contributions besides ketchup are assembly-line standardization, in American Taylorist and managerial traditions, and marketing. Thus, it would make more sense to consider McDonaldization as a form of intercultural hybridization, partly in its origins and certainly in its present globally localizing variety of forms.

McDonaldization has sparked growing resistance and wide debate. In its home country, McDonald's is past its peak, its shares declining and franchises closing. Obesity as a national disease and changing diets, saturation of the fast food market, resistance, and litigation contribute to the decline. Beyond "rationalization" this takes us to the shifting shapes of contemporary capitalism. Is contemporary capitalism a homogenizing force? A stream of studies examines the cultures of late capitalism, a problematic often structured by world system thinking or at least vocabulary. The commodification of labor, services, and information takes myriad forms, under headings each of which are another lament: McJobs, McInformation, McCitizens, McUniversity, McTourism, McCulture, McPrisons, McCourts. One study seeks "to intervene in discourses on transnational capitalism whose tendency is to totalize the world system," but in the process finds that "capitalism has proceeded not through global homogenization but through differentiation of labor markets, material resources, consumer markets, and production operations." The economist Michael Storper finds a combined effect of homogenization and diversification across the world:

> The loss of "authentic" local culture in these places [smaller US cities] is a constant lament. But on the other hand, for the residents of such places – or of Paris, Columbus, or Belo Horizonte, for that matter – there has been an undeniable increase in the variety of material, service, and cultural outputs. In short, the perceived loss of diversity would appear to be attributable to a certain rescaling of territories: from a world of more internally homogeneous localities, where diversity was found by traveling between places with significantly different material cultures to a world where one travels between more similar places but finds increasing variety within them.

Most studies of capitalism and culture find diverse and hybrid outcomes. This suggests that capitalism itself hosts more diversity than is usually assumed – so the appropriate analytic would rather be capitalisms; and its cultural intersections are more diverse than is generally assumed. The rhizome of capitalism twins then with the rhizome of culture, which brings us to the theme of hybridization.

Hybridization: the Rhizome of Culture

Mixing has been perennial as a process but new as an imaginary. As a perspective, it differs fundamentally from the previous two paradigms. It does not build on an older theorem but opens new windows. It is fundamentally excluded from the other two paradigms. It springs from the taboo zone of race thinking because it refers to that which the doctrines of racial purity and cultural integrism could not bear to acknowledge the existence of: the half-caste, mixed-breed, métis. If it was acknowledged at all, it was cast in diabolical terms. Nineteenth-century race thinking abhorred mixing because, according to Comte de Gobineau and many others, in any mixture the "lower" element would predominate. The idea of mixing goes against all the doctrines of *purity* as strength and sanctity, ancient and classical, of which "race science" and racism have been modern, biologized versions.

Hybridization is an antidote to the cultural differentialism of racial and nationalist doctrines because it takes as its point of departure precisely those experiences that have been banished, marginalized, tabooed in cultural differentialism. It subverts nationalism because it privileges border-crossing. It subverts identity politics such as ethnic or other claims to purity and authenticity because it starts out from the fuzziness of boundaries. If modernity stands for an ethos of order and neat separation by tight boundaries, hybridization reflects a postmodern sensibility of cut'n'mix, transgression, subversion. It represents, in Foucault's terms, a "resurrection of subjugated knowledges" because it foregrounds those effects and experiences which modern cosmologies, whether rationalist or romantic, would not tolerate.

Hybridization goes under various aliases such as syncretism, creolization, métissage, mestizaje, crossover. Related notions are global ecumene, global localization, and local globalization. [...] Hybridization may conceal the asymmetry and unevenness in the process and the elements of mixing. Distinctions need to be made between different times, patterns, types, and styles of mixing; besides mixing carries different meanings in different cultural settings.

Hybridization occurs of course also among cultural elements and spheres *within* societies. In Japan, "Grandmothers in kimonos bow in gratitude to their automated banking machines. Young couples bring hand-held computer games along for romantic evenings out." Is the hybridization of cultural styles then typically an urban phenomenon, a consequence of urbanization and industrialization? If we look into the countryside virtually anywhere in the world, we find traces of cultural mixing: the crops planted, planting methods and agricultural techniques, implements and inputs used (seeds, fertilizer, irrigation methods, credit) are usually of translocal origin. Farmers and peasants throughout the world are wired, direct or indirect, to the fluctuations of global commodity prices that affect their economies and decision-making. The ecologies of agriculture may be local, but the cultural resources are translocal. Agriculture is a prime site of globalization.

An interesting objection to the hybridization argument is that what are actually being mixed are cultural *languages* rather than *grammars*. The distinction runs between surface and deep-seated elements of culture. It is, then, the folkloric, superficial elements of culture – foods, costumes, fashions, consumption habits, arts and crafts, entertainments, healing methods – that travel, while deeper attitudes and values, the way elements hang together, the structural ensemble of culture, remain contextually bound. There are several implications to this argument. It would imply that contemporary "planetarization" is a surface phenomenon only because "deep down" humanity remains divided in historically formed cultural clusters. Does this also imply that the new social technologies of telecommunication – from jet aircraft to electronic media – are surface phenomena only that don't affect deep-seated attitudes? If so, the implications would be profoundly conservative. A midway position is that the new technologies are profound in themselves while each historically framed culture develops its own takes on the new spaces of commonality.

Another issue is immigrant and settler societies where intermingling over time represents a historical momentum profound enough to engage cultural grammar and not just language. A prime example is North America. Probably part of the profound and peculiar appeal of American popular culture is precisely its mixed and "traveling" character, its "footloose" lightness, unhinged from the feudal past. In this culture, the grammars of multiple cultures mingle, and this intercultural density may be part of the subliminal attraction of American popular media, music, film, television: the encounter, and often enough the clash, but an intimate clash, of ethnicities, cultures, histories. The intermingling of cultural grammars then makes up the deeply human appeal of American narratives and its worldly character, repackaging elements that came from other shores, in a "Mississippi Massala."

Intercultural mingling itself is a deeply creative process not only in the present phase of accelerated globalization but stretching far back in time. Cees Hamelink notes: "The richest cultural traditions emerged at the actual meeting point of markedly different cultures, such as Sudan, Athens, the Indus Valley, and Mexico."

This sheds a different light on the language/grammar argument: presumably, some grammars have been mingling all along. Thus, a mixture of cultural grammars is part of the intrinsic meaning of the world religions (as against tribal, national religions). More fundamentally, the question is whether the distinction between cultural language and cultural grammar can be maintained at all, as a distinction between surface and depth. Certainly we know that in some spheres nothing has greater depth than the surface. This is the lesson taught by art and aesthetics. Superficial mingling then may have deep overtones. Even so we have been so trained and indoctrinated to think of culture in territorial packages of assorted "imagined communities" that to seriously address the windows opened and questions raised by hybridization in effect requires a decolonization of imagination.

A schematic précis of the three paradigms of cultural difference is in table 1.

Futures

The futures evoked by these three paradigms are dramatically different. McDonaldization evokes both

Table 1 Three ways of seeing cultural difference

Dimensions	Differentialism	Convergence	Mixing
Cosmologies	Purity	Emanation	Synthesis
Analytics	Territorial culture	Cultural centers and diffusion	Translocal culture
Lineages	Differences in language, religion, region. Caste.	Imperial and religious universalisms. Ancient "centrisms."	Cultural mixing of technologies, languages, religions
Modern times	Romantic differentialism. Race thinking, chauvinism. Cultural relativism.	Rationalist universalism. Evolutionism. Modernization. Coca-colonization.	Métissage, hybridization, creolization, syncretism
Present	"Clash of civilizations." Ethnic cleansing. Ethnodevelopment.	McDonaldization, Disneyfication, Barbiefication. Homogenization.	Postmodern views of culture, cultural flows, crossover, cut'n'mix
Futures	A mosaic of immutably different cultures and civilizations	Global cultural homogeneity	Open-ended ongoing mixing

a triumphalist Americanism and a gloomy picture of a global "iron cage" and global cultural disenchantment. The clash of civilizations likewise offers a horizon of a world of iron, a deeply pessimistic politics of cultural division as a curse that dooms humanity to lasting conflict and rivalry; the world as an archipelago of incommunicable differences, the human dialogue as a dialogue of war, and the global ecumene as an everlasting battlefield. The political scientist Benjamin Barber in *Jihad vs. McWorld* presents the clash between these two perspectives without giving a sense of the third option, mixing. Mixing or hybridization is open-ended in terms of experience as well as in a theoretical sense. Its newness means that its ramifications over time are not predictable because it doesn't fit an existing matrix or established paradigm but itself signifies a paradigm shift.

Each paradigm represents a different politics of *multiculturalism*. Cultural differentialism translates into a policy of closure and apartheid. If outsiders are let in at all, they are preferably kept at arm's length in ghettos, reservations, or concentration zones. Cultural communities are best kept separate, as in colonial "plural society" in which communities are not supposed to mix except in the marketplace, or as in gated communities that keep themselves apart. Cultural convergence translates into a politics of assimilation with the dominant group as the cultural center of gravity. Cultural mixing refers to a politics of integration without the need to give up cultural identity while cohabitation is expected to yield new cross-cultural patterns of difference. This is a future of ongoing mixing, ever-generating new commonalities and new differences.

At a deeper level, each paradigm resonates with particular sensibilities and cosmologies. The paradigm of differentialism follows the principle of *purity*, as in ritual purity in the caste system, the *limpieza de sangre* in Spain after the Reconquest, and the preoccupation with purity of blood and lineage among aristocracies, a concern that was subsequently translated into thinking about "race" and class. The paradigm of convergence follows the theory of *emanation*, according to which phenomena are the outward expressions of an ultimate numinous realm of being. In its sacred version, this reflects a theology and cosmogony of emanation outward from a spiritual center of power (as in Gnosticism). What follows upon the cycle of emanation, dissemination, and divergence is a cycle of "in-gathering," or a process of convergence. A temporal reflection of this cosmology is the ancient imperial system in which the empire is the circumference of the world and the emperor its center (as in the case of the Pharaoh, the emperor of China as the "middle of the middle kingdom," and imperial Rome) and divine kingship, in which the king embodies the land and the people. Western imperialism and its *mission civilisatrice* or White Man's Burden was a variation on this perspective. Since decolonization, the principle of radiation outward from an imperial center has retained its structure but changed its meaning, from positive to negative, as in dependency theory and the critique of cultural imperialism and Eurocentrism.

The third view is the synthesis that acts as the solvent between these polar perspectives. As such, it owes its existence to the previous two principles and is meaningful only in relation to them. It resolves the tension between purity and emanation, between the local and the global, in a dialectic according to which the local is in the global and the global is in the local. An example in which we see this synthetic motion in operation is Christmas: "The ability of this festival to become potentially the very epitome of globalization derives from the very same quality of easy syncretism which makes Christmas in each and every place the triumph of localism, the protector and legitimation for specific regional and particular customs and traditions."

Each paradigm involves a different take on *globalization*. According to cultural differentialism, globalization is a surface phenomenon only: the real dynamic is regionalization, or the formation of regional blocs, which tend to correspond with civilizational clusters. Therefore, the future of globalization is interregional rivalry. According to the convergence principle, contemporary globalization is westernization or Americanization writ large, a fulfillment in installments of the classical imperial and the modernization theses. According to the mixing approach, the outcome of globalization processes is open-ended and current globalization is as much a process of easternization as of westernization, as well as of many interstitial influences.

In the end it turns out that the two clashing trends noted at the beginning, growing awareness of cultural difference and globalization, are not simply contradictory but interdependent. Growing awareness of cultural difference is a function of globalization. Increasing cross-cultural communication, mobility, migration, trade, investment, tourism, all generate awareness of cultural difference. The other side of the politics of difference is that the very striving for recognition implies a claim to equality, equal rights, same treatment: in other words, *a common universe of difference*. Accordingly, the clash between cultural diversity and globalization may well be considered a creative clash.

These views find adherents in each setting and their dispute echoes in every arena. Arguably, cultural self-understandings and empirical evidence confirm the third perspective more than the others do. Through most of Asia, ideas of East-West fusion are a dominant motif. In Africa, recombinations of local and foreign practices are a common notion. Latin America and the Caribbean are steeped in syncretism and creolization. But the imprint of other paradigms runs deep, disputes over identity and meaning are ubiquitous, and besides there is disagreement over the meaning and dynamics of hybridity.

Creolization, Hybridity, and Glocalization

While we present three key, interrelated ideas in this chapter, there are many others that present much the same idea, including "collage, mélange, hotchpotch, montage, synergy, bricolage . . . mestizaje, mongrelization, syncretism, transculturation, third cultures."[1]

Glocalization can be defined as the interpenetration of the global and the local resulting in unique outcomes in different geographic areas. Based on the work of Roland Robertson,[2] the essential elements of the perspective on globalization adopted by those who emphasize glocalization are that the world is growing more pluralistic (glocalization theory is exceptionally alert to differences within and between areas of the world); individuals and local groups have great power to adapt, innovate, and maneuver within a glocalized world (glocalization theory sees individuals and local groups as important and creative agents); social processes are relational and contingent (globalization provokes a variety of reactions – ranging from nationalist entrenchment to cosmopolitan embrace – that produce glocalization); and commodities and the media are seen *not* as (totally) coercive, but rather as providing material to be used in individual and group creation throughout the glocalized areas of the world.

A discussion of some closely related terms (and related examples) will be of considerable help in getting a better sense of glocalization, as well as the broader issue of

cultural hybridization.[3] Of course, *hybridization* itself is one such term, emphasizing increasing diversity associated with the unique mixtures of the global and the local as opposed to the tendency toward *uniformity* often associated with globalization. A cultural hybrid involves the combination of two or more elements from different cultures and/or parts of the world. Among the examples of hybridization (and heterogenization, glocalization) are Ugandan tourists visiting Amsterdam to watch Moroccan women engage in Thai boxing, Argentinians watching Asian rap performed by a South American band at a London club owned by a Saudi Arabian, and the more mundane experiences of Americans eating such concoctions as Irish bagels, Chinese tacos, Kosher pizza, and so on. Obviously, the list of such hybrids is long and growing rapidly with increasing hybridization. The contrast, of course, would be such uniform experiences as eating hamburgers in the United States, quiche in France, or sushi in Japan.

Yet another concept that is closely related to glocalization is *creolization*. The term "creole" generally refers to people of mixed race, but it has been extended to the idea of the creolization of language and culture involving a combination of languages and cultures that were previously unintelligible to one another.[4]

Given this general overview, we turn to a summary of the seminal works on this topic presented in this chapter. The popularity of the term "creolization" is traceable to Ulf Hannerz's 1987 essay, "The World in Creolisation." As he defines it, "creole cultures . . . are those which draw in some way on two or more historical sources, often originally widely different."[5] In some cases creole cultures can be based on internal differences such as rural–urban differences, the division of labor, the division of knowledge, and so on.

In developing this idea, Hannerz takes on ideas associated with cultural homogeneity. Instead of producing homogeneity, Hannerz sees the world system as a new source of diversity. Foreign cultural influence can be destructive, but it can also "give people access to technological and symbolic resources for dealing with their own ideas, managing their own culture, in new ways."[6] Hannerz puts this in the context of "conversations" among cultures. The creolist perspective leads to the view "that the different cultural streams engaging one another in creolisation may all be actively involved in shaping the resultant forms . . . active handling of meanings of various local and foreign derivations."[7]

Jan Nederveen Pieterse's work is closely related to the idea of hybridization. One definition of the term, with a focus on culture, is "the ways in which forms become separated from existing practices and recombine with new forms in new practices."[8] In a later work, Nederveen Pieterse associates the term with that which "denotes a wide register of multiple identity, crossover, pic-'n'-mix, boundary-crossing experiences and styles, matching a world of growing migration and diaspora lives, intensive intercultural communication, everyday multiculturalism and erosion of boundaries."[9]

Nederveen Pieterse also distinguishes between cultural and structural hybridization. In terms of the latter, this involves the concept of hybridization "extended to structural forms of social organization."[10] Structural hybridization can lead to an alteration in, even a weakening of, the nation-state and the national economy.[11] For example, migrants to a given nation can live in it, but engage in "long distance nationalism"[12] with their country of origin (and even those in other nations who come from that country). This weakens the nation-state in which they live since the migrants do not owe their allegiance, or at least some of it, to that country.

Such structural hybridization can be compared to the period between the 1840s and the 1960s when the nation-state predominated as a structural form. In recent decades we have seen a proliferation in the modes of organization including "transnational, international, macro-regional, national, micro-regional, municipal, local."[13] Each of these may exist on its own, or they may exist in a multitude of permutations and combinations without any single one having priority or exercising a monopoly over the others.

Beyond the structures, the informal open spaces that emerge within the interstices between these structures are also important. These interstices are inhabited by "diasporas, migrants, nomads, exiles, stateless people, etc."[14] Also relevant here are border zones, world cities, and ethnic mélange neighborhoods. They are all meeting points of structures of a variety of different types. Thus, multiple cultures and identities are paralleled by a multitude of organizational forms. This focus on structures demonstrates Nederveen Pieterse's more sociological approach to hybridity in contrast to Hannerz's more anthropological approach to creolization focusing on culture and cultural differences.

Nederveen Pieterse also deals with the impact of non-Western cultures on the West and the production in the latter of both hybrid culture and structures (a global mélange). In this context, he sees the emergence of global crossover cultures including examples such as the Irish bagels mentioned above. All of this indicates, of course, that hybridity applies to culture just as it does to structure.

Nederveen Pieterse explicitly relates the idea of hybridity to creolization which he sees as offering a Caribbean view of the world. It, like hybridity, emphasizes "the mixed and in-between" as well as "boundary crossing." It also stands in contrast to Westernization and communicates the view that the West itself is involved in the process of creolization. More recently Nederveen Pieterse has offered a contrast between old and new hybridity (see table opposite).[15]

Nederveen Pieterse also sees the study of hybridity as proliferating. It first entered the social sciences through the anthropology of religion and the idea of "syncretism," or "uniting pieces of the mythical history of two different traditions in one that continued to be ordered by a single system."[16] It then found its way into linguistics as the study of creole languages and creolization more generally. Nederveen Pieterse also details work in cultural hybridization (e.g. art), structural and institutional hybridization (e.g. forms of

Varieties of hybridity

New hybridity: recent combinations of cultural and/or institutional forms	Existing or old hybridity: existing cultural and institutional forms are translocal and crosscultural combinations already
Dynamics: migration, trade, ICT, multiculturalism, globalization	Dynamics: crosscultural trade, conquest and contact
Analytics: new modernities	Analytics: history as collage
Examples: Punjabi pop, Mandarin pop, Islamic fashion shows	Examples: too many
Objective: as observed by outsiders	Subjective: as experience and self
As process: hybridization As outcome: hybrid phenomena	As discourse and perspective: hybridity consciousness

governance), hybrid organization forms, and diverse cultural influences (e.g. American, Japanese) on management, science (e.g. ecological economics), food (eclectic menus), and most commonly on "identities, consumer behaviour, lifestyle, etc."[17] Finally, even the newly popular "hybrid car" is a reflection of hybridization.

This chapter closes with one of Roland Robertson's essays on glocalization. Some of the basic characteristics of this concept have been outlined at the beginning of the introduction to this chapter. We will also have more to say about it in the following chapter.

NOTES

1 Ulf Hannerz, "Flows, Boundaries and Hybrids: Keywords in Transnational Anthropology." Working Paper Series WPTC-2K-02, Transnational Communities Programme, University of Oxford, 1997, 13.

2 Roland Robertson, "Globalization Theory 2000 Plus: Major Problematics." In George Ritzer and Barry Smart, eds., *Handbook of Social Theory*. London: Sage, 2001, 458–71; see also Jonathan Friedman, *Culture Identity and Global Processes*. London: Sage, 1994, 102ff.

3 Nestor Garcia Canclini, *Hybrid Cultures: Strategies for Entering and Leaving Modernity*. Minneapolis: University of Minnesota Press, 1995; Jan Nederveen Pieterse, *Globalization and Culture: Global Mélange*. Lanham, MD: Rowman and Littlefield, 2004.

4 Robin Cohen, "Creolization and Cultural Globalization: The Soft Sounds of Fugitive Power." *Globalizations* 4, 3, 2007: 369–84.

5 Ulf Hannerz, "The World in Creolisation." *Africa* 57, 1987: 552.

6 Ibid., 555.

7 Ibid., 555.

8 William Rowe and Vivian Schelling, *Memory and Modernity: Popular Culture in Latin America*. London: Verso, 1991. Cited in Jan Nederveen Pieterse, "Globalization as Hybridization." In M. Featherstone, S. Lash and R. Robertson, eds., *Global Modernities*. Thousand Oaks, CA: Sage, 1995, 49.

9 Jan Nederveen Pieterse, "Hybridity, So What? The Anti-Hybridity Backlash and the Riddles of Recognition." *Theory, Culture and Society* 18, 2001: 221.

10 Pieterse, "Globalization as Hybridization," 49.

11 It can also lead to a strengthening of these structures.

12 Pieterse, "Globalization as Hybridization," 49.

13 Ibid., 50.

14 Ibid., 50.

15 Pieterse, "Hybridity, So What?," 222, table 1.

16 R. Bastide, "Mémoire collective et sociologie du bricolage." *L'Année Sociologique* 21: 65–108. Cited in Pieterse, "Hybridity, So What?," 223.

17 Pieterse, "Hybridity, So What?," 223.

 **READING 47**

The World in Creolisation

Ulf Hannerz

[. . .]

So cultural studies could well benefit from a fresh start in this area, one that sees the world as it is in the late twentieth century. Scattered here and there in anthropology recently, there have been intimations that this world of movement and mixture is a world in creolisation; that a concept of creole culture with its congeners may be our most promising root metaphor. Moving from the social and cultural history of particular colonial societies (where they have tended to apply especially to particular racial or ethnic categories) to the discourse of linguists, creole concepts have become more general in their applications. And it is with a usage along such lines that they are now being retrieved. Drummond thus moves from a consideration of internal variability and change in the symbolic processes of ethnicity in Guyana to a general view that there are now no distinct cultures, only intersystemically connected, creolising Culture. Fabian suggests that the colonial system in Africa – 'frequently disjointed, hastily thrown together for the purpose of establishing political footholds' – produced pidgin contact cultures. In the following period there was creolisation, the emergence of viable new syntheses. In Zaire he finds this represented in popular painting, such as in the *mamba muntu* genre of mermaid images; in the Jamaa religious movement, based on a Belgian missionary's interpretation of Bantu philosophy; and in Congo jazz. Graburn sees new creole art forms, anchored in the reformulated consciousness of Third and Fourth World peoples, expanding beyond the restricted codes of tourist art.

Current creolist linguistics probably has enough theoretical diversity and controversy to allow for rather varied borrowings into cultural theory. As I see it myself, creole cultures like creole languages are those which draw in some way on two or more historical sources, often originally widely different. They have had some time to develop and integrate, and to become elaborate

and pervasive. People are formed from birth by these systems of meaning and largely live their lives in contexts shaped by them. There is that sense of a continuous spectrum of interacting forms, in which the various contributing sources of the culture are differentially visible and active. And, in relation to this, there is a built-in political economy of culture, as social power and material resources are matched with the spectrum of cultural forms. A number of important points seem to come together here.

If the 'Standard', the officially approved language of the metropolis, stands at one end of the creole continuum of language, metropolitan culture in some prestige variant occupies the corresponding position on the cultural spectrum. But what are the mechanisms which place it there, on the range of variations of a national culture, and how do the members of the society come to be arranged in some fashion along that range on the basis of their personal cultural repertoires? I sketched such a spectrum above in spatial terms, from city to village, but this tends not to explain much in itself. If we should look for the mechanisms which are more directly involved in the distributive ordering of culture, we must note first of all that in Third World societies, as elsewhere, the division of labour now plays a major part in generating cultural complexity. Anthropological thinking about culture seems too often to disregard this fact. On the one hand, the division of labour entails a division of knowledge, bringing people into interaction precisely because they do not share all understandings. By not sharing, of course, they can increase their collective cultural inventory. On the other hand, as people are differently placed within the division of labour, they develop varied perspectives going beyond that knowledge which is in some sense commoditised, involved in material transactions.

[. . .]

I believe there is room for a more optimistic view of the vitality of popular expressive forms in the Third

World, at least if the Nigerian example is anything at all to go by. But, of course, these forms are by no means pure traditional Nigerian culture. The world system, rather than creating massive cultural homogeneity on a global scale, is replacing one diversity with another; and the new diversity is based relatively more on inter-relations and less on autonomy. Yet meanings and modes of expressing them can be born in the inter-relations. We must be aware that openness to foreign cultural influences need not involve only an impover-ishment of local and national culture. It may give people access to technological and symbolic resources for dealing with their own ideas, managing their own culture, in new ways. Very briefly, what is needed to understand the transforming power of media tech-nology, from print to electronics, on cultures generally is a subtle understanding of the interplay between ideas, symbolic modalities with their varied potentialities, and the ability of the media to create new social rela-tionships and contexts (as well as to alter old ones). Of that subtle understanding there is as yet little in the anthropology of complex cultures, at least in any systematic form.

Along the entire creolising spectrum, from First World metropolis to Third World village, through education and popular culture, by way of missionaries, consultants, critical intellectuals and small-town story-tellers, a conversation between cultures goes on. One of the advantages of a creolist view of contemporary Third World cultural organisation, it seems to me, is that it suggests that the different cultural streams engaging one another in creolisation may all be actively involved in shaping the resultant forms; and that the merger of quite different streams can create a particular intensity in cultural processes. The active handling of meanings of various local and foreign derivations can allow them to work as commentaries on one another, through never-ending intermingling and counterpoint. Fela Anikulapo-Kuti, or Fela for short, the creator of Afro beat music, political radical and hero of Nigerian popular culture, tells his biog-rapher that he was Africanised by a black American girlfriend in California who gave him a consciousness-raising working-over. Third World intellectuals gener-ally – writers, artists or academics – may be close to the point of entry of the international flow of meaning into national cultures, but, like intellectuals in most

places, they are to some extent counter-cultural, carriers of an adversary culture. While far from immune to the charms of the metropolis, they respond to them critic-ally as well, self-consciously making themselves the spokesmen and guardians of Third World cultures (at least some of the time). What they may broadcast about metropolitan culture through the channels of communication reaching into their society, then, is not necessarily that culture itself, in either a pure or a somehow diluted form. It is their report on the dia-logue between the metropolitan culture and them-selves – as they have heard it. Back in the provincial town a schoolteacher may speak admiringly of the classic ethnography of his people, from the heyday of colonialism, although he may be critical at points on the basis of the oral history he has collected himself. Receding into the past, the 'serial polyandry' of their forefathers and foremothers now seems as titillating to the sophisticates in Kafanchan as Mormon polygamy may be to many Americans. They cannot take the subject as seriously as the missionaries and the first generation of Christian converts did.

The dominant varieties of world system thought which have developed in recent times seem mostly to leave anthropologists uninterested, ambivalent or hostile. This may in part be due to the tradition of anthropological practice, with its preference for the small-scale, the face-to-face, the authentically alien. Another reason, however, would seem to be our dis-trust of approaches which seem too determined not to let small facts get in the way of large issues, too sure that the dominant is totally dominant, too little con-cerned with what the peripheries do both for themselves and to the centre. World system thought sometimes indeed breeds its own rhetorical oversimplifications, its own vulgarities. It seems a little too ready to forget that the influences of any one centre on the peripheries may not be wholly monolithic, but may be varied, unco-ordinated and possibly contradictory. In its typical figures of speech there may be no room for recognising that there may be several centres, conflicting or com-plementary, and that certain of them may not be the products of colonial or post-colonial periods. (For Ahmadu Bello, the northern Nigerian politician, the real Mecca was not London; Mecca was Mecca.) And, last but not least, too often in world system thinking there simply seems to be no room for culture.

A macro-anthropology of culture which takes into account the world system and its centre–periphery relation appears to be well served by a creolist point of view. It could even be the most distinctive contribution anthropology can make to world system studies. It identifies diversity itself as a source of cultural vitality; it demands of us that we see complexity and fluidity as an intellectual challenge rather than as something to escape from. It should point us to ways of looking at systems of meaning which do not hide their connections with the facts of power and material life.

We can perhaps benefit from it, too, because an understanding of the world system in cultural terms can be enlightening not only in Third World studies but also as we try to make of anthropology a truly gen-eral and comparative study of culture. Creole cultures are not necessarily only colonial and post-colonial cultures. I spend most of my time in a small country which for the last half-millennium or so has been nobody's colony, at least not as far as politics goes. Yet we are also drawn into the world system and its centre–periphery relations, and the terms of debate in these 1980s seem to be those of creolisation. What is really Swedish culture? In an era of population movements and communication satellites will it survive, or will it be enriched? And the questions are perhaps just slightly changed in the real centres of the world. What would life be like there without swamis and without reggae, without Olympic Games and 'the Japanese model'? In the end, it seems, we are all being creolised.

 READING 48

Flows, Boundaries and Hybrids: Keywords in Transnational Anthropology

Ulf Hannerz

[. . .]

Anyway, here we are now, with hybridity, collage, mélange, hotchpotch, montage, synergy, bricolage, creolization, mestizaje, mongrelization, syncretism, transculturation, third cultures and what have you; some terms used perhaps only in passing as summary metaphors, others with claims to more analytical status, and others again with more regional or thematic strongholds. Mostly they seem to suggest a concern with cultural form, cultural products (and conspicu-ously often, they relate to domains of fairly tangible cultural materials, such as language, music, art, ritual, or cuisine); some appear more concerned with process than others.

It seems hybridity is at present the more favored general term; no doubt drawing strength, like "flow", from easy mobility between disciplines (but then several of the other terms are also fairly footloose). Despite its biologistic flavor, it has a strength not least in literary scholarship, due in large part to its presence in the work of Mikhail Bakhtin. For Bakhtin, I take it, hybridity was above all the coexistence of two languages, two linguistic consciousnesses, even within a single utterance; commenting on one another, unmasking each other, entailing contradiction, ambiguity, irony; again, the trickster theme may seem not far away. As Homi Bhabha takes the notion into the cultural critique of colonialism, it comes to draw attention to the subversion, the destabilization, of colonial cultural authority. But as different commentators, from a range of disciplines, have taken it in different directions, with varied analytical objectives, hybridity is by now itself a term which is far from unambiguous.

Let us have a quick look at some of the other words for mixture. "Synergy" may not have much of a past in anthropology; it has been pointed out that the concept shows up in some of Ruth Benedict's lecture notes, from 1941. But Benedict used it for situations under-stood as internal to cultures, where an "act or skill that advantages the individual at the same time advantages the group". At present, too, the term seems less popular in anthropology than among professionals in the growing field of intercultural communication, who use it to refer to the dynamic advantages of contacts and mergers between cultures. And of course, these interculturalists themselves often move in the border-lands of the world of business, where the idea of synergy tends to lend an attractive aura to mergers and takeovers. "Synergy", that is to say, has distinctly celebratory overtones built into it.

Going back about equally far in anthropology is "transculturation", a term coined by the Cuban social historian Fernando Ortiz in his book *Cuban Counter-point*. Bronislaw Malinowski, who met Ortiz in Havana in 1939, wrote an introduction (dated 1940) to the book, stating that he had promised the author to "appropriate the new expression for /his/ own use, acknowledging its paternity, and use it constantly and loyally". It was, Malinowski felt, a term much prefer-able to acculturation, which he thought fell upon the ear in an unpleasant way – "sounds like a cross between a hiccup and a belch" – and which, as he understood it, suggested a more one-sided cultural change. Trans-culturation, he agreed with Ortiz, was a system of give and take, "a process from which a new reality emerges, transformed and complex, a reality that is not a mechanical agglomeration of traits, nor even a mosaic, but a new phenomenon, original and independent". It hardly seems that at least some of Malinowski's American colleagues actually understood acculturation very differently. In recent times, "transculturation" may have been made more popular again especially by Pratt's use of it in her study of travel writing. And in postcolonial times, one of the attractions of this concept may be that it is in itself an example of counterflow, from periphery to center.

Perhaps, despite their somewhat different histories and emphases, it does not matter much which of these concepts one chooses, but that to which I have been most strongly drawn myself, primarily on the basis of my field experience in Nigeria, is "creolization". While I believe that the others mostly denote cultural mixture as such, and although "creolization" is no doubt some-times also so used, I think this concept can be used in a more precise, and at the same time restricted, way.

The origins of the idea of "creole" people and cultural phenomena are in the particular culture-historical context of New World plantation societies, and some might feel that the notion should be left there; one could have a debate over this much like those over other concepts which have been taken out of particular areas to be used for more comparative purposes (caste, totem, taboo . . .). In any case, the more expansive use has been an established fact for some time, particularly in sociolinguistics, and in analogy with creolist under-standings there, I would argue that a creolist view is particularly applicable to processes of cultural conflu-ence within a more or less open continuum of diversity, stretched out along a structure of center-periphery relationships which may well extend transnationally, and which is characterized also by inequality in power, prestige and material resource terms. Along such lines it appears to me possible to integrate cultural with social analysis, in a way not equally clearly suggested by many of the other concepts in this cluster, and thus also to pursue a more macroanthropological vision. But again, this also means that creolization becomes a less general term, by referring to a more elaborated type. (And it may also suggest a social landscape which is rather more structured, not so much a frontier or a borderland.)

The identification of creole cultures draws attention to the fact that some cultures are very conspicuously *not* "bounded", "pure", "homogeneous", and "timeless", as in the anthropological tradition cultures have often been made to seem; and to the extent that the celebra-tory stance toward hybridity recurs here as well, it is also suggested that these cultures draw some of their vitality and creativity precisely from the dynamics of mixture (although the celebration here may be somewhat tempered by the recognition that the cul-tures are also built around structures of inequality). One objection occasionally raised against the cre-olization concept – and other related notions may be confronted with it as well – is that such an iden-tification of creole cultures as a particular category might simply push those features of essentialism a step back, implying that the cultural currents joined through creolization were pure, bounded, and so forth, until they were thus joined.

I do not find this implication inevitable. Drawing on the linguistic parallel again, there are a number of English-based creole languages in the world, but nobody would seriously argue that the English lan-guage is historically pure. (Remember 1066, and all that.) The claim need only be that in one particular period, some cultures are more creole than others, to the extent that the cultural streams coming together, under the given conditions and with more or less dramatic results, are historically distinct from another, even as they themselves may have resulted from other confluences. At some point or other, we or our forefathers may all have been creolized, but we are not forever engaged in it to the same degree.

Finally, syncretism; again an old idea, although perhaps not a continuously highly visible one, used in and out of anthropology, but especially in the field of comparative religion, for example in the study of how, in Afro-American cultures, West African deities have merged with Catholic saints. Recently there appears to have been some revival of interest, coupled with an interest in "anti-syncretism" – in a world where academics study non-academic lives and non-academics read academic texts, the leaders and adherents of some of the faiths involved are not particularly pleased with scholarship which appears to deny the authenticity and purity of their beliefs and practices.

[. . .]

READING 49

Globalization as Hybridization

Jan Nederveen Pieterse

Global Mélange: Windows for Research on Globalization

How do we come to terms with phenomena such as Thai boxing by Moroccan girls in Amsterdam, Asian rap in London, Irish bagels, Chinese tacos and Mardi Gras Indians in the United States, or 'Mexican schoolgirls dressed in Greek togas dancing in the style of Isidora Duncan'? How do we interpret Peter Brook directing the Mahabharata, or Ariane Mânouchkine staging a Shakespeare play in Japanese Kabuki style for a Paris audience in the Théâtre Soleil? Cultural experiences, past or present, have not been simply moving in the direction of cultural uniformity and standardization. This is not to say that the notion of global cultural synchronization is irrelevant – on the contrary – but it is fundamentally incomplete. It overlooks the counter-currents – the impact non-Western cultures have been making on the West. It downplays the ambivalence of the globalizing momentum and ignores the role of local reception of Western culture – for example the indigenization of Western elements. It fails to see the influence non-Western cultures have been exercising on one another. It has no room for crossover culture – as in the development of 'third cultures' such as world music. It overrates the homogeneity of Western culture and overlooks the fact that many of the standards exported by the West and its cultural industries themselves turn out to be of culturally mixed character if we examine their cultural lineages. Centuries of South–North cultural osmosis have resulted in an intercontinental crossover culture. European and Western culture are *part* of this global mélange. This is an obvious case if we reckon that Europe until the fourteenth century was invariably the recipient of cultural influences from 'the Orient'. The hegemony of the West dates only from very recent times, from around 1800, and, arguably, from industrialization.

One of the terms offered to describe this interplay is the *creolization* of global culture. This approach is derived from creole languages and linguistics. 'Creolization' itself is an odd, hybrid term. In the Caribbean and North America it stands for the mixture of African and European (the Creole *cuisine* of New Orleans, etc.), while in Hispanic America 'criollo' originally denotes those of European descent born in the continent. 'Creolization' means a Caribbean window on the world. Part of its appeal is that it goes against the grain of nineteenth-century racism and the accompanying abhorrence of *métissage* as miscegenation, as in Comte de Gobineau's view that race mixture leads to decadence and decay for in every mixture the lower element is bound to predominate. The doctrine of racial purity involves the fear of and *dédain* for the half-caste. By stressing and foregrounding the *mestizo* factor, the mixed and in-between, creolization highlights what has been hidden and valorizes boundary crossing. It also implies an argument with Westernization: the West itself may be viewed as a mixture and Western culture as a creole culture.

The Latin American term *mestizaje* also refers to boundary-crossing mixture. Since the early part of the century, however, this has served as a hegemonic élite ideology, which, in effect, refers to 'whitening' or Europeanization as the overall project for Latin American countries: while the European element is supposed to maintain the upper hand, through the gradual 'whitening' of the population and culture, Latin America is supposed to achieve modernity. A limitation of both creolization and *mestizaje* is that they are confined to the experience of the post-sixteenth-century Americas.

Another terminology is the 'orientalization of the world', which has been referred to as 'a distinct global process'. In Duke Ellington's words, 'We are all becoming a little Oriental'. It is reminiscent of the theme of 'East wind prevailing over West wind', which runs through Sultan Galiev, Mao and Abdel-Malek. In the setting of the 'Japanese challenge' and the development model of East Asian Newly Industrialized Countries, it evokes the Pacific Century and the twenty-first century as the 'Asian century'.

Each of these terms – 'creolization', '*mestizaje*', 'orientalization' – opens a different window on the global mélange. In the United States 'crossover culture' denotes the adoption of black cultural characteristics by European-Americans and of white elements by African-Americans. As a general notion, this may aptly describe global intercultural osmosis and interplay. Global 'crossover culture' may be an appropriate characterization of the long-term global North–South mélange. Still, what is not clarified are the *terms* under which cultural interplay and crossover take place. Likewise in terms such as 'global mélange', what is missing is acknowledgement of the actual unevenness, asymmetry and inequality in global relations.

Politics of Hybridity

Given the backdrop of nineteenth-century discourse, it's no wonder that arguments that acknowledge

hybridity often do so on a note of regret and loss – loss of purity, wholeness, authenticity. Thus, according to Hisham Sharabi, neo-patriarchical society in the contemporary Arab world is 'a new, hybrid sort of society/culture', 'neither modern nor traditional'. The 'neopatriarchal petty bourgeoisie' is likewise characterized as a 'hybrid class'. This argument is based on an analysis of 'the political and economic conditions of distorted, dependent capitalism' in the Arab world, in other words, it is derived from the framework of dependency theory.

In arguments such as these hybridity functions as a negative trope, in line with the nineteenth-century paradigm according to which hybridity, mixture, mutation are regarded as negative developments which detract from prelapsarian purity – in society and culture, as in biology. Since the development of Mendelian genetics in the 1870s and its subsequent adoption in early twentieth-century biology, however, a revaluation has taken place according to which crossbreeding and polygenic inheritance have come to be positively valued as enrichments of gene pools. Gradually this theme has been seeping through to wider circles; the work of Bateson, as one of the few to reconnect the natural sciences and social sciences, has been influential in this regard.

In post-structuralist and postmodern analysis, hybridity and syncretism have become keywords. Thus hybridity is the antidote to essentialist notions of identity and ethnicity. Cultural syncretism refers to the methodology of montage and collage, to 'cross-cultural plots of music, clothing, behaviour, advertising, theatre, body language, or [. . .] visual communication, spreading multi-ethnic and multi-centric patterns'. Interculturalism, rather than multiculturalism, is the keynote of this kind of perspective. But it also raises different problems. What is the political *portée* of the celebration of hybridity? Is it merely another sign of perplexity turned into virtue by those grouped on the consumer end of social change? According to Ella Shohat, 'A celebration of syncretism and hybridity per se, if not articulated in conjunction with questions of hegemony and neo-colonial power relations, runs the risk of appearing to sanctify the *fait accompli* of colonial violence'. Hence a further step would be not merely to celebrate but to theorize hybridity.

A theory of hybridity would be attractive. We are so used to theories that are concerned with establishing boundaries and demarcations among phenomena – units or processes that are as neatly as possible set apart from other units or processes – that a theory which instead would focus on fuzziness and mélange, cut-and-mix, crisscross and crossover, might well be a relief in itself. Yet, ironically, of course, it would have to prove itself by giving as neat as possible a version of messiness, or an unhybrid categorization of hybridities.

By what yardstick would we differentiate hybridities? One consideration is in what context hybridity functions. At a general level hybridity concerns the mixture of phenomena which are held to be different, separate; hybridization then refers to a *cross-category* process. Thus with Bakhtin hybridization refers to sites, such as fairs, which bring together the exotic and the familiar, villagers and townsmen, performers and observers. The categories can also be cultures, nations, ethnicities, status groups, classes, genres, and hybridity, by its very existence, blurs the distinctions among them. Hybridity functions, next, as part of a power relationship between centre and margin, hegemony and minority, and indicates a blurring, destabilization or subversion of that hierarchical relationship.

One of the original notions of hybridity is *syncretism*, the fusion of religious forms. Here we can distinguish, on the one hand, syncretism as *mimicry* – as in Santería, Candomblé, Vodûn, in which Catholic saints are adapted to serve as masks behind which non-Christian forms of worship are practised. The Virgin of Guadeloupe as a mask for Pacha Mama is another example. On the other hand, we find syncretism as a mélange not only of forms but also of beliefs, a merger in which *both* religions, Christian and native, have changed and a 'third religion' has developed (as in Kimbangism in the Congo).

Another phenomenon is hybridity as migration mélange. A common observation is that second-generation immigrants, in the West and elsewhere, display mixed cultural patterns – for example, a separation between and, next, a mix of a home culture and language (matching the culture of origin) and an outdoor culture (matching the culture of residence), as in the combination 'Muslim in the daytime, disco in the evening'.

In postcolonial studies hybridity is a familiar and ambivalent trope. Homi Bhabha refers to hybrids as intercultural brokers in the interstices between nation and empire, producing counter-narratives from the nation's margins to the 'totalizing boundaries' of the nation. At the same time, refusing nostalgic models of precolonial purity, hybrids, by way of mimicry, may conform to the 'hegemonized rewriting of the Eurocentre'. Hybridity, in this perspective, can be a condition tantamount to alienation, a state of homelessness. Smadar Lavie comments: 'This is a response-oriented model of hybridity. It lacks agency, by not empowering the hybrid. The result is a fragmented Otherness in the hybrid'. In the work of Gloria Anzaldúa and others, she recognizes, on the other hand, a community-oriented mode of hybridity, and notes that 'reworking the past exposes its hybridity, and to recognize and acknowledge this hybrid past in terms of the present empowers the community and gives it agency'.

An ironical case of hybridity as intercultural crossover is mentioned by Michael Bérubé, interviewing the African American literary critic Houston Baker, Jr: 'That reminds me of your article in *Technoculture*, where you write that when a bunch of Columbia-graduate white boys known as Third Bass attack Hammer for not being black enough or strong enough [. . .] *that's* the moment of hybridity'.

Taking in these lines of thought, we can construct a *continuum of hybridities*: on one end, an assimilationist hybridity that leans over towards the centre, adopts the canon and mimics the hegemony, and, at the other end, a destabilizing hybridity that blurs the canon, reverses the current, subverts the centre. Hybridities, then, may be differentiated according to the components in the mélange. On the one hand, an assimilationist hybridity in which the centre predominates – as in V.S. Naipaul, known for his trenchant observations such as there's no decent cup of coffee to be had in Trinidad. A posture which has given rise to the term Naipaulitis. And on the other hand, an hybridity that blurs (passive) or destabilizes (active) the canon and its categories. Perhaps this spectrum of hybridities can be summed up as ranging from Naipaul to Salman Rushdie, Edward Said, Gayatri Spivak. Still, what does it mean to destabilize the canon? It's worth reflecting on the politics of hybridity.

Politics of Hybridity: towards Political Theory on a Global Scale

Relations of power and hegemony are inscribed and reproduced *within* hybridity for wherever we look closely enough we find the traces of asymmetry in culture, place, descent. Hence hybridity raises the question of the *terms* of mixture, the conditions of mixing and mélange. At the same time it's important to note the ways in which hegemony is not merely reproduced but *refigured* in the process of hybridization. Generally, what is the bearing of hybridization in relation to political engagement?

> At times, the anti-essentialist emphasis on hybrid identities comes dangerously close to dismissing all searches for communitarian origins as an archaeological excavation of an idealized, irretrievable past. Yet, on another level, while avoiding any nostalgia for a prelapsarian community, or for any unitary and transparent identity predating the 'fall', we must also ask whether it is possible to forge a collective resistance without inscribing a communal past.

Isn't there a close relationship between political mobilization and collective memory? Isn't the remembrance of deeds past, the commemoration of collective itineraries, victories and defeats – such as the Matanza for the FMLN in El Salvador, Katipunan for the NPA in the Philippines, Heroes Day for the ANC – fundamental to the symbolism of resistance and the moral economy of mobilization? Still, this line of argument involves several problems. While there may be a link, there is no necessary symmetry between communal past/collective resistance. What is the basis of bonding in collective action – past or future, memory or project? While communal symbolism may be important, collective symbolism and discourse merging a heterogeneous collectivity in a common project may be more important. Thus, while Heroes Day is significant to the ANC (16 December is the founding day of Umkhonto we Sizwe), the Freedom Charter, and more specifically, the project of non-racial democracy (non-sexism has been added later) has been of much greater importance. These projects are not of a 'communal' nature: part of their strength is precisely that they transcend communal boundaries. Generally, emancipations may be thought of in the plural, as a project or ensemble of projects that in itself is diverse, heterogeneous, multivocal. The argument linking communal past/collective resistance imposes a unity and transparency which in effect reduces the space for *critical* resistance, for plurality *within* the movement, diversity within the

process of emancipation. It privileges a communal view of collective action, a primordialist view of identity, and ignores or downplays the importance of *intra*group differences and conflicts over group representation, demands and tactics, including reconstructions of the past. It argues as if the questions of whether demands should be for autonomy or inclusion, whether the group should be inward or outward looking, have already been settled, while in reality these are political dilemmas. The nexus between communal past/collective engagement is one strand in political mobilization, but so are the hybrid past/plural projects, and in actual everyday politics the point is how to negotiate these strands in round-table politics. This involves going beyond a past to a future orientation – for what is the point of collective action without a future? The lure of community, powerful and prevalent in left as well as right politics, has been questioned often enough. In contrast, hybridity when thought of as a politics may be subversive of essentialism and homogeneity, disruptive of static spatial and political categories of centre and periphery, high and low, class and ethnos, and in recognizing multiple identities, widen the space for critical engagement. Thus the nostalgia paradigm of community politics has been contrasted to the landscape of the city, along with a reading of 'politics as relations among strangers'.

What is the significance of this outlook in the context of global inequities and politics? Political theory on a global scale is relatively undeveloped. Traditionally political theory is concerned with the relations between sovereign and people, state and society. It's of little help to turn to the 'great political theorists' from Locke to Mill for they are all essentially concerned with the state–society framework. International relations theory extrapolates from this core preoccupation with concepts such as national interest and balance of power. Strictly speaking international relations theory, at any rate neo-realist theory, precludes global political theory. In the absence of a concept of 'world society', how can there be a notion of a world-wide social contract or global democracy? This frontier has opened up through concepts such as global civil society, referring to the transnational networks and activities of voluntary and non-governmental organizations: 'the growth of global civil society represents an ongoing project of civil society to reconstruct, re-imagine, or re-map world politics'. Global society and postinternational politics are other relevant notions. A limitation to these reconceptualizations remains the absence of legal provisions that are globally binding rather than merely in interstate relations.

The question remains as to what kind of conceptual tools we can develop to address questions such as the double standards prevailing in global politics: perennial issues such as Western countries practising democracy at home and imperialism abroad; the edifying use of terms such as self-determination and sovereignty while the United States are invading Panama or Grenada. The term 'imperialism' may no longer be adequate to address the present situation. It may be adequate in relation to US actions in Panama or Grenada, but less so to describe the Gulf War. Imperialism is the policy of establishing or maintaining an empire, and empire is the control exercised by a state over the domestic and foreign policy of another political society. This is not an adequate terminology to characterize the Gulf War episode. If we consider that major actors in today's global circumstance are the IMF and World Bank, transnational corporations and regional investment banks, it is easy to acknowledge their influence on the domestic policies of countries from Brazil to the Philippines, but the situation differs from imperialism in two ways: the actors are not states and the foreign policy of the countries involved is not necessarily affected. The casual use of terms such as recolonization or neocolonialism to describe the impact of IMF conditionalities on African countries remains just that, casual. The situation has changed also since the emergence of regional blocs which can potentially exercise joint foreign policy (for example, the European Community) or which within themselves contain two or more 'worlds' (for example, NAFTA, APEC). Both these situations differ from imperialism in the old sense. Current literature on international political economy shows a shift from 'imperialism' to 'globalization'. The latter may be used with critical intent but is more often used in an open-ended sense. I've used the term 'critical globalism' as an approach to current configurations. According to Tomlinson,

> the distribution of global power that we know as 'imperialism' [. . .] characterised the modern period up to, say, the 1960s. What replaces 'imperialism' is 'globalisation'. Globalisation may be distinguished

from imperialism in that it is a far less coherent or culturally directed process [. . .] The idea of 'globalisation' suggests interconnection and interdependency of all global areas which happens in a less purposeful way.

This is a particularly narrow interpretation in which globalization matches the epoch of late capitalism and flexible accumulation; still, what is interesting is the observation that the present phase of globalization is less coherent and less purposeful than imperialism. That does not mean the end of inequality and domination, although domination may be more dispersed, less orchestrated, more heterogeneous. To address global inequalities and develop global political theory a different kind of conceptualization is needed. We are not without points of reference but we lack a theory of global political action. Melucci has discussed the 'planetarization' of collective action. Some of the implications of globalization for democracy have been examined by Held. As regards the basics of a global political consensus, the UN Declaration of Human Rights, and its subsequent amendments by the Movement of Non-Aligned Countries, may be a point of reference.

Post-Hybridity?

Cultural hybridization refers to the mixing of Asian, African, American, European cultures: hybridization is the making of global culture as a global mélange. As a category hybridity serves a purpose on the basis of the assumption of *difference* between the categories, forms, beliefs that go into the mixture. Yet the very process of hybridization shows the difference to be relative and, with a slight shift of perspective, the relationship can also be described in terms of an affirmation of *similarity*. Thus, the Catholic saints can be taken as icons of Christianity, but can also be viewed as holdovers of pre-Christian paganism inscribed in the Christian canon. In that light, their use as masks for non-Christian gods is less quaint and rather intimates transcultural pagan affinities.

Ariane Mânouchkine's use of Kabuki style to stage a Shakespeare play leads to the question, which Shakespeare play? The play is Henry IV, which is set in the context of European high feudalism. In that light, the use of Japanese feudal Samurai style to portray European feudalism makes a point about transcultural historical affinities.

'Mexican schoolgirls dressed in Greek togas dancing in the style of Isidora Duncan', mentioned before, reflects transnational bourgeois class affinities, mirroring themselves in classical European culture. Chinese tacos and Irish bagels reflect ethnic crossover in employment patterns in the American fast food sector. Asian rap refers to cross-cultural stylistic convergence in popular youth culture.

An episode that can serve to probe this more deeply is the influence of Japanese art on European painting. The impact of *Japonisme* is well known: it inspired impressionism which in turn set the stage for modernism. The colour woodcuts that made such a profound impression on Seurat, Manet, Van Gogh, Toulouse Lautrec, Whistler belonged to the Ukiyo school – a bourgeois genre that flourished in Japan between the seventeenth and nineteenth centuries, sponsored by the merchant class. Ukiyo-e typically depicted urban scenes of ephemeral character, such as streetlife, entertainments, theatre, or prostitution, and also landscapes. It was a popular art form which, unlike the high art of aristocracy, was readily available at reasonable prices in book stores (rather than cloistered in courts or monasteries) and therefore also accessible to Europeans. This episode, then, is not so much an exotic irruption in European culture, but rather reflects the fact that bourgeois sensibilities had found iconographic expression in Japan earlier than in Europe. In other words, Japanese popular art was modern before European art was. Thus what from one angle appears as hybridity to the point of exoticism, from another angle, again, reflects transcultural class affinities in sensibilities *vis à vis* urban life and nature. In other words, the other side of cultural hybridity is transcultural convergence.

What makes it difficult to discuss these issues is that two quite distinct concepts of *culture* are generally being used indiscriminately. The first concept of culture (culture 1) views culture as essentially territorial; it assumes that culture stems from a learning process that is, in the main, localized. This is culture in the sense of *a culture*, that is the culture of a society or social group. A notion that goes back to nineteenth-century romanticism and that has been elaborated in twentieth-century anthropology, in particular cultural relativism – with the notion of cultures as a whole, a Gestalt, configuration. A related idea is the organic or 'tree' model of culture.

A wider understanding of culture (culture 2) views culture as a general human 'software', as in nature/culture arguments. This notion has been implicit in theories of evolution and diffusion, in which culture is viewed as, in the main, a *translocal* learning process. These understandings are not incompatible: culture 2 finds expression in culture 1, cultures are the vehicle of culture. But they do reflect different emphases in relation to historical processes of culture formation and hence generate markedly different assessments of cultural relations. Divergent meta-assumptions about culture underlie the varied vocabularies in which cultural relations are discussed.

Assumptions about culture	
Territorial culture	*Translocal culture*
endogenous	exogenous
orthogenetic	heterogenetic
societies, nations, empires	diasporas, migrations
locales, regions	crossroads, borders, interstices
community-based	networks, brokers, strangers
organic, unitary	diffusion, heterogeneity
authenticity	translation
inward looking	outward looking
community linguistics	contact linguistics
race	half-caste, mixed-breed, métis
ethnicity	new ethnicity
identity	identification, new identity

Culture 2 or translocal culture is not without place (there is no culture without place), but it involves an *outward-looking* sense of place, whereas culture 1 is based on an *inward-looking* sense of place. Culture 2 involves what Doreen Massey calls 'a global sense of place': 'the specificity of place which derives from the fact that each place is the focus of a distinct *mixture* of wider and more local social relations'.

The general terminology of cultural pluralism, multicultural society, intercultural relations, etc. does not clarify whether it refers to culture 1 or culture 2. Thus, relations among cultures can be viewed in a static fashion (in which cultures retain their separateness in interaction) or a fluid fashion (in which cultures interpenetrate).

Cultural relations	
Static	*Fluid*
plural society (Furnivall)	pluralism, melting pot
multiculturalism (static)	multiculturalism (fluid), interculturalism
global mosaic	cultural flow in space (Hannerz)
clash of civilizations	third cultures

Hybridization as a perspective belongs to the fluid end of relations between cultures: it's the mixing of cultures and not their separateness that is emphasized. At the same time, the underlying assumption about culture is that of culture/place. Cultural forms are called hybrid/syncretic/mixed/creolized because the elements in the mix derive from different cultural contexts. Thus Ulf Hannerz defines creole cultures as follows: 'creole cultures like creole languages are those which draw in some way on two or more historical sources, often originally widely different. They have had some time to develop and integrate, and to become elaborate and pervasive'. But, in this sense, would not every culture be a creole culture? Can we identify any culture that is *not* creole in the sense of drawing on one or more different historical sources? A scholar of music makes a similar point about world music: 'all music is essentially world music'.

A further question is: are cultural elements different merely because they originate from different cultures? More often what may be at issue, as argued above, is the *similarity* of cultural elements when viewed from the point of class, status group, life-style sensibilities or function. Hence, at some stage, towards the end of the story, the notion of cultural hybridity itself unravels or, at least, needs reworking.

To explore what this means in the context of globalization, we can contrast the vocabularies and connotations of globalization-as-homogenization and globalization-as-hybridization.

Globalization/homogenization	*Globalization/diversification*
cultural imperialism	cultural planetarization
cultural dependence	cultural interdependence
cultural hegemony	cultural interpenetration
autonomy	syncretism, synthesis, hybridity
modernization	modernizations
Westernization	global mélange
cultural synchronization	creolization, crossover
world civilization	global ecumene

What is common to some perspectives on both sides of the globalization/homogenization/heterogenization axis is a territorial view of culture. The territoriality of culture, however, itself is not constant over time. For some time we have entered a period of accelerated globalization and cultural mixing. This also involves an overall tendency towards the 'deterritorialization' of culture, or an overall shift in orientation from culture 1 to culture 2. Introverted cultures, which have been prominent over a long stretch of history and which overshadowed translocal culture, are gradually receding into the background, while translocal culture made up of diverse elements is coming into the foreground. This transition and the hybridization processes themselves unleash intense and dramatic nostalgia politics, of which ethnic upsurges, ethnicization of nations, and religious revivalism form part.

Hybridization refers not only to the crisscrossing of cultures (culture 1) but also and by the same token to a transition from the provenance of culture 1 to culture 2. Another aspect of this transition is that due to advancing information technology and biotechnology, different *modes* of hybridity emerge on the horizon: in the light of hybrid forms, such as cyborgs, virtual reality and electronic simulation, intercultural differences may begin to pale to relative insignificance – although of great local intensity. Biotechnology opens up the perspective of 'merged evolution', in the sense of the merger of the evolutionary streams of genetics, cultural evolution and information technology, and the near prospect of humans intervening in genetic evolution, through the matrix of cultural evolution and information technologies.

Conclusion: towards a Global Sociology

Globalization/hybridization makes, first, an empirical case: that processes of globalization, past and present, can be adequately described as processes of hybridization. Secondly, it is a critical argument: against viewing globalization in terms of homogenization, or of modernization/Westernization, as empirically narrow and historically flat.

The career of sociology has been coterminous with the career of nation-state formation and nationalism, and from this followed the constitution of the object of sociology as society and the equation of society with the nation. Culminating in structural functionalism and modernization theory, this career in the context of globalization is in for retooling. A global sociology is taking shape, around notions such as social networks (rather than 'societies'), border zones, boundary crossing and global society. In other words, a sociology conceived within the framework of nations/societies is making place for a post-inter/national sociology of hybrid formations, times and spaces.

Structural hybridization, or the increase in the range of organizational options, and cultural hybridization, or the doors of erstwhile imagined communities opening up, are signs of an age of boundary crossing. Not, surely, of the erasure of boundaries. Thus, state power remains extremely strategic, but it is no longer the only game in town. The tide of globalization reduces the room of manoeuvre for states, while international institutions, transnational transactions, regional co-operation, sub-national dynamics and non-governmental organizations expand in impact and scope.

In historical terms, this perspective may be deepened by writing diaspora histories of global culture. Due to nationalism as the dominant paradigm since the nineteenth century, cultural achievements have been routinely claimed for 'nations' – that is, culture has been 'nationalized', territorialized. A different historical record can be constructed on the basis of the contributions to culture formation and diffusion by diasporas, migrations, strangers, brokers. A related project would be histories of the hybridization of metropolitan cultures, that is a counter-history to the narrative of imperial history. Such historical inquiries may show that hybridization has been taking place all along but over time has been concealed by religious, national, imperial and civilizational chauvinism. Moreover, they may deepen our understanding of the temporalities of hybridization: how certain junctures witness downturns or upswings of hybridization, slowdowns or speed-ups. At the same time it follows that, if we accept that cultures have been hybrid *all along*, hybridization is in effect a tautology: contemporary accelerated globalization means the hybridization of hybrid cultures.

As such, the hybridization perspective remains meaningful only as a critique of essentialism. Essentialism will remain strategic as a mobilizational device as long as the units of nation, state, region, civilization, ethnicity remain strategic: and for just as long hybridization remains a relevant approach. Hybridity unsettles the introverted concept of culture which underlies romantic nationalism, racism, ethnicism, religious revivalism, civilizational chauvinism, and culturalist essentialism. Hybridization, then, is a perspective that is meaningful as a counterweight to the introverted notion of culture; at the same time, the very process of hybridization unsettles the introverted gaze, and accordingly, hybridization eventually ushers in post-hybridity, or transcultural cut and mix.

Hybridization is a factor in the reorganization of social spaces. Structural hybridization, or the emergence of new practices of social co-operation and competition, and cultural hybridization, or new translocal cultural expressions, are interdependent: new forms of co-operation require and evoke new cultural imaginaries. Hybridization is a contribution to a sociology of the in-between, a sociology from the interstices. This involves merging endogenous/exogenous understandings of culture. This parallels the attempt in international relations theory to overcome the dualism between the nation-state and international system perspectives. Other significant perspectives are Hannerz' macro-anthropology and his concern with mapping micro–macro linkages and contemporary work in geography and cultural studies.

In relation to the global human condition of inequality, the hybridization perspective releases reflection and engagement from the boundaries of nation, community, ethnicity, or class. Fixities have become fragments as the kaleidoscope of collective experience is in motion. It has been in motion all along and the fixities of nation, community, ethnicity and class have been grids superimposed upon experiences more complex and subtle than reflexivity and organization could accommodate.

 **READING 50**

Glocalization: Time–Space and Homogeneity–Heterogeneity

Roland Robertson

[. . .]

The need to introduce the concept of glocalization firmly into social theory arises from the following considerations. Much of the talk about globalization has tended to assume that it is a process which overrides locality, including large-scale locality such as is exhibited in the various ethnic nationalisms which have seemingly arisen in various parts of the world in recent years. This interpretation neglects two things. First, it neglects the extent to which what is called local is in large degree constructed on a trans- or super-local basis. In other words, much of the promotion of locality is in fact done from above or outside. Much of what is often declared to be local is in fact the local expressed in terms of generalized recipes of locality. Even in cases where there is apparently no concrete recipe at work – as in the case of some of the more aggressive forms of contemporary nationalism – there is still, or so I would claim, a translocal factor at work. Here I am simply maintaining that the contemporary assertion of ethnicity and/or nationality is made within the global terms of identity and particularity.

Second, while there has been increasing interest in spatial considerations and expanding attention to the intimate links between temporal and spatial dimensions of human life, these considerations have made relatively little impact as yet on the discussion of globalization and related matters. In particular there has been little attempt to connect the discussion of time-and-space to the thorny issue of universalism-and-particularism. Interest in the theme of postmodernity has involved much attention to the supposed weaknesses of mainstream concern with 'universal time' and advancement of the claim that 'particularistic space' be given much greater attention; but in spite of a few serious efforts to resist the tendency, universalism has been persistently counterposed to particularism (in line with characterizations in the old debate about societal modernization in the 1950s and 1960s). At this time the emphasis on space is frequently expressed as a diminution of temporal considerations.

[. . .]

The leading argument in this discussion is thus centred on the claim that the debate about global homogenization versus heterogenization should be transcended. It is not a question of *either* homogenization or heterogenization, but rather of the ways in which both of these two tendencies have become features of life across much of the late-twentieth-century world. In this perspective the problem becomes that of spelling out the ways in which homogenizing and heterogenizing tendencies are mutually implicative. This is in fact much more of an empirical problem than might at first be thought. In various areas of contemporary life – some of which are discussed in the following pages – there are ongoing, calculated attempts to combine homogeneity with heterogeneity and universalism with particularism.

In this respect we may well speak of the way in which academic disciplines have lagged behind 'real life'. At the same time, we need, of course, to provide analyses and interpretations of these features of 'reality' (recognizing that the distinction between theory and reality is extremely problematic and, I believe, ultimately untenable). I hope to show that outside academic/intellectual discourse there are many who take it for granted that the universal and particular can and *should* be combined. The question for them is: how and in what form should these be synthesized? It is not whether they *can* be interrelated. In order to comprehend the 'how' rather than the 'whether' we need to attend more directly to the question as to what is actually 'going on'. Asking that question does not, as some might well think, involve a disinterest in issues of a 'critical' nature concerning, for example, the interests served by strategies of what I here call glocalization;

not least because, as I will intermittently emphasize, strategies of glocalization are – at least at this historical moment and for the foreseeable future – themselves grounded in particularistic frames of reference. There is no viable and practical Archimedean point from which strategies of glocalization can be fully maintained. Nevertheless, we appear to live in a world in which the expectation of uniqueness has become increasingly institutionalized and globally widespread.

Glocalization

According to *The Oxford Dictionary of New Words* the term 'glocal' and the process noun 'glocalization' are 'formed by telescoping *global* and *local* to make a blend'. Also according to the *Dictionary* that idea has been 'modelled on Japanese *dochakuka* (deriving from *dochaku* "living on one's own land"), originally the agricultural principle of adapting one's farming techniques to local conditions, but also adopted in Japanese business for *global localization*, a global outlook adapted to local conditions' (emphasis in original). More specifically, the terms 'glocal' and 'glocalization' became aspects of business jargon during the 1980s, but their major locus of origin was in fact Japan, a country which has for a very long time strongly cultivated the spatio-cultural significance of Japan itself and where the general issue of the relationship between the particular and the universal has historically received almost obsessive attention. By now it has become, again in the words of *The Oxford Dictionary of New Words*, 'one of the main marketing buzzwords of the beginning of the nineties'.

The idea of glocalization in its business sense is closely related to what in some contexts is called, in more straightforwardly economic terms, micro-marketing: the tailoring and advertising of goods and services on a global or near-global basis to increasingly differentiated local and particular markets. Almost needless to say, in the world of capitalistic production for increasingly global markets the adaptation to local and other particular conditions is not simply a case of business responses to existing global variety – to civilizational, regional, societal, ethnic, gender and still other types of differentiated consumers – as if such variety or heterogeneity existed simply 'in itself'. To a considerable extent micromarketing – or, in the more comprehensive

phrase, glocalization – involves *the construction* of increasingly differentiated consumers, the 'invention' of 'consumer traditions' (of which tourism, arguably the biggest 'industry' of the contemporary world, is undoubtedly the most clear-cut example). To put it very simply, diversity sells. From the consumer's point of view it can be a significant basis of cultural capital formation. This, it should be emphasized, is not its only function. The proliferation of, for example, 'ethnic' supermarkets in California and elsewhere does to a large extent cater not so much to difference for the sake of difference, but to the desire for the familiar and/or to nostalgic wishes. On the other hand, these too can also be bases of cultural capital formation.

It is not my purpose here to delve into the comparative history of capitalistic business practices. Thus the accuracy of the etymology concerning 'glocalization' provided by *The Oxford Dictionary of New Words* is not a crucial issue. Rather I want to use the general idea of glocalization to make a number of points about the global–local problematic. There is a widespread tendency to regard this problematic as straightforwardly involving a polarity, which assumes its most acute form in the claim that we live in a world of local assertions *against* globalizing trends, a world in which the very idea of locality is sometimes cast as a form of opposition or resistance to the hegemonically global (or one in which the assertion of 'locality' or *Gemeinschaft* is seen as the pitting of subaltern 'universals' against the 'hegemonic universal' of dominant cultures and/or classes). An interesting variant of this general view is to be found in the replication of the German culture–civilization distinction at the global level: the old notion of ('good') culture is pitted against the ('bad') notion of civilization. In this traditional German perspective local culture becomes, in effect, national culture, while civilization is given a distinctively global, world-wide colouring.

We have, in my judgement, to be much more subtle about the dynamics of the production and reproduction of difference and, in the broadest sense, locality. Speaking in reference to the local–cosmopolitan distinction, Hannerz has remarked that for locals diversity 'happens to be the principle which allows all locals to stick to their respective cultures'. At the same time, cosmopolitans largely depend on 'other people' carving out 'special niches' for their cultures. Thus 'there can be no cosmopolitans without locals'. This point has some bearing on the particular nature of the intellectual interest in and the approach to the local–global issue. In relation to Hannerz's general argument, however, we should note that in the contemporary world, or at least in the West, the current counter-urbanization trend, much of which in the USA is producing 'fortress communities', proceeds in terms of the standardization of locality, rather than straightforwardly in terms of 'the principle of difference'.

[. . .]

Thus the notion of glocalization actually conveys much of what I myself have previously written about globalization. From my own analytic and interpretative standpoint the concept of globalization has involved the simultaneity and the interpenetration of what are conventionally called the global and the local, or – in more abstract vein – the universal and the particular. (Talking strictly of my own position in the current debate about and the discourse of globalization, it may even become necessary to substitute the term 'glocalization' for the contested term 'globalization' in order to make my argument more precise.) I certainly do not wish to fall victim, cognitive or otherwise, to a particular brand of current marketing terminology. Insofar as we regard the idea of glocalization as simply a capitalistic business term (of apparent Japanese origin) then I would of course reject it as, *inter alia*, not having sufficient analytic-interpretative leverage. On the other hand, we are surely coming to recognize that seemingly autonomous economic terms frequently have deep cultural roots. In the Japanese and other societal cases the cognitive and moral 'struggle' even to recognize the economic domain as relatively autonomous has never really been 'won'. In any case, we live in a world which increasingly acknowledges the quotidian conflation of the economic and the cultural. But we inherited from classical social theory, particularly in its German version in the decades from about 1880 to about 1920, a view that talk of 'culture' and 'cultivation' was distinctly at odds with 'materialism' and the rhetoric of economics and instrumental rationality.

My deliberations in this chapter on the local–global problematic hinge upon the view that contemporary conceptions of locality are largely produced in something like global terms, but this certainly does not mean that all forms of locality are thus substantively homogenized (notwithstanding the standardization, for example, of relatively new suburban, fortress communities). An important thing to recognize in this connection is that there is an increasingly globe-wide discourse of locality, community, home and the like. One of the ways of considering the idea of *global culture* is in terms of its being constituted by the increasing interconnectedness of many local cultures both large and small, although I certainly do not myself think that global culture is entirely constituted by such interconnectedness. In any case we should be careful *not to equate the communicative and interactional connecting of such cultures* – including very asymmetrical forms of such communication and interaction, as well as 'third cultures' of mediation – *with the notion of homogenization of all cultures*.

I have in mind the rapid, recent development of a relatively autonomous discourse of 'intercultural communication'. This discourse is being promoted by a growing number of professionals, along the lines of an older genre of 'how to' literature. So it is not simply a question of social and cultural theorists talking about cultural difference and countervailing forces of homogenization. One of the 'proper objects' of study here is the phenomenon of 'experts' who specialize in the 'instrumentally rational' promotion of intercultural communication. These 'experts' have in fact a vested interest in the promotion and protection of variety and diversity. Their jobs and their profession depend upon the expansion and reproduction of heterogeneity. The same seems to apply to strong themes in modern American business practice.

We should also be more interested in the conditions for the production of cultural pluralism – as well as geographical pluralism. Let me also say that the idea of locality, indeed of globality, is very relative. In spatial terms a village community is of course local relative to a region of a society, while a society is local relative to a civilizational area, and so on. Relativity also arises in temporal terms. Contrasting the well-known pair consisting of locals and cosmopolitans, Hannerz has written that 'what was cosmopolitan in the early 1940s may be counted as a moderate form of localism by now'. I do not in the present context get explicitly involved in the problem of relativity (or relativism). But sensitivity to the problem does inform much of what I say.

There are certain conditions that are currently promoting the production of concern with the local–global problematic within the academy. King has addressed an important aspect of this. In talking specifically of the spatial compression dimension of globalization he remarks on the increasing numbers of 'protoprofessionals from so-called "Third World" societies' who are travelling to 'the core' for professional education. The educational sector of 'core' countries 'depends increasingly on this input of students from the global periphery'. It is the experience of 'flying round the world and needing schemata to make sense of what they see' on the one hand, and encountering students from all over the world in the classroom on the other, which forms an important experiential basis for academics of what King calls totalizing and global theories. I would maintain, however, that it is *interest in 'the local'* as much as the 'totally global' which is promoted in this way.

The Local in the Global? The Global in the Local?

In one way or another the issue of the relationship between the 'local' and the 'global' has become increasingly salient in a wide variety of intellectual and practical contexts. In some respects this development hinges upon the increasing recognition of the significance of space, as opposed to time, in a number of fields of academic and practical endeavour. The general interest in the idea of postmodernity, whatever its limitations, is probably the most intellectually tangible manifestation of this. The most well known maxim – virtually a cliché – proclaimed in the diagnosis of 'the postmodern condition' is of course that 'grand narratives' have come to an end, and that we are now in a circumstance of proliferating and often competing narratives. In this perspective there are no longer any stable accounts of dominant change in the world. This view itself has developed, on the other hand, at precisely the same time that there has crystallized an increasing interest in the world as a whole as a single place. As the sense of temporal unidirectionality has faded so, on the other hand, has the sense of 'representational' space within which all kinds of narratives may be inserted expanded. This of course has increasingly raised in recent years

the vital question as to whether the apparent collapse – and the 'deconstruction' – of the heretofore dominant social-evolutionist accounts of implicit or explicit world history are leading rapidly to a situation of chaos or one in which, to quote Giddens, 'an infinite number of purely idiosyncratic "histories" can be written'. Giddens claims in fact that we *can* make generalizations about 'definite episodes of historical transition'. However, since he also maintains that 'modernity' on a global scale has amounted to a rupture with virtually all prior forms of life he provides no guidance as to how history or histories might actually be done.

In numerous contemporary accounts, then, globalizing trends are regarded as in tension with 'local' assertions of identity and culture. Thus ideas such as the global *versus* the local, the global *versus* the 'tribal', the international *versus* the national, and the universal *versus* the particular are widely promoted. For some, these alleged oppositions are simply puzzles, while for others the second part of each opposition is seen as a reaction against the first. For still others they are contradictions. In the perspective of contradiction the tension between, for example, the universal and the particular may be seen either in the dynamic sense of being a relatively progressive source of overall change or as a modality which preserves an existing global system in its present state. We find both views in Wallerstein's argument that the relation between the universal and the particular is basically a product of expanding world-systemic capitalism. Only what Wallerstein calls anti-systemic movements – and then only those which effectively challenge its 'metaphysical presuppositions' – can move the world beyond the presuppositions of its present (capitalist) condition. In that light we may regard the contemporary proliferation of 'minority discourses' as being encouraged by the presentation of a 'world-system'. Indeed, there is much to suggest that adherents to minority discourses have, somewhat paradoxically, a special liking for Wallersteinian or other 'totalistic' forms of world-systems theory. But it must also be noted that many of the enthusiastic participants in the discourse of 'minorities' describe their intellectual practice in terms of the *singular*, minority discourse. This suggests that there is indeed a potentially *global* mode of writing and talking on behalf of, or at least about, minorities.

Barber argues that 'tribalism' and 'globalism' have become what he describes as the two axial principles of our time. In this he echoes a very widespread view of 'the new world (dis)order'. I chose to consider his position because it is succinctly stated and has been quite widely disseminated. Barber sees these two principles as inevitably in tension – a 'McWorld' of homogenizing globalization *versus* a 'Jihad world' of particularizing 'lebanonization'. (He might well now say 'balkanization'.) Barber is primarily interested in the bearing which each of these supposedly clashing principles have on the prospects for democracy. That is certainly a very important matter, but I am here only directly concerned with the global–local debate.

Like many others, Barber defines globalization as the opposite of localization. He argues that 'four imperatives make up the dynamic of McWorld: a market imperative, a resource imperative, an information-technology imperative, and an ecological imperative'. Each of these contributes to 'shrinking the world and diminishing the salience of national borders' and together they have 'achieved a considerable victory over factiousness and particularism, and not least over their most virulent traditional form – nationalism'. Remarking that 'the Enlightenment dream of a universal rational society has to a remarkable degree been realized', Barber emphasizes that that achievement has, however, been realized in commercialized, bureaucratized, homogenized and what he calls 'depoliticized' form. Moreover, he argues that it is a very incomplete achievement because it is 'in competition with forces of global breakdown, national dissolution, and centrifugal corruption'. While notions of localism, locality and locale do not figure explicitly in Barber's essay they certainly diffusely inform it.

There is no good reason, other than recently established convention in some quarters, to define globalization largely in terms of homogenization. Of course, anyone is at liberty to so define globalization, but I think that there is a great deal to be said against such a procedure. Indeed, while each of the imperatives of Barber's McWorld appear superficially to suggest homogenization, when one considers them more closely, they each have a local, diversifying aspect. I maintain also that it makes no good sense to define the global as if the global excludes the local. In somewhat technical terms, defining the global in such a way suggests that the global lies beyond all localities, as having systemic properties over and beyond the attributes of units within a global system. This way of talking flows along the lines suggested by the macro–micro distinction, which has held much sway in the discipline of economics and has recently become a popular theme in sociology and other social sciences.

Without denying that the world-as-a-whole has some systemic properties beyond those of the 'units' within it, it must be emphasized, on the other hand, that such units themselves are to a large degree constructed in terms of extra-unit processes and actions, in terms of increasingly global dynamics. For example, nationally organized societies – and the 'local' aspirations for establishing yet more nationally organized societies – are not simply units within a global context or texts within a context or intertext. Both their existence, and particularly the form of their existence, is largely the result of extra-societal – more generally, extra-local – processes and actions. If we grant with Wallerstein and Greenfeld that 'the national' is a 'prototype of the particular' we must, on the other hand, also recognize that the nation-state – more generally, the national society – is in a crucial respect a *cultural idea* (as Greenfeld herself seems to acknowledge). Much of the apparatus of contemporary nations, of the national-state organization of societies, including *the form* of their particularities – the construction of their unique identities – is very similar across the entire world, in spite of much variation in levels of 'development'. This is, perhaps, the most tangible of contemporary sites of the interpenetration of particularism and universalism.

Before coming directly to the contemporary circumstance, it is necessary to say a few words about globalization in a longer, historical perspective. One can undoubtedly trace far back into human history developments involving the expansion of chains of connectedness across wide expanses of the earth. In that sense 'world formation' has been proceeding for many hundreds, indeed thousands, of years. At the same time, we can undoubtedly trace through human history periods during which the consciousness of the potential for world 'unity' was in one way or another particularly acute. One of the major tasks of students of globalization is, as I have said, to comprehend *the form* in which the present, seemingly rapid shifts towards a highly interdependent world was structured. I have specifically argued that that form has been centred upon four main elements of the global-human condition: societies, individuals, the international system of societies, and humankind. It is around the changing relationships between, different emphases upon and often conflicting interpretations of these aspects of human life that the contemporary world as a whole has crystallized. So in my perspective the issue of what is to be included under the notion of the global is treated very comprehensively. The global is not in and of itself counterposed to the local. Rather, what is often referred to as the local is essentially included within the global.

In this respect globalization, defined in its most general sense as the compression of the world as a whole, involves the linking of localities. But it also involves the 'invention' of locality, in the same general sense as the idea of the invention of tradition, as well as its 'imagination'. There is indeed currently something like an 'ideology of home' which has in fact come into being partly in response to the constant repetition and global diffusion of the claim that we now live in a condition of homelessness or rootlessness; as if in prior periods of history the vast majority of people lived in 'secure' and homogenized locales. Two things, among others, must be said in objection to such ideas. First, the form of globalization has involved considerable emphasis, at least until now, on the cultural homogenization of nationally constituted societies; but, on the other hand, prior to that emphasis, which began to develop at the end of the eighteenth century, what McNeill calls polyethnicity was normal. Second, the phenomenological diagnosis of the generalized homelessness of modern man and woman has been developed as if 'the same people are behaving and interpreting at the same time in the same broad social process'; whereas there is in fact much to suggest that it is increasingly global expectations concerning the relationship between individual and society that have produced both routinized and 'existential' selves. On top of that, the very ability to identify 'home', directly or indirectly, is contingent upon the (contested) construction and organization of interlaced categories of space and time.

But it is not my purpose here to go over this ground again, but rather to emphasize the significance of certain periods prior to the second half of the twentieth century when the possibilities for a single world seemed at the time to be considerable, but also problematic. Developing research along such lines will undoubtedly emphasize a variety of areas of the world and different periods. But as far as relatively recent times are concerned, I would draw attention to two arguments, both of which draw attention to rapid extension of communication across the world as a whole and thematize the central issue of changing conceptions of time-and-space. Johnson has in his book, *The Birth of the Modern*, argued that 'world society' – or 'international society in its totality' – largely crystallized in the period 1815–30. Here the emphasis is upon the crucial significance of the Congress of Vienna which was assembled following Bonaparte's first abdication in 1814. According to Johnson, the peace settlement in Vienna, following what was in effect the first world war, was 'reinforced by the powerful currents of romanticism sweeping through the world'. Thus was established 'an international order which, in most respects, endured for a century'. Regardless of its particular ideological bent, Johnson's book is important because he does attempt not merely to cover all continents of the world but also to range freely over many aspects of life generally, not just world politics or international relations. He raises significant issues concerning the development of consciousness of the world as a whole, which was largely made possible by the industrial and communicative revolution on the one hand, and the Enlightenment on the other.

Second (and, regardless of the issue of the periodization of globalization, more important), Kern has drawn attention to the crucial period of 1880–1918, in a way that is particularly relevant to the present set of issues. In his study of the *Culture of Time and Space* Kern's most basic point is that in the last two decades of the nineteenth century and the first twenty years or so of the twentieth century very consequential shifts took place with respect to both our sense of space and time. There occurred, through international negotiations and technological innovations, a standardization of time–space which was inevitably both universal and particular: world time organized in terms of particularistic space, in a sense the co-ordination of objectiveness and subjectiveness. In other words, homogenization went hand in hand with heterogenization. They made each other possible. It was in this period that 'the world' became locked into a particular *form* of a strong shift to unicity. It was during this time that the four major 'components' of globalization which I have previously specified were given formidable concreteness. Moreover, it was in the late nineteenth century that there occurred a big spurt in the organized attempts to link localities on an international or ecumenical basis.

An immediate precursor of such was the beginning of international exhibitions in the mid nineteenth century, involving the internationally organized display of particular national 'glories' and achievements. The last two decades of the century witnessed many more such international or cross-cultural ventures, among them the beginnings of the modern religious ecumenical movement, which at one and the same time celebrated difference and searched for commonality within the framework of an emergent culture for 'doing' the relationship between the particular and the, certainly not uncontested, universal. An interesting example of the latter is provided by the International Youth Hostel movement, which spread quite rapidly and not only in the northern hemisphere. This movement attempted on an organized international, or global, basis to promote the cultivation of communal, 'back to nature' values. Thus at one and the same time particularity was valorized but this was done on an increasingly globe-wide, pan-local basis.

The present century has seen a remarkable proliferation with respect to the 'international' organization and promotion of locality. A very pertinent example is provided by the current attempts to organize globally the promotion of the rights and identities of native, or indigenous, peoples. This was a strong feature, for example, of the Global Forum in Brazil in 1992, which, so to say, surrounded the official United Nations 'Earth Summit'. Another is the attempt by the World Health Organization to promote 'world health' by the reactivation and, if need be, the invention of 'indigenous' local medicine. It should be stressed that these are only a few examples taken from a multifaceted trend.

Glocalization and the Cultural Imperialism Thesis

Some of the issues which I have been raising are considered from a very different angle in Appiah's work on the viability of Pan-Africanism. Appiah's primary theme is 'the question of how we are to think about Africa's contemporary cultures in the light of the two main external determinants of her recent history – European and Afro-New World conceptions of Africa – and of her own endogenous cultural traditions'. His contention is that the 'ideological decolonization' which he seeks to effect can only be made possible by what he calls finding a 'negotiable middle way' between endogenous 'tradition' and 'Western' ideas, both of the latter designations being placed within quotation marks by Appiah himself. He objects strongly to what he calls the racial and racist thrusts of much of the Pan-African idea, pointing out that insofar as Pan-Africanism makes assumptions about the racial unity of all Africans, then this derives in large part from the experience and memory of non-African ideas about Africa and Africans which were prevalent in Europe and the USA during the latter part of the nineteenth century. Speaking specifically of the idea of the 'decolonization' of African literature, Appiah insists, I think correctly, that in much of the talk about decolonization we find what Appiah himself calls (again within quotation marks) a 'reverse discourse':

> The pose of repudiation actually presupposes the cultural institutions of the West and the ideological matrix in which they, in turn, are imbricated. Railing against the cultural hegemony of the West, the nativists are of its party without knowing it [...] [D]efiance is determined less by 'indigenous' notions of resistance than by the dictates of the West's own Herderian legacy – its highly elaborated ideologies of national autonomy, of language and literature as their cultural substrate. Native nostalgia, in short is largely fueled by that Western sentimentalism so familiar after Rousseau; few things, then, are less native than nativism in its current form.

Appiah's statement facilitates the explication of a particularly important point. It helps to demonstrate that much of the conception of contemporary locality and indigeneity is itself historically contingent upon *encounters* between one civilizational region and another. Within such interactions, many of them historically imperialistic, has developed a sense of particularistic locality. But the latter is in large part a consequence of the increasingly global 'institutionalization' of the expectation and construction of local particularism. Not merely is variety continuously produced and reproduced in the contemporary world, that variety is *largely an aspect of the very dynamics which a considerable number of commentators interpret as homogenization.* So in this light we are again required to come up with a more subtle interpretation than is usually offered in the general debate about locality and globality.

Some important aspects of the local–global issue are manifested in the general and growing debate about and the discourse of cultural imperialism. There is of course a quite popular intellectual view which would have it that the entire world is being swamped by Western – more specifically, American – culture. This view has undoubtedly exacerbated recent French political complaints about American cultural imperialism, particularly within the context of GATT negotiations. There are, on the other hand, more probing discussions of and research on this matter. For starters, it should be emphasized that the virtually overwhelming evidence is that even 'cultural messages' which emanate directly from 'the USA' are *differentially* received and interpreted; that 'local' groups 'absorb' communication from the 'centre' in a great variety of ways. Second, we have to realize that the major alleged producers of 'global culture' – such as those in Atlanta (CNN) and Los Angeles (Hollywood) – increasingly tailor their products to a differentiated global market (which they partly construct). For example, Hollywood attempts to employ mixed, 'multinational' casts of actors and a variety of 'local' settings when it is particularly concerned, as it increasingly is, to get a global audience. Third, there is much to suggest that seemingly 'national' symbolic resources are in fact increasingly available for differentiated global interpretation and consumption. For example, in a recent discussion of the staging of Shakespeare's plays, Billington notes that in recent years Shakespeare has been subject to wide-ranging cultural interpretation and staging. Shakespeare no longer belongs to England. Shakespeare has assumed a universalistic significance; and we have to distinguish in this respect between Shakespeare as representing Englishness and Shakespeare as of 'local-cum-global' relevance. Fourth, clearly many have seriously underestimated the flow of ideas and practices from the so-called Third World to the seemingly dominant societies and regions of the world.

Much of global 'mass culture' is in fact impregnated with ideas, styles and genres concerning religion, music, art, cooking, and so on. In fact the whole question of what will 'fly' globally and what will not is a very important question in the present global situation. We know of course that the question of what 'flies' is in part contingent upon issues of power; but we would be very ill-advised to think of this simply as a matter of the hegemonic extension of Western modernity. As Tomlinson has argued, 'local cultures' are, in Sartre's phrase, *condemned to freedom.* And their global participation has been greatly (and politically) underestimated. At this time 'freedom' is manifested particularly in terms of the social construction of identity-and-tradition, by the appropriation of cultural traditions. Although, as I have emphasized, this reflexiveness is typically undertaken along relatively standardized global-cultural lines. (For example, in 1982 the UN fully recognized the existence of indigenous peoples. In so doing it effectively established *criteria* in terms of which indigenous groups could and should identify themselves and be recognized formally. There are national parallels to this, in the sense that some societies have legal criteria for ethnic groups and cultural traditions.)

Then there is the question of diversity at the local level. This issue has been raised in a particularly salient way by Balibar, who talks of *world spaces.* The latter are places in which the world-as-a-whole is potentially inserted. The general idea of world-space suggests that we should consider the local as a 'micro' manifestation of the global – in opposition, *inter alia*, to the implication that the local indicates enclaves of cultural, ethnic, or racial homogeneity. Where, in other words, is *home* in the late-twentieth century? Balibar's analysis – which is centred on contemporary Europe – suggests that in the present situation of global complexity, the idea of home has to be divorced analytically from the idea of locality. There may well be groups and categories which equate the two, but that doesn't entitle them or their representatives to project their perspective onto humanity as a whole. In fact there is much to suggest that the senses of home and locality are contingent upon alienation from home and/or locale. How else could one have (reflexive) consciousness of such? We talk of the mixing of cultures, of polyethnicity, but we also often underestimate the significance of what Lila Abu-Lughod calls 'halfies'. As Geertz has said, 'like nostalgia, diversity is not what it used to be'. One of the most significant aspects of contemporary diversity is indeed the complication it raises for conventional notions of culture. We must be careful not to remain in thrall to the old and rather well established view that cultures are organically binding and sharply bounded. In fact Lila Abu-Lughod opposes the very idea of culture because it seems to her to deny the importance of 'halfies', those who combine in themselves as individuals a number of cultural, ethnic and genderal features. This issue is closely related to the frequently addressed theme of global hybridization, even more closely to the idea of creolization.

Conclusion: Sameness and Difference

My emphasis upon the significance of the concept of glocalization has arisen mainly from what I perceive to be major weaknesses in much of the employment of the term 'globalization'. In particular, I have tried to transcend the tendency to cast the idea of globalization as inevitably in tension with the idea of localization. I have instead maintained that globalization – in the broadest sense, the compression of the world – has involved and increasingly involves the creation and the incorporation of locality, processes which themselves largely shape, in turn, the compression of the world as a whole. Even though we are, for various reasons, likely to continue to use the concept of globalization, it might well be preferable to replace it for certain purposes with the concept of glocalization. The latter concept has the definite advantage of making the concern with space as important as the focus upon temporal issues. At the same time emphasis upon the global condition – that is, upon globality – further constrains us to make our analysis and interpretation of the contemporary world both spatial and temporal, geographical as well as historical.

Systematic incorporation of the concept of glocalization into the current debate about globalization is of assistance with respect to the issue of what I have called form. The form of globalization has specifically to do with the way in which the compression of the world is, in the broadest sense, structured. This means that the issue of the form of globalization is related to the ideo-logically laden notion of world order. However, I want to emphasize strongly that insofar as this is indeed the case, my own effort here has been directed only at making sense of two *seemingly* opposing trends: homogenization and heterogenization. These simultaneous trends are, in the last instance, complementary and interpenetrative; even though they certainly can and do collide in concrete situations. Moreover, glocalization can be – in fact, is – used strategically, as in the strategies of glocalization employed by contemporary TV enterprises seeking global markets (MTV, then CNN, and now others). Thus we should realize that in arguing that the current form of globalization involves what is best described as glocalization I fully acknowledge that there are many different modes of practical glocalization. Thus, even though much of what I said in this chapter has been hinged upon the Japanese conception of glocalization, I have in fact generalized that concept so as, in principle, to encompass the world as a whole. In this latter perspective the Japanese notion of glocalization appears as a *particular version* of a very general phenomenon.

An important issue which arises from my overall discussion has to do with the ways in which, since the era of the nation-state began in the late eighteenth century, the nation-state itself has been a major agency for the production of diversity and hybridization. Again, it happens to be the case that Japan provides the most well-known example of what Westney calls cross-societal emulation, most clearly during the early Meiji period. I would, however, prefer the term, selective incorporation in order to describe the very widespread tendency for nation-states to 'copy' ideas and practices from other societies – to engage, in varying degrees of systematicity, in projects of importation and hybridization. So, even though I have emphasized that the cultural idea of the nation-state is a 'global fact', we also should recognize that nation-states have, particularly since the late nineteenth century, been engaged in selective learning from other societies, each nation-state thus incorporating a different mixture of 'alien' ideas.

There is still another factor in this brief consideration of 'hybridized national cultures'. This is the phenomenon of cultural nationalism. Yet again, this concept has emerged in particular reference to Japan. On the basis of a discussion of *nihonjinron* (the discourse on and of Japanese uniqueness), Yoshino argues that *nihonjinron* has, in varying degrees, been a common practice. Specifically, modern nations have tended to promote discourses concerning their own unique difference, a practice much encouraged in and by the great globalizing thrusts of the late nineteenth and early twentieth centuries. In this respect what is sometimes these days called strategic essentialism – mainly in reference to liberation movements of various kinds – is much older than some may think. It is in fact an extension and generalization of a long drawn-out process.

Finally, in returning to the issue of form, I would argue that no matter how much we may speak of global disorder, uncertainty and the like, generalizations and theorizations of such are inevitable. We should not entirely conflate the empirical issues with the interpretative-analytical ones. Speaking in the latter vein we can conclude that the form of globalization is currently being reflexively reshaped in such a way as to increasingly make projects of glocalization the constitutive features of contemporary globalization.

Critiquing Creolization, Hybridity, and Glocalization

In the preceding chapter we offered some of the key works on several central concepts in the cultural aspects of globalization – creolization, hybridity, and glocalization. These ideas, like all ideas in the field of globalization, have elicited a variety of critiques and led to a number of debates.

We begin with an excerpt from an essay by Jan Nederveen Pieterse, the leading exponent of the concept of hybridity, on what he calls "the anti-hybridity backlash." While this is too strong an expression, and his enumeration of the criticisms of hybridity is more about a defense of the concept, it nonetheless gives us a handy overview of those criticisms. They are that:

- Hybridity is meaningful only as a critique of essentialism.
- It is doubtful that colonialist times were really so essentialist.
- Hybridity is dependent on essentialism rather than combating it.
- It is a trivial concept.
- It is only hybrid self-identification that is important.
- All the talk of hybridity is a function of the decline of Western hegemony.
- Cosmopolitans like the idea of hybridity; it helps in their quest for hegemony.

- Intellectuals celebrate border-crossing, but real border-crossers fear the border.

Keith Nurse looks at carnival as a hybrid phenomenon. He looks at carnival in one of its "homes" in Trinidad, as well as overseas in places like Britain. However, rather than the normal focus on flows from the core (e.g. Britain) to the periphery (Trinidad), he looks at "colonization in reverse," the "extraordinary process of periphery-induced creolization in the cosmopolis."[1] While the carnival in Trinidad is a hybrid in form and influence, this is at least as much the case in carnival overseas:

- Masks from Jamaica and the Bahamas, not used in the Trinidad carnival, are evident in many overseas carnivals which are more pan-Caribbean in character.
- Carnivalesque traditions from other immigrant communities have been integrated including those from South America (Brazilian samba dancers), Africa and Asia.
- Local whites have become participants.
- Carnivals overseas have become "multicultural or poly-ethnic festivals."[2]
- Overseas carnivals have, over time, become more contained and controlled as they have become larger and more commercial.

- Other carnivals, especially in Europe, have come to be affected ("colonized") by carnival.
- Conflict has arisen between ethnic groups (Trinidadians vs Jamaicans) over, for example, the preferred music (reggae vs calypso).

Nurse concludes that Trinidad carnival needs to be seen as both the *"localization of global influences"* and the *"globalization of local impulses."*[3] In terms of the former, the Trinidad carnival is the "historical outcome of the hybridization of multiple ethnicities and cultures," while in terms of the latter the carnivals have come "to embrace, if not 'colonize,' the wider community in the respective host societies."[4] Overall, Nurse concludes: "The Trinidad carnival and its overseas offspring is a popular globalized celebration of hybridity and cultural identity, a contested space and practice, a ritual of resistance which facilitates the centring of the periphery."[5]

Kraidy examines hybrid practices among Maronite youth in Lebanon. He finds that they form hybrid identities largely out of the consumption of US and Lebanese television programs. While these were accepted, they dismissed Egyptian soap operas and Latin American *telenovelas* because of their poor quality. Hybrid identities were formed out of both that which was accepted and that which was rejected. Identities were both Western and Arab, but also different from both. Kraidy studied his subjects using "native ethnography" which itself requires a hybrid ability to deal with different cultural systems. Using this methodology, Kraidy is able to show that "hybridity is not a negation of identity, but its quotidian, vicarious, and inevitable condition."[6]

This research shows the importance not only of the local but also of the West, of its popular culture, and of cultural imperialism. Rather than looking at the conflict between the global and the local, Kraidy prefers the term "glocalization" which "takes into account the local, national, regional, and global contexts of intercultural communicative processes."[7] In this context, identity can be seen "as a process which is simultaneously assimilationist and subversive, restrictive and liberating."[8]

Kraidy's conclusion on the utility of the concept of glocalization feeds nicely into Thornton's work on the glocal, as well as the limitations of the concept of glocalization. Thornton gets at a key weakness in the idea of glocalization in that it "may operate at the expense of more 'revolting' but ultimately more resistant strains of difference . . . may too easily resolve the critical tension between global and local values."[9] Thus, Thornton critiques the concept of glocalization, especially as it was articulated by Robertson, as having "no teeth" and as not being able to consider "dissonant rumblings."[10] Thornton also sees Robertson's notion of glocalization as "an inoculation against further resistance" and as serving "capitalist globalization by naturalizing it."[11]

Ritzer expands on this critique of the glocal by arguing that it needs to be complemented by the idea of the "grobal" and more generally by "grobalization" (as a complement to glocalization), defined as "the imperialistic ambitions of nations, corporations, organizations, and other entities and their desire – indeed their need – to impose themselves on various geographic areas."[12] This obviously carries with it negative implications, and when employed in combination with glocalization gives globalization both a positive and a negative orientation. Thus, we need to look at the relationship between grobalization and glocalization.

Further nuances are given to globalization theory when we add the something/nothing dimension. Something is that which is indigenously conceived, indigenously controlled, and rich in distinctive content, while nothing is that which is centrally conceived, centrally controlled, and lacking in distinctive content. Something is often, but *not* always, positive, while nothing is often, but *not* always, negative. Putting the two sets of concepts together, the most negative is the grobalization of nothing, especially because it poses a threat to the glocalization of something. A critical impulse is also added when we recognize that the glocalization of nothing is also a largely negative process. Rounding out the theoretical alternatives, the grobalization of something is largely positive in nature. Taken together, all of this gives globalization theory far greater balance in terms of its ability, unlike glocalization theory, to deal with both the positive and the negative aspects of globalization.

Finally, Ritzer's ideas have themselves come under the critical gaze of Douglas Kellner. The heart of his critique lies in Ritzer's focus on consumption and Kellner's argument that the theoretical approach used by Ritzer must include production as well. That is, production is being globalized and the ideas of grobalization/glocalization and something/nothing can be applied usefully to it as well.

Kellner also critiques Ritzer for a number of the examples he offers of nothing. Kellner argues that brands, logos, audio guides, and soaps should not be considered as examples of nothing, but rather are something because people (including Kellner in some cases) value them. While this is true, it involves slipping into a conventional usage of the terms "something" (and "nothing") rather than adhering to Ritzer's definition. By that definition, all of the examples cited by Kellner *are* nothing – centrally conceived, controlled, lacking in distinctive content.

Kellner recognizes the central tension made clear in Ritzer's work in both globalization theory and in the globalization process itself. However he sees a need for more concrete goals and actions and he sees far more evidence of the local than Ritzer does in his discussion of the "death of the local."

NOTES

1   Louise Bennett, and Orlando Patterson, cited in Keith Nurse, "Globalization and Trinidad Carnival: Diaspora, Hybridity and Identity in Global Culture." *Cultural Studies* 13, 4, 1999: 663.
2   Ibid., 675.
3   Ibid., 683, italics in original.
4   Ibid., 683.
5   Ibid., 685.
6   Marwan M. Kraidy, "The Global, the Local, and the Hybrid: a Native Ethnography of Glocalization." *Critical Studies in Mass Communications* 16, 1999: 471.
7   Ibid., 472.
8   Ibid., 473.
9   William H. Thornton, "Mapping the 'Glocal' Village: The Political Limits of 'Glocalization.'" *Continuum: Journal of Media and Cultural Studies* 14, 1, 2000: 81.
10 Ibid., 81.
11 Ibid., 82.
12 George Ritzer, "Rethinking Globalization: Glocalization/Grobalization and Something/Nothing." *Sociological Theory* 21, 3, 2003: 194.

 **READING 51**

Hybridity, So What? The Anti-Hybridity Backlash and the Riddles of Recognition

Jan Nederveen Pieterse

[. . .]

Criticisms of particular versions of hybridity arguments and quirks in hybridity thinking are familiar. The most conspicuous shortcoming is that hybridity skips over questions of power and inequality: 'hybridity is not parity.' Some arguments make no distinction between different levels: 'The triumph of the hybrid is in fact a triumph of neo-liberal multiculturalism, a part of the triumph of global capitalism.' These wholesale repudiations of hybridity thinking belong in a different category: this is the anti-hybridity backlash, which this article takes on. In the discussion below most arguments against hybridity thinking have been taken from Friedman as representative of a wider view. A précis of anti-hybridity arguments and rejoinders is in Table 1.

Hybridity Is Meaningful Only as a Critique of Essentialism

There is plenty of essentialism to go round. Boundary fetishism has long been, and in many circles continues to be, the norm. After the nation, one of the latest forms of boundary fetishism is 'ethnicity'. Another reification is the 'local'. Friedman cites the statement

Table 1 Arguments for and against hybridity

Contra hybridity	Pro hybridity
Hybridity is meaningful only as a critique of essentialism.	There is plenty of essentialism around.
Were colonial times really so essentialist?	Enough for hybrids to be despised.
Hybridity is a dependent notion.	So are boundaries.
Asserting that all cultures and languages are mixed is trivial.	Claims of purity have long been dominant.
Hybridity matters to the extent that it is a self-identification.	Hybrid self-identification is hindered by classification boundaries.
Hybridity talk is a function of the decline of Western hegemony.	It also destabilizes other hegemonies.
Hybridity talk is carried by a new cultural class of cosmopolitans.	Would this qualify an old cultural class of boundary police?
'The lumpenproletariat real border-crossers live in constant fear of the border.'	Crossborder knowledge is survival knowledge.
'Hybridity is not parity.'	Boundaries don't usually help either.

above and then concludes that 'hybridization is a political and normative discourse.' Indeed, but so of course is essentialism and boundary fetishism. 'In a world of multiplying diasporas, one of the things that is not happening is that boundaries are disappearing.' That, on the other hand, is much too sweeping a statement to be meaningful. On the whole, cross-boundary and cross-border activities have been on the increase, as a wide body of work in international relations and international political economy testifies, where the erosion of boundaries is one of the most common accounts of contemporary times and globalization.

Were Colonial Times Really So Essentialist?

This is a question raised by Young. Here we can distinguish multiple levels: actual social relations, in which there was plenty of border-crossing, and discourse, which is differentiated between mainstream and marginal discourses. Discourse and representation were also complex and multilayered, witness for instance the mélange of motifs in Orientalism. While history, then, is a history of ambivalence, attraction and repulsion, double takes and zigzag moves, nevertheless the 19th and early 20th-century colonial world was steeped in a Eurocentric pathos of difference, *dédain*, distinction. All the numerous countermoves in the interstices of history do not annul the *overall* pathos of the White Man's Burden and the *mission civilisatrice*, nor its consequences.

But the imperial frontiers are not only geographical frontiers, where the 'civilized' and the 'barbarians' confront and contact one another; they are also frontiers of status and ethnicity which run through imperialized societies, as in the form of the colonial 'colour bar'. Here colonizers and colonized are segregated and meet, here slave masters and slaves face one another and here, where imperial posturing is at its most pompous and hatred is most intense, the imperial house of cards folds and paradox takes over. For this frontier is also the locus of a *genetic dialectic*, a dialectic which, in the midst of the most strenuous contradictions, gives rise to that strangest of cultural and genetic syntheses – the *mulatto, mestizo*, half-caste. The mestizo is the personification of the dialectics of empire and

emancipation. No wonder that in the age of empire the mestizo was dreaded as a monster, an infertile hybrid, an impossibility: subversive of the foundations of empire and race. The mestizo is the living testimony of an attraction that is being repressed on both sides of the frontier. The mestizo is proof that East and West *did* meet and that there is humanity on either side.

Hybridity Is a Dependent Notion

'In the struggle against the racism of purity, hybridity invokes the dependent, not converse, notion of the mongrel. Instead of combating essentialism, it merely hybridizes it.' The mongrel, half-caste, mixed race, *métis, mestizo* was a taboo figure in the colonial world. When so much pathos was invested in boundaries, boundary crossing involved dangerous liaisons. In an era of thinking in biological terms, boundaries were biologized ('race'), and by extension so was boundary crossing. Status, class, race, nation were all thought of as *biological* entities in the lineage from Comte de Boulainvilliers and Gobineau to Houston Stewart Chamberlain and Hitler.

By the turn of the century, genetics had gone through a paradigm shift from a dominant view that gene mixing was weakening and debilitating (decadence) to the view in Mendelian genetics that gene mixing is invigorating and that combining diverse strains creates '*hybrid vigour*'. This principle still guides plant-breeding companies now. Social and cultural hybridity thinking takes this further and revalorizes the half-castes. The gradual emergence of hybrid awareness (in 19th-century novels, psychoanalysis, modernism, bricolage) and its articulation in the late 20th century can be sociologically situated in the rapid succession of waning aristocracy (as represented in the theme of *décadence*), bourgeois hegemony and its supersession and reworking from the second half of the 20th century.

Hybridity as a point of view is meaningless *without* the prior assumption of difference, purity, fixed boundaries. Meaningless not in the sense that it would be inaccurate or untrue as a description, but that, without an existing regard for boundaries, it would not be a point worth making. Without reference to a prior cult of purity and boundaries, a pathos of hierarchy and gradient of difference, the point of hybridity would be moot.

Asserting That All Cultures and Languages are Mixed Is Trivial

Trivial? When since time immemorial the dominant idea has been that of pure origins, pure lineages? As in perspectives on language, nation, race, culture, status, class, gender. The hieratic view was preoccupied with divine or sacred origins. The patriarchal view posited strong gender boundaries. The aristocratic view cultivated blue blood. The philological view saw language as the repository of the genius of peoples, as with Herder and the subsequent 'Aryan' thesis. The racial view involved a hierarchy of races. The Westphalian system locked sovereignty within territorial borders. Next came the nation and chauvinism. All these views share a preoccupation with pure origins, strong boundaries, firm borders. The contemporary acknowledgement of mixture in origins and lineages indicates a sea change in subjectivities and consciousness that correlates, of course, with sea changes in social structures and practices. It indicates a different ethos that in time will translate into different institutions. To regard this as trivial is to misread history profoundly.

Hybridity Matters to the Extent That It Is Self-Identification

Hybridity only exists as a social phenomenon when it is identified as such by those involved in social interaction. This implies that where people do not so identify, the fact of cultural mixture is without social significance [. . .] hybridity is in the eyes of the beholder, or more precisely in the practice of the beholder.

Hybrid self-identification *is* in fact common: obvious instances are second-generation immigrants and indeed hyphenated identities. Tiger Woods, the champion golfer, describes himself as 'Cablinasian': 'a blend of Caucasian, black, Indian and Asian.' Donald Yee, who is part black, part Asian and part Native American, can sympathize. 'When Mr Yee fills out racial questionnaires, he frequently checks "multiracial". If that is not an option, he goes with either black or Asian. "Nothing bothers me", he said. "It is just that it doesn't capture all of me".'
Creolization in the Caribbean, *mestizaje* in Latin America and fusion in Asia are common self-definitions.

In some countries national identity is overtly hybrid. Zanzibar is a classic instance. Mexico and Brazil identify themselves as hybrid cultures. Nepal is a mélange of Tibetan, Chinese and Indian culture of the Gangetic plains and the same applies to Bhutan. Singapore's identity is often referred to as Anglo-Chinese.

Even so, the view that, in relation to hybridity, only self-identification matters presents several problems. (1) The obvious problem is how to monitor hybrid self-identification since most systems of classification and instruments of measurement do not permit multiple or in-between identification. In the United States, 'Until 1967 states were constitutionally permitted to ban mixed-race marriages. More than half the states had anti-miscegenation statutes in 1945; 19 still had them in 1966.' The US census is a case in point. The 2000 census is the first that, after much resistance and amid ample controversy, permits multiple self-identification, i.e. as being white as well as African American, Hispanic, etc. (2) What about the in-betweens? The point of hybridity thinking is that the in-betweens have been numerous all along and because of structural changes have been growing in number. (3) Only the eye of the beholder counts? Going native as epistemological principle? Because most people in the Middle Ages thought the earth is flat, it was flat? Because between 1840 and 1950 many people were racist, there are races? Or, there were as long as most people thought so? Jews were bad when most Germans under National Socialism thought so? *Vox populi, vox dei* – since when? This is unacceptable in principle and untenable in practice.

Hybridity Talk Is a Function of the Decline of Western Hegemony

This is true in that the world of Eurocentric colonialism, imperialism and racism is past. It is only partially true because hybridity talk can refer just as much to the passing of other centrisms and hegemonies, such as China the middle kingdom, Japan and the myth of the pure Japanese race, Brahmins in India, Sinhala Buddhists in Sri Lanka and their claim to 'Aryan' origins, Israel the Jewish state, Kemalist Turkism centred on Anatolia, Greekness among the Greeks. For all hegemonies, the claim to purity has served as part of a claim to power. This applies to all status boundaries, not just those of nation, ethnicity or race. The Church clamped down on heresies; the aristocracy and then the bourgeoisie despised mésalliance. Status requires boundaries and with boundaries come boundary police.

Hybridity Talk Is Carried by a New Cultural Class of Cosmopolitans Who Seek to Establish Hegemony

Hybridity represents 'a new "elite" gaze', 'a new cosmopolitan elite'. Here innuendo comes in. *Ad hominem* reasoning, casting aspersions on the motives of the advocates of an idea, rather than debating the idea, is not the most elevated mode of debate. Then, should we discuss the motives of those who talk homogeneity? Of those who talk of boundaries allegedly on behalf of the working class and 'redneck' virtues? Of those who create a false opposition between working-class locals and cosmopolitan airheads? According to Friedman, 'Cosmopolitans are a product of modernity, individuals whose shared experience is based on a certain loss of rootedness [. . .] Cosmopolitans identify with the urban, with the "modern" [. . .] They are the sworn enemies of national and ethnic identities.'

Aversion to cosmopolitanism and the decadence of city life was part of Hitler's outlook and the Nazi ideology of blood and soil. With it came the Nazi idealization of the German peasant and, on the other hand, anti-Semitism. According to a German source in 1935: 'Dangers threaten the nation when it migrates to the cities. It withers away in a few generations, because it lacks the vital connection with the earth. The German must be rooted in the soil, if he wants to remain alive.'

It is odd to find this combination of elements restated. For one thing, it is an ideological and not an analytical discourse. Brief rejoinders are as follows. (1) The specific discourse of cosmopolitanism does not really belong in this context; there is no necessary relationship. But if it is brought in, one would rather say that humanity is a cosmopolitan species. Adaptability to a variety of ecological settings is inherent in the species. (2) Also if this view is not accepted, cosmopolitanism still pre-dates modernity and goes back to the intercivilizational travel of itinerant craftsmen, traders and pilgrims. (3) The stereotype that is implicitly invoked here echoes another stereotype, that of the wandering Jew. (4) Why or by which yardstick would or should

'rootedness' be the norm? Have nomadism and itineracy not also a long record? (5) Why should affinity with the urban (if it would apply at all) necessarily involve animosity to national and ethnic identities? The Romantics thought otherwise. Cities have been central to national as well as regional identities. (6) According to Friedman, 'Modernist identity as an ideal type is anti-ethnic, anti-cultural and anti-religious.' 'Anti-cultural' in this context simply does not make sense. Apparently this take on modernism excludes Herder and the Romantics and assumes a single ideal-type modernity.

While Intellectuals May Celebrate Border-Crossing, the Lumpenproletariat Real Border-Crossers Live in Constant Fear of the Border

Experiences with borders and boundaries are too complex and diverse to be captured under simple headings. Even where boundaries are strong and fences high, knowing what is on either side is survival knowledge. This is part of the political economy of mobility. Geographical mobility is an alternative key to social mobility. In negotiating borders, hybrid bicultural knowledge and cultural shape-shifting acquire survival value. 'Passing' in different milieus is a survival technique. This applies to the large and growing transborder informal sector in which migrant grassroots entrepreneurs turn borders to their advantage.

Friedman sees it otherwise.

> But for whom, one might ask, is such cultural transmigration a reality? In the works of the post-colonial border-crossers, it is always the poet, the artist, the intellectual, who sustains the displacement and objectifies it in the printed word. But who reads the poetry, and what are the other kinds of identification occurring in the lower reaches of social reality?

(Elsewhere: 'This author, just as all hybrid ideologues, takes refuge in literature.') This is deeply at odds with common experience. Thus, research in English and German major cities finds that it is precisely lower-class youngsters, second-generation immigrants, who now develop new, mixed lifestyles. Friedman recognizes this among Turks in Berlin but then neutralizes this finding by arguing that 'the internal dynamics of identification and world-definition aim at coherence.' Why not? Hybridity is an argument against homogeneity, not against coherence. The point is precisely that homogeneity is not a requirement for coherence.

When Friedman does acknowledge hybridity he shifts the goalposts. 'Now this combination of cultural elements might be called hybridization, but it would tell us nothing about the processes involved.' The processes involved indeed may vary widely. And probably there is something like a stereotyping of hybridity – of world music stamp.

Friedman's argument against hybridity is inconsistent, contradictory and at times far-fetched, so it is not worth pursuing far. Friedman argues that all cultures are hybrid but that boundaries are not disappearing: these two statements alone are difficult to put together. He argues that hybridity talk is trivial unless it is self-identification, but if hybridity is part of self-identification it is overruled by coherence, and we should examine the processes involved. However, if all cultures are hybrid all along, then the problem is not hybridity but boundaries: how is it that boundaries are historically and socially so significant? How come that while boundaries continuously change shape in the currents and tides of history, boundary fetishism remains, even among social scientists? If hybridity is real but boundaries are prominent, how *can* hybridity be a self-identification: in a world of boundaries, what room and legitimacy are there for boundary-crossing identities, politically, culturally?

How to situate the anti-hybridity argument? At one level it is another instalment of the critique of 'postmodernism', which in these times recurs with different emphases every 10 years or so. In the present wave, the polemical emphasis is 'Marxism versus cultural studies', which is obviously a broad-stroke target. At another level the argument reflects unease with multiculturalism. When these two lines coincide we get the novel combination of redneck Marxism. In this view multiculturalism is a fad that detracts from, well, class struggle. A positive reading is that this refocuses the attention on political economy, class, social justice and hard politics, which is surely a point worth making in relation to Tinkberbell postmodernism. At the same time, this is an exercise in symbolic politics, unfolding on a narrow canvas, for it mainly concerns positioning within academia. Would this explain why so much is missing from the debate? Among the fundamental considerations that are missing in the anti-hybridity backlash is the historical depth of hybridity viewed in the *longue durée*. More important still is the circumstance that boundaries and borders can be issues of life and death; and the failure to recognize and acknowledge hybridity is then a political point that may be measured in lives.

[...]

 ## READING 52

The Global, the Local, and the Hybrid: A Native Ethnography of Glocalization

Marwan M. Kraidy

[...]

Enacting Hybrid Identities: Consumption, Mimicry, and Nomadism

Hybridity as consumption

After Baudrillard defined consumption as "an active mode of relationships [...] a mode of systematic activity and global response upon which the entirety of our cultural system is founded," thinkers like Bourdieu, de Certeau, and the active audience formation gave consumption its *lettres de noblesse* as the prime meaning-making everyday life activity. Oddly, interlocutors began with literary examples to explain how they gravitated towards hybrid television and musical genres. They revealed a predilection for consuming ostensibly hybrid publications. Citing Milan Kundera and Tahar Ben Jalloun, Fuad said that he "love[d] and identified with border-crossing writers," living "between two or more worlds" and "perpetually looking for an identity of their own." Antoun, Maha, Adib and Peter also mentioned Lebanese-French author Amin Maalouf and the anti-colonial *négritude* formation in Africa as favorite writers.

Some claimed admiring Rushdie as a typical "in-between" (Fuad) writer. Because of the controversy caused by Rushdie's *Satanic Verses* and the outrage of Muslim clerics throughout the Arab and Muslim worlds, a symbolic alliance with the West via Rushdie's books ostensibly serves to differentiate the Maronites from their Muslim neighbors. However, some interlocutors criticized the *Satanic Verses* for its offending content to Muslims while at the same time praising Rushdie's other books, thus assuming an "in-between" position, once again symptomatic of hybridity. On yet another level, the fact that interlocutors claimed that they had access to Rushdie's books, banned in Lebanon since the publication of *Satanic Verses*, reflects another tactic of cultural poaching through the acquisition and consumption of prohibited material.

The Lebanese television industry has historically shunned local dramatic productions and favored less costly Egyptian, French, and American imports. The few locally produced television dramas focused on village life or historical events. *The Storm Blows Twice*, a 1996–7 Lebanese dramatic series, marked a break with that tradition in its daring treatment of contemporary social issues. The series depicted a society caught between tradition and modernity, with characters, including women, struggling to keep a balance between family and career, conservative social norms, and individual freedoms. Religious restrictions are questioned, social taboos broken, and controversial issues tackled in the program. Characters explicitly discussed premarital sex and divorced women were positively depicted pursuing successful careers. This is unusual in conservative Arab societies. Young Maronites especially appreciated that *The Storm Blows Twice* broke social taboos in a daring but not offensive manner. In doing so, interlocutors said, characters in the *Storm Blows Twice* picked the best from tradition and modernity but did not completely embrace either of the two. Young Maronites closely identified with the daily negotiation of the two worldviews.

The enactment of hybridity is strongly manifest in my interlocutors' infatuation with the music of Fairuz and the Rahbanis, who are one of Lebanon's most famous cultural exports, enjoying a nearly mythical status in Lebanon and the Arab world, and an appreciation in Europe and North America. Fairuz and the Rahbanis' monumental oeuvre blended Lebanese folk melodies with modern music. Ziad, the son of Fairuz and Assi Rahbani, introduced jazz to Lebanese music. Revealing hybridity's dual assimilationist and subversive thrusts, Peter describes Ziad's music as "very homogenous" and yet "pluralistic" but "not fragmented," stressing that the "harmonious mélange" of Rahbani's "Oriental jazz" was "the greatest music ever." Whereas Elham described the music as a "unique mixture of [. . .] conflicting cultural legacies," like Hebdige's cut'n'mix Caribbean musics, Fuad and Antoun agreed with Peter that Ziad's music was "influenced by so many musical forms and currents, but [. . .] [was] different from all of them." In Fuad's words:

> You cannot discern different structural musical elements in his music. You cannot say this part is jazz, this other Arabic. It is a unique and innovative blending. Just like his father was influenced by classical music but never let it dominate his music, Ziad is very subtle in mixing differences. Others have been trying to blend Western and Eastern music, but the result is artificial. It has no genius and no creativity.

Thus the resonance of Stuart Hall's rhetorical question: "Are there any musics left that have not heard some other music?"

The assertion predominant in interlocutors' narratives that Fairuz and Ziad Rahbani were "typically Lebanese," and Elham's description of their music as "more Lebanese than the cedar," underscores how important hybrid texts were to young Maronites. Since the cedar is the quintessential symbol of Lebanon, such a hierarchical reversal posits Fairuz and the Rahbanis as the paramount cultural text, indeed the only cultural matrice that all young Maronites I spoke with identified with unconditionally. It also posited Lebanon itself as a hybrid national space. This preference for hybrid cultural products reveals the importance of "cultural proximity" in audience tastes and choices.

Hybridity as mimicry

Early in my fieldwork, I noticed that many young Maronites mimicked snapshots of Western lifestyles. My interlocutors validated my observations and made several unsolicited remarks about mass media's perceived role in the phenomenon of imitation. Antoun claimed that young Maronites liked to live "the European way, or the American way," in his own words, "maybe because of all [those] programs on television." As an example, Antoun invoked the "torn jeans fashion" which he imputed to the influence of *Music Television (MTV)*. Using the same example of torn jeans, Peter spoke of a "tremendous phenomenon of imitation of everything Western, particularly from the United States." Claiming that fads took "phenomenal proportions" among young Maronites," he argued that "things [were] swallowed rapidly, snatched up, as if [young people were] waiting for something new to swallow in order *to fill an unbearable void*" (emphasis mine). Invoking "this urge to imitate," Serge told me how sentences from *Beverly Hills 90210* became "leitmotifs, repeated over and over again: the word 'man,' for instance. Also 'hi guys,' 'I've had it' and others." Serge concluded that *90210* had become a cult series in Lebanon because "young people really [identified] with that bright picture of happy shiny boys and girls."

Interlocutors, in a somewhat self-criticizing tone, stated that the phenomenon of imitation of Western fashion and lifestyles among young Maronites was mostly on the superficial level of appearances rather than mentalities and actions. In other words, it is a phenomenon of simulation. Baudrillard established an interesting connection between dissimulation and simulation. "[T]o dissimulate," he wrote, "is to feign not to have what one has," while "to simulate is to feign to have what one has not." According to Baudrillard, simulation means concealment of the non-existence of something; in other words, it is the display of a simulacrum, a copy with no original. Interlocutors' adoption of simulative strategies reflected a perceived lack of cultural identity wherein simulated action masks the absence of such an identity. Hence, mimicking Western popular culture served to symbolically fill a void. Elham explained, first in Arabic: "*We have a fragmented identity lost between two or three languages,*

between different world views. This leads to a crisis. An identity crisis" (emphasis mine). She proceeded in French: "*Nous sommes à cheval entre deux cultures* [we are straddling two cultures]. We do not really have any identity; *the stronger your feeling of not having an identity, the more you want to pretend to have one*" (emphasis mine).

Young Maronites thus constructed their identity by using hybridizing acts of mimicry and simulation. Simulation, because "it is simulacrum and because it undergoes a metamorphosis into signs and is invented on the basis of signs," serves to hide that a void exists and to project the impression that the void does not exist. As such, simulation helps young Maronites to navigate a cultural realm whose matrices irrevocably slip into hybridity. Baudrillard claimed that resorting to simulation is a manifestation of deterritorialization which is "no longer an exile at all [...] [but rather] a deprivation of meaning and territory." Nomadic tactics in my interlocutors" everyday life underscored that deprivation of meaning and territory.

Hybridity as nomadism

Media audiences have been theorized as nomadic communities of "impossible subjects," inhabiting no physical space, only discursive positions. In Grossberg's words, audiences are "located within varying multiple discourses which are never entirely outside of the media discourses themselves." In order to understand how young Maronites weave their hybrid identities, we need to articulate their media consumption with a variety of social, political, and cultural factors – local and global. More precisely, we need to look at the quotidian tactics young Maronites mobilize to make sense of these manifold factors. Mouffe embraces Derrida's notion of the "constitutive outside" which sees every identity as "irremediably destabilized by its exterior" and argues that identity is relational. From this perspective, the relationality of Maronite identity with its "Western" and "Arab" dialogical counterpoints is manifest in nomadic identity postures. In this context, Peter expressed his reluctance to identify himself as an Arab when he is among Westerners because of his weariness of being associated with Western stereotypes of Arabs. Antoun strongly expressed this context-bound nomadism when he said:

Sometimes yes, I am an Arab, but only sometimes. It depends. If a Christian asks me "are you Arab?" I will say "yes." If a Muslim asks me the same question, my answer will be "no." Why? Because if you are a Christian in an Arab country, you lose your rights and freedom.

This sweeping statement underscores the insecurity felt by a member of a minority whose apprehensions are expressed differently depending on the context. The multitude of competing identities and worldviews living cheek-by-jowl in Lebanon imposes on young Maronites nomadic tactics of identity construction and display. Thus young Maronites are *cultural chameleons*, nomads who blend with the different settings they cross.

Etymologically, the term "nomad" stems from the Greek "nomos," meaning "an occupied space without limits," and the Greek "nemo," which means "to pasture." Thus, a "nomad" is someone who lives in an open space without restrictions. Furthermore, "pasture" connotes a temporary sojourn in a particular location which the nomad leaves after having used what it had to offer. The term nomad does not necessarily imply physical movement from one place to the other. In *Nomadology: The War Machine*, Deleuze and Guattari explicate differences between nomads and migrants:

> The nomad is not at all the same as the migrant; for the migrant goes principally from one point to another, even if the second point is uncertain, unforeseen and not very well localized. But the nomad only goes from point to point as a consequence and as a factual necessity: in principle, points for him are relays along a trajectory.

Conflating Maronite and Lebanese identities, Fuad suggested that following nomadic identity construction strategies reflected the fact that Maronite Lebanese "*roam [. . .] in search of several identities*" (emphasis mine). Fuad lamented how slippery and blurred Lebanese, and especially Maronite, identity was. Of it he said:

> It is impossible to paint a portrait and point to it and say "this is the Lebanese." It is the Lebbedeh [traditional head dress] and the Sherwel [traditional pants] now, jeans and T-shirt some other time, and (smiling) maybe the [Indian] sari at some other occasion.

Fuad appeared to suggest that nomadic itineraries of self-definition were triggered by an absence of identity, resulting in a perpetual, circuitous, and never satisfied, search for an identity to adopt. This constant change of territory following peripatetic trajectories reflects the continuous evacuation of meaning inherent in the construction of hybrid cultural identities.

Conclusion: Glocalization, Hybridity, Hegemony

Departing from theoretical formulations of international interactions converging on hybridity, this paper explored the intersection of global and local media and cultural spheres in terms of the hybrid cultural identities enacted by young Maronites in Lebanon. I focused on the quotidian practices by which young middle-class Maronites develop and maintain a cultural identity located on the faultline between Western and Arab worldviews. In consuming media and popular culture, young Maronites use tactics of consumption, mimicry, and nomadism to weave the hybrid fabric of their cultural identities. In doing so, they identify with key elements from the cultural capital made available by a plethora of media. These constitutive elements were mainly US and Lebanese television programs, with the exception of the music and songs of Fairuz and Ziad Rahbani who emerged as favorite cultural texts. On the other hand, other programs such as Egyptian soap operas and Latin American *telenovelas* were harshly dismissed for their perceived poor dramatic and production qualities. By setting their own rules of inclusion and exclusion, young Maronite audience members used favorite and unpopular programs as dialogical counterpoints between which symbolic codes and cultural discourses were harnessed to construct, preserve, and defend hybrid identities. These identities were articulated as being part of both Western and Arab discourses but simultaneously different from both.

Responding to Appadurai's call for "ethnographic cosmopolitanism," I designed and used the method of native ethnography. A branch of critical ethnography, native ethnography occupies an intermediary position on the border between different worldviews. Because of their hybrid ability to negotiate a variety of traditions and contexts, native ethnographers are uniquely positioned to understand and conciliate these different cultural systems. As such, this study is an "ethnography of the particular," concerned with the explication of ways in which extralocal events and processes are articulated locally by people making do in their everyday life. As an enunciative modality, native ethnography demonstrates that hybridity is not a negation of identity, but its quotidian, vicarious, and inevitable condition. Native ethnography can thus be a significant contributor to the true internationalization of media and cultural studies.

The empirical data generated in this study in the form of personal narratives suggests entangled articulations of global and local discourses. However, we need to retain an important lesson from the literature on cultural imperialism, for despite its perceived unsubtlety, this perspective has unraveled inequities and power imbalances in international communication. The fact that "the West" was one of two overriding cultural worldviews revealed by young Maronites is witness to the ubiquity of Western popular culture. Besides, the hegemonic overtones that colored some young Maronites' perspective on "the Arabs" merit further empirical investigation and theoretical contemplation in addition to class and gender issues. Nevertheless, instead of looking at power in unifying terms of two distinct poles (the global and the local) locked in an unequal relationship in which the former dominates the latter, we need to trace, map, and study what Grossberg referred to as the "messy reality" of power in society. Rather than perceiving global/local interaction in terms of oppression and resistance, we should focus on power differences as they are manifested in everyday life "modalities of action" and "formalities of practices." If we are to understand local/global encounters, the discussion should focus on axiological *and* ontological grounds, adding "how" questions to "*why*" and "*in whose interest*" questions. We need to recognize with Murdock that "although [global] arenas circumscribe options for [local] action, they do not dictate them. There is always a repertoire of choices." More empirical cross-cultural research is needed to tackle these local options and to ground the underlying threads of and to better grasp the experiential manifestations of cultural hybridity.

Furthermore, we should perhaps adopt terms that better reflect global/local encounters than the now cliché "globalization." The term "*glocalization*" obtained

by telescoping "globalization" and "localization" is a more heuristic concept that takes into account the local, national, regional, and global contexts of intercultural communicative processes. The term has already been used in marketing, sociology, and geography. The communication discipline, more specifically international and intercultural communication research, could benefit from conceptual inroads made in other fields especially when these inroads carry a reinvigorating interdisciplinary potential. It is with this potential in mind that I propose a conceptualization of hybridity as *glocalization*, at the intersection of globalization and localization.

A deeper understanding of global/local interfaces can be achieved if empirical investigation departs from the following theoretical stances. First, we need to commit to the recognition that cultural hybridity is the rule rather than the exception in that what we commonly refer to as "local" and "global" have been long hybridized. Although historians have for years offered competing theories about the origins of the Maronites, young Maronites are more concerned with understanding and preserving their hybrid identities than eager to seek untraceable and mythical origins. This offers additional evidence to Stuart Hall and others' argument that intercultural contacts and their manifestations testify that "it is hybridity all the way down." Therefore, hybridity needs to be understood as a tautology rather than as a causation, hence the reading of globalization itself as hybridization.

Second, we need to acknowledge that hybridity is not a mere summation of differences whereby eclectic symbolic elements cohabitate. Rather, hybridity is the dialogical re-inscription of various codes and discourses in a spatio-temporal zone of signification. As such, conceptualizing hybridity entails re-formulating intercultural and international communication beyond buoyant models of resistance and inauspicious patterns of domination. The articulation of hybridity with hegemony is a step towards exiting the material/symbolic, political economy/cultural studies impasse. Such a leap would entail moving beyond an understanding of local/global interactions in strictly dialectical terms where the mingling of a variety of foreign cultural elements allegedly neutralizes differences. We need to theoretically establish and empirically investigate the quotidian tactics of hybridity as a knotty articulation of the dialectical and the dialogical. Articulating the poetics of meaning construction and the politics of consent formation, such a perspective looks at hybridity as an assertion of differences coupled with an enactment of identity, as a process which is simultaneously assimilationist and subversive, restrictive and liberating. In this endeavor, it may be helpful to remember Trinh Minh Ha's remark that "no matter how desperate our attempts to mend, categories will always leak."

 READING 53

Globalization and Trinidad Carnival: Diaspora, Hybridity and Identity in Global Culture

Keith Nurse

[. . .]

In the current debate about globalization and the growth of a global culture the main tendency is to focus on the recent acceleration in the flow of technology, people and resources in a North to South or centre to periphery direction. In this sense much of the literature on globalization is really a depoliticized interpretation of the long-standing process of Westernization and imperialism, terms that have become very unfashionable in these so-called postmodern times. Alternatively, the article is premised on the view that 'culturally, the periphery is greatly influenced by the society of the center, but the reverse is also the case'. Therefore, the aim of the study is to examine the counter-flow, the periphery-to-centre cultural flows, or what Patterson calls the 'extraordinary process of periphery-induced creolization in the cosmopolis'. In this respect it is a case study of 'globalization in reverse', a take on what Jamaican poet Louise Bennett calls 'colonization in reverse'.

The argument here is that the Trinidad carnival and its overseas or diasporic offspring are both products of and responses to the processes of globalization as well as 'intercultural and transnational formations' that relate to the concept of a Black Atlantic. Carnival is theorized as a hybrid site for the ritual negotiation of cultural identity and practice between and among various social groups. Carnival employs an 'esthetic of resistance' that confronts and subverts hegemonic modes of representation and thus acts as a counter-hegemonic tradition for the contestations and conflicts embodied in constructions of class, nation, 'race', gender, sexuality and ethnicity.

[. . .]

The Overseas Caribbean Carnivals

It is estimated that there are over sixty overseas Caribbean carnivals in North America and Europe. No other carnival can claim to have spawned so many offspring. These are festivals that are patterned on the Trinidad carnival or borrow heavily from it in that they incorporate the artistic forms (pan, mas and calypso) and the Afro-creole celebratory traditions (street parade/theatre) of the Trinidad carnival. Organized by the diasporic Caribbean communities, the overseas carnivals have come to symbolize the quest for 'psychic, if not physical return' to an imagined ancestral past and the search for a 'pan-Caribbean unity, a demonstration of the fragile but persistent belief that "All o' we is one"'. In the UK alone, there are as many as thirty carnivals that fall into this category. They are held during the summer months rather than in the pre-Lenten or Shrovetide period associated with the Christian calendar. The main parade routes are generally through the city centre or within the confines of the immigrant community – the former is predominant, especially with the larger carnivals.

Like its parent, the overseas carnival is hybrid in form and influence. The Jonkonnu masks of Jamaica and the Bahamas, not reflected in the Trinidad carnival, are clearly evident in many of these carnivals, thereby making them pan-Caribbean in scope. The carnivals have over time incorporated carnivalesque traditions from other immigrant communities: South Americans (e.g. Brazilians), Africans and Asians. For instance, it is not uncharacteristic to see Brazilian samba drummers and dancers parading through the streets of London, Toronto or New York during Notting Hill, Caribana or Labour Day. The white population in the respective locations have also become participants, largely as spectators, but increasingly as festival managers, masqueraders and pan players. Another development is that the art-forms and the celebratory traditions of the overseas Caribbean carnivals have been borrowed, appropriated or integrated into European carnivals to enhance them. Indeed, in some instances, the European carnivals have been totally transformed. Examples of this are the Barrow-in-Furness and Luton carnivals where there is a long tradition of British carnival. One also finds a similar trend taking place in carnivals in France, Germany, the Netherlands, Switzerland and Sweden, as they draw inspiration from the success of the Notting Hill carnival.

The first overseas Caribbean carnival began in the 1920s in Harlem, New York. This festival was later to become the Labour Day celebrations in 1947, the name that it goes by today. The major overseas Caribbean carnivals, for example, Notting Hill and Caribana, became institutionalized during the mid- to late 1960s at the peak in Caribbean migration. Nunley and Bettleheim relate the timing to the rise in nationalism in the Caribbean with the independence movement of the 1950s and 1960s. The emergence of the carnivals can also be related to the rise of black power consciousness. The growth in the number and size of the overseas Caribbean carnivals came in two waves. The first involved the consolidation of the early carnivals during the 1960s until the mid 1970s. From the mid 1970s, two parallel developments took place: the early carnivals expanded in size by broadening the appeal of the festival, for example, playing reggae music; and, through demonstration effect, a number of smaller carnivals emerged as satellites to the larger, older ones.

The carnivals have developed to be a means to promote cultural identity and sociopolitical integration within the Caribbean diasporic community as well as with the host society. The diversity in participation suggests that the overseas Caribbean carnivals have become multicultural or poly-ethnic festivals. For instance, Manning argues that the overseas Caribbean carnivals provide:

> a kind of social therapy that overcomes the separation and isolation imposed by the diaspora and restores to West Indian immigrants both a sense of community with each other and sense of connection to the culture that they claim as a birthright. Politically, however, there is more to these carnivals than cultural nostalgia. They are also a means through which West Indians seek and symbolize integration into the metropolitan society, by coming to terms with the opportunities, as well as the constraints, that surround them.

Manning's explanation of the significance of carnivals to the Caribbean diaspora is supported by the observations of Dabydeen:

> For those of us resident in Britain, the Notting Hill carnival is our living link with this ancestral history, our chief means of keeping in touch with the ghosts of 'back home'. In a society which constantly threatens or diminishes black efforts, carnival has become an occasion for self-assertion, for striking back – not with bricks and bottles but by beating pan, by conjuring music from steel, itself a symbol of the way we can convert steely oppression into celebration. We take over the drab streets and infuse them with our colours. The memory of the hardship of the cold winter gone, and that to come, is eclipsed in the heat of music. We regroup our scattered black communities from Birmingham, Manchester, Glasgow and all over the kingdom to one spot in London: a coming together of proud celebration.

Dabydeen goes on to illustrate that the carnivals are an integrative force in an otherwise segregated social milieu:

> We also pull in crowds of native whites, Europeans, Japanese, Arabs, to witness and participate in our entertainment, bringing alien peoples together in a swamp or community of festivity. Carnival breaks down barriers of colour, race, nationality, age, gender. And the police who would normally arrest us for doing those things (making noise, exhibitionism, drinking, or simply being black) are made to smile and

be ever so courteous, giving direction, telling you the time, crossing old people over to the other side, undertaking all manner of unusual tasks. They fear that bricks and bottles would fly if they behaved as normal. Thus the sight of smiling policemen is absorbed into the general masquerade.

From another perspective it is argued that the overseas carnivals reflect rather than contest institutionalized social hierarchies. In each of the major overseas carnivals the festival has been represented in ways which fit into the colonialist discourse of race, gender, nation and empire. The festival has suffered from racial and sexual stigmas and stereotypes in the media which are based on constructions of 'otherness' and 'blackness'. This situation became heightened as the carnivals became larger and therefore more threatening to the prevailing order. In the early phase, from the mid-1960s to the mid-1970s, the carnivals were viewed as exotic, received little if any press and were essentially tolerated by the state authorities. From the mid-1970s, as attendance at the festivals enlarged, the carnivals became more menacing and policing escalated, resulting in a backlash from the immigrant Caribbean community. Violent clashes between the British police and the Notting Hill carnival came to the fore in the mid- to late 1970s. Similar confrontations occurred at the other major overseas carnivals in New York and Toronto. Through a gendered lens 'black' male participants in the festivals have been portrayed as 'dangerous' and 'criminal'. Female participants, on the other hand, are viewed as 'erotic' and 'promiscuous'.

These modes of representation have come in tandem with heightened surveillance mechanisms from the state and the police. In the case of London, the expenditure by the state on the policing of the festival is several times larger than its contribution to the staging of the festival. The politics of cultural representation has negatively affected the viability of the overseas carnivals. The adverse publicity and racialized stigmas of violence, crime and disorder has allowed for the blockage of investments from the public and private sectors in spite of the fact that the carnivals have proved to be violence-free relative to other large public events or festivals. In the case of the UK, for instance, official figures show that Notting Hill, which attracts two million people, has fewer reported incidents of crime than the Glastonbury rock festival which attracts

60,000 people. Yet the general perception is that Notting Hill is more violence-prone.

Under increased surveillance the carnivals became more contained and controlled during the 1980s. The perspective of governments, business leaders and the media began changing when it was recognized that the carnivals were major tourist attractions and generated significant sums in visitor expenditures. For example, the publication of a 1990 visitor survey of Caribana, which showed that the festival generated Cnd$96 million from 500,000 attendees, resulted in the Provincial Minister of Tourism and Recreation visiting Trinidad in 1995 to see how the parent festival operated. Provincial funding for the festival increased accordingly. In 1995, for the first time, London's Notting Hill carnival was sponsored by a large multinational corporation. The Coca-Cola company, under its product Lilt, a 'tropical' beverage, paid the organizers £150,000 for the festival to be called the 'Lilt Notting Hill Carnival' and for exclusive rights to advertise along the masquerade route and to sell its soft drinks. That same year the BBC produced and televised a programme on the thirty-year history of the Notting Hill carnival. By the mid 1990s, as one Canadian analyst puts it, the carnivals were reduced to a few journalistic essentials: 'the policing and control of the crowd, the potential for violence, the weather, island images, the size of the crowd, the city economy and, most recently, the great potential benefit for the provincial tourist industry'. These developments created concern among some analysts. For example, Amkpa argues that:

strategies for incorporating and neutralizing the political efficacies of carnivals by black communities are already at work. Transnational corporations are beginning to sponsor some of the festivals and are contributing to creating a mass commercialized audience under the guise of bogus multiculturalisms.

Another analyst saw the increasing role of the state in these terms:

The funding bodies appear to treat it as a social policy as part of the race relations syndrome: a neutralised form of exotica to entertain the tourists, providing images of Black women dancing with policemen, or failing this, footage for the media to construct distortions and mis(sed)representations. Moreover, this view also sees that, if not for the problems it causes the police, courts, local authorities, and auditors, Carnival could be another enterprising venture.

In this respect one can argue that the sociopolitical and cultural conflicts, based on race, class, gender, ethnicity, nation and empire that are embedded in the Trinidad carnival were transplanted to the metropolitan context. In many ways the overseas carnivals, like the Trinidad parent, have become trapped between the negative imagery of stigmas and stereotypes, the co-optive strategics of capitalist and state organizations and the desires of the carnivalists for official funding and validation.

[. . .]

Trinidad Carnival and Globalization Theory

The foregoing analysis of the historical and global significance of Trinidad carnival presents some challenges to globalization theory. It suggests that the globalization of Trinidad carnival needs to be viewed as a dual process: the first relates to the *localization of global influences* and the second involves the *globalization of local impulses*. Drawing from the case of Trinidad carnival one can therefore argue that the formation of carnival in Trinidad is based upon the localization of global influences. The Trinidad carnival is the historical outcome of the hybridization of multiple ethnicities and cultures brought together under the rubric of colonial and capitalist expansion. New identities are forged and negotiated in the process. On the other hand, the exportation of carnival to overseas diasporic communities refers to the globalization of the local. The overseas Caribbean carnivals have grown in scale and scope beyond the confines of the immigrant population to embrace, if not 'colonize', the wider community in the respective host societies. This is what is referred to as 'globalization in reverse'. In sum, the overseas carnivals have become a basis for pan-Caribbean identity, a mechanism for social integration into metropolitan society and a ritual act of transnational, transcultural, transgressive politics.

Another observation is that historically, core societies are the ones most involved in the globalization of their local culture. For example, in most developed economies

cultural industry exports are seen as part of foreign economic policy. They recognize that perpetuating or transplanting one's culture is a critical factor in influencing international public opinion, attitude and value judgement. Peripheral societies are those that are more subject to importing cultural influences as opposed to exporting them. It is also the case that when peripheral societies export their culture they often lack the organizational capability and the political and economic leverage to control or maximize the commercial returns. This is in marked contrast to the capabilities of core societies where there is not only an ability to maximize on exports but also to co-opt imported cultures. What it comes down to is who is globalizing whom. In this business there are 'globalizers' and 'globalizees', those who are the producers and those who are just consumers of global culture. In this regard, it is far too premature to argue, as Appadurai has suggested, that centre periphery theories lack explanatory capability when it comes to transformations in the global cultural economy.

From this perspective one can argue that Trinidad, like other peripheral countries, has been on the receiving end of globalization except in the case of its carnival. This is to say that in an evaluation of globalization an appreciation for the resultant political hierarchies and asymmetries must be evident and caution should be employed so as not to construct new mythologies of change that depoliticize the systemic properties of the capitalist world system. In this regard, it is critical that the relevant historical period is conceptualized. The case of the Trinidad carnival suggests that the growth of historical capitalism in the past five hundred years is pivotal to understanding the causal relations and social forces that shaped and have evolved from the festival, both locally and globally, both in the recent past and the *longue durée*.

Another critical methodological issue is the conceptualization of space. Because of the heavy reliance on statecentric and nationalist analyses in the social sciences a wide array of activities and structures have escaped mainstream thought. The argument here is that the world has not changed as much as some make out, rather, it is that our awareness of change has been sharpened by the inadequacy of conventional thought. For example, one of the major contributions of postcolonial theory has been to introduce diaspora as a unit of analysis. This approach is particularly applicable to the case of Trinidad carnival, given the dual processes of globalization identified. The Trinidad carnival and its overseas offspring fits into Gilroy's concept of a Black Atlantic where 'double consciousness' and transnationalism are focal processes in the Caribbean's experience with globalization.

The study of the Trinidad carnival and its overseas offspring illustrates that globalization presents opportunities for some reversal in hegemonic trends. However, the case study shows that globalization is not a benign process and that there are limited possibilities for transformation, given the strictures and rigidities in the global political economy. The limitations are systemic in nature in that they relate to large-scale, long-term processes such as colonialist discourse and imperialism. In peripheral societies the political and economic elite are generally insecure and view the social protest in popular culture with much trepidation. They are therefore loath to acknowledge, far more invest in, the globalizing potential of the local popular culture. They are more likely to denigrate and marginalize it, and failing that, to co-opt it. Consequently, the tendency is for local capabilities not to be fully maximized at home. This suggests that the future contribution of Trinidad carnival to global culture may begin to move outside the control of the parent carnival and the home territory if a localized global strategy is not developed.

Historically, the carnivalesque spirit of festivity, laughter and irreverence feeds off the enduring celebration of birth, death and renewal and the eternal search for freedom from the strictures of official culture. From this perspective the Trinidad carnival confronts and unmasks sociohierarchical inequalities and hegemonic discourses at home and in the diaspora. Aesthetic and symbolic rituals operate as the basis for critiquing the unequal distribution of power and resources and a mode of resistance to colonialist and neocolonialist cultural representations and signifying practices. The Trinidad carnival and its overseas offspring is a popular globalized celebration of hybridity and cultural identity, a contested space and practice, a ritual of resistance which facilitates the centring of the periphery.

 READING 54

Mapping the "Glocal" Village: The Political Limits of "Glocalization"

William H. Thornton

[. . .]

'Glocalization' – a word that tellingly has its roots in Japanese commercial strategy – erases the dividing line between universalism and particularism, modernity and tradition. The resulting hybrid demythologizes locality as an independent sphere of values and undermines the classic Tonniesian antithesis of benign culture versus malign civilization. It operates, for example, in micro-marketing strategies that 'invent' (g)local traditions as needed – needed for the simple reason that diversity sells [. . .] In the case of Massey's 'global sense of place', this predilection for locational invention is flowing over into academic discourse, and particularly into cultural studies.

The danger is that this 'glocal' invention of difference may operate at the expense of more 'revolting' but ultimately more resistant strains of difference. Glocal theory, that is, may too easily resolve the critical tension between global and local values, thus abetting global commercial interests. For many on the Left, most notably David Harvey and Fredric Jameson, postmodernism is quite simply a solvent for global capitalism. From this perspective modernism arose out of an incomplete modernization and remained at least partially at odds with capitalistic 'logic'. Postmodernism, by contrast, issues from the triumphant completion of modernization and has no use for 'Pazian' resistance.

This study shares the wariness of Harvey and Jameson toward International Postmodernism, yet is equally wary of any Marxist solution to the problem. So too it is wary of some geocultural correctives, which replace the global anti-globalism of the Left with a hybrid (g)localism that, on closer examination, has no teeth.

Robertson, more than anyone, has made a signal contribution to the new cultural studies by countering the reductionist logic that allows Immanuel Wallerstein to treat religion, for example, as a pure epiphenomenon. One senses, however, a nascent rift between Robertson's anti-reductionism and Featherstone's. The latter contains, as will be shown, an agonistic current that saves it from the tension-dissolving synthesis of the 'glocal'. Robertson's optimism towards glocalization invites comparison with the global imperative he locates at the core of Elias's civilizational project. This obscures the significance of reactive cultural forms such as Pazian localism and/or Huntingtonesque (cultural or civilizational) regionalism, as well as reactive readings of culture in general. These dissonant rumblings simply do not register with Robertson.

[. . .]

Since Robertson rejects this pejorative view of the global, he has no need of the reactive concept of culture that would contest glocalization. Whether we are dealing with the retreatist localism of modernism or the resurgent localism of postmodernism, Robertson's 'glocal' amounts to an inoculation against further resistance. However inadvertently, this version of the glocal serves capitalist globalization by naturalizing it, rendering it acceptable by rendering it numbingly familiar. This puts the wolf in sheep's clothing, albeit a designer brand.

Robertson is well aware that his thesis runs counter to reactive views of culture such as that of Anthony Giddens, which he correctly sees as part of the global/local duality that 'glocalization' would expel. Vincent Leitch ties such reaction to the fact that every global,

virtualizing force is met by a stubborn alterity, such that the postmodern condition involves a dialectical intensification of both globalism and localism. Featherstone expands that dialectic to the plane of nation-states, while Huntington carries it all the way to the level of civilizational clash.

Try as he will, Robertson cannot escape the pull of this global/local duality, unless of course he gives up on resistance altogether. We have seen that he allows for the 'glocal' construction of diversity, if only as a tourist attraction. Can there be any doubt that a big part of that attraction is its place in an action–reaction dialectic? The category of (g)locality lives on in the global imaginary as the locus of all we feel to be missing in our social lives. It springs to life from the same horror vacui that Michel Maffesoli sees as the dialectical source of tribalism in mass society. It thrives, that is, in inverse proportion to the nostalgia that gives it life.

In a recent essay on glocalization, Robertson himself cites nostalgia as a prime source of cultural formation, and in a prior essay he underscores the continuing force of nostalgic resistance to globalization. Obviously his difference with Giddens is not founded on an empirical rejection of nostalgia as an element in the global/local dialectic. It derives, rather, from his normative judgement that the proper role of the theorist is 'positive'. By that he means 'analytic and critical' as opposed to the nostalgic negativity that he locates in Giddens. This preference rests on the assumption that negativity inherently voids critical analysis; yet negativity is sometimes (as was the case with Frankfurt School negative dialectics) the best available medium of social critique.

[. . .]

 READING 55

Rethinking Globalization: Glocalization/Grobalization and Something/Nothing

George Ritzer

This essay seeks to offer a unique theoretical perspective by reflecting on and integrating some well-known ideas in sociology (and the social sciences) on globalization

and a body of thinking, virtually unknown in sociology, on the concept of nothing (and, implicitly, something). The substantive focus will be on consumption, and all of the examples will be drawn from it. However, the implications of this analysis extend far beyond that realm, or even the economy more generally.

It is beyond the scope of this discussion to deal fully with globalization, but two centrally important processes – glocalization and grobalization – will be of focal concern. Glocalization (and related ideas such as hybridity and creolization) gets to the heart of what many – perhaps most – contemporary globalization theorists think about the nature of transnational processes. *Glocalization* can be defined as the interpenetration of the global and the local, resulting in unique outcomes in different geographic areas. This view emphasizes global heterogeneity and tends to reject the idea that forces emanating from the West in general and the United States in particular are leading to economic, political, institutional, and – most importantly – cultural homogeneity.

One of the reasons for the popularity of theories of glocalization is that they stand in stark contrast to the much hated and maligned *modernization theory* that had such a wide following in sociology and the social sciences for many years. Some of the defining characteristics of this theory were its orientation to issues of central concern in the West, the preeminence it accorded to developments there, and the idea that the rest of the world had little choice but to become increasingly like it (more democratic, more capitalistic, more consumption-oriented, and so on). While there were good reasons to question and to reject modernization theory and to develop the notion of glocalization, there are elements of that theory that remain relevant to thinking about globalization today.

In fact, some of those associated with globalization theory have adhered to and further developed perspectives that, while rejecting most of modernization theory, retain an emphasis on the role of Westernization and Americanization in global processes. Such concerns point to the need for a concept – *grobalization* – coined here for the first time as a much-needed companion to the notion of glocalization. While it does *not* deny the importance of glocalization and, in fact, complements it, grobalization focuses on the imperialistic ambitions of nations, corporations, organizations, and other entities and their desire – indeed, their need – to impose themselves on various geographic areas. Their main interest is in seeing their power, influence, and (in some cases) profits *grow* (hence the term "*gro*balization") throughout the world. It will be argued that grobalization tends to be associated with the proliferation of nothing, while glocalization tends to be tied more to something and therefore stands opposed, at least partially (and along with the local itself), to the spread of nothing. Globalization as a whole is not unidirectional, because these two processes coexist under that broad heading and because they are, at least to some degree, in conflict in terms of their implications for the spread of nothingness around the world.

Having already begun to use the concepts of nothing and something, we need to define them as they will be used here. Actually, it is the concept of nothing that is of central interest here (as well as to earlier scholars); the idea of something enters the discussion mainly because nothing is meaningless without a sense of something. However, nothing is a notoriously obscure concept: "Nothing is an awe-inspiring yet essentially undigested concept, highly esteemed by writers of a mystical or existentialist tendency, but by most others regarded with anxiety, nausea, or panic."

While the idea of nothing was of concern to ancient (Parmenides and Zeno) and medieval philosophers (St Augustine) and to early scientists (Galileo and Pascal) who were interested in the physical vacuum, the best-known early work was done by Shakespeare, most notably in *Much Ado About Nothing*. Of more direct interest is the work of some of the leading philosophers of the last several centuries, including Immanuel Kant, Georg Hegel, Martin Heidegger, and Jean-Paul Sartre. However, this is neither a work in philosophy nor the place to offer a detailed exposition of the recondite thoughts of these thinkers. Overall, the following generalizations can be offered about the contributions of the philosophical literature on nothing. First, it confirms a widespread and enduring interest in the topic, at least outside of sociology. Second, it fails to create a sense of nothing (and something) that applies well to and is usable in this analysis. Third, especially in the work of Kant and, later, Simmel, it leads us in the direction of thinking about form and content as central to conceptualizing nothing/something. Finally, it suggests issues such as loss as related to any consideration of nothing and its spread.

Conceptualizing Nothing (and Something)

Nothing is defined here as *a social form that is generally centrally conceived, controlled, and comparatively devoid of distinctive substantive content.* This leads to a definition of *something* as *a social form that is generally indigenously conceived, controlled, and comparatively rich in distinctive substantive content.* This definition of nothing's companion term makes it clear that neither nothing nor something exists independently of the other: *each makes sense only when paired with and contrasted to the other.* While presented as a dichotomy, this implies a *continuum* from something to nothing, and that is precisely the way the concepts will be employed here – as the two poles of that continuum.

A major and far more specific source of the interest here in nothing – especially conceptually – is the work in social geography by anthropologist Marc Auge on the concept of nonplaces (see also Morse on "nonspaces"; Relph). To Augé, nonplaces are "the real measure of our time." This can be generalized to say that nothing is, in many ways, the true measure of our time! The present work extends the idea of nonplaces to nonthings, nonpeople, and nonservices and, following the logic used above, none of these make sense without their polar opposites – places, things, people, and services. In addition, they need to be seen as the poles of four subtypes that are subsumed under the broader heading of the something/nothing continuum. Figure 1 offers an overview of the overarching something/nothing continuum and these four subtypes, as well as an example of each.

Following the definition of nothing, it can be argued that a credit card is nothing (or at least lies toward that end of the something/nothing continuum) because it is centrally conceived and controlled by the credit card company and there is little to distinguish one credit card (except a few numbers and a name) from any other (they all do just about the same things). Extending this logic, a contemporary credit card company, especially its telephone center, is a nonplace, the highly programmed and scripted individuals who answer the phones are nonpeople, and the often automated functions can be thought of as nonservices. Those entities that are to be found at the something end of each continuum are locally conceived and controlled forms that are rich in distinctive substance. Thus, a traditional line of credit negotiated by local bankers and personal clients is a thing; a place is the community bank to which people can go and deal with bank employees in person and obtain from them individualized services.

Nothing/Something and Grobalization/Glocalization

We turn now to a discussion of the relationship between grobalization/glocalization and something/nothing. Figure 2 offers the four basic possibilities that emerge when we cross-cut the grobalization/glocalization and something/nothing continua (along with representative examples of places/nonplaces, things/nonthings, people/nonpeople, and services/nonservices for each of the four possibilities and quadrants). It should be noted that while this yields four "ideal types," there are no hard and fast lines between them. This is reflected in the use of both dotted lines and multidirectional arrows in Figure 2.

Quadrants one and four in Figure 2 are of greatest importance, at least for the purposes of this analysis.

SOMETHING	NOTHING
Place (community bank)....................	Nonplace (credit card company)
Thing (personal loan)........................	Nothing (credit card loan)
Person (personal banker).................	Nonperson (telemarketer)
Service (individualized assistance)....	Nonservice (automated, dial-up aid)

Figure 1 The four major subtypes of something/nothing (with examples) presented as subcontinua under the broad something/nothing continuum

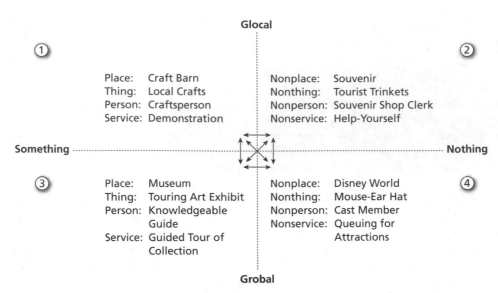

Figure 2 The relationship between glocal–grobal and something–nothing with exemplary (non-)places, (non-)things, (non-)persons, and (non-)services

They represent a key point of tension and conflict in the world today. Clearly, there is great pressure to grobalize nothing, and often all that stands in its way in terms of achieving global hegemony is the glocalization of something. We will return to this conflict and its implications below.

While the other two quadrants (two and three) are clearly residual in nature and of secondary importance, it is necessary to recognize that there is, at least to some degree, a glocalization of nothing (quadrant two) and a grobalization of something (quadrant three). Whatever tensions may exist between them are of far less significance than that between the grobalization of nothing and the glocalization of something. However, a discussion of the glocalization of nothing and the grobalization of something makes it clear that grobalization is not an unmitigated source of nothing (it can involve something) and that glocalization is not to be seen solely as a source of something (it can involve nothing).

The close and centrally important relationship between (1) grobalization and nothing and (2) glocalization and something leads to the view that there is an *elective affinity* between the two elements of each of these pairs. The idea of elective affinity, derived from the historical comparative sociology of Max Weber, is meant to imply that there is *not* a necessary, law-like causal relationship between these elements. That is, neither in the case of grobalization and nothing nor that of glocalization and something does one of these elements "cause" the other to come into existence. Rather, the development and diffusion of one tends to go hand in hand with the other. Another way of putting this is that grobalization/nothing and glocalization/something tend to mutually favor one another; they are inclined to combine with one another. Thus, it is far easier to grobalize nothing than something: the development of grobalization creates a favorable ground for the development and spread of nothing (and nothing is easily grobalized). Similarly, it is far easier to glocalize something than nothing: the development of glocalization creates a favorable ground for the development and proliferation of something (and something is easily glocalized).

However, the situation is more complex than this, since we can also see support for the argument that grobalization can, at times, involve something (e.g., art exhibits that move among art galleries throughout the world; Italian exports of food such as Parmigiano-Reggiano and Culatello ham; touring symphony orchestras and rock bands that perform in venues throughout the world) and that glocalization can sometimes involve nothing (e.g., the production of local souvenirs and trinkets for tourists from around the world). However, we would *not* argue that there is an elective affinity between grobalization and something and between glocalization and nothing. The existence of examples of the grobalization of something and the glocalization of nothing makes it clear why we need to think in terms of elective affinities and not law-like relationships.

The Grobalization of Something

Some types of something have been grobalized to a considerable degree. For example, gourmet foods, handmade crafts, custom-made clothes, and Rolling Stones concerts are now much more available throughout the world, and more likely to move transnationally, than ever in history. In a very specific example in the arts, a touring series of "Silk Road" concerts recently brought together Persian artists and music, an American symphony orchestra, and Rimsky-Korsakov's (Russian) "Scheherezade."

Returning to Figure 2, we have used as examples of the grobalization of something touring art exhibitions (thing) of the works of Vincent van Gogh, the museums throughout the world in which such exhibitions occur (place), the knowledgeable guides who show visitors the highlights of the exhibition (person), and the detailed information and insights they are able to impart in response to questions from gallery visitors (service).

In spite of the existence of examples like these, why is there comparatively little affinity between grobalization and something? First, there is simply far less demand throughout the world for most forms of something, at least in comparison to the demand for nothing. One reason for this is that the distinctiveness of something tends to appeal to far more limited tastes than nothing, be it gourmet foods, handmade crafts, or Rolling Stones or Silk Road concerts. Second, the complexity of something, especially the fact that it is likely to have many different elements, means that it is more likely that it will have at least some characteristics that will be off-putting for or will even offend large numbers of people in many different cultures. For example, a Russian audience at a Silk Road concert might be bothered by the juxtaposition of Persian music with that of Rimsky-Korsakov. Third, the various forms of something are usually more expensive – frequently much more expensive – than competing forms of nothing (gourmet food is much more costly than fast food). Higher cost means, of course, that far fewer people can afford something. As a result, the global demand for expensive forms of something is minuscule in comparison to that for the inexpensive varieties of nothing. Fourth, because the prices are high and the demand is comparatively low, far less can be spent on the advertising and marketing of something, which serves to keep demand low. Fifth, something is far more difficult to mass-manufacture and, in some cases (Silk Road concerts, van Gogh exhibitions), impossible to produce in this way. Sixth, since the demand for something is less price-sensitive than nothing (the relatively small number of people who can afford it are willing, and often able, to pay almost any price), there is less need to mass-manufacture it (assuming it could be produced in this way) in order to lower prices. Seventh, the costs of shipping (insurance, careful packing and packaging, special transports) of something (gourmet foods, the van Gogh paintings) are usually very high, adding to the price and thereby reducing the demand.

It could also be argued that the fact that the grobalization of something (compared to nothing) occurs to a lesser degree helps to distinguish something from nothing. Because it is relatively scarce, something retains its status and its distinction from nothing. If something came to be mass-produced and grobalized, it is likely that it would move toward the nothing end of the continuum. This raises the intriguing question of what comes first – nothing, or grobalization and the associated mass production. That is, does a phenomenon start out as nothing? Or is it transformed into nothing by mass production and grobalization? We will return to this issue below.

The Grobalization of Nothing

The example of the grobalization of nothing in Figure 2 is a trip to one of Disney's worlds. Any of Disney's worlds is a nonplace, awash with a wide range of nonthings (such as mouse-ear hats), staffed largely by nonpeople (the "cast members," in costume or out), who offer nonservices (what is offered is often dictated by rules, regulations, and the scripts followed by employees).

The main reasons for the strong elective affinity between grobalization and nothing are basically the inverse of the reasons for the lack of such affinity between grobalization and something. Above all, there is a far greater demand throughout the world for nothing than something. This is the case because nothing tends to be less expensive than something (although this is not always true), with the result that more people can afford the former than the latter. Large numbers of people are also far more likely to want the various forms of nothing, because their comparative simplicity and lack of distinctiveness appeals to a wide range of tastes. In addition, as pointed out earlier, that which is nothing – largely devoid of distinctive content – is far less likely to bother or offend those in other cultures. Finally, because of the far greater potential sales, much more money can be – and is – devoted to the advertising and marketing of nothing, thereby creating a still greater demand for it than for something.

Given the great demand, it is far easier to mass-produce and mass-distribute the empty forms of nothing than the substantively rich forms of something. Indeed, many forms of something lend themselves best to limited, if not one-of-a-kind, production. A skilled potter may produce a few dozen pieces of pottery and an artist a painting or two in, perhaps, a week, a month, or even (a) year(s). While these craft and artworks may, over time, move from owner to owner in various parts of the world, this traffic barely registers in the total of global trade and commerce. Of course, there are the rare masterpieces that may bring millions of dollars, but in the main these are small-ticket items. In contrast, thousands, even many millions, and sometimes billions of varieties of nothing are mass-produced and sold throughout the globe. Thus, the global sale of Coca-Cola, Whoppers, Benetton sweaters, Gucci bags, and even Rolex watches is a far greater factor in grobalization than is the international sale of pieces of high art or of tickets to the Rolling Stones' most recent world tour. Furthermore, the various forms of nothing can range in cost from a dollar or two to thousands, even tens of thousands of dollars. The cumulative total is enormous and infinitely greater than the global trade in something.

Furthermore, the economics of the marketplace demands that the massive amount of nothing that is produced be marketed and sold on a grobal basis. For one thing, the economies of scale mean that the more that is produced and sold, the lower the price. This means that, almost inevitably, American producers of nothing (and they are, by far, the world leaders in this) must become dissatisfied with the American market, no matter how vast it is, and aggressively pursue a world market for their consumer products. The greater the grobal market, the lower the price that can be charged. This, in turn, means that even greater numbers of nothing can be sold and farther reaches of the globe in less-developed countries can be reached. Another economic factor stems from the demand of the stock market that corporations that produce and sell nothing (indeed, all corporations) increase sales and profits from one year to the next. Those corporations that simply meet the previous year's profitability or experience a decline are likely to be punished in the stock market and see their stock prices fall, sometimes precipitously. In order to increase profits continually, the corporation is forced, as Marx understood long ago, to continue to search out new markets. One way of doing that is constantly to expand globally. In contrast, since something is less likely to be produced by corporations – certainly by the large corporations listed in the stock market – there is far less pressure to expand the market for it. In any case, as we saw above, given the limited number of these things that can be produced by artisans, skilled chefs, artists, and so on, there are profound limits on such expansion. This, in turn, brings us back to the pricing issue and relates to the price advantage that nothing ordinarily has over something. As a general rule, the various types of nothing cost far less than something. The result, obviously, is that nothing can be marketed globally far more aggressively than something.

Also, nothing has an advantage in terms of transportation around the world. These are things that generally can be easily and efficiently packaged and moved, often over vast areas. Lunchables, for example, are compact, prepackaged lunch foods, largely for schoolchildren, that require no refrigeration and have a long shelf life. Furthermore, because the unit cost of such items is low, it is of no great consequence if they go awry, are lost, or are stolen. In contrast, it is more difficult and expensive to package something – say, a piece of handmade pottery or an antique vase – and losing such things or having them stolen or broken is a disaster. As a result, it is far more expensive to insure something than nothing, and this difference is another reason for the cost advantage that nothing has over something. It is these sorts of things that serve to greatly limit the global trade in items that can be included under the heading of something.

It is important to remember that while most of our examples in this section are nothings, it is the case that nonplaces (franchises), nonpeople (counterpeople in fast-food chains), and nonservices (automatic teller machines – ATMs) are also being grobalized.

While the grobalization of nothing dominates in the arena of consumption as it is generally defined, we find domains – medicine, science, pharmaceuticals, biotechnology, education, and others – in which the grobalization of something is of far greater importance. While these areas have experienced their share of the grobalization of nothing, they are also characterized by a high degree of the grobalization of something. For example, the worldwide scientific community benefits from the almost instantaneous distribution of important scientific findings, often, these days, via new journals on the Internet. Thus, our focus on the grobalization of nothing should not blind us to the existence and importance – especially in areas such as these – of the grobalization of something.

The Glocalization of Nothing

Just as there has historically been a tendency to romanticize and glorify the local, there has been a similar trend in recent years among globalization theorists to overestimate the glocal. It is seen by many as not only the alternative to the evils of grobalization, but also a key source of much that is worthwhile in the world today. Theorists often privilege the glocal something over the grobal nothing (as well as over the glocal nothing, which rarely appears in their analyses). For example, Jonathan Friedman associates cultural pluralism with "a dehegemonizing, dehomogenizing world incapable of a formerly enforced politics of assimilation or cultural hierarchy." Later, he links the "decline of hegemony" to "a liberation of the world arena to the free play of already extant but suppressed

projects and potential new projects." Then there are the essays in James Watson's *McDonald's in East Asia*, which, in the main, focus on glocal adaptations (and generally downplay grobal impositions) and tend to describe them in positive terms.

While most globalization theorists are not postmodernists, the wide-scale acceptance of various postmodern ideas (and rejection of many modern positions) has helped lead to positive attitudes toward glocalization among many globalization theorists. Friedman is one who explicitly links "cultural pluralism" and the "postmodernization of the world." The postmodern perspective is linked to glocalization theory in a number of ways. For example, the work of de Certeau and others on the power of the agent in the face of larger powers (such as grobalization) fits with the view that indigenous actors can create unique phenomena out of the interaction of the global and the local. De Certeau talks of actors as "unrecognized producers, poets of their own affairs, trailblazers in the jungles of functionalist rationality." A similar focus on the local community gives it the power to create unique glocal realities. More generally, a postmodern perspective is tied to hybridity, which, in turn, is "subversive" of such modern perspectives as essentialism and homogeneity.

While there are good reasons for the interest in and preference for glocalization among globalization theorists, such interest is clearly overdone. For one thing, grobalization (especially of nothing) is far more prevalent and powerful than glocalization (especially of something). For another, glocalization itself is a significant source of nothing.

One of the best examples of the glocalization of nothing is to be found in the realm of tourism, especially where the grobal tourist meets the local manufacturer and retailer (where they still exist) in the production and sale of glocal goods and services (this is illustrated in quadrant two of Figure 2). There are certainly instances – perhaps even many of them – in which tourism stimulates the production of something: well-made, high-quality craft products made for discerning tourists; meals lovingly prepared by local chefs using traditional recipes and the best of local ingredients. However, far more often – and increasingly, as time goes by – grobal tourism leads to the glocalization of nothing. Souvenir shops are likely to be bursting at the seams with trinkets reflecting a bit of the local culture. Such souvenirs are increasingly likely to be mass-manufactured – perhaps using components from other parts of the world – in local factories. If demand grows great enough and the possibilities of profitability high enough, low-priced souvenirs may be manufactured by the thousands or millions elsewhere in the world and then shipped back to the local area to be sold to tourists (who may not notice, or care about, the "made in China" label embossed on their souvenir replicas of the Eiffel Tower). The clerks in these souvenir shops are likely to act like nonpeople, and tourists are highly likely to serve themselves. Similarly, large numbers of meals slapped together by semiskilled chefs to suggest vaguely local cooking are far more likely than authentic meals that are true to the region, or that truly integrate local elements. Such meals are likely to be offered in "touristy" restaurants that are close to the nonplace end of the continuum and to be served by nonpeople who offer little in the way of service.

Another major example involves the production of native shows – often involving traditional costumes, dances, and music – for grobal tourists. While these could be something, there is a very strong tendency for them to be transformed into nothing to satisfy grobal tour operators and their clientele. Hence these shows are examples of the glocalization of nothing, because they become centrally conceived and controlled empty forms. They are often watered down, if not eviscerated, with esoteric or possibly offensive elements removed. The performances are designed to please the throngs of tourists and to put off as few of them as possible. They take place with great frequency, and interchangeable performers often seem as if they are going through the motions in a desultory fashion. For their part, this is about all the grobal tourists want in their rush (and that of the tour operator) to see a performance, to eat an ersatz local meal, and then to move on to the next stop on the tour. Thus, in the area of tourism – in souvenirs, performances, and meals – we are far more likely to see the glocalization of nothing than of something.

The Glocalization of Something

The example of the glocalization of something in Figure 2 (quadrant Í) is in the realm of indigenous crafts such as pottery or weaving. Such craft products

are things, and they are likely to be displayed and sold in places such as craft barns. The craftperson who makes and demonstrates his or her wares is a person, and customers are apt to be offered a great deal of service.

Such glocal products are likely to remain something, although there are certainly innumerable examples of glocal forms of something that have been transformed into glocal – and in some cases grobal – forms of nothing (see below for a discussion of Kokopelli figures and matryoshka dolls). In fact, there is often a kind of progression here, from glocal something to glocal nothing as demand grows, and then to grobal nothing if some entrepreneur believes that there might be a global market for such products. However, some glocal forms of something are able to resist this process.

Glocal forms of something tend to remain as such for various reasons. For one thing, they tend to be costly, at least in comparison to mass-manufactured competitors. High price tends to keep demand down locally, let alone globally. Second, glocal forms of something are loaded with distinctive content. Among other things, this means that they are harder and more expensive to produce and that consumers, especially in other cultures, find them harder to understand and appreciate. Furthermore, their idiosyncratic and complex character make it more likely that those in other cultures will find something about them they do not like or even find offensive. Third, unlike larger manufacturers of nothing, those who create glocal forms of something are not pushed to expand their business and increase profits to satisfy stockholders and the stock market. While craftspeople are not immune to the desire to earn more money, the pressure to do so is more internal than external, and it is not nearly as great or inexorable. In any case, the desire to earn more money is tempered by the fact that the production of each craft product is time-consuming and only so many of them can be produced in a given time. Further, craft products are even less likely to lend themselves to mass marketing and advertising than they are to mass manufacture.

Which Comes First: Nothing, or Its Grobalization?

At this point, we need to deal with a difficult issue: is it possible to determine which comes first – nothing or its grobalization? The key components of the definition of nothing – central conception and control, lack of distinctive content – tend to lead us to associate nothing with the modern era of mass production. After all, the system of mass production is characterized by centralized conception and control, and it is uniquely able to turn out large numbers of products lacking in distinctive content. While there undoubtedly were isolated examples of nothing prior to the Industrial Revolution, it is hard to find many that fit our basic definition of nothing.

Thus, as a general rule, nothing requires the prior existence of mass production. However, that which emanates from mass-production systems need not necessarily be distributed and sold globally. Nevertheless, as we have discussed, there are great pressures on those who mass-produce nothing to market it globally. Thus, there is now a very close relationship between mass production and grobalization; the view here is that *both* precede nothing and are prerequisites to it.

Take, for example, such historic examples of something in the realm of folk art as Kokopellis from the southwestern United States and matryoshka dolls from Russia. At their points of origin long ago in local cultures, these were clearly hand-made products that one would have had to put close to the something end of the continuum. For example, the Kokopelli, usually depicted as an arch-backed flute player, can be traced back to at least 800 AD and to rock art in the mountains and deserts of the southwestern United States. Such rock art is clearly something. But in recent years, Kokopellis have become popular among tourists to the area and have come to be produced in huge numbers in innumerable forms (figurines, lamps, keychains, light-switch covers, Christmas ornaments, and so on), with increasingly less attention to the craftsmanship involved in producing them. Indeed, they are increasingly likely to be mass-produced in large factories. Furthermore, offending elements are removed in order not to put off potential consumers anywhere in the world. For example, the exposed genitals that usually accompanied the arched back and the flute have been removed. More recently, Kokopellis have moved out of their locales of origin in the Southwest and come to be sold globally. In order for them to be marketed globally at a low price, much of the distinctive character

and craftsmanship involved in producing the Kokopelli is removed. That is, the grobalization of Kokopellis has moved them even closer to the nothing end of the continuum.

A similar scenario has occurred in the case of the matryoshka doll (from five to as many as 30 dolls of increasingly small size nested within one another), although its roots in Russian culture are not nearly as deep (little more than a century) as that of the Kokopelli in the culture of the southwestern United States. Originally hand-made and hand-painted by skilled craftspeople and made from seasoned birch (or lime), the traditional matryoshka doll was (and is) rich in detail. With the fall of communism and the Soviet Union, Russia has grown as a tourist destination, and the matryoshka doll has become a popular souvenir. In order to supply the increasing demand of tourists, and even to distribute matryoshka dolls around the world, they are now far more likely to be machine-made: automatically painted; made of poor quality, unseasoned wood; and greatly reduced in detail. In many cases, the matryoshka doll has been reduced to the lowest level of schlock and kitsch in order to enhance sales. For example, the traditional designs depicting precommunist nobles and merchants have been supplemented with caricatures of global celebrities such as Bill Clinton, Mikhail Gorbachev, and – post-September 11 – Osama bin Laden. Such mass-produced and mass-distributed matryoshka dolls bear little resemblance to the folk art that is at their root. The mass production and grobalization of these dolls has transformed that which was something into nothing. Many other products have followed that course, and still more will do so in the future.

While we have focused here on nonthings that were things at one time, much the same argument can be made about places, people, and services. That is, they, too, have come to be mass-manufactured and grobalized, especially in the realm of consumption. This is most obvious in virtually all franchises for which settings are much the same throughout the world (using many mass-manufactured components), people are trained and scripted to work in much the same way, and the same "services" are offered in much the same way. They all have been centrally conceived, are centrally controlled, and are lacking in distinctive content.

Grobalization and Loss

Grobalization has brought with it a proliferation of nothing around the world. While it carries with it many advantages (as does the grobalization of something), it has also led to a loss, as local (and glocal) forms of something are progressively threatened and replaced by grobalized (and glocalized) forms of nothing.

This reality and sense of loss are far greater in much of the rest of the world than they are in the United States. As the center and source of much nothingness, the United States has also progressed furthest in the direction of nothing and away from something. Thus, Americans are long accustomed to nothing and have fewer and fewer forms of something with which to compare it. Each new form of or advance in nothing barely creates a ripple in American society.

However, the situation is different in much of the rest of the world. Myriad forms of something remain well entrenched and actively supported. The various forms of nothing – often, at least initially, imports from the United States – are quickly and easily perceived as nothing, since alternative forms of something, and the standards they provide, are alive and well. Certainly, large numbers of people in these countries demand and flock to nothing in its various forms, but many others are critical of it and on guard against it. The various forms of something thriving in these countries give supporters places, things, people, and services to rally around in the face of the onslaught of nothing. Thus, it is not surprising that the Slow Food Movement, oriented to the defense of "slow food" against the incursion of fast food, began in Italy (in fact, the origin of this movement was a battle to prevent McDonald's from opening a restaurant at the foot of the Spanish Steps in Rome) and has its greatest support throughout Europe.

The Increase in Nothing!
The Decline in Something?

A basic idea – even a grand narrative – in this essay is the idea that there is a long-term trend in the social world in general, and in the realm of consumption in particular, in the direction of nothing. More specifically, there is an historic movement from something to nothing. Recall that this is simply an argument about the increase in forms that are centrally conceived and controlled and are largely devoid of distinctive content. In other words, we have witnessed a long-term trend *from* a world in which indigenously conceived and controlled forms laden with distinctive content predominated *to* one where centrally conceived and controlled forms largely lacking in distinctive content are increasingly predominant.

There is no question that there has been an increase in nothing and a relative decline in something, but many forms of something have not experienced a decline in any absolute sense. In fact, in many cases, forms of something have increased; they have simply not increased at anything like the pace of the increase in nothing. For example, while the number of fast-food restaurants (nonplaces) has increased astronomically since the founding of the McDonald's chain in 1955, the number of independent gourmet and ethnic restaurants (places) has also increased, although at not nearly the pace of fast-food restaurants. This helps to account for the fact that a city such as Washington, DC (to take an example I know well) has, over the last half century, witnessed a massive increase in fast-food restaurants *at the same time* that there has been a substantial expansion of gourmet and ethnic restaurants. In fact, it could be argued that there is a dialectic here – that the absolute increase in nothing sometimes serves to spur at least some increase in something. That is, as people are increasingly surrounded by nothing, at least some are driven to search out or create something. However, the grand narrative presented here is more about the relative ascendancy of nothing and the relative decline in something than about absolute change.

Nonetheless, at least some forms of something (e.g., local groceries, cafeterias) have suffered absolute declines and may have disappeared or be on the verge of disappearance. It could be argued that all of these have been victims of what Joseph Schumpeter called "creative destruction." That is, while they have largely disappeared, in their place have arisen successors such as the fast-food restaurant, the supermarket, and the "dinner-house" (e.g., the Cheesecake Factory). While there is no question that extensive destruction of older forms has occurred, and that considerable creativity has gone into the new forms, one must question Schumpeter's one-sidedly positive view of this process.

Perhaps some things – even some measure of creativity – have been lost with the passing of these older forms. It may be that the destruction has not always been so creative.

However, no overall value judgment needs to be made here; forms laden with content are not inherently better than those devoid of content, or vice versa. In fact, there were and are many forms rich in content that are among the most heinous of the world's creations. We could think, for example, of the pogroms that were so common in Russia, Poland, and elsewhere. These were largely locally conceived and controlled and were awash in distinctive content (anti-Semitism, nationalism, and so on). Conversely, forms largely devoid of content are not necessarily harmful. For example, the bureaucracy, as Weber pointed out, is a form (and ideal type) that is largely lacking in content. As such, it is able to operate in a way that other, more content-laden forms of organization – those associated with traditional and charismatic forms of organization – could not. That is, it was set up to be impartial – to *not* (at least theoretically) discriminate against anyone.

There is very strong support for the argument, especially in the realm of consumption, that we are in the midst of a long-term trend away from something and in the direction of nothing. By the way, this implies a forecast for the future: we will see further increases in nothing and further erosions of something in the years to come.

The Economics of Nothing

Several points can be made about the economics of nothing. First, it is clear that, in general, there is an inverse relationship between income and nothing. That is, those with money can still afford to acquire various forms of something, whereas those with little money are largely restricted to nothing. Thus, only the affluent can afford expensive bottles of complex wine, or gourmet French meals with truffles. Those with little means are largely restricted to Coca-Cola, Lunchables, microwave meals, and McDonald's fries.

Second, there is an economic floor to this: those below a certain income level cannot even afford much of that which is categorized here as nothing. Thus, there are those near or below the poverty line in America who often cannot afford a meal at McDonald's or a six-pack of Coca-Cola. More importantly, there are many more people in the less-developed parts of the world who do not have access to and cannot afford such forms of nothing. Interestingly, extreme poverty relegates people to something – homemade meals and home brews made from whatever is available. However, in this case it is hard to make the argument for something. These forms of something are often meager, and those who are restricted to them would love to have access to that which has been defined here, as well as by many people throughout the world, as nothing.

Third, thinking of society as a whole, some minimum level of affluence and prosperity must be reached before it can afford nothing. That is, there are few ATMs, fast-food restaurants, and Victoria's Secret boutiques in the truly impoverished nations of the world. There simply is not enough income and wealth for people to be able to afford nothing; people in these societies are, ironically, doomed – at least for the time being – to something. Thus, they are more oriented to barter, preparing food at home from scratch, and making their own nightgowns. It is not that they would not readily trade their something for the forms of nothing described above, but that they are unable to do so. It seems clear that as soon as the level of wealth in such a country reaches some minimal level, the various forms of nothing will be welcomed and, for their part, the companies that produce them will enter eagerly.

Fourth, even the wealthiest of people often consume nothing. For one thing, as has been pointed out previously, nothing is not restricted to inexpensive (non)places, (non)things, (non)people, and (non)services. Some forms of nothing – a Four Seasons hotel room, a Dolce and Gabbana frock, the salesperson at Gucci, and the service of a waiter at a Morton's steakhouse – are very costly, but they still qualify as nothing as that term is used here: relatively empty forms that are centrally conceived and controlled. The consumption of these very expensive forms of nothing is obviously restricted to the uppermost reaches of the economic ladder.

Fifth, the wealthy are drawn to many of the same low-priced forms of nothing that cater to the mass of the population, even those who would be considered poor or very close to it. A credit card knows no income barriers – at least at the high end of the spectrum – and the same is true of ATMs. The wealthy, especially wealthy teenagers, are just as likely to be attracted to fast-food restaurants as are those from virtually every other income group.

There is no simple relationship between wealth and nothingness.

Grobalization versus Glocalization

Returning to the issue with which we began this discussion, one of the key contributions here is the argument that the/a key dynamic under the broad heading of globalization is the conflict between grobalization and glocalization. This is a very different view than *any* of the conventional perspectives on global conflict. For example, I think a large number of observers have tended to see the defining conflict, where one is seen to exist, as that between globalization and the local. However, the perspective offered here differs from that perspective on several crucial points.

First, globalization does not represent one side in the central conflict. It is far too broad a concept, encompassing as it does all transnational processes. It needs further refinement to be useful in this context, such as the distinction between grobalization and glocalization. When that differentiation is made, it is clear that the broad process of globalization already encompasses important conflicting processes. Since globalization contains the key poles in the conflict, it therefore is not, and cannot be, one position in that conflict.

Second, the other side of the traditional view of that conflict – the local – is relegated to secondary importance in this conceptualization. That is, to the degree that the local continues to exist, it is seen as increasingly insigni-

ficant and a marginal player in the dynamics of globalization. Little of the local remains that has been untouched by the global. Thus, much of what we often think of as the local is, in reality, the glocal. As the grobal increasingly penetrates the local, less and less of the latter will remain free of grobal influences. That which does will be relegated to the peripheries and interstices of the local community. The bulk of that which remains is much better described as glocal than local. In community after community, the real struggle is between the more purely grobal versus the glocal. One absolutely crucial implication of this is that *it is increasingly difficult to find anything in the world untouched by globalization.* Ironically, then, the hope for those opposed to globalization, especially the grobalization of nothing, seems to lie in an alternative form of globalization – glocalization. This is hardly a stirring hope as far as most opponents of grobalization are concerned, but it is the most realistic and viable one available. The implication is that those who wish to oppose globalization, and specifically grobalization, must support and align themselves with the other major form of globalization – glocalization.

Yet glocalization does represent some measure of hope. For one thing, it is the last outpost of most lingering (if already adulterated by grobalization) forms of the local. That is, important vestiges of the local remain in the glocal. For another, the interaction of the grobal and the local produces unique phenomena that are not reducible to either the grobal or the local. If the local alone is no longer the source that it once was of uniqueness, at least some of the slack has been picked up by the glocal. It is even conceivable that the glocal and the interaction among various glocalities are – or at least can be – a significant source of uniqueness and innovation.

 READING 56

Dialectics of Something and Nothing: Critical Reflections on Ritzer's Globalization Analysis

Douglas Kellner

George Ritzer's *The Globalization of Nothing* provides a highly original take on globalization that illuminates

aspects of globalization neglected in many standard works. Ritzer produces a wide range of categories, some original, to delineate how globalization produces massification, homogenization, and standardization of consumer products and practices. Thus, his recent book is a worthy successor to *The McDonaldization of Society*, *Expressing America*, and *Enchanting a Disenchanted World: Revolutionizing the Means of Consumption* as well as his other recent work on McDonaldization.

In addition, Ritzer's *The Globalization of Nothing* articulates the dialectic between the global and the local, between its empty forms, or nothing in his terminology, and its specific forms of something, of particularity and difference. His recent studies of globalization have many of the virtues of his earlier books in providing a wealth of sociological insight and analysis to a popular audience. The text particularly illuminates and helps develop Ritzer's earlier concepts of McDonaldization, Americanization, and delineation of the new means of consumption, and it adds a wide range of important insights into globalization, whilst providing useful categories and distinctions to describe globalization itself.

In these comments, first, I want to critically engage with an issue that Ritzer might have addressed, that in my view would have substantially strengthened his conceptual optic. Then, I will make some comments on things I like and find important in the book, and will signal some disagreements.

Globalization and Nothing: the Missing Dialectic

Ritzer sets out his definition of the globalization of nothing as "generally centrally conceived and controlled social forms that are comparatively devoid of distinctive content," such as the form of Mills corporation shopping malls, airports, chain hotels, credit cards, and of course McDonald's and fast-food restaurants. He presents a dialectic of something and nothing in a continuum of social forms with "something" presented as "a social form that is generally indigenously conceived, controlled, and comparatively rich in distinctive substantive content; a form that is to a large degree substantively unique." Both presuppose each other and make "sense only when paired with, and contrasted to, the other."

The dialectic of something and nothing is fleshed out with a series of conceptual contrasts between places and non-places, things and non-things, persons and non-persons, and services and non-services, encompassing, as examples, credit card companies, telemarketing, fast food production, and global branding (I will provide further examples and explication as I proceed). He also develops a set of other categories like "glocalization" (building on Roland Robertson), through which global and local forces hybridize, and "grobalization" through which global processes absorb and in some cases destroy local artifacts, customs, and culture.

Ritzer says he will offend fans of many "somethings," such as products or forms of consumption that he critically analyzes, but I am not in the least offended by this critique, and would be happy to see Ritzer and others develop the analysis of nothing and the destruction of something(s) further. Indeed, this brings me to my central critique of Ritzer's book.

In the Preface, Ritzer states "My focal interest in these pages is in the globalization of nothing within the realm of consumption," and here I wish that Ritzer had embraced the dialectic of production and consumption and critically engaged both, as he does to some extent in his analysis of McDonaldization, which is both a form of production and consumption. Ritzer does have a short section at the end of chapter 1 on "the production of nothing" where he mentions that he will not engage with the "developing" world; whose inhabitants often cannot afford, or do not have access to the nothings of globalization; and also, will not engage with global production, such as Nike shoe factories, that have received a lot of attention and criticism. Ritzer says that there has been a "productivist bias" in social theory and that he wants to compensate for what he sees as a one-sidedness in this direction. But, while there was perhaps once a problem of a production bias in fields of social theory and consumption studies that needs correction, I would assert that production and consumption are so tightly and importantly linked that one needs a dialectic of production and consumption to adequately grasp the general processes of globalization.

In fact, within cultural studies and a lot of social theory, there has been a booming field of consumption studies, of which Ritzer is an important part, so I am not sure that we need to worry about a productivist bias in social theory and cultural studies, but should rather worry about the production deficit (this has been one of my worries and themes in cultural studies for some years now and is reappearing here in the context of the sociology of consumption). But, I would also argue that it is imperative to analyze the dialectic of production and consumption which is absolutely central to grasping, and engaging with globalization in order to conceptualize its key dynamics – as important, I would argue and perhaps more so, than the dialectic of something and nothing that Ritzer takes on (in fact, I will argue that they go together).

To make this point, let me take an example from Ritzer's earlier study of McDonald's, surely a sociological classic of our time. One key insight of this text was the analysis of McDonaldization as a mode of production and consumption. McDonald's provides an entire business model (the franchise) and a model of fast-food production and consumption marked by the features of efficiency, speed, predictability, calculability, and rationalization. This model spread to many other fields of production and consumption, as Ritzer points out. Indeed, it is McDonaldization as a dialectic of production and consumption, that makes the corporation so paradigmatic for corporate globalization.

Now, extending Ritzer's argument of the dialectic of production and consumption to the sphere of labor, I would argue that the spread, diffusion, and the impact of the forms of production described as postFordism, McDonaldization, technocapitalism, or the networked society, range from the global spread of assembly-line labor described by Harry Braverman and other, mostly, Marxists as contributing to a deskilling of labor to the forms of labor described by Dan Schiller and other critics of digital capitalism. It is true, however, that there are a couple of mentions of production in Ritzer's book, such as a passage on page 177, where Ritzer notes that his analyses of the grobalization of nothing:

> Certainly applies as well to consumption's other face – production. We literally could not have the grobalization of, for example, non-things without the existence of systems that produce massive numbers of the non-things that are to be sold and distributed worldwide. But even production, or the production-consumption nexus, is too narrow a domain for examining the grobalization of nothing. Nothing spreads globally within politics, or the church, or the criminal justice system, for myriad reasons, many of them specific to each of those domains, that have nothing to do with production or consumption.

Far from it being for me to deny the relative autonomy of politics, the legal system, or culture, but all of these things are centrally related to production and, increasingly, to consumption. There is also another phenomenon of immense importance that Ritzer's analysis suggests, but does not critically engage with, and that is the replacement of human labor power by machines. In terms of one of Ritzer's sets of categorical distinctions involving non-places, non-things, non-persons, and non-services (encompassing as examples credit card companies, telemarketing, and computerized services of various sorts), this proliferation of nullities, to use Ritzer's terms, involves a rather substantial global restructuring of labor, which both eliminates a lot of jobs and creates a wealth of "McJobs" that could serve as paradigms of contemporary alienated labor (consider telemarketing, or all the clerical work that credit cards, airline reservations, sales of many sorts and the like involve). Now, as Marx argued in the *Grundrisse*, replacement of human labor power by technology can be progressive, but as we have seen, it can also be disastrous for certain categories of labor, in the sense that it eliminates more creative, unionized and well-paid and secure jobs and creates more deadening, alienating, lower-paid and insecure ones.

This is an immense world-historical phenomenon that lies at the heart of current concerns about globalization, and I think that Ritzer's dialectic of something and nothing could have interestingly illuminated and critically engaged this phenomenon. There is one passage where Ritzer mentions that Marx's analysis of alienation, while not especially useful in talking about consumption (although some might contest this), "is probably more relevant than ever to the less-developed world where much of the kind of production-oriented work analyzed by Marx is increasingly done." I would agree with this, but would suggest that alienated labor is also wide-spread in the kind of domains that Ritzer is analyzing, such as telemarketing, computerized services, and most clerical and other office work needed to sustain global production.

Parenthetically, I might mention that the film *One Hour Photo* that Ritzer uses to illustrate the empty forms of consumption is about empty forms of production and labor as much as consumption and that the Robin Williams character illustrates the dehumanizing and alienating effects that doing nothing, i.e.

laboring in a completely prescribed, impersonalized, and uncreative way, can have distorting effects on the personality. Yet, the film can also be read as suggesting that, even in the most dehumanizing matrixes of production and consumption, there are attempts to create human relationships and creative work – that is something.

As a hopelessly Hegelian dialectician, I appreciate the dialectic of nothing and something in Ritzer's book, as well as the dialectic of glocalization and grobalization, but would have liked to see him engage more with the dialectic of production and consumption. I would note, also, that there is one important passage and concept thrown out, but not developed, where Ritzer mentions the "double affliction" of those workers in extremely low paying jobs who are not able to afford the very products that they are producing. Both afflictions are heart wrenching, but I fear that they are a widespread global phenomenon, whose development and documentation could provide a sharp critical edge to how we view globalization.

I suspect part of Ritzer's answer would be his statement:

> It is worth remembering that it was not too long ago that the United States was the world leader in production. In many ways, consumption has replaced production as the focus of the American economy and it has become the nation's prime export to the rest of the world. It is interesting to ponder the implications of what it means to have gone from the world leader in the production of steel to, say, the world leader in the exportation of fast-food restaurants and the shopping mall.

I would agree with Ritzer that to some extent consumption has replaced production as the US's prime export, but I think that, globally, production is as important as consumption. As postFordist theory makes clear, production is increasingly moving from one place to another and, to some extent, this process embodies Ritzer's analysis, in that the forms of production are increasingly similar, whether sneakers, for instance, are produced in Los Angeles, Indonesia, Vietnam, or China.

In general, I would agree with Marx's model in the *Grundrisse*, that there is a circuit of capital that involves production, exchange, distribution, and consumption, and while one could debate whether production is or is not the primary moment in this circuit, as Marx claims, I think it is clear that, taking globalization as a whole, the dialectic of production and consumption, and circuits of capital are crucial to the process (i.e. that there is no consumption without production and that they are linked in circuits including exchange and distribution, much of which Ritzer engages with, so he might as well take on production as well to complete the circuit).

Another criticism of Ritzer's McDonald's analysis that could be leveled against *The Globalization of Nothing* is that he does not have enough on creative consumption, or the ways that something and nothing produce hybrids, or local variants of global products, or forms like McDonald's. Hybridization has been taken as a key form of the construction of local cultures within globalization that postmodernists, and others including Stuart Hall and the studies of McDonald's in *Golden Arches East*, positively valorize as a cultural synthesis of local and global, and traditional and modern. While hybridizations have been exaggerated and many of the celebrations of hybridization, or local inflection of global phenomena, such as the *Golden Arches East* studies cited above, overlook the elements of cultural imperialism (if I may use an old-fashioned term), of destruction of the traditional, and of loss, as Ritzer repeatedly stresses, nonetheless, more global forms can always be inflected globally and creative hybrids can be produced of the global and the local.

Yet, Ritzer focuses on the form of consumption and nothing, and downplays creative use and active audience appropriation of commodities, cultural forms, or globalized phenomena of various sorts. British cultural studies highlights the active audience as constitutive of the popular and, while this emphasis can overplay subjectivity and the power of the consumer, I think highlights a potential production of difference, meaning and creative practice (i.e. something) that Ritzer does not adequately address. He might, thus, add a dialectic of nothing and something to the activity of the consumer in the process of consumption, in which one class or pole of consumers is ideal-type characterized as largely passive and consumes in a standardized way, whereby another class or pole can consume in highly creative and idiosyncratic ways that can transform nothing into something (to use Ritzer's dialectic). Ritzer does have a section on "Making something out of nothing" in his internet chapter and he valorizes the slow food movement in a concluding chapter, but I think he needs more on active and creative uses of consumption, or globalized technology like the internet.

Both Andrew Feenberg and I, in developing theories of technology, stress how technologies can be reconstructed in ways that people can make something out of nothing, to use Ritzer's terminology; that is, use technology for their own self-valorization, projects, and purposes, and not just those of capital or whoever produces the technology. For example, people use traditional medicines, or natural child-bearing, instead of the standardized forms of corporate medicine, and have constructed the internet as a decommodified realm of communication, cultural dissemination, and political organization, often going beyond the purposes of the creators of the technology.

Further, whereas I find many of Ritzer's concepts and distinctions in the book valuable, like his analysis "Meet the nullities," where he analyzes the forms of non-place, non-things, non-persons, and non-services of corporate globalization, I have a conceptual problem with his analysis of non-things where he writes:

> Our bodies are covered by an array of non-things and even when we go to bed at night, we are likely to be surrounded by non-things (Sealy Posturepedic mattresses, Martha Stewart sheets and pillow cases, Chanel perfumes or colognes, and so on).

While many consumer brands are nullities, and so in Ritzer's vocabulary are arguably "no things," many brands are important things to many people. Reducing so many consumer brands to nullities downplays the importance of logos and branding that Naomi Klein and others claim is at the very heart of globalization. While Ritzer provides a powerful critique of current modes of branding, I am just not convinced that some of the brands Ritzer cites in his text are "nothings." Such a concept of brands neglects the kind of sign value and system of difference in consumption stressed by Baudrillard, and worked out more concretely by sociologists like Robert Goldman and Stephen Papson. Now, some of the brands like Visa and MasterCard that Ritzer engages with are paradigms of brands that

are pure forms, where there is little if any material difference in the way the cards work, but some people strongly identify with brands of airlines, autos, clothing lines, and other commodities. Admittedly, the distinction is often hard to make regarding whether brands are something or nothing: while arguably a Gucci bag can be seen as a nothing, as Ritzer claims, in which pure form dominates, there are genuine differences in some fashion houses and clothing lines that have passionate detractors and fans. Certainly, as pirating and simulation of original products indicates, replication is big business, but the fact that many products are run off indicates precisely that they are "something" with commodity sign-value.

And although there may be some pure models of strip malls, or even mega malls and other sites of consumption, that appear as "nothing" (i.e. not distinctive, interesting, locally-inflected, and so on), it is precisely the differences that make some malls stand out, like the Grove and Fairfax Farmer's Market in Los Angeles, or Edmonton Mall in Canada. Likewise, when Ritzer cites the Ford Edsel as an example of nothing in the appendix, this just seems wrong: Edsel is symbolic of something different, a product line that flopped in a spectacular way (as did Classic Coke).

Another problem with Ritzer's categorization is that he appears sometimes to be too loose with his application of nothing or, at least, one could raise questions whether certain phenomena are something or nothing. I would question, for example, Ritzer's citing of the use of audio guides in museums as an example of nothing:

> An interesting example of the trend toward nothingness is the increasing use of audio guides and rented tape players at such shows and at museums more generally.

While it is true that more and more museums are using similar types of audio guides to accompany their art shows, they are uneven in quality, but more important, facilitate qualitatively different aesthetic experiences and uses. I personally avoided these audio guides at first, as I thought that they distracted from the aesthetic experience. I found, however, that some were very informative and could, if used properly, enhance the overall experience of the art show. Some, indeed, strike me as quite something. For instance, the audio guide that accompanied the 2003 Kandinsky-Schönberg show at the Jewish Museum in New York, not only had very informative and intelligent commentary, but large sections of music by Schönberg and others, so that one could enjoy Schönberg's music while looking at his paintings, or just take a break, close one's eyes, and imagine one was at a concert.

Another place where one could contest Ritzer's overly loose use of "nothing" is his claim that "the media are, themselves [...] purveyors of nothing (for example, the 'soaps,' *CNN Headline News*, sitcoms)." Whereas there are rather empty forms of global news and entertainment (reality TV, headline news, and maybe at least some US sitcoms), other forms like "soaps" are arguably quite varied, diverse, local, and thus, presumably, something. I have been to telenovela panels at conferences, and read papers on the topic, that insist on the major differences between Latin American telenovelas and American soaps, and the differences between programs of this genre in, say, Mexico, Argentina, Cuba, and Brazil, and even within these countries (aficionados can discourse for great length on varieties of Brazilian telenovela, and one friend loaned me tapes of a Cuban soap opera that was a quite interesting political drama, using certain formats of American soaps but producing something significantly different, and thus I would conclude, something).

Parenthetically, I might note that Ritzer uses nothing, nothingness, nullity, and such cognates interchangeably and while I have learned to live with, and even appreciate, "the globalization of nothing" and find the "nullities" concept amusing and illuminating, I cringe a bit when I read "nothingness," no doubt because of my early immersion in Sartre and Heidegger and association of nothingness with anxiety, death, and disturbing forms of non-being. Hence, I would question Ritzer as to whether there is a difference between nothing and nothingness in his categorizations, and why he uses the latter term when it carries a lot of conceptual baggage from existential philosophy. In other words, "nothing" is an empty enough concept to serve Ritzer's purposes, but "nothingness" is to me branded heavily in terms of Heidegger's and Sartre's existential philosophy, and does not readily serve as a cognate for Ritzer's nothing.

Consequently, while Ritzer is using a flexible model of ideal types ranging from something to nothing, I think there is room for contestations of at least some

of his presentations of nothing, and hope at least that more varied and diverse somethings might proliferate in a global economy, as opposed to the undeniable proliferation of nothing, the grobalization of the local, and general tendencies toward standardization, exchangeability, massification, that it is the virtue of Ritzer's analysis to warn us about.

Globalization and the Contemporary Moment

Indeed, Ritzer is telling a very dramatic story that comes most alive, for me at least, in the titanic battle between the glocalization of something and grobalization of something and nothing that takes place in the middle of his book. He concludes chapter 5 by stating:

> Thus, we live in an era in which a variety of its basic characteristics have led to a tremendous expansion in the grobalization of nothing. Furthermore, current trends lead to the view that the future will bring with it an even greater proliferation of nothing throughout the globe.

This is a rather ominous prospect concerning the growing hegemony of the "grobalization of nothing," of pure forms or models of production and consumption that could obliterate the local, singularity, heterogeneity, and difference. Of course, there are countervailing tendencies that varieties of postmodern theory and Roland Robertson extol, but I think Ritzer provides an important cautionary warning that major trends of globalization are destroying individuality and particularity, and producing standardization and homogeneity.

To some extent this is a familiar story, told by various neo-Marxists, Weberians and other critics of modernity, but it is salutary to rehear the story as a warning against too enthusiastic globophilac embraces of a globalization that is producing, according to its postmodern champions, bountiful heterogeneity, hybridity, and difference. Ritzer claims, near the end of the book, that his major conceptual contribution to this story, and to theorizing globalization, is his account of the growing conflict between glocalization and grobalization. This optic helps balance tendencies to celebrate and overrate the local and catches the fact that the anti-corporate globalization movement that wants to protect the local and the human from corporate domination, or grobalization in Ritzer's vocabulary, is itself global in nature and thus represents a form of glocalization.

But, I think more concrete goals need to be attached and defended via the anti-corporate globalization movement (that is not any longer, strictly speaking, anti-globalization *tout court*, but anti-corporate or anti-capitalist globalization). In particular, the anti-corporate globalization (or social justice movement) is not just for preserving the local over global appropriation and control, but also for specific goals like human rights, labor rights, the rights of specific groups, like women, gays, the otherly-abled, or animals, as well as for goals like environmental preservation, safe food, democratization, and social justice. These goals are at least somewhat universal in many conceptualizations, so there is something of a synthesis of the global and local in the anti-corporate globalization movement. Hence in my view, these universal values and goals are valuable somethings, and the anti-corporate and social justice movement is important for defending important universal values, preserving local sites, cultures, and values, and providing innovative alternatives and political strategies and practices (though as Ritzer warns, they may erode into nothings if they merely repeat the same slogans and actions time after time).

I am not sure that one can quite as easily or cavalierly dismiss the local as Ritzer does, suggesting it has largely disappeared and cannot be resuscitated, as you cannot have glocals without locals and there are still many places, cuisines, products, peoples, cultures, and the like that have not yet been largely glocalized (or so I would imagine, though here Ritzer may be right in the long term). For instance, the day before a panel on Ritzer's book at the Eastern Sociology Association conference in New York in February 2004, I took a walk down Lexington Avenue and encountered on one block the stores "Good Old Things," "Fine Antiques," and other specialty shops. The next block had Indian vegetarian restaurants next to one that read "Non-Vegetarian Indian" and even "Kosher Vegetarian Indian," as well as a variety of other foreign restaurants. I passed the Armory that had the famous 1913 modern art exposition and was having an antiques fair

that weekend. At Union Square there was a market that was selling fresh bison meat, ostrich burgers, and freshly brewed hot apple cider that I tasted. Beyond the Square, the Strand bookstore still exists along with a few other surviving used bookstores in the neighborhood. And best of all, I found on the way back that the Grammercy Cinema was now the home of the MOMA Cinematique and was showing, for a six dollar a day pass, films by major Iranian and Korean directors, as well as a pair of Godard classics.

So, while somethings and the local are clearly under attack through corporate globalization (and one could give a detailed analysis of the grobalization of New York starting with the Disneyfication of 42nd Street and corporatization of Times Square), nonetheless, there are some locales still existent and they should be treasured, defended, and supported.

Shifting the register, I would also quibble about Ritzer's interpretation of the 9/11 attacks and, more broadly, why a certain breed of fundamentalist terrorism is anti-US. Ritzer rightly calls attention to a growing anti-Americanism and growing hostility to the grobalization, to use his term, of American culture, values, politics, and the military, but he does not mention George W. Bush, and I would argue that much of the skyrocketing anti-Americanism evident in the PEW polls, that Ritzer cites as evidence of growing anti-Americanism, is a specific reaction to the Bush administration's militarist unilateralism, nationalistic chauvinism, and just plain arrogance.

While the 9/11 and other Jihadist attacks might have happened no matter who was president, and while many parts of the world resent American grobalization, as Ritzer suggests, I think these resentments and reaction have been greatly intensified, perhaps dangerously so, by the Bush administration. Another caveat, in presenting Ben Barber's *Jihad vs McWorld*, Ritzer saliently presents McWorld as an example of grobalization, or nothing, but wrongly, in my view, presents Jihad as something. There is little so formulaic as bin Laden's anti-west ravings, and I suggest that terrorism has been extremely formulaic and repetitive (look at suicide bombings in Israel or Iraq), much more so than the anti-corporate globalization movement that Ritzer claims is repeating empty forms of internet connections and protest, rather than creating new and original forms of protest (only partly true, in my opinion, but a salutary warning to be creative, innovative, and surprising in constructing forms of global protest and oppositional politics). Finally, in regard to Jihad, I would argue that the Islamic schools, or *madrassa*, are as formulaic as the textbooks and McSchools that Ritzer rightly complains about.

And so in conclusion, I find George Ritzer's *The Globalization of Nothing* highly provocative, useful in its dialectic of something and nothing and glocalization vs grobalization in terms of theorizing globalization. As noted, I would have liked to see more of a dialectic of production and consumption, which I think would have enriched the project. Someone still needs to rewrite Marx's account of capitalism and the alienation of labor, in terms of global and hi-tech production and labor and new forms of culture and consumption. Nick Dyer-Witheford in *CyberMarx* has begun this enterprise, and those wishing to continue this thematic could well use many of Ritzer's categories applied to production and labour. Hence, whereas Ritzer's text is useful for illuminating aspects of consumption and globalization, the dialectic of production and consumption on local, national, and global scales still needs to be taken up.

McDonaldization

McDonaldization is the major example used by Nederveen Pieterse to illustrate the cultural convergence paradigm. Indeed, as we will see in this chapter, much of the debate surrounding the "McDonaldization thesis" deals with the issue of whether the model that is associated with the creation of the McDonald's chain in 1955 is accepted and practiced uniformly throughout the world.

We begin with "An Introduction to McDonaldization" from the fifth edition of *The McDonaldization of Society*. The basic definition of McDonaldization makes clear the fact that it is seen as a global phenomenon. That is, McDonaldization is defined as "the process by which the principles of the fast-food restaurant are coming to dominate more and more sectors of American society *as well as the rest of the world*."[1] It is the latter, italicized, phrase that makes it clear that McDonaldization is a global phenomenon. It is a global phenomenon in several obvious senses: McDonald's and other McDonaldized chains (both inside and outside the fast-food industry) have proliferated around the globe, other nations have developed their own McDonaldized chains, and now other nations are exporting their McDonaldized chains to the US. McDonald's, itself, has become a global icon that at least some consider more iconic than the US itself, or at least its ambassadors and embassies. However, we should bear in mind that

McDonaldization is not restricted to McDonald's, the fast-food industry, and even food. Rather it is seen as a wide-ranging process affecting many sectors of society (e.g. religion, education, and criminal justice).

The key to McDonaldization, as is made clear in the above definition, is its dimensions: efficiency, calculability, predictability, control, and, paradoxically, the irrationality of the seemingly highly rational process of McDonaldization. The key point is that the heart of McDonaldization is these principles and the system or structure that they represent and create. The issue from the point of view of globalization is the degree to which these principles and systems/structures have been globalized. As we will see, the critics of McDonaldization from the point of view of globalization tend to focus on things like the differences in the food in McDonald's in different parts of the world. While this is an issue, it does not get to the heart of whether McDonaldization has been globalized or whether it has tended to lead to at least some degree of homogeneity throughout the world. The central issue is whether McDonald's, and other McDonaldized systems, wherever they are in the world, adhere to the basic principles outlined above; whether they are based on the same system or structure.

Malcolm Waters contends that Ritzer argues that globalization must be seen as homogenization.

However, Ritzer does not equate McDonaldization with globalization; globalization is clearly a much broader process of which McDonaldization is but one component. Waters's second point is much more interesting and provocative. He recognizes that while McDonaldization may have homogenizing effects, it also can be used by local communities throughout the world in ways that are unanticipated by the forces that push it. That is, McDonaldization may be used in ways that further heterogeneity rather than homogeneity.

James Watson draws a number of conclusions that tend to support Waters's position on, and the critique of, McDonaldization as being inextricably linked to homogenization. Although Watson recognizes that McDonaldization has led to small and influential changes in East Asia that have made it and its dietary patterns more like those in the West, his most general conclusion is that "East Asian consumers have quietly, and in some cases stubbornly, transformed their neighborhood McDonald's into local institutions."[2] This represents not only a lack of global homogeneity, but resistance to it.

One of Watson's most interesting contentions is that East Asian cities are being reinvented so rapidly that it is hard even to differentiate between what is local and what is global. That is, the global is adopted and adapted so rapidly that it becomes part of the local. Thus, many Japanese children are likely to think that Ronald McDonald is Japanese.

Watson also does not see McDonald's as a typical transnational corporation with headquarters in the first world. Rather, to him, McDonald's is more like "a federation of semiautonomous enterprises" with the result that local McDonald's are empowered to go their separate ways, at least to some degree. Thus, locals have accepted some of McDonald's "standard operating procedures," but they have also modified or rejected others. McDonald's undergoes a process of localization whereby the locals, especially young people, no longer see it as a "foreign" entity.

While Watson takes the process of localization as a positive development, it can also be seen as more worrisome from the perspective of those who are concerned about the growing McDonaldization of the world. If McDonaldization remains a "foreign" presence, it is easy to identify and oppose, at least by those concerned about it. However, if it worms its way into the local culture and comes to be perceived as a local phenomenon, it becomes virtually impossible to identify and to oppose.

Bryan Turner surveys the ways in which McDonald's has modified itself in order to fit into various regions of the world: Russia, Australia, Asia, and the Middle East. He demonstrates the global power and reach of McDonald's and McDonaldization. He, like most other critics, focuses on the food – *not* the principles – and concludes that McDonald's has made major modifications in its menu in many locales. He sees this as compromising the basic McDonald's model – burgers and fries – at least as far as food is concerned. Turner's limited perspective is shaped by his view that: "At the end of the day, McDonald's simply is a burger joint."[3] This, of course, stands in contrast to the view in *The McDonaldization of Society* that in spawning McDonaldization, McDonald's is far more than a burger joint. Rather it is a framework with a basic set of principles that has served as a model for the creation and orchestration of a wide range of social structures and social institutions in the United States and throughout the world.

Bryman understands that McDonaldization is really about systems for accomplishing various tasks and achieving various goals. In fact, such systems define not only McDonaldization but also Disneyization. The key is the basic principles of McDonaldization (and Disneyization) that lie at the base of these systems. And those principles remain essentially the same whatever products and/or services are being proffered and wherever in the world they are on offer. This perspective reduces the import of the critiques offered by analysts like Waters, Watson, and especially Turner because their focus is largely limited to the foods and the ways in which they are adapted to different cultures.

Uri Ram understands this fact and demonstrates it in a case study of McDonald's in Israel. Although McDonald's has been successful there, it has not destroyed the local falafel industry. Rather, one part of the falafel business has been McDonaldized, while another has been "gourmetized." Depicted is a complex mix of the global and the local rather than one winning out over the other. Ram puts this in the context of the debate between one-way (e.g. McDonaldization, although now that process is multidirectional and not just running from the US to the rest of the world) and two-way (e.g. Appadurai's "landscapes"[4]) models of globalization. Ram responds creatively that *both* approaches are correct but on different levels. Structurally, he sees a one-way model as predominant, but symbolically it is a two-way street. So, much of the falafel industry in Israel has been transformed structurally into an industrial standardized system – a McDonaldized system. Symbolically, a two-way system is operant, with the falafel and the McDonald's hamburger coexisting and mutually affecting one another. Thus, although Israel is characterized by considerable structural uniformity, symbolically Israel remains internally differentiated as well as different from other societies, including the US. However, Ram seems to betray this perspective by arguing that Israeli differences have only "managed to linger on." Such phrasing seems to indicate that even to Ram, symbolic differences, like structural differences, may disappear, leading to increasing McDonaldization in both realms.

NOTES

1   George Ritzer, *The McDonaldization of Society*, 5th edn. Thousand Oaks, CA: Pine Forge, 2008, 1, italics added.
2   James L. Watson, "Transnationalism, Localization, and Fast Foods in East Asia." In James L. Watson, ed., *Golden Arches East: McDonald's in East Asia*. Stanford, CA: Stanford University Press, 1997, 6.
3   Bryan S. Turner, "McDonaldization: Linearity and Liquidity in Consumer Cultures." *American Behavioral Scientist* 47, 2, 2003: 151. What is most surprising about this assertion is that Turner, a Weberian scholar, should know better. Such an assertion is akin to critiquing Weber's work by saying that "a bureaucracy is simply an organization." Bureaucracy plays the same paradigmatic role in Weber's work as McDonald's does in Ritzer's.
4   Arjun Appadurai, *Modernity at Large: Cultural Dimensions of Globalization*. Minneapolis: University of Minnesota Press, 1996.

READING 57

An Introduction to McDonaldization

George Ritzer

[. . .]

McDonald's has been a resounding success in the international arena. Over half of McDonald's restaurants are outside the United States (in the mid 1980s, only 25% of McDonald's were outside the United States). The majority (233) of the 280 new restaurants opened in 2006 were overseas (in the United States, the number of restaurants increased by only 47). Well over half of the revenue for McDonald's comes from its overseas operations. McDonald's restaurants are now found in 118 nations around the world, serving 50 million customers a day. The leader by far, as of the beginning of 2007, is Japan with 3,828 restaurants, followed by Canada with over 1,375 and Germany with over 1,200. There are currently 780 McDonald's restaurants in China (but Yum! Brands operates over 2,000 KFCs – the Chinese greatly prefer chicken to beef – and 300 Pizza Huts in China). McDonald's will add 100 new restaurants a year in China with a goal of 1,000 restaurants by the opening of the 2008 Beijing Olympics (but KFC will add 400 a year!). As of 2006, there were 155 McDonald's in Russia, and the company plans to open many more restaurants in the former Soviet Union and in the vast new territory in eastern Europe that has been laid bare to the invasion of fast-food restaurants. Although there have been recent setbacks for McDonald's in Great Britain, that nation remains the "fast-food capital of Europe," and Israel is described as "McDonaldized," with its shopping malls populated by "Ace Hardware, Toys 'R' Us, Office Depot, and TCBY."

Many highly McDonaldized firms outside the fast-food industry have also had success globally. Although most of Blockbuster's 9,000-plus sites are in the United States, about 2,000 of them are found in 24 other countries. Wal-Mart is the world's largest retailer with 1.8 million employees and over $312 billion in sales. There are almost 4,000 of its stores in the United States (as of 2006). It opened its first international store (in Mexico) in 1991; it now has more than 2,700 units in Puerto Rico, Canada, Mexico, Argentina, Costa Rica, El Salvador, Guatemala, Honduras, Nicaragua, Brazil, China, Korea, Japan, Germany, and the United Kingdom. In any given week, more than 175 million customers visit Wal-Mart stores worldwide.

Other nations have developed their own variants on the McDonald's chain. Canada has a chain of coffee shops called Tim Hortons (merged with Wendy's in 1995), with 2,711 outlets (336 in the United States). It is Canada's largest food service provider with nearly twice as many outlets as McDonald's in that country. The chain has 62% of the coffee business (Starbucks is a distant second with just 7% of that business). Paris, a city whose love for fine cuisine might lead you to think it would prove immune to fast food, has a large number of fast-food croissanteries; the revered French bread has also been McDonaldized. India has a chain of fast-food restaurants, Nirula's, that sells mutton burgers (about 80% of Indians are Hindus, who eat no beef) as well as local Indian cuisine. Mos Burger is a Japanese chain with over 1,600 restaurants that, in addition to the usual fare, sell Teriyaki chicken burgers, rice burgers, and "Oshiruko with brown rice cake." Perhaps the most unlikely spot for an indigenous fast-food restaurant, war-ravaged Beirut of 1984, witnessed the opening of Juicy Burger, with a rainbow instead of golden arches and J. B. the Clown standing in for Ronald McDonald. Its owners hoped it would become the "McDonald's of the Arab world." In the immediate wake of the 2003 invasion of Iraq, clones of McDonald's (sporting names like "MaDonal" and "Matbax") opened in that country complete with hamburgers, french fries, and even golden arches.

And now McDonaldization is coming full circle. Other countries with their own McDonaldized institutions have begun to export them to the United States. The Body Shop, an ecologically sensitive British cosmetics chain, had, as of 2006, over 2,100 shops in

55 nations, 300 of them in the United States. American firms have followed the lead and opened copies of this British chain, such as Bath & Body Works. Pret A Manger, a chain of sandwich shops that also originated in Great Britain (interestingly, McDonald's purchased a 33% minority share of the company in 2001), has over 150 company-owned and -run restaurants, mostly in the United Kingdom but now also in New York, Hong Kong, and Tokyo. Pollo Campero was founded in Guatemala in 1971 and by mid-2006 had more than 200 restaurants in Latin America and the United States. In the latter, 23 restaurants were in several major cities, and the company planned to open 10 more in such cities by the end of 2006. (Jollibee, a Philippine chain, has 10 US outlets.) Though Pollo Campero is a smaller presence in the United States than the American-owned Pollo Tropical chain (which has 80 U.S. outlets), Pollo Campero is more significant because it involves the invasion of the United States, the home of fast food, by a foreign chain.

IKEA (more on this important chain later), a Swedish-based (but Dutch-owned) home furnishings company, did about 17.6 billion euros of business in 2006, derived from the over 410 million people visiting their 251 stores in 34 countries. Purchases were also made from the 160 million copies of their catalog printed in over 44 languages. In fact, that catalog is reputed to print annually the second largest number of copies in the world, just after the Bible. IKEA's web site features over 12,000 products and reported over 125 million "hits" in 2006. Another international chain to watch in the coming years is H&M clothing, founded in 1947 and now operating 1,345 stores in 24 countries with plans to open another 170 stores by the end of 2007. It currently employs over 60,000 people and sells more than 500 million items a year. Based in Spain, Inditex Group, whose flagship store is Zara, overtook H&M in March 2006 to become Europe's largest fashion retailer with more than 3,100 stores in 64 countries.

[. . .]

At the opening of the McDonald's in Moscow, one journalist described the franchise as the "ultimate icon of Americana." When Pizza Hut opened in Moscow in 1990, a Russian student said, "It's a piece of America." Reflecting on the growth of fast-food restaurants in Brazil, an executive associated with Pizza Hut of Brazil said that his nation "is experiencing a passion for things American." On the popularity of Kentucky Fried Chicken in Malaysia, the local owner said, "Anything Western, especially American, people here love [. . .] They want to be associated with America."

One could go further and argue that in at least some ways McDonald's has become more important than the United States itself. Take the following story about a former US ambassador to Israel officiating at the opening of the first McDonald's in Jerusalem wearing a baseball hat with the McDonald's golden arches logo:

An Israeli teen-ager walked up to him, carrying his own McDonald's hat, which he handed to Ambassador Indyk with a pen and asked: "Are you the Ambassador? Can I have your autograph?" Somewhat sheepishly, Ambassador Indyk replied: "Sure. I've never been asked for my autograph before."

As the Ambassador prepared to sign his name, the Israeli teen-ager said to him, "Wow, what's it like to be the ambassador from McDonald's, going around the world opening McDonald's restaurants everywhere?"

Ambassador Indyk looked at the Israeli youth and said, "No, no. I'm the American ambassador – not the ambassador from McDonald's!" Ambassador Indyk described what happened next: "I said to him, 'Does this mean you don't want my autograph?' And the kid said, 'No, I don't want your autograph,' and he took his hat back and walked away."

Two other indices of the significance of McDonald's (and, implicitly, McDonaldization) are worth mentioning. The first is the annual "Big Mac Index" (part of "burgernomics"), published, tongue-in-cheek, by a prestigious magazine, the *Economist*. It indicates the purchasing power of various currencies around the world based on the local price (in dollars) of the Big Mac. The Big Mac is used because it is a uniform commodity sold in many different nations. In the 2007 survey, a Big Mac in the United States cost an average of $3.22; in China it was $1.41; in Switzerland it cost $5.5; the costliest was $7.44 in Iceland. This measure indicates, at least roughly, where the cost of living is high or low, as well as which currencies are undervalued (China) and which are overvalued (Switzerland). Although the *Economist* is calculating the Big Mac Index only half-seriously, the index represents the ubiquity and importance of McDonald's around the world.

The second indicator of the global significance of McDonald's is the idea developed by Thomas Friedman that "no two countries that both have a McDonald's have ever fought a war since they each got McDonald's." Friedman calls this the "Golden Arches Theory of Conflict Prevention." Another tongue-in-cheek idea, it implies that the path to world peace lies through the continued international expansion of McDonald's. Unfortunately, it was proved wrong by the NATO bombing of Yugoslavia in 1999, which had McDonald's at the time (as of 2007, there are 16 McDonald's there).

To many people throughout the world, McDonald's has become a sacred institution. At that opening of the McDonald's in Moscow, a worker spoke of it "as if it were the Cathedral in Chartres [. . .] a place to experience 'celestial joy.'" Kowinski argues that indoor shopping malls, which almost always encompass fast-food restaurants, are the modern "cathedrals of consumption" to which people go to practice their "consumer religion." Similarly, a visit to another central element of McDonaldized society, Walt Disney World, has been described as "the middle-class hajj, the compulsory visit to the sunbaked holy city."

[. . .]

The Dimensions of McDonaldization

Why has the McDonald's model proven so irresistible? Eating fast food at McDonald's has certainly become a "sign" that, among other things, one is in tune with the contemporary lifestyle. There is also a kind of magic or enchantment associated with such food and its settings. The focus here, however, is on the four alluring dimensions that lie at the heart of the success of this model and, more generally, of McDonaldization. In short, McDonald's has succeeded because it offers consumers, workers, and managers efficiency, calculability, predictability, and control. [. . .]

Efficiency

One important element of the success of McDonald's is *efficiency*, or the optimum method for getting from one point to another. For consumers, McDonald's (its drive-through is a good example) offers the best available way to get from being hungry to being full. The fast-food model offers, or at least appears to offer, an efficient method for satisfying many other needs, as well. Woody Allen's orgasmatron offered an efficient method for getting people from quiescence to sexual gratification. Other institutions fashioned on the McDonald's model offer similar efficiency in exercising, losing weight, lubricating cars, getting new glasses or contacts, or completing income tax forms. Like their customers, workers in McDonaldized systems function efficiently by following the steps in a pre-designed process.

Calculability

Calculability emphasizes the quantitative aspects of products sold (portion size, cost) and services offered (the time it takes to get the product). In McDonaldized systems, quantity has become equivalent to quality; a lot of something, or the quick delivery of it, means it must be good. As two observers of contemporary American culture put it, "As a culture, we tend to believe deeply that in general 'bigger is better.'" People can quantify things and feel that they are getting a lot of food for what appears to be a nominal sum of money (best exemplified by the McDonald's current "Dollar Menu," which played a key role in recent years in leading McDonald's out of its doldrums and to steadily increasing sales). In a recent Denny's ad, a man says, "I'm going to eat too much, but I'm never going to pay too much." This calculation does not take into account an important point, however: the high profit margin of fast-food chains indicates that the owners, not the consumers, get the best deal.

People also calculate how much time it will take to drive to McDonald's, be served the food, eat it, and return home; they then compare that interval to the time required to prepare food at home. They often conclude, rightly or wrongly, that a trip to the fast-food restaurant will take less time than eating at home. This sort of calculation particularly supports home delivery franchises such as Domino's, as well as other chains that emphasize saving time. A notable example of time savings in another sort of chain is LensCrafters, which promises people "Glasses fast, glasses in one hour." H&M is known for its "fast fashion."

Some McDonaldized institutions combine the emphases on time and money. Domino's promises pizza delivery in half an hour, or the pizza is free. Pizza Hut will serve a personal pan pizza in 5 minutes, or it, too, will be free.

Workers in McDonaldized systems also emphasize the quantitative rather than the qualitative aspects of their work. Since the quality of the work is allowed to vary little, workers focus on things such as how quickly tasks can be accomplished. In a situation analogous to that of the customer, workers are expected to do a lot of work, very quickly, for low pay.

Predictability

McDonald's also offers *predictability*, the assurance that products and services will be the same over time and in all locales. The Egg McMuffin in New York will be, for all intents and purposes, identical to those in Chicago and Los Angeles. Also, those eaten next week or next year will be identical to those eaten today. Customers take great comfort in knowing that McDonald's offers no surprises. People know that the next Egg McMuffin they eat will not be awful, although it will not be exceptionally delicious, either. The success of the McDonald's model suggests that many people have come to prefer a world in which there are few surprises. "This is strange," notes a British observer, "considering [McDonald's is] the product of a culture which honours individualism above all."

The workers in McDonaldized systems also behave in predictable ways. They follow corporate rules as well as the dictates of their managers. In many cases, what they do, and even what they say, is highly predictable.

Control

The fourth element in the success of McDonald's, *control*, is exerted over the people who enter the world of McDonald's. Lines, limited menus, few options, and uncomfortable seats all lead diners to do what management wishes them to do – eat quickly and leave. Furthermore, the drive-through (in some cases, walk-through) window invites diners to leave before they eat. In the Domino's model, customers never enter in the first place.

The people who work in McDonaldized organizations are also controlled to a high degree, usually more blatantly and directly than customers. They are trained to do a limited number of things in precisely the way they are told to do them. This control is reinforced by the technologies used and the way the organization is set up to bolster this control. Managers and inspectors make sure that workers toe the line.

A Critique of McDonaldization: the Irrationality of Rationality

McDonaldization offers powerful advantages. In fact, efficiency, predictability, calculability, and control through nonhuman technology (that is, technology that controls people rather than being controlled by them) can be thought of as not only the basic components of a rational system but also as powerful advantages of such a system. However, rational systems inevitably spawn irrationalities. The downside of McDonaldization will be dealt with most systematically under the heading of the irrationality of rationality; in fact, paradoxically, the irrationality of rationality can be thought of as the fifth dimension of McDonaldization [. . .]

Criticism, in fact, can be applied to all facets of the McDonaldizing world. As just one example, at the opening of Euro Disney, a French politician said that it will "bombard France with uprooted creations that are to culture what fast food is to gastronomy." Although McDonaldization offers many advantages [. . .], this book will focus on the great costs and enormous risks of McDonaldization. McDonald's and other purveyors of the fast-food model spend billions of dollars each year detailing the benefits of their system. Critics of the system, however, have few outlets for their ideas. For example, no one sponsors commercials between Saturday-morning cartoons warning children of the dangers associated with fast-food restaurants.

Nonetheless, a legitimate question may be raised about this critique of McDonaldization: is it animated by a romanticization of the past, an impossible desire to return to a world that no longer exists? Some critics do base their critiques on nostalgia for a time when life was slower and offered more surprises, when at least some people (those who were better off economically) were freer, and when one was more likely to deal with a human being than a robot or a computer. Although they have a point, these critics have undoubtedly exaggerated the positive aspects of a world without McDonald's, and they have certainly tended to forget the liabilities associated with earlier eras. As an example of the latter, take the following anecdote about a visit to a pizzeria in Havana, Cuba, which in some respects is decades behind the United States:

> The pizza's not much to rave about – they scrimp on tomato sauce, and the dough is mushy.
>
> It was about 7:30 p.m., and as usual the place was standing-room-only, with people two deep jostling for a stool to come open and a waiting line spilling out onto the sidewalk.
>
> The menu is similarly Spartan [. . .] To drink, there is tap water. That's it – no toppings, no soda, no beer, no coffee, no salt, no pepper. And no special orders.
>
> A very few people are eating. Most are waiting [. . .] Fingers are drumming, flies are buzzing, the clock is ticking. The waiter wears a watch around his belt loop, but he hardly needs it; time is evidently not his chief concern. After a while, tempers begin to fray.
>
> But right now, it's 8:45 p.m. at the pizzeria, I've been waiting an hour and a quarter for two small pies.

Few would prefer such a restaurant to the fast, friendly, diverse offerings of, say, Pizza Hut. More important, however, critics who revere the past do not seem to realize that we are not returning to such a world. In fact, fast-food restaurants have begun to appear even in Havana (and many more are likely after the death of Fidel Castro). The increase in the number of people crowding the planet, the acceleration of technological change, the increasing pace of life – all this and more make it impossible to go back to the world, if it ever existed, of home-cooked meals, traditional restaurant dinners, high-quality foods, meals loaded with surprises, and restaurants run by chefs free to express their creativity.

It is more valid to critique McDonaldization from the perspective of a conceivable future. Unfettered by the constraints of McDonaldized systems, but using the technological advances made possible by them, people could have the potential to be far more thoughtful, skillful, creative, and well-rounded than they are now.

In short, if the world was less McDonaldized, people would be better able to live up to their human potential.

We must look at McDonaldization as both "enabling" and "constraining." McDonaldized systems enable us to do many things we were not able to do in the past; however, these systems also keep us from doing things we otherwise would do. McDonaldization is a "double-edged" phenomenon. We must not lose sight of that fact, even though this book will focus on the constraints associated with McDonaldization – its "dark side."

Illustrating the Dimensions of McDonaldization: the Case of IKEA

An interesting example of McDonaldization, especially since it has its roots in Sweden rather than the United States, is IKEA. Its popularity stems from the fact that it offers at very low prices trendy furniture based on well-known Swedish designs. It has a large and devoted clientele throughout the world. What is interesting about IKEA from the point of view of this book is how well it fits the dimensions of McDonaldization. The similarities go beyond that, however. For example, just as with the opening of a new McDonald's, there is great anticipation over the opening of the first IKEA in a particular location. Just the rumor that one was to open in Dayton, Ohio, led to the following statement: "We here in Dayton are peeing our collective pants waiting for the IKEA announcement." IKEA is also a global phenomenon – it is now in 34 countries (including China and Japan) and sells in those countries both its signature products as well as those more adapted to local tastes and interests.

In terms of *efficiency*, IKEA offers one-stop furniture shopping with an extraordinary range of furniture. In general, there is no waiting for one's purchases, since a huge warehouse is attached to each store (one often enters through the warehouse), with large numbers of virtually everything in stock.

Much of the efficiency at IKEA stems from the fact that customers are expected to do a lot of the work:

- Unlike McDonald's, there are relatively few IKEA's in any given area; thus, customers most often spend many hours driving great distances to get to a store. This is known as the "IKEA road trip."

- On entry, customers are expected to take a map to guide themselves through the huge and purposely maze-like store (IKEA hopes, like Las Vegas casinos, that customers will get "lost" in the maze and wander for hours, spending money as they go). There are no employees to guide anyone, but there are arrows painted on the floor that customers can follow on their own.
- Also upon entry, customers are expected to grab a pencil and an order form and to write down the shelf and bin numbers for the larger items they wish to purchase; a yellow shopping bag is to be picked up on entry for smaller items. There are few employees and little in the way of help available as customers wander through the stores. Customers can switch from a shopping bag to a shopping cart after leaving the showroom and entering the marketplace, where they can pick up other smaller items.
- If customers eat in the cafeteria, they are expected to clean their tables after eating. There is even this helpful sign: "Why should I clean my own table? At IKEA, cleaning your own table at the end of your meal is one of the reasons you paid less at the start."
- Most of the furniture sold is unassembled in flat packages, and customers are expected to load most of the items (except the largest) into their cars themselves. After they get home, they must break down (and dispose) of the packaging and then put their furniture together; the only tool supposedly required is an Allen wrench.
- If the furniture does not fit into your car, you can rent a truck on site to transport it home or have it delivered, although the cost tends to be high, especially relative to the price paid for the furniture.
- To get a catalog, customers often sign up online.

Calculability is at the heart of IKEA, especially the idea that what is offered is at a very low price. Like a McDonald's "Dollar Menu," one can get a lot of furniture – a roomful, even a houseful – at bargain prices. As with value meals, customers feel they are getting value for their money. (There is even a large cafeteria offering low-priced food, including the chain's signature Swedish meatballs and 99-cent breakfasts.) However, as is always the case in McDonaldized settings, low price generally means that the quality is inferior, and it is often the case that IKEA products fall apart in relatively short order. IKEA also emphasizes the huge size of its stores, which often approach 300,000 square feet or about four to five football fields. This mammoth size leads the consumer to believe that there will be a lot of furniture offered (and there is) and that, given the store's reputation, most of it will be highly affordable.

Of course, there is great *predictability* about any given IKEA – large parking lots, a supervised children's play area (where IKEA provides personnel, but only because supervised children give parents more time and peace of mind to shop and spend), the masses of inexpensive, Swedish-design furniture, exit through the warehouse and the checkout counters, boxes to take home with furniture requiring assembly, and so on.

An IKEA is a highly *controlled* environment, mainly in the sense that the maze-like structure of the store virtually forces the consumer to traverse the entire place and to see virtually everything it has to offer. If one tries to take a path other than that set by IKEA, one is likely to become lost and disoriented. There seems to be no way out that does not lead to the checkout counter, where you pay for your purchases.

There are a variety of *irrationalities* associated with the rationality of IKEA, most notably the poor quality of most of its products. Although the furniture is purportedly easy to assemble, many are more likely to think of it as "impossible-to-assemble." Then there are the often long hours required to get to an IKEA, to wander through it, to drive back home, and then to assemble the purchases.

[. . .]

McDonaldization and the Global Culture of Consumption

Malcolm Waters

On the face of it [...] Ritzer offers a persuasive case that McDonaldization is an influential globalizing flow. The imperatives of the rationalization of consumption appear to drive McDonald's and like enterprises into every corner of the globe so that all localities are assimilated. The imperatives of such rationalization are expressed neatly:

[C]onsumption is work, it takes time and it competes with itself since choosing, hauling, maintaining and repairing the things we buy is so time-consuming that we are forced to save time on eating, drinking, sex, dressing, sleeping, exercising and relaxing. The result is that Americans have taught us to eat standing, walking, running and driving – and, above all, never to finish a meal in favour of the endless snack [...] we can now pizza, burger, fry and coffee ourselves as quickly as we can gas our autos.

[...] The globalization of "McTopia," a paradise of effortless and instantaneous consumption, is also underpinned by its democratizing effect. It democratizes by de-skilling, but not merely by de-skilling McWorkers but also by de-skilling family domestic labor. The kitchen is invaded by frozen food and microwaves so that domestic cooks, usually women, can provide McDonaldized fare at home. In the process, non-cooks, usually men and children, can share the cooking. Meals can become "de-familized" (i.e., de-differentiated) insofar as all members can cook, purchase, and consume the same fatty, starchy, sugary foods. Consequently, while "America is the only country in the world where the rich eat as badly as the poor," the appeal of such "gastronomic leveling" can serve as a magnet for others elsewhere.

However, we can put in perspective the alarmist in Ritzer's neo-Weberian suggestions that globalization will lead to a homogenized common culture of consumption if we expose them to the full gamut of globalization theory. Globalization theory normally specifies that a globalized culture is chaotic rather than orderly – it is integrated and connected so that the meanings of its components are "relativized" to one another but it is not unified or centralized. The absolute globalization of culture would involve the creation of a common but hyperdifferentiated field of value, taste, and style opportunities, accessible by each individual without constraint for purposes either of self-expression or consumption. Under a globalized cultural regime, Islam would not be linked to particular territorially based communities in the Middle East, North Africa, and Asia but would be universally available across the planet and with varying degrees of "orthodoxy." Similarly, in the sphere of the political ideology, the apparently opposed political values of private property and power sharing might be combined to establish new ideologies of economic enterprise. In the sphere of consumption, cardboard hamburgers would be available not only in Pasadena but anywhere in the world, just as classical French cuisine would be available not only in Escoffier's in Paris but anywhere. A globalized culture thus admits a continuous flow of ideas, information, commitment, values, and tastes mediated through mobile individuals, symbolic tokens, and electronic simulations. Its key feature is to suggest that the world is one place not because it is homogenized but because it accepts only social differentiation and not spatial or geographical differentiation.

These flows give a globalized culture a particular shape. First, it links together previously encapsulated and formerly homogeneous cultural niches. Local developments and preferences are ineluctably shaped by similar patterns occurring in very distant locations. Second, it allows for the development of genuinely transnational cultures not linked to any particular nation-state-society, which may be either novel or syncretistic. Appadurai's increasingly influential argument about the global cultural economy identifies several of the important fields in which these developments take place. The fields are identified by the suffix "-scape"; that is, they are globalized mental pictures of the social world perceived from the flows of cultural objects. The flows include ethnoscapes, the distribution of mobile individuals (tourists, migrants, refugees, etc.); technoscapes, the distribution of technology; finanscapes, the distribution of capital; mediascapes, the distribution of information; and ideoscapes, the distribution of political ideas and values (e.g., freedom, democracy, human rights).

McDonaldization infiltrates several of these flows, including ethnoscapes, technoscapes, finanscapes, and ideoscapes. However, its effects are by no means universally homogenizing. The dynamics that are at work center on processes of relativization, reflexivity, and localization that operate against the assumed capacity of McDonaldization to regiment consumer behavior into uniform patterns. The return of agency that many authors have identified is not simply a series of isolated and individualized coping reactions of the type advocated by Ritzer in *McDonaldization* but a generalized feature of contemporary society that arises from the intersection of these globalizing flows. Indeed, such developments might be called the dysfunctions of McDonaldization in much the way that post-Weberian organizational theorists wrote of the dysfunctions of bureaucracy [. . .]

The term "relativization" [. . .] implies that globalizing flows do not simply swamp local differences. Rather, it implies that the inhabitants of local contexts must now make sense of their lifeworlds not only by reference to embedded traditions and practices but by reference to events occurring in distant places. McDonaldization is such an intrusive, neonistic development that it implies decisions about whether to accept its modernizing and rationalizing potential or to reject it in favor of a reassertion of local products and traditions. In some instances, this may involve a reorganization of local practices to meet the challenge. If we remain at the mundane level of hamburgers to find our examples, there is a story about the introduction of McDonald's in the Philippines that can illustrate the point:

Originally, Filipino hamburger chains marketed their product on the basis of its "Americanness." However, when McDonald's entered the field and, as it were, monopolized the symbols of "Americanness," the indigenous chains began to market their product on the basis of local taste.

The relativization effect of McDonaldization goes of course much further than this because it involves the global diffusion not only of particular products but of icons of American capitalist culture. Relativizing reactions can therefore encompass highly generalized responses to that culture, whether positive or negative.

As people increasingly become implicated in global cultural flows they also become more reflexive. [. . .] Participation in a global system means that one's lifeworld is determined by impersonal flows of money and expertise that are beyond one's personal or even organizational control. If European governments cannot even control the values of their currencies against speculation, then individual lifeworlds must be highly vulnerable. Aware of such risk, people constantly watch, seek information about, and consider the value of money and the validity of expertise. Modern society is therefore specifically reflexive in character. Social activity is constantly informed by flows of information and analysis that subject it to continuous revision and thereby constitute and reproduce it. "Knowing what to do" in modern society, even in such resolutely traditional contexts as kinship or child rearing, almost always involves acquiring knowledge about how to do it from books, or television programs, or expert consultations, rather than relying on habit, mimesis, or authoritative direction from elders. McDonaldization is implicated in this process precisely because it challenges the validity of habit and tradition by introducing expertly rationalized systems, especially insofar as its capacity to commercialize and to commodify has never been in doubt.

The concept of localization is connected with the notions of relativization and reflexivity. The latter imply that the residents of a local area will increasingly come to want to make conscious decisions about which values and amenities they want to stress in their communities and that these decisions will increasingly be referenced against global scapes. Localization implies a reflexive reconstruction of community in the face of the dehumanizing implications of such rationalizing and commodifying forces as McDonaldization. The activist middle classes who mobilize civic initiatives and heritage preservation associations often stand in direct opposition to the expansion of McDonaldized outlets and hark back to an often merely imagined prior golden age.

Returning to more abstract issues, these three processes can assure us that a globalized world will not be a McWorld. It is a world with the potential for the displacement of local homogeneity not by global homogeneity but by global diversity. Three developments can confirm this hopeful prognosis.

First, one of the features of Fordist mass-production systems, of which McDonaldization might be the ultimate example, is that they sought to standardize at the levels of both production and consumption. Ultimately, they failed not only because they refused to recognize that responsible and committed workers would produce more in quantity and quality than controlled and alienated ones but because markets for standardized products became saturated. The succeeding paradigm of "flexible specialization" involved flexibly contracted workers using multiple skills and computerized machinery to dovetail products to rapidly shifting market demand. So consumer products took on a new form and function. Taste became the only determinant of their utility, so it became ephemeral and subject to whim. Product demand is determined by fashion, and unfashionable products are disposable. Moreover, taste and fashion became linked to social standing as product-based classes appeared as central features of social organization.

The outcome has been a restless search by producers for niche-marketing strategies in which they can multiply product variation in order to match market demand. In many instances, this has forced a downscaling of enterprises that can maximize market sensitivity. Correspondingly, affluent consumers engage in a restless search for authenticity. The intersection of these trends implies a multiplication of products and production styles. The world is becoming an enormous bazaar as much as a consumption factory. One of the most impressive examples of consumer and producer resistance to rationalization is the French bread industry, which is as non-McDonaldized as can be. [. . .] Consumers and producers struggled collec-tively against invasions by industrialized bakers, the former to preserve the authenticity of their food, the latter to maintain independent enterprises. Bread-baking is an artisanal form of production that reproduces peasant domestic traditions. About 80 percent of baking (Ritzer's *Croissanteries* notwithstanding) is still done in small firms. The product, of course, is the envy of global, middle-class consumers.

This diversification is accelerated by an aestheticization of production. As is well known, the history of modern society involves an increasing production of mass-cultural items. For most of this century, this production has been Fordist in character, an obvious example being broadcasting by large-scale private or state TV networks to closed markets. Three key features in the current period are the deregulation of markets by the introduction of direct-satellite and broadband fiber-optic technology; the vertical disintegration of aesthetic production to produce "a transaction-rich nexus of markets linking small firms, often of one self-employed person"; and the tendency of de-differentiation of producer and consumer within emerging multimedia technologies associated with the Internet and interactive television. The implication is that a very rapidly increasing proportion of consumption is aesthetic in character, that aesthetic production is taking place within an increasingly perfectionalized market, and that these aesthetic products are decreasingly susceptible to McDonaldization. An enormous range of individualized, unpredictable, inefficient, and irrational products can be inspected simply by surfing the Internet.

The last development that can disconfirm the thesis of a homogenized global culture is the way in which globalization has released opposing forces of opinion, commitment, and interest that many observers find threatening to the fabric of society and indeed to global security. One of these is the widespread religious revivalism that is often expressed as fundamentalism. Globalization carries the discontents of modernization and postmodernization (including McDonaldization) to religious traditions that might previously have remained encapsulated. [. . .] Religious systems are obliged to relativize themselves to these global post-modernizing trends. This relativization can involve an embracement of postmodernizing patterns, an abstract and humanistic ecumenism, but it can also take the form of a rejective search for original traditions. It is this latter that has given rise to both Islamic fundamentalism and [...] the New Christian Right.

Globalization equally contributes to ethnic diversity. It pluralizes the world by recognizing the value of cultural niches and local abilities. Importantly, it weakens the putative nexus between nation and state releasing absorbed ethnic minorities and allowing the reconstitution of nations across former state boundaries. This is especially important in the context of states that are confederations of minorities. It can actually alter the mix of ethnic identities in any nation-state by virtue of the flow of economic migrants from relatively disadvantaged sectors of the globe to relatively advantaged ones. Previously homogeneous nation-states have, as a consequence, moved in the direction of multiculturalism.

Conclusion

The paradox of McDonaldization is that in seeking to control consumers it recognizes that human individuals potentially are autonomous, a feature that is notoriously lacking in "cultural dupe" or "couch potato" theories of the spread of consumer culture. As dire as they may be, fast-food restaurants only take money in return for modestly nutritious and palatable fare. They do not seek to run the lives of their customers, although they might seek to run their diets. They attract rather than coerce so that one can always choose not to enter. Indeed, advertising gives consumers the message, however dubious, that they are exercising choice.

It might therefore be argued, *contra* Ritzer, that consumer culture is the source of the increased cultural effectivity that is often argued to accompany globalization and postmodernization. Insofar as we have a consumer culture, the individual is expected to exercise choice. Under such a culture, political issues and work can equally become items of consumption. A liberal-democratic political system might only be possible where there is a culture of consumption precisely because it offers the possibility of election – even if such a democracy itself tends to become McDonaldized, as leaders become the mass-mediated images of photo opportunities and juicy one-liners, and issues are drawn in starkly simplistic packages. Equally, work

can no longer be expected to be a duty or a calling or even a means of creative self-expression. Choice of occupation, indeed choice of whether to work at all, can be expected increasingly to become a matter of status affiliation rather than of material advantage.

Ritzer is about right when he suggests that McDonaldization is an extension, perhaps the ultimate extension, of Fordism. However, the implication is that just as one now has a better chance of finding a Fordist factory in Russia or India than in Detroit, it should not surprise us to find that McDonaldization is penetrating the furthest corners of the globe, and there is some indication that, as far as the restaurant goes, there is stagnation if not yet decline in the homeland. McDonaldization faces post-Fordist limits and part of the crisis that these limits imply involves a transformation to a chaotic, taste- and value-driven, irrational, and possibly threatening global society. It will not be harmonious, but the price of harmony would be to accept the predominance of Christendom, or Communism, or Fordism, or McDonaldism.

This chapter, then, takes issue with the position taken by Ritzer. [...] First, there is a single globalization–localization process in which local sensibilities are aroused and exacerbated in fundamentalist forms by such modernizing flows as McDonaldization. Even in the fast-food realm, McDonaldization promotes demands for authenticity, even to the extent of the fundamentalism of vegetarianism. Second, the emerging global culture is likely to exhibit a rich level of diversity that arises out of this intersection. Globalization exposes each locality to numerous global flows so that any such locality can accommodate, to use food examples once again, not only burgers but a kaleidoscope of ethnically diverse possibilities hierarchically ordered by price and thus by the extent to which the meal has been crafted as opposed to manufactured. Thus while it is not possible to escape the ubiquity of McDonald's in one sense, the golden arches are indeed everywhere, in another it certainly is, one can simply drive by and buy either finger food from a market stall or haute cuisine at a high priced restaurant. Ritzer is not wrong then to argue that McDonaldization is a significant component of globalization. Rather, he is mistaken in assuming first that globalization must be understood as homogenization and second that McDonaldization only has homogenizing effects.

The McDonald's Mosaic: Glocalization and Diversity

Bryan S. Turner

There is considerable ethnographic evidence that McDonald's outlets have adjusted to local circumstances by incorporating local cuisines and values into their customer services. The success of global McDonald's has been to organize and present itself as a local company, where it specifically aims to incorporate local taste and local dishes – the curry potato pie from Hong Kong, the Singapore Loveburger (grilled chicken, honey, and mustard sauce), and the Teriyaki burger (sausage patty) and the Tukbul burger with cheese for the Korean market. Let us take the Russian example. The Russian experience of Western culture in the last decade has been intensely ambiguous. The obvious seduction of Western consumerism that had begun in the 1970s continued into the early 1990s, and young people in particular rushed to embrace the latest Western consumer goods and habits. Yet unsurprisingly, the promise of a widespread democratic consumer culture has not been fulfilled. Among older Russians, there has been a growing nostalgia for a putative Russian "way of doing things" and a concomitant suspicion of Western cultural institutions.

In this context of disappointed ambitions and expectations, one would expect McDonald's to be an obvious target of Russian hostility. Even in Western countries themselves, McDonald's is often seen as representative of the detrimental, exploitative, and pervasive reach of global capitalism. For many critics, McDonald's exploits and poisons workers. Its culture of fast and unimaginative food is symbolic of the worst aspects of consumerism. From a Russian perspective, the characteristics of McDonald's, including its style – such as its particular forms of graphic design and its presentation of food – its emphasis on customer service and training, and its standardized global presence are decidedly Western. Russia is a society in which, as a result of its communist legacy, personal service,

friendliness, and helpfulness are still corrupt bourgeois customs.

Of interest, however, Russians have a decidedly ambivalent view of McDonald's, in part because they are pragmatic in their responses to Western influences. Seventy years of Soviet rule has taught them to be judicious in their use of principle because they have learned to live with inconsistency and contradiction. McDonald's offers a surfeit of cultural contradiction because, notwithstanding the overtly Western style of McDonald's, there are also numerous forms of convergence with Russian habits and values.

First, there is the compatibility of the Fordist labor process, food process, and purchasing protocols in McDonald's with those that were developed during the Soviet period in Russia and that have continued under postcommunism. These processes and protocols, although often different in content, are consistently Fordist in form and structure. In both a McDonald's and postcommunist setting, there are clear expectations of standardized and predictable products, delivery of products, staff and their uniform dress, and consumer protocols. In both settings, production and social interaction are rule driven and steered through authoritarian decision-making processes.

Second, the formal standardized structure and method of operation of a McDonald's restaurant is underpinned by an egalitarian ethos. In particular, the egalitarian ethos in Russia has been manifested in disdain for the external trappings of a service culture (as a sign of inequality) and is currently manifested in popular contempt for the ostentatious consumption of "the new Russians." McDonald's presents its food as sustenance for the "common people." In addition, the way of eating the food, using hands rather than knives and forks, appeals to ordinary people in a country where haute cuisine has been seen as, and continues to be defined as, a form of cultural pretension. The service culture of McDonald's is based on a commitment to a formal equality between customer and service assistant.

Finally, the actual content of McDonald's food has a definite appeal to Russian taste. For example, McDonald's food, such as the buns, sauces, and even the meat, tends to be sweeter than the average European or Asian cuisine. Desserts are generally based on dairy produce and include exceedingly sweet sauces. Potato chips and fried chicken appeal to the Russian preference for food fried in saturated fat rather than food that is grilled or uncooked. Thus, although McDonald's might be seen as a harbinger of the worst of Western cultural imperialism, the pragmatic Russian will usually be prepared to frequent McDonald's restaurants because of the quality and compatibility of the food with Russian taste and the familiarity of the setting and delivery process. However, the cost of McDonald's food in Russia is prohibitive and for many is a luxury item for which the average family must save.

In Australia, by contrast, McDonald's culture is highly compatible with a society that has embraced egalitarianism to such an extent that cultural distinction is explicitly rejected in such popular expressions as "to cut down tall poppies" and by the emphasis on mateship. Historically, the Australian food consumption has contained a high level of meat, especially lamb and beef. Dietary innovations such as replacing lard by canola resulted in a 50% cut in sales in Sydney stores. McDonald's has been particularly successful down under, where it is claimed by the *Weekend Australian* that a million Australians consume more than $4.8 million burgers, fries, and drinks at the 683 McDonald's stores each day. McDonald's arrived in Australia in 1971, opening 118 stores in its first year. The company had an important impact on services in Australia, where it led the way in modernizing work practices, corporate culture, and philanthropy. Their business strategy involved the development of community and educational links through Rotary clubs and churches. McDonald's successfully survived much local criticism against American cultural imperialism and developed educational programs that have been addressed to kindergartens and schools. McDonald's built playgrounds and distributed toys. Through the development of McHappy Day, it donates generously to hospitals and charities. It also developed Ronald McDonald

House Charities that in 2001 raised $2.4 million for charity. Ray Kroc's four commandments – quality, service, cleanliness, and value – have been adopted as core elements in a two-unit educational diploma that can be taken in certain Australian high schools as components of their educational experience.

Although it has been a significant commercial success and now controls 42% of the fast-food market, the high-water mark was achieved in the mid 1990s when 145 stores were opened in the space of 2 years. Sales figures have become static, customer satisfaction is declining, and McDonald's has been the subject of public criticism. McDonald's suffered economically when the Liberal Government of John Howard introduced the GST (General Sales Tax) and McDonald's hamburgers were not exempt. The result was 10% decline in sales, and they failed to achieve their target of 900 stores by the year 2000. McDonald's has responded to this decline in several ways, including the diversification of their products into McCafes and by moving upmarket into Mexican-style restaurants and sandwich bars.

In Asia, McDonald's outlets have been successful in penetrating local markets. In the process, however, McDonald's products have been changing. The doctrine that societies that are connected by trade do not go to war is being tested in the case of China and Taiwan. For example, Taiwan has 341 and the People's Republic of China has 326 McDonald's restaurants. The new Chinese elite in its drive to industrialize and modernize society has accepted McDonald's outlets because McDonald's is seen to epitomize healthy food based on nutritious ingredients and scientific cooking. Although the Party is still in control and formally promotes communist ideals of loyalty and dedication, young people have adopted the Ronald McDonald backpack as a sign of modernist consumerism. McDonald's entered Taiwan in 1984, where it now sells 92 million hamburgers and 60 million McNuggets to a population of 22.2 million. McDonald's has become ubiquitous partly by adding corn soup to its regular menu once it was realized that no meal is complete without soup. McDonald's in Taiwan also abandoned its antiloitering policy once it accepted the fact that students saw the air-conditioned McDonald's as an attractive and cool venue for study. Other changes in this densely populated society followed, such as building three-storey outlets that can seat more than 250 people at a time.

South Korea is another society that enthusiastically embraced McDonald's. The first outlet was opened in Apkujong-dong in Seoul during the 1988 Olympic Games and expanded rapidly to become the second largest fast-food service retailer after Lotteria. The World Cup provided important marketing opportunities for McDonald's, and the company sought to increase its outlets, adding another 100 restaurants. The company initiated a "Player Escort" scheme to select Korean children to participate by escorting soccer players to the football dome. The current McDonald's president Kim Hyung soo has adopted the sociological expression "glocalization" to describe the customization of McDonald's menus to satisfy the demands of local customers by developing Korean-style burgers such as Bulgogi Burger and Kimchi Burger. Another promotional strategy has been to make Internet available in its restaurants located in famous hang-out places for Korean youth, such as the ASEM mall and Shinohon.

The market in Asia is also diversifying as further Westernized commodities and lifestyles are imported. [. . .] The growing demand for coffee in Asia, where it is now beginning to challenge the cultural hegemony of tea. [. . .] in the last 5 years, Starbucks has become as widespread as McDonald's. [. . .]

McDonald's has responded by creating McSnack. [. . .] It offers chicken and beef curry rice, bagels and English muffin sandwiches, and waffles. It also offers nine different hot and cold coffee drinks. The important feature of the coffee craze is that Korean customers expect to loiter in the outlets, which are used as meeting places and spaces for study. McDonald's staff tolerate customers who sit for hours inside the restaurant or on chairs outside hardly buying anything. During their university examinations period, students are packed into McSnack and so actual customers often find it difficult to secure a seat. Customers also bring food into McSnack from other restaurants to eat at the nice, clean, air-conditioned outlets.

These national case studies show us how McDonald's fast-food outlets interact with local cultures. Perhaps the best illustration of these local tensions is in the Middle East, where 300 McDonald's have opened, mainly following the Gulf war. McDonald's has been successful in Saudi Arabia, where McDonald's has spread rapidly, despite periodic fundamentalist boycotts, and where its stores are closed five times a day for prayers. The company now intends to open McDonald's in Afghanistan. In Turkey, McDonald's started to open branches in the 1980s in Istanbul and Ankara. Although McDonald's has expanded to around 100 outlets, almost half of these are in Istanbul. There is a McDonald's in Kayseri, the center of the Islamist vote in Istanbul. The only remarkable protest against McDonald's was held at the Middle East Technical University when it tried to open a branch there in the 1990s, but this protest came from socialists not Islamic students. Ironically, Muslim couples often use McDonald's as a place to meet because they know that their traditionalist parents would not dine there. McDonald's in Turkey also has been sensitive to Islamic norms and it offers *iftar*, an evening meal served during Ramadan. In Egypt, McDonald's has also become popular and serves sandwiches, Egyptian boulettes, and other local items. Although Egyptian intellectuals condemn Kentucky Fried Chicken and McDonald's as examples of Western corruption of local taste and cuisine, McDonald's now exists without conflict alongside street vendors and local cafes.

McDonald's outlets have paradoxically been popular in many Muslim societies, despite strong anti-American sentiments, because parents recognize them as places where alcohol will not be served. In addition, the mildly exotic Western taste of a burger and fries is an alternative to local fare. Indonesian youth use McDonald's in the same way that Western youth gravitate toward shopping malls. With temperatures consistently in the 30°C range (90°F) and humidity often more than 80%, McDonald's is simply a convenient, clean, and cool place to be. The company has once more adapted to local taste by introducing sweet iced tea, spicy burgers, and rice. The economic crisis in early 1998 forced McDonald's to experiment with a cheaper menu as the price of burgers exploded. McDonald's customers remained with the company to consume McTime, PaNas, and Paket Nasi. For many years, McDonald's has advertised its products as *halal*, reassuring its Muslim customers that its products are religiously clean. Similar to Egyptian McDonald's, in Indonesia, a postsunset meal is offered as a "special" during Ramadan. To avoid any criticism of Americanization, McDonald's is a local business that is owned by a Muslim, whose advertising banners proclaim in Arabic that McDonald's Indonesia is fully owned by an indigenous Muslim. Proprietors also will proudly boast their Muslim status by the use of post-pilgrimage titles such as *Haji*.

Conclusions: Cultural Liquidity

These local case studies show how the rational model of McDonald's adjusts to local cultural preferences, but the result is a diminution of the original McDonald's product (the burger and fries). In fact, the more the company adjusts to local conditions, the more the appeal of the specifically American product may be lost. At the end of the day, McDonald's simply is a burger joint. Therefore [. . .] we need to distinguish between specific studies of McDonald's and macro-studies of McDonaldization as rationalization. [. . .] The global reach of McDonald's is hardly at issue, and I have attempted to illustrate some of the complexity

of that reach through several vignettes of McDonald's in Russia, Australia, the Middle East, and Asia. The spread of McDonald's clearly illustrates the fact that McDonaldization has been a powerful force behind the administrative rationalism of modern societies. With globalization, rationalization has become a global dimension of the basic social processes of any modern society. In this sense, the McDonaldization thesis is also a potent defense of the continuing relevance of Weber's general sociology of modernity.

More fundamentally, the diversification of McDonald's through its interaction with local cultures has produced new management strategies, consumer cultures, and product range that depart radically from the Fordist linearity of the original model. McDonald's is slowly disappearing under the weight of its fragmentation, differentiation, and adaptation. [. . .] The unstoppable march of McDonald's through urban society has come to an end.

READING 60

Transnationalism, Localization, and Fast Foods in East Asia

James L. Watson

Does the spread of fast food undermine the integrity of indigenous cuisines? Are food chains helping to create a homogeneous, global culture better suited to the needs of a capitalist world order?

[. . .] We do not celebrate McDonald's as a paragon of capitalist virtue, nor do we condemn the corporation as an evil empire. Our goal is to produce ethnographic accounts of McDonald's social, political, and economic impact on five local cultures. These are not small-scale cultures under imminent threat of extinction; we are dealing with economically resilient, technologically advanced societies noted for their haute cuisines. If McDonald's can make inroads in these societies, one might be tempted to conclude, it may indeed be an irresistible force for world culinary change. But isn't another scenario possible? Have people in East Asia conspired to change McDonald's, modifying

this seemingly monolithic institution to fit local conditions?

[. . .] The interaction process works both ways. McDonald's *has* effected small but influential changes in East Asian dietary patterns. Until the introduction of McDonald's, for example, Japanese consumers rarely, *if* ever, ate with their hands [. . .] this is now an acceptable mode of dining. In Hong Kong, McDonald's has replaced traditional teahouses and street stalls as the most popular breakfast venue. And among Taiwanese youth, French fries have become a dietary staple, owing almost entirely to the influence of McDonald's.

At the same time, however, East Asian consumers have quietly, and in some cases stubbornly, transformed their neighborhood McDonald's into a local institution. In the United States, fast food may indeed imply fast consumption, but this is certainly not the case everywhere. In Beijing, Seoul, and Taipei, for instance, McDonald's restaurants are treated as leisure centers, where people can retreat from the stresses of urban life. In Hong Kong, middle school students often sit in McDonald's for hours – studying, gossiping, and picking over snacks; for them, the restaurants are the equivalent of youth clubs. [. . .] Suffice it to note here that McDonald's does not always call the shots.

Globalism and Local Cultures

[. . .] The operative term is "local culture," shorthand for the experience of everyday life as lived by ordinary people in specific localities. In using it, we attempt to capture the feelings of appropriateness, comfort, and correctness that govern the construction of personal preferences, or "tastes." Dietary patterns, attitudes toward food, and notions of what constitutes a proper meal [. . .] are central to the experience of everyday life and hence are integral to the maintenance of local cultures.

Readers will note [. . .] class, gender, and status differences, especially in relation to consumption practices. One surprise was the discovery that many McDonald's restaurants in East Asia have become sanctuaries for women who wish to avoid male-dominated settings. In Beijing and Seoul, new categories of yuppies treat McDonald's as an arena for conspicuous consumption. Anthropologists who work in such settings must pay close attention to rapid changes in consumer preferences. Twenty years ago, McDonald's catered to the children of Hong Kong's wealthy elite; the current generation of Hong Kong hyperconsumers has long since abandoned the golden arches and moved up-market to more expensive watering holes (e.g., Planet Hollywood). Meanwhile, McDonald's has become a mainstay for working-class people, who are attracted by its low cost, convenience, and predictability.

One of our conclusions [. . .] is that societies in East Asia are changing as fast as cuisines – there is nothing immutable or primordial about cultural systems. In Hong Kong, for instance, it would be impossible to isolate what is specifically "local" about the cuisine, given the propensity of Hong Kong people to adopt new foods. [. . .] Hong Kong's cuisine, and with it Hong Kong's local culture, is a moving target.

Hong Kong is the quintessential postmodern environment, where the boundaries of status, style, and taste dissolve almost as fast as they are formed. What is "in" today is "out" tomorrow.

Transnationalism and the Multilocal Corporation

It has become an academic cliché to argue that people are constantly reinventing themselves. Nevertheless, the speed of that reinvention process in places like Hong Kong, Taipei, and Seoul is so rapid that it defies description. In the realm of popular culture, it is no longer possible to distinguish between what is "local" and what is "foreign." Who is to say that Mickey Mouse is not Japanese, or that Ronald McDonald is not Chinese? To millions of children who watch Chinese television, "Uncle McDonald" (alias Ronald) is probably more familiar than the mythical characters of Chinese folklore.

We have entered here the realm of the transnational, a new field of study that focuses on the "deterritorialization" of popular culture. [. . .] The world economy can no longer be understood by assuming that the original producers of a commodity necessarily control its consumption. A good example is the spread of "Asian" martial arts to North and South America, fostered by Hollywood and the Hong Kong film industry. Transnationalism describes a condition by which people, commodities, and ideas literally cross – transgress – national boundaries and are not identified with a single place of origin. One of the leading theorists of this new field argues that transnational phenomena are best perceived as the building blocks of "third cultures," which are "oriented beyond national boundaries."

Transnational corporations are popularly regarded as the clearest expressions of this new adaptation, given that business operations, manufacturing, and marketing are often spread around the globe to dozens of societies.

At first glance, McDonald's would appear to be the quintessential transnational. On closer inspection, however, the company does not conform to expectations; it resembles a federation of semiautonomous enterprises. James Cantalupo, former President of McDonald's International, claims that the goal of McDonald's is to "become as much a part of the local culture as possible."

He objects when "[p]eople call us a multinational. I like to call us *multilocal*," meaning that McDonald's goes to great lengths to find local suppliers and local partners whenever new branches are opened.

[...] McDonald's International retains at least a 50 percent stake in its East Asian enterprises; the other half is owned by local operators.

Modified Menus and Local Sensitivities: McDonald's Adapts

The key to McDonald's worldwide success is that people everywhere know what to expect when they pass through the Golden Arches. This does not mean, however, that the corporation has resisted change or refused to adapt when local customs require flexibility. [...] McDonald's restaurants in India serve Vegetable McNuggets and a mutton-based Maharaja Mac, innovations that are necessary in a country where Hindus do not eat beef, Muslims do not eat pork, and Jains (among others) do not eat meat of any type. In Malaysia and Singapore, McDonald's underwent rigorous inspections by Muslim clerics to ensure ritual cleanliness; the chain was rewarded with a *halal* ("clean," "acceptable") certificate, indicating the total absence of pork products.

Variations on McDonald's original, American-style menu exist in many parts of the world: chilled yogurt drinks (*ayran*) in Turkey, espresso and cold pasta in Italy, teriyaki burgers in Japan (also in Taiwan and Hong Kong), vegetarian burgers in the Netherlands, McSpaghetti in the Philippines, McLaks (grilled salmon sandwich) in Norway, frankfurters and beer in Germany, McHuevo (poached egg hamburger) in Uruguay. [...]

Irrespective of local variations (espresso, McLaks) and recent additions (carrot sticks), the structure of the McDonald's menu remains essentially uniform the world over: main course burger/sandwich, fries, and a drink – overwhelmingly Coca-Cola. The keystone of this winning combination is *not*, as most observers might assume, the Big Mac or even the generic hamburger. It is the fries. The main course may vary widely (fish sandwiches in Hong Kong, vegetable burgers in Amsterdam), but the signature innovation of McDonald's – thin, elongated fries cut from russet potatoes – is everpresent and consumed with great gusto by Muslims, Jews, Christians, Buddhists,

Hindus, vegetarians (now that vegetable oil is used), communists, Tories, marathoners, and armchair athletes. [...]

Conclusion: McDonaldization versus Localization

McDonald's has become such a powerful symbol of the standardization and routinization of modern life that it has inspired a new vocabulary: McThink, McMyth, McJobs, McSpiritually, and, of course, McDonaldization. George Ritzer, author of a popular book titled *The McDonaldization of Society* [...] treats McDonald's as the "paradigm case" of social regimentation and argues that "McDonaldization has shown every sign of being an inexorable process as it sweeps through seemingly impervious institutions and parts of the world."

Is McDonald's in fact the revolutionary, disruptive institution that theorists of cultural imperialism deem it to be? Evidence [...] could be marshaled in support of such a view but only at the risk of ignoring historical process. There is indeed an initial, "intrusive" encounter when McDonald's enters a new market – especially in an environment where American-style fast food is largely unknown to the ordinary consumer. In five cases [...] McDonald's was treated as an exotic import – a taste of Americana – during its first few years of operation. Indeed, the company drew on this association to establish itself in foreign markets. But this initial euphoria cannot sustain a mature business.

Unlike Coca-Cola and Spam, for instance, McDonald's standard fare (the burger-and-fries combo) could not be absorbed into the preexisting cuisines of East Asia. [...] Spam quickly became an integral feature of Korean cooking in the aftermath of the Korean War; it was a recognizable form of meat that required no special preparation. Coca-Cola, too, was a relatively neutral import when first introduced to Chinese consumers. During the 1960s, villagers in rural Hong Kong treated Coke as a special beverage, reserved primarily for medicinal use. It was served most frequently as *bo ho la*, Cantonese for "boiled Cola," a tangy blend of fresh ginger and herbs served in piping hot Coke – an excellent remedy for colds. Only later was the beverage consumed by itself, first at banquets (mixed with brandy) and later for special events such as a visit by relatives. There was nothing particularly revolutionary about Coca-Cola or Spam; both products were quickly adapted to suit local needs and did not require any radical adjustments on the part of consumers.

McDonald's is something altogether different. Eating at the Golden Arches is a total experience, one that takes people out of their ordinary routines. One "goes to" a McDonald's; it does not come to the consumer, nor is it taken home. [. . .]

From this vantage point it would appear that McDonald's may indeed have been an intrusive force, undermining the integrity of East Asian cuisines. On closer inspection, however, it is clear that consumers are not the automatons many analysts would have us believe they are. The initial encounter soon begins to fade as McDonald's loses its exotic appeal and gradually gains acceptance (or rejection) as ordinary food for busy consumers. The hamburger-fries combo becomes simply another alternative among many types of ready-made food.

The process of localization is a two-way street: it implies changes in the local culture as well as modifications in the company's standard operating procedures. Key elements of McDonald's industrialized system – queuing, self-provisioning, self-seating – have been accepted by consumers throughout East Asia. Other aspects of the industrial model have been rejected, notably those relating to time and space. In many parts of East Asia, consumers have turned their local McDonald's into leisure centers and after school clubs. The meaning of "fast" has been subverted in these settings: it refers to the *delivery* of food, not to its consumption. Resident managers have had little choice but to embrace these consumer trends and make virtues of them: "Students create a good atmosphere which is good for our business," one Hong Kong manager told me as he surveyed a sea of young people chatting, studying, and snacking in his restaurant.

The process of localization correlates closely with the maturation of a generation of local people who grew up eating at the Golden Arches. By the time the children of these original consumers enter the scene, McDonald's is no longer perceived as a foreign enterprise. Parents see it as a haven of cleanliness and predictability. For children, McDonald's represents fun, familiarity, and a place where they can choose their own food – something that may not be permitted at home.

[. . .] Localization is not a unilinear process that ends the same everywhere. McDonald's has become a routine, unremarkable feature of the urban landscape in Japan and Hong Kong. It is so local that many younger consumers do not know of the company's foreign origins. The process of localization has hardly begun in China, where McDonald's outlets are still treated as exotic outposts, selling a cultural experience rather than food. At this writing, it is unclear what will happen to expansion efforts in Korea; the political environment there is such that many citizens will continue to treat the Golden Arches as a symbol of American imperialism. In Taiwan, the confused, and exhilarating, pace of identity politics may well rebound on American corporations in ways as yet unseen. Irrespective of these imponderables, McDonald's is no longer dependent on the United States market for its future development. [. . .]

As McDonald's enters the 21st century, its multilocal strategy, like its famous double-arches logo, is being pirated by a vast array of corporations eager to emulate its success. In the end, however, McDonald's is likely to prove difficult to clone.

 READING 61

Global Implications of McDonaldization and Disneyization

Alan Bryman

One way in which Disneyization and McDonaldization can be viewed as parallel processes is that both can legitimately be viewed as signals of globalization. Ritzer makes this point in relation to McDonaldization in his more recent work, and it is apparent that the dimensions of Disneyization [. . .] are similarly spreading throughout the globe. But what is striking about the two concepts is that they do not refer specifically to the global diffusion of products. Much of the writing on globalization is full of hyperbole about the global spread and recognizability of prominent brands: Nike, Coca-Cola, Pepsi-Cola, Pizza Hut, KFC, Benetton, Body Shop, and so on. And, of course, one could hardly disregard the golden arches of McDonald's or Mickey's ears and Walt's signature as involved in the global travels of brand names. But that is *not* what McDonaldization and Disneyization are about: they are concerned essentially with the diffusion of *modes of delivery* of goods and services. McDonaldization relates primarily to a mode of delivery in the sense of the *production* of goods and services. It is a means of providing an efficient and highly predictable product in a manner that would have appealed to people such as Ford and Taylor. It belongs to an era of mass consumption that is not disappearing but whose emphases are becoming less central to modern society with the passage of time. Disneyization is a mode of delivery in the sense of the *staging* of goods and services for consumption. It provides a context for increasing the allure of goods and services. Indeed, it may be that one of the reasons for the growing use of theming in the form of external narratives in some McDonald's restaurants has to do with the limitations of McDonaldization itself. McDonaldization's emphasis on standardization sits uneasily in an increasingly post-Fordist era of choice and variety. Theming becomes a means of reducing the sense of sameness and thereby enhancing the appeal of its products.

What is important about such a suggestion is that it is crucial to appreciate that McDonaldization and Disneyization are both *systems*, that is, they are ways of producing or presenting goods and services. One of the problems with tying the names of these systems to well-known icons of popular culture – McDonald's and Disney – is that it is easy to make the mistake of lapsing into a discussion of just McDonald's and Disney. This is an error because the two companies are merely emblems of the underlying processes associated with their respective systems.

By emphasizing processes associated with Disneyization and McDonaldization as systems, it is possible to get away from the shrill but not always revealing accounts of the global reach of prominent brands. It can hardly be doubted that there is a clutch of high-profile brands that have spread through much of the globe, but systems such as Disneyization and McDonaldization are in a sense more significant than that. For one thing, their presence is perhaps less immediately obvious than the arrival of McDonald's restaurants or the impending arrival of a new Disney theme park in Hong Kong. Focusing on the products obscures the more fundamental issue of the diffusion of underlying principles through which goods and services are produced and then put into people's mouths and homes. Although McDonald's restaurants have been the focus of anti-globalization campaigners and Disney was given a decidedly gallic cold shoulder among intellectuals in France when Disneyland Paris was in the planning stage, occasioning the famous "cultural Chernobyl" comment, the spread of the fundamental principles that can be divined from an examination of what McDonald's and the Disney theme parks exemplify is much less frequently, and perhaps less likely to be, a focus of comment.

When considered in this way, it is striking how poorly Disneyization and McDonaldization fit into Appadurai's influential delineation of different forms of "-scape," that is, contexts for the flow of goods, people, finance, and other items around the globe. Appadurai distinguished between five scapes; ethnoscapes (the movement of people), technoscapes (the movement of technology), financescapes (the movement of capital), mediascapes (the movement of information), and ideoscapes (the movement of ideas and ideals). Waters has argued that "McDonaldization infiltrates several of these flows." However, such a view does not do justice to the significance of McDonaldization and by implication Disneyization. In a sense, we need a new conceptual term for them, which we might call "system-scapes," to refer to the flow of contexts for the production and display of goods and services. Although they incorporate elements of the five scapes, as Waters suggests, McDonaldization and Disneyization are somewhat more than this. They represent important templates for the production of goods and services and their exhibition for sale.

Of course, we must give due consideration to the charge that we are subscribing here to a simplistic globalization or Americanization thesis that depicts icons of American culture spreading by design across the globe and riding roughshod over local conditions and practices. Research on McDonald's which can be treated as the *locus classicus* of McDonaldization, suggests that it is dangerous to think of a simple process of subsuming foreign cultures. Not only has McDonald's accommodated to local tastes and dietary requirements and preferences but it is also used in different ways in different cultures. It is sometimes regarded as a sophisticated eating environment for special occasions or dating couples, as a meeting place, as an area for study, and so on. Similar remarks can be made in relation to the Disney theme parks when they have been transported abroad. Raz observes in relation to Tokyo Disneyland that although it is invariably claimed to be a copy of the American original, it has in fact been Japanized. Thus, the Mystery Tour in the castle in Tokyo Disneyland is a Disney version of the Japanese ghost house. The Meet the World show is [. . .] as "a show about and for the Japanese." Similar adaptation can be seen in Disneyland Paris, where after a disappointing beginning, the company was forced to adapt the park to European tastes. The alcohol ban, in particular, had to be dropped. Such local adaptations and accommodations are frequently and quite rightly latched on to by the critics of a simple globalization thesis. They are also reassuring that the world is not becoming a single homogenized realm because there are signs of resistance even in the face of the momentum of two revered representatives of popular culture.

However, although reassuring, these indications of the continued relevance of the local for McDonald's and the Disney theme parks should not blind us to the fact that although McDonald's may be used differently in Taipei and that Tokyo Disneyland has adapted many attractions to the Japanese sensibility, this is not what McDonaldization and Disneyization are about. As previously argued, they are about *principles* to do with the production and delivery of goods and services. What the researchers who tell us about the different ways that McDonald's has adapted to or been differentially appropriated by diverse cultures is how McDonald's has been adapted to and appropriated, not McDonaldization as such. In a sense, Disneyization and McDonaldization are more worrying for the critics of

globalization as a homogenizing force than the arrival of golden arches in far reaches of the globe or the transplanting of Disney theme parks abroad. They are more worrying because Disneyization and McDonaldization are potentially more insidious processes because they are far less visible and immediately obvious in their emergence than the appearance of golden arches or of magic kingdoms on nations' doorsteps. As Ritzer points out in relation to McDonald's, "The fundamental operating procedures remain essentially the same everywhere in the globe," a view that is largely endorsed by company representatives. Robert Kwan, at the time managing director of McDonald's in Singapore, is quoted by Watson as saying, "McDonald's sells [. . .] a system, not products." In other words, finding adaptations to and local uses of McDonald's and Disney theme parks should not make us think that this means or even necessarily entails adaptations to and local uses of McDonaldization and Disneyization.

Turning more specifically to Disneyization, particularly in relation to McDonald's, none of what has been said previously should be taken to imply that there are likely to be no processes of local adaptation or resistance or culturally specific uses in relation to Disneyization. Emotional labor has been a particularly prominent site for resistance, as studies of the local reception of McDonald's demonstrate. Watson has observed that during the early period of the restaurant's arrival in Moscow, people standing in queues had to be given information about such things as how to order. In addition, they had to be told, "The employees inside will smile at you. This does not mean that they are laughing at you. We smile because we are happy to serve you." Watson also remarks on the basis of his fieldwork in Hong Kong that people who are overly congenial are regarded with suspicion, so that a smile is not necessarily regarded as a positive feature. Also, consumers did not display any interest in the displays of friendliness from crew personnel. It is not surprising, therefore, that the display of emotional labor is not a significant feature of the behavior and demeanor of counter staff in McDonald's in Hong Kong. Watson says, "Instead, they project qualities that are admired in the local culture: competence, directness, and unflappability. [. . .] Workers who smile on the job are assumed to be enjoying themselves at the consumer's (and the management's) expense."

A somewhat different slant is provided by Fantasia's account of the reception of McDonald's in France. There, the attraction of McDonald's for young people was what he calls the "American ambience." Insofar as the display of emotional labor is an ingredient of this ambience, it may be that it is not that the French enthusiasts respond positively to emotional labor per se but that in the context of McDonald's they respond positively to the total package, of which smiling counter staff is a component. In other words, as the writers who emphasize local adaptations to global processes point out, local consumers frequently make their own culturally bespoke uses of the forces of globalization.

Clearly, there are risks with the foregoing argument. At a time when writers on globalization prefer to emphasize "glocalization" or "creolization" as ways of coming to terms with the varied ways in which global forces have to run the gauntlet of local cultural conditions and preferences, it is unfashionable to suggest that impulses emanating from the United States are tramping over the globe. Indeed, as the previously cited evidence concerned with emotional labor implies, we do need to take into account the ways such global influences are working their way into and are being incorporated into local cultures. But Disneyization is a more invisible process than the arrival of brand names on foreign shores. It is designed to maximize consumers' willingness to purchase goods and services that in many cases they might not otherwise have been prompted to buy. Theming provides the consumer with a narrative that acts as a draw by providing an experience that lessens the sense of an economic transaction and increases the likelihood of purchasing merchandise. Dedifferentiation of consumption is meant to give the consumer as many opportunities as possible to make purchases and therefore to keep them as long as possible in the theme park, mall, or whatever. Emotional labor is the oil of the whole process in many ways: in differentiating otherwise identical goods and services, as an enactment of theming, and as a milieu for increasing the inclination to purchase merchandise. It may be that, as in Russia and Hong Kong, emotional labor is ignored or not effective. However, these are fairly small responses to the diffusion of these instruments of consumerism. And insofar as we can regard McDonald's as a Disneyized institution, the process of Disneyization has a high-profile partner that is likely to enhance the global spread of its underlying principles.

 READING 62

Glocommodification: How the Global Consumes the Local – McDonald's in Israel

Uri Ram

One of the more controversial aspects of globalization is its cultural implications: does globalization lead to universal cultural uniformity, or does it leave room for particularism and cultural diversity? The global–local encounter has spawned a complex polemic between 'homogenizers' and 'heterogenizers.' This article proposes to shift the ground of the debate from the homogeneous–heterogeneous dichotomy to a structural-symbolic construct. It is argued here that while both homogenization and heterogenizations are dimensions of globalization, they take place at different societal levels: homogenization occurs at the structural-institutional level; heterogenization, at the expressive-symbolic. The proposed structural-symbolic model facilitates a realistic assessment of global–local relations. In this view, while global technological, organizational and commercial flows need not destroy local habits and customs, but, indeed, may preserve or even revive them, the global does tend to subsume and appropriate the local, or to consume it, so to say, sometimes to the extent that the seemingly local, symbolically, becomes a specimen of the global, structurally.

The starting point for this analysis is the McDonaldization of Israeli culture. McDonald's opened its first outlet in Israel in 1993. Since then, it has been involved in a variety of symbolic encounters [. . .] [in] the encounter between McDonald's, as the epitome of global fast food, and the local version of fast food, namely the falafel [. . .] local idioms have thrived, though only symbolically. On the structural level, they have been subsumed and appropriated by global social relationships.

Global Commerce Encounters the Local Eating Habitus: McDonald's and the Falafel

The industrialized hamburger first arrived on Israel's shores back in the late 1960s, although the chains involved at the time did not make much of an impression. In 1972, Burger Ranch (BR) opened a local hamburger joint that expanded into a chain only in the 1980s. It took the advent of McDonald's, however, for the 'great gluttony' of the fast hamburger to begin. McDonald's opened its first branch in October 1993. It was followed by Burger King (BK), the world's second largest hamburger chain, which opened its first branch in Israel in early 1994. Between McDonald's arrival and the year 2000, sales in the hamburger industry soared by 600 percent. By 2000, annual revenues from fast-food chains in Israel reached NIS 1 billion (about US$200 million according to the 2002 exchange rate). McDonald's is the leading chain in the industry, with 50 percent of the sales, followed by BR with 32 percent, and BK with 18 percent. In 2002 the three chains had a total of 250 branches in place: McDonald's, 100; BR, 94 and BK, 56.

McDonald's, like Coca-Cola – both flagship American brands – conquered front-line positions in the war over the Israeli consumer. The same is true of many other American styles and brands, such as jeans, T-shirts, Nike and Reebok footwear, as well as megastores, such as Home Center Office Depot, Super-Pharm, etc. [. . .] As for eating habits, apart from the spread of fast-food chains, other Americanisms have found a growing niche in the Israeli market: frozen 'TV dinners,' whether in family or individual packs, and an upsurge in fast-food deliveries. These developments stem from the transformation of the familial lifestyle as an increasing number of women are no longer (or not only) housewives, the growth of singles

households, and the rise in family incomes. All this, along with accelerated economic activity, has raised the demand for fast or easy-to-prepare foods. As has happened elsewhere, technological advancements and business interests have set the stage for changes in Israeli eating habits. Another typical development has been the mirror process that accompanies the expansion of standardized fast foods, namely, the proliferation of particularist cuisines and ethnic foods as evinced by the sprouting of restaurants that cater to the culinary curiosity and open purses of a new Yuppie class in Tel Aviv, Herzliya and elsewhere.

As in other countries, the 'arrival' of McDonald's in Israel raised questions and even concern about the survival of the local national culture. A common complaint against McDonald's is that it impinges on local cultures, as manifested primarily in the local eating habitus both actual and symbolic. If Israel ever had a distinct national equivalent to fast food, it was unquestionably the falafel – fried chick-pea balls served in a 'pocket' of pita bread with vegetable salad and tahini (sesame) sauce. The falafel, a Mediterranean delicacy of Egyptian origin, was adopted in Israel as its 'national food.' Although in the 1930s and 1940s the falafel was primarily eaten by the young and impecunious, in the 1950s and 1960s a family visit to the falafel stand for a fast, hot bite became common practice, much like the visit paid nowadays to McDonald's. The falafel even became an Israeli tourist symbol, served as a national dish at formal receptions of the Ministry of Foreign Affairs. Indeed, one kiosk in Tel Aviv advertises itself as a "'mighty' falafel for a mighty people."

Despite the falafel's fall from glory in the 1970s and 1980s vis-à-vis other fast foods, such as *shawarma* (lamb or turkey pieces on a spit), pizza and the early hamburger stands, and notwithstanding the unwholesome reputation it developed, an estimated 1200 falafel eateries currently operate in Israel. Altogether, they dish up about 200,000 portions a day to the 62 percent of Israelis who are self-confessed falafel eaters. The annual industry turnover is some NIS 600 million – not that far short of the hamburger industry. Thus, surprisingly enough, in the late 1990s, McDonald's presence, or rather the general McDonaldization of Israeli food habits, led to the falafel's renaissance, rather than to its demise.

The falafel's comeback, vintage 2000, is available in two forms: gourmet and fast-food. The clean, refined, gourmet Tel-Avivian specimen targets mainly yuppies and was launched in 1999 – five years after McDonald's landed in the country – in a prestigious restaurant owned by two women, famed as Orna and Ella. Located in the financial district, which is swiftly being gentrified, it is known as 'The Falafel Queens' – a hip, ironic feminist version of the well-known 'Falafel King' – one of the most popular designations for Israeli falafel joints, which always take the masculine form. The new, 'improved' gourmet model comes in a variety of flavors. Apart from the traditional 'brown' variety, the Queens offer an original 'red' falafel, based on roasted peppers, as well as 'green' falafel, based on olive paste. Beverages are a mixed bag, including orange-Campari and grapefruit-arrack ice. Owner Ella Shein rightly notes that the falafel's revival reflects a composite global–local trend:

> We have opened up to the world culinarily speaking, we have been exposed to new raw materials, new techniques, a process that occurs simultaneously with a kind of return to one's origins, to one's roots.

Apart from its 'gourmetization,' the falafel has simultaneously undergone 'McDonaldized' standardization. The Israeli franchise of Domino's Pizza inaugurated a new falafel chain, setting itself a nationwide target of 60 branches. Furthermore, its reported intention is to 'take the tidings of Israeli fast-food abroad.' The falafel has thus been rescued from parochialism and upgraded to a world standard-bearer of 'Israeli fast food,' or, as one observer put it, it has been transformed from 'grub' into 'brand.' In fact, the Ma'oz chain already operates 12 falafel eateries in Amsterdam, Paris and Barcelona and, lately, also in Israel. The new chains have developed a 'concept' of 'clean, fresh, and healthy,' with global implications, because: 'if you are handed an inferior product at "Ma'oz" in Amsterdam, you won't set foot in the Paris branch' either. In contrast to the traditional falafel stand, which stands in the street and absorbs street fumes and filth, the new falafel is served indoors, at spruce, air-conditioned outlets, where portions are wrapped in designer bags and sauces flow out of stylized fountains. At Falafels, the balls are not moulded manually, but dispensed by a mechanical implement at the rate of 80 balls/minute. There are two kinds – the Syrian Zafur and the Turkish

Baladi. And as befits an industrial commodity, the new falafel is 'engineered' by food technicians and subjected to tastings by focus groups.

Like any self-respecting post-Fordist commodity, the falafel of the new chains is not only a matter of matter but, as stated above, of concept or, more precisely, of fantasy, rendering the past as nostalgia or retro. Branches are designed in a nostalgic style – in order to evoke yearning within the primary target sector – and they carry, in the name of 'retro,' old-fashioned soda pops. This is the local Israeli habitus dusted off, 'branded' and 'designed' so as to be marketed as a mass standardized commodity. Another trendy aspect of the new falafel is its linkage to the new discourses on the environment or nutrition. The proprietor of Ma'oz notes that 'salads, tehini, and falafel are healthy foods, and we have taken the health issue further by offering also whole-wheat pita bread. The health issue is becoming so central that we are now considering establishing a falafel branch that would serve only organic vegetables.' To sum up, the distinction between the old falafel and the new, post-McDonald's falafel, is identified in a local newspaper report as follows:

> If in the past every Falafel King took pride in the unique taste [of his own product, the secret of] which was sometimes passed down from father to son, and which acquired a reputation that attracted customers from far and wide, in the [new] chains, the taste would always be the same. Uniqueness and authenticity would be lost for the sake of quality and free market rules.

One major change in Israel's culinary habitus as a result of its McDonaldization, therefore, is the demise of the old 'authentic' falafel and the appearance of the new commodified 'falafel 2000.'

But McDonald's had to surmount another – no less challenging – culinary hurdle: the Israeli carnivorous palate. [. . .] Given this hankering for meat, especially of the grilled variety, the McDonald's hamburger appeared rather puny, and the Israeli consumer tended to favour the Burger King broiled product. In 1998, McDonald's bowed to the Israeli appetite, changing both the preparation and size of its hamburger. It shifted to a combined technique of fire and charcoal, and increased portion size by 25 percent. The Israeli customer now has the distinction of being served the largest hamburger (120 grams) marketed by McDonald's worldwide. But the most striking fast-food modification to the Israeli habitus is the 'Combina' (the Hebrew equivalent of 'combo'), launched in 2001 by Burger Ranch – a packaged meal for four eaters that taps into the local custom of 'sharing' and, to quote the marketing blurb, allows for 'a group experience while retaining individual dining expression.'

It may thus be concluded that the interrelations of McDonald's and the falafel are not simply a contrast between local decline and global rise. Rather, they are a complex mix, though certainly under the banner of the global. Indeed, the global (McDonald's) contributed somewhat to the revival of the local (the falafel). In the process, however, the global also transformed the nature and meaning of the local. The local, in turn, caused a slight modification in the taste and size of the global, while leaving its basic institutional patterns and organizational practices intact. The 'new falafel' is a component of both a mass-standardized consumer market, on the one hand, and a post-modern consumer market niche, on the other. This sort of relationship between McDonald's and the falafel, in which the global does not eliminate the local symbolically but rather restructures or appropriates it structurally, is typical of the global–local interrelations epitomized by McDonald's.

Discussion I: 'One-Way' or 'Two-Way'?

Based on this case analysis, how, then, are we to conceive the relations between global commerce and local idioms?

The literature on relations between the global and the local presents a myriad of cases. Heuristically, the lessons from these may be condensed into two competing – contrasting, almost – approaches: the one gives more weight to globalization, which it regards as fostering cultural uniformity (or homogeneity); the other gives more weight to localization, which it regards as preserving cultural plurality, or cultural 'differences' (or heterogeneity). [. . .] the former is known also as cultural imperialism and McDonaldization [. . .] The latter is known also as hybridization. [. . .] For the sake

of simplicity we shall call the former the 'one-way' approach, i.e., seeing the effect as emanating from the global to the local; and the latter, as the 'two-way' approach, i.e., seeing the effect as an interchange between the global and the local.

The most prominent exponent of the one-way approach is George Ritzer, in his book *The McDonaldization of Society*. Ritzer, more than anyone else, is responsible for the term that describes the social process of McDonaldization. [. . .]

Contrary to this one-way approach [. . .] the literature offers another view, which we call here the two-way approach. This view considers globalization only a single vector in two-way traffic, the other vector being localization. The latter suspends, refines, or diffuses the intakes from the former, so that traditional and local cultures do not dissolve; they rather ingest global flows and reshape them in the digestion.

Arjun Appadurai, for one, asserts that it is impossible to think of the processes of cultural globalization in terms of mechanical flow from center to periphery. Their complexity and disjunctures allow for a chaotic contest between the global and the local that is never resolved. [. . .]

One typical significant omission of the two-way perspective is its disregard for imbalances of power. [. . .] Positing 'localization' as a counterbalance to globalization, rather than as an offshoot, some of the cultural studies literature is indeed rich in texture and subtlety when depicting the encounters of global commerce with local popular cultures and everyday life. This literature is at its best when acknowledging that its task is to 'twist the stick in the other direction,' from the top-down political-economic perspective to a bottom-up cultural perspective. It falters, however, when it attempts to replace, wholesale, the top-down approach with a bottom-up one, without weighting the relative power of the top and the bottom.

The latter move is evident in an ethnographic study of McDonaldization conducted in Southeast Asia by a team of anthropologists. They argue overall that even though McDonald's transformed local customs, customers were nonetheless able to transform McDonald's in their areas into local establishments; this led them to conclude that McDonald's does not always call the shots. They claim that, in the realm of popular culture, it is no longer possible to distinguish between the 'local' and the 'external.' Who, they protest, is to say whether or not Mickey Mouse is Japanese, or Ronald McDonald, Chinese; perhaps, this attests to a 'third culture' that belongs neither to one nationality nor the other, but constitutes rather a transnational culture.

This ethnographic discussion stresses the variety of supplemental dishes McDonald's has included on its menu in order to accommodate various local cultures. Applying this approach to our case study, the new falafel, for instance, can be considered a manifestation of [. . .] hybridization of McDonald's. The new falafel assimilated some of McDonald's practices, but accommodated them to local traditions and tastes.

The two-way approach to the global–local encounter is usually portrayed as critical and espoused by radical social scientists, because it 'empowers' the sustainability of local cultures and fosters local identities. [. . .]

Discussion II: 'Both Ways'

[. . .] To the question of homogenization vs heterogenization in global–local relationships, we suggest here the following resolution: (1) both perspectives are valid; (2) yet they apply to discrete societal levels; and (3) the one-way approach is restricted to one level of social reality, the structural-institutional level, i.e., patterns and practices which are inscribed into institutions and organizations; the two-way approach is restricted to the symbolic-expressive level of social reality, i.e., the level of explicit symbolization. Finally, (4) we suggest a global–local structural-symbolic model, in which the one-way structural homogenization process and the two-way symbolic heterogenization process are combined. Thus, heuristically speaking, our theoretical resolution is predicated on the distinction between two different levels, the structural-institutional level and the expressive-symbolic level.

While each of the rival perspectives on the global–local encounter is attuned to only one of these levels, we propose that globalization be seen as a process that is simultaneously one-sided and two-sided but in two distinct societal levels. In other words, on the structural level, globalization is a one-way street; but on the symbolic level, it is a two-way street. In Israel's case, for instance, this would mean that, symbolically,

the falafel and McDonald's coexist side by side; structurally, however, the falafel is produced and consumed as if it were an industrialized-standardized (McDonaldized) hamburger, or as its artisan-made 'gourmet' counterpart. [. . .]

The two-way approach to globalization, which highlights the persistence of cultural 'difference,' contains more than a grain of empirical truth. On the symbolic level, it accounts for the diversity that does not succumb to homogeneity – in our case, the falafel once again steams from the pita; the Israeli hamburger is larger than other national McDonald's specimens (and kosher for Passover [. . .]). On the symbolic level, the 'difference' that renders the local distinctive has managed to linger on. At the same time, on the structural level, that great leveller of 'sameness' at all locales prevails: the falafel has become McDonaldized. [. . .]

A strong structuralist argument sees symbolic 'differences' not merely as tolerated but indeed as functional to structural 'sameness,' in that they are purported to conceal the structure's underlying uniformity and to promote niches of consumer identity. In other words, the variety of local cultural identities 'licensed' under global capitalist commercial expansion disguises the unified formula of capital, thereby fostering legitimacy and even sales.

[. . .] A variety of observers – all with the intention of 'giving voice' to the 'other' and the 'subaltern' – may unwittingly be achieving an opposite effect. [. . .] Exclusive attention to explicit symbolism may divert attention from implicit structures.

Transnational corporations are quick to take advantage of multiculturalism, postcolonialism and ethnography, and exploit genuine cultural concerns to their benefit. It is worth quoting at some length a former Coca-Cola marketing executive:

> We don't change the concept. What we do is maybe change the music, maybe change the execution, certainly change the casting, but in terms of what it sounds like and what it looks like and what it is selling, at a particular point in time, we have kept it more or less patterned. [. . .] [our activity] has been all keyed on a local basis, overlaid with an umbrella of the global strategy. We have been dealing with various ethnic demographic groups with an overall concept. Very recently [. . .] the company has moved to a more

fragmented approach, based on the assumption that the media today is fragmented and that each of these groups that are targeted by that media core should be communicated to in their own way with their own message, with their own sound, with their own visualization. [. . .]

The case study presented here has shown a number of instances of the process whereby global commodities appropriate local traditions. To recap with the example of the 'new falafel,' McDonaldization did not bring about its demise, but, indeed, contributed to its revival, vindicating, as it were, the two-way perspective. The falafel's new lease on life, however, is modelled after McDonald's, that is, a standardized, mechanical, mass-commodified product, on the one hand; or responds to it in a commercial 'gourmetized' and 'ethnicitized' product, on the other hand. In both cases, global McDonaldization prevails structurally, while it may give a symbolic leeway to the local. [. . .] Indeed, from the end-user's or individual consumer's perspective, the particular explicit symbolic 'difference' may be a source of great emotional gratification; but from the perspective of the social structure, the system of production and consumption, what matters is the exact opposite – namely, the implicit structural homogenization.

Thus, the question of global homogenization vs. local heterogenization cannot be exhausted by invoking symbolic differences, as is attempted by the two-way approach. 'McDonaldization' is not merely or mainly about the manufactured objects – the hamburgers – but first and foremost about the deep-seated social relationships involved in their production and consumption – i.e., it is about commodification and instrumentalization. In its broadest sense here, McDonaldization represents a robust commodification and instrumentalization of social relations, production and consumption, and therefore an appropriation of local cultures by global flows. This study [. . .] proposes looking at the relations between the global and the local as a composite of the structural and symbolic levels, a composite in which the structural inherently appropriates the symbolic but without explicitly suppressing it. [. . .]

This is what is meant by glocommodification – global commodification combining structural uniformity with symbolic diversity.

World Culture

The idea of world culture revolves around the work of John Meyer and a group of sociologists, some of whom were Meyer's students. They include a wide variety of phenomena under the heading of world culture, ranging all the way from a growing global consensus against genocide, to similar educational systems, to local chess clubs where the game is played in accord with global rules. In contrast to Marxists and neo-Marxists (like Wallerstein; see chapter 8), world culture theorists focus, in Marxian terms, on the superstructure (culture) rather than on the base (the material, the economic). This chapter opens with several excerpts from *World Culture: Origins and Consequences* by Frank Lechner and John Boli.

In the first excerpt they outline several dimensions of world culture. First, world culture is global, at least in its potential reach, although of course some parts of the globe may not (yet) be affected. Second, world culture is distinct, although it does not overwhelm or replace local culture. Third, world culture is complex: it is not unidimensional. Fourth, world culture is seen as an entity "with its own content and structure,"[1] but it is *not* a reified entity with tight boundaries clearly separating it from other cultural phenomena. Fifth, it is cultural in the sense that it involves "socially constructed and socially shared symbolism."[2] Sixth, it is dynamic and tends to grow over time; it is "open to new ideas, vulnerable to

new conflicts, and subject to continual reinterpretation."[3] Finally, world culture is significant; it "matters for the world as a whole and for the world in all its varied parts."[4]

Under the heading "World Culture as Ontology[5] of World Society," Lechner and Boli argue that "organizations in a particular field experience the same institutional pressure, they are likely to become more similar over time."[6] Organizations are especially likely to feel pressed to become increasingly rational. The latter means, in the case of education, that school systems around the world, to take one example, are likely to implement certain "procedures and curricula, certain styles of teaching and studying," use professional teachers and textbooks, and so on.[7] This "institutionalist view" (education is an institution that experiences institutional pressure to be like other educational systems around the world) has several components:

- World culture is the culture of many nation-states; of a decentralized world polity.
- "It contains rules and assumptions, often unstated and taken for granted, that are built into global institutions and practices."[8]
- It can be seen as a "script" that is the joint product of many different people (e.g. professionals and organizational leaders) from many different parts of the world.

- World culture is "universalistic": "the same assumptions, the same models are relevant, indeed valid, across the globe."[9] This does not mean that they are the same throughout the world, but local practices depend on global norms to at least some extent.

Much of world culture today has its origin in the West. It includes ideas about "individual value and autonomy, the importance of rationality in the pursuit of secular process, and the status of states as sovereign actors."[10] However, world culture has now become global "because its main structural elements are similar across the globe and because they are deemed to be universally applicable."[11] The result is global isomorphism, "the increasing institutional similarity of differently situated societies"[12] in such domains as organized science and women's rights.

In spite of increasing similarities throughout the world, world culture theory recognizes that differences exist throughout the world due to incomplete institutionalization, resistance to world culture in some quarters, its acceptance and practice primarily by powerful societies, and the disparities, even contradictions, in its basic principles (e.g. between equality and liberty).

Under "Differentiating World Culture," Lechner and Boli focus on the issues of the degree to which world culture is feared in many parts of the world (especially France) and the threat it poses to global differences. In this context, they deal with many other issues dealt with in this book including McWorld (chapter 12), McDonaldization (chapter 15), and MNCs and TNCs (chapter 7). While not rejecting these views, Lechner and Boli make the point that these processes lead not only to global similarity but also to diversity and to "cultural cross-fertilization";[13] that locals react creatively to these global processes; and that world culture is not of one piece. Indeed, they go further to argue that diversity is fundamental to, built into, world culture. Also in this context, Lechner and Boli argue against the idea that the nation-state is being eroded or destroyed by globalization. Rather, they see world culture and the nation-state as intertwined; indeed, characteristics of the nation-state have come to be part of world culture.

Finnemore critiques the world culture perspective on several grounds. First, it focuses the effects of world culture, but tells us little about either the causes or the mechanisms of its spread. Second, it tends to emphasize the spread of an internally harmonious Western culture, especially its rational systems. However, what is ignored are the conflicts and tensions within that culture, especially those between progress and justice and between markets and bureaucracies. Third, the world culture perspective is silent on agency. Finally, it overlooks the role of power and coercion, that is politics, in the spread of world culture.

NOTES

1 Frank Lechner and John Boli, *World Culture: Origins and Consequences.* Malden, MA: Blackwell, 2005, 27.
2 Ibid., 27.
3 Ibid., 28.
4 Ibid., 28.
5 It is an ontology because it is a "deep structure underlying global practices." Included in this structure are rules, principles, institutions, etc.
6 Lechner and Boli, *World Culture*, 43.
7 Ibid., 44.
8 Ibid., 44.
9 Ibid., 44.
10 Ibid., 46.
11 Ibid., 46.
12 Ibid., 46.
13 Ibid., 141.

 READING 63

World Culture: Origins and Consequences

Frank J. Lechner and John Boli

[. . .]

The Case for World Culture

This [reading] proposes a view of world culture as a global, distinct, complex, and dynamic phenomenon and supports this view by analyzing its different dimensions with concrete examples. As prelude to our substantive chapters, we now summarize our perspective on world culture.

World culture as global

In speaking of "world" culture, we have in effect treated it as global, as the globe-spanning culture of actual world society. Though the distinction between "world" phenomena, as properties of large geographical areas, and "global" ones, of true planetary scope, once may have mattered, world and global in these senses have practically converged. [. . .] [W]hat matters for our purposes is that certain ideas and principles are presented as globally relevant and valid, and are seen as such by those who absorb them. At any rate, the claim does not have to be wholly correct as an empirical matter (for example, not all parts of the globe need to be equally enamored of chess [. . .]) to be useful as a working hypothesis (for example, because the chess subculture works on common assumptions [. . .]).

World culture as distinct

Arguing that the world has a culture might seem to slight the diversity that still prevails today. However, our point is not that world culture obliterates all others, supersedes the local, or makes the world one in the sense of being utterly similar. To be sure, from our

analytical point of view, it does have a coherence and content of its own, but this does not imply empirically that the world is on a long slide toward Turning Point's monoculture. Nor does it rule out the possibility of a "clash of civilizations." [. . .] We suggest that world culture grows alongside of, and in complex interaction with, the more particularistic cultures of the world. In relating to world culture the more particularistic ones also change. For example [. . .] the civilizations central to Huntington's argument are always already embedded in an encompassing global civilization, which to some extent constrains their interactions and bridges their differences. Within world culture, civilizations cannot be self-centered, taken-for-granted practices, if they ever were. Actual cultural practices in particular places, as well as the thinking of particular individuals, are likely to exhibit mixtures of "world" and more local symbolism. In treating world culture as distinct, we do not claim to capture the full range of those practices. As our argument about how to distinguish world culture implies, world culture is not the sum of all things cultural.

World culture as complex

From another angle, our analysis of world culture might seem too complex, too focused on teasing out tension and difference. The monocultural scenario, after all, has numerous supporters. According to the popular "McDonaldization" argument, for instance, institutional forces pressing for efficiency and control threaten to impose one way of life everywhere. We think the direction sketched by this argument is partly correct: rationalization is powerful, and in fact a certain kind of rationality has become an influential cultural model. But even on the culinary scene, rationalization is not a cul-de-sac. The fast-food experience takes many forms, single models of food production come in multiple versions, foods and tastes mix around the world. From our perspective, the McDonaldization thesis is not so much wrong as one-sided. World culture encompasses different domains and contains tensions among its different components. Global consciousness does not come in one styrofoam package.

World culture as an entity

We have already ascribed several characteristics to world culture. Whenever we say that world culture "does" X, the specter of reification lurks. In some instances, of course, talking of world culture as an active whole is a matter of convenience, sparing us the need to unpack it into components or into the actions of people using the symbolic resources at their disposal. Treating it in this way does not entail seeing it divorced from other realms of human activity. As we have already hinted in our discussion of "real world" institutions, we think the analytical move to distinguish the cultural from, say, the political and economic, should actually enable us to see how those aspects of human activity are mutually constitutive. However, we do not want to grant critics of reification too much. In the final analysis, we do claim that a distinct and recognizable world culture is crystallizing as a phenomenon with its own content and structure. At the same time, we do not draw tight boundaries. In exploring what issues reasonably fit under the heading of world culture, we err on the side of inclusion.

World culture as culture

As we explained, we hold a particular view of culture. We regard it as socially constructed and socially shared symbolism. Our position is "holistic" and "constructionist." This rules out subjective or purely textual views of culture – it is neither (just) in people's heads nor (just) in esoteric documents. It also leaves aside popular grab-bag notions of culture as a way of life. However, it incorporates many other perspectives, from which we borrow liberally. Our holistic constructionism directs attention to the way in which culture is created and consciousness is formed. It suggests that, once created, cultural forms do have a dynamic of their own. It requires analysis of how cultural elements come to be shared, notably through the work of institutions that

carry abstract ideas into practice. It points to the fault lines and tectonic stresses that may become sources of change. We argue [elsewhere] that this perspective builds on and complements much previous work on world culture. We apply this perspective heuristically. Our purpose [. . .] is to marshal available resources to illuminate our problems, not to engage in scholarly polemics by advocating one theory to the exclusion of others. We hope that our view of culture is sufficiently ecumenical to be useful to a wide range of readers.

World culture as dynamic

Our opening example of global sports showed how rules, ideas, and symbolism surrounding this transnational practice have grown over the years. The world culture of sports is always being constructed and reconstructed. The point applies more generally. World culture is not simply a finished structure, a done deal. Certainly, some world-cultural patterns display continuity over many decades, as the global commitment to the nation-state form illustrates. But world culture is open to new ideas, vulnerable to new conflicts, and subject to continual reinterpretation. Even the apparent convergence of people and countries from many regions on the merits of liberal democracy as a model for organizing societies hardly counts as the "end of history." Much as we appreciate the value of the model itself, we lack the Hegelian confidence to think of contemporary world culture as the fully formed end point of humanity's ideological evolution, or as the irreversible progress of reason that has achieved a system immune to future contradictions.

World culture as significant

Needless to say, we think world culture is significant in many ways. We argue against the view that it is a veneer, a set of fairly abstract notions only variably relevant in real people's lives. Examples such as the globalization backlash, one could argue, still refer to the concerns of a relatively small elite. Models such as neoliberalism or even the nation-state would seem irrelevant in West African states on the verge of collapse. We agree that the relevance of world culture can vary in this way, but this does not diminish its significance as a feature of world society. Without grasping world culture we could not understand the direction of world affairs, as we have already suggested. However, it is also vastly more pervasive in particular places than ever before. Anti-globalization discourse affects African dealings with international organizations, neoliberalism shapes development strategies even of countries with few resources, and the nation-state has become the operative model for groups not naturally hospitable to living within one political system. Even more concretely, as our earlier examples show, many regular activities now embody world culture in some way. World culture matters for the world as a whole and for the world in all its varied parts. [. . .]

   [. . .]

World Culture as Ontology of World Society

In the 1970s, John Meyer and his colleagues faced a puzzle about the spread of formal education around the globe. Why, they wondered, did states with very different needs and resources adopt very similar educational institutions and methods, even when these did not obviously suit their particular situations? Meyer's previous work on educational organizations suggested a way to address the issue. He had argued that in modern societies organizations are not so much tools deliberately designed to solve problems as institutions driven by outside pressure to implement practices defined as "rational." Organizations are "dramatic enactments" of rules that pervade a particular sector of society. By adopting these rules, ceremoniously as it were, by operating according to the official "myths" of rationality, organizations increase their legitimacy. Because all organizations in a particular field experience the same institutional pressure, they are likely to become more similar over time.

The insight Meyer and his colleagues brought to bear first on education, and ultimately on world culture as a whole, is that this "institutionalist" account also works at the global level. What, then, are the rules and assumptions built into the globalization of formal education? First of all, education has become the obligatory work of states. States themselves are constrained by global rules to act in rational fashion for the sake of progress: according to prevalent global models, states have ultimate authority in many areas of life, and they must exercise that authority by building "rational" institutions that promote "growth." Formal public education is one such institution. Any modern state must have it, even if, as in the case of Malawi and similar countries, the country has few resources to sustain it and its people have basic needs not served by this foreign import. Second, education seems so compelling in part because it is inextricably linked to great collective goals. According to the global script, learning increases human capital, educational investment raises growth, the spread of knowledge is the road to progress. Third, education has to take a certain form. A "rational" system is not one specifically designed to produce growth and literate citizens in a way that suits a particular country, but rather one that implements certain kinds of procedures and curricula, certain styles of teaching and studying. Thus, Malawi strives to implement a modern curriculum with professional teachers who exercise authority in their classroom, however difficult this may be when books and pencils are lacking. In globalizing education, form trumps function. Fourth, education reflects particular ideas about the people involved in it, especially the students. They are to be treated as individuals capable of learning, entitled to opportunity, eager to expand their horizons. Education must foster individual growth, but it must also connect students to their country: both implicitly and explicitly, it is always a kind of citizenship training. Around the world, formal education is one large civics lesson. Here again, Malawi is a case in point, even if individuality is unlikely to be fostered through mass teaching in drafty classrooms.

The example shows several characteristics of world culture as institutionalists view it. It is the culture of a decentralized "world polity," in which many states are legitimate players but none controls the rules of the game (this account is therefore often called "world polity theory"). It contains rules and assumptions, often unstated and taken for granted, that are built into global institutions and practices. When we illustrated [elsewhere] how many features of world society are "deeply cultural," for instance in the case of world chess, we already were applying an institutionalist insight. Moreover, no single person, organization, or state chooses the rules it follows; these are, to a large extent, exogenous – features of the world polity as a whole. In part for this reason, institutionalists sometimes describe world culture as composed of "scripts." Of course, a script does not simply create itself. It is the joint product of teachers and administrators, ministry officials and consultants, UNESCO representatives and NGO advocates. Like many aspects of world culture, it is the focus of much specialized professional activity, notably in international organizations. Finally, world culture is universalistic: the same assumptions, the same models are relevant, indeed valid, across the globe. To return to our example, this is not to say that actual educational practice exactly lives up to a single global model, but, institutionalists claim, the power of world culture is evident in the extent to which local practice depends on global norms.

Because these scholars view world culture as a deep structure underlying global practices, they have described it as a kind of "ontology." In using this term, they do not imply that global actors routinely speculate philosophically about the nature of being, but they do think there are now powerful, globally shared ideas about what is "real" in world society. Ontology, in their sense, comprises a set of rules and principles that define, among other things, the very actors that can legitimately participate in world affairs. "Culture has both an ontological aspect, assigning reality to actors and action, to means and ends; and it has a significatory aspect, endowing actor and action, means and ends, with meaning and legitimacy." It "includes the institutional models of society itself." It specifies what the constituent parts of world society are and what kinds of things are to be considered valuable in the first place. This culture constitutes the array of authoritative organizations carrying out its mandates. Because the world cultural order shapes not only the nation-state system but also other organizations and even human identities, Meyer and his colleagues ultimately present the world as the enactment of culture. Of course, this implies that world culture is not simply made by actors, the product of contending groups in a given system; it does not necessarily sustain a particular type of political economy or justify the position of actors within it, as the materialist account would have it. World culture cuts deeper.

What is the content of this ontology? As the education example shows, one prime tenet of world culture is that the world consists of states – corporate actors in control of territory and population, endowed with sovereignty, charged with numerous tasks, and expected to operate rationally in pursuit of globally defined progress. Though states encounter many difficulties, the idea has a powerful grip on global practice. But states are not the only actors, for the second main tenet of world culture, again evident in the education example, is that the world also consists of individuals – human actors endowed with rights and needs, possessing a distinct subjective consciousness, moving through a common life course, and acting as choosers and decision makers. Of course, Meyer does not mean that world culture somehow creates flesh-and-blood persons. However, how we understand and express ourselves as persons, the way we assert our rights and needs, does depend on globally relevant ideas. States and individuals are inextricably linked through a third tenet, the global principle of citizenship, which requires the cultivation of individual capacities as a basis for societal growth, respect for the equal rights and status of all members of society, and the creation of commonality among individuals as a way to integrate society. In short, the way we belong to a society is not simply an accident of birth or a result of personal choice; to some extent, belonging fits a global mold. Yet individuals are not merely citizens of states: since in principle all have the same rights and duties, may pursue their own interests freely, and can contribute to solving collective problems, they are construed as citizens of the world polity as a whole.

The origins of this world culture clearly lie in the core Western cultural account, itself derived from medieval Christendom. Notions of individual value and autonomy, the importance of rationality in the pursuit of secular progress, and the status of states as sovereign actors, have deep roots in European history. Even in the nineteenth century, such basic ideas were still applied first and foremost by and for Westerners. However, this culture is now effectively global, both because its main structural elements are similar across the globe and because they are deemed to be universally applicable. It has become global due to a decades-long process of institutionalization. Intergovernmental organizations enshrined many of the tenets we have described, for example in international conventions and declarations [. . .]. After the Second World War, state building proceeded largely according to global scripts, resulting in a world of sovereign, rational, nominally equal states. Institutions focused on cultivating individuals have expanded rapidly. These include, of course, the educational institutions we have referred to in this chapter, but many others as well [. . .]. International nongovernmental organizations – voluntary associations of interested individuals – have assumed increased influence in articulating global principles. Many people, groups, and institutions, in short, have done the work of world culture. A key consequence of that work is global isomorphism, the increasing institutional similarity of differently situated societies. Where materialist accounts of the capitalist world-system would expect variation by economic status and historical trajectory, institutionalists find homogeneity, for example, in the way organized science spreads to all corners of the globe or in the way women's rights gain recognition within many states.

Since institutionalists treat world culture as constitutive of reality, as a symbolic structure that shapes the ways people act and feel, they do not need to assume any widespread, explicit agreement on the fundamentals of world culture. They would suggest that even ostensible critics of existing world culture, such as environmentalists or feminists, ultimately conform to important tenets. However, this is not to say that world culture is a seamless web. For one thing, institutionalization is always incomplete, due to numerous local constraints, as we already noted in the case of formal education in Malawi, which resembles the supposed global script in only some respects. World culture also provokes genuine conflict. Thus, the assertion of equal rights for women has been challenged by Islamic groups as incompatible with their tradition. The notion that world culture is now global, universally shared and applicable, is itself subject to challenge in practice, insofar as it is disproportionately the product of powerful states. A case in point is the expansion of education, at least in part a consequence of America's exercise of hegemony. World culture could not be seamless in any case, since many of its principles are contradictory, as is evident in the well-known tensions between equality and liberty, efficiency and individuality, and expectations for states to "be themselves" and "act alike."

World culture thus creates a culturally dynamic world: "Ironically, world-cultural structuration produces more mobilization and competition among the various types of similarly constructed actors than would occur in a genuinely segmental world. Increasing consensus on the meaning and value of individuals, organizations, and nation-states yields more numerous and intense struggles to achieve independence, autonomy, progress, justice, and equality."

The institutionalists agree with the Wallersteinians on several empirical features of the modern world-system, but they account for the origins and reproduction of that system in different ways. As our brief summary has shown, the institutionalists give culture much greater weight. It becomes, so to speak, base rather than superstructure. As an analytical standpoint, this carries its own risks. For example, it is tempting to find evidence of deep culture at work in the activities of various institutions and then use the understanding of culture thus acquired to explain the evidence that served to generate the independent cultural variable in the first place. While avoiding such circular reasoning, we will partly rely on the institutionalist argument for guidance.

[. . .]

Differentiating World Culture: National Identity and the Pursuit of Diversity

[. . .]

World culture, in this view, is simply the globalization of the West. Like deterritorialization, McDonaldization, Coca-Colanization, and Americanization, the cultural imperialism argument portends a single world with a single culture.

While this scenario contains important kernels of truth, it is far too stark. In many ways, globalization itself is a "motor of diversity." For example, McDonaldization, much derided by French activists, captures only part of global food trends. As Asia takes to hamburgers and Cokes, Europe and North America adopt Eastern cuisines; though a less standardized product, sushi is as global as the golden arches. Globalization thus fosters many kinds of cultural cross-fertilization. Within particular countries, it usually expands the "menu of choice" for individuals by liberating them from the constraints of place, as French consumers of jeans and jazz and Japanese electronics – as well as hamburgers and "French fries" – can attest. This applies to language as well. Through translation, particular languages and literatures increasingly build "bridgeheads" to other places, as illustrated by the success of Latin-American novelists in France. Immigration and cultural contact introduce impure innovations, a form of linguistic diversity that official French opposition to franglais ironically attempts to stifle. Among global audiences, globalization thus fosters cultural experimentation. Even in the industry that most provoked French ire, the feared homogeneity brought about by American dominance is by no means absolute: Hollywood must still compete with other centers of film production, such as India's Bollywood; its global success depends in part on its ability to attract non-American talents and adapt to non-American tastes; and it encourages the development of home-grown niche productions. As these examples suggest, the "creative destruction" of global competition also has diversifying consequences. Even if we implausibly assume that all place-based culture is doomed, there is no reason why supposedly deterritorialized communities should be culturally uniform. Underground dance club aficionados are distinct from professional soccer players or fruit fly researchers; the proliferation of their multifarious ties stimulates new kinds of transnational diversity. While multiplying empirical examples of diversity "on the ground" would take us too far afield in examining the intricacies of the much-feared global-ization process, our first response to the doomsayers is simply that the world is still a very diverse and surprising place, unlikely to be smothered in one cultural goulash.

Our second rejoinder is more closely related to the agenda of this book. As we have described world culture, some of its fundamental substance is quite abstract. McDonald's-style rationality, American neo-liberalism, and Western universalism provide only very general models for social action. At least two of the theories we [have] discussed argue that the implica-tions of such models impel creative adaptation by particular groups in particular places. Robertson calls this glocalization, Hannerz creolization. World-cultural precepts become socially real by being incorporated in locally situated practices. Thus, Hong Kong becomes

McDonaldized rather differently than Peoria; neoliberal privatization proceeds differently in India than in the UK; democracy takes hold differently in Mexico than in the Czech Republic. Certain kinds of pop music may be transnationally popular, but their vibrancy still depends on the way musicians make such music part of their own traditions. The upshot of the Robertson/ Hannerz line of thought is that groups and societies mix and match, borrow and adapt, learn and revise. By its very generality, world culture gives an impetus toward highly varied interpretation.

What is more, world culture is not of a piece. Even the elements in the homogenization scenario are not identical: though McDonald's does serve Coke, McDonaldization and Coca-Colanization metaphoric-ally capture different forms of homogenization. While the "America" that has left its imprint on world culture is a reality, "the West" is a vague community of values. McDonald's and Coke, America and "the West" – the forces evoked by those terms are themselves quite dif-ferent. However close the affinities of these components, a culture that is rationalized à la McDonald's, dependent on lowbrow consumer taste, influenced by popular culture, and infused with falsely universal aspirations is not an internally consistent whole. The components define different aspects of global reality; they vary in strength and scope. More generally, as we [have seen], Appadurai has described the "disjunctures" between the different dimensions of world culture. World culture contains different sets of universally applicable and influential ideas that operate at different rhythms, creating multiple tensions and unpredictable intersec-tions. For example, two of the core ideas identified with American-style cultural imperialism – namely a fully liberalized market and democratic governance – may have contradictory implications when applied globally: one encourages the unregulated pursuit of self-interest, the other stresses deliberate collective control of social affairs. World culture therefore does not, and could not, prescribe any single course of action to be followed by everyone everywhere. Indeed, the very process of globalization itself is molded by contending views of how it ought to be structured. [. . .] Disjuncture and contention preserve diversity.

We [have seen] that such varied interpretations and disjunctures are a common theme in much recent scholarship. Recall Breidenbach and Zukrigl's book on the "dance of cultures," which shows with a wealth of ethnographic examples how people around the world incorporate global products and practices into their own world-views, adapt new categories such as "feminism" according to their own needs, "talk back" to the supposed sources of cultural flows, and engage in all manner of resistance. Or recall the Berger and Huntington volume on "many globalizations," which shows how a supposedly universal process takes different paths according to local cultural contexts. Both in his book on cultural globalization and even more systematically in an earlier work on cultural imperialism, Tomlinson has subjected the cultural imperialism argument to criticism, arguing against the idea that the cultural "synchronization" produced by the spread of modern institutions is a destructive imposition. As we pointed out, not all these scholars fully share our view of world culture. But we detect in their work a convergence on a basic point, namely that any emerging world culture is bound to be refracted in complex ways by the prisms of specific groups and societies and that diversity is bound to flourish through the multiple ways in which they relate to such an overarching world culture.

We can go a step further. Difference is flourishing not just in the way a nascent world culture "plays out" in practice, but also as an organizing principle of world culture. To return to the example we started with, while the tone of the francophone conference may have been defensive, it also emphatically called for recognition of diversity as a value in its own right. It advocated the mutual recognition of cultures and their right to participate on an equal basis in the "concert of nations." Though its cause was French, its appeal was framed in universal terms, referring specifically to the support of international organizations such as UNESCO. Indigenous peoples and movements working on their behalf similarly claim the right to maintain their particularistic heritages. The importance of "cultural survival" as such, to cite the name of one advocacy group, has become conventional wisdom. The very concept of indigeneity points to a burgeoning global respect for the heritages of minority groups. Both national and indigenous defenders of difference have vested some of their hopes in UNESCO, and that organization has become a linchpin in the globalization of diversity as a value. Thus far, it has done its share in fostering difference by issuing reports charting cultural

diversity within world culture, celebrating diversity as a goal for the world community, and instituting programs to protect the world's cultural heritage. Among state leaders, movement activists, and IGO officials, the cause of diversity thus has been gaining strength. In Western academic circles, such trends have been bolstered by the discourse of "multiculturalism," which itself has swept across the globe, assigning equal value to different cultures and promoting coexistence rather than dominance. However justified the fear of indigenous groups may be as a practical matter, the globalizing diversity industry indicates that world culture is more complex than the imperialist scenario allows. Diversity has been enshrined as a counterpoint to homogenization. Particularism is universalized, as Robertson has suggested. This is not to say, of course, that "mere" rhetoric will help the French build a bulwark against Hollywood blockbusters. It is to suggest, though, that world culture itself nurtures the seeds of difference.

In other words, difference is built in. This more differentiated view of world culture follows straightforwardly from the work of Robertson, who makes contrasting definitions of the global situation the hallmark of world culture, and from the work of anthropologists like Hannerz, who treat world culture as the organization of diversity. However, the same idea also appears in world-system theory, which assumes that the geographical division of labor within a single world market depends on competition among culturally distinct units within the system. By comparison with these perspectives, Meyer and his colleagues put more emphasis on the way in which similar institutions are enacted across the globe, though they also portray the world polity as internally differentiated. The scholarly pendulum is thus swinging away from the kind of anxiety that dominates much public discourse. But rather than dismissing the fears of cultural loss, our picture of world culture helps to put them in perspective: as world culture grows, some differences may fall by the wayside, others require redefinition, still others are constantly created. To summarize, while fears of a world-cultural goulash are understandable, there are at least three reasons to be skeptical of the scenario such fears assume: the globalization process, regarded as pushing homogenization, actually has varied effects; the process takes place in the context of an existing world culture to which individual groups and societies relate in varied ways; and world culture itself fosters difference through the principles it contains and the institutions it legitimates. This argument entails that fears of American hegemony, so common in French responses to world culture, are overstated. To make our case more concrete, we now turn to a particular form of difference often regarded as endangered, the kind at issue in the francophone example with which we started, namely national difference. Our argument implies that nations can flourish as distinct entities under the canopy of world culture. We support this argument by showing how, in one instance, the reproduction of national difference occurs. However, we do not aim for a Panglossian conclusion that for nations this is the best of all possible worlds. While world culture fosters national difference, it also embeds nations in a transnational framework that constrains and homogenizes them.

The Difference Nations Make

Is there still a place for nations in the world? Those who fear global uniformity believe the answer may be negative, for the reasons we discussed above. Deterritorialization implies that the control of the nation-state over its own affairs diminishes. The upshot of McDonaldization, Coca-Colanization, and Americanization is the accelerating demise of the national, as national distinctions are undermined by transnational rules, tastes, and institutions. Not surprisingly, then, influential authors foresee the end of the nation: "Too remote to manage the problems of our daily life, the nation nevertheless remains too constrained to confront the global problems that affect us." Another student of nationalism concludes that today's world "can no longer be contained within the limits of 'nations' and 'nation-states' as these used to be defined [. . .] It will see [these] primarily as retreating before, resisting, adapting to, being absorbed or dislocated by, the new supranational restructuring of the globe." As the "isomorphism of people, territory, and legitimate sovereignty that constitutes the normative character of the nation-state" has eroded, the nation-state itself "has become obsolete and other formations for allegiance and identity have taken its place." Under conditions of globalization, "[t]he centrality of national cultures,

national identities and their institutions is challenged." While Appadurai and Held et al. do not infer from the nation's dire straits a picture of a homogeneous world culture, the more common diagnosis remains that a one-size-fits-all culture leaves little room for national difference.

We argue against this common diagnosis. As we have suggested, the "one-size-fits-all" view of world culture is itself misleading. Focusing on the nation allows us to elaborate our main points about world culture, namely that it produces difference in practice and contains difference in principle. Addressing the demise-of-nations scenario further enables us to refine our position by showing how the fear of uniformity rests on questionable assumptions about static national cultures confronting an oppressive, alien force. For illustrations we return to the French example we have cited before. This case is especially pertinent because France has played a major role in the history of nationalism as the "archetype" of a nation-state and because, as we have seen, many influential figures have championed France as a nation in the global debate about difference. This championing is rich with ironies. In discussing these ironies, we aim not to convey all the ways in which nations reproduce their identity but only to focus on the extent to which the reproduction of difference revolves around the operations of world culture.

The first irony in presenting the nation as a bulwark of cultural difference is that historically the drive toward nationhood itself has often obliterated differences. In most places, nations were forged out of previously distinct regions and peoples. The unity they possessed often sprang from visions of coherence pursued by elites in control of states who deliberately created "imagined" communities. According to one interpretation, these visions themselves first gained plausibility in industrial societies that placed a premium on a shared high culture, fostered by formal education, that facilitated communication among large populations. Historically, then, nation is to difference what single-crop agriculture is to biodiversity. The French state, for example, has itself been relentlessly homogenizing, not least by requiring the use of standard French throughout its territory. To Bretons, the idea of French as a carrier of diversity can seem far-fetched. This implies that one assumption underlying the common scenario is implausible. Since nations are relatively recent creations, it is misleading to think of them as fully formed cultural wholes suddenly confronted by a stream of global cultural material that invades them. Though it is now conventional wisdom to think of nations as "constructed" and "imagined" rather than "primordial," defenders of difference have an ironically primordial view of national identity, insofar as they treat it as something deeply rooted and unchanging. However, even in seemingly old nations, national identity is always in flux. In the case of France, that identity was established by turning "peasants into Frenchmen," as the title of one prominent study put it, in the late nineteenth century, when "long didactic campaigns" taught inhabitants of France to speak French and to think of themselves as French. The Third Republic of that period built a new nation by means of "coercive elimination" of regional diversity and languages, notably by instituting a nationalized system of free public schools. Applied to France itself, the "defense of difference" advocated by the public figures discussed above risks locking into place a fairly recent version of national identity at the expense of further experimentation and the "intertemporal diversity" that might result.

Another irony in the defense of national difference against a global cultural juggernaut is that critics of homogeneity appear to have so little faith in actually existing difference. It is as if uniformly hapless countries await a common fate. However, even a cursory glance across the globe shows that nations vary greatly in their understanding of what it means to be a nation. "Is there in fact any one thing called a nation?" asks one scholar, explaining that "[e]ach nation-state now on earth could supply a slightly different meaning for the word 'nation,' a different official account (perhaps more than one), not only of its own origins and development, but of the idea of national identity that it supposedly embodies." In many instances, these "different accounts" were deliberately created by elites attempting to draw distinctions between their own nation and foreign counterparts, thereby charting distinct paths toward nationhood. Pecora's point applies to France as well, where at least two conceptions of the nation – one rooted in visions of the traditional Catholic monarchy, the other in the revolutionary vision of a secular republic – have been at odds for two centuries, perhaps to be replaced by a third vision more attuned

to new social realities. To infer from the enormous variety of national situations that a single world-view or way of life will prove uniformly devastating is simply implausible. The case for national difference against global homogeneity depends on a far too homogeneous view of national identity.

Scenarios that oppose nation to world culture portray them as somehow separate. However, far from being unrelated adversaries, world culture and national cultures evolved together. [. . .] nineteenth-century world culture was in part made by and for nations. From the outset, nationalism was itself a *trans*national movement, important first in Latin America and Europe, later in Asia and Africa. The creation of nations was always accompanied by claims to universal respect of politically organized but culturally distinct communities. Of course, world culture could only become "transnational" when the form and legitimacy of nations were largely taken for granted. In this entwinement of world and national culture, France in fact played a pivotal role. France took shape as a nation-state when its revolutionary elite articulated a new creed with universal aspirations. Liberty, equality, and brotherhood have been ideological elements of world culture ever since. By organizing itself dramatically as a nation-state at the time of its revolution, France created a model for others to follow. Ironically, the world culture French intellectuals bemoan is therefore, at least in part, of their predecessors' making. By presenting this particular nation as the embodiment of universal values, France also created an influential, nonethnic or "civic" version of nationhood, which competed with others such as the "ethnic" German version. This variety in the ways nations formed and asserted themselves has itself become entrenched in world culture. National difference has long been built into world culture.

World-system theorists would modify this point about entwinement of the global and the national. As we have seen, they regard the existence of politically and culturally distinct units as critical to the system. The worst-case scenario for world capitalism, their argument implies, is the transformation of a differentiated market system into a single world empire. A more successful Napoleon might have wrecked that system. Differences are therefore functional, but they hardly produce the kind of tolerant diversity current critics of world culture envision. Historically, differences fueled competition and conflict. The rise of nation-states amounted to the reorganization of previously existing regional differences into more politically organized and internally homogeneous units fit for global competition. France's rise as a nation-state was therefore less a matter of spreading a revolutionary faith than of positioning it for that competition. Yet that faith had consequences as well. For all its universalism, it also set up a hierarchy among nations. Some countries could fully live up to France's standards, others only partly so, while still other groups could not even aspire to nationhood. Nationalism, Wallerstein has noted, "first emerged as the response to the universalizing imperialism of the revolutionary power, France." This form of "popular antisystemic mobilization" subsequently "received sustenance particularly from the successive waves of struggle taking place in the semiperipheral areas of the world-economy." As the embodiment of enlightened principles, France helped to create a world culture legitimating Euro-American dominance throughout the colonial age. The irony here is that for most of two centuries world culture actually resembled the hegemonic kind of culture French commentators now oppose, a hegemony to which Frenchmen actively contributed. In world culture, then, not all differences are created equal. Given its involvement in establishing a hierarchical version of world culture, France's more ecumenical defense of difference today rings a bit hollow to world-system theorists.

World polity theorists amplify the point about the historical entwinement of the global and the national in a slightly different way. They are most impressed with the way in which the trappings of the nation have become truly global norms, applying equally to all properly constituted societies. In the nineteenth century, even European nation-states' capacity to control their territory and their people was actually quite limited. For a long time, nationalism was more vision than reality, but the nationalist definition of the global situation was real in its consequences. Once the model was defined, its content expanded greatly, as we have seen in earlier chapters. In some ways, of course, this reduced global pluralism. By the year 2000, more countries looked more alike. Yet the very success of the nation-state model now also provides global standards for what nations must do to reproduce themselves, and globally legitimated tools to satisfy those standards. Nation-states cannot be passive. They have work to

do in upholding their identity. We have already seen examples of that work in the French case. The media policy that protects France's cultural "exception" depends on global norms authorizing state responsibility in this area. The same goes for its educational policies. France's effort to teach children across the country in the same way, striving for closely coordinated teaching in a single system designed to turn individuals into good citizens, is a particularly energetic way of discharging a global responsibility. The broader point here is that locally distinctive policy processes such as these are ways to reproduce national identities in keeping with world-cultural standards. Upholding national identity through national institutions is the world-cultural thing to do.

Robertson's globalization theory also complements our analysis. As we have seen, this theory portrays world culture as stimulating rather than suppressing difference. With regard to national culture, this works in at least two ways. National and world culture stand in a kind of dialectical relationship. To Robertson, the generalization of a partly French model of nation-states to globally legitimate status is an instance of the "universalization of the particular." But such universalization always provokes the opposite trend of "particularization of the universal," in the French case an increasingly anxious attempt to define more actively and precisely what makes France stand out as a nation among others. Nations are therefore always caught in the interplay of standardizing uniformity and diversifying particularity. Worrying about how-to-be-national is inherent in the rules of the world-cultural game. To some extent, nations have always been part of a single "game," identifying their position relative to certain universal rules and principles. Relativization, to use Robertson's term, is nothing new. However, as world culture has grown along with other forms of global integration, this burden of distinct identification has increased as well. The common notion that many French lamentations stem from a loss of former great-power status is relevant here, since this relativization especially hits home in a society that was so instrumental in building up the world-cultural edifice within which it now must find a new place. The French concern about the viability of their national identity is thus rooted in the key world-cultural process Robertson has identified. But the Robertsonian argument also suggests that the French can be sanguine about their prospects, since relativization with regard to world culture, from the historically varied standpoints of nations, will lead to a great variety of outcomes. Through relativization, world culture actually drives differentiation. By redefining their role as defenders of difference, the French are thus playing out a differentiated scenario.

Even more emphatically than Robertson, anthropologists like Hannerz stress the highly variable entwinement of world and national culture. To push their point a bit with French metaphors, national culture becomes a *bricolage* or *mélange* of world-cultural elements through creolization. This is not a case, however, of world culture bearing down on hapless nations. Creolization refers to continuous, critical interaction. When France sticks to its media quotas while also enjoying Hollywood fare, when French-speakers adopt franglais, when Disney icons rival the Eiffel Tower, when adherents of a secular universal faith discover the value of diversity, the result is a national culture less pristine than its leading intellectuals prefer but more distinctive than they are prepared to admit. Further, because creolization is a form of interaction, world culture is affected as well. In practice, it is a composite of the ways nations make sense of it. The French way of "doing" world culture contributes to the overall organization of diversity. By their actions, ironically, French critics of homogeneity disprove their point.

We have argued that world culture contains and fosters difference, but with regard to nations it has not always done so in the same way or to the same degree. Until recently, leading nation-states were more intent on spreading their influence than on guaranteeing difference. France, for instance, has had little compunction about globalizing its own culture, including the use of its own language as an international lingua franca. Would the French be as worried about uniformity if it were expressed in French? Would French politicians lead La Francophonie in defense of difference if France's once universal aspirations had been universally accepted? The irony here is that our argument, as well as the position of the French intellectuals itself, depends in part on the outcome of struggles against former French dominance in world culture. It is the success of movements asserting their right to political and cultural independence from colonizing powers – movements that themselves took different directions – that has helped to entrench the right to self-determination and

distinctive identities as universal principles in a world culture less tainted by hegemony.

Conclusions

As our extended illustration shows, the global and the local/national are thoroughly intertwined in the reproduction of difference. In carrying out its identity work through public policy, as France has done in enforcing its cultural exception, the nation-state firmly rests on world-cultural principles of great legitimacy. The very task of defining the nation is a standard responsibility of the state, taken especially seriously in France, and in this sense any definition of national identity is always more-than-national. Insofar as the forces of globalization undermine a nation's settled forms of self-understanding, which is certainly true of France, the magnitude of that task increases. Where the capacity of a nation to respond is in question, the salience of national identity as a project may well be even greater, especially if, as illustrated by the strong sentiments of many French public figures cited above, the relevant cultural elite is deeply invested in it. Nations can show resilience precisely in becoming embattled, as the French example shows. In fact, the components of French identity – universalistic culture, a strong state, a quest for a world role – may make it especially suitable as a defender of difference, and its defensive actions may well enrich and expand French culture.

Of course, the degree to which a nation's identity becomes embattled and the particular way in which it shows resilience are shaped by the sediments in its cultural foundation, by its global exposure and vantage point, and by its own historical trajectory in relation to the interplay between globalizing forces and national sediments. The global–national dialectic is clearly path-dependent. We showed, for example, how France's own involvement in the history of world culture now shapes its critical posture. Other factors we can only mention here will further shape the way France deals with the "crisis" of its national identity. Will the growing presence of Muslim immigrants and their offspring lead to a gradual loosening of national attachment or trigger strong reaffirmations of "traditional" national identity? Will European integration further erode the sovereignty and domestic control of the French state?

Will the relatively low proportion of the French people who consider themselves "very proud" of their nationality, around 40 percent in 1999–2000, increase or decrease? Since histories vary, depending in part on such "local" factors, no single case such as the one we have discussed can fully illuminate the dynamics of what is now a global experience. The very fact that it is difficult to generalize supports our argument against the homogenization scenario. But even if France no longer serves as a global model, the French are not unique in the predicament they face and the response they have fashioned. The upshot of our analysis is, once again, that expectations of cultural doom or the demise of diversity are simplistic. However, our argument should not dispel such notions entirely. The world-cultural legitimation of difference depends for its efficacy on the practical identity work of distinct groups, work that is contingent on "local" factors. From a general picture of world culture, even combined with strong assumptions about globalization, we therefore cannot derive clear-cut local predictions. Because the observable diversity of world culture stems from the multiplicity of particular identity projects, however much relativized and implicated in the global circumstance, that diversity is, so to speak, always up for grabs.

We also cannot say that the nation and national identity are secure as the defining form of difference in the twenty-first century. For all the current focus on the national in France, national identity may well become less salient over time. This is by no means to forecast a happy cosmopolitan future; rather, it is to suggest that national distinction may lose out in competition with other forms of collective identity, other claims on particularized loyalty. The rise of indigenous movements raises this possibility. On the horizon are more forceful claims for recognition of groups that differ in their sexual orientation and practices. World culture in principle legitimates alternative forms of particularism and therefore allows for such a pluralism of differences. We therefore need not fear homogeneity, little comfort though it may offer to defenders of any specific, uniquely cherished kind of difference, such as the French exception. In any case, a world in which a hundred differences bloom is not necessarily peaceful or pleasant. It may not sustain the kind of difference, such as the national, for which many have given their lives. When it comes to difference, contemporary world culture offers no guarantees.

[. . .]

 READING 64

Norms, Culture, and World Politics: Insights from Sociology's Institutionalism

Martha Finnemore

[. . .]

First, institutionalist research has been more concerned with documenting the effects of world cultural structure than investigating its causes or the mechanisms of change in the cultural structure itself. Institutionalists tend to produce global correlative studies whose structure and logic follow from Meyer and Rowan's initial insights about isomorphism in the face of dissimilar task demands. Institutionalist studies generally proceed by collecting quantitative data on a large number of units (usually states) and demonstrating that rather than correlating with local task demands, attributes or behavior of the units correlate with attributes or behavior of other units or with worldwide phenomena (international conferences and treaties or world historical events, for example). These analyses are often quite sophisticated, using event history analysis and other techniques that look exotic to most political scientists. However, once correlation is established, world cultural causes are assumed. Detailed process-tracing and case study analysis to validate and elaborate the inferences based on correlation are missing. Research to uncover the processes and mechanisms whereby world cultural norms spread and evolve would have at least two effects. The first would be to enrich the institutionalist argument. Such research would open up a more truly dialectical relationship between agency and structure and enable more persuasive accounts of the origins and dynamics of the world cultural structure.

Detailed case studies about the mechanisms by which cultural norms evolve and spread are also likely to call into question the cognitive basis of institutionalist theory. Institutionalists ground their arguments about the ways in which culture operates in social psychology. Meyer credits Erving Goffman, Guy Swanson, and C. Wright Mills with providing a connection between this social psychological literature and institutions. Detailed examination of cases of spreading Western culture is likely to reveal that its triumph is not due only or even primarily to cognition. The picture painted by institutionalist studies is one in which world culture marches effortlessly and facelessly across the globe. Little attention is paid either to contestation or coercion. To any political scientist (or historian) an account of the rise of the modern state in the West and its expansion across Africa, Asia, and the Americas that omits conflict, violence, and leadership is grossly incomplete. Similarly, the implication that human rights or citizen rights or even market economies become established and spread in a peaceful, orderly fashion through cognition alone is untenable to anyone who has detailed knowledge of cases.

The lack of case study analysis or on-the-ground investigation of the mechanisms whereby world culture produces isomorphism obscures the roles of politics and power in world history and normative change. The cognitive processes to which institutionalists point are important, but they are by no means the only processes at work in international life. Destroying cultural competitors, both figuratively and literally, is a time-honored way of establishing cultural dominance. Treatment of the native populations in North America is one example. Attempts at ethnic cleansing in Nazi Germany, Bosnia, Rwanda, and elsewhere are another. Cultural rules are often established not by persuasion or cognitive processes of institutionalization but by force and fiat. Over time, cultural norms established by force indeed may become institutionalized in the sense that they come to have a "taken-for-granted" quality that shapes action in the ways institutionalists describe. But emphasizing the institutionalized quality of sovereignty, for example, and its effects in world politics

should not obscure the role played by force and coercion in imposing sovereignty rules and in arbitering their ongoing evolution.

One instance where force and military power may be particularly important to institutionalist concerns involves the Reformation and eventual Protestant domination of the West. Institutionalists trace their Western cultural norms back to medieval Christendom without a word about the Reformation or Protestantism's effect on these cultural rules. This is a startling omission given the intellectual debt these scholars owe Max Weber. Many of the cultural rules institutionalists emphasize – individualism and markets, for example – arguably have strong ties to Protestantism specifically, not Christianity generally. One could argue that the Western culture that is expanding across the globe is really a Protestant culture. Protestantism did not come to dominate Europe through cognition and persuasion alone, as centuries of religious wars make clear. Western culture may look the ways it does because of three centuries of Anglo-American (i.e., Protestant) power and domination of the West, domination that was secured through repeated military conquest of France.

The second feature of institutionalist research that should concern political scientists is their specification of the content of world culture. Institutionalists focus on Western rationality as the means to both progress and equality. Progress is defined as wealth accumulation, justice is defined as equality, and rational means, in institutionalist research, are usually bureaucracies and markets. Institutionalists tend to treat these elements of Western modernity as at least loosely compatible. Equality, in the form of individual rights, expands together with markets and bureaucracies across the globe, and institutionalist research documents the collective and interrelated spread of these cultural norms.

The implication, which will be suspect to all political scientists, is that all "good" things (in the Western cultural frame) can and do go together. Institutionalists may not intend this implication, but both their research and their theorizing consistently underscore the mutually reinforcing nature of these Western cultural rules.

In fact, there are good reasons to believe that the elements of world culture, even as the institutionalists have specified it, contain deep tensions and contradictions that constrain isomorphism and limit the stability of behavioral convergence. Most obvious is the tension between the two "ends" of Western world culture – progress, defined as economic accumulation, and justice, defined as equality. The trade-off between equity and growth in development economics is well-known. In making decisions about economic policies, the two pillars of the normative structure often pull in opposite directions. Partisans of redistributionist policies have invoked equality norms in their defense. Those pushing for more and faster growth will evoke progress norms. Policymakers often have to make explicit and controversial trade-offs between the two.

Similarly, the two rational means to justice and progress – markets and bureaucracies – may be in tension. Market arrangements may be justified normatively by their efficient contributions to progress (wealth accumulation) and by equality defined as opportunity or access, but they often create outcomes that offend other definitions of equality, notably equality of outcomes. Markets tend to produce unequal distributional outcomes. The common solution is to bring in bureaucracy, in the form of the state, to remedy the equality offenses of markets. However bureaucracies may compromise the efficiency of markets and so compromise progress. Again, progress (wealth) conflicts with justice (equality). And, again, no obvious or equilibrium set of arrangements can resolve this.

Contradictions among dominant cultural norms mean that social institutions are continually being contested, albeit to varying degrees at different times. Unresolved normative tensions in a set of social compromises at one time may be the mobilizing basis for attacks on that set of social arrangements later as people articulate normative claims that earlier were pushed aside. Further, compromises among competing world normative principles may be contingent on local circumstances and personalities and are likely to reflect local norms and customs with which international norms have had to compromise. Thus, after World War II Japan was forced (note the process was not cognitive) to accept a set of Western economic and political arrangements that had been forged elsewhere, in the United States. Over time, those arrangements became institutionalized in Japan but in unique ways that reflected non-Western local cultural norms. The subsequent success of Japan in Western terms (a great deal of economic accumulation with relative equality) has prompted Western firms and Asian states to adopt a number of Japanese practices, policies, and norms. This kind of cultural feedback, from periphery to core, is neglected by the unidirectional institutionalist model.

These contestation processes for normative dominance are political. In fact, normative contestation is in large part what politics is all about; it is about competing values and understandings of what is good, desirable, and appropriate in our collective communal life. Debates about civil rights, affirmative action, social safety nets, regulation and deregulation, and the appropriate degree of government intrusion into the lives of citizens are all debates precisely because there is no clear stable normative solution. Further, they are all debates involving conflict among the basic normative goods identified by the institutionalists. Civil rights, affirmative action, and to some extent social safety nets are debates about the nature of equality – who attains equality and how that equality is measured. Since the solutions all involve bureaucratic intervention, these debates are also about the relationship of bureaucracies and the state to equality. Debates about social safety nets raise specific issues about the relationship between bureaucracies and markets and the degree to which the latter may be compromised by the former in the service of equality. Debates over regulation and government intrusion are both about the degree to which bureaucracy can compromise markets, on the one hand, or equality and individual rights that derive from equality, on the other.

If one takes seriously the tensions and contradictions among elements of culture, research must focus on politics and process. If cultural elements stand in paradoxical relations such that equilibrium arrangements are limited or constrained, the interesting questions become, which arrangements are adopted where – and why? Institutionalists may be right. Common global norms may create similar structures and push both people and states toward similar behavior at given times, but if the body of international norms is not completely congruent, then those isomorphisms will not be stable. Further, people may adopt similar organizational forms but show little similarity in behavior beyond that. Botswana and the United States may both be organized in the form of a modern state, but the content of those forms and the behavior within them are very different. Isomorphism is not homogeneity; it does not create identical behavioral outcomes. Without a specification of culture that attends to oppositions within the overall structure, institutionalists will not be able to account for either diversity or change in that structure.

Conclusions

Institutionalist arguments emphasize structure at the expense of agency. Doing so has important intellectual benefits. It allows institutionalists to ask questions about features of social and political life that other perspectives take for granted – ubiquitous sovereign statehood and expanding claims by individuals, for example. Further, from an IR theory perspective, institutionalists' emphasis on structure allows for system-level explanations that compete with other dominant paradigms and so enrich the body of theory available to tackle puzzles in the field.

If the neglect of agency were only an omission, there would be little cause for concern. No theory explains everything. One can always explain a few more data points by adding a few more variables and increasing the complexity of the model. But the institutionalists' inattention to agency leads them into more serious errors. It leads them to misspecify both the mechanisms by which social structure produces change and the content of the social structure itself.

Cognitive processes may dominate organizational change in many empirical domains, but they compete with and often are eclipsed by coercion in many of the empirical domains that concern IR scholars. Educational curricula may change in peaceful ways driven by cognitive decision-making processes; state authority structures often do not. Violence is a fundamentally different mechanism of change than cognition. Both mechanisms may operate in a given situation. Often there are choices to be made even within the constraints imposed by force, but outcomes imposed externally through violence are not captured by a cognitive theoretical framework.

Institutionalists are not alone in this tendency to overlook power and coercion in explaining organizational outcomes. Much of organization theory shares this characteristic. Terry Moe has noted the failure of the new economics of organization to incorporate considerations of power, but even Moe, a political scientist, is not particularly concerned with issues of violence since these occur rarely in his own empirical domain – US bureaucracy.

Institutionalist models imply a world social structure made up of norms that are largely congruent. Their emphasis is on the mutually reinforcing and expansive nature of these norms. They stress the consensus that arises around various cultural models – of citizenship, of statehood, of education, of individual rights – to the point that these norms and institutions are taken for granted in contemporary life. The implication is that the spread of world culture is relatively peaceful. Institutionalists specify no sources of instability, conflict, or opposition to the progressive expansion of world culture. Yasemin Soysal's work is perhaps the most attuned to contradictions among the cultural elements of citizenship she studies. However, even in her work these contradictions result only in paradoxical arrangements with which people seem to live reasonably peacefully.

The result of this specification is that all of politics becomes problematic in an institutionalist framework. If the world culture they specify is so powerful and congruent, the institutionalists have no grounds for explaining value conflicts or normative contestation – in other words, politics. A research design that attended to agency and the processes whereby isomorphic effects are produced would have prevented institutionalists from falling into this trap. Focusing more closely on process would draw attention to the contradictions among normative claims and force institutionalists to rethink both the specification of world culture and its likely effects.

These problematic features of institutionalist theory lie squarely on the turf of political scientists. Politics and process, coercion and violence, value conflict and normative contestation are our business. Institutionalism would benefit greatly from a dialogue with political scientists. Likewise, political scientists could learn a great deal from institutionalists. Thus far, IR scholars interested in norms have lacked a substantive systemic theory from which to hypothesize and carry out research. Institutionalism provides this. Taking its claims seriously may produce radical revisions to the existing sociologists' theories. It may also produce opposing theoretical arguments. Either outcome would advance research in both disciplines and enrich our understanding of world politics.

SOURCES AND CREDITS

Mauro F. Guillén. "Is Globalization Civilizing, Destructive or Feeble? A Critique of Five Key Debates in the Social Science Literature." *Annual Review of Sociology* 27, 2001: 235–60. Reprinted with permission from the *Annual Review of Sociology*, Volume 27 © 2001 by Annual Reviews, www.annualreviews.org.

Samuel P. Huntington. "The Clash of Civilizations?" *Foreign Affairs* 72, 3, Summer 1993: 22–49. Reprinted by permission of *Foreign Affairs*, 72, 3, Summer 1993. Copyright © 1993 by the Council on Foreign Relations, Inc. www.ForeignAffairs.com.

John Gray. "Global Utopias and Clashing Civilizations: Misunderstanding the Present." *International Affairs* 74, 1, 1998: 149–63.

Jack F. Matlock, Jr. "Can Civilizations Clash?" *Proceedings of the American Philosophical Society* 143, 3, September 1999: 428–39.

Chris Brown. "History Ends, Worlds Collide." *Review of International Studies* 25, 1999: 41–57. © 1999 by British International Studies Association. Published and reproduced by permission of Cambridge University Press.

Samuel P. Huntington. "If Not Civilizations, What? Paradigms of the Post-Cold War World." *Foreign Affairs* 72, 5, Fall 1993: 186–94. Reprinted by permission of *Foreign Affairs*, 72, 5, Fall 1993. Copyright © 1993 by the Council on Foreign Relations, Inc. www.ForeignAffairs.com.

Edward W. Said. "Introduction." In *Orientalism*. New York: Vintage Books, 1979/1994. "Introduction" from *Orientalism* by Edward W. Said, copyright © 1978 by Edward W. Said. Used by permission of Pantheon Books, a division of Random House, Inc.

Sadik Jalal al-'Azm. "Orientalism and Orientalism in Reverse." In A.L. Macfie ed. *Orientalism: a Reader*. New York: New York University Press, 2001: chapter 24. Reprinted with permission from Edinburgh University Press, www.euppublishing.com.

Ali Rattansi. "Postcolonialism and Its Discontents." *Economy and Society* 26, 4, November 1997: 480–500. © 1997 by Routledge. Reprinted with permission from Taylor & Francis Group. www.informaworld.com.

Peter Marcuse. "Said's Orientalism: a Vital Contribution Today." *Antipode* 2004: 809–17. © 2004 by Editorial Board of *Antipode*.

William Easterly. "Freedom versus Collectivism in Foreign Aid." In *Economic Freedom of the World: 2006 Annual Report*: chapter 2. © 2006 by the Fraser Institute. Reprinted with permission from The Fraser Institute, www.freetheworld.com.

Karl Polanyi. *The Great Transformation: the Political and Economic Origins of Our Time* (1944). Boston: Beacon Press, 2001. © 2001 by Karl Polanyi. Reprinted by permission of Kari Polanyi-Levitt.

David Harvey. "Freedom's Just Another Word . . . ." In *A Brief History of Neoliberalism*. Oxford: Oxford University Press, 2005: chapter 1. © 2005 by David Harvey. Reprinted with permission from Oxford University Press.

Aiwha Ong. " Neoliberalism as Exception, Exception to Neoliberalism." In *Neoliberalism as Exception: Mutations in Citizenship and Sovereignty*. Durham: Duke University Press, 2006: Introduction. © 2006 Duke University Press. All rights reserved. Used by permission of the publisher.

Jim Glassman and Pádraig Carmody. "Structural Adjustment in East and Southeast Asia: Lessons from Latin America." *Geoforum* 32, 2001: 77–90. © 2001 Elsevier Science Ltd. All rights reserved. Reprinted with permission from Elsevier.

Sarah Babb. "The Social Consequences of Structural Adjustment: Recent Evidence and Current Debates." *Annual Review of Sociology* 31, 2005: 199–222. Reprinted with permission from the *Annual Review of Sociology*, Volume 31 © 2005 by Annual Reviews, www.annualreviews.org.

M. Rodwan Abouharb and David L. Cingranelli. "The Human Rights Effects of World Bank Structural Adjustment, 1981–2000." *International Studies Quarterly* 50, 2006: 233–62.

Vincent Lloyd and Robert Weissman. "How International Monetary Fund and World Bank Policies Undermine Labor Power and Rights." *International*

Journal of Health Services 32, 3, 2002: 433–42. Multinational Monitor, Online by Vincent Lloyd and Robert Weissman. Copyright © 2001 by Essential Information. Reproduced with permission of Essential Information in the format Textbook via Copyright Clearance Center.

Gerald Scott. "Who Has Failed Africa? IMF Measures or the African Leadership?" *Journal of Asian and African Studies* XXXIII, 3, 1998: 265–74. © 1998 by Brill Academic Publishers. Reproduced with permission of Brill Academic Publishers in the format Textbook via Copyright Clearance Center.

Donald N. Levine. "Sociology and the Nation-State in an Era of Shifting Boundaries." *Sociological Inquiry* 66, 3, 1996: 253–66.

Susan Strange. "The Westfailure System." *Review of International Studies* 25, 1999: 345–54. © 1999 by British International Studies Association. Published and reproduced by permission of Cambridge University Press.

Linda Weiss, "Globalization and the Myth of the Powerless State." *New Left Review* I/225, September–October 1997: 3–27. © 1997 by *New Left Review*. Reprinted with permission from *New Left Review*.

Daniel Béland. *States of Global Insecurity: Policy, Politics, and Society*. New York: Worth, 2008: 47–54.

William I. Robinson. "Beyond Nation-State Paradigms: Globalization, Sociology, and the Challenge of Transnational Studies." *Sociological Forum* 13, 4, 1998: 561–94.

Leslie Sklair. *Globalization: Capitalism and Its Alternatives*. Oxford: Oxford University Press, 2002. © 2002 by Leslie Sklair. Reprinted with permission from Oxford University Press.

William I. Robinson. "Social Theory and Globalization: the Rise of a Transnational State." *Theory and Society* 30, 2001: 157–200. © 2001, Springer Netherlands. With kind permission from Springer Science & Business Media.

Philip McMichael. "Revisiting the Question of the Transnational State: a Comment on William Robinson's 'Social Theory and Globalization'." *Theory and Society* 30, 2001: 201–10. © 2001, Springer Netherlands. With kind permission from Springer Science & Business Media.

Immanuel Wallerstein. "Theoretical Reprise." In *The Modern World-System*. Vol. 1. New York: Academic Press, 1974: chapter 7.

Leslie Sklair. "Competing Conceptions of Globalization." *Journal of World Systems Research* 5, 2, Summer 1999: 143–63. © 1999 by Leslie Sklair. Reprinted with permission from the author.

Michael Hardt and Antonio Negri. *Empire*. Cambridge, MA: Harvard University Press, 2000. © 2000 by the President and Fellows of Harvard College. Reprinted with permission from Harvard University Press.

Michael Hardt and Antonio Negri interviewed by Nicholas Brown and Imre Szeman. "The Global Coliseum: on *Empire*." *Cultural Studies* 16, 2, 2002: 177–92. © 2002 by Routledge. Reprinted with permission from Taylor & Francis Group. www.informaworld.com.

Tarak Barkawi and Mark Laffey. "Retrieving the Imperial: *Empire* and International Relations." *Millennium: Journal of International Studies* 31, 1, 2002: 109–27. © 2002, *Millennium: Journal of International Studies*. Reprinted by permission of Sage.

David Moore. "Africa: the Black Hole at the Middle of *Empire*?" *Rethinking Marxism* 13, 3/4, 2001: 100–18. © 2001 by Routledge. Reprinted with permission from Taylor & Francis Group. www.informaworld.com.

Stanley Aronowitz. "The New World Order (They Mean It)." *The Nation* July 17, 2000: 25–8.

Michael Hardt and Antonio Negri. "Adventures of the Multitude: Response of the Authors." *Rethinking Marxism* 13, 3/4, Fall/Winter 2001: 236–43. © 2001 by Routledge. Reprinted with permission from Taylor & Francis Group. www.informaworld.com.

Manuel Castells. "Toward a Sociology of the Network Society." *Contemporary Sociology* 29, 5, September 2000: 693–9. © 2000 by Manuel Castells. Reprinted with permission from the author.

Peter Marcuse. "Depoliticizing Globalization: From Neo-Marxism to the Network Society of Manuel Castells." In John Eade and Christopher Mele eds. *Understanding the City: Contemporary and Future Perspectives*. Oxford: Blackwell, 2002: chapter 7. © 2002 by Blackwell Publishers Ltd.

Ulrich Beck. "The Terrorist Threat: World Risk Society Revisited." *Theory, Culture & Society* 19, 4, 2002: 39–55. © 2002, *Theory, Culture & Society* Ltd. Reprinted by permission of Sage.

Darryl S. L. Jarvis. "Risk, Globalisation and the State: a Critical Appraisal of Ulrich Beck and the World Risk Society Thesis." *Global Society* 21, 1, January 2007: 23–46. © 2007 by Routledge. Reprinted with permission from Taylor & Francis Group. www.informaworld.com.

Ulrich Beck and Natan Sznaider. "Unpacking Cosmopolitanism for the Social Sciences: a Research Agenda." *The British Journal of Sociology* 57, 1, 2006: 1–23.

Craig Calhoun. "Cosmopolitanism and Nationalism." *Nations and Nationalism* 14, 3, 2008: 427–48. © 2008 by Craig Calhoun.

Benjamin R. Barber. "Jihad vs McWorld." *Atlantic Monthly* March 1992: 53–63. © 1992 by *The Atlantic Monthly*. Reproduced with permission of *The Atlantic Monthly* in the format Textbook via Copyright Clearance Center.

Fareed Zakaria. "Paris Is Burning." *The New Republic* January 22, 1996: 27–30. © 1996 by Fareed Zakaria. Reprinted with permission from the author.

Bryan S. Turner. "Sovereignty and Emergency: Political Theology, Islam and American Conservatism." *Theory, Culture & Society* 19, 4, 2002: 103–19. © 2002, *Theory, Culture & Society* Ltd. Reprinted by permission of Sage.

Benjamin R. Barber. "On Terrorism and the New Democratic Realism." *The Nation* January 22, 2002: 17–18.

Jan Nederveen Pieterse. "Globalization and Culture: Three Paradigms." In *Globalization and Culture: Global Mélange.* Rowman and Littlefield, 2003: chapter 3. © 2003 by Rowman & Littlefield Publishing Group, Inc. Reproduced with permission of Rowman & Littlefield Publishing Group, Inc., in the format Textbook via Copyright Clearance Center.

Ulf Hannerz. "The World in Creolisation." *Africa* 57, 4, 1987: 546–59. © 1987, International African Institute. Reprinted with permission.

Ulf Hannerz. "Flows, Boundaries and Hybrids: Keywords in Transnational Anthropology." Working Paper Series WPTC-2K-02, Transnational Communities Programme, University of Oxford. © Ulf Hannerz. Reprinted with permission from the author.

Jan Nederveen Pieterse. "Globalization as Hybridization." In M. Featherstone, S. Lash and R. Robertson eds. *Global Modernities.* Thousand Oaks, CA: Sage, 1995: chapter 3.

Roland Robertson. "Glocalization: Time–Space and Homogeneity–Heterogeneity." In M. Featherstone, S. Lash, and R. Robertson eds. *Global Modernities.* Thousand Oaks, CA: Sage, 1995: chapter 2.

Jan Nederveen Pieterse. "Hybridity, So What? The Anti-Hybridity Backlash and the Riddles of Recognition." *Theory, Culture & Society* 18, 2–3, 2001: 219–45. © 2001, *Theory, Culture & Society* Ltd. Reprinted by permission of Sage.

Marwan M. Kraidy. "The Global, the Local and the Hybrid: a Native Ethnography of Glocalization." *Critical Studies in Mass Communications* 16, 4,

1999: 456–76. © 1999 by Routledge. Reprinted with permission from Taylor & Francis Group. www.informaworld.com.

Keith Nurse. "Globalization and Trinidad Carnival: Diaspora, Hybridity and Identity in Global Culture." *Cultural Studies* 13, 4, 1999: 661–90. © 1999 by Routledge. Reprinted with permission from Taylor & Francis Group. www.informaworld.com.

William H. Thornton. "Mapping the 'Glocal' Village: the Political Limits of 'Glocalization'." *Continuum: Journal of Media & Cultural Studies* 14, 1, 2000: 78–89. © 2000 by Routledge. Reprinted with permission from Taylor & Francis Group. www.informaworld.com.

George Ritzer. "Rethinking Globalization: Glocalization/ Grobalization and Something/Nothing." *Sociological Theory* 21, 3, 2003: 193–209. © 2003, American Sociological Association. Reprinted with permission of the publisher.

Douglas Kellner. "Dialectics of Something and Nothing: Critical Reflections on Ritzer's Globalization Analysis." *Critical Perspectives on International Business* 1, 4, 2005: 263–72. © 2005, Emerald Group Publishing Limited. All rights reserved. Reprinted by permission.

George Ritzer. "An Introduction to McDonaldization." In *The McDonaldization of Society* 5th edn. Thousand Oaks, CA: Pine Forge Press, 2008: chapter 1. © 2008 by Sage Publications, Inc. Books. Reproduced with permission of Sage Publications, Inc. Books in the format Textbook via Copyright Clearance Center.

Malcolm Waters. "McDonaldization and the Global Culture of Consumption." *Sociale Wetenschappen* 39, 1996: 17–28.

Bryan S. Turner. "The McDonald's Mosaic: Glocalization and Diversity." *American Behavioral Scientist* 47, 2, 2003: 137–53. © 2003 by Sage Publications, Inc. Journals. Reproduced with permission of Sage Publications, Inc. Journals in the format Textbook via Copyright Clearance Center.

James L. Watson, "Transnationalism, Localization, and Fast Foods in East Asia." In J. L. Watson ed. *Golden Arches East: McDonald's in East Asia.* Stanford: Stanford University Press, 1997. © 1997, 2006 by the Board of Trustees of the Leland Stanford Junior University. All rights reserved. Used with the permission of Stanford University Press, www.sup.org.

Alan Bryman. "Global Implications of McDonaldization and Disneyization." *American Behavioral Scientist* 47, 2, 2003: 154–67. © 2003 by Sage Publications, Inc. Journals. Reproduced with permission of Sage

INDEX